Dear Fire Fighter Instructor and Student:

For over 100 years, the National Fire Protection Association (NFPA) and the International Association of Fire Chiefs (IAFC) have had a profound impact on the fire service. Through research, education, and vision, each organization has helped reduce the burden of fire and other hazards on the quality of life. At your fingertips now is an innovation resulting from the partnership of the NFPA and IAFC that will improve fire fighter education for years to come.

As today's fire fighters respond to all types of incidents from fire calls and medical emergencies to hazardous materials incidents and terrorist attacks, it is imperative that lack of knowledge not compromise fire fighters' effectiveness or safety. *Fundamentals of Fire Fighter Skills* thoroughly covers the 2002 edition of NFPA 1001, *Standard for Fire Fighter Professional Qualifications,* while also including a variety of enhancements to enrich fire fighter education. *Fundamentals of Fire Fighter Skills* is an advanced learning system that will have a profound impact on the field of fire fighter education.

Instructors and students alike will find this text current, clear in its presentation, and complete in its content. A complete array of resources has been designed to better support instructors and to prepare future fire fighters for the job. Experienced educators have contributed to and reviewed these training materials to ensure that they are both comprehensive and engaging.

On behalf of the NFPA and IAFC, we applaud the efforts of all of the contributors and reviewers who have participated in the development of the *Fundamentals of Fire Fighter Skills* materials. Together, we are changing the face of the fire service.

Chief Ernest Mitchell	**James M. Shannon**
IAFC President	NFPA President

Fundamentals of Fire Fighter Skills

JONES AND BARTLETT PUBLISHERS
Sudbury, Massachusetts
BOSTON TORONTO LONDON SINGAPORE

Jones and Bartlett Publishers
World Headquarters
40 Tall Pine Drive
Sudbury, MA 01776
978-443-5000
www.jbpub.com

Jones and Bartlett Publishers Canada
2406 Nikanna Road
Mississauga, ON L5C 2W6
CANADA

Jones and Bartlett Publishers International
Barb House, Barb Mews
London W6 7PA
United Kingdom

National Fire Protection Association
1 Batterymarch Park
Quincy, MA 02169-7471
www.NFPA.org

International Association of Fire Chiefs
4025 Fair Ridge Drive
Fairfax, VA 22033
www.IAFC.org

Production Credits
Chief Executive Officer: Clayton E. Jones
Chief Operating Officer: Donald W. Jones, Jr.
President: Robert W. Holland, Jr.
V.P. of Sales and Marketing: William J. Kane
V.P. of Production and Design: Anne Spencer
V.P. of Manufacturing and Inventory Control: Therese Bräuer
Publisher, Public Safety Group: Kimberly Brophy
Associate Managing Editor: Jennifer Reed
Associate Managing Editor: Erin Roberts
Senior Production Editor: Linda S. DeBruyn
Production Editor: Scarlett Stoppa
Production Assistant: Carolyn F. Rogers
Photo Researcher: Kimberly Potvin
Director of Interactive Technology: Adam Alboyadjian
Director of Marketing: Alisha Weisman
Design Manager: Kristin Ohlin
Production Services Manager: Bret Kerr
Interactive Technology Manager: Dawn Mahon Priest
Design and Composition: Studio Montage
Cover Photograph: © Michael Heller; 911 Pictures
Text Printing and Binding: The Courier Company
Cover Printing: Lehigh Press

Library of Congress Cataloging-in-Publication Data
National Fire Protection Association.
Fundamentals of fire fighter skills / National Fire Protection Association.-- 1st ed.
p. cm.
Includes index.
ISBN 0-7637-2233-2 (pbk. : alk. paper)
1. Fire prevention. 2. Fire extinction. I. Title.
TH9146.N29 2003
628.9'2--dc22

Printed in the United States of America
08 07 06 05 04 10 9 8 7 6 5 4 3 2 1

Brief Contents

Table of Contents

Chapter 1

Chapter 2

Chapter 3

Chapter 4

Chapter 5

Chapter 6

Chapter 7

Chapter 8

Chapter 9

Chapter 10

Chapter 11

Chapter 12

Chapter 13

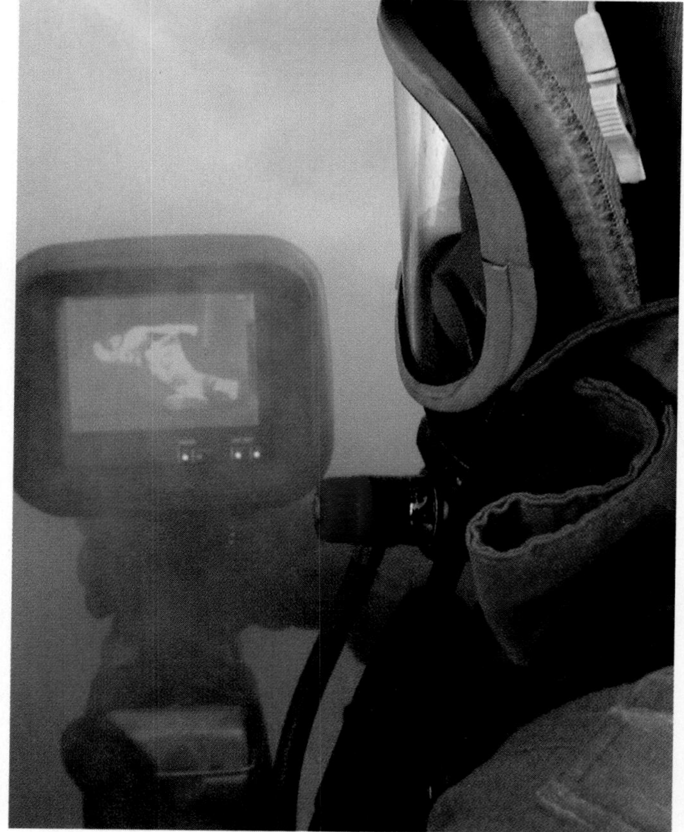

Chapter 14

Chapter 15

Chapter 16

Chapter 17

Chapter 18

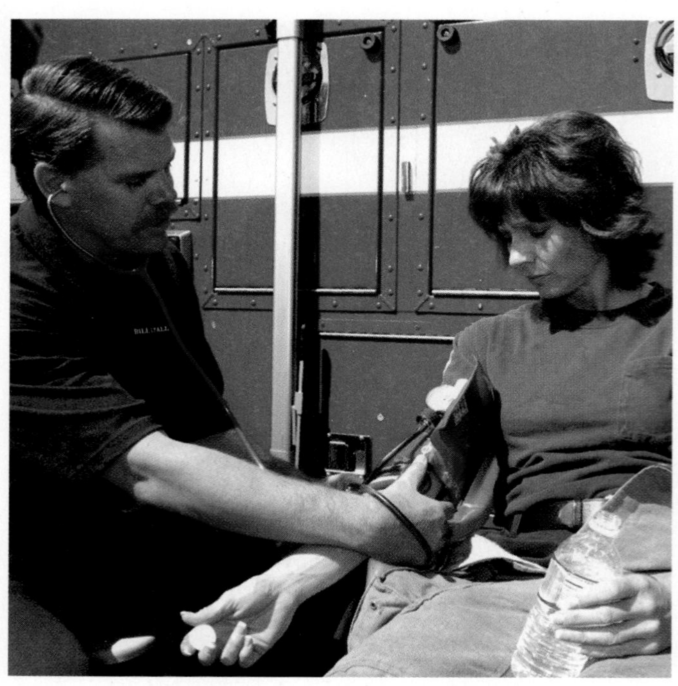

Chapter 19

Chapter 20

Chapter 21

Chapter 22

Chapter 23

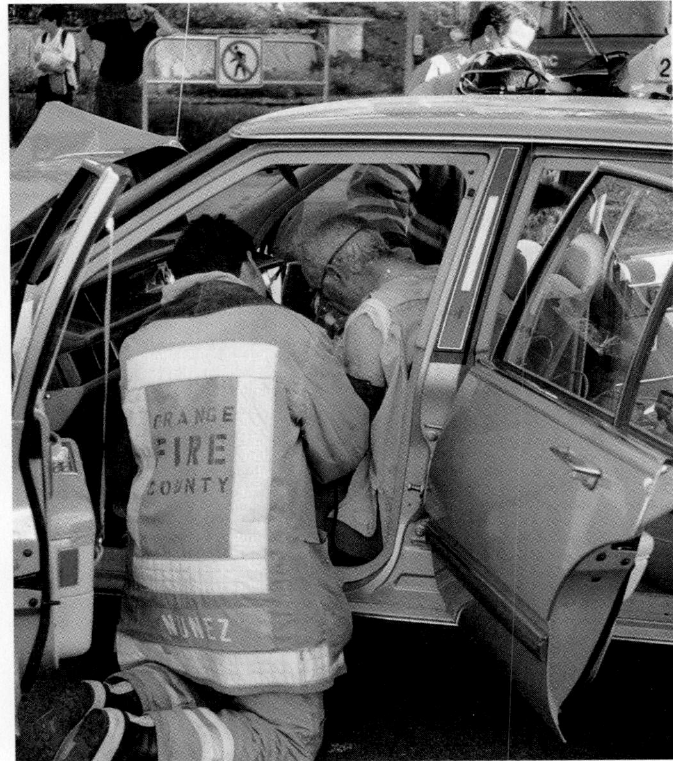

Chapter 24

Chapter 25

Chapter 26

Chapter 27

Chapter 28

Chapter 29

Chapter 30

Chapter 31

Chapter 32

SPARKY® is a trademark of NFPA.

Chapter 33

Chapter 34

Chapter 35

Chapter 36

Chapter 37

Appendix A

Appendix B

Skill Drills

Chapter 12 (continued)

Chapter 13

Chapter 14

Chapter 15

Chapter 16

Skill Drills

Chapter 16 (continued)

Chapter 18

Chapter 21

Chapter 24

Chapter 25

Chapter 33

Resource Preview

Fundamentals of Fire Fighter Skills

The National Fire Protection Association (NFPA) and the International Association of Fire Chiefs (IAFC) are pleased to bring you *Fundamentals of Fire Fighter Skills*, a modern integrated teaching and learning system for the Fire Fighter I and II levels. It combines current content with dynamic features, interactive technology, and both instructor and student resources.

Today's fire fighters respond to all types of crises from fire calls to medical emergencies to hazardous materials incidents to terrorist attacks. As they answer these calls, it is imperative that no gaps in knowledge compromise their effectiveness or safety. Fire fighters need up-to-date, comprehensive training materials to thoroughly prepare for any situation that may arise. *Fundamentals of Fire Fighter Skills* will set the standard for fire fighter education for years to come.

Chapter Resources

Fundamentals of Fire Fighter Skills thoroughly supports instructors and prepares students for the job. The text covers the entire spectrum of the 2002 Edition of NFPA 1001, *Standard for Fire Fighter Professional Qualifications* from firefighting and hazardous materials incidents to emergency medical care in one volume, eliminating the needs for additional texts.

The text is the core of the teaching and learning system with features that will reinforce and expand on the essential information and make information retrieval a snap. These features include:

Navigational Toolbar

Found at the beginning of each chapter, the navigational toolbar will guide you through the technology resources and text features available for that chapter.

Chapter Objectives

NFPA 1001 Standard, Additional NFPA Standards, Knowledge Objectives, and Skills Objectives are listed at the beginning of each chapter.

- Portions of the NFPA 1001 Standard that are highlighted in red are applicable to the chapter.
- Additional NFPA Standards that apply to the chapter are listed for reference.
- Knowledge Objectives outline the most important topics covered in the chapter.
- Skills Objectives map the skill drills provided in the chapter.

Resource Preview

Chapter Resources (continued)

You Are the Fire Fighter

Each chapter opens with a case study that will stimulate classroom discussion, capture students' attention, and provide an overview of the chapter. An additional case study is included in the end-of-chapter Wrap-Up.

Voices of Experience

In the Voices of Experience essays, veteran fire fighters share accounts of memorable incidents while offering advice and encouragement. These essays highlight what it is truly like to be a fire fighter.

You Are the Fire Fighter

As part of your training as a new fire fighter, you have been assigned to a ride along with Captain Lewis at Station #3. After greeting you and introducing you to the rest of the crew, Captain Lewis shows you the equipment on the engine. Just as you are walking back to the kitchen, you hear the tones go off. "Engine 403 and Medic 433, respond to 11285 Main Street for a 63-year-old male with chest pain. The patient is reporting shortness of breath."

As you strap yourself into your assigned seat, you wonder why fire fighters are responding to a call for medical assistance for someone having a heart attack. Then you start thinking of questions you could ask the Captain later.

1. *Why does the fire department respond to emergency medical calls?*
2. *Will you have a role in delivering emergency medical services?*
3. *Will you need additional training?*
4. *How does the delivery of emergency medical services relate to your duties as a fire fighter?*

Introduction

This chapter will help you understand the role played by the fire department in providing emergency medical services (EMS). This role varies from one department to another. By learning the components of the EMS system, you will have a better understanding of the relationship between fire departments and EMS, including:

- The importance of EMS as part of the fire department mission of saving lives and protecting property
- The difference between basic life support (BLS) and advanced life support (ALS) levels of emergency medical care
- The different types of EMS training and certification
- The importance of EMS training for fire fighters
- The different types of EMS systems provided by fire departments
- The variety of interactions between EMS providers and other members of the health care team

The material presented in this chapter is not designed to make you an EMS provider. To achieve that goal, you will need to take an approved course leading to certification or registration as a First Responder, an Emergency Medical Technician-Basic (EMT-Basic), an Emergency Medical Technician-Intermediate (EMT-Intermediate), or an Emergency Medical Technician-Paramedic (EMT-Paramedic) (▶ Table 23-1). Instead, this material introduces you to the different levels of care that EMS systems provide, the types of training available in EMS, the types of EMS systems found in fire departments, and the other health care providers who are also part of the EMS system.

Table 23-1
EMS Certification Levels
• First Responder
• Emergency Medical Technician-Basic (EMT-Basic)
• Emergency Medical Technician-Intermediate (EMT-Intermediate)
• Emergency Medical Technician-Paramedic (EMT-Paramedic)

The Importance of EMS to the Fire Service and the Community

The mission of the fire service is to save lives and protect property. When many people think of their fire department, they think of fire suppression first. However, in some fire departments, more than 80% of the emergency calls are requests for emergency medical services (▶ Figure 23-1). Many fire departments have added "EMS" or "rescue" to their names to reflect this change. As a new fire fighter, you will have an important role in providing emergency medical services to the citizens of your community.

Most fire departments provide some level of emergency medical services, although the degree of involvement varies. Some departments provide only extrication and medical assistance for vehicle collisions. Others are the first responders for emergency medical calls, or provide basic or advanced life support until another agency arrives to transport the patient. Still other fire departments provide both medical first response and patient transportation services.

Voices of Experience

"If this kind of breakdown can take place in a controlled environment with physically fit fire fighters, imagine the potential when dealing with fire fighters in a chaotic atmosphere full of unknowns."

On a hot August day in 1996, the Salt Lake City Fire Department recruits spent the morning donning and doffing Class A hazardous materials suits (heavy, encapsulating suits that envelop both the wearer and SCBA totally) and conducting decontamination drills. After a short lunch break, the afternoon was filled with ladder drills. In the late afternoon, one of the recruits began having difficulty completing the evolutions. As her condition worsened, she was unable to think clearly and demonstrate basic skills. She became confused, and one of the training officers asked if she was all right. When she was unable to give a clear answer, she was told to take a break and rehydrate. A fire fighter approached her to see how she was feeling and quickly realized that something was not right. The fire fighter took the recruit's vital signs and discovered that her temperature, pulse, and blood pressure were all elevated. Her skin was now bright red and she showed no sign of sweating. The decision was made immediately to start an IV and transport her to the hospital. The assessment in the emergency room revealed that the recruit was dehydrated and had low blood sugar levels. She was treated for her condition and released when her vital signs returned to normal.

This incident involved a recruit, in training, who was in above average physical condition. If this kind of breakdown can take place in a controlled environment with physically fit fire fighters, imagine the potential when dealing with fire fighters in a chaotic atmosphere full of unknowns. Rehabilitation on the fire ground, or any other physically demanding scene, should be as much a part of actual fire ground activities as putting water on the fire. Fire fighter fitness levels and events leading up to the incident, such as recent workouts, heavy meals, or illness, create a level of unpredictability when it comes to crew capabilities. Salt Lake City Fire Department has made many advances in ensuring proper rehabilitation for all members in training or on the emergency scene.

It is interesting to note that this fire fighter did not stop working without being approached. Fire fighters need to take responsibility for themselves and their crewmates. There are many distractions on the fire ground. A heightened awareness of your own body's signals, and paying close attention to your crewmates' actions, can allow for early intervention. A simple 10- to 15-minute break could make the difference between being a functional, contributing fire fighter, or becoming part of the problem.

Martha Ellis
Salt Lake City Fire Department
Salt Lake City, Utah

270 FUNDAMENTALS OF FIRE FIGHTER SKILLS

Skill Drill

Hoisting a Ladder

The team that needs the ladder should lower a rope with enough extra rope to tie onto the ladder.

Tie a figure eight on a bight to make a loop 3' or 4' in diameter.

Pass the rope between the rungs of the ladder, three or four rungs from the top.

Pull the loop under the rungs toward the top of the ladder.

Place the loop around the top of the ladder.

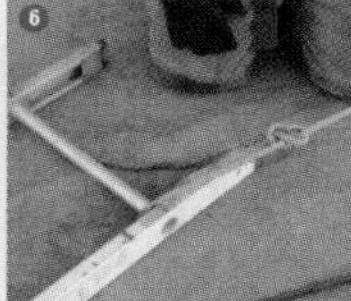

Remove the slack from the rope and allow the loop to slide down the ladder.

Attach a tag line from below to control the ladder as it is hoisted. Prepare to raise the ladder.

Skill Drills

Skill Drills provide written step-by-step explanations and visual summaries of important skills and procedures. Skill Drills are provided in a format that enhances student comprehension of complex procedures.

56 FUNDAMENTALS OF FIRE FIGHTER SKILLS

Face pieces for various brands and models of SCBAs are slightly different. Some have the regulator mounted on the face piece; others have it mounted on the harness straps. Fire fighters must learn about the face pieces used by their departments.

Follow the steps in Skill Drill 2-7 to don a face piece:

1. Make sure you have donned your protective hood. Remove your helmet and pull the hood down over your neck.
2. Fully extend the straps on the face piece. **(Step 1)**
3. Rest your chin in the chin pocket at the bottom of the mask. **(Step 2)**
4. Fit the face piece to your face, bringing the straps or webbing over your head. **(Step 3)**
5. Tighten the lowest two straps. To tighten, pull the straps straight back, not out and away from your head. **(Step 4)**
6. Tighten the pair of straps at your temple, if any.
7. If your model has additional straps, tighten the top strap(s) last. **(Step 5)**
8. Check for a proper seal. This process depends on the model and type of face piece you use. **(Step 6)**
9. Pull the protective hood up so it covers all bare skin. Be sure it does not get under your face piece or obscure your vision.
10. Replace your helmet and secure the chin strap. **(Step 7)**
11. Install the regulator on your face piece or attach the low-pressure air supply hose to the regulator. **(Step 8)**

2-7 Skill Drill

Donning a Face Piece

Fully extend the straps on the face piece.

Place your chin in the chin pocket.

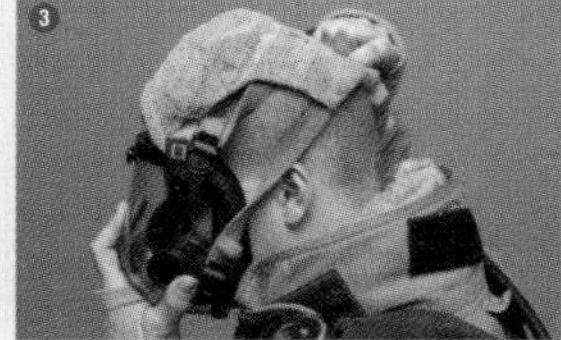

Fit the face piece to your face, bringing the straps or webbing over your head.

Tighten the lowest two straps.

Resource Preview

Chapter Resources (continued)

Fire Marks

This boxed feature offers history, lore, legends, and other interesting information.

Teamwork Tips

Teamwork Tips offer practical advice and information on teamwork and communication.

Teamwork Tips

At a minimum, one member of every company must remain oriented to their location in the building.

way into an apartment while the other members perform a systematic search inside. The searchers work their way around the apartment in one direction until they get back to the fire fighter at the door.

A guideline can be used for orientation when inside a structure. This is a rope attached to an object on the exterior or a known fixed location. The guideline is stretched out as a team enters the structure. They can follow the guideline back out or another team can follow the guideline in to find them. Some fire departments use a series of knots to mark distances along the rope to indicate how far it is stretched. The guideline technique requires intense practice.

Training sessions in conditions with no visibility will build confidence. This type of training can be conducted inside any building, while using an SCBA facemask with the lens covered. More realistic conditions can be simulated by using smoke machines inside training facilities. Fire fighters should practice navigating around furniture, following charged hose lines, climbing and descending stairs, and fitting through narrow spaces. Distracting noises, such as operating fans, can

larly so that you will be ready to perform them without hesitation if you are ever in a situation that requires them.

Disentanglement is an important skill that needs to be learned and practiced. When crawling in a smoke filled building, a fire fighter can easily become entangled in wires, cables, debris, or whatever may be hidden in the smoke. Many fire fighters carry small tools in the pockets of their PPE that can be used to cut through wires or small cables. This can be very difficult if visibility does not allow the entangling material to be seen and identified.

Some additional self-rescue methods involve using tools and equipment in manners that they were not designed for, such as the head-first "bail-out" technique onto a ground ladder. These are considered last resort methods. They should only be taught by experienced and qualified instructors and practiced with strict safety measures in place. There is controversy over several of these methods. Some fire departments consider them unacceptable due to the high risk of injury when they are used or even practiced. Other fire departments train in using them because they could save the life of a fire fighter in an extreme situation. This is a policy decision that must be made by each fire department.

Safe Havens

A safe haven is a temporary location that provides refuge while awaiting rescue or finding a method of self-rescue

Figure 23-1 In some fire departments, more than 80% of emergency calls are for emergency medical services.

Figure 23-2 EMS providers need to be supportive of the patient in all types of EMS calls.

Fire Marks

According to a survey of the International Association of Fire Chiefs (IAFC), 94% of their member's agencies provide EMS to their community. Of that, 56% provide patient transportation services.

Delivering emergency medical service fits directly into the fire department mission of protecting lives and property. Strict fire codes and effective public education efforts have been successful in reducing the numbers of fires. This has enabled the fire service to take a greater role in providing emergency medical care.

The same fire stations, emergency vehicles, and fire fighters that are available to respond quickly to fires can also be used to deliver assistance to people who need urgent medical care. Relatively small investments in EMS training and equipment can increase the fire department's ability to deliver life-saving services to the public.

In most communities, much of the funding for the fire department's operations comes from tax dollars paid by the citizens of that community. In this sense, the citizens of your community are your customers, and the services you provide should be customer-oriented. Their continued support is vital for the department's fire suppression and EMS activities. EMS functions deliver both an essential service to the citizens and positive public relations for the department. The community responds positively, and its support enables your department to maintain its effectiveness in all areas of operation.

When you respond to an emergency, you should realize that the citizen who called 9-1-1 probably had little or no training in identifying a medical emergency or knowing what to do to help a person in distress. Sometimes, what the caller perceives as an emergency is actually a minor incident. At other times, the call may not be made until the patient's condition is literally life-threatening. A competent and caring EMS provider will always remain supportive of the patient and the caller, even when the call was unnecessary or delayed. Remember, from the caller's perspective, this is an emergency (▲ Figure 23-2).

Levels of Service

Emergency medical services are provided at the basic life support (BLS) level and at the advanced life support (ALS) level.

Basic Life Support

Basic life support is the level of medical care that can be provided by persons trained to perform a limited set of emergency medical skills. These skills include:

- Scene control
- Evaluating conditions for responder safety
- Patient assessment
- Basic airway management techniques
- Cardiopulmonary resuscitation (CPR)
- Administering oxygen
- Splinting
- Controlling external bleeding and bandaging
- Treating for shock
- Lifting and moving patients (► Figure 23-3)
- Transporting patients to an appropriate medical facility

Figure 9-5 Utility ropes are used for hoisting and lowering tools.

Figure 9-6 Some utility ropes are made from natural fibers.

Because these escape ropes are so important, they can be used only once. After they are used once, they are discarded. The rope may have been damaged in some way and might fail if it ever has to be used again. You cannot take that chance; your life may depend on the quality and strength of your personal escape rope.

Utility Rope

Utility rope is used when it is NOT necessary to support the weight of a person. Fire department utility rope is used for hoisting or lowering tools or equipment, for ladder halyards (rope used on extension ladders to raise a fly section), for marking off areas, and for stabilizing objects (▲ Figure 9-5). Utility ropes also require regular inspection.

Utility ropes must not be used in situations where life safety rope is required. Conversely, life safety rope must not be used for utility applications. A fire fighter must be able to instantly recognize the category of a rope from its appearance and markings.

Fire Fighter Safety Tips

Many fire departments use color-coding or other visible markings to identify different types of rope. This allows a fire fighter to very quickly determine if a rope is a one-person or two-person life safety rope or a utility rope. The length of each rope should also be clearly marked by a tag or a label on the rope bag.

Rope Materials

Ropes can be made from many different types of materials. The earliest ropes were made from naturally occurring vines or fibers that were woven together. Ropes are now made of synthetic materials such as nylon or polypropylene. Because ropes have many different uses, different materials may work better than others in various situations.

Natural Fibers

In the past, fire departments used ropes made from natural fibers, such as manila, because there were no alternatives (▲ Figure 9-6). The natural fibers are twisted together to form strands. A strand may contain hundreds of individual fibers of different lengths. Today, ropes made from natural fibers are still used as utility ropes, but are no longer acceptable as life safety ropes (► Table 9-1). Natural fiber ropes can be weakened by mildew and deteriorate with age, even when properly stored. A wet manila rope can absorb 50% of its weight in water, making it very susceptible to deterioration. A wet natural fiber rope is very difficult to dry.

Synthetic Fibers

Since nylon was first manufactured in 1938, synthetic fibers have been used to make ropes. In addition to nylon, several newer synthetic materials such as polyester, polypropylene, and polyethylene are used in rope construction (► Figure 9-7). Synthetic fibers have several advantages over natural fibers (► Table 9-2). Synthetic fibers are generally stronger than natural fibers, so it may be possible to use a smaller diameter rope without sacrificing strength. Synthetic materials can also pro-

Fire Fighter Safety Tips

Fire Fighter Safety Tips reinforce safety concerns for both fire fighters and victims.

Fire Fighter Tips

Fire Fighter Tips provide advice from masters of the trade.

Canadian Perspectives

This boxed feature offers information that is specific to Canada.

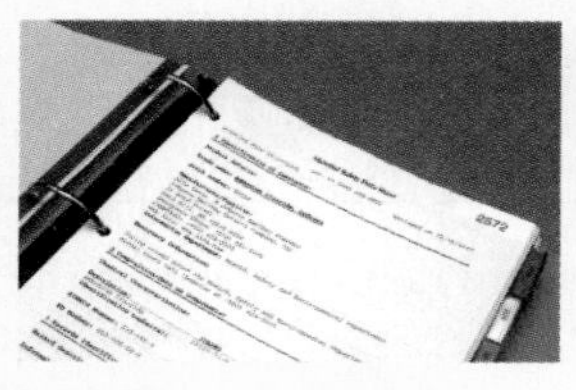

Figure 27-5 A material safety data sheet (MSDS) details the properties of a specific chemical.

hazardous wastes should have a minimum amount of training. Additionally, this law laid the foundation that ultimately allowed local fire departments and the community at large to obtain information on how and where hazardous materials were stored in their community. The Emergency Planning and Community Right to Know Act (EPCRA) requires a business that handles chemicals to report storage type, quantity, and storage methods to the fire department and the local emergency planning committee.

Fire Fighter Tips

You can't do it all! Different members of your department have different skill levels and capabilities. All personnel must use their collective talents and training at a hazardous materials incident.

Difference Between Hazardous Materials Incidents and Other Types of Emergencies

A critical distinction must be made between a hazardous materials event and other emergencies, such as a structure fire. Fire fighters cannot approach a hazardous materials incident with the same mindset used for a structure fire. Structural firefighting employs a fairly basic strategy, determined during the size-up phase, with the ultimate goal of extinguishing the fire. Hazardous materials incidents are more complicated. The strategies used must be thought out and planned in advance.

For example, most of the time, a hazardous materials emergency moves more slowly than a structure fire. If a rescue is required, or if the situation is imminently dangerous and requires rapid intervention, events may move quickly, but for the most part, a hazardous materials emergency requires time, forethought, and planning to ensure that the incident is han-

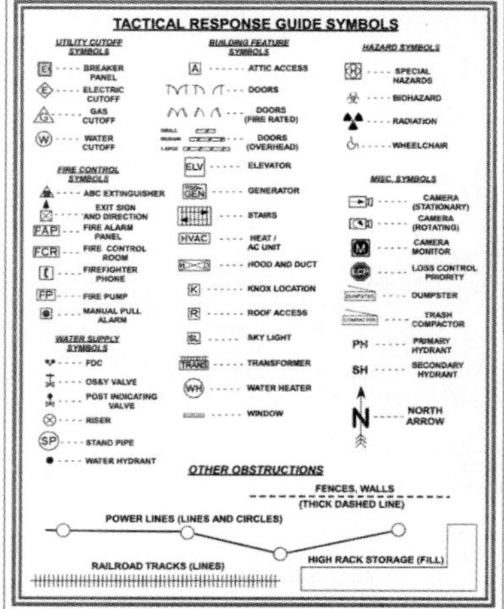

Figure 22-5 Preincident plan symbols.

mation to a final drawing. In some cases, the building owner can provide the survey team with a copy of a plot plan or a floor plan. Many fire departments also use digital cameras to record information during the survey.

The completed drawing should use standard, easily understood map symbols (▲ Figure 22-5 and ► Figure 22-6). Many fire departments use computer-assisted drawing software to create and store these diagrams.

Preincident Planning for Response and Access

Building layout and access information is particularly important during the response phase of an emergency incident. A preincident plan should provide information that would be valuable to units en route to an incident. For instance, the plan could identify the most efficient route to the fire building and also note an alternate route if the time of the day and local traffic patterns would affect the primary route. Alternate routes should also be

Canadian Perspectives

In Canadian communities, the Fire Underwriters' Survey, a division of the Insurers' Advisory Organization, determines the Public Fire Protection Classification (PFPC). The PFPC rating is then used by insurance companies in setting premiums. Forty percent of the PFPC grading is based on the fire department's effectiveness. A key element of this rating is the thoroughness and completeness of preincident planning within the community. The Insurance Services Office (ISO) rating in the United States is very similar, although pre-planning is not heavily weighted.

established if primary routes require crossing railroad tracks, drawbridges, or other potentially blocked routes.

When conducting a preincident survey, fire fighters should ensure that the building address is easily visible to save time in locating an emergency incident. If the building is part of a complex, the best route to each individual building, section, or apartment should be clearly indicated on a map. The locations of hydrants and fire department connections for sprinkler and standpipe systems should always be indicated.

Points of access into the building must also be noted, particularly if there are multiple entrances or if access points are not easily seen from the street. In some cases, different entrances should be used, depending on the location of the emergency within the building. In other cases, the first unit should always respond to a designated entrance where the fire alarm annunciator panel is located or to meet a security guard who can act as a guide.

Diagrams should clearly indicate where gates, fences, or other barriers block access to parts of a building. The preincident survey should also note whether the topography makes one or more sides of a building inaccessible to fire apparatus or if an underground parking area will not support the weight of apparatus.

Access to the Exterior of the Building

The preincident plan should address access to the exterior of the building. For example, you should ask the following questions during your survey:

- Are there several roads that lead to the building, or just a few?
- Where are hydrants located?
- Where are the fire department connections for automatic sprinkler and standpipe systems located?

Figure 7-12 Light hazard areas include offices, churches, and classrooms.

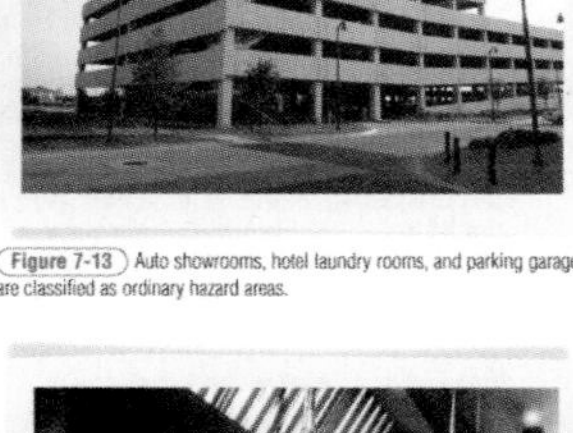

Figure 7-13 Auto showrooms, hotel laundry rooms, and parking garages are classified as ordinary hazard areas.

Ordinary or Moderate Hazard

The ordinary (or moderate) hazard locations contain more Class A and Class B materials than light hazard locations. Typical examples of ordinary hazard locations include retail stores with on-site storage areas, light manufacturing facilities, auto showrooms, parking garages, research facilities, and workshops or service areas that support light hazard locations, such as hotel laundry rooms or restaurant kitchens (► Figure 7-13).

Ordinary hazard areas also include warehouses that contain Class I and Class II commodities. Class I commodities include noncombustible products stored on wooden pallets or in corrugated cartons that are shrink-wrapped or wrapped in paper. Class II commodities include noncombustible products stored in wooden crates or multilayered corrugated cartons.

Extra or High Hazard

Extra (or high) hazard locations contain more Class A combustibles and/or Class B flammables than ordinary hazard locations. Typical examples of extra hazard areas include woodworking shops; service or repair facilities for cars, aircraft, or boats; and many kitchens and other cooking areas that have deep fryers, flammable liquids, or gases under pressure (► Figure 7-14). In addition, areas used for manufacturing processes such as painting, dipping, or coating, and facilities used for storing or handling flammable liquids are classified as extra hazard environments. Warehouses containing products that do not meet the definitions of Class I and Class II commodities are also considered extra hazard locations.

Figure 7-14 Kitchens, woodworking shops, and auto repair shops, are considered possible extra hazard locations.

Hot Terms

Hot Terms are easily identifiable within the chapter and define key terms the student must know. A comprehensive glossary of Hot Terms is in the end-of-chapter Wrap-Up. The Hot Term Explorer on **www.FireFighter.jbpub.com** provides interactivites for students.

Resource Preview

Chapter Resources (continued)

Wrap-Up

End-of-chapter activities reinforce important concepts and improve students' comprehension. Additional instructor support and answers to all activities are contained in the Instructor's Resource Manual.

Ready for Review summarizes chapter content to help students prepare for exams.

Hot Terms provide key terms and definitions from the chapter.

Wrap-Up

Ready for Review

This chapter examined the entire spectrum of a model Incident Management System. IMS is useful not only for large-scale operations, but also for any incident, regardless of size or type. All of the functions outlined in IMS must be addressed at every incident. It is the size and/or type of incident that dictates the degree to which each function is addressed. IMS allows the IC to delegate responsibilities in a standard incremental manner. The IC must perform any function that is not delegated. The IC's ultimate responsibility is to ensure that all incident requirements are met.

All fire fighters must understand the basic structure, concepts, and terminology of the IMS used by their departments and their role in its implementation. Company officers must be thoroughly trained in IMS and continually practice and refine the related skills.

Chief Concepts

- Several characteristics that are critical to an Incident Management System:
  - Organized approach: IMS imposes "order on chaos" on the fireground and enables a safer and more efficient operation than would be possible if personnel and units worked independently of each other.
  - Terminology: IMS uses a standard terminology for effective communications.
  - All-risk: IMS can be used at any type of emergency incident.
  - Jurisdictional authority: IMS enables different jurisdictions, agencies, and organizations to work cooperatively on a single incident.
  - Span of control: IMS maintains the desired span of control through flexible levels of organization.
  - Unity of command: Everyone reports to only one supervisor to avoid conflicts in giving orders.
  - Everyday applicability: IMS can and should be used on every single incident, every single time.
  - Modular: IMS is based on standard modules that are activated as needed to manage an incident.
  - Integrated communications: Everyone on the incident can communicate up and down the chain of command as needed.
  - Incident Action Plan: Every incident has a plan that outlines the strategic objectives. Large incidents will have formal plans.
  - Facilities: Standardized facilities, such as a command post, staging area, or rehabilitation area, are established as needed.
- Five major functions are part of IMS:
  - Command: Responsible for the entire incident; this is the only function that is always staffed.
  - Operations: Responsible for most fireground functions including suppression, search and rescue, and ventilation
  - Planning: Responsible for developing the Incident Action Plan
  - Logistics: Responsible for obtaining the resources needed to support the incident
  - Finance/Administration: Responsible for tracking expenditures and managing the administrative functions at the incident
- The Command Staff assists the Incident Commander at the incident:
  - Safety Officer: Responsible for overall safety of the incident; has the authority to stop any action or operation if it creates a safety hazard on the scene.
  - Liaison: Responsible for coordinating operations between the fire department and other agencies that may be involved in the incident
  - Information: Responsible for coordinating media activities and providing the necessary information to the various media organizations
- An example of a single resource would be an engine company or a ladder company.
- Single resources can be combined into task forces or strike teams.
- Other organizational units that can be established under IMS include divisions, groups, and sectors.
- IMS can be expanded infinitely to accommodate any size incident. Branches can be established to group similar functions, such as suppression, EMS, or hazardous materials.
- At every incident, someone must always assume command. As the incident grows or continues, it may be necessary to transfer command to another officer. This has to be done in a seamless manner to ensure continuity of command.

Hot Terms

Branch A supervisory level established to manage the span of control above the division, group, or sector level; usually applied to operations or logistics functions.

Branch director Officer in charge of all resources operating within a specified branch, responsible to the next higher level in the incident organization (either a Section Chief or the Incident Commander).

Command The first component of the IMS system. This is the only position in the IMS system that must always be staffed.

Command post The location at the scene of an emergency where the Incident Commander is located and where command, coordination, control, and communications are centralized.

Command staff Staff positions established to assume responsibility for key activities in the incident management system; individuals at this level report directly to the IC. Command staff include the Safety Officer, Public Information Officer, and Liaison Officer.

Company officer The officer in charge of a fire department company or station.

Crew An organized group of fire fighters under the leadership of a company officer, crew leader, or other designated official.

Designated incident facilities Assigned locations where specific functions are always performed.

Division An organizational level within IMS that divides an incident into geographic areas of operational responsibility.

Division supervisor The officer in charge of all resources operating within a specified division, responsible to the next higher level in the incident organization, and the point-of-contact for the division within the organization.

Finance/Administration Section The command-level section of IMS responsible for all costs and financial aspects of the incident, as well as any legal issues that arise.

Fireground Command System (FCS) An incident management system developed in the 1970s for day-to-day fire department incidents (generally handled with fewer than 25 units or companies).

FIRESCOPE (**FI**re **RES**ources of **C**alifornia **O**rganized for **P**otential **E**mergencies) An organization of agencies established in the early 1970s to develop a standardized system for managing fire resources at large-scale incidents such as wildland fires.

Group An organization level within IMS that divides an incident according to functional areas of operation.

Group supervisor The officer in charge of all resources operating within a specified group, responsible to the next higher level in the incident organization, and the point-of-contact for the group within the organization.

IMS General Staff The chiefs of each of the four major sections of IMS: Operations, Planning, Logistics, and Finance/Administration.

Incident Action Plan (IAP) The objectives for the overall incident strategy, tactics, risk management, and member safety that are developed by the IC. Incident Action Plans are updated throughout the incident.

Incident Commander (IC) The person in charge of the incident site who is responsible for all decisions relating to the management of the incident.

Incident Command System (ICS) The first standard system for organizing large, multi-jurisdictional and multi-agency incidents involving more than 25 resources or operating units; eventually developed into the Incident Management System (IMS).

Incident Management System (IMS) The combination of facilities, equipment, personnel, procedures, and communications under a standard organizational structure to manage assigned resources effectively to accomplish stated objectives for an incident. Also known as Incident Command System (ICS).

Integrated communications The ability of all appropriate personnel at the emergency scene to communicate with their supervisor and their subordinates.

Liaison Officer The position within IMS that establishes a point of contact with outside agency representatives.

Logistics Section The section within IMS responsible for providing facilities, services, and materials for the incident.

Logistics Section Chief The General Staff position responsible for directing the logistics function; generally assigned on complex, resource-intensive, or long-duration incidents.

Operations Section The section within IMS responsible for all tactical operations at the incident.

Operations Section Chief The general staff position responsible for managing all operations activities; usually assigned when complex incidents involve more than 20 single resources or when the IC cannot be involved in the details of tactical operations.

Passing command Option that can be used by the first-arriving company officer to direct the next arriving unit to assume command.

Planning Section The section within IMS responsible for the collection, evaluation, and dissemination of tactical information related to the incident and for preparation and documentation of incident management plans.

Chief Concepts provide the most important information from the chapter in a bulleted form to help students prepare for exams.

Fire Fighter in Action promotes critical thinking through the use of case studies and provides you with discussion points for your classroom presentation.

Wrap-Up

Ready for Review

This chapter addressed the phases of response and size-up. Preparations for an emergency response begin long before an alarm is received. Each fire fighter must ensure that all personal protective equipment and tools are ready for response. Adhering to simple procedures enables crews to arrive at an emergency quickly and safely.

Size-up is the ongoing mental evaluation of an emergency situation. The size-up process begins when information about an emergency incident is received. Size-up involves both an evaluation of the known facts of the situation, including the incident location, time, weather, type structure, exposures, available resources, and life hazard, and a consideration of any probabilities that could alter the situation. The incident commander processes this information and determines a plan of action. The incident action plan must address the fireground priorities of rescue, exposure protection, confinement, extinguishment, salvage, and overhaul.

Generally, a fire fighter is not expected to assess and take command of an emergency incident. However, understanding the size-up process will enable fire fighters to provide needed information to the fire officer and incident commander.

Chief Concepts

- Fire fighters must understand the need to receive dispatch information properly.
- As a fire fighter, your job is to respond safely to a scene. Follow your department's SOPs when riding in apparatus to the scene.
- Carefully mount and dismount all apparatus.
- Be aware of your surroundings at all times, especially when responding to an incident on the highway. Motorists might be distracted and not see you in the road.
- All information received at the communication center must be relayed to fire department units. What may seem insignificant to the dispatcher may be essential to the incident commander.
- Size-up is critical to the successful outcome of an emergency incident.

Hot Terms

Balloon-frame construction An older style of wood-frame construction that contains channels in the wall extending from the basement to the attic and providing a path for fire to spread.

Defensive attack Exterior fire suppression operations directed at protecting exposures.

Exposure An area adjacent to the fire that may become involved if not protected.

Extension Fire that moves into areas not originally involved, including walls, ceilings, and attic spaces; also the movement of fire into uninvolved areas of a structure.

Freelancing A fire fighter or unit operating without direction from the Incident Commander.

Offensive attack An advance into the fire building by fire fighters with hose lines or other extinguishing agents to overpower the fire.

Overhaul Examination of all areas of the building and contents involved in a fire to ensure that the fire is completely extinguished.

Personal alert safety system (PASS) Device worn by a fire fighter that sounds an alarm if the fire fighter is motionless for a period of time.

Personnel accountability system Procedure that tracks personnel on an emergency incident.

Personnel accountability tag (PAT) Identification card used to track the location of a fire fighter on an emergency incident.

Personal protective equipment (PPE) Gear worn by fire fighters that includes helmet, gloves, hood, coat, pants, SCBA, and boots. The personal protective equipment provides a thermal barrier for fire fighters against intense heat.

Preincident plan A written document based on general and detailed information and used by public emergency response agencies and private industry for determining the response to reasonable, anticipated emergency incidents at a specific facility.

Reconnaissance report The inspection and exploration of a specific area in order to gather information for the incident commander.

Rekindle A situation where a fire, which was thought to be completely extinguished, re-ignites.

Response Activities that occur in preparation for an emergency and continue until the arrival of emergency apparatus at the scene.

Salvage Removing or protecting property that could be damaged during fire fighting or overhaul operations.

Seat of the fire The main area of the fire

Size-up The ongoing mental evaluation of an emergency incident.

Thermal imaging devices Electronic devices that detect differences in temperature based on infrared energy and then generate images based on that data.

Response and Size-Up 295

Fire Fighter in Action

You have just arrived for your shift and received your riding assignment on the engine. While completing your morning inspections you are dispatched to a grease fire at a small drive-through hamburger restaurant. You prepare to mount the apparatus.

1. What preparations should you make at the beginning of each shift to make sure you are ready for emergency response?
   - A. Clean the station, and ensure you uniform is pressed and you boots are polished.
   - B. Have a shift meeting, then discuss meal plans and the time for physical fitness.
   - C. Inspect and place PPE in a designated location, inspect your SCBA and PASS device, and check apparatus and tools assigned to you.
   - D. Eat a good breakfast, wash the apparatus, and check the reserve apparatus.
2. What is the proper procedure for donning PPE and mounting the apparatus?
   - A. Jump on the apparatus as quickly as you can and don PPE in the cab to save time.
   - B. Place your PPE on the apparatus and don it after you arrive on the scene.
   - C. Don all of your PPE, always have at least one foot firmly placed on a foot surface, fasten your seat belt, and don your hearing protection.
   - D. Put your helmet on and jump on the apparatus. If you have a closed cab, you can don your PPE in the cab, fasten your seat belt once your PPE is on correctly, and, if provided, wear hearing protection.

You arrive on the scene and your lieutenant is performing size-up. The building is a small wooD framed structure that is fully involved. Lieutenant Hansen tells you to shut off the natural gas, pull a 1 $^{1}/_{2}$" preconnected hose line, and attack the fire.

3. To turn off the natural gas you should:
   - A. force the locking device and turn the valve so the handle is at a right angle to the pipe.
   - B. force the locking device and turn the valve so the handle is in line to the pipe.
   - C. turn off the main disconnection switch.
   - D. turn the underground valve counterclockwise with a wrench.
4. What are the five basic fire ground objectives, in order of priority?
   - A. Rescue any victims, confine the fire, extinguish the fire, protect exposures, salvage the property and overhaul the fire.
   - B. Rescue any victims, protect exposures, confine the fire, extinguish the fire, salvage the property and overhaul the fire.
   - C. Rescue any victims, extinguish the fire, confine the fire, protect exposures, salvage the property and overhaul the fire.
   - D. Rescue any victims, confine the fire, extinguish the fire, salvage the property and overhaul the fire, protect exposures.

www.FireFighter.jbpub.com

- Chapter Pretests
- Interactivities
- Hot Term Explorer
- Web Links
- Review Manual
- FireLearn

www.FireFighter.jbpub.com

Technology Resources guide exploration of topics online at **www.FireFighter.jbpub.com**, where additional activities reinforce and expand on important information from the chapter.

Resource Preview

Instructor Resources

A complete teaching and learning system developed by educators with an intimate knowledge of the obstacles you face each day supports *Fundamentals of Fire Fighter Skills*. The supplements provide practical, hands-on, time-saving tools like PowerPoint™ presentations, customizable test banks, and web-based distance learning resources to better support you and your students. This system includes the text plus the following resources:

Instructor's ToolKit CD-ROM

ISBN: 0-7637-2491-2

Preparing for class is easy with the resources found on this CD-ROM, including:

- **PowerPoint™ Presentations,** providing you with a powerful way to make presentations that are both educational and engaging to your students. The slides can be edited and modified to meet your needs.
- **Lecture Outlines,** providing you with complete, ready-to-use lesson plans that outline all of the topics covered in the text. The lesson plans can be modified and customized to fit your course.
- **Electronic Test Bank,** containing multiple-choice and scenario-based questions, allows you to originate tailor-made classroom tests and quizzes quickly and easily by selecting, editing, organizing, and printing a test along with an answer key, that includes page references to the text.
- **Image and Table Bank,** providing you with many of the images and tables found in the text. You can use them to incorporate more images into the PowerPoint™ presentations, make handouts, or enlarge a specific image for further discussion.

The resources found on the Instructor's ToolKit CD-ROM have been formatted so that you can seamlessly integrate them onto the most popular course administration tools. Please contact Jones and Bartlett technical support at any time with questions.

Instructor's Resource Manual

ISBN: 0-7637-2512-9

The Instructor's Resource Manual is your guide to the entire teaching and learning system. This indispensable manual contains:

- Detailed lesson plans that are keyed to the PowerPoint™ presentations with sample lectures, lesson quizzes, and teaching strategies.
- Teaching tips and ideas to enhance your classroom presentation.
- Answers to all end-of-chapter student questions found in the text.

Instructor's Test Bank (print)

ISBN: 0-7637-2554-4

This is the printed version of the Test Bank available on the Instructor's ToolKit CD-ROM. It contains multiple-choice and scenario-based questions with page references.

Student Resources

To help students retain the most important information and to assist them in preparing for exams, we have developed the following resources:

Student Workbook

ISBN: 0-7637-2556-0

This resource is designed to encourage critical thinking and aid comprehension of the course material through:

- Case studies and corresponding questions
- Figure labeling exercises
- Crossword puzzles
- Matching, fill-in-the-blank, short answer, and multiple-choice questions
- Skill Drill activities

Student Review Manual

Print – ISBN: 0-7637-2559-5

CD-ROM – ISBN: 0-7637-2558-7

Online – ISBN: 0-7637-2560-9

This Review Manual has been designed to prepare students for exams by including the same type of questions that they are likely to see on classroom and certification examinations. The manual contains multiple-choice question exams with an answer key and page references. It is available in print, on CD-ROM, and online.

Resource Preview

Technology Resources

A key component to the teaching and learning system are interactivities and simulations to help students become great fire fighters.

www.FireFighter.jbpub.com

Make full use of today's teaching and learning technology with **www.FireFighter.jbpub.com.** This site has been specifically designed to compliment *Fundamentals of Fire Fighter Skills* and is regularly updated. Some of the resources available include:

- **Chapter Pretests** to prepare students for training. Each chapter has a pretest and provides instant results, feedback on incorrect answers, and page references for further study.
- **Interactivities** allow your students to experience with the most important skills and procedures in the safety of a virtual environment.
- **Hot Term Explorer** is your virtual dictionary. Here, students can review key terms, test their knowledge of key terms through quizzes and flashcards, and complete crossword puzzles.
- **Web Links** provide up-to-date information for instructors and students including current trends in the fire service community and new equipment.
- **Review Manual** allows students to prepare for classroom and certification examinations providing instant results, feedback, and page references.

FireLearn

ISBN: 0-7637-3145-5

Reinforce your skills and knowledge online!

An eLearning component to *Fundamentals of Fire Fighter Skills*, FireLearn gives students the opportunity to study knowledge and critical thinking skills online. It brings Fire Fighter I and II material to life with interactive audio, video, Flash™ animation, photographs, illustrations, and case-based scenarios.

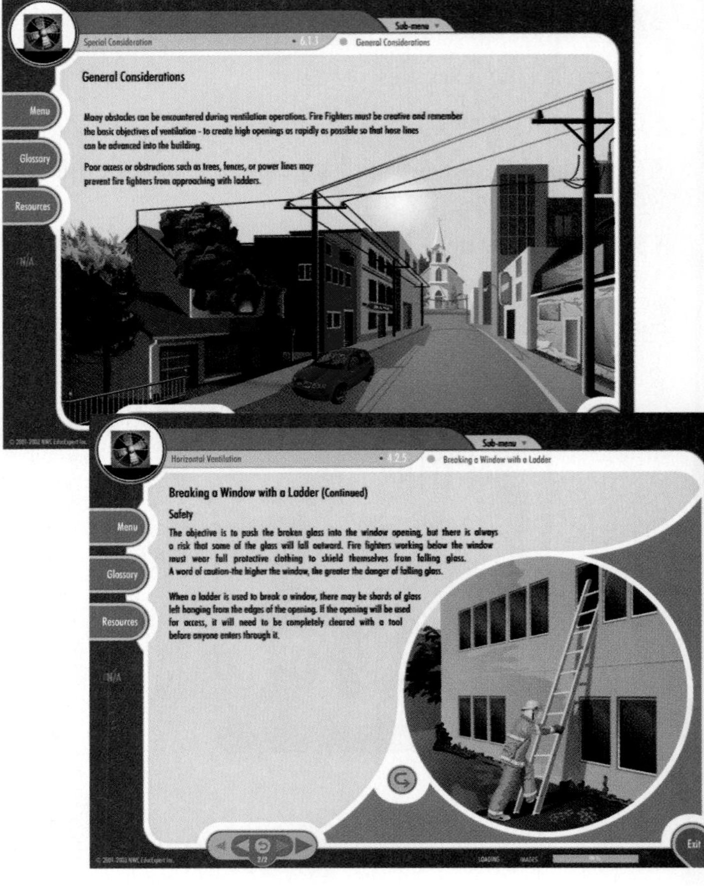

Acknowledgments

Editorial Board

Contributors

Christopher Armstrong
Safety International, Inc.
Pembroke Pines, Florida

Richard Avallone
Elevator Safety Training Programs Inc.
Coconut Creek, Florida

David Bentley
Fairfax County Fire and Rescue
Fairfax, Virginia

Ed Comeau
writer-tech
Belchertown, Massachusetts

Carl Cox
Sedgwick County Fire Department
Narragansett, Rhode Island

Richard Davis, PE
FM Global
Norwood, Massachusetts

Mike DeVirgillo
City of Tempe Fire Department
Tempe, Arizona

Dan Donahue
Paradise Valley Community College
Phoenix Fire Department
Phoenix, Arizona

Mike Estep
Fire Tech Metro
Cameron Park, California

Cathleen Connor Goetz
Eastern Connecticut Fire School
Willimantic, Connecticut

David Hall
Springfield Missouri Fire Department
Rogersville, Missouri

Timothy A. Hawthrone
Cranston Fire Prevention
Cranston, Rhode Island

Attlia J. Hertelendy
University of Mississippi Medical Center
Brandon, Mississippi

Judith Hoffmann
St. Joseph County Airport Authority (retired)
South Bend, Indiana

John E. Jackson
Lake Mohegan Fire District
Mohegan Lake, New York

Glenn P. Jirka
Energy Management
Columbia, Missouri

David King
Chesapeake Fire Department
Chesapeake, Virginia

Dennis Krebs
Baltimore County Fire Department
Baltimore, Maryland

Peter J. Lamb
Massachusetts Fire Academy
Stow, Massachusetts

Thomas E. Loraine, CHMM
Indiana Board of Fire Fighting Personnel Standards and Education
Rome City, Indiana

Loren Marshall
Tree Line Writer's Group
Anchorage, Alaska

John McPhee
Iowa Fire Service Training Bureau
Ames, Iowa

Scott J. Murray
Central Coventry Fire District
Coventry, Rhode Island

Ted J. Pagels
FireProse, LLC
De Pere, Wisconsin

Anthony J. Piontek
North East Wisconsin Technical College
City of Green Bay Fire Department
Green Bay, Wisconsin

Contributors (continued)

Jake Rixner
Richmond Fire Bureau
Richmond, Virginia

Gordon M. Sachs
Fairfield Fire and EMS
Fairfield, Pennsylvania

Robb Schnepp
Alameda County Fire Department
San Leandro, California

Ron Shaw
Extrication.com
Plymouth, Massachusetts

James B. Stevens
Dayton, Ohio

Lindsay Stromgen
Amherst Fire Department
Amherst, New Hampshire

Thomas B. Sturtevant, CFPS
ESTI-TEEX
Texas A&M University System
College Station, Texas

Francis "Sandy" Sullivan
Pinellas Park Fire Department
Pinellas Park, Florida

Mark J. Thiry
Ashwaubenon Department of Public Safety
Ashwaubenon, Wisconsin

Greg Thompson
South Bay Fire Department
Olympia, Washington

Sean Tracy
National Fire Protection Association
Ottawa, Ontario, Canada

Steering Committee

Terry Allen
Cambridge Fire Department
Cambridge, Ontario, Canada

Mary L. Corso
Office of the State Marshal
Olympia, Washington

Art Cota
California Department of Forestry and Fire Protection
Sacramento, California

Luther Fincher
Charlotte Fire Department
Charlotte, North Carolina

Tim Fuller
Minnesota State Fire Chiefs Association
St. Paul, Minnesota

Jimmy Hall
Bureau of Fire Prevention and Public Safety
Los Angeles Fire Department
Los Angeles, California

Maureen T. Hennessy
Prince George's County Fire/EMS Department
Largo, Maryland

Jerry Laughlin
Alabama Fire College
Tuscaloosa, Alabama

Jack Peltier
International Society of Fire Service Instructors
Stafford, Virginia

Bill Perrin
Stevensville Fire Department
Stevensville, Montana

Gary Tokle
National Fire Protection Association
Quincy, Massachusetts

Reviewers

Richard Guyton
Alabama Fire College
Tuscaloosa, Alabama

Paul Sanchez
State of Alaska, Fire Service Training
Fort Richardson, Alaska

Steve Schreck
State of Alaska, Fire Service Training
Anchorage, Alaska

Mike DeVirgilio
City of Tempe Fire Department
Tempe, Arizona

Dan Donahue
Paradise Valley Community College
Phoenix, Arizona

James Cantrell
Little Rock Fire Department
Little Rock, Arkansas

Richard Keller
Santa Ana College Basic Fire Academy
Santa Ana, California

John Linstrom
Victor Valley College
Apple Valley, California

Tracy E. Rickman, MPA
Rio Hondo College Fire Academy
Whittier, California

Todd R. Gilgren
Arvada Fire Protection Agency
Arvada, Colorado

Al Bassett
Connecticut Fire Academy
Windsor Locks, Connecticut

William M. Higgins
Connecticut Fire Academy
Windsor Locks, Connecticut

John F. McDonald, III
Submarine Base Fire Department
Groton, Connecticut

Joseph L. Murabito
Delaware State Fire School
Dover, Delaware

Michael Dashosh
Indian River Community College
Fort Pierce, Florida

Frank DeFrancesco
Largo Fire Rescue
Spring Hill, Florida

Bruce A. Grabbe
Department of Defense Firefighter Certification Program
Panama City, Florida

Ray Hansen
City of Pinellas Park Fire Department
Pinellas Park, Florida

Hugh Pike
United States Air Force
Fountain, Florida

Matt Scott
Sanibel Fire Rescue
Sanibel, Florida

Joey Hartley
Monroe County Emergency Services
Forsyth, Georgia

Edward Kalinowski
University of Hawaii
Honolulu, Hawaii

Brian Bauer
University of Illinois Fire Service Institute
Champaign, Illinois

Dan Hite
University of Illinois Fire Service Institute
Champaign, Illinois

John S. Lehman
Illinois Fire Service Institute
City of Aurora Illinois Fire Department
Aurora, Illinois

Paul Martin
Chicago Fire Department
Chicago, Illinois

Randy Brown
Angola Fire Department
Angola, Indiana

Tony Clark
Farmland Volunteer Fire Department
Farmland, Indiana

Rhonda Healy
Chandler, Indiana

Brett Healy
Chandler, Indiana

Judith Hoffmann
St. Joseph County Airport Authority (retired)
South Bend, Indiana

Jason Lewis
Oolitic Fire Department
Oolitic, Indiana

Thomas E. Loraine, CHMM
Indiana Board of Fire Fighting Personnel Standards and Education
Rome City, Indiana

Maynard D. Masters
Emergency Books and Training, Inc.
Indianapolis, Indiana

John D. McPhee
Iowa Fire Service Training Bureau
Ames, Iowa

Randy E. Novak
Iowa Fire Service Training Bureau
Ames, Iowa

Raymond M. Palczynski, Jr.
Davenport Iowa Fire Department
University of Illinois Fire Service Institute
Davenport, Iowa

Carl Cox
Kansas Fire and Rescue Training Institute
Sedgwick County Fire District #1
Lawrence, Kansas

William D. Hicks Jr.
Eastern Kentucky University
Richmond, Kentucky

Mark H. Cogswell
Massachusetts Firefighting Academy
Stow, Massachusetts

Donald A. Dominick
Henry Ford Community College
Northfield Township Fire Department
Whitmore Lake, Michigan

Michael F. Greis
Michigan Fire Fighters Training Council
Lansing, Michigan

Raymond C. Hoff
Illinois Fire Service Institute
Topinabee Michigan Fire Department
Topinabee, Michigan

Roger Land
Lake Superior University
Sault Ste. Marie, Michigan

Lawrence M. Morabito
Wayne County Community College District
Taylor, Michigan

Steve Morelan
Little Canada Fire Department
Little Canada, Minnesota

William H. Clark
Escatawpa Volunteer Fire Department
Moss Point, Mississippi

Chris Simpson
Choctaw Fire Department
Carthage, Mississippi

Michael Arnhart
High Ridge Fire Protection District
High Ridge, Missouri

David A. Hall
Springfield Missouri Fire Department
Rogersville, Missouri

Reviewers (continued)

Gary L. Wilson
Fire and Rescue Training Institute
University of Missouri
Columbia, Missouri

Bob Stober
Big Sky Fire Department (retired)
Big Sky, Montana

Curt Rohling
Southeast Community College
Lincoln, Nebraska

Bruce Evans
Community College of Southern Nevada
Henderson, Nevada

Gary M. Courtney
New Hampshire Technical College
Laconia, New Hampshire

Robert E. Ward
Essex County College Police Academy
Cedar Grove, New Jersey

Joseph D'Agosto
New York City Fire Department
Rescue Training Institute, Inc.
Bayside, New York

Paul R. Gerardi
Fairview Fire Department
White Plains, New York

Mark D. Gregory
New York City Fire Department
Suffolk County Fire Academy
New York, New York

S. Paul Iarocci
New Rochelle Fire Department
New Rochelle, New York

John E. Jackson
Lake Mohegan Fire District
Mohegan Lake, New York

Matthew B. Thorpe
City of King Fire Department
King, North Carolina

Sandy Waggoner
EHOVE Ghrist Adult Career Center
Milan, Ohio

Mark Ayers
Department of Public Safety, Standards and Training
Monmouth, Oregon

Gene Fisher
Chemeketa Community College
Salem, Oregon

Jim Whelan
Stanfield Fire Department
Stanfield, Oregon

William Shouldis
Philadelphia Fire Department
Philadelphia, Pennsylvania

Richard A. Blaine, Jr.
Rhode Island Fire Academy
Warwick Fire Department
Warwick, Rhode Island

Paul E. Calderwood
Providence College
Providence, Rhode Island

John Nicholas Carnegis
Rhode Island Fire Academy
North Providence Fire Department
North Providence, Rhode Island

Timothy A. Hawthorne
Cranston Fire Department
Cranston, Rhode Island

Scott J. Murray
Central Coventry Fire District
Coventry, Rhode Island

Kenneth J. Scandariato
Rhode Island Fire Academy
North Providence, Rhode Island

J. Curtis Varone
Providence College
Rhode Island Fire Academy
Exeter, Rhode Island

Dwayne Duncan
City of Green Fire Department
Green, South Carolina

James Rex Smith
Midway Fire Rescue
Pawleys Island, South Carolina

Barron D. Kennedy, III
Tennessee Fire Service and Codes Enforcement Academy
Tazewell, Tennessee

Robert Havens
Lamar Regional Fire Academy
Lamar Institute of Technology
Beaumont, Texas

Gary Kistner
San Antonio College
San Antonio, Texas

Jerry Ashbridge
Salt Lake City Fire Department
Salt Lake City, Utah

Stephen W. Carrier
Windsor Vermont Fire Department
White River Junction, Vermont

Peter Lynch
Vermont Fire Service Training
Pittsford, Vermont

James M. Armstrong
Roanoke City Fire Department
Roanoke, Virginia

Jeffrey H. Bailey
Loudoun County Fire and Rescue Recruit Training
Leesburg, Virginia

Thomas C. Fuqua
Roanoke Regional Fire and EMS Training Center
Roanoke, Virginia

James T. Ghi
Fire Station 25
Fredericksburg, Virginia

David I. King, Jr.
City of Chesapeake Fire Department
Chesapeake, Virginia

Michael M. Athey, Ed.D.
West Virginia University
Shepherdstown, West Virginia

Michael Caravasos
West Virginia University, Fire Service Extension
Morgantown Fire Department
Morgantown, West Virginia

Larry E. Huber
Lakeshore Technical College
Cleveland, Wisconsin

Ted Pagels
FireProse, LLC
DePene, Wisconsin

Mark J. Thiry
Northeast Wisconsin Technical College
Ashwaubenon Department of Public Safety
Green Bay, Wisconsin

Peter Sells
Toronto Fire Academy
Toronto, Ontario, Canada

Ray Unrau
Saskatoon Fire and Protective Services
Saskatoon, Saskatchewan, Canada

Photographic Contributors

Jim Brown, MA, EMT-B
Maryland Institute for Emergency Medical Services Systems
Baltimore, Maryland

Brian Slack, BA
Maryland Institute for Emergency Medical Services Systems
Baltimore, Maryland

Berta A. Daniels, BA
Maynard, Massachusetts

We would also like to thank the following people and organizations for their collaboration on the photo shoots for this project. Their assistance is greatly appreciated.

Tempe Fire Department
Tempe, Arizona

Plano Fire Department
Plano, Texas

Massachusetts Fire Fighting Academy
Stow, Massachusetts

Eastern Connecticut Fire Training School
Willimantic, Connecticut

Mohegan Tribe of Indians of Connecticut
Uncasville, Connecticut

Windham Center Fire Department
Windham, Connecticut

Margate Fire Rescue Department
Margate, Florida

Cascade Fire Equipment Company

Sterling Rope

Ver Sales, Inc.

The History and Orientation of the Fire Service

Chapter 1

NFPA 1001 Standard

Fire Fighter I

5.1.1 For certification at Level 1, the fire fighter candidate shall meet the general knowledge requirements in 5.1.1.1, the general skill requirements in 5.1.1.2, and the job performance requirements defined in Sections 5.2 through 5.5 of this standard and the requirements defined in Chapter 4, Competencies for the First Responder at the Awareness Level, of NFPA 472, Standard for Professional Competence of Responders to Hazardous Materials Incidents.

5.1.1.1 *General Knowledge Requirements.* The organization of the fire department; the role of the Fire Fighter I in the organization; the mission of fire service; the fire department's standard operating procedures and rules and regulations as they apply to the Fire Fighter I; the role of other agencies as they relate to the fire department; aspects of the fire department's member assistance program; the critical aspects of NFPA 1500, *Standard on Fire Department Occupational Safety and Health Program,* as they apply to the Fire Fighter I; knot types and usage; the difference between life safety and utility rope; reasons for placing rope out of service; the types of knots to use for given tools, ropes, or situations; hoisting methods for tools and equipment; and using rope to support response activities.

5.1.1.2 *General Skill Requirements.* The ability to don personal protective clothing within one minute; doff personal protective clothing and prepare for reuse; hoist tools and equipment using ropes and the correct knot; tie a bowline, clove hitch, figure eight on a bight, half hitch, becket or sheet bend, and safety knots; and locate information in departmental documents and standard or code materials.

5.5.1 (A) *Requisite Knowledge.* Organizational policy and procedures, common causes of fire and their prevention, the importance of fire safety survey and public education programs to fire department public relations and community, and referral procedures.

Fire Fighter II

6.1.1 For certification at Level II, the Fire Fighter I shall meet the general knowledge requirements in 6.1.1.1, the general skill requirements in 6.1.1.2, and the job performance requirements defined in Sections 6.2 through 6.5 of this standard and the requirements defined in Chapter 5, Competencies for the First Responder at the Operational Level, of NFPA 472, Standard for Professional Competence of Responders to Hazardous Materials Incidents.

6.1.1.1 *General Knowledge Requirements.* Responsibilities of the Fire Fighter II in assuming and transferring command within an incident management system, performing assigned duties in conformance with applicable NFPA and other safety regulations and authority having jurisdiction procedures, and the role of a Fire Fighter II within the organization.

6.2.2 Communicate the need for team assistance, given fire department communications equipment, standard operating procedures (SOPs), and a team, so that the supervisor is consistently informed of team needs, departmental SOPs are followed, and the assignment is accomplished safely.

Additional NFPA Standards

National Fire Incident Reporting System

NFPA 1500, *Standard on Fire Department Occupational Safety and Health Program*

Knowledge Objectives

After studying this chapter, you will be able to:

- Describe changes in the fire department from the colonial days to the present.
- Describe the four basic principles of organization of the fire department.
- Define the chain of command as it applies to a fire department.
- List the different types of fire department companies and describe their functions.
- Describe the roles of fire fighters within the fire department.
- Describe the fire department's regulations, policies, and standard operating procedures, and how they apply to the fire fighter.
- Locate information in departmental documents and standard operating procedures.
- Define the roles and responsibilities of Fire Fighter I and Fire Fighter II.
- List five guidelines for successful fire fighter training.

Skills Objectives

There are no skills objectives for this chapter.

You Are the Fire Fighter

As a newly hired fire fighter, you are asked to research practices of earlier fire departments. Your grandfather was a fire fighter in a large city in the 1950s, so you decide to visit him. He describes his early days as a new fire fighter, including life in the fire station and firefighting techniques. Some procedures have changed drastically since he joined the service, but others have not changed at all.

1. ***What aspects of fighting fires do you think have changed the most in the past 50 years?***
2. ***What things do you think have not changed in the past 50 years?***

Introduction

Training to become a fire fighter is challenging. The art and science of extinguishing fires is much more complex than most people imagine. You will be challenged—both physically and mentally—during this course. You must keep your body in excellent condition so you can complete your assignments, and you must remain mentally alert to cope with the various conditions you will encounter. Fire fighter training will expand your understanding of fire suppression, the activities involved in controlling and extinguishing fires.

By the time you complete this course, you will be equipped to continue a centuries-old tradition of preserving lives and property threatened by fires. You will have an understanding of the many duties and responsibilities of the fire service, and a feeling of personal satisfaction.

This chapter reviews the history and development of today's fire service. It explains the organization of the fire service and describes the chain of command that connects all members of a department. It also presents the responsibilities of a Fire Fighter I and a Fire Fighter II. Finally, this chapter offers five guidelines for you to remember as you progress through this course and take your place as a fire fighter in your community.

Fire Service in the United States

According to the National Fire Protection Association (NFPA), there are approximately 1.1 million fire fighters in the United States. Of this total, 27% are full-time, career fire fighters and 73% are volunteers. Three out of four career fire fighters work in communities with populations of 25,000 or more. More than half of the volunteer fire fighters work in fire departments that protect small, rural communities with populations of 2,500 or less.

There are approximately 30,000 fire departments throughout the United States. Fire departments may be composed of all career fire fighters, all volunteer fire fighters, or a mix of career and volunteer fire fighters, depending upon the community's needs and available resources (▼ Figure 1-1).

- All career—All of the members of the department are paid, full-time fire fighters.
- Combination departments:
  - Mostly career—These departments include both paid, full-time fire fighters and either on-call fire fighters or volunteers. More than half of the members are paid, full-time fire fighters.
  - Mostly volunteer—More than half of the members in these departments are volunteers, although they do include paid, full-time fire fighters as well.
- All volunteer—All of the members of the fire department are volunteer fire fighters.

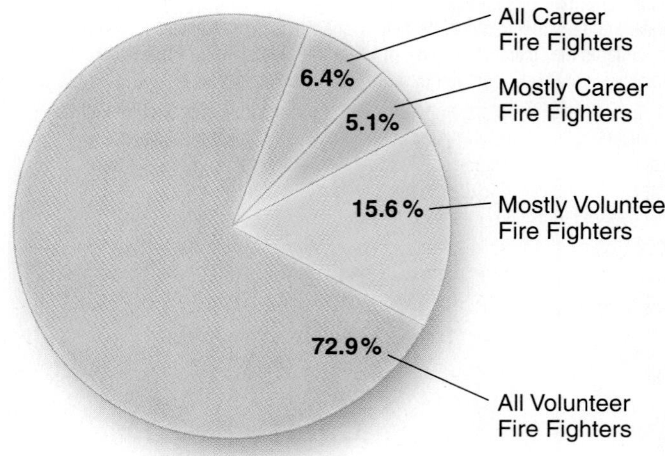

Figure 1-1 The majority of fire departments consist of fire fighters who are volunteers.

Fire Fighter Safety Tips

Fire Statistics for the United States

- There were 401,000 residential fires in the United States in 2002—an average of 46 residential fires every hour.
- A total of 2,695 people died in residential fires in 2002—one every 195 minutes.
- The total number of people killed in fires in 2002 was 3,380.
- Every 19 seconds, a fire department responds to a fire somewhere in the United States.
- An estimated 44,500 intentionally set structure fires occurred in 2002.
- There were 839,000 outside fires in 2002.

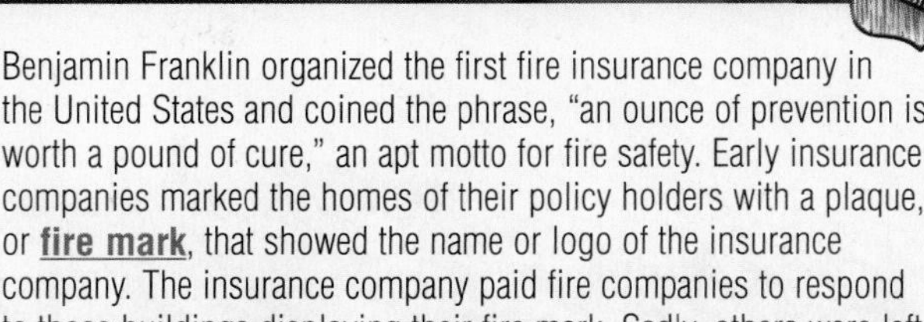

Fire Marks

Benjamin Franklin organized the first fire insurance company in the United States and coined the phrase, "an ounce of prevention is worth a pound of cure," an apt motto for fire safety. Early insurance companies marked the homes of their policy holders with a plaque, or **fire mark**, that showed the name or logo of the insurance company. The insurance company paid fire companies to respond to those buildings displaying their fire mark. Sadly, others were left to burn.

Canadian Perspectives

Statistical information on the composition of the fire service in Canada has been based upon information provided by provincial associations. Recently, however, the NFPA has begun to compile statistics. A study conducted by NFPA in December 2000 showed that there were a total of 101,535 fire fighters in Canada. Of these, 22,308 (22%) are career fire fighters and 79,227 (78%) are volunteers. Throughout the country, there are approximately 3,700 fire brigades.

The History of the Fire Service

Since prehistoric times, controlled fire has been a source of comfort and warmth, but uncontrolled fire has brought death and destruction. Historic accounts from the ancient Roman empire describe community efforts to suppress uncontrolled fire. In 24 BC, the Roman emperor Augustus Caesar created what was probably the first fire department. Called the *Familia Publica,* it was composed of about 600 slaves who were stationed around the city to watch for and fight fires. But because the *Familia Publica* was slaves, they had little interest in preserving the homes of their masters and little desire to take risks, so fires continued to be a problem. By about 60 AD, under the emperor Nero, the *Corps of Vigiles* was the fire protectors. This group of 7,000 free men was responsible for firefighting, fire prevention, and building inspections. The *Corps* adopted the formal rank structure of the Roman military, which is still used by fire departments.

The first documented fire in North America was in Jamestown, Virginia, in 1607. The fire started in the community blockhouse and almost burned down the entire settlement. At that time, most structures were built entirely of combustible materials such as straw and wood. Local ordinances soon required the use of less flammable building materials and mandated that fires be **banked**, or covered over, throughout the night. In 1630, Boston, Massachusetts, established the first fire regulations in North America when it banned wood chimneys and thatched roofs. In the Dutch colony of New York in 1647, Governor Peter Stuyvesant not only banned wood chimneys and thatched roofs but also required that chimneys be swept out regularly. **Fire wardens** imposed fines on those who did not obey the regulations; the money collected was used to pay for firefighting equipment.

The first paid fire department in the United States was established in 1679 in Boston, which also had the first fire stations and fire engines. The first volunteer fire company began in Philadelphia in 1735, under the leadership of Benjamin Franklin. There, citizens were required to keep buckets filled with water outside their doors, available to fight fires. Franklin recognized the many dangers of fire and continually sought ways to prevent it. For example, he developed the lightning rod to help draw lightning strikes away from homes, a common cause of fires. Another early volunteer fire fighter, George Washington, imported one of the first fire engines from England, which he donated to the Alexandria Fire Department in 1765.

In 1871, two major fires significantly affected the development of both the fire service and fire codes. At the time, the city of Chicago was a "boom town" with 60,000 buildings—40,000 of which were constructed wholly of wood, with roofs of tar and felt or wooden shingles. With the lax construction regulations and no rain for three weeks, the city was extremely dry. On October 8, 1871, a fire started in a barn on the west side of the city, which some say was caused by a cow kicking over a lantern. The fire department was already exhausted from fighting a four-block fire earlier in the day. Errors in judging the location of the fire and in signaling the alarm resulted in a delayed response time. The Great Chicago Fire burned through the city for three days (► **Figure 1-2**). When it was over, more than 2,000 acres and 17,000 homes had been destroyed, the city had suffered over $200 million in damage, and 300 people were dead and 90,000 homeless.

Figure 1-2 The Great Chicago Fire of 1871 caused the deaths of 300 people and led to changes in firefighting operations in America. © Chicago Historical Society #IC Hi-02808.

At the same time, another major fire was raging just 262 miles north of Chicago in Peshtigo, Wisconsin. Although not as highly publicized as the Chicago Fire, the Peshtigo Fire would become the deadliest fire in United States history. Throughout the summer, the north woods of Wisconsin had experienced drought-like conditions. Logging operations had left pine branches carpeting the forest floor. A flash forest fire created a "tornado of fire" more than 1,000′ high and five miles wide. More than 2,400 square miles of forest land burned, several small communities were destroyed, and more than 2,200 people lost their lives. The Peshtigo fire storm even jumped the 60-mile-wide Green Bay to destroy several hundred more square miles of land and settlements on Wisconsin's northeast peninsula.

Particularly as a result of the Great Chicago Fire, but also in response to the Peshtigo Fire, communities began to enact strict building and fire codes. The development of water pumping systems, advances in firefighting equipment, and improvements in communications and alarm systems helped ensure that such tragedies did not recur.

Fire Equipment

Today's equipment and apparatus developed over the years as new inventions were adapted to the needs of the fire service. Colonial fire fighters, for example, had only buckets, ladders, and **fire hooks** (tools used to pull down burning structures). Homeowners were required to keep buckets filled with water and to bring them to the scene of the fire. Some towns also required that ladders be available so that fire fighters could access the roof to extinguish small fires. If all else failed, the fire hook was used to pull down a burning building and prevent the fire from spreading to nearby structures. The "hook-and-ladder truck" evolved from this early equipment.

Buckets gave way to hand-powered pumpers in 1720, when Richard Newsham developed the first such pumper in London, England. Several strong men powered the pump, making it possible to propel a steady stream of water from a safe distance. In 1829, more powerful steam-powered pumpers replaced the hand-powered pumpers. Many volunteer fire fighters felt threatened by the steam engines and fought against their use. Steam engines were heavy machines that were pulled to the fire by a trained team of horses. They required constant attention, which limited their use to larger cities that could meet the costs of maintaining the horses and the steamers.

The advent of the internal combustion engine in the early 1900s greatly changed the fire service and enabled even small towns to have machine-powered pumpers. Today, both staffed and unstaffed firehouses have fire engines ready to respond at any hour of the day or night. Although they require regular maintenance, current equipment does not require the constant attention that horses or steam engines did. Modern fire apparatus carry water, a pumping mechanism, hoses, equipment, and personnel. A single apparatus has replaced several vehicles.

The progress in fire protection equipment extends beyond trucks. Without an adequate water supply, modern day apparatus would be helpless. The advent of municipal water systems provides large quantities of water to extinguish major fires.

Fire Marks

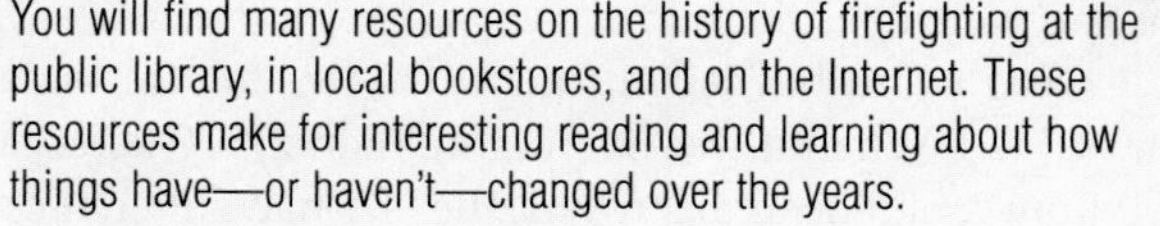

You will find many resources on the history of firefighting at the public library, in local bookstores, and on the Internet. These resources make for interesting reading and learning about how things have—or haven't—changed over the years.

Fire Marks

Buckets made of leather could not hold water for any length of time, so in some parts of the country, fire buckets were kept filled with sand. After the sand was thrown on the fire, the bucket would be refilled with water and used in a bucket brigade.

Romans developed the first municipal water systems, just as they had developed the first fire companies. But it wasn't until the 1800s that water distribution systems were applied to fire suppression efforts. George Smith, a fire fighter in New York City, developed the first fire hydrants in 1817. He realized that using a valve to control access to the water in the pipes would enable fire fighters to tap into the system when there was a fire. These valves, or **fireplugs**, were used with both above-ground and below-ground piping systems.

Because small fires are more easily controlled, the sooner a fire department is notified, the more likely it will be able to extinguish the fire and minimize losses. The introduction of public call boxes in Washington, DC in 1860 was a major advance. The call boxes, located around the city, enabled citizens to send a coded telegraph signal to the fire department in a series of bells. The fire department would know the location of the fire alarm box by the number of bells in the signal. When fire fighters arrived at the alarm box, the caller could direct them to the exact location of the fire. Similar units are still used in many areas, but are being replaced by more immediate and effective communications systems.

Communications

Communications are vital for effective firefighting. When a fire is discovered, fire fighters must be summoned and citizens alerted to the danger. During the firefight, officers must be able to communicate with fire fighters or summon additional resources. Improvements in communication systems are tied to the history of the fire service.

During the colonial period, a fire warden or night watchman patrolled neighborhoods and sounded the alarm if a fire was discovered. Some towns, including Charleston, South Carolina, built a series of fire towers where wardens would watch for fires. In many towns, ringing the community fire bell or church bells alerted citizens to a fire.

In the late 1800s, telegraph fire alarm systems were installed in large cities. These systems enabled more rapid reporting of fires and made it possible for officers to request additional resources or to let dispatchers know that the fire was extinguished. Telegraph fire alarm systems were gradually replaced by community sirens mounted on poles or tall rooftops to signal a fire.

These early communications systems have been replaced by hardwired and cellular telephones that enable citizens to report an emergency from almost anywhere (◄ Figure 1-3). Telephones have reduced the time and difficulty to report a fire. The introduction of computer-aided dispatch facilities has improved response times, because the closest available fire units can quickly be sent to the emergency.

Communications during the firefight also have improved over the years, from simply shouting loud enough to be heard over the chaos to using two-way radios. Before electronic amplification was available, the chief officer shouted commands through his trumpet. The **chief's trumpet**, or bugles, eventually became a symbol of authority (▼ Figure 1-4).

Figure 1-3 Modern technology allows citizens to report an emergency from almost any location.

Figure 1-4 The chief's trumpet was once used to amplify his voice. Today it is a symbol of authority in the fire service.

Fire Marks

The historic symbol of the chief officer's trumpet is still used on a chief's badge. This series of crossed trumpets is one of the cherished traditions of the fire service.

Although chief officers no longer use trumpets for communicating, the use of multiple trumpets symbolizing the rank of chief signifies the need to communicate as well as lead. Today's two-way radios enable fire units and individual fire fighters to remain in contact with each other at all times.

Building Codes

Throughout history, fires have served as an impetus for communities to establish building codes. Although the first building codes, developed in ancient Egypt, focused on preventing building collapse, building codes were quickly recognized as an effective means of preventing, limiting, and containing fires.

Colonial communities had few building codes. The first settlers had a difficult time erecting even primitive shelters, which often were constructed of wood with straw-thatched roofs. The fireplaces used for cooking and heating may have had chimneys constructed of smaller logs. The all-wood construction and open fires meant that fires were a constant threat. As communities developed, they enacted codes restricting the hours during which open fires were permitted and the materials that could be used for roofs and chimneys. In 1678, Boston required that "tyle" or slate be used for all roofs. After the British burned Washington, DC in 1814, codes prohibited the building of wooden houses. Some building codes required the construction of a fire-resistive wall, or firewall, of brick or mortar between two buildings.

Today's building codes not only govern construction materials, they also frequently require built-in fire prevention and safety measures. Required fire detection equipment notifies both building occupants and the fire department. Built-in fire suppression or sprinkler systems help contain fires to a small area and prevent small fires from becoming major fires. Fire escapes, stairways, and doors that unlock when the alarms sound and open outward enable occupants to escape a burning building safely. Without modern building code requirements, high-rise buildings and large shopping centers could not be built safely.

Code Development

At first, each community established its own building codes. As larger government jurisdictions were established, a basic, uniform code was adopted, reflecting a minimum standard. Local communities could make the code stricter as needed.

Today, codes and standards are written by national organizations such as the NFPA. Volunteer committees of citizens and representatives of businesses, insurance companies, and government agencies research and develop proposals that are debated and reviewed by various groups. The final document, or the **consensus document**, is then presented to the public. Many states adopt selected NFPA codes and standards as law.

Paying for Fire Service

As previously noted, volunteer fire departments were common in colonial America and are still used in many areas. The first fire wardens were employed by communities and paid from community funds. The introduction of hand-powered pumpers hastened the development of a permanent fire service. The question of who would pay for the equipment and the fire fighters, however, was not settled for many years.

Fire insurance companies were established in England soon after the Great Fire of London in 1666, to help cope with the financial loss from fires. The first fire insurance company in America was established in 1736 in Charleston, South Carolina, but it folded shortly after a major fire in 1740. The companies would collect fees (premiums) from homeowners and businesses, and pledged to repay the owner for any losses due to fire.

Because the insurance companies could save money if the fire was put out before much damage was done, they agreed to pay a fire company for trying to extinguish the fire. Houses that had insurance were designated with a fire mark (▶ Figure 1-5). Most fire companies were loosely governed and organized, and more than one company might show up to fight a fire. If two fire companies arrived at a fire, however, a dispute might arise over which company would collect the money. Consequently, municipalities began assuming the role of providing fire protection.

Today almost all the fire protection in this country is funded directly or indirectly through tax dollars. Some jurisdictions fund the fire department as a public service, while others contract with an independent fire department. Volunteer fire departments frequently conduct fund-raising activities. Ultimately, however, all funds for operating fire departments come from the citizens of the community.

Training and Education

Fire fighter training and education has also come a long way over the years. The first fire fighters required simply muscular strength and endurance to pass buckets or operate a hand pumper. As equipment became more complex, the importance of formalized training and good judgment increased.

Today's fire fighters operate high-tech, costly equipment, including apparatus, radios, **thermal imaging device** (a high-tech device that uses infrared technology to find

Figure 1-5 A fire mark indicated the homeowner had insurance that would pay the fire company for extinguishing the blaze.

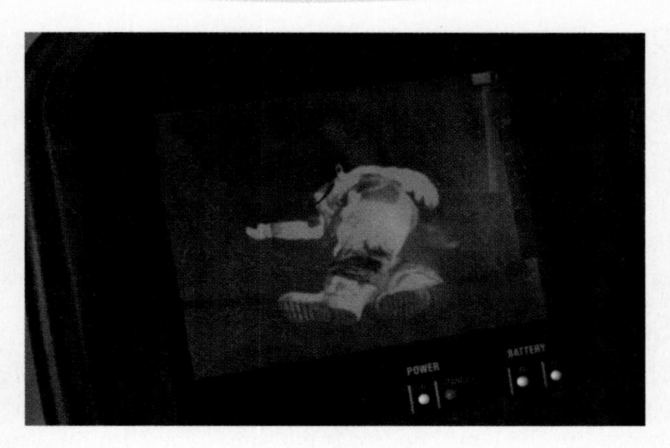

Figure 1-6 A thermal imaging device is one of the many tools available to modern fire fighters.

objects giving off a heat signature), and **self-contained breathing apparatus or SCBA** (air packs used by fire fighters to enter a hazardous atmosphere) ▲ Figure 1-6. These tools, as well as better fire detection devices, increase the safety and effectiveness of modern fire fighters. The most important "machines" on the fire scene, however, remain the knowledgeable, well-trained, physically capable fire fighters who have the ability and determination to attack the fire. A thermal imaging camera may be able to find someone trapped in a burning building, but it takes a smart, able-bodied fire fighter to remove the victim safely.

The increasing complexity of the world and the science of firefighting requires that fire fighters continually sharpen their skills and increase their knowledge of various hazards. That is why training courses such as this are just as important as good physical fitness to today's fire fighters.

The Organization of the Fire Service

Source of Authority

Governments—whether municipal, state, provincial, or national—are charged with protecting the welfare of the public against common threats. Fire is one such peril; an uncontrolled fire threatens everyone in the community. Citizens accept certain restrictions on their behavior and pay taxes to protect themselves and the common good. People charged with protecting the public may be given certain privileges to enable them to perform effectively. For example, fire departments can legally enter a locked home without permission to extinguish a fire and protect the public.

The fire service draws its authority from the governing entity responsible for protecting the public from fire—whether it is a town, a city, a county, a township, or a special fire district. Federal and state governments also grant authority to fire departments. In some states, private corporations have government contracts to provide fire services. The head of the fire department, the fire chief, is accountable to the leader of the governing body, such as the city council, the county commission, the mayor, or the city manager. Because of the relationship between a fire department and local government, fire fighters should consider themselves as civil servants, working for the tax-paying citizens who fund the fire department.

Basic Principles of Organization

The fire department uses a paramilitary style of leadership. Fire fighters operate under a rank system, which establishes a chain of command with the fire chief as head of the department. Most fire departments use four basic management principles:

1. Unity of command
2. Span of control
3. Division of labor
4. Discipline

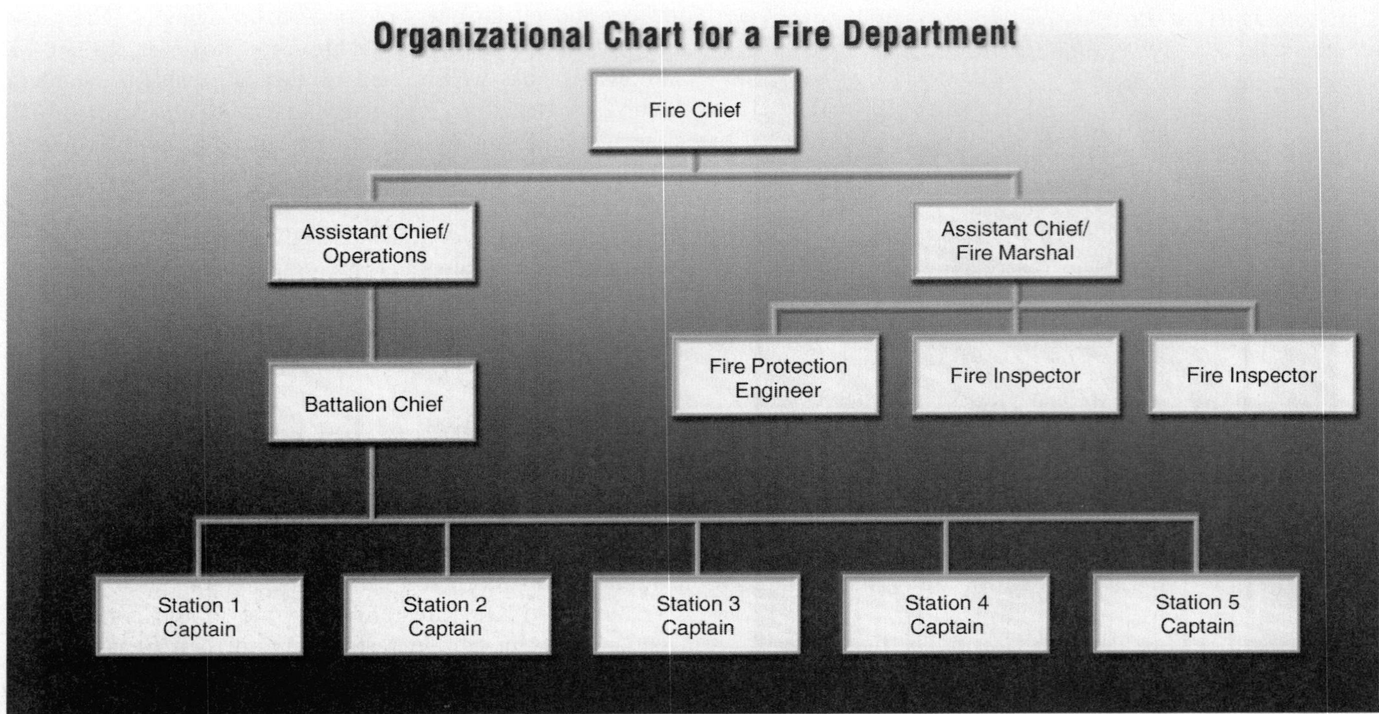

Figure 1-7 The organization of a typical fire department.

Unity of Command

Unity of command is the theory that each fire fighter answers to only one supervisor (▲ **Figure 1-7**). Each supervisor answers to only one boss, and so on. In this way, the chain of command ensures that everyone is answerable to the fire chief and establishes a direct route of responsibility from the chief to the fire fighter.

At a fire ground, all functions are assigned according to incident priorities. A fire fighter with more than one supervising officer during an emergency may be overwhelmed with various assignments and the incident priorities may not get accomplished in a timely and efficient manner.

Span of Control

Span of control is the number of people that one person can supervise effectively. Most experts believe that span of control should extend to no more than five people in a complex or rapidly changing environment, but this number can change, depending on the assignment or task to be completed.

Division of Labor

Division of labor is a way of organizing an incident by breaking down the overall strategy into smaller tasks. Some fire departments are divided into units based upon function. For example, the functions of engine companies are to establish water supplies and flow water; truck companies perform forcible entry and rescue functions. Each of these functions can be divided into multiple assignments, which can then be assigned to individual fire fighters. With division of labor, the specific assignment of a task to an individual makes that person responsible for completing the task and prevents duplication of job assignments.

Discipline

Discipline is the set of guidelines that a department establishes for fire fighters. Standard operating procedures, suggested operating guidelines, policies, and procedures are all forms of discipline, because they outline how things are to be done, and usually how far a person can go without requesting further guidance. Firefighting needs strong discipline, both positive and corrective, to operate safely and effectively.

Chain of Command

The organizational structure of a fire department consists of a **chain of command**. The ranks may vary in different departments, but the basic concept is the same. The chain of command creates a structure for managing the department and the fire ground operations. Fire fighters usually report to a **lieutenant**, who is responsible for a single fire company (such as an engine company) on a single shift. Lieutenants can provide a number of practical skills and tips to new recruits.

The next level in the chain of command is the **captain**. Captains are not only responsible for a fire company on their shift, but are also responsible for coordinating the company's activities with other shifts. A captain may be in charge of the activities of a fire station.

Captains report directly to chiefs. There are several levels of chiefs. Battalion chiefs, or district chiefs, are responsible for coordinating the activities of several fire companies in a defined geographic area such as a station or district. A battalion chief is usually the officer in charge of a single-alarm working fire.

Above battalion chiefs are assistant or division chiefs. Assistant or division chiefs are usually in charge of a functional area, such as training, within the department. These officers report directly to the chief of the department.

The chief of the department has overall responsibility for the administration and operations of the department. The chief can delegate responsibilities to other members of the department, but is still responsible for assuring that these activities are properly carried out.

The chain of command is used to implement department policies. This organizational structure enables a fire department to determine the most efficient and effective way to fulfill its mission and to communicate this information to all members of the department (► Figure 1-8). Using the chain of command ensures that a given task is carried out in a uniform manner. Several different documents, including regulations, policies, procedures, and standard operating procedures, are used to achieve this goal and will be reviewed later in the chapter.

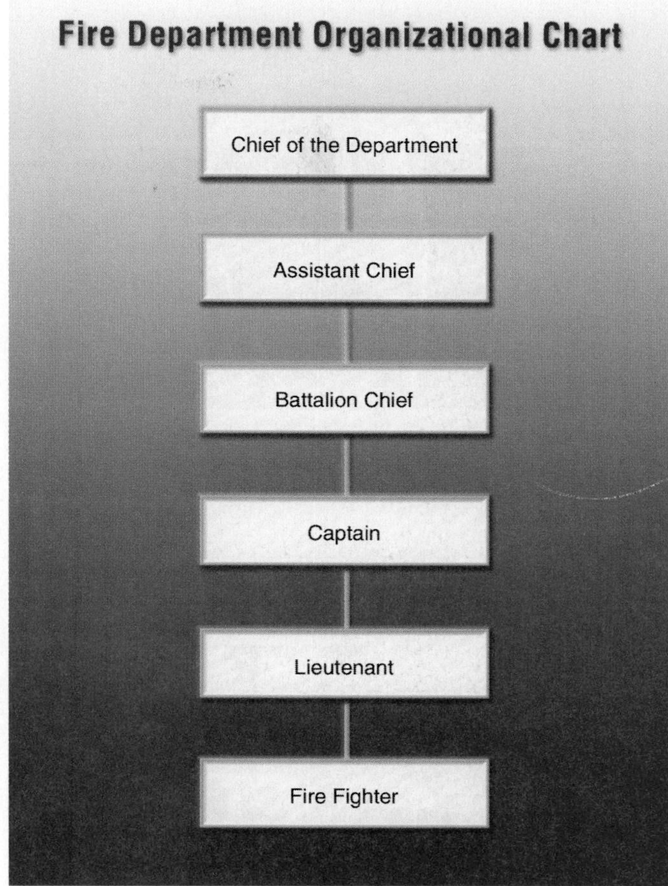

Figure 1-8 The chain of command ensures that the department's mission is carried out efficiently and effectively.

Company Types

There are many different types of companies in the fire department, each with its own job at the scene of an emergency. Companies may be composed of various combinations of people and equipment; in smaller departments, a company may fill many roles. The most common types of companies and their roles include:

- Engine company: An engine company is responsible for securing a water source, deploying handlines, conducting search-and-rescue operations, and putting water on the fire (► Figure 1-9). The fire engines have a pump, carry hoses, and a booster tank of water. Engines also carry a limited quantity of ladders and hand tools.
- Truck company: A truck company specializes in forcible entry, ventilation, roof operations, search-and-rescue operations above the fire, and deployment of ground ladders (► Figure 1-10). Trucks carry several ground ladders, ranging from 8′ to 50′ in length and an extensive quantity of tools. Trucks are also equipped with an aerial device, such as an aerial ladder, tower ladder, or ladder/platform. These aerial devices can be raised and positioned above a roof to provide fire fighters with a stable, safe work zone. They are also called ladder companies.

Figure 1-9 An engine company secures a water source and extinguishes the fire.

Figure 1-10 A truck company provides aerial support at a fire.

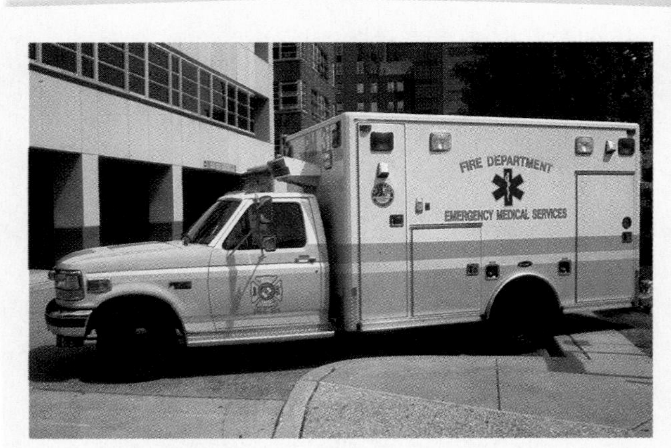

Figure 1-11 An EMS company delivers medical services.

- Rescue company: A rescue company usually is responsible for rescuing victims from fires, confined spaces, trenches, and high-angle situations. Rescue companies carry an extensive quantity of regular and specialized tools.
- Wildland/brush company: A wildland/brush company is dispatched to wildland and brush fires that larger engines cannot reach. Wildland/brush companies use four-wheel drive vehicles and carry special firefighting equipment such as portable pumps, rakes, shovels, and other specialized tools.
- Hazardous materials company: A hazardous materials company responds to and controls scenes involving spilled or leaking hazardous materials. They have special equipment, personal protective equipment (PPE), and training to handle most chemical emergencies.
- Emergency medical services (EMS) company: An EMS company may include medical units such as ambulances or first-response vehicles. They respond to and assist in transporting medical and trauma patients to medical facilities for further treatment. They often have medications, defibrillators, and other equipment that can stabilize a critical patient during transport. Engine or truck companies may be staffed with EMS providers as first responders until a transport ambulance arrives (◄ Figure 1-11).

Other Views of Organization

There are several different ways to look at the organization of a fire department.

Function

Fire departments can be organized along functional lines. For example, the training division is responsible for leading and coordinating department-wide training activities. Truck companies have certain functional responsibilities at a fire. Hazardous material squads have different functional responsibilities that support the overall mission of the fire department.

Geography

Each fire department is responsible for a specific geographic area, and each station in the department is responsible for a geographic area within the community. Fire stations are located throughout a community to ensure a rapid response time to every location in that community. This also enables the fire department to distribute and use specialized equipment efficiently throughout the community.

Staffing

A fire department must have sufficient trained personnel available to respond to a fire at any hour of the day, every day of the year. Staffing issues affect all fire departments—career departments, combination departments, and volunteer departments.

In volunteer departments, it is particularly important to ensure that there are enough responders available at all times, particularly during the day. In the past, when many people worked in or near the community where they lived, volunteer response time was not an issue. Today, people have longer commutes to work and work longer hours, so the number of people available to respond during the day may be limited. Some volunteer departments have had to hire full-time fire fighters during daytime hours to ensure sufficient personnel to respond to an incident.

Roles Within the Department

General Roles

A fire fighter may have many roles during a tenure, particularly in smaller departments. Some of the more common positions that fire fighters must assume include:

- Fire fighter: The fire fighter may be assigned any task from placing hose lines to extinguishing fires. Generally, the fire fighter is not responsible for any command functions and does not supervise other personnel, except on a temporary basis when promoted to an acting officer.
- Driver/operator: Often called an engineer, the driver is responsible for getting the fire apparatus to the scene safely, setting up, and running the pump or operating the aerial ladder once it arrives on the scene. This is a full-time role in some departments, but is rotated among the fire fighters in other departments.
- Company officer: This is usually a lieutenant or captain in charge of an apparatus. This person is in charge of the company both on scene and at the station. The company officer is responsible for initial firefighting strategy, personnel safety, and the overall activities of the fire fighters on their apparatus. Once command is established, the company officer focuses on tactics.
- Safety officer: The safety officer watches the overall operation for unsafe practices. He or she has the authority to stop any firefighting activity until it can be done safely and correctly. The senior ranking officer may act as safety officer until the appointed safety officer arrives or until another officer is delegated those duties.
- Training officer: The training officer is responsible for updating the training of current fire fighters and for training new fire fighters. He or she must be aware of the most current techniques of firefighting and EMS.
- Incident Commander: The incident commander is the individual responsible for the management of all the incident operations. This position focuses on the overall strategy of the incident and is often assumed by the Battalion/District Chief.
- Fire marshal/fire inspector/fire investigator: Fire marshals inspect businesses and enforce public safety laws and fire codes. They may respond to fire scenes to help investigate the cause of a fire. Investigators may have full police powers to investigate and arrest suspected arsonists and people causing false alarms.
- Fire and life safety education specialist: These individuals educate the public about fire safety and injury prevention and presents juvenile fire safety programs.
- 9-1-1 dispatcher/telecommunicator: From the communications center, dispatchers take calls from the public, send appropriate units to the scene, assist callers with emergency medical information (treatment that can be performed until the medical unit arrives), and assist the incident commander with needed resources.
- Fire apparatus maintenance personnel: Apparatus mechanics repair, service, and keep fire and EMS vehicles ready to respond to emergencies. They are usually trained by equipment manufacturers to repair vehicle engines, lights, and all parts of the fire pump and aerial ladders.
- Fire police: Fire police are usually fire fighters who control traffic and secure the scene from public access. Many fire police are also sworn peace officers.
- Information management: "Info techs" are fire fighters or civilians who take care of a department's computer and networking systems. Computerized fire reports, e-mail, and a functional computer network all require properly trained computer personnel.
- Public information officer: The public information officer serves as a liaison between the incident commander and the news media.
- Fire protection engineer: The fire protection engineer usually has an engineering degree, reviews plans, and works with building owners to ensure that their fire suppression and detection systems will meet code and function as needed. Some fire protection engineers actually design these systems.

Specialized Response Roles

Many assignments require specialized training. Most large departments have teams of fire fighters who respond to specific types of calls. Specialist positions include:

- Aircraft/crash rescue fire fighter: Aircraft rescue fire fighters are based on military and civilian airports and receive specialized training in aircraft fires, extrication, and extinguishing agents. They wear special personal protective equipment and respond in specialized fire apparatus that protects them from high-temperature fires caused by substances such as jet fuel.
- Hazardous materials technician: "Hazmat" technicians have training and certification in chemical identification, leak control, decontamination, and clean-up procedures.

Smallville Fire Department

Standard Operating Procedure

Date: 1/1/00
Section: 1 – Administration SOP 01-01 Page 1 of 2
Maintaining Station Logbooks

Purpose

This guideline is provided to ensure that information documented in station logbooks is maintained in a consistent manner, station to station, shift to shift, throughout the department. The station commander shall have discretion in formatting the information.

Scope

This guideline shall be followed by all authorized department personnel entering information into the station logbooks. Any deviations from this guideline will be the responsibility of the individual making the deviation.

Policy

The station log is an important component of the fire department's record-keeping system. All entries must be legible, written in black ink, and the writer identified by name and computer identification number. All members should treat the logbook as a legal document and record all of the station's activities as well as other information deemed important by the station commander.

Procedure

The following guidelines should be used when making logbook notations:

1. The day, date, and shift noted on the top line of the page.
2. A list of each person assigned to that shift to include their computer identification numbers, and the apparatus to which they are assigned for the shift. Personnel on leave should be identified with the type of leave (ie, annual, sick, leave without pay, funeral, military, training) noted. A notation should be made by the name of any member temporarily assigned to another station during that shift.
3. Daily safety talks should be recorded prior to the section dedicated to emergency responses. For more information on safety talks, refer to SOP 31.3.
4. The section of the log dedicated to recording chronological activity should have two (2) columns on the left side of the book: the far left should note time and the second column the incident number. A column to the far right should be used to note the identification number of the person making the entry.
5. Incident numbers for medic unit responses should be noted in blue ink.
6. Incident numbers for fire suppression responses should be noted in red ink.
7. Units responding, the address to which the response is being made, and the type of situation found should be noted behind the message number. (ie, E-5 responded to 304 Albemarle Drive/Gas Leak)
8. The term fill-in should be used in the logbook to denote one station standing by in another station to cover a company that has responded to an incident.
9. Entries for medic units should include the hospital to which the patient was transported.
10. Station maintenance/repairs and apparatus maintenance should be documented in the section of the logbook dedicated to that purpose. This documentation may also be maintained in a separate logbook dedicated to those types of entries.

Although the information requirements provided here must be maintained in the station's logbook, the format used for ensuring that documentation is left to the station commander's discretion.

Approved: ______________________________ Date: ______________________

Fire Chief

Figure 1-12 A sample Standard Operating Procedure.

- **Technical rescue technician**: A "tech rescue" technician is trained in special rescue techniques for incidents involving structural collapse, trench rescue, swiftwater rescue, confined-space rescue, high-angle rescue, and other unusual situations. They are sometimes called Urban Search and Rescue (USAR) teams.
- **SCUBA dive rescue technician**: Many fire departments, especially those around waterways, lakes, or an ocean, use SCUBA technicians who are trained in rescue, recovery, and search procedures in both water and under-ice situations. (SCUBA stands for self-contained underwater breathing apparatus.)
- **Emergency Medical Services (EMS) personnel**: EMS personnel administer prehospital care to people who are sick or injured. Prehospital calls account for the majority of responses in many departments, so fire fighters are often cross-trained with EMS personnel. EMS training levels are normally divided into three categories: Emergency Medical Technician-Basic, Emergency Medical Technician-Intermediate, and Emergency Medical Technician-Paramedic.
  - **Emergency Medical Technician – Basic (EMT-Basic)**: Most EMS providers are EMT-Basics. They have training in basic emergency care skills, including oxygen therapy, bleeding control, cardiopulmonary resuscitation (CPR), automated external defibrillation, basic airway devices, and assisting patients with certain medications.
  - **Emergency Medical Technician – Intermediate (EMT-Intermediate)**: EMT-Intermediates can perform more procedures than EMT-Basics, but they are not yet EMT-Paramedics. They have training in specific aspects of advanced life support, such as intravenous (IV) therapy, interpretation of cardiac rhythms, defibrillation, and airway intubation.
  - **Emergency Medical Technician – Paramedic (EMT-Paramedic)**: An EMT-Paramedic is the highest level of training in EMS. EMT-Paramedics have extensive training in advanced life support, including IV therapy, administering drugs, cardiac monitoring, inserting advanced airways (endotracheal tubes), manual defibrillation, and other advanced assessment and treatment skills.

Regulations, Policies, and Standard Operating Procedures

Regulations are developed by various government or government-authorized organizations to implement a law that has been passed by a government body. For example, federal occupational health and safety laws are adopted as regulations by some states. These regulations may apply to activities within the fire department.

Policies are developed to provide definite guidelines for present and future actions. Fire department policies outline what is expected in stated conditions. Policies often require personnel to make judgments and to determine the best course of action within the stated policy. Policies governing parts of a fire department's operations may be enacted by other government agencies, such as personnel policies that cover all employees of a city or county.

Standard operating procedures (SOPs) provide specific information on the actions that should be taken to accomplish a certain task (◄ Figure 1-12). SOPs are developed within the fire department, approved by the chief of the department, and ensure that all members of the department perform a given task in the same manner. SOPs provide a uniform way to deal with emergency situations, enabling fire fighters from different stations or companies to work together smoothly, even if they have never worked together before. They are vital because they enable everyone in the department to function properly and know what is expected for each task. Fire fighters must learn and frequently review departmental SOPs.

In some states, Suggested Operating Guidelines (SOGs) is the preferred terminology, because conditions may dictate that the fire fighter or officer use personal judgment in completing the procedure. This allows the responder to deviate from a set procedure and still be held accountable for that action.

A practical way to organize a department's SOP manual is with removable pages in a three-ring binder so that regular updates can be added easily. Each department member should have an SOP manual and must update it as needed. The SOP manual should be organized in sections—such as administration, safety, scene operations, apparatus and equipment, station duties, uniforms, and miscellaneous—and use a simple numbering system based on section and policy numbers.

Working with Other Organizations

Fire departments are part of the structure of the community. In order to fulfill its mission, the fire department must interact with other organizations in the community, particularly in situations such as motor vehicle crashes. In an area with a centralized 9-1-1 call center, the fire department, EMS, law enforcement officials, and tow truck operators will all be notified (► Figure 1-13). If the crash affects utility service, the utility company will be notified. If EMS is not part of the fire department, the EMS provider will be contacted. Fire fighters frequently interact with hospital personnel who assume care for the ill or injured patients who are transported to emergency rooms.

When multiple agencies work together, a unified command system must be established. This unified com-

Figure 1-13 Many agencies work together at a motor vehicle crash.

mand system is called the **Incident Management System (IMS)** and will be outlined later in this text. A unified command system eliminates multiple command posts, establishes a single set of incident goals and objectives, and ensures mutual communication and cooperation. There must be only one incident commander on scene, but each agency must have input in handling an emergency.

Other organizations may be involved in different types of incidents. Utility companies, the Salvation Army, and the Red Cross may be part of the team for structure fires. Wildland fires may involve representatives from various government agencies and jurisdictions. Large-scale incidents may call upon several agencies including:

- Public works
- School administrators
- Funeral directors
- Government officials (mayor, city manager)
- Federal Bureau of Investigation (FBI)
- Military
- Federal Emergency Management Agency (FEMA)
- Search and rescue teams
- Fire investigators
- Various state agencies

The communications center should keep a current list of contact names and phone numbers so they may be notified without delay.

Roles and Responsibilities of Fire Fighter I and Fire Fighter II

New fire fighters—whether they are training to become a Fire Fighter I or Fire Fighter II—must understand their roles and responsibilities. As you continue through this text, you will learn what to do and how to do it so that you can take your place confidently among the ranks.

Roles and Responsibilities for Fire Fighter I

- **Don** (put on) and **doff** (take off) personal protective equipment properly.
- Hoist hand tools using appropriate ropes and knots.
- Understand and correctly apply appropriate communication protocols.
- Use self-contained breathing apparatus (SCBA).
- Respond on apparatus to an emergency scene.
- Force entry into a structure.
- Exit a hazardous area safely as a team.
- Set up ground ladders safely and correctly.
- Attack a passenger vehicle fire, an exterior Class A fire, and an interior structure fire.
- Conduct search and rescue in a structure.
- Perform ventilation of an involved structure.
- Overhaul a fire scene.
- Conserve property with salvage tools and equipment.
- Connect a fire department engine to a water supply.
- Extinguish incipient Class A, Class B, Class C, and Class D fires.
- Illuminate an emergency scene.
- Turn off utilities.
- Perform fire safety surveys.
- Clean and maintain equipment.
- Present fire safety information to station visitors, community groups, or schools.

Every member of the fire service will interact with the public. In addition to dealing with citizens at the scene of incidents, people may visit the fire station, requesting a tour or asking questions on specific fire safety issues (► Figure 1-14). Fire fighters should be prepared to assist these visitors and use this opportunity to provide them with additional fire safety information. Use every contact with the public to deliver positive public relations and an educational message.

Additional Roles and Responsibilities for Fire Fighter II

- Coordinate an interior attack line team.
- Extinguish an ignitable liquid fire.
- Control a flammable gas cylinder fire.
- Protect evidence of fire cause and origin.
- Assess and disentangle victims from motor vehicle accidents.
- Assist special rescue team operations.
- Perform annual service tests on fire hose.
- Test the operability of and flow from a fire hydrant.

Figure 1-14 Any contact with the public should be used as an educational opportunity.

Fire Fighter Guidelines

During this course of study, you will need to practice and work hard. Do your best. The five guidelines listed below will help to keep you on target to become a proud and accomplished fire fighter.

1. **Be safe.** Safety should always be uppermost in your mind. Keep yourself safe. Keep your teammates safe. Keep the public you serve safe.
2. **Follow orders.** Your supervisors have more training and experience than you do. If you can be counted on to follow orders, you will become a dependable member of the team.
3. **Work as a team.** Fighting fires requires the coordinated efforts of each department member. Teamwork is essential to success.
4. **Think!** Lives will depend on the choices you make. Put your brain in gear. Think about what you are studying.
5. **Follow the Golden Rule.** Treat each person, patient, or victim as an important person or as a member of your family. Everyone is an important person or family member to someone and deserves your best efforts.

Wrap-Up

Ready for Review

The history of the fire service dates as far back as Roman civilization, and spans the entire history of the United States, from colonial days to the present. Advances in firefighting procedures, equipment, communications, building codes, payment systems, and training have helped shape today's fire service. The organization of a fire department includes different ranks, ranging from fire fighter to chief. The chain of command stretches from the chief to the newest member of the department.

Regulations, policies, and SOPs provide uniformity and consistent performance. Department personnel are often required to perform various functions to fulfill the mission of the fire department. Departments must be located to meet the geographic needs of the community and staffed appropriately to provide continuous protection. Many incidents will require the fire department to work smoothly with other organizations. The roles and responsibilities of Fire Fighter I and Fire Fighter II include both knowledge areas and technical skills.

Remember the five guidelines throughout your training and during your service as a fire fighter: be safe, follow orders, work as a team, think, and follow the Golden Rule.

Chief Concepts

- Although the mission of the fire service has remained the same, the equipment and means of achieving that mission have changed greatly since colonial days.
- Fire departments use an organizational chain of command.
- Fire departments are organized to perform multiple functions.
- You should know the roles and responsibilities of Fire Fighter I and Fire Fighter II.
- Each part of your fire fighter training builds upon the materials you have already learned. A team effort involving simultaneous actions is necessary for successful fire suppression.
- The five fire fighter guidelines will help you focus on becoming a good fire fighter.

Hot Terms

9-1-1 dispatcher/telecommunicator From the communications center, the dispatcher takes the calls from the public, sends appropriate units to the scene, assists callers with treatment instructions until the medic unit arrives, and assists the incident commander with needed resources.

Aircraft/crash rescue fire fighter Individuals with specialized training in aircraft fires, extrication, and extinguishing agents; they wear special types of turnout gear and respond in fire apparatus that protect them from high-temperature fires.

Assistant or division chief Midlevel chief who often has a functional area of responsibility, such as training, and answers directly to the fire chief.

Banked Covering a fire to ensure low burning.

Battalion chief Usually the first level of fire chief; also called district chief. These chiefs are often in charge of running calls and supervising multiple stations or districts within a city. A battalion chief is usually the officer in charge of a single-alarm working fire.

Captain The second rank of promotion, between the lieutenant and battalion chief. Captains are responsible for a fire company and for coordinating the activities of that company among the other shifts.

Chain of command A rank structure, from the fire fighter through the fire chief, for managing a fire department and fireground operations.

Chief of the department The top position in the fire department. The fire chief has ultimate responsibility for the fire department and usually answers directly to the mayor or designated public official.

Chief's trumpet An obsolete amplification device that enabled a chief officer to give orders to fire fighters during an emergency; precursor to a bullhorn and portable radios.

Company officer Usually a lieutenant or captain in charge of a team of fire fighters, both on scene and at the station. The company officer is responsible for firefighting strategy, safety of personnel, and the overall activities of the fire fighters on their apparatus.

Consensus document A code document developed through agreement between people representing different organizations and interests. NFPA codes and standards are consensus documents.

Discipline The guidelines that a department sets for fire fighters to work within.

Division of labor Breaking down an incident or task into smaller, more manageable tasks and assigning personnel to complete those tasks.

Doff To take off an item of clothing or equipment.

Don To put on an item of clothing or equipment.

Driver/operator Often called an engineer, this person operates the fire apparatus. The driver/operator is responsible for getting the apparatus safely to the scene, setting up, and operating the apparatus on scene.

Emergency medical services (EMS) company EMS companies may be made up of medical units and first-response vehicles. They respond to and assist in the transport of medical and trauma patients to medical facilities. They often have medications, defibrillators, and paramedics who can stabilize a critical patient.

Emergency medical services (EMS) personnel EMS personnel are those responsible for administering prehospital care to people who are sick and injured. Prehospital calls make up the majority of responses in most fire departments and EMS personnel are cross-trained as fire fighters.

Emergency Medical Technician (EMT)–Basic EMT-Basics account for most of the EMS providers in the country. EMT-Basics have training in basic emergency care skills, including oxygen therapy, bleeding control, cardiopulmonary resuscitation (CPR), automated external defibrillation, use of basic airway devices, and assisting patients with certain medications.

Emergency Medical Technician (EMT)–Intermediate EMT-Intermediate personnel can perform limited procedures that usually fall between those provided by an EMT- Basic and those provided by an EMT-Paramedic, including intravenous (IV) therapy, interpretation of cardiac rhythms, defibrillation, and airway intubation.

Emergency Medical Technician (EMT)–Paramedic An EMT-Paramedic has the highest level of training in EMS, including cardiac monitoring, administering drugs, inserting advanced airways, manual defibrillation, and other advanced assessment and treatment skills.

Engine company Engine companies are responsible for securing a water source, deploying hose lines, conducting search-and-rescue operations, and putting water on the fire.

Fire and life safety education specialist This person deals with the public on education, fire safety, and juvenile fire safety programs.

Fire apparatus maintenance personnel The people who repair and service the fire and EMS vehicles so that they are always ready to respond to emergencies.

Fire fighter This person is tasked with anything from hose line placement to extinguishing fires. Generally, the fire fighter is not responsible for any command functions or supervising other personnel.

Fire hook A tool used to pull down burning structures.

Fire mark Historically, an identifying symbol on a building to let fire fighters know that the building was insured by a company that would pay them for extinguishing the fire.

Fire marshal/fire inspector/fire investigator These people inspect businesses and enforce laws that deal with public safety and fire codes. Fire investigators may respond to fire scenes to help incident commanders investigate the cause of a fire. Investigators may have full police powers of arrest and deal directly with investigations and arrests.

Fire plug A valve installed to control water accessed from wooden pipes.

Fire police Fire police protect fire fighters by controlling traffic and securing the scene from public access. Many fire police are sworn peace officers as well as fire fighters.

Fire protection engineer The person in this position has received a college degree in one of several engineering disciplines. He or she is responsible for reviewing plans and working with building owners to ensure that the design of and systems for fire detection and suppression will meet code and function as needed.

Fire wardens Individuals designated to enforce fire regulations in colonial America.

Hazardous materials company Hazardous materials companies respond to and control scenes where hazardous materials spilled or leaked. Responders wear special suits and are trained to deal with most chemicals.

Hazardous materials technician "Hazmat" technicians are fire fighters who receive training in chemical identification, leak control, decontamination, and clean-up procedures. They would be called to handle the threat of weapons of mass destruction.

Incident commander The person in charge of the incident site who is responsible for all decisions relating to the management of the incident.

Incident management system (IMS) The combination of facilities, equipment, personnel, procedures, and communications under a standard organizational structure to manage assigned resources effectively to accomplish stated objectives for an incident. Also known as Incident Command System (ICS).

Wrap-Up

Information management Fire fighters or civilians who take care of all the computer and networking systems that a fire department needs to operate.

Lieutenant A company officer who is usually responsible for a single fire company on a single shift; the first in line of company officers.

Policies Formal statements that provide guidelines for present and future actions; policies often require personnel to make judgments.

Public information officer The position within IMS responsible for gathering and releasing incident information to the media and other appropriate agencies.

Regulations Rules, usually issued by a government or other legally authorized agency, that dictate how something must be done; regulations are often developed to implement a law.

Rescue company Rescue companies usually are tasked with rescue of victims from fires, confined spaces, trenches, and high-angle situations.

Safety officer The position within IMS responsible for identifying and evaluating hazardous or unsafe conditions at the scene of an incident. Safety officers have the authority to stop any activity that is deemed unsafe.

SCUBA dive rescue technician A responder trained to handle water rescues and emergencies, including recovery and search procedures, in both water and under-ice situations. (SCUBA stands for self-contained underwater breathing apparatus.)

Self-contained breathing apparatus (SCBA) Respirator with independent air supply used by fire fighters to enter toxic and otherwise dangerous atmospheres.

Span of control The number of people that a single person supervises. The maximum number of people that one person can effectively supervise is about five.

Standard operating procedures (SOPs) Written rules, policies, regulations, and procedures enforced to structure the normal operations of most fire departments.

Technical rescue technician Someone trained in special rescue techniques for incidents involving structural collapse, trench rescue, swiftwater rescue, confined space rescue, and other unusual rescue situations.

Thermal imaging device Electronic devices that detect differences in temperature based on infrared energy and then generate images based on that data. Commonly used in obscured environments to locate victims.

Training officer Responsible for updating the training of current employees and for training new fire fighters in the current techniques of firefighting and EMS.

Truck company Truck companies specialize in forcible entry, ventilation, roof operations, search-and-rescue operations above the fire, and deployment of ground ladders.

Unity of command A characteristic of the IMS structure that has each individual reporting to a single supervisor and everyone reporting to the IC directly or through the chain of command.

Wildland/brush company Wildland/brush companies are dispatched to woods and brush fires where larger engines cannot gain access. Wildland/brush companies have four-wheel drive vehicles and special firefighting equipment.

Fire Fighter in Action

You are excited about becoming a part of the fire service family. As you look around the fire station, you see many antiques and display cases illustrating the history and tradition of the fire service. You notice the fire service has changed to keep up with technology and current culture, but also is proud of its past and history. You feel that it is important to know the history of the fire service to better understand its mission and goals.

1. The Great Chicago Fire and the Peshtigo Fire lead to what changes in the fire operations of America?
   - A. Fire departments ensured that all homes have a fire bucket outside of their doors.
   - B. Insurance companies introduced the use of fire marks.
   - C. Communities enacted strict building and fire codes.
   - D. Legislation banned the use of wood chimneys.

2. The chief's trumpet is a symbol of:
   - A. authority.
   - B. communication center.
   - C. pull station location.
   - D. teamwork.

You have read your textbook on the history of the fire service, but want to learn more. Your uncle is a retired fire chief, so you decide to talk with him. He tells you about his experience and how much his fire department has progressed. You ask him what you should study next, and without hesitation he tells you to learn the chain of command. He says it is important to understand how you and all of the personnel fit in the organization, and what your responsibilities are to fulfill the mission.

3. What position is responsible for a functional area in the department and directly reports to the chief?
   - A. Lieutenant
   - B. Captain
   - C. Battalion chief
   - D. Assistant or division chief

4. What position is responsible for the overall management of incident operations?
   - A. Fire chief
   - B. Incident commander
   - C. Battalion chief
   - D. Safety officer

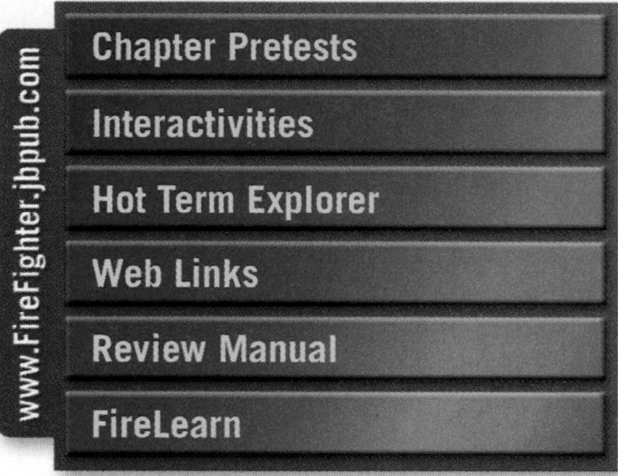

Fire Fighter Qualifications and Safety

Chapter 2

NFPA 1001 Standard

Fire Fighter I

5.1.1.2 *General Skill Requirements.* The ability to don personal protective clothing within one minute; doff personal protective clothing and prepare for reuse; hoist tools and equipment using ropes and the correct knot; tie a bowline, clove hitch, figure eight on a bight, half hitch, becket or sheet bend, and safety knots; and locate information in departmental documents and standard or code materials.

5.3.1 Use SCBA during emergency operations, given SCBA and other personal protective equipment, so that the SCBA is correctly donned and activated within one minute, the SCBA is correctly worn, controlled breathing techniques are used, emergency procedures are enacted if the SCBA fails, all low-air warnings are recognized, respiratory protection is not intentionally compromised, and hazardous areas are exited prior to air depletion.

5.3.1 (A) *Requisite Knowledge.* Conditions that require respiratory protection, uses, and limitations of SCBA, components of SCBA, donning procedures, breathing techniques, indications for and emergency procedures used with SCBA, and physical requirements of the SCBA wearer.

5.3.1 (B) *Requisite Skills.* The ability to control breathing, replace SCBA air cylinders, use SCBA to exit through restricted passages, initiate and complete emergency procedures in the event of SCBA failure or air depletion, and complete donning procedures.

5.5.3 Clean and check ladders, ventilation equipment, self-contained breathing apparatus (SCBA), ropes, salvage equipment, and hand tools, given cleaning tools, cleaning supplies, and an assignment, so that equipment is clean and maintained according to the manufacturer's or departmental guidelines, maintenance is recorded, and equipment is placed in a ready state or reported otherwise.

Fire Fighter II

6.1.1 For certification at Level II, the Fire Fighter I shall meet the general knowledge requirements in 6.1.1.1, the general skill requirements in 6.1.1.2, and the job performance requirements defined in Sections 6.2 through 6.5 of this standard and the requirements defined in Chapter 5, Competencies for the First Responder at the Operational Level of NFPA 472, *Standard for Professional Competence of Responders to Hazardous Materials Incidents.*

Additional NFPA Standards

NFPA 1404, *Standard for Fire Service Respiratory Protection Training*

NFPA 1500, *Standard on Fire Department Occupational Safety and Health Program*

NFPA 1582, *Standard on Medical Requirements for Fire Fighters and Information for Fire Department Physicians*

NFPA 1971, *Standard on Protective Ensemble for Structural Firefighting*

NFPA 1975, *Standard on Station/Work Uniforms for Fire and Emergency Services*

NFPA 1976, *Standard on Protective Ensemble for Proximity Firefighting*

NFPA 1977, *Standard on Protective Clothing and Equipment for Wildland Firefighting*

NFPA 1981, *Standard on Open-Circuit Self-Contained Breathing Apparatus for Fire and Emergency Services*

NFPA 1982, *Standard on Personal Alert Safety Systems (PASS)*

Knowledge Objectives

After completing this chapter, you will be able to:

- Discuss the educational, age, medical, physical fitness, and emergency medical care requirements for becoming a fire fighter.
- Describe how standards and procedures, personnel, training, and equipment are related to the prevention of fire fighter injuries and deaths.
- List safety precautions you need to take during training, during emergency responses, at emergency incidents, at the fire station, and outside your workplace.
- Describe the protection provided by personal protective equipment (PPE).
- Explain the importance of standards for PPE.
- Describe the limitations of PPE.
- Describe how to properly maintain PPE.
- Describe the hazards of smoke and other toxic environments.
- Explain why respiratory protection is needed in the fire service.
- Describe the differences between open-circuit breathing apparatus and closed-circuit breathing apparatus.
- Describe the limitations associated with self-contained breathing apparatus (SCBA).
- List and describe the major components of SCBA.
- Explain the skip-breathing technique.
- Explain the safety precautions you should remember when using SCBA.
- Describe the importance of daily, monthly, and annual SCBA inspections.
- Explain the procedures for refilling SCBA cylinders.
- List the steps for donning a complete PPE ensemble.

Skills Objectives

After completing this chapter, you will be able to perform the following skills:

- Don approved personal protective clothing.
- Doff approved personal protective clothing.
- Don an SCBA from a seat-mounted bracket.
- Don an SCBA from a side-mounted compartment.
- Don an SCBA from a storage case using the over-the-head method.
- Don an SCBA from a storage case using the coat method.
- Don a face piece.
- Doff an SCBA.
- Perform daily SCBA inspections.
- Perform monthly SCBA inspections.
- Replace an SCBA cylinder.
- Clean and sanitize an SCBA.

You Are the Fire Fighter

Before leaving the fire station to respond to a house fire, you don your personal protective clothing, board the apparatus, and fasten your seat belt. The fire is on the second floor. Your officer tells you and your partner to mount an interior attack, so you put on your self-contained breathing apparatus (SCBA) and stretch a hose line to the house. Smoke fills the downstairs as you make your way to the seat of the fire. You can feel the heat through your face piece and hear the sounds of breaking glass and crackling flames. You open the nozzle and direct a stream of water onto the fire to extinguish it. You continue to use your SCBA until the Safety Officer approves working without it.

1. ***How does your personal protective equipment keep you safe in this hostile environment?***
2. ***What are some of the limitations of your personal protective equipment?***

Fire Fighter Qualifications

Not everyone can become a fire fighter. Those who do understand the vital mission of the fire department: to save lives and protect property. A fire fighter must be healthy and in good physical condition, assertive enough to enter a dangerous situation, but mature enough to work as a member of a team (► Figure 2-1). The job requires a person who has the desire to learn, the will to practice, and the ability to apply the skills of the trade. A fire fighter is constantly learning as the body of knowledge about fires increases and the technology used in fighting fires develops.

The training and performance qualifications for fire fighters are specified in NFPA 1001, *Standard for Fire Fighter Professional Qualifications*. Age, education requirements, medical requirements, and other criteria are established locally.

Age Requirements

Most career fire departments require that candidates be at least 18 years of age, although some require a minimum age of 21. Candidates should possess a valid driver's license, have a clean driving record, have no criminal record, and be drug-free.

Volunteer fire departments usually have different age requirements, which often depend on insurance considerations. Some volunteer fire departments allow "junior" members to join at age 16, but restrict their activities until they reach age 18.

Education Requirements

Most career fire departments require that applicants have at least a high school diploma or equivalent. Some require candidates to have more advanced college-level courses in a fire- or EMS-related field. All departments will require fire fighters who wish to be considered for promotion or extra responsibility to take additional courses.

Medical Requirements

Because firefighting is both stressful and physically demanding, fire fighters must have a medical evaluation before training begins. The medical evaluation will identify any medical condition or physical limitation that could increase the risk of injury or illness to the candidate or other fire fighters. Medical requirements for fire fighters are specified in NFPA 1582, *Standard on Medical Requirements for Fire Fighters and Information for Fire Department Physicians.*

Physical Fitness Requirements

Physical fitness requirements are established to ensure that fire fighters have the strength and stamina needed to perform the tasks associated with firefighting and emergency operations. Several performance-testing scenarios have been validated by qualified organizations. NFPA 1001 allows indi-

Figure 2-1 Firefighting requires a special person.

vidual fire departments to choose the fitness-testing method that will be used for fire fighter candidates.

Emergency Medical Care Requirements

Delivering emergency medical care is an important function of most fire departments. NFPA 1001 allows individual fire departments to specify the level of emergency medical care training required for entry-level personnel. At a minimum, you will need to understand infection control procedures, perform CPR, control bleeding, and manage shock. These skills are discussed in more detail in Chapter 23, Fire and Emergency Medical Care. Many departments require fire fighters to become certified at the first responder, EMT-Basic, or higher levels.

Fire Fighter Safety

Firefighting, by its very nature, is dangerous. Each individual fire fighter must learn safe methods of confronting the risks presented during training exercises, on the fireground, and at other emergency scenes.

Every fire department must do what it can to reduce the hazards and dangers of the job and help prevent fire fighter injuries and deaths. Each fire department must have a strong commitment to fire fighter safety and health with designated personnel to oversee these programs. Safety must be fully integrated into every activity, procedure, and job description.

Appropriate safety measures must be applied routinely and consistently. During serious incidents, safety officers are responsible for evaluating the hazards of various situations and recommending appropriate safety measures to the Incident Commander (IC). Each accident or injury must be thoroughly investigated to learn why it happened and how it can be avoided in the future.

Advances in technology and equipment require fire departments to review and revise their safety policies and procedures regularly. Information reviews and research by designated safety personnel can identify new hazards as well as appropriate risk-management measures. Reports of accidents and fatalities from other fire departments can help identify problems and develop preventive actions.

Causes of Fire Fighter Deaths and Injuries

Each year about 100 fire fighters are killed in the line of duty in the United States. These deaths occur not only at emergency incident scenes, but also in the station, during training, and while responding to or returning from emergency situations. Approximately the same number of fire fighter deaths occur on the fireground or emergency scene as during training or while performing other nonemergency duties. The remainder, approximately 24%, occur while responding to or returning from alarms (► Figure 2-2). The leading cause of fire fighter deaths is heart attacks, both on and off the fireground.

Vehicle collisions are a major cause of fire fighter fatalities. For every 1,000 emergency responses, there is one vehicle collision involving an emergency vehicle (▼ Figure 2-3). One study found that 27% of the fire fighters who died in those incidents were ejected from the vehicle, which suggests that they were not using seatbelts. Fire fighters should never overlook basic safety procedures, such as always fastening seat belts, especially during emergency responses.

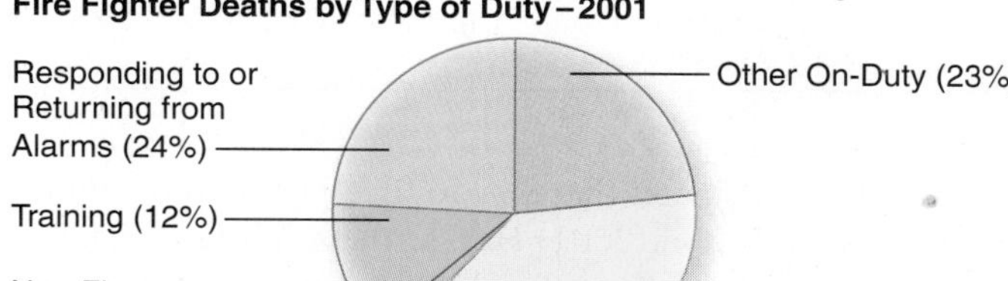

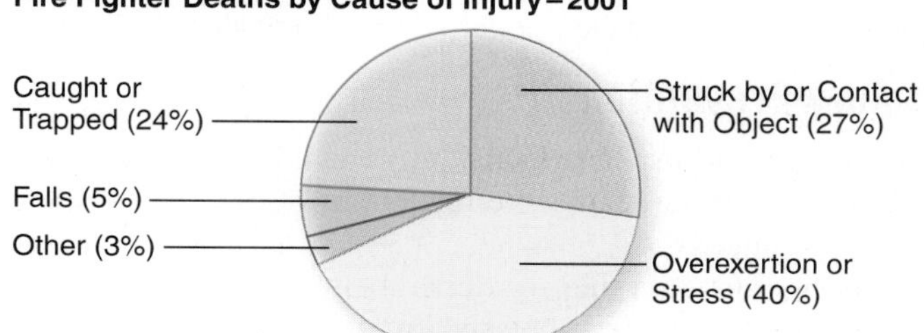

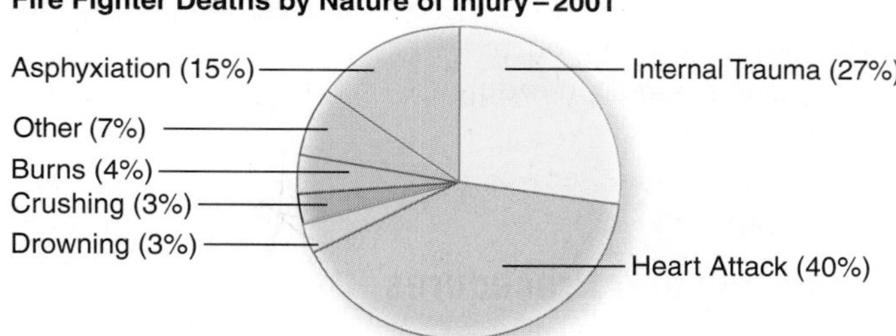

Figure 2-2 Fire fighter deaths in the United States.

Figure 2-3 Motor vehicle collisions are a leading cause of death for fire fighters.

Table 2-1 Fire Fighter Injuries

Type of Injury	% of total
Strains and sprains	40%
Soft-tissue injuries	22%
Burns	7.9%
Smoke and gas inhalation	6.2%

The NFPA estimates that 82,250 fire fighters were injured in the line of duty in 2001. Half of these injuries occurred while fighting fires and another 17% occurred at other emergencies. The rest occurred during other on-duty activities. The most common injuries were strains, sprains, and soft-tissue injuries. Burn injuries and smoke and gas inhalation made up only a small percentage of total injuries (▲ Table 2-1).

Injury Prevention

Injury prevention is a responsibility shared by each member of the firefighting team. Fire fighters must always consider:

- Personal safety
- The safety of other team members
- The safety of everyone present at an emergency scene

To reduce the risks of accidents, injuries, occupational illnesses, and fatalities, a successful safety program must have four major components:

- Standards and procedures
- Personnel
- Training
- Equipment

Standards and Procedures

Because safety is such a high priority, several organizations set standards for a safe working environment for the fire service. NFPA 1500, *Standard on Fire Department Occupational Safety and Health Program* provides a template for implementing a comprehensive health and safety program. Other NFPA standards focus on specific subjects directly related to health and safety. The Federal **Occupational Safety and Health Administration (OSHA)**, as well as various state and provincial health and safety agencies, develops and enforces government regulations on workplace safety. NFPA standards often are incorporated by reference in government regulations.

Every fire department should have a set of **standard operating procedures (SOPs)** or standard operating guidelines (SOGs), which outline how to perform various functions and operations. SOPs and SOGs cover a range of topics from uniform and grooming standards to emergency scene operations. These procedures should incorporate safe practices and policies. Each fire fighter is responsible for understanding and following these procedures.

The fire department chain of command also enforces safety goals and procedures. The command structure keeps everyone working toward common goals in a safe manner. The **Incident Management System (IMS)** is a nationally recognized plan to establish command and control of emergency incidents. Flexible enough to meet the needs of any emergency situation, IMS should be implemented at every emergency scene, from a routine auto accident to a major disaster involving numerous agencies.

Many fire departments have a health and safety committee responsible for establishing policies on fire fighter safety. Members of the committee should include representatives from every area, component, and level within the department, from fire fighters to chief officers. The safety officer and the fire department physician also should be members of the committee.

Personnel

A safety program is only as effective as the individuals who implement it. Personnel selection and training in the science of safe and effective fire suppression is a significant part of fire department operations and budgets.

Teamwork is an essential element of safe emergency operations. On the fireground and during any hazardous activity, fire fighters must work together to get the job done. The lives of citizens, as well as the lives of other members of the department, depend on compliance with basic safety concepts and principles of operation.

An overall plan also is essential to coordinate the activities of every team, crew, or unit involved in the operation. The IMS coordinates and tracks the location and function of every individual or work group involved in an operation.

Freelancing is acting independently of a superior's orders or the fire department's SOPs. Freelancing has no place on the fireground; it is a danger to both the fire fighter who acts independently and every other fire fighter. A fire fighter who freelances can easily get into trouble by being in the wrong place at the wrong time or by doing the wrong thing. For example, a fire fighter who enters a burning structure without informing a superior may be trapped by rapidly changing conditions. By the time the fire fighter is missed, it may be too late to perform a rescue. Searching for a missing fire fighter exposes others to unnecessary risk.

Safety officers are designated members of the fire department whose primary responsibility is safety. At the emergency scene, a safety officer reports directly to the IC and has the authority to stop any part of an action that is judged to be unsafe. Safety officers observe operations and conditions, evaluate risks, and work with the IC to identify hazards and ensure the safety of all personnel. Safety officers also determine when fire fighters can work without self-contained breathing apparatus (SCBA) after a fire is extinguished.

Safety officers contribute to safety in the workplace, at emergency incidents, and at training exercises. However, each member of the department shares the responsibility for safety, as an individual and as a member of the team.

Training

Adequate training is essential for fire fighter safety. The initial fire fighter training covers the potential hazards of each skill and evolution and outlines the steps necessary to avoid injury. Fire fighters must avoid sloppy practices or shortcuts that can contribute to injuries and learn how to identify hazards and unsafe conditions.

The knowledge and skills developed during training classes are essential for safety. The initial training course is only the beginning. Fire fighters must continually seek out additional courses and work to keep their skills current to ensure personal and team safety.

Equipment

A fire fighter's equipment ranges from power and hand tools to **personal protective equipment (PPE)** and electronic instruments. Fire fighters must know how to use equipment properly and operate it safely. Equipment also must be properly maintained. Poorly maintained equipment can create additional hazards to the user or fail to operate when needed.

Manufacturers usually supply operating instructions and safety procedures. Instructions cover proper use, limitations, and warnings of potential hazards. Fire fighters must read and heed these warnings and instructions. New equipment must meet applicable standards to ensure that it can perform under difficult and dangerous conditions on the fireground.

Safety and Health

Safety and well-being are directly related to personal health and physical fitness. Although fire departments regularly monitor and evaluate the health of fire fighters, each department member is responsible for personal conditioning and nutrition. Fire fighters should eat a healthy diet, maintain a healthy weight, and exercise regularly.

All fire fighters, whether paid or volunteer, should spend at least an hour a day in physical fitness training. Fire fighters should be examined by either a personal or departmental physician before beginning any new workout routine. An exercise routine that includes weight training, cardiovascular workouts, and stretching with a concentration on job-related exercises is ideal. For example, many fire fighters will use a stair-climbing machine and focus on the muscle groups used for firefighting. This builds strength and endurance on the fireground, but other muscle groups should not be neglected.

Hydration is an important part of every workout. A good guideline is to consume 8 to 10 ounces of water for every 5 to 10 minutes of physical exertion. Do not wait until you feel thirsty to start rehydrating. Fire fighters should drink up to a gallon of water each day to keep properly hydrated. Proper hydration enables muscles to work longer and reduces the risk of injuries at the emergency scene.

Diet is another important aspect of physical fitness. A healthy menu includes fruits, vegetables, low-fat foods, whole grains, and lean protein. Pay attention to portion sizes; most people eat larger portions than their bodies need. Substitute healthy choices such as fruit for high-calorie desserts.

Heart disease is the leading cause of death in the United States and among fire fighters. A healthy lifestyle that includes a balanced diet, weight training, and cardiovascular exercises helps reduce many risk factors for heart disease and enables fire fighters to meet the physical demands of the job (▼ Figure 2-4).

Many fire departments have adopted policies that prohibit the use of tobacco products by fire fighters, on-duty or off-duty. Smoking is a factor in cardiovascular disease, reduces the efficiency of the body's respiratory system, and increases the risk of lung cancer. Fire fighters should avoid tobacco products entirely for both health and insurance reasons.

Alcohol is another substance that fire fighters should avoid. Alcohol is a mood-altering substance that can be abused. Excessive alcohol use can damage the body and affect performance. Fire fighters who have consumed alcohol must not be permitted to engage in emergency operations.

Figure 2-4 Regular exercise will help you to stay healthy and perform your job.

Drug use has absolutely no place in the fire service. Many fire departments have drug-testing programs to ensure that fire fighters do not use or abuse drugs. The illegal use of drugs endangers your life, the lives of your team members, and the public you serve.

Everyone is subject to an occasional illness or injury. Fire fighters should not try to work when ill or injured. Operating safely as a member of a team requires fitness and concentration. Do not compromise the safety of the team or your personal health by trying to work while ill or injured.

Employee Assistance Programs

Employee assistance programs (EAPs) provide confidential help with a wide range of problems that might affect performance. Many fire departments have EAPs so that fire fighters can get counseling, support, or other assistance in dealing with a physical, financial, emotional, or substance abuse problem. An officer may refer a fire fighter to an EAP if the problem begins to affect job performance. Fire fighters who use an EAP can do so with complete confidentiality and without fear of retribution.

Safety During Training

During training, fire fighters learn the actual skills that are later used under emergency conditions. The patterns that develop during training will continue during actual emergency incidents. Developing the proper working habits during training courses helps ensure safety later.

Many of the skills covered during training can be dangerous if they are not performed correctly. According to the NFPA, an average of nine fire fighters are fatally injured during training exercises every year. Proper protective gear and teamwork are as important during training as they are on the fireground.

Instructors and veteran fire fighters are more than willing to share their experiences and advice. They can explain and demonstrate every skill and point out the safety hazards involved because they have performed these skills hundreds of times and know what to do. But here, too, safety is a shared responsibility. Do not attempt anything you feel is beyond your ability or knowledge. If you see something that you feel is an unsafe practice, bring it to the attention of your instructors or a designated safety officer.

Do not freelance on the training ground. Wait for specific instructions or orders before beginning any task. Do not assume that something is safe and act independently. Follow instructions and learn to work according to the proper procedures.

Teamwork is also important during training exercises. Assignments are given to firefighting teams during most live fire exercises. Teams must stay together. If any member of the team becomes fatigued, is in pain or discomfort, or needs to leave the training area for any reason, notify the instructor or safety officer. EMS personnel will be nearby to perform an examination and refer for further treatment if necessary. A fire fighter injured during training should not return until medically cleared for duty.

Safety During Emergency Response

When a company is dispatched to an emergency, fire fighters need to get to the apparatus and don appropriate PPE quickly before mounting the vehicle and proceeding to the incident. Walk quickly to the apparatus; do not run. Be careful not to slip and become injured before reaching the apparatus.

Personal protective gear should be properly positioned so you can don it quickly before getting into the apparatus (▼ Figure 2-5). Be sure that seat belts are properly fastened before the apparatus begins to move. All personnel responding on fire apparatus must be seated with seat belts fastened. Seat belts should remain fastened until the apparatus comes to a complete stop. Fire fighters can don SCBA while seated in some vehicles. Learn how to do this without compromising safety.

Drivers have a great responsibility. They must get the apparatus and the crew members to the emergency scene without having or causing a traffic accident en route. They must know the streets in their first-due area and any target hazards. They must be able to operate the vehicle skillfully and keep it under

Figure 2-5 Protective clothing should be properly positioned so you can quickly don it.

control at all times. They must anticipate all responses from other drivers who might not see or hear an approaching emergency vehicle or know what to do. Prompt response is a goal, but safe response is a much higher priority.

Volunteer fire department members must exercise extreme care when responding in private vehicles. Other volunteers will be responding at the same time, so you must be aware of their vehicles. Excessive speed is a major factor in many fire fighter fatalities involving private vehicles. The few seconds gained by speeding are not worth the risk created. A fire fighter responding in a private vehicle is legally required to comply with all traffic laws.

Safety at Emergency Incidents

At the emergency scene, fire fighters should never charge blindly into action. The officer in command will "size-up" the situation, carefully evaluating the conditions to determine if the burning building is safe to enter. Wait for instructions and follow directions for the specific tasks to be performed. Do not freelance or act independently of command.

Teamwork

On the fireground, a firefighting team should always consist of at least two fire fighters who work together and are in constant communication with each other. Some departments call this the **buddy system** (▼ Figure 2-6). In some cases, fire fighters work directly with the company officer, and all crew members function as a team. In other situations, two individual fire fighters may be a team assigned to perform a specific task. In either case, the company officer must always know where teams are and what they are doing.

Partners or assigned team members should enter together, work together, and leave together. If one member of a team must leave the fire building for any reason, the entire team must leave together, regardless if it is a two-person team or an entire crew working as a team.

Figure 2-6 A firefighting team should consist of at least two members who work together.

Before entering a burning building to perform interior search and rescue or fire suppression operations, fire fighters must be properly equipped with approved PPE. Partners should check each other's PPE to ensure it is on and working correctly before they enter a hazardous area.

Team members working in a hazardous area should maintain visual, vocal, or physical contact with each other at all times. At least one member of each team should have a portable **two-way radio** to maintain contact with the IC or a designated individual in the chain of command who remains outside the hazardous area. The radio can be used to relay pertinent information and to summon help if the team becomes disoriented, trapped, or injured. The IC can contact the team with new instructions or an evacuation order.

Fire fighters operating in a hazardous area require back-up personnel. The back-up team must be able to communicate with the entry team, either by sight or by radio, and ready to provide assistance. During the initial stages of an incident, at least two fire fighters must remain outside the hazardous area, properly equipped to respond immediately if the entry team has to be rescued.

As the incident progresses and additional crews are assigned to work in the hazardous area, a designated **rapid intervention company/crew (RIC)** should be established and positioned outside the hazardous area. This team's sole responsibility is to be prepared to provide emergency assistance to crews working inside the hazardous area.

Accountability

Every fire department should have a **personnel accountability system** to track personnel and assignments on the emergency scene. The system should record the individuals assigned to each company, crew, or entry team; the assignments for each team; and the team's current activities. Several kinds of accountability systems are acceptable, ranging from paper assignments or display boards to laptop computers and electronic tracking devices.

Some departments use a "passport" system. Each company officer carries a small magnetic board called a passport. Each crew member on duty with that company has a magnetic name tag on the board (► Figure 2-7). At the incident scene, the company's passport is given to a designated individual who uses it to track the assignment and location of every company at the incident.

An "accountability tag" system is often used by many fire departments. Each fire fighter carries a name tag and turns it in or places it in a designated location on the apparatus when the individual is on the scene. The tags are collected and used to track crews working together as a unit. This system works well when crews are organized based on the available personnel.

Both systems provide an up-to-date accounting of everyone who is working at the incident and how they are organized. At set intervals, an accountability check is performed

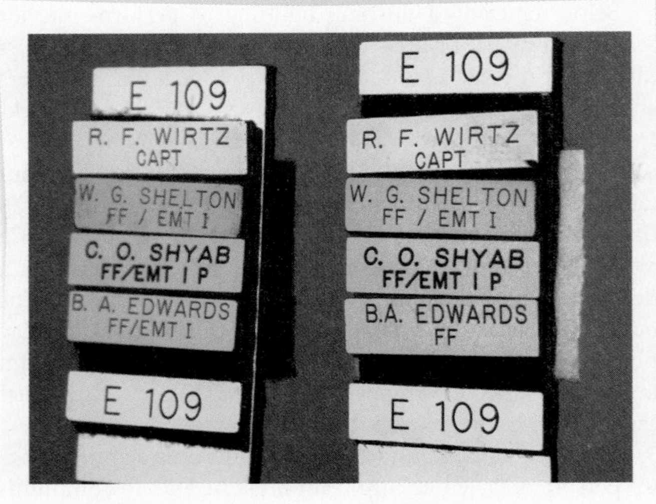

Figure 2-7 A passport lists each fire fighter assigned to a crew.

to account for everyone. Usually, a company officer reports on the status of each crew. The company officer should always know exactly where each crew is and what it is doing. If a crew splits into two or more teams, the company officer should be in contact with at least one member of each team.

An accountability check is also performed when there is a change in operational strategy or when a situation occurs that could endanger fire fighters. If an accountability check is needed, the list of personnel and assignments is available at the command post.

Fire fighters must learn their department's accountability system, how to work within it, and how it works within the IMS. Fire fighters are responsible for complying with the system and staying in contact with a company officer or assigned supervisor at all times. Teams must stay together.

Incident Scene Hazards

Fire fighters must be aware of their surroundings when performing their assigned tasks at an emergency scene. At an incident, make a safe exit from the apparatus and look at the building or situation for safety hazards such as traffic, downed utility wires, and adverse environmental conditions. An incident on a street or highway must first be secured with proper traffic- and scene-control devices. Flares, traffic cones, or barrier tape can keep the scene safe and the public at a safe distance. Always operate within established boundaries and protected work areas.

Changing fire conditions will also affect safety. During the overhaul phase and while picking up equipment, watch out for falling debris, smoldering areas of fire, and sharp objects. If a safety officer is not on the scene, another qualified person should be assigned to monitor the atmosphere for the presence of **carbon monoxide** (CO). CO is an odorless, colorless, tasteless gas that can cause asphyxiation, resulting in unconsciousness or death. Because the chance for injury increases when you are tired, do not let down your safety guard even though the main part of the fire is over.

Using Tools and Equipment Safely

Fire fighters must learn how to use tools and equipment properly and safely before using them at an emergency incident. Follow the proper procedures and safety precautions in training and at an incident scene. Use protective gear such as PPE, safety glasses, and hearing protection when they are required.

Proper maintenance includes sharpening, lubricating, and cleaning each tool. Equipment should always be in excellent condition and ready for use. Fire fighters should be able to do basic repairs such as changing a saw blade or a handlight battery. Practice these tasks at the fire station until you can perform them quickly and safely on the emergency scene.

Electrical Safety

Electricity is an emergency scene hazard that must always be respected. Many fires are caused by electricity, such as those ignited by faulty wiring or involving electric-powered equipment. Energized power lines may be present on the fireground. Fire fighters must always check for overhead power lines when raising ladders. During any fire, the electric power supply to the building should be turned off. This is part of the fireground task called "controlling the utilities."

Fire departments are often called to electrical emergencies, such as downed power lines, fires or arcing in transformers and switchgear, and stuck elevators. Always disconnect the power to any electrical equipment involved in an emergency incident.

Park apparatus outside the area and away from power lines when responding to a call for an electrical emergency. A downed power line should be considered energized until the power company confirms that it is dead. Secure the area around the power line and keep the public at a safe distance. Never drive fire apparatus over a downed line or attempt to move it using tools. If a sparking power line causes a brush fire, attempt to contain the fire but do not use water near the line.

Lifting and Moving

Lifting and moving objects are part of a fire fighter's daily duties. Do not try to move something that is too heavy alone—ask for help. Never bend at the waist to lift an object; always bend at the knees and use the legs to lift. Use equip-

Fire Fighter Safety Tips

Never be afraid to ask for help when moving a heavy object.

ment such as handcarts, hand trucks, and wheelbarrows to move objects a long distance.

Fire fighters must often move sick or injured patients. Discuss and evaluate the options before moving a patient, then proceed very carefully. If necessary, request help. Never be afraid to call for additional resources, such as an additional engine or truck company, to assist in lifting and moving a heavy patient.

Working in Adverse Weather Conditions

In adverse weather conditions, fire fighters must dress appropriately. A turnout coat and helmet can keep you warm and dry in rain, snow, or ice. Firefighting gloves and knit caps will also help retain body heat and keep you warm. If conditions are icy, make smaller movements, watch your step, and keep your balance.

During the hot summer months, many departments allow their members to wear cotton T-shirts and short pants. Check your department's SOPs.

Rehabilitation

Rehabilitation is a systematic process to provide periods of rest and recovery for emergency workers during an incident. Rehabilitation is usually conducted in a designated area away from the hazards of the emergency scene. The rehabilitation area, or "rehab," is usually staffed by EMS personnel.

A fire fighter who is sent to rehab should be accompanied by the other members of the crew. The company officer should inform the IC of their change in location. While in rehab, fire fighters should take advantage of the opportunity to rest, rehydrate, have their vital signs checked by EMS personnel, and have minor injuries treated (► **Figure 2-8**). Rehab gives fire fighters the chance to cool off in hot weather and to warm up in cold weather.

Rehabilitation time can be used to replace SCBA cylinders, obtain new batteries for portable radios, and make repairs or adjustments to tools or equipment. Firefighting teams can discuss recently completed assignments and plan their next work cycle. When a crew is released from rehab, they should be rested, refreshed, and ready for another work cycle. If the crew is too exhausted or unable to return to work, they should be replaced and released from the incident.

Never be afraid or embarrassed to admit you need a break when on the emergency scene. Heat exhaustion is a common condition, characterized by profuse sweating, dizziness, confusion, headache, nausea, and cramping. If a fire fighter shows the signs or symptoms of heat exhaustion, the company officer should be notified immediately. The company officer will request approval for rehabilitation so the problem can be treated.

Heat exhaustion is usually not life-threatening and, if identified early, can be remedied by rehydration, cooling, and rest. Left untreated, heat exhaustion can progress quickly to heat stroke, which is a life-threatening emergency. A lack of sweating, low blood pressure, shallow breathing, and seizures are some of the signs of heat stroke. If you or a member of your team experience these symptoms during training or on the fireground, place a "mayday" or "fire fighter in trouble" call. You must leave the fireground or training area and seek immediate medical attention.

A fire fighter who experiences chest pain or discomfort should stop and seek medical attention immediately. Heart attacks are the leading cause of death among fire fighters.

Violence at the Scene

Fire fighters are often dispatched to situations involving domestic disputes, injuries from an assault, or other violent scenes. The staging area and the apparatus should be some distance from the scene until the police arrive, investigate, and declare the area safe. Only then should fire fighters proceed into the emergency scene. A fire fighter's personal safety should always be paramount. If there is any threat to personal safety, slowly back away from the emergency scene to a safe distance and request the police to secure the scene. Do not become a victim.

If you are confronted with a potentially violent situation, do not respond violently. Remain calm, speak quietly, and attempt to gain the person's trust. You may consider taking additional classes to increase your understanding and develop appropriate skills for these situations.

Figure 2-8 In the rehabilitation area, fire fighters can rest and rehydrate.

Fire Fighter Safety Tips

Remember these guidelines to stay safe—on and off the job.

- **You are personally responsible for safety.** Keep yourself safe. Keep your teammates safe. Keep the citizens—your customers—safe.
- **Work as a team.** The safety of the firefighting unit depends on the efforts of each member. Become a dependable member of the team.
- **Follow orders.** Freelancing can endanger other fire fighters as well as yourself.
- **Think!** Before you act, think about what you are doing. Many people are depending on you.

Critical Incident Stress Debriefing

Some calls are particularly difficult and emotionally traumatic. Afterwards, fire fighters who were involved may be required to attend a critical incident stress debriefing (CISD). Usually, a stress debriefing is held as soon as possible after a traumatic call. It provides a forum for firefighting and EMS personnel to discuss the anxieties, stress, and emotions triggered by a difficult call. Follow-up sessions can be arranged for individuals who continue to experience stressful or emotional responses after a challenging incident.

Most fire departments have qualified, designated CISD staff available 24 hours a day. The initial CISD is usually a group session for all fire fighters and rescuers. It can also be done on a one-on-one basis or in smaller groups. Everyone handles stress differently, and new fire fighters need time to develop the personal resources to deal with difficult situations. If any call is emotionally disturbing, ask your superior officer for a referral to a qualified CISD counselor.

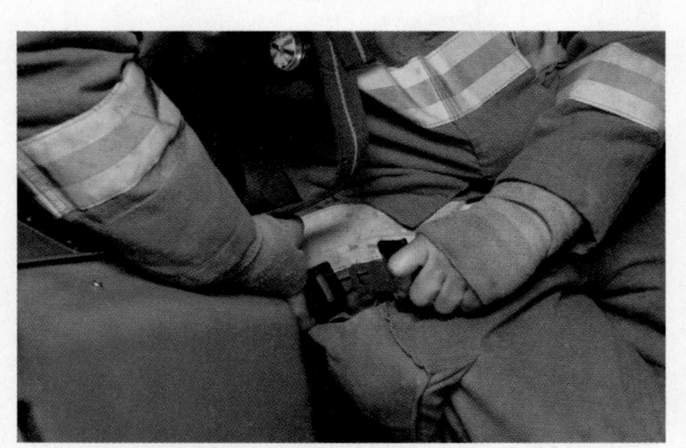

Figure 2-9 Always use your seat belt.

Safety at the Fire Station

The fire station is just as much a workplace as the fireground. Fire fighters will spend much of the time during a 12- or 24-hour shift at the fire station. Be careful when working with power tools, ladders, electrical appliances, pressurized cylinders, and hot surfaces. Injuries that occur at the firehouse can be just as devastating as those that occur at an emergency incident scene.

Safety Outside Your Workplace

Continue to follow safe practices when you are off-duty as well. An accident or injury, regardless of where it happens, can end your career as a fire fighter. For example, if you are using a ladder while off duty, follow the same safety practices that you would use on duty. Use the seat belts in your personal vehicle, just as you are required to do when you are on duty (◄ Figure 2-9).

Personal Protective Equipment

Personal protective equipment (PPE) is an essential component of a fire fighter's safety system. It enables a person to survive under conditions that would otherwise result in death or serious injury. Different PPE ensembles are designed for specific

Figure 2-10 A protective ensemble for structural firefighting provides protection from multiple hazards.

Fire Fighter Safety Tips

A structural firefighting ensemble is designed only for structural firefighting. It is not designed for other functions such as hazardous materials, water rescue, or wildland firefighting. Each of these specialized functions requires specific PPE. If your department performs specialized operations, you need the appropriate PPE for that activity.

hazardous conditions, such as structural firefighting, wildland firefighting, airport rescue and firefighting, hazardous materials operations, and emergency medical operations.

PPE ensembles provide specific protections, so an understanding of their designs, applications, and limitations is critical. For example, a structural firefighting ensemble will protect the wearer from the heat, smoke, and toxic gases present in building fires (◄ Figure 2-10). It cannot provide long-term protection from extreme weather conditions and it limits range of motion. The more you know about the protection your PPE can provide, the better you will be able to judge conditions that exceed its limitations.

A fire fighter's PPE must provide full body coverage and protection from a variety of hazards. To be effective, the entire ensemble must be worn whenever potential exposure to those hazards exists. PPE must be cleaned, maintained, and inspected regularly to ensure that it will provide the intended degree of protection when it is needed. Worn or damaged articles must be repaired or replaced.

Structural Firefighting Ensemble

Structural firefighting PPE enables fire fighters to enter burning buildings and work in areas with high temperatures and concentrations of toxic gases. Without PPE, fire fighters would be unable to conduct search-and-rescue operations or perform fire suppression activities. A structural firefighting ensemble is designed to cover every inch of the body. It provides protection from the fire, keeps water away from the body, and helps reduce trauma from cuts or falls. Structural firefighting PPE is designed to be worn with **self-contained breathing apparatus (SCBA)**, which provides respiratory protection.

The structural firefighting ensemble consists of a protective coat, trousers or coveralls, a helmet, a hood, boots, and gloves. The helmet must have a face shield, goggles, or both. The clothing is worn with SCBA and a personal alert safety system (PASS) device. All of these elements must be worn together to provide the necessary level of protection (► Figure 2-12).

Protection Provided

A structural firefighting ensemble is designed for full body coverage and provides several different types of protection

Standards for Personal Protective Equipment

Personal protective equipment for fire fighters must be manufactured according to exacting standards. Each item must have a permanent label verifying that the particular item meets the requirements of the standard (▼ Figure 2-11). Usage limitations as well as cleaning and maintenance instructions should also be provided. The requirements for firefighting PPE are outlined in:

- NFPA 1971, *Standard on Protective Ensemble for Structural Firefighting*
- NFPA 1976, *Standard on Protective Ensemble for Proximity Firefighting*
- NFPA 1977, *Standard on Protective Clothing and Equipment for Wildland Firefighting*

The requirements for self-contained breathing apparatus and personal alert safety systems are outlined in:

- NFPA 1981, *Standard on Open-Circuit Self-Contained Breathing Apparatus for Fire and Emergency Services*
- NFPA 1982, *Standard on Personal Alert Safety Systems (PASS)*

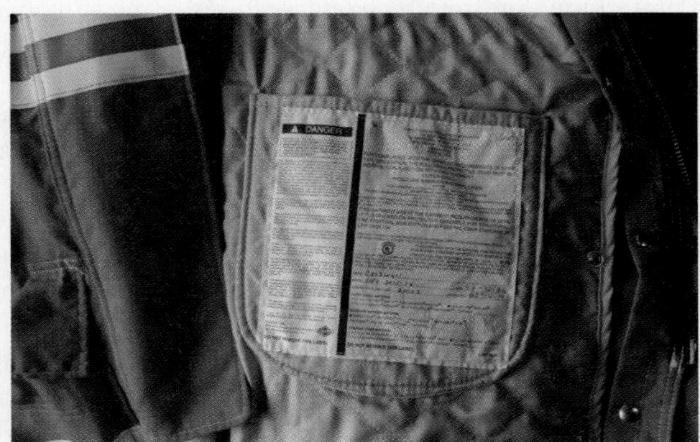

Figure 2-11 The label provides important information about each item of PPE.

Figure 2-12 The complete structural firefighting ensemble consists of a helmet, coat, trousers, protective hood, gloves, boots, SCBA, and PASS device.

Table 2-2 Types of Protection Furnished by PPE

Personal protective equipment:

- Provides thermal protection
- Repels water
- Provides impact protection
- Provides protection against cuts and abrasions
- Furnishes padding against injury
- Increases your visibility
- Provides respiratory protection

Figure 2-13 A helmet is constructed with multiple layers and components.

▲ Table 2-2. The coat and trousers have tough outer shells that can withstand high temperatures, repel water, and provide protection from abrasions and sharp objects. The knees may be reinforced with pads for greater protection when crawling. Fluorescent/reflective trim adds visibility in dark or smoky environments. Insulating layers of fire-resistant materials protect the skin from high temperatures. A moisture barrier between the layers keeps liquids and vapors, such as hot water or steam, from reaching the skin.

The helmet provides protection from trauma to the head and includes ear coverings. The face shield helps protect the eyes. A fire-retardant hood covers any exposed skin between the coat collar and the helmet. Gloves protect the hands from heat, cuts, and abrasions. Boots protect the feet and ankles from the fire, keep them dry, prevent puncture injuries, and protect the toes from crushing injuries.

Self-contained breathing apparatus provides respiratory protection. An SCBA gives the fire fighter an independent air supply. This protects the respiratory system from toxic products and hot gases present in the atmosphere.

Helmet

Fire helmets are manufactured in several designs and shapes using different materials. Each design must meet the requirements specified in NFPA 1971, *Standard on Protective Ensemble for Structural Firefighting*. The hard outer shell is lined with energy-absorbing material and has a suspension system to provide impact protection against falling objects ► Figure 2-13. The helmet shell also repels water, protects against steam, and creates a thermal barrier against heat and cold. The shape of the helmet helps to deflect water away from the head and neck.

Face and eye protection can be provided by a face shield, goggles, or both. These components must be attached to the helmet and are used when SCBA is not needed or when the SCBA face piece is not in place. A chin strap is also required and must be worn to keep the helmet in the proper position. The chin strap also helps to keep the helmet on the fire fighter's head during an impact.

Fire helmets have an inner liner for added thermal protection. This liner also provides protection for the ears and neck. When entering a burning building, the fire fighter should pull down the ear tabs for maximum protection.

Helmets are manufactured with adjustable inner suspension systems that hold the head away from the shell and cushion it against impacts. This suspension system must be adjusted to fit the individual, with the SCBA face piece and hood in place. Helmet shells also come in various sizes. Fire fighters must be sure to wear the correct size shell.

Helmet shells are often color-coded according to the fire fighter's rank and function. They often carry company numbers, rank insignia, or other markings. Some fire departments use a shield mounted on the front of the helmet to identify the fire fighter's rank and company. Bright, reflective materials are applied to make the fire fighter more visible in all types of lighting conditions.

NFPA 1971 requires that helmets approved for structural firefighting have a label permanently attached to the inside of the shell. This label lists the manufacturer, model, date of manufacture, weight, size, and recommended cleaning procedures.

Protective Hood

Although the helmet's ear tabs cover the ears and neck, this area is still at risk for burns when the head is turned or the neck is flexed. **Protective hoods** provide additional thermal protection for these areas. The hood, which is constructed of flame-resistant materials such as **Nomex®** or **PBI®**, covers the whole head and neck, except for that part of the face protected by the SCBA face piece ► Figure 2-14. The lower part of the hood, which is called the bib, drapes down inside the turnout coat.

Figure 2-14 A protective hood.

Protective hoods are worn over the face piece but under the helmet. After securing the face piece straps, carefully fit the hood around the face piece so that no areas of bare skin are left exposed. The hood must fit snugly around the clear area of the face piece so that vision is not compromised and hot gases cannot leak between the face piece and the hood.

Turnout Coat

Coats used for structural firefighting are generally called **bunker coats** or **turnout coats** (▼ Figure 2-15). Only coats that meet NFPA 1971 should be used for structural firefighting. Turnout coats have three layers. The outer layer or shell is constructed of a sturdy, flame-resistant, water-repellant material such as Nomex, **Kevlar®**, or PBI. Fluorescent reflective material applied to the light-colored outer shell makes the fire fighter more visible in smoky conditions and at night. The light-colored fabric also makes it easier to identify contaminants such as hydrocarbons, blood, and body fluids on the coat.

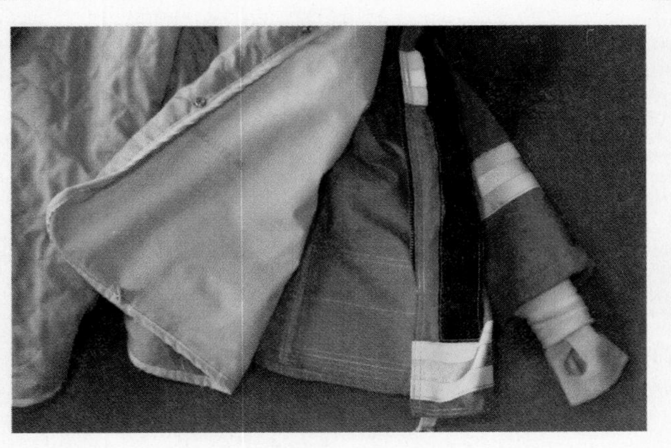

Figure 2-15 A bunker or turnout coat.

Fire Marks

Before hoods were introduced, the skin on the neck and ears was often exposed. Fire fighters "measured" the temperature in a burning building by how hot their ears felt. This often resulted in burns to the ears. Fire fighters still must be careful to avoid situations where the temperature exceeds the protection provided by PPE.

The second layer of the coat is the moisture barrier, which is usually a flexible membrane attached to a thermal barrier material (the third layer). The moisture barrier helps prevent the transfer of water, steam, and other fluids to the skin. Water applied to a fire generates large amounts of superheated steam, which can engulf fire fighters and burn unprotected skin.

The thermal barrier is a multilayered or quilted material that insulates the body from external temperatures. It enables fire fighters to operate in the high temperatures generated by a fire and keeps the body warm during cold weather.

The front of the turnout coat has an overlapping flap to provide a secure seal. The inner closure is secured first, and then the outer flap is secured, creating a double seal. Several different combinations of D-rings, snaps, zippers, and Velcro can be used to secure the inner and outer closures.

The collar of the coat works with the hood to protect the neck. The collar has snaps or a Velcro closure system in front to keep it in a raised position. The coat's sleeves have wristlets that prevent liquids or hot embers from getting between the sleeves and the skin. They also prevent the sleeves from riding up the wrists, so there is no exposed skin between the gloves and the sleeves, which could result in wrist burns.

Bunker coats come in two different lengths—long and short. Both will protect the body as long as the matching style of pants or coveralls are also worn. The coat must be long enough that you can raise your arms over your head without exposing your midsection. The sleeve length should not hinder arm movement, and the coat should be large enough that it does not interfere with breathing or other movements.

Fire Fighter Safety Tips

The inner thermal liner of most turnout coats can be removed while the outer shell is being cleaned, but the turnout coat should never be used without the thermal liner. Severe injury could occur if the coat is used without the liner.

Fire Marks

Bunker coats were originally made of rubber and designed primarily to repel water. Today's coats provide protection from flames, heat, abrasions, and cuts as well as water.

Pockets in the coat can be used for carrying small tools or extra gloves. Additional pockets or loops can be installed to hold radios, microphones, flashlights, or other accessories.

Bunker Pants

Protective trousers are also called **bunker pants** or **turnout pants** (▼ Figure 2-16). They can be constructed in a waist-length design or bib-overall configuration. Bunker pants also must meet the NFPA 1971 and are constructed with multiple layers, just like bunker coats. The outer shell resists abrasion and repels water. The second layer is a moisture barrier to protect the skin from liquids and steam burns, and the inner layer is a quilted, thermal barrier to protect the body from elevated temperatures. Bunker pants are reinforced around the ankles and knees with leather or extra padding.

Figure 2-16 Bunker pants.

Bunker pants are manufactured with a double fastener system at the waist, similar to the front flap of a turnout coat. Florescent or reflective stripes around the ankles provide added visibility. Suspenders hold the pants up. Pants should be large enough to allow you to don them quickly. They should be big enough to allow you to crawl and bend your knees easily, but they should not be bigger than necessary.

Boots

Structural firefighting boots can be constructed of rubber or leather and come in different lengths. Rubber firefighting boots come in a step-in style without laces (▼ Figure 2-17). Leather firefighting boots are available in a knee-length, pull-on style or in a shorter version with laces (▼ Figure 2-18). Many fire fighters install a zipper on the laced boots to aid in quick donning and doffing.

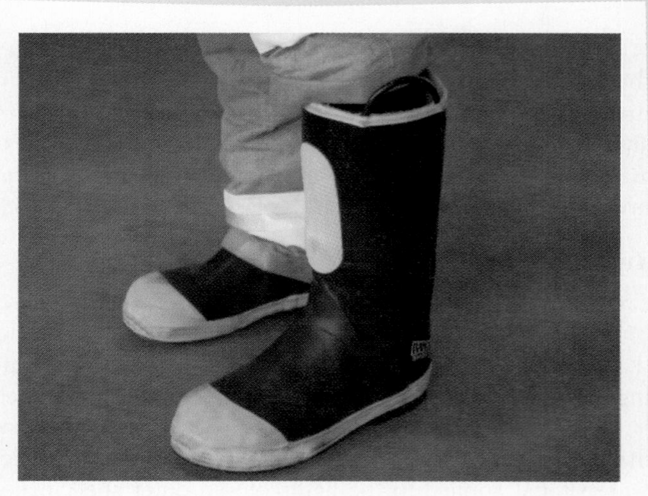

Figure 2-17 Rubber firefighting boots.

Figure 2-18 Leather firefighting boots.

Fire Fighter Safety Tips

Fire boots must be worn with approved bunker pants. Boots worn without bunker pants leave the legs unprotected and exposed to injury.

Fire Fighter Safety Tips

Plain leather work gloves or plastic-coated gloves should never be used for structural firefighting. Gloves used for structural firefighting must be labeled and meet NFPA 1971.

Both boot styles must meet the same test requirements specified in NFPA 1971. The outer layer repels water and must be both flame- and cut- resistant. The boots must have a heavy sole with a slip-resistant design. Boots must have a puncture-resistant sole and a reinforced toe to prevent injury from falling objects. An inner liner constructed of materials such as Nomex or Kelvar adds thermal protection.

Boots must be the correct size for the foot. The foot should be secure within the boot to prevent ankle injuries and enable secure footing on ladders or uneven surfaces. Improperly sized boots will cause blisters and other problems.

Gloves

Gloves are an important part of the firefighting ensemble, because most fire suppression tasks require the use of the hands (▼ Figure 2-19). Gloves must provide adequate protection and still enable the manual dexterity needed to accomplish tasks. NFPA 1971 specifies that gloves must be resistant to heat, liquid absorption, vapors, cuts, and penetration. Gloves must have a wristlet to prevent skin exposure during normal firefighting activities.

Firefighting gloves are usually constructed of heat-resistant leather. The wristlets are usually made of knitted Nomex or Kevlar. The liner adds thermal protection and serves as a moisture barrier.

Many fire fighters carry a second set of gloves in their bunker gear or on the fire apparatus so they can change gloves if one pair gets wet or damaged. Do not wring or twist wet gloves because this can tear or damage the inner liners.

Although gloves furnish needed protection, they reduce manual dexterity. Fire fighters need to practice manual skills while wearing gloves to become accustomed to them and to adjust movement accordingly.

Respiratory Protection

Self-contained breathing apparatus is an essential component of the PPE used for structural firefighting. Without adequate respiratory protection, fire fighters would be unable to mount an attack. The design, parts, operation, use, and maintenance of this complex equipment are covered later in this chapter.

Personal Alert Safety System (PASS)

A personal alert safety system (PASS) is an electronic device that sounds a loud audible signal when a fire fighter becomes trapped or injured. A PASS will sound automatically if a fire fighter is motionless for a set time period. The PASS can also be manually activated to notify other fire fighters that the user needs assistance.

A PASS can be separate from or integrated into the SCBA unit. (▼ Figure 2-20) The integrated PASS devices automatically turn on when the SCBA is activated. The separate PASS devices are often worn on the SCBA harness and must be turned on manually. Fire fighters must check that their PASS

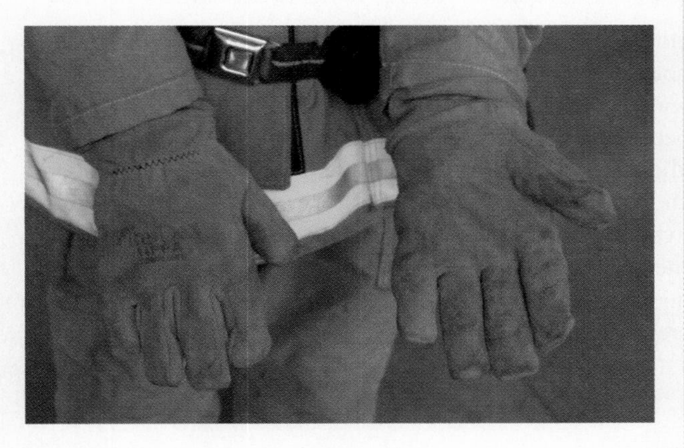

Figure 2-19 Firefighting gloves.

Figure 2-20 A PASS can save a fire fighter's life.

Voices of Experience

“Despite advances in protective clothing and equipment, we are not invincible on the job.”

On a late winter evening early in my career, the fire tone activated, reporting a fire in a two-story residential structure in the number 1 district. Arriving on scene, we encountered a woman in the front yard who was screaming that her children were upstairs. She said the fire started when she fell asleep while waiting for food to cook on the stove. She stated that the stairwell was right off the kitchen. The officer instructed us to enter and attack the fire using a “fog” pattern because he believed the fire had already vented sufficiently and a quick “knock-down” could provide relief for those trapped upstairs.

As we entered the front door, we could see heavy fire in the kitchen, approximately twenty feet in front of us. Visibility was good and we could see approximately five feet above the floor. Suddenly, the conditions began to change. The smoke banked down quickly, heat build-up was now intense, and we could see the flame rolling over toward us from the kitchen. What we could not see was the fire development in the living room immediately behind us and to the left. I opened the nozzle and began to knock down the fire ahead of us. Within seconds, we were hit with a tremendous blast of super-heated gas from behind and then in front. Crews outside later reported a flash of bright orange light with flame rolling out the front door, and with us silhouetted in the middle. We continued to flow water, but this seemed to make conditions worse. As we backed our line out to the door, we could see the fire from the living room pushing toward us. We turned on it and proceeded to knock down the fire, and then made our withdrawal.

Once we returned to the street, we performed a quick check of our team and our gear. We had been fully protected, with the latest personal protective clothing, SCBA, gloves, and hoods. Everyone appeared to be okay, except for me. My helmet shield was melted, my earflaps were burned, and my hood was stuck to my ears. I had received second- and third-degree burns to both ears, and first-degree burns to the sides of my face.

Looking back on this incident, I still consider the options and the choices we made that night and what we could have done differently. All the signs of the impending flashover were present—rapid heat build-up, banking smoke conditions, and rollover—but we were focused on the stairwell off the kitchen and the children upstairs, whom we found out later were not at home. We let the emotion of the incident drive our attack without regard for our own safety. We believed if we were aggressive, we could knock down the fire. However, aggressiveness is not a substitute for patience and planning when attacking a fire. Despite advances in protective clothing and equipment, we are not invincible on the job. Plan your interior attack carefully and move methodically with regard to safety.

Paul E. Ricci
Sandusky Fire Division
Sandusky, Ohio

Fire Fighter Tips

Fire fighters have personal preferences about the tools and equipment they carry in their pockets. Observe what seasoned members of your department carry in their pockets. This will help you decide whether you need to carry similar equipment.

is on and working properly before they enter a burning building or hazardous area.

The PASS combines an electronic motion sensor with an alarm system. If the user is motionless for 30 seconds, the PASS will sound a low warning tone before sounding a full alarm. The user can reset the device by moving during this warning period. Newer PASS devices also have a radio transmitter that sends a signal to the command post when the alarm sounds.

Additional Personal Protective Equipment

Eye protection is provided by the face shield mounted on fire helmets. When additional eye protection is needed, such as when using power saws or hydraulic rescue tools, fire fighters can use approved goggles. Goggles can be carried easily in turnout coat pockets.

Fire fighters can also be exposed to loud noises such as sirens and engines. Because hearing loss is cumulative, it is important to limit exposure to loud sounds. An intercom system on the apparatus can provide hearing protection (▼ Figure 2-21). A small microphone is incorporated into large earmuffs located at each riding and operating position. Fire fighters don the earmuffs, which reduce engine and siren noise. The microphone enables crew members to talk to each other in a normal tone of voice and to hear the apparatus radio.

Flexible ear plugs are useful in other situations involving loud sounds. Fire fighters should use the hearing protection supplied by their departments to prevent hearing loss.

A fire fighter should always carry a **hand light**. Most interior firefighting is done in near-dark, zero-visibility conditions. A good working hand light can illuminate your surroundings, mark your location, and help you find your way under difficult conditions.

Two-way radios link the members of a firefighting team. At least one member of each team working inside a burning building or in any hazardous area should always have a radio. Some fire departments provide a radio for every on-duty fire fighter. Follow your department's SOP on radio use. A radio should be considered part of PPE and carried with you whenever it is appropriate.

Limitations of the Structural Firefighting Ensemble

The structural firefighting ensemble protects a fire fighter from the hostile environment of a fire. Each component must be properly donned and worn to provide complete protection. But even today's advanced PPE has drawbacks and limitations. Understanding those limitations will help you avoid situations that could result in serious injury or death.

First, this ensemble is not easy to don. There are several components that must be put on in the proper order and correctly secured. You must be able to don your equipment quickly and correctly, either at the firehouse before you respond to an emergency, or after you arrive at the scene. Properly donning and doffing PPE takes practice (▼ Figure 2-22).

PPE is also heavy—nearly 50 pounds of extra weight. This increased weight means that everything you do—even walking—requires more energy and strength. Tasks such as advancing an attack line up a stairway or using an axe to ventilate a roof can be difficult, even for a fire fighter in excellent physical condition.

Figure 2-21 Some fire apparatus are equipped with an intercom system.

Figure 2-22 It takes practice to don the full PPE.

PPE retains body heat and perspiration, making it difficult for the body to cool itself. Perspiration is retained inside the protective clothing rather than released through evaporation to cool you. Fire fighters in full protective gear can rapidly develop elevated body temperatures, even when the ambient temperature is cool. The problem of overheating is more acute when surrounding temperatures are high. This is one reason that fire fighters must undergo regular rehabilitation and fluid replacement.

Another drawback to PPE is that it limits mobility. Full turnout gear not only limits the range of motion, it also makes movements awkward and difficult. This increases energy expenditure and adds stress.

Wearing PPE also decreases normal sensory abilities. The sense of touch is reduced by wearing heavy gloves. Turnout coats and pants protect skin but reduce its ability to determine the temperature of hot air. Sight is restricted when you wear SCBA. The plastic face piece reduces peripheral vision and the helmet, hood, and coat make turning the head difficult. Earflaps and the protective hood over the ears limit hearing. Speaking becomes muffled and distorted by the SCBA face piece, unless it is equipped with a special voice amplification system.

For these reasons, fire fighters must become accustomed to wearing and using PPE. Practicing skills while wearing PPE will help you become comfortable with its operation and limitations.

Work Uniforms

A fire fighter's work uniform is also part of the personal protective package. Certain synthetic fabrics can melt at relatively low temperatures and cause severe burns, even when worn under a complete personal protective ensemble. NFPA 1975, *Standard on Station/Work Uniforms for Fire and Emergency Services,* defines criteria for selecting appropriate fabrics for work uniforms.

Clothing containing nylon or polyester, even if they are blended with natural fibers, should not be worn in a firefighting environment. Clothing made of natural fibers, such as cotton or wool, is generally safer to wear. Nylon and polyester will melt, but cotton and wool may char or even burn without melting. Special synthetic fibers such as Nomex and PBI, which are used in turnout clothing, have excellent resistance to high temperatures. Volunteer fire fighters, who often wear PPE over their normal clothing, should consider these fabric properties when selecting their wardrobe.

Fire Fighter Tips

NFPA 1001 requires fire fighters to don personal protective clothing in one minute. Practice donning your protective clothing ensemble until you can do it quickly and smoothly. Remember that each piece of the ensemble must be in place for maximum protection.

Fire Fighter Safety Tips

DO NOT don protective clothing inside the apparatus while en route to an emergency incident. Stay in your assigned seat, properly secured by a seat belt or safety harness while the vehicle is in motion. Don protective clothing in the firehouse before mounting the apparatus, or on the scene after you arrive.

Donning Personal Protective Clothing

Donning protective clothing must be done in a specific order to obtain maximum protection. It also must be done quickly. Fire fighters should be able to **don**, or put on, personal protective clothing in one minute or less. This requires considerable practice, but remember that your goals should first be to don PPE properly, and second to do it consistently in 60 seconds or less.

Following a set pattern of donning PPE can help reduce the time it takes to dress. Many fire fighters follow the pattern described in ▶ Skill Drill 2-1. This exercise does not include donning SCBA. First become proficient in donning personal protective clothing, and then add the SCBA. Follow the steps in Skill Drill 2-1 to don personal protective clothing:

1. Place equipment in a logical order for donning. **(Step 1)**
   - Place the legs of your bunker pants over your boots and fold the pants down around them.
   - Place your protective hood over the top of your boots to remind you to put on your protective hood.
   - Lay out your coat, helmet, and gloves.
2. Place your protective hood over your head and down around your neck. **(Step 2)**
3. Step into your boots and pull up your bunker pants. Place the suspenders over your shoulders and secure the front of the pants using the closure system. **(Step 3)**
4. Put on your turnout coat and secure the inner and outer closures. **(Step 4)**
5. Place your helmet on your head with ear tabs extended and adjust the chin strap securely. Turn up your coat collar and secure it in front. **(Step 5)**
6. Put on your gloves. **(Step 6)**
7. Check all clothing to be sure it is properly secured and in the correct position. Have your partner check your clothing. **(Step 7)**

2-1 Skill Drill

Donning Personal Protective Clothing

Place your equipment in a logical order for donning.

Place your protective hood over your head and down around your neck.

Put on boots and pull up bunker pants. Place the suspenders over your shoulders and secure the front of the pants.

Put on your turnout coat and close the front of the coat.

Place your helmet on your head and adjust the chin strap securely. Turn up your coat collar and secure it in front.

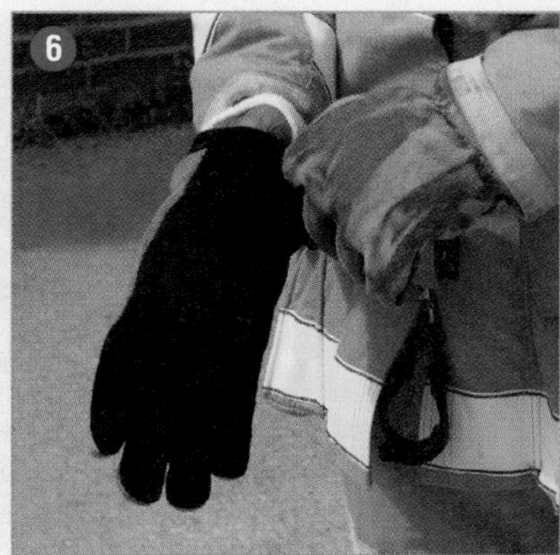

Put on your gloves.

Have your partner check your clothing.

Fire Fighter Safety Tips

In hot weather, remove as much of your PPE as possible when you arrive at rehab. This will help you to cool down quickly. Some fire departments even place fans and misting equipment in the rehab area to provide additional cooling.

Doffing Personal Protective Clothing

To doff, or remove, your personal protective clothing, reverse the procedure used in getting dressed. Follow the steps in ▶ Skill Drill 2-2 to doff personal protective clothing:

1. Remove your gloves. **(Step 1)**
2. Open the turnout coat collar. **(Step 2)**
3. Release the helmet chin strap and remove your helmet. **(Step 3)**
4. Remove your turnout coat. **(Step 4)**
5. Remove your protective hood. **(Step 5)**
6. Remove your bunker pants and boots. **(Step 6)**

If necessary, PPE should be cleaned after it is used, and then kept in a convenient location for the next response. PPE may be kept close to the apparatus, on the apparatus, or in an equipment locker. Personal protective clothing must be properly maintained, organized, and ready for the next emergency response.

Care of Personal Protective Clothing

Approved personal protective clothing is built to exacting standards but requires proper care to continue to afford maximum protection. Avoid unnecessary wear on turnout clothing. A complete set of approved turnout clothing (excluding SCBA) costs more than $1,000. Keep this expensive equipment in good shape for its intended use—structural firefighting.

Check the condition of PPE on a regular basis. Clean it when necessary; repair worn or damaged PPE at once. PPE that is worn or damaged beyond repair must be replaced immediately because it will not be able to protect you.

Avoid unnecessary cuts or abrasions on the outer material. This material already meets NFPA standards; do not look for opportunities to test its effectiveness. If the fabric is damaged, it must be properly repaired to retain its protective qualities. Follow the manufacturer's instructions for repairing or replacing PPE.

Personal protective clothing must be kept clean to maintain its protective properties. Dirt will build up in the clothing fibers from routine use and exposure to fire environments. Smoke particles will also become embedded in the outer shell material. The interior layers will frequently be soaked with perspiration. Regular cleaning should remove most of these contaminants.

Other contaminants are formed from the by-products of burnt plastics and synthetic products. These residues are flammable and can be trapped between the fibers or build up on the outside of PPE, damaging the materials and reducing their protective qualities. A fire fighter who is wearing contaminated PPE is actually bringing additional fuel into the fire on the clothing.

PPE that has been badly soiled by exposure to smoke, other products of combustion, melted tar, petroleum products, or other contaminants needs to be cleaned as soon as possible. Items that have been exposed to chemicals or hazardous materials may have to be impounded for decontamination or disposal.

Cleaning instructions are listed on the tag attached to the clothing. Follow the manufacturer's cleaning instructions. Failure to do so may reduce the effectiveness of the garment and create an unsafe situation for the wearer.

Some fire departments have special washing machines that are approved for cleaning turnout clothing. Other departments contract with an outside firm to clean and repair protective clothing. In either case, the manufacturer's instructions for cleaning and maintaining the garment must be followed.

Other PPE also require regular cleaning and maintenance. The outer shell of your helmet should be cleaned with a mild soap as recommended by the manufacturer. The inner parts of helmets should be removed and cleaned according to the manufacturer's instructions. The chin strap and suspension system must be properly adjusted and all parts of the helmet kept in good repair.

Protective hoods and gloves get dirty quickly and should be cleaned according to the manufacturer's instructions. Most hoods can be washed with appropriate soaps or detergents. Repair or discard gloves or hoods that have holes in them; do not use them. A small cut or opening can result in a burn injury.

Boots should be maintained according to the manufacture's instructions. Rubber boots need to be kept in a place that does not result in damage to the boot. Leather boots must be properly maintained to keep them supple and in good repair. Boots need to be repaired or replaced if the outer shell is damaged.

Fire Fighter Safety Tips

Make sure your PPE is dry before using it on the fireground. If wet protective clothing is exposed to the high temperatures of a structural fire, the water trapped in the liner materials will turn into steam and be trapped inside the moisture barrier. This can result in painful steam burns.

Skill Drill

Doffing Personal Protective Clothing

Remove your gloves.

Open the collar of your turnout coat.

Release the helmet chin strap and remove your helmet.

Remove your turnout coat.

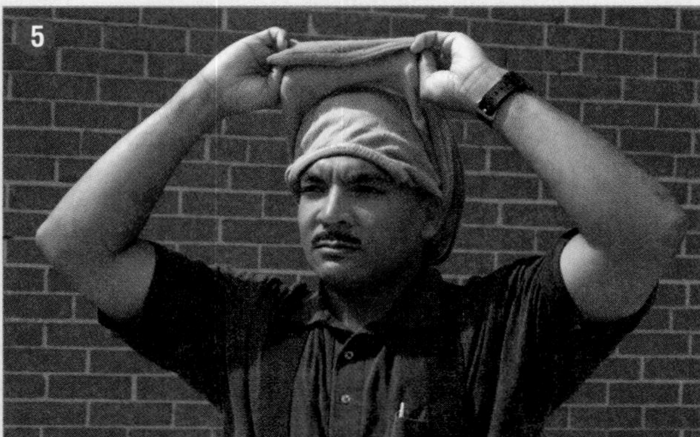

Remove your protective hood.

Remove your bunker pants and boots.

Fire Fighter Safety Tips

Turnout clothing must be properly cleaned and maintained to provide maximum protection. Dirty turnout clothing is a sign of carelessness, not of experience.

Specialized Protective Clothing

Vehicle Extrication

Due to the risk of fire at the scene of a vehicle extrication incident, most members of the emergency team will wear full turnout gear. The officer in charge may also designate one or more members to don SCBA and stand by with a charged hoseline. A structural firefighting ensemble protects against many of the hazards present at a vehicle extrication incident, such as broken glass and sharp metal objects.

There is protective clothing, such as special gloves and coveralls or jumpsuits, specifically designed for vehicle extrication. These items are generally lighter in weight and more flexible than structural firefighting PPE, although they may use the same basic materials.

Fire fighters performing a vehicle extrication must always be aware of the possibility of contact with blood or other body fluids. Latex gloves should be worn when providing patient treatment. Eye protection also should be worn, due to the possibilities of breaking glass, contact with body fluids, metal debris, and accidents with tools.

Wildland Fires

Firefighting gear designed specially for fighting wildland or brush fires must meet NFPA 1977, *Standard on Protective Clothing and Equipment for Wildland Firefighting*. The jacket and pants are made of fire-resistant materials, such as Nomex or specially treated cotton, which are designed for comfort and maneuverability while working in the wilderness. Wildland fire fighters wear a helmet of a thermo-resistant plastic, eye protection, and pigskin or leather gloves. Their boots are designed for comfort and sure footing while hiking to a remote fire scene, and for protection against crush and puncture injuries.

Respiratory Protection

Respiratory protection equipment is an essential component of the structural firefighting personal protection ensemble. A self-contained breathing apparatus is both expensive and complicated; using one confidently requires practice. Fire fighters must be proficient in using SCBA before they engage in interior fire suppression activities.

The interior atmosphere of a burning building is considered to be **IDLH (immediately dangerous to life and health)**. Attempting to work in this atmosphere without proper respiratory protection can cause serious injury or death. Never enter or operate in a fire atmosphere without appropriate respiratory protection.

Respiratory Hazards of Fires

Fire fighters need respiratory protection for several reasons. A fire involves a complex series of chemical reactions that can rapidly affect the atmosphere in unpredictable ways.

The most evident by-product of a fire is smoke. The visible smoke produced by a fire contains many different substances, some of which are dangerous if inhaled. In addition, smoke contains invisible, highly toxic products of combustion. The process of combustion consumes oxygen and can lower the oxygen concentration in the atmosphere below the level necessary to support life. The atmosphere of a fire may become so hot that one unprotected breath can result in fatal respiratory burns.

These respiratory hazards require fire fighters to use respiratory protection in all fire environments, regardless of whether the environment is known to be contaminated, suspected of being contaminated, or could possibly become contaminated without warning. The use of SCBA allows fire fighters to enter and work in a fire atmosphere with a safe, independent air supply.

Smoke

Most fires do not have an adequate supply of oxygen to consume all of the available fuel. This results in **incomplete combustion** and produces a variety of by-products, which are released into the atmosphere. Many of these by-products are extremely toxic. Collectively, the airborne products of combustion are called smoke, which has three major components: particles, vapors, and gases.

Smoke Particles

Smoke particles consist of unburned, partially burned, and completely burned substances. These particles are lifted in the thermal column produced by the fire and are usually very visible. The completely burned particles are primarily ash; the unburned and partially burned smoke particles can

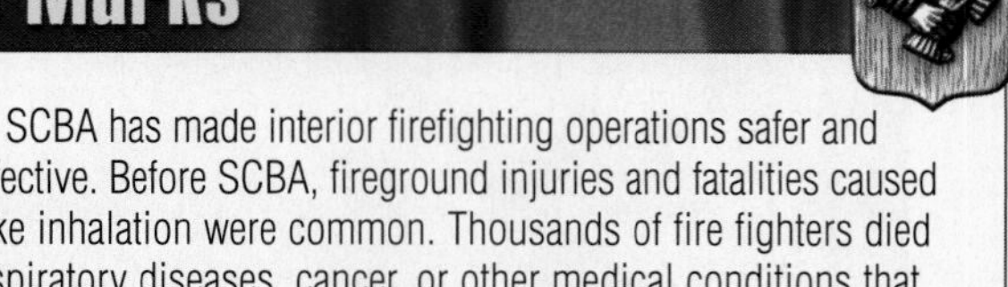

Fire Marks

Modern SCBA has made interior firefighting operations safer and more effective. Before SCBA, fireground injuries and fatalities caused by smoke inhalation were common. Thousands of fire fighters died from respiratory diseases, cancer, or other medical conditions that were direct consequences of repeated, unprotected exposure to smoke and other products of combustion. These exposures often had a cumulative, sometimes delayed effect. Each time fire fighters inhaled poisonous compounds, their lungs became more damaged, even though the damage was not evident until years later.

Fire Fighter Safety Tips

Smoky environments and oxygen-deficient atmospheres are deadly! It is essential to use SCBA at all times when operating in a smoky environment.

include various substances. The concentration of unburned or partially burned particles depends on the amount of oxygen that was available to the fire.

Many smoke particles are so small that they can pass through the natural protective mechanisms of the respiratory system and enter the lungs. Some of these particles can be toxic to the body and result in severe injuries or death if they are inhaled. They also can be extremely irritating to the eyes and digestive system.

Smoke Vapors

Smoke also contains small droplets of liquids. These smoke vapors are similar to fog, which consists of small water droplets suspended in the air. When oil-based compounds burn, they produce small hydrocarbon droplets that become part of the smoke. If inhaled or ingested, these compounds can affect the respiratory and circulatory systems. Some of the toxic droplets in smoke can cause poisoning if they are absorbed through the skin.

Water applied to a fire creates steam and water droplets that also become part of the smoke. These water droplets can absorb some of the toxic substances contained in the smoke.

Toxic Gases

A fire also produces several gases. The amount of oxygen available to the fire and the type of fuel being burned determine which gases are produced. A fire fueled by wood produces a different mixture of gases than one fueled by petroleum-based products. (Remember, many common products such as plastics are made from petroleum-based compounds.) The concentrations of these gases can change rapidly as the oxygen supply is consumed or as fresh oxygen is introduced to the combustion process.

Many of the gases commonly produced by residential or commercial fires are very toxic. Carbon monoxide, hydrogen cyanide, and phosgene are three of many gases often present in smoke.

Carbon monoxide is deadly in small quantities. When inhaled, carbon monoxide quickly replaces the oxygen in the bloodstream because it combines with the hemoglobin in the blood 200 times more readily than oxygen. A small concentration of carbon monoxide can quickly disable and kill a fire fighter. Carbon monoxide is odorless, colorless, and tasteless, which also makes it dangerous. Because carbon monoxide cannot be detected without special instruments, fire fighters must always assume that it is present in the atmosphere around a fire.

Hydrogen cyanide is formed when plastic products, such as the PVC pipe used in residential construction, burn. It is a poisonous gas that is quickly absorbed by the blood and interferes with cellular respiration. A small amount of hydrogen cyanide can easily render a person unconscious. **Phosgene** gas is formed from incomplete combustion of many common household products, including vinyl materials. Phosgene gas can affect the body in several ways. At low levels, it causes itchy eyes, a sore throat, and a burning cough. At higher levels, phosgene gas can cause pulmonary edema (fluid retention in the lungs) and death. Phosgene gas was used as a weapon in World War I to disable and kill soldiers.

Fires produce other gases that have different effects on the human body. Among these toxic gases are hydrogen chloride and several compounds containing different combinations of nitrogen and oxygen. Carbon dioxide, which is produced when there is sufficient oxygen for complete combustion, is not toxic, but it can displace oxygen from the atmosphere and cause asphyxiation.

Oxygen Deficiency

Oxygen is required to sustain human life. Normal outside or room air contains about 21% oxygen. A decrease in the amount of oxygen in the air will drastically affect an individual's ability to function as shown in ▼Table 2-3. Atmosphere with an oxygen concentration of 19.5% or less is considered oxygen-deficient. If the oxygen level drops below 17%, people can experience disorientation, an inability to control their muscles, and irrational thinking, which can make escaping a fire much more difficult.

During compartment fires (fires burning within enclosed areas), **oxygen deficiency** occurs in two ways. First, the fire consumes large quantities of the available oxygen, decreasing the concentration of oxygen in the atmosphere. Second, the fire produces large quantities of other gases, which decrease

Table 2-3 Physiological Effects of Reduced Oxygen Concentration

Oxygen Concentration	Effect
21%	Normal breathing air
17%	Judgment and coordination impaired; lack of muscle control
12%	Headache, dizziness, nausea, fatigue
9%	Unconsciousness
6%	Respiratory arrest, cardiac arrest, death

the oxygen concentration by displacing the oxygen that would otherwise be present inside the compartment.

Increased Temperature

Heat is also a respiratory hazard. The temperature of the gases generated during a fire varies, depending on fire conditions and the distance traveled by the hot gases. Inhaling superheated gases produced by a fire can cause severe burns of the respiratory tract. If the gases are hot enough, a single inhalation can cause fatal respiratory burns. More information about fire behavior and products of combustion is presented in Chapter 5, Fire Behavior.

Other Toxic Environments

Not all hazardous atmospheric conditions are caused by fires. Fire fighters will encounter toxic gases or oxygen-deficient atmospheres in many emergency situations. Respiratory protection is just as important in these situations as in a fire suppression operation.

Toxic gases can be released at hazardous materials incidents from leaking storage containers or industrial equipment, from chemical reactions, or from the normal decay of organic materials. Carbon monoxide can be produced by internal combustion engines or improperly adjusted heating appliances. Exposure to carbon monoxide or some other toxic gas should be suspected whenever fire fighters find one or more unconscious patients with no evident cause.

Toxic gases can quickly fill confined spaces or below-grade structures. Any confined space or below-grade area must be treated as a hazardous atmosphere until it has been tested to ensure that an adequate concentration of oxygen and no hazardous or dangerous gases are present.

Conditions that Require Respiratory Protection

We have looked at several factors that contribute to the dangers that exist in or near a fire atmosphere. Fires produce huge quantities of smoke, which contains unburned, partially burned, and completely burned poisonous particles, as well as toxic compounds and gases. Most fire deaths are caused by smoke inhalation rather than burns.

Fine liquid droplets, suspended in the smoke, contain highly toxic compounds formed from the breakdown of fuels. Smoke contains a wide variety of highly toxic gases resulting from incomplete combustion. Fires consume huge quantities of oxygen and generate huge quantities of poisonous gases, which can displace oxygen, causing an oxygen-deficient environment. Superheated gases are also produced. All of these factors contribute to classifying the atmosphere in a fire environment as IDLH.

Adequate respiratory protection is essential to your safety. The products of combustion from house fires and commercial fires are so toxic that a few breaths can result in death. Most fire deaths are caused by smoke inhalation rather than burns. As you arrive at the scene of a fire you do not have any way to measure the immediate danger to your life and your health posed by that fire. You must use approved breathing apparatus if you are going to enter and operate within this atmosphere.

Anytime you are in an area where there is smoke, SCBA must be used. This includes outside fires such as vehicle and dumpster fires. Utilize your SCBA at the fire scene until the air has been tested and proven to be safe by the safety officer. Do not remove your SCBA just because a fire has been knocked down. SCBA should be worn during overhaul until the air has been tested and deemed safe by your safety officer.

SCBA must also be used in any situation where there is a possibility of toxic gases being present or oxygen deficiency, such as a confined space. Always assume that the atmosphere is hazardous until it has been tested and proven to be safe.

Fire Fighter Safety Tips

Consider all below-grade areas or confined spaces to be IDLH–immediately dangerous to life and health—until they have been tested.

Types of Breathing Apparatus

The basic respiratory protection hardware used by the fire service is the self-contained breathing apparatus or SCBA.

Figure 2-23 Open-circuit SCBA.

Fire Fighter Safety Tips

Always put on SCBA before entering a confined space or a below-ground structure. Continue to use respiratory protection until the air is tested and deemed safe by the safety officer. Do not attempt confined space operations without special training.

Fire Fighter Tips

Do not confuse SCBA and SCUBA. Self-contained breathing apparatus, SCBA, is used by fire fighters during fire suppression activities. Divers use **self-contained underwater breathing apparatus (SCUBA)** while swimming underwater.

The term self-contained refers to the requirement that the apparatus is the sole source of the fire fighter's air supply. It is an independent air supply that will last for a predictable duration.

The two main types of self-contained breathing apparatus are **open-circuit breathing apparatus** and **closed-circuit breathing apparatus**. Open-circuit breathing apparatus is usually used for structural firefighting. A tank of compressed air provides the breathing air supply for the user (◄ Figure 2-23). Exhaled air is released into the atmosphere through a one-way valve. Approved open-circuit SCBA comes in several different models, designs, and options.

Closed-circuit breathing apparatus recycles the user's exhaled air. The air passes through a mechanism that removes carbon dioxide and adds oxygen within a closed system (► Figure 2-24). The oxygen is generated from a chemical canister. Many closed-circuit SCBAs have a small oxygen tank as well as the chemically-generated oxygen. Closed-circuit breathing apparatus is more often used for extended operations, such as mine rescue work, where breathing apparatus must be worn for a long time.

A **supplied-air respirator (SAR)** uses an external source for the breathing air (► Figure 2-25). A hose line is connected to a breathing-air compressor or to compressed air cylinders located outside the hazardous area. The user breathes air through the line and exhales through a one-way valve, just as with an open-circuit SCBA.

Although supplied-air respirators are commonly used in industrial settings, they are not used by fire fighters for structural firefighting. Hazardous materials teams and confined space rescue teams sometimes use supplied-air respirators for specialized operations. Some fire service SCBA units can be adapted for use as supplied-air respirators.

Figure 2-24 Closed-circuit SCBA.

Figure 2-25 A supplied-air respirator may be needed for special rescue operations.

SCBA Standards and Regulations

In the United States, the **National Institute for Occupational Safety and Health (NIOSH)** sets the design, testing, and certification requirements for SCBA. NIOSH is a federal agency that researches, develops, and implements occupational safety and health programs. It also investigates fire fighter fatalities and serious injuries, and makes recommendations on how to prevent accidents from recurring.

The U.S. Occupational Safety and Health Administration (OSHA) and state agencies are responsible for establishing and enforcing regulations for respiratory protection programs. In some states and in Canadian provinces, there are

Fire Marks

Many of the occupational health and safety regulations for the fire service and other workers use the term **respirator**. A respirator is a device that provides respiratory protection for the user. There are several different types of respirators for different applications. A self-contained breathing apparatus is a particular type of respirator.

individual occupational safety and health agencies to establish and enforce regulations.

The NFPA has developed three standards directly related to SCBA. NFPA 1500, *Standard for a Fire Department Occupational Safety and Health Program,* includes the basic requirements for SCBA use and program management. NFPA 1404, *Standard for Fire Service Respiratory Protection Training,* sets requirements for an SCBA training program within a fire department. NFPA 1981, *Standard on Open-Circuit Self-Contained Breathing Apparatus for Fire and Emergency Services,* includes requirements for the design, performance, testing, and certification of open-circuit SCBA for the fire service. Each fire department must follow applicable standards and regulations to ensure safe working conditions for all personnel.

Uses and Limitations of Self-Contained Breathing Apparatus

SCBA is designed to provide pure breathing air to fire fighters working in the hostile environment of a fire. It must meet rigid manufacturing specifications so that it can function in the increased temperature and smoke-filled environments that fire fighters encounter. When properly maintained, it will provide sufficient quantities of air for fire fighters to perform rigorous tasks.

Using SCBA requires that fire fighters develop unique skills, including different breathing techniques. SCBA limits normal sensory awareness; scent, hearing, and sight are all affected by the apparatus. Proficiency in the use of SCBA and other PPE requires ongoing training and practice.

SCBA also has its limitations, as with any type of equipment. Some of these limitations apply to the equipment; others apply to the user's physical and psychological abilities.

Limitations of the Equipment

Because an SCBA carries its own air supply in a pressurized cylinder, its use is limited by the amount of air in the cylinder. SCBA for structural firefighting must carry enough air for a minimum of 30 minutes; cylinders rated for 45 minutes and 60 minutes are also available. These duration ratings, however, are based on ideal laboratory conditions. The realistic useful life of an SCBA cylinder for firefighting operations is usually much less than the rated duration, and actual use time will depend on the activities being performed and on the person using the apparatus. An SCBA cylinder will generally have a realistic useful life of 50% of the rated time. For example, an SCBA cylinder rated for 30 minutes can be expected to last for a maximum of 15 minutes of interior fire fighting.

Figure 2-26 SCBA expands a user's profile, making it more difficult to get through tight spaces.

Fire fighters must manage their working time while using SCBA so they have enough time to exit from the hazardous area before exhausting the air supply. If it takes 5 minutes to reach the work area and 5 minutes to return to fresh air, a 15-minute air supply only provides a maximum of 5 minutes of working time.

The weight of an SCBA varies, based on the manufacturer and the type and size of cylinder. Generally, an SCBA weighs at least 25 pounds. The size of the unit also makes it more difficult for the user to fit into small places (▲ Figure 2-26). The added weight and bulk decrease the user's flexibility and mobility, and shift the user's center of gravity.

The design of the SCBA face piece limits visibility, particularly peripheral vision. The face piece may fog up under some conditions, further limiting visibility. SCBA also may affect the user's ability to communicate, depending on the type of face piece and any additional hardware provided, such as voice amplification and radio microphones. The equipment is noisy during inhalation and exhalation, which may limit the user's hearing.

Physical Limitations of the User

Conditioning is important for SCBA users. An out-of-shape fire fighter will consume the air supply more quickly and will have to exit the fire building long before a well-conditioned fire fighter. Overweight or poorly conditioned fire fighters are also at greater risk for heart attacks due to physical stress.

The protective clothing and SCBA that must be worn when fighting fires weigh more than 50 pounds. Moving with this extra weight requires additional energy, which increases air consumption and body temperature. This places additional stress on a fire fighter's body. A person with ideal body weight will be able to perform more work per cylinder of air than a person who is overweight.

The weight and bulk of the complete PPE ensemble limits a fire fighter's ability to walk, climb ladders, lift weight, and crawl through restricted spaces. Fire fighters must become accustomed to these limitations and learn to adjust their movements accordingly. Practice and conditioning are required to become proficient in wearing and using PPE while fighting fires.

Psychological Limitations of the User

In addition to the physical limitations, the user must also make mental adjustments when wearing an SCBA. Breathing through an SCBA is different from normal breathing and can be very stressful. Covering your face with a face piece, hearing the air rushing in, hearing valves open and close, and exhaling against a positive pressure are all foreign sensations. The surrounding environment, which is often dark and filled with smoke, is foreign as well.

Fire fighters must adjust to these stressful conditions. Practice in donning PPE, breathing through SCBA, and performing firefighting tasks in darkness helps to build confidence, not only in the equipment, but also in personal skills. Training generally introduces one skill at a time. Practice each skill as it is introduced and try to become proficient in that skill. As your skills improve, you will be able to tackle increasing levels of difficulty.

Components of Self-Contained Breathing Apparatus

SCBA consists of four main parts: the backpack and harness, the air cylinder assembly, the regulator assembly, and the face piece assembly. There are several different manufacturers and models with varying features and operations. Although the basics are similar, you need to become familiar with the specific SCBA used by your department.

Backpack and Harness

The **backpack** provides the frame for mounting the other working parts of the SCBA (► Figure 2-27). It is usually constructed of a lightweight metal or composite material. The **harness** consists of the straps and fasteners used to attach the SCBA to the fire fighter. Most harnesses have two adjustable shoulder straps and a waist belt. Depending on the specific model of SCBA, the waist belt and shoulder straps will carry different proportions of the weight. The procedures for tightening and adjusting the straps also vary based on the model.

The harness must be secure enough to keep the SCBA firmly fastened to the user, but not so tight that it interferes with breathing or movements. The waist strap must be tight enough to keep the SCBA from moving from side to side or getting caught on obstructions.

Air Cylinder Assembly

A compressed **air cylinder** holds the breathing air for an SCBA. This removable cylinder is attached to the backpack harness and can be changed quickly in the field. An experienced fire fighter should be able to remove and replace the cylinder in complete darkness.

Fire fighters should be familiar with the type of cylinders used in their departments. Cylinders are marked with the materials used in construction, the working pressure, and the rated duration.

The air pressure in filled SCBA cylinders ranges from 2200 to 4500 **pounds per square inch (psi)**. The greater the air pressure, the more air that can be stored in the cylinder.

Low-pressure cylinders, rated for 2200 psi, can be constructed of steel or aluminum and are usually rated for 30 minutes of use. Composite cylinders are generally constructed of an aluminum shell wrapped with carbon, Kevlar, or glass fibers. They are significantly lighter in weight, can be pressurized up to 4500 psi, and are rated for 30, 45, or 60 minutes of use.

As previously noted, the rated duration times are established under laboratory conditions. A working fire fighter can quickly use up the air because of exertion. Generally, the

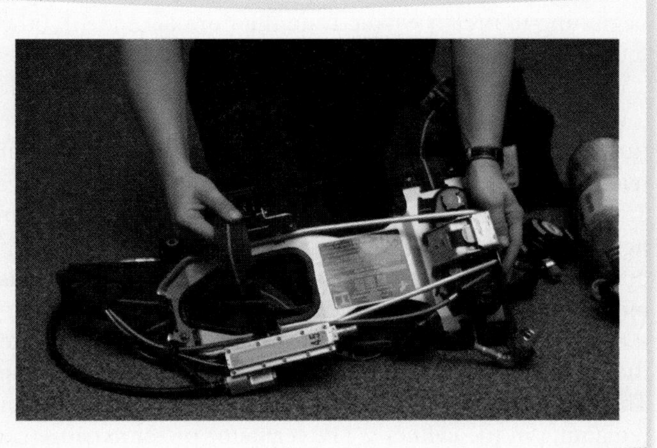

Figure 2-27 SCBA backpacks come in different models.

working time available for a particular cylinder is half the rated duration.

The neck of an air cylinder is equipped with a hand-operated shut-off valve. The **pressure gauge** is located near the shut-off valve and shows the pressure of the cylinder when full and shows the amount of pressure currently in the cylinder. Be careful not to damage the threads or let any dirt get into the outlet of the cylinder.

Regulator Assembly

SCBA regulators may be mounted on the waist belt or shoulder strap of the harness or attached directly to the face piece (▶ Figure 2-28). The regulator controls the flow of air to the user. Inhaling decreases the air pressure in the face piece. This opens the regulator, which releases air from the cylinder into the face piece. When inhalation stops, the regulator shuts off the air supply. Exhaling opens a second valve, the exhalation valve, to exhaust used air into the atmosphere. SCBA regulators are capable of delivering large volumes of air to support the strenuous activities required in firefighting.

SCBA regulators will maintain a slightly positive air pressure in the face piece. This feature prevents hazardous outside air from leaking into the face piece during inhalation. If there is any leakage around the face piece, the positive pressure will let breathing air seep out and prevent contaminated air from leaking in. Breathing with this slight positive pressure may require some practice. New fire fighters often say that it takes more energy to breathe when first using positive-pressure SCBA. This sensation gradually decreases.

Regulators are equipped for two modes of operation—the normal mode and the emergency by-pass mode. In the normal mode, described above, the regulator supplies breathing air during inhalation, stops when inhalation stops, then opens an exhalation valve to exhaust used air into the atmosphere. To activate the normal mode, some SCBA models require that the user turn a yellow-colored valve. Simply attaching the regulator to the face piece and beginning to breath will activate the normal mode in other models.

The **emergency by-pass mode** is used only if the regulator malfunctions (▶ Figure 2-29). It is activated when the user turns on the red-colored emergency by-pass valve. This releases a constant flow of breathing air into the face piece. The emergency by-pass mode uses more air, but it enables fire fighters to exit a hazardous environment if the regulator stops operating. A fire fighter who must use the emergency by-pass mode must leave the hazardous area IMMEDIATELY.

Like the cylinder, the regulator has a gauge that indicates the pressure of the breathing air remaining in the cylinder. This gauge enables the user to monitor the amount of air remaining in the cylinder. The regulator pressure gauge can be mounted directly on the regulator or on a separate hose so it can be attached to a shoulder strap for easier viewing.

Figure 2-28 SCBA regulator.

The regulator and cylinder pressure gauges should read within 100 psi of each other.

NFPA standards require that SCBA have a low-air alarm that activates when one quarter of the air supply remains. This alarm may be a bell or whistle, a vibration, or a flashing **light-emitting diode (LED)**. Early SCBAs (those manufactured before the 1997 edition of NFPA Standard 1981 became effective) have a single low-air supply alarm. Newer models have two different types of low-air alarms that operate independently of each other and activate different senses. For example, one alarm might ring a bell, and the second alarm might vibrate or flash an LED. Fire fighters should never ignore the low-air alarm.

Some SCBAs are manufactured with integrated PASS devices. Turing on the air supply automatically activates these PASS devices. This ensures that a fire fighter doesn't forget to turn the PASS device on when entering a hazardous

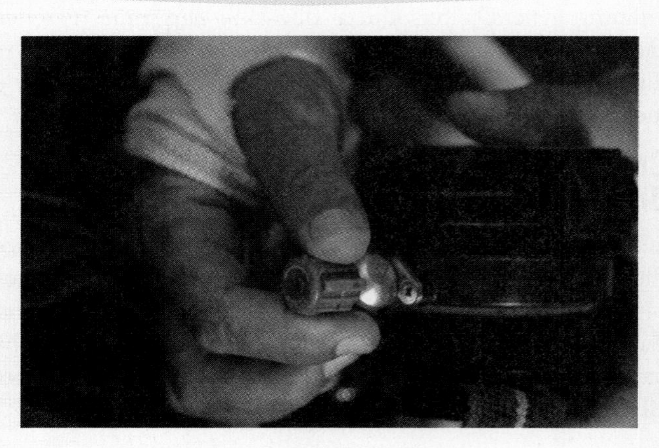

Figure 2-29 The SCBA emergency by-pass valve is used if the regulator malfunctions.

Figure 2-30 A PASS device can be integrated into an SCBA.

Figure 2-31 SCBA face pieces come in several sizes.

area. If your SCBA is equipped with a PASS device, learn how to activate it and how to turn it off (▲ Figure 2-30).

Although there are many different models of SCBA regulators, fire fighters must learn how to operate the particular model that is used in their department. Fire fighters should be able to operate the regulator in the dark and with gloves on.

Face Piece Assembly

The face piece delivers breathing air to the fire fighter (► Figure 2-31). The face piece also protects the face from high temperatures and smoke. The face piece consists of a facemask with a clear lens, an exhalation valve, and—on models with a harness-mounted regulator—a flexible low-pressure supply hose. Other models will have the regulator attached directly to the face piece.

The face piece should cover the entire face. The part in contact with the skin is made of special rubber or silicon to provide a tight seal. The clear lens allows for vision. Exhaled air is exhausted from the face piece through the one-way exhalation valve, which has a spring mechanism to maintain positive pressure inside the face piece. Because it is hard to communicate through a face piece, some models have a voice amplification device.

Face pieces may be equipped with nose cups to help prevent fogging of the clear lens. Fogging is a greater problem in colder climates. The compressed air you breathe is dry, but the air you exhale is moist. The flow of dry air helps to prevent fogging of the lens.

The face piece is held in place with a web-like series of straps, or a net and straps. Face pieces should be stored with the straps in the longest position to make them easier to don. Pull the end of the straps toward the back of the head (not out to the sides) to tighten them and ensure a snug fit.

A leak in the face piece seal can be caused by an improperly sized face piece, improper donning, or facial hair around the edge of the face piece. Leaks are dangerous for two reasons. First, a large leak could overcome the positive pressure in the face piece and allow contaminated air to enter the face piece. Second, a leak of any size will deplete the breathing air and reduce the amount of time available for fire fighting.

Face pieces are manufactured in several sizes. NFPA 1500 requires that all fire fighters must have their face pieces fit-tested annually to ensure that they are wearing the proper size. Some departments issue individual face pieces to each fire fighter; others provide a selection of sizes on each apparatus. NFPA 1500 also requires that the sealing surface of the face piece must be in direct contact with the user's skin. There must not be hair or a beard in the seal area.

Pathway of Air through an SCBA

The breathing air is stored under pressure in the cylinder. The air passes through the cylinder shut-off valve into the high-pressure air line, or hose, that takes it to the regulator. The regulator reduces the high pressure to low pressure.

The regulator opens when the user inhales, reducing the pressure on the downstream side. In an SCBA unit with a face piece-mounted regulator, the air next goes directly into the face piece. In units with a harness-mounted regulator, the air travels from the regulator through a low-pressure hose into the face piece. From the face piece, the air is inhaled through the user's air passages and into the lungs.

When the user exhales, used air is returned to the face piece. The exhaled air is exhausted from the face piece through the exhalation valve. This cycle repeats with every

Fire Fighter Safety Tips

Be sure that your face piece is properly fitted and the correct size for your face.

Fire Fighter Tips

Controlled-Breathing Technique

Controlled breathing helps extend the SCBA air supply. Controlled breathing is a conscious effort to inhale naturally through the nose and to force exhalation from the mouth. Practicing controlled breathing during training will help you to maximize the efficient use of air while you are working.

inhalation. As the pressure in the face piece drops, the exhalation valve closes and the regulator opens.

Skip-Breathing Technique

The skip-breathing technique helps conserve air while using an SCBA in a firefighting situation. The technique is to take a short breath, hold, take a second short breath (do not exhale in between breaths), and then relax with a long exhale. Each breath should take 5 seconds.

A simple drill can demonstrate the benefits of skip breathing. One fire fighter dons PPE and an SCBA with a full air cylinder, and walks in a circle around a set of traffic cones, around the track at the local school, or, if safety permits, around the parking lot at the fire station. A second fire fighter times how long it takes for the fire fighter to completely deplete the air in the SCBA. After the first fire fighter is completely rested, replace the air cylinder, and repeat the same drill using the skip-breathing technique. Compare times after completion of both evolutions.

Mounting Breathing Apparatus

SCBA should be located so that fire fighters can don it quickly when they arrive at the scene of a fire. Seat-mounted brackets enable fire fighters to don SCBA en route to an emergency scene, without unfastening their seat belts or otherwise endangering themselves. This enables fire fighters to begin work as soon as they arrive.

There are several types of apparatus seat-mounting brackets. Some hold the SCBA with the friction of a clip. Others are equipped with a mechanical hold-down device that must be released to remove the SCBA. Regardless of the mounting system used, it must hold the SCBA securely in the bracket. A collision or sudden stop should not dislodge the SCBA from the brackets. A loose SCBA can be a dangerous projectile. The fire fighter who dons SCBA from a seat-mounted bracket should not tighten the shoulder straps while seated, so as not to dislodge the SCBA in a sudden stop situation. The fire fighter should be secured by a seat belt or combination seat belt and shoulder harness.

Compartment-mounted SCBA units also can be donned quickly. These units are used by fire fighters who arrive in vehicles without seat-mounted SCBA or whose seats were not equipped with them. Apparatus drivers and fire fighters who arrive in their private vehicles often use compartment-mounted SCBA units. The mounting brackets should be positioned high enough for easy donning. Some mounting brackets allow the fire fighter to lower the SCBA without removing it from the mounting bracket. Older apparatus may have the brackets mounted on the exterior of the vehicle. An exterior-mounted SCBA should be protected from weather and dirt by a secure cover.

SCBA also may be kept in a storage case. This method is most appropriate for transporting extra SCBA units. It should not be used to transport SCBA that will be used during the initial phase of operations at a fire scene.

The SCBA should be stored on the apparatus in ready-for-use condition, with the main cylinder valve closed. After checking the SCBA, close the cylinder valve and slowly open the by-pass valve to release pressure in the system. The low-pressure alarm should sound as the pressure bleeds down. After releasing the pressure, close the by-pass valve. The SCBA is now ready to be placed in a bracket for immediate use.

Donning Self-Contained Breathing Apparatus

Donning SCBA is an important skill. Fire fighters must be able to don and activate SCBA in one minute. Personal safety and the effectiveness of the firefighting operation depend on this skill. Fire fighters must be wearing full PPE before donning SCBA. Before beginning the actual donning process, fire fighters must carefully check the SCBA to ensure it is ready for operation.

- Check to be sure the air cylinder has at least 90% of its rated pressure.
- If the SCBA has a donning/doffing switch, be sure that it is activated.
- Open the cylinder valve two or three turns, listen for the low-air alarm to sound, and then open the valve fully.
- Check the pressure gauges on the regulator and on the cylinder. Both gauges should read within 100 psi of each other.
- Check all harness straps to be sure they are fully extended.
- Check all valves to be sure they are in the correct position. (An open by-pass valve will waste air.)

Donning SCBA from an Apparatus Seat Mount

Donning SCBA while en route to an emergency can save valuable time. However, this requires that you don all of your protective clothing before mounting the apparatus. Place your arms through the shoulder straps as you sit down, and then fasten your seat belt. Or, you can fasten your seat belt first, and then slide one arm at a time through the shoulder straps of the SCBA harness. You can partially tighten the shoulder straps while you are seated.

When you arrive at the emergency scene, release your seat belt, activate the bracket release, and exit the apparatus. Be sure to take a face piece with you. Face pieces should be kept in a storage bag close to each seat-mounted SCBA or attached to the harness. After exiting from the apparatus, attach the waist strap, and then tighten and adjust the shoulder and waist straps.

Follow the steps in ► **Skill Drill 2-3** to don SCBA from a seat-mounted bracket. Before beginning this skill drill, inspect the SCBA to ensure it is ready for service.

1. Don full PPE ensemble prior to mounting the fire apparatus. Safely mount the apparatus and sit in the seat, placing arms through SCBA shoulder straps. **(Step 1)**
2. Fasten your seat belt. Partially tighten the shoulder straps. Do not fully tighten at this time. When the apparatus comes to a complete stop at the emergency scene, release your seat belt and release the SCBA from the mounting bracket. Carefully exit the apparatus. **(Step 2)**
3. Attach the waist belt and cinch down. **(Step 3)**
4. Adjust shoulder straps until they are tight. **(Step 4)**
5. Open the main cylinder valve. **(Step 5)**
6. Remove or loosen your helmet and pull back the protective hood. Don the face piece and check for leaks. Pull the protective hood up over the head, put the helmet back in place, and secure the chin strap. **(Step 6)**
7. If necessary, connect the regulator to the face piece. **(Step 7)**
8. Activate the airflow and PASS alarm. **(Step 8)**

Donning SCBA from a Compartment Mount

To don a compartment-mounted SCBA, slide one arm through the shoulder harness strap. Slide the other arm through the other shoulder strap. Release the SCBA from the mounting bracket. Adjust the shoulder straps to carry the SCBA fairly high on your back. Attach the ends of the waist strap and tighten the waist strap.

Follow the steps in **Skill Drill 2-4** to don SCBA from a side-mounted compartment or bracket. Before starting, check to be sure that the SCBA has been inspected and is ready for service. Examine the SCBA to see how it is mounted in the compartment. Check the release mechanism and how it operates. If the SCBA is mounted on an exterior bracket, remove the protective cover before beginning the donning sequence.

1. Stand in front of the SCBA bracket and fully open the main cylinder valve.
2. Turn your back toward the SCBA, slide your arms through the shoulder straps, and partially tighten the straps.
3. Release the SCBA from the bracket and step away from the apparatus.
4. Attach the waist belt and tighten.
5. Adjust the shoulder straps.
6. Remove your helmet and pull the hood back.
7. Don the face piece and check for adequate seal.
8. Pull the protective hood into position, replace your helmet, and secure the chin strap.
9. If necessary, connect the regulator to the face piece.
10. Activate the airflow and PASS alarm.

Donning SCBA from the Ground, the Floor, or a Storage Case

Fire fighters must sometimes don an SCBA that is stored in a case or on the ground. Two methods can be used: the over-the-head method and the coat method.

Over-the-Head Method

To don an SCBA using the over-the-head method, place the SCBA on the ground or on the floor with the cylinder valve facing away from you. Lay the shoulder straps out to each side of the backpack. Grasp the backplate with both hands and lift the SCBA over your head. Let the backpack slide down your back. The straps will slide down your arms. Balance the unit on your back. Attach and tighten the waist strap and then tighten the shoulder straps. Follow the steps in ► **Skill Drill 2-5** to don SCBA using the over-the-head method. Before starting, ensure that the SCBA has been inspected and is ready for service.

1. If necessary, open the protective case and lay out the SCBA so that the cylinder valve is away from you and the shoulder straps are to the sides. **(Step 1)**
2. Fully open the main cylinder valve. **(Step 2)**
3. Bend down and grasp the SCBA backplate with both hands. Using your legs, lift the SCBA over your head. Once the SCBA clears your head, rotate it 180°, so the waist straps are pointed toward the ground. **(Step 3)**
4. Slowly slide the pack down your back. Make sure that your arms slide into the shoulder straps. Once the SCBA is in place, tighten the shoulder straps and secure the waist strap. **(Step 4)**
5. Remove your helmet and pull the hood back. Don the face piece and check for an adequate seal. Pull your protective hood into position, replace your helmet, and secure the chin strap. **(Step 5)**
6. If necessary, connect the regulator to the face piece. Activate the airflow and PASS alarm. **(Step 6)**

Coat Method

To don an SCBA using the coat method, place the SCBA on the ground or on the floor with the cylinder valve facing toward you. Spread out and extend the shoulder straps. Use your left hand to grasp the left shoulder strap close to the backplate. Use your right hand to grasp the right shoulder strap farther away from the backplate. Swing the SCBA over your left shoulder. Release your right arm and slide it

2-3 Skill Drill

Donning SCBA from a Seat-Mounted Bracket

1

Don full PPE ensemble prior to mounting fire apparatus. Safely mount the apparatus and sit in the seat, placing arms through SCBA shoulder straps.

2

Fasten your seat belt. Partially tighten the shoulder straps. When the apparatus stops, release the seat belt and release the SCBA from its brackets. Exit apparatus.

3

Attach waist belt and cinch down.

4

Adjust shoulder straps until they are tight.

through the right shoulder harness strap. Tighten both shoulder straps. Attach and tighten the waist belt.

Follow the steps in ► Skill Drill 2-6 to don SCBA using the coat method. Before starting, ensure that the SCBA has been inspected and is ready for service.

1. If necessary, open the protective case and lay out the SCBA so that the cylinder valve is facing you and the straps are laid out to the sides. Fully open the main cylinder valve. Place your dominant hand on the opposite shoulder strap. For safety reasons be sure to grasp the strap as close to the backplate as possible. **(Step 1)**
2. Lift the SCBA and swing it over your dominant shoulder, being careful of people or objects around you. **(Step 2)**
3. Slide your other hand between the SCBA cylinder and the corresponding shoulder strap. **(Step 3)**
4. Tighten the shoulder straps. **(Step 4)**
5. Attach the waist belt and adjust tightness. **(Step 5)**
6. Remove your helmet and pull your hood back. Don the face piece and check for an adequate seal. **(Step 6)**
7. Pull the protective hood into position, replace the helmet, and secure the chin strap. If necessary, con-

2-3 **Skill Drill**

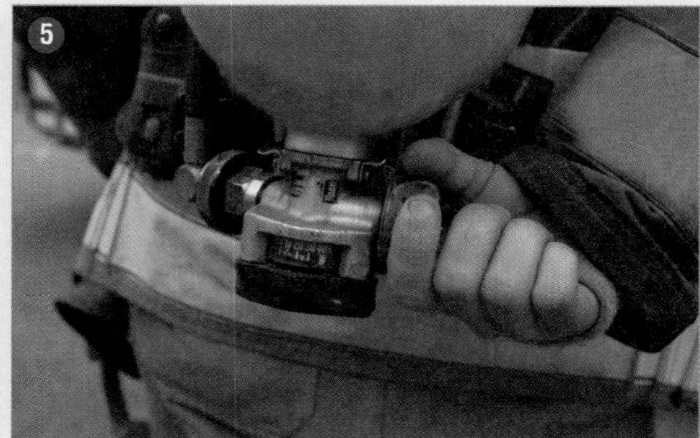

5 Open the main cylinder valve.

6 Loosen or remove helmet and pull hood back. Don face piece and check for leaks. Replace protective hood and helmet and secure chin strap.

7 If necessary, connect regulator to face piece.

8 Activate airflow and PASS alarm.

nect the regulator to the face piece. Activate the airflow and PASS alarm. **(Step 7)**

These instructions will have to be modified for different SCBA units. The sequence for adjusting shoulder straps and waist belts varies with different models. Modifications must also be made for SCBAs with waist-mounted regulators. Refer to the specific manufacturer's instructions supplied with each unit. Follow the standard operating procedures for your department.

Donning the Face Piece

Your face piece keeps contaminated air outside and pure breathable air inside. To perform properly, it must be the correct size and it must be adjusted to fit your face. Be sure you have been tested to determine your proper size.

There must be no facial hair in the seal area. Eyeglasses that pass through the seal area cannot be worn with a face piece, because they can cause leakage between the face piece and your skin. Your face piece must match your SCBA. You cannot interchange a face piece from a different SCBA model.

2-5 Skill Drill

Donning SCBA Using the Over-the-Head Method

Open the case and lay out the SCBA with the cylinder valve away from you and the shoulder straps out to the sides.

Fully open the main cylinder valve.

Bend down and grasp the SCBA backplate with both hands. Using your legs, lift the SCBA over your head. Rotate the SCBA 180°, so the waist straps are pointed to the ground.

Slide the SCBA down your back while your arms slide into the shoulder straps. Tighten the shoulder straps and secure the waist belt.

Remove your helmet and pull the hood back. Don your face piece and check for an adequate seal. Pull your protective hood into position, replace your helmet, and secure the chin strap.

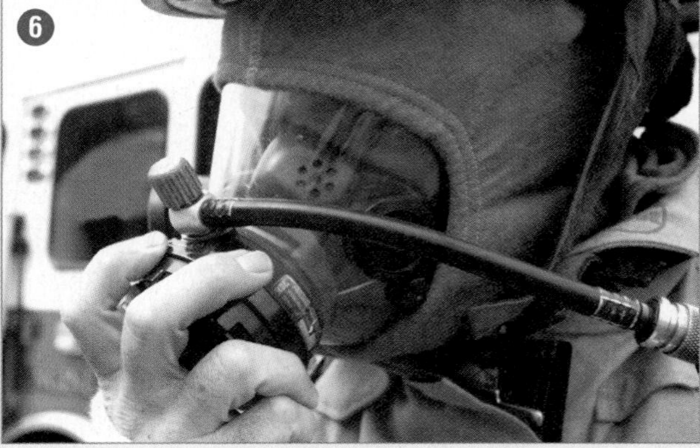

If necessary, connect the regulator to the face piece. Activate the airflow and PASS alarm.

2-6 Skill Drill

Donning SCBA Using the Coat Method

1 Open the case and lay out the SCBA with the cylinder valve away from you and the shoulder straps out to the sides. Fully open the main cylinder valve. Place your dominant hand on the opposite shoulder strap.

2 Lift the SCBA and swing it over your dominant shoulder.

3 Slide your other hand between the SCBA cylinder and the corresponding shoulder strap.

4 Tighten the shoulder straps.

5 Attach the waist belt and adjust tightness.

6 Remove your helmet and pull your hood back. Don the face piece and check for an adequate seal.

7 Pull the hood into position, replace the helmet, and secure the chin strap. If necessary, connect the regulator to the face piece. Activate the airflow and PASS alarm.

Face pieces for various brands and models of SCBAs are slightly different. Some have the regulator mounted on the face piece; others have it mounted on the harness straps. Fire fighters must learn about the face pieces used by their departments.

Follow the steps in **Skill Drill 2-7** to don a face piece:

1. Make sure you have donned your protective hood. Remove your helmet and pull the hood down over your neck.
2. Fully extend the straps on the face piece. **(Step 1)**
3. Rest your chin in the chin pocket at the bottom of the mask. **(Step 2)**
4. Fit the face piece to your face, bringing the straps or webbing over your head. **(Step 3)**
5. Tighten the lowest two straps. To tighten, pull the straps straight back, not out and away from your head. **(Step 4)**
6. Tighten the pair of straps at your temple, if any.
7. If your model has additional straps, tighten the top strap(s) last. **(Step 5)**
8. Check for a proper seal. This process depends on the model and type of face piece you use. **(Step 6)**
9. Pull the protective hood up so it covers all bare skin. Be sure it does not get under your face piece or obscure your vision.
10. Replace your helmet and secure the chin strap. **(Step 7)**
11. Install the regulator on your face piece or attach the low-pressure air supply hose to the regulator. **(Step 8)**

2-7 **Skill Drill**

Donning a Face Piece

1 Fully extend the straps on the face piece.

2 Place your chin in the chin pocket.

3 Fit the face piece to your face, bringing the straps or webbing over your head.

4 Tighten the lowest two straps.

Safety Precautions for Self-Contained Breathing Apparatus

As you practice using your SCBA, remember that this equipment is your protection against serious injury or death in hazardous conditions. Practice safe procedures from the beginning.

Learn to recognize the low-air alarm on your SCBA. As soon as your alarm goes off, you must exit the hazardous environment before your air supply is depleted. Never get into a situation from which you cannot escape when your low-air alarm goes off.

Before you enter a hazardous environment, make sure your PASS device is activated. Be sure you are properly logged into your accountability system. Always work in teams of two in hostile environments. Always have at least two fire fighters outside at the ready whenever two fire fighters are working in a hostile environment.

Fire Fighter Safety Tips

NFPA 1001 requires you to don and activate your SCBA in one minute. Practice donning your SCBA until you can do it quickly and smoothly.

2-7 Skill Drill

5 If there are more straps, tighten the top straps last.

6 Check for a proper seal.

7 Pull your protective hood up so it covers all bare skin. Don your helmet and secure the chin strap.

8 Install the regulator on your face piece or attach the low-pressure air supply hose to the regulator.

Fire Fighter Tips

Restricted Spaces

The size and shape of SCBA may make it difficult for you to fit through tight openings. Several techniques may help you navigate these spaces.

- Change your body position: Rotate your body 45° and try again.
- Loosen one shoulder strap and change the location of the SCBA on your back.
- If you have no other choice, you may have to remove your SCBA. In this case, do not let go of the backpack and harness for any reason. Keep the unit in front of you as you navigate through the tight space. Re-attach the harness as soon as you are through the restricted space. **This is an absolutely "last resort" procedure!**

Preparing for Emergency Situations

Because hostile environments are often unpredictable, fire fighters must be prepared to react if an emergency situation occurs while they are using SCBA. In emergencies, follow simple guidelines. First, keep calm, stop, and think. Panic increases air consumption. Try to control your breathing by maintaining a steady rate of respirations. A calm person has a greater chance of surviving an emergency.

If the problem is with your SCBA, try to exit the hostile environment. Use the emergency by-pass valve so you can breathe if your regulator malfunctions.

If you are in danger, activate your PASS device. Use your hand light to attract attention. If you have a portable radio, call for help. These are simple but effective steps; additional emergency techniques are covered in Chapter 17, Fire Fighter Survival.

Doffing Self-Contained Breathing Apparatus

The procedure for doffing your SCBA depends on the model and whether it has a face piece-mounted regulator or a harness-mounted regulator. Follow the procedures recommended by the manufacturer and your department's SOPs.

In general, you should reverse the steps used to don your SCBA. Follow the steps in **► Skill Drill 2-8** to doff your SCBA:

1. Remove your gloves. Remove the regulator from your face piece or disconnect the low-pressure air supply hose from the regulator. **(Step 1)**
2. Shut off the air-supply valve.
3. Remove your helmet and pull your protective hood down around your neck. **(Step 2)**
4. Loosen the straps on your face piece. **(Step 3)**
5. Remove your face piece. **(Step 4)**
6. Release your waist belt. **(Step 5)**
7. Loosen the shoulder straps and remove the SCBA. **(Step 6)**
8. Shut off the air-cylinder valve. **(Step 7)**
9. Bleed the air pressure from the regulator by opening the emergency by-pass valve. **(Step 8)**
10. If you have an integrated PASS device, turn it off.
11. Place the SCBA in a safe location where it will not get dirty or damaged. **(Step 9)**

Putting It All Together: Donning the Entire PPE Ensemble

The complete PPE ensemble consists of both personal protective clothing and respiratory protection (SCBA). Although donning personal protective clothing and donning and operating SCBA can be learned and practiced separately, you must be able to integrate these skills to have a complete PPE ensemble. Each part of the complete ensemble must be in the proper place to provide whole-body protection.

The steps for donning a complete PPE are listed below:

- Place the protective hood over your head.
- Put on your bunker pants and boots. Adjust the suspenders and secure the front flap of the pants.
- Put on your turnout coat and secure the front.
- Open the air-cylinder valve on your SCBA and check the air pressure.
- Put on your SCBA.
- Tighten both shoulder straps.
- Attach the waist belt and tighten it.
- Fit the face piece to your face.
- Tighten the straps, beginning with the lowest straps.
- Check the face piece for a proper seal.
- Pull the protective hood up so that it covers all bare skin, but does not obscure vision.
- Place your helmet on your head with the ear tabs extended and secure the chin strap.
- Turn up your coat collar and secure it in front.
- Put on your gloves.
- Check your clothing to be sure it is properly secured.
- Be sure your PASS device is turned on.
- Attach your regulator or turn it on to start the flow of breathing air.
- Work safely!

SCBA Inspection and Maintenance

SCBA must be properly serviced and prepared for the next use each time it is used, whether it is an actual emergency incident or a training exercise. The air cylinder must be changed or refilled, the face piece and regulator must be sanitized according to the manufacturer's instructions, and the unit must be cleaned, inspected and checked for proper operation. It is the

2-8 Skill Drill

Doffing SCBA

Remove your gloves. Remove the regulator from the face piece or disconnect the low-pressure hose from the regulator.

Remove your helmet and pull the protective hood down around your neck.

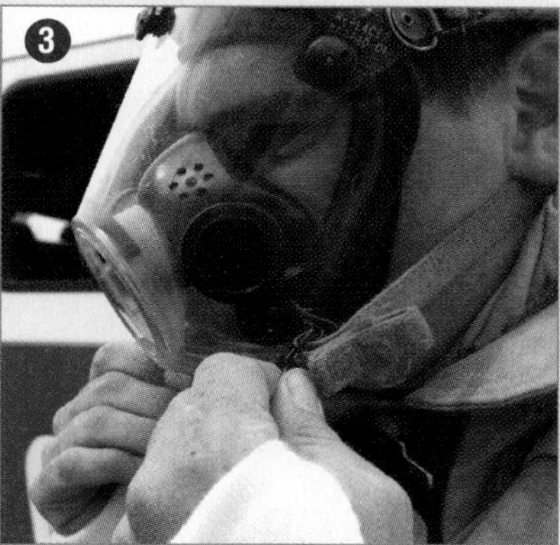

Loosen the face piece straps.

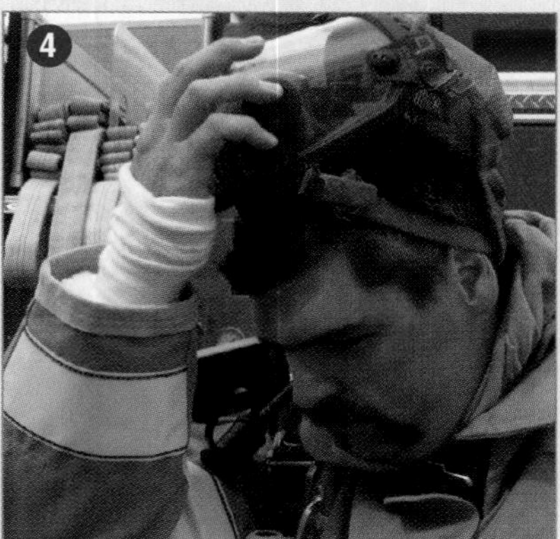

Remove your face piece.

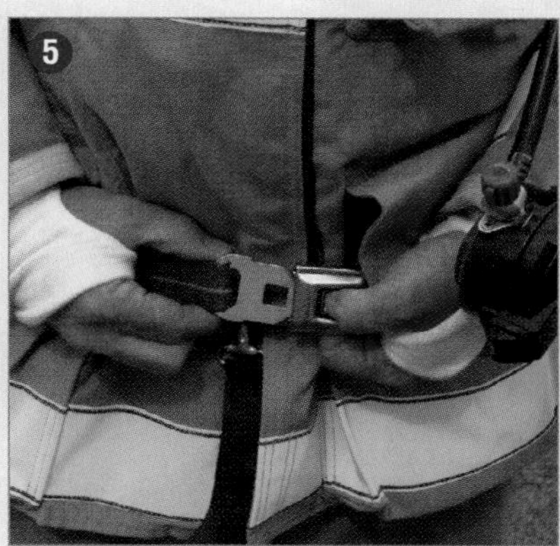

Release your waist belt.

Loosen the shoulder straps and remove the SCBA.

Shut off the air-cylinder valve.

Bleed the air pressure from the regulator.

Place the SCBA in a safe location.

user's responsibility to ensure that the SCBA is in ready condition before it is returned to the fire apparatus.

Each SCBA must be checked on a regular basis to ensure that it is ready for use. There are different procedures for daily, monthly, and annual inspections. The daily inspection procedure should be used when restoring a unit to service after it has been used.

If an SCBA inspection reveals any problems that cannot be remedied by routine maintenance, the SCBA must be removed from service for repair. Only properly trained and certified personnel are authorized to repair SCBA.

Daily Inspection

Each SCBA unit should be inspected daily or at the beginning of each shift. When fire stations are not staffed, SCBA should be inspected at least once a week. Follow the steps in ▼ **Skill Drill 2-9** for daily SCBA inspection:

1. Check the backpack and harness straps. Make sure these components are intact and the straps are kept lengthened. **(Step 1)**
2. Check the air-cylinder pressure. Make sure it is full. Turn on the air-cylinder valve. Check the regulator gauge or remote gauge. It should read within 100 psi of the cylinder gauge. **(Step 2)**

2-9 **Skill Drill**

Daily SCBA Inspection

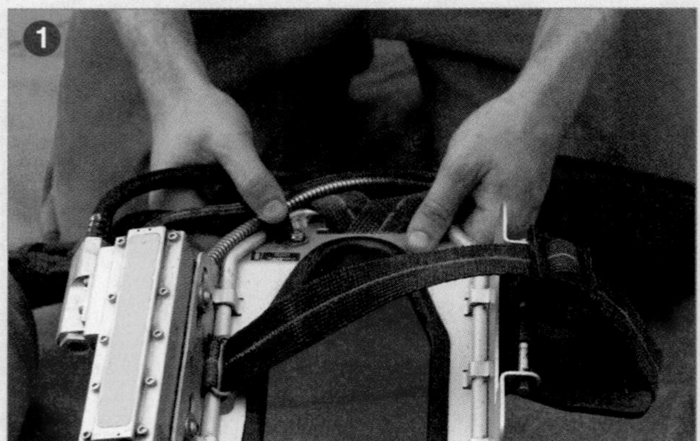

1 Check backpack and harness straps.

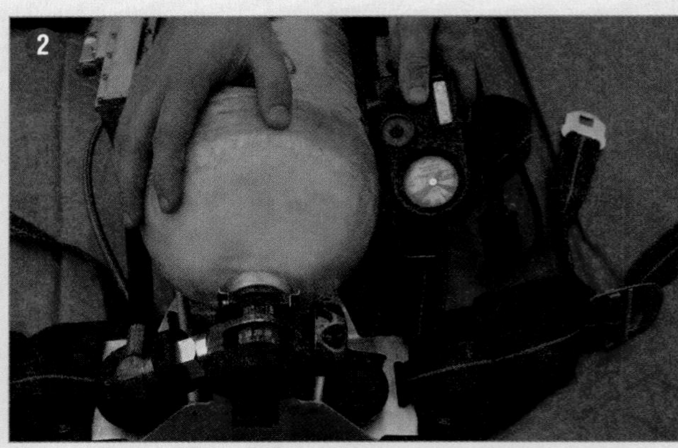

2 Check air-cylinder pressure. Turn on the air-cylinder valve and check gauge pressure.

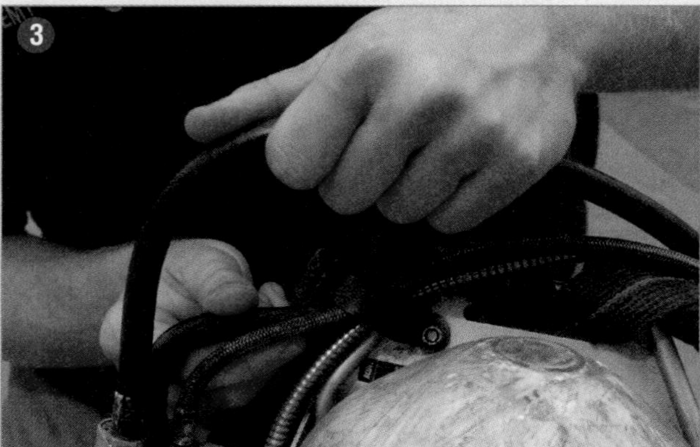

3 Check condition of all hoses while pressurized.

4 Activate integrated PASS device, if present.

3. Check the condition of all hoses while they are pressurized. **(Step 3)**
4. Activate the integrated PASS device, if there is one. **(Step 4)**
5. Check the face piece. It should be clean and undamaged. Check the operation of the exhalation valve. **(Step 5)**
6. Connect the regulator to the face piece and take one or two test breaths. **(Step 6)**
7. Close the cylinder valve and open the emergency by-pass valve to bleed the pressure. **(Step 7)**
8. Check to ensure that the low-pressure alarm(s) activate at the proper pressure. Close the emergency by-pass valve and restore the SCBA to ready condition. **(Step 8)**

Monthly Inspection

SCBA should be completely checked each month for proper operation, for leaks, and for any deterioration. Follow the steps in **Skill Drill 2-10** for monthly SCBA inspection:

1. Remove the SCBA from the apparatus and place it on the floor or on a workbench.
2. Inspect the mounting bracket for damage or wear. Lubricate it if this is recommended by manufacturer.
3. Look at the overall condition of the SCBA and note any damage.

2-9 Skill Drill

Check the face piece.

Connect the regulator to the face piece and take test breaths.

Close the cylinder valve and open the emergency by-pass valve to bleed the pressure.

Check function and activation pressure of low-air alarm. Close by-pass valve and restore unit to ready condition.

4. Remove the air cylinder from the SCBA harness and check the hydrostatic test date.
5. Check the air cylinder for damage or wear.
6. Inspect the SCBA shoulder straps and waist belt for damage, cuts, burns, or wear.
7. Check all buckles and fasteners to ensure they work properly.
8. Examine the backplate for damage, cracks, or rust.
9. Make sure all connection points between the cylinder and the SCBA harness operate properly and are free of damage or corrosion. Lubricate them if this is recommended by the manufacturer's literature.
10. Reattach the air cylinder to the SCBA harness.
11. Check all hoses and connection points for wear, cuts, or damage.
12. Open the air-cylinder valve. Compare readings on the cylinder and regulator gauges to ensure they match.
13. Attach the face piece and check the regulator for proper operation.
14. Activate the PASS alarm. Allow the SCBA to sit idle until the PASS alarm sounds.
15. Shut off the air-cylinder valve and open the by-pass valve to bleed the pressure. Check the low-pressure alarm as the pressure bleeds down.
16. Return the SCBA to the mounting bracket.
17. Complete all necessary paperwork.

Annual Inspection

A complete annual inspection and maintenance must be performed on each SCBA. The annual inspection must be performed by a certified manufacturer's representative or a person who has been trained and certified to perform this work. SCBA requires regular inspection and maintenance to ensure that it will perform as intended.

Servicing SCBA Cylinders

A pressurized SCBA cylinder contains a tremendous amount of potential energy. Not only does the air within the cylinder exert considerable pressure on its walls, the cylinder itself is used under extreme conditions on the fireground. If the cylinder ruptures and suddenly releases this energy, it can cause serious injury or death. Cylinders must be regularly inspected and tested to ensure they are safe. Cylinders must be visually inspected during daily and monthly inspections. More detailed inspection is required if a cylinder has been exposed to excessive heat, come into contact with flame, exposed to chemicals, or dropped.

The U.S. Department of Transportation requires hydrostatic testing for SCBA cylinders on a periodic basis and limits the number of years that a cylinder can be used. Hydrostatic testing identifies defects or damage that render the cylinder unsafe. Any cylinder that fails a hydrostatic test is immediately taken out of service and cannot be used.

Cylinders constructed of different materials have different testing requirements. Aluminum, steel, and carbon-fiber cylinders must be hydrostatically tested every five years. Cylinders constructed of composite materials such as Kevlar-aramid or fiberglass fibers must be tested every three years. Fire fighters must know what type of cylinders are used by their departments and must check each cylinder for a current hydrostatic test date before filling it.

Replacing SCBA Cylinders

A used air cylinder can be quickly replaced with a full cylinder in the field to enable you to continue firefighting activities. A single fire fighter must doff SCBA to replace the air cylinder; two fire fighters working together can change cylinders without removing SCBA. The steps listed below outline how a single person makes a cylinder change. These procedures will change depending on the model of SCBA being used. Follow the procedures recommended by the manufacturer and by department SOPs.

Practice changing air cylinders until you become proficient. A fire fighter should be able to change cylinders in the dark and while wearing gloves. Follow the steps in ▶ **Skill Drill 2-11** to replace an SCBA cylinder:

1. Place the SCBA on the floor or a bench. **(Step 1)**
2. Turn off the cylinder valve. **(Step 2)**
3. Bleed off the pressure by opening the by-pass valve. **(Step 3)**
4. Disconnect the high-pressure supply hose. Keep the ends clean. **(Step 4)**
5. Release the cylinder from the backpack. **(Step 5)**
6. Slide a full cylinder into the backpack. Align the outlet to connect the supply hose. Lock the cylinder in place. **(Step 6)**
7. Check that the "O" ring is present and in good shape. **(Step 7)**
8. Connect the high-pressure hose to the cylinder. Hand tighten only. **(Step 8)**
9. Open the cylinder valve. Check the regulator gauge or remote gauge. It should read within 100 psi of the cylinder gauge. **(Step 9)**

To save time, someone else can replace the air cylinder while you are wearing the SCBA harness. You should not overtax yourself, however, by replacing the cylinder and going back to work without adequate rest when you need it.

Refilling SCBA Cylinders

Compressors and cascade systems are used to refill SCBA cylinders. A compressor or a cascade system can be permanently located at a maintenance facility or at a firehouse, or they can be mounted on a truck or a trailer for mobile use. Mobile filling units are often brought to the scene of a large fire.

2-11 Skill Drill

Replacing an SCBA Cylinder

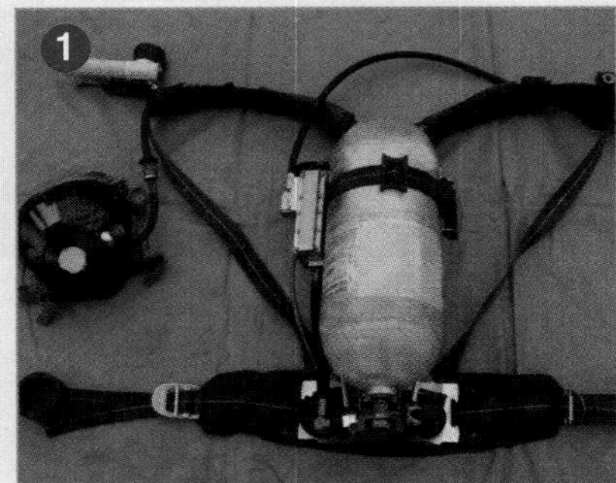
1 Place the SCBA on the floor or a bench.

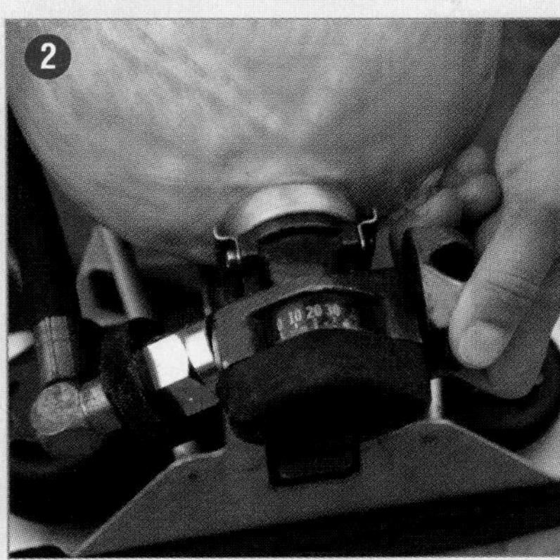
2 Turn off the cylinder valve.

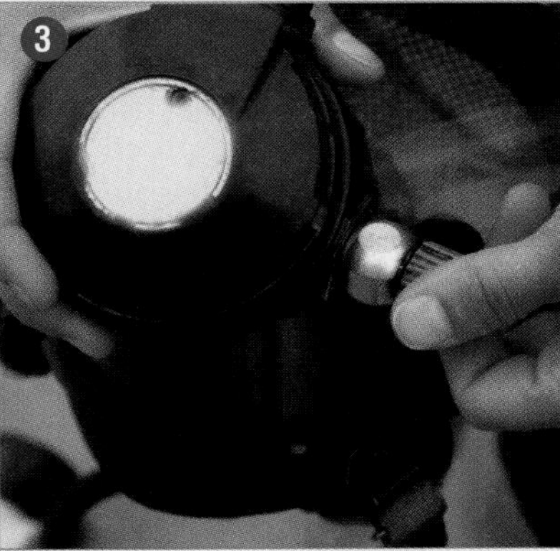
3 Open the by-pass valve to bleed off pressure.

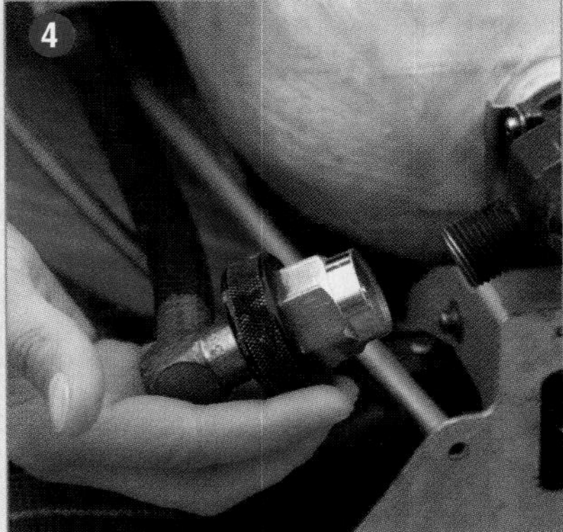
4 Disconnect the high-pressure supply hose.

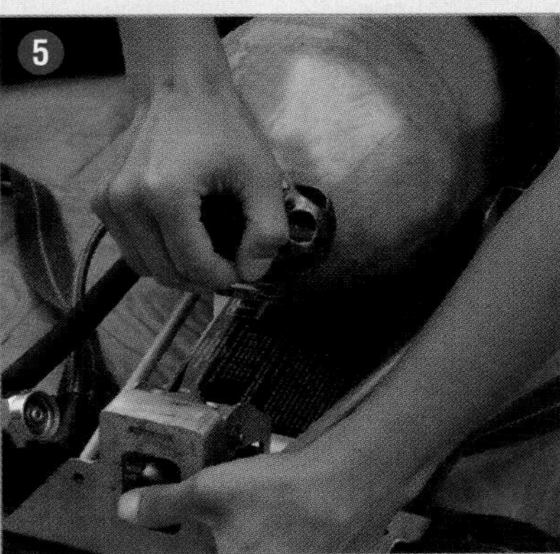
5 Release the cylinder from the backpack.

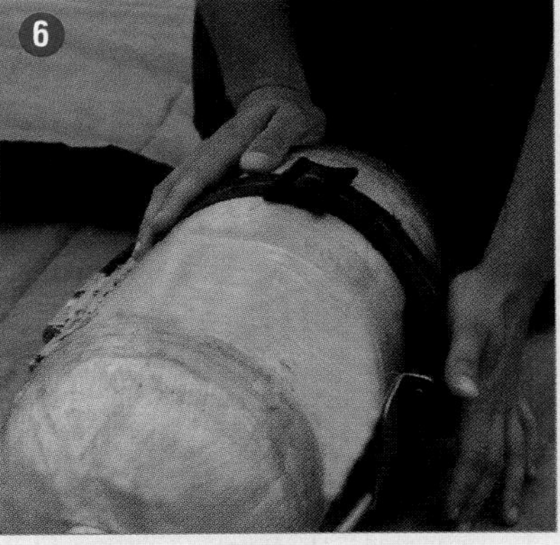
6 Slide a full cylinder into the backpack. Align the outlet to the supply hose. Lock the cylinder in place.

7 Check that the "O" ring is present and in good shape.

8 Connect the high-pressure hose to the air cylinder.

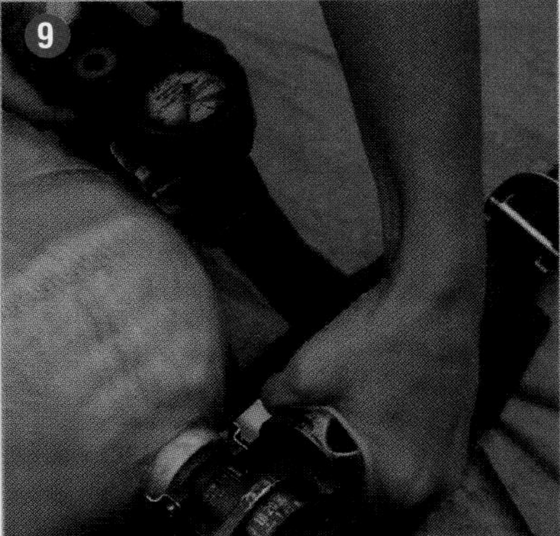
9 Open the cylinder valve. Check gauge reading.

Compressor systems filter atmospheric air, compress it to a high pressure, and transfer it to the SCBA cylinders (► Figure 2-32). Cascade systems have several large storage cylinders of compressed breathing air connected by a high-pressure manifold system. The empty SCBA cylinder is connected to the cascade system, and compressed air is transferred from the storage tanks to the cylinder. The storage cylinder valves must be opened and closed, one at a time, to fill the SCBA cylinder to the recommended pressure.

Proper training is required to fill SCBA cylinders. Whether your department has an air compressor or a cascade system, only those fire fighters who have been trained on the equipment should use it to refill air cylinders.

Figure 2-32 Air compressors are installed at many fire stations to refill SCBA cylinders.

Cleaning and Sanitizing SCBA

Most SCBA manufacturers will provide specific instructions for the care and cleaning of their models. The first step in cleaning the SCBA is to rinse the entire unit using a hose with clean water. The harness assembly and cylinder can be cleaned with a mild soap and water solution. If additional cleaning is needed, the unit can be scrubbed with a stiff brush. After scrubbing, the SCBA harness and cylinder should be rinsed with clean water.

After a fire, face pieces and regulators can be cleaned with a mild soap and warm water or a disinfectant cleaning solution. The face piece should be fully submerged in the cleaning solution. If additional cleaning is needed, a soft brush can be used to scrub the face piece. Avoid scratching the lens or damaging the exhalation valve. The regulator can be cleaned with the same solution, but should not be submerged. The face piece and regulator should then be rinsed with clean water.

Allow the SCBA time to dry completely before returning it to service. Check for any damage before returning the SCBA to service. Follow the steps in Skill Drill 2-12 to clean and sanitize an SCBA:

1. Inspect the SCBA for any damage that may have occurred before cleaning.
2. Remove the face piece from the regulator. On some models, the regulator also can be removed from the harness.
3. Detach the SCBA cylinder from the harness.
4. Rinse all parts of the SCBA with clean water. Water from a garden hose can be used for this step.
5. Using a stiff brush, along with a soap-and-water solution, scrub the SCBA cylinder and harness. Rinse and set aside to dry.
6. In a 5-gallon bucket make a mixture of mild soap and water or use the manufacturer's recommended cleaning and disinfecting solution and water.
7. Submerge the SCBA face piece into the soapy water or cleaning solution. For heavier cleaning, allow the face piece to soak.
8. Clean the regulator with the soapy water or cleaning solution, following the manufacturer's instructions.
9. Use a soft brush, if necessary, to scrub contaminants from the face piece and regulator.
10. Completely rinse the face piece and the regulator with clean water. Set them aside and allow them to dry
11. Reassemble and inspect the entire SCBA before placing it back in service.

Fire Fighter Safety Tips

Refilling SCBA cylinders requires special precautions because of the high pressures that are involved. The SCBA cylinder must be in a shielded container while it is being refilled (▼ Figure 2-33). The container is designed to prevent injury if the cylinder ruptures. The hydrostatic test date must be checked before the cylinder is refilled to ensure that its certification has not expired. Special procedures must be followed to ensure that the air used to fill the SCBA cylinder is not contaminated.

Figure 2-33 SCBA cylinders are refilled in a protective enclosure.

Wrap-Up

Ready for Review

This chapter discussed the qualifications required to become a fire fighter. Safety is a critical part of a fire fighter's job. Preventing injuries is always preferable to treating them. Injury prevention includes standards and procedures, personnel, training, and the department's equipment. It is important to exercise good safety practices during training, during responses, at emergency incidents, at the fire station, and outside the workplace.

Personal protective equipment must meet NFPA standards to ensure your safety. The personal protective ensemble includes a helmet, a protective hood, a turnout coat, bunker pants, boots, gloves, SCBA, and a PASS device. All parts of the ensemble must be donned and in place for maximum protection. Fire fighters must be able to don protective clothing in one minute or less. Fire fighters also must care for and maintain PPE properly so that it works as it should.

Fires produce smoke particles, smoke vapors, toxic gases, oxygen-deficient atmospheres, and high temperatures. These conditions require fire fighters to use respiratory protection. The primary respiratory protection equipment used by the fire service is SCBA. This chapter described the limitations of SCBA, the physical limitations of the user, the psychological limitations of the user, and the relationship of these factors to the safe use of SCBA.

Fire fighters must understand the major parts of SCBA and be able to don the equipment quickly. You should be able to don your SCBA in one minute or less. This chapter described how to don SCBA from an apparatus seat mount, from a compartment mount, and from a storage case. It discussed the safety precautions to follow when using SCBA and how to prepare for emergency conditions.

Regular inspection and proper maintenance of SCBA is vital to your safety. This chapter outlined the steps to change air cylinders. It described the importance of hydrostatic testing. It discussed the difference between a cascade filling system and a compressor filling system for air cylinders.

Chief Concepts

- Qualifications for becoming a fire fighter include age requirements, medical requirements, physical fitness requirements, and emergency medical care requirements.
- Good safety practices must be followed during training, during response, at emergency incidents, at the firehouse, and outside the workplace.
- The PPE ensemble for structural firefighting consists of a helmet, a protective hood, a turnout coat, bunker pants, boots, gloves, SCBA, and a PASS device.
- Respiratory hazards from fires include smoke particles and vapors, toxic gases, an oxygen-deficient environment, and high temperatures.
- Understanding the uses and limitations of SCBA is essential for your safety at fire scenes.

Wrap-Up

Hot Terms

Air cylinder The component of the SCBA that stores the compressed air supply.

Air line The hose through which air flows, either within an SCBA or from an outside source to a supplied air respirator.

Backpack The harness of the SCBA that supports the components worn by a fire fighter.

Buddy system A system in which two fire fighters always work as a team for safety purposes.

Bunker coat The protective coat worn by a fire fighter for interior structural firefighting; also called a turnout coat.

Bunker pants The protective trousers worn by a fire fighter for interior structural firefighting; also called turnout pants.

Carbon monoxide A toxic gas produced through incomplete combustion.

Cascade system An apparatus consisting of multiple tanks used to store compressed air and fill SCBA cylinders.

Closed-circuit breathing apparatus SCBA designed to recycle the user's exhaled air. The system removes carbon dioxide and generates fresh oxygen.

Compressor A mechanical device that increases the pressure and decreases the volume of atmospheric air, used to refill SCBA cylinders.

Critical incident stress debriefing (CISD) A confidential group discussion among those who served at a traumatic incident to address emotional, psychological, and stressful issues; usually occurs within 24 to 72 hours of the incident.

Doff To take off an item of clothing or equipment.

Don To put on an item of clothing or equipment.

Emergency by-pass mode Operating mode that allows an SCBA to be used even if part of the regulator fails to function properly.

Employee assistance program (EAP) Program adopted by many departments for fire fighters to receive confidential help with problems such as substance abuse, stress, depression, or burn out that can affect their work performance.

Face piece Component of SCBA that fits over the face.

Fire helmet Protective head covering worn by fire fighters to protect the head from falling objects, blunt trauma, and heat.

Freelancing Dangerous practice of acting independently of command instructions.

Hand light Small, portable light carried by fire fighters to improve visibility at emergency scenes, often powered by rechargeable batteries.

Hydrogen cyanide Toxic gas produced by combustion of materials containing cyanide.

Hydrostatic testing Periodic certification test performed on pressure vessels, including SCBA cylinders.

Immediate Danger to Life and Health (IDLH) An atmospheric concentration of any toxic, corrosive, or asphyxiant substance that poses an immediate threat to life or could cause irreversible or delayed adverse health effects. There are three general IDLH atmospheres: toxic, flammable, and oxygen-deficient.

Incident Commander (IC) The person in charge of the incident site who is responsible for all decisions relating to the management of the incident.

Incident Management System (IMS) The combination of facilities, equipment, personnel, procedures, and communications under a standard organizational structure to manage assigned resources effectively to accomplish stated objectives for an incident. Also known as Incident Command System (ICS).

Incomplete combustion A burning process in which the fuel is not completely consumed, usually due to a limited supply of oxygen.

Kevlar® Strong, synthetic material used in the construction of protective clothing and equipment.

Light-emitting diode (LED) An electronic semiconductor that emits a single-color light when activated.

National Institute for Occupational Safety and Health (NIOSH) A U.S. Federal agency responsible for research and development on occupational safety and health issues.

Nomex® A fire-resistant synthetic material used in the construction of personal protective equipment for fire fighting.

Nose cups An insert inside the face piece of an SCBA that fits over the user's mouth and nose.

Occupational Safety and Health Administration (OSHA) The federal agency that regulates worker safety and, in some cases, responder safety. OSHA is part of the Department of Labor.

Open-circuit breathing apparatus SCBA in which the exhaled air is released into the atmosphere and is not reused.

Oxygen deficiency Any atmosphere where the oxygen level is below 19.5%. Low oxygen levels can have serious effects on people, including adverse reactions such as poor judgment and lack of muscle control.

PBI® A fire-retardant synthetic material used in the construction of personal protective equipment.

Personnel accountability system A method of tracking the identity, assignment, and location of fire fighters operating at an incident scene.

Personal alert safety system (PASS) Device worn by a fire fighter that sounds an alarm if the fire fighter is motionless for a period of time.

Personal protective equipment (PPE) Gear worn by fire fighters that includes helmet, gloves, hood, coat, pants, SCBA, and boots. The personal protective equipment provides a thermal barrier for fire fighters against intense heat.

Phosgene A chemical agent that causes severe pulmonary damage.

Pounds per square inch (psi) Standard unit used in measuring pressure.

Pressure gauge A device that measures and displays pressure readings. In an SCBA, the pressure gauges indicate the quantity of breathing air that is available at any time.

Protective hood A part of a fire fighter's PPE designed to be worn over the head and under the helmet to provide thermal protection for the neck and ears.

Rapid intervention company/crew (RIC) A minimum of two fully equipped personnel on site, in a ready state, for immediate rescue of injured or trapped fire fighters. In some departments, this is also known as Rapid Intervention Team.

Rehabilitation A systematic process to provide periods of rest and recovery for emergency workers during an incident; usually conducted in a designated area away from the hazardous area.

Respirator A protective device used to provide safe breathing air to a user in a hostile or dangerous atmosphere.

Safety Officer The position within IMS responsible for identifying and evaluating hazardous or unsafe conditions at the scene of the incident. Safety officers have the authority to stop any activity deemed unsafe.

SCBA harness Part of SCBA that allows fire fighters to wear it as a "backpack."

SCBA regulators Part of the SCBA that reduces the high pressure in the cylinder to a usable lower pressure and controls the flow of air to the user.

Self-contained breathing apparatus (SCBA) Respirator with independent air supply used by fire fighters to enter toxic and otherwise dangerous atmospheres.

Self-contained underwater breathing apparatus (SCUBA) Respirator with independent air supply used by underwater divers.

Smoke particles Airborne solid material consisting of ash and unburned or partially burned fuel released by a fire.

Standard operating procedures (SOPs) Written rules, policies, regulations, and procedures enforced to structure the normal operations of most fire departments.

Supplied-air respirator (SAR) A respirator that gets its air through a hose from a remote source, such as a compressor or storage cylinder.

Turnout coat Protective coat that is part of a protective clothing ensemble for structural firefighting; also called a bunker coat.

Turnout pants Protective trousers that are part of a protective clothing ensemble for structural firefighting; also called bunker pants.

Two-way radio A portable communication device used by fire fighters. Every firefighting team should carry at least one radio to communicate distress, progress, changes in fire conditions, and other pertinent information.

Fire Fighter in Action

You have just completed Fire Fighter I and II training and have begun your probationary period at the fire department. The importance of safety has been reinforced throughout your training. Fire Fighter Rogers is assigned as your mentor, and you are going to be her shadow for the next six months. On your first day at the fire station, Fire Fighter Rogers shows you around the station and introduces you to daily duties, and she familiarizes you with the apparatus and equipment. You are assigned to Engine 4 and are ready for your first emergency call. You have just sat down for lunch when your engine is dispatched to a garage fire. You feel a rush of adrenaline—your first call is a working structure fire! You and your crew must now get ready to respond to the call.

1. What is the proper order of donning your PPE?

- **A.** Hood, turnout coat, bunker gear pants and boots, SCBA, face piece, helmet, and gloves
- **B.** Hood, bunker gear pants and boots, turnout coat, SCBA, face piece, helmet, and gloves
- **C.** Hood, bunker gear pants and boots, turnout coat, SCBA, helmet, face piece, and gloves
- **D.** Hood, bunker gear pants and boots, SCBA, turnout coat, helmet, face piece, and gloves

2. While riding in the apparatus you should:

- **A.** finish donning your personal protective gear before fastening your seat belt to save time.
- **B.** not fasten your seat belt while donning your seat-mounted SCBA so that you do not inadvertently fasten yourself to the seat.
- **C.** not wear a seat belt. As long as you ride in the back, seat belts are not required.
- **D.** ensure your personal protective gear is donned properly prior to entering the apparatus and fasten you seat belt before the apparatus moves. Remain seated until the apparatus comes to a complete stop.

When you arrive on the scene, you observe smoke and fire rolling out of the garage windows. Your lieutenant orders you and Fire Fighter Rogers to do a quick attack with a 2 $^{1}/_{2}$" hose line.

3. You and Fire Fighter Rogers are a team, which means you should:

- **A.** check each other's PPE to ensure it is on and working correctly.
- **B.** enter together, work together, and leave together.
- **C.** maintain visual, vocal, or physical contact with each other.
- **D.** all of the above

4. Fire Fighter Rogers tells you to turn on your PASS device. What is a PASS device?

- **A.** Primary Alert Sounding System—an electronic device that alerts fire fighters when toxins are in the air
- **B.** Personal Alert Safety System—an electronic device that sounds a loud audible signal when a fire fighter becomes trapped or injured
- **C.** Probationary Accountability Safety System—an electronic device that tracks new probationary fire fighters to ensure accountability at all times
- **D.** Prevent Asbestos Safety System—an electronic device that detects asbestos so that interior crews will limit the disruption of materials containing asbestos

www.FireFighter.jbpub.com

www.FireFighter.jbpub.com

Chapter Pretests
Interactivities
Hot Term Explorer
Web Links
Review Manual
FireLearn

Fire Service Communications

Chapter 3

NFPA 1001 Standard

Fire Fighter I

5.2 *Fire Department Communications.* This duty involves initiating responses, receiving telephone calls, and using fire department communications equipment to correctly relay verbal or written information, according to the following job performance requirements.

5.2.1 Initiate the response to a reported emergency, given the report of an emergency, fire department standard operating procedures, and communications equipment, so that all necessary information is obtained, communications equipment is operated correctly, and the information is promptly and accurately relayed to the dispatch center.

5.2.1 (A) *Requisite Knowledge.* Procedures for reporting an emergency, departmental standard operating procedures for taking and receiving alarms, radio codes or procedures, and information needs of dispatch center.

5.2.1 (B) *Requisite Skills.* The ability to operate fire department communications equipment, relay information, and record information.

5.2.2 Receive a business or personal telephone call, given a fire department business phone, so that procedures for answering the phone are used and the caller's information is relayed.

5.2.2 (A) *Requisite Knowledge.* Fire department procedures for answering nonemergency telephone calls.

5.2.2 (B) *Requisite Skills.* The ability to operate fire station telephone and intercom equipment.

5.2.3 Transmit and receive messages via the fire department radio, given a fire department radio and operating procedures, so that the information is accurate, complete, clear, and relayed within the time established by the AHJ.

5.2.3 (A) *Requisite Knowledge.* Departmental radio procedures and etiquette for routine traffic, emergency traffic, and emergency evacuation signals.

5.2.3 (B) *Requisite Skills.* The ability to operate radio equipment and discriminate between routine and emergency traffic.

Fire Fighter II

6.2 *Fire Department Communications.* This duty involves performing activities related to initiating and reporting responses, according to the following job performance requirements.

6.2.1 Complete a basic incident report, given the report forms, guidelines, and information, so that all pertinent information is recorded, the information is accurate, and the report is complete.

6.2.1 (A) *Requisite Knowledge.* Content requirements for basic incident reports, the purpose and usefulness of accurate reports, how to obtain necessary information, and required coding procedures.

6.2.1 (B) *Requisite Skills.* The ability to determine necessary codes, proof reports, and operate fire department computers or other equipment necessary to complete reports.

6.2.2 Communicate the need for team assistance, given fire department communications equipment, standard operating procedures (SOPs), and a team, so that the supervisor is consistently informed of team needs, departmental SOPs are followed, and the assignment is accomplished safely.

6.2.2 (A) *Requisite Knowledge.* SOPs for alarm assignments and fire department radio communication procedures.

6.2.2 (B) *Requisite Skills.* The ability to operate fire department communications equipment.

Additional NFPA Standards

NFPA 901, *Standard Classifications for Incident Reporting and Fire Protection Data,* 2001 edition

NFPA 902, *Fire Reporting Field Incident Guide*

NFPA 1061, *Standard for Professional Qualifications for Public Safety Telecommunicator,* 2003 edition

NFPA 1221, *Standard for the Installation, Maintenance, and Use of Emergency Services Communications Systems,* 2002 edition

Knowledge Objectives

After studying this chapter, you will be able to:

- Describe the roles of the telecommunicator and dispatch.
- Describe how to receive an emergency call.
- Describe how to initiate a response.
- Describe fire department radio communications.
- Describe radio codes.
- Describe emergency traffic and emergency evacuation signals.
- Define the content requirements for basic incident reports.
- Define how to obtain necessary information, required coding procedures, and the consequences of incomplete and inaccurate reports.
- Describe fire department procedures for answering nonemergency business and personal telephone calls.

Skills Objectives

After studying this chapter, you will be able to:

- Transmit and receive messages via the fire department radio.
- Complete a basic incident report accurately and completely.
- Operate and answer a fire station telephone.

You Are the Fire Fighter

You are the senior fire fighter at the station, staffing a reserve engine while the on-duty crew is covering a fire station across town. When the station's business phone rings, you answer it according to your department's standard operating procedure.

"Good evening. Fire Station 4, Fire Fighter Smith speaking. May I help you?"

The woman caller sounds a little nervous. "I didn't want to call 9-1-1, because I don't know if there's a fire; but I smell smoke in my house."

1. ***How do you handle this situation?***
2. ***What information do you get from the caller? What questions do you ask? To whom do you relay this information, and how?***
3. ***What instructions do you give the caller?***

Introduction

Rapidly developing technology and advanced communications systems are having a tremendous impact on fire department communications. As a fire fighter, you must be familiar with the communications systems, equipment, and procedures used in your department. This chapter provides a basic guide to help you understand how fire department communications systems work and how common systems are configured.

A fire department depends on a functional communications system. When a citizen requests assistance or an alarm sounds, the communications center dispatches the appropriate units to the incident, and maintains communication with those units throughout the incident. The communications system is the link between fire fighters on the scene and the rest of the organization. The communications center monitors everything that happens at the incident scene and processes all requests for assistance or special resources.

At the scene, fire fighters need to communicate with each other, so that the incident commander (IC) can manage the operation efficiently based on progress reports or requests for assistance. The Incident Management System depends on a functional on-site communications system. During large-scale incidents, fire fighters must be able to communicate not only with each other but also with other emergency response agencies.

The communications center does more than simply dispatch units and communicate with them during emergency operations. It also must track the location and status of every other fire department unit. The communications center must always know which units can be dispatched to an incident and it must be able to contact those units. The communications center is responsible for redeploying units to maintain adequate coverage for all areas.

In addition to these special communications requirements, a fire department must have a communications infrastructure that allows it to function as an effective organization. Basic administration and day-to-day management require an efficient communications network, including telephone and data links with every fire station and work site.

The Communications Center

The **communications center** is the hub of the fire department emergency response system. It is the central processing point for all information relating to an emergency incident and all of the information relating to the location, status, and activities of fire department units. It connects and controls all of the department's communications systems. The communications center functions like the human brain. Information comes in via the nerves, it is processed and sent back out to be acted upon in all of the different parts of the body.

The communications center is a physical location. Its size and complexity may vary depending on the needs of the department. The communications center for one department may be a small room in the fire station; another department may have a specially designed, highly sophisticated facility with advanced technological equipment. The fire department in a small community may need a simple system, while the public safety agencies in a metropolitan area may require a large facility. Regardless of the size, all communications centers perform the same basic functions (► **Figure 3-1**).

Many fire departments operate their own "stand-alone" communications centers, serving only a single agency. Others operate or are served by a communications center that serves several fire departments, sometimes all of the fire departments in a county or region. In many areas, the fire

Figure 3-1 Regardless of the size, all communications centers perform the same basic functions.

department communications center is co-located with other public safety agencies, including law enforcement and/or separate emergency medical service providers. In a joint facility there may be separate personnel and independent systems for each service delivery agency or the entire operation may be integrated, with all employees cross-trained to receive calls and dispatch any type of emergency incident.

Telecommunicators

The employees who staff a communications center are known as dispatchers or **telecommunicators**. Telecommunicators have been professionally trained to work in a public safety communications environment. Advanced training and professional certification programs ensure that skilled, competent telecommunicators can perform their critical role in the public safety system.

Just as there are special qualities that can make an individual a good fire fighter, there are qualities that a telecommunicator must possess in order to be successful. The job of a telecommunicator can be complicated, demanding, and extremely stressful. The successful telecommunicator must be able to understand and follow complicated procedures, perform multiple tasks effectively, memorize information, and make decisions quickly.

One of the telecommunicator's most important skills is an ability to communicate effectively with citizens to obtain critical information, even when the citizen is highly stressed or in extreme personal danger. An emotional caller may criticize or insult the telecommunicator. The telecommunicator must respond professionally and focus on obtaining the essential information. Voice control and the ability to maintain composure under pressure are important qualities; the telecommunicator must be clear, calm, and in control.

A telecommunicator must be skilled in operating all of the systems and equipment in the communications center. He or she must understand and follow the fire department's operational procedures, particularly those relating to dispatch policies and protocols, radio communications, and incident management. The telecommunicator must keep track of the status and location of each unit at all times and monitor the overall deployment and availability of resources throughout the system. NFPA 1061, *Standard for Professional Qualification for Public Safety Telecommunicator,* contains a complete list of qualifications for telecommunicator candidates.

Communications Facility Requirements

The fire department communications center must be designed and operated to ensure that its critical mission can be performed with a very high degree of reliability. The performance requirements in NFPA 1221, *Standard for the Installation, Maintenance, and Use of Emergency Services Communications Systems,* should be used to design and construct a fire department or public safety communications center. These requirements apply whether the communications center serves a small community with only one or two fire stations or a metropolitan area with several hundred stations.

The communications center should be well-protected against natural threats such as floods, and it should be constructed to withstand predictable damaging forces such as severe storms and earthquakes. The communications center should be able to operate at maximum capacity, without interruption, even when other community services are severely affected. The building should be equipped with emergency generators and other systems so that it can continue to operate for several days in the most challenging conditions.

NFPA 1221 also requires back-up systems for all of the critical equipment in a communications center, so that the failure of a single component or system will not disable the operation. For example, there must be more than one way of transmitting a dispatch message from the communications center to each fire station. There should be a back-up radio transmitter, and the telephone system must be able to receive calls, even if part of the system is damaged. The facility should be designed to minimize its vulnerability to a fire originating inside the building as well as to nearby fires. It must be secured to prevent unauthorized entry.

Finally, there should be a back-up communications center at a different location. If some unanticipated situation makes it impossible to operate from the primary location, the back-up location can be activated.

Communications Center Equipment

A communications center is usually equipped with several different types of communications systems and equipment. Although specific requirements depend on the size of the operation and the configuration of the local communications systems, most centers will have the following equipment:

- Dedicated 9-1-1 telephones
- Public telephones
- Direct-line phones to other agencies
- Equipment to receive alarms from public or private fire alarm systems
- Computers and/or hard copy files and maps to locate addresses and select units to dispatch
- Equipment for alerting and dispatching units to emergency calls
- Two-way radio system(s)
- Recording devices to record phone calls and radio traffic
- Back-up electrical generators

Computer-Aided Dispatch (CAD)

Computers are used in almost all communications centers. Most large communications centers use **computer-aided dispatch (CAD)** systems and many smaller centers have smaller scale versions of CAD systems.

As the name suggests, a CAD system is designed to assist a telecommunicator by performing specific functions more quickly and efficiently than they can be done manually. There are many variations in CAD systems, from very simple versions to highly sophisticated systems that perform many different functions. All of the functions performed by a CAD system can be performed manually by trained and experienced telecommunicators. The computer assists the telecommunicator by performing these functions more quickly and accurately (▼ Figure 3-2).

A well-designed CAD system can shorten the time it takes for a telecommunicator to receive and dispatch calls. A CAD system helps meet the most important objective in processing an emergency call: sending the appropriate units to the correct location as quickly as possible. The increased efficiency provided by a CAD system is particularly important in large communications centers that process a high volume of calls.

Because a computer can manage large amounts of information efficiently and accurately, a CAD system can be used to track the status of a large number of units. The CAD system will know which units are available to respond to a call, which units are assigned to incidents, and which units are temporarily assigned to cover different areas. The most advanced CAD system has global positioning system (GPS) devices in the department's vehicles, so it can track the exact location of each unit. A CAD system can also look up addresses and determine the closest fire stations in order of response for any location. The CAD system can then select the units that can respond quickly to an alarm, even if some of the units that would normally respond are unavailable.

Some CAD systems transmit dispatch information directly to **mobile data terminals (MDT)**, computers that are located in the fire station or on the apparatus (► Figure 3-3). CAD systems can also provide immediate access to information such as preincident plans and hazardous materials lists for an address. By linking the CAD system to other data files, fire fighters have access to additional useful information such as travel route instructions, maps, and reference materials.

Figure 3-2 A CAD system enables a telecommunicator to work more quickly and efficiently.

Voice Recorders and Activity Logs

Almost everything that happens in a communications center is recorded, either by a **voice recording system** or by an **activity logging system**. Most communications centers can automatically record everything that is said over the telephone or radio, 24 hours a day. Most centers also have an instant playback unit that allows the telecommunicator to replay conversations for the previous 10 to 15 minutes at the touch of a button. This feature is particularly valuable if the caller talks very quickly, has an unusual accent, hangs up, or is disconnected. The telecommunicator can replay the message several times, if necessary, to understand exactly what was said.

The logging system keeps a detailed record of every incident and activity that occurs. The records include every call that is entered, every unit that is dispatched, and every significant event that occurs in relation to an emergency incident. Times are recorded when a call is received, when the units are dispatched, when they report that they are en route, when they arrive at the scene, when the incident is under control, and when the last unit leaves the scene. One of the advantages of a CAD system is that it automatically captures and stores every event as it occurs. Before CAD systems were developed, someone had to write down and time stamp every transaction, including every radio transmission. All hard-copy logs had to be retained for future reference.

There are several reasons for keeping voice recorders and activity logs. They serve as legal records of the official delivery of a government service by a public agency, the fire department. These records may be required for legal proceedings, sometimes years after the incident occurred. They may be needed to defend the fire department's actions when questions are raised about an unfortunate outcome. The records provided by voice recorders and incident logs accurately document the events and can often demonstrate that the organization and its employees performed ethically, responsibly, and professionally. They also make it difficult to hide an error, if a mistake was made.

Records are also valuable in reviewing and analyzing information about department operations. Good recordkeeping makes it possible to examine what happened on a particular call as well as to measure workloads, system performance, activity trends, and other factors as part of planning and budget preparation. Analysts and planners can use data from CAD systems to study deployment strategies and to make the most efficient use of fire department resources.

Call Response and Dispatch

Critical functions performed by most CAD systems include verifying an address and determining which units should respond to an alarm. This is usually a simple task in a small fire department with only two or three fire stations, but it is much more complicated in a large urban system with hundreds of fire stations and individual units. When the telecommunicator enters an address and an incident description code into a CAD system, the system can make a recommendation on the appropriate units to dispatch in less than a second.

Data links that provide the location of the caller and automatically enter the address can also save time. If there are potential duplicate addresses or if the address entered is not a valid location, the CAD system will prompt the telecommunicator to ask the caller for more information. Hard copy files of the most essential information should always be maintained for back-up, so that calls can be dispatched if the system is down (out-of-service).

The telecommunicator's first responsibility is to obtain the information that is required to dispatch the appropriate units to the correct location (▼ Figure 3-4). Then the incident must be processed according to standard protocols. The

Figure 3-3 Some CAD systems transmit dispatch information directly to terminals in fire stations and mobile data terminals in the apparatus.

Figure 3-4 Telecommunicators must obtain information and relay it accurately to the appropriate responders.

telecommunicator must decide which agencies and/or units should respond and transmit the necessary information to them. The generally accepted performance objective, from the time a call reaches the communications center until the units are dispatched, is one minute or less.

Most requests for fire department response are made by telephone. Although some communities use a regular seven-digit telephone number or "0" to report an emergency, most areas in the United States and Canada have implemented the 9-1-1 system. According to the National Emergency Number Association (NENA), 93% of the population in the United States and Canada uses some type of 9-1-1 system to report an emergency.

All calls to 9-1-1 are automatically directed to a designated **public safety answering point (PSAP)** for that community or jurisdiction. If all public safety communications are in the same facility, the call can be answered, processed, and dispatched immediately. If fire and/or EMS communications have a separate facility, those calls must be transferred to a fire/EMS-trained telecommunicator. The call-taking process is described in more detail later in this chapter.

Communications Center Operations

Several different functions are performed in a fire department or public safety communications center. All of these activities must be performed accurately and efficiently, even in the most challenging circumstances. For example, the activity level in a communications center can increase from calm to extreme in less than a minute during an emergency, but the chaos outside must not affect the operations inside. If the communications center fails to perform its mission, the fire department will not be able to deliver much-needed emergency services.

The basic functions performed in a communications center include:

- Receiving calls for emergency incidents and dispatching fire department units
- Supporting the operations of fire department units delivering emergency services
- Coordinating fire department operations with other agencies
- Keeping track of the status of each fire department unit at all times
- Monitoring the level of coverage and managing the deployment of available units
- Notifying designated individuals and agencies of particular events and situations
- Maintaining records of all emergency-related activities
- Maintaining information required for dispatch purposes

Receiving and Dispatching Emergency Calls

Most fire department responses begin when a call comes in to the communications center, which then dispatches one or more units. Even if a citizen reports an emergency directly to a fire station or a crew discovers a situation, the communications center is immediately notified to initiate the emergency response process.

The major steps in processing an emergency incident include:

- Call receipt
- Location validation
- Classification and prioritization
- Unit selection
- Dispatch

Call receipt refers to the process of receiving an initial call and obtaining the necessary information to initiate a response. This includes telephone calls from the general public, as well as other notification methods, such as automatic fire alarm systems, public fire alarm boxes, requests from other agencies, calls reported directly to fire stations, and calls initiated via radio from fire department units or other public safety agencies.

Location validation ensures that the information received is adequate to dispatch units to the correct location. Potential duplicate addresses, such as two streets with the same or very similar names, must be eliminated. The information must point to a valid location on a map or in a street index system. It must be within the geographic jurisdiction of the potential dispatch units. Without a valid address, a communications center will be unable to send units to the proper location.

Classification and prioritization is the process of assigning a response category, based on the nature of the reported problem. Most fire departments respond to many kinds of situations, ranging from outside vegetation fires and fires in high-rise buildings to heart attacks or multi-casualty incidents. The nature of the call dictates which units or combinations of units should be dispatched. Dispatch policies are usually spelled out in the department's SOPs and dispatch protocols, which state the numbers and types of units to dispatch for each type of situation.

Unit selection is the process of determining exactly which unit or units to dispatch, based on the location and classification of the incident. The usual policy is to dispatch the closest available unit that can provide the necessary assistance. When all units are available and in their fire stations, this can usually be determined quickly from pre-programmed information. Unit selection becomes more complicated when some units are out of position or unavailable. Unit selection often requires quick decision-making skills, even with a CAD system that is programmed to follow set policies for various situations.

Dispatch is the important step of actually alerting the selected units to respond and transmitting the information to them. Fire departments use different dispatch systems, ranging from telephone lines to radio systems. The communications center must have at least two separate ways of notifying each fire station.

Call Receipt

Telephones

The public generally uses telephones to report emergency incidents. In most communities, calls to 9-1-1 connect the caller with a public safety answering point (PSAP). Some communities have implemented a 3-1-1 system for non-emergency calls. The PSAP can take the information immediately, or transfer the call to the appropriate agency, based on the nature of the emergency. In some systems, 9-1-1 connects the caller directly with a telecommunicator, who obtains the required information.

The communications center usually has a seven-digit telephone number as well. People who remember the seven-digit number used before their communities introduced the 9-1-1 system may continue to call that number. Even when every effort is made to encourage people to use 9-1-1 to report an emergency, some people may be reluctant to use the system because they are not sure that their particular situation is serious enough to be considered a "true emergency." Any number that is published as a fire department telephone number should be answered at all times, including nights and weekends. The emergency number should always be pronounced as "nine-one-one," not "nine-eleven."

The telecommunicator who takes the call must conduct a **telephone interrogation**, asking the caller questions to obtain the required information. Initially, the telecommunicator will need to know the location of the emergency and the nature of the situation. Many 9-1-1 systems can automatically provide the location of the telephone where a call originates. But this information may be unavailable or inaccurate if the call was made on a wireless phone or from someplace other than the location of the emergency incident. The exact location must be obtained so that units can respond directly to the incident scene.

The telecommunicator must also interpret the nature of the problem from the caller's description. The caller may be distressed, excited, and unable to organize his or her thoughts. There may be language barriers; the caller may speak a different language or may not know the right words to explain the situation. Telecommunicators must follow SOPs and use active listening to interpret the information. Many communications centers use a structured set of questions to obtain and classify information on the nature of the situation. The telecommunicator must remember that the caller thinks the situation is an emergency, and must treat every call as such until it is determined that no emergency exists.

Telecommunicators cannot allow gaps of silence to occur while questioning the caller. If the caller suddenly becomes silent, something may have happened to him or her; the caller may be in personal danger or extremely upset. If the telecommunicator is silent, the caller may think that the telecommunicator is no longer on the line or no longer listening. Disconnects are another problem. Callers to 9-1-1 might hang up accidentally or be disconnected after initially reaching the communication center. A telecommunicator who is unable to return the call and reach the original caller will usually dispatch a police officer to the location to determine if more help is needed.

The telecommunicator should never argue with a caller. Although callers may raise their voices, scream, or shout, the telecommunicator or fire fighter must continue to speak calmly and remain professional at all times. The telecommunicator should remember that the caller is simply reacting to the emergency situation.

With just two critical pieces of information—location and nature of the problem—the telecommunicator can initiate a response. However, local SOPs may require the telecommunicator to obtain additional information. Obtaining the caller's name and contact phone number is useful in case it is necessary to call back for additional information. If the caller is in danger or distress, the telecommunicator will try to keep the line open and remain in contact with the caller until help arrives. In many communities, telecommunicators are trained to provide self-help instructions for callers, advising them on what to do until the fire department arrives. For example, if a building fire is being reported, the telecommunicator will advise the occupants to evacuate and wait outside; if the caller is reporting a medical incident, the telecommunicator may provide first aid instructions, based on the patient's symptoms.

Location Validation

Enhanced 9-1-1 Systems

Most enhanced 9-1-1 systems have features that can help the telecommunicator obtain identifying information. **Automatic Number Identification (ANI)** shows the telephone number where the call originated. **Automatic Location Identification (ALI)** queries a database to show the location of the telephone, the subscriber's name, and other details (▼ Figure 3-5). These

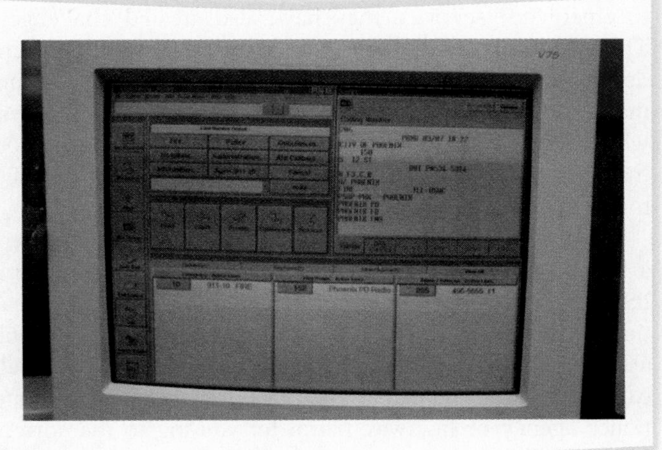

Figure 3-5 An ALI system provides information about the origin of a 9-1-1 call.

Fire Fighter Tips

Sometimes, the communications center receives calls about issues that are not handled by the fire department or other participating agencies. In these cases, the telecommunicator or fire fighter should make every effort to accommodate the caller. If possible, transfer the caller to the proper agency. In addition, provide the caller with the proper contact information in case the transfer fails. The caller will appreciate the assistance and retain a positive image of the department's service.

features, combined with an accurate database, will generate a computer display listing all of the information when the call is received. Another feature in an enhanced 9-1-1 system ensures that each call will be directed to the appropriate PSAP for that location.

Using enhanced 9-1-1, the telecommunicator often can pinpoint the location of an incident, even if the caller is unable to provide that information. The telecommunicator should always confirm that the information is correct and refers to the actual location of the emergency. If the caller has recently moved, the database may contain the old address; sometimes, a billing address for the telephone service is provided rather than the physical location of the telephone.

These enhanced 9-1-1 features can save both time and lives, but they require an accurate, up-to-date database. Telephone service providers, state regulatory agencies, and local emergency service agencies must cooperate to keep the database current and accurate.

Wireless telephone technology means that people no longer have to go to a phone to make a call. Instead, they carry their phones with them. This enables citizens to call for assistance whenever they encounter an emergency situation. Today, calls from wireless telephones are as common as those from hard-wired telephones.

Although wireless phones have enhanced public access to emergency services, they have also created challenges. The ANI and ALI enhancements to 9-1-1 systems were developed for hard-wired telephone systems. These systems may have difficulty accommodating cellular telephones, which have no fixed location. Often, the best information that ALI can provide is the physical location of the radio tower that relays the call.

Because radio waves can cross jurisdictional boundaries, a 9-1-1 call from a wireless phone can easily be sent to a PSAP in a different jurisdiction. Determining the actual location of the caller so that the call can be redirected to the appropriate jurisdiction may be difficult. In some areas, all calls to 9-1-1 from wireless phones are directed to a state police agency or highway patrol for routing to the correct communications center.

Determining the location of the incident is also problematic. People who call 9-1-1 on their cellular telephones may not know their exact location. When an incident occurs on a busy highway or other highly visible location, the local PSAP may receive dozens of calls from individuals who can provide only a vague approximation of their own location or the location of the emergency.

Figure 3-6 TDD/TTY/Text Phone systems enable the hearing-impaired to communicate over phone lines.

Fortunately, a technological solution to these problems is available. Wireless telephone systems are being updated to use GPS technology to pinpoint the geographic coordinates of a 9-1-1 call from a cellular phone.

TDD/TTY/Text Phones

Speech and/or hearing-impaired citizens can communicate by telephone using special devices that display text rather than voice (▲ **Figure 3-6**). Three such systems are **TDD** (Telecommunications Device for the Deaf), **TTY** (Teletype), and **Text Phones**. The Americans with Disabilities Act (ADA) requires that communications centers be able to receive calls via text messages as well as by voice communication. Telecommunicators must know how to use this equipment to communicate with hearing-impaired persons. Many communications centers have a special telephone line with a seven-digit number set aside for such calls, but the 9-1-1 lines must also include text messaging systems. Any incoming calls without voice contact at the other end should be checked for TDD/TTY/Text Phone systems.

Tracing Calls

The ANI/ALI displays have practically eliminated the need to trace calls through the telephone company. Many agencies without enhanced 9-1-1 service have installed caller ID services, which can immediately identify the source of a call. Where enhanced 9-1-1 or caller ID services are not available, the telecommunicator may need to request that the phone company trace the call to help identify the source of a call. If a telecommunicator needs to trace an active call, a second telecommunicator must contact the local phone company to

Canadian Perspectives

The 9-1-1 systems used in Canada and the United States are essentially the same.

request an immediate call trace on a specific line. The phone company supervisor will usually ask a series of questions before authorizing the trace.

Call Receipt

Direct Line Telephones

A direct line (or ring down) telephone connects two predetermined points. Picking up the phone at one end causes an immediate ring at the other end. Direct line connections often link police and fire communications centers or two fire communications centers that serve adjacent areas. Direct lines also may connect hospitals, private alarm companies, utility companies, airports, and similar facilities with the fire department communications center. These lines are often used in both directions, so that the communications center can receive calls for assistance or send notifications and requests for response.

The communications center may have direct lines to each fire station in its jurisdiction. A direct line also can be linked to the station's public address speakers to announce dispatch messages.

Walk-Ins

Although most emergencies are reported by telephone, people may actually come to a fire station seeking assistance (► Figure 3-7). When a walk-in occurs, the station should contact and advise the communications center of the situation immediately. The communications center will create an incident record and dispatch any needed additional assistance. Even if the units at the station can handle the situation, they should notify the communications center that they are occupied with an incident.

Citizens should be able to come to a fire station and report an emergency at all times, even when the station is unoccupied. Many departments install a direct line telephone to the communications center just outside each fire station. These phones should be marked with a simple sign stating, "If station is closed, pick up telephone in red box to report an emergency."

Municipal Fire Alarm Systems

Many communities have fire alarm boxes or emergency telephones located on street corners or in public places (► Figure 3-8). A fire alarm box transmits a coded signal to the communications center. The signal indicates an alarm and identifies the location of the box, but does not indicate

Figure 3-7 A citizen might walk into the fire station and report an emergency.

Figure 3-8 Public fire alarm boxes are often placed at intersections or in front of public buildings.

what type of emergency is occurring. The major drawbacks to this type of reporting system is often a high rate of false alarms and not knowing the nature of the call. Public fire alarm boxes may also be connected to automatic fire alarm systems in buildings. When the building's alarm system is activated, the public fire alarm box transmits an alarm directly to the communications center.

Public fire alarm boxes were once connected by networks of dedicated cables that transmitted the alarm signals to a fire department facility and to individual fire stations. These hard-wired boxes often have been replaced by radio-operated fire alarm boxes. Today, many communities have eliminated their public fire alarm systems due to increasing numbers of false alarms, high maintenance costs, and the development of alternative notification systems, including private and public hard-wire and cellular telephones. Other communities have converted their public fire alarm systems to call box systems.

Fire Marks

The Boston, Massachusetts Fire Department placed the first emergency call box into service on April 28, 1852.

A **call box** connects a person directly to a telecommunicator. The caller can request a full range of emergency assistance—from police, fire, or emergency medical assistance to a tow truck. Call boxes can be directly connected to a wireless or hard-wired telephone network or operate on a radio system. Emergency call boxes are often located along major highways, in bridges and tunnels, and in other locations without nearby phones. In locations without electric or telephone service, call boxes are solar-powered (► Figure 3-9).

Figure 3-9 Wireless call boxes are often found in places without other telephone service.

Private and Automatic Fire Alarm Systems

Private and automatic fire alarm systems use several different arrangements to transmit alarms to the local fire department communications center. Many commercial, industrial, and residential buildings have automatic fire alarm systems, which use heat detectors, smoke detectors, or other devices to initiate an alarm. Other buildings may have manual pull-station fire alarms. These alarms may be connected to the fire department or monitored from a remote location. Water flow alarms on automatic sprinkler systems also can be connected to local fire alarm systems or monitored at a remote location. Chapter 36, Fire Protection, Suppression, and Detection Systems discusses these systems in detail.

The connection used to transmit an alarm from a private system to a public fire communications center depends on many factors. Private companies may monitor alarm systems and relay alarms to the local police or fire department. Many private systems have automatic telephone dialers programmed to call an emergency number with a recorded message when the system is activated. These systems may be monitored by a privately operated central station alarm service, which relays the alarm to the fire communications center. There may be a direct line between a central alarm service and the fire department communications center. Some communications centers provide monitoring services. Finally, private alarm systems may be connected to a public fire alarm box. No matter how the alarm reaches the communications center, the result is the creation of an incident report and the dispatch of fire department apparatus.

Call Classification and Prioritization

The next step in processing an emergency incident is classifying and prioritizing each call as it is received. The nature of the problem—what is happening at the scene—determines the urgency of the call and its priority. Although most fire department calls are dispatched immediately, it is sometimes necessary to delay dispatch to a lower priority call if a more urgent situation arises or if several calls come in at the same time. Some calls qualify for emergency response (red lights and siren) while others are considered nonemergencies, depending on local SOPs.

Call classification determines the number and types of units that are dispatched. The standard response to different incidents is established by the fire department's SOPs and can range from a single unit to an initial assignment of 10 or more units and dozens of fire fighters.

Unit Selection

After verifying the location and classifying the situation, a telecommunicator must select the specific units to dispatch to an incident. Generally, the standard assignment to each type of incident in each geographic area is stored on a system of **run cards**. Run cards are prepared in advance and stored in hard copy form or in computer files. Run cards list units in the proper order of response, based on response distance or estimated response time, and often specify the units that would be dispatched through several levels of multiple alarms.

The run card assignments can be used as long as all of the units that would normally respond on a call are available. If any of the units are unavailable, the telecommunicator must adjust the assignments by selecting substitute units in their order of response. In a large, busy system with many simultaneous incidents, the status of individual units is constantly changing and complicating the assignment process.

Most CAD systems are programmed to select the units for an incident automatically, based on the location, call classification, and actual status of all units. The CAD system will recommend a dispatch assignment, which the telecommunicator can accept or adjust, based on circumstances or special information. The same process is used to dispatch any additional units to an incident, whether it is a multiple alarm or a request for particular units or capabilities.

Dispatch

The telecommunicator's next step is to transmit the dispatch information to the assigned units quickly and accurately. The information can be transmitted in many different ways, but every department must have at least two separate methods for sending a dispatch message from the communications center to each fire station. The primary connection can be a hard-wired circuit, a telephone line, a data line, a microwave link, or a radio system.

Most fire departments dispatch verbal messages to the appropriate fire stations. The dispatch message is broadcast over speakers in the fire station, so that everyone immediately knows the location and nature of the incident and the units that should respond. Radio transmissions are used to contact a unit that is out of the station. The fire station or each vehicle must confirm that the message was received and the units are responding. If the communications center does not receive the confirmation within a set time period, it must dispatch substitute units.

A CAD system can be programmed to alert the appropriate fire stations automatically. The system can send the dispatch information to computer terminals or printers, sound distinctive tones, turn on lights and public address speakers, turn off the stove, and perform additional functions. The dispatch message also can go directly to each individual vehicle that has a computer terminal as well as to the station. The CAD notification often is accompanied by a verbal announcement over the radio and the fire station speakers.

Fire departments with volunteer responders must be able to reach them with a dispatch message. Some departments issue pagers to their individual volunteers. Some pagers can receive a vocal dispatch message; others display an alphanumeric message. Volunteer fire departments in many communities rely on outdoor sirens, horns, or whistles to notify their members of an emergency. These audible devices usually can be activated by remote control from the communications center. Volunteers call the communications center by radio or telephone to receive specific instructions.

Operational Support and Coordination

After the communications center dispatches the units, it begins to provide incident support and coordination. Someone in the communications center must remain in contact with the responding units throughout the incident. The telecommunicator must confirm that the dispatched units actually received the alarm, record their enroute times, provide any additional or updated information, and record their on-scene arrival times.

Operational support and coordination encompass all communications between the units and the communications center during an entire incident. Two-way radio or mobile computer terminals are used to exchange information. Many fire departments also issue cellular telephones to command officers and/or individual units.

Generally, the incident commander will communicate with a telecommunicator operating a radio in the communications center. Progress and incident status reports, requests for additional units or release of extra units, notifications, and requests for information or outside resources are examples of incident communications. The communications center closely monitors the radio and provides any needed support for the incident.

Each part of the public safety network—fire, EMS, and police—must be aware of what other agencies are doing at the incident. The communication center is the hub of the network that supports units operating at emergency incidents. It coordinates the fire department's activities and requirements with other agencies and resources. For example, the telecommunicator may need to notify nonemergency resources such as gas, electric, or telephone companies and request that they take certain actions. The communications center should have accurate, current telephone numbers and contact information for every relevant agency.

Status Tracking and Deployment Management

The communications center must know the location and status of every fire department unit at all times. Units should never get "lost" in the system, whether they are available for dispatch, assigned to an incident, or in the repair shop. As previously stated, the communications center must always know which units are available and which units are not available for dispatch. The changing conditions at an incident may require frequent reassignment of units, including ambulances. In addition, units may be needed from outside the normal response area or from other districts. Status tracking is difficult if the telecommunicator must rely on radio reports and colored magnets or tags on a map or status board. CAD systems make this job much easier, because status changes can be entered through digital status units or computer terminals. GPS devices, which accurately track the location of each unit, are also helpful on large incidents.

Communications centers also must continually monitor the availability of units in each geographic area and redeploy units when there is insufficient coverage in an area. Many fire departments list both unit relocations and multiple response units on the run cards. This information is only valid, however, if major incidents occur one at a time and when all units are available. If a department has a large volume of routine incidents, it may need to redeploy units to balance coverage, even when there are no major incidents in progress.

Usually, a supervisor in the communications center is responsible for determining when and where to redeploy units, as well as for requesting coverage from surrounding jurisdictions. Units from many different jurisdictions can be redeployed to respond to large-scale incidents under regional or statewide plans. These plans must include a system for tracking every unit and a designated communications center for maintaining contact with all units.

Radio Systems

Fire department communications systems depend on two-way radio systems. Radios link the communications center and individual units; they also link units at an incident scene. A radio system is an integral component of the Incident Management System because it links all of the units on an incident—both up and down the chain of command and across the organization chart. Usually, every fire department vehicle has a mobile radio, and at least one—if not every—member of a team carries a portable radio during an emergency incident. Often, a radio is the fire fighter's only link to the incident organization and the only means to call for help in a dangerous situation. Radios also are used to transmit dispatch information to fire stations, to page volunteer fire fighters, and to link mobile computer terminals.

Figure 3-10 A portable radio can be carried by an individual fire fighter.

Fire departments use many different types of radios and radio systems. Technological advances are rapidly adding new features and system configurations, making it impossible to describe all of the possible features, principles, and systems. This section describes common systems and operating features, but as a fire fighter, you must know how to operate your assigned radio and learn your department's radio procedures.

Radio Equipment

There are three types of fire service radios: the portable radio, the mobile radio, and the base station. A **portable radio** is a hand-held two-way radio small enough for a fire fighter to carry at all times (◄ Figure 3-10). The radio body contains an integrated speaker and microphone, an on/off switch or knob, a volume control, and a "push-to-talk" (PTT) button. Most radios also have a knob or switch so the user can move between a radio channel and a talk group. A knob or button on the radio adjusts the sensitivity of the radio's signal reception. This enables the user to **squelch** or eliminate weak transmissions.

A portable radio must have an antenna to receive and transmit signals. A frequently used optional attachment is an extension microphone/speaker unit that can be clipped to a collar or shoulder strap, while the radio remains in a pocket or pouch.

A portable radio is usually powered by a rechargeable battery, which should be checked at the beginning of each shift or prior to each use. Because the battery has a limited capacity, it must be recharged or replaced after extended operations. Battery-operated portable radios also have limited transmitting power. The signal can be heard only within a certain range and is easily blocked or overpowered by a stronger signal.

Mobile radios are more powerful two-way radios permanently mounted in vehicles and powered by the vehicle's electrical system (▼ Figure 3-11). Both mobile and portable

Figure 3-11 A mobile radio is permanently mounted in a vehicle.

Figure 3-12 Base station radios are installed at fixed locations.

radios have similar features, but mobile radios usually have a fixed speaker and an attached, hand-held microphone on a coiled cord. The "push-to-talk" button is on the microphone, while the antenna is usually mounted on the exterior of the vehicle. Fire apparatus often include headsets with a combined intercom/radio system that enable crew members to talk to each other and hear the radio.

Base station radios are permanently mounted in a building, such as a fire station, communications center, or remote transmitter site (▲ Figure 3-12). Base station radios are more powerful than portable or mobile radios. The antenna is often mounted on a radio tower so the transmissions have a wide coverage area (► Figure 3-13). Public safety radio systems often use multiple base stations at different locations to cover large geographic areas. The communications center operates these stations by remote control.

Mobile data terminals are computer devices that transmit data by radio. Some departments use them to track unit status, transmit dispatch messages, and exchange different types of information. Some terminals simply use buttons to send status messages such as "responding," "on the scene," and "back in service" while others are advanced laptop computers (► Figure 3-14).

Radio Systems

The design, installation, and operation of two-way radio systems is closely regulated by the Federal Communications Commission (FCC). The FCC has strict limitations governing the assignment of frequencies to ensure that all users have adequate access. Every system must be licensed and operated within established guidelines.

Radios work by broadcasting electronic signals on certain frequencies. A frequency is an assigned space on the radio spectrum; only those radios tuned to that specific frequency can hear the message. The FCC licenses an agency to operate on one or more specific frequencies. A particular agency may be licensed to operate on several frequencies. The frequencies are programmed into the radio and can be adjusted only by a qualified technician.

Figure 3-13 A base station radio antenna is often mounted on a radio tower to provide maximum coverage.

Figure 3-14 Mobile computers exchange data over a radio channel.

A radio channel uses either one frequency or two frequencies. A simplex channel uses only one frequency. Each radio transmits signals and receives signals on the same frequency, so a message goes directly from one radio to every

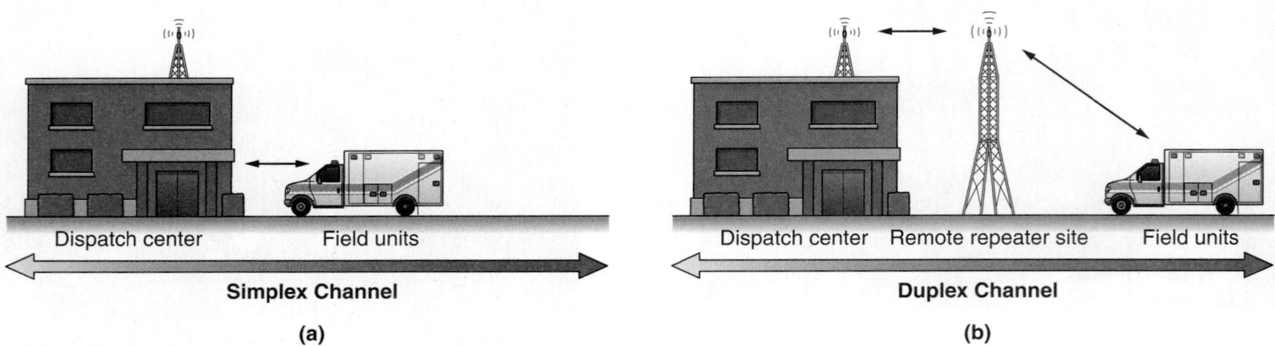

Figure 3-15 Simplex and duplex channels send and receive transmissions in different ways.

other radio that is set to receive that frequency. A **duplex channel** uses two frequencies, with each radio transmitting signals on one frequency and receiving on the other frequency (▲ **Figure 3-15**). Duplex channels are used with repeater systems, which are described later.

U.S. fire service frequencies are in several different ranges, including 33 to 46 MHz (VHF low band); 150 to 174 MHz (VHF high band); 450 to 460 MHz (UHF band); and the 800 MHz band. Additional groups of frequencies are allocated in specific geographic areas with a high demand for public safety agency radio channels. Each band has certain advantages and disadvantages relating to geographic coverage, topography (hills and valleys), and penetration into structures. Some bands have a large number of different users, which can cause interference problems, particularly in densely populated metropolitan areas.

Generally, one radio can be programmed to operate on several frequencies in a particular band, but cannot be used across different bands. This is a problem if neighboring fire departments or police, fire, and EMS services within the same jurisdiction are on different bands. When different agencies working at the same incident are on different bands, they must make complicated arrangements with cross band repeaters and multiple radios in command vehicles to communicate with each other.

Canadian Perspectives

Industry Canada is the federal regulator for radio communications systems in Canada. Canadian and U.S. rules and regulations are similar, although not identical.

Repeater Systems

Radio messages can be broadcast over a limited distance. The signal weakens as it travels farther from the source. Buildings, tunnels, and topography create interference. These problems are most significant for hand-held portable radios, which have limited transmitting power. Mobile radios that operate in systems covering large geographic areas face similar problems. To compensate, fire departments may use **radio repeater systems**.

In a repeater system, each radio channel uses two separate frequencies—one to transmit and the other to receive. When a low-power radio transmits over the first frequency, the signal is received by a repeater unit that automatically rebroadcasts it on the second frequency over a more powerful radio. All radios set on the designated channel receive the boosted signal on the second frequency. This enables the transmission to reach a wider coverage area.

Many public safety radio systems have multiple receivers (voter receivers), which are geographically distributed over the service area to capture weak signals. The individual receivers forward their signals to a device that selects the strongest signal and rebroadcasts it over the system's base station radio(s) and transmission tower(s). With this configuration, messages that originate from a mobile or hand-held portable radio have as strong a signal as a base station radio at the communications center. As long as the original signal is strong enough to reach one of the voter receivers, the system is effective. If the signal does not reach a repeater, no one will hear it.

Some fire departments switch to a simplex radio channel for on-scene communications. This is sometimes called a **talk-around channel**, because it bypasses the repeater system. In this configuration, the radios transmit and receive on the same frequency, and the signal is not repeated. This often works well for short-distance communications, such as from the command post to inside crews at a house fire or from one unit to another inside a building. Conversations on a talk-around channel cannot be monitored by the communications center. The incident commander will use a more powerful radio on a repeater channel to maintain contact with the communications center.

Mobile repeater systems also can boost signals at the incident scene by creating a localized, on-site repeater system. The mobile repeater can be permanently mounted in a vehicle or set up at the command post. It captures the weak signal from a portable radio for rebroadcast. Some fire departments use cross-band repeaters, which boost the power from a weak signal and transmit it over a different radio band. This enables a fire fighter using a UHF portable inside a building to communicate with the IC who is outside on a VHF radio. Similar systems may be used in large buildings or underground structures so crews working inside can communicate with units on the outside or with the communications center.

Trunking Systems

Radio systems using new technology and digital communications are being developed to take advantage of the additional frequencies being allocated for public safety use. Unfortunately, the new systems are generally incompatible with older radios and require expensive infrastructure changes. They also can be more complicated to operate because the radio settings and characteristics are quite different from older systems (► Figure 3-16). Although the new systems will have tremendous operational advantages over conventional radio systems, certain basic requirements still must be met. The new systems demand well-designed software support, an adequate number of frequencies and transmitter/receiver sites, and well-trained users.

The new radio technologies use a **trunking system**, a group of shared frequencies controlled by a computer. The computer allocates the frequencies for each transmission. The radio operator sets the radio on a talk group and communicates with the computer on a control frequency. When a user presses the transmit button, the computer assigns a frequency for that message and directs all of the radios in the talk group to receive the message on that frequency. A trunk system makes more efficient use of available frequencies and can provide coverage over extensive networks. Using digital rather than analog technology also is more efficient and allows more users to communicate at the same time in a limited portion of the radio spectrum.

Using a Radio

As a fire fighter, you must learn how to operate any radio assigned to you, and how to work with the particular radio

Figure 3-16 New technology (trunking) radios are much more advanced than conventional (VHF) radios.

system(s) used by your fire department. You must know when to use "Channel 3" or "Tac 5" or "Charlie 2," which buttons to press, and which knobs to turn. Training materials and the SOPs used by your department should cover this information.

When a radio is assigned to you, make sure you know how to operate it and check that the battery is fully charged. On the fireground, your radio is your link to the outside world. It must work properly because your life depends on it. To use a radio, follow the steps in (► Skill Drill 3-1).

1. Before transmitting, determine that the channel is clear of any other traffic. Depress the "push-to-talk" (PTT) button and wait at least two seconds before speaking. This enables the system to capture the channel without cutting off the first part of the message. Some systems sound a distinctive tone when the channel is ready. **(Step 1)**
2. When you speak into the microphone, always speak across the microphone at a 45° angle and hold the microphone 1" to 2" from the mouth. Never speak on the radio if you have anything in your mouth. **(Step 2)**
3. Always know what you are going to say before you start talking. Speak clearly and keep the message brief and to the point.
4. Do not release the PTT button until you have finished speaking. This will avoid cutting off the final part of the message. **(Step 3)**

If you hold a portable radio perpendicular to the ground with the antenna pointing toward the sky, you will get better transmission and reception. Range and transmission quality also will improve if you remove the radio from the radio pocket or belt clip before you use it.

3-1 Skill Drill

Using a Radio

Before transmitting, determine that the channel is clear of any other traffic. Depress the "push-to-talk" (PTT) button and wait at least two seconds before speaking.

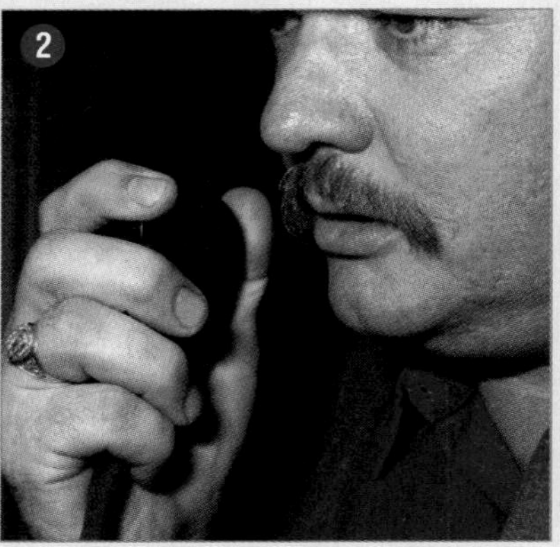

Speak across the microphone at a 45° angle and hold the microphone 1" to 2" from the mouth.

Speak clearly and keep the message brief. Do not release the PTT button until you have finished speaking.

You should be familiar with all departmental SOPs governing the use of radios. All radio transmissions should be pertinent to the situation, without any unnecessary radio traffic. Remember that radio communications are automatically recorded. They also may be heard by everyone with a radio scanner, including the news media and the general public. Avoid saying anything of a sensitive nature or anything you might later regret. Radio tapes provide a complete record of everything that is said and can be admissible as evidence in a legal case.

Most fire departments use clear speech (plain English) for radio communications, but some use codes for standard messages. A few departments have an intricate system of radio codes to fit any situation. **Ten-codes**, a system of coded messages that begin with the number 10, were once widely used. They can be problematic, however, when units from different jurisdictions have to communicate with each other. To be effective, codes must be understood by all and mean the same thing to all users. NFPA standards recommend using clear speech rather than codes. If your department uses radio codes, you must know all the codes and how they are used.

Even clear speech may include some generally accepted terminology within a system, such as "code blue" for a cardiac arrest. The objective is to be clearly and easily understood, without having to explain every message in detail. If your department has a few codes or special words for special situations, they should be specified in the SOPs.

Arrival and Progress Reports

Fire fighters and telecommunicators must know how to transmit clear, accurate size-up and progress reports. The first-arriving unit at an incident—whether a fire fighter, a company officer, or a chief—should always give a brief initial radio report and establish command as described in the chapter on the incident management system. This initial report should convey a preliminary assessment of the situation. This report gives other units a sense of what is happening so they can anticipate what their assignments may be when they arrive at the scene. The communications center should repeat the initial size-up information. For example, on arrival at the scene, a company may report:

"Engine 41 to Communications. Engine 41 is on location at the intersection of High Street and Chambersburg Road. We have a two-story, wood-frame dwelling with fire showing from two windows on the second floor. Engine 41 is assuming High Street Command. This will be an offensive operation."

The communications center would respond by announcing:

"Engine 41 is on the scene assuming High Street Command. Fire showing from the second floor of a two-story dwelling. Offensive operation."

This initial report should be followed by a more detailed report after the IC has had time to gather additional information and make initial assignments. An example of such a report follows:

IC: *"High Street Command to Communications."*

Communications center: *"Go ahead, High Street Command."*

IC: *"At 176 High Street, we are using all companies on the first alarm for a fire on the second floor of a two-story, wood-frame dwelling, approximately 30' × 40'. Exposure Alpha is the street, Exposure Baker is a similar dwelling, Exposure Charlie is an alley, and Exposure Delta is a vacant lot. Companies are engaged in offensive operations and the fire appears to be confined. We have an all clear at this time. Battalion Chief 6 is in Command."*

Communications center: *"Copied High Street Command. You have an all-clear for a fire in a dwelling at 176 High Street. The fire appears to be confined to the second floor. Battalion 6 is IC."*

For the duration of an incident, the IC should provide regular updates to the communications center. Information about certain events, such as command transfer, "all clear," and "fire under control" should be provided when they occur. Reports also should be made when the situation changes significantly and at least every 10 to 20 minutes during a working incident.

Fire Fighter Safety Tips

Verifying the information received over the radio is vital. Always restate an important message or instruction to confirm that it has been received and understood.

IC: *"Command to Ladder 7. We need you to open the roof for ventilation, directly over the fire area."*

Ladder 7: *"Ladder 7 copied. We are going to the roof to ventilate directly over the fire."*

In some fire departments, the communications center prompts the IC to report at set intervals, known as **time marks**. Time marking allows the IC to assess the progress of the incident and decide if any changes should be made in strategy or tactics. All progress reports are noted in the activity log to document the incident.

If additional resources, such as apparatus, equipment, or personnel, are needed, the IC transmits the request to the communications center. The communications center will advise the IC of additional resources that are dispatched.

Emergency Messages

There are times when the telecommunicator must interrupt normal radio transmissions for **emergency traffic**. Emergency traffic is an urgent message that takes priority over all other communications. When a unit needs to transmit emergency traffic, the telecommunicator generates a distinctive alert tone to notify everyone on the frequency to stand by, so the channel is available for the emergency communication. Once the emergency message is complete, the telecommunicator notifies all units to resume normal radio traffic.

The most important emergency traffic is a fire fighter's call for help. Most departments use "**mayday**" to indicate that a fire fighter is lost, missing, or requires immediate assistance. If a mayday call is heard on the radio, all other radio traffic should stop immediately. The fire fighter making the mayday call should describe the situation, location, and help needed. Fire fighters should study and practice the procedure for responding to a mayday call. An example of a mayday call follows:

Fire fighter: *"MAYDAY.....MAYDAY.....MAYDAY."*

All radio traffic stops.

IC: *"Unit calling MAYDAY, go ahead."*

Fire fighter: *"This is Engine 4. We are on the second floor and running out of air. Fire has cut off our escape route. We request a ladder to the window on the "Charlie" side of the building so we can evacuate."*

IC: *"Command copied. Engine 4, your escape route cut off by fire. I am sending the rapid intervention crew to Charlie side with a ladder."*

Voices of Experience

"Documentation is an important task of fire fighting, and you must develop the ability to write down the events of a fire detail-by-detail, step-by-step."

Emotion and stress can cloud memory. As a fire fighter, it is critical to remain calm and clear-headed so that you can later recall events logically and thoroughly. Documentation is an important task of fire fighting, and you must develop the ability to write down the events of a fire detail-by-detail, step-by-step. Consider jotting down the events of the fire quickly in an outline form following the progression of the fire. After you have reviewed your outline for accuracy, fill in the details of the event.

Proper documentation is especially crucial in court cases. Both sides will scrutinize your report. Any inconsistencies will be questioned in court, which is why your documentation of the event needs to be perfect. Practice your writing skills and work on developing your recall skills. Remember, if it is not written down, you cannot prove that you and your company followed the proper procedures during the fire.

John Carnegis
North Providence Fire Department
Program Chair Rhode Island Fire Academy
Providence, Rhode Island

Teamwork Tips

When using the radio at an incident involving multiple jurisdictions, use clear speech to avoid confusion.

The procedure for responding to a mayday call should be studied and practiced frequently. It is used when fire fighters are in imminent danger.

Another emergency traffic message is "abandon the building" which warns all units inside a structure to evacuate immediately. Most fire departments have a standard evacuation signal to warn all personnel to pull back to a safe location. The evacuation signal could be a sequence of three blasts on an apparatus air horn, repeated several times, or sirens sounded on "high-low" for 15 seconds. An evacuation warning should be announced at least three times to ensure that everyone hears it and announced on the radio by the IC. Because there is no universal evacuation signal, fire fighters must learn their department's SOP for emergency evacuation.

After the evacuation, the radio airwaves should remain clear so the IC can do a roll call of all units. This ensures that all personnel have safely evacuated. After the roll call, the IC will allow the resumption of normal radio traffic.

Records and Reporting

The final communications step in an emergency situation is the incident report. Every time a fire department unit responds to an incident, it must complete the proper reports. The incident report includes where and when the incident occurred, who was involved, and what happened. Incident reports for fires should include details about how the fire started, the extent of damage, and any injuries or fatalities. Incident reports can be completed and submitted on paper; however, many fire departments enter the reports on computers and store the information in a database. Several different incident-reporting software packages are available, and many large fire departments use custom-designed software.

In the United States, the National Fire Incident Reporting System (NFIRS) is used to compile and analyze incident reports at the local, state, and/or national levels. The information is used to help reduce the loss of life and property by fire. NFIRS 5.0 was developed by the United States Fire Administration (USFA) in partnership with the National Fire Information Council (NFIC). It has been offered to all states and territories to develop a more usable database of fire statistics. The NFIRS 5.0 system includes code improvements and additional modules such as Emergency Medical Services (EMS), Department Apparatus, Wildland, and Personnel that address the increasingly diverse demands faced by fire departments. It expands data collection beyond fires to the full range of fire department activities.

Canadian Perspectives

In Canada, the provinces and territories are responsible for fire reporting and the development of fire statistics. Legislation empowers the provincial and territorial fire commissioners and/or fire marshals (whichever applies) to collect fire incident data.

How to Obtain Necessary Information

After each incident, the officer in charge will need to complete an incident report. The property owner and/or occupant is a primary source of information for the report. Any bystanders or eyewitnesses should also be questioned on what they observed. The officer preparing the report should share information with any authorities such as a fire marshal or insurance investigator who also are looking into the cause of the fire. The model number and serial number of any equipment damaged by or involved in the fire should be recorded.

Required Coding Procedures

Codes are often used in incident reports to indicate incident type, actions taken, and property use. The *NFIRS Reference Guide* includes the information codes, as does the *Quick Reference Guide,* a condensed version that does not include graphics or examples. The *Coding Questions and Answers Guide* also provides coding information for NFIRS 5.0 incident reports. Fire administrators in all states have been invited to participate in NFIRS 5.0, although participation is not mandated. If your department has its own incident reports and coding procedures, you should follow department SOPs.

Consequences of Incomplete and Inaccurate Reports

From a legal standpoint, records and reports are vital parts of the emergency process. Information must be complete, clear, and concise because these records can become admissible evidence in a court case. Improper or inadequate documentation can have long-term negative consequences. Fire reports are considered as public records under the Freedom of Information Act, and may be viewed by an attorney, an insurance company, the news media, or the public. If a fatality or loss occurs, incomplete or inaccurate reports may be used to prove that the fire department was negligent. The department, the Fire Chief, and others may be held accountable. To use the NFIRS Data Entry Tool, follow the steps in ► Skill Drill 3-2.

1. Connect to the Internet. To start the NFIRS Data Entry Tool, go to the Windows Start Menu and move your cursor to Programs, then NFIRSv521, then Data Entry Tool. **(Step 1)**
2. A connection will be established and the NFIRS Data Entry Tool will open. Beside the red fire fighter's hat is your FDID (Fire Department Identification) and department name, which is displayed within your state or county level. Under the Fire Dept menu, select "Open Fire Dept" and enter your department, personnel, and apparatus information. Click "OK" when you are done. **(Step 2)**
3. To enter a new incident, select "New Incident" under the Incident menu, or click on the "New" button under Incident on the right-hand menu. Type in your FDID number, the incident date (mm/dd/yyyy), and the incident number, if this information does not appear by default. (Note that yellow fields must be completed.) Click "OK" when you are done. **(Step 3)**
4. Now that you've entered the key information, you must complete the required basic module. Click "Basic Module" to select it, then click the "Open" button under the module on the right-hand menu. **(Step 4)**
5. Sections B-E of the basic module will open. After you complete them, click on the "Sections F-J" tab at the top of your screen, or click on the "Next Tab" key on the lower right. **(Step 5)**
6. Continue through the forms, completing each one. Use the tabs at the top or the tabs to the lower-right to advance through the forms. When you are finished, save the information by selecting "Save Incident" under the Incident menu or click on the "Save" button under Incident on the right-hand menu. **(Step 6)**
7. If you have not completed all required fields or have entered an invalid code, the "Critical Validation Errors Exist" prompt will alert you. The Validation Errors and Warning screen gives details of the form, field, error message, and error level (critical or warning). You must correct critical errors to save the incident as valid (V), but you can save the incident as invalid (I) and edit it at a later date. **(Step 7)**
8. To search for incidents, select "Open Incident" under the Incident menu, or click on the "Open" button under Incident on the right-hand menu, then click "Search." **(Step 8)**
9. For a general search, complete the fields under "Address Search Criteria." For a more detailed search, complete the incident number, exposure number, status (valid or invalid), incident type, incident date range, and/or property use under "Key Search Criteria." **(Step 9)**
10. To exit the NFIRS Data Entry Tool, select Exit NFIRS from the Incident menu. Click "Yes" when the pop-up asks "Exit NFIRS Data Entry Tool?" **(Step 10)**

3-2 Skill Drill

Using the NFIRS Data Entry Tool

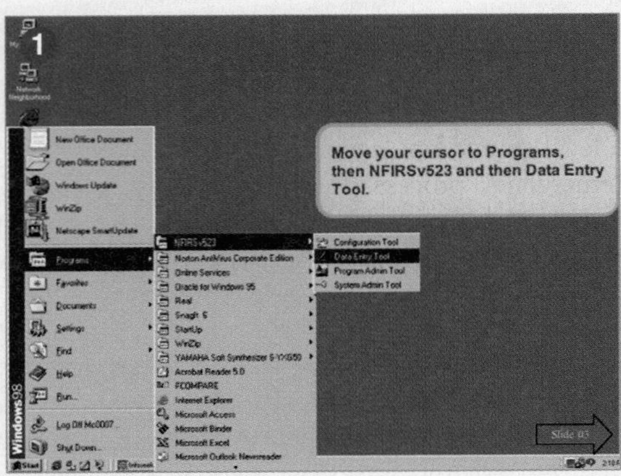

Connect to the Internet. To start the NFIRS Data Entry Tool, go to the Windows Start Menu and move your cursor to Programs, then NFIRSv521, and then Data Entry Tool.

Skill Drill

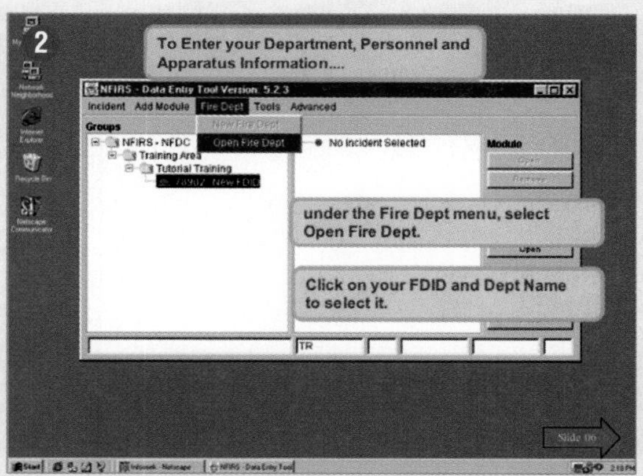

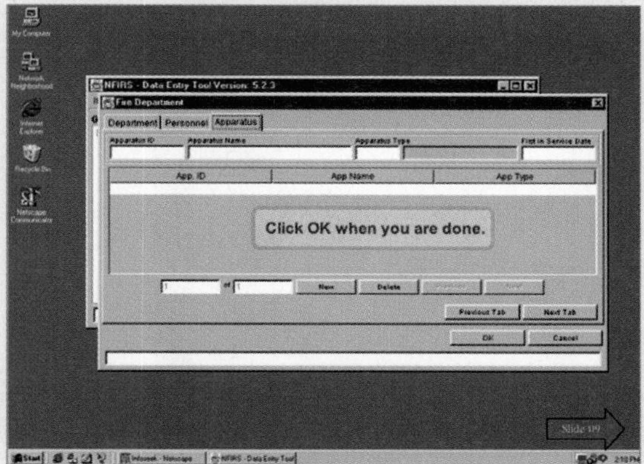

A connection will be established and the NFIRS Data Entry Tool will open. Beside the red fire fighter's hat is your FDID (Fire Department Identification) and department name, which is displayed within your state or county level. Under the Fire Dept menu, select "Open Fire Dept" and enter your department, personnel, and apparatus information. Click "OK" when you are done.

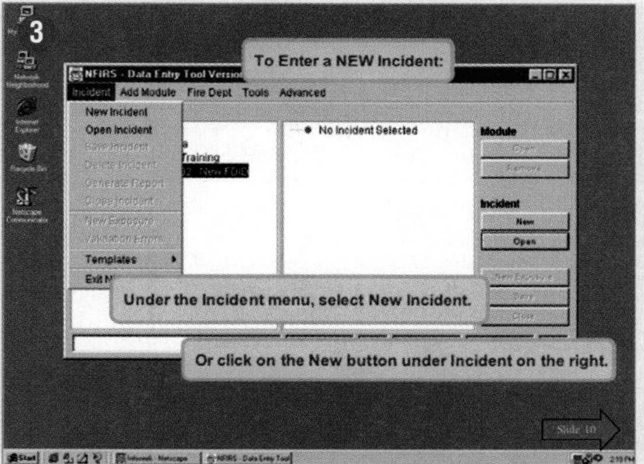

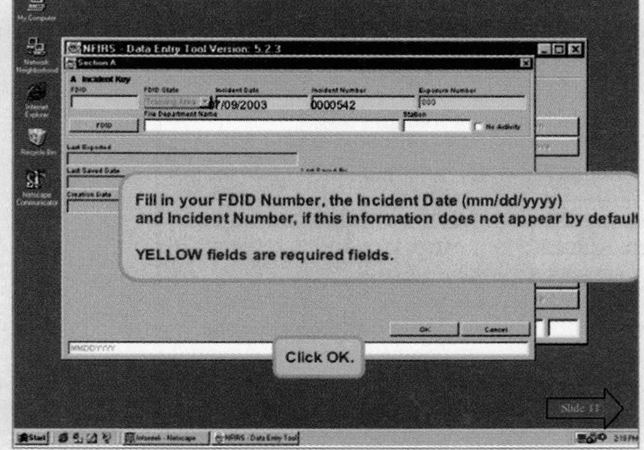

To enter a new incident, select "New Incident" under the Incident menu, or click on the "New" button under Incident on the right-hand menu and fill in your FDID number, the incident date (mm/dd/yyyy), and the incident number, if this information does not appear by default. (Yellow fields are required fields.) Click "OK" when you are done.

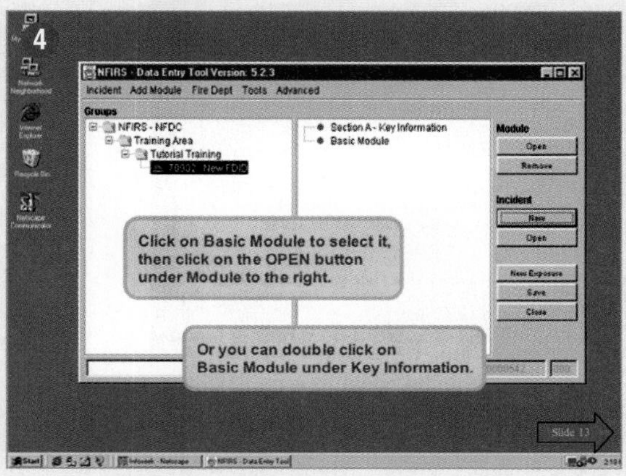

After you have entered Key Information, you are required to complete the Basic Module. Click on "Basic Module" to select it, then click on the "Open" button under the module on the right-hand menu.

Continued

Skill Drill

Using the NFIRS Data Entry Tool

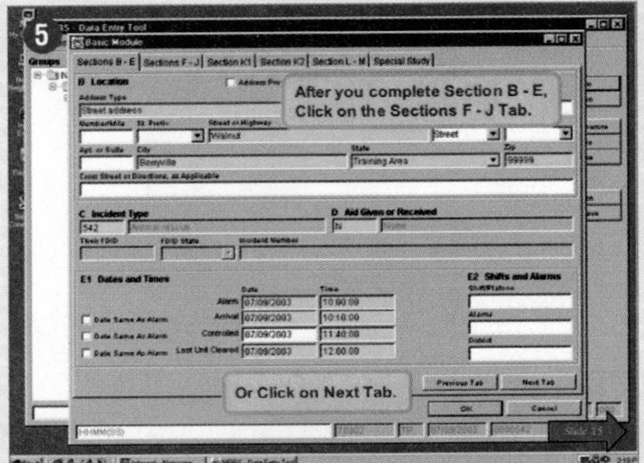

Sections B-E of the basic module will open. After you complete them, click on the "Sections F-J" tab at the top of your screen, or click on the "Next Tab" key on the lower right.

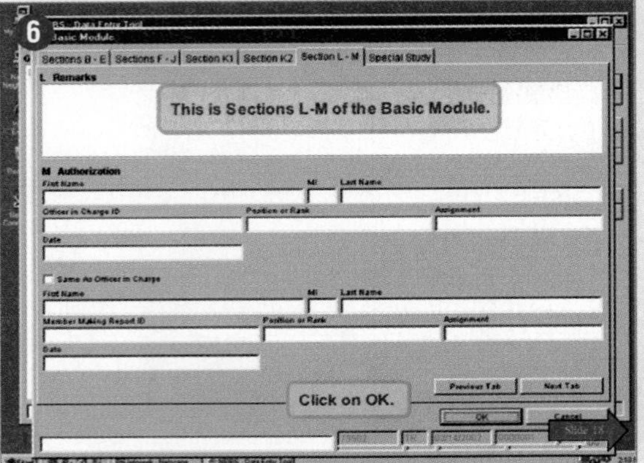

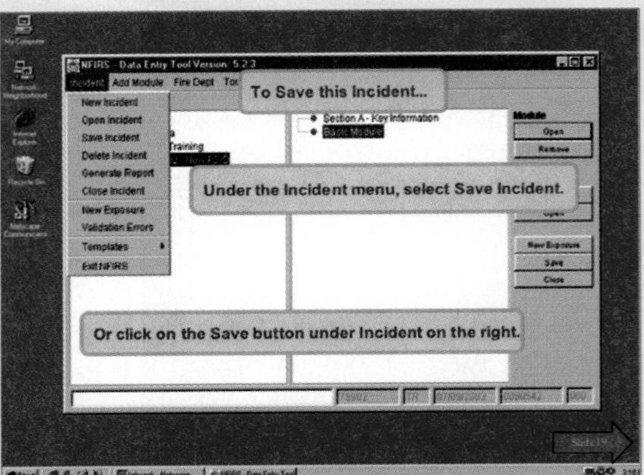

Continue through the forms, completing each one. Use the tabs at the top or the tabs to the lower right to advance through the forms. When you are finished, save the information by selecting "Save Incident" under the Incident menu or click on the "Save" button under Incident on the right-hand menu.

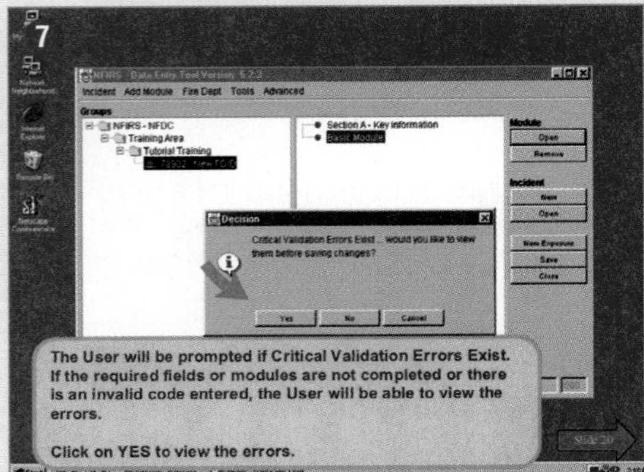

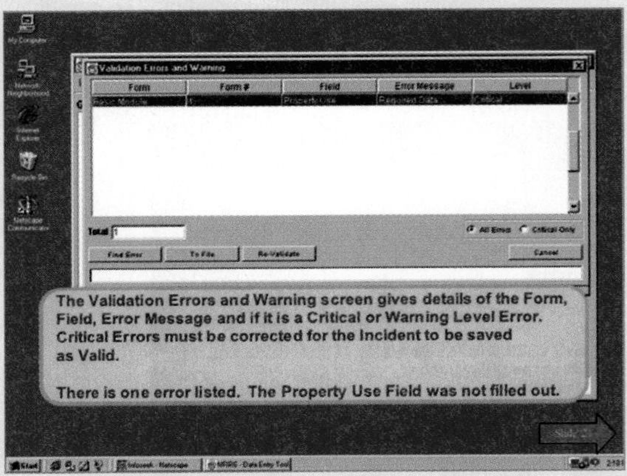

If required fields are incomplete or there is an invalid code, the "Critical Validation Errors Exist" prompt will alert you. The Validation Errors and Warning screen gives details of the form, field, error message, and error level (critical or warning). You must correct critical errors to save the incident as valid (V), but you can save the incident as invalid (I) and edit it at a later date.

Skill Drill

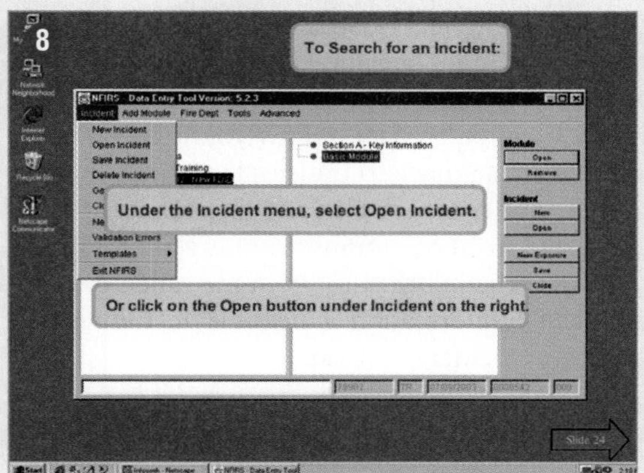

To search for incidents, select "Open Incident" under the Incident menu, or click on the "Open" button under Incident on the right-hand menu, then click "Search."

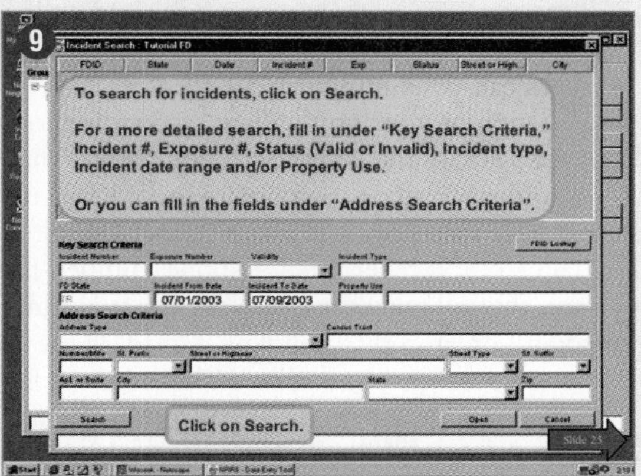

For a general search, complete the fields under "Address Search Criteria." For a more detailed search, fill in the incident number, exposure number, status (valid or invalid), incident type, incident date range, and/or property use under the "Key Search Criteria" section.

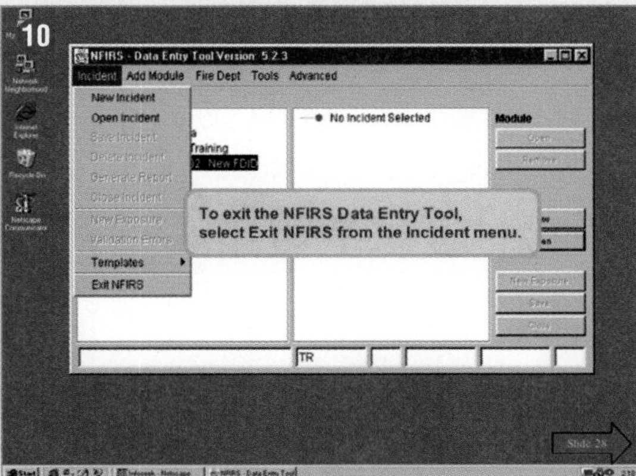

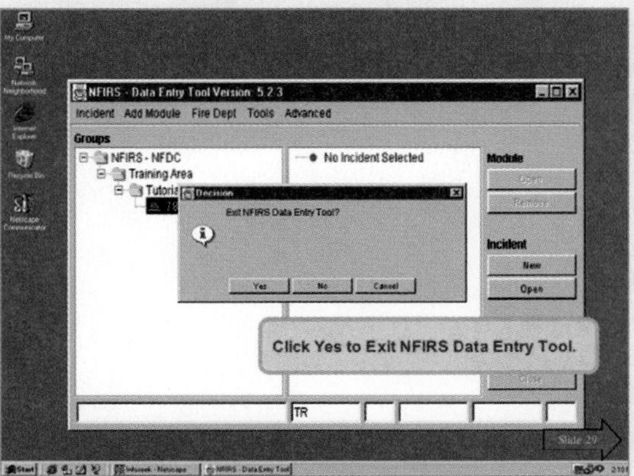

To exit the NFIRS Data Entry Tool, select "Exit NFIRS" from the Incident menu. Click "Yes" when the pop-up asks "Exit NFIRS Data Entry Tool?"

Taking Calls: Emergency, Nonemergency, and Personal Calls

One of the first things you should learn when assigned to a fire station is how to use the telephone and intercom systems. You must be able to use them to answer a call or to announce a response. You should keep personal calls to a minimum, so incoming phone lines are open to receive emergency calls.

A fire fighter who answers the telephone in a fire station, fire department facility, or communications center is a representative of the fire department. Use your department's standard greeting when you answer the phone: "Good afternoon, Pleasant Town Fire Department. Captain Smith speaking, how may I help you?" Be prompt, polite, professional, and concise.

Detailed SOPs for obtaining information and processing calls should be provided to any fire fighter assigned to answer incoming emergency telephone lines. An emergency call can come in on any fire department telephone line. Someone may call your fire station directly instead of dialing 9-1-1 or another published emergency number. If this happens, you are responsible for ensuring that the caller receives the appropriate emergency assistance. Your department's SOPs should outline exactly what steps you should take in this situation, such as whether you should take the information yourself or connect the caller directly to the communications center. **► Skill Drill 3-3** lists the steps in obtaining essential information from a caller for an emergency response.

1. When you answer a call on any incoming line, follow your department's SOPs. Answer promptly and professionally. Identify yourself and the department or location. For example, "Good afternoon. Fire Station 4, Fire Fighter Jones speaking."
2. Determine immediately if the caller has an emergency.
3. If there is an emergency, follow your department's SOPs. You may have to transfer the call to the communications center or take the information yourself and relay it to the communications center. **(Step 1)**
4. If you take the information from the caller, be sure to get as much information as possible, including:
   a. Type of incident/situation
   b. Location of the incident
   c. Cross streets or identifying landmarks
   d. Indication of scene safety
   e. Caller's name
   f. Caller's location, if different from the incident location
   g. Caller's call-back number
5. If your station or your unit will be responding to the call, always advise the communications center immediately before responding. Be prepared to take accurate information or messages on all calls. Include the date, time, name of the caller, caller's phone number, message, and the call taker's name. Have questions already formed and organized to maintain control of the conversation. Never leave someone on hold for any longer than necessary. Always terminate the call in a courteous manner. Let the caller hang up first. **(Step 2)**

3-3 Skill Drill

Operating and Answering Fire Station Telephone and Intercom Systems

1 Determine immediately if the caller has an emergency. If there is an emergency, follow your department's SOPs.

2 If you take the information from the caller, focus on obtaining vital information. If your station or your unit will be responding to the call, advise the communications center immediately. Always be prepared to take accurate information or messages for emergency, nonemergency, and personal calls. Never leave someone on hold for a long time. Always let the caller hang up first.

Wrap-Up

Ready for Review

This chapter discussed the role and qualifications of the telecommunicator, who may or may not be a fire fighter. The telecommunicator works in the communications center, which may be a fire station or a separate facility. Communications equipment, including radio systems and computers, was discussed. Computers are used for storing data and for assisting the telecommunicator with dispatch. The chapter also outlined how to receive emergency reports from the public, how to dispatch the appropriate units, how to alert personnel, and how to use the radio. Various radio systems and examples of arrival and progress reports were explained.

To understand the steps in dispatching communications and alarms, a fire fighter needs a working knowledge of the telecommunicator's job. If possible, make arrangements to observe a fire fighter or telecommunicator working a dispatch shift; it will give you a greater appreciation and a better understanding of the difficulties they face.

Chief Concepts

- Fire fighters must know how to use communication equipment properly, whether they are taking a request for routine or emergency assistance or using radio equipment at an emergency incident.
- The primary responsibilities of the telecommunicator are to receive requests from the public, determine the appropriate level of response, notify the proper agency, and record the proper information.
- A call for help is always viewed as an emergency by the person making the call.
- Emergencies may be reported in several ways, including by telephone, wireless/cellular telephone, emergency call boxes, public alarm systems, radio, or walk-ins.
- The telecommunicator can use several systems, ranging from manual run cards to computer-aided dispatch (CAD) systems, to determine which units should be deployed.
- The first-arriving unit on the scene—whether it is a fire fighter, an engine crew, an ambulance crew, or a chief—should establish command and communicate a brief initial report that conveys the initial size-up of the incident.
- In some instances, emergency traffic (an urgent message) must take precedence over normal radio communications.

Hot Terms

Activity logging system Device that keeps a detailed record of every incident and activity that occurs.

Automatic Location Identification (ALI) An enhanced 9-1-1 service feature that displays where the call originated or where the phone service is billed.

Automatic Number Identification (ANI) An enhanced 9-1-1 service feature that shows the calling party's phone number on the display screen at the telecommunicator's terminal.

Base station Radios are permanently mounted in a building, such as a fire station, communications center, or remote transmitter site.

Call box System of telephones connected by phone lines, radio equipment, or cellular technology to communicate with a communications center or fire department.

Communications center Facility that receives emergency or non-emergency reports from citizens. Many communications centers are responsible for dispatching fire department units as well.

Computer-Aided Dispatch (CAD) Computer-based, automated systems used by telecommunicators to obtain and assess dispatch information. The system recommends the type of response required.

Direct line Telephone that connects two predetermined points.

Dispatch A summons to fire department units to respond to an emergency. Also known as alerting, dispatch is performed by the telecommunicator at the communications center.

Duplex channel A radio system that uses two frequencies per channel. One frequency transmits and the other receives messages. The system uses a repeater site to transmit messages over a greater distance than a simplex system.

Emergency traffic An urgent message, such as a call for help or evacuation, transmitted over a radio that takes precedence over all normal radio traffic.

Evacuation signal Warn all personnel to pull back to a safe location.

Federal Communications Commission (FCC) United States federal regulatory authority that oversees radio communications.

Mayday Code that indicates that a fire fighter is lost, missing, or requires immediate assistance.

Mobile data terminals (MDT) Technology that allows fire fighters to receive information in the apparatus or at the station.

Mobile radio Two-way radio that is permanently mounted in a fire apparatus.

National Fire Incident Reporting System (NFIRS) System used by fire departments to report and maintain computerized records of fires and other fire department incidents in a uniform manner.

Portable radio A battery-operated, hand-held transceiver.

Public safety answering point (PSAP) The community's emergency response communications center.

Radio repeater system A radio system that automatically retransmits a radio signal on a different frequency.

Run cards Cards used to determine a predesignated response to an emergency.

Simplex channel Radio system that uses one frequency to transmit and receive all messages.

Squelch An electric circuit designed to cut off weak radio transmissions that are only capable of generating noise.

Talk around channel A simplex channel used for on-site communications

TDD/TTY/Text Phone system User devices that allow speech- and/or hearing-impaired citizens to communicate over a telephone system. TDD stands for Telecommunications Device for the Deaf; TTY stands for teletype, and Text Phones visually display text. The displayed text is the equivalent of a verbal conversation between two hearing persons.

Telecommunicator A trained individual responsible for answering requests for emergency and nonemergency assistance from citizens. This individual assesses the need for response and alerts responders to the incident.

Telephone interrogation Phase in a 9-1-1 call when a telecommunicator asks questions to obtain vital information such as the location of the emergency.

Ten-codes System of predetermined, coded messages, such as "What is your 10-20?" used by responders over the radio.

Time marks Status updates provided to the communications center every 10 to 20 minutes. This update should include the type of operation, the progress of the incident, the anticipated actions, and the need for additional resources.

Trunking system A radio system that uses a shared bank of frequencies to make the most efficient use of radio resources.

Voice recording system Recording devices or computer equipment connected to telephone lines and radio equipment in a communications center to record telephone calls and radio traffic.

Fire Fighter in Action

Your captain has scheduled a training session at the communications center. The purpose of the drill is to better understand how the communications center works and to build a better relationship between dispatchers and department personnel. The training includes a tour of the facility and an opportunity to sit with a dispatcher while they are taking calls and dispatching units.

1. You learn that the major steps in processing an emergency incident are:

- **A.** call receipt, location validation, classification and prioritization, unit selection, and dispatch.
- **B.** telephone interrogation, assign incident number, and dispatch.
- **C.** call receipt, unit selection, finding the nature of the problem, dispatch.
- **D.** dispatch, giving a short report to the unit, repeating the scene size-up.

2. What kind of equipment can you expect to find in a communications center?

- **A.** Dedicated 911 telephones
- **B.** Computers and/or hard copy files and maps to locate addresses and select units to dispatch
- **C.** Equipment for alerting and dispatching units to emergency calls
- **D.** Recording devices to record phone calls and radio traffic
- **E.** All of the above

You sit down with a dispatcher and he explains his console and CAD system. As you watch him work, you realize how complicated, demanding, and extremely stressful the dispatcher's job is.

3. Successful dispatchers must be able to:

- **A.** understand and follow complicated procedures, perform multiple tasks effectively, memorize information, and make decisions quickly.
- **B.** maintain composure under pressure and be clear, calm, and in control.
- **C.** lift their own body weight while bench pressing, eat nutritional meals, and perform cardiovascular workouts at least three times a week.
- **D.** both a and b

4. What type of fire service radio is the most powerful?

- **A.** Portable radio
- **B.** Mobile radio
- **C.** Base station
- **D.** CB radio

Incident Management System

Technology Resources

www.FireFighter.jbpub.com

- Chapter Pretests
- Interactivities
- Hot Term Explorer
- Web Links
- Review Manual
- FireLearn

Chapter Features

- Skill Drills
- Voices of Experience
- Fire Marks
- Teamwork Tips
- Fire Fighter Safety Tips
- Fire Fighter Tips
- Canadian Perspectives
- Hot Terms
- Wrap-Up

Chapter 4

NFPA 1001 Standard

Fire Fighter I

NFPA 1001 contains no Fire Fighter I Job Performance Requirements for this chapter.

Fire Fighter II

6.1.1.1 *General Knowledge Requirements.* Responsibilities of the Fire Fighter II in assuming and transferring command within an incident management system, performing assigned duties in conformance with applicable NFPA and other safety regulations and authority having jurisdiction procedures, and the role of a Fire Fighter II within the organization.

6.1.1.2 *General Skill Requirements.* The ability to determine the need for command, to organize and coordinate an incident management system until command is transferred, and to function within an assigned role in the incident management system.

Additional NFPA Standards

NFPA 1500, *Standard on Fire Department Occupational Safety and Health Program*

NFPA 1521, *Standard for Fire Department Safety Officer*

NFPA 1561, *Standard on Emergency Services Incident Management System*

Knowledge Objectives

After studying this chapter, you will be able to:

- Describe the characteristics of the incident management system.
- Explain the organization of the incident management system.
- Function within an assigned role within the incident management system.
- Organize and coordinate an incident management system until command is transferred.
- Transfer command within an incident management system.

Skills Objectives

There are no skills objectives for this chapter.

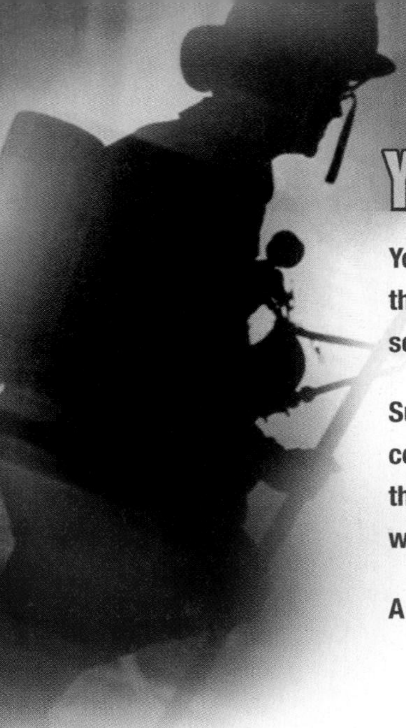

You Are the Fire Fighter

You are in the station, listening to radio traffic about a three-alarm warehouse fire. Because of the size of the incident and the number of crews working, the command structure now includes several Divisions and Branches.

Suddenly, you hear the Incident Commander order a fourth alarm. That's the summons for your company. The dispatch tone sounds, and a voice announces the location of the staging area and the companies that should respond. The watch desk printer provides the same information, as well as a description of the burning building and instructions for the fastest response route.

A few blocks from the scene, you hear the following exchange on your headset:

"Staging—from University Drive Command."

"University Drive Command, this is Staging."

"Staging, have the next two engines and one ladder report to Division C for assignment."

"University Drive Command from Staging—Engine 7 and Ladder 2 will be reporting to Division C. The second engine will be assigned as soon as they arrive at Staging."

Based on your understanding of the Incident Management System and its terminology, you anticipate that your company will probably be assigned to work at the back of the building. You're glad that the IMS is in place and organizing equipment and crews. It will mean less confusion and safer operating conditions for you and your crew.

1. ***Why is it important to establish a well-organized command structure at a major incident?***
2. ***Why are arriving engines dispatched to a staging area instead of directly to the incident location?***
3. ***Where would you expect to find the Incident Commander?***

Introduction

The Incident Management System (IMS) is a management structure that utilizes only the components that are needed at each emergency incident. The structure of IMS is similar to that of a company. A company is organized with similar components, but different titles. The company always has a head, the CEO, who makes the final decisions and creates the general plan of action for the company. The company is then broken down into departments (Production, Purchasing, Finance, Scheduling) to implement the CEO's plan. Each department has a support staff to help accomplish the department's responsibilities. Like a company, IMS is also broken down into components. There is an overall person in charge, the Incident Commander, and each department (Command, Operations, Planning, Logistics, and Finance/Administration) has specific responsibilities.

All fire department emergency operations and training exercises should be conducted within the framework of an Incident Management System (IMS). Using an IMS ensures that operations are coordinated and conducted safely and effectively. An IMS provides a standard approach, structure, and operational procedure to organize and manage any operation, from emergencies to training sessions. The same principles apply, whether the situation involves a single engine company or hundreds of emergency responders from dozens of different agencies. They also can be applied to many nonemergency events, such as large-scale public events, that require planning, communications, and coordination.

Planning, supervision, and communications are key components of an incident management system. Model procedures for incident management have been developed and widely adopted to provide a standard approach that can be used by many different agencies. Within a single fire department or a group of fire departments that routinely respond together, the use of a standard system for managing an incident is essential. Enhanced coordination results when every organization that could be involved in a major incident is familiar with and has practiced with the same incident management system.

Some fire departments use procedures that vary from the IMS model or other model systems in different respects. As a fire fighter, it is essential for you to become intimately familiar with the system that is used within your jurisdiction. Where local variations exist, the overall system usually remains similar in structure and application to the IMS model.

History of the Incident Management System

Prior to the 1970s, each individual fire department had its own method for commanding and managing incidents. Often, the organization established to direct operations at an incident scene depended on the style of the chief on duty. However, this individualized approach did not work well when units from different districts or mutual aid companies responded to a major incident. What was adequate for routine incidents was ineffective and confusing with large-scale incidents, rapidly changing situations, and units that did not normally work together.

Such a fragmented approach to managing emergency incidents is no longer accepted. Over the past 30 years, formal incident management systems (or incident command systems) have been developed and refined. Today's incident management systems include an organized system of roles, responsibilities, and standard operating procedures that are widely used to manage and direct emergency operations. The same basic approaches, organization structures, and terminology are used by thousands of fire departments and emergency response agencies.

The move to develop a standard system began about 40 years ago, after several large-scale wildland fires in Southern California proved disastrous for both the fire service and residents. A number of fire-related agencies at the local, state, and federal levels decided that better organization was necessary to effectively combat these costly fires. These agencies established an organization known as FIRESCOPE (**FI**re **RES**ources of **C**alifornia **O**rganized for **P**otential **E**mergencies) to develop solutions for a variety of problems associated with large, complex emergency incidents during wildland fires. These problems included:

- Command and control procedures
- Resource management
- Terminology
- Communications

FIRESCOPE developed the first standard Incident Command System (ICS). Originally, ICS was intended only for large, multijurisdictional/multiagency incidents involving more than 25 resources or operating units. However, it was so successful that it was applied to structural firefighting and eventually became an accepted system for managing all emergency incidents (► Figure 4-1).

At the same time, the Fireground Command (FGC) System was developed and adopted by many fire departments. The concepts behind the FGC system were similar to the ICS, although there were some differences in terminology and organization structure. The FGC system was initially designed for day-to-day fire department incidents involving fewer than 25 fire suppression companies, but it could be expanded to meet the needs of larger scale incidents.

During the 1980s, the ICS developed by FIRESCOPE was adopted by all federal and most state wildland firefighting agencies. The National Fire Academy (NFA) also used the FIRESCOPE ICS as the model fire service incident management system for all of its courses. Additional federal agencies, including the Federal Emergency Management Agency (FEMA) and the Federal Bureau of Investigation (FBI), adopted the same model for use during major disasters or terrorist events. All federal agencies could now learn and use one basic system.

Several federal regulations and consensus standards, including NFPA 1500, *Standard on Fire Department Occupational Safety and Health Program*, were adopted during the 1980s. These standards mandated the use of an incident command system at emergency incidents. NFPA 1561, *Standard on a Fire Department Incident Management System*, released in 1990, identified the key components of an effective system and the importance of using such a system at all emergency incidents. Fire departments could use either ICS or FGC to meet the requirements of NFPA 1561.

In the following years, users of different systems from across the country formed the National Fire Service Incident Management System Consortium to develop "model procedure guides" for implementing effective incident management systems at various types of incidents. The resulting system, which blended the best aspects of both ICS and FGC, is now known formally as IMS—the Incident Management System. The IMS can be used at any type or size of emergency incident and by any type or size department or agency. NFPA 1561 is now called *Standard on Emergency Services Incident Management System.*

Figure 4-1 ICS was first developed to coordinate efforts during large-scale wildland fires.

Characteristics of the Incident Management System

An IMS provides a standard, professional, organized approach to managing emergency incidents. The use of an IMS enables a fire department and any other type of emergency response agency to operate more safely and effectively. A standardized approach facilitates and coordinates the use of resources from multiple agencies, working toward common objectives. It also eliminates the need to develop a special approach for each situation.

All incident management systems require an organizational structure to provide both a hierarchy of authority and responsibility and formal channels for communications. The specific responsibility and authority of everyone in the organization is clearly stated and the relationships are well-defined. Important characteristics of an incident management system include:

- Recognized jurisdictional authority and responsibility
- Applicable to all risk and hazard situations
- Applicable to day-to-day operations as well as major incidents
- Unity of command
- Span of control
- Modular organization
- Common terminology
- Integrated communications
- Consolidated incident action plans
- Designated incident facilities
- Resource management

Jurisdictional Authority

The identification of one individual as the ultimate authority with overall responsibility for managing an incident is a key concept for the success of IMS. The issue of jurisdictional authority can become a serious concern at large-scale incidents involving multiple agencies and different levels of government. Each agency has specific responsibilities and legal authority for different situations, depending on the nature and scale of the incident, the geographic location, and other factors. Someone has to be "in charge" to ensure that operations are coordinated and safe.

Jurisdictional authority is usually not a problem at a structure fire, where the highest ranking or designated officer from the local fire department is clearly the **Incident Commander (IC)**. It can be more complicated when several jurisdictions are involved. For example, a military aircraft crashes in a national park and starts a wildland fire that spreads across the park's boundaries into an adjoining state. There, the fire causes a hazardous material to leak into a navigable waterway and approaches a county jail located inside the limits of an incorporated city. Such a situation involves both military and civilian agencies, as well as multiple levels of government. Chaos would prevail if each affected jurisdiction claimed to be in charge and attempted to issue orders.

An effective incident management system clearly defines the agency that will be in charge of each situation. When there are overlapping responsibilities, the IMS may employ a **unified command**. This brings representatives of different agencies together to work on one plan and ensures that all actions are fully coordinated. Each agency has command authority for specific, well-defined areas of responsibility.

All-Risk and All-Hazard System

IMS has evolved into an all-risk, all-hazard system that can be applied to manage resources at fires, floods, tornadoes, plane crashes, earthquakes, hazardous materials incidents, or any other type of emergency situation (► Figure 4-2). IMS has also been used to manage many nonemergency events, such as large-scale public events, that have similar requirements for command, control, and communications. The flexibility of the system enables the management structure to expand as needed, using whatever components are needed. Multiple agencies and organizations can be integrated in the management of the incident.

Everyday Applicability

IMS can and should be used for everyday operations as well as major incidents. Someone is in command at every incident, whether it is a trash bin fire, an automobile accident, or a building fire (► Figure 4-3). Regular use of the system builds familiarity with standard procedures and terminology. It also increases the users' confidence in the system. Frequent use of IMS for routine situations makes it easier to apply to larger incidents.

Unity of Command

Unity of command is a management concept in which each person has only one direct supervisor. All orders and assignments come directly from that supervisor, and all reports are made to the same supervisor. This eliminates the confusion that can result when a person receives orders from more than one boss. Unity of command reduces delays in solving problems, the potential for life and property losses, and safety risks for fire fighters.

IMS is not necessarily a rank-oriented system. The best qualified person should be assigned at the appropriate level for each situation, even if a lower ranking individual is temporarily assigned to a higher level position. This concept is critical for the effective application of the system and must be embraced by all participants.

Span of Control

Span of control refers to the number of subordinates who report to one supervisor at any level within the organization. Span of control relates to all levels of IMS, from the strategic level to the operational/tactical level and to the task level (individual companies or crews).

In most situations, one person can effectively supervise only three to seven people. Because of the dynamic nature of

Figure 4-2 IMS can be used to manage different types of emergency incidents involving several agencies.

Figure 4-3 IMS can be used effectively during day-to-day operations.

emergency incidents, an individual who has command or supervisory responsibilities in an IMS normally should not directly supervise more than five people. The actual span of control should depend on the complexity of the incident and the nature of the work being performed. For example, at a complex hazardous materials incident, the span of control might be only three; during less intense operations, the span of control could be as high as seven.

Modular Organization

IMS is designed to be flexible and modular. The IMS modules—Command, Operations, Planning, Logistics, and Finance/Administration—are predefined, ready to be staffed and made operational as needed. IMS has often been characterized as an organizational toolbox—only the tools needed for the specific incident are used. In IMS, the tools are position titles, job descriptions, and an organization structure that defines the relationships between positions. Some positions and functions are used frequently; others are only needed for complex or unusual situations. Any position can be activated simply by assigning someone to it.

For example, a small structure fire can usually be managed by an Incident Commander, who directly supervises four or five company officers. Each company officer supervises their respective company fire fighters. At a larger fire, with more companies, the IMS would expand to include divisions of up to five companies. At the same time, officers would be assigned to perform specific functions, such as safety and planning. At more complex incidents, additional levels and positions within the IMS structure would be filled in the same manner. The organization can expand as much as needed to manage the incident.

Common Terminology

IMS promotes the use of common terminology both within an organization and among all of the agencies involved in emergency incidents. Common terminology means that each word has a single definition, and no two words used in managing an emergency incident have the same definition. Everyone uses the same terms to communicate the same thoughts, so everyone understands what is meant. Each job comes with one set of responsibilities, and everyone knows who is responsible for each duty.

Common terminology is particularly important for radio communications. For example, in many fire departments, the term "tanker" refers to a mobile water supply apparatus. But to wildland firefighting agencies, it means a fixed-wing aircraft used to drop extinguishing agent on a fire. Without common terminology, an Incident Commander who called for a "tanker" could get an airplane instead of the truck he or she expected.

Integrated Communications

Integrated communications ensure that everyone at an emergency can communicate with both their supervisors and their subordinates. The IMS must support communication up and down the chain of command at every level. A message must be able to move efficiently through the system from the Incident Commander down to the lowest level and from the lowest level up to the Incident Commander (▶ Figure 4-4). Within a fire department, the primary means of communicating at the incident scene is by radio.

Consolidated Incident Action Plans

An IMS ensures that everyone involved in the incident is following one overall plan. Different components of the organization may perform different functions, but all of their efforts contribute to the same goals and objectives. Everything that occurs is coordinated with everything else.

At smaller incidents, the Incident Commander develops an action plan and communicates the incident priorities, objectives, strategies, and tactics to all of the operating units. Representatives from all participating agencies meet regularly

Figure 4-4 Integrated communications are essential for a successful operation.

to develop and update the plan. In both large and small incidents, those involved in the incident understand their specific roles and how they fit into the overall plan.

Designated Incident Facilities

Designated incident facilities are assigned locations where specific functions are always performed. For example, the Incident Commander will always be based at the incident command post. The staging area, rehabilitation area, casualty collection point, treatment area, base of operations, and helispot are all designated areas where particular functions take place. The facilities required for the incident are established according to the IMS plan.

Resource Management

Resource management is a standard system of assigning and keeping track of the resources involved in the incident. In structural firefighting, the basic units are companies. Engine companies, ladder companies, and other units are dispatched to the incident and assigned a specific function. In some cases, groups from several different companies are assigned to task forces and strike teams. The incident management system keeps track of company assignments.

At large-scale incidents, units are often dispatched to a staging area and not directly to the incident location. A staging area is a location close to the incident scene where a number of units can be held in reserve, ready to be assigned if they are needed.

The IMS Organization

The IMS structure identifies a full range of duties, responsibilities, and functions that are performed at emergency incidents. It defines the relationships among all these different components. Some components are used on almost every incident, while others apply to only the largest and most complex situations. The five major components of an IMS organization are Command, Operations, Planning, Logistics, and Finance/Administration.

The IMS organization chart may be quite simple or very complex. Each block on an IMS organization chart refers to a function area or a job description. Positions are staffed as they are needed. The only position that must be filled at every incident is the Incident Commander (IC). The IC decides which additional components are needed for the situation and activates those positions by assigning someone to perform those tasks.

Fire fighters must understand the overall structure of IMS, as well as the basic roles and responsibilities of each position. As an emergency develops, a fire fighter could start in logistics, move to operations, and eventually assume a command position. Knowing how IMS works enables fire fighters to see how different roles and responsibilities work together and relate to each other. This makes it easier to focus on a particular assignment, instead of being overwhelmed by the incident.

Command

On an IMS organization chart, the first component is Command (► Figure 4-5). The Incident Commander or IC is the only position in the IMS that must **always** be filled because there must always be someone in charge. Command is established when the first unit arrives on the scene and is maintained until the last unit leaves the scene. The IC is equivalent to the CEO of a company.

In the IMS structure, one person, the IC, is ultimately responsible for managing an incident and has the necessary authority to direct all activities at the incident scene. The IC is directly responsible for:

- Determining strategy
- Selecting incident tactics
- Setting the action plan
- Developing the IMS organization
- Managing resources
- Coordinating resource activities
- Providing for scene safety
- Releasing information about the incident
- Coordinating with outside agencies

Unified Command

Most incidents have one IC, who is directly responsible for all of the command functions. When multiple agencies with overlapping jurisdictions or legal responsibilities are involved in the same incident, a unified command is used. Representatives from each agency cooperate to share command authority. They work together and are directly involved in the decision-making process. Unified command helps ensure cooperation, avoids confusion, and guarantees agreement on goals and objectives (► Figure 4-6).

Command Post

The command post is the headquarters location for the incident. Command functions are centered in the command post. The IC and all direct support staff should always be located at the command post.

The command post should be in a nearby, protected location. Often, the command post for a major incident is located in a special vehicle or building. This enables the command staff to function without needless distractions or interruptions.

Command Staff

Individuals on the command staff perform functions that report directly to the IC and cannot be delegated to other major sections of the organization (► Figure 4-7). The Safety Officer, Liaison Officer, and Information Officer are always part of the command staff. In addition, aides, assistants, and advisors may be assigned to work directly for the IC. An aide is a fire fighter (sometimes an officer) who serves as a direct assistant to a command officer.

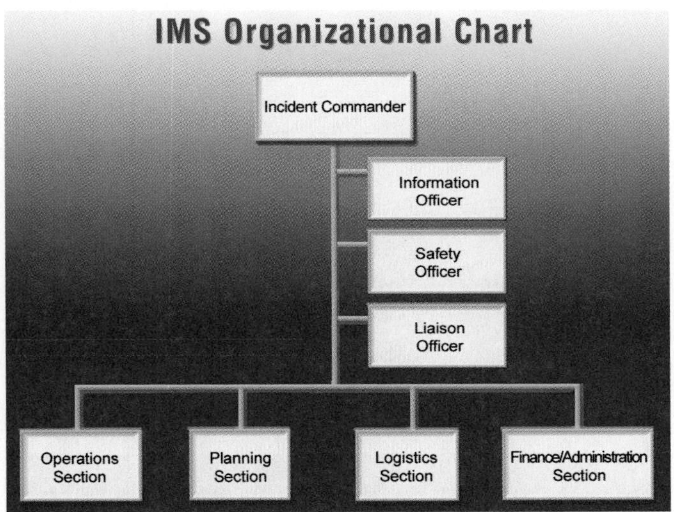

Figure 4-5 The IMS Organization Chart.

Figure 4-6 A unified command involves many agencies directly in the decision-making process for a large incident.

Safety Officer

The Safety Officer is responsible for ensuring that safety issues are managed effectively at the incident scene. The Safety Officer is the eyes and ears of the IC for safety, identifying and evaluating hazardous conditions, watching out for unsafe practices, and ensuring that safety procedures are followed. The Safety Officer is appointed early during an incident. As the incident becomes more complex and the number of resources increase, additional qualified personnel can be assigned as assistant safety officers.

The Safety Officer is an advisor to the IC, but has the authority to stop or suspend operations when unsafe situations occur. This authority is clearly stated in national standards, including NFPA 1500, *Standard on Fire Department Occupational Safety and Health Program,* NFPA 1521, *Standard for Fire Department Safety Officer,* and NFPA 1561, *Standard on Emergency Services Incident Management System.* Several state and federal regulations require the assignment of a Safety Officer at hazardous materials incidents and certain technical rescue incidents.

The Safety Officer should be a qualified individual who is knowledgeable in fire behavior, building construction and collapse potential, firefighting strategy and tactics, hazardous materials, rescue practices, and departmental safety rules and regulations. A Safety Officer should also have considerable experience in incident response and specialized training in occupational safety and health. Many fire departments have full-time Safety Officers who perform administrative functions relating to health and safety when they are not responding to emergency incidents. The National Fire Academy's *Incident Safety Officer* and *Advanced Safety Operations and Management* courses are excellent resources for people interested in serving in this capacity.

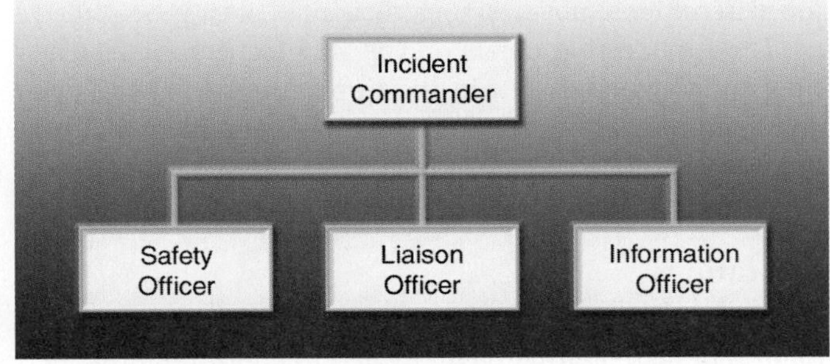

Figure 4-7 The command staff report directly to the IC.

Liaison Officer

The Liaison Officer is the IC's representative to a point of contact for representatives from outside agencies. The Liaison Officer is responsible for exchanging information with representatives from those agencies. During an active incident, the IC may not have time to meet directly with everyone who comes to the command post. The Liaison Officer position takes the IC's place, obtaining and providing information, or directing people to the proper location or authority. The Liaison area should be adjacent to, but not inside, the command post.

Public Information Officer

The Public Information Officer is responsible for gathering and releasing incident information to the news media and other appropriate agencies (► Figure 4-8). At a major incident, the public will want to know what is being done. Because the IC must make managing the incident his or her top priority, the Public Information Officer serves as the contact person for media requests. This frees the IC to concentrate on the incident. A media headquarters should be established near, but not in, the command post.

Figure 4-8 The Public Information Officer is responsible for gathering and releasing incident information to the media and other appropriate agengies.

General Staff Functions

The IC has the overall responsibility for the entire incident management organization, although some elements of the IC's responsibilities can be handled by the command staff. When the incident is too large or too complex for just one person to manage effectively, the IC may appoint someone to oversee parts of the operation. Everything that occurs at an emergency incident can be divided among the four major functional components within IMS (► Figure 4-9):

- Operations
- Planning
- Logistics
- Finance/Administration

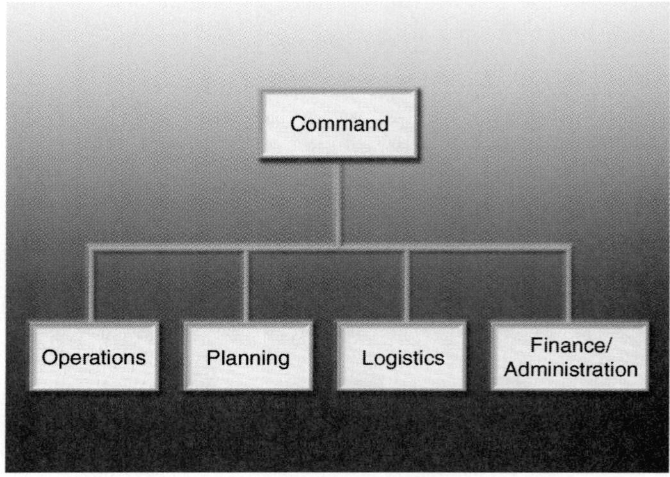

Figure 4-9 Major functional components of IMS.

The chiefs of these four sections are known as the IMS General Staff. The IC decides which (if any) of these four positions need to be activated, when to activate them, and who should be placed in each position. (Remember that the blocks on the IMS organization chart refer to function areas or job descriptions, not positions that must always be staffed.)

The four section chiefs on the IMS General Staff, when they are assigned, may run their operations from the main command post, but this is not required. At a large incident, the four functional organizations may operate from different locations, but will always be in direct contact with the IC.

Operations

The Operations Section is responsible for the management of all actions that are directly related to controlling the incident (► Figure 4-10). The Operations Section fights the fire, rescues any trapped individuals, treats the patients, and does whatever else is necessary to alleviate the emergency situation.

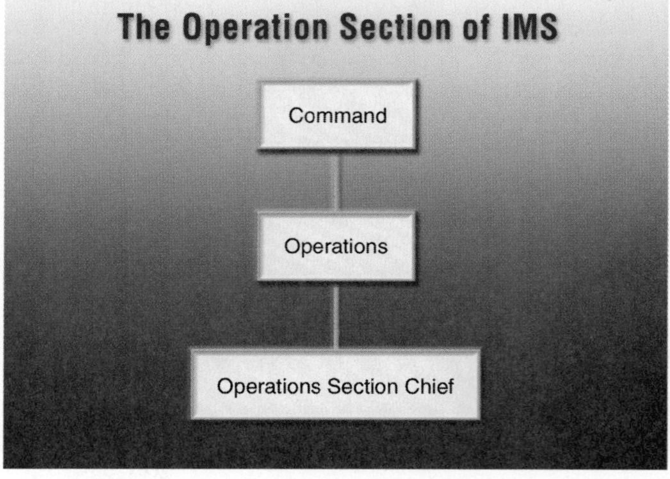

Figure 4-10 The organizational structure of the Operations Section.

Fire Fighter Tips

Strategy is the "big picture" plan of what has to be done. "Stop the fire from extending into Exposure D" is a strategic objective. Tactics are the steps that are taken to achieve the strategic objectives, such as "Make a trench cut in the roof and get ahead of the fire with hose lines on the top floor." Tasks are the specific assignments that will get the job done. "Ladder 1 will go to the roof to make the trench cut; Engines 3 and 5 will take the hose lines to the third floor, and Ladder 4 will pull ceilings to provide access into the cockloft" are all tasks.

The Operations sections is equivalent to the production department of a company. This is the department that produces results.

For most structure fires, the IC directly supervises the functions of the Operations Section. A separate **Operations Section Chief** is used at complex incidents so that the IC can focus on overall strategy while the Operations Section Chief focuses on the tactics that are required to get the job done.

Operations are conducted in accordance with an **Incident Action Plan (IAP)** that outlines what the strategic objectives are and how emergency operations will be conducted. At most incidents, the IAP is relatively simple and can be expressed in a few words or phrases. The IAP for a large-scale incident can be a lengthy document that is regularly updated and used for daily briefings of the command staff.

Planning

The **Planning Section** is responsible for the collection, evaluation, dissemination, and use of information relevant to the incident (► Figure 4-11). The Planning Section works with preincident plans, building construction drawings, maps, aerial photographs, diagrams, reference materials, and status boards. The Planning Section is also responsible for developing and updating the IAP. The Planning section is similar to the scheduling department at a company. It plans what needs to be done by whom and what resources are needed.

The IC activates the Planning Section when information needs to be obtained, managed, and analyzed. The **Planning Section Chief** reports directly to the IC. Individuals assigned to planning examine the current situation, review available information, predict the probable course of events, and prepare recommendations for strategies and tactics. The Planning Section also keeps track of resources at large-scale incidents and provides the IC with regular situation and resource status reports.

Logistics

The **Logistics Section** is responsible for providing supplies, services, facilities, and materials during the incident (▲ Figure 4-12). The **Logistics Section Chief** reports directly to the IC and serves as the supply officer for the incident. Among the responsibilities of this section would be keeping apparatus fueled, providing food and refreshments for the fire fighters, obtaining the foam concentrate needed to fight a flammable liquids fire, and arranging for a bulldozer to remove a large pile of debris. The Logistics section is similar to the purchasing department of a company. It ensures that adequate resources are always available.

In many fire departments, these logistical functions are routinely performed by support services personnel. These groups work in the background to ensure that the members of the Operations Section have whatever they need to get the job done. Resource-intensive or long-duration situations may require assignment of a Logistics Section Chief because service and support requirements are so complex or extensive that they need their own management component.

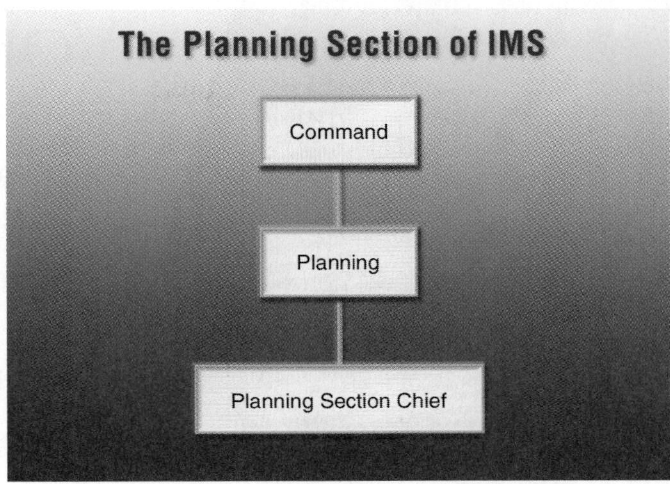

Figure 4-11 The organizational structure of the Planning Section.

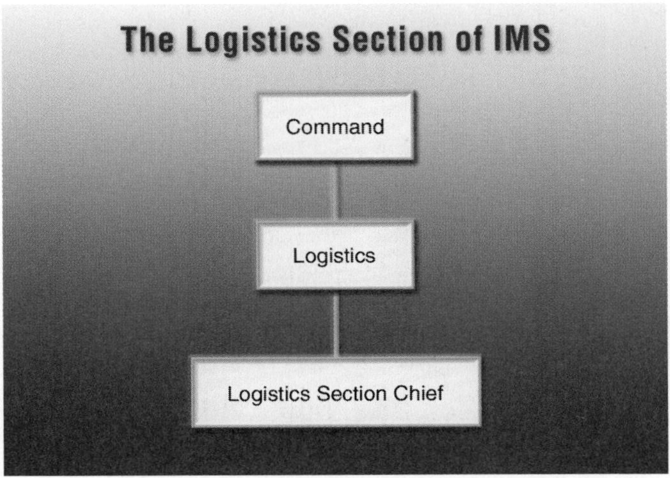

Figure 4-12 The organizational structure of the Logistics Section.

Voices of Experience

"Working together under the IMS system, we rescued the injured workers, transported them to the hospital, suppressed the fire, and protected the surrounding area from environmental damage."

It was snowy in Wisconsin on April 3, 2001. A call came in from dispatch and the information was vague. Our crew was sent to, "an unknown type of explosion in the area of the intersection of Ridge Road and Glory Road." This was unusual and everyone wondered what we were getting into. As we drove to the scene, we could see a large black plume of smoke rising in the southwest while we were still several miles away from the scene. I turned to the crew and said, "I hope you're ready to work, because I think we're going to have our hands full."

As we learned later, a small private plane had crashed shortly after taking off from the local airport. The plane encountered a problem in flight and had attempted to return to Austin Straubel International Airport. The plane lost altitude and crashed into the warehouse of Morning Glory Dairy, a milk processing plant. The plane had a full load of fuel, so the ensuing fire was large. The crash occurred during a shift change at the plant, so it was unknown how many workers were in the plant.

On arrival, the scene was chaos. There were badly burned victims and a large structure fire to combat. I met with the Incident Commander and asked, "Sir, where do you want us to start?"

He explained, "Mark, EMS has begun triage and treatment. They are transporting the first of the injured to the hospital. I need your crew to start on the fire operation and search the interior for more victims." We then lay in our supply line, advanced a 2 $^1/_2$" attack hose, and entered the burning warehouse.

At that point the incident command structure was simple. It only consisted of a Medical sector for the injured and a Fire Operation sector to suppress the fire. Within two hours, the command structure rapidly grew to accommodate the large and complex incident. Law Enforcement provided for security. The hazardous materials team dealt with leaking ammonia coolant from warehouse refrigerators and the runoff of ammonia, fuel, water, and dairy products. Public Works solved water and sewer issues. Logistical support services provided food, beverages, shelter, and sanitary needs for the personnel working at the scene. The Public Information section provided information to the news media. The Liaison sector interacted with local, county, state, and federal government officials.

In the end, the fire was suppressed. The pilot of the plane died and seven workers from the Morning Glory Dairy plant were injured. Working together under the IMS system, we rescued the injured workers, transported them to the hospital, suppressed the fire, and protected the surrounding area from environmental damage. The IMS system brought order to a scene of chaos and allowed us to do our jobs.

Mark Thiry
Ashwaubenon Department of Public Safety
Ashwaubenon, Wisconsin

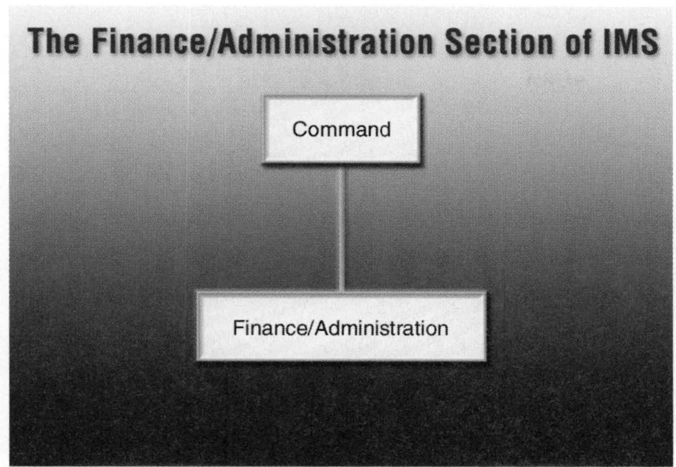

Figure 4-13 The organizational structure of the Finance/Administration Section.

Figure 4-14 A single resource is an individual company and the personnel that arrive on that unit.

Finance/Administration

The Finance/Administration Section is the fourth major IMS component under the IC (▲ Figure 4-13). This section is responsible for the accounting and financial aspects of an incident, as well as any legal issues that may arise. This function is not staffed at most incidents, because cost and accounting issues are usually addressed after the incident. A Finance/Administration Section may be needed at large-scale and long-term incidents that require immediate fiscal management, particularly when outside resources must be procured quickly. A Finance/Administration Section may also be established during a natural disaster or during a hazardous materials incident where reimbursement may come from the shipper, carrier, chemical manufacturer, or insurance company. The Finance/Administration Section is equivalent to the finance department in a company. It accounts for all activities of the company and ensures that there is enough money to keep the company running.

Standard IMS Concepts and Terminology

Fire departments and other emergency response agencies implement IMS to organize emergency scene activities in a standard manner. Emergency scenes tend to be chaotic, and organizing operations is often a serious challenge, particularly if the agencies involved use different terms to describe certain concepts and resources. One of the strengths of IMS is standard terminology. In IMS, specific terms apply to various parts of an incident organization. Understanding these basic concepts and terminology is the first step in understanding the system.

Some fire departments may use slightly different terminology. Fire fighters must know the terminology used by their department, as well as the standard IMS terminology. This section defines important IMS terms and examines their use in organizing and managing an incident.

Single Resources and Crews

A single resource is an individual vehicle and its assigned personnel (▲ Figure 4-14). For example, an engine and its crew would be a single resource; a ladder company would be a second single resource. A company officer is the individual in charge of a company. A company operates as a work unit, with all crew members working under the supervision of the company officer. Companies are assigned to perform tasks such as search and rescue, attacking the fire with a hose line, forcible entry, or ventilation.

A crew is a group of fire fighters who are working without apparatus. For example, members of an engine company or ladder company assigned to operate inside a building would be considered a crew. Additional personnel at the scene of an incident who are assembled to perform a specific task may also be called a crew. A crew must have an assigned leader or company officer.

Divisions, Groups, and Sectors

Organizational units such as divisions, groups, and sectors are established to group single resources and/or crews under one supervisor. The primary reason for establishing divisions, groups, and sectors is to maintain an effective span of control.

- A division usually refers to companies and/or crews working in the same geographic area.
- A group usually refers to companies and/or crews working on the same task or objective, although not necessarily in the same location.
- A sector can refer to companies and/or crews that have been assigned on the basis of either geography or function.

The flexibility of the IMS enables organizational units to be created as needed, depending upon the size and scope of the incident. In the early stages of an incident, individual companies are often assigned to work in different areas or perform different tasks. As the incident grows and more companies are assigned to it, the IC can establish divisions, groups, and sectors. An individual is assigned to supervise each division, group, or sector as it is created.

These organizational units are particularly useful when several resources are working near each other, such as on the same floor inside a building, or are doing similar tasks, such as ventilation. The assigned supervisor can directly observe and coordinate the actions of several crews.

A division is comprised of the resources responsible for operations within a defined geographic area. This area could be a floor inside a building, the rear of the fire building, or a section of a wildland/brush fire. Divisions are most often employed during routine fire department emergency operations. By assigning all of the units in one area to one supervisor, the IC is better able to coordinate their activities and ensure that all resources are working towards a similar strategy.

The division structure provides effective coordination of the tactics being employed by different companies working in the same area. For example, the division supervisor would coordinate the actions of a crew that is advancing a hose line into the area, a crew that is conducting search-and-rescue operations in the same area, and a crew that is performing horizontal ventilation (► Figure 4-15).

An alternative way of organizing resources is by function rather than location. A group is comprised of resources assigned to a specific function, such as ventilation, search and rescue, or water supply. Groups are responsible for performing an assignment, wherever it may be required, and often work across division lines. For example, the officer assigned to supervise the ventilation group uses the radio designation "Ventilation Group."

An example of a local terminology preference is the term sector. A sector can be either geographically-based (used instead of a division) or functionally-based (used instead of a group). However, departments that use the term sector must understand the meaning of division and group as well, because they may have to work with departments that use those terms.

Division, group, and sector supervisors all have the same rank within IMS. Divisions do not report to groups, and groups do not report to divisions. The supervisors are required to coordinate their actions and activities with each other. For example, a **group supervisor** must coordinate with the **division supervisor** when the group enters the division's geographic area, particularly if the group's assignment will affect the division's personnel, operations, or safety. The division supervisor must be aware of everything that is happening within that area. Effective communication among divisions, groups, and sectors is critical during emergency operations.

Branches

A **branch** is a higher level of combined resources than divisions, groups, and sectors. At a major incident, several different activities may occur in separate geographic areas or involve distinct functions. The span of control might still be a problem, even after the establishment of divisions, groups, and sectors. In these situations, the IC can establish branches to place a higher level supervisor (a **branch director**) in charge of a number of divisions, groups, and/or sectors.

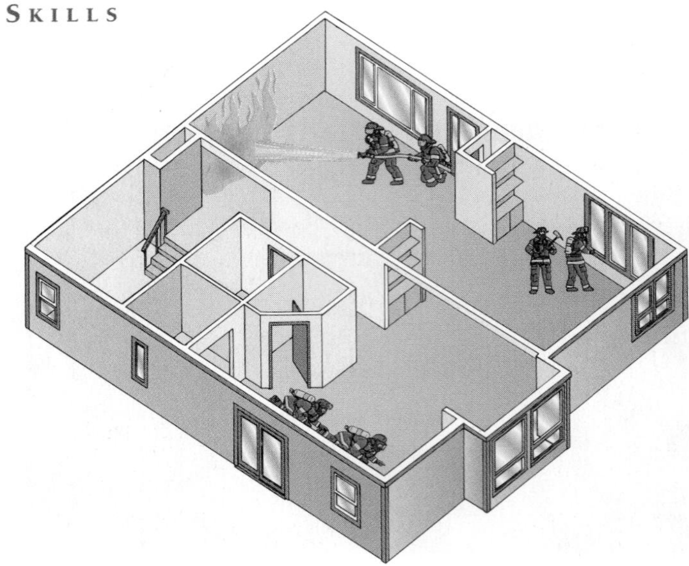

Figure 4-15 The organization of a division.

Location Designators

IMS uses a standard system to identify the different parts of a building or a fire scene. Every fire fighter must be familiar with this terminology.

The exterior sides of a building are generally known as sides A, B, C, and D. The front of the building is side A, with sides B, C, and D following in a clockwise direction around the building. The companies working in front of the building are assigned to Division A and the radio designation for their supervisor is "Division A." Similar terminology is used for the sides and rear of the building.

The areas adjacent to a burning building are called exposures. Exposures take the same letter as the adjacent side of the building. A fire fighter facing Side A can see the adjacent building on the left (Exposure B), and the building to the right (Exposure D). If the burning building is in a row of buildings, the buildings to the left are called Exposures B, B1, B2, and so on. The buildings to the right are Exposures D, D1, D2, and so on.

Within a building, divisions commonly take the number of the floor on which they are working. For example, fire fighters working on the fifth floor would be in Division 5, and the radio designation for the chief assigned to that area would be "Division 5." Crews doing different tasks on the fifth floor would all be part of this division.

(► Figure 4-16) illustrates the location designators for different parts of a burning structure, exposures, or floors of a building.

Task Forces and Strike Teams

Task forces and strike teams are groups of single resources that have been assigned to work together for a specific purpose or

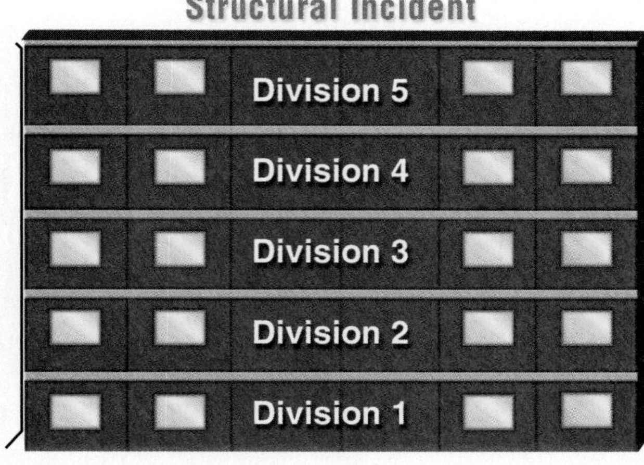

Figure 4-16 Location designators in IMS.

Figure 4-17 A strike team is five units of the same type under one leader.

for a period of time. Grouping resources reduces the span of control by placing several units under a single supervisor.

A task force includes two to five single resources, such as different types of units assembled to accomplish a specific task. For example, a task force may be composed of two engines and one truck company, two engines and two brush units, or one rescue company and four ambulances. A task force operates under the supervision of a task force leader. All communications for the separate units in the task force are directed to the task force leader.

Task forces are often part of a fire department's standard dispatch philosophy. In the Los Angeles City Fire Department, for example, a task force consists of two engines and one ladder truck, staffed by a total of 10 fire fighters. Some departments create task forces of one engine and one brush unit for response during wildland fire season. The brush unit responds with the engine company wherever it goes.

A strike team is five units of the same type with an assigned leader. A strike team could be five engines (engine strike team), five trucks (truck strike team), or five ambulances (EMS strike team) (► Figure 4-17). A strike team operates under the supervision of a strike team leader.

Strike teams are commonly used for wildland fires, where dozens or hundreds of companies may respond. During wildland fire seasons, many departments establish strike teams of five engine companies that will be dispatched and work together on major wildland fires (► Figure 4-18). The assigned companies rendezvous at a designated location and then respond to the scene together. Each engine has an officer and firefighters, but only one officer is designated as the strike team leader. All communications for the strike team are directed to the strike team leader.

EMS strike teams, consisting of five ambulances and a supervisor, are often organized to respond to multi-casualty incidents or disasters. Rather than requesting 15 ambulances and establishing an organizational structure to supervise 15 single resources, the IC can request three EMS strike teams and can coordinate with three strike team leaders.

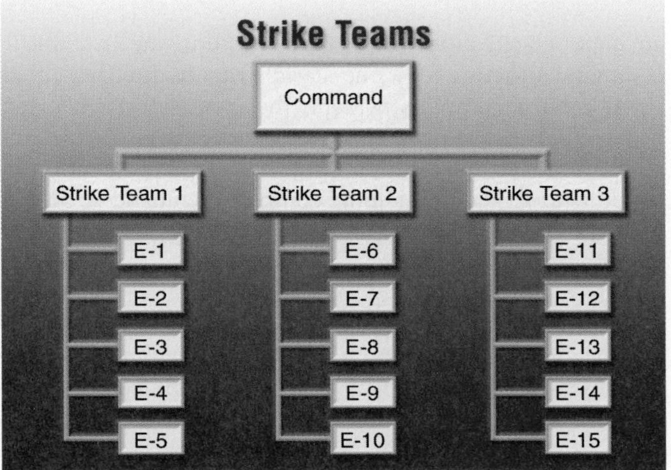

Figure 4-18 The organization of a strike team.

Implementing IMS

IMS helps to organize every incident scene in a standard, consistent manner. Fire departments develop standard operating procedures, then train and practice using IMS to ensure consistency at emergency incidents. This approach increases safety and efficiency. As an incident escalates in size or complexity, the IMS organization expands to fit the situation. The same system can be used at a house fire or a major wildland fire.

A small-scale incident can often be handled successfully by one company or a first-alarm assignment. The organizational structure implemented at an emergency incident starts small, with the arrival of the first unit or units. When only a limited number of resources are involved, an IC can often manage the organization personally or with the assistance of

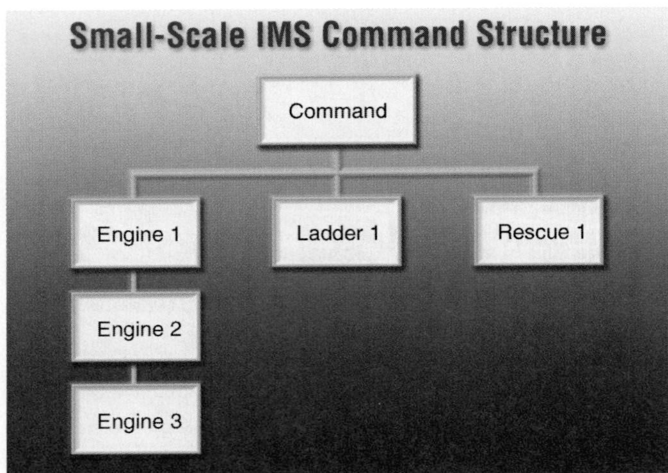

Figure 4-19 At a small-scale incident, the typical IMS command structure may consist solely of the IC and reporting resources.

an experienced aide. ▲ Figure 4-19 illustrates a typical command structure for a one-alarm structure fire.

At a more complex incident, the increasing number of problems and resources places greater demands on the IC and can quickly exceed the IC's effective span of control. A larger incident requires a more complex command structure to ensure that no details are overlooked or personnel safety compromised.

The modular design of IMS allows the organization to expand, based on the needs of the incident, by activating predesignated components. The IC can delegate specific responsibilities and authority to others, creating an effective incident organization. An individual who gets an assignment knows the basic responsibilities of the job, because they are defined in advance.

The most frequently used IMS components in structural firefighting are divisions, groups, and sectors ► Figure 4-20. These components place several single resources under one supervisor, effectively reducing the IC's span of control. The IC can also assign individuals to special jobs, such as Safety Officer and Liaison Officer, to establish a more effective organization for the incident.

In the largest and most complex incidents, other IMS components can be activated. The IC can delegate responsibility for managing major components of the operation, such as Operations or Logistics, to General Staff positions. Each of these positions has a wide range of responsibilities for a major component of the organization.

If necessary, branches can be established, creating an additional level within the organization ► Figure 4-21. For example, a structural collapse during a major fire may result in several trapped victims and many patients who need medical treatment and transportation. In such a situation, the Operations Section Chief would have multiple responsibilities that could exceed his span of control. Activating a Fire Suppression Branch, a Rescue Branch, and a Medical

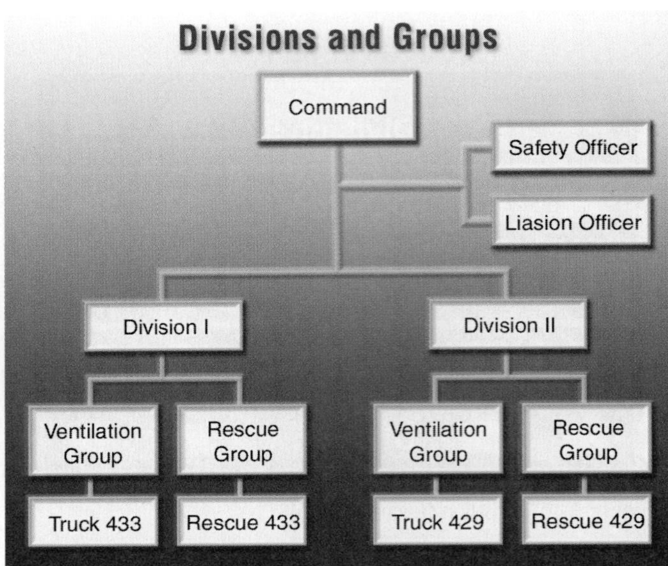

Figure 4-20 Divisions and groups are organized to manage the span of control and to supervise and coordinate units working together.

Branch would address this problem. One officer would be responsible for each branch and report to the Operations Chief. Within each branch, several divisions, groups, or sectors would report to the branch director.

Standard Position Titles

To help clarify roles within the IMS organization, standard position titles are used. Each level of the organization has a different designator for the individual in charge. The position title typically includes the functional or geographic area of responsibility, followed by a specific designator. ▼ Table 4-1 shows the standard IMS levels, functions, and position designators.

Table 4-1 Levels of an IMS Organization

IMS Level	IMS Function/ Location	Position Designator
Command	Command and control	Incident Commander
Command Staff	Safety, Liaison, Information	Officer
General Staff	Operations, Planning, Logistics, Finance/ Administration	Section Chief
Branch	Varies (e.g., EMS)	Director
Division/Group	Varies (e.g., Div. A)	Supervisor
Unit/Crew/Strike Team/Task Force	Varies (e.g., Rehab)	Leader (Company Officer)

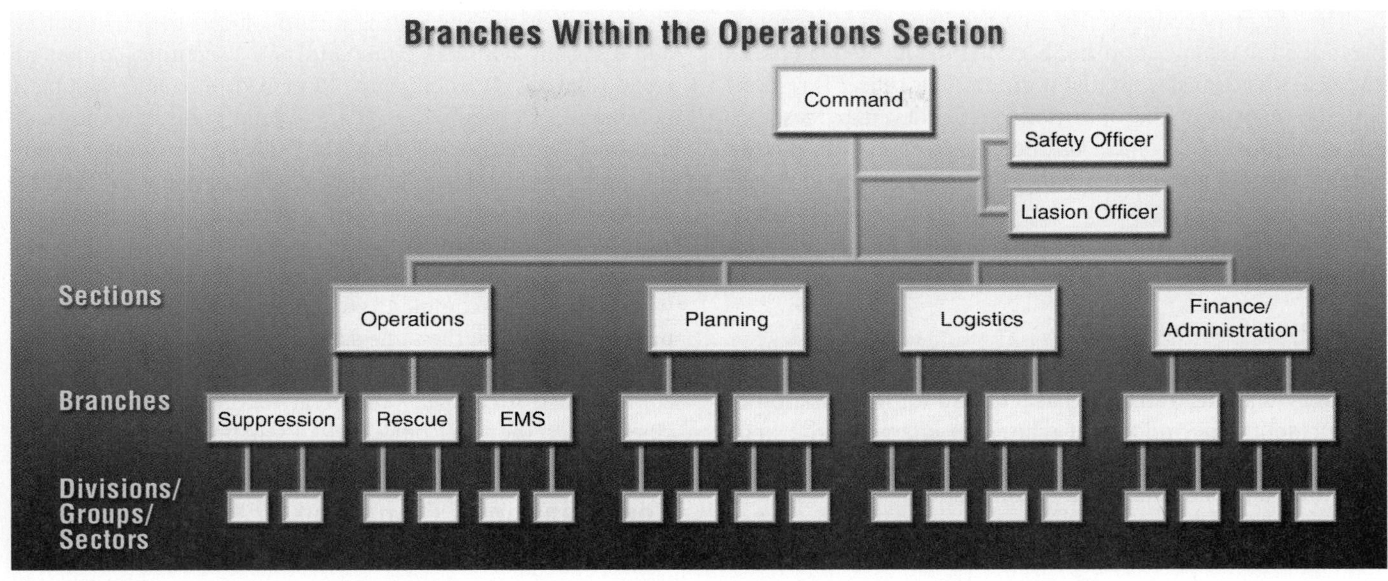

Figure 4-21 Creating branches within the Operations Section is one way to manage span of control during a large incident.

Division and group supervisors are usually chief officers, but company officers may be assigned to these positions as well. Typically, the company officer from the first unit assigned to a division or group will serve as the supervisor until a chief officer is available. A chief officer with a strong background in operations is usually assigned to the position of Operations Section Chief. The other section chief positions (Planning, Logistics, and Finance/Administration) are staffed as needed by the Incident Commander. These individuals should understand their roles and be able to meet the responsibilities of these positions.

Working within the Incident Management System

To an outsider, the IMS appears to be a large, complicated organization model that involves a complex set of standard operating procedures. But to the individual fire fighter working within the system, IMS is really quite simple and uncomplicated. Fire fighters should understand what IMS is and how it works. More importantly, each individual fire fighter must understand his or her place and role in the system.

This section discusses how every emergency incident is conducted. Certain components of the IMS are used on every incident and at every training exercise. In addition, fire fighters generally have specific responsibilities and procedures to follow in most situations.

The three basic components that always apply are:

- ***Someone is always in command of every incident,*** from the time that the first unit arrives until the last unit leaves. The identity of the Incident Commander may change, but there is always one IC in charge of the operation and responsible for everything that happens. Standard operating procedures dictate who will be the IC at any time.
- ***You always report to one supervisor.*** A fire fighter's supervisor will usually be a company officer. The company officer directly supervises a small group of fire fighters, such as an engine company or ladder company, who work together. At an incident scene, the officer provides instructions and must always know where each fire fighter is and what he or she is doing. If the company officer assigns two fire fighters to work together away from the rest of the company, both fire fighters are still under the supervision of the company officer. The company officer could be an acting officer (a fire fighter temporarily designated as a "fill-in" officer) or a fire fighter could be assigned to work temporarily under the supervision of a different officer.
- ***The company officer reports to the IC.*** If there is only one company on the scene, the company officer is the IC, until someone else arrives and assumes that role. At a small incident, the company officer may report directly to the IC, while at a large incident, there may be several layers of supervision between a company officer and the IC.

Responsibilities of the First-Arriving Fire Fighters

The first fire fighters to arrive at an emergency incident are the foundation of the IMS organization structure. Rarely is a high-ranking chief sent to a fire to evaluate the situation, design the organization structure, and order the resources that will be needed. IMS builds its organization from the bottom up, around the units that take initial action.

The officer in charge of the first-arriving unit is responsible for taking initial action and becomes the IC, with all of the authority that comes with that position. This officer is responsible for managing the operation until relieved by a senior officer. If there is no officer on the first-arriving unit, the fire fighter with the greatest seniority is in charge until an officer arrives and assumes command. The position of Incident Commander must have an unbroken line of succession from the moment the first unit arrives on the scene.

Assuming Command

The IMS assumes that command is initiated when the first unit arrives and is transferred as required for the duration of the incident. The individual who is in charge of the first-arriving unit automatically assumes command of the incident until a superior takes over command.

The individual who initially assumes command must formally announce that fact over the radio. This announcement eliminates any possible confusion over who is in charge. The initial report should include the following information:

- Command designation
- Unit or individual who is assuming command
- An initial situation report
- Initial action being taken

For example, a radio report from the first-arriving unit at a single-family dwelling fire might be:

"Engine 4 is on the scene at 309 Central Avenue where smoke is showing from a single-family dwelling. Engine 4 is laying a supply line and will make an offensive attack through the front door. Engine 4 is assuming Central Avenue Command."

Usually, the first-arriving unit at an incident scene is an engine company or some other unit that takes direct action. Regardless of the type of emergency or the type of unit, the first officer on the scene is in charge. This officer must assess the situation, determine incident priorities, and provide direction for his or her own crew as well as any other units that are arriving. Any units coming in behind the first unit know that they will be taking their orders from the initial IC until a higher level officer assumes command.

The officer who initially assumes command must decide whether to take action directly supervising the initial attack crew or to concentrate on managing the incident as the Incident Commander. This decision depends on many factors, including the nature of the situation, the resources that are on the scene or expected to arrive quickly, and the ability of the crew to work safely without direct supervision.

If the incident is large and complicated, the best option for the IC is to establish a command post and focus on sizing up the situation, directing incoming units, and requesting additional resources. The IC's own crew can be assigned to work with an acting officer or to join forces with another company officer and crew. If the situation is less critical, the IC might be able to function both as a company officer and as the incident commander at the same time, at least temporarily.

If the first-arriving unit is a chief officer, the chief officer automatically assumes command and executing command responsibilities. If a company officer had previously assumed command, the company officer would transfer command authority to the first-arriving chief officer. The officer relinquishing command should provide an assessment of the situation to the new Incident Commander, as described under Transfer of Command.

Most fire departments have written procedures that specify who will assume command in certain situations. If multiple units respond from the same station and arrive together, there should be an established protocol for which officer assumes command. If a company officer arrives seconds ahead of a chief officer, the chief officer should assume command from the outset.

Confirmation of Command

The initial announcement of command definitely confirms that command has been established at an incident. If no one announces that he or she is in command, the entire system realizes that there is no IC in place. The announcement also reinforces the IC's personal commitment to the position through a conscious personal act and a standard organizational act.

Identification of the Incident

Individual fire department procedures may vary on the specific protocol for naming an incident. The first officer to assume command should establish an identity that clearly identifies the location of the incident, such as, *"Engine 10 will be 7th Avenue IC."* This reduces confusion on the radio and establishes a continuous identity for the IC, regardless of who holds that position during the incident.

There can only be one "7th Avenue IC" at a time, so there is no confusion about who is in charge of the incident. When anyone needs to talk to the IC, a call to "7th Avenue Command" should be answered by the individual who is in command.

Passing Command

Passing command is an option that can be used by a first-arriving company officer, if there is a compelling reason that prevents that officer from assuming command of the incident. Passing command directs the next-arriving unit, whether it is a company or chief, to assume command.

A company officer is allowed to pass command only under a precise set of guidelines, in situations where his or her direct involvement in operations will have a significant impact on the outcome of the incident. For example, a four-person company might arrive at a fire and immediately need to rescue someone from a second-floor window. Because the officer must help the fire fighters get the ladder into position to rescue the victim, he or she would pass command to the officer in charge of the next-arriving unit. The notice of passing command is simply stated: *"Engine 1 passing command to Engine 2."* Engine 2 must acknowledge this message. As soon as Engine

2 arrives, the company officer will assume command as though he or she was the first-arriving company officer.

Transfer of Command

Transfer of command occurs whenever one person relinquishes command of an incident and another individual becomes the IC. For example, the first-arriving company officer would transfer command to the first-arriving chief officer who would later transfer command to a higher ranking officer during a major incident. Some departments require transfer of command when a higher ranking officer arrives at an incident; others give the higher ranking officer the option of assuming command or leaving the existing IC in charge.

When a higher ranking officer arrives at the scene of an emergency incident and takes charge, that officer assumes the moral and legal responsibility for managing the overall operation. Established procedures must be followed whenever command is transferred. One of the most important requirements of command transfer is the accurate and complete exchange of incident information (▶ Figure 4-22). The officer who is relinquishing command needs to give the new IC a current situation status report that includes:

- Tactical priorities
- Action plans
- Hazardous or potentially hazardous conditions
- Accomplishments
- Assessment of effectiveness of operations
- Current status of resources
  - Assigned or working
  - Available for assignment
  - En route
- Additional resource requirements

If transfer of command occurs very early during an incident, the transfer of information may be brief. For example, the first-arriving company officer might have been in command for only a few minutes and have little information to report when command transfers to the first-arriving chief officer. The chief may have heard all of the exchanges over the radio and know what the current situation is. However, if the company officer has been IC for several minutes, there may be a significant amount of information to report, such as the current assignments of all first-alarm companies. Whether the information is minimal or substantial, the transfer must be accurate and complete.

Each fire department establishes specific procedures for transferring command. In some cases, the transfer of command may be done via radio, but the most effective method is face-to-face communication. Standard command worksheets and a status board are valuable tools for tracking and transferring information at a command post.

Command is always maintained for the entire duration of an incident. In the later stages of an incident, after the situation is under control, command might be transferred to a lower-level commander. A downward transfer of command requires the same type of briefing and exchange of information as an upward command transfer. The officer in charge of the last company remaining on the scene would be the last IC. When that company leaves the scene, command is terminated.

Figure 4-22 The new IC needs to be briefed before assuming command.

Command Transfer Rationale

There are important reasons for transferring command at different points during an emergency incident. A first-arriving company officer can usually direct the initial operations of two or three additional companies, but as situations become more complex, the problems of maintaining control increase rapidly. A company officer's primary responsibility is to supervise one crew and ensure that they operate safely. When three or more companies are operating at an incident, it is better to have a chief officer assume command.

As more companies are assigned to an incident, the command structure must also expand. The organization must grow to maintain an effective span of control. Additional chief officers may be assigned to the incident, and a higher ranking officer may assume command. A command transfer may be required if the situation is beyond the training and experience of the current IC. A more experienced officer may have to assume command to ensure proper management of the incident.

If multiple agencies are involved, overall command of an incident may be transferred to a different agency. At a terrorist incident, for example, the fire department could have the command responsibility until victims have been removed and fires extinguished. At that point, a law enforcement agency might assume overall command. Fire department units operating at the scene would maintain their internal command structure, but the IC could be a law enforcement officer.

Wrap-Up

Ready for Review

This chapter examined the entire spectrum of a model Incident Management System. IMS is useful not only for large-scale operations, but also for any incident, regardless of size or type. All of the functions outlined in IMS must be addressed at every incident. It is the size and/or type of incident that dictates the degree to which each function is addressed. IMS allows the IC to delegate responsibilities in a standard incremental manner. The IC must perform any function that is not delegated. The IC's ultimate responsibility is to ensure that all incident requirements are met.

All fire fighters must understand the basic structure, concepts, and terminology of the IMS used by their departments and their role in its implementation. Company officers must be thoroughly trained in IMS and continually practice and refine the related skills.

Chief Concepts

- Several characteristics that are critical to an Incident Management System:
    - Organized approach: IMS imposes "order on chaos" on the fireground and enables a safer and more efficient operation than would be possible if personnel and units worked independently of each other.
    - Terminology: IMS uses a standard terminology for effective communications.
    - All-risk: IMS can be used at any type of emergency incident.
    - Jurisdictional authority: IMS enables different jurisdictions, agencies, and organizations to work cooperatively on a single incident.
    - Span of control: IMS maintains the desired span of control through flexible levels of organization.
    - Unity of command: Everyone reports to only one supervisor to avoid conflicts in giving orders.
    - Everyday applicability: IMS can and should be used on every single incident, every single time.
    - Modular: IMS is based on standard modules that are activated as needed to manage an incident.
    - Integrated communications: Everyone on the incident can communicate up and down the chain of command as needed.
    - Incident Action Plan: Every incident has a plan that outlines the strategic objectives. Large incidents will have formal plans.
    - Facilities: Standardized facilities, such as a command post, staging area, or rehabilitation area, are established as needed.
- Five major functions are part of IMS:
    - Command: Responsible for the entire incident; this is the only function that is always staffed.
    - Operations: Responsible for most fireground functions including suppression, search and rescue, and ventilation
    - Planning: Responsible for developing the Incident Action Plan
    - Logistics: Responsible for obtaining the resources needed to support the incident
    - Finance/Administration: Responsible for tracking expenditures and managing the administrative functions at the incident
- The Command Staff assists the Incident Commander at the incident:
    - Safety Officer: Responsible for overall safety of the incident; has the authority to stop any action or operation if it creates a safety hazard on the scene.
    - Liaison: Responsible for coordinating operations between the fire department and other agencies that may be involved in the incident
    - Information: Responsible for coordinating media activities and providing the necessary information to the various media organizations
- An example of a single resource would be an engine company or a ladder company.
- Single resources can be combined into task forces or strike teams.
- Other organizational units that can be established under IMS include divisions, groups, and sectors.
- IMS can be expanded infinitely to accommodate any size incident. Branches can be established to group similar functions, such as suppression, EMS, or hazardous materials.
- At every incident, someone must always assume command. As the incident grows or continues, it may be necessary to transfer command to another officer. This has to be done in a seamless manner to ensure continuity of command.

Hot Terms

Branch A supervisory level established to manage the span of control above the division, group, or sector level; usually applied to operations or logistics functions.

Branch director Officer in charge of all resources operating within a specified branch, responsible to the next higher level in the incident organization (either a Section Chief or the Incident Commander).

Command The first component of the IMS system. This is the only position in the IMS system that must always be staffed.

Command post The location at the scene of an emergency where the Incident Commander is located and where command, coordination, control, and communications are centralized.

Command staff Staff positions established to assume responsibility for key activities in the incident management system; individuals at this level report directly to the IC. Command staff include the Safety Officer, Public Information Officer, and Liaison Officer.

Company officer Usually a lieutenant or captain in charge of a team of fire fighters, both on scene and at the station. The company officer is responsible for firefighting strategy, safety of personnel, and the overall activities of the fire fighters on their apparatus.

Crew An organized group of fire fighters under the leadership of a company officer, crew leader, or other designated official.

Designated incident facilities Assigned locations where specific functions are always performed.

Division An organizational level within IMS that divides an incident into geographic areas of operational responsibility.

Division supervisor The officer in charge of all resources operating within a specified division, responsible to the next higher level in the incident organization, and the point-of-contact for the division within the organization.

Finance/Administration Section The command-level section of IMS responsible for all costs and financial aspects of the incident, as well as any legal issues that arise.

Fireground Command System (FCS) An incident management system developed in the 1970s for day-to-day fire department incidents (generally handled with fewer than 25 units or companies).

FIRESCOPE (**FI**re **RES**ources of **C**alifornia **O**rganized for **P**otential **E**mergencies) An organization of agencies established in the early 1970s to develop a standardized system for managing fire resources at large-scale incidents such as wildland fires.

Group An organization level within IMS that divides an incident according to functional areas of operation.

Group supervisor The officer in charge of all resources operating within a specified group, responsible to the next higher level in the incident organization, and the point-of-contact for the group within the organization.

IMS General Staff The chiefs of each of the four major sections of IMS: Operations, Planning, Logistics, and Finance/Administration.

Incident Action Plan (IAP) The objectives for the overall incident strategy, tactics, risk management, and member safety that are developed by the IC. Incident Action Plans are updated throughout the incident.

Incident Command System (ICS) The first standard system for organizing large, multi-jurisdictional and multi-agency incidents involving more than 25 resources or operating units; eventually developed into the Incident Management System (IMS).

Incident Commander (IC) The person in charge of the incident site who is responsible for all decisions relating to the management of the incident.

Incident Management System (IMS) The combination of facilities, equipment, personnel, procedures, and communications under a standard organizational structure to manage assigned resources effectively to accomplish stated objectives for an incident. Also known as Incident Command System (ICS).

Integrated communications The ability of all appropriate personnel at the emergency scene to communicate with their supervisor and their subordinates.

Liaison Officer The position within IMS that establishes a point of contact with outside agency representatives.

Logistics Section The section within IMS responsible for providing facilities, services, and materials for the incident.

Logistics Section Chief The General Staff position responsible for directing the logistics function; generally assigned on complex, resource-intensive, or long-duration incidents.

Operations Section The section within IMS responsible for all tactical operations at the incident.

Operations Section Chief The general staff position responsible for managing all operations activities; usually assigned when complex incidents involve more than 20 single resources or when the IC cannot be involved in the details of tactical operations.

Passing command Option that can be used by the first-arriving company officer to direct the next arriving unit to assume command.

Planning Section The section within IMS responsible for the collection, evaluation, and dissemination of tactical information related to the incident and for preparation and documentation of incident management plans.

Wrap-Up

Planning Section Chief The general staff position responsible for planning functions; assigned when the IC needs assistance in managing information.

Public Information Officer The position within IMS responsible for gathering and releasing incident information to the media and other appropriate agencies.

Resource management A standard system of assigning and keeping track of the resources involved in the incident.

Safety Officer The position within IMS responsible for identifying and evaluating hazardous or unsafe conditions at the scene of the incident. Safety officers have the authority to stop any activity that is deemed unsafe.

Sector Alternate terminology used for an organizational level defined by either a geographic or functional assignment; comparable to a division or group.

Single resource An individual vehicle and the personnel that arrive on that unit.

Span of control The number of people that a single person supervises. The maximum number of people that one person can effectively supervise is about five.

Staging area A prearranged, strategically placed area where support personnel, vehicles, and other equipment can be held in an organized state of readiness for use during an emergency.

Strike team Five units of the same resource category, such as engines or ambulances, with a leader.

Strike team leader The person in charge of a strike team, responsible to the next higher level in the incident organization, and the point-of-contact for the strike team within the organization.

Task force Any combination of single resources assembled for a particular tactical need; has common communications and a leader.

Task force leader The person in charge of a task force, responsible to the next higher level in the incident organization, and the point-of-contact for the task force within the organization.

Transfer of command Reassignment of command authority and responsibility from one individual to another.

Unified command IMS option that allows representatives from multiple jurisdictions and/or agencies to share command authority and responsibility, working together as a "joint" incident command team.

Unity of command A characteristic of the IMS structure that has each individual reporting to a single supervisor and everyone reporting to the IC directly or through the chain of command.

Fire Fighter in Action

It is a Saturday morning in February when your engine is dispatched to a single family residential dwelling for a chimney fire. Dispatch reports that fire is discharging from the top of the chimney with light smoke in the building and has advised the occupants to evacuate the structure. You arrive on the scene and Lieutenant Harn performs size-up and establishes command. The house is a medium size 3-story balloon framed structure with the fireplace located on the first floor. Your first alarm response includes three engines, one ladder truck, and a battalion chief.

1. What should the command structure at this incident look like?
   - **A.** IC, operations section chief, and division A supervisor
   - **B.** IC, fire branch director, and strike team leader
   - **C.** IC, PIO, rescue group supervisor
   - **D.** IC and team leaders
2. While working in the IMS you abide by a management concept called unity of command. Unity of command states:
   - **A.** a command officer can effectively manage three to seven personnel at one time.
   - **B.** each person has only one direct supervisor.
   - **C.** a command officer must use an accountability system in conjunction with IMS.
   - **D.** each person can effectively work for three to seven managers at one time.

Lieutenant Harn tells you to enter the building and extinguish the fire in the fireplace with a pressurized water extinguisher. After you extinguish the fire, you hear crackling and popping from within the wall. Your partner examines the wall with a thermal imaging camera and states that it appears there is substantial fire in the wall. You report your findings to your lieutenant. Lieutenant Harn reports seeing smoke coming from the attic, calls for a second alarm, and tells you to back out and prepare for an offensive attack. Due to the balloon frame construction, the fire has reached the attic. The second alarm consists of three engines, one heavy rescue, and an air/rehabilitation vehicle. The incident commander is considering breaking the incident into divisions and groups to maintain an effective span of control.

3. What is the difference between a division and group?
   - **A.** A division oversees a group.
   - **B.** A division is geographical and a group is functional.
   - **C.** A division consists of five of the same type units and a group consists of two to five single resources.
   - **D.** A group oversees a division.
4. Due to the second alarm, it would be pertinent to establish a staging area. Where should staging be located?
   - **A.** Next to the command post
   - **B.** Close to the incident scene where units can be assigned if they are needed
   - **C.** In a division
   - **D.** Next to the first in engine so that waiting fire fighters can decontaminate fire fighters exiting the fire

www.FireFighter.jbpub.com

Chapter Pretests
Interactivities
Hot Term Explorer
Web Links
Review Manual
FireLearn

Fire Behavior

Chapter 5

NFPA 1001 Standard

Fire Fighter I

5.3.10 (A) *Requisite Knowledge.* Principles of fire streams; types, design, operation, nozzle pressure effects, and flow capabilities of nozzles; precautions to be followed when advancing hose lines to a fire; observable results that a fire stream has been properly applied; dangerous building conditions created by fire; principles of exposure protection; potential long-term consequences of exposure to products of combustion; physical states of matter in which fuels are found; common types of accidents or injuries and their causes; the application of each size and type of attack line; the role of the backup team in fire attack situations; attack and control techniques for grade level and above and below grade levels; and exposing hidden fires.

5.3.11 (A) *Requisite Knowledge.* The principles, advantages, limitations, and effects of horizontal, mechanical, and hydraulic ventilation; safety considerations when venting a structure; fire behavior within a structure; the products of combustion found within a structure fire; the signs, causes, effects, and prevention of backdrafts; and the relationship of oxygen concentration to life safety and fire growth.

5.3.12 (A) *Requisite Knowledge.* The methods of heat transfer; the principles of thermal layering within a structure on fire; the techniques and safety precautions for venting flat roofs, pitched roofs, and basements; basic indicators of potential collapse or roof failure; the effects of construction type and elapsed time under fire conditions on structural integrity; and the advantages and disadvantages of vertical and trench/strip ventilation.

Fire Fighter II

There are no job performance requirements for this chapter.

Knowledge Objectives

After studying this chapter, you will be able to:

- Discuss the fire tetrahedron.
- Identify the physical states of matter in which fuels are found.
- Describe the methods of heat transfer.
- Define flash point, flame point, and ignition temperature as they relate to liquid fuel fires.
- Define the relationship of vapor density and flammability limits to gas fuel fires.
- Define Class A, B, C, D, and K fires.
- Describe the phases of fire.
- Describe the characteristics of an interior structure fire.
- Describe rollover and flashover.
- Describe backdrafts.
- Describe the principles of thermal layering within a structure.

Skills Objectives

There are no skill objectives for this chapter.

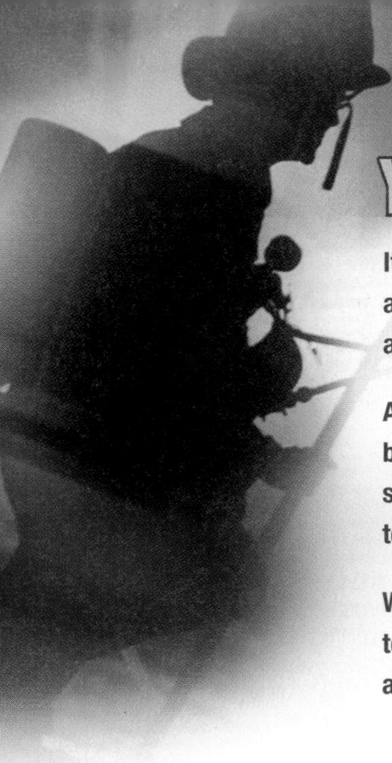

You Are the Fire Fighter

It is a quiet Sunday night, and your engine company sits down to eat its evening meal. Suddenly a dispatch comes out for your company to respond to a reported fire on the third floor of an older apartment building.

As your company arrives, you encounter light smoke on the ground floor. People are leaving the building walking upright and seem to be only mildly concerned about the situation. You climb the stairs to the second floor where the smoke is heavier. People are coughing and choking as they try to come down the stairs. Visibility is partially obscured, but you have no problem finding your way.

When you reach the third floor, it is suddenly hot and dark. You need to use your hand light just to see the reflective material on your partner's turnout gear. After dropping down to your hands and knees, you find a layer of air that allows you to see along the floor.

1. ***Why is visibility better near the ground?***
2. ***What concerns should you have upon reaching the third floor and it suddenly becoming hot and dark?***

Introduction

Before learning the methods and tactics for extinguishing fires, it is important to understand what goes on physically and chemically to make a fire occur. It is also important to understand how a fire behaves in different situations. A fire is a complex chemical process that converts one combination of substances into a different combination of substances and, at the same time, releases energy in the forms of light and heat. The understanding of fire behavior is the basis for all firefighting principles and actions.

Fire Fighter Tips

Units of Measure

When studying fire behavior, some basic units of measure must be used. In the United States, the British system of units is used, while in Canada and most other nations, the International System of Units, also known as the SI or metric system, is used. Either system can be used; however, it is important to use one system consistently. This book uses the British system and provides metric conversion factors or equivalent values as necessary.

In the British system, distance is measured in feet and inches, liquid volume is measured in gallons, temperature is measured in degrees Fahrenheit, and pressure is measured in pounds per square inch.

In the metric system, distance is measured in meters, liquid volume is measured in liters, temperature is measured in degrees Celsius, and pressure is measured in pascals or kilopascals.

Fire Triangle and Tetrahedron

In order to understand fire behavior, we will begin by identifying the conditions that are necessary for fire (also known as combustion) to occur. Three basic ingredients are required to create a fire: fuel, oxygen, and heat. We refer to these three essential components as the fire triangle. First, a combustible fuel must be present. The fuel is the material that burns. Second, oxygen must be available in sufficient quantities. Third, a source of heat must be present. The heat is essential to initiate and sustain the reaction between the fuel and the oxygen.

Heat is required initially to raise the temperature of the fuel and the oxygen to a point at which they will react together. After the fire is ignited, the process of combustion releases heat energy, which is usually sufficient to keep the reaction going. If nothing interferes with the combustion process, the fuel and the oxygen will continue to burn until the supply of one or the other is exhausted.

A fourth factor is also required to maintain a self-sustaining fire. A self-sustaining series of chemical chain reactions has to occur in order to keep the fire burning. When this factor is included, the process can be represented by a four-sided geometric shape called the fire tetrahedron (► Figure 5-1). Each side of the tetrahedron depicts one of the four factors that must be present for fire to take place. A fire cannot occur without all four components.

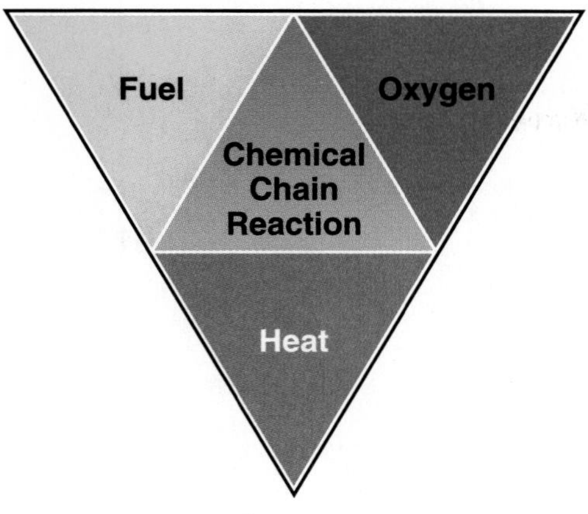

Figure 5-1 The tetrahedron model of fire (the fire tetrahedron) includes chemical chain reactions as an essential part of the combustion process.

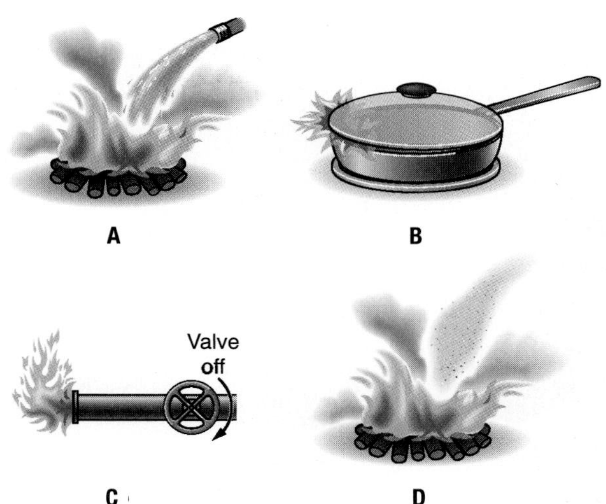

Figure 5-2 Each of the four basic methods of fire extinguishment (illustrated in relation to the fire tetrahedron). **A.** Cool the burning material. **B.** Exclude oxygen from the fire. **C.** Remove fuel from the fire. **D.** Break the chemical reaction with a flame inhibitor.

Methods of Extinguishment

The concepts of the fire triangle and the fire tetrahedron explain the conditions that are necessary for a fire to occur. Our objective as fire fighters is to extinguish fires. Although there are many different methods that can be used to extinguish fires, they can be summarized by four main methods. The four basic methods of extinguishing fires are as follows:

- Cool the burning material.
- Exclude oxygen from the fire.
- Remove fuel from the fire.
- Break the chemical reaction.

Each method of extinguishment can be understood as removing one of the essential components of the fire triangle or the fire tetrahedron. The first three methods relate to the fire triangle. When the tetrahedron model of fire is considered, all four extinguishing methods can be applied (► Figure 5-2).

The most common method used to extinguish fires is to cool the burning material with water. The water absorbs the heat of the fire as it is turned into steam. The second method is to exclude oxygen from the fire. One example of this is placing a lid over a skillet containing flaming food. Another example is applying foam to a petroleum fire, which creates a blanket that traps the vapors and separates the oxygen from the burning fuel. The third method is to remove the fuel from a fire. The easiest way to extinguish a natural gas fire is to close the gas supply valve. The fourth method of extinguishing a fire is to interrupt the chemical chain reactions with a chemical agent.

Fuel

Fuel is the actual material that is being consumed by a fire, allowing the fire to take place. Any material or matter that

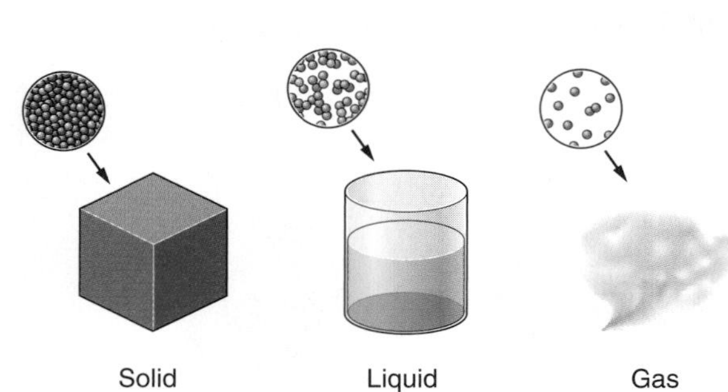

Figure 5-3 Fuels can be classified as solids, liquids, or gases.

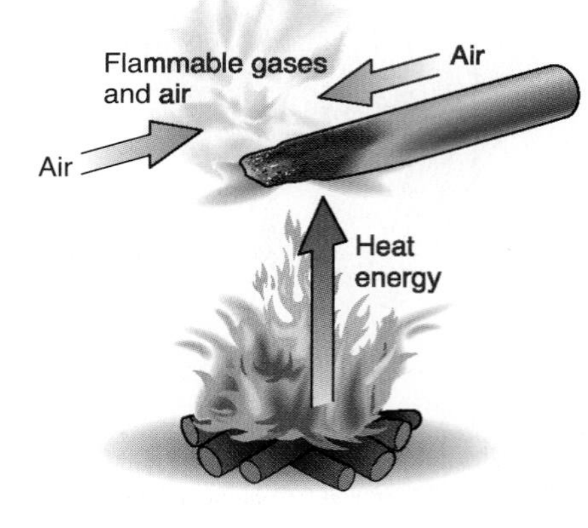

Figure 5-4 Pyrolysis.

will burn is classified as a fuel. Individual fuels can be classified as solids, liquids, or gases (▲ Figure 5-3). The ability of a material to burn is regulated by its composition and its current physical state (that is, whether it is a solid, liquid, or gas). Additional factors, such as the surface to mass ratio, have a strong influence on the ease of ignition and the rate of combustion for a particular fuel.

When substances are heated, they tend to change from a solid to a liquid state, and then to a gas or vapor state. When they are cooled, the reverse tends to take place. When substances are exposed to an increase in pressure, they tend to change from a gas or vapor state to a liquid, and then to a solid state.

The combustion reaction actually occurs when the fuel is in the gas or vapor state. Individual fuel molecules must be released from a solid or a liquid fuel in order to react with oxygen from the air.

Solids

Solids have a definite shape. Most of the fuels we encounter in fires are solids. In a structure fire, the building materials and most of the contents are solids. Some of the characteristics of solids include the following:

- Solids have the ability to resist forces and retain a definite size and shape under ordinary conditions.
- Most solids expand slightly when heated and contract slightly when cooled.
- Most solids become more brittle when cooled and more flexible when heated.

When a solid fuel is burned, the actual combustion reaction does not occur within the solid itself. As the solid fuel heats up, it decomposes in a process known as **pyrolysis** and releases individual molecules into the atmosphere (▲ Figure 5-4). The combustion occurs when oxygen molecules in the air react with the individual fuel molecules, at the surface or slightly above the surface of the solid material.

The **surface to mass ratio (STMR)** is an important factor when looking at a solid fuel. Wood is one of the most common solid fuels. A large solid piece of wood, such as a log, will burn, but it is not easy to ignite. A log has a low surface to mass ratio, because it has a relatively small surface area in relation to its large mass. A large amount of energy is needed to heat a log before it will release enough molecules to sustain a fire.

If a log is cut into 2 × 4s, the STMR increases, because each individual piece of wood has more surface area in relation to its total mass. Less heat energy is required to heat a 2 × 4, and there is more surface area to release molecules of fuel. Therefore, a 2 × 4 is easier to ignite and will burn more rapidly than a log.

If the wood is reduced down to sawdust, the STMR increases further. Each individual particle of sawdust has a small mass and a very large surface area in relation to its mass. A sawdust particle has a high STMR. Little energy is required to ignite the sawdust particles, and they will burn extremely quickly. Under the right conditions, the rate of combustion can be so fast that an explosion occurs.

Liquid

A **liquid** does not have a definite shape; it assumes the shape of the container in which it is placed. Most liquids contract when cooled and expand when heated. When sufficiently heated, most liquids will turn into gases. Liquids, for all practical purposes, are not very compressible; their volumes do not change significantly with changes in pressure. This characteristic allows us to pump water for long distances through pipelines or hoses.

Liquid fuels are one stage closer than solids to being in the optimum state for combustion. Vaporization occurs as a liquid fuel is heated, releasing molecules of fuel from the surface into the gaseous state. The vaporized fuel can then mix with oxygen, allowing combustion to occur. A fire involving a liquid fuel burns at the surface of the liquid.

The **surface to volume ratio** is an important factor for liquids, just as STMR is an important factor for solid fuels. As the surface area of the liquid increases, more molecules can be vaporized. The larger the surface area of a spill or container, the easier it is for combustion to occur and the more quickly the fuel is consumed.

Gas

A **gas** has neither independent shape nor volume and tends to expand indefinitely. The gas we most commonly encounter is air. Air is a mix of invisible, odorless, tasteless gases that surround the earth. It consists of approximately 21% oxygen and 78% nitrogen, with small amounts of other gases like carbon dioxide making up the remaining 1%. Some fuels exist in the form of a gas such as propane or natural gas.

In the gaseous state the oxygen molecules and fuel molecules can mix freely and come into direct contact with each other. Fuels in the gaseous state need very little energy to ignite, since they are already in the optimum state for combustion. The important factor for a gaseous fuel is the ratio of fuel to air in the mixture. The mixture of fuel and air must be within a certain range for combustion to occur. The combustible range is different for each fuel. If the mixture is too lean (too much air and not enough fuel), it will not burn. If the mixture is too rich (too much fuel and not enough air), it will not burn.

Oxygen and Oxidizing Agents

Oxygen is required to combine with the fuel for combustion to occur. The amount of oxygen required for combustion depends on the chemical composition of the solid fuel. If there is too little oxygen, the fuel will not burn. If there is an excess of oxygen, the solid fuel will generally burn faster and hotter.

An **oxidizing agent** can be used instead of oxygen itself in the combustion process. In a chemical reaction, an oxidizing agent behaves in the same manner as oxygen. Materials that are classified as oxidizers will support the combustion of other materials, even if no oxygen is present.

Heat

Heat energy is required to ignite a fire. Once the fire is ignited, the combustion process produces more heat. If the fuel is in the form of a solid or a liquid, heat is required to raise its temperature in order to begin to liberate fuel molecules to a gaseous state before the fuel can burn. Additional heat is required for the gaseous fuel molecules to reach their ignition temperature, at which point the reaction between the fuel molecules and the oxygen molecules begins. Once the combustion is initiated, the extra heat energy that is produced will increase the temperature and accelerate the entire process.

Heat is one of many different forms of energy. The energy that is required to produce an ignition can come from a variety of sources. These include other forms of energy, such as mechanical, electrical, and chemical energy, that are easily converted to heat.

Types of Energy

Mechanical Energy

Mechanical energy is converted to heat when two materials rub against each other and create friction. For example, a fan belt rubbing against a seized pulley produces heat. Heat is also produced when mechanical energy is used to compress air in a compressor.

Chemical Energy

Chemical energy is the energy created by a chemical reaction. Some chemical reactions produce heat (exothermic) and some absorb heat (endothermic). The combustion process is an exothermic reaction, because it releases heat energy.

Heat is produced whenever oxygen combines with a combustible material. If the reaction occurs slowly in a well-ventilated area, the heat is released harmlessly into the air. If the reaction occurs very rapidly or within an enclosed space, the mixture can be heated to its ignition temperature and can begin to burn. A bundle of rags soaked with linseed oil will begin to burn spontaneously because of the heat produced by oxidation that occurs within the mass of rags.

Electrical Energy

Electrical energy is converted to heat energy in several different manners. Electricity produces heat when it flows through a wire, or any other conductive material. The greater the flow of electricity and the greater the resistance of the material, the greater the amount of heat produced. Examples of electrical energy that can produce enough heat to start a fire include heating elements, overloaded wires, electrical arcs, and lightning.

Chemical Chain Reaction

The three basic ingredients of a fire are fuel, oxygen, and heat. When these three factors are present, under most circumstances, combustion will occur. The actual chemical processes of combustion involve a very complicated series of chain reactions at the molecular level. Normally, these chain reactions will continue to occur as long as there is an adequate supply of fuel and oxygen and sufficient heat. One way to extinguish a fire is to interrupt the sequence of chemical chain reactions by introducing other chemicals.

Chemistry of Combustion

As materials decompose and burn, there are many different chemical reactions and physical changes that occur. In many cases, the chemical constituents that are present will break down and recombine several times. As described previously, reactions that result in the release of heat energy are referred to as exothermic reactions (▶ Figure 5-5). Reactions that absorb heat or require heat to be added are called endothermic reactions (▶ Figure 5-6).

The reactions that occur within most fires are complex. In addition to large quantities of heat and light, the combustion process also produces a wide variety of other by-products. The by-products change depending on the ratio of the amount of oxygen to the amount of fuel that is available to burn at any instant. This ratio can change often and quickly.

Many of the fires we encounter as fire fighters burn with a limited amount of oxygen. During the early stages of a fire there is usually much more oxygen available than the fire can consume. As a fire grows inside a building, it will often consume most of the available oxygen until the rate of combustion is forced to slow down due to the shortage of oxygen. The fire can only consume oxygen at the rate that more oxygen can enter the space where the fire is burning. This tends to result in incomplete combustion, which produces large quantities of deadly toxic gases and compounds.

It is helpful to differentiate between oxidation, combustion, and pyrolysis. Oxidation is the process of chemically combining oxygen with another substance to create a new compound. The process of oxidation can be slow or fast. For example, steel that is exposed to oxygen will rust. Rusting is a very slow oxidation process.

Combustion is a rapid, self-sustaining process that combines oxygen with another substance and results in the release of heat and light. In effect, combustion is rapid oxidation. For our purposes, the terms combustion and fire can be used interchangeably.

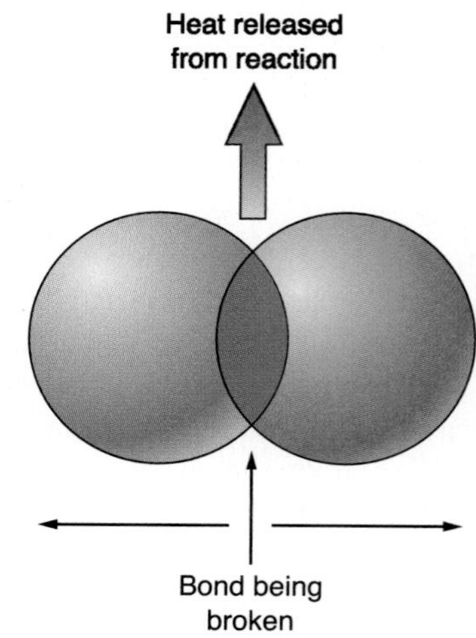

Figure 5-5 Exothermic reaction.

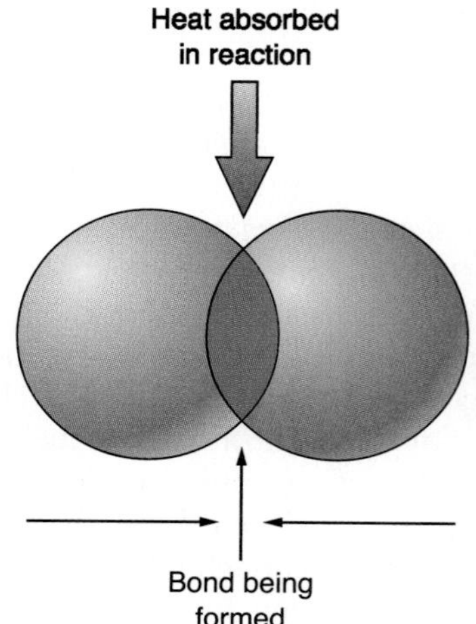

Figure 5-6 Endothermic reaction.

Pyrolysis is the decomposition of a material caused by external heating. Some of the products of decomposition are gases, which can burn in the atmosphere. Pyrolysis can eventually produce enough heat to cause combustion.

Products of Combustion

Products of combustion are the substances that are produced by the combustion reaction. Fires produce a tremendous variety of products of combustion, depending on the fuel that is being burned, the temperature of the fire, and the amount of oxygen that is available to combine with the fuel. Very few fires consume all of the available fuel with maximum efficiency. This usually results in incomplete combustion and produces a variety of by-products.

We refer to the airborne products of combustion as **smoke**. There are three major components of smoke: particles, vapors, and gases. Many of the constituents of smoke are toxic. Inhalation of these particles, droplets, and gases can cause severe respiratory injuries. Because smoke is the direct result of fire, it is generally hot when it is created. The inhalation of superheated gases in smoke can cause severe burns of the respiratory tract. The temperature of smoke will vary depending on the conditions of the fire and the distance the smoke has traveled from the fire.

Smoke Particles

Smoke particles are solid matter consisting of unburned, partially burned, or completely burned substances. These particles are lifted in the thermal column produced by the fire. In some cases the unburned and partially burned particles are hot enough to burn and can ignite if they come into contact with a supply of oxygen. The completely burned particles are primarily ash.

Most of the particles found in smoke are toxic. Some of these particles are small enough to get past the protective mechanisms of the respiratory system and enter the lungs.

Liquids

Smoke often contains small droplets of liquids. When oil-based compounds burn, they produce small droplets that become part of the smoke. Oil-based or lipid compounds can be toxic and dangerous if inhaled. Some toxic droplets also cause poisoning if absorbed through the skin. Known examples of poisonous carcinogens that are readily absorbed through the skin include aromatic hydrocarbons such as xylene, toluene, and benzene.

When water is applied to a fire, small water droplets are also suspended in the smoke or haze that forms. This is similar to the fog that is formed on a cool night. The water droplets often absorb particulate matter from the smoke.

Fire Fighter Safety Tips

Smoky environments are deadly! It is essential to use self-contained breathing apparatus any time you are operating in a smoky environment. This applies whether it is a structure fire, dumpster fire, car fire, or any other type of fire that generates smoke. Common toxic gases in smoky environments:

- Carbon monoxide
- Hydrogen chloride
- Hydrogen cyanide
- Carbon dioxide
- Nitrogen dioxide
- Phosgene
- Ammonia
- Chlorine

Gases

Smoke contains a wide variety of gases. The composition of gases in smoke varies greatly, depending on the substance being burned, the temperature, and the amount of oxygen available to the fire at a given instant. Burning wood will produce a different combination of gases than a fire fueled by petroleum-based products. (Remember that plastics are made from petroleum-based compounds.)

Most of the gases produced by a fire are toxic. Carbon monoxide, hydrogen cyanide, and phosgene are three of many toxic gases that are often present in smoke. Carbon monoxide is deadly in small quantities. Hydrogen cyanide was used to kill convicted criminals in gas chambers. Phosgene was used in World War I as a poisonous gas to disable soldiers. Fires also produce carbon dioxide, which is an inert gas. Although it is not toxic, carbon dioxide can displace oxygen and cause **hypoxia**.

Heat Transfer

Heat transfer is important to a fire fighter, because heat is essential for a fire to be initiated and for combustion to occur. Fires also produce large quantities of heat, which causes the fire to spread and involve additional fuel. Heat transfer is particularly important in relation to protecting the fire fighter from the heat of the fire. Protective clothing is a shield that helps limit the heat that is transferred from the fire to the fire fighter.

There are three mechanisms of heat transfer. Heat energy can be transferred from a hotter mass to a colder mass by any one of the mechanisms or by any combination of the three mechanisms working together. The fire spread we observe as fire fighters is often a combination of conduction, convection, and radiation, all occurring simultaneously. In most situations one of the three mechanisms is more powerful than the other two mechanisms.

Conduction

Conduction is the process of transferring heat energy within matter by transferring the energy directly from one molecule to another (▶ Figure 5-7). Objects vary in their ability to conduct energy. Metals generally conduct heat very easily. Insulating materials, such as fiberglass, are very poor conductors of heat.

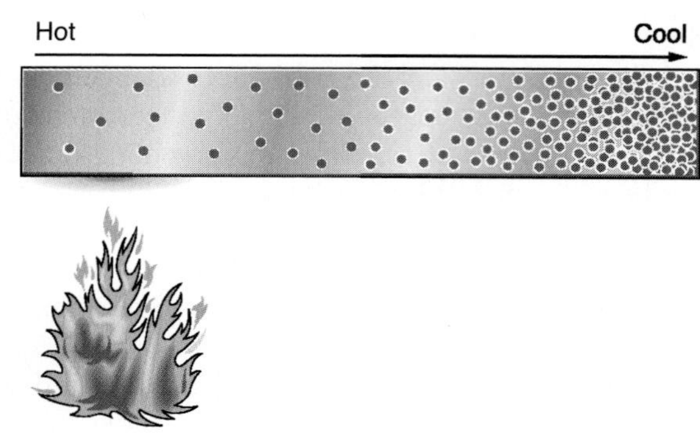

Figure 5-7 Conduction.

Wood is a relatively poor conductor of heat. When wood is heated, the energy tends to remain in the area where the heat is applied, and the temperature in that area increases. Since wood has a fairly low ignition temperature, if the heat source is sustained, it will begin to burn in that localized area. The localized combustion will release more heat and cause the fire to spread to a larger area.

The same amount of heat energy applied to a steel beam will result in lower temperatures at the point where the heat is applied, because much of the heat will be dissipated by conduction to other parts of the beam. The temperature of the entire beam will increase, although the spot where the heat is applied will still be hotter than any other part of the beam.

Steel conducts heat so well that it can cause another material to ignite. Consider a steel beam that is in contact with a flame at one end and wood at the other end. By conduction, the steel could eventually transfer enough heat to the wood to cause the wood to ignite.

Convection

Convection is the movement of heat through a fluid medium such as air or a liquid (▶ Figure 5-8). Heated molecules within a fluid medium will always rise. Convection currents can be observed above hot soup or a cup of coffee.

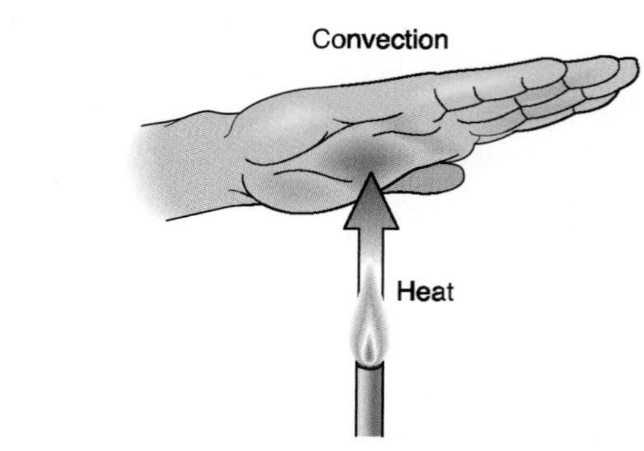

Figure 5-8 Convection.

The convection currents created by a fire primarily involve gases generated by the fire. The heat of the fire warms the gases and particles in the smoke. A large fire burning in the open can generate a **plume** or column of heated gases and smoke that rises high into the air. This convection stream can carry the smoke and sometimes large bands of burning fuel several blocks before the gases cool and fall back to the earth. If there are winds present during a large fire, the winds can push the convection currents in different directions, causing the fire to spread.

When a fire is burning inside a room, the hot gases will rise to the ceiling. They will then travel horizontally along the ceiling until they hit a wall. If there are no openings in the room that allow the gases to escape, a layer of heated air will build up at the ceiling level and then move downward. The highest temperatures will always occur at the ceiling.

If there are openings in the fire room, convection currents can carry the fire gases outside the room of origin to other parts of the building. The superheated gases can be hot enough to ignite other materials. The hot gases will flow

Figure 5-9 Radiation.

toward the highest level of the building, using stairways or any open shaft, and collect under the roof. Opening a ventilation hole in the roof releases the hot gases and allows energy to escape from the building by convection. This can rapidly lower temperatures inside the building, making it easier for fire fighters to enter and attack the fire. We will explore the effects of convection currents on structure fires by studying how a fire spreads beyond the room of origin.

Radiation

Radiation is the transfer of heat energy in the form of invisible waves (▲ Figure 5-9). The sun radiates energy to the earth through outer space. The electromagnetic radiation from the sun travels through the vacuum of space until it reaches the earth. The earth in turn absorbs the radiant heat energy from the sun. The human body absorbs radiant heat energy in the same manner, whether it comes from the sun, from a space heater, or from an open fire. Some materials reflect radiant energy more than they absorb it. A sheet of shiny aluminum foil will reflect the rays from the sun and send them in another direction.

Radiated heat energy from a fire travels in all directions. The effect of radiation is not seen or felt until the radiation strikes an object and heats the surface of the object. A large building that is fully involved in fire can radiate a tremendous amount of heat energy to the surrounding buildings. That energy can travel several hundred feet and ignite an exposed building. The radiated heat can even pass through glass windows and ignite objects inside other buildings.

VOICES OF EXPERIENCE

"Just because you do not see the danger signs, does not mean that there is no danger."

"It must be the coal stove in the basement," the homeowner told me as we arrived at the scene of a house fire. I asked the homeowner how to get into the basement and he pointed to the garage. A member of my crew had stretched a hand line to my position. I motioned for him to follow me and we moved to an entry door next to the overhead garage door.

A large window showed that smoke was in the garage but no fire was present. There were no pulsing windows, the smoke was not dark, and it was not sucking or puffing around the windows and doors. Since I did not see the usual backdraft danger signs, I decided to open the door to get a look at the stairway. I opened the door and smoke pushed out but did not suck back in. I could hear the cackling of fire and turned my back to request a second alarm. Suddenly, I felt that something was wrong.

I felt the hair stand up on the back of my neck. I heard a loud roar, my back became warm, and a blast of hot gas and smoke pushed me away from my position. I was dazed for a few seconds. I stood up and shook my head to clear it. I realized I had just been in a backdraft. From the horrified looks of the bystanders, it must have been quite a site. I quickly made a head count of my crew. Thankfully, everyone was accounted for and uninjured.

The force of the backdraft sent window glass flying across a street and cut the tires on the pick-up truck of a responding fire fighter. Entire window and door frames lay in the yard. The garage was now several inches from the house. The overhead garage door was blown completely off and twisted like a ribbon.

My hair was singed on the back of my head where my helmet was not covering it. I recovered quickly, but I will never forget the effects of the backdraft. My department now runs drills on the causes of backdraft and the preventative measures that should always be taken. Just because you do not see the danger signs, does not mean that there is no danger.

Adrian Borry
Lincoln Fire Company
Ephrata, Pennsylvania
Lebanon Bureau of Fire
Lebanon, Pennsylvania

Characteristics of Liquid Fuel Fires

Fires involving liquid fuels have some different characteristics from fires that involve solid fuels. As discussed previously, solid fuels do not burn in a solid state. A fuel must be decomposed and the flammable compounds must be vaporized before the fuel can burn. The same characteristic applies to liquids. A liquid must be converted to a vapor before it will burn. In addition, the vaporized fuel molecules must be mixed with air in the proper concentration. There is a minimum and a maximum concentration that is necessary for any vapor to be ignited. These two concentrations are called the flammability limits.

There are three conditions that must be present in order for a vapor and air mixture to ignite:

- The fuel and air mixture must be within the flammable limits.
- There must be an ignition source with enough energy to ignite the vapors.
- There must be enough sustained contact between the ignition source and the fuel–air mixture to transfer the ignition energy.

As a liquid is heated, the molecules in the liquid become more active, and some of them begin to break away from the surface as vapor. The process of molecules being released from the liquid's surface is called evaporation. The rate of evaporation increases as the temperature of the liquid increases. At the boiling point, all of the liquid is converted to vapor. The amount of liquid that will be vaporized is also related to the **volatility** of the liquid. A more volatile liquid evaporates more quickly than a less volatile liquid.

As more of the liquid vaporizes, the mixture will reach a point where there is enough vapor in the air to create a flammable fuel–air mixture. Most flammable liquids can generate enough vapors to be ignited at temperatures well below their boiling point. For example, gasoline can generate flammable vapors at temperatures as low as –45°F. If the vapors are ignited, the fire will warm the liquid surface and cause vapors to be released more rapidly. This increases the rate of burning.

Because most liquid fuels are mixtures of different compounds, they do not have single boiling points. For example, gasoline is a mixture of about 100 different compounds and the individual compounds evaporate at different rates. Gasoline vapors are composed of the individual vapors of all of the compounds mixed together.

The flammability of a liquid is defined in terms of three temperatures. These terms are used to compare characteristics of different fuels: The **flash point** is the lowest temperature at which a liquid produces a flammable vapor. This is measured by determining the lowest temperature at which the liquid will produce enough vapor to support a small flame for a short period of time. At this concentration the flame may go out quickly. The **fire point** or **flame point** is the lowest temperature at which a liquid produces enough vapor to sustain a continuous fire. The **ignition temperature** is the temperature at which the fuel–air mixture produced by a liquid will spontaneously ignite. The ignition temperature of the mixture is actually determined by the compound with the lowest ignition temperature.

Characteristics of Gas Fuel Fires

There are two terms that help to describe the characteristics of flammable gases and vapors. These are vapor density and flammability limits.

Vapor Density

Vapor density refers to the weight of a gaseous fuel. Vapor density measures the weight of the gas molecules compared to air. The weight of air is assigned the value of 1.0. A gas with a vapor density of less than 1.0 will rise to the top of a confined space or rise into the open atmosphere. A gas with a vapor density greater than 1.0 is heavier than air and will settle close to the ground.

In situations where a flammable gas is present, knowing the vapor density of the product allows you to predict whether the danger of ignition is at a high level or at a low level within an area. Hydrogen gas has a density of 0.07, so it is very light and will rise in the atmosphere. Propane gas has a vapor density of 1.51, so it will settle to the ground. Carbon monoxide has a density of 0.97, which is very close to air. Carbon monoxide mixes readily with all layers of the air since it has almost the same density.

Flammability Limits

Mixtures of flammable gases and air will burn only when they are mixed in certain concentrations. If there is too little fuel (vapor) present in the mixture, there will not be enough fuel to support the combustion process. This mixture is known as being too lean. If too much fuel vapor is present in the mixture, there will not be enough oxygen present to support the combustion. This mixture is known as being too rich.

The range of mixtures that will burn varies from one fuel to another. Natural gas must be mixed with air in concentrations of between 4.5% and 15.0% in order to burn. These values are known as the lower and upper flammability limits. The terms **flammability limits** and **explosive limits** are used interchangeably in this book. Under most conditions, if the flammable gas and air mixture can be ignited, it has the ability to explode. Test instruments are available to measure the percent of fuels in gas and air mixtures and to determine when an emergency scene is safe.

The lower explosive limit (LEL) refers to the minimum amount of gaseous fuel that must be present in a gas and air mixture for the mixture to be flammable or explosive. In the case of carbon monoxide, the LEL is 12.5%.

The upper explosive limit (UEL) of carbon monoxide is 74%. These two values tell us that carbon monoxide can burn or explode when the concentration is at least 12.5% and no greater than 74% in air.

Boiling-Liquid, Expanding-Vapor Explosion (BLEVE)

You must understand one potentially deadly set of circumstances involving liquid and gas fuels. This is a boiling-liquid, expanding-vapor explosion or BLEVE. A BLEVE can occur when a liquid fuel is stored in a vessel under pressure. A propane tank is an example; the vessel is partly filled with the liquid propane and the rest of the vessel is taken up by propane in the form of a vapor.

The most common cause of a BLEVE is a fire that is exposed to the tank. The fire heats up the liquid in the tank, causing it to generate more vapors. This increases the internal pressure of the tank to a point where the tank can rupture catastrophically. When this happens, large pieces of the tank can be propelled significant distances, injuring and even killing fire fighters. In addition, any remaining flammable liquid is immediately released from the tank. Because the temperature of the liquid is at or above its boiling point, the liquid immediately turns into vapor and creates a rapidly expanding cloud. The fire can then ignite the escaping vapors, creating a fireball. All of this can happen extremely rapidly, in a matter of a second or two.

Over the years dozens of fire fighters have been killed by BLEVEs that occurred while they were trying to fight fires that involved tanks of liquefied gaseous fuels. By understanding the characteristics of flammable gas fuels and the mechanism of a BLEVE, we can help to prevent injuries or deaths in emergency situations.

Fire Fighter Safety Tips

When water is applied to certain combustible metals, such as magnesium, the water reacts with the metal to produce hydrogen and oxygen. The hydrogen can burn with an explosive force.

Classes of Fire

Fires are classified according to the type of fuel that is burning. The classification system is designed to simplify the decisions that must be made when choosing extinguishing agents and methods for different types of fires. Fires are placed into one of five classes: Class A, Class B, Class C, Class D, and Class K. For each class of fire there are corresponding classes of extinguishing agents. Class A extinguishing agents are used to fight Class A fires and so on. We will consider the methods of attacking specific types of fires in more detail in subsequent chapters.

A fire may fit into more than one class. For example, a fire could involve a wood building (Class A) as well as petroleum products (Class B). A fire involving energized electrical circuits (Class C), might also involve Class A or Class B materials.

Class A Fires

Class A fires involve ordinary solid combustible materials such as wood, paper, and cloth. The best method to extinguish a Class A fire is to cool the fuel to a temperature that is below its ignition temperature. Water is the most frequently used extinguishing agent for Class A fires, because it is plentiful and efficient in absorbing large quantities of heat.

A Class A fire should not be considered extinguished until the entire mass has been thoroughly cooled. If a Class A fire is not completely cooled, the remaining hot embers may rekindle, and the remaining fuel may start to burn again. Smothering a Class A fire to exclude oxygen may extinguish the fire temporarily, but it will not cool the fuel. The remaining hot embers can reignite the fire if they are exposed to fresh air.

Class B Fires

Class B fires involve flammable or combustible liquids such as gasoline, kerosene, oils, paints, and tar. Most Class B fires are best extinguished by creating a barrier between the fuel and the oxygen. A layer of foam is applied to the surface of the fuel to stop the production of vapors. If the liquid can no longer release any vapors to mix with the air, the fire goes out. Foam also cools the surface of the liquid and heated objects in the vicinity of the fire. Different types of foam are required for different liquid fuels.

Some liquid fires can be extinguished by cooling them with water, similar to a Class A fire. This is most effective on liquids that have high flash points, in which cases the water can cool the surface of the liquids below their respective flash points. A water mist is generally used in these cases.

Alternative extinguishing agents are available for Class B fires, including dry chemicals and carbon dioxide. Dry chemical fire extinguishers attack a fire by interrupting the fire's chain reactions. Carbon dioxide (CO_2) is applied as a smothering agent to exclude oxygen from the area where the fire is burning.

Class C Fires

Class C fires involve energized electrical equipment. Electricity itself doesn't burn, but it provides energy that can ignite a fire. The most common extinguishing agents for electrical fires are carbon dioxide or dry chemical agents because they do not conduct electricity. Halogenated agent extinguishers are used for special Class C fires involving electrical equipment such as computers. The fuel that is burning is often Class A or Class B, however the special classification is required due to the particular hazards of electricity. If the electrical source is disconnected it becomes a Class A or Class B fire.

The use of water on a Class C fire is very dangerous because water can conduct electricity. Whenever possible, the electrical supply should be shut-off prior to applying water in any form.

Class D Fires

Class D fires involve burning metals. Combustible metals include sodium, potassium, lithium, titanium, zirconium, magnesium, aluminum, and many of their alloys. These fires are less common, but they can be very difficult to extinguish. Most cars contain numerous components made from such alloys. The greatest ignition hazard exists when the combustible metals are present as shavings or in a molten form.

Fighting a Class D fire with water can cause a chemical reaction, and it can generate explosive hydrogen gas. Special extinguishing powders based on sodium chloride or other salts are available. Extinguishing with clean, dry sand is another option.

Class K Fires

Class K fires involve combustible cooking media, such as oils and grease, commonly found in commercial kitchens. The Class K designation is new and coincides with a new classification of Class K extinguishing agents recognized by NFPA 10, *Standard for the Installation of Portable Fire Extinguishers*.

Phases of Fire

When oxygen, fuel, heat, and the necessary chain reactions are all present under adequate circumstances, a fire will occur. As a typical fire progresses, it will pass through four distinct phases, unless the process is interrupted. The four phases of fire are as follows:

- Ignition
- Growth
- Fully developed
- Decay

Ignition Phase

The ignition phase of the fire is the starting point (▼ Figure 5-10). When the four parts of the fire tetrahedron are present and the fuel is heated to its ignition temperature, ignition occurs.

To study an example, consider a fire that begins when a burning cigarette ignites the arm of a sofa. Further consider that the sofa is sitting between an end table and a coffee table in a living room, and that there is a good supply of fuel, and the concentration of oxygen is 21% within the room. The fire begins small, with a localized flame at the point where the cigarette is in contact with the fabric.

Gradually, more of the fabric is ignited and a small plume of hot gases starts to rise from the sofa arm. This plume ignites even more of the fabric, increasing the size and intensity of the flame. At this point the convection of hot gases is the primary means of fire growth, with some additional energy being radiated to the area close to the flame.

Within a minute the sofa starts to ignite. Oxygen is drawn into the fire at floor level. At this point the fire could probably be extinguished with a portable fire extinguisher.

Figure 5-10 Ignition phase.

Figure 5-11 Growth phase.

Growth Phase

During the **growth phase** of a fire, additional fuel becomes involved in the fire. A combination of convection and radiation ignites more surfaces of the sofa and the table closest to it (▲ Figure 5-11). As more fuel is ignited, the size of the fire increases and the visible plume reaches the ceiling.

The convection flow begins to draw additional air into the fire. With this additional air and more fuel becoming involved, the fire grows in intensity. Hot gases from the plume flow across the ceiling toward the four walls of the room.

This is the point where **thermal layering** begins. The hotter air collects at the top of the room and banks down, while the cooler air stays closer to the floor. During this growth phase, the fire will continue to grow, as long as there is enough oxygen available and the fuel source is maintained.

Flashover

Flashover is a the point between the growth phase and the fully developed phase where all of the combustible materials in a room become ignited.

Most fires begin small. As a fire grows in size, the combined forces of radiation, convection, and conduction cause the surface temperatures of all other combustible materials in the room or space to approach their ignition temperatures. The surface temperatures reach a point where individ-

Fire Fighter Safety Tips

Fire fighters must learn to recognize the warning signs of flashover and use the reach of their hose stream to cool the atmosphere. A fire fighter even in full turnout gear has little chance of avoiding severe injury or even death if caught in a flashover.

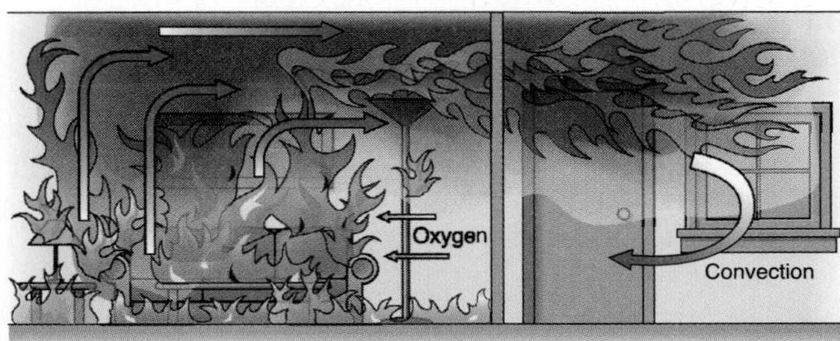

Figure 5-12 Fully developed phase.

Figure 5-13 Decay phase.

ual items begin to ignite in rapid succession. As each additional item ignites, it adds even more heat energy to the room. At the flashover point, all of the remaining items reach their ignition temperatures at virtually the same time. Suddenly everything in the room begins to burn, releasing energy at a much faster rate. Temperatures reach the vicinity of 1000°F in just a few seconds.

Flashover is often the most deadly to fire fighters and victims because conditions can change radically in an instant. The details of a flashover are discussed in more detail in the following section.

Fully Developed Phase

During the **fully developed phase**, the fire has progressed beyond flashover (◄ Figure 5-12). All combustibles have ignited and heat is being produced at the maximum rate. The fire at this point has everything it needs, and produces large amounts of fire gases.

During this phase the available oxygen is consumed rapidly. The fire can only burn at this rate for as long as the oxygen supply is adequate. If the fire consumes oxygen more rapidly than it can be replaced by fresh air, the rate of combustion will slow down.

Decay Phase

The final phase of a fire is the **decay phase** (◄ Figure 5-13). At this point the fire is running out of fuel. The atmosphere is still hot and heat energy is still being released; however, the rate of combustion is slowing down. The intensity of the fire decreases to the point where there is only smoldering fuel. Eventually, all of the fuel will be consumed and the fire will go out.

Characteristics of an Interior Structure Fire

The greatest part of the material in this book is directly related to fires that occur in buildings. Interior fires have some special characteristics, because the fire is fully or at least partially contained within the building. The building acts as a box, keeping heat and products of combustion inside. The structure can also limit the flow of clean air that can reach the fire, which can change the rate of combustion and the products of combustion. In addition, the structure limits the ability of fire fighters to reach the fire in order to extinguish it.

On a smaller scale, each room within a building is also a compartment where a fire can originate and grow. Sometimes the fire stays within a room and sometimes it spreads to other rooms or spaces. The following sections describe some special considerations that are directly related to fires in rooms and within buildings.

Room Contents

Most fires in buildings involve the building's contents. A large proportion of fires in buildings do not involve the structure itself, but only burn the contents within individual rooms and spaces. These fires originate in the contents, and most of the heat and products of combustion are produced by the contents. Buildings themselves can and do burn; however, the building often functions as a container, and most of the fire occurs inside it.

Fifty years ago, a typical room in a residential occupancy contained many products made of wood and natural fibers. Today most rooms are heavily loaded with plastics, and many of the synthetic items in our normal environment are made from petroleum products. This change in room contents has changed the behavior of fires. When heated, most plastics release volatile products that are both flammable and toxic. Burning plastics generate dense smoke that is rich in flammable vapors, which makes fire suppression more difficult and more dangerous. Some of these products melt and drip at high temperatures. If the droplets are on fire, they can contribute to the spread of the fire.

The upholstered furniture that is manufactured today is generally more resistant to ignition from glowing sources such as cigarettes than in the past. However, it has very little resistance to ignition from flaming sources. After it is ignited, such furniture can be completely involved in fire in 3 to 5 minutes and reduced to a burning frame in less than 10 minutes.

Walls and ceilings are often painted using emulsions of latex, acrylics, or polyvinyl materials. When these paints are applied, they form a plastic-like coating. This coating has the characteristics of the plastic from which it was formed and can burn readily. Varnishes and lacquers are also combustible. Most paints and coatings add to the fire load in a building and can aid in the spread of the fire from one area to another. It should be apparent that the increased use of plastics in furnishings and finishes has set the stage for fires that are different from a simple campfire.

Fuel Load and Fire Spread

Fuel load refers to the total quantity of all of the combustible products that are within a room or a space. The fuel load determines how much heat and smoke will be produced by a fire, assuming that all of the combustible fuel in that space is consumed. The size and shape of interior objects and the types of materials used to create them have a tremendous impact on the objects' ability to burn and their rate of combustion. The same factors influence the rate of fire spread to other objects. Even the arrangement of the furnishings within a room can make a difference in the way a fire burns.

Take the example of a couch on fire in a living room that is furnished in the following typical fashion: in addition to the couch, the room contains two upholstered chairs, an entertainment center, a coffee table, several small end tables, and a wastebasket. The walls are painted and the floor is covered with synthetic carpeting. Almost everything in the room is highly combustible. This situation involves many different factors that will trigger events simultaneously or in rapid succession:

- Most of the materials used to construct a couch are typically plastics. As the couch burns it will release tremendous quantities of heat energy within the room. If nothing is done to extinguish the fire, it will burn until all of the fuel is consumed.
- The heat will cause the plastic materials in the couch and the other pieces of furniture in the room to release toxic and flammable gases, even before they are ignited.
- The burning plastics will release large quantities of smoke and other products of combustion. These products are both flammable and toxic, and they will completely obscure the visibility in the room, making the atmosphere extremely hazardous to anyone who remains in the room.
- The burning plastics in the couch will begin to drip onto the carpet. As a result, the carpet will also catch fire and burn.
- As the fire continues to burn, all of the contents of the room as well as the ceiling and wall finishes will become involved in the fire.

This is a relatively simple example, considering only the contents of a typical living room. A fire in a building becomes much more complicated when the heat and products of combustion begin to spread from room to room. All of these factors and many others influence the behavior of the fire.

Special Considerations in Structure Fires

As a fire fighter, you must understand the behavior of fires and work to control the power of a fire rather than try to attack it in a reckless manner. There are four special conditions you must understand in order to work safely to extinguish structure fires. These special conditions are flashover, rollover (flameover), backdraft, and the thermal layering of gases.

Flashover and Rollover

Flashover was defined previously as the sudden ignition of all of the combustible objects within a room or a confined space. Flashovers can occur with explosive force, but in a different way than with a backdraft, which is described on the next page. A flashover is a much more frequent occurrence than a backdraft.

A flashover occurs within a room or space when a growing fire has sufficiently heated everything around it. The hot gases produced by the fire spread across the ceiling and reach out to the walls. These gases are so hot that heat energy is radiated down from the smoke layer to the contents of the room. The ceiling and the walls are also hot and radiate heat back into the space. This intense buildup of heat causes all of the combustible objects in the space to reach their ignition temperatures almost simultaneously. As soon as the first objects begin to ignite, more heat energy is released and everything else in the room is heated even faster. The flashover occurs very quickly.

A fire fighter in full protective gear cannot survive flashover conditions for more than a few seconds. The sudden increase in temperature will overwhelm the protective capabilities of protective clothing and equipment.

Fire fighters often arrive at the scene of a fire when it is close to the flashover point. It is very important to be able to recognize a situation in which flashover conditions are present. The signs and conditions that indicate that a flashover may be imminent are as follows:

- Dense black smoke pushes out of a doorway or window opening under pressure. The smoke moves several feet from the opening as it is being pushed by high heat. This dense smoke with tightly packed curls is sometimes referred to as black fire.
- Dense smoke fills over one half of a door or window opening into the fire area. Inside the room the smoke will be banked down to doorknob height. Below this level the atmosphere is often clear and the fire can be seen within the room.

Rollover or **flameover** is a term used to describe burning fire gases. This condition occurs in one of two manners. It can be produced when ignited fire gases leave the fuel, rise to the ceiling, and then spread out horizontally. The second way it is produced is when the fuel produces unburned combustible fire gases that rise to the ceiling and spread out horizontally. The fire gases have sufficient heat to ignite, but are deficient in oxygen. Once the combustible gases reach an opening such as a door or window, they mix with oxygen-rich air and ignite into an open flame. In both situations, it is the fire gases that are burning. In order to stop rollover, the fire must be controlled. Left uncontrolled, this condition will likely lead to flashover.

The only thing that fire fighters can do to prevent flashover at this point is to immediately cool the atmosphere. This situation calls for aggressive action to cool the atmosphere, to make an immediate exit, or for immediate ventilation.

Backdraft

A **backdraft** is best described as an explosion that occurs when oxygen is suddenly admitted to a confined area that is very hot and contains large amounts of combustible vapors and smoke (► Figure 5-14). A backdraft is sometimes called a smoke explosion, because, in effect, it is the smoke produced by the fire that explodes.

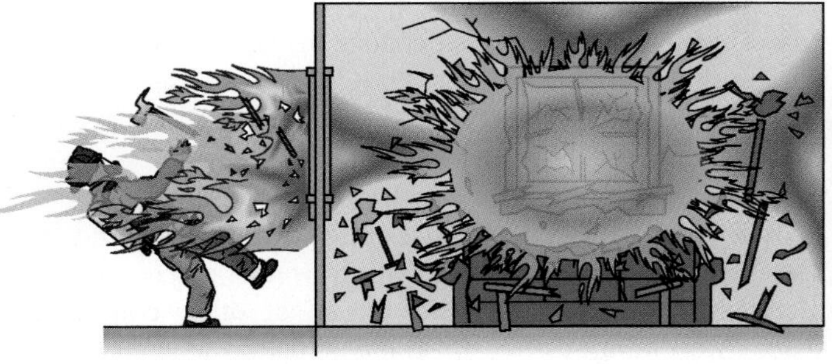

Figure 5-14 Backdraft.

A backdraft usually occurs when a fire is smoldering or when the intensity of the fire has been reduced by a lack of oxygen. The typical conditions that lead to a backdraft begin with a fire inside a building that produces large amounts of smoke and carbon monoxide. These products of combustion have collected within a confined space or throughout the entire building. Because the building has relatively few openings, the products of combustion have difficulty escaping and the flow of fresh air into the building is restricted. As the fire continues to burn, it consumes most of the available oxygen, and the rate of combustion slows down. Sometimes all flaming combustion stops and the fire smolders.

The situation at this point is a building or a space within a building that is filled with carbon monoxide, which is a flammable gas, and other products of combustion. The smoke contains large quantities of material that is only partially burned. This mixture is hot enough to burn and the only thing that is keeping it from bursting into flames is the lack of oxygen. If oxygen is suddenly introduced, the mixture will ignite with explosive force. This can occur if fire fighters open a door or window in a way that allows oxygen to enter the space. There have been cases where large explosions have sent glass, portions of walls, and even buildings towards fire fighters.

Fire fighters should learn to recognize the signs of an impending backdraft. These signs are as follows:

- Little or no flame is visible from the exterior of the building. (The fire is smoldering, not flaming.)
- Smoke is emanating under pressure from cracks around doors, windows, and eaves.
- Typically, there are no large openings in the building, such as open doors or windows.
- A "living fire" is visible, where smoke is puffing from the building and being drawn back in so that it looks like it is breathing.
- There is an unexplained change in the color of the smoke.
- Glass is smoke stained and blackened due to heavy carbon deposits from the smoke.
- Signs of extreme heat conditions are present.

It is important to note that these conditions could be present in an entire building or only within a space inside the building such as an attic, closet, or other concealed space. A backdraft can occur in any situation where oxygen is suddenly introduced into an enclosed space that is filled with heated products of combustion.

The best way to prevent a backdraft from occurring is to make a ventilation opening at a high level, so that the hot gases can escape from the interior without allowing fresh air to enter. This is not a simple task. It requires very close coordination between ventilation and fire attack. The most important consideration for fire fighters is to recognize when potential backdraft conditions are present and to avoid triggering the backdraft by making an opening at the wrong place at the wrong time.

Thermal Layering and Thermal Balance

During a fire in a closed room or space, the fire gases are said to be in thermal balance when they are allowed to seek their own level. Prior to flashover, the hottest gases always rise to the upper levels, and the temperature at the floor level is relatively cool compared to the ceiling (although it still could be hot enough at the floor level to burn anyone without full protective clothing and equipment). By keeping low in a normal thermal balance situation, you stay in the area where the temperature is the lowest. Your chance of survival is increased with the proper personal protective gear and by staying low.

If this normal thermal balance is upset, severe injury can occur to fire fighters. When water is sprayed into the upper part of a room, the energy in the superheated gases at the top of the room converts the water to steam. Steam takes up many times the space of the same amount of liquid water, so the steam expands rapidly. The freshly produced steam becomes superheated as it absorbs more heat energy from the fire gases.

The expanding superheated steam can fill the whole room and will also displace superheated fire gases from the ceiling down toward the floor. This mixture of hot steam and fire gases can result in severe burns to fire fighters, even when they are close to floor level and wearing full personal protective gear. This action also seriously jeopardizes the chances of survival for any victims trapped.

It is important that you understand the concept of thermal layering so you can avoid creating a thermal imbalance. Two actions are important. First, actions among fire fighters should be coordinated so that, whenever possible, the superheated gases are being vented from the fire room as you are attacking the fire. Second, it is important to use the proper fire stream for the situation. Opening a fog stream into a heated enclosed area will produce much more steam than a straight stream. Selecting a straight stream will allow more of the water to reach the seat of the fire where it can have the greatest effect. Use proper fire suppression techniques to avoid thermal imbalance.

Wrap-Up

Ready for Review

In order to be a successful fire fighter you must learn as much as possible about the characteristics and behavior of fire, which were covered in this chapter. Within this chapter the different characteristics of solids, liquids, and gases were described. The conditions necessary for combustion to occur and the products of combustion were described. The mechanisms of heat transfer and the propagation of fires due to conduction, convection, and radiation were covered. The four principal methods of fire extinguishment were covered. Class A, B, C, D, and K fires were defined. The characteristics of a solid fuel fire were covered through the ignition phase, the growth phase, the fully developed phase, and the decay phase. The characteristics of each phase of a room and contents fire were described. The causes, prevention, and characteristics of rollover, flashover, thermal layering, and backdraft were illustrated. The characteristics of liquid fuel fires were described. The three conditions that are necessary for a vapor and air mixture to ignite were listed. The concepts of ignition temperature, flash point, and flame point were outlined. The characteristics of gas fuel fires were described. The concepts of vapor density and flammability limits were described. The causes and conditions necessary for a BLEVE were outlined.

Chief Concepts

- Matter exists in three states: solid, liquid, and gas.
- Energy exists in many forms including chemical, mechanical, and electrical.
- Conditions necessary for a fire include fuel, oxygen, heat and a self-sustaining reaction
- Fires spread by conduction, convection, and radiation.
- The four principal methods of fire extinguishment are cooling the fuel, excluding the oxygen, removing the fuel, and inhibiting the chemical reaction.
- Fires are categorized as Class A, Class B, Class C, Class D, and Class K. These classes reflect the type of fuel that is burning and the type of hazard the fire represents.
- Solid fuel fires develop through four phases: the ignition phase, the growth phase, the fully developed phase, and the decay phase.
- The growth of room and contents fires is dependent on the characteristics of the room and the contents of the room.
- Special considerations related to room and contents fires include flameover, flashover, backdraft, and the thermal layering of gases.
- Liquid fuel fires require the proper mixture of fuel and air, an ignition source, and contact between the fuel mixture and the ignition.
- The characteristics of gas fuel fires are different from other fires.
- Vapor density reflects the weight of a gas compared to air.
- Flammability limits vary widely for different fuels.
- A BLEVE is a catastrophic explosion in a vessel containing a boiling liquid and a vapor.

Wrap-Up

Hot Terms

Backdraft The sudden explosive ignition of fire gases when oxygen is introduced into a superheated space previously deprived of oxygen.

Boiling-liquid, expanding-vapor explosion (BLEVE) An explosion that can occur when a tank containing a volatile liquid is heated.

Chemical energy Energy created or released by the combination or decomposition of chemical compounds.

Class A fires Fires involving ordinary combustible materials, such as wood, cloth, paper, rubber, and many plastics.

Class B fires Fires involving flammable and combustible liquids, oils, greases, tars, oil-based paints, lacquers, and flammable gases.

Class C fires Fires that involve energized electrical equipment where the electrical conductivity of the extinguishing media is of importance.

Class D fires Fires involving combustible metals such as magnesium, titanium, zirconium, sodium, and potassium.

Class K fires Fires involving combustible cooking media such as vegetable oils, animal oils, and fats.

Combustion A chemical process of oxidation that occurs at a rate fast enough to produce heat and usually light in the form of either a glow or flames.

Conduction Heat transfer to another body or within a body by direct contact.

Convection Heat transfer by circulation within a medium such as a gas or a liquid.

Decay phase The phase of fire development where the fire has consumed either the available fuel or oxygen and is starting to die down.

Electrical energy Heat produced by electricity.

Endothermic reactions Reactions that absorb heat or require heat to be added.

Exothermic reactions Reactions that result in the release of energy in the form of heat.

Fire point (flame point) The lowest temperature at which a substance releases enough vapors to ignite and sustain combustion.

Fire tetrahedron A geometric shape used to depict the four components required for a fire to occur: fuel, oxygen, heat, and chemical chain reactions.

Flashover The condition where all combustibles in a room or confined space have been heated to the point at which they release vapors that will support combustion, causing all combustibles to ignite simultaneously.

Flash point The minimum temperature at which a liquid or solid releases sufficient vapor to form an ignitable mixture with the air.

Flammability limits (explosive limits) The upper and lower concentration limits (at a specified temperature and pressure) of a flammable gas or vapor in air that can be ignited, expressed as a percentage of fuel by volume.

Fuel All combustible materials. The actual material that is being consumed by a fire, allowing the fire to take place.

Fully developed phase The phase of fire development where the fire is free-burning and consuming much of the fuel.

Gas One of the three phases of matter. A substance that will expand indefinitely and assume the shape of the container that holds it.

Growth phase The phase of fire development where the fire is spreading beyond the point of origin and beginning to involve other fuels in the immediate area.

Hypoxia A state of inadequate oxygenation of the blood and tissue.

Ignition phase The phase of fire development where the fire is limited to the immediate point of origin.

Ignition temperature The minimum temperature at which a fuel, when heated, will ignite in air and continue to burn.

Liquid One of the three phases of matter. A nongaseous substance that is composed of molecules that move and flow freely and assumes the shape of the container that holds it.

Lower explosive limit (LEL) The minimum amount of gaseous fuel that must be present in the air mixture for the mixture to be flammable or explosive.

Mechanical energy Heat energy created by friction.

Oxidation A chemical reaction initiated by combining an element with oxygen, resulting in the form of the element or one of its compounds.

Oxidizing agent A substance that will release oxygen or act in the same manner as oxygen in a chemical reaction.

Plume The column of hot gases, flames, and smoke that rises above a fire, also called a convection column, thermal updraft, or thermal column.

Pyrolysis The chemical decomposition of a compound into one or more other substances by heat alone; pyrolysis often precedes combustion.

Radiation The combined process of emission, transmission, and absorption of energy traveling by electromagnetic wave propagation between a region of higher temperature and a region of lower temperature.

Rollover (flameover) The condition where unburned products of combustion from a fire have accumulated in the ceiling layer of gas to a sufficient concentration (i.e., at or above the lower flammable limit) that they ignite momentarily.

Smoke An airborne particulate product of incomplete combustion suspended in gases, vapors, or solid and liquid aerosols.

Solid One of the three phases of matter. A substance that has three dimensions and is firm in substance.

Surface to mass ratio (STMR) The surface area of a material in proportion to the mass of that material.

Surface to volume ratio The surface area of a liquid in proportion to the mass.

Thermal column A cylindrical area above a fire in which heated air and gases rise and travel upward.

Thermal layering The stratification, or heat layers, that occur in a room as a result of a fire.

Upper explosive limit (UEL) Defines the boundaries of a fuel/air mixture necessary for a combustible material to burn properly.

Vapor density The weight of an airborne concentration (vapor or gas) as compared to an equal volume of dry air.

Volatility The ready ability of a substance to produce combustible vapors.

Fire Fighter in Action

You are in your last weeks of training and the day has arrived to do your first live fire exercise. You are wearing all of your PPE including SCBA when you are instructed to enter a large burn room. There are two small piles of wood pallets on fire in the front corners of the room. The fires are growing, and hot gases are gathering at the ceiling. Smoke is starting to bank down towards the floor.

1. What is the class of this fire?

A. Class A

B. Class B

C. Class C

D. Class D

E. Class K

2. What phase is this fire in?

A. Ignition

B. Growth

C. Flashover

D. Fully developed

E. Decay

You start to feel the heat coming down from the ceiling. Directly over the fire you can see flames igniting in the upper layers of smoke. The instructor directs straight streams at the base of both fires.

3. The flames that you see within the upper layers of smoke indicate a:

A. backdraft.

B. fire point.

C. flameover.

D. flashover.

4. What method of extinguishment did the instructor use?

A. Cooling

B. Excluding oxygen

C. Removing fuel

D. Breaking the chemical chain reaction

www.FireFighter.jbpub.com

www.FireFighter.jbpub.com

Chapter Pretests

Interactivities

Hot Term Explorer

Web Links

Review Manual

FireLearn

Building Construction

Chapter 

NFPA 1001 Standard

Fire Fighter I

5.3.12 (A) *Requisite Knowledge.* The methods of heat transfer; the principles of thermal layering within a structure on fire; the techniques and safety precautions for venting flat roofs, pitched roofs, and basements; basic indicators of potential collapse or roof failure; the effects of construction type and elapsed time under fire conditions on structural integrity; and the advantages and disadvantages of vertical and trench/strip ventilation.

Additional NFPA Standards

NFPA 80, *Standard for Fire Doors and Fire Windows*

NFPA 220, *Standard on Types of Building Construction*

Knowledge Objectives

After studying this chapter, you will be able to:

- Describe the characteristics of the following building materials: masonry, concrete, steel, glass, gypsum board, and wood.
- List the characteristics of each of the following types of building construction: fire-resistive construction, noncombustible construction, ordinary construction, heavy timber construction, and wood-frame construction.
- Describe how each of the five types of building construction react to fire.
- Describe the function of each of the following building components: foundations, floors, ceilings, roofs, trusses, walls, doors, windows, interior finishes, and floor coverings.

Skills Objectives

There are no skills objectives for this chapter.

You Are the Fire Fighter

Your engine company is dispatched for a building fire at an unfamiliar address. The building, on a residential street, is a two-story house with smoke coming out of a second-floor window. As the engine pulls to a stop, you quickly identify the building construction and materials. As you dismount, you mentally run through the hazards presented by this construction.

1. ***Why must fire fighters understand building construction?***
2. ***What is the best way for fire fighters to learn about the types of construction in their response area?***

Introduction

Knowing the basic types of building construction is vital for fire fighters because building construction affects how fires grow and spread. Fire fighters must be able to identify different types of building construction quickly so that they can anticipate the fire's behavior and respond accordingly. An understanding of building construction will help determine when it is safe to enter a burning building and when it is necessary to evacuate a building. Your safety, the safety of your team members, and the safety of the building's occupants depend on a knowledge of building construction.

But construction is only one component of a complex relationship. Fire fighters also must consider the occupancy of the building and the building contents. Fire risk factors vary depending on how a building is used. To establish a course of action at a fire, it is necessary to consider the interactions between three factors: the building construction, the occupancy of the building, and the contents of the building. The following sections relate to occupancy and contents, while the rest of this chapter focuses on those aspects of building construction that are important to a fire fighter.

Occupancy

The term **occupancy** refers to how a building is used. Based on a building's occupancy classification, a fire fighter can predict who is likely to be inside the building, how many people, and what they are likely to be doing. Occupancy also suggests the types of hazards and situations that may be encountered in the building.

Occupancy classifications are used with building and safety codes to establish regulatory requirements, including the types of construction that can be used to construct buildings of a particular size, use, or location. Regulations divide building occupancies into major categories—such as residential, health care, business, and industrial—based on common characteristics.

Occupancy classifications can be used to predict the number of occupants that are likely to be at risk in a fire. Hospitals and nursing homes are occupied 24 hours a day by persons who will probably need assistance to evacuate. An office building probably has a large population during the day, but most of the workers should be able to evacuate without assistance. A day care center may be filled with young children during the day and empty at night. A nightclub is probably empty during the day and crowded at night.

Fire hazards in different types of occupancies also vary greatly. A computer repair shop has different characteristics than an auto transmission shop. A factory could be used to produce cast-iron pipe fittings or airplanes. The materials used could be flammable, toxic, or fire resistant. Fire fighters must consider each of these factors when responding to a particular building.

Building codes require that certain types of building construction be used for specific types of occupancies. In most communities, however, buildings that were built for one reason are now being used for some other purpose. Buildings constructed before modern codes were adopted were often grandfathered in during inspections. This presents hazards for fire fighters.

Contents

Contents also must be considered when responding to a building. Building contents vary widely, but are usually closely related to the occupancy of a building. Some buildings have noncombustible contents that would not feed a fire, while others could be so dangerous that fire fighters could not safely attack any fire. For example, an automobile repair shop would contain many toxic and flammable substances. Whenever possible, fire fighters should prepare an advance plan for buildings that present special hazards.

Similar occupancies can pose different levels of risk. For example, three warehouses, identical from the exterior, could have very different contents that increase or reduce the risks to fire fighters. One might contain ceramic floor tiles, the second might be filled with wooden furniture, and the third might be a storehouse for swimming pool chemicals. The tiles will not burn, the furniture will, and the chemicals could create toxic products of combustion and contaminated runoff.

Types of Construction Materials

An understanding of building construction begins with the materials used. Building components are usually made of different materials. The properties of these materials and the details of their construction determine the basic fire characteristics of the building itself.

Function, appearance, price, and compliance with building and fire codes are all considerations when selecting building materials and construction methods. Architects often place a priority on functionality and aesthetics when selecting materials; builders are often more concerned about price and ease of construction, and building owners are usually interested in durability and maintenance expenses, as well as initial cost and aesthetics.

Fire fighters, however, use a different set of factors to evaluate construction methods and materials. Their chief concern is the behavior of the building under fire conditions. A building material or construction method that is attractive to architects, builders, and building owners can create serious problems or deadly hazards for fire fighters.

The most common building materials are wood, masonry, concrete, steel, aluminum, glass, gypsum board, and plastics. Within these basic categories are hundreds of variations. The key factors that affect the behavior of each of these materials under fire conditions include:

- **Combustibility**: Whether or not a material will burn determines its combustibility. Materials such as wood will burn when they are ignited, releasing heat, light, and smoke, until they are completely consumed by the fire. Concrete, brick, and steel are noncombustible materials that cannot be ignited and are not consumed by a fire.
- **Thermal conductivity**: This describes how quickly a material will conduct heat. Heat flows very quickly through metals such as steel and aluminum. Brick, concrete, and gypsum board are poor conductors of heat.
- Decrease in strength at elevated temperatures: Many materials lose strength at elevated temperatures. Steel loses strength and will bend or buckle when exposed to fire temperatures. Aluminum melts in a fire. Bricks and concrete can generally withstand high temperatures for extended periods of time.
- Thermal expansion when heated: Some materials, steel in particular, expand significantly when they are heated. A steel beam exposed to a fire will stretch (elongate); if it is restrained so that it cannot elongate, it will sag, warp, or twist.

Fire Fighter Tips

The four characteristics of building materials under fire are:

- Combustibility
- Thermal conductivity
- Decrease in strength with increased temperature
- Rate of thermal expansion

Masonry

Masonry includes stone, concrete blocks, and brick. The individual components are usually bonded together into a solid mass with mortar, which is produced by mixing sand, lime, water, and Portland cement. A concrete-masonry unit, or CMU, is a pre-assembled wall section of brick or concrete blocks that contains steel reinforcing rods. A complete CMU can be delivered to a construction side, ready to be erected.

Masonry materials are inherently fire resistive (▼ Figure 6-1). They do not burn or deteriorate at the temperatures normally encountered in building fires. Masonry is also a poor conductor of heat, so it will limit the passage of heat from one side through to the other side. For these reasons, masonry is often used to construct **fire walls** to

Figure 6-1 Masonry materials are inherently fire resistive.

protect vulnerable materials. A fire wall helps prevent the spread of a fire from one side to the other side of the wall.

All masonry walls are not necessarily fire walls. A single layer of masonry may be used over a wood-framed building to make it appear more substantial. If there are unprotected openings in a masonry wall, a fire can often spread through them. If the mortar has deteriorated or the wall has been exposed to fire for a prolonged time, a masonry wall can collapse during a fire.

A masonry structure also can collapse under fire conditions if the roof or floor assembly collapses. The masonry falls because of the mechanical action of the collapse. Aged, weakened mortar can also contribute to a collapse. Regardless of the cause, a collapsing masonry wall is a deadly hazard to fire fighters.

Concrete

Concrete is also a naturally fire-resistive material. It does not burn or conduct heat well, so it is often used to insulate other building materials from fire. Concrete does not have a high degree of thermal expansion—that is, it does not expand greatly when exposed to heat and fire. Nor will it lose strength when exposed to heat and fire.

Concrete is made from a mixture of Portland cement and aggregates such as sand and gravel. Different formulations of concrete can be produced for specific building purposes. Concrete is inexpensive and easy to shape and form. It can be used for foundations, columns, floors, walls, roofs, and exterior pavement.

Under compression, concrete is strong and can support a great deal of weight, but under tension, it is weak. When used in building construction, concrete usually has embedded steel reinforcing rods to strengthen it under tension. In turn, the concrete insulates the steel reinforcing rods from heat.

Although it is inherently fire resistive, concrete can be damaged by exposure to a fire. A fire can convert trapped moisture in the concrete to steam. As the steam expands, it creates internal pressure that can cause sections of the concrete surface to break off, a process called **spalling**. Severe spalling in reinforced concrete can expose the steel reinforcing rods to the heat of the fire. If the fire is hot enough, the steel might weaken, resulting in a structural collapse. However, this rarely occurs.

Steel

Steel is the strongest building material in common use. It is very strong in both tension and compression and can be produced in a wide range of shapes and sizes, from heavy beams and columns to thin sheets. Steel is often used in the structural framework of a building to support floor and roof assemblies. It is resistant to aging and does not rot; however, most types of steel will rust unless they are protected from exposure to air and moisture.

Steel is an alloy of iron and carbon; additional metals may be added to produce steel with special properties, such as stainless steel or galvanized steel.

Steel by itself is not fire resistive. Steel will melt at extremely high temperatures, but these temperatures are not normally encountered at structure fires (► Figure 6-2). Steel conducts heat well, and will expand and lose strength as the temperature increases. Other materials, such as masonry, concrete, or layers of gypsum board, are often used to protect steel from the heat of a fire. Sprayed-on coatings of mineral or cement-like materials are also used to insulate steel members. The amount of heat absorbed by steel depends on the mass of the object and the amount of protection surrounding it. Smaller, lighter pieces of steel heat more easily than larger and heavier pieces.

Steel will expand as well as lose strength as it is heated. An unprotected steel roof beam directly exposed to a fire can elongate sufficiently to cause a supporting wall to collapse. Heated steel beams will often sag and twist, while columns tend to buckle as they lose strength. The bending and distortion are caused by the uneven heating that occurs in actual fire situations.

Failure of a steel structure is dependent on three factors: the mass of the steel components, the loads placed upon them, and the methods used to connect the steel pieces. There are no accurate indicators that enable fire fighters to predict when a steel beam will fail. Any sign of bending, sagging, or stretching of steel structural members is a warning of an immediate risk of failure.

Other Metals

A variety of other metals including aluminum, copper, and zinc are used in building construction. Aluminum is often used for siding, window frames, door frames, and roof panels. Copper is used primarily for electrical wiring and piping; it is sometimes used for decorative roofs, gutters, and down spouts. Zinc is used primarily as a coating to protect metal parts from rust and corrosion.

Aluminum is occasionally used as a structural material in building construction. It is more expensive and not as strong as steel, so its use is generally limited to light-duty applications such as awnings and sunshades. Aluminum expands more than steel when heated and loses strength quickly when exposed to a fire. Aluminum has a lower melting point than steel, so it will often melt and drip in a fire.

Glass

Almost all buildings contain glass—in windows, doors, skylights, and sometimes walls. Ordinary glass breaks very easily, but glass can be manufactured to resist breakage and to withstand impacts or high temperatures.

Glass is noncombustible, but it is not fire resistive. Ordinary glass will usually break when exposed to fire but specially formulated glass can be used as a fire barrier in par-

Figure 6-2 Steel will melt at extremely high temperatures.

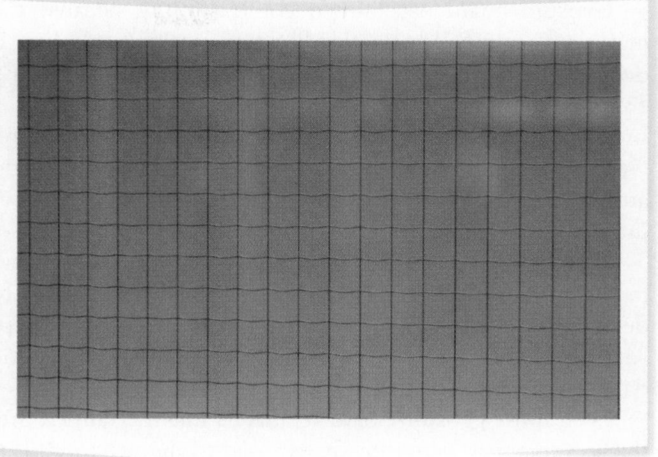

Figure 6-3 Wired glass.

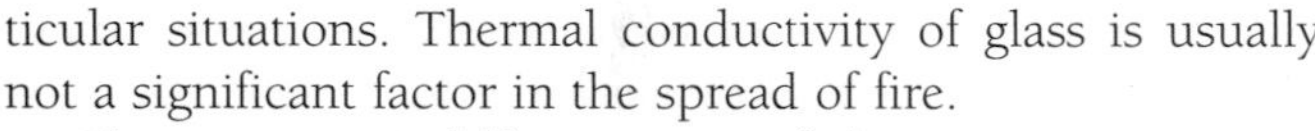

ticular situations. Thermal conductivity of glass is usually not a significant factor in the spread of fire.

There are many different types of glass:

- Ordinary window glass will usually break with a loud pop when heat exposure to one side causes it to expand and creates internal stresses that fracture the glass. The broken glass forms large shards, which usually have sharp edges.
- **Tempered glass** is much stronger than ordinary glass and harder to break. Some tempered glass can be broken with a spring-loaded center punch, and will shatter into small pieces that do not have the sharp edges of ordinary glass.
- **Laminated glass** is manufactured with a thin sheet of plastic between two sheets of glass. It is much stronger than ordinary glass, difficult to break with ordinary hand tools, and will usually deform instead of breaking. When exposed to a fire, laminated glass windows are likely to crack and remain in place. Laminated glass is sometimes used in buildings to help soundproof areas.
- **Glass blocks** are thick pieces of glass similar to bricks or tiles. They are designed to be built into a wall with mortar so that light can be transmitted through the wall. Glass blocks have limited strength and cannot be used as part of a load-bearing wall, but they can usually withstand a fire. Some glass blocks are approved for use with fire-rated masonry walls.
- **Wired glass** is made by molding tempered glass with a reinforcing wire mesh (► Figure 6-3). When wired glass is subjected to heat, the wire holds the glass together and prevents it from breaking. It is often used in fire doors and windows designed to prevent fire spread.

Gypsum Board

Gypsum is a naturally occurring mineral composed of calcium sulfate and water molecules, used to make plaster of Paris. Gypsum is a good insulator and noncombustible; it will not burn even in atmospheres of pure oxygen.

Gypsum board, which is also called drywall, sheetrock, or plasterboard, is commonly used to cover the interior walls and ceilings of residential living areas and commercial spaces (▼ Figure 6-4). Gypsum board is manufactured in large sheets consisting of a layer of compacted gypsum sandwiched between two layers of specially produced paper. The sheets are nailed or screwed in place on a framework of wood or metal studs, and the edges are secured with a special tape. The nail or screw heads and tape are then covered with a thin layer of plaster.

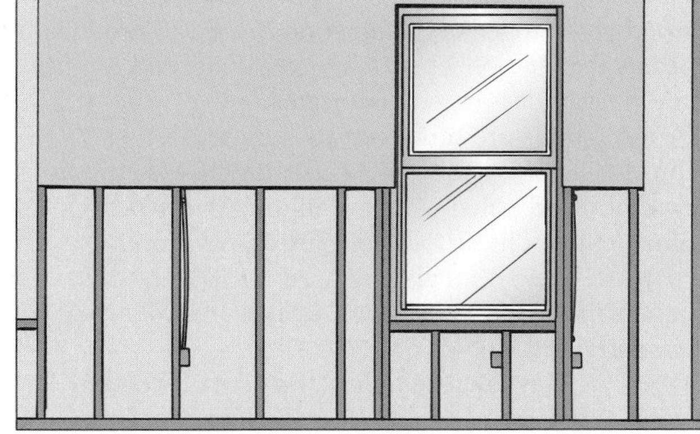

Figure 6-4 Gypsum board is widely used in many buildings.

Gypsum board has limited combustibility, because the paper covering will burn slowly when exposed to a fire. It does not conduct or release heat to an extent that would contribute to fire spread, and is often used to create a firestop or to protect building components from fire. Gypsum blocks are sometimes used as protective insulation around steel members or to create fire-resistive enclosures or interior fire walls.

When gypsum board is heated, some of the water found in the calcium sulfate will evaporate, causing the board to deteriorate. If it is exposed to a fire for a long time, the gypsum board will fail. Water will also weaken and permanently damage gypsum board.

Even if the gypsum board retains its integrity after a fire, the sections that were directly exposed to the fire should be replaced. The sections that were heated will probably be unable to serve as effective fire barriers in the future.

Although gypsum board is a good finishing material and an effective fire barrier, it is not a strong structural material. It must be properly mounted on and supported by wood or steel studs. Gypsum board mounted on wood framing will protect the wood from fire for a limited time.

Wood

Wood is probably the most commonly used building material in our environment. It is inexpensive to produce, easy to use, and can be shaped into many different forms, from heavy structural supports to thin strips of exterior siding. Both soft woods such as pine and hard woods such as oak are used for building construction.

A wide variety of wood products are used in building construction:

- Solid lumber is squared and cut into uniform lengths. Examples of solid lumber include the heavy timbers used in mills and barns and the lightweight boards used for siding and decorative trim.
- **Laminated wood** consists of individual pieces of wood glued together. Lamination is used to produce beams that are longer and stronger than solid lumber and to manufacture curved beams.
- **Wood panels** are produced by gluing together thin sheets of wood. Plywood is the most common type of wood panel used in building construction. Small chips (chip board) or particles of wood (particle board) can also be used to make wood panels that are usually much weaker than plywood or solid lumber.
- **Wood trusses** are assemblies of pieces of wood or wood and metal, often used to support floors and roofs. The structure of a truss enables a limited amount of material to support a heavy load.
- **Wooden beams** are efficient load-bearing members assembled from individual wood components. The shape of a wooden I-beam or box beam enables it to support the same load that a solid wood beam could support.

For fire fighters, the most important characteristic of wood is combustibility. Wood adds fuel to a structural fire and can provide a path for the fire to spread. It ignites at fairly low temperatures and is gradually consumed by the fire, weakening and eventually collapsing the structure. Great quantities of heat and hot gases are created until all that remains is a small quantity of residual ash.

Wood components weaken as they are consumed by fire. A burning wood structure gets weaker every minute, making it imperative that fire fighters always remember to operate on the side of safety.

How fast wood ignites, burns, and decomposes depends on several factors:

- **Ignition:** A small ignition source contains less energy and will take a longer time to ignite a fire. Using an accelerant will greatly speed up the process.
- **Moisture:** Damp or moist wood takes longer to ignite and burn. New lumber usually contains more moisture than lumber that has been in a building for many years. A higher relative humidity in the atmosphere also makes wood more difficult to ignite.
- **Density:** Heavy, dense wood is harder to ignite than lighter or less dense wood.
- **Preheating:** The more the wood is preheated, the faster it ignites.
- **Size and form:** The rate of combustion is directly related to the surface area of the wood. The combustion process occurs as high temperatures release flammable gases from the surface of the wood. A large solid beam is difficult to ignite and burns relatively slowly. A lightweight truss is more easily ignited and rapidly consumed.

Exposure to the high temperatures generated by a fire can also decrease the strength of wood through the process of **pyrolysis**. This chemical change occurs when wood or other materials are heated to a temperature high enough to release some of the volatile compounds in the wood, without igniting these gases. The wood begins to decompose without combustion.

Fire-Retardant-Treated Wood

Wood cannot be treated to make it completely noncombustible. But impregnating wood with mineral salts makes it more difficult to ignite and slows the rate of burning. Fire-retardant treatment can significantly reduce the fire hazards of wood construction, although the treatment process can also reduce the strength of the wood.

In some cases, fire-retardant-treated wood can pose a danger for fire fighters. Some fire-retardant chemicals used to treat plywood roofing panels will cause the wood to deteriorate

Fire Fighter Safety Tips

Avoid standing on roofs that have been constructed with fire-retardant-treated wood. Atmospheric decomposition of the wood could cause these roofs to collapse, even without fire damage.

and weaken. The plywood can fail, necessitating the replacement of these panels. Fire fighters should know whether fire-retardant-treated plywood is used in the community and avoid standing on roofs made with these panels during a fire. The extra weight could cause the plywood to collapse.

Plastics

Plastics are synthetic materials used in many products today ▶ **Figure 6-5**. Plastics may be transparent or opaque, stiff or flexible, tough or brittle. They are often combined with other materials to produce building products.

Although plastics are rarely used for structural support, they can be found throughout a building. Building exteriors may include vinyl siding, plastic window frames, and plastic panel skylights. Foam plastic materials can be used as exterior or interior insulation. Plastic pipe and fittings, plastic tub and shower enclosures, and plastic lighting fixtures are commonly used today. Even carpeting and floor coverings often contain plastics.

The combustibility of plastics varies greatly. Some plastics ignite easily and burn quickly, while others will burn only while an external source of heat is present. Some plastics will withstand high temperatures and fire exposure without igniting.

Many plastics produce quantities of heavy, dense, dark smoke and release high concentrations of toxic gases as they burn. The smoke resembles smoke from a petroleum fire because most plastics are made from petroleum products.

Thermoplastic materials melt and drip when exposed to high temperatures, even those as low as 500°F. Heat is used to shape these materials into different forms, but the dripping, burning plastic can rapidly spread a fire. **Thermoset materials**, however, are fused by heat and will not melt. They will lose strength as the plastic burns, but will not melt.

Types of Construction

Fire fighters need to understand and recognize five different types of building construction. These classifications are directly related to fire protection and fire behavior in the building.

Buildings are classified based on the combustibility of the structure and the fire resistance of its components. Buildings using Type I or Type II construction are assembled with primarily noncombustible materials and limited amounts of wood and other materials that will burn. This does not mean that Type I and Type II buildings cannot be damaged by a fire. If the contents of the building burn, the fire could seriously damage or destroy the structure.

Figure 6-5 Plastic building materials come in many different forms.

In buildings using Type III, Type IV, or Type V construction, both the structural components and the building contents will burn. Wood and wood products are used to varying degrees in these buildings. If the wood ignites, the fire will weaken and consume the structure as well as the contents. Structural elements of these building can be damaged or destroyed in a very short time.

Fire resistance refers to the length of time that a building or building component can withstand a fire before igniting. Fire resistance ratings are stated in hours, based on the time that a test assembly withstands or contains a standard test fire. For example, walls are rated based on whether they stop the progress of the test fire for periods from 20 minutes to 4 hours. A floor assembly could be rated based on whether it supports a load for 1 hour or 2 hours. Because ratings are based on a standard test fire, and an actual fire could be more or less severe than the test fire, ratings are only guidelines; a 2-hour rated assembly will not necessarily withstand a fire for 2 hours.

Building codes specify the type of construction to be used, based upon the height, area, occupancy classification, and location of the building. Additional factors, such as the presence of automatic sprinklers, will also affect the required construction classification. NFPA 220, *Standard on Types of Building Construction,* provides the detailed requirements for each type of building construction.

Type I Construction: Fire Resistive

Type I construction is the most fire-resistive category of building construction. It is used for buildings designed for

Figure 6-6 Type I construction is commonly found in schools, hospitals, and high-rise buildings.

Figure 6-7 Sprayed-on fireproofing materials are often used to protect structural steel.

large numbers of people, buildings with a high life-safety hazard, tall buildings, large-area buildings, and buildings containing special hazards. Type I construction is commonly found in schools, hospitals, and high-rise buildings (▲ Figure 6-6).

Building with Type I construction can withstand and contain a fire for a specified period of time. The fire resistance and combustibility of all construction materials are carefully evaluated. Each building component must be engineered to contribute to fire resistance of the entire building.

All of the structural members and components used in Type I construction must be made of noncombustible materials, such as steel, concrete, or gypsum board. In addition, the structure must be constructed or protected so that there is at least two hours of fire resistance. Building codes specify the fire resistance requirements for different components. For example, columns and load-bearing walls in multi-story buildings could have a fire resistance requirement of 3 or 4 hours, while a floor might be required to have a fire resistance rating of only 2 hours. Some codes allow Type I buildings to use a limited amount of combustible material as interior finish.

If a Type I building exceeds specific height and area limitations, codes generally require the use of fire-resistive walls and/or floors to subdivide it into compartments. A compartment could be a single floor in a high-rise building or a part of a floor in a large-area building. A fire in one compartment should not spread to any other parts of the building. Stairways, elevators shafts, and utility shafts should be enclosed in construction that prevents fire from spreading from floor to floor or compartment to compartment.

Type I buildings usually use reinforced concrete and protected steel-frame construction. Concrete is noncombustible and provides thermal protection around the steel reinforcing rods. Reinforced concrete can fail, however, if it is subjected to a fire for a long period of time or if the building contents create an extreme fire load. Structural steel framing must be protected from the heat of a fire. In Type I construction, the structural steel members are generally encased in concrete, shielded by a fire-resistive ceiling, covered with multiple layers of gypsum board, or protected by a sprayed-on insulating material (▲ Figure 6-7). An unprotected steel beam exposed to a fire could fail and cause the entire building to collapse.

Type I building materials should not provide enough fuel by themselves to create a serious fire. It is the contents of the building that determine the severity of a Type I building fire. In theory, a fire could consume all of the combustible contents of a Type I building and leave the structure basically intact.

Although Type I construction may provide the highest degree of safety, it does not eliminate all of the risks to occupants and fire fighters. Serious fires can occur in fire-resistive buildings, because the burning contents can produce copious quantities of heat and smoke. Even though the structure is designed to give fire fighters time to conduct an interior attack, it could be very difficult to extinguish the fire. A fire that is fully contained within fire-resistive construction can be very hot and difficult to ventilate. For this reason, automatic sprinklers, as well as fire-resistive construction, are used in modern high-rise buildings.

Sometimes, the burning contents provide sufficient fuel and generate enough heat to overwhelm the fire-resistant properties of the construction. In these cases, the fire can escape from its compartment and damage or destroy the structure. Similarly, inadequate or poorly maintained construction can also undermine the fire-resistive properties of the building materials. Under extreme fire conditions, a Type I building can collapse, but these buildings have a much lower potential for structural failure than other buildings.

Type II Construction: Noncombustible

Type II construction is also referred to as noncombustible construction (► Figure 6-8). All of the structural components in a Type II building must be made of noncombustible materials. The fire-resistive requirements, however, are less stringent for Type II construction. In some cases, there are no fire resistance requirements. In other cases, known as protected noncombustible construction, a fire resistance rating of 1 or 2 hours may be required for certain elements.

Noncombustible construction is most common in single-story warehouse or factory buildings, where vertical fire spread is not an issue. Unprotected noncombustible construction is generally limited to a maximum of two stories. Some multi-story buildings are constructed using protected noncombustible construction.

Steel is the most common structural material in Type II buildings. Insulating materials can be applied to the steel when fire resistance is required. A typical example of Type II construction is a large-area, single-story building with a steel frame, metal or concrete block walls, and a metal roof deck. Fire walls are sometimes used to subdivide these large-area buildings and prevent catastrophic losses. Undivided floor areas must be protected with automatic sprinklers to limit the fire risk.

Fire severity in a Type II building is determined by the contents because the structural components contribute little or no fuel and interior finish materials are limited. If the building contents provide a high fuel load, a fire could destroy the structure. Automatic sprinklers should be used to protect combustible and valuable contents.

Type III Construction: Ordinary

Type III construction is also referred to as ordinary construction because it is used in a wide variety of buildings, ranging from commercial strip malls to small apartment buildings (► Figure 6-9). Ordinary construction is usually limited to buildings of no more than four stories, but it can sometimes be found in buildings as tall as six or seven stories.

Type III construction buildings have masonry exterior walls, which support the floors and the roof structure. The interior structural and nonstructural members, including the walls, floor, and roof, are all constructed of wood. In most ordinary construction buildings, gypsum board or plaster is used as an interior finish material, covering the wood framework and providing minimal fire protection.

Fire resistance requirements for interior construction of Type III buildings are limited. The gypsum or plaster coverings over the interior wood components provide some fire resistance. Key interior structural components may be required to have fire resistance ratings of 1 or 2 hours, or there may be no requirements at all. Some Type III buildings use interior masonry load-bearing walls to meet requirements for fire-resistant structural support.

A building of Type III construction has two separate fire loads: the contents and the combustible building materials used. Even a vacant building will contain a sufficient quan-

Fire Fighter Tips

Fire-resistive structures with sprinklers are a great asset in assisting fire fighters in controlling the burning contents. However, when the sprinklers activate, steam is produced when the water hits the fire. As the sprinklers continue to operate, the steam is cooled and can fill an enclosed area from floor to ceiling. Without proper ventilation fire fighters may find themselves with virtually no visibility.

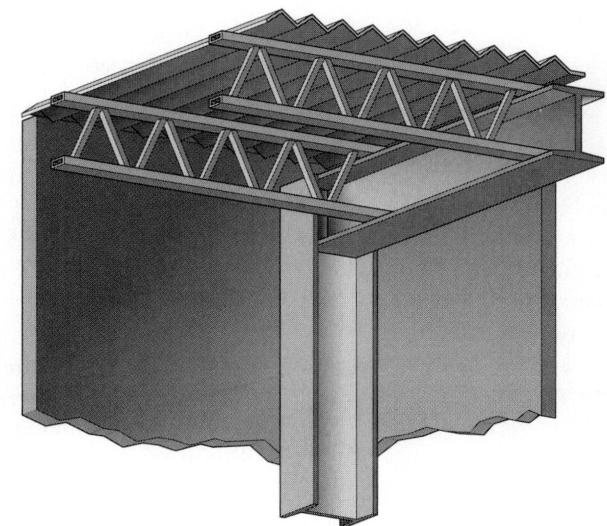

Figure 6-8 Type II construction is also referred to as noncombustible construction.

Figure 6-9 Type III construction is also referred to as ordinary construction because it is used in a wide variety of buildings.

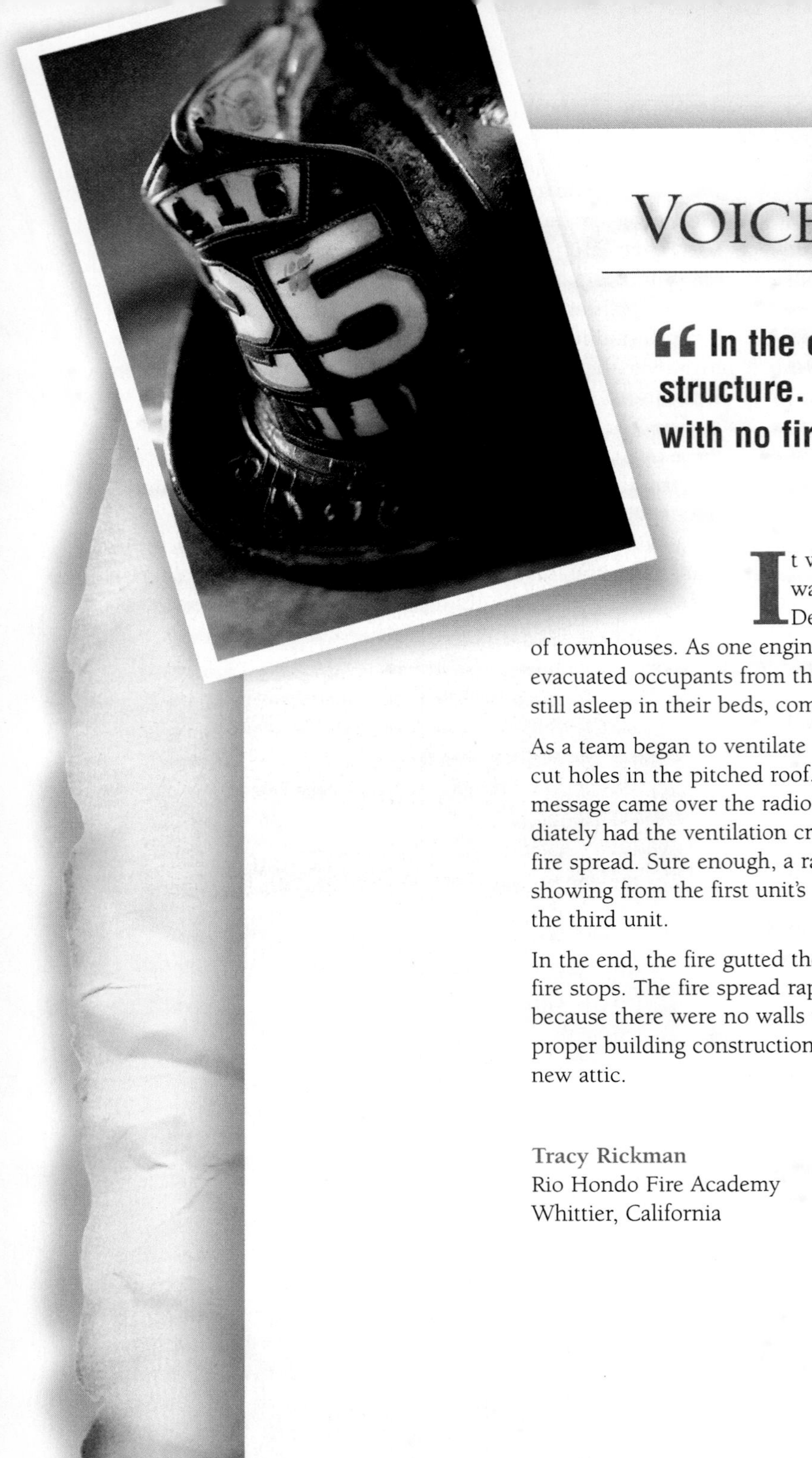

Voices of Experience

"In the end, the fire gutted the entire structure. The culprit was a common attic with no fire stops."

It was about 20°F one Sunday morning in January. I was a new fire fighter at Grand Forks Air Force Base Fire Department responding to a fire in the sixth unit in a row of townhouses. As one engine company attacked the fire, another engine company evacuated occupants from their homes. The fire fighters found most of the occupants still asleep in their beds, completely unaware of the fire.

As a team began to ventilate using windows on the upper floor, additional resources cut holes in the pitched roof. Within minutes of opening the first hole on the roof, a message came over the radio that the fire was spreading to the fifth unit. The IC immediately had the ventilation crews make a cut in the roof of the first unit to determine fire spread. Sure enough, a radio message back to the IC indicated that the fire was showing from the first unit's attic space. Fire was also noticed in the attic space in the third unit.

In the end, the fire gutted the entire structure. The culprit was a common attic with no fire stops. The fire spread rapidly in a horizontal direction through the common attic because there were no walls to stop its advance. When the structure was rebuilt, the proper building construction codes were followed and fire stops were built into the new attic.

Tracy Rickman
Rio Hondo Fire Academy
Whittier, California

tity of wood and other combustible components to produce a large fire. A fire involving both the contents and the structural components can quickly destroy the building.

Fortunately, most fires originate with the building contents and are extinguished before the flames ignite the structure itself. Once a fire extends into the structure itself and begins to consume the fuel within the walls and above the ceilings, it becomes much more difficult to control.

Type III construction presents several problems for fire fighters. For example, an electrical fire can begin inside the void spaces within the walls, floors, and roof assemblies and extend to the contents. The void spaces also allow a fire to extend vertically and horizontally, spreading from room to room and from floor to floor. Fire fighters will have to open the void spaces to fight the fire. An uncontrolled fire within the void spaces is likely to destroy the building.

The fire resistance of interior structural components often depends on the age of the building and on local building codes. Older Type III buildings were built from solid lumber, which can contain or withstand a fire for a limited time. Newer or renovated buildings may have lightweight assemblies that can be damaged much more quickly and are prone to early failure. Floors and roofs in these buildings can collapse suddenly and without warning.

Exterior walls also could collapse if a fire causes significant damage to the interior structure. Because the exterior walls, the floors, and the roof are all connected in a stable building, the collapse of the interior structure could make the freestanding masonry walls unstable and likely to collapse.

Type IV Construction: Heavy Timber

Type IV construction is also known as heavy timber construction (► Figure 6-10). A heavy timber building has exterior walls of masonry construction, and interior walls, columns, beams, floor assemblies, and roof structure of wood. The exterior walls are usually brick and extra thick to support the weight of the building and its contents.

The wood used in Type IV construction is much heavier than that used in Type III construction. It is more difficult to ignite and will withstand a fire for a much longer time before it collapses. A typical heavy timber building could have 8"-square wood posts supporting 14"-deep wood beams and 8" floor joists. The floors could be constructed of solid wood planks, 2" or 3" thick, with top layer of wood as the finished floor surface.

Heavy timber construction has no concealed spaces or voids. This helps reduce the horizontal and vertical fire spread that often occurs in ordinary construction buildings. Many heavy timber buildings do have vertical openings for elevators, stairs, or machinery, which can provide a path for a fire to travel from one floor to another.

Heavy timber construction has been used to construct buildings as tall as six to eight stories with open spaces suitable for manufacturing and storage occupancies. Similar

Figure 6-10 Type IV construction is also known as heavy timber construction.

modern buildings are usually built with either fire resistive (Type I) or noncombustible (Type II) construction. New buildings of Type IV construction are rare, except for special structures that feature the construction components as architectural elements.

The heavy, solid wood columns, support beams, floor assemblies, and roof assemblies used in heavy timber construction will withstand a fire much longer than the smaller wood members used in ordinary and lightweight combustible construction. But once involved in a fire, the structure of a heavy timber building can burn for many hours. A fire that ignites the combustible portions of heavy timber construction is likely to burn until it runs out of fuel and the building is reduced to a pile of rubble. As the fire consumes the heavy timber support members, the masonry walls will become unstable and collapse.

Mill construction was common during the 1800s, especially in northeastern states. Large mill buildings were built as factories and warehouses. In many communities, dozens of these large buildings were clustered together in industrial areas, creating the potential for huge fires involving multiple multi-story buildings. The radiant heat from a major fire in one of these buildings could spread the fire to nearby buildings, resulting in the loss of several surrounding structures.

Mill construction was state of the art for its time. Automatic sprinkler systems were developed to protect buildings of mill construction; as long as the sprinklers are properly maintained, these buildings have a good safety record. Without sprinklers, a heavy timber mill building is a major fire hazard.

Only a few of the original mill buildings are still used for manufacturing. Many have been converted to small shops, galleries, office buildings, and residential occupancies. The conversions divide the open spaces into smaller compartments and create void spaces within the structure. Appropriate fire protection and life safety features, such as modern sprinkler systems, must be built into the conversions to ensure the safety of occupants.

Type V Construction: Wood Frame

In Type V construction, all of the major components are constructed of wood or other combustible materials (► Figure 6-11). Type V construction is often called wood-frame construction and is the most common type of construction used today. Wood frame construction is used not only in one- and two-family dwellings and small commercial buildings, but in larger structures such as apartment and condominium complexes and office buildings up to four stories in height.

Many wood frame buildings do not have any fire-resistive components. Some codes require a 1-hour fire resistance rating for limited parts of wood frame construction buildings, particularly those of more than two stories. Plaster or gypsum board barriers are often used to achieve this rating, but fire fighters should not assume that these barriers will be present or effective in all Type V construction.

Because all of the structural components are combustible, wood frame buildings can rapidly become totally involved in a fire. In addition, Type V building construction usually contains voids and channels that allow a fire within the structure to spread quickly. A fire that originates in a Type V building can easily extend to other nearby buildings.

Fire Fighter Safety Tips

Many buildings may have a brick veneer on the exterior walls of wood frame construction. A thin layer of brick or stone might be applied to enhance the appearance of the building or to reduce the risk of fire spread. But this practice can give the impression that a building is Type III (ordinary) construction when it is really Type V (wood frame) construction.

During a structural fire, the brick veneer is likely to collapse or peel away as the wall behind it burns or collapses. A single thickness of bricks is not usually strong enough to stand independently. Fire fighters should be aware of buildings constructed in this manner and maintain a safe distance from potential falling bricks.

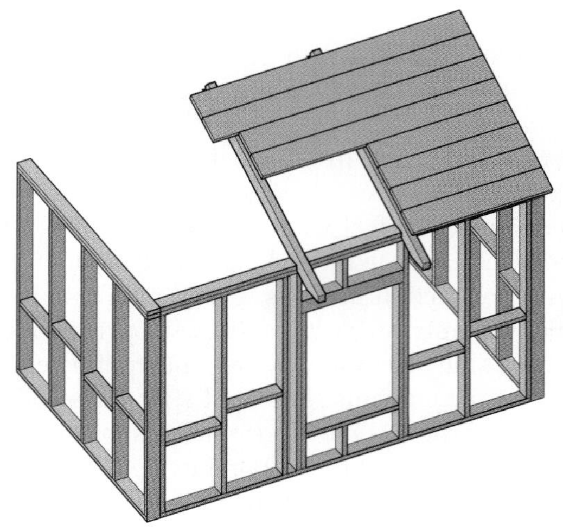

Figure 6-11 Type V, or wood frame, construction is the most common type of construction used today.

Wood frame buildings often collapse and suffer major destruction from fires. Smoke detectors are essential to warn building residents early if a fire occurs. Although compartmentalization can help limit the spread of a fire, automatic sprinklers are the most effective way of protecting lives and property in Type V buildings.

Wood frame buildings are constructed in various ways. Older wood frame construction was assembled from solid lumber, which relied on its mass to provide strength. To reduce costs and create the largest building with the least material, lightweight construction makes extensive use of wooden I-beams and wooden trusses. Structural assemblies are engineered to be just strong enough to carry the required load. As a result, there is no built-in safety margin, and these buildings can collapse early, suddenly, and completely during a fire.

Two systems are used to assemble wood frame buildings: balloon-frame construction and platform construction. Because these systems developed in different eras, fire fighters can anticipate the type of construction based on the age of a building.

Balloon-Frame Construction

Balloon-frame construction was popular between the late 1800s and the mid-1900s. A balloon-frame building has the exterior walls assembled with wood studs that are continuous from the basement to the roof (► Figure 6-12). In a two-story building, the floor joists that support the first and second floors are nailed to these continuous studs. As a result, there is an open channel between each pair of studs that extends from the foundation to the attic. Each of these channels provides a path that enables a fire to spread from the basement to the attic without being visible on the first- or second-floor levels. Fire fighters must anticipate that a fire originating on a lower level will quickly extend through

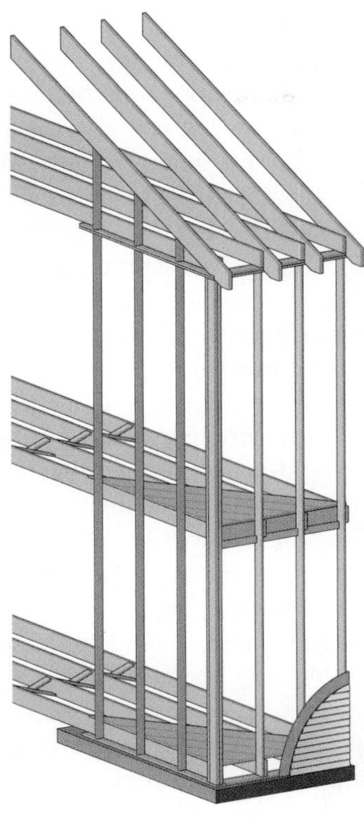

Figure 6-12 In a balloon-frame building, the exterior walls create channels that enable a fire to spread from the basement to the attic.

Figure 6-13 The foundation supports the entire weight of the building.

these voids, and should open the void spaces to check for hidden fires and to prevent rapid vertical extension.

Platform-Frame Construction

Platform-frame construction is used for almost all modern wood frame construction. In a building with platform construction, the exterior wall studs are not continuous. The first floor is constructed as a platform, and the studs for the exterior walls are erected on top of it. The first set of studs extends only to the underside of the second-floor platform, which blocks any vertical void spaces. The studs for the second-floor exterior walls are erected on top of the second-floor platform. At each level, the floor platform blocks the path of any fire rising within the void spaces. Platform framing prevents fire spread from one floor to another through continuous stud spaces. A fire can eventually burn through the wood platform, but the platform will slow the fire spread.

Building Components

The construction classification system gives fire fighters an idea of the materials used in building and an indication of how the materials will react to fire. But each building also has several different components. A fire fighter who understands how these components function will have a better understanding of the risks involved with a burning building. Some building construction features are safer for fire fighters, as this section shows.

The major components of a building discussed in this section are:

- Foundations
- Floors and ceilings
- Roofs
- Trusses
- Walls
- Doors and windows
- Interior finishes and floor coverings

Foundations

The primary purpose of a building foundation is to transfer the weight of the building and its contents to the ground (▲ Figure 6-13). The weight of the building itself is called the **dead load**. The weight of the building's contents is called the **live load**. The foundation ensures that the base of the building is planted firmly in a fixed location, which helps keep all other components connected.

Modern foundations are usually constructed of concrete or masonry, although wood may be used in some areas. Foundations can be shallow or deep, depending on the type of building and soil composition. Some buildings are built on concrete footings or piers; others are supported by steel piles or wooden posts driven into the ground.

As long as the foundation is stable and intact, it is usually not a major concern for fire fighters. If the foundation shifts or is in poor repair, however, it can become a very critical concern. A burning building with a weak foundation or inadequate lateral bracing could be in imminent danger of collapse.

Most foundation problems are caused by circumstances other than a fire, such as improper construction, shifting soil conditions, or earthquakes. Fires can damage timber post and other wood foundations, but most foundations remain intact even after a severe fire has damaged the rest of the building.

When examining a building, take a close look at the foundation. Look for cracks in the foundation that indicate movement. If the building has been modified or remodeled, look for areas where the support could be compromised.

Floors and Ceilings

Floor construction is very important to fire fighters for three major reasons. First, fire fighters who are working inside a building must rely on the integrity of the floor to support their weight. A floor that fails could drop the fire fighters into a fire on a lower level. Second, in a multi-story structure, fire fighters may be working below a floor (or roof), which would fall on them if it collapsed. Finally, the floor system influences whether a fire spreads vertically from floor to floor within a building or is contained on a single level. Some floor systems are designed to resist fires, while others have no fire-resistant capabilities.

Fire-Resistive Floors

In multi-story buildings, floors and ceilings are generally considered a combined structural system. The structure that supports a floor also supports the ceiling of the story below. In a fire-resistive building, this system is designed to prevent a fire from spreading vertically and to prevent a collapse when a fire occurs in the space below the floor-ceiling assembly. Fire-resistive floor-ceiling systems are rated in hours of fire resistance based on a standard test fire.

Concrete floors are common in fire-resistive construction. The concrete can be cast in place or assembled from panels or beams of precast concrete, which are made at a factory and transported to the construction site. The thickness of the concrete floor depends on the load that the floor needs to support.

Concrete floors can be self-supporting or supported by a system of steel beams or trusses. The steel can be protected by sprayed-on insulating materials or covered with concrete or gypsum board. If the ceiling is part of the fire-resistive rating, it provides a thermal barrier to protect the steel members from a fire in the area below the ceiling.

The ceiling below a fire-resistant floor can be constructed with plaster or gypsum board, or it can be a system of tiles suspended from the floor structure. In many cases, a void space between the ceiling and the floor above contains building systems and equipment such as electrical or telephone wiring, heating and air conditioning ducts, and plumbing and fire sprinkler system pipes (▶ **Figure 6-14**). If the space above the ceiling is not subdivided by fire-resistant partitions or protected by automatic sprinklers, a fire can quickly extend horizontally across a large area.

Wood-Supported Floors

Wood floor structures are common in nonfire-resistive construction. Wooden floor systems range from heavy timber construction, found in old mill buildings, to modern, lightweight construction.

Heavy timber construction can provide a huge fuel load for a fire, but it can also contain and withstand a fire for a considerable length of time without collapsing. Heavy timber construction uses posts and beams that are at least 8″ on the smallest side and often as large as 14″ in depth. The floor decking is often assembled from solid wood boards, 2″ or 3″ thick, which are covered by an additional 1″ of finished wood flooring. The depth of the wood in this system will often contain a fire for an hour or more before the floor fails or burns through.

Conventional wood flooring, which was widely used for many decades, is much lighter than heavy timber, but used solid lumber as beams, floor joists, decking, and finished flooring. It burned readily when exposed to a fire, but generally took about 20 minutes to burn through or reach structural failure. This time factor is only a general, unscientific guideline that depends on many circumstances.

Modern construction uses structural elements that are much less substantial than conventional lumber. Lightweight wooden trusses or wooden I-beams are often used as supporting structures. Thin sheets of plywood are used as decking, and the top layer is often a thin layer of concrete or wood covered by carpet. This floor construction provides little resistance to fire. The lightweight structural elements can fail or the fire can burn through the decking quickly.

Unfortunately, it is impossible to tell how a floor is constructed by looking at it from above. The important information about a floor can only be observed from below, if it is visible at all. The building's age and local construction methods can provide significant clues to a floor's stability, but many older buildings have been renovated using mod-

Figure 6-14 The space between the ceiling and the underside of the floor above is often used for electrical and communications wiring and other building systems.

ern lightweight systems and materials. Fire fighters should use preincident surveys and other planning activities to gather essential structural information about buildings.

Roofs

Safe interior firefighting operations depend on a stable roof. If the roof collapses, the fire fighters inside the building may be injured or killed. Interior fires generate hot gases that accumulate under the roof. Fire fighters performing ventilation activities on the roof to release the heated gases also must depend on the structural integrity of the roof.

The primary purpose of a roof is to protect the inside of a building from the weather. Often, roofs are not designed to be as strong as floors, especially in warmer climates. If the space under the roof is used for storage, or if extra heating, ventilating, or air conditioning equipment is mounted on the roof, the load could exceed the designer's expectations. Adding the weight of several fire fighters to an overloaded roof could be disastrous.

Several methods and materials are used for roof construction. Roofs are constructed in three primary designs: pitched, curved, and flat. The major components of a roof assembly are the supporting structure, the roof deck, and the roof covering.

Pitched Roofs

A pitched roof has sloping or inclined surfaces. Pitched roofs are used on many houses and some commercial buildings. The pitch or angle of the roof can vary depending on local architectural styles and climate. Variations of pitched roofs include gable, hip, mansard, gambrel, and lean-to roofs (▼ Figure 6-15).

The slope of a pitched roof can present either a minor inconvenience or a complete lack of secure footing. Fire fighters working on a pitched roof are always in danger of falling, particularly when the roof is wet or icy, when the roof covering is not secure, or when smoke or darkness obscures

A

B

C

D

Figure 6-15 Examples of pitched roofs include: **A.** A gable roof. **B.** A hip roof. **C.** A mansard roof. **D.** A gambrel roof.

vision. Roof ladders are used to provide a secure working platform for fire fighters working on a pitched roof.

Pitched roofs are usually supported by either rafters or trusses. Rafters are solid wood joists mounted in an inclined position. Pitched roofs supported by rafters usually have solid wood boards as the roof decking. Fire fighters can detect the impending failure of a roof with solid decking supported by rafters by a spongy feeling underfoot.

Modern lightweight construction uses manufactured wood trusses to construct most pitched roofs. The decking is usually thin plywood or a sheeting material such as wood particleboard. The trusses and decking material can burn through quickly. Fire fighters cannot work safely on this roof if the supporting structures become involved in a fire.

Steel trusses also are used to support pitched roofs. The fire resistance of steel trusses is directly related to how well the steel is protected from the heat of the fire.

Several roof-covering materials are used on pitched roofs, usually in the form of shingles or tiles (► Figure 6-16). Shingles are usually made from felt or mineral fibers impregnated with asphalt, although metal and fiberglass shingles are also used. Wood shingles, often made from cedar, are popular in some areas. In older construction, wood shingles were often mounted on individual wood slats instead of continuous decking, and would burn rapidly in dry weather conditions.

Shingles are generally durable, economical, and easily repaired. A shingle roof that has aged and deteriorated should be removed completely and replaced. Some older buildings may have newer layers of shingles on top of older layers, which can make it difficult to cut an opening for ventilation.

Roofing tiles are usually made from clay or concrete products. Clay tiles, which have been used for roofing since ancient times, can be flat or rounded. Rounded clay tiles are sometimes called mission tiles. Clay tiles are both durable and fire resistant.

Slate tiles are produced from thin sheets of rock. Slate is an expensive, long-lasting roofing material, but very brittle and slippery when wet.

Metal panels of galvanized steel, copper, and aluminum also are used on pitched roofs. Expensive metals such as copper are often used for their decorative appearance, while lower-cost galvanized steel is used on barns and industrial buildings. Corrugated, galvanized metal panels are strong enough to be mounted on a roof without roof decking. Metal roof coverings will not burn, but they can conduct heat to the roof decking.

Curved Roofs

Curved roofs are often used for supermarkets, warehouses, industrial buildings, arenas, auditoriums, bowling alleys, churches, airplane hangars, and other large buildings that

Figure 6-16 Several roof-covering materials are used on pitched roofs.

require large, open interiors. Curved roofs are usually supported by steel or wood bowstring trusses or arches.

The decking on curved roof buildings can range from solid wooden boards or plywood to corrugated steel sheets. The decking material must be identified before ventilation openings can be made. Often, the roof covering consists of layers that include felt, mineral fibers, and asphalt, although some curved roofs are covered with foam plastics or plastic panels.

Flat Roofs

Flat roofs are found on houses, apartment buildings, shopping centers, warehouses, factories, schools, and hospitals (▶ Figure 6-17). Most flat roofs have a slight slope so water can drain off. If the roof does not have the proper slope or the drains are not maintained, water can pool on the roof, overloading the structure and causing a collapse.

The support systems for most flat roofs are constructed of either wood or steel. A wood support structure uses solid wood beams and joists; laminated wood beams may be used for extra strength or to span long distances. Lightweight construction uses wood trusses or wooden I-beams as the supporting members. Lightweight assemblies are much less fire resistant and collapse more quickly than solid wood systems.

Steel also is used to support flat roofs. Open-web steel trusses, sometimes called bar joists, can span long distances and remain stable during a fire if they are not subjected to excessive heat. Automatic sprinklers can protect the steel from the heat, but without such protection, the steel over a fire will soon weaken and sag. Heated steel support members can even elongate enough to push out the exterior walls of the building. Cracks in the outside walls at the roof level are a warning indicator.

Flat roof decks can be constructed of wood planking, plywood, corrugated steel, gypsum, or concrete. The material used depends on the building's age and size, the climate, the cost of materials, and the type of roof covering.

The roof covering is applied on top of the deck. Most flat roofs are covered with multiple-layer, built-up roofing systems that help prevent leakage. A typical built-up roof covering will have five layers: a vapor barrier, thermal insulation, a waterproof membrane, a drainage layer, and a wear course. The outside of the flat roof reveals only the top layer, which is often gravel, that serves as the wear course and protects the underlying layers from wear and tear. Cutting through all of the layers to open a ventilation hole can be a challenge.

Most flat roof coverings contain highly combustible materials, including asphalt, roofing felt, tarpaper, rubber or plastic membranes, and plastic insulation layers. These materials can be difficult to ignite, but once lit, they burn readily, releasing great quantities of heat and black smoke.

Flat roofs can present unique problems during vertical ventilation operations. A roof with a history of leaks could have several patches where additional layers of roofing material make the covering extra-thick. Sometimes a whole new roof is constructed on top of the old one. After opening a hole in the new roof, fire fighters might discover the old roof and have to cut into that as well. Fire fighters might discover fire burning in the space between the two roofs as well.

Figure 6-17 Most flat roofs have a slight slope so water can drain off.

A similar problem may be encountered in dealing with remodeled buildings or additions. There could be an old flat roof under a new pitched section, resulting in two separate void spaces. The old roof provides a large supply of fuel for the fire. Such a situation is difficult to predict unless the incident commander (IC) has the building plan in advance.

Trusses

Trusses are used extensively in support systems for both floors and roofs. A truss is a structural component composed of smaller components in a triangular configuration or a system of triangles. Trusses are common in modern construction for several reasons. The triangular geometry creates a strong, rigid structure that can support a load much greater than its own mass. For example, both a solid beam and a simple truss with the same overall dimensions can support the same load. The truss requires much less material than the beam, is much lighter, and can span a long distance without supports. Trusses are often prefabricated and transported to the building location.

Fire Fighter Safety Tips

When working on a flat roof, avoid skylights and other openings that will not support your weight. If you are temporarily blinded by smoke, use a tool to test the surface in front of you and to your sides. Stay away from the edges of the roof.

Trusses are used in residential construction, apartment buildings, small office buildings, commercial buildings, warehouses, fast food restaurants, airplane hangars, churches, and even firehouses. They may be clearly visible, or concealed within the construction. They are widely used in new construction and often replace heavier solid beams and joists in renovated or modified older buildings.

The strength of a truss depends on both its members and the connections between them. A properly assembled and installed truss is strong. If the members or connectors begin to fail, however, the strength of the truss is compromised.

Trusses can be used for many different purposes. In building construction, trusses made of wood and steel or combinations of wood and steel are used primarily to support roofs and floors.

A **parallel chord truss** has two parallel horizontal members connected by a system of diagonal and sometimes vertical members. The top and bottom members are called the chords, and the connecting pieces are the web members. Parallel chord trusses are often used to support flat roofs or floors. In lightweight construction, an engineered wood truss is often assembled with wood chords and either wood or light steel web members. A steel bar joist is another example of a parallel chord truss.

A **pitched chord truss** is typically used to support a sloping roof. Most modern residential construction uses a series of prefabricated wood pitched chord trusses to support the roof. The roof deck is supported by the top chords and the ceilings of the occupied rooms are attached to the bottom chords. In this way, the trusses define the shape of the attic.

A **bowstring truss** has the same shape as an archery bow. The top chord represents the curved bow and the bottom chord represents the straight bowstring. Bowstring trusses are usually quite large and widely spaced. They were popular in warehouses, supermarkets, and similar structures with large, open floor areas. The roof of a building with bowstring trusses has a distinctive curved shape.

Fire Fighter Safety Tips

The United States Fire Administration (USFA) has partnered with the American Forest & Paper Association (AF&PA) to develop educational materials to enhance fire fighter awareness regarding the fire performance of different types of lightweight construction components. The fire fighter educational material developed under this partnership will be available in December 2004, free of charge from http://www.FireFighter.jbpub.com and http://usfa.fema.gov.

Walls

Walls are the most visible parts of a building because they shape the exterior and define the interior. Walls may be constructed of masonry, wood, steel, aluminum, glass, and many other materials.

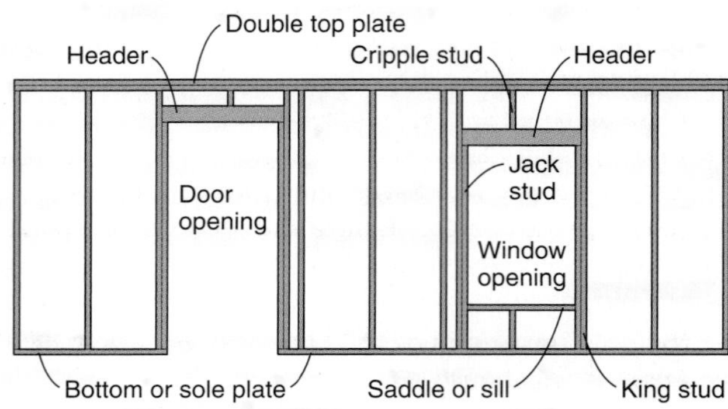

Figure 6-18 A load-bearing wall provides structural support.

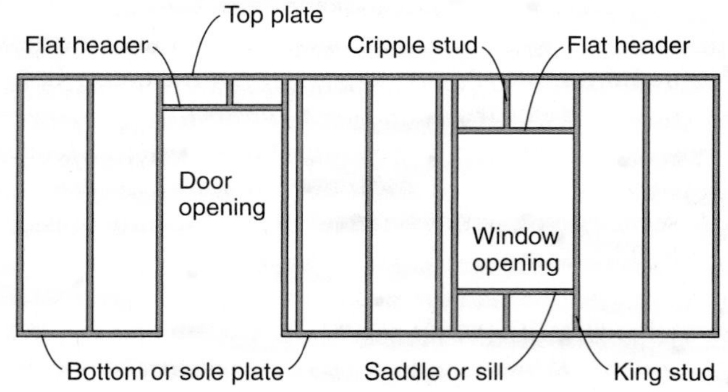

Figure 6-19 A nonbearing wall supports only its own weight.

Walls are either load-bearing or nonbearing. **Load-bearing walls** provide structural support (► Figure 6-18). A load-bearing wall supports a portion of both the building's weight (dead load) and its contents (live load), transmitting that load down to the building's foundation. Damaging or removing a load-bearing wall can result in a partial or total collapse of the building. Load-bearing walls can be either exterior or interior walls.

Nonbearing walls support only their own weight (► Figure 6-19). Most nonbearing walls can be breached or removed without compromising the structural integrity of the building. Many nonbearing walls are interior partitions that divide the building into rooms and spaces. The exterior walls of a building can also be nonbearing, particularly when a system of columns supports the building.

In addition to load-bearing and nonbearing walls, there are several specialized walls.

- **Party walls** are constructed on the line between two properties, and are shared by a building on each side of the line. They are almost always load-bearing walls. A party wall is often, but not always, constructed as a fire wall between the two properties.
- Fire walls are designed to limit the spread of fire from one side of the wall to the other side. A fire wall might divide a large building into sections or separate two attached buildings. Fire walls usually extend from the foundation up to and through the roof of a building. They are constructed of fire-resistant materials and may be fire-rated.
- **Fire partitions** are interior walls that extend from a floor to the underside of the floor above. Fire partitions often enclose fire-rated interior corridors or divide a floor area into separate fire compartments.
- **Fire enclosures** are fire-rated assemblies that enclose interior vertical openings, such as stairwells, elevator shafts, and chases for building utilities. A fire enclosure prevents fire and smoke from spreading from floor to floor via the vertical opening. In multi-story buildings, fire enclosures also protect the occupants using the exit stairways.
- **Curtain walls** are nonbearing exterior walls attached to the outside of the building. Curtain walls often serve as the exterior skin on a steel-framed high-rise building.

Solid, load-bearing masonry walls, at least 6" to 8" thick, can be used for buildings up to six stories high. Nonbearing masonry walls can be almost any height. Masonry walls provide a durable, fire-resistant outer covering for a building, and are often used as fire walls. A well-designed masonry fire wall is often completely independent of the structures on either side. Even if the building on one side burns completely and collapses, the fire wall should prevent the fire from spreading to the building on the other side.

Older buildings often had masonry load-bearing walls several feet thick at the bottom that decreased in thickness as height increased. Modern masonry walls are often reinforced with steel rods or concrete to provide a more efficient and more durable structural system.

When properly constructed and maintained, masonry walls are strong and can withstand a vigorous assault by fire. But if the interior structure begins to collapse and exerts unanticipated forces on the exterior walls, solid masonry walls can fail during a fire.

Even though a building has an outer layer of brick or stone, fire fighters should never assume that the exterior walls are masonry. Many buildings that look like masonry are actually constructed using wood frame techniques and materials. A single layer of brick or stone, called a veneer layer, is applied to the exterior walls to give the appearance of a durable and architecturally desirable outer covering. If the wood structure is damaged during a fire, the veneer is likely to collapse.

Wood framing is used to construct the walls in most houses and many small commercial buildings. The wood framing used for exterior walls can be covered with various materials, including wooden siding, vinyl siding, aluminum siding, stucco, and masonry veneer. Moisture barriers, wind barriers, and other types of insulation are usually applied to the outside of the vertical studs before the outer covering is applied.

Vertical wooden studs support the walls and partitions inside the building. If fire resistance is critical, steel studs are used to frame walls. The wood framing is usually covered with gypsum board and a variety of interior finish materials. The space between the two wall surface coverings may be empty, it may contain thermal and sound insulating materials, or it may contain electrical wiring, telephone wires, and plumbing. These spaces often provide pathways through which fire can spread.

> **Fire Fighter Tips**
>
> **Manufactured (mobile) homes** use lightweight building components throughout the structure to reduce weight. As a result, most parts of manufactured (mobile) homes are combustible. Manufactured (mobile) homes typically have few doors and small windows, making entry for fire suppression or rescue difficult. Once a fire starts, especially in an older manufactured (mobile) home, it can destroy the entire structure within a few minutes. Often all that is left is the frame and the wheels.

Doors and Windows

Doors and windows are important components of any building. Although they generally have different functions—doors provide entry and exit while windows provide light and ventilation—in an emergency, doors and windows are almost interchangeable. A window can serve as an entry or an exit while a door can provide light and ventilation.

There are hundreds of door and window designs, with many different applications. Of particular concern to fire fighters are fire doors and fire windows.

Door Assemblies

Most doors are constructed of either wood or metal. Hollow-core wooden doors are often used inside buildings. A typical hollow-core door has an internal framework with its outer surfaces covered by thin sheets of wood. Because hollow-core doors can be easily opened with simple hand tools, they should not be used where security is a concern. A fire can usually burn through a hollow-core door in just a few minutes.

Solid-core wooden doors are used where a more substantial door is required. Solid-core doors are manufactured from solid panels or blocks of wood, and are more difficult to force open. A solid-core door provides some fire resistance and can often keep a fire contained within a room for 20 minutes or more, giving fire fighters a chance to arrive and mount an effective attack.

Metal doors are more durable and fire resistant than wooden doors. Some metal doors have a solid wood core covered on both sides by a thin sheet of metal. Others are constructed entirely of metal and reinforced for added security. The interior of a metal door can be hollow, or filled with wood, sound-deadening material, or thermal insulation. Although most approved fire doors are metal, not all metal doors should be considered fire doors.

Fire Fighter Safety Tips

Buildings under construction or demolition and buildings that are being renovated are extremely hazardous for fire fighters.

Window Assemblies

Fire fighters frequently use windows as entry points to attack a fire and as emergency exits. Windows also provide ventilation during fires, allowing smoke and heat to escape and cooler, fresh air to enter.

Windows come in many shapes, sizes, and designs for different buildings and occupancies (▼ Figure 6-20). Fire fighters must become familiar with the specific types of windows found in local occupancies, learn to recognize them, and understand how they operate. It is especially important to know if a particular type of window is very difficult to open or cannot be opened. The chapter on forcible entry contains information on window construction.

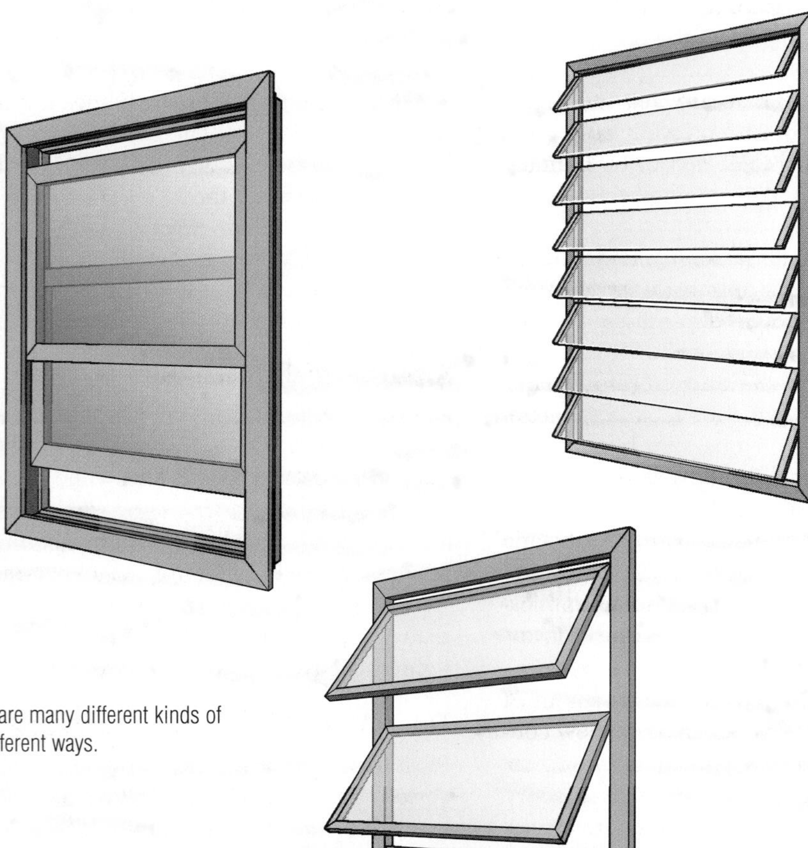

Figure 6-20 There are many different kinds of windows that open in different ways.

Fire Doors and Fire Windows

Fire doors and fire windows are constructed to prevent the passage of flames, heat, and smoke through an opening during a fire. They must be tested and meet the standards set in NFPA 80, *Standard for Fire Doors and Fire Windows.* Fire doors and fire windows come in different shapes and sizes, and provide different levels of fire resistance. For example, fire doors can swing on hinges, slide down or across an opening, or roll down to cover an opening.

The fire rating on a door or window covers the actual door or window, the frame, the hinges or closing mechanism, the latching hardware, and any other equipment that is required to operate the door or window. All of these items must be tested and approved as a combined system. All fire doors must have a mechanism that keeps the door closed or automatically closes the door when a fire occurs. Doors that are normally open can be closed by the release of a fusible link, by a smoke or heat detector, or by activation of the fire alarm system.

Fire doors and fire windows are rated for a particular duration of fire resistance to a standard test fire. This is similar to the fire resistance rating system used for building construction assemblies. A 1-hour rating, however, does not guarantee that the door will resist any fire for 60 minutes; it only establishes that the door will resist the standard test fire for 60 minutes. In any given fire situation, a door rated at 1 hour will probably last twice as long as a door rated 30 minutes.

Fire doors and windows are labeled and assigned the letters A, B, C, D, or E, based on their approved-use locations (► Figure 6-21). For example, doors designated "A" are approved for use as part of a fire wall. Doors designated "B" are approved for use in stair shafts and elevator shafts. (► Table 6-1) shows the approved uses for each of the letter designations.

Fire windows are used when a window is needed in a required fire-resistant wall (► Figure 6-22). Fire windows are often made of wired glass, which is designed to withstand exposures to high temperatures without breaking. Fire-resistant glass without wires is available for some applications; special steel window frames are required to keep the glass firmly in place.

Wired glass is also used to provide vision panels in or next to fire doors. Vision panels allow a person to view conditions on the opposite side of a door without opening the door, or to make sure no one is standing in front of the door before opening it. In these configurations, the entire assembly—including the door, the window, and the frame—must all be tested and approved together.

When a window is required only for light passage, glass blocks can sometimes be used instead of wired glass. Glass blocks will resist high temperatures and remain in place if properly installed. The size of the opening is limited and depends on the fire-resistive rating of the wall.

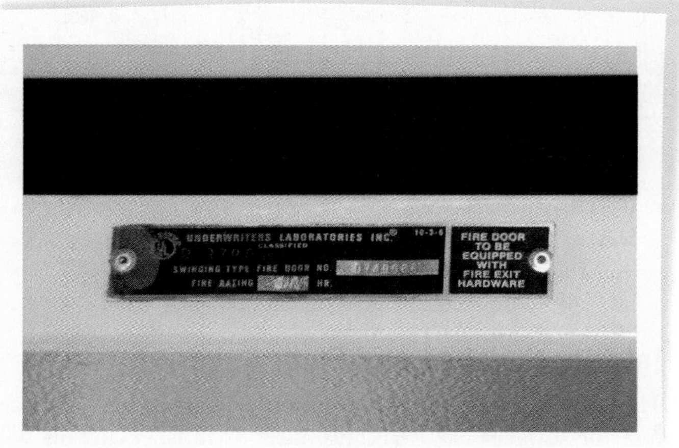

Figure 6-21 An approved fire door has a label indicating its classification and rating.

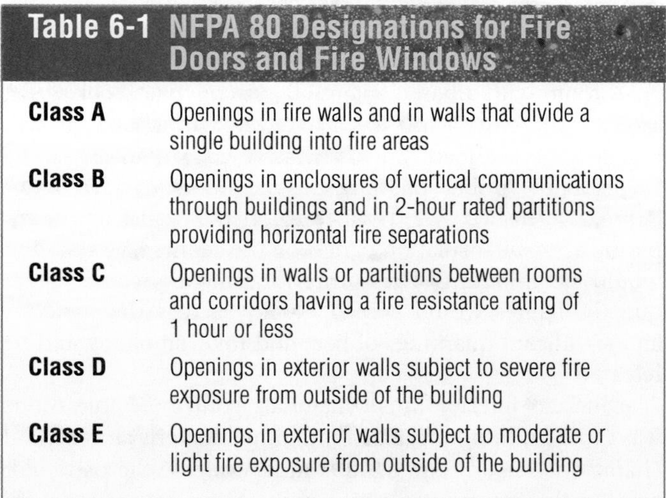

Table 6-1 NFPA 80 Designations for Fire Doors and Fire Windows

Class A	Openings in fire walls and in walls that divide a single building into fire areas
Class B	Openings in enclosures of vertical communications through buildings and in 2-hour rated partitions providing horizontal fire separations
Class C	Openings in walls or partitions between rooms and corridors having a fire resistance rating of 1 hour or less
Class D	Openings in exterior walls subject to severe fire exposure from outside of the building
Class E	Openings in exterior walls subject to moderate or light fire exposure from outside of the building

Figure 6-22 Fire windows are used to protect openings in walls that are required to be fire resistant.

Fire Fighter Safety Tips

Beware of materials attached to the walls or ceilings of a building to deaden sound. Some of these materials are highly flammable and can provide the fuel for a very intense, smoky fire. Check to determine if the material is approved for use.

Interior Finishes and Floor Coverings

The term **interior finish** commonly refers to the exposed interior surfaces of a building. Interior finish materials will affect how a particular building or occupancy reacts when a fire occurs. Interior finish considerations include whether a material ignites easily or resists ignition, how quickly a flame will spread across the material surface, how much energy the material will release when it burns, how much smoke it will produce, and what the smoke will contain.

A room with a bare concrete floor, concrete block walls, and a concrete ceiling has no interior finishes that will increase the fire load. But if the same room had an acrylic carpet with rubber padding on the floor, wooden baseboards, varnished wood paneling on the walls, and foam plastic acoustic insulation panels on the ceiling, the situation would be different. These interior finishes would ignite quickly, flames would spread quickly across the surfaces, and significant quantities of heat and toxic smoke would be released.

Different interior finish materials contribute in various ways to a building fire. Each individual material has certain characteristics, and fire fighters must evaluate the particular combination of materials in a room or space. Typical wall coverings include painted plaster or gypsum board, wallpaper or vinyl wall coverings, wood paneling, and many other surface finishes. Floor coverings might include different types of carpet, vinyl floor tiles, or finished wood flooring. All of these products will burn to some extent, and each one involves a different set of fire risk factors. Fire fighters must understand the hazards posed by different interior finish materials to operate safely at the scene of a fire.

Buildings Under Construction or Demolition

Buildings that are under construction or renovation or in the process of demolition present a variety of problems and extra hazards for fire fighters. In many cases, the fire protection features found in finished buildings are missing. Automatic sprinklers may not yet be installed; smoke detectors may have been removed. There may be no coverings on the walls, leaving the entire wood framework fully exposed. Missing doors and windows could provide an unlimited supply of fresh air to feed the fire and allow the flames to spread rapidly. Fire-resistive enclosures could be missing, leaving critical structural components unprotected. Sprinkler and standpipe systems might be inoperative. All of these factors can contribute to a fire that spreads rapidly, burns intensely, and causes structural failure quickly.

Many fires at construction and demolition sites are inadvertently caused by workers using torches to weld or take apart pieces of the structure. Tanks of flammable gases and piles of highly combustible construction materials may be left in locations where they could add even more fuel to a fire. Buildings under construction or demolition are often unoccupied for many hours, resulting in delayed discovery and reporting of fires. It may be difficult for fire apparatus to approach the structure or for fire fighters to access working hydrants. All of these problems must be anticipated when considering the fire risks of a construction or demolition site.

On-Duty Fire Fighter Fatalities

In 2000, two fire fighters died during a fast-food restaurant fire that involved a structure with a lightweight, wood truss roof. This incident prompted a review of on-duty fire fighter fatalities in the ten years preceding 2000. During the period 1991 through 2000, 962 fire fighters died while on-duty. Of those 962 fire fighter deaths, 19 died in roof collapses involving 15 structures. Sufficient details are not available to determine the materials used in fabrication of the roof structures used in the 15 structures.

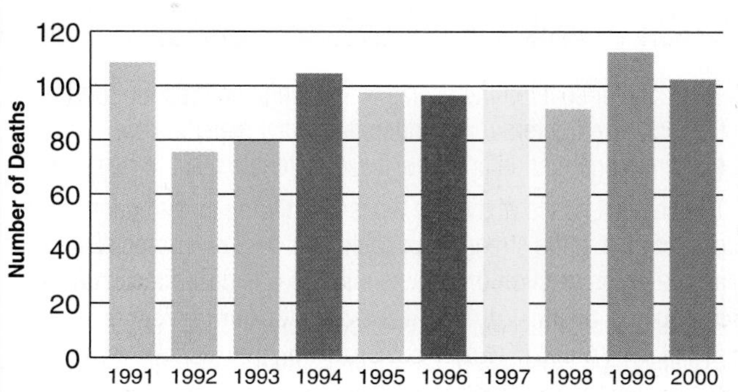

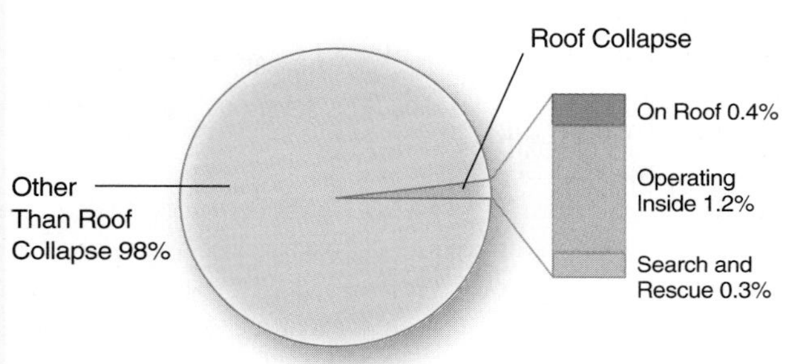

Of the 19 fire fighter deaths due to roof collapse, 4 occurred while operating on the roof of the structure; 12 occurred while operating inside the structure; and 3 occurred while conducting search and rescue operations.The threat of roof collapse during fire incidents needs to constantly be on the mind of all fire fighters.

Source: Fahy, Rita and LeBlanc, Paul, "U.S. Firefighter Fatalities," *NFPA J.*, July/August 2001.

Wrap-Up

Ready for Review

This chapter covered the basics of building construction for fire fighters. Fire fighters must understand what materials are used in building construction and how these materials react to heat and fire.

The characteristics of the five types of building construction were presented and the strengths of each type were discussed. The resistance and the reaction of each type to fire was described.

Buildings contain certain parts or components. This chapter explained the functions of foundations, floors, ceilings, roofs, trusses, walls, doors, windows, interior finishes, and floor coverings. It discussed the various materials that can be used for each component, and the reaction of each component to fire.

Chief Concepts

- Building materials under fire will combust, conduct heat, expand, or lose strength.
- Most construction uses masonry, concrete, steel, glass, gypsum board, wood, and plastics.
- Fire-resistive construction, noncombustible construction, ordinary construction, heavy timber construction, and wood frame construction are the five major types of building construction.
- Each type of building construction has different strengths and hazards during a fire.
- Foundations, floors, ceilings, roofs, trusses, walls, doors, windows, interior finishes, and floor coverings are the major components of buildings.

Hot Terms

Balloon-frame construction An older type of wood frame construction in which the wall studs extend vertically from the basement of a structure to the roof without any fire stops.

Bowstring trusses Trusses that are curved on the top and straight on the bottom.

Combustibility Determines whether or not a material will burn.

Curtain walls Nonbearing walls used to separate the inside and outside of the building, but not part of the support structure for the building.

Curved roofs Roofs that have a curved shape.

Dead load The weight of a building; the dead load consists of the weight of all materials of construction incorporated into a building, including but not limited to walls, floors, roofs, ceilings, stairways, built-in partitions, finishes, cladding, and other similarly incorporated architectural and structural items, as well as fixed service equipment, including the weight of cranes.

Fire enclosures Fire-rated assemblies used to enclose vertical openings such as stairwells, elevator shafts, and chases for building utilities.

Fire partitions Interior walls extending from the floor to the underside of the floor above.

Fire wall A wall with a fire-resistive rating and structural stability that separates buildings or subdivides a building to prevent the spread of fire.

Fire window A window or glass block assembly with a fire-resistive rating.

Flat roofs Horizontal roofs often found on commercial or industrial occupancies.

Glass blocks Thick pieces of glass similar to bricks or tiles.

Gypsum A naturally occurring material composed of calcium sulfate and water molecules

Gypsum board The generic name for a family of sheet products consisting of a noncombustible core primarily of gypsum with paper surfacing.

Interior finish Any coating or veneer applied as a finish to a bulkhead, structural insulation, or overhead, including the visible finish, all intermediate materials, and all application materials and adhesives.

Laminated glass Glass manufactured with a thin vinyl core covered by glass on each side of the core.

Laminated wood Pieces of wood that are glued together.

Live load The weight of the building contents.

Load-bearing wall Walls designed for structural support.

Masonry Built-up unit of construction or combination of materials such as brick, clay tiles, or stone set in mortar.

Manufactured (mobile) homes A factory-assembled structure or structures transportable in one or more sections that is built on a permanent chassis and designed to be used as a dwelling without a permanent foundation when connected to the required utilities, including the plumbing, heating, air-conditioning, and electric systems contained therein.

Nonbearing wall Wall designed to support only the weight of the wall itself.

Occupancy The purpose for which a building or other structure, or part thereof, is used or intended to be used.

Parallel chord trusses A truss in which the top and bottom chords are parallel.

Party walls Walls constructed on the line between two properties.

Pitched chord truss Type of truss typically used to support a sloping roof.

Pitched roof A roof with sloping or inclined surfaces.

Platform-frame construction Construction technique for building the frame of the structure one floor at a time. Each floor has a top and bottom plate that acts as a fire stop.

Pyrolysis The destructive distillation of organic compounds in an oxygen-free environment that converts the organic matter into gases, liquids, and char.

Rafters Joists that are mounted in an inclined position to support a roof.

Spalling Chipping or pitting of concrete or masonry surfaces.

Steel joints A steel truss used as a horizontal beam.

Tempered glass Glass that is much stronger and harder to break than ordinary glass.

Thermal conductivity Describes how quickly a material will conduct heat.

Thermoplastic material Plastic material capable of being repeatedly softened by heating and hardened by cooling and, that in the softened state, can be repeatedly shaped by molding or forming.

Thermoset material Plastic material that, after having been cured by heat or other means, is substantially infusible and cannot be softened and formed.

Wrap-Up

Truss A collection of lightweight structural components joined in a triangular configuration that can be used to support either floors or roofs.

Type I construction Buildings with structural members made of noncombustible materials that have a specified fire resistance.

Type II construction Buildings with structural members made of noncombustible materials without fire resistance.

Type III construction Buildings with the exterior walls made of noncombustible or limited-combustible materials, but interior floors and walls made of combustible materials.

Type IV construction Buildings constructed with noncombustible or limited-combustible exterior walls, and interior walls and floors made of large dimension combustible materials.

Type V construction Buildings with exterior walls, interior walls, floors, and roof structures made of wood.

Wired glass Glass made by molding glass around a special wire mesh.

Wood panels Thin sheets of wood glued together.

Wood trusses Assemblies of small pieces of wood or wood and metal.

Wooden beams Load-bearing members assembled from individual wood components.

Fire Fighter in Action

It is 1:00 p.m. on a Monday afternoon when your engine company is dispatched to a two-story wood frame residential structure with a pitched roof. Upon arrival you find fire in two bedrooms on the second floor. Your lieutenant tells you to complete an interior attack with a preconnected hose line.

1. What type of construction is this structure?
   - A. Type I
   - B. Type II
   - C. Type III
   - D. Type IV
   - E. Type V

2. What type of roof truss system is this structure likely to have?
   - A. Parallel cord truss
   - B. Pitched cord truss
   - C. Bowstring truss
   - D. None of the above

3. You should be cautious when damaging or removing load-bearing walls because doing so:
   - A. divides the building into rooms or spaces and may encourage fire spread.
   - B. can result in a partial or total collapse of the building.
   - C. might disrupt ventilation.
   - D. may cause the veneer to collapse.

4. Balloon-frame constructed buildings are of concern because:
   - A. they used a lot of brick veneer, which collapses easily.
   - B. the interior wall creates channels that allow fire to spread horizontally.
   - C. the exterior walls create channels that enable vertical fire spread.
   - D. both A and B

www.FireFighter.jbpub.com

Portable Fire Extinguishers

Technology Resources

www.FireFighter.jbpub.com

- Chapter Pretests
- Interactivities
- Hot Term Explorer
- Web Links
- Review Manual
- FireLearn

Chapter Features

- Skill Drills
- Voices of Experience
- Fire Marks
- Teamwork Tips
- Fire Fighter Safety Tips
- Fire Fighter Tips
- Canadian Perspectives
- Hot Terms
- Wrap-Up

Chapter 7

NFPA 1001 Standard

Fire Fighter I

5.3.16 Extinguish incipient Class A, Class B, and Class C fires, given a selection of portable fire extinguishers, so that the correct extinguisher is chosen, the fire is completely extinguished, and proper extinguisher-handling techniques are followed.

5.3.16 (A) *Requisite Knowledge.* The classifications of fire; the types of, rating systems for, and risks associated with each class of fire; and the operating methods and limitations of portable fire extinguishers.

5.3.16 (B) *Requisite Skills.* The ability to operate portable fire extinguishers, approach fire with portable fire extinguishers, select an appropriate extinguisher based on the size and type of fire, and safely carry portable fire extinguishers.

Fire Fighter II

NFPA 1001 contains no Fire Fighter II Job Performance Requirements for this chapter.

Additional NFPA Standards

NFPA 10, *Standard for Portable Fire Extinguishers*

NFPA 11, *Standard for Low-, Medium-, and High-Expansion Foam*

Knowledge Objectives

After studying this chapter, you will be able to:

- State the primary purposes of fire extinguishers.
- Define Class A fires.
- Define Class B fires.
- Define Class C fires.
- Define Class D fires.
- Define Class K fires.
- Explain the classification and rating system for fire extinguishers.
- Describe the types of agents used in fire extinguishers.
- Describe the types of operating systems in fire extinguishers.
- Describe the basic steps of fire extinguisher operation.
- Explain the basic steps of inspecting, maintaining, recharging, and hydrostatic testing of fire extinguishers.

Skills Objectives

After completing this chapter, you will be able to perform the following skills:

- Transport the extinguisher to the location of the fire.
- Select and operate a portable fire extinguisher safely to effectively extinguish an incipient fire.
- Attack a Class A fire with a stored-pressure water-type fire extinguisher.
- Attack a Class A fire with a multipurpose dry chemical fire extinguisher.
- Attack a Class B flammable liquid fire with a dry chemical fire extinguisher.
- Attack a Class B flammable liquid fire with a stored-pressure foam fire extinguisher.
- Use a wet chemical fire extinguisher.
- Use a halogenated agent-type extinguisher.
- Use dry powder fire extinguishing agents.

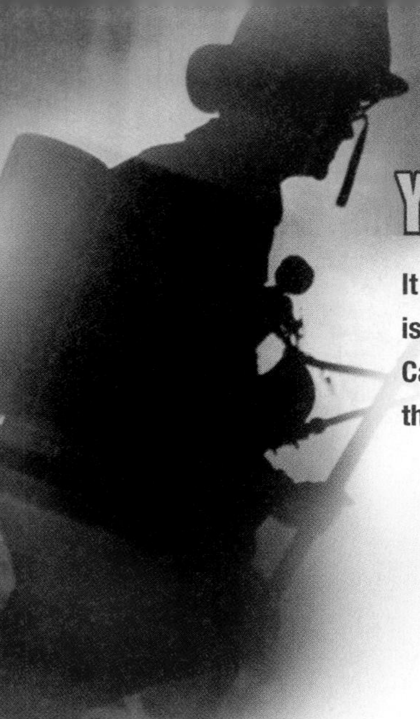

You Are the Fire Fighter

It is your first shift as a fire fighter with Captain Jones at Station 34. The first alarm of the day is a reported car fire in a parking lot. The fire is in the engine compartment of a midsized sedan. Captain Jones directs you to don your SCBA gear, get the dry chemical extinguisher, and back up the other crew members who are pulling a 1¾" hose line to fight the fire. As you obey, you wonder:

1. ***Are there any advantages to using a portable fire extinguisher to initially attack a car fire?***
2. ***Does it have to be a dry chemical extinguisher, or would another type of extinguisher work just as well?***

Introduction

Portable fire extinguishers are required in many types of occupancies, as well as in commercial vehicles, boats, aircraft, and various other locations. Fire prevention efforts encourage citizens to keep fire extinguishers in their homes, particularly in their kitchens. Fire extinguishers are used successfully to put out hundreds of fires every day, preventing millions of dollars in property damage as well as saving lives. Most fire extinguishers are easy to operate and can be used effectively by an individual with only basic training.

Fire extinguishers range in size from models that can be operated with one hand to large, wheeled models that contain several hundred pounds of extinguishing agent (a material used to stop the combustion process) (► Figure 7-1 A & B). Extinguishing agents include water, water with different additives, dry chemicals, dry powders, and gaseous agents. Each agent is suitable for specific types of fires.

Fire extinguishers are designed for different purposes and involve different operational methods. As a fire fighter, you must know which is the most appropriate kind of extinguisher to use for different types of fires, which kinds must not be used for certain fires, and how to use any fire extinguisher you may encounter, especially those used by your department and carried on your apparatus. The selection of the proper fire extinguisher builds on the information presented in Chapter 5, Fire Behavior.

Fire fighters use fire extinguishers to control small fires that do not require the use of a hose line. A portable backpack-type fire extinguisher can be used to control and overhaul a wildland fire located beyond the reach of hoses. Special types of fire extinguishers are appropriate for situations where the application of water would be dangerous, ineffective, or undesirable. For example, applying water to a fire that involves expensive electronic equipment is both dangerous and costly. Using the correct fire extinguisher could control the fire without causing additional damage to the equipment.

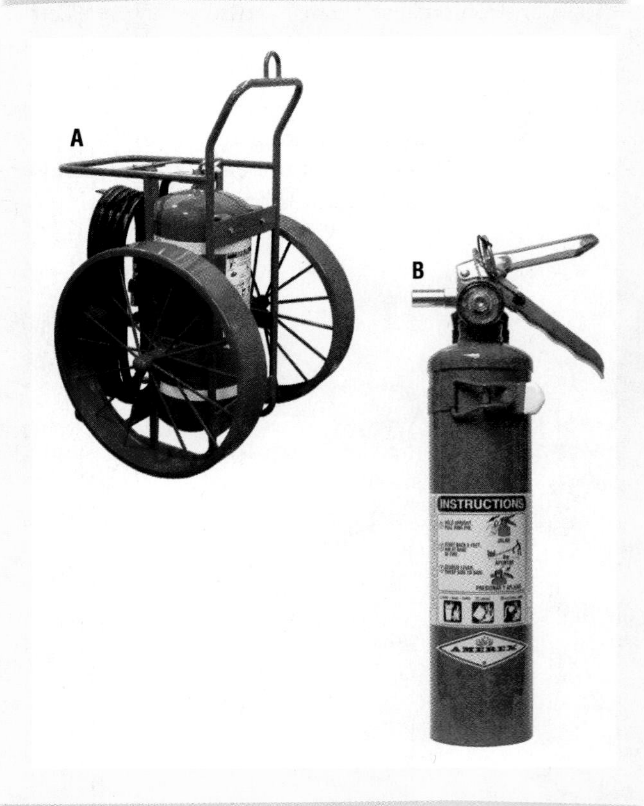

Figure 7-1 Portable fire extinguishers can be large or small. **A.** A wheeled extinguisher. **B.** A one hand fire extinguisher.

This chapter also covers the operation and maintenance of the most common types of portable fire extinguishers. These principles will enable you to use fire extinguishers correctly and effectively to reduce the risk of personal injury and property damage.

Purposes of Fire Extinguishers

Portable fire extinguishers have two primary uses: to extinguish incipient fires (those that have not spread beyond the area of origin), and to control fires where traditional methods of fire suppression are not recommended.

Fire extinguishers are placed in many locations so that they will be available for immediate use on small, incipient-stage fires, such as a fire in a wastebasket. A trained individual with a suitable fire extinguisher could easily control this type of fire (▼ Figure 7-2). As the flames spread beyond the wastebasket to other contents of the room, the fire becomes increasingly difficult to control with only a portable fire extinguisher.

Fire extinguishers are also used to control fires where traditional extinguishing methods are not recommended. For example, using water on fires that involve energized electrical equipment increases the risk of electrocution to fire fighters. Applying water to a fire in a computer or electrical control room could cause extensive damage to the electrical equipment. In these cases, it would be better to use a fire extinguisher with the appropriate extinguishing agent. Special extinguishing agents are also required for fires that involve flammable liquids, cooking oils, and combustible metals. The appropriate type of fire extinguisher should be available in areas containing these hazards.

Figure 7-2 A trained individual with a suitable fire extinguisher can easily control an incipient fire.

Fire Fighter Safety Tips

Do not place yourself in a dangerous situation by trying to fight a large fire with a small fire extinguisher. You can't fight a fire or protect yourself with an empty extinguisher.

Fire Fighter Safety Tips

Proper technique in using a portable fire extinguisher is important for both safety and effectiveness. Practice using a portable fire extinguisher under the careful supervision of a trained instructor.

Incipient Fires

Most fire department vehicles carry at least one fire extinguisher; many vehicles carry two or more extinguishers of different types. Fire fighters often use these portable extinguishers to control incipient-stage fires quickly. At times, a fire fighter may even use an extinguisher from the premises to control an incipient fire.

One advantage of fire extinguishers is their portability. It may take less time to control a fire with a portable extinguisher than it would to advance and charge a hose line. Fire department vehicles that are not equipped with water or fire hoses usually carry at least one multipurpose fire extinguisher. If you arrive at an incipient-stage fire in one of these vehicles, you might be able to control the flames with the portable extinguisher.

The primary disadvantage of fire extinguishers is that they are "one-shot" devices. Once the contents of a fire extinguisher have been discharged, it is no longer effective in fighting fires until it is recharged. If the extinguisher does not control the fire, some other device or method will have to be employed. This is a serious limitation when compared to a fire hose with a continuous water supply.

Special Extinguishing Agents

Some types of fires require special extinguishing agents. As a fire fighter, you must know which fires require special extinguishing agents, what type of extinguisher should be used, and how to operate the different types of special-purpose extinguishers.

Special extinguishing agents are used for kitchen fires that involve combustible cooking oils and fats, for combustible metal fires, and for fires in electronic equipment. Using water or an unsuitable fire-extinguishing chemical to fight these types of fires can cause more damage than the fire.

Figure 7-3 Portable extinguishers containing specific extinguishing agents are sometimes helpful in overhauling a fire.

Portable extinguishers are sometimes used in combination with other techniques. For example, with a pressurized gas fire, water may be used to cool hot surfaces and prevent re-ignition, while a dry chemical agent is used to extinguish the flames. Certain types of portable extinguishers can be helpful in overhauling a fire (▲ Figure 7-3). The extinguishing agents contained in these devices break down the surface tension of the water, so the water can penetrate the materials and reach deep-seated fires.

Classes of Fires

It is essential to match the appropriate type of extinguisher to the type of fire. Fires and fire extinguishers are grouped into classes according to their characteristics. Some extinguishing agents work more efficiently than others on certain types of fires. In some cases, selecting the proper extinguishing agent will mean the difference between extinguishing a fire and being unable to control it.

More importantly, in some cases it is dangerous to apply the wrong extinguishing agent to a fire. Using a water extinguisher on an electrical fire can cause an electrical shock as well as a short circuit in the equipment. A water extinguisher should never be used to fight a grease fire. Burning grease is generally hotter than 212°F (100°C), so the water converts to steam, which expands very rapidly. If the water penetrates the surface of the grease, the steam is produced within the grease. As the steam expands, the hot grease erupts like a volcano and splatters over everything and everyone nearby, resulting in burns or injuries to people and spreading the fire.

Before selecting a fire extinguisher, ask yourself, "What class of fire am I fighting?" Remember, there are five classes of fire and each will affect the choice of extinguishing equipment.

Fire Fighter Tips

Most residential fires involve Class A materials. Even though the source of the ignition could be Class B or Class C in nature, the resulting fire frequently involves normal combustibles. A fire that begins with an unattended pot of grease on the stove might ignite the wooden cabinets and other kitchen contents. A short circuit in an electrical outlet might ignite ordinary combustibles in the immediate vicinity. In both of these cases, the fire would involve primarily Class A materials.

Class A Fires

Class A fires involve ordinary combustibles such as wood, paper, cloth, rubber, household rubbish, and some plastics (► Figure 7-4). Natural vegetation, such as grass and trees, is also Class A material. Water is the most commonly used extinguishing agent for Class A fires, although several other agents can be used effectively.

Class B Fires

Class B fires involve flammable or combustible liquids, such as gasoline, oil, grease, tar, lacquer, oil-based paints, and some plastics (► Figure 7-5). Fires involving flammable gases, such as propane or natural gas, are also categorized as Class B fires. Several different types of extinguishing agents are approved for Class B fires.

Examples of Class B fires include a fire in a pot of molten roofing tar, a fire involving splashed fuel on a hot lawnmower engine, and burning natural gas that is escaping from a gas meter struck by a vehicle.

Class C Fires

Class C fires involve energized electrical equipment, which includes any device that uses, produces, or delivers electrical energy (► Figure 7-6). A Class C fire could involve building wiring and outlets, fuse boxes, circuit breakers, transformers, generators, or electric motors. Power tools, lighting fixtures, household appliances, and electronic devices such as televisions, radios, and computers could be involved in Class C fires. The equipment must be plugged in or connected to an electrical source, but not necessarily operating.

Electricity does not burn, but electrical energy can generate tremendous heat that could ignite nearby Class A or B materials. As long as the equipment is energized, it must be treated as a Class C fire. Agents that will not conduct electricity, such as dry chemicals or carbon dioxide, must be used on Class C fires.

Figure 7-4 Most ordinary combustible materials are included in the definition of Class A fires.

Figure 7-5 Class B fires involve flammable liquids and gases.

Figure 7-6 Class C fires involve energized electrical equipment or appliances.

Figure 7-7 Combustible metals in Class D fires require special extinguishing agents.

Class D Fires

Class D fires involve combustible metals such as magnesium, titanium, zirconium, sodium, lithium, and potassium. Special techniques and extinguishing agents are required to fight combustible metals fires (▶ Figure 7-7). Normal extinguishing agents can react violently, even explosively, if they come in contact with burning metals. Violent reactions also can occur when water strikes burning combustible metals.

Class D fires are most often encountered in industrial occupancies, such as machine shops and repair shops, as well as in fires involving aircraft and automobiles. Magnesium and titanium, both combustible metals, are used to produce automotive and aircraft parts because they combine high strength with light weight. Sparks from cutting, welding, or grinding operations could ignite a Class D fire, or the metal items could become involved in a fire that originated elsewhere.

Because of the chemical reactions that could occur during a Class D fire, it is important to select the proper extinguishing agent and application technique. Choosing the correct fire extinguisher for a Class D fire requires expert knowledge and experience.

Class K Fires

Class K fires involve combustible cooking oils and fats (▶ Figure 7-8). This is a relatively new classification; cooking oil fires were previously classified as Class B combustible liquid fires. The use of high-efficiency modern cooking equipment and the trend toward using vegetable oils instead of animal fats to fry foods required the develop-

Figure 7-8 Class K fires involve cooking oils and fats.

ment of a new class of extinguishing agents. Many restaurants are still using extinguishing agents that were approved for Class B fires.

Classification of Fire Extinguishers

Portable fire extinguishers are classified and rated based on their characteristics and capabilities. This information is important for selecting the proper extinguisher to fight a particular fire (► Table 7-1). It is also used to determine what type or types of fire extinguishers should be placed in a given location so that incipient fires can be quickly controlled.

In the United States, Underwriters Laboratories Inc. (UL) is the organization that developed the standards, classification, and rating system for portable fire extinguishers. This system rates fire extinguishers for both safety and effectiveness. Each fire extinguisher has a specific rating that identifies the class or classes of fires for which it is both safe and effective.

The classification system for fire extinguishers uses letters and numbers. The letters indicate the class or classes of fire for which the extinguisher can be used, and the numbers indicate its effectiveness. Fire extinguishers that are safe and effective for more than one class will be rated with multiple letters. For example, an extinguisher that is safe and effective for Class A fires will be rated with an "A;" one that is safe and effective for Class B fires will be rated with a "B;" and one that is safe and effective for both Class A and Class B fires will be rated with both an "A" and a "B."

Table 7-1 Types of Fires

Class A	Ordinary combustibles
Class B	Flammable or combustible liquids
Class C	Energized electrical equipment
Class D	Combustible metals
Class K	Combustible cooking media

Canadian Perspectives

In Canada, the Underwriters Laboratories of Canada (ULC) develops the standards and conducts the certification and testing of portable fire extinguishers under Standard CAN/ULC-S508-M. The rating system is almost identical to the UL ratings in the United States. NFPA 10 *Standard for Portable Fire Extinguishers* is also widely used in Canada.

Class A and Class B fire extinguishers also include a number, indicating the relative effectiveness of the fire extinguisher in the hands of a nonexpert user.

On Class A extinguishers, the number is related to an amount of water. An extinguisher that is rated 1-A contains the equivalent of 1.25 gallons of water. A typical Class A extinguisher contains 2.5 gallons of water and has a 2-A rating. The higher the number, the greater the extinguishing capability of the extinguisher. An extinguisher that is rated 4-A should be able to extinguish approximately twice as much fire as one that is rated 2-A.

The effectiveness of Class B extinguishers is based on the approximate area (measured in square feet) of burning fuel they are capable of extinguishing. A 10-B rating indicates that a nonexpert user should be able to extinguish a fire in a pan of flammable liquid that is 10 square feet in surface area. An extinguisher rated 40-B should be able to control a flammable liquid pan fire with a surface area of 40 square feet.

Numbers are used to rate an extinguisher's effectiveness only for Class A and Class B fires. If the fire extinguisher can also be used for Class C fires, it contains an agent proven to be nonconductive to electricity and safe for use on energized electrical equipment. For instance, a fire extinguisher that carries a 2-A:10-B:C rating can be used on Class A, Class B, and Class C fires. It has the extinguishing capabilities of a 2-

Fire Fighter Tips

The safest and surest way to extinguish a Class C fire is to turn off the power and treat it like a Class A or B fire. If you are unable to turn off the power, you should be prepared for re-ignition. The electricity could re-ignite the fire after it is extinguished.

Fire Fighter Tips

Fire fighters must understand both the Underwriters Laboratories classification and rating system and the labeling system for fire extinguishers.

A extinguisher when applied to Class A fires, the capabilities of a 10-B extinguisher for Class B fires, and can be used safely on energized electrical equipment.

Standard test fires are used to rate the effectiveness of fire extinguishers. The testing may involve different agents, amounts, application rates, and application methods. Fire extinguishers are rated for their ability to control a specific type of fire as well as for the extinguishing agent's ability to prevent rekindling. Some agents can successfully suppress a fire, but are unable to prevent the material from re-igniting. A rating is only given if the extinguisher completely extinguishes the standard test fire and prevents rekindling.

Labeling of Fire Extinguishers

Fire extinguishers that have been tested and approved by an independent laboratory are labeled to clearly designate the class or classes of fire the unit is capable of extinguishing safely. The traditional lettering system has been used for many years and is still found on many fire extinguishers. Recently, however, a universal pictograph system, which does not require the user to be familiar with the alphabetic codes for the different classes of fires, has been developed.

Traditional Lettering System (▶ Figure 7-9)

The traditional lettering system uses the following labels:

- Extinguishers suitable for use on Class A fires are identified by the letter A on a solid green triangle. The triangle has a graphic relationship to the letter A.
- Extinguishers suitable for use on Class B fires are identified by the letter B on a solid red square. Again, the shape of the letter mirrors the graphic shape of the box.
- Extinguishers suitable for use on Class C fires are identified by the letter C on a solid blue circle, which

(Figure 7-9) Traditional letter labels on fire extinguishers often incorporated a shape as well as a letter.

(Figure 7-10) The icons for Classes A, B, C, and K fires.

also incorporates a graphic relationship between the letter C and the circle.
- Extinguishers suitable for use on Class D fires are identified by the letter D on a solid yellow five-pointed star.
- Extinguishers suitable for use on Class K (combustible cooking oil) fires are identified by a pictograph showing a fire in a frying pan. Because the Class K designation is new, there is no traditional-system alphabet graphic for it.

Pictograph Labeling System

The pictograph system, such as described for Class K fire extinguishers, uses symbols rather than letters on the labels. This system also clearly indicates if an extinguisher is inappropriate for use on a particular class of fire. The pictographs are all square icons that are designed to represent each class of fire (▲ Figure 7-10). The icon for Class A fires is a burning trash can beside a wood fire. The Class B fire extinguisher icon is a flame and a gasoline can; the Class C icon is a flame and an electrical plug and socket. There is no pictograph for Class D extinguishers. Extinguishers rated for fighting Class K fires are labeled with an icon showing a fire in the frying pan.

Under this pictograph labeling system, the presence of an icon indicates that the extinguisher has been rated for that class of fire. A missing icon indicates that the extinguisher has not been rated for that class of fire. A red slash across an icon indicates that the extinguisher must not be used on that type of fire, because doing so would create additional risk.

An extinguisher rated for Class A fires only would show all three icons, but the icons for Class B and Class C would have a red diagonal line through them. This three-icon array signifies that the extinguisher uses a water-based extinguishing agent, making it unsafe to use on flammable liquid or electrical fires.

Certain extinguishers labeled for Class B and Class C fires do not include the Class A icon, but may be used to put out small Class A fires. The fact that they have not been rated for Class A fires indicates that they are less effective in extinguishing a common combustible fire than a comparable Class A extinguisher would be.

Fire Extinguisher Placement

Fire codes and regulations require the installation of fire extinguishers in many areas so that they will be available to fight incipient fires. NFPA 10 *Standard for Portable Fire Extinguishers* lists the requirements for placing and mounting portable fire extinguishers as well as the appropriate mounting heights.

The regulations for each type of occupancy specify the maximum floor area that can be protected by each extinguisher, the maximum travel distance from the closest extinguisher to a potential fire, and the types of fire extinguishers that should be provided. Two key factors must be considered when determining which type of extinguisher should be placed in each area: the class of fire that is likely to occur and the potential magnitude of an incipient fire.

Extinguishers should be mounted so they are readily visible and easily accessed (► Figure 7-11). Heavy extinguishers should not be mounted high on a wall. If the extinguisher is mounted too high, a smaller person might be unable to lift it off its hook or could be injured in the attempt.

Figure 7-11 Extinguishers should be mounted in locations with unobstructed access and visibility.

According to NFPA 10, the recommended mounting heights for the placement of fire extinguishers are:

- Fire extinguishers weighing up to 40 lb (18.14 kg) should be mounted so that the top of the extinguisher is not more than 5′ (1.53 m) above the floor.
- Fire extinguishers weighing more than 40 lb (18.14 kg) should be mounted so that the top of the extinguisher is not more than 3′ (1.07 m) above the floor.
- The bottom of an extinguisher should be at least 4″ (10.2 cm) above the floor.

Classifying Area Hazards

Areas are divided into three risk classifications—light, ordinary, and extra hazard—according to the amount and type of combustibles that are present, including building materials, contents, decorations, and furniture. The quantity of combustible materials present is sometimes called a building's fire load and is measured as the average weight of combustible materials per square foot or per square meter of floor area. The larger the fire load, the larger the potential fire.

Occupancy use category does not necessarily determine the appropriate hazard classification. The recommended hazard classifications for different types of occupancies are guidelines based on typical situations. The hazard classification for each area should be based on the actual amount and type of combustibles that are present.

Light or Low Hazard

Light (or low) hazard locations are areas where the majority of materials are noncombustible or arranged so that a fire is not likely to spread. Light hazard environments usually contain limited amounts of Class A combustibles, such as wood, paper products, cloth, and similar materials. A light hazard environment might also contain some Class B combustibles (flammable liquids and gases), such as copy machine chemicals or modest quantities of paints and solvents, but all Class B materials must be kept in closed containers and stored safely. Examples of common light hazard environments are most offices, classrooms, churches, assembly halls, and hotel guest rooms (► Figure 7-12).

Figure 7-12 Light hazard areas include offices, churches, and classrooms.

Figure 7-13 Auto showrooms, hotel laundry rooms, and parking garages are classified as ordinary hazard areas.

Ordinary or Moderate Hazard

The ordinary (or moderate) hazard locations contain more Class A and Class B materials than light hazard locations. Typical examples of ordinary hazard locations include retail stores with on-site storage areas, light manufacturing facilities, auto showrooms, parking garages, research facilities, and workshops or service areas that support light hazard locations, such as hotel laundry rooms or restaurant kitchens (► Figure 7-13).

Ordinary hazard areas also include warehouses that contain Class I and Class II commodities. Class I commodities include noncombustible products stored on wooden pallets or in corrugated cartons that are shrink-wrapped or wrapped in paper. Class II commodities include noncombustible products stored in wooden crates or multilayered corrugated cartons.

Extra or High Hazard

Extra (or high) hazard locations contain more Class A combustibles and/or Class B flammables than ordinary hazard locations. Typical examples of extra hazard areas include woodworking shops; service or repair facilities for cars, aircraft, or boats; and many kitchens and other cooking areas that have deep fryers, flammable liquids, or gases under pressure (► Figure 7-14). In addition, areas used for manufacturing processes such as painting, dipping, or coating, and facilities used for storing or handling flammable liquids are classified as extra hazard environments. Warehouses containing products that do not meet the definitions of Class I and Class II commodities are also considered extra hazard locations.

Figure 7-14 Kitchens, woodworking shops, and auto repair shops, are considered possible extra hazard locations.

Determining the Appropriate Class of Fire Extinguisher

Several factors must be considered when determining the number and types of fire extinguishers that should be placed in each area of an occupancy. Among these factors are the types of fuels found in the area and the quantities of those materials.

Some areas may need extinguishers with more than one rating or more than one type of fire extinguisher. Environments that include Class A combustibles require an extinguisher rated for Class A fires; those with Class B combustibles require an extinguisher rated for Class B fires; and areas that have both Class A and Class B combustibles require either an extinguisher that is rated for both types of fires or a separate extinguisher for each class of fire.

Most buildings require extinguishers that are suitable for fighting Class A fires because ordinary combustible materials, such as furniture, partitions, interior finish materials, paper, and packaging products, are so common. Even where other classes of products are used or stored, there is still a need to defend the facility from a fire involving common combustibles.

A single multipurpose extinguisher is generally less expensive than two individual fire extinguishers and eliminates the problem of selecting the proper extinguisher for a particular fire. However, it is sometimes more appropriate to install Class A extinguishers in general use areas and to place extinguishers that are especially effective in fighting Class B or Class C fires near those hazards.

Some facilities present a variety of conditions. In these occupancies, each area must be individually evaluated so that extinguisher installation is tailored to the particular circumstances. A restaurant is a good example of this situation. The dining areas contain common combustibles, such as furniture, tablecloths, and paper products, that would require an extinguisher rated for Class A fires. In the restaurant's kitchen, where the risk of fire involves cooking oils, a Class K extinguisher would provide the best defense.

Similarly, within a hospital, extinguishers for Class A fires would be appropriate in the hallways, offices, lobbies, and patient rooms. Class B extinguishers should be mounted in laboratories and areas where flammable anesthetics are stored or handled. Electrical rooms should have extinguishers that are approved for use on Class C fires. Hospital kitchens would need Class K extinguishers.

Methods of Fire Extinguishment

Understanding the nature of fire is key to understanding how extinguishing agents work and how they differ from each other. All fires require three basic ingredients: fuel, heat, and oxygen. Scientifically, burning is called **rapid oxidation**. It is a chemical process that occurs when a fuel is combined with oxygen, resulting in the formation of ash or other waste products and the release of energy as heat and light.

Figure 7-15 Covering a pan of burning food with a lid will extinguish a fire by cutting off the supply of oxygen.

The combustion process begins when the fuel is heated to its **ignition point**—the temperature at which it begins to burn. The energy that initiates the process can come from many different sources, including a spark or flame, friction, electrical energy, or a chemical reaction. Once a substance begins to burn, it will generally continue burning as long as there are adequate supplies of oxygen and fuel to sustain the chemical reaction, unless something interrupts the process.

Most extinguishers stop the burning by cooling the fuel below its ignition point or kindling temperature, by cutting off the supply of oxygen, or by combining these two techniques. Some extinguishing agents interrupt the complex system of molecular chain reactions that occur between the heated fuel and the oxygen. Modern portable fire extinguishers contain agents that use one or more of these methods.

Cooling the Fuel

If the temperature of the fuel falls below its kindling temperature, the combustion process will stop. Water extinguishes a fire using this method.

Cutting Off the Supply of Oxygen

Creating a barrier that interrupts the flow of oxygen to the flames will also extinguish a fire. Putting a lid on a pan of burning food is an example of this technique (▲ Figure 7-15). Applying a blanket of foam to the surface of a burning liquid is another example. Surrounding the fuel with a layer of carbon dioxide can also cut off the supply of oxygen necessary to sustain the burning process.

Interrupting the Chain of Reactions

Some extinguishing agents work by interrupting the molecular chain reactions required to sustain combustion. In some cases, a very small quantity of the agent can accomplish the objective.

Types of Extinguishing Agents

An extinguishing agent is the substance contained in a portable fire extinguisher that puts out a fire. Various different chemicals, including water, are used in portable fire extinguishers. The best extinguishing agent for a particular hazard depends on several factors, including the types of materials involved and the anticipated size of the fire. Portable fire extinguishers use seven basic types of extinguishing agents:

- Water
- Dry chemicals
- Carbon dioxide
- Foam
- Wet chemicals
- Halogenated agents
- Dry powder

Water

Water is an efficient, plentiful, and inexpensive extinguishing agent. When water is applied to a fire, it quickly converts from liquid into steam, absorbing great quantities of heat in the process. As the heat is removed from the combustion process, the fuel cools below its ignition temperature and the fire stops burning.

Water is an excellent extinguishing agent for Class A fires. Many Class A fuels will absorb liquid water, which further lowers the temperature of the fuel. This also prevents rekindling.

Water is a much less effective extinguishing agent for other fire classes. Applying water to hot cooking oil can cause splattering, which can spread the fire and possibly endanger the extinguisher operator. Many burning flammable liquids will float on top of water. Because water conducts electricity, it is dangerous to apply a stream of water to any fire that involves energized electrical equipment. If water is applied to a burning combustible metal, a violent reaction can occur. Because of these limitations, plain water is only used in Class A fire extinguishers.

One disadvantage of water is that it freezes at 32°F (0°C). In areas that are subject to freezing, **loaded-stream extinguishers** can be used. These extinguishers combine an alkali metal salt with water. The salt lowers the freezing point of water, so the extinguisher can be used in much colder areas.

Wetting agents can also be added to the water in a fire extinguisher. These agents reduce the surface tension of the water, allowing it to penetrate more effectively into many fuels, such as baled cotton or fibrous materials.

Dry Chemical

Dry chemical fire extinguishers deliver a stream of very finely ground particles onto a fire. Different chemical compounds are used to produce extinguishers of varying capabilities and characteristics. The dry chemical extinguishing agents work in two ways. First, the dry chemicals interrupt the chemical chain reactions that occur within the combustion process. In addition, the tremendous surface area of the finely ground particles allows them to absorb large quantities of heat.

Figure 7-16 Multipurpose dry chemical extinguishers can be used for Class A, B, and C fires.

Dry chemical extinguishing agents offer several advantages over water extinguishers:

- They are effective on Class B (flammable liquids and gases) fires.
- They can be used on Class C (energized electrical equipment) fires, because the chemicals are nonconductive.
- They are not subject to freezing.

The first dry chemical extinguishers were introduced during the 1950s and were rated only for Class B and C fires. The industry term for these B:C-rated units is "ordinary dry chemical" extinguishers.

During the 1960s "**multipurpose dry chemical**" extinguishers were introduced. These extinguishers are rated for Class A, B, and C fires. The chemicals in these extinguishers form a crust over Class A combustible fuels to prevent rekindling (▲ Figure 7-16).

Multipurpose dry chemical extinguishing agents are in the form of fine particles and are treated with other chemicals to help maintain an even flow when the extinguisher is being used. Additional additives prevent them from absorbing moisture, which could cause packing or caking and interfere with the discharge.

One disadvantage of dry chemical extinguishers is that the chemicals, particularly the multipurpose dry chemicals, are corrosive and can damage electronic equipment, such as computers, telephones, and copy machines. The fine particles are carried by the air and settle like a fine dust inside the equipment. Over a period of months, the residue can corrode metal parts, causing considerable damage. If electronic equipment is exposed to multipurpose dry chemical extinguishing agents, it should be cleaned professionally within 48 hours after exposure.

The five primary compounds used as dry chemicals extinguishing agents are:

1. Sodium bicarbonate (rated for Class B and C fires only)
2. Potassium bicarbonate (rated for Class B and C fires only)
3. Urea-based potassium bicarbonate (rated for Class B and C fires only)
4. Potassium chloride (rated for Class B and C fires only)
5. Ammonium phosphate (rated for Class A, B, and C fires)

Sodium bicarbonate is often used in small household extinguishers. Potassium bicarbonate, potassium chloride, and urea-based potassium bicarbonate all have greater fire-extinguishing capabilities (per unit volume) for Class B fires than sodium bicarbonate. Potassium chloride is more corrosive than the other dry chemical extinguishing agents.

Ammonium phosphate is the only dry chemical extinguishing agent rated as suitable for use on Class A fires. Although ordinary dry chemical extinguishers can also be used against Class A fires, a water dousing is also needed to extinguish any smoldering embers and prevent rekindling.

Which dry chemical extinguisher to use depends on the compatibility of different agents with each other and with products they might contact. Some dry chemical extinguishing agents cannot be used in combination with particular types of foam.

Carbon Dioxide

Carbon dioxide (CO_2) is a gas that is 1.5 times heavier than air. When carbon dioxide is discharged on a fire, it forms a dense cloud that displaces the air surrounding the fuel. This interrupts the combustion process by reducing the amount of oxygen that can reach the fuel. A blanket of carbon dioxide over the surface of a liquid fuel can also disrupt the fuel's ability to vaporize.

In portable **carbon dioxide fire extinguishers**, carbon dioxide is stored under pressure as a liquid. It is colorless and odorless. It is discharged through a hose and expelled on the fire through a horn. When it is released, the carbon dioxide is very cold and forms a visible cloud of "dry ice" because moisture in the air will freeze when it comes into contact with the carbon dioxide.

Figure 7-17 Carbon dioxide extinguishers are heavy due to the weight of the container and the quantity of agent needed. They also have a large discharge nozzle, making them easily identifiable.

Carbon dioxide is rated for Class B and C fires only. It does not conduct electricity and has two significant advantages over dry chemical agents: it is not corrosive and it does not leave any residue.

Carbon dioxide also has several limitations and disadvantages. These include:

- Weight: Carbon dioxide extinguishers are heavier than similarly rated extinguishers that use other extinguishing agents (▲ Figure 7-17).
- Range: Carbon dioxide extinguishers have a short discharge range, which requires the operator to be close to the fire, increasing the risk of personal injury.
- Weather: Carbon dioxide does not perform well at temperatures below 0°F (-18°C) or in windy or drafty conditions, because it dissipates before it reaches the fire.
- Confined spaces: When used in confined areas, carbon dioxide dilutes the oxygen in the air. If enough oxygen is displaced, people in the space can begin to suffocate.
- Suitability: Carbon dioxide extinguishers are not suitable for use on fires involving pressurized fuel or on cooking grease fires.

Foam

Foam fire extinguishers discharge a water-based solution with a measured amount of foam concentrate added. The nozzles on foam extinguishers are designed to introduce air into the discharge stream, thus producing a foam blanket. Foam extinguishing agents are formulated for use on either Class A or Class B fires.

Class A foam extinguishers for ordinary combustible fires extinguish fires in the same way that water extinguishes fires. This type of extinguisher can be produced by adding Class A foam concentrate to the water in a standard, 2.5 gallons, stored-pressure extinguisher. The foam concentrate reduces the surface tension of the water, allowing for better penetration into the burning materials.

Class B foam extinguishers discharge a foam solution that floats across the surface of a burning liquid and prevents the fuel from vaporizing. The foam blanket forms a barrier between the fuel and the oxygen, extinguishing the flames and preventing re-ignition. These agents are not suitable for Class B fires that involve pressurized fuels or cooking oils.

The most common Class B additives are aqueous film-forming foam (AFFF) and film-forming fluoroprotein (FFFP) foam (► Figure 7-18). Both concentrates produce very effective foams. Which one should be used depends on the product's compatibility with a particular flammable liquid and other extinguishing agents that could be used on the same fire.

Some Class B foam extinguishing agents are approved for use on polar solvents, which are water-soluble flammable liquids such as alcohols, acetone, esters, and ketones. Only extinguishers that are specifically labeled for use with polar solvents should be used if these products are present.

Although they are not specifically intended for Class A fires, most Class B foams can also be used on ordinary combustibles. The reverse is not true, however; Class A foams are not effective on Class B fires. Foam extinguishers are not suitable for use on Class C fires and cannot be stored or used at freezing temperatures.

Figure 7-18 An AFFF extinguisher produces an effective foam for use on Class B fires.

Wet Chemical

Wet chemical extinguishers are the only type of extinguisher to qualify under the new Class K rating requirements. They use wet chemical extinguishing agents, which are chemicals applied as water solutions. Before Class K extinguishing agents were developed, most fire extinguishing systems for kitchens used dry chemicals. The minimum requirement for a commercial kitchen was a 40-B-rated sodium bicarbonate or potassium bicarbonate extinguisher. These systems required extensive clean-up after their use, which often result in serious business interruptions.

All new fixed extinguishing systems in restaurants and commercial kitchens now use wet chemical extinguishing agents. These agents are specifically formulated for use in commercial kitchens and food-product manufacturing facilities, especially where food is cooked in a deep fryer. The fixed systems discharge the agent directly over the cooking surfaces. There is no numeric rating of their efficiency in portable fire extinguishers.

The Class K wet chemical agents include aqueous solutions of potassium acetate, potassium carbonate, and potassium citrate, either singly or in various combinations. The wet agents convert the fatty acids in cooking oils or fats to a soap or foam, a process known as saponification.

When wet chemical agents are applied to burning vegetable oils, they create a thick blanket of foam that quickly smothers the fire and prevents it from re-igniting while the hot oil cools. The agents are discharged as a fine spray, which reduces the risk of splattering. They are very effective at extinguishing cooking oil fires, and clean-up afterward is much easier, allowing a business to reopen sooner.

Halogenated Agents

Halogenated extinguishing agents are produced from a family of liquified gases, known as halogens, that includes fluorine, bromine, iodine, and chlorine. Hundreds of different formulations can be produced from these elements with many different properties and potential uses. Although several of these formulations are very effective for extinguishing fires, only a few of them are commonly used as extinguishing agents.

Halogenated extinguishing agents are called **clean agents** because they leave no residue and are ideally suited for areas that contain computers or sensitive electronic equipment. Per pound, they are approximately twice as effective at extinguishing fires as carbon dioxide.

There are two categories of halogenated extinguishing agents: halons and halocarbons. A 1987 international agreement, known as the Montreal Protocol, limits halon production because these agents damage the earth's ozone layer. Halons have been replaced by a new family of extinguishing agents, halocarbons.

The halogenated agents are stored as liquids and are discharged under relatively high pressure. They release a mist of vapor and liquid droplets that disrupts the molecular chain reactions within the combustion process to extinguish a fire.

These agents dissipate rapidly in windy conditions, as does carbon dioxide, so their effectiveness is limited in outdoor locations. Because these agents also displace oxygen, they should be used with care in confined areas.

Halon 1211 (bromochlorodifluoromethane) should be used judiciously and only in situations where its clean properties are essential, due to its environmental impact. Small Halon 1211 extinguishers are rated for Class B and C fires, but are unsuited for use on fires involving pressurized fuels or cooking grease. Larger halon extinguishers are also rated for Class A fires.

Currently, four types of halocarbon agents are used in portable extinguishers: hydrochlorofluorocarbon (HCFC), hydrofluorocarbon (HFC), perfluorocarbon (PFC), and fluoroiodocarbon (FIC).

Dry Powder

Dry powder extinguishing agents are chemical compounds used to extinguish fires involving combustible metals (Class D fires). These agents are stored in fine granular or powdered form and are applied to smother the fire. They form a solid crust over the burning metal to exclude oxygen and absorb heat.

The most commonly used dry powder extinguishing agent is formulated from finely ground sodium chloride (table salt) plus additives to help it flow freely over a fire. A thermoplastic material mixed with the agent binds the sodium chloride particles into a solid mass when they come into contact with a burning metal.

Another dry powder agent is produced from a mixture of finely granulated graphite powder and compounds containing phosphorus. This agent cannot be expelled from fire extinguishers; it is produced in bulk form and applied by hand, using a scoop or a shovel. When applied to a metal fire, the phosphorus compounds release gases that blanket the fire and cut off its supply of oxygen. The graphite absorbs heat from the fire, allowing the metal to cool below its ignition point. Other specialized dry powder extinguishing agents are available for fighting specific types of metal fires. For details, see NFPA's *Fire Protection Handbook*.

Class D agents must be applied very carefully so that the molten metal does not splatter. No water should come in contact with the burning metal. Even a trace quantity of moisture can cause a violent reaction.

Fire Extinguisher Design

All portable fire extinguishers use pressure to expel their contents onto a fire. Many portable fire extinguishers rely on pressurized gas to expel the extinguishing agent. The gas can be stored with the extinguishing agent in the body of the extinguisher, or externally, in a separate cartridge or cylinder. With external storage, the extinguishing agent is put under pressure only as it is used.

Some extinguishing agents, such as carbon dioxide, are called **self-expelling** agents. Most of these agents are normally gases, which are stored as liquids under pressure. When the confining pressure is released, the agent rapidly expands, causing it to self-discharge. Hand-operated pumps are used to expel the agent when water or water with additives is the extinguishing agent.

Portable Fire Extinguisher Components

Most hand-held portable fire extinguishers have six basic parts (► Figure 7-19):

- A cylinder or container that holds the extinguishing agent
- A carrying handle
- A nozzle or horn
- A trigger and discharge valve assembly
- A locking mechanism to prevent accidental discharge
- A pressure indicator

Cylinder or Container

The body of the extinguisher, known as the **cylinder** (fire extinguisher) or container, holds the extinguishing agent. Nitrogen, compressed air, or carbon dioxide can be used to pressurize the cylinder to expel the agent. **Stored-pressure extinguishers** store both the extinguishing agent, in wet or dry form, and the expeller gas under pressure in the cylinder. **Cartridge/cylinder extinguishers** rely on an external cartridge of pressurized gas, which is only released when the extinguisher is to be used.

Handle

The **handle** is used to carry a portable fire extinguisher and, in many cases, to hold it during use. The actual design of the handle varies from model to model, but all extinguishers that weigh more than 3 lb have handles. In many cases, the handle is located just below the trigger mechanism.

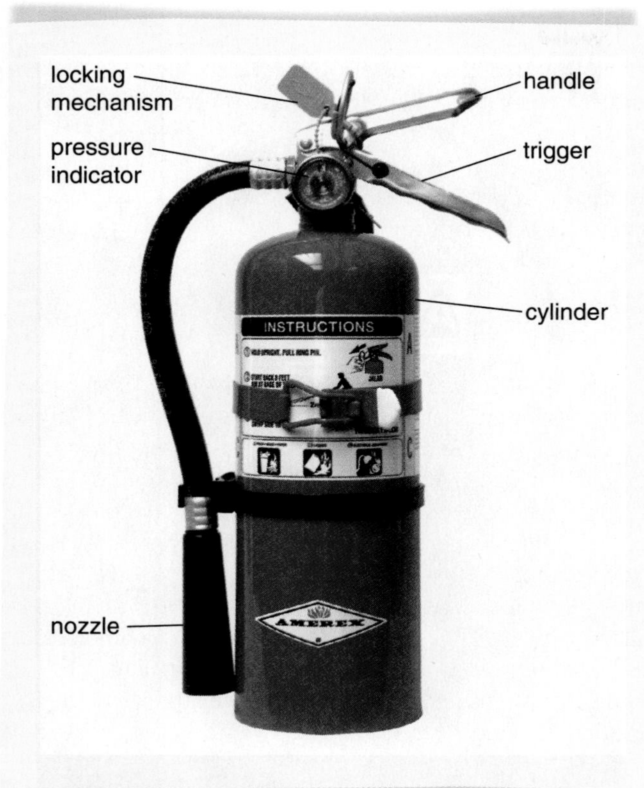

Figure 7-19 The portable fire extinguisher.

Nozzle or Horn

The extinguishing agent is expelled through a nozzle (fire extinguisher) or horn. In some extinguishers, the nozzle is attached directly to the valve assembly at the top of the extinguisher. In other models, the nozzle is at the end of a short hose.

Foam extinguishers have a special aspirating nozzle that introduces air into the extinguishing agent, creating the foam. Carbon dioxide extinguishers have a tubular or conical horn, which is often mounted at the end of a short hose.

Pump tank extinguishers, which are nonpressurized, manually operated water extinguishers, usually have a nozzle at the end of a short hose. The manually operated pump may be mounted directly on the cylinder or it may be part of the nozzle assembly.

Trigger

The trigger is the mechanism that is squeezed or depressed to discharge the extinguishing agent. In some models, the trigger is a button positioned just above the handle. Most of the time, however, the trigger is a lever located above the handle. The operator lifts the extinguisher by the handle and simultaneously squeezes down on the discharge lever.

Cartridge/cylinder extinguisher models usually have a two-step operating sequence. First, a handle or lever is pushed to pressurize the stored agent, then a trigger-type mechanism incorporated in the nozzle assembly is used to control the discharge.

Locking Mechanism

The locking mechanism is a simple quick-release device that prevents accidental discharge of the extinguishing agent. The simplest form of locking mechanism is a stiff pin, which is inserted through a hole in the trigger to prevent it from being depressed. The pin usually has a ring at the end so that it can be removed quickly.

A special plastic tie, called a tamper seal, is used to secure the pin. The tamper seal is designed to break easily when the pin is pulled. Removing the pin and tamper seal is best accomplished with a twisting motion. The tamper seal makes it easy to see whether the extinguisher has been used and not recharged. The seal also discourages people from playing or tinkering with the extinguisher.

Pressure Indicator

The pressure indicator or gauge shows whether a stored-pressure extinguisher has sufficient pressure to operate properly. Over time, the pressure in an extinguisher may dissipate. Checking the gauge first will tell you whether the extinguisher is ready for use.

Pressure indicators vary in design and sophistication. Most extinguishers use a needle gauge. Pressure may be shown in pounds per square inch (psi) or on a three-step scale (too low, proper range, too high). Pressure gauges are usually color-coded; a green area indicates the proper pressure zone. CO_2 extinguishers do not have pressure indicators or gauges, but are weighted to determine the remaining agent.

Some disposable fire extinguishers intended for home use have an even simpler pressure indicator. A plastic pin is built into the cylinder. Pressing on the pin tests the pressure within the extinguisher. If the pin pops back up, the extinguisher has enough pressure to operate; if it remains depressed, the pressure has dropped below an acceptable level. Carbon dioxide extinguishers do not have pressure indicators or gauges. They are weighed to determine the amount of remaining agent.

Wheeled Fire Extinguishers

Wheeled fire extinguishers are large units mounted on wheeled carriages. Wheeled extinguishers typically contain between 150 and 350 lb of extinguishing agent. The wheeled design lets one person transport the extinguisher to the fire. If a wheeled extinguisher is intended for indoor use, doorways and aisles must be wide enough to allow passage to every area where it could be needed.

Wheeled fire extinguishers usually have long delivery hoses, so the unit can stay in one spot as the operator moves around to attack the fire from more than one side. Usually, a separate cylinder containing nitrogen or some other compressed gas provides the pressure necessary to operate the extinguisher.

Wheeled fire extinguishers are most often installed in special hazard areas. They are often found at military bases and airports. Models with rubber tires or wide-rimmed wheels are available for outdoor installations.

Fire Extinguisher Characteristics

Portable fire extinguishers vary according to their extinguishing agent, capacity, effective range, and the time it takes to completely discharge their agent. They also have different mechanical designs. This section describes the basic characteristics of several types of extinguishers, organized by type of extinguishing agent.

The seven types of extinguishers described include:

- Water extinguishers
- Dry chemical extinguishers
- Carbon dioxide extinguishers
- Class B foam extinguishers
- Halogenated-agent extinguishers
- Dry powder extinguishing agents
- Wet chemical extinguishers

Water Extinguishers

Water extinguishers are used to cool the burning fuel below its ignition temperature. Water extinguishers are intended for use primarily on Class A fires. Class B foam extinguishers, which are a specific type of water extinguisher, are intended for flammable liquids fires. Water extinguishers include stored-pressure, loaded-stream, and wetting-agent models.

Stored-Pressure Water-Type Extinguishers

The most popular kind of stored-pressure water-type extinguisher is the 2.5 gallons model with a 2-A rating (► Figure 7-20). Many fire department vehicles carry this type of fire extinguisher for use on incipient Class A fires. This extinguisher expels water in a solid stream with a range of 35′ to 40′ through a nozzle at the end of a short hose. The discharge time is approximately 55 seconds if the extinguisher is used continuously. A full extinguisher weighs about 30 lb.

Because the contents of these extinguishers can freeze, they should not be installed in areas where the temperature is expected to drop below 32°F (0°C). Antifreeze models of stored-pressure water-type extinguishers, called loaded-stream extinguishers, are available. The loaded-stream agent will not freeze at temperatures as low as -40°F (-40°C).

The recommended procedure for operating a stored-pressure water extinguisher is to set it on the ground, grasp the handle with one hand, and pull out the ring pin or release the locking latch with the other hand. Now the extinguisher can be lifted and used to douse the fire. Use one hand to aim the stream at the fire, and squeeze the trigger with the other hand. The stream of water can be made into a spray by putting a thumb at the end of the nozzle; this technique is often used after the flames have been extinguished to thoroughly soak the fuel.

Figure 7-20 Stored-pressure water-type extinguishers are used by most fire departments.

Stored-pressure water-type extinguishers can be recharged at any location that provides water and a source of compressed air. Follow the manufacturer's instructions to ensure proper and safe recharging.

Loaded-Stream Water-Type Extinguishers

Loaded-stream water-type extinguishers discharge a solution of water containing an alkali metal salt that prevents freezing at temperatures as low as -40°F (-40°C). The most common model is the 2.5 gallons unit, which is identical to a typical stored-pressure water extinguisher. Hand-held models are available with capacities of 1 to 2.5 gallons of water, and are rated from 1-A to 3-A.

Larger units, including a 17 gallons unit rated 10-A and a 33 gallons unit rated 20-A, are also available. Pressure for these extinguishers is supplied by a separate cylinder of carbon dioxide.

Wetting-Agent and Class A Foam Water-Type Extinguishers

Wetting-agent water-type extinguishers expel water that contains a solution to reduce its surface tension (the physical property that causes water to bead or form a puddle on a flat surface). Reducing the surface tension allows water to spread over the fire and penetrate more efficiently into Class A fuels.

Fire Marks

Stored-pressure water extinguishers have replaced soda-acid extinguishers, which had to be inverted (turned upside down) to trigger their discharge. The soda-acid models are no longer manufactured and should have been replaced in all installations.

Class A foam extinguishers contain a solution of water and Class A foam concentrate. This agent has foaming properties as well as the ability to reduce surface tension.

Both wetting-agent and Class A foam extinguishers are available in the same configurations as water extinguishers, including hand-held stored-pressure models and wheeled units. These extinguishers should not be exposed to temperatures below 40°F (4°C).

Pump Tank Water-Type Extinguishers

Pump tank water-type extinguishers come in sizes ranging from 1-A rated, 1.5 gallons units to 4-A rated, 5 gallons units. The water in these units is not stored under pressure. The pressure to expel the water is provided by a hand-operated, double-acting, vertical piston pump, which moves water out through a short hose on both the up and the down strokes. This type of extinguisher sits upright on the ground during use. A small bracket at the bottom allows the operator to steady the extinguisher with one foot while pumping.

Pump tank extinguishers can be used with antifreeze. The manufacturer should be consulted for details because some antifreezes (such as common salt) can corrode the extinguisher or damage the pump. Extinguishers with steel shells corrode more easily than those with copper or nonmetallic shells.

Backpack Water-Type Extinguishers

Backpack water-type extinguishers are used primarily outdoors for fighting brush and grass fires (► Figure 7-21). Most of these units have a tank capacity of 5 gallons and weigh about 50 lb when full. Backpack extinguishers are listed by Underwriters Laboratories but do not carry numeric ratings.

The water tank can be made of fiberglass, stainless steel, galvanized steel, or brass. Backpack extinguishers are designed to refill easily in the field, from a lake or a stream, through a wide-mouth opening at the top. A filter keeps dirt, stones, and other contaminates from entering the tank.

Antifreeze agents, wetting agents, or other special water-based extinguishing agents can be used with backpack water-type extinguishers.

Most backpack extinguishers have hand pumps. The most common design has a trombone-type double-acting piston pump located at the nozzle, which is attached to the tank by a short rubber hose. The operator holds the pump in both hands and moves the piston back and forth.

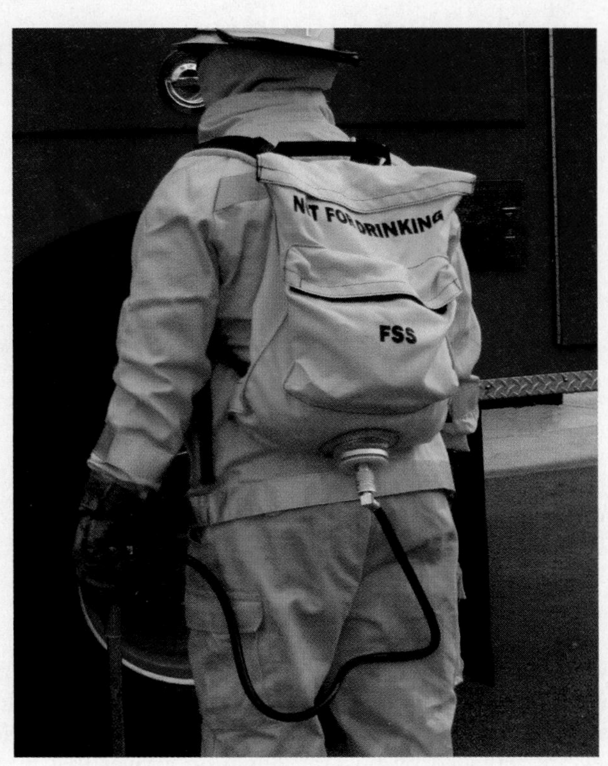

Figure 7-21 A backpack extinguisher can be easily refilled from a natural water source such as a lake or river.

Some models have a compression pump built into the side of the tank. It takes about 10 strokes of the pump handle to build the initial pressure, which is maintained through continuous slow strokes. The operator uses the other hand to control the discharge. A lever-operated shut-off nozzle is provided at the end of a short hose.

Dry Chemical Extinguishers

Dry chemical extinguishers contain a variety of chemical extinguishing agents in granular form. Hand-held dry chemical extinguishers are available with capacities ranging from 1 to 30 lbs of agent. Wheeled fire extinguishers are available with capacities up to 350 lbs of agent. Ordinary dry chemical models can be used to extinguish Class B and C fires. Multipurpose dry chemical models are rated for use on Class A, B, and C fires.

All dry chemical extinguishing agents can be used on Class C fires that involve energized electrical equipment; however, the residue left by the dry chemical can be very damaging to computers, electronic devices, and electrical equipment.

Stored-pressure units expel the dry chemical agent in the same manner as a stored-pressure water extinguisher. The

Fire Fighter Safety Tips

Discharging a dry chemical fire extinguisher in a confined space can create a cloud of very fine dust that can impair vision and cause difficulty breathing. SCBA should be used to protect fire fighters from both toxic gases from the fire and the dry chemical dust discharged from the fire extinguisher.

Fire Fighter Tips

A dry chemical extinguisher will often continue to lose pressure after a partial discharge. Pressure loss can occur even when only a very small amount of agent has been discharged and the pressure gauge still indicates that the extinguisher is properly charged. This occurs because the agent can leave residue in the valve assembly and allow the stored pressure to leak out slowly.

dry chemical agent in a cartridge/cylinder extinguisher is not stored under pressure. These extinguishers have a sealed, pressurized cartridge connected to the storage cylinder. They are activated by pushing down on a lever that punctures the cartridge and pressurizes the cylinder.

Most small, hand-held dry chemical extinguishers are designed to discharge completely in as few as 8 to 20 seconds. Larger units may discharge for as long as 30 seconds. All hand-held dry chemical extinguishers are designed to be carried and operated simultaneously.

Depending on the extinguisher's size, the horizontal range of the discharge stream can be from 5′ to 30′. Some models have special nozzles that allow for a longer range. The long-range nozzles are useful when the fire involves burning gas or a flammable liquid under pressure or working in a strong wind.

The trigger allows the extinguisher to be discharged intermittently, starting and stopping the agent flow. Releasing the trigger stops the flow of agent; however, this does not mean that the extinguisher can be put aside and used again later. These extinguishers do not retain their internal pressure for extended periods, because the granular dry chemical causes the valve to leak.

Anytime a dry chemical extinguisher has been activated or partially used, the extinguisher must be serviced and recharged to replenish the extinguishing agent and restore the unit's pressure. Disposable models are not refillable. Dry chemical fire extinguishers can be stored and used in areas with temperatures below freezing.

Ordinary Dry Chemical Extinguishers

Ordinary dry chemical extinguishers are available in hand-held models with ratings up to 160-B:C. Larger, wheeled units carry ratings up to 640-B:C.

Multipurpose Dry Chemical Extinguishers

Ammonium phosphate is commonly called a multipurpose dry chemical agent because it can be used on Class A, B, and C fires. When it is used on Class A fires, the chemical coats the surface of the fuel to prevent continuing combustion.

Multipurpose dry chemical extinguishers are available in hand-held models with ratings ranging from 1-A to 20-A and from 10-B:C to 120-B:C. Larger, wheeled models have ratings ranging from 20-A to 40-A and from 60-B:C to 320-B:C.

Multipurpose dry chemical extinguishers should never be used on cooking oil (Class K) fires. The ammonium phosphate-based extinguishing agent is acidic and will not react with cooking oils to produce the smothering foam needed to extinguish the fire. Worse, the acid will counteract the foam-forming properties of any alkaline extinguishing agent that is applied to the same fire.

Carbon Dioxide Extinguishers

Carbon dioxide (CO_2) extinguishers are rated to fight Class B and C fires. Carbon dioxide extinguishes a fire by enveloping the fuel in a cloud of inert gas, which reduces the oxygen content of the surrounding atmosphere and smothers the flames. Because the carbon dioxide discharge is very cold, it also helps to cool the burning materials and surrounding areas.

Carbon dioxide gas is 1.5 times heavier than air, colorless, odorless, non-conductive, and inert. It is also noncorrosive and does not leave any residue. These factors are important where costly electronic components or computer equipment must be protected. Carbon dioxide is also used around food preparation areas and in laboratories.

Carbon dioxide is both an expelling agent and an extinguishing agent. The agent is stored under a pressure of 823 psi, which keeps the carbon dioxide in liquid form at room temperatures. When the pressure is released, the liquid carbon dioxide rapidly converts to a gas. The expanding gas forces the agent out of the container.

The carbon dioxide is discharged through a siphon tube that reaches to the bottom part of the storage cylinder. It is forced through a hose to a horn or cone-shaped applicator that is used to direct the flow of the agent. When discharged, the agent is very cold and contains a mixture of carbon dioxide gas and solid carbon dioxide, which quickly converts to gas.

Compared to other types of extinguishers, carbon dioxide extinguishers have relatively short discharge ranges, 3′ to 8′. This presents a safety issue because the operator must be close to the fire. Depending on size, carbon dioxide extinguishers can discharge completely in 8 to 30 seconds.

Carbon dioxide fire extinguishers are not recommended for outdoor use or for locations with strong air currents. The carbon dioxide will be rapidly dissipated and will not be effective in smothering the flames.

Carbon dioxide extinguishers have a trigger mechanism that can be operated intermittently to preserve the remaining agent. The pressurized carbon dioxide will remain in the extinguisher, but the extinguisher will have to be recharged after use. The extinguisher is weighed to determine how much agent is left in the storage cylinder.

The smaller carbon dioxide extinguishers contain from 2 to 5 lb of agent. These units are designed to be operated with one hand. The horn is attached directly to the discharge valve on the top of the extinguisher by a hinged metal tube. In larger models, the horn is attached at the end of a short hose. These models require two-handed operation.

The horns of some older carbon dioxide fire extinguishers were constructed of metal, which conducts electricity. These extinguishers do not carry a Class C rating and must not be used around fires involving energized electrical equipment. Metal horns are no longer made for carbon dioxide fire extinguishers, but some of these units may still be in service.

Class B Foam Extinguishers

Class B foam extinguishers are very similar in appearance, and operation to water extinguishers. Instead of plain water, they discharge a solution of water and either AFFF or FFFP foam concentrate. The agent is discharged through an aspirating nozzle, which mixes air into the stream. They create a foam blanket that will float over the surface of a flammable liquid.

Class B foam agents are also very effective in fighting Class A fires. They are not suitable for Class C fires or for fires involving flammable liquids or gases under pressure. They are not intended for use on cooking oil fires, and only certain foam extinguishers can be used on fires involving polar solvents. Detailed information on the use of AFFF and FFFP is available in NFPA 11 *Standard for Low-, Medium-, and High-Expansion Foam*.

AFFF and FFFP hand-held stored-pressure extinguishers are available in two sizes: 1.6 gallons, rated 2-A:10-B, and 2.5 gallons, rated 3-A:20-B. A wheeled model with a 33 gallons capacity is rated 20-A:160-B.

Foam extinguishing agents are not effective at freezing temperatures. Consult the extinguisher manufacturer for information on using foam agents effectively at low temperatures.

Wet Chemical Extinguishers

Wet chemical extinguishers are used to protect Class K installations, which include cooking oils, deep fryers, and grills. Many commercial cooking installations use fixed, automatic fire extinguishing systems as their first line of defense. Portable Class K wet chemical extinguishers are currently available in two sizes, 1.5 gallons and 2.5 gallons. There are no numerical ratings for these extinguishers.

Halogenated-Agent Extinguishers

Halogenated-agent extinguishers include both halon agents and halocarbon agents. Because halon agents can destroy the earth's protective ozone layer, their use is strictly controlled.

Fire Fighter Safety Tips

Do not aim carbon dioxide extinguisher discharge at anyone or allow it to come in contact with exposed skin. Frostbite could result. Carbon dioxide discharged into a confined space will reduce the oxygen level in that space. Self-contained breathing apparatus must be used by anyone entering the confined area.

The halocarbons are not subject to the same environmental restrictions. Both types of agents are available in hand-held extinguishers rated for Class B and C fires. Larger capacity models are also rated for use on Class A fires.

The agent is discharged as a streaming liquid, which can be directed at the base of a fire. The discharges from these extinguishers have a horizontal stream range of 9′ to 15′.

The halogenated agents are nonconductive and leave no residue that can damage electrical equipment. These agents are relatively expensive but they perform better than carbon dioxide models in most applications, particularly in windy conditions.

Halon 1211 (bromochlorodifluoromethane) is available in hand-held stored-pressure extinguishers with capacities that range from 1 lb, rated 1-B:C, to 22 lb, rated 4-A:80-B:C. Wheeled Halon 1211 models are available with capacities up to 150 lb with a rating of 30-A:160-B:C. The wheeled fire extinguishers use a nitrogen booster charge from an auxiliary cylinder to expel the agent.

Dry Powder Extinguishing Agents

Dry powder extinguishing agents are intended for fighting Class D fires involving combustible metals. The extinguishing agents and the techniques required to extinguish Class D fires vary greatly, and depend on the specific fuel, the quantity involved, and the physical form of the fuel, such as grindings, shavings, or solid objects. The agent and the application method must be suited to the particular situation.

Each dry powder extinguishing agent is listed for use on specific combustible metal fires. This information and recommended application methods are printed on the container. Consult the manufacturer's recommendations for information about Class D agents and extinguishers, as well as NFPA's *Fire Protection Handbook*.

Dry Powder Fire Extinguishers

Dry powder fire extinguishers using sodium chloride-based agents are available with 30 lb capacity in either stored-pressure or cylinder/cartridge models. Wheeled models are available with 150 lb and 350 lb capacities.

Dry powder extinguishers have adjustable nozzles that allow the operator to vary the flow of the extinguishing agent. When the nozzle is fully opened, the hand-held models have a range of 6′ to 8′. Extension wand applicators are available to direct the discharge from a more distant position.

Fire Fighter Tips

Dry chemical agents and dry powder agents have very different meanings in relation to fire extinguishers. Dry powder fire extinguishers are designed for use in suppressing Class D fires. Dry chemical fire extinguishers are rated for Class B and C fires or for Class A, B, and C fires. These two terms must not be confused.

Fire Fighter Safety Tips

Before deciding to use a fire extinguisher, size up the fire to ensure that the extinguisher is adequately sized and has the proper extinguishing agent.

Bulk Dry Powder Agents

Bulk dry powder agents are available in 40 lb and 50 lb pails and 350 lb drums. The same sodium chloride-based agent that is used in portable fire extinguishers can be stored in bulk form and applied by hand.

Another dry powder extinguishing agent for Class D fires, graded granular graphite mixed with compounds containing phosphorus, cannot be expelled from a portable fire extinguisher. This agent must be applied manually from a pail or other container using a shovel or scoop.

Use of Fire Extinguishers

Fire extinguishers should be simple to operate. An individual with only basic training should be able to use most fire extinguishers safely and effectively. Every portable extinguisher should be labeled with printed operating instructions.

There are six basic steps in extinguishing a fire with a portable fire extinguisher. They are:

1. Locate the fire extinguisher.
2. Select the proper classification of extinguisher.
3. Transport the extinguisher to the location of the fire.
4. Activate the extinguisher to release the extinguishing agent.
5. Apply the extinguishing agent to the fire for maximum effect.
6. Ensure your personal safety by having an exit route.

Although these steps are not complicated, practice and training are essential for effective fire suppression. Tests have shown that the effective use of Class B portable fire extinguishers depends heavily on user training and expertise. A trained expert can extinguish a fire up to twice as large as a non-expert can, using the same extinguisher.

As a fire fighter, you should be able to operate any fire extinguisher that you might be required to use, whether it is carried on your fire apparatus, hanging on the wall of your firehouse, or placed in some other location.

Locating a Fire Extinguisher

Fire fighters should know what types of fire extinguishers are carried on department apparatus and where each type of extinguisher is located. You should also know where fire extinguishers are located in and around the fire station and other work places. You should have at least one fire extinguisher in your home and another in your personal vehicle, and you should know exactly where they are located. Knowing the exact locations of extinguishers can save valuable time in an emergency.

Selecting the Proper Fire Extinguisher

Selecting the proper extinguisher requires an understanding of the classification and rating system for fire extinguishers. Knowing the different types of agents, how they work, the ratings of the fire extinguishers carried on your fire apparatus, and which extinguisher is appropriate for a particular fire situation is also important.

Fire fighters should be able to assess a fire quickly, determine if the fire can be controlled by an extinguisher, and identify the appropriate extinguisher. Using an extinguisher with an insufficient rating may not completely extinguish the fire, which can place the operator in danger of being burned or otherwise injured. If the fire is too large for the extinguisher, you will have to consider other options such as obtaining additional extinguishers or making sure that a charged hose line is ready to provide back-up.

Understanding the fire extinguisher rating system and the different types of agents will enable a fire fighter who must use an unfamiliar extinguisher to determine if it is suitable for a particular fire situation. A quick look at the label should be all that is needed.

Fire fighters should also be able to determine the most appropriate type of fire extinguisher to place in a given area, based on the types of fires that could occur and the hazards that are present. In some cases, one type of extinguisher might be preferred over another. An extinguishing agent such as carbon dioxide or a halogenated agent is better for a fire involving electronic equipment because it leaves no residue. A dry chemical extinguisher is generally more appropriate than a carbon dioxide extinguisher to fight an outdoor fire, because wind will quickly dissipate carbon dioxide. A dry chemical extinguisher would also be the best choice for a fire involving a flammable liquid leaking under pressure from a pipe, but foam is better for a liquid spill on the ground.

▶ Table 7-2 provides information on different types of extinguishers.

Transporting a Fire Extinguisher

The best method of transporting a hand-held portable fire extinguisher depends on the size, weight, and design of the extinguisher. Hand-held portable fire extinguishers can weigh as little as 1 lb to as much as 50 lb. The ability to handle the heavier extinguishers depends on an individual operator's personal strength.

Extinguishers with a fixed nozzle should be carried in the favored or stronger hand. This enables the operator to depress the trigger and direct the discharge easily. Extinguishers that have a hose between the trigger and the nozzle should be carried in the weaker or less-favored hand so that the favored hand can grip and aim the nozzle.

Heavier extinguishers may have to be carried as close as possible to the fire and placed upright on the ground. The operator can depress the trigger with one hand, while holding the nozzle and directing the stream with the other hand.

Follow the steps in ► **Skill Drill 7-1** to transport a fire extinguisher:

1. Locate the extinguisher and remove it from the mounting bracket. **(Step 1)**
2. When lifting an extinguisher from a low bracket or off the floor, always lift with your legs and not your back. **(Step 2)**
3. If the extinguisher is very heavy, use both hands to carry it. Grasp the handle with your less-favored hand and support the bottom of the extinguisher in your favored or stronger hand. **(Step 3)**
4. Walk briskly toward the fire. Never run when carrying a fire extinguisher.
5. If the extinguisher has a fixed nozzle, hold the extinguisher at arm's length in your favored or stronger hand.
6. If the extinguisher has a hose with a nozzle, carry the extinguisher in your less-favored hand and grip the nozzle with your other hand. **(Step 4)**

Basic Steps of Fire Extinguisher Operation

Activating a fire extinguisher to apply the extinguishing agent is a single operation in four steps. The P-A-S-S acronym is a helpful way to remember these steps:

- Pull the safety pin.
- Aim the nozzle at the base of the flames.
- Squeeze the trigger to discharge the agent.
- Sweep the nozzle across the base of the flames.

Most fire extinguishers have very simple operation systems. Practice discharging different types of extinguishers in training situations to build confidence in your ability to use them properly and effectively.

Ensure Your Personal Safety

When using a fire extinguisher, always approach the fire with an exit behind you. If the fire suddenly expands or the extinguisher fails to control it, you must have a planned escape route. Never let the fire get between you and a safe exit. After suppressing a fire, do not turn your back on it. Always watch and be prepared for a rekindle until the fire has been fully overhauled.

When fire extinguishers are used by ordinary civilians on incipient fires, the operators are probably wearing their normal clothing. As a fire fighter, however, you should wear your personal protective clothing and use appropriate personal protective equipment (PPE). Take advantage of the protection they provide.

If you must enter an enclosed area where an extinguisher has been discharged, wear full PPE and use SCBA. The atmosphere within the enclosed area will probably contain a mixture of combustion products and extinguishing agents. The oxygen content within the space may be dangerously depleted.

Follow the steps in ► **Skill Drill 7-2** to operate a carbon dioxide extinguisher:

1. Size up the fire to determine what is burning, if there is energized electrical equipment involved or nearby, and if there are any other hazards present. Select the proper extinguisher.
2. Be sure the rating of the extinguisher matches the size of fire. If the fire is too large for a portable fire extinguisher, back away until other suppression methods are available. If possible, close off the area where the fire is located to limit the spread of smoke and fire.
3. Check the pressure gauge on the extinguisher to ensure that it is properly charged. Remember, carbon dioxide fire extinguishers do not have pressure gauges.
4. Pull the pin on the handle. **(Step 1)**
5. Remove the horn or nozzle from the secured position on the extinguisher and aim in the direction of your approach. **(Step 2)**
6. Give the trigger a quick squeeze to ensure that the extinguisher is operational and the agent discharges properly. **(Step 3)**
7. Approach the fire with an exit to your back. Never let the fire get between you and the exit. Always have a safe exit path in case you have to evacuate. **(Step 4)**
8. Aim the nozzle of the fire extinguisher at the base of the fire and squeeze the trigger. **(Step 5)**
9. Sweep the extinguishing agent from side to side, continuing to aim at the base of the flames. Continue to use the extinguisher until the fire is out or the extinguisher is empty. **(Step 6)**
10. If the extinguisher empties before the fire is completely suppressed, back away to a safe location and wait for assistance. If possible, close off the area where the fire is located to limit the spread of the fire until assistance arrives.
11. Back away from the fire. Never turn your back on the fire. Be prepared in case it re-ignites.
12. Overhaul the fire to ensure that it is completely extinguished. Have additional fire extinguishers or a hose line available in case the fire flares up again. **(Step 7)**

Follow the steps of ► **Skill Drill 7-3** to attack a Class A fire with a stored-pressure water-type fire extinguisher.

1. Begin to attack the fire from a safe distance to take advantage of the reach of the stream. **(Step 1)**

Table 7-2 Fire Extinguisher Types

Extinguishing Agent	Method of Operation	Capacity	Horizontal Range of Stream	Approximate Time of Discharge	Protection Required below 40°F (4°C)	UL or ULC Classifications[a]
Water	Stored-pressure	6L	30 to 40 ft	40 sec	Yes	1-A
	Stored-pressure or pump	2 1/2 gal	30 to 40 ft	1 min	Yes	2-A
	Pump	4 gal	30 to 40 ft	2 min	Yes	3-A
	Pump	5 gal	30 to 40 ft	2 to 3 min	Yes	4-A
Water (wetting agent)	Stored-pressure	1 1/2 gal	20 ft	30 sec	Yes	2-A
		25 gal (wheeled)	35 ft	1 1/2 min	Yes	10-A
		45 gal (wheeled)	35 ft	2 min	Yes	30-A
		60 gal (wheeled)	35 ft	2 1/2 min	Yes	40-A
Loaded stream	Stored-pressure	2 1/2 gal	30 to 40 ft	1 min	No	2 to 3-A:1-B
		33 gal (wheeled)	50 ft	3 min	No	
AFFF, FFFP	Stored-pressure	2 1/2 gal	20 to 25 ft	50 sec	Yes	3-A:20 to 40-B
	Stored-pressure	6L	20 to 25 ft	50 sec	Yes	2A:10B
	Nitrogen cylinder	33 gal	30 ft	1 min	Yes	20-A:160-B
Carbon dioxide[b]	Self-expelling	2 1/2 to 5 lb	3 to 8 ft	8 to 30 sec	No	1 to 5-B:C
	Self-expelling	10 to 15 lb	3 to 8 ft	8 to 30 sec	No	2 to 10-B:C
	Self-expelling	20 lb	3 to 8 ft	10 to 30 sec	No	10-B:C
	Self-expelling	50 to 100 lb (wheeled)	3 to 10 ft	10 to 30 sec	No	10 to 20-B:C
Regular dry chemical (sodium bicarbonate)	Stored-pressure	1 to 2 1/2 lb	5 to 8 ft	8 to 12 sec	No	2 to 10-B:C
	Cartridge or stored-pressure	2 3/4 to 5 lb	5 to 20 ft	8 to 25 sec	No	5 to 20-B:C
	Cartridge or stored-pressure	6 to 30 lb	5 to 20 ft	10 to 25 sec	No	10 to 160-B:C
	Stored-pressure	50 lb (wheeled)	20 ft	35 sec	No	160-B:C
	Nitrogen cylinder or stored-pressure	75 to 350 lb (wheeled)	15 to 45 ft	20 to 105 sec	No	40 to 320-B:C
Purple K dry chemical (potassium bicarbonate)	Cartridge or stored-pressure	2 to 5 lb	5 to 12 ft	8 to 10 sec	No	5 to 30-B:C
	Cartridge or stored-pressure	5 1/2 to 10 lb	5 to 20 ft	8 to 20 sec	No	10 to 80-B:C
	Cartridge or stored-pressure	16 to 30 lb	10 to 20 ft	8 to 25 sec	No	40 to 120-B:C
	Cartridge or stored-pressure	48 to 50 lb (wheeled)	20 ft	30 to 35 sec	No	120 to 160-B:C
	Nitrogen cylinder or stored-pressure	125 to 315 lb (wheeled)	15 to 45 ft	30 to 80 sec	No	80 to 640-B:C
Super K dry chemical (potassium chloride)	Cartridge or stored-pressure	2 to 5 lb	5 to 8 ft	8 to 10 sec	No	5 to 10-B:C
	Cartridge or stored-pressure	5 to 9 lb	8 to 12 ft	10 to 15 sec	No	20 to 40-B:C
	Cartridge or stored-pressure	9 1/2 to 20 lb	10 to 15 ft	15 to 20 sec	No	40 to 60-B:C
	Cartridge or stored-pressure	19 1/2 to 30 lb	5 to 20 ft	10 to 25 sec	No	60 to 80-B:C
	Cartridge or stored-pressure	125 to 200 lb (wheeled)	15 to 45 ft	30 to 40 sec	No	160-B:C

Table 7-2—continued

Extinguishing Agent	Method of Operation	Capacity	Horizontal Range of Stream	Approximate Time of Discharge	Protection Required below 40°F (4°C)	UL or ULC Classifications[a]
Multipurpose/ABC dry chemical (ammonium phosphate)	Stored-pressure	1 to 5 lb	5 to 12 ft	8 to 10 sec	No	1 to 3-A[c] and 2 to 10-B:C
	Stored-pressure or cartridge	2 1/2 to 9 lb	5 to 12 ft	8 to 15 sec	No	1 to 4-A and 10 to 40-B:C
	Stored-pressure or cartridge	9 to 17 lb	5 to 20 ft	10 to 25 sec	No	2 to 20-A and 10 to 80-B:C
	Stored-pressure or cartridge	17 to 30 lb	5 to 20 ft	10 to 25 sec	No	3 to 20-A and 30 to 120-B:C
	Stored-pressure or cartridge	45 to 50 lb (wheeled)	20 ft	25 to 35 sec	No	20 to 30-A and 80 to 160-B:C
	Nitrogen cylinder or stored-pressure	110 to 315 lb (wheeled)	15 to 45 ft	30 to 60 sec	No	20 to 40-A and 60 to 320-B:C
Dry chemical (foam compatible)	Cartridge or stored-pressure	4 3/4 to 9 lb	5 to 20 ft	8 to 10 sec	No	10 to 20-B:C
	Cartridge or stored-pressure	9 to 27 lb	5 to 20 ft	10 to 25 sec	No	20 to 30-B:C
	Cartridge or stored-pressure	18 to 30 lb	5 to 20 ft	10 to 25 sec	No	40 to 60-B:C
	Nitrogen cylinder or stored-pressure	150 to 350 lb (wheeled)	15 to 45 ft	20 to 150 sec	No	80 to 240-B:C
Dry chemical (potassium bicarbonate urea based)	Stored-pressure	5 to 11 lb	11 to 22 ft	18 sec	No	40 to 80-B:C
	Stored-pressure	9 to 23 lb	15 to 30 ft	17 to 33 sec	No	60 to 160-B:C
		175 lb (wheeled)	70 ft	62 sec	No	480-B:C
		3L	8 to 12 ft	30 sec	No	K
Wet chemical	Stored-pressure	6L (2 1/2 gal)	8 to 12 ft	35 to 45 sec	No	2A:1-B:C:K
			8 to 12 ft	75 to 85 sec	No	2-A:1-B:C:K
Halon 1211 (bromochlorodi-fluoromethane)	Stored-pressure	0.9 to 2 lb	6 to 10 ft	8 to 10 sec	No	1 to 2-B:C
		2 to 3 lb	6 to 10 ft	8 to 10 sec	No	5-B:C
		5 1/2 to 9 lb	9 to 15 ft	8 to 15 sec	No	1-A:10-B:C
		13 to 22 lb	14 to 16 ft	10 to 18 sec	No	2 to 4-A and 20 to 80-B:C
		50 lb	35 ft	30 sec	No	10-A:120-B:C
		150 lb (wheeled)	20 to 35 ft	30 to 44 sec	No	30-A:160 to 240-B:C
Halon 1211/1301 (bromochlorodi-fluoromethane bromotrifluoro-methane) mixtures	Stored-pressure or self-expelling	0.9 to 5 lb	3 to 12 ft	8 to 10 sec	No	1 to 10-B:C
	Stored-pressure	9 to 20 lb	10 to 18 ft	10 to 22 sec	No	1-A:10-B:C to 4-A:80-B:C
Halocarbon type	Stored-pressure	1.4 to 150 lb	6 to 35 ft	9 to 23 sec	No	1B:C to 10A:80-B:C

Note: Halon should be used only where its unique properties are deemed necessary.

[a]UL and ULC ratings checked as of July 24, 1987. Readers concerned with subsequent ratings should review the pertinent lists and supplements issued by these laboratories: Underwriters' Laboratories Inc., 333 Pfingsten Road, Northbrook, IL 60062, or Underwriters' Laboratories of Canada, 7 Crouse Road, Scarborough, Ontario, Canada M1R 3A9.

[b]Carbon dioxide extinguishers with metal horns do not carry a C classification.

[c]Some small extinguishers containing ammonium phosphate-based dry chemical do not carry an A classification.

Source: National Fire Protection Association. *Fire Protection Handbook* Vol. 2, 19th ed. Quincy, MA, 2003, pp 11-121–11-122.

2. Aim the stream directly at the base of the flames.
3. Sweep the nozzle back and forth, moving closer as the fire goes out. **(Step 2)**
4. After the flames are out, position your finger in front of the nozzle to create a spray and soak the fuel. **(Step 3)**
5. If the fire is deep-seated (burning below the surface, as in a tightly packed trash barrel or a brush pile), break the fuel apart with a stick or long-handled tool. Apply the extinguishing agent to any smoldering, smoking, or glowing surfaces. **(Step 4)**
6. Apply additional water spray to prevent the fire from re-igniting. **(Step 5)**

Follow the steps of **Skill Drill 7-4** to attack a Class A fire with a multipurpose dry chemical fire extinguisher.

1. Begin to attack the fire from a safe distance taking advantage of the reach of the dry chemical discharge stream.
2. Aim the stream directly at the base of the flames.
3. Sweep the nozzle back and forth, moving closer as the fire is extinguished.
4. After the flames are out, apply additional agent in short bursts to ensure that all hot surfaces are coated with dry chemical.
5. If the fire is deep-seated (burning below the surface, as in a tightly packed trash barrel or a brush pile), break the fuel apart with a stick or long-handled tool. Apply additional extinguishing agent to any smoldering, smoking, or glowing surfaces.
6. Watch for indications of re-ignition and apply additional agent in short bursts as required. If necessary, use a fine water spray to soak the fuel.

Follow the steps in **► Skill Drill 7-5** to attack a Class B flammable liquid spill or pool fire with a dry chemical extinguisher.

1. Check the pressure gauge to ensure that it is properly charged **(Step 1)** and pull the pin on the handle. **(Step 2)**
2. Begin fighting the fire from a safe distance, as specified on the extinguisher's label. **(Step 3)**
3. Do not aim the initial discharge directly into the liquid from close range. The high-velocity stream of extinguishing agent could splash and spread the burning fuel. **(Step 4)**
4. Discharge the stream at the base of the flame, starting at the near edge of the fire and working toward the back. **(Step 5)**
5. Sweep the nozzle back and forth across the surface of the flammable liquid. **(Step 6)**
6. After the flames are extinguished, watch for indications of re-ignition and be prepared to apply additional agent. Look for hot or smoldering objects that could provide a source of re-ignition. **(Step 7)**

Fire Fighter Safety Tips

Always test a fire extinguisher before using it. Activate the trigger long enough to ensure that the agent discharges properly.

Follow the steps in **► Skill Drill 7-6** to attack a Class B flammable liquid spill or pool fire with a stored-pressure fire extinguisher containing AFFF or FFFP extinguishing agent.

1. Begin fighting the fire from a safe distance, as specified on the extinguisher's label. **(Step 1)**
2. Aim the discharge stream high over the top of the fire. Aiming directly into the liquid could cause the burning fuel to splash and spread the fire. **(Step 2)**
3. Discharge the stream in an arc over the top of the fire, so that the foam drops gently onto the surface of the burning liquid. Lob the foam to create a blanket that floats on the surface. **(Step 3)**
4. Slowly sweep the stream back and forth above the flammable liquid to widen the foam blanket. Use the stream to carefully push the foam blanket toward the back of the liquid surface.
5. If the fire is a spill on the ground, aim the stream at the ground in front of the fire and let the foam bounce onto the front part of the fire. Let the foam blanket flow across the surface of the burning liquid.
6. After the flames are extinguished, continue to apply agent until the surface of the flammable liquid is fully covered by a foam blanket. Look for hot or smoldering objects that could re-ignite the liquid and apply the agent directly on them. **(Step 4)**
7. Apply additional agent as needed to maintain the foam blanket and control any re-ignition. **(Step 5)**

Follow the steps in **► Skill Drill 7-7** to apply a wet (Class K) extinguishing agent on a fire in a deep fryer.

1. Begin to apply the agent from a safe distance. **(Step 1)**
2. Do not direct the agent stream directly into the burning liquid. Avoid splashing the burning liquid.
3. Lob the extinguishing agent, expelled as a fine spray, lightly onto the burning surface. It will form a thick foam blanket to smother the fire. **(Step 2)**
4. Continue discharging the extinguisher until the foam blanket has extinguished all flames and secured the entire surface of the oil. **(Step 3)**
5. Do not disturb the foam blanket after the flames die down. The surrounding metal and other materials may still be hot enough to re-ignite the oil if the foam blanket is broken.
6. Be prepared for a possible flare up until the cooking oil and the surrounding area have adequately cooled. If re-ignition occurs, repeat these steps. **(Step 4)**

7-1 Skill Drill

Transporting a Fire Extinguisher

Locate the extinguisher and remove it from its mounting bracket.

When lifting an extinguisher, always lift with your legs and not your back.

Grasp the handle with the less-favored hand and support the bottom of the extinguisher in the favored or stronger hand.

If the extinguisher has a fixed nozzle, hold the extinguisher at arm's length in your favored or stronger hand. If the extinguisher has a hose with a nozzle, carry the extinguisher in your less-favored hand and use the other hand to grip the nozzle.

7-2 Skill Drill

Operating a Carbon Dioxide Extinguisher

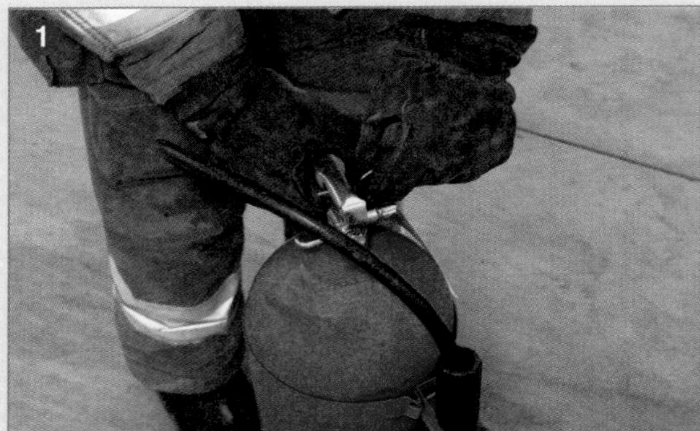

1 Pull the pin on the handle.

2 Remove the horn or nozzle from the secured position on the extinguisher and aim in the direction of your approach.

3 Give the trigger a quick squeeze to ensure that the extinguisher is operational and the agent discharges properly.

4 Approach the fire with an exit to your back. Never let the fire get between you and the exit.

5 Aim the nozzle of the fire extinguisher at the base of the fire and squeeze the trigger.

6 Sweep the extinguishing agent from side to side, continuing to aim at the base of the flames. Continue to use the extinguisher until the fire is out or the extinguisher is empty.

7 Back away from the fire. Overhaul the fire to ensure that it is completely extinguished.

Skill Drill

Attacking a Class A Fire with a Stored-Pressure Water-Type Fire Extinguisher

Begin to attack the fire from a safe distance.

Aim the stream directly at the base of the flames. Sweep the nozzle back and forth, moving closer as the fire goes out.

After the flames are out, position your finger in front of the nozzle to create a spray and soak the fuel.

Break apart the fuel with a stick and apply the extinguishing agent to any smoldering, smoking, or glowing surfaces.

Apply additional water spray to prevent the fire from re-igniting.

Skill Drill

Attacking a Class B Flammable Liquid Fire with a Dry Chemical Fire Extinguisher

1 Check the pressure gauge to ensure that the extinguisher is properly charged.

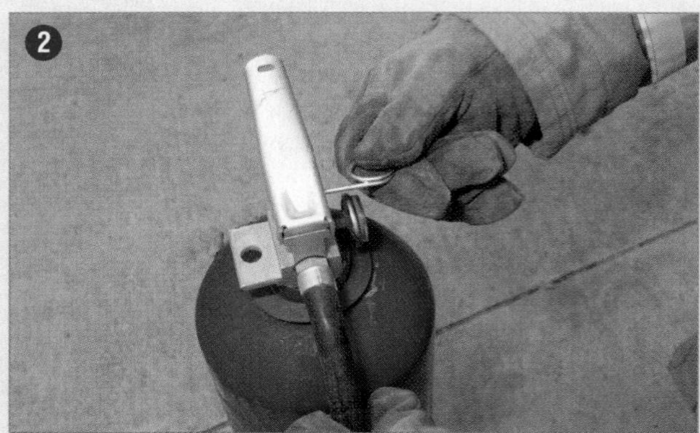

2 Pull the pin on the handle.

3 Begin fighting the fire from a safe distance.

4 Do not aim the initial discharge into the liquid at close range.

5 Discharge the stream at the base of the flame, starting at the near edge of the fire and working toward the back.

6 Sweep the nozzle back and forth across the surface of the flammable liquid.

7 Look for hot or smoldering objects that could re-ignite the liquid.

Attacking a Class B Flammable Liquid Fire with a Stored-Pressure Foam Fire Extinguisher (AFFF or FFFP)

Begin fighting the fire from a safe distance.

Do not discharge stream directly into the liquid.

Lob the foam in an arc over the fire to create a blanket that floats on the surface.

Slowly sweep the stream back and forth above the flaming surface to widen the foam blanket. Let the foam blanket flow across the surface of the burning liquid. After the flames are extinguished, continue to apply agent until the liquid surface is fully covered by the foam blanket.

Apply additional agent as needed to maintain the foam blanket and control any re-ignition.

7-7 Skill Drill

Use of Wet Chemical Fire Extinguishers

Begin to apply the agent from a safe distance.

Do not direct the agent stream directly into the burning liquid. Lob the extinguishing agent, expelled as a fine spray, lightly onto the burning surface to create a foam blanket.

Continue to discharge the extinguisher until the foam blanket has extinguished all flames.

Do not disturb the foam blanket even after all flames have died down. If re-ignition occurs, repeat these steps.

Follow the steps in **Skill Drill 7-8** to apply a halogenated extinguishing agent to a fire in an electrical equipment room.

1. If possible, turn off or disconnect the electrical power from a remote location before attacking the fire.
2. Stand back at least 8′ (2.4 m) when you begin to discharge the agent.
3. Direct the stream at the base of the flames, sweeping the agent slowly from side to side.
4. Continue to apply the extinguishing agent for a short time after the flames have gone out. This allows the fuel to cool, preventing re-ignition.
5. Watch the fire area for re-ignition. Repeat the agent application, if necessary

Follow the steps in **▶ Skill Drill 7-9** to apply a dry powder extinguishing agent to an incipient-stage fire involving combustible metal filings or shavings (Class D fire).

1. If the fire is very hot, begin to discharge the agent with the nozzle fully opened at the maximum range of 6′ to 7′ (1.7 to 2.4 m) away from the fire. Direct the agent so that it falls onto the top of the burning material. **(Step 1)**
2. As the fire comes under control, close the nozzle valve to produce a soft, heavy flow. Move in closer and cover the fire area completely. **(Step 2)**
3. If you are using a scoop or a shovel instead of an extinguisher, apply a thick, even coat of extinguishing agent over the entire fire to smoother it.

7-9 Skill Drill

Use of Dry Powder Fire Extinguishing Agents

Open the nozzle completely and direct the agent so that it falls onto the top of the burning material.

Close the nozzle valve to produce a soft, heavy flow and move closer to cover the fire area completely.

Do not disturb the blanket of extinguishing agent until the fire has cooled completely.

Voices of Experience

"The staff member and the twenty-dollar portable fire extinguisher prevented the incipient fire from spreading to the grass and to the Class 1 explosive magazine."

While I was working for a United States Army fire department at an ammunitions plant, the decision was made to supply the complex with stored pressure extinguishers. These fire extinguishers were also placed on every vehicle on the site. Because of this massive acquisition, we held portable fire extinguisher training for the entire staff to teach them how to properly use the new equipment.

This training paid off one hot summer day. A staff member was riding a tractor and mowing the lawn near an ammunition magazine. The tractor experienced a mechanical failure and caught on fire. Thanks to his training, the staff member jumped into action. He retrieved the 2-A:10-B:C extinguisher, pulled the pin, and quickly suppressed the fire. The staff member saved the tractor from serious damage.

Without the portable fire extinguisher, the fire would have surely spread because the staff member was alone in a field without any means of communication. Also, the response time for the fire department would have been an additional 10 minutes. The staff member and the twenty-dollar portable fire extinguisher prevented the incipient fire from spreading to the grass and to the Class 1 explosive magazine. When used for the appropriate situation and in the hands of a properly trained operator, fire extinguishers can prevent serious incidents.

Gary Wilson
University of Missouri, Fire and Rescue Training Institute
Columbia, Missouri

4. If the burning metal is on a combustible surface, first cover the fire with the dry powder extinguishing agent. Then spread a 1" to 2" (2.5 to 5.0 cm) layer of dry powder extinguishing agent nearby. Carefully shovel the burning metal onto this layer of powder and top with more extinguishing agent as needed.
5. If hot spots develop, apply more extinguishing agent to cover them.
6. Allow the remains of the fire to cool undisturbed.
7. Do not disturb the blanket of extinguishing agent until the fire has cooled completely. **(Step 3)**
   NOTE: Do not use water or any other agent to cool the fuel or the area around it! Combustible metals react violently with water and many other agents.

The Care of Fire Extinguishers: Inspection, Maintenance, Recharging, and Hydrostatic Testing

Fire extinguishers must be regularly inspected and properly maintained so they are available for use in an emergency. Records must be kept to ensure that the required inspections and maintenance have been performed on schedule. The individuals assigned to perform these functions must be properly trained and must always follow the manufacturer's recommendations for inspecting, maintaining, recharging, and testing the equipment.

Inspection

According to NFPA 10, an inspection is a "quick check" to verify that a fire extinguisher is available and ready for immediate use. Fire extinguishers on fire apparatus should be inspected as part of the regular equipment checks mandated by your department. The fire fighter charged with inspecting the extinguishers should:

- Ensure that tamper seals are intact.
- Determine fullness by weighing or "hefting" the extinguisher.
- Examine all parts for signs of physical damage, corrosion, or leakage.
- Check the pressure gauge to be sure it is in the operable range.
- Ensure that the extinguisher is properly identified by type and rating.
- Shake dry chemical extinguishers to mix or redistribute the agent.
- Check the nozzle for damage or obstruction by foreign objects.

If an inspection reveals any problems, the extinguisher should be removed from service until the required maintenance procedures are performed. Spare extinguishers should be used until the problem is corrected.

The pressure gauge on a stored-pressure extinguisher indicates whether the pressure is sufficient to expel the entire agent. Some stored-pressure extinguishers use compressed air while others use compressed nitrogen. The weight of the extinguisher and the presence of an intact tamper seal should indicate that it is full of extinguishing agent.

Cartridge-type extinguishers contain a predetermined quantity of a pressurizing gas that expels the agent. The gas is only released when the cartridge is punctured. A properly charged extinguisher will be full of extinguishing agent and will have a cartridge that has not been punctured.

Self-expelling agents, such as carbon dioxide and some halogenated agents, do not require a separate gas cartridge. The only way to determine if a carbon dioxide extinguisher is properly charged is to weigh it. The proper weight should be indicated on the outside of the extinguisher.

Maintenance

The maintenance requirements and intervals for various types of fire extinguishers are outlined in NFPA 10. Maintenance includes an internal inspection as well as any repairs that may be required. Maintenance procedures must be performed periodically, depending on the type of extinguisher. An inspection may also reveal the need to perform maintenance procedures.

Maintenance procedures must always be performed by a qualified person. Some procedures can only be performed at a properly licensed facility. The specific qualifications and training requirements are determined by the manufacturer and the jurisdictional authority. Untrained personnel should never be allowed to perform fire extinguisher maintenance.

Common indications that an extinguisher needs maintenance include:

- The pressure gauge reading is outside the normal range.
- The inspection tag is out-of-date.
- The tamper seal is broken, especially in extinguishers with no pressure gauge.
- There is any indication that the extinguisher is not full of extinguishing agent.
- The hose and/or nozzle assembly is obstructed.
- There are signs of physical damage, corrosion, or rust.
- Signs of leakage around the discharge valve or nozzle assembly can be seen.

Recharging

A fire extinguisher must be recharged after each and every use, even if it was not completely discharged. The only exceptions are non-rechargeable extinguishers, which should be replaced after any use. All performance standards assume that an extinguisher will be fully charged when it has to be used. Immediately after any use, an extinguisher should be taken out of service until it has been properly recharged. This also applies to any extinguisher that leaks or has a pressure gauge reading above or below the proper operating range. Most rechargeable extinguishers must be recharged by qualified and licensed personnel.

When an extinguisher is recharged, the extinguishing agent is refilled and the system that is provided to expel the agent is fully charged. Both the quantity of the agent and the pressurization must be verified. A tamper seal is installed after recharging. The tamper seal provides assurance that the extinguisher has not been fully or even partially discharged since the last time it was recharged.

Typical 2.5 gallons stored-pressure water extinguishers can be recharged by fire fighters using water and a source of compressed air. Before the extinguisher is refilled, all remaining stored pressure must be discharged so that the valve assembly can be safely removed. Water is added up to the water-level indicator and the valve assembly is replaced. Compressed air is then introduced to raise the pressure to the level indicated on the gauge.

Hydrostatic Testing

Most fire extinguishers are pressure vessels, designed to hold a steady internal pressure. The ability of an extinguisher to withstand this internal pressure is measured by periodic hydrostatic testing. The hydrostatic testing requirements for fire extinguishers are established by the U.S. Department of Transportation and can be found in NFPA 10 (► Table 7-3). The test is conducted in a special test facility and involves filling the extinguisher with water and applying above-normal pressure.

Each extinguisher has an assigned maximum interval between hydrostatic tests, usually 5 or 12 years, depending on the construction material and vessel type (► Table 7-4). The date of the most recent hydrostatic test must be indicated on the outside of the extinguisher. An extinguisher may not be refilled if the date of the most recent hydrostatic test is not within the prescribed limit. Any extinguisher that is out of date should be removed from service and sent to the appropriate maintenance facility for hydrostatic testing.

Table 7-3 Hydrostatic Test Pressure Requirements

Carbon dioxide extinguishers Carbon dioxide and nitrogen cylinders (used with wheeled extinguishers)	$^5/_3$ service pressure stamped on cylinder
Carbon dioxide extinguishers with cylinder specification ICC3	3000 psi (20 685 kPa)
All stored-pressure and bromo-chlorodifluoromethane (1211)	Factory test pressure not to exceed 3 times the service pressure
Carbon dioxide hose assemblies	1,250 psi (8619 kPa)
Dry chemical and dry powder hose assemblies	300 psi (2068 kPa) or at service pressure, whichever is higher

Note: The factory test pressure is the pressure at which the shell was tested at time of manufacture. This pressure is shown on the nameplate.

The service pressure is the normal operating pressure as indicated on the guage and nameplate.

Table 7-4 Hydrostatic Test Interval for Extinguishers

Extinguisher Type	Test Interval (Years)
Stored-pressure water-loaded stream and/or antifreeze	5
Wetting agent	5
AFFF (aqueous film-forming foam)	5
FFFP (film-forming fluoroprotein foam)	5
Dry chemical with stainless-steel shells	5
Carbon dioxide	5
Wet chemical	5
Dry chemical, stored-pressure, with mild steel shells, brazed brass shells, or aluminum shells	12
Dry chemical, cartridge- or cylinder-operated, with mild steel shells	12
Halogenated agents	12
Dry powder, stored-pressure, cartridge- or cylinder-operated, with mild steel shells	12

Wrap-Up

Ready for Review

This chapter covered the two main uses of fire extinguishers: to control incipient-stage fires and to apply special extinguishing agents. It covered the rating and classification system for fire extinguishers, the basic steps for operating different types of fire extinguishers, and the requirements for inspecting, testing, and maintaining fire extinguishers. Information on water, dry chemical, carbon dioxide, halogenated agent, dry powder, foam, and wet chemical fire extinguishers was provided.

The chapter described how to select the best fire extinguisher based on the type of burning material, the size of the fire, the location of the fire, the effectiveness of the extinguisher, and the suitability of different extinguishing agents for different situations. Safety concerns relating to the use of fire extinguishers were also presented. The importance of regular inspection, maintenance, recharging, and hydrostatic testing of extinguishers was discussed.

Chief Concepts

- Portable fire extinguishers are effective for fighting fires only if the extinguisher is suitable for the class of fire.
- Fires are classified by their fuels (the material that is burning).
- Extinguishers are rated by independent testing laboratories according to the class or classes of fire the units can extinguish.
- Portable fire extinguishers that are rated for use on Class A and Class B fires are also rated numerically according to the size of the fire they can extinguish.
- The environment where a fire is likely to take place is an important factor in choosing the proper extinguisher.
- Environments are classified in three major hazard groups: light, ordinary, and extra.
- The basic steps of fire extinguisher operation are: locate, select, transport, activate, and apply.
- The common types of fire extinguishers are: water, dry chemical, carbon dioxide, foam, halogenated agents, dry powder, and wet chemical.
- Selecting the best fire extinguisher depends on the type of material burning, the size of the fire, the location of the fire, the effectiveness of the extinguisher, the ambient temperature, and wind conditions.
- Extinguishers differ by how their extinguishing agents are stored and expelled.
- Fire extinguishers must be regularly inspected, maintained, recharged, and hydrostatically tested.

Wrap-Up

Hot Terms

Ammonium phosphate An extinguishing agent used in dry chemical fire extinguishers that can be used on Class A, B, and C fires.

Aqueous film-forming foam (AFFF) A water-based extinguishing agent used on Class B fires that forms a foam layer over the liquid and stops the production of flammable vapors.

Carbon dioxide (CO_2) extinguisher A fire extinguisher that uses carbon dioxide gas as the extinguishing agent.

Cartridge/cylinder fire extinguisher A fire extinguisher that has the expellant gas in a separate container from the extinguishing agent storage container. The storage container is pressurized by a mechanical action that releases the expellant gas.

Class A fires Fires involving ordinary combustible materials such as wood, cloth, paper, rubber, and many plastics.

Class B fires Fires involving flammable and combustible liquids, oils, greases, tars, oil-based paints, lacquers, and flammable gases.

Class C fires Fires that involve energized electrical equipment where the electrical conductivity of the extinguishing media is of importance.

Class D fires Fires involving combustible metals such as magnesium, titanium, zirconium, sodium, and potassium.

Class K fires Fires involving combustible cooking media such as vegetable oils, animal oils, and fats.

Clean agent Volatile or gaseous fire extinguishing agent that does not leave a residue when it evaporates. Also known as halogenated agents.

Cylinder (fire extinguisher) The body of the fire extinguisher where the extinguishing agent is stored.

Dry chemical fire extinguisher An extinguisher that uses a mixture of finely divided solid particles to extinguish fires. The agent is usually sodium bicarbonate-, potassium bicarbonate-, or ammonium phosphate-based, with additives to provide resistance to packing and moisture absorption and to promote proper flow characteristics.

Dry powder extinguishing agent Extinguishing agent used in putting out Class D fires. The common dry powder extinguishing agents include sodium chloride and graphite-based powders.

Dry powder fire extinguisher A fire extinguisher that uses an extinguishing agent in powder or granular form, designed to extinguish Class D combustible metal fires by crusting, smothering, or heat-transferring means.

Extinguishing agent Material used to stop the combustion process; extinguishing agents may include liquids, gases, dry chemical compounds, and dry powder compounds.

Extra hazard locations Occupancies where the total amounts of Class A combustibles and Class B flammables are greater than expected in occupancies classed as ordinary (moderate) hazard.

Film-forming fluoroprotein (FFFP) A water-based extinguishing agent used on Class B fires that forms a foam layer over the liquid and stops the production of flammable vapors.

Fire load The weight of combustibles in a fire area or on a floor in buildings and structures, including either contents or building parts, or both.

Halogenated-agent extinguisher An extinguisher that uses halogenated extinguishing agents.

Halogenated extinguishing agent A liquefied gas extinguishing agent that puts out fires by chemically interrupting the combustion reaction between the fuel and oxygen.

Halon 1211 Bromochlorodifluoromethane ($CBrClF_2$), a halogenated agent that is effective on Class A, B, and C fires.

Handle The grip used for holding and carrying a portable fire extinguisher.

Horn The tapered discharge nozzle of a carbon dioxide-type fire extinguisher.

Hydrostatic testing Periodic testing of an extinguisher to verify it has sufficient strength to withstand internal pressures.

Ignition point The minimum temperature at which a substance will burn.

Incipient The initial stage of a fire.

Light hazard locations Occupancies where the total amount of combustible materials is less than expected in an ordinary hazard location.

Loaded-stream extinguisher A water-based fire extinguisher that uses an alkali metal salt as a freezing point depressant.

Lob A method of discharging extinguishing agent in an arc to avoid splashing or spreading the burning fuel.

Locking mechanism A device that locks an extinguisher's trigger to prevent accidental discharge.

Multipurpose dry chemical extinguishers Extinguishers rated to fight Class A, B, and C fires.

Nozzle (fire extinguisher) The discharge orifice of a portable fire extinguisher.

Ordinary hazard location Occupancies that contain more Class A and Class B materials than are found in light hazard locations.

P-A-S-S Acronym used for operating a portable fire extinguisher: Pull pin, Aim nozzle, Squeeze trigger, Sweep across burning fuel.

Polar solvent One of the water-soluble flammable liquids such as alcohol, acetone, ester, and ketone.

Pressure indicator A gauge on a pressurized portable fire extinguisher that indicates the internal pressure of the expellant.

Pump tank extinguishers Nonpressurized, manually operated water extinguishers, usually have a nozzle at the end of a short hose.

Pump tank water-type fire extinguisher A nonpressurized portable water fire extinguisher. Discharge pressure is provided by a hand-operated double-acting piston pump.

Rapid oxidation Chemical process that occurs when a fuel is combined with oxygen, resulting in the formation of ash or other waste products and the release of energy as heat and light.

Saponification The process of converting the fatty acids in cooking oils or fats to soap or foam.

Self-expelling A fire extinguisher in which the agents have sufficient vapor pressure at normal operating temperatures to expel themselves.

Stored-pressure extinguisher A fire extinguisher in which both the extinguishing agent and the expellant gas are kept in a single container; generally equipped with a pressure indicator or gauge.

Stored-pressure water-type fire extinguisher A fire extinguisher in which water or a water-based extinguishing agent is stored under pressure.

Tamper seal A retaining device that breaks when the locking mechanism is released.

Trigger The button or lever used to discharge the agent from a portable fire extinguisher.

Underwriters Laboratories Inc. (UL) The U.S. organization that tests and certifies that fire extinguishers (among many other products) meet established standards. The Canadian equivalent is Underwriters Laboratories of Canada (ULC).

Wet chemical extinguisher A fire extinguisher for use on Class K fires that contains wet chemical extinguishing agents.

Wet chemical extinguishing agent An extinguishing agent for Class K fires; commonly uses solutions of water and potassium acetate, potassium carbonate, potassium citrate, or any combination thereof.

Wetting-agent water-type extinguisher An extinguisher that expels water combined with a chemical or chemicals to reduce its surface tension.

Wheeled fire extinguisher A portable fire extinguisher equipped with a carriage and wheels intended to be transported to the fire by one person.

Fire Fighter in Action

Your inventory of all of the fire extinguishers carried on your station's apparatus includes:

Vehicle	Agent	Quantity	Rating
Engine 34	Water	2.5 gallons	2-A
	Dry Chemical	30 lb	80-B:C
Ladder 34	Water	2.5 gallons	2-A
	Dry Chemical	30 lb	80-B:C
Rescue Squad 34	AFFF	2.5 gallons	3-A:40-B
	Dry Chemical	30 lb	80-B:C
Hazardous Material 34	CO_2	(2) X 20 lb	10-B:C
	AFFF	(2) X 2.5 gallons	3-A:40-B
	Halon 1211	20 lb	4-A:80-B:C
	Dry Powder	(2) X 30 lb	D
Ambulance 34	Dry Chemical	10 lb	2-A:10-B:C
Chief 34	Dry Chemical	10 lb	2-A:10-B:C
Utility 34	Dry Chemical	10 lb	2-A:10-B:C

When you present your list to Captain Jones, he tests your understanding of fire extinguishers by asking you the following questions.

1. Engine 34 responds to an automatic fire alarm in a high-rise apartment building. The alarm panel indicates that the activated smoke detector is in the elevator machinery room. Which fire extinguisher would you take when you check the elevator machinery room?
   - **A.** Water extinguisher
   - **B.** Dry chemical extinguisher
   - **C.** Both extinguishers
   - **D.** Neither one
2. On the way back from another alarm, Ladder 34 is flagged down by a motorist whose car is on fire. The driver says that passengers were smoking cigarettes in the back seat area. There is light-colored smoke inside the passenger compartment. How would you attack this fire?
   - **A.** Water extinguisher
   - **B.** Dry chemical extinguisher
   - **C.** Both extinguishers
   - **D.** Neither one—this situation requires an engine company
3. Engine 34, Ladder 34, Rescue Squad 34, and Chief 34 are all dispatched to a fire in a restaurant. Moderate smoke is coming from the kitchen, and flames from a deep fryer reach the stainless steel vent hood at the ceiling level. The restaurant owner explains that the automatic fire extinguishing system is being serviced and is not activated. How would you plan to attack this fire?
   - **A.** Water extinguisher
   - **B.** AFFF extinguisher
   - **C.** Dry chemical
   - **D.** Hose line extinguisher
   - **E.** None of the above

4. Rescue Squad 34 is on the scene of an automobile accident with a driver trapped in a vehicle. During the scene size-up, a pool of gasoline is noted under the rear part of the car. The gasoline is dripping from a small split in the fuel tank. How would you use the fire extinguishers on Rescue Squad 34 to deal with this situation?

A. Stand by with the dry chemical extinguisher in case the gasoline ignites.

B. Stand by with the AFFF extinguisher in case the gasoline ignites.

C. Use the AFFF extinguisher to cover the pool of gasoline with a blanket of foam and stand by with the dry chemical extinguisher in case a fire ignites.

D. Discharge the dry chemical extinguisher to cover the pool of gasoline and stand by with the AFFF in case a fire ignites.

5. Engine 34 responds to an outside trash fire at the rear of an industrial building. As you approach, you observe a very bright fire and large quantities of billowing gray-white smoke. An employee tells you that a barrel of metal shavings containing magnesium was placed there yesterday to be picked up by a recycling company. How would you attack this fire?

A. Lob water into the top of the barrel from a safe distance.

B. Overturn the barrel and spread the contents out on the ground, then apply dry chemical.

C. Shovel dirt or sand into the barrel to bury the burning material.

D. Call for Hazardous Materials 34 to respond with the Class D extinguishers. Stand by until they arrive.

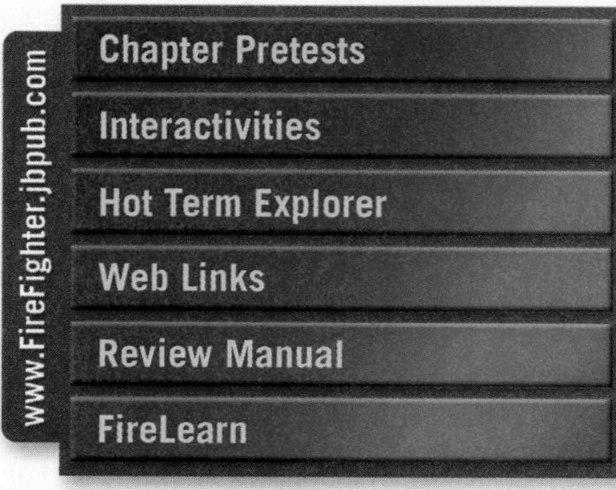

Fire Fighter Tools and Equipment

Chapter 8

NFPA 1001 Standard

Fire Fighter I

5.3.4 Force entry into a structure, given personal protective equipment, tools, and an assignment, so that the tools are used as designed, the barrier is removed, and the opening is in a safe condition and ready for entry.

5.3.13 (A) *Requisite Knowledge.* Types of fire attack lines and water application devices most effective for overhaul, water application methods for extinguishments that limit water damage, types of tools and methods used to expose hidden fire, dangers associated with overhaul, obvious signs of area of origin or signs of arson, and reasons for protection of fire scene.

5.5.3 Clean and check ladders, ventilation equipment, self-contained breathing apparatus (SCBA), ropes, salvage equipment, and hand tools, given cleaning tools, cleaning supplies, and an assignment, so that equipment is clean and maintained according to manufacturer's or departmental guidelines, maintenance is recorded, and equipment is placed in a ready state or reported otherwise.

5.5.3 (A) *Requisite Knowledge.* Types of cleaning methods for various tools and equipment, correct use of cleaning solvents, and manufacturers' guidelines or departmental guidelines for cleaning equipment and tools.

5.5.3 (B) *Requisite Skills.* The ability to select correct tools for various parts and pieces of equipment, follow guidelines, and complete recording and reporting procedures.

Fire Fighter II

6.4.2 Assist rescue operation teams, given standard operating procedures, necessary rescue equipment, and an assignment, so that procedures are followed, rescue items are recognized and retrieved in the time as prescribed by the AHJ, and the assignment is completed

6.4.2 (A) *Requisite Knowledge.* The fire fighter's role at a special rescue operation, the hazards associated with special rescue operations, types and uses for rescue tools, and rescue practices and goals.

6.4.2 (B) *Requisite Skills.* The ability to identify and retrieve various types of rescue tools, establish public barriers, and assist rescue teams as a member of the team when assigned.

6.5.2 Maintain power plants, power tools, and lighting equipment, given tools and manufacturers' instructions, so that equipment is clean and maintained according to manufacturer and departmental guidelines, maintenance is recorded, and equipment is placed in a ready state or reported otherwise.

6.5.2 (A) *Requisite Knowledge.* Types of cleaning methods, correct use of cleaning solvents, manufacturer and departmental guidelines for maintaining equipment and its documentation, and problem-reporting practices.

6.5.2 (B) *Requisite Skills.* The ability to select correct tools; follow guidelines, complete recording and reporting procedures, and operate power plants, power tools, and lighting equipment.

Knowledge Objectives

After studying this chapter, you will be able to:

- Describe the general purposes of tools and equipment.
- Describe the safety considerations for the use of tools and equipment.
- Describe why it is important to use tools and equipment effectively.
- Describe why it is important for you to know where tools are stored.
- List and describe tools and equipment that are used for rotating.
- List and describe tools and equipment that are used for pushing or pulling.
- List and describe tools and equipment that are used for prying or spreading.
- List and describe tools and equipment that are used for striking.
- List and describe tools and equipment that are used for cutting.
- Describe the tools used in response and scene size-up activities.
- Describe the tools used in a forcible entry.
- Describe the tools used during an interior attack.
- Describe the tools used in search-and-rescue operations.
- Describe ventilation tools.
- Describe the hand tools needed during an overhaul assignment.
- Describe the importance of properly maintaining tools and equipment.
- Describe how to clean and inspect hand tools.
- Describe how to maintain power plants and power tools.

Skills Objectives

There are no skills objectives for this chapter.

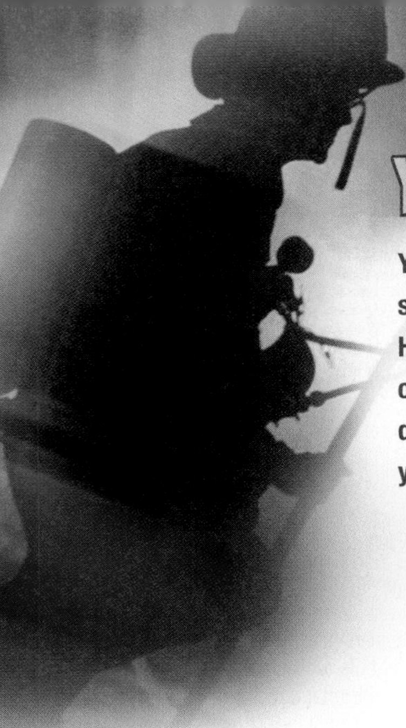

You Are the Fire Fighter

You are on a call, and when the engine comes to a stop, you can see fire coming out of a second-story window of a brick house. As you unbuckle your seatbelt, your captain tells you to bring a Halligan tool and a flat-head axe and follow him. Meanwhile, the other fire fighters start pulling off a preconnected hose line to advance to the front door. You and the captain try to open the door, but it is locked. Everyone dons an SCBA face mask, the hose line is charged with water, and you position the Halligan tool to force open the door.

1. *Where were these tools stored?*
2. *How will you use these tools?*
3. *Without these tools, what would you have done?*

Introduction

Fire fighters use tools and equipment to perform a wide range of activities. A fire fighter must know how to use tools effectively, efficiently, and safely, even when it is dark or visibility is limited. This chapter provides an overview of the general functions of the most commonly used tools and equipment and discusses how they are used during fire suppression and rescue operations. The same tools may be used in different ways during each phase of fire suppression or rescue operations.

General Considerations

Tools and equipment are used in almost all fire suppression and rescue operations. As you progress through your training, you will learn how to use and operate the different types of tools and equipment carried by your department.

Hand tools are used to extend or multiply the actions of your body and increase your effectiveness in performing specific functions. Most hand tools operate using simple machine principles. A pike pole extends your reach, allows you to penetrate through a ceiling, and enables you to apply force to pull down ceiling material. An axe multiplies the cutting force you can exert on a given area. Power tools and equipment use an external source of power, such as an electric motor or an internal combustion engine, and are faster and more efficient than hand tools.

Safety

Safety is a prime consideration when using tools and equipment. You need to operate the equipment so that you, fellow fire fighters, victims, and bystanders are not accidentally injured. Personal safety also requires the use of proper personal protective equipment (PPE) that includes:

- Approved helmet
- Firefighting hood
- Eye protection
- Face shield
- Approved firefighting gloves
- Turnout coat
- Bunker pants
- Boots
- Self-contained breathing apparatus (SCBA)
- Personal alert safety system

Conditions of Use/Operating Conditions

The best way to learn how to use tools and equipment properly is under optimal conditions of visibility and safety. In the beginning, you must be able to see what you are doing and practice without endangering yourself and others. As you become more proficient, you should practice using tools and equipment under more realistic working conditions. Eventually, you must be able to use tools and equipment safely and effectively when darkness or smoke decreases visibility. You must be able to work safely in hazardous areas, where you are surrounded by noise and other activities, while wearing all of your protective clothing and using your SCBA. Many departments require fire fighters to practice certain skills and evolutions in total darkness or with their face masks covered to simulate the darkness of actual fires.

Effective Use

Effective, efficient use of tools and equipment means using the least amount of energy to accomplish the task. Being effective means you achieve the desired goal. Being efficient means that you produce the desired effect without wasting time or energy. When assigned a task on the fireground, your objective is to complete that task safely and quickly. If you waste energy by working inefficiently, you will not be able to perform additional tasks. However, if you know which tools and equipment are needed for each phase of firefighting, you will be able to achieve the desired objective quickly and have the energy needed to complete the remaining tasks.

New fire fighters are often surprised by the strength and energy required to perform many tasks. An aggressive, continuous program of physical fitness will enable you to maintain your body in the optimal state of readiness. It does little good to practice for hours if you are not in prime condition.

As your training continues, you will learn which tools and equipment are used during different phases of fireground operations. For example, the tools needed for forcible entry are different from the tools usually needed for overhaul. Knowing which tools are needed for the work that must be done will help you prepare for the different tasks that unfold on the scene of a fire. The specific steps for properly using tools and equipment will be presented in this chapter, throughout the textbook, and by your fire department.

Most fire departments have standard operating procedures or guidelines that specify the tools and equipment needed in various situations. As a fire fighter, you must know where every tool and piece of equipment is carried on your apparatus. Knowing how to use a piece of equipment does you no good if you cannot find it quickly. Your company officer is responsible for telling you which tools to bring along for different situations.

Some fire fighters carry a selection of small tools and equipment in the pockets of their turnout coats or bunker pants. Check to see if your department requires you to carry certain tools and equipment at all times. Ask senior fire fighters for recommendations about what tools and equipment you should carry.

Functions

An engine or truck company carries a number of tools and different types of equipment. Often, the easiest way to learn and remember these tools is to group them by the function each performs (► Tables 8-1, 8-2). Most of the tools used by fire departments fit into the following functional categories:

- Rotating (assembly or disassembly)
- Pushing or pulling
- Prying or spreading
- Striking
- Cutting
- Multiple use

Rotating Tools

Rotating tools apply a rotational force to make something turn. The most common rotating tools are screwdrivers, wrenches, and pliers, which are used to assemble (fit together) or disassemble (take apart) parts that are connected with threaded fasteners (► Table 8-3).

Table 8-1 Tool Functions
Rotate, assemble, or disassemble (► Figure 8-1).
Pry or spread (► Figure 8-2).
Push or pull (► Figure 8-3).
Strike (► Figure 8-4).
Cut (► Figure 8-5).
A combination of functions (► Figure 8-6).

Assembling and disassembling are basic mechanical skills that are routinely employed by fire fighters to solve problems. Most fire apparatus carry a tool kit with a selection of screwdrivers with different heads, open-end wrenches, box wrenches, socket wrenches, pliers, adjustable wrenches, and pipe wrenches.

There are various sizes and types of screw heads, including slotted head, Phillips head, Roberts head, and others. A screwdriver with interchangeable heads is sometimes more useful than a selection of different screwdrivers. Pliers and wrenches also come in various shapes and sizes for different applications.

A spanner wrench is used to tighten or loosen fire hose **couplings** (one set or pair of connection devices that allow a fire hose to be interconnected with additional lengths of hose). If the couplings are not too tight, they can be loosened by hand, without the use of a spanner wrench. A hydrant wrench is used to open or close a hydrant by rotating the valve stem, or to remove the caps from the hydrant outlets.

Common examples of rotating tools include (► Figure 8-7):

- **Box-end wrench** A hand tool with a closed end that is used to tighten or loosen bolts; some styles of box-end wrenches have ratchets for easier use
- **Gripping pliers** A hand tool with a pincer-like working end that can also be used to bend wire or hold smaller objects
- **Hydrant wrench** A hand tool used to operate the valves on a hydrant; some are plain wrenches and others have a ratchet (a mechanism that will only rotate one way) feature (► Figure 8-8)

Fire Fighter Safety Tips

Learn to operate tools and equipment in areas without hazards and with adequate light. Gradually, increase the amount of PPE worn and decrease the light to levels typically found on the fireground. Eventually, you should be so familiar with your tools and equipment that you can use them under adverse conditions while wearing PPE.

- Open-end wrench A hand tool with an open end that is used to tighten or loosen bolts
- Pipe wrench A wrench with one fixed grip and one moveable grip that can be adjusted to fit securely around pipes and other tubular objects
- Screwdriver A tool that is used to turn screws
- Socket wrench A wrench that fits over a nut or bolt and uses the action of an attached handle to tighten or loosen the nut or bolt
- Spanner wrench A special wrench used to tighten or loosen hose couplings; spanners come in several sizes

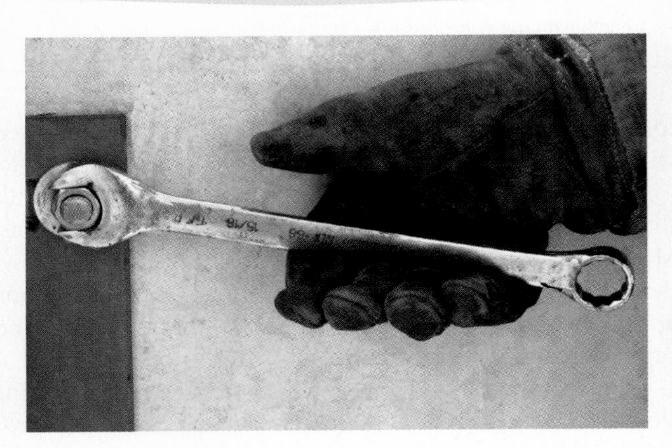

Figure 8-1 Rotate, assemble, or disassemble.

Figure 8-2 Pry or spread.

Figure 8-3 Push or pull.

Figure 8-4 Strike.

Figure 8-5 Cut.

Figure 8-6 A combination of functions.

Figure 8-7 Hydrant wrench, spanner wrench, pipe wrench, open-end wrench, box-end wrench, gripping pliers, screwdriver, socket wrench.

Table 8-2 Tool Functions

Function	Tool
Rotate	Box-end wrench Gripping pliers Hydrant wrench Open-end wrench Pipe wrench Screwdriver Socket wrench Spanner wrench
Push or pull	Ceiling hook Clemens hook Drywall hook K tool Multipurpose hook Pike pole Plaster hook Roofman's hook San Francisco hook
Pry or spread	Claw bar Crowbar Flat bar Halligan tool Hux bar Hydraulic spreader Kelly tool Pry bar Rabbet tool
Strike	Battering ram Chisel Flat-head axe Hammer Mallet Maul Pick-head axe Sledgehammer Spring-loaded center punch
Cut	Axe Bolt cutter Chain saw Cutting torch Hacksaw Handsaw Hydraulic shears Reciprocating saw Rotary saw Seatbelt cutter

Table 8-3 Common Tools for Assembly and Disassembly

Box-end wrenches	Pipe wrenches
Gripping pliers	Screwdrivers
Hydrant wrenches	Socket wrenches
Open-end wrenches	Spanner wrenches

Figure 8-8 Hydrant wrench in action.

Fire Marks

The word "pike" means sharp point or spike. In ancient times, a pike was the tip of a spear, and pike poles were used for very different purposes than current pike poles.

Figure 8-9 A pike pole in action.

Pushing/Pulling Tools

A second type of tool is used for pushing or pulling. These tools can extend the reach of the fire fighter as well as increase the power that can be exerted upon an object (▼ Table 8-4). These tools have many different uses in fire department operations.

An example of a tool that extends reach is a **pike pole** (► Figure 8-9). A pike pole consists of a wood or fiberglass pole with a metal head attached to one end. A pike pole is used primarily to pull down a ceiling to get to the seat of a fire burning above. The metal head has a sharpened point that can be punched through the ceiling and a hook that can grab and pull it down. The proper use of a pike pole will be discussed in depth in Chapter 18, Salvage and Overhaul.

Pike poles come in several different sizes and with a variety of heads (► Figure 8-10). The most common length of 4′ to 6′ enables a fire fighter to stand on a floor and pull down a 10′-high ceiling. **Closet hooks**, intended for use in tight spaces, are commonly 2′ to 4′ long. Some pike poles are equipped with handles as long as 12′ or 14′ for use in rooms with very high ceilings; others may have a "D"-type handle for better pulling power.

Table 8-4 Common Tools for Pushing and Pulling

Ceiling hook	Multipurpose hook	Roofman's hook
Clemens hook	Pike pole	San Francisco hook
Drywall hook	Plaster hook	

The different head designs are intended for different types of ceilings and come in a variety of configurations. Many fire departments use one type of pike pole for plaster ceilings and another for drywall ceilings.

Common types of pulling poles and hooks include:

- **Ceiling hook** A tool with a long, wood or fiberglass pole and a metal point with a spur at right angles that can be used to probe ceilings and pull down plaster lath material
- **Clemens hook** A multipurpose tool that can be used for forcible entry and ventilation applications because of its unique head design
- **Drywall hook** A specialized version of a pike pole designed to remove drywall
- **Multipurpose hook** A long pole with a wooden or fiberglass handle and a metal hook
- **Plaster hook** A long pole with a pointed head and two retractable cutting blades on the side
- **Roofman's hook** A long pole with a solid metal hook
- **San Francisco hook** A multipurpose tool used for

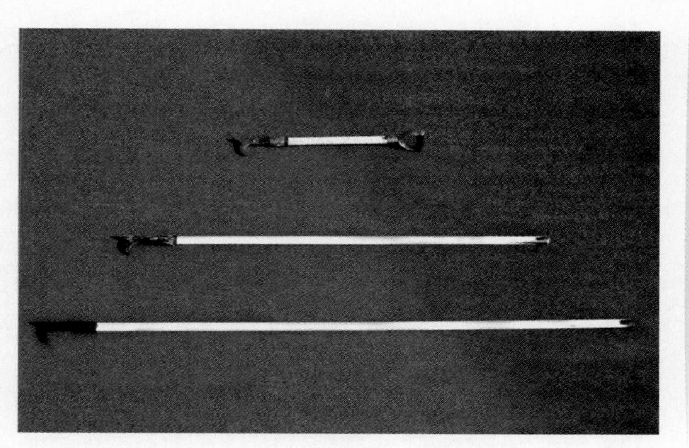

(Figure 8-10) Pike poles come in several different sizes.

(Figure 8-11) Components of a K tool.

forcible entry and ventilation applications because it has a built-in gas shut-off and directional slot

A K tool is another type of pushing or pulling tool (► Figure 8-11). This tool is used to pull the lock cylinder out of a door, exposing the locking mechanism so it can be unlocked easily.

Prying/Spreading Tools

A third type of tool is used for prying or spreading. These tools may be as simple as a pry bar or as mechanically complex as a hydraulic spreader (► Table 8-5). They also come in several variations for different applications.

A simple pry bar consists of a hardened steel rod with a tapered end that can be inserted into a small area. The bar acts as a lever to multiply the force that a person can exert to bend or pry objects apart. A properly positioned pry bar can apply an enormous amount of force.

Fire departments use a wide range of prying tools in different sizes and with different features. One of the most popular is a Halligan tool, which was designed by a New York City fire fighter. This tool incorporates a sharp pick, a flat prying surface, and a forked claw. It can be used for forcible entry applications. Several other prying tools can also be used for forcible entry. The specific uses and applications of these tools will be covered in the chapters on the phases of fire suppression.

Tools used for prying include (► Figure 8-12):

- Claw bar A tool with a pointed claw-hook on one end and a forked- or flat-chisel pry on the other end that can be used for forcible entry
- Crowbar A straight bar made of steel or iron with a forked-like chisel on the working end
- Flat bar A specialized prying tool made of flat steel with prying ends suitable for performing forced entry
- Halligan tool A prying tool designed for use in the

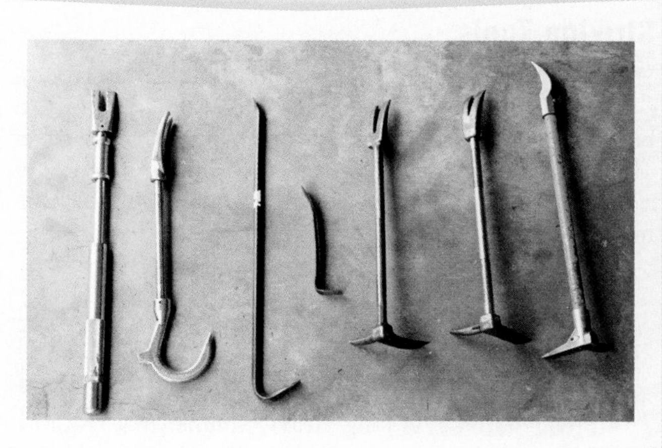

(Figure 8-12) Common tools used for prying and spreading.

Table 8-5 Common Tools for Prying and Spreading

Claw bar	Halligan tool	Kelly tool
Crowbar	Hux bar	Pry bar
Flat bar	Hydraulic spreader	Rabbet tool

Teamwork Tips

Always try before you pry! One person should check to see if a door or window is already unlocked while another readies the right forcible entry tool or tools.

fire service that also incorporates a pick and a fork; sometimes known as a Hooligan tool

- Hux bar A multipurpose tool that can be used for forcible entry and ventilation applications because of its unique design; also can be used as a hydrant wrench
- Kelly tool A steel bar with two main features: a large pick and a large chisel or fork
- Pry bar A specialized prying tool made of a hardened steel rod with a tapered end that can be inserted into a small area (► Figure 8-13)

Hydraulic spreaders are an example of machine-powered rescue tools (► Figure 8-14). The use of hydraulic power enables you to apply several tons of force on a very small area. You need to have special training to operate these machines safely. Fire and rescue departments most commonly use them for extrication from vehicles (► Figure 8-15).

Hand-powered hydraulic tools are also used for prying and spreading. One of these, called a rabbet tool, is designed for quickly opening doors (► Figure 8-16).

Striking Tools

Striking tools are used to apply an impact force to an object (► Table 8-6). They are often used to gain entrance to a building or a vehicle or to make an opening in a wall or roof. They can also be used to force the end of a prying tool into a small opening. Specific use of these tools will be covered in Chapter 11, Forcible Entry. Striking tools include:

- Hammer A hand tool constructed of solid material with a long handle and a head affixed to the top of the handle with one side of the head used for striking and the other side used for prying
- Mallet A short-handled hammer with a round head
- Sledgehammer A long, heavy hammer that requires the use of both hands
- Maul A specialized striking tool (weighing six pounds or more) with an axe on one end and a sledgehammer on the other end (► Figure 8-17)
- Flat-head axe A tool with a head that has an axe blade on one end and a flat head on the opposite end (► Figure 8-18 A)
- Pick-head axe A tool with a head that has an axe blade on one end and a pointed "pick" on the opposite end (► Figure 8-18 B)

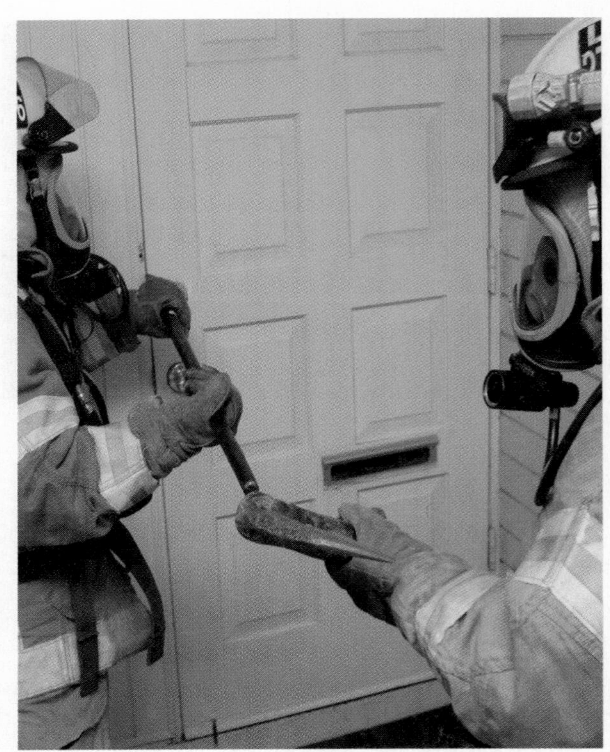

Figure 8-13 A pry bar in action.

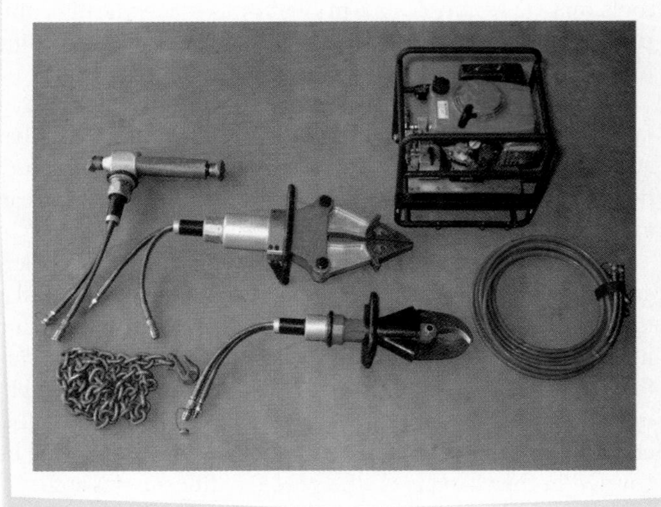

Figure 8-14 The components of a hydraulic rescue tool.

Fire Fighter Tips

Many fire departments carry a striking tool and a prying bar strapped together. This combination of tools is sometimes referred to as the irons. A flat-head axe and a Halligan tool are carried together on many vehicles and are always taken into a fire building by one of the crew members.

Table 8-6 Common Striking Tools

Battering ram	Hammer	Pick-head axe
Chisel	Mallet	Sledgehammer
Flat-head axe	Maul	Spring-loaded center punch

Figure 8-15 A hydraulic spreader is often used for vehicle extrication.

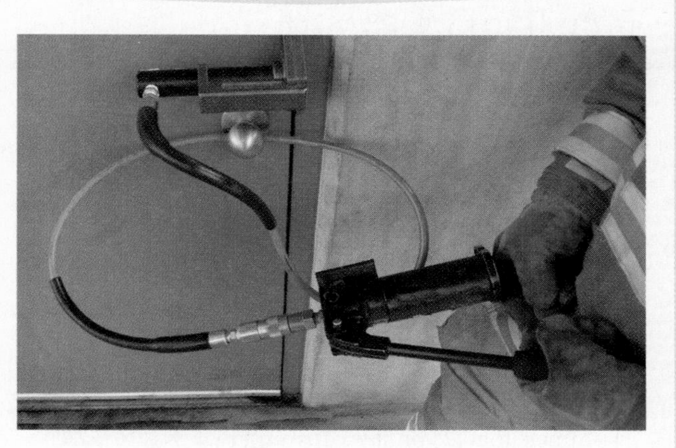

Figure 8-16 A rabbet tool can open doors quickly.

Fire Fighter Safety Tips

Some fire fighters carry knives, combination tools, a small length of rope, or wooden wedges in their turnout gear. The tools and equipment that you carry will depend on your department requirements, the type of area served by your department, and your personal preference.

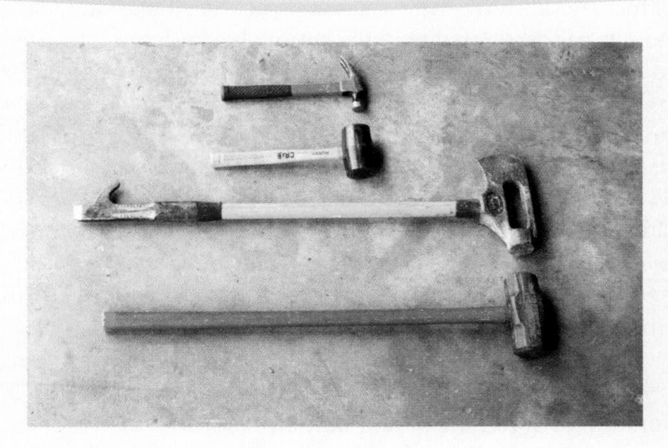

Figure 8-17 Striking tools (from top): hammer, mallet, maul, sledgehammer.

A

B

Figure 8-18 Two types of axes: **A.** Flat-head axe and **B.** A pick-head axe.

Fire Fighter Safety Tips

Some tools require a considerable amount of movement and room to operate. Always check to be sure no one is in danger of being injured before you use these tools. Look around and make sure you can operate an axe or sledgehammer safely and effectively.

- Battering ram A heavy metal bar used to break down doors (► Figure 8-19)
- Chisel A metal tool with one sharpened end that can be used to break apart material in conjunction with a hammer, mallet, or sledgehammer (► Figure 8-20)
- Spring-loaded center punch A spring-loaded punch used to break tempered automobile glass

One of the most frequently used tools in the fire service is the axe. Both flat-head axes and pick-head axes are used. Both types of axes have a wide cutting blade that can be used to chop into a wall, roof, or door. A flat-head axe also can be used as a striking tool for forcible entry, usually in combination with a prying tool, such as a Halligan tool. Together, the flat-head axe and the Halligan tool are sometimes referred to as "the irons" and are very effective for most forcible entry situations (► Figure 8-21). A pick-head axe has a point or pick that can be used for puncturing, pulling, and prying (► Figure 8-22). Forcible entry will be covered in depth in Chapter 11, Forcible Entry.

The spring-loaded center punch is a striking tool used primarily on cars (► Figure 8-23). This tool can exert a large amount of force on a pinpoint-size portion of tempered automobile glass. This disrupts the integrity of the glass and causes the window to shatter into small, uniform-sized pieces. A spring-loaded center punch is often used in vehicular crashes to gain access to a patient who needs care. Vehicle extrications will be covered in depth in Chapter 25, Vehicle Rescue and Extrication.

Cutting Tools

Cutting tools have a sharp edge that severs an object. They come in several forms and are used to cut a wide variety of substances (► Figure 8-24 A–D). Cutting tools used by fire fighters range from knives or wire cutters carried in the pockets of turnout coats to seatbelt cutters, bolt cutters (a scissors-like tool used to cut through items such as chains or padlocks), saws, cutting torches (a torch that produces a high temperature flame capable of melting through metal), and hydraulic shears (► Table 8-7). Each is designed to work on certain types of materials. Fire fighters can be injured and cutting tools can be ruined if tools are used incorrectly.

Bolt cutters are most often used to cut through chains or padlocks to open doors or gates. By concentrating the cut-

Figure 8-19 Using a battering ram.

Figure 8-20 Using a chisel with a hammer.

Figure 8-21 A flat-head axe can be used with a Halligan tool to force open a door.

Figure 8-22 A pick-head axe can be used to pry up boards.

Figure 8-23 A spring-loaded center punch can be used to break a car window safely.

A

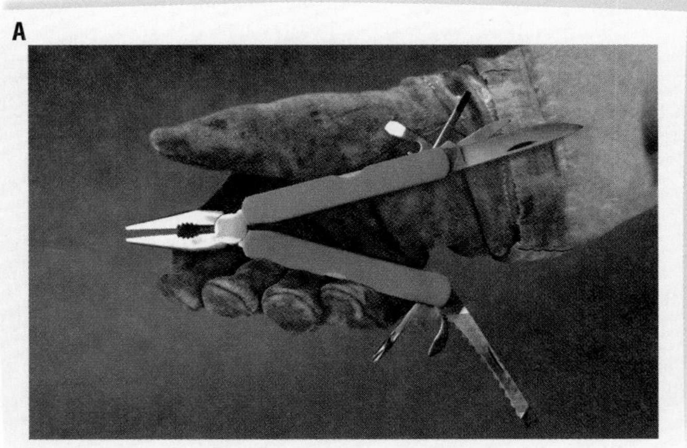

B

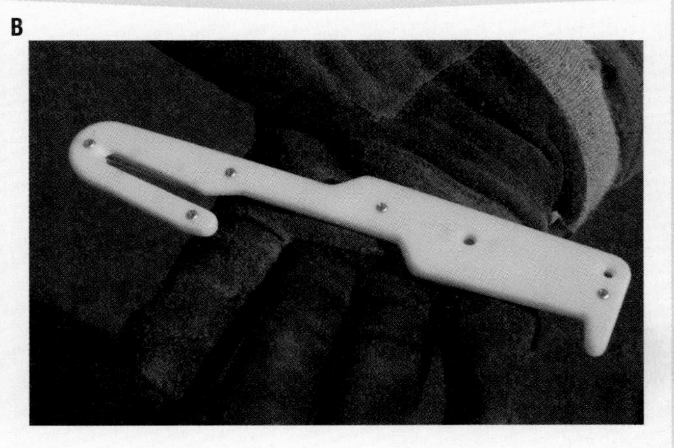

C

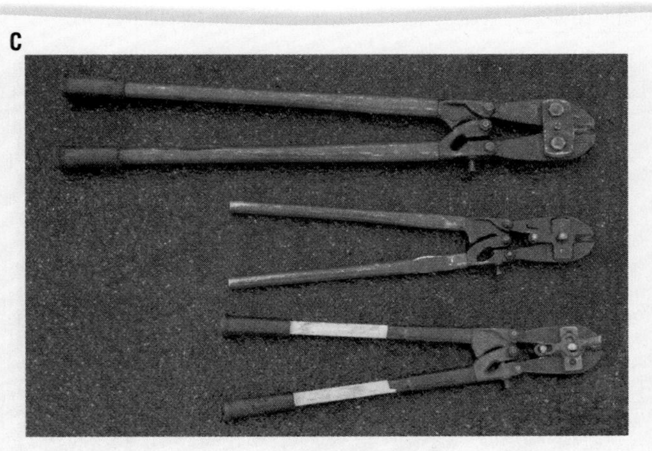

D

Figure 8-24 Cutting tools. **A.** Combination tool. **B.** Seatbelt cutter. **C.** Bolt cutters. **D.** Various handsaws.

Table 8-7 Common Cutting Tools

Axes	Bolt cutters
Chain saws	Cutting torches
Hacksaws	Handsaws
Hydraulic shears	Reciprocating saws
Rotary saws	Seatbelt cutter

ting force on a small area, you can break through a chain in just a few seconds.

Fire departments often carry several different types of saws. Saws can be divided into two main categories, based on the power source. Handsaws are manually powered and mechanical saws are usually powered by electric motors or gasoline-powered engines. Handsaws include hacksaws, carpenter's handsaws, keyhole saws, and coping saws.

Hacksaws are designed to cut metal (▼ Figure 8-25). Different blades can be used, depending on the type of metal being cut. Hacksaws are useful when metal needs to be cut under closely controlled conditions.

Carpenter's handsaws are designed for cutting wood (▼ Figure 8-26). Saws with large teeth are effective in cutting large timbers or tree branches. These saws are useful at auto crashes where tree limbs may hamper the rescue effort. Saws with finer teeth are designed for cutting finished lumber. A coping saw is used to cut curves in wood. It is a handsaw with a narrow blade set between the ends of a U-shaped frame. A keyhole saw, a specialty saw, is narrow and slender and can be used to cut keyholes in wood.

There are three primary types of mechanical saws: chain saws, rotary saws, and reciprocating saws. Although handsaws have a valuable role, power saws have the advantage of accomplishing more work in a shorter period of time. They also enable fire fighters to conserve energy, resulting in less

Figure 8-25 A fire fighter using a hacksaw to cut through metal.

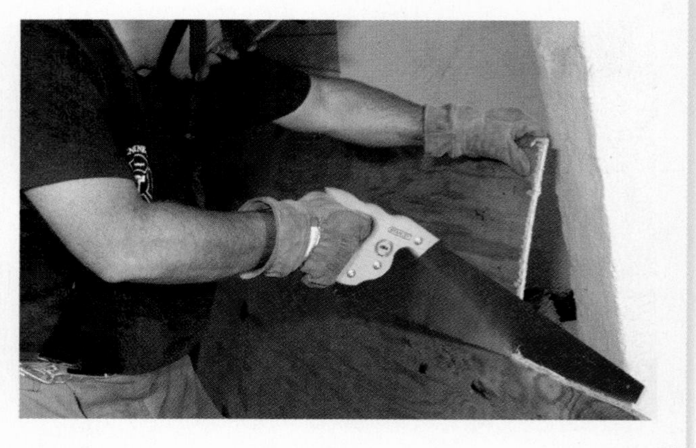

Figure 8-26 A carpenter's handsaw is designed for cutting wood.

Figure 8-27 A fire fighter can use a chain saw to cut through a roof.

Figure 8-28 Fire fighter using a rotary saw.

Fire Fighter Safety Tips

When operating any mechanical saw, it is important to use a saw blade or saw disk designed specifically for the material being cut. Always use the proper eye protection as well as a helmet shield when using power saws.

fatigue. Because mechanical saws are powerful, they must only be used by trained operators. But there are some disadvantages to using power saws. They are heavy to carry and sometimes can be difficult to start. They may also require an electrical connection, although cordless models are becoming more available.

Most people are familiar with gasoline-powered or electric chain saws commonly used to cut wood, particularly trees. Fire fighters often use saws with special chains to cut ventilation openings in roofs constructed of wood, metal, tar, gravel, or insulating materials (◄ Figure 8-27).

Rotary saws are powered by electric motors or gasoline engines (◄ Figure 8-28). In some rotary saws, the cutting part of the saw is a round metal blade with teeth. Different blades are used depending on the type of material being cut. Other rotary saws use a flat, abrasive disk for cutting. The disks are made of composite materials and are designed to wear down as they are used. It is important to match the appropriate saw blade or saw disk to the material being cut.

Reciprocating saws are powered by an electric or battery motor that rapidly pulls a saw blade back and forth (► Figure 8-29 A & B). As with rotary saws, reciprocating saws use different blades to cut different materials. Reciprocating saws are most commonly used to cut metal during a vehicle extrication.

The cutting tools that require the most training are hydraulic shears and the cutting torch. Hydraulic shears are often used with hydraulic spreaders and rams in automobile extrication. The same hydraulic power source can be used with all three types of tools. Hydraulic shears can cut quickly through metal posts and bars.

Cutting torches produce an extremely high temperature flame and are capable of heating steel until it melts, thereby cutting through the object (► Figure 8-30). Because these torches produce such high temperatures (5,700°F), operators must be specially trained before using this tool. Cutting torches are sometimes used for rescue situations and for cutting through heavy steel objects.

Multiple Function Tools

Certain tools are designed to perform multiple functions, thus reducing the total number of tools needed to achieve a goal. One example is a flat-head axe because it can be used as either a cutting or a striking tool. There are combination tools that can be used to cut, to pry, to strike, and to turn off utilities.

A

B

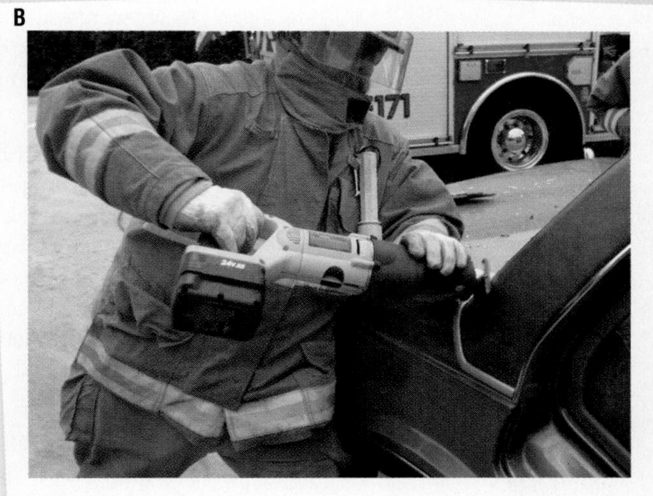

Figure 8-29 **A.** The components of a reciprocating saw. **B.** A reciprocating saw being used during a vehicle extrication.

Figure 8-30 A cutting torch can be used to cut through a metal door.

Voices of Experience

"I leapt out of the way just as the entire 4′ x 8′ section of the ceiling came crashing down—insulation, water, and the lost axe included!"

When I was a rookie fire fighter, I was dispatched to a structure fire at a local retail store. Upon arrival at the scene, my company was assigned to ventilation. As we were ventilating the roof, another fire fighter, wearing wet gloves, accidentally dropped the pick-head axe. The axe was on the down swing and the drywall ceiling below the roof caught it. We decided that we would recover the axe later during overhaul operations.

Once the fire had been extinguished, my crew was reassigned to overhaul. My partner and I grabbed a pike pole and began pulling down the ceiling in the most heavily damaged section of the store. Because the interior attack crew had fought the fire from the attic through an opening in the ceiling, the drywall was wet and heavy. As I dug the pike pole into a section of the ceiling, I suddenly heard my partner shout, "Hugh, look out!" I leapt out of the way just as the entire 4′ × 8′ section of the ceiling came crashing down—insulation, water, and the lost axe included! The axe hit my helmet. Fortunately, I had all of my PPE properly in place, and was only a little dizzy from the impact of the axe.

Knowing where all your equipment is, and how to use it properly, is critical for ensuring safety on the fireground. Had I forgotten to secure my helmet that day, the impact of the falling axe would have caused a very serious injury, possibly even death. Fire fighters must always keep track of the tools and equipment they use throughout emergency operations—it can mean the difference between life and death.

Hugh Pike
United States Air Force, Retired
Fountain, Florida

Special Use Tools

Some fire situations require special use tools that perform other functions. For example, fire departments located in areas where brush and ground fires occur frequently may need to carry rakes, brooms, shovels, and combination tools that can be used for raking, chopping, cutting, and leaf blowing.

Fire department rescue squads also use specialized equipment such as jacks and air bags for lifting heavy objects, come alongs or lever blocks (small, hand-operated winches) for dragging heavy objects, and tripods (► Figure 8-31). You can learn more about the proper use of this special equipment by taking special rescue courses or during in-service training.

Figure 8-31 Heavy-duty air bags can be used to lift vehicles in rescue situations.

Phases of Use

The process of extinguishing a fire usually involves a sequence of steps or stages. Each phase of a fireground operation may require the use of certain types of tools and equipment. The basic steps of fire suppression include:

- Response/Size-Up: This phase begins when the emergency call is received and continues as the units travel to the incident scene. The last part of this phase involves the initial observation and evaluation of factors used to determine the strategy and tactics that will be employed.
- Forcible Entry: This phase begins when entry to buildings, vehicles, aircraft, or other confined areas are locked or blocked, requiring fire fighters to use special techniques to gain access.
- Interior Attack: During this phase, a team of fire fighters is assigned to enter a structure and attempt fire suppression.
- Search and Rescue: As its name suggests, this phase involves a search for any victims trapped by the fire and their rescue from the building.
- Rapid Intervention: When a rapid intervention company/crew (RIC) provides immediate assistance to injured or trapped fire fighters.
- Ventilation: This step involves changing air within a compartment by natural or mechanical means.
- Overhaul: The final phase is to ensure that all hidden fires are extinguished after the main fire has been suppressed.

Response/Size-Up

The response and size-up phase enables you to anticipate emergency situations. At this time, you should consider the information from the dispatcher along with preincident plan information about the location. This information can provide you with an idea of the nature and possible gravity of the situation, as well as the types of problems that might arise. For example, an automobile fire on the highway may present different problems and require different tools than a call for smoke coming from a single-family house. A different thinking process occurs when you are dispatched at midnight to a house fire that may have trapped a family inside than when you respond to a report of a kitchen stove fire at suppertime. Even though information is limited, this is the time to start thinking about the types of tools and equipment that you might need.

Most fire departments have standard operating procedures or guidelines that specify the tools and equipment required for different types of fires. Each crew member is expected to bring specific tools and equipment from the apparatus. These requirements take into account the roles that different units have within the fire department. A member of an engine company will not bring the same tools that a member of a truck company will bring. The tools carried for an interior fire attack are different from those carried for an outside or defensive attack.

Upon arrival at the scene, the company officers will size-up the situation and develop the action plans for each company, following standard operating procedures and guidelines. They will use their senses of sight, sound, and smell, as well as any other available information, to make a determination of how to best attack the fire. As a crew member, you are responsible for following the directions of your company officer.

Forcible Entry

Gaining entrance to a locked building or structure can present a challenge to even the most seasoned fire fighter. Buildings are often equipped with security devices designed to keep unwanted people out. However, these same devices can make it very difficult for fire fighters to gain access to the building. Forcible entry is the process of entering a building by overcoming these barriers. The skills involved will be discussed in depth in Chapter 11, Forcible Entry.

Several types of tools can be used for forcible entry, including an axe, a prying tool, or a K tool. A flat-head axe and a Halligan tool are often used in combination to quickly pry open a door, although they may permanently damage both the door and the frame. Prying tools used for gaining access include pry bars, crowbars, Halligan tools, hux bars, and the hydraulic-powered rabbit tool.

A K tool can be used to pull out a cylinder lock mounted in a wood or heavy metal door, so that the lock can be released (► Figure 8-32). This is a comparatively nondestructive process that leaves the door and most of the locking mechanism undamaged. The building owner can have the lock cylinder replaced at a relatively low cost.

Various striking tools can be used for forcible entry when brute force is needed to break into a building. These include flat-head axes, hammers, sledgehammers, and battering rams.

Sometimes the easiest or only way to gain access is to use cutting tools. An axe can be used to cut out a door panel. A power saw can be used to cut through a wood wall. Bolt cutters can be used to remove a padlock. Cutting torches or power saws can be used to cut through metal security bars. Although cutting is a destructive process, it is justified to save lives or property.

Many techniques and types of tools can be used to gain entry into secured structures (▼ Table 8-8). The exact tool needed will depend upon the method of entry and the type of obstacle. Because experience usually determines the best way to gain entry in each situation, rely on the orders and advice of your captain and coworkers.

Figure 8-32 Use a K tool to pull out and release a cylinder lock.

Interior Firefighting Tools and Equipment

The process of fighting a fire inside a building involves several tasks that are usually performed simultaneously or in rapid succession by teams of fire fighters. While one crew is advancing a hose line to attack the fire, another crew may be searching for occupants and another may be performing ventilation tasks. An additional company may be standing by as a rapid intervention crew, in case a fire fighter needs to be rescued.

Some basic tools and equipment should be carried by every crew working inside a burning building. Crews may

Table 8-8 Forcible Entry Tools

Type of tool	Use in forcible entry
Prying tools	These can include Halligan tools, a flat bar, a crowbar, and other prying tools. They can be used to break windows or force open doors.
Axe, flat-head axe, pick-head axe	An axe can be used to cut through a door or break a window. A flat-head axe can be used in conjunction with a Halligan tool to force open a door.
Sledgehammer	A sledgehammer can be used to breach walls or break a window. It can also be used in conjunction with other tools, such as a Halligan tool, to force open a door or break off a padlock.
Hammer, mallet	These tools may be used in conjunction with other tools such as chisels or punches to force entry through windows or doors.
Chisel, punch	These tools can be used to make small openings through doors or windows.
K tool	The K tool provides a "through the lock" method of opening a door, minimizing damage to the door.
Rotary saw, chain saw, reciprocating saw	Power saws can cut openings through various obstacles, including doors, walls, fences, gates, security bars, and other barriers.
Bolt cutter	A bolt cutter can be used to cut off padlocks or cut through obstacles such as fences.
Battering ram	A battering ram can be used to breach walls.
Hydraulic door opener	A hydraulic door opener can be used to force open a door.
Hydraulic rescue tool	A hydraulic rescue tool can be used in a variety of situations to break, breach, or force openings in doors, windows, walls, fences, or gates.

Figure 8-33 A thermal imaging device.

also carry specialized tools and equipment needed for their particular assignment. The basic tools enable them to solve problems they may encounter while performing interior operations. For example, the crew may encounter obstacles such as locked doors, or they may need to open an emergency escape route. They may need to establish **horizontal ventilation** by forcing, opening, or breaking a window. They may have to gain access to the space above the ceiling by using a pike pole or making a hole in a wall or floor with an axe. A powerful light is important, because smoke can quickly reduce interior visibility to just a few inches.

The basic set of tools for interior firefighting includes:

- A prying tool, such as a Halligan tool
- A striking tool, such as a flat-head axe or sledgehammer
- A cutting tool, such as an axe
- A pushing/pulling tool, such as a pike pole
- A strong **hand light** or portable light

The specific tools that must be carried by each crew are usually defined in a fire department's training manuals and standard operating procedures. The requirements are based on local conditions and preferences and may differ, depending on the type of company and the assignment.

The interior attack team is responsible for advancing a hose line, finding the fire, and applying water to extinguish the flames. They need the basic tools that will allow them to reach the seat of the fire.

Figure 8-34 Fire fighter using a thermal imaging device.

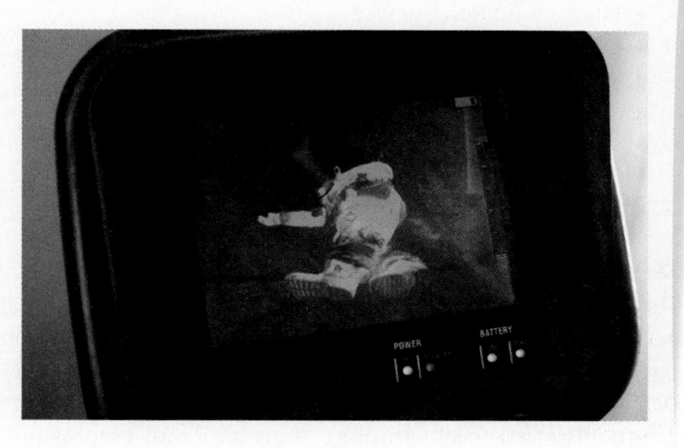

Figure 8-35 An image produced by a thermal imaging device.

Search and Rescue Tools and Equipment

Search and rescue needs to be carried out quickly, shortly after arrival on the fire ground. A search team should carry the same basic hand tools as the interior attack team:

- Pushing tool (short pike pole)
- Prying tool (Halligan tool)
- Striking tool (sledgehammer or flat-head axe)
- Cutting tool (axe)
- Hand light

In addition to being equipped for forcible entry and emergency exit, a search-and-rescue team may also use tools to probe under beds for unconscious victims. A short pike pole or closet hook is relatively light and reduces the time needed to search an area by extending the fire fighter's reach. An axe handle can also be used for this purpose.

Other types of tools used for search and rescue include **thermal imaging devices** (▲ Figures 8-33, ▲ 8-34, ▲ 8-35), portable lights, and **lifelines**. The techniques of search and rescue are discussed in Chapter 13, Search and Rescue.

Rapid Intervention Tools and Equipment

A rapid intervention company or crew (RIC) is designated to stand by to provide immediate assistance to any fire fighters who become lost, trapped, or injured during an incident or training exercise. The RIC team should have the standard set of tools for interior firefighting as well as extra tools and equipment particularly important for search-and-rescue tasks. The extra tools and equipment should help them find and gain access to a fire fighter who is in trouble, extricate a fire fighter who is trapped under debris, provide breathing air for a fire fighter who has experienced an SCBA failure or run out of air, and remove an injured or unconscious fire fighter from the building. All of this equipment should be gathered and staged with the rapid intervention crew where it will be immediately available if it is needed.

The special equipment that a rapid intervention team should carry includes:

- Thermal imaging device
- Additional portable lighting
- Lifelines
- Prying tools
- Striking tools
- Cutting tools, including a power saw
- SCBA and spare air cylinders

Ventilation Tools and Equipment

The objective of ventilation is to provide openings so that fresh air can enter and hot gases and products of combustion can escape. Many of the same tools used for forcible entry are also used to provide ventilation. Power saws and axes are commonly used to cut through roofs and vent combustion by-products.

Fans are often used either to remove smoke from a building or to introduce fresh air into a structure. With **positive pressure ventilation**, fresh air is blown into a building through selected openings to force contaminated air out through other openings. Positive pressure ventilation is often used to allow fire fighters to work quickly, safely, and effectively to locate and suppress the fire.

Negative pressure ventilation uses fans placed at selected openings to draw contaminated air out of a building. It is used when there are no suitable openings for positive pressure ventilation or when introducing pressurized air could accelerate the fire. Negative pressure ventilation may also be the best option when positive pressure could spread the products of combustion.

Ventilation fans can be powered either by electric or gasoline motors or by water pressure (▶ Figure 8-36). A gasoline-powered fan may not be suitable in some situations, because it can introduce carbon monoxide into a structure if an exhaust hose is not available. If there are potentially dangerous levels of gases in the building atmosphere, a water-powered fan may be the best choice.

Horizontal ventilation usually involves opening outer doors and windows to allow fresh air to enter and to remove contaminated air. When ventilation fans are used, they are placed at these openings. Positive pressure fans blow in fresh air, and negative pressure fans draw out smoke and contaminants. Unlocked or easily released windows and doors should be opened normally. Locked or jammed windows and doors may have to be broken or forced open using basic interior firefighting tools.

It may also be necessary to make interior openings within the building so that contaminated air can reach the exterior openings. Opening interior doors is the easiest, quickest, and often most effective way to ventilate interior spaces. A series of electrically powered fans may be placed throughout a large structure to help move smoke in a desired direction.

Vertical ventilation requires openings in the roof or the highest part of a building to allow smoke and hot gases to escape. Whenever possible, existing openings such as doors, windows, roof hatches, and skylights should be used for vertical ventilation. It may be necessary to force them open or to break them using forcible entry tools.

In some circumstances it may be necessary to cut through a roof to make an effective vertical ventilation opening. Cutting tools such as axes and power saws are used to make these openings. Pike poles will also be needed to pull down ceilings after the roof covering is opened.

The special equipment needed for ventilation includes:

- Positive pressure fans
- Negative pressure (exhaust) fans
- Pulling and pushing tools (long pike poles)
- Cutting tools (power saws and axes)

Overhaul Tools and Equipment

The purpose of overhaul is to examine the fire scene carefully and ensure that all hidden fires are extinguished.

Figure 8-36 Different types of fans used in ventilation.

Fire Fighter Safety Tips

There is a valuable saying in the fire service. "Hand tools—never leave your vehicle without them."

Burned debris must be removed and potential hot spots in enclosed spaces behind walls, above ceilings, and under floors must be exposed. Both tasks can be accomplished using simple hand tools.

Pike poles are commonly used for pulling down ceilings and opening holes in walls. Axes, and sometimes power saws, are used to open walls and floors. Prying and striking tools are also used to open closed spaces. Shovels, brooms, and rakes are used to clear away debris.

The increasing use of infrared thermal imaging devices has made it possible to "see" hot spots behind walls without physically cutting into them (► Figure 8-37). This technology has reduced the risk of missing dangerous hot spots as well as the time and effort it takes to overhaul a fire scene. The process of overhauling a fire scene and the tools used to accomplish this will be covered more fully in Chapter 18, Salvage and Overhaul.

Figure 8-37 Thermal imaging devices can "see" hot spots.

Tools used during overhaul operations include:

- Pushing tools (pike poles of varying lengths)
- Prying tools (Halligan tool)
- Striking tools (sledgehammer, flat-head axe, hammer, mallet)
- Cutting tools (axes, power saws)
- Debris-removal tools (shovels, brooms, rakes, carryalls)
- Water-removal equipment (water vacuums)
- Ventilation equipment (electric, gas, or water-powered fans)
- Portable lighting
- Thermal imaging device

Tool Staging

Many fire departments have standard operating procedures for staging necessary equipment nearby during a fire or rescue operation. This often involves placing a salvage cover on the ground at a designated location and laying out commonly used tools and equipment where they can be accessed readily. A similar procedure may be used for rescue operations, where tools that are likely to be needed can be laid out ready for use. This saves valuable time because fire fighters do not have to return to their own apparatus or search several different vehicles to find a particular tool.

A department's standard operating procedure usually specifies the types of tools and equipment to be staged. The tool-staging area could be outside the building or, in the case of a high rise or large building, at a convenient location inside. Additional personnel may be directed to bring particular items to the tool staging location at working fires.

Maintenance

Tools and equipment must be properly maintained so that they will be ready for use when they are needed. Keep equipment clean and free from rust. Keep cutting blades sharpened and fuel tanks filled. Every tool and piece of equipment must be ready for use before you respond to an emergency incident.

Use power equipment only after you have been instructed on its use. Read and heed the instructions supplied with the equipment. Test power equipment frequently and have it serviced regularly by a qualified shop. Keep records of all inspections and maintenance performed on power tools. Preventive maintenance will help ensure that equipment will operate properly when it is needed. Treat tools and equipment as if someone's life depends on them, because it will!

Use equipment only for its intended purposes. For example, a pike pole is made for pushing and pulling; it is not a lever and will break if used inappropriately. Use the right tool for the job.

Fire Fighter Safety Tips

Keep the manufacturers' manuals to all of the department's tools and equipment in a safe and easily accessible location.

Cleaning and Inspecting Salvage, Overhaul, and Ventilation Equipment and Tools

Clean tools used in ventilation according to the manufacturer's instructions. Power tools are used frequently during overhaul and should also be cleaned according to the manufacturer's instructions. After returning from a fire, clean and inspect, and record all of your overhaul tools to ensure that they are in a "ready state."

Cleaning and Inspecting Hand Tools

All hand tools should be completely cleaned, inspected, and recorded after use, as shown in ► Table 8-9. Remove all dirt and debris. If appropriate, use water streams to remove the debris and soap to clean the equipment thoroughly. To prevent rust, metal tools must be dried completely, either by hand or by air, before being returned to the apparatus. Cutting tools should be sharpened after each use. Before any tool is placed back into service, it should be inspected for damage.

Avoid painting tools, because this will hide any possible defects or visible damage. Keep the number of markings on a tool to a minimum.

Cleaning and Inspecting Power Tools

All power equipment should be left in a "ready state" for immediate use at the next incident. This means that:

- All debris should be removed and the tool should be clean and dry.
- All fuel tanks should be filled completely with fresh fuel.
- Any dull or damaged blades/chains should be replaced.
- Belts should be inspected to ensure they are tight and undamaged.
- All guards should be securely in place.
- All hydraulic hoses should be cleaned and inspected.
- All power cords should be inspected for damage.
- All hose fittings should be cleaned, inspected, and tested to ensure tight fit.
- The tools should be started to ensure that they operate properly.
- Tanks on water vacuums should be emptied, washed, cleaned, and dried.
- Hoses and nozzles on water vacuums should be cleaned and dried.

It is very important to read the manufacturer's manual and follow all instructions on the care and inspection of power tools and equipment. Keep all manufacturer's manuals in a safe and easily accessible location. Refer to them when cleaning and inspecting the tools and equipment. Remember, your safety depends on the quality of your tools and equipment.

Table 8-9 Cleaning and Inspecting Hand Tools

Metal parts	All metal parts should be clean and dry. Remove rust with steel wool. Do not oil the striking surface of metal tools because this may cause them to slip.
Wood handles	Inspect for damage such as cracks, splinters, etc. Sand if necessary. Do not paint or varnish. Apply a coat of boiled linseed oil. Check that the tool head is tightly fixed to the handle.
Fiberglass handles	Clean with soap and water. Inspect for damage. Check that the tool head is tightly fixed to the handle.
Cutting edges	Inspect for nicks or other damage. File and sharpen as needed. *Note:* Power grinding may weaken some tools. Hand sharpening may be required.

Wrap-Up

Ready for Review

This chapter covered many of the tools and equipment used by a fire department. It is important to understand the purpose of each tool and piece of equipment on the apparatus. Tools and equipment may be used in all phases of the fire suppression sequence. They are used in conditions of smoke, darkness, decreased visibility, and limited motion. A fire fighter must know where tools and equipment are stored and be able to use them safely and effectively.

Most tools and equipment perform one or more of the following functions: rotating (assembling or disassembling), pushing or pulling, prying or spreading, striking, or cutting. A fire fighter should know which tools and equipment are most often used in each phase of the fire suppression sequence. By evaluating tool and equipment needs during response and size-up, a fire fighter can help realize the goals of forcible entry, search and rescue, rapid intervention, ventilation, and overhaul.

Selecting the appropriate tools and properly maintaining them will help ensure that they work properly during an emergency situation. This chapter provided an overview of the function and uses of selected tools and equipment. Later chapters will discuss these applications in greater detail.

Chief Concepts

- Tools and equipment extend your reach and multiply the force you can apply to an object.
- Your safety and the safety of others are of paramount importance when using tools and equipment.
- Learn to use tools and equipment under adverse conditions, while wearing motion-limiting PPE gear.
- Strive for effective and efficient use of tools and equipment.
- Know where tools and equipment are stored in the fire station and on the apparatus.
- The primary functions of tools and equipment are:
  - Rotate—assemble or disassemble
  - Push or pull
  - Pry or spread
  - Strike
  - Cut
  - A combination
- Use dispatch information and size-up information to anticipate the types of tools you may need.
- Learn which tools and equipment are used during the following phases of fire suppression:
  - Forcible Entry
  - Interior Attack
  - Search and Rescue
  - Rapid Intervention Company/Crew
  - Ventilation
  - Overhaul
- Proper selection and maintenance of tools and equipment are essential.

Hot Terms

Battering ram A tool made of hardened steel with handles on the sides used to force doors and to breach walls. Larger versions are used by up to four people, smaller versions are made for one or two people.

Bolt cutter A cutting tool used to cut through thick metal objects such as bolts, locks, and wire fences.

Box-end wrench A hand tool used to tighten or loosen bolts. The end is enclosed, as opposed to an open-end wrench. Each wrench is a specific size and most have ratchets for easier use.

Carpenter's handsaw A saw designed for cutting wood.

Ceiling hook A tool with a long wooden or fiberglass pole that has a metal point with a spur at right angles at one end. It can be used to probe ceilings and pull down plaster lath material.

Chain saw A power saw that uses the rotating movement of a chain equipped with sharpened cutting edges. Typically used to cut through wood.

Chisel A metal tool with one sharpened end used to break apart material in conjunction with a hammer, mallet, or sledgehammer.

Claw bar A tool with a pointed claw-hook on one end and a forked- or flat-chisel pry on the other end. It is often used for forcible entry.

Clemens hook A multipurpose tool that can be used for several forcible entry and ventilation applications because of its unique head design.

Closet hook A type of pike pole intended for use in tight spaces, commonly 2′ to 4′ in length.

Come along A hand-operated tool used for dragging or lifting heavy objects. Sometimes known as lever blocks.

Coping saw A saw designed to cut curves in wood.

Coupling One set or a pair of connection devices attached to a fire hose that allows the hose to be interconnected to additional lengths of hose.

Crowbar A straight bar made of steel or iron with a forked-like chisel on the working end suitable for performing forcible entry.

Cutting torch A torch that produces a high temperature flame capable of heating metal to its melting point, thereby cutting through an object. Because of the high temperatures (5,700°F) that these torches produce, the operator must be specially trained before using this tool.

Drywall hook A specialized version of a pike pole that can remove drywall more effectively because of its hook design.

Flat bar A specialized type of prying tool made of flat steel with prying ends suitable for performing forcible entry.

Flat-head axe A tool that has a head with an axe on one side and a flat head on the opposite side.

Forcible entry Techniques used by fire fighters to gain entry into buildings, vehicles, aircraft, or other areas when normal means of entry are locked or blocked.

Gripping pliers A hand tool with a pincer-like working end that can be used to bend wire or hold smaller objects.

Hacksaw A cutting tool designed for use on metal. Different blades can be used for cutting different types of metal.

Halligan tool A prying tool that incorporates a pick and a fork, designed for use in the fire service. Sometimes known as a Hooligan tool.

Hammer A striking tool.

Hand light Small, portable light carried by fire fighters to improve visibility at emergency scenes, often powered by rechargeable batteries.

Handsaw A manually powered saw designed to cut different types of materials. Examples include hacksaws, carpenter's handsaws, keyhole saws, and coping saws.

Horizontal ventilation The process of making openings so that smoke, heat, and gases can escape horizontally from a building through openings such as doors and windows.

Hux bar A multipurpose tool that can be used for several forcible entry and ventilation applications because of its unique design. Also can be used as a hydrant wrench.

Hydrant wrench A hand tool is used to operate the valves on a hydrant; may also be used as a spanner wrench. Some are plain wrenches, and others have a ratchet feature.

Hydraulic shears A lightweight hand-operated tool that can produce up to 10,000 pounds of cutting force.

Hydraulic spreader A lightweight hand-operated tool that can produce up to 10,000 pounds of prying and spreading force.

Interior attack The assignment of a team of fire fighters to enter a structure and attempt fire suppression.

Irons A combination tool, normally the Halligan tool and the flat-head axe.

K tool Used to remove lock cylinders from structural doors so the locking mechanism can be unlocked.

Kelly tool A steel bar with two main features: a large pick and a large chisel or fork.

Keyhole saw A saw designed to cut keyholes in wood.

Lifeline A rope secured to a fire fighter that enables him or her to retrace his or her steps out of a structure.

Mallet A short-handled hammer.

Maul A specialized striking tool, weighing six pounds or more, with an axe on one end and a sledgehammer on the other end.

Mechanical saw Usually powered by electric motors or gasoline engines. The three primary types are chain saws, rotary saws, and reciprocating saws.

Multipurpose hook A long pole with a wooden or fiberglass handle and a metal hook on one end used for pulling.

Negative pressure ventilation Ventilation that relies upon electric fans to pull or draw the air from a structure or area.

Open-end wrench A hand tool used to tighten or loosen bolts. The end is open, as opposed to a box-end wrench. Each wrench is a specific size.

Overhaul Examination of all areas of the building and contents involved in a fire to ensure that the fire is completely extinguished.

Personal protective equipment (PPE) Gear worn by fire fighters that includes helmet, gloves, hood, coat, pants, SCBA, and boots. The personal protective equipment provides a thermal barrier for fire fighters against intense heat.

Pick-head axe A tool that has a head with an axe on one side and a pointed end, or "pick" on the opposite side.

Pike pole A pole with a sharp point, or pike, on one end coupled with a hook. Used to make openings in ceilings, walls, etc.

Pipe wrench A wrench having one fixed grip and one moveable grip that can be adjusted to fit securely around pipes and other tubular objects.

Plaster hook A long pole with a pointed head and two retractable cutting blades on the side.

Positive pressure ventilation Ventilation that relies upon fans to push or force clean air into a structure.

Pry bar A specialized prying tool made of a hardened steel rod with a tapered end that can be inserted into a small area.

Wrap-Up

Rabbet tool Hydraulic spreading tool designed to pry open doors that swing inward.

Rapid intervention company/crew (RIC) A minimum of two fully equipped personnel on site, in a ready state, for immediate rescue of injured or trapped fire fighters. In some departments, this is also known as a Rapid Intervention Team.

Reciprocating saw Powered by electric or battery motors, a reciprocating saw's blade moves back and forth.

Response Activities that occur in preparation for an emergency and continue until the arrival of emergency apparatus at the scene.

Roofman's hook A long pole with a solid metal hook used for pulling.

Rotary saw Powered by electric motors or gasoline engines, a rotary saw uses a large rotating blade to cut through material. The blades can be changed depending upon the material that is being cut.

San Francisco hook A multipurpose tool that can be used for several forcible entry and ventilation applications because of its unique design, which includes a built-in gas shut-off and directional slot.

Screwdriver A tool used for turning screws.

Search and rescue The process of searching a building for a victim and extricating the victim from the building.

Seatbelt cutter A specialized cutting device that cuts through seatbelts.

Size-up The ongoing observation and evaluation of factors that are used to develop objectives, strategy, and tactics for fire suppression.

Sledgehammer A long, heavy hammer that requires the use of both hands.

Socket wrench A wrench that fits over a nut or bolt and uses the ratchet action of an attached handle to tighten or loosen the nut or bolt.

Spanner wrench A type of tool used in coupling or uncoupling hoses by turning the rocker lugs on the connections.

Spring-loaded center punch A spring-loaded punch used to break automobile glass.

Thermal imaging device High-tech device that uses infrared technology to find objects giving off a heat signature.

Ventilation The process of removing smoke, heat, and toxic gases from a burning structure and replacing them with clean air.

Vertical ventilation The process of making openings so that the smoke, heat, and gases can escape vertically from a structure.

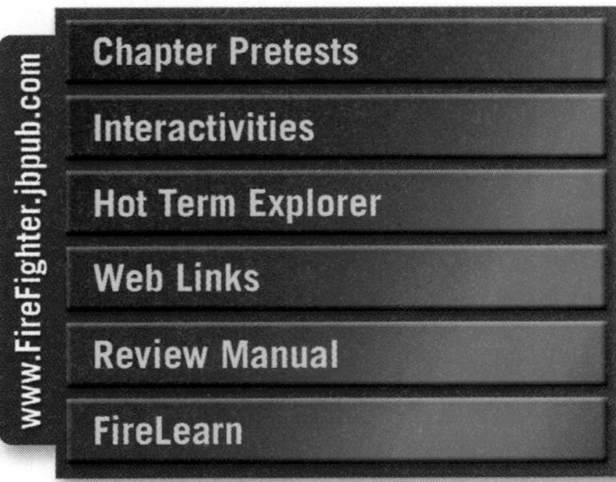

Fire Fighter in Action

You have recently completed your basic training and you are being assigned to a series of different companies for 30-day periods. You have just been assigned to a ladder company for the first time.

1. One of the first things that you should do is:
   A. Check the warranty for all of the power tools carried on the apparatus.
   B. Make a mental note of where each tool is carried so you can find it quickly when you need it.
   C. Sharpen the blades on all of the cutting tools.
   D. Make sure that the striking tools and the prying tools are kept in separate compartments.
2. The fire officer tells you that your assignment for the day is to carry "the irons." You understand from this assignment that your first responsibility at the scene of a fire will be:
   A. Search and rescue
   B. Forcible entry
   C. Ventilation
   D. Rapid intervention
3. "The irons" refers to two specific tools that are usually used together. These tools are:
   A. Pick-head axe and pike pole
   B. K tool and rabbit tool
   C. Flat-head axe and Halligan tool
   D. Battering ram and bolt cutters
4. You respond on the second alarm for a fire in a three-story garden apartment building. While the fire officer reports to the Incident Commander, you observe that smoke is coming from the top floor and escaping from under the eaves of the building. When the officer returns, he tells you and the other crew members to bring pike poles inside the building. Your assignment will most likely be:
   A. Vertical ventilation
   B. Rapid intervention crew
   C. Horizontal ventilation
   D. Opening ceilings to expose hidden fire
5. While working with a pike pole inside an apartment on the third floor, you observe smoke coming from behind the wooden baseboards, just above the floor. The fire officer tells you to open this area and see if there is fire in the wall. You should:
   A. Use your pike pole to pry the baseboard away from the wall
   B. Use a sledgehammer to break the baseboard
   C. Get a screwdriver from the tool kit and remove the screws that secure the baseboard to the wall
   D. Get a pick-head axe and use it to cut the baseboard away from the wall
6. The fire officer tells you and another fire fighter to set up positive pressure ventilation to clear the smoke out of the area where you will be working. You should place a positive pressure ventilation fan:
   A. Outside a doorway to blow fresh air into the building
   B. On the roof of the building to suck smoke out through a hole
   C. In an open window to blow smoke out
   D. Inside the building to blow smoke toward an open door
7. Late one night, your ladder company responds to a call for smoke coming from a store in a shopping center. When you arrive, you see that all of the stores are closed, but there is a light smoke inside a coffee shop. The coffee shop has a cylinder lock in the metal frame of a glass door. Which tool are you most likely to use to gain entry?
   A. Pick-head axe
   B. Spring-loaded center punch
   C. K tool
   D. Rabbit tool

Ropes and Knots

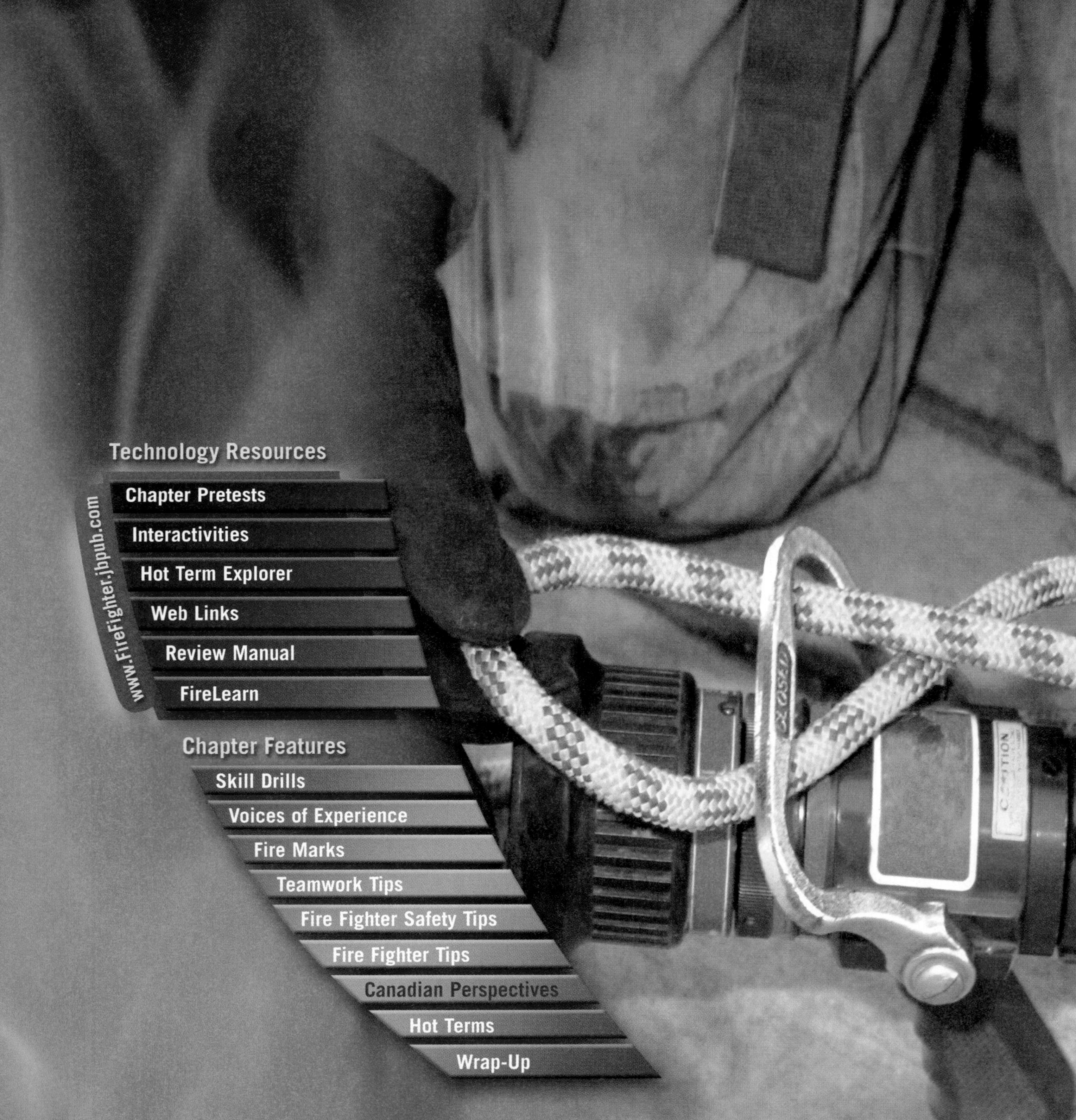

Chapter 9

NFPA 1001 Standard

Fire Fighter I

5.1.1.1 *General Knowledge Requirements.* The organization of the fire department; the role of the Fire Fighter I in the organization; the mission of fire service; the fire department's standard operating procedures and rules and regulations as they apply to the Fire Fighter I; the role of other agencies as they relate to the fire department; aspects of the fire department's member assistance program; the critical aspects of NFPA 1500, *Standard on Fire Department Occupational Safety and Health Program,* as they apply to the Fire Fighter I; knot types and usage; the difference between life safety and utility rope; reasons for placing rope out of service; the types of knots to use for given tools, ropes, or situations; hoisting methods for tools and equipment; and using rope to support response activities.

5.1.1.2 *General Skill Requirements.* The ability to don personal protective clothing within one minute; doff personal protective clothing and prepare for reuse; hoist tools and equipment using ropes and the correct knot; tie a bowline, clove hitch, figure eight on a bight, half hitch, becket or sheet bend, and safety knots; and locate information in departmental documents and standard or code materials.

5.5.3 Clean and check ladders, ventilation equipment, self-contained breathing apparatus (SCBA), ropes, salvage equipment, and hand tools, given cleaning tools, cleaning supplies, and an assignment, so that equipment is clean and maintained according to manufacturer's or departmental guidelines, maintenance is recorded, and equipment is placed in a ready state or reported otherwise.

Fire Fighter II

6.4.2 Assist rescue operation teams, given standard operating procedures, necessary rescue equipment, and an assignment, so that procedures are followed, rescue items are recognized and retrieved in the time as prescribed by the AHJ, and the assignment is completed.

Additional NFPA Standards

NFPA 1983, *Standard on Fire Service Life Safety Rope and System Components*

Knowledge Objectives

After studying this chapter, you will be able to:

- Describe the differences between life safety rope and utility rope.
- List the three most common synthetic fiber ropes used for fire department operations.
- Describe the construction of a kernmantle rope.
- Describe how to use rope to support response activities.
- Describe how to clean and check ropes.
- Describe how to record rope maintenance.
- List the reasons for placing a life safety rope out of service.
- Describe the knot types and their usage in the fire service.
- Describe how to tie safety, half hitch, clove hitch, figure eight, figure eight on a bight, figure eight with a follow-through, bowline, and sheet bend or Becket bend knots.
- Describe the types of knots to use for given tools, ropes, or situations.
- Describe hoisting methods for tools and equipment.

Skills Objectives

After completing this chapter, you will be able to perform the following skills:

- Place a life safety rope into a rope bag.
- Tie the following knots:
  Safety (overhand)
  Half hitch
  Clove hitch
  Figure eight
  Figure eight on a bight
  Figure eight with a follow-through
  Bowline
  Sheet bend or Becket bend
- Hoist the following tools using the correct knots:
  Axe
  Pike pole
  Ladder
  Charged hose line
  Uncharged hose line
  Exhaust fan

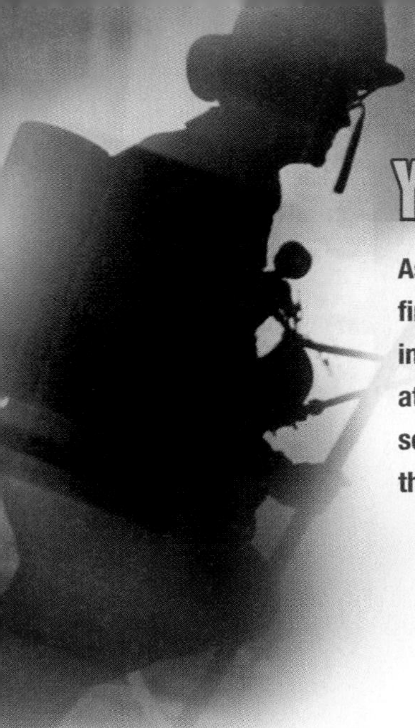

You Are the Fire Fighter

As a new recruit, you are riding as the fourth member of a truck company to the scene of your first fire. Flames are showing from the third floor in the rear of the building. Two fire fighters go into the building to perform search-and-rescue operations while the engine company sets up an attack line to begin fire suppression activities. As the engine company enters the building, you see the rescue team at an upstairs window. The captain tells you that they need another axe and that they are going to drop a rope so you can send it up to them.

1. ***As the rescue team lowers a rope from the upstairs window, what factors do you need to consider?***
2. ***How do you safely tie the axe to the rope?***
3. ***What kind of knot will you use?***
4. ***How can you be sure that the axe will not slip out of the rope?***

Introduction

In the fire service, ropes are widely used to hoist or lower tools, appliances, or people; to pull a person to safety; or to serve as a lifeline in an emergency. A rope may be your only means of accessing a trapped person or your only way of escaping from a fire.

Learning about ropes and knots is an important part of your training as a fire fighter. This chapter will give you a basic understanding of the importance of ropes and knots in the fire service. You can then build on this foundation as you develop skills in handling ropes and tying knots. You must be able to tie simple knots accurately without hesitation or delay.

This chapter discusses different types of rope construction and the materials used in making ropes. It covers the care, cleaning, inspection, and storage of ropes. It also shows how to tie eight essential knots and how to secure tools and equipment so they can be raised or lowered using ropes. Finally, it discusses additional uses for ropes in various rescue situations.

Types of Rope

There are two primary types of rope used in the fire service, each dedicated to a distinct function. Life safety rope is used solely for supporting people. Life safety rope must be used anytime a rope is needed to support a person, whether during training or during firefighting, rescue, or other emergency operations (▶ Figure 9-1). Utility rope is used in most other cases, when it is NOT necessary to support the weight of a person, such as when hoisting or lowering tools or equipment.

Life Safety Rope

The life safety rope is a critical tool used only for life-saving purposes. It must never be used for utility purposes. Life safety rope must be used in every situation where the rope must support the weight of one or more persons. In these situations, rope failure could result in serious injury or death.

Because a fire fighter's equipment must be extremely reliable, the criteria for design, construction, and performance of life safety rope and related equipment are specified in NFPA 1983, *Standard on Fire Service Life Safety Rope and System Components*. Life safety ropes are rated for either one

Figure 9-1 A life safety rope is a critical tool for fire fighters.

person or two persons. A two-person rope must be used in rescue operations where both the rescued individual and the rescuer require support.

NFPA 1983 lists very specific standards for the construction of life safety rope. The *Standard* also requires the rope manufacturer to include detailed instructions for the proper use, maintenance, and inspection of the life safety rope, including the conditions for removing the rope from service. The manufacturer must also supply a list of criteria that must be reviewed before a life safety rope that has been used in the field can be used again. If the rope does not meet all of the criteria, it must be retired from service.

Types of Life Safety Ropes

The two primary types of life safety ropes are the one-person rope and the two-person rope. In NFPA 1983, the one-person rope is classified as a light duty life safety rope and the two-person rope is classified as a general duty life safety rope.

Figure 9-2 **A.** A one-person rope is a light duty life safety rope. **B.** A two-person rope is a general duty life safety rope.

Figure 9-3 An personal escape rope is designed to be used only once.

A one-person life safety rope is designed to bear the weight of a single person (300 lb) (◂ Figure 9-2A). A two-person life safety rope is designed to bear the weight of two people (600 lb) (◂ Figure 9-2B). After each use, these ropes must be inspected according to the criteria provided by the manufacturer before they can be used again. If a life safety rope has been damaged or overstressed, or if it does not meet the inspection criteria, it cannot be reused as a life safety rope.

Personal Escape Ropes

An personal escape rope is intended to be used by a fire fighter only for self-rescue from an extreme situation. This rope is designed to carry the weight of only one person and to be used only one time (◂ Figure 9-3). Its purpose is to provide the fire fighter with a method of escaping from a life-threatening situation. After one use, the personal escape rope should be replaced by a new rope.

When you are fighting a fire, you should always have a safe way to get out of a situation and to a safe location. You may be able to go back through the door that you entered, or you may have another exit route, such as through a different door, through a window, or down a ladder. If conditions suddenly change for the worse, having an escape route can save your life.

Sometimes, however, you may find yourself in a situation where conditions deteriorate so quickly that you cannot use your planned exit route. For example, the stairway you used collapses behind you, or the room you are in suddenly flashes over (a phase in the development of a contained fire in which exposed surfaces reach ignition temperature more or less simultaneously and fire spreads rapidly throughout the space), blocking your planned route out. In such a situation, you may need to take extreme measures to get out of the building. The personal escape rope was developed specifically for this type of emergency self-rescue situation. A personal escape rope can support the weight of one person and fits easily in a small packet or pouch (▾ Figure 9-4).

Figure 9-4 A personal escape rope can be easily carried by a fire fighter.

Figure 9-5 Utility ropes are used for hoisting and lowering tools.

Figure 9-6 Some utility ropes are made from natural fibers.

Because these personal escape ropes are so important, they can be used only once. After they are used once, they are discarded. The rope may have been damaged in some way and might fail if it ever has to be used again. You cannot take that chance; your life may depend on the quality and strength of your personal escape rope.

Utility Rope

Utility rope is used when it is NOT necessary to support the weight of a person. Fire department utility rope is used for hoisting or lowering tools or equipment, for ladder halyards (rope used on extension ladders to raise a fly section), for marking off areas, and for stabilizing objects (▲ Figure 9-5). Utility ropes also require regular inspection.

Utility ropes must not be used in situations where life safety rope is required. Conversely, life safety rope must not be used for utility applications. A fire fighter must be able to instantly recognize the category of a rope from its appearance and markings.

Fire Fighter Safety Tips

Many fire departments use color-coding or other visible markings to identify different types of rope. This allows a fire fighter to very quickly determine if a rope is a one-person or two-person life safety rope or a utility rope. The length of each rope should also be clearly marked by a tag or a label on the rope bag.

Rope Materials

Ropes can be made from many different types of materials. The earliest ropes were made from naturally occurring vines or fibers that were woven together. Ropes are now made of synthetic materials such as nylon or polypropylene. Because ropes have many different uses, different materials may work better than others in various situations.

Natural Fibers

In the past, fire departments used ropes made from natural fibers, such as manila, because there were no alternatives (▲ Figure 9-6). The natural fibers are twisted together to form strands. A strand may contain hundreds of individual fibers of different lengths. Today, ropes made from natural fibers are still used as utility ropes, but are no longer acceptable as life safety ropes (► Table 9-1). Natural fiber ropes can be weakened by mildew and deteriorate with age, even when properly stored. A wet manila rope can absorb 50% of its weight in water, making it very susceptible to deterioration. A wet natural fiber rope is very difficult to dry.

Synthetic Fibers

Since nylon was first manufactured in 1938, synthetic fibers have been used to make ropes. In addition to nylon, several newer synthetic materials such as polyester, polypropylene, and polyethylene are used in rope construction (► Figure 9-7). Synthetic fibers have several advantages over natural fibers (► Table 9-2). Synthetic fibers are generally stronger than natural fibers, so it may be possible to use a smaller diameter rope without sacrificing strength. Synthetic materials can also pro-

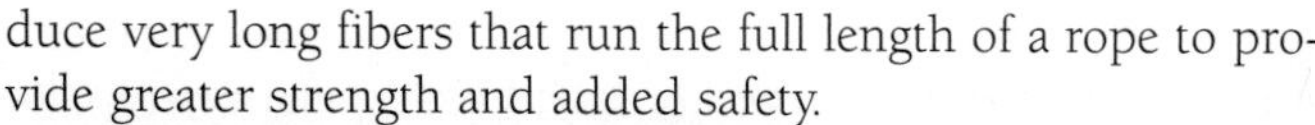

Figure 9-7 Synthetic fibers are generally stronger than natural fibers.

Figure 9-8 Polypropylene rope is often used in water rescues.

duce very long fibers that run the full length of a rope to provide greater strength and added safety.

Synthetic ropes are more resistant to rotting and mildew than natural fiber ropes and do not age or degrade as quickly. Depending on the material, synthetic ropes may provide more resistance to melting and burning than natural fiber ropes. They also absorb much less water when wet and can be washed and dried. Some types of synthetic rope can float on water, which is a major advantage in water rescue situations.

However, ropes made from synthetic fibers do have some drawbacks (► Table 9-3). Prolonged exposure to ultraviolet light as well as exposure to strong acids or alkalis can damage a synthetic rope and decrease its life expectancy. In addition, synthetic materials may be highly susceptible to abrasion or cutting.

Life safety ropes are always made of synthetic fibers. Before any rope can be used for life safety purposes, it must meet the requirements outlined in the current *NFPA 1983*. These standards specify that life safety rope must be made of continuous filament virgin fiber and woven of **block creel construction** (without knots or splices in the yarns, ply yarns, strands, braids, or rope). Rope of any other material or construction may not be used as a life safety rope.

The most common synthetic fiber used in life safety ropes is nylon. It has a high melting temperature with good abrasion resistance and is strong and lightweight. Nylon ropes are also resistant to most acids and alkalis. Polyester is the second most common synthetic fiber used for life safety ropes. Some life safety ropes are made of a combination of nylon and polyester or other synthetic fibers.

Polypropylene is the lightest of the synthetic fibers. Because it does not absorb water and floats, polypropylene rope is often used for water rescue situations (► Figure 9-8). However, it is not as suitable as nylon for fire department life safety uses because it is not as strong, it is hard to knot, and it has a low melting point.

Table 9-1 Drawbacks to Using Natural Fiber Ropes
• Lose their load-carrying ability over time
• Subject to mildew
• Absorb 50% of their weight in water
• Degrade quickly

Table 9-2 Advantages to Using Synthetic Fiber Ropes
• Thinner without sacrificing strength
• Less absorbent than natural fiber ropes
• Greater resistance to rotting and mildew
• Longer-lasting than natural fiber ropes
• Greater strength and added safety
• More fire-retardant than natural fiber ropes

Table 9-3 Drawbacks to Using Synthetic Fiber Ropes
• Can be damaged by prolonged exposure to ultraviolet light
• Can be damaged by exposure to strong acids or alkalis
• Susceptible to abrasion

Table 9-4 Properties of Rope Materials

Type	Material	Properties: Positive	Properties: Negative	Application
Natural	Manila	No real advantages over synthetic rope	Absorbs water easily Cannot bear as much weight as synthetic rope Noncontinuous fibers Easily degraded	Utility rope
Synthetic	Nylon Polyester	High melting temperature Good abrasion resistance Less absorbent than natural ropes Greater resistance to rotting and mildew Lasts longer than natural ropes	Can be damaged from exposure to sunlight, oils, gas, acids, bases, or fumes	Life safety rescue rope Utility rope
	Polypropylene	Does not absorb water Floats	Hard to knot Low melting point	Water rescue rope Utility rope

Rope Construction

There are several different types of rope construction. The best choice of rope construction depends on the specific application (▲ Table 9-4).

Twisted and Braided Rope

Twisted ropes, which are also called laid ropes, are made of individual fibers twisted into strands. The strands are then twisted together to make the rope (► Figure 9-9 A). This method of rope construction has been used for hundreds of years. Both natural and synthetic fibers can be used to make twisted rope.

This method of construction exposes all of the fibers to the outside of the rope where they are subject to abrasion. Abrasion can damage the rope fibers and may reduce rope strength. Twisted ropes tend to stretch and are prone to unraveling when a load is applied.

Braided ropes are constructed by weaving or intertwining strands together in the same way that hair is braided (► Figure 9-9 B). This method of construction also exposes all of the strands to the outside of the rope where they are subject to abrasion. Most braided rope is constructed from synthetic fibers. Braided rope will stretch under a load, but it is not prone to twisting. A double braided rope has an inner braided core covered by a protective braided sleeve, so that only the fibers in the outer sleeve are exposed to the outside. The inner core is protected from abrasion.

Kernmantle Rope

Kernmantle rope consists of two distinct parts: the kern and the mantle. The kern is the center or core of the rope; it provides about 70% of the strength of the rope. The mantle or sheath is a braided covering that protects the core from dirt and abrasion. Only about 30% of the strength of the rope comes from the mantle. Both parts of a kernmantle rope are made with synthetic fibers, but different fibers may be used for the kern and the mantle.

Each fiber in the kern extends for the entire length of the rope without knots or splices. This block creel construction is required under NFPA 1983 for all life safety ropes. The continuous filaments produce a core that is stronger than one constructed of shorter fibers that are twisted or braided together.

Fire Fighter Tips

To understand how a kernmantle rope gets its strength, imagine a small, monofilament fishing line rated at 15 pounds. Putting 100 identical strands of this filament side-by-side would create a cable capable of supporting 1500 pounds. When covered with a sheath, the fishing line cable would create a static kernmantle rope.

Figure 9-9 **A.** Twisted rope. **B.** Braided rope.

Figure 9-10 A. Suspension bridge cables use the same type of construction as kernmantle ropes. B. This close-up shows the core and mantle construction.

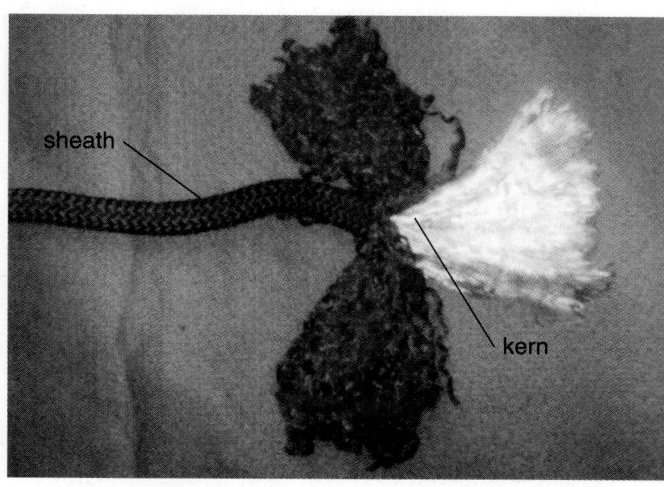

Figure 9-11 The parts of a kernmantle rope.

Kernmantle construction is also used for the cables that support a suspension bridge (▲ Figure 9-10 A & B). Thin strands of steel are laid from one end of the bridge to the other. Although each strand only supports a small weight, multiple strands are strong enough to support both the bridge and the load carried by the bridge. The cables are then wrapped with a protective covering, similar to the mantle portion of a rope.

Kernmantle construction produces a very strong and flexible rope that is relatively thin and lightweight (► Figure 9-11). This construction is well suited for rescue work and is very popular for life safety rope.

Dynamic and Static Rope

A rope can be either dynamic or static, depending on how it reacts to an applied load. A dynamic rope is designed to be elastic and will stretch when it is loaded. A static rope will not stretch under load. The differences between dynamic and static ropes result from both the fibers used and the construction method.

Dynamic rope is usually used in safety lines for mountain climbing, because it will stretch and cushion the shock if a climber falls a long distance. A static rope is more suitable for most fire rescue situations, where falls from great heights are not anticipated. Teams that specialize in rope rescue often carry both static and dynamic ropes for use in different situations.

Dynamic and Static Kernmantle Ropes

Kernmantle ropes can be either dynamic or static. A dynamic kernmantle rope is constructed with overlapping or woven fibers in the core. When the rope is loaded, the core fibers are pulled tighter, which gives the rope its elasticity.

The core of a static kernmantle rope has all of the fibers laid parallel to each other. A static kernmantle rope has very little elasticity and limited elongation under an applied load. Most fire department life safety ropes use static kernmantle construction. It is well suited for lowering a person and can be used with a pulley system for lifting individuals. It can also be used to create a bridge between two structures.

Rope Strength

Life safety ropes are rated to carry a specific amount of weight under the minimum requirements of NFPA 1983 (▼ Table 9-5). The required minimum breaking strength for a life safety

Table 9-5 Required Strength of Life Safety Ropes

Classification	Rated Load (Persons)	Rated Load (Weight)	Minimum Breaking Strength	Safety Factor
Personal escape rope	One	300 lb	3,000 lbf (13.34 kN)	10:1
Light use life safety rope	One	300 lb	4,500 lbf (20 kN)	15:1
General use life safety rope	Two	600 lb	9,000 lbf (40 kN)	15:1

SOURCE: NFPA 1983, *Standard on Fire Service Life Safety Rope and System Components*

Fire Marks

Although synthetic fiber ropes were introduced to the fire service in the 1950s, they were not widely adopted until the 1980s.

rope is based on an assumed loading of 300 lb per person with a safety factor of 15:1. The safety factor allows for reductions in strength due to knots, twists, abrasion, or any other cause. The safety factor also allows for shock loading if a weight is applied very suddenly. For example, shock loading could occur if the person who is tied to the rope falls and then is stopped by the rope. A personal escape rope is also expected to support a weight of 300 lb, representing one person, with a safety factor of 10:1.

The actual breaking strength of a rope depends on the material, the diameter, and the type of construction. The rope manufacturer should be consulted for detailed specifications.

Figure 9-12 A carabiner.

Technical Rescue Hardware

During technical rescue incidents, ropes are often used to access and extricate individuals. In addition to the rope itself, several hardware components may also be used. The one most commonly used by fire fighters is a carabiner, or a snap link (► Figure 9-12). This device is used to connect one rope to another rope, to a harness, or to itself. There are different types of carabiners, and you should know how to operate the type used by your department.

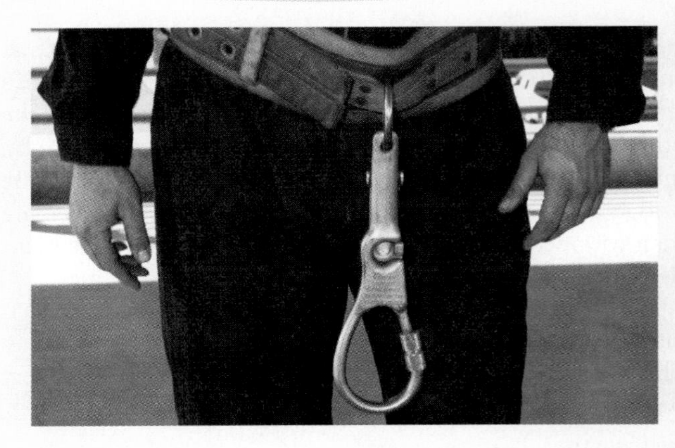

Figure 9-13 Class I harness (ladder belt harness).

Harnesses

A harness is a piece of rescue or safety equipment made of webbing and worn by a person. It is used to secure the person to a rope or to a solid object. Three different types of harnesses—belts, seats, and chest harnesses—are used by rescuers, depending upon the type of circumstances encountered.

- The ladder belt harness (Class I harness) is used to keep a fire fighter in place on a ladder (► Figure 9-13).
- The seat harness (Class II harness) is used to support a fire fighter, particularly in rescue situations (► Figure 9-14).
- The chest harness (Class III harness) is the most secure type of harness and is often used to support a fire fighter who is being raised or lowered on a life safety rope (► Figure 9-15).

Harnesses need to be cleaned and inspected regularly, just as you do for life safety ropes. Follow the manufacturer's instructions for cleaning and inspecting harnesses.

Figure 9-14 Class II harness (seat harness).

Rope Rescue

Rope rescue involves raising and lowering rescuers to access injured or trapped individuals, as well as raising or lowering those rescued so they can be given appropriate medical treatment. An approved rope rescue course is required to attain proficiency in rope rescue skills. However, this chapter will provide you with the basics of rope rescue and give you a foundation for learning the more complex parts of rope rescue.

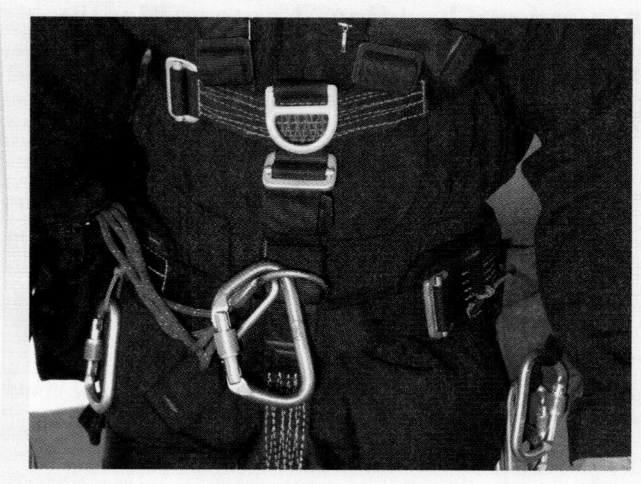

Figure 9-15 Class III harness (chest harness).

Rope rescue courses cover the technical skills needed to raise or lower people using mechanical advantage systems and to remove someone from a rock ledge or a confined space (▼ Figure 9-16). They also cover the equipment and skills needed to accomplish these rescues safely.

Rope Rescue Incidents

Most rope rescue incidents involve people who are trapped in normally inaccessible locations such as a mountainside or the outside of a building (▼ Figure 9-17). Rescuers often have to lower themselves using a system of anchors, webbing, ropes, carabiners, and other devices to reach the trapped person. Once rescuers reach the person, they then have to stabilize him or her and determine how to get the person to safety. Sometimes the person will have to be lowered or raised to a safe location. Extreme cases could involve more complicated operations, such as transporting the person in a basket lowered by a helicopter.

The type and number of ropes used in a rope rescue will depend upon the situation. There is almost always a primary rope that will bear the weight of the rescuer (or rescuers) in order to reach the person. The rescuers will often

Figure 9-16 Rope rescues require intense technical training.

Figure 9-17 Ropes are invaluable when a person is trapped in an inaccessible area, like a high building.

Voices of Experience

"That night, when it came time to tie those lifelines around my buddies, I automatically remembered the knot that secured them to their rescue lines."

One cold winter night, we received a report from dispatch that two children had fallen through the ice at a retention pond. The temperature that night was around 30°F. We quickly mobilized and headed toward the pond. When we arrived at the scene, I could see where the children had gone through the ice, but there was no sign of them. We immediately broke into teams and went into the water two at a time. The water was about 8′ deep and the water temperature was around 20 to 25°F. We could only stay in the water for about two minutes at a time.

My initial role was rope tender, so I tied lifelines to my buddies going into the water. Most fire fighters do not practice tying knots once they get out of the academy. After I left the academy, I purchased a small length of kernmantle rope for about $4.00 from a fire supply store. Whenever it was slow around the station, I would practice tying different knots. That night, when it came time to tie those lifelines around my buddies, I automatically remembered the knot that secured them to their rescue lines. I tied bowlines that night and kept them safe.

Don't take your training for granted. Practice the skills that you learned at the academy during downtime. You never know when you will need to tie a bowline or a clove hitch. Your life or a friend's life just may depend on that knot.

David King
Chesapeake Fire Department
Chesapeake, Virginia

have a second line attached to them, known as a belay line, which serves as a backup if the main line fails. Additional lines may be needed to raise or lower the trapped individual, depending upon the circumstances.

Trench Rescue

Rescues in collapsed trenches often are complicated and involve a number of different skills, such as shoring, air quality monitoring, confined space operations, and rope rescue. Ropes are often used to remove the trapped person. After the rescuers shore the walls of the trench and remove the dirt covering the person, they will place the person in a Stokes basket or on a backboard and lift him or her to the surface. If the trench is deep, ropes may be used to raise the patient to the surface.

Confined Space Rescue

A confined space rescue can take place in locations such as tanks, silos, underground electrical vaults, storm drains, and similar structures. It is often very difficult to extricate an unconscious or injured person from these locations because of the poor ventilation and limited entry or exit area. For this reason, ropes are often used to remove an injured or unconscious person (◄ Figure 9-18).

Figure 9-18 Ropes are often used to remove an injured or unconscious person from a confined space.

Water Rescue

Ropes can be used in a variety of ways during water rescue operations. The simplest situation involves a rescuer on the shore throwing a rope to a person in the water and pulling the person to shore (◄ Figure 9-19). A more complicated situation may involve a rope stretched across a stream or river. A boat is tethered to the rope, and rescuers on shore maneuver the boat using a series of ropes and pulleys.

Figure 9-19 Ropes ensure rescuer safety during water rescues.

Rope Maintenance

All ropes, especially life safety ropes, need proper care to perform in an optimal manner. Maintenance is necessary for all kinds of equipment and all types of rope, and it is absolutely essential for life safety ropes. Your life and the lives of others depend on the proper maintenance of your life safety ropes.

There are four parts to the maintenance formula:

- Care
- Clean
- Inspect
- Store

Care

You must follow certain principles to preserve the strength and integrity of rope (▼ Table 9-6):

- Protect the rope from sharp and abrasive surfaces. Use edge protectors when the rope must pass over a sharp or unpadded surface.
- Protect the rope from heat, chemicals, and flames.
- Protect the rope from rubbing against another rope or webbing. Friction generates heat, which can damage or destroy the rope.
- Protect the rope from prolonged exposure to sunlight. Ultraviolet radiation can damage rope.

Table 9-6 Principles to Preserve Strength and Integrity of Rope

- Protect from sharp abrasive surfaces.
- Protect from heat, chemicals, and flames.
- Protect from rubbing against another rope.
- Protect from prolonged exposure to sunlight.
- Never step on a rope.
- Follow manufacturer's recommendations.

Fire Fighter Safety Tips

A **shock load** can occur when a rope is suddenly placed under unusual tension. This could occur if someone attached to a life safety rope falls until the length of the rope or another rescuer stops the drop. A utility rope can be shock-loaded in a similar manner if a piece of equipment that is being raised or lowered suddenly drops.

Any rope that has been shock-loaded should be inspected and may have to be removed from service. Although there may not be any visible damage, shock-loading may cause damage that is not immediately apparent. Repeated shock loads can severely weaken a rope so that it can no longer be used safely. Accurate rope records will help identify potentially damaged rope.

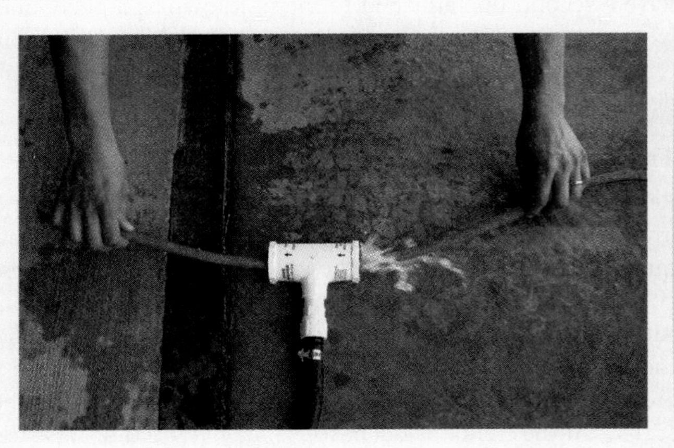

Figure 9-20 Some fire departments use a rope washer to clean their ropes.

- Never step on a rope! Your footstep could force shards of glass, splinters, or abrasive particles into the core of the rope, damaging the rope fibers.
- Follow the manufacturer's recommendations for rope care.

Clean

Many ropes made from synthetic fibers can be washed with a mild soap and water. A special rope washer can be attached to a garden hose (► Figure 9-20). Some manufacturers recommend placing the rope in a mesh bag and washing it in a frontloading washing machine.

Use a mild detergent. Do not use bleach because it can damage rope fibers. Follow the manufacturer's recommendations for specific care of your rope. Do not pack or store wet or damp rope. Air-drying is usually recommended, but rope should not be dried in direct sunlight. The use of mechanical drying devices is not usually recommended.

Inspect

Life safety ropes must be inspected after each use, whether the rope was used for an emergency incident or in a training exercise (▼ Table 9-7). Unused rope should be inspected on a regular schedule. Some departments inspect all rope, including life safety and utility ropes, every three months. Obtain the inspection criteria from the rope manufacturer.

Inspect the rope visually, looking for cuts, frays, or other damage, as you run it through your fingers. Because you cannot see the inner core of a kernmantle rope, feel for any **depressions** (flat spots or lumps on the inside). Examine the sheath for any discolorations, abrasions, or flat spots (► Figure 9-21). If you have any doubt about whether the rope has been damaged, consult with your company officer (▼ Table 9-8).

A life safety rope that is no longer useable must be destroyed. In some cases, a used life safety rope can be downgraded and used as a utility rope. A downgraded rope must be clearly marked so that it cannot be confused with a life safety rope.

Rope Record

Each piece of rope must be marked for identification. A **rope record** must be kept for each piece of life safety rope. This record should include a history of when the rope was purchased, each time it was used, how it was used, and the types of loads applied to it. Each inspection should also be recorded. Many fire departments maintain records for both utility ropes and life safety ropes.

Table 9-7 Questions to Consider When Inspecting Life Safety Ropes

- Has the rope been exposed to heat or flame?
- Has the rope been exposed to abrasion?
- Has the rope been exposed to chemicals?
- Has the rope been exposed to shock loads?
- Are there any depressions, discoloration, or lumps in the rope?

Table 9-8 Signs of Possible Rope Deterioration

- Discoloration
- Shiny markings from heat or friction
- Damaged sheath
- Core fibers poking through the sheath

Figure 9-21 Rope inspection is a critical step.

Figure 9-23 Ropes may be coiled for storage.

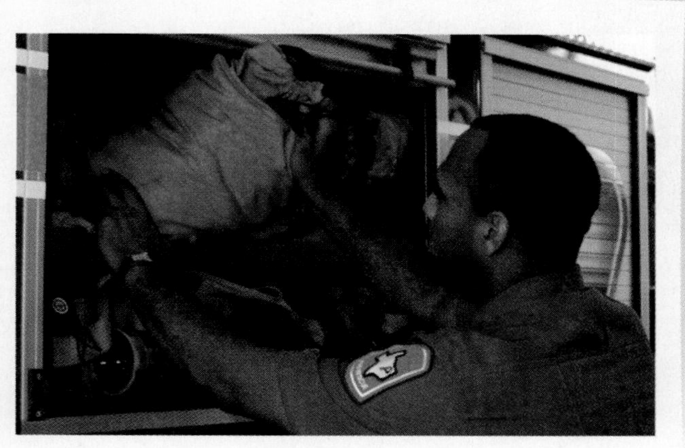

Figure 9-22 Ensure that ropes are stored safely.

Figure 9-24 Long sections of rope are sometimes stored on reels.

Store

Proper care will assure a long life for your rope and reduce the chance of equipment failure and accident. Store ropes away from temperature extremes, out of sunlight, and in areas where there is some air circulation. Avoid placing ropes where fumes from gasoline, oils, or hydraulic fluids can damage the rope. Apparatus compartments used to store ropes should be separated from compartments used to store any oil-based products or machinery powered by gasoline or diesel fuel. Do not place any heavy objects on top of the rope (▲ Figure 9-22).

Rope bags are used to protect and store ropes. Each bag should contain only one rope. Rope may also be coiled for storage (► Figure 9-23). Very long pieces of rope are sometimes stored on reels (► Figure 9-24). Follow the steps in (► Skill Drill 9-1) to place a life safety rope into a rope bag:

1. Tie a figure eight knot on a bight in the first end of the life safety rope to be placed into the rope bag. **(Step 1)**
2. Load the life safety rope into the rope bag carefully. **(Step 2)**
3. Do not try to coil the rope in the bag, because this will cause it to kink and become tangled when it is pulled out. **(Step 3)**

Knots

Knots are prescribed ways of fastening lengths of rope or webbing to objects or to each other. As a fire fighter, you must know how to tie and when to use certain knots. Knots can be used for one or more particular purposes. Hitches, such as the clove hitch, are used to attach a rope around an

9-1 Skill Drill

Placing a Life Safety Rope into a Rope Bag

Tie a figure eight knot on a bight in the first end of the rope to go into the bag.

Load the rope into the bag.

Do not coil the rope in the bag.

object. Knots, such as the figure eight and the bowline, are used to form loops. Bends, such as the sheet bend or becket bend, are used to join two ropes together. Safety knots, such as the overhand knot, are used to secure the ends of ropes to prevent them from coming untied.

Any knot will reduce the load-carrying capacity of the rope by a certain percentage (► Table 9-9). You can avoid an unnecessary reduction in rope strength if you know what type of knot to use and how to tie it correctly.

Terminology

Specific terminology is used to refer to the parts of a rope in describing how to tie knots (► Figure 9-25).

- The **working end** is the part of the rope used for forming the knot.
- The **running end** is the part of the rope used for lifting or hoisting.
- The **standing part** is the rope between the working end and the running end.
- A **bight** is formed by reversing the direction of the rope to form a "U" bend with two parallel ends (► Figure 9-26).
- A **loop** is formed by making a circle in the rope (► Figure 9-27).
- A **round turn** is formed by making a loop and then bringing the two ends of the rope parallel to each other (► Figure 9-28).

Teamwork Tips

In a rescue situation, always have your partner check your knots. Taking a few extra seconds to be sure knots are tied correctly can save lives.

Table 9-9 Effect of Knots on Rope Strength

Group	Knot	Reduction in Strength
Loop knots	Figure eight on a bight	20%
	Figure eight with a follow-through	19%
	Bowline	33%

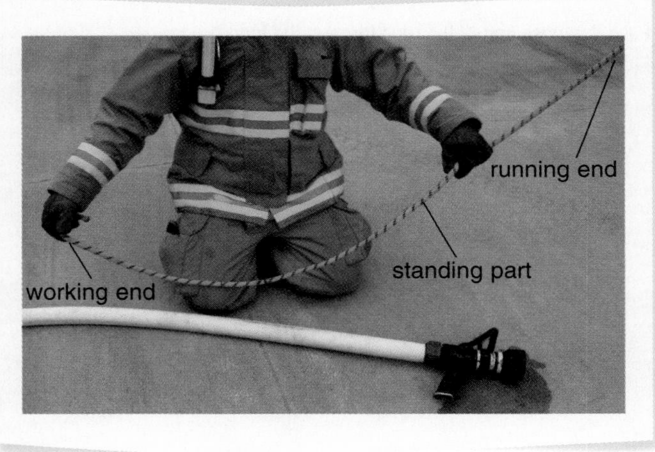

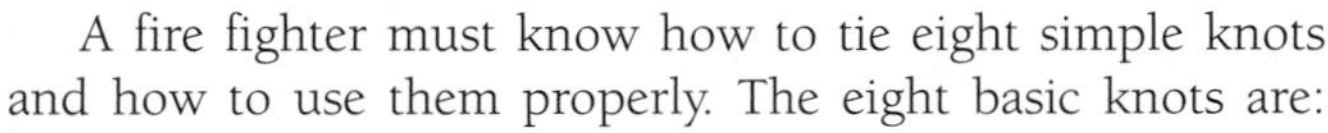
Figure 9-25 The sections of a rope used in tying knots.

Figure 9-26 A bight.

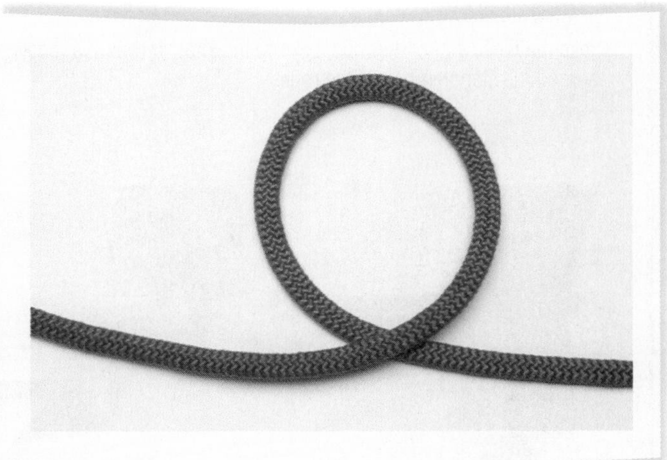
Figure 9-27 A loop.

Figure 9-28 A round turn.

A fire fighter must know how to tie eight simple knots and how to use them properly. The eight basic knots are:

- Safety knot (overhand knot)
- Half hitch
- Clove hitch
- Figure eight
- Figure eight on a bight
- Figure eight with a follow-through
- Bowline
- Bend (sheet or becket bend)

Safety Knot

A safety knot (also referred to as an overhand knot or a keeper knot) is used to secure the leftover working end of the rope. It provides a degree of safety to ensure that the primary knot will not become undone. A safety knot should always be used to finish the other basic knots.

9-2 Skill Drill

Safety Knot

Take the loose end of the rope, beyond the knot, and form a loop around the standing part of the rope.

Pass the loose end of the rope through the loop.

Tighten the safety knot by pulling on both ends at the same time.

A safety knot is simply an overhand knot in the loose end of the rope that is made around the standing part of the rope. This secures the loose end and prevents it from slipping back through the primary knot. Follow the steps in ▲ **Skill Drill 9-2** to tie a safety knot:

1. Take the loose end of the rope, beyond the knot, and form a loop around the standing part of the rope. **(Step 1)**
2. Pass the loose end of the rope through the loop. **(Step 2)**
3. Tighten the safety knot by pulling on both ends at the same time. To test whether you've tied a safety knot correctly, try sliding it on the standing part of the rope. A knot that is tied correctly will slide. **(Step 3)**

Hitches

Hitches are knots that wrap around an object, such as a pike pole or a fencepost. They are used to secure the working end of a rope to a solid object or to tie a rope to an object before hoisting it.

Half Hitch

The half hitch is not a secure knot by itself. It is used only in conjunction with other knots. For example, when hoisting

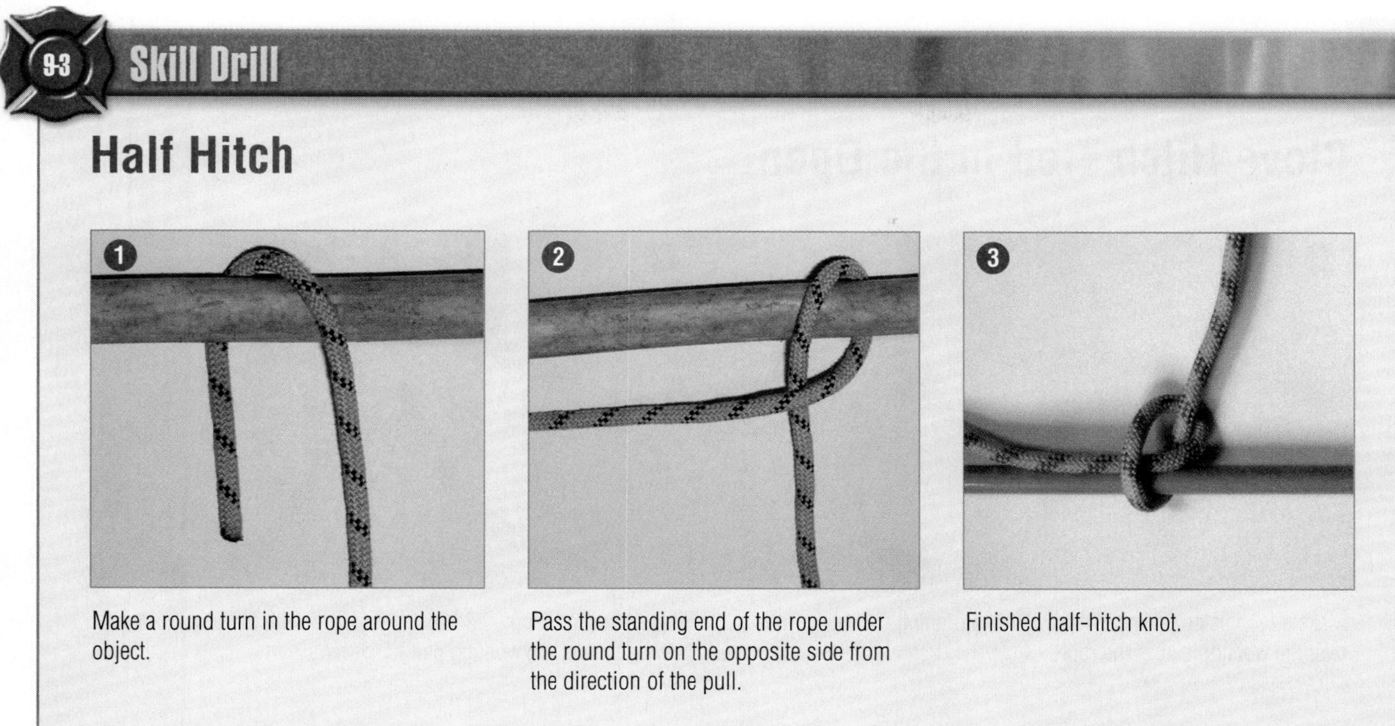

Make a round turn in the rope around the object.

Pass the standing end of the rope under the round turn on the opposite side from the direction of the pull.

Finished half-hitch knot.

an axe or pike pole, you will use the half hitch to keep the hoisting rope aligned with the handle. On long objects, you may need to use several half hitches. Follow the steps in **▲ Skill Drill 9-3** to tie a half hitch:

1. Make a round turn in the rope around the object you are hoisting. **(Step 1)**
2. Pass the standing end of the rope under the round turn on the opposite side from the direction of the pull. **(Step 2)**

Clove Hitch

A clove hitch is used to attach a rope firmly to a round object, such as a tree or a fence post. It can also be used to tie a hoisting rope around an axe or pike pole. A clove hitch can be tied anywhere in a rope and will hold equally well if tension is applied to either end of the rope or both ends simultaneously. There are two different methods of tying this knot.

A clove hitch tied in the open is used when the knot can be formed and then slipped over the end of an object, such as an axe or pike pole. It is tied by making two consecutive loops in the rope. Follow the steps in **► Skill Drill 9-4** to tie a clove hitch in the open:

1. The first loop is made with the left hand and has the running part of the rope passing over the working part. **(Step 1)**
2. The second loop is made with the right hand and has the running part of the rope passing under the working part. **(Step 2)**
3. The right-hand loop is then placed on top of the left-hand loop so that the openings are aligned. **(Step 3)**
4. The two loops are held together, and the knot is slipped over the object. **(Step 4)**
5. Tighten the clove hitch by simultaneously pulling both ends of the rope in opposite directions. **(Step 5)**
6. Tie a safety knot in the working end of the rope. **(Step 6)**

If the object is too large or too long to slip the clove hitch over one end, the same knot can be tied around the object. Follow the steps in **► Skill Drill 9-5** to tie a clove hitch around an object:

1. First, loop the rope completely around the object with the working end below the running end. **(Step 1)**
2. Loop the working end around the object a second time to form a second loop, slightly above the first loop. **(Step 2)**
3. Pass the working end under the second loop, just above the point where the second loop crosses over the first loop. **(Step 3)**
4. Secure the knot by pulling on both ends. **(Step 4)**
5. Tie a safety knot in the working end of the rope. **(Step 5)**

Loop Knots

Loop knots are used to form a loop in the end of a rope. These loops may be used for hoisting tools, for securing a person during a rescue, for securing a rope to a fixed object, or for identifying the end of a rope stored in a rope bag. When tied properly, these knots will not slip and are easy to untie.

Skill Drill

Clove Hitch Tied in the Open

Make a loop using your left hand, with the running part of the rope over the working part of the rope.

Make a second loop using your right hand, with the running part of the rope under the working part of the rope.

Bring the right-hand loop on top of the left-hand loop.

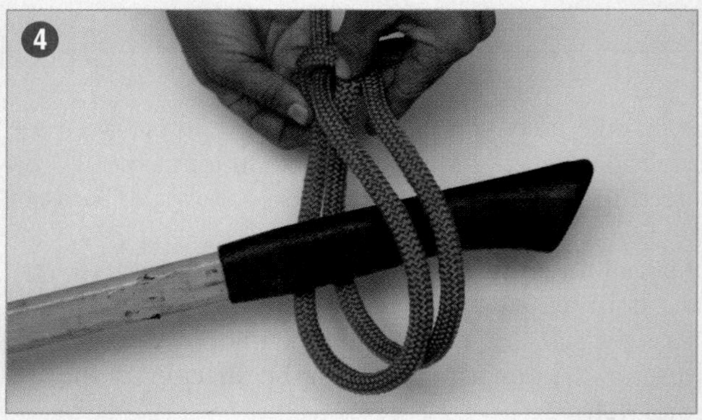

Slide both loops over the object.

Pull in opposite directions to tighten the clove hitch.

Tie a safety knot in the working end of the rope.

9-5 Skill Drill

Clove Hitch Tied Around an Object

Make a complete loop around the object, working end down.

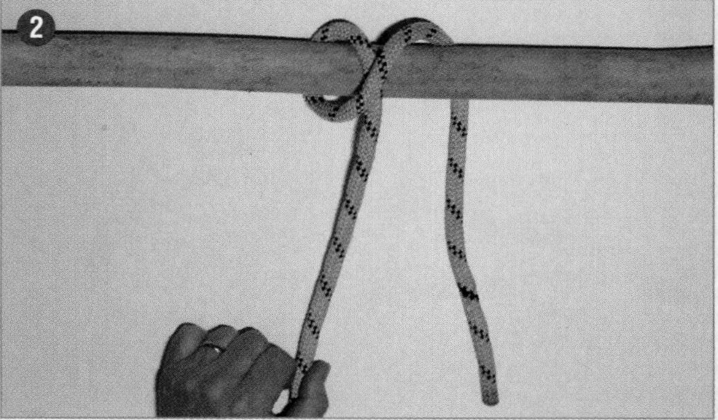

Make a second loop around the object a short distance above the first loop.

Now pass the working end of the rope under the second loop, above the point where the second loop crosses over the first loop.

Tighten the knot and secure it by pulling on both ends.

Tie a safety knot in the working end of the rope.

9-6 Skill Drill

Figure Eight Knot

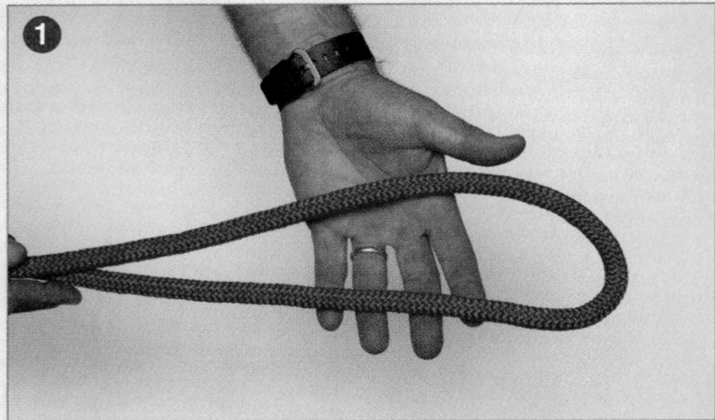
Form a bight in the rope.

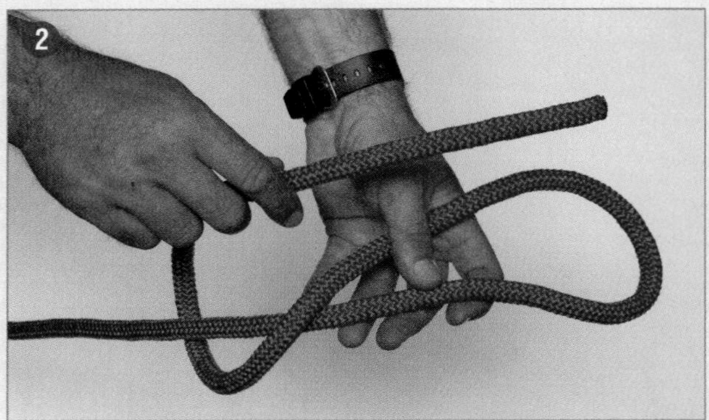
Loop the working end of the rope completely around the standing part of the rope.

Thread the working end back through the bight.

Tighten the knot by pulling on both ends simultaneously.

Figure Eight Knot

A figure eight is a basic knot used to produce a family of other knots, including the figure eight on a bight and the figure eight with a follow-through. A simple figure eight knot is seldom used alone. Follow the steps in ▲ **Skill Drill 9-6** to tie a figure eight knot:

1. Form a bight in the rope. **(Step 1)**
2. Loop the working end of the rope completely around the standing end of the rope. **(Step 2)**
3. Thread the working end through the opening of the bight. **(Step 3)**
4. Tighten the knot by pulling on both ends simultaneously. When you pull the knot tight, it will have the shape of a figure eight. **(Step 4)**

Figure Eight on a Bight

The figure eight on a bight knot creates a secure loop at the working end of a rope. The loop can be used to attach the end of the rope to a fixed object or a piece of equipment, or to tie a life safety rope around a person. The loop may be any size—from an inch to several feet in diameter. Follow the steps in ► **Skill Drill 9-7** to tie a figure eight on a bight:

1. The figure eight on a bight is tied in a section of the rope that has been doubled over to form a bight. The closed end of the bight becomes the working end of the rope. **(Step 1)**
2. Hold the two sides of the bight as if they were one rope. Form a loop in the doubled section of the rope. **(Step 2)**

9-7 Skill Drill

Figure Eight on a Bight

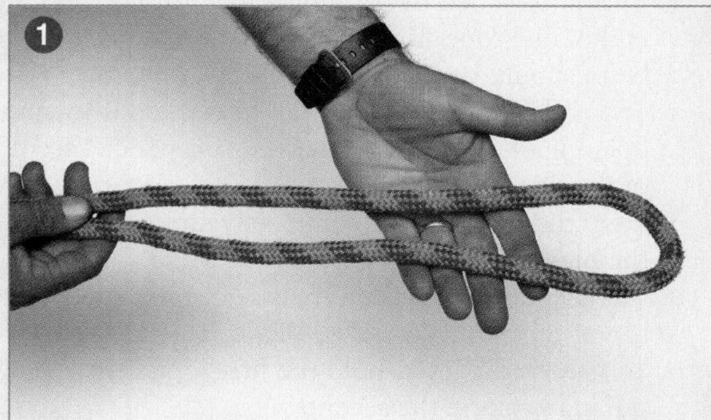

1 Form a bight and identify the end of the bight as the working end.

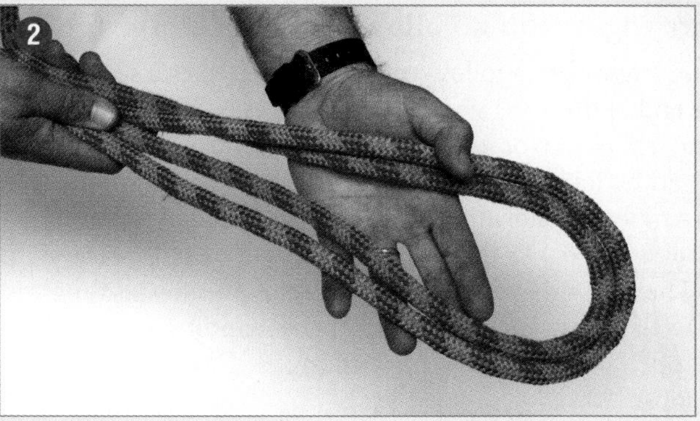

2 Holding both sides of the bight together, form a loop.

3 Feed the working end of the bight back through the loop.

4 Pull the knot tight.

5 Secure the loose end of the rope with a safety knot.

3. Pass the working end of the bight through the loop. **(Step 3)**
4. Pull the knot tight. Pulling the knot tight locks the neck of the bight and forms a secure loop. **(Step 4)**
5. Use a safety knot to tie the loose end of the rope to the standing part. **(Step 5)**

Figure Eight with a Follow-Through

A figure eight with a follow-through knot creates a secure loop at the end of the rope when the working end must be wrapped around an object or passed through an opening before the loop can be formed. It is very useful for attaching a rope to a fixed ring or a solid object with an "eye." Follow the steps in ► **Skill Drill 9-8** to tie a figure eight with a follow-through:

1. The first step in a figure eight with a follow-through is to tie a simple figure eight in the standing part of the rope, about 2′ from the working end. Leave this knot loose. **(Step 1)**
2. Thread the working end through the opening or around the object and bring it back to the figure eight knot. **(Step 2)**
3. Secure the working end by threading it back through the figure eight, tracing the path of the original knot in the opposite direction. **(Step 3)**
4. Once the working end has been threaded through the knot, you can pull the knot tight. **(Step 4)**
5. Secure the loose end with a safety knot. **(Step 5)**

This knot can also be used to tie the ends of two ropes together securely. To do this:

1. Tie the simple figure eight near the end of one rope.
2. Thread the end of the second rope completely through the knot in the opposite direction.
3. Pull the knot tight.
4. Secure the loose end of each rope to the standing part of the second rope.

Bowline

A bowline knot also can be used to form a loop. It is frequently used to secure the end of a rope to an object or anchor point. Follow the steps in ► **Skill Drill 9-9** to tie a bowline:

1. To tie a bowline, form a loop and bring the working end of the rope back to the standing part. **(Step 1)**
2. Make a small loop in the standing part of the rope so that the section closer to the working end passes on top of the section closer to the running end. **(Step 2)**
3. Thread the working end of the rope through the opening in the small loop from below. **(Step 3)**
4. Pass the working end over the loop, around and under the standing part of the rope, and back down through the same opening. **(Step 4)**
5. Tighten the knot by holding the working end and pulling the standing part of the rope backward. **(Step 5)**
6. Use a safety knot to secure the working end of the rope. **(Step 6)**

Bends

Bends are used to join two ropes together. The sheet bend or Becket bend can be used to join two ropes of unequal size. A sheet bend knot also can be used to join rope to a chain. Follow the steps in ► **Skill Drill 9-10** to tie a sheet or becket bend:

1. This knot is tied by forming a bight in the working end of one rope. If the ropes are of unequal size, the bend should be made in the larger rope. **(Step 1)**
2. The end of the second rope is passed up through the opening of the bight, under the two parallel sections of the first rope. **(Step 2)**
3. Loop the second rope completely around both sides of the bight. **(Step 3)**
4. The second rope is then threaded under itself and on top of the two sides of the bight. **(Step 4)**
5. To tighten the knot, hold the first rope firmly while pulling back on the second rope. **(Step 5)**
6. A safety knot should be used to secure the loose ends of both ropes. **(Step 6)**

There are many ways to tie each of these knots. Find one method that works for you and use it all the time. In addition, your department may require that you learn how to tie other knots. It is important to become proficient in tying knots. With practice, you should be able to tie these knots in the dark, with heavy gloves on, and behind your back.

A knot should be properly "dressed" by tightening and removing twists, kinks, and slack from the rope. The finished knot is firmly fixed in position. The configuration of a properly dressed knot should be evident so that it can be easily inspected. All loose ends should be secured by safety knots to ensure that the primary knot cannot be released accidentally.

Knot-tying skills can be quickly lost without practice. Practice tying knots while you are on the telephone or watching TV ▼ **Figure 9-29**. You never know when you will need to use your skills in an emergency situation.

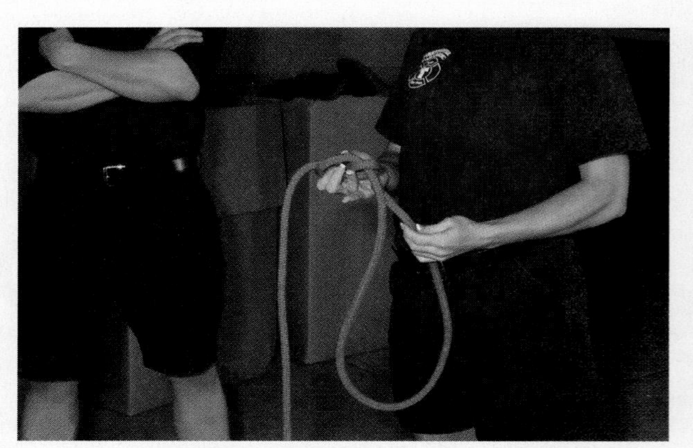

Figure 9-29 To maintain your knot-tying skills, practice tying different knots frequently.

Skill Drill

Figure Eight with a Follow-Through

Tie a regular figure eight leaving approximately 2′ of rope at the working end.

Loop the working end around or through the object to be secured.

Thread the working end back through the original figure eight in the opposite direction.

Pull the knot tight.

Tie a safety knot in the working end.

9-9 Skill Drill

Bowline

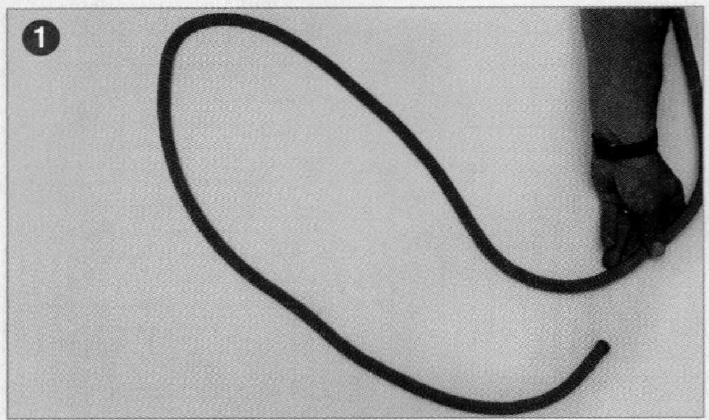

Make the desired size loop and bring the working end back to the standing part.

Form another small loop in the standing part of the rope with the section closest to the working end on top.

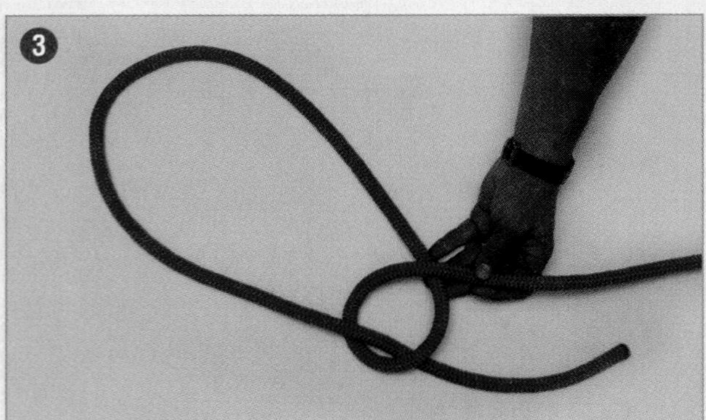

Thread the working end up through this loop from the bottom.

Pass the working end over the loop, around and under the standing part, and back down through the same opening.

Tighten the knot by holding the working end and pulling the standard part of the rope backward.

Tie a safety knot in the working end of the rope.

Skill Drill

Sheet or Becket Bend

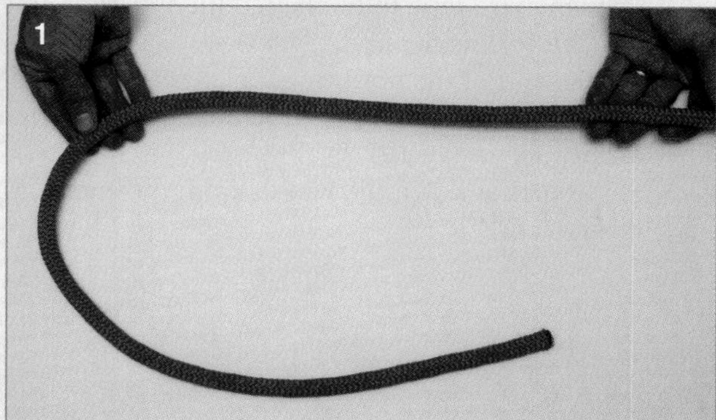

Using your left hand, form a bight at the working end of the first (larger) rope.

Thread the working end of the second (smaller) rope up through the bight.

Loop the second (smaller) rope completely around both sides of the bight.

Pass the working end of the second (smaller) rope between the original bight and under the second rope.

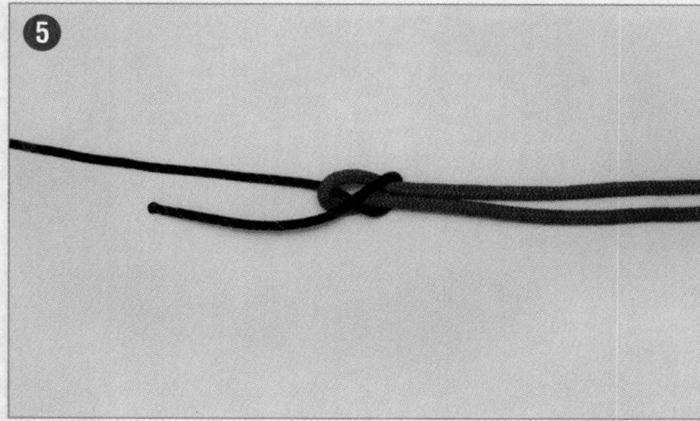

Tighten the knot.

Tie a safety knot in the working end of each rope.

Hoisting

Tying knots is not an idle exercise but a practical skill that you will use on the job. In emergency situations, you may have to raise or lower a tool to other fire fighters. It is important for you to learn how to raise and lower an axe, a pike pole, a ladder, a hose line, and an exhaust fan.

You must ensure that the rope is tied securely to the object being hoisted so the tool does not fall. Additionally, your coworkers must be able to remove and place the tool into service quickly. When you are hoisting or lowering a tool, make sure no one is standing under the object. Keep the scene clear of people to avoid any chance for accident.

Hoisting an Axe

An axe should be hoisted in a vertical position with the head of the axe down. Follow the steps in ▼ Skill Drill 9-11 to hoist an axe:

1. The team that needs the axe should lower a rope with enough extra rope to tie the required knot around the axe and for the tag line. **(Step 1)**
2. Tie the end of the hoisting rope around the handle near the head using either a figure eight on a bight or a clove hitch. **(Step 2)**
3. Slip the knot down the handle from the end to the head. **(Step 3)**

9-11 **Skill Drill**

Hoisting an Axe

The team that needs the axe should lower a rope with enough extra rope to tie the required knot around the axe.

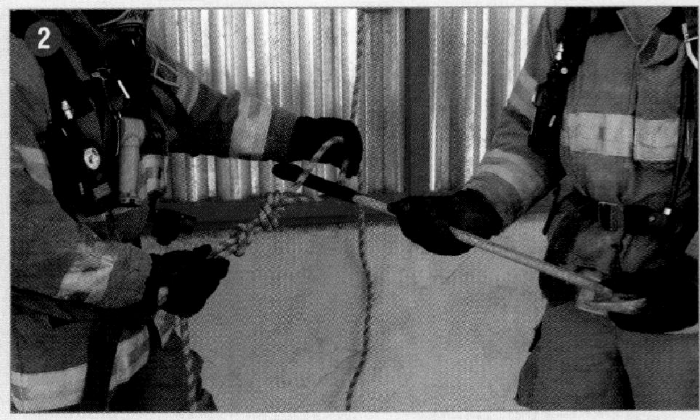

Tie a figure eight on a bight to make a small loop.

Place the loop over the axe handle near the head.

Pass the standing part of the rope around the head of the axe.

4. Loop the standing part of the rope under the head. **(Step 4)**
5. Place the standing part of the rope parallel to the axe handle. **(Step 5)**
6. You can use one or two half hitches along the axe handle to keep the handle parallel to the rope. **(Step 6)**
7. Prepare to raise the axe. **(Step 7)**

To release the axe, hold the middle of the handle, release the half hitches, and slip the knot up and off.

Hoisting a Pike Pole

A pike pole should be hoisted in a vertical position with the head at the top, so it can be used immediately. Follow the steps in ► Skill Drill 9-12 to hoist a pike pole:

1. The team that needs the pike pole should lower a rope with enough extra rope available for the required knot and the tag line. **(Step 1)**
2. Place a clove hitch over the end of the handle, slip along the handle, and secure it close to the head. **(Step 2)**
3. Depending on the length of the handle, one or two half hitches can be placed around the handle below the clove hitch to keep the rope parallel to the handle. **(Step 3)**
4. Slip a second clove hitch over the handle and secure it near the bottom of the pike pole. Leave an additional length of rope below the second clove hitch to be used as a tag line when hoisting the pike pole. **(Step 4)**
5. Prepare to raise the pike pole. **(Step 5)**

Skill Drill 9-11

5 Place the standing part of the rope parallel to the axe handle.

6 Tie one or two half hitches along the axe handle.

7 Prepare to raise the axe.

9-12 Skill Drill

Hoisting a Pike Pole

1. The team that needs the pike pole should lower a rope with enough extra rope available for the required knot and the tag line.

2. Tie a clove hitch in the open and slip it over the handle. Secure it near the head.

3. Place a half hitch around the handle below the clove hitch.

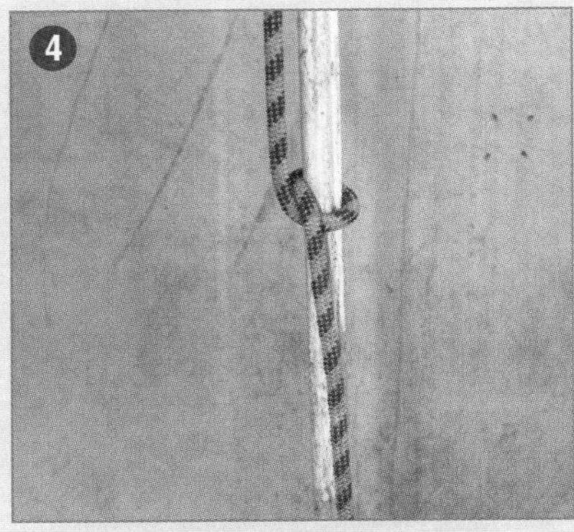

4. Place a second clove hitch around the handle near the bottom of the pike pole.

5. Prepare to raise the pike pole.

Hoisting a Ladder

A ladder should be hoisted in a vertical position. A tag line should be attached to the bottom to keep it under control as it is hoisted. If it is a roof ladder, the hooks should be in the retracted position. Follow the steps in ► **Skill Drill 9-13** to hoist a ladder:

1. The team that needs the ladder should lower a rope with enough extra rope to tie onto the ladder. **(Step 1)**
2. A figure eight on a bight should be tied to create a loop, approximately 3′ or 4′ in diameter, that is large enough to fit around both ladder beams. **(Step 2)**
3. The rope should be passed between two rungs of the ladder, three or four rungs from the top. **(Step 3)**
4. The end of the loop is then pulled under the rungs toward the top of the ladder. **(Step 4)**
5. Place the loop around the top of the ladder. **(Step 5)**
6. Remove the slack from the rope and allow the loop to slide down the ladder. **(Step 6)**
7. Attach a tag line from below to control the ladder as it is hoisted. The size and weight of a ladder require that a tag line be attached to the lower part of the ladder. This enables a fire fighter on the ground to keep the ladder under control as it is raised. Prepare to raise the ladder. When the hoisting rope is pulled up, the ladder will be securely tied to the rope. **(Step 7)**

The ladder can be easily released by pulling the loop back over the top. When lowering a ladder, reverse the loop down through the ladder. This will pull the bottom of the ladder away from the building, preventing damage.

Fire Fighter Tips

Raise a pike pole in the same orientation it will be used, pick first. This eliminates the need to turn the pole in close confines.

Hoisting a Charged Hose Line

It is almost always preferable to hoist a dry hose line, because water adds considerable weight to a charged line. Water weighs 8.33 pounds per gallon, which can make hoisting much more difficult. Follow the steps in ▶ Skill Drill 9-14 to hoist a charged hose line:

1. The team that needs the hose should lower a rope with enough extra rope available to tie onto the hose. **(Step 1)**
2. Make sure that the nozzle is completely closed and secure. A charged hose line should have the nozzle secured in a closed position as it is hoisted. An unsecured handle (known as the bale) could get caught on something as the hose is being hoisted. The nozzle could open suddenly, resulting in an out-of-control hose line suspended in mid-air, and the potential for serious injuries and damage. **(Step 2)**
3. Use a clove hitch, 1′ or 2′ behind the nozzle, to tie the end of the hoisting rope around a charged hose line. A safety knot should be used to secure the loose end of the rope below the clove hitch. **(Step 3)**
4. A bight should then be made in the rope even with the nozzle shut-off handle, which must be in the forward (completely off) position. **(Step 4)**
5. Insert the bight through the handle opening and slip it over the end of the nozzle. When the bight is pulled tight, it will create a half hitch and secure the handle in the off position while the charged hose line is hoisted. **(Step 5)**
6. Prepare to hoist the hose. **(Step 6)**

The knot can be released after the line is hoisted by removing the tension from the rope and slipping the bight back over the end of the nozzle.

Hoisting an Uncharged Hose Line

Before hoisting a dry hose line, you should fold the hose back on itself and place the nozzle on top of the hose. This ensures that water will not reach the nozzle if the hose is accidentally charged while being hoisted. It also eliminates an unnecessary stress on the couplings by ensuring that the rope pulls on the hose and not directly on the nozzle. Follow the steps in ▶ Skill Drill 9-15 to hoist an uncharged hose line:

1. The team that needs the hose should lower enough rope so the ground crew can tie on the hose. **(Step 1)**
2. Fold about 3′ of hose back on itself and place the nozzle on top of the hose. **(Step 2)**
3. Make a half hitch in the rope and slip it over the nozzle. Move the half hitch along the hose and secure it about 6″ from the fold. **(Step 3)**
4. Tie a clove hitch near the end of the rope, wrapping the rope securely around both the nozzle and the hose. The clove hitch should hold both the nozzle and the hose. **(Step 4)**
5. Hoist the hose with the fold at the top and the nozzle pointing down. Before releasing the rope, the fire fighters at the top must pull up enough hose so that the weight of the hanging hose does not drag down the hose. **(Step 5)**

Hoisting an Exhaust Fan or Power Tool

Several different types of tools and equipment, including an exhaust fan, a chain saw, or circular saw, or any other object that has a strong closed handle, can be hoisted using the same technique. The hoisting rope is secured to the object by passing the rope through the opening in the handle. A figure eight with a follow-through knot is used to close the loop.

Some types of equipment require that you use additional half hitches to balance the object in a particular position while it is being hoisted. Power saws are hoisted in a level position to prevent the fuel from leaking out.

You should practice hoisting the actual tools and equipment used in your department. You should be able to do them automatically and in adverse conditions.

Remember, you always use utility rope for hoisting tools. You do not want to get oil or grease on designated life safety ropes. If a life safety rope gets oily or greasy, it should be taken out of service and destroyed so that it will not be mistakenly used again as a life safety rope. It can be cut into short lengths and used for utility rope.

Follow the steps in ▶ Skill Drill 9-16 to hoist an exhaust fan:

1. The team that needs the equipment lowers enough rope so the ground crew can tie on the exhaust fan. **(Step 1)**
2. Tie a figure eight knot in the rope about 3′ from the working end of the rope. **(Step 2)**
3. Loop the working end of the rope around the fan handle and back to the figure eight knot. **(Step 3)**
4. Secure the rope by tying a figure eight with a follow-through. Thread the working end back through the first figure eight in the opposite direction. **(Step 4)**
5. Attach a tag line to the fan for better control. **(Step 5)**
6. Prepare to hoist the fan. **(Step 6)**

Fire Fighter Safety Tips

When hoisting tools:

- Wear a helmet and gloves.
- Keep the scene clear.
- Use the hand-over-hand method.
- Use an **edge roller** (a device used to prevent damage to a rope from jagged edges or friction) to protect the rope.
- Avoid electric wires.

Skill Drill

Hoisting a Ladder

The team that needs the ladder should lower a rope with enough extra rope to tie onto the ladder.

Tie a figure eight on a bight to make a loop 3′ or 4′ in diameter.

Pass the rope between the rungs of the ladder, three or four rungs from the top.

Pull the loop under the rungs toward the top of the ladder.

Place the loop around the top of the ladder.

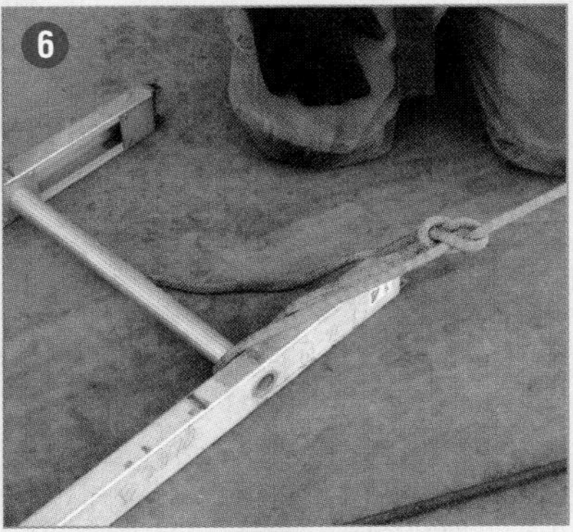

Remove the slack from the rope and allow the loop to slide down the ladder.

Attach a tag line from below to control the ladder as it is hoisted. Prepare to raise the ladder.

Skill Drill

Hoisting a Charged Hose Line

The team that needs the hose should lower a rope with enough extra rope available to tie onto the hose.

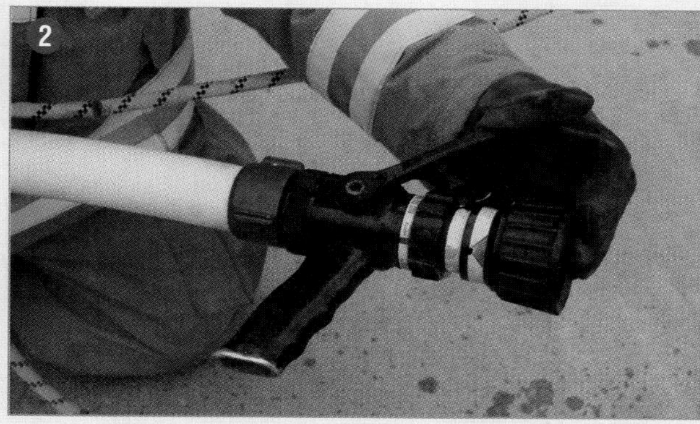

Make sure that the nozzle is completely closed and secure.

Tie a clove hitch around the hose, 1′ or 2′ behind the nozzle.

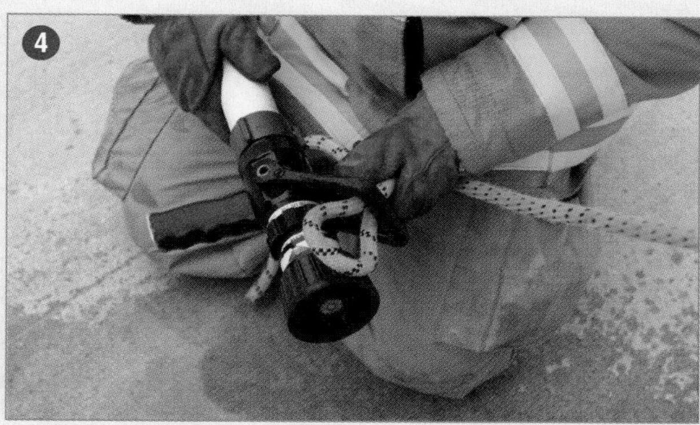

Make a bight in the rope.

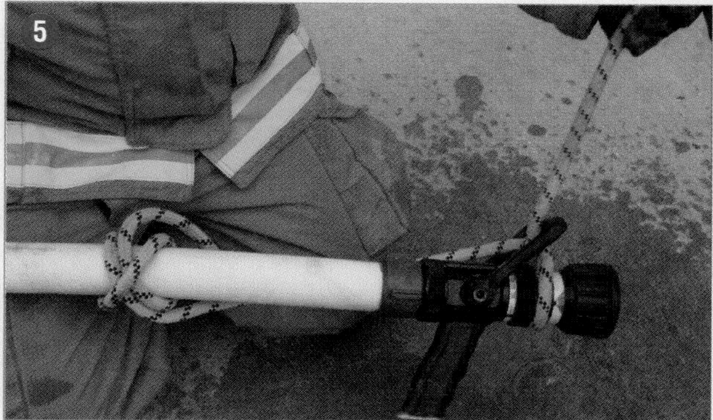

Pass the bight through the nozzle handle (bale) and slip the bight over the nozzle tip. This creates a half hitch that will help keep the nozzle closed while the hose is being raised.

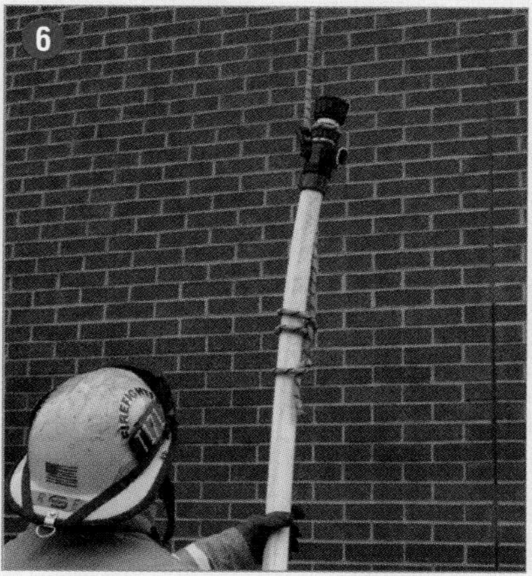

Prepare to hoist the hose.

Skill Drill

Hoisting an Uncharged Hose Line

The team that needs the hose should lower enough rope so the ground crew can tie on the hose.

Fold about 3′ of hose back on itself and place the nozzle on top of the hose.

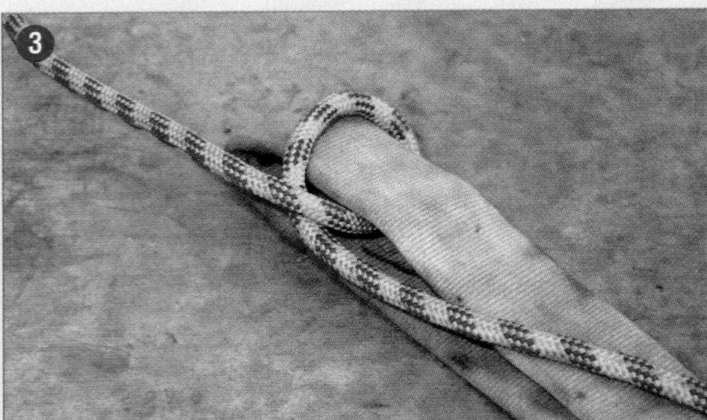

Tie a half hitch and slip it over the nozzle. Move the half hitch along the hose and secure it about 6″ from the fold.

Tie a clove hitch near the end of the rope, wrapping the rope around both the nozzle and the hose.

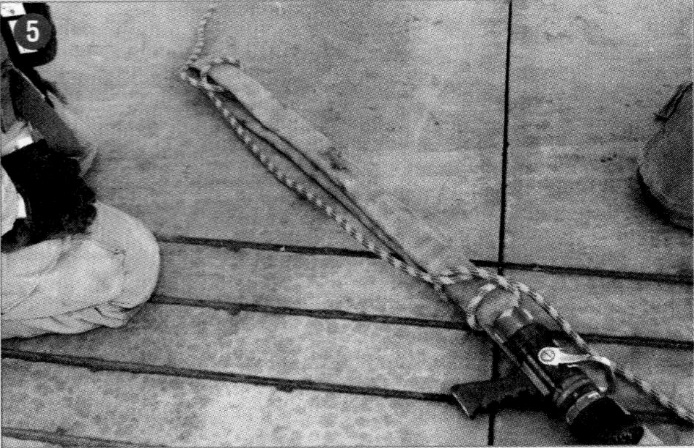

Prepare to hoist the hose line with the fold at the top and the nozzle pointing down.

9-16 Skill Drill

Hoisting an Exhaust Fan

The team that needs the equipment should lower enough rope so the ground crew can tie on the exhaust fan.

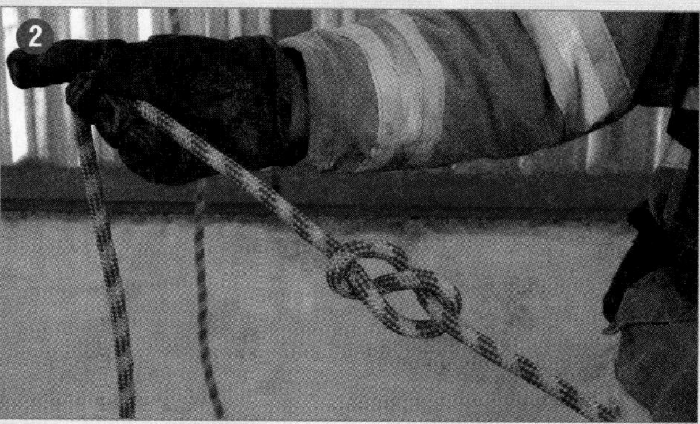

Tie a figure eight knot in the rope about 3′ from the working end of the rope.

Loop the working end of the rope around the fan handle and back to the figure eight knot.

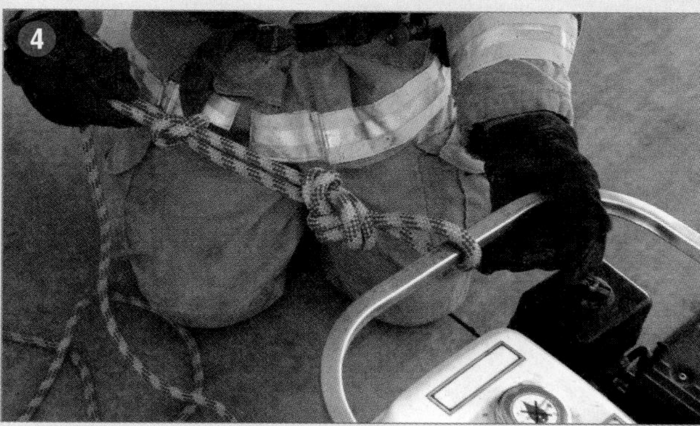

Secure the rope by tying a figure eight with a follow-through. Thread the working end back through the first figure eight in the opposite direction.

Attach a tag line to the fan for better control.

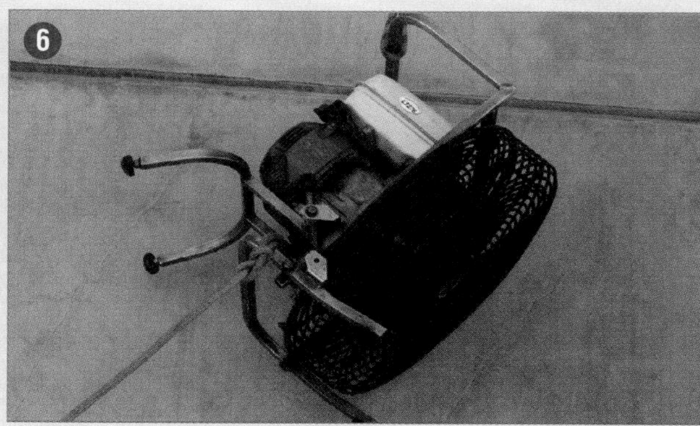

Prepare to hoist the fan.

Wrap-Up

Ready for Review

This chapter covered the basic skills required to work with ropes and knots. It addressed the requirements of NFPA 1983 and described the differences between life safety ropes, personal escape ropes, and utility ropes. It explained the differences between dynamic and static ropes, between different rope materials, and between different types of rope construction. The advantage of synthetic fibers for life safety ropes and the reasons that static kernmantle is generally preferred for fire department rescue rope were discussed.

Also covered in this chapter were the requirements for regular inspection and maintenance of life safety ropes as well as record-keeping procedures. Proper procedures for care, cleaning, inspection, and storage of ropes were discussed. The common types of hardware used with ropes were described, including the three different classes of harnesses.

The characteristics of a good knot were presented as well as the importance of regular practice tying the basic knots. The proper procedures for tying an overhand safety knot, a half hitch, a clove hitch, figure eight knots, a bowline, and a sheet or becket bend were presented and illustrated. The proper methods of using these knots to hoist an axe, a pike pole, a ladder, a hose line, and an exhaust fan were presented.

Chief Concepts

- Life safety ropes are used for supporting a person (or persons) during rescue operations and should be used only for that purpose.
- Life safety ropes must be made of continuous filament virgin fiber and woven of block creel construction.
- Proper maintenance consists of preventing damage through proper care, proper cleaning, regular inspection, and proper storage.
- Knots have characteristics that make them good for selected uses. Basic knots used by fire fighters include the overhand safety knot, hitches, loop knots, and bends.
- Fire fighters must be able to select and tie the appropriate knots to hoist axes, pike poles, ladders, hose lines, and exhaust fans.

Hot Terms

Bends Knots used to join two ropes together.

Bight A U-shape created by bending a rope with the two sides parallel.

Block creel construction Rope constructed without knots or splices in the yarns, ply yarns, strands, braids, or rope.

Braided rope Rope constructed by intertwining strands in the same way that hair is braided.

Carabiner A piece of metal hardware used extensively in rope rescue operations. It is generally an oval-shaped device with a spring-loaded clip that can be used for connecting together pieces of rope, webbing, or other hardware. Also referred to as a "snap link."

Depressions Indentations felt on a kernmantle rope that indicate damage to the interior, or kern, of the rope.

Dynamic rope A rope generally made out of synthetic materials that is designed to be elastic and stretch when loaded. Used often by mountain climbers.

Edge roller A device used to prevent damage to a rope from jagged edges or friction.

Harness A piece of equipment worn by a rescuer that can be attached to a life safety rope.

Hitches Knots that attach to or wrap around an object.

Kernmantle rope Rope made of two parts—the kern (interior component) and the mantle (the outside sheath).

Knots A fastening made by tying together lengths of rope or webbing in a prescribed way, used for a variety of purposes.

Ladder halyards Rope used on extension ladders to raise a fly section.

Life safety rope Rope used solely for the purpose of supporting people during firefighting, rescue, other emergency operations, or during training exercises.

Loop A piece of rope formed into a circle.

One-person rope A rope rated to carry the weight of a single person (300 lbs).

Personal escape rope An emergency use rope designed to carry the weight of only one person and to be used only once.

Rope bag A bag used to protect and store rope so that the rope can be easily and rapidly deployed without kinking.

Rope record A record for each piece of rope that includes a history of when the rope was placed in service, when it was inspected, when and how it was used, and what types of loads were placed on it.

Round turn A piece of rope looped to form a complete circle with the two ends parallel.

Running end The part of a rope used for lifting or hoisting.

Safety knot A knot used to secure the leftover working end of the rope. Also known as an overhand knot or keep knot.

Shock load An instantaneous load that places a rope under extreme tension, such as when a falling load is suddenly stopped when the rope becomes taut.

Standing part The part of a rope between the working end and the running end.

Static rope A rope generally made out of synthetic material that stretches very little under load.

Twisted rope Rope constructed of fibers twisted into strands, which are then twisted together.

Two-person rope A rope rated to carry the weight of two people (600 lbs).

Utility rope Rope used for securing objects, for hoisting equipment, or for securing a scene to prevent bystanders from being injured. It is never to be used in life safety operations.

Working end The part of the rope used for forming the knot.

Fire Fighter in Action

You are assigned to a ladder company. On your first day, you are given an inventory list and told to look in all of the compartments on the apparatus and identify all of the equipment found in each location. This will help you learn where things are kept so that you will be able to go to the proper location and find any piece of equipment needed at the scene of an incident.

The inventory list for one of the compartments includes the following items under the category "ropes":

Life Safety	**two-person x 100'**	**2**
Life Safety	**one-person x 200'**	**2**
Personal Escape	**50'**	**4**
Utility	**5/8" x 100'**	**2**

All of the ropes are stored in nylon bags with one end protruding. A tag is attached to a loop at the end of each rope.

1. Later that day you respond to a working fire in a three-story apartment building. Your company is assigned to work on the roof. Captain Landry tells you to bring a rope in case you have to hoist up additional equipment. Which rope will you bring?
   - **A.** A personal escape rope
   - **B.** A utility rope
   - **C.** A one-person life safety rope
   - **D.** A two-person life safety rope
2. Once on the roof, you are told to drop the rope down to a fire fighter on the ground and hoist up a hose line. The hose line is not charged. The fire fighter who attaches the rope should:
   - **A.** Tie a clove hitch around the hose and a half hitch through the shutoff handle.
   - **B.** Pass the rope through the shutoff handle and tie a figure eight with a follow-through.
   - **C.** Fold the hose and tie a clove hitch around the hose and nozzle and a half hitch near the fold.
   - **D.** Tell the pump operator to charge the line before hoisting.
3. While hoisting the hose up to the roof, you should:
   - **A.** Raise the rope with a hand-over-hand motion.
   - **B.** Wrap the rope around your waist and walk away from the edge of the roof.
   - **C.** Look out for electrical wires to avoid shock loading the rope.
   - **D.** Have someone stand directly below to ensure that the hose does not get snagged on an obstacle.
4. While you were working on the roof, you used both a chain saw and a circular saw to cut ventilation holes. Now that the fire is under control, you plan to lower the saws to the ground by rope rather than carry them down the ladder. What type of knot will you use to attach the saws to the rope?
   - **A.** Figure eight with a follow-through
   - **B.** Clove hitch
   - **C.** Becket bend
   - **D.** Bowline
5. After using the rope you will:
   - **A.** Coil it carefully and place the entire rope back into the bag.
   - **B.** Open the top of the bag and simply stuff the entire rope inside.
   - **C.** Feed the rope progressively into the bag without coiling.
   - **D.** Take the rope back to the fire station to be cleaned and inspected before putting it back into the bag.

The following day, you attend a company training session at the fire department drill tower. In one of the exercises, a fire fighter is trapped at a third-floor window and has to be rescued by another fire fighter who rappels down from the roof.

6. The rope that will be used for this exercise is:

A. A personal escape rope

B. A utility rope

C. A one-person life safety rope

D. A two-person life safety rope

7. The type of rope most likely to be used for this exercise is:

A. Braided polyethylene

B. A blend of nylon and manila fibers

C. Dynamic kernmantle

D. Static kernmantle

8. After using this rope, it must be:

A. Cut into short lengths to be used for other purposes

B. Immediately coiled and put back into its bag

C. Inspected according to the manufacturer's instructions

D. Tested by dropping a 300-lb weight from the roof

Response and Size-Up

Technology Resources

www.FireFighter.jbpub.com

- Chapter Pretests
- Interactivities
- Hot Term Explorer
- Web Links
- Review Manual
- FireLearn

Chapter Features

- Skill Drills
- Voices of Experience
- Fire Marks
- Teamwork Tips
- Fire Fighter Safety Tips
- Fire Fighter Tips
- Canadian Perspectives
- Hot Terms
- Wrap-Up

Chapter 10

NFPA 1001 Standard

Fire Fighter I

5.3.2 Respond on apparatus to an emergency scene, given personal protective clothing and other necessary personal protective equipment, so that the apparatus is correctly mounted and dismounted, seat belts are used while the vehicle is in motion, and other personal protective equipment is correctly used.

5.3.2 (A) *Requisite Knowledge.* Mounting and dismounting procedures for riding fire apparatus; hazards and ways to avoid hazards associated with riding apparatus; prohibited practices; types of department personal protective equipment and the means for usage.

5.3.2 (B) *Requisite Skills.* The ability to use each piece of provided safety equipment.

5.3.3 Operate in established work areas at emergency scenes, given protective equipment, traffic and scene control devices, structure fire and roadway emergency scenes, traffic hazards and downed electrical wires, so that procedures are followed, protective equipment is worn, protected work areas are established as directed using traffic and scene control devices, and the fire fighter performs assigned tasks only in established, protected work areas.

5.3.3 (A) *Requisite Knowledge.* Potential hazards involved in operating on emergency scenes including vehicle traffic, utilities, and environmental conditions; proper procedures for dismounting apparatus in traffic; procedures for safe operation at emergency scenes; and the protective equipment available for members' safety on emergency scenes and work zone designations.

5.3.3 (B) *Requisite Skills.* The ability to use PPC, the deployment of traffic and scene control devices, dismount apparatus, and operate in the protected work areas as directed.

5.3.18 Turn off building utilities, given tools and an assignment, so that the assignment is safely completed.

5.3.18 (A) *Requisite Knowledge.* Properties, principles, and safety concerns for electricity, gas, and water systems; utility disconnect methods and associated dangers; and use of required safety equipment.

5.3.18 (B) *Requisite Skills.* The ability to identify utility control devices, operate control valves or switches, and assess for related hazards.

Fire Fighter II

NFPA 1001 contains no Fire Fighter II Job Performance Requirements for this chapter.

Additional NFPA Standards

NFPA 1500, *Standard on Fire Department Occupational Safety and Health Program*

Knowledge Objectives

After studying this chapter, you will be able to:

- Describe your role in assuring safe and efficient response to an emergency scene.
- Describe how to ride an emergency vehicle safely.
- Describe how to dismount an emergency vehicle safely.
- Describe how to shut off utilities.
- Define and describe size-up.

Skills Objectives

After studying this chapter, you will be able to:

- Mount an apparatus safely.
- Dismount from an apparatus safely.

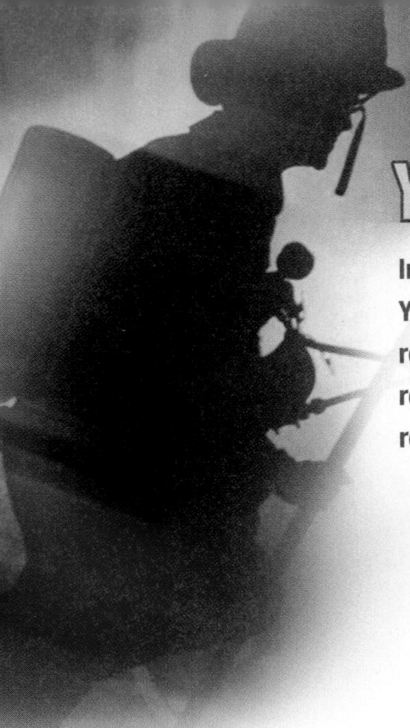

You Are the Fire Fighter

In the middle of the night, your company is dispatched as the third-due engine to a structure fire. You don your PPE, strap yourself into your seat on the apparatus, and put on your headphone. En route to the scene, you listen to the initial reports from the first-arriving company officer. The reports help you develop a mental picture of the structure and conditions you can expect. You realize how important initial size-up is to the success of your mission.

1. *As a new fire fighter, how do you contribute to the process of incident size-up?*
2. *How does incident size-up increase the safety and effectiveness of firefighting?*

Introduction

Response involves a series of actions that begin when a crew is dispatched to an alarm and end with their arrival at the emergency incident. Response actions include receiving the alarm, donning protective clothing and equipment, mounting the apparatus, and transporting equipment and personnel to the emergency incident quickly and safely.

Because fire fighters must be ready to react immediately to an alarm, preparations for response should begin even before the alarm is sounded. These preparations include checking personal equipment, ensuring the fire apparatus is ready, and making sure that all equipment carried on the apparatus is ready for use. Fire fighters also should be familiar with their response district, know the buildings under their protection, and understand their department's standard operating procedures (SOPs).

Response actions for the apparatus driver also include considering road and traffic conditions, determining the best route to the incident, identifying nearby hydrant locations or water sources, and selecting the best position for the apparatus at the incident scene.

Size-up is a systematic process of information gathering and situation evaluation that begins when an alarm is received. Size-up continues during response and includes the initial observations made upon arrival at the incident scene. Size-up information is essential for determining the appropriate strategy and tactics for each situation.

The Incident Commander (IC) and company officers are ultimately responsible for obtaining the necessary information to manage the emergency incident. Each fire fighter should also be involved in the process of gathering and processing information. The observations made by individual fire fighters will help them anticipate necessary actions and provide company officers with needed information.

Response

Trained fire fighters must be ready to respond to an emergency at any time during their tour of duty. This process begins by ensuring that **personal protective equipment (PPE)** is complete, ready for use, and in good condition. This includes checking that gloves, protective hood, and flashlight are ready for immediate use. At the beginning of each tour of duty, place PPE in its designated location, which depends on your assigned riding position on the apparatus (► Figure 10-1).

Conduct a daily inspection of the self-contained breathing apparatus (SCBA) for your riding position at the beginning of each tour of duty (► Figure 10-2). The air supply should be adequate, the face piece clean, and the **personal alert safety system (PASS)** operable. Also check the availability and operation of your hand light and any hand tools you might require, based on your assigned position on the apparatus (► Figure 10-3). Recheck personal protective equipment and tools thoroughly when you return from each emergency response. This will help ensure that your gear and equipment will be functional for the next alarm.

Alarm Receipt

The emergency response process begins when an alarm is received at the fire station. Fire fighters should be familiar with the dispatch method or methods used by their departments. In many fire departments, a local or regional communications center dispatches individual units, although in smaller departments, the dispatcher is often located at the fire station. Most departments have both a primary and a back-up method of transmitting alarms to stations.

Radio, telephone, or public address systems are often used to transmit information to fire stations, and the use of computer terminals and printers to transmit dispatch messages is

Figure 10-1 Check your PPE and place it in the designated location for your riding position.

increasing. Some fire departments still use a system of bells to transmit alarms; volunteer or rural departments may use outdoor sirens or horns to summon fire fighters to an emergency. Most volunteer fire fighters, however, receive dispatch messages over pagers.

Dispatch messages contain varying amounts of useful information, based on what the dispatcher learns from the caller. At a minimum, the dispatch information will include the location of the incident, the type of emergency (vehicle fire, structure fire, medical call), and the units that are due to respond. Computer-aided dispatch systems often provide additional information about the building or premises on the dispatch printout.

As additional information becomes available, the telecommunicator will include it in dispatch messages to later-responding units or advise the responding units by radio while they are en route. Each additional piece of information can help responding fire fighters plan strategies and prepare themselves for the incident.

When an alarm is received, the response should be prompt and efficient. Responding fire fighters should walk briskly to the apparatus. There is no need to run; the objective is to respond quickly, without injuring anyone or causing any damage. Follow established procedures to ensure that

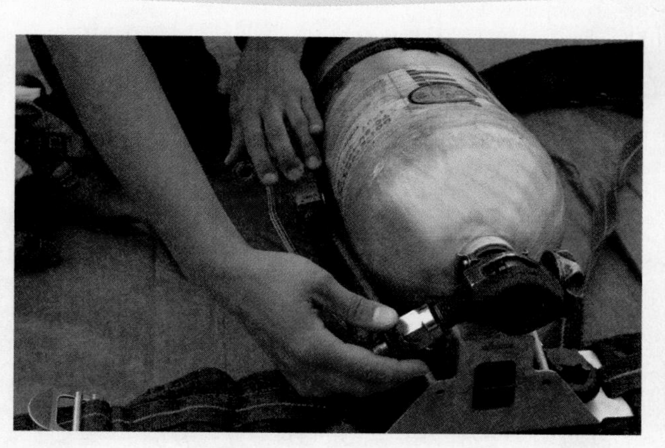

Figure 10-2 Conduct the daily check of your SCBA at the beginning of each tour of duty.

Figure 10-3 Check the tools assigned to your riding position.

Fire Fighter Tips

Volunteer fire fighters who are not assigned to specific tours of duty or riding positions should check PPE, SCBA, and associated tools and equipment on a regular basis so that they are ready for use whenever there is an emergency response. After each use, all PPE should be carefully checked before it is put away.

stoves, faucets, and other appliances at the station are shut off. Wait until the apparatus doors are fully open before leaving the station.

Riding the Apparatus

Don PPE before mounting the apparatus. Do not attempt to dress while the apparatus is on the road. Wait until you

10-1 Skill Drill

Mounting Apparatus

When mounting (climbing aboard) fire apparatus, always have at least one hand firmly grasping a handhold and at least one foot firmly placed on a foot surface. Maintain the one hand and one foot placement until you are seated.

Fasten your seatbelt and then don any other required safety equipment for response, such as hearing protection. Eye and face protection are required for seating areas that are not fully enclosed.

dismount at the incident scene to don any protective clothing that is not donned prior to mounting the apparatus. Don SCBA only after the apparatus stops at the scene, unless the SCBA is seat-mounted.

All equipment should be properly mounted, stowed, or secured on the fire apparatus. Any unsecured equipment in the crew compartment can be dangerous if the apparatus must stop or turn quickly. A flying tool, map book, or PPE can seriously injure a fire fighter.

Be careful when mounting and dismounting apparatus. The steps on fire apparatus are often high and can be slippery. Follow the steps in ▲ Skill Drill 10-1 to mount an apparatus properly.

1. When mounting (climbing aboard) fire apparatus, always have at least one hand firmly grasping a handhold and at least one foot firmly placed on a foot surface. Maintain the one hand and one foot placement until you are seated. **(Step 1)**
2. Fasten your seatbelt and then don any other required safety equipment for response, such as hearing protection. Eye and face protection are required for seating areas that are not fully enclosed. **(Step 2)**

All fire fighters must be seated in their assigned riding positions with seatbelts and/or harnesses fastened before the apparatus begins to move. NFPA 1500, *Standard on Fire Department Occupational Safety and Health Program,* requires all fire fighters to be in their seats, with seatbelts secured, whenever the vehicle is in motion. Do not unbuckle a seatbelt to don any clothing or equipment while the apparatus is en route to an incident.

The noise produced by sirens and air horns can have long-term, damaging effects on a fire fighter's hearing. If your department provides hearing protection for personnel riding on fire apparatus, use it. These devices often include radio and intercom capabilities so that fire fighters can talk to each other and hear information from the dispatcher or IC.

During transport, limit conversation to the exchange of pertinent information. Listen for instructions from the IC, your company officer, and for additional information about the incident over the radio. The vehicle operator's attention should be focused on driving the apparatus safely to the scene of the incident.

The ride to the incident is a good time to consider any relevant factors that could affect the situation. These factors could include the time of day or night, the temperature, the presence of precipitation or wind, as well as the location and type of incident. Using the time to think ahead will help you mentally prepare for various possibilities.

Emergency Response

The fire apparatus operator must always exercise caution when driving to an incident. Fire apparatus can be very large, heavy, and difficult to maneuver. Operating an emergency vehicle without the proper regard for safety can endanger the lives of both the fire fighters on the vehicle and civilian drivers and pedestrians. Although the impulse to respond as speedily as possible is understandable, never compromise safety for a faster response time.

Fire fighters who drive emergency vehicles must have special driver training and know the laws and regulations that apply to emergency response. Many jurisdictions require a special driver's license to operate fire apparatus. The rules that apply to emergency vehicles are very specific, and the driver is legally responsible for the safe operation of the vehicle at all times.

An emergency vehicle must always be operated with due regard for the safety of everyone on the road. Although most states and provinces permit drivers of emergency vehicles to disregard specific traffic regulations when responding to emergency incidents, apparatus operators must consider the actions of other drivers before making such a decision. For example, traffic laws require other drivers to yield the right of way to an emergency vehicle. There is no assurance, however, that other drivers will do so when an emergency vehicle approaches. The apparatus driver must also anticipate what routes other units responding to the same incident will take.

Fire fighters who respond to emergency incidents in personal vehicles must follow the specific laws and regulations of their state or province and follow departmental standard operating procedures (SOPs). Some jurisdictions allow fire fighters to equip their privately-owned vehicles with warning devices and to operate them as emergency vehicles; others grant no special status to privately-owned vehicles. In some areas, volunteer fire fighters responding to an emergency incident can use colored lights to request the right of way from other drivers.

Prohibited Practices

For safety's sake, SOPs often prohibit specific actions during response. As noted previously, fire fighters must remain seated with seatbelts securely fastened while the emergency vehicle is in motion. Never unfasten your seatbelt to retrieve or don equipment. Do not dismount the apparatus until the vehicle comes to a complete stop.

Never stand up while riding on apparatus. Do not hold on to the side of a moving vehicle or stand on the rear step. When a vehicle is in motion, everyone aboard must be seated and belted in an approved riding position.

Dismounting a Stopped Apparatus

When the apparatus arrives at the incident scene, the driver will park it in a location that is both safe and functional. Wait until the vehicle comes to a complete stop before dismounting. Always check for traffic before opening doors or stepping out of the apparatus. During the dismount, watch for other hazards, such as ice and snow, downed power lines, or hazardous materials, that could be present.

Be careful when dismounting apparatus. The increased weight of PPE and adverse conditions can contribute to slips, strains, and sprains. Use handrails when mounting or dismounting the apparatus. Follow the steps in ► Skill Drill 10-2 to dismount an apparatus safely.

1. Become familiar with your riding position and the safest way to dismount. **(Step 1)**
2. One hand should always be grasping a handhold, and one foot should always be placed firmly on a flat surface when leaving the apparatus, especially on wet or potentially icy roadway surfaces. **(Step 2)**

Traffic Safety on the Scene

An emergency incident scene presents several risks to fire fighters in addition to the hazards of fighting fires and performing other duties. One of these dangers is traffic, particularly when the incident scene is on a street or highway. Traffic safety should be a major concern for the first-arriving units because approaching drivers might not see emergency workers or realize how much room fire fighters need to work safely.

The first unit or units to arrive at the incident scene have a dual responsibility. Not only must the fire fighters focus on the emergency situation facing them, they must also consider approaching traffic, including other emergency vehicles, and other, less obvious hazards. Always check for traffic before opening doors and dismounting from the apparatus. Watch out for traffic when working in the street. Follow departmental SOPs to close streets quickly and to block access to areas where operations are being conducted.

One of the most dangerous work areas for fire fighters is on the scene of a highway incident, where traffic can be

Fire Fighter Safety Tips

- Do not attempt to mount or dismount a moving vehicle.
- Do not remove your seatbelt until the apparatus comes to a complete stop.
- Do not stand directly behind an apparatus that is backing up. Stand off to one side where the driver can see you in the rear-view mirror. All fire apparatus should have working, audible, alarms when in reverse gear.

10-2 Skill Drill

Dismounting Apparatus

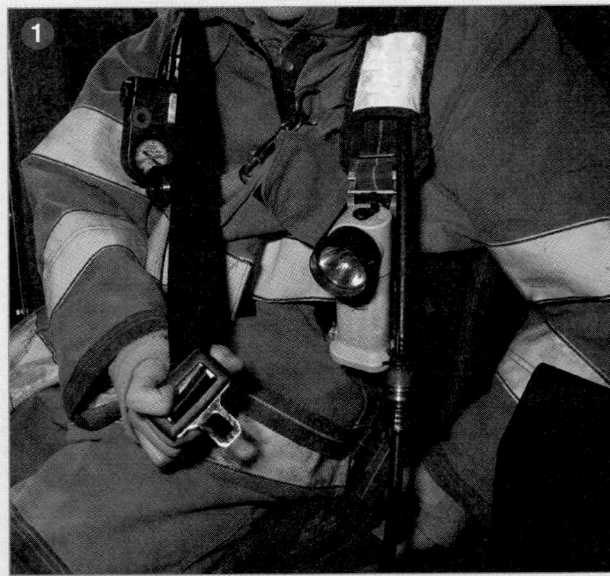

1 Become familiar with your riding position and the safest way to dismount.

2 Maintain the one hand and one foot placement when leaving the apparatus, especially on wet or potentially icy roadway surfaces.

approaching at high speeds. Place traffic cones, flares, and other warning devices far enough away from the incident to slow approaching traffic and direct it away from the work area. All fire fighters working at highway incidents should wear high-visibility safety vests, as well as their normal PPE. Many fire departments have specific SOPs covering required safety procedures for these incidents.

Arrival at the Incident Scene

From the moment that fire fighters arrive at an emergency incident scene, SOPs and the structured incident management system must guide all actions. Fire fighters should always work in assigned teams (companies or crews) and be guided by a strategic plan for the incident. Teamwork and disciplined action are essential for the safety of all fire fighters and the effective, efficient conduct of operations.

The command structure plans, coordinates, and directs operations. Fire fighters responding to an incident on an apparatus comprise the crew assigned to that vehicle and take direction from their company officer, who ensures that their actions coordinate with the overall plan. A fire fighter who arrives at the scene independent of an apparatus must report to the IC and receive an assignment to a company or crew under the supervision of a company officer.

Fire fighters who take action on their own, without regard to SOPs, the command structure, or the strategic plan, are **freelancing**. This type of activity, whether it is done by individual fire fighters, groups of fire fighters, or full companies, is unacceptable. Freelancing cannot be tolerated at any emergency incident. The safety of every person on the scene can be compromised by fire fighters who do not work within the system.

Personnel should not respond to an emergency incident scene unless they have been dispatched or have an assigned duty to respond. Unassigned units and individual personnel arriving on the scene can overload the IC's ability to manage the incident effectively. Individuals who simply show up and find something to do are likely to compromise their own safety and create more problems for the command staff. All personnel must operate within the established system, reporting to a designated supervisor under the direction of the IC.

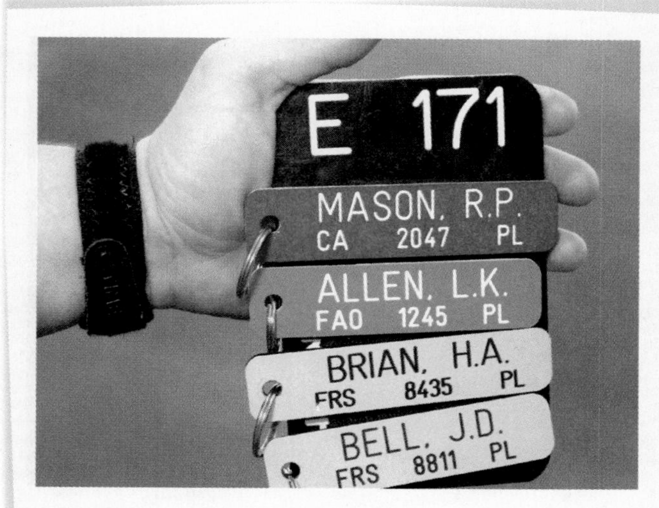

Figure 10-4 There should be a designated location on the fire apparatus for fire fighters to deposit their PATs.

Personnel Accountability System

A **personnel accountability system** should be used to track every fire fighter at every incident scene. The system maintains an updated list of the fire fighters assigned to each vehicle or crew and tracks each crew's assignment at the fire scene. If any fire fighters are reported missing, lost, or injured, the accountability system can identify who is missing and their last assignment.

Different fire departments use various types of accountability systems. Consult your agency's SOPs for more specific information about the system that is being used and always follow the required procedures.

Many fire departments use **personnel accountability tags (PATs)** to track individual fire fighters. An accountability tag generally includes the fire fighter's name, ID number, and photograph and may include additional information, such as an individual's medical history, medications, allergies, and other factors. Fire fighters who respond to an incident on an apparatus deposit their PATs in a designated location on the vehicle (▲ **Figure 10-4**). The PATs are then collected from each vehicle and taken to the command post. Fire fighters who respond directly to the scene must report to the command post to receive an assignment and deposit their PATs.

Teamwork Tips

Teamwork is a hallmark of effective fireground operations. Training and teamwork produce effective, coordinated operations. Unassigned individual efforts and freelancing can disrupt the operations and endanger the lives of both fire fighters and civilians.

Controlling Utilities

Controlling utilities is one of the first tasks that must be accomplished at many working structure fires. Most departments have written SOPs that define when the utilities are to be shut off. Although this responsibility is often assigned to a particular company or crew, all fire fighters should know how to shut off a building's electrical, gas, and water service.

There are several reasons to disconnect electrical and gas utilities to a burning building. If faulty electrical equipment or a gas appliance caused the fire, shutting off the supply will help alleviate the problem. Disconnecting electrical service can also prevent problems such as short circuits and electrical arcing that could result from fire or water damage. Shutting down gas service eliminates the potential for explosion due to damaged, leaking, or ruptured gas piping.

Controlling utilities is particularly important if fire fighters need to open walls or ceilings to look for fire in the void spaces, to cut ventilation holes in roofs, or to penetrate through floors. Because these spaces often contain both electrical lines and gas pipes, the danger of electrocution or explosion exists unless these utilities are disconnected.

For example, a fire fighter using an axe to open a void space could be electrocuted if the tool contacts energized electrical wires. Fire fighters operating a hose could be injured if the water conducts electricity back to the nozzle. Fire fighters can also be electrocuted by coming in contact with wires that have been pulled loose or exposed by a structural collapse.

Gas pipes in walls, ceilings, roofs, or floors can also be inadvertently damaged by fire fighters. A power saw slicing through a gas pipe can cause a rapid release of gas and create the potential for an explosion. A structural collapse can rupture gas lines or create leaks in the stressed piping.

Controlling utilities also alleviates the risk of an additional fire or explosion. Shutting off electrical service eliminates potential ignition sources that could cause an explosion if leaking gas has accumulated inside the building. Shutting off the gas supply will prevent any further leakage. The electrical supply must be disconnected at a location outside the area where gas might be present. Interrupting the power at a remote location alleviates the risk of an electrical arc that could cause a gas explosion.

If there is a serious water leak inside the building, shut off the water supply to prevent electrical problems and to help minimize additional water damage to the structure and contents. Shutting off the water supply to a fire-damaged building is often necessary to control leaking pipes, which could cause more water damage than the water that was used to fight the fire.

Electrical Service

Electric delivery service arrangements depend on the providing utility company and the age of the system. The most common installation is a service drop from above-ground utility wires to the electric meter, which is mounted on the outside of the building (▶ Figure 10-5). Other electric service installations include underground connections, inside meters, and multiple services or separate electric meters for different tenants or areas inside a building.

Fire departments must work with their local electric utility companies to identify the different types of electric services and arrangements used and the proper procedures for dealing with each type. Often, a main disconnect switch can be operated to interrupt the power. Most utility companies will train fire fighters to identify and operate shut-off devices on typical installations. Other utility companies request that the fire department call a company representative to shut off the power. Shutting off large systems that involve high-voltage equipment should be done by a company technician or a trained individual from the premises.

A utility company representative should be called to interrupt power from a remote location, such as a utility pole. This may be necessary if the outside wires have been damaged by fire, if fire fighters are working with ladders or aerial apparatus, or if an explosion is possible. In urban areas, the electric utility company can usually dispatch a qualified technician or crew to respond quickly to a fire or emergency incident. In many cases, qualified technicians are dispatched automatically to all working fires. Some utility companies can interrupt the service to an area from a remote location in response to a fire department request.

Figure 10-5 The electric service usually has an exterior meter connected to above-ground utility wires.

Gas Service

Natural gas or liquefied propane (LP) gas are used for heating and cooking. Generally, natural gas is delivered through a network of underground pipes. LP gas is usually delivered by a tank truck and stored in a container on the premises. In some areas, LP gas is distributed from one large storage tank through a local network of underground pipes to several customers.

A single valve usually controls the natural gas supply to a building. This valve is generally located outside the building at the entry point of the gas piping. Natural gas service has a distinctive piping arrangement. In older buildings, the shut-off valve for a natural gas system may be in the basement.

The shut-off valve for a natural gas system is usually a quarter-turn valve with a locking device so it can be secured in the off position (▶ Figure 10-6). When the handle is in line with the pipe, the valve is open; when the handle is at a right angle to the pipe, the valve is closed. A special key or an adjustable wrench is needed to turn the handle.

The valve for an LP gas system is usually located at the storage tank. The more common type of LP gas valve, however, has a distinctive handle that indicates the proper rotation direction to open or close the valve. To shut off the flow of gas, rotate the handle to the fully closed position.

A gas valve that has been shut off must not be reopened until the system piping has been inspected by a qualified person. Air must be purged from the system and any pilot lights must be re-ignited to prevent gas leaks.

Water Service

Water service to a building can usually be shut off by closing one valve at the entry point. Many communities permit

Figure 10-6 A typical natural gas service includes a pressure-reducing valve and a gas meter along with the shut-off valve.

the water service to be turned off at the connection between the utility pipes and the building's system. This underground valve is outside the building and can be operated with a special wrench or key. In most cases, there is also a valve inside the building, usually in the basement (if there is one), where the water line enters.

Size-up

Size-up is the process of evaluating an emergency situation to determine what actions need to be taken. It is always the first step in making plans to bring the situation under control. The initial size-up of an incident is conducted when the first unit arrives on the scene and determines the appropriate actions for that unit or units.

Initial size-up is often conducted by the first-arriving company officer, who serves as the IC until a higher-ranking officer arrives and assumes command. The IC uses the size-up to develop an initial plan and to set the stage for the actions that follow.

As more complete and detailed information becomes available, the IC will improve both the initial size-up and the initial plan. At a major incident, the size-up process might continue through several stages. The ongoing size-up must consider the effectiveness of the initial plan, the impact fire fighters are having on the problem, and any changing circumstances at the incident.

Although officers usually perform size-up, fire fighters must understand how to formulate an operational plan, how to gather and process information, and how this information can change plans during the operation. If there is no officer on the first arriving unit, a fire fighter could be responsible for assuming command and conducting the preliminary size-up until an officer arrives. Individual fire fighters are often asked to obtain information or to report their observations for ongoing size-up. Fire fighters should routinely make observations during incidents to maintain their personal awareness of the situation and to develop their personal competence.

Managing Information

The size-up process requires a systematic approach to managing information. Emergency incidents are often complicated, chaotic, and rapidly changing. The IC must look at a complicated situation, identify the key factors that apply, and develop an action plan based on known facts, observations, realistic expectations, and certain assumptions. As the operation unfolds, the IC must continually revise, update, supplement, and reprocess the size-up information to ensure that the plan is still valid or to identify when the plan needs to be changed.

Size-up is based on two basic categories of information: facts and probabilities. Facts are data elements that are accurate and based on prior knowledge, a reliable source of information, or an immediate, on-site observation. Probabilities are factors that can be reasonably assumed, predicted, or expected to occur, but are not necessarily accurate.

Often, the initial size-up is based on a limited amount of factual information and a larger amount of probable information. It is refined as more factual information is obtained and as probable information is either confirmed or revised. Effective size-up requires a combination of training, experience, and good judgment.

Facts

Facts are bits of accurate information obtained from various sources. For example, the communications center will provide some facts about the incident during dispatch. A preincident plan will contain many facts about the structure. Maps, manuals, and other references provide additional information. Individuals with specific training, such as the building engineer or a utility representative, can add specific information. A seasoned officer has built up a bank of information based on experience, training, and direct observations.

The initial dispatch information will contain facts such as the location and nature of the situation. The dispatch information could be very general (*"a building reported on fire in the vicinity of Central Avenue and Main Street"*) or very specific (*"a smoke alarm activation in room 3102 of the First National Bank Building"*).

The time of day, temperature, and weather conditions are other factors that can easily be determined and incorporated into the initial size-up. Based on these basic facts, an officer might have certain expectations about the incident. For example, whether a building is likely to be occupied or unoccupied, whether the occupants are likely to be awake or

Voices of Experience

“Your direction can serve as a prompt for response and size-up by incoming companies.”

“Engine 46 on the scene. Southbound, 8517 Burley. Occupied, 2 story residential frame, 25 by 60. Fire, 2nd floor rear.”

Verbally, the above statement would take only a few seconds for you to transmit on the radio. Its value is not only in its brevity, but also in the wealth of information that it transmits to the dispatcher and the additional responding units. With less than twenty words, you have described the incident scene and set in motion the preparation for extinguishment and other fireground actions.

By repeating the address, you have verified to the dispatcher the correct location and, depending on the municipality, you may have also indicated the location of the nearest hydrant. Your direction can serve as a prompt for response and size-up by incoming companies.

The quick, brief description of the building and fire location prepare the responders for what they may encounter. Dimensions, floor, and fire location can help responders estimate the required hose stretch and determine what portable ground ladders may be required. Location also provides a starting point for search, ventilation, and potential salvage activities.

Construction type and occupancy provide important information regarding fire spread, potential hazards, and the need for additional assistance.

The radio protocol for every fire department differs and individuals must comply with their respective department's regulations. If allowable, a brief, concise radio report that paints a picture of the fireground provides immeasurable benefits for the dispatchers and responding units.

Paul R. Martin
Chicago Fire Department
Chicago, Illinois

Fire Fighter Tips

Maintenance personnel can often provide valuable information about a structure and its mechanical systems.

sleeping, and whether traffic will delay the arrival of additional units may be inferred from the time of day. Weather conditions such as snow and ice will delay the arrival of fire apparatus, create operational problems with equipment, and require additional fire fighters to perform basic functions such as stretching hose lines and raising ladders. Strong winds can cause rapid **extension** or spread of a fire to exposed buildings. High heat and humidity will affect fire fighters' performance and may cause heat casualties.

An experienced fire officer will also have a basic knowledge of the community and the department's available resources. The officer may remember the types of occupancies, construction characteristics of typical buildings, and specific information about particular buildings from previous incidents or preincident planning visits.

A **preincident plan** can be especially helpful during size-up because it contains significant information about a structure. A preincident plan generally provides details about a building's construction, layout, contents, special hazards, and fire protection systems. Chapter 22, Preincident Planning, contains specific information about the contents and development of such plans.

Basic facts about a building's size, layout, construction, and occupancy can often be observed upon arrival, if they are not known in advance. The officer must consider the size, height, and construction of the building during the size-up. The action plan for a single-story, wood-frame dwelling on an acre of land will be quite different than that for a steel-frame high-rise tower on a crowded urban street.

The age of the building is often an important consideration in size-up because building and fire safety codes change over time. Older wooden buildings may have **balloon-frame construction**, which can provide a path for fire to spread rapidly into uninvolved areas. Floor and roof systems in newer buildings use trusses for support, which are susceptible to early failure and collapse. The weight of air conditioning or heating units on the roof may also contribute to structural collapse.

A plan for rescuing occupants and attacking a fire must consider information about the building layout, such as the number, locations, and construction of stairways. The plan for a building with an open stairway that connects several floors might have keeping the fire out of the stairway as a priority. This enables fire fighters to use the stairway for rescue and prevents the fire from spreading to other floors. Ladder placement, use of aerial or ground ladders, and emergency exit routes all depend on the building layout.

Any special factors that will assist or hinder operations must be identified during size-up. For example, bars on the windows will limit access and complicate the rescue of trapped victims, but firewalls and sprinkler systems will help to confine or extinguish a fire.

The occupancy of a building is also critical information. An apartment building, a restaurant, an automotive repair garage, an office building, a warehouse filled with concrete blocks, and a chemical manufacturing plant all present fire fighters with different sets of problems that must be addressed.

The size of the fire and its location within the building help determine hose line placement and ventilation sites. The fire's location is also critical in determining which occupants are in immediate need of rescue. For example, someone directly above a burning apartment should be rescued immediately, but a person at the opposite end of the building, far from the fire, is probably not in immediate danger.

Direct visual observations will give the best information about the size and location of a fire, particularly when combined with information about the building. Visible flames indicate where the fire is located and how intensely it is burning but might not tell the whole story. Flames issuing from only one window suggest that the fire is confined to that room, but it could also be spreading through void spaces to other parts of the building.

Often, the location of a fire within a building cannot be determined from the exterior, particularly when visibility is obscured by smoke (► Figure 10-7). An experienced fire fighter will observe where smoke is visible, how much and what color it is, how it moves, and even how it smells. A burning pot of food, an overheated electrical motor, or burning wood all have distinctive odors.

Inside a building, fire fighters can use their observations and sensations to help them work safely and effectively. If smoke limits visibility, a crackling sound or the sensation of heat coming from one direction may indicate the **seat of the fire**, which is the main area of the fire. Blistering paint and smoke seeping through cracks can lead fire fighters to a hidden fire burning within a wall. Advanced technology, such as thermal imaging cameras, can be even more effective in identifying hidden fires.

The IC needs to gather as much factual information as possible about a fire. Because the IC often is located at a command post outside the building, company officers are requested to report their observations from different locations. A company that is operating a hose line inside the building can report on interior conditions, while the ventilation crew on the roof will have a different perspective. The IC may request that an officer or a fire fighter prepare a **reconnaissance report**. The reconnaissance report is the

Figure 10-7 It can be difficult to determine the location and extent of the fire inside a building.

inspection and exploration of a specific area to gather information for the IC. The IC assembles, interprets, and bases decisions on this information.

Regular progress reports from companies working in different areas update information about the situation. Progress reports enable an IC to judge whether an operational plan is effective or needs to be changed.

Probabilities

Probabilities refer to events and outcomes that can be predicted or anticipated, based on facts, observations, common sense, and previous experiences. Fire fighters frequently use probabilities to anticipate or predict what is likely to happen in various situations. The attack plan is also based on probabilities, predicting where the fire is likely to spread and anticipating potential problems.

An IC must be able to identify the probabilities that apply to a given situation quickly. For example, an apartment building fire in the middle of the night will probably involve occupants who need to be rescued. Therefore, the IC would assign additional crews to search for potential victims, even if no factual information indicates that any occupants need to be rescued. Similarly, a fire burning on the top floor of a structure in a row of attached buildings has a high probability of spreading to adjoining buildings through the cockloft. In this case, the IC's plan would include opening the roof above the fire and sending crews into the exposed occupancies to check for fire extension in the cockloft.

The concepts of convection, conduction, and radiation enable an IC to predict how a fire will extend in a particular situation. By observing a particular combination of smoke and fire conditions in a particular type of building, the IC can identify a range of possibilities for fire extension within the structure or to other exposed buildings. Using these probabilities, the IC can predict what is likely to happen and develop a plan to control the situation effectively.

The IC must also evaluate the potential for collapse of a burning structure. The building's construction, the location and intensity of the fire, and the length of time the structure has been burning are all factors that must be considered. If the possibility of collapse exists, risk management dictates that the IC order fire fighters out of the building before it occurs.

Resources

Resources include all of the means that are available to fight a fire or conduct emergency operations at any other type of emergency incident. Resource requirements depend on the size and type of incident. Resource availability depends on the capacity of a fire department to deliver fire fighters, fire apparatus, equipment, water, and other items that can be used at the scene of an incident.

A fire department's basic resources are personnel and apparatus. Firefighting resources are usually defined as the numbers of engine companies, ladder companies, special units, and command officers required to control a particular fire. An IC should be able to request the required number of companies and know that each unit will arrive with the appropriate equipment and the necessary fire fighters to perform a standard set of functions at the emergency scene. The IC usually will have a good idea of the number and types of companies available to respond, how they are staffed and equipped, and how long they should take to arrive. This information might have to be updated at the incident, particularly in areas served by volunteer fire fighters, because the number of fire fighters available to respond may vary at different times.

Water supply is another critical resource. Water supply is generally not a problem if the fire is in a district that has hydrants and a strong, reliable municipal water system. Water supply could limit operations in an area without hydrants. It takes time to establish a water supply from a static source, and the amount of water that can be delivered by a tanker shuttle is limited. In this situation, the IC would need to call for additional tankers or engine companies with large-diameter hoses.

Resources include more than fire fighters, fire apparatus, and a water source. A fire in a flammable liquids storage facility will require large quantities of foam and the equipment to apply it effectively. A hazardous materials incident might require special monitoring equipment, chemical protective clothing, and bulk supplies to neutralize or absorb a spilled product. A building collapse might require heavy equipment to move debris and a structural engineer to determine where fire fighters can work safely. The fire department must have these supplies available or be able to obtain them quickly when they are needed. Resources for a large-scale incident must include food and fluids for the rehydration of fire fighters, fuel for the apparatus, and other supplies. The Red Cross and law enforcement agencies often provide support resources.

The size-up process enables the IC to determine what resources will be needed to control the situation and to ensure their availability. An action plan to control an incident can be effective only if the necessary resources can be assembled on a timely basis. If there is a delay, the IC must anticipate how much the fire will grow and where it will spread. If the desired resources cannot be obtained, the IC must develop a realistic plan using available resources to gain eventual control of the situation.

Ideally, a fire department will be able to dispatch enough fire fighters and apparatus to control any situation within its jurisdiction. Most departments have mutual aid agreements with surrounding jurisdictions to assist each other if a situation requires more resources than the local community can provide. In some areas, hundreds of fire fighters and apparatus can be assembled for a large-scale incident. In more remote areas, the resources available to fight a fire can be very limited.

If resources are insufficient or delayed, a fire can become too large to be controlled by available personnel. For example, if only 20 fire fighters and four apparatus are available, they will have to bring the fire under control before it gets too big or wait until the fire burns itself down to a manageable size.

Resources must be organized to support efficient emergency operations. Fire fighters must be properly equipped and organized in companies. Equipment and procedures must be standardized. The incident management system must be designed to manage all of the resources that could be used at a large-scale incident, and the communications system must enable the IC to coordinate operations effectively.

Incident Action Plan

The IC develops an incident action plan that outlines the steps needed to control the situation. The incident action plan is based on information gathered during size-up. The initial IC develops a basic plan for beginning operations. If the situation expands or becomes more complicated, this plan is revised and expanded as additional information is obtained, more resources become available, and the incident management structure grows.

The incident action plan should be based on the five basic fireground objectives. In order of priority, these objectives are:

1. Rescue any victims.
2. Protect exposures.
3. Confine the fire.
4. Extinguish the fire.
5. Salvage property and overhaul the fire.

This system of priorities clearly establishes that the highest priority in any emergency situation is saving lives. The remaining priorities involve saving property. Making sure that the fire does not spread to any **exposure** (an area adjacent to the fire that may become involved if not protected) is a higher priority than confining the fire within the burning building. After the fire is confined, the next priority is to extinguish it. The final priority is to protect property from additional damage and make sure the fire is completely out.

These priorities are not separate and exclusive. One objective does not have to be accomplished before fire fighters tackle the next one. Often, more than one objective can be addressed simultaneously, and certain activities help achieve more than one objective. For example, if a direct attack on the fire will bring it under control very quickly, the objectives of protecting exposures, confining the fire, and extinguishing the fire might all be conducted together. Extinguishing the fire might also be the best way to protect the lives of building occupants. Frequently, salvage crews may be working on lower floors while fire suppression crews are still attacking a fire on an upper floor.

These priorities guide the IC in making difficult decisions, particularly if not enough resources are available to address every priority. If the decision is between saving lives and saving property, saving lives comes first. After rescue is completed and if the fire is still spreading, the IC should place exposure protection ahead of salvage and overhaul.

A similar set of priorities can be established for any emergency situation. Saving lives is always more important than protecting property. The IC must always place a higher priority on bringing the problem under control than on cleaning up after the problem.

Rescue

Protecting lives is the first consideration at a fire or any other emergency incident. The need for rescue depends on many circumstances. The number of people in danger varies based on the type of occupancy and the time of day. A commercial building that is crowded during the workday might have few, if any, occupants at night. At night, rescue is more likely needed in a residential occupancy, such as a house, an apartment building, or a hotel.

The degree of risk to the lives of occupants must also be evaluated. A fire that involves one apartment on the 10th floor of a high-rise apartment building could threaten the lives of the residents of that floor and those directly above the fire. However, residents below the fire are probably not in any danger, and residents several floors above the fire might be safer staying in their apartments until the fire is extinguished, instead of walking down smoke-filled stairways.

Often, the best way to protect lives is to extinguish the fire quickly, so efforts to control the fire are usually initiated at the same time as rescue. Hose lines may have to be used to protect exit paths and keep the fire away from victims during search-and-rescue operations. The chapter on search and rescue discusses specific tactics and techniques for these operations.

Exposure Protection

Within the secondary objective of protecting property, there are multiple priorities. The first priority in protecting property is to keep the fire from spreading beyond the area of origin or involvement when the fire department arrives. The IC must start by making sure the fire is not expanding. If the fire is burning in only one room, the objective should be to ensure that it does not spread beyond that room. If more than one room is involved, the objective might be to contain the fire to one apartment or one floor level. Sometimes the objective is to confine the fire to one building, particularly if multiple buildings are attached or closely spaced.

In some cases, the IC has to look ahead of the fire and identify a place to stop its spread. If flames are extending quickly through the cocklofts in a row of attached buildings, the IC will place companies with hose lines ahead of the fire to stop its progress.

The IC must sometimes weigh potential losses when deciding where to attack a fire. If a fire in a vacant building threatens to spread to an adjacent occupied building, the IC will usually assign companies to protect the exposure before attacking the main body of fire. If, however, a fire in an occupied building could spread to a vacant building, the IC's decision might be to attack the fire first and worry about controlling the spread later.

Confinement

After ensuring that the fire is not actively extending into any exposed areas, the IC will focus on confining it to a specific area. The IC will define a perimeter and plan operations so that the fire does not expand beyond that area. Fire fighters on the perimeter must be alert to any indications that the fire is extending.

Thermal imaging devices are electronic cameras that can detect sources of heat. They are valuable tools for finding fires in void spaces. The principles and use of thermal imaging devices are covered more fully in Chapter 13, Search and Rescue.

Extinguishment

Depending on the size of the fire and the risk involved, the IC will mount either an offensive attack or a defensive strategy to extinguish a fire. An **offensive attack** is used with most small fires. Fire fighters advance into the fire building with hose lines or other extinguishing agents and overpower the fire. If the fire is not too large and the attacking fire fighters can apply enough extinguishing agent, the fire can usually be extinguished quickly and efficiently. Extinguishing the fire in this way often resolves several priorities at the same time, including exposure protection and confinement.

At regular intervals during an offensive attack, the IC must evaluate the progress being made. If not enough progress is being made, the IC must either adjust the tactics or alter the overall strategy. For example, the fire may produce more heat than the water from hose lines can absorb or fire fighters may be unable to penetrate far enough to make a direct attack on the fire.

Sometimes the problem can be resolved by adding resources and intensifying the offensive attack. Larger hose lines or more fire fighters with additional hose lines might be able to extinguish the fire successfully. Coordinating ventilation with an offensive attack might enable fire fighters to get close enough to extinguish the fire. Special extinguishing agents may be used to extinguish the fire.

When the fire is too large or too dangerous to extinguish with an offensive attack, the IC will implement a **defensive attack**. In these situations, all fire fighters are ordered out of the building, and heavy streams are operated from outside the fire building. A defensive strategy is required when the IC determines that the risk to fire fighters' lives is excessive, as in situations where structural collapse is possible. The IC who adopts a defensive strategy has determined that there is no property left to save or that the potential for saving property does not justify the risk to fire fighters. Sometimes a defensive strategy is effective in extinguishing the fire; at other times, it simply keeps the fire from spreading to exposed properties (► Figure 10-8).

There are some situations in which all fire fighters are withdrawn from the area and the fire is allowed to burn itself out. These situations generally involve potentially explosive or hazardous materials that pose an extreme danger to fire fighters.

Salvage and Overhaul

Salvage operations are conducted to save property by preventing avoidable property losses. **Salvage** is the removal or protection of property that could be damaged during firefighting or overhaul operations (► Figure 10-9). Salvage operations are often aimed at reducing smoke and water

Figure 10-8 A defensive strategy involves an exterior fire attack with heavy streams that emphasizes protecting the exposures to the fire building.

Figure 10-9 Salvage operations are conducted to save property by preventing avoidable property losses.

damage to the structure and contents once the fire is under control.

The **overhaul** process is conducted after a fire is under control to completely extinguish any remaining pockets of fire (► Figure 10-10). The IC is responsible for ensuring that the fire is completely out before terminating operations. Floors, walls, ceilings, and attic spaces should be checked for signs of heat, smoke, or fire. Window casings, wooden door jambs, baseboards, electrical outlets, and heating/air conditioning vents can often hide small, smoldering fires. Debris from the burned contents of the structure should be removed and thoroughly doused to reduce the potential for **rekindle**, or a reignition of the fire. For large fires, a fire department unit may be assigned to return to the scene every few hours to check for indications of residual fire. These operations are covered in depth in Chapter 18, Salvage and Overhaul.

Figure 10-10 Overhaul is conducted after a fire is under control to completely extinguish any remaining pockets of fire.

Wrap-Up

Ready for Review

This chapter addressed the phases of response and size-up. Preparations for an emergency response begin long before an alarm is received. Each fire fighter must ensure that all personal protective equipment and tools are ready for response. Adhering to simple procedures enables crews to arrive at an emergency quickly and safely.

Size-up is the ongoing mental evaluation of an emergency situation. The size-up process begins when information about an emergency incident is received. Size-up involves both an evaluation of the known facts of the situation, including the incident location, time, weather, type structure, exposures, available resources, and life hazard, and a consideration of any probabilities that could alter the situation. The incident commander processes this information and determines a plan of action. The incident action plan must address the fireground priorities of rescue, exposure protection, confinement, extinguishment, salvage, and overhaul.

Generally, a fire fighter is not expected to assess and take command of an emergency incident. However, understanding the size-up process will enable fire fighters to provide needed information to the fire officer and incident commander.

Chief Concepts

- Fire fighters must understand the need to receive dispatch information properly.
- As a fire fighter, your job is to respond safely to a scene. Follow your department's SOPs when riding in apparatus to the scene.
- Carefully mount and dismount all apparatus.
- Be aware of your surroundings at all times, especially when responding to an incident on the highway. Motorists might be distracted and not see you in the road.
- All information received at the communication center must be relayed to fire department units. What may seem insignificant to the dispatcher may be essential to the incident commander.
- Size-up is critical to the successful outcome of an emergency incident.

Hot Terms

Balloon-frame construction An older style of wood-frame construction in which the wall studs extend vertically from the basement to the roof without any fire stops.

Defensive attack Exterior fire suppression operations directed at protecting exposures.

Exposure Any person or property that may be endangered by flames, smoke, gases, heat, or runoff from a fire.

Extension Fire that moves into areas not originally involved, including walls, ceilings, and attic spaces; also the movement of fire into uninvolved areas of a structure.

Freelancing Dangerous practice of acting independently of command instructions.

Offensive attack An advance into the fire building by fire fighters with hose lines or other extinguishing agents to overpower the fire.

Overhaul Examination of all areas of the building and contents involved in a fire to ensure that the fire is completely extinguished.

Personal alert safety system (PASS) Device worn by a fire fighter that sounds an alarm if the fire fighter is motionless for a period of time.

Personal protective equipment (PPE) Gear worn by fire fighters that includes helmet, gloves, hood, coat, pants, SCBA, and boots. The personal protective equipment provides a thermal barrier for fire fighters against intense heat.

Personnel accountability system A method of tracking the identity, assignment, and location of fire fighters operating at an incident scene.

Personnel accountability tag (PAT) Identification card used to track the location of a fire fighter on an emergency incident.

Preincident plan A written document resulting from the gathering of general and detailed information to be used by public emergency response agencies and private industry for determining the response to reasonable anticipated emergency incidents at a specific facility.

Reconnaissance report The inspection and exploration of a specific area in order to gather information for the incident commander.

Rekindle A situation where a fire, which was thought to be completely extinguished, reignites.

Response Activities that occur in preparation for an emergency and continue until the arrival of emergency apparatus at the scene.

Salvage Removing or protecting property that could be damaged during firefighting or overhaul operations.

Seat of the fire The main area of the fire

Size-up The ongoing observation and evaluation of factors that are used to develop objectives, strategy, and tactics for fire suppression.

Thermal imaging devices Electronic devices that detect differences in temperature based on infrared energy and then generate images based on that data. Commonly used in obscured environments to locate victims.

Fire Fighter in Action

You have just arrived for your shift and received your riding assignment on the engine. While completing your morning inspections you are dispatched to a grease fire at a small drive-through hamburger restaurant. You prepare to mount the apparatus.

1. What preparations should you make at the beginning of each shift to make sure you are ready for emergency response?
   - **A.** Clean the station, and ensure you uniform is pressed and you boots are polished.
   - **B.** Have a shift meeting, then discuss meal plans and the time for physical fitness.
   - **C.** Inspect and place PPE in a designated location, inspect your SCBA and PASS device, and check apparatus and tools assigned to you.
   - **D.** Eat a good breakfast, wash the apparatus, and check the reserve apparatus.
2. What is the proper procedure for donning PPE and mounting the apparatus?
   - **A.** Jump on the apparatus as quickly as you can and don PPE in the cab to save time.
   - **B.** Place your PPE on the apparatus and don it after you arrive on the scene.
   - **C.** Don all of your PPE, always have at least one foot firmly placed on a foot surface, fasten your seat belt, and don your hearing protection.
   - **D.** Put your helmet on and jump on the apparatus. If you have a closed cab, you can don your PPE in the cab, fasten your seat belt once your PPE is on correctly, and, if provided, wear hearing protection.

You arrive on the scene and your lieutenant is performing size-up. The building is a small wood framed structure that is fully involved. Lieutenant Hansen tells you to shut off the natural gas, pull a 1 $^{1}/_{2}$" preconnected hose line, and attack the fire.

3. To turn off the natural gas you should:
   - **A.** force the locking device and turn the valve so the handle is at a right angle to the pipe.
   - **B.** force the locking device and turn the valve so the handle is in line to the pipe.
   - **C.** turn off the main disconnection switch.
   - **D.** turn the underground valve counterclockwise with a wrench.
4. What are the five basic fire ground objectives, in order of priority?
   - **A.** Rescue any victims, confine the fire, extinguish the fire, protect exposures, salvage the property and overhaul the fire.
   - **B.** Rescue any victims, protect exposures, confine the fire, extinguish the fire, salvage the property and overhaul the fire.
   - **C.** Rescue any victims, extinguish the fire, confine the fire, protect exposures, salvage the property and overhaul the fire.
   - **D.** Rescue any victims, confine the fire, extinguish the fire, salvage the property and overhaul the fire, protect exposures.

www.FireFighter.jbpub.com

- Chapter Pretests
- Interactivities
- Hot Term Explorer
- Web Links
- Review Manual
- FireLearn

www.FireFighter.jbpub.com

Forcible Entry

Technology Resources

www.FireFighter.jbpub.com

- Chapter Pretests
- Interactivities
- Hot Term Explorer
- Web Links
- Review Manual
- FireLearn

Chapter Features

- Skill Drills
- Voices of Experience
- Fire Marks
- Teamwork Tips
- Fire Fighter Safety Tips
- Fire Fighter Tips
- Canadian Perspectives
- Hot Terms
- Wrap-Up

Chapter 11

NFPA 1001 Standard

Fire Fighter I

5.3.4 Force entry into a structure, given personal protective equipment, tools, and an assignment, so that the tools are used as designed, the barrier is removed, and the opening is in a safe condition and ready for entry.

5.3.4 (A) *Requisite Knowledge.* Basic construction of typical doors, windows, and walls within the department's community or service area; operation of doors, windows, and locks; and the dangers associated with forcing entry through doors, windows, and walls.

5.3.4 (B) *Requisite Skills.* The ability to transport and operate hand and power tools and to force entry through doors, windows, and walls using assorted methods and tools.

5.3.14 (A) The purpose of property conservation and its value to the public, methods used to protect property, types of and uses for salvage covers, operations at properties protected with automatic sprinklers, how to stop the flow of water from the sprinkler head, identification of the main control valve on an automatic sprinkler system, and forcible entry issues related to salvage.

Fire Fighter II

6.3.2 (A) *Requisite Knowledge.* Selection of the nozzle and hose for fire attack given different fire situations; selection of adapters and appliances to be used for specific fire ground situations; dangerous building conditions created by fire and suppression activities; indicators of building collapse; the effects of fire and fire suppression activities on wood, masonry (brick, block, stone), cast iron, steel, reinforced concrete, gypsum wall board, glass, and plaster on lath; search-and-rescue and ventilation procedures; indicators of structural instability; suppression approaches and practices for various types of structural fires; and the association between specific tools and special forcible entry needs.

Knowledge Objectives

After studying this chapter, you will be able to:

- Understand the association between specific tools and special forcible entry needs.
- Describe the basic construction of typical doors, windows, and walls.
- Know the dangers associated with forcing entry through doors, windows, and walls.
- Describe how to force entry through doors, windows, locks, and walls.
- Know how forcible entry relates to salvage.

Skills Objectives

After studying this chapter, you will be able to:

- Force entry through an inward-opening door.
- Force entry through an outward-opening door.
- Use the triangle method to open an overhead garage door.
- Force entry through a double-hung window.
- Force entry through a casement window.
- Force entry through a projected window.
- Force entry using a K tool.
- Force entry using an A tool.
- Force entry by unscrewing the lock.
- Breach a wall frame.
- Breach a masonry wall.
- Breach a metal wall.
- Breach a wood floor.

You Are the Fire Fighter

It's late and you're tired, but you respond at once when your ladder company is dispatched to a reported structure fire with people trapped. Your ladder company is generally assigned to perform forcible entry if needed, ventilation operations, and a primary search for victims. En route, you think about the forcible entry tools and techniques you know and begin to consider what might be the best spot to force entry. The neighborhood has seen a rise in crime in the past couple of years, so there could be bars on the windows and multiple locks on the doors.

1. ***What tools will you need to force entry in this situation?***
2. ***What obstacles might you counter inside the building?***
3. ***What are some of the dangers associated with forcible entry?***

Introduction

One of a fire fighter's most dynamic and challenging tasks, forcible entry is defined as gaining access to a structure when the normal means of entry are locked, secured, obstructed, blocked, or unable to be used for some other reason. The term forcible entry usually refers to structures; extrication is the term used for entry to a vehicle. This chapter examines forcible entry into structures.

Forcible entry requires strength, knowledge, proper techniques, and skill. Because forcible entry often causes damage to the property, fire fighters must consider the results of using different forcible entry methods, and select the one appropriate for the situation. If rapid entry is needed to save a life or prevent a more serious loss of property, it is appropriate to use maximum force. When the situation is less urgent, fire fighters can take more time and use less force.

Another factor to consider when making forcible entry is the need to secure the premises after operations are completed. Fire fighters must never leave the premises in a condition that would allow unauthorized entry. If a door or a lock is destroyed during an urgent forcible entry, arrangements should be made to board up or repair the opening afterward. When the situation is not urgent, consider using entry methods that result in less damage and can be more easily repaired.

A fire fighter who is skilled in forcible entry should be able to get the job done quickly, with as little damage as possible. The fire fighter must consider the type of building construction, possible entry points, the types of securing device(s) that are present, and the best tools and techniques for each situation. Selecting the right tool and using the proper technique initially saves valuable time and could save lives.

Doors, windows, locks, and security devices can be combined in countless variations to prevent easy entry. Fire fighters must keep up with technology, including the new styles of windows, doors, locks, and security devices that are common in the local response area and how they operate. Learn to recognize different types of doors, windows, and locks. The best time to examine these components is during inspection and preincident planning visits. Touring buildings under construction or renovation is also an excellent way to learn about building construction and to examine different devices. Talking with construction workers and locksmiths can provide valuable information about how to best attack a particular lock, window, or door.

This chapter will review door, window, and lock construction, as well as discuss the tools and techniques used for forcible entry in several types of situations. The skill drills cover the most common forcible entry techniques.

Forcible Entry Situations

Forcible entry is usually required at emergency incidents where time is a critical factor. A rapid forcible entry, using the right tools and methods, might result in a successful rescue or allow an engine company to make an interior attack and control a fire before it extends. Fire fighters must study and practice forcible entry techniques so that proficiency increases with every experience.

Company officers usually select both the point of entry and the method to be used. They ensure that the efforts of different companies are properly coordinated for safe, effective operations. For example, forcible entry actions must be coordinated with hose teams because the entry must be made before engine companies can get inside to attack a fire. Opening a door before the hose lines are in place and ready to advance could allow clean air to enter, resulting in a possible fire spread or backdraft. Making an opening at the wrong location could undermine a well-planned attack.

Before beginning forcible entry, remember to "try before you pry." Always check doors or windows to ensure that forcible entry is needed. An unlocked door requires no force. A window that can be opened does not need to be broken. Checking first takes only a few seconds, and could save several minutes of effort and unnecessary property damage. It is equally important to look for alternative entry points. Do not spend time working on a locked door when a nearby window provides easy access to the same room.

Unusually difficult forcible entry situations may require the expertise of specialty companies such as technical rescue teams. These companies have experience working with specialized tools and equipment to gain access to almost any area.

Forcible Entry Tools

Choosing the right forcible entry tool can be a very important decision. Fire departments use several forcible entry tools, ranging from basic cutting, prying, and striking tools to sophisticated mechanical and hydraulic equipment. Fire fighters must know:

- What tools the department uses
- The uses and limitations of each tool
- How to select the proper tool for the job
- How to safely operate each tool
- How to carry the tools safely
- How to inspect and maintain each tool, to ensure readiness for service and safe operation

General Tool Safety

Any tool that is used incorrectly or is maintained improperly can be dangerous to both the operator and other persons. Anyone who uses a tool should understand the proper operating and maintenance procedures before beginning the task. Always follow the manufacturer's recommendations for operation and maintenance.

General safety tips for using tools include:

1. Always wear the appropriate protective equipment. When conducting forcible entry during fire suppression operations, this includes a full set of structural firefighting personal protective equipment (PPE).
   - Goggles or approved eye protection are required when working with cutting or striking tools to prevent eye injuries from pieces of metal, glass, or other materials.
   - Gloves provide protection from sharp cutting blades as well as sharp edges and broken glass. A good glove can easily make the difference between a minor injury and a severed finger.
   - A helmet provides essential protection from falling debris.
   - Turnout coat and pants help to protect skin from sharp metal edges or broken glass.
   - Boots protect the feet from nails or other sharp objects.
2. Learn to recognize the materials used in building and lock construction and the appropriate tools and techniques for each. For example, many locks are made of **case-hardened steel**, which combines carbon and nitrogen to produce a very hard outer coating that cannot be penetrated by most ordinary cutting tools. Using the wrong tool to cut case-hardened steel could break the tool and potentially injure the user and other fire fighters.
3. Keep all tools clean, properly serviced according to the manufacturer's guidelines, and ready for use. Immediately report any tool that is damaged or broken and take it out of service.
4. Do not leave tools lying on the ground or floor. Return them to a tool staging area or the apparatus. Tools that are left lying around could cause injuries.

General Carrying Tips

Tools can cause injuries even when not in use. Carrying a tool improperly can result in muscle strains, abrasions, or lacerations. General carrying tips that apply to all tools include:

- Do not try to carry a tool or piece of equipment that is too heavy or designed to be used by more than one person (▼ Figure 11-1). Request assistance from another fire fighter.
- Always use your legs—not your back—when lifting heavy tools.
- Keep all sharp edges and points away from your body at all times. Cover or shield them with a gloved hand to protect those around you.
- Carry long tools with the head down toward the ground. Be aware of overhead obstructions and wires, especially when using pike poles.

Figure 11-1 Request assistance to help carry a heavy tool.

General Maintenance Tips

All tools should be kept in a "ready state." This means that the tool is in proper working order, in its proper storage place, and ready for immediate use. Forcible entry tools are generally stored in a designated compartment on the apparatus. Hand tools should be clean, and cutting blades should be sharp. Power tools should be completely fueled and treated with a fuel stabilizing product, if necessary, to ensure easy starting. Every crew member should be able to locate the right tool immediately, confident that it is ready for use (▶ Figure 11-2).

All tools require regular maintenance and cleaning to ensure that they will be ready for use in an emergency. Thorough, conscientious daily or weekly checks should be performed, particularly with infrequently used tools. Always follow the manufacturer's instructions and guidelines, and store maintenance manuals in an easily accessible location. Keep proper records to track maintenance, repairs, and any warranty work that is performed.

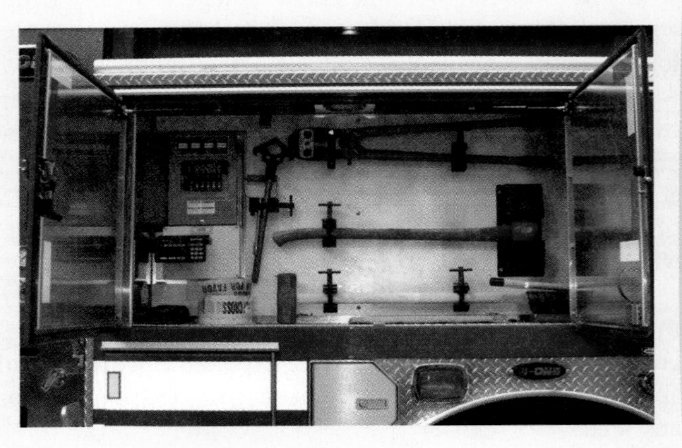

Figure 11-2 All tools should be kept in a ready state.

Types of Forcible Entry Tools

Forcible entry tools include hand tools and power tools. They can be categorized as:

- Striking tools
- Prying/Spreading tools
- Cutting tools
- Lock tools

Many of the basic fire fighting tools were described and pictured in Chapter 8, Fire Fighter Tools and Equipment.

Striking Tools

Striking tools are used to generate an impact force directly on an object or another tool. Striking tools are generally hand tools powered by human energy. The head of a striking tool is usually made of hardened steel.

Flat-head Axe

The flat-head axe was one of the first tools developed by humans and is still widely used. One side of the axe head is a cutting blade and the other side is a flat striking surface. Fire fighters often use the flat side to strike a Halligan tool and drive a wedge into an opening. Most fire apparatus carry both flat-head axes and pick-head axes. The pick-head axe is a cutting tool, not a striking tool.

Battering Ram

The **battering ram** is another tool used to force doors and breach walls. Originally, it was a large log used to smash through enemy fortifications. Today's battering rams are usually made of hardened steel and have handles; two to four people are needed to use a battering ram.

Sledgehammer or Maul

Sledgehammers (sometimes called mauls) come in various weights and sizes. The head of the hammer can weigh from two pounds to 20 pounds. The handle may be short like a carpenter's hammer or long like an axe handle. A sledgehammer can be used by itself to break down a door or with other striking tools such as the Halligan.

Prying/Spreading Hand Tools

Hand tools designed for prying and spreading are often used by fire fighters to force entry into buildings. This section describes the most commonly used forcible entry tools.

Halligan Tool (Bar)

The Halligan tool or bar is widely used by the fire service (▶ Figure 11-3). Pairing a Halligan tool with a flat-head axe creates a tool often referred to as "the irons." These two tools are commonly used to perform forcible entry. The Halligan tool incorporates three different tools—the **adz**, **pick**, and

Fire Fighter Tips

The forcible entry tools covered in this chapter are not the only ones available. Different fire departments use different tools, and there are often regional preferences for a particular piece of equipment. Become familiar with the tools used in your department.

Fire Fighter Safety Tips

Before swinging a striking tool, make sure there is a clear area of at least the length of the tool handle. Station another fire fighter outside the swing area as a safety person to keep others from entering the zone.

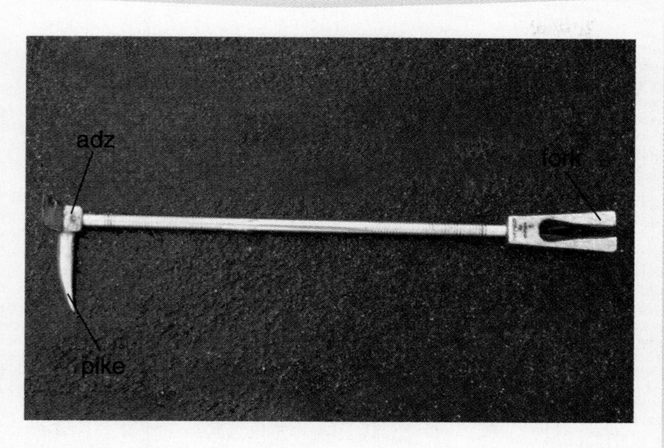

Figure 11-3 A Halligan tool has multiple purposes.

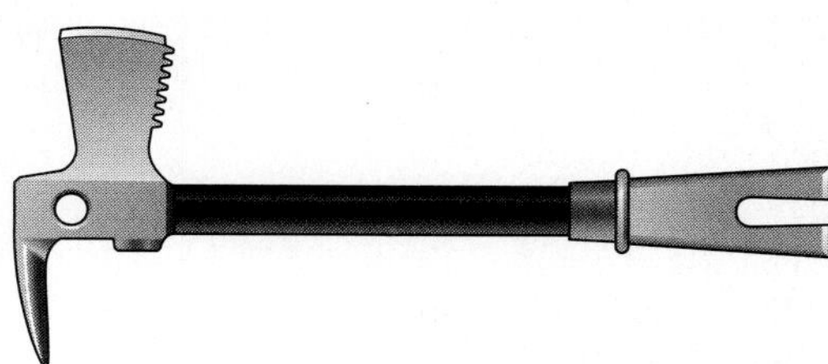

Figure 11-4 The pry axe is a multipurpose tool that can be used to cut and pry.

claw. The adz end is used to pry open doors and windows; the pick end is used to make holes or break glass; and the claw is used to pull nails and to pry apart wooden slats. A fire fighter can use a flat-head axe to strike the Halligan into place and create a better bite (a small opening that allows better tool access in forcible entry) between a door or window and its frame.

Pry Bar/Hux Bar/Crow Bar

Pry bars, Hux bars, and crow bars are made from hardened steel in a variety of shapes and sizes. They are most commonly used to force doors and windows, to remove nails, or to separate building materials. The various shapes allow fire fighters to exert different amounts of leverage in diverse situations.

Pry Axe

The pry axe is a multipurpose tool that can be used both to cut and to force open doors and windows (► Figure 11-4). The pry axe includes an adz, a pick, and a claw. The tool consists of two parts, the body, which has the adz and pick and looks similar to a miniature pick axe. It also includes a handle with a claw at the end, which slides into the body. The handle can be extended to provide extra leverage when prying or it may be removed and inserted into the head of the adz to provide rotational leverage. Extreme caution should be used when handling this tool. Over time the mechanism that locks the handle into position may become worn, allowing the handle to slip out and potentially injure fire fighters.

Fire Marks

The Halligan bar, designed by New York City fire fighter Hugh Halligan, is one of the most innovative tools ever created by a fire fighter. It is used by thousands of fire departments.

Figure 11-5 A hydraulic spreader.

Hydraulic Tools

Hydraulic-powered tools such as spreaders, cutters, and rams are often used for forcible entry. These tools require hydraulic pressure, which can be provided by a high-pressure, motor-operated pump or a hand pump.

Hydraulic cutters and spreaders are usually used in vehicle extrication, but can also be used in some forcible entry situations (▲ Figure 11-5). Hydraulic rams come in different lengths and sizes and can be used to apply a powerful force in one direction. The operation of these tools is covered in the chapter on vehicle rescue and extrication.

A rabbet tool is a small hydraulic spreader operated by a hand-powered pump. This tool is designed with teeth that

Fire Fighter Safety Tips

If hydraulic fluid under pressure begins to leak, it can cause serious injuries. Do not try to disconnect a pressurized hose because this can cause serious injury or splash hydraulic fluid into the eyes. If the fluid enters the body, it can cause tissue necrosis (death). Always relieve hydraulic system pressure before disconnecting a hose or fitting. Avoid wiping your face or eyes if there is hydraulic fluid on your hands.

will fit into a door jamb or rabbet (a type of door frame that has the door stop cut into the frame). As the spreader opens, it applies a powerful force that can open many doors.

Cutting Tools

Cutting tools are primarily used for cutting doors, roofs, walls, and floors. Although not as fast as power tools, hand-operated cutting tools are proven and reliable in many situations. Because they do not require a power source, they often can be deployed more quickly than power tools.

Several power cutting tools are available for different applications. These tools can be powered by batteries, electricity, gasoline, or hydraulics, depending on the amount of power required. Each type of tool and power source has advantages and disadvantages. For example, battery-powered tools are portable and can be placed in operation quickly, but have limited power and operating times.

Axe

There are many different types of axes, including flat-head, pick-head, pry, and multipurpose axes. The cutting edge of an axe is used to break into plaster and wood walls, roofs, and doors (► Figure 11-6).

Although a flat-head axe has been described as a striking tool, it is generally classified as a cutting tool. The pick-head axe is similar to a flat-head axe, but it has a pick instead of a striking surface opposite the blade. This pick can be used to make an entry point or a small hole if needed.

Specialty axes such as the pry axe and the multipurpose axe can be used for purposes other than cutting. The multipurpose axe, for example, can be used for cutting, striking, or prying, and includes a pick, a nail puller, a hydrant wrench, and a gas main shut-off wrench.

Bolt Cutters

Bolt cutters are used to cut metal components such as bolts, padlocks, chains, and chain-link fences. Bolt cutters are available in several different sizes, based on the blade opening and the handle length. The longer the handles, the greater the cutting force that can be applied. Bolt cutters may not be able to cut into some heavy-duty padlocks that are made with case-hardened metal.

Figure 11-6 The cutting edge of an axe is used to break into plaster and wood walls, roofs, and doors.

Circular Saw

Gasoline-powered circular saws are used by most fire departments both for forcible entry and for cutting ventilation holes. They are light, powerful, and easy-to-use, with blades that can be changed quickly. The different blades enable the saw to cut several materials.

Fire departments generally carry several different circular saw blades so that the proper blade can be used in different situations.

- Carbide-tipped blades are specially designed to cut through hard surfaces or wood. They stay sharp for long periods, so more cuts can be made before the blade needs to be changed.
- Metal-cutting blades are a composite material made with aluminum oxide. They are used to cut metal doors, locks, or gates.
- Masonry-cutting blades are abrasive and made of a composite material that includes silicon carbide or steel. They can cut concrete, masonry, and similar materials. Because they resemble metal-cutting blades, the operator must check the label on the blade before using it. Blades with missing labels should be discarded. Do not store masonry-cutting blades near gasoline because the gasoline vapors will cause the

Fire Fighter Tips

Almost all cutting tools have built-in safety features. Always take advantage of and use the safety features on every tool. To avoid injury, never remove safety guards from tools, not even for "just one cut."

composite materials in the blade to decompose. When the blade is later used, it could disintegrate.

Lock Tools/Specialty Tools

The lock and specialty tools are used to disassemble the locking mechanism on a door. These devices will cause minimal damage to the door and the door frame. If they are used properly, the door and frame should be undamaged, although the lock will generally need to be replaced. An experienced user can usually gain entry in less than a minute with these tools.

K Tool, A Tool, and J Tool

The K tool is designed to cut into a lock cylinder (► Figure 11-7 A). To shear the cylinder from a lock, place the cutting edge of the K tool on the top of the lock, insert a pry bar into the K tool, and strike the pry bar with a flathead axe. If the lock has a protective ring, the K tool may not be able to cut through it. Once the lock cylinder is removed, another tool is then used to open the locking mechanism.

The A tool is similar to the K tool except that it has the pry bar built into the cutting part of the tool (► Figure 11-7 B). This tool gets its name from its A-shaped cutting edges. To use an A tool, put the cutting head on the lock cylinder and use a striking tool to force it down into the cylinder until the lock can be forced out of the door. Once the lock cylinder is removed, use another tool to open the locking mechanism.

A J tool will fit between double doors that have panic bars (► Figure 11-7 C). Slide the J tool between the doors and pull to engage the panic bars.

Shove Knife

A shove knife is an old tool frequently used. Slipping the knife between the door and the frame at the latch forces the latch back and opens the door. Most new doors have latches that will not respond to a tool like this.

Duck-Billed Lock Breakers

To open a padlock using duck-billed lock breakers, drive the point into the shackles (the U-shaped part) of the lock. The increasing size of the point forces the shackles apart until they break. This causes no damage to the lock.

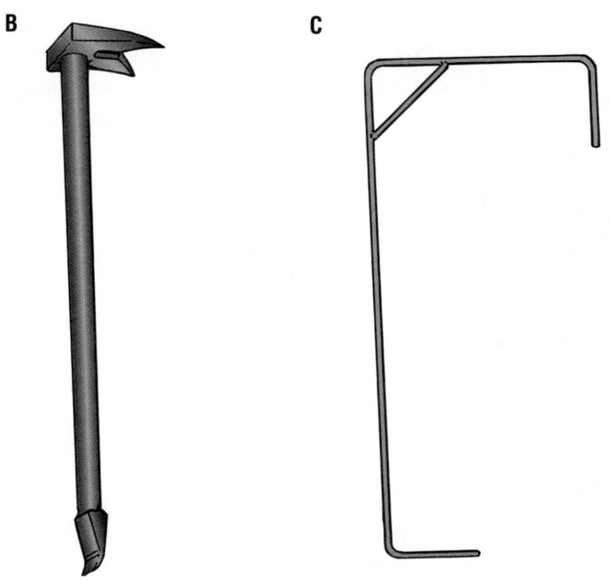

Figure 11-7 A–C A. K tool. B. A tool. C. J tool.

Locking Pliers and Chain

The locking pliers and chain are used to clamp a padlock securely in place so that the shackles can be cut safely with a circular saw or cutting torch. Clamp the pliers to the lock body while a second fire fighter maintains a steady tension on the chain as the lock is being cut.

Bam-Bam Tool

The bam-bam tool has a case-hardened screw, which is inserted and secured in the keyway of a lock. After the screw is in place, drive the handle of the tool into the keyway and pull the tumbler out of the lock.

Doors

Usually, the best point to attempt forcible entry to a structure is the door or window. Doors and windows are constructed as entry points and are generally of weaker materials than walls or roofs. An understanding of the basic construction of doors will help you select the proper tool and increase the likelihood of successfully gaining entry.

Basic Door Construction

Doors can be categorized by their construction material, and by the way they open. Both interior and exterior doors have the same basic components (▶ Figure 11-8):

- Door (the entryway itself)
- Jamb (the frame)
- Hardware (the handles, hinges, etc)
- Locking device

Construction Material

Doors are generally constructed of wood, metal, or glass. The design of the door and the construction material used will determine the difficulty in forcing entry.

Wood

Wood doors are commonly used in residences and may be found in some commercial buildings. There are three types of wood swinging doors: slab, ledge, and panel. Slab doors come in solid-core or hollow-core designs and are attached to wood frame construction with normal hardware (▶ Figure 11-9).

Solid-core doors are constructed of solid wood core blocks covered by a face panel. Solid-core doors may have an Underwriters Laboratories (UL) or Factory Mutual fire rating. They are usually used for entrance doors. Solid-core doors are heavy and may be difficult to force but their construction enables them to contain fire better than hollow-core doors.

Hollow-core doors have a lightweight, honeycomb interior, covered by a face panel. They are often used as interior doors, such as for bedrooms. Hollow-core doors are easy to force and will burn through quickly.

Ledge doors are simply wood doors with horizontal bracing. These doors, which are often constructed of tongue-and-groove boards, may be found on warehouses, sheds, and barns.

Panel doors are solid wood doors that are made from solid planks to form a rigid frame with solid wood panels set into the frame. Panel doors are used as both exterior and interior doors and may be made from a variety of types of wood. These doors resist fire longer than hollow-core slab doors and are typically easier to breach than solid-core slab doors if entry is attempted at the panels.

Figure 11-8 The parts of a door.

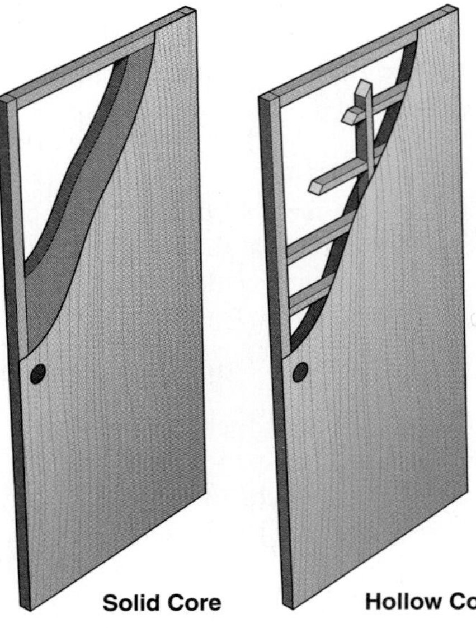

Figure 11-9 A slab door may have either solid-core or hollow-core construction.

Fire Fighter Safety Tips

Opening a door can allow clean air to enter, resulting in a possible fire spread or backdraft. Be careful and listen to instructions.

Metal

Metal doors may be decorative for residential use or utilitarian for warehouses and factories. They also may have either a hollow-core or solid-core construction. Hollow-core metal doors have a metal framework interior so they are as lightweight as possible. Solid-core metal doors have a foam or wood interior to reduce weight without affecting strength. Metal doors may be set in either a wood frame or a metal frame. Residential metal doors may appear to be panel doors and are usually used as entry doors.

Glass

Glass doors generally have a steel frame with tempered glass; or are simply tempered glass, and do not require a frame, but have metal supports to attach hardware. Glass doors are easy to force, but can be dangerous due to the large amount of small broken pieces that are produced when glass is broken.

Types of Doors

Doors also can be classified by how they open. The five most common ways that doors open are: inward, outward, sliding, revolving, and overhead. Overhead, sliding, and revolving doors are the most easily identifiable. Inward-opening and outward-opening doors can be differentiated by whether or not the hinges are visible. If you can see the hardware, the door will swing toward you (outward); if the hinges are not visible, the door will swing away from you (inward) (► **Figure 11-10 A & B**).

Door frames may be constructed of either wood or metal. Wood-framed doors come in two styles: stopped and rabbet. Stopped door frames have a piece of wood attached to the frame to stop the door from swinging past the latch. Rabbeted door frames are constructed with the stop cut built into the frame so it cannot be removed.

Metal-framed doors are more difficult to force than wood-framed doors. Metal frames have little flexibility; when a metal frame is used with a metal door, forcing entry will be very difficult. Metal frames look like rabbeted door frames.

Inward-Opening Doors

Design

Inward-opening doors of wood, steel, or glass can be found in most structures. They have an exterior frame with a stop or rabbet that keeps the door from opening past the latch. The locking mechanisms may range from standard door knob locks to deadbolt locks or sliding latches.

A

B

Figure 11-10 A & B **A.** An outward-opening door will have hinges showing. **B.** A door with the hinges not showing will open inward.

Skill Drill

Forcing Entry into an Inward-Opening Door

Size-up the door looking for any safety hazards. Inspect the door for the location and number of locks and mechanisms.

Place the forked end of the Halligan tool into the door frame between the door jamb and the door stop. Insert the tool near the lock, with the beveled end of the tool against the door.

Once the Halligan tool is in position, have your partner, on your command, drive the tool further into the gap between the rabbeted jamb or stop and the door. Make sure that the tool is not driven into the door jamb itself.

Once the tool is past the stop, and between the door and the jamb, push the Halligan toward the door to force it open. If more leverage is needed, your partner can slide the axe head between the bevel of the Halligan bar and the door. It may be necessary to push in on the door. Secure the door to prevent it from closing behind you.

Forcing Entry

A simple solution to gaining entry through these doors may be to break a small window on the door or adjacent to it, reach inside, and operate the locking mechanism. Remember to "try before you pry."

If no window is available and the door is locked, stronger measures might be required. To force an inward-opening door, first ascertain what type of frame it has. If the door has a stopped frame, use a prying tool near the locking mechanism to pry the stop away from the frame. After removing the stop, reinsert the prying tool near the latch and pry the door away from the frame. Once the latch clears the strike plate, push the door inward. Use a striking tool to force the prying tool further into the jamb. To force entry into an in-swinging door, follow the steps in (◄ Skill Drill 11-1).

1. Look for any safety hazards as you evaluate the door. Inspect the door for the location and number of locks and their mechanisms. **(Step 1)**
2. Place the forked end of the Halligan tool into the door frame, between the door jamb and the door stop. Place the tool near the lock, with the beveled end against the door. **(Step 2)**
3. Once the Halligan tool is in position, have your partner, on your command, drive the tool further into the gap between the rabbeted jamb or stop and the door. Make sure that the tool is not driven into the door jamb itself. **(Step 3)**
4. Once the tool is past the stop, and between the door and the jamb, push the Halligan toward the door to force it open. If more leverage is needed, your partner can slide the axe head between the bevel of the Halligan bar and the door. It may be necessary to push in the door. **(Step 4)**
5. Secure the door to prevent it from closing behind you. **(Step 5)**

Outward-Opening Doors

Design

Outward-opening doors are used in commercial occupancies and for most exits (► Figure 11-11). They are designed so that people can leave a building quickly during an emergency. Outward-opening doors may be constructed of wood, metal, or glass. They usually have exposed hinges, which may present an entry opportunity. More frequently, however, these hinges will be sealed so that the pins cannot be removed. Several types of locks, including handle-style locks and dead bolts may be used with these doors.

Forcing Entry

Before forcing entry to an outward-opening door, check the hinges to see if they can be disassembled or the pins removed. If that would take too long or cannot be done, place the adz end of a prying tool into the door frame near the locking mechanism. Use a striking tool to drive it further into the door jamb and get a good bite on the door. Then leverage the tool to force the door outward away from the jamb. To force entry into an out-swinging door, follow the steps in (► Skill Drill 11-2).

Figure 11-11 An outward-opening door enables rapid exit.

1. Size-up the door looking for any safety hazards. Determine the number and location of locks and the locking mechanism.
2. Place the adz end of the Halligan tool between the door and the frame, near the locking mechanism, or between the mechanism and a secondary lock. **(Step 1)**
3. Once the Halligan tool is in position, give the command to your partner to strike the Halligan and drive the adz end further into the gap. **(Step 2)**
4. Pry down with the fork end of the tool and then force the door outward.
5. Always secure the door to prevent it from closing behind you. **(Step 3)**

Sliding Doors

Design

Most sliding doors are constructed of tempered glass in a wooden or metal frame. They are commonly found in residences and hotel rooms that open onto balconies or patios (► Figure 11-12). Sliding doors generally have two sections and a double track; one side is fixed in place while the other side slides. A weak latch on the frame of the door secures the

11-2 Skill Drill

Forcing Entry into an Outward-Opening Door

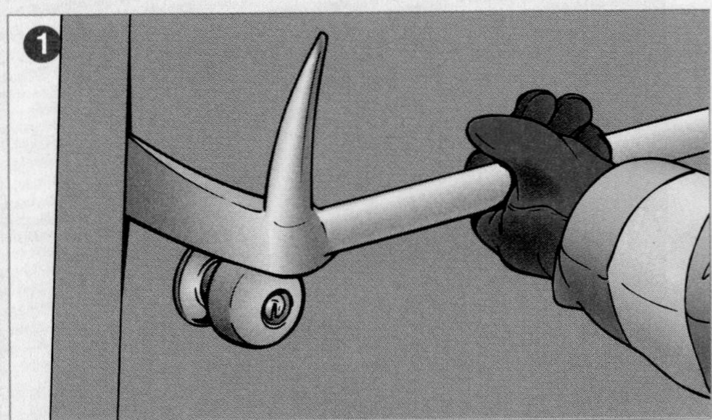

1 Size-up the door looking for any safety hazards. Place the adz end of the Halligan tool between the door and the frame either near the locking mechanism, or between the mechanism and a secondary lock.

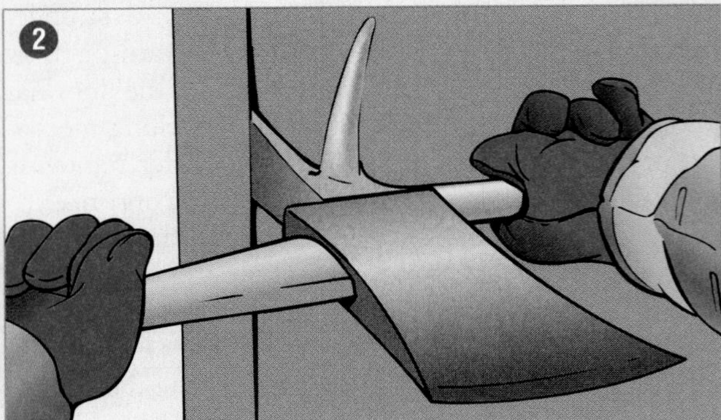

2 Once the Halligan tool is in position, have your partner strike the Halligan on your command and drive the adz end further into the gap.

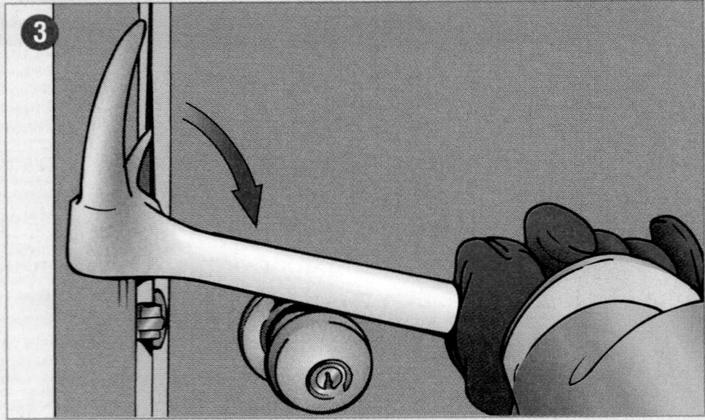

3 Pry in a downward direction with the fork end of the tool and then force the door outward. Always secure the door to prevent it from closing behind you.

moveable side. Many people prop a wood or metal rod in the track to provide additional security. If this has been done, look for another means of entry if possible to avoid destroying the door and leaving the premise open.

Forcing Entry

Forcing open sliding glass doors may be either very easy or very difficult. If the doors are not reinforced with a rod in the track, they should be easy to force. The locking mechanisms are not strong and any type of prying tool can be used. Place the tool into the door frame near the locking mechanism and force the door away from the mechanism. If there is a rod in the track that prevents the door from moving and there is no other way to enter, the only forcible entry choice is to break the glass. Remember to clean all broken glass from the opening.

Revolving Doors

Design

Revolving doors are most commonly found in upscale buildings and buildings in large cities (▶ Figure 11-13). They are usually made of four glass panels with metal frames. The panels are designed to collapse outward with a certain amount of pressure to allow rapid emergency escape. Revolving doors are generally secured by a standard cylinder lock or slide latch lock.

Figure 11-12 A reinforced sliding glass door may be difficult to open without breaking the glass.

Figure 11-13 Building codes may require outward-opening doors next to a revolving door.

Because of *Life Safety Code*®, fire prevention, and building code requirements, there are often standard, outward-opening doors adjacent to the revolving doors. It may be easier to attempt to force entry through these doors instead of through the revolving doors. Repairing any damage to the side doors will be less expensive for the building owner, too.

Forcing Entry

Forced entry through revolving doors should be avoided whenever possible. Even if the doors can be forced open, the opening created will not be large enough to allow many people to exit easily. Forced entry through a revolving door can be done by attacking the locking mechanism directly or by breaking the glass.

Overhead Doors

Design

Overhead doors come in many different designs—from standard residential garage doors to high-security commercial roll-up doors (▼ Figure 11-14 A & B). Most residential garage

A

B

Figure 11-14 A & B **A.** Commercial overhead doors are made of metal panels or steel rods. **B.** Residential overhead garage doors may tilt or roll up.

VOICES OF EXPERIENCE

"Sure enough, my predictions of a multi-family dwelling were correct—we found four separate apartments inside the structure, each of which had to be forced into."

Early one morning, my crew and I were dispatched to a reported structure fire involving a single-family dwelling. Our engine company arrived on the scene and performed a brief size-up. Suppression crews located the nearest hydrant and began stretching the attack line. My partner and I were assigned forcible entry and search duties.

As my partner and I approached the structure, I noticed that all of the first floor windows had bars bolted to their casings. Three cars were parked in the driveway, and there were four separate mailboxes next to the front door. This didn't seem right to me—why would a single-family home have four separate mailboxes? When my partner informed me that he found four gas meters on the side of the structure, I knew that we were dealing with a multi-family dwelling, and not a single-family dwelling as the dispatcher had reported. I immediately notified the incident commander of my suspicions.

The inward-swinging front door was made of wood and set in a wood frame, with six locking mechanisms. I placed the Halligan tool against the door with the bevel of the fork facing the door. "Strike, strike!" I shouted. The fork sank into the frame. I pushed forward and the door began to move. My partner tied a piece of webbing around the knob to make sure that I did not lose control of the door. I pushed a little harder and the door was freed of the locks. My partner and I backed out of the way to let the suppression crews in. As the suppression crews entered with their attack line, my partner and I followed to conduct a search. Sure enough, my predictions of a multi-family dwelling were correct—we found four separate apartments inside the structure, each of which had to be forced into. I advised the incident commander of our findings, and he then instructed the next-in company to begin removing the bars from the first floor windows to speed up forcible entry operations. Due to our quick thinking and effective forcible entry operations, crews were able to successfully attack the fire, vent the structure, and rescue all occupants.

Forcible entry size-up is done both before the incident and during. As part of a forcible entry team, it is imperative for you to recognize external clues about a structure's occupants, construction, and potential hazards. You must also be able to manipulate your arsenal of tools to meet the needs of the incident. These skills will help you to get the job done quickly and safely, with the least amount of damage possible.

Matt Thorpe
City of King Fire Department
King, North Carolina

Fire Fighter Safety Tips

Always secure doors in the open position before entering a building. Prop open overhead doors with a pike pole. Overhead doors could come down on fire fighters or their attack lines, trapping them or cutting off their water supply. This could complicate the attack and lead to injuries or deaths.

doors have three or four panels that may include windows. Some residential overhead garage doors come in a single section and tilt rather than roll up. These doors may be made of wood or metal. They usually have a hollow-core filled with insulation or foam. Commercial security overhead doors are made of metal panels or hardened steel rods. They may be solid-core or hollow-core, depending on the amount of security needed. Overhead doors can be secured with cylinder-style locks, padlocks, or automatic garage door openers.

Forcing Entry

Before forcing entry through an overhead door, make a careful size-up of the door. Most residential garage doors are not very sturdy; breaking a window or panel and manually operating the door lock or pulling the emergency release on the automatic opener from the inside may be all that is needed. If the fire is behind the overhead door, it may have weakened the door springs, making it impossible to raise the door. In such a case, the door must be cut or another means of entry found.

Secure the opened door with a pike pole or support to ensure that it does not close. Put the prop under the door near the track.

The quickest way to force entry through a security roll-up door is to cut the door with a torch or saw. Make the cut in the shape of a triangle, with the point at the top. Pad the opening to prevent injury to those entering and exiting through the opening. To open an overhead garage door using the triangle method, follow the steps in Skill Drill 11-3.

1. Before cutting, check for any safety hazards during the size-up of the garage door.
2. Select the appropriate tool to make the cut. The best choice will probably be a power saw with a metal-cutting blade.
3. Wearing full turnout gear and eye protection, start the saw and ensure it is in proper working order.
4. Be aware of the environment behind the door. If necessary, cut a small inspection hole, large enough to insert a hose nozzle.
5. Starting at a center high point in the door, make a diagonal cut to the right, down to the bottom of the door.
6. From the same starting point, make a second diagonal cut to the left, down to the bottom of the door, forming a large triangle.
7. Pad or protect the cut edges of the triangle and the bottom panel to prevent injuries as fire fighters enter or leave.

Windows

Windows provide airflow and light to the inside of buildings; they also can provide emergency entry or exits. Windows are often easier to force than doors. Understanding how to force entry into a window requires an understanding of both window-frame construction and glass construction. Window frames are made of the same materials used in doors—wood, metal, vinyl, or combinations of these materials—and will often match the door construction of a building.

This section reviews window construction, glass construction, and forcible entry techniques. As always, the goal is to force entry with minimal or no damage to the window. As with doors, try to open the window before using any force. Although breaking the glass is the easiest way to force entry, it is also dangerous. Glass may be the least expensive part of the window, but replacement costs are increasing because energy-efficient windows are being used more widely.

Safety

Always take proper protective measures, including wearing full PPE, when forcing entry through windows. Ensure that the area around the window, both inside and outside the building and below the window, is clear of other personnel. Broken pieces of wood, shattered glass, and sharp metal can cause serious injury.

Remember that forcing entry through windows during a fire situation may cause the fire to spread. Do not attempt forcible entry through a window unless a proper fire attack is in place. Always stand to the windward side, with your hands higher than the breaking point, when breaking windows. This ensures that broken glass will fall away from the hands and body. Placing the tip of the tool in the corner of the window will give you more control in breaking the window. After the window is broken, clear glass from the entire frame, so that no glass shards will stick out and cause injury. This also allows safe passage for those entering or exiting through the window.

Glass Construction

The **glazed** (transparent) part of the window is most commonly made of glass. Window glass comes in several configurations: regular glass, double-pane glass, plate glass (for large windows), laminated glass, and tempered glass. Other substances such as Plexiglas may also be used. The window may contain one or more panes of glass; insulated glass usually has two or more pieces of glass in the window and will be discussed in more detail later.

Regular or Annealed Glass

Single-pane, regular, or **annealed** glass is normally used in construction because it is relatively inexpensive; larger pieces are called plate glass. It is easily broken with a pike pole.

When broken, plate glass creates long, sharp pieces called shards, which can penetrate helmets, boots, and other protective gear, causing severe lacerations and other injuries.

Double-Pane Glass (Insulated windows)

Double-pane glass is being used in many homes because it improves home insulation by using two panes of glass with an air pocket between them. Some double-pane windows may have an inert gas such as argon between the panes for additional insulation value. These windows are sealed units, and are more expensive to replace. However, replacing the glass alone is less expensive than replacing the entire window assembly. Forcing entry through these windows is basically the same as for single-pane windows except that the two panes may need to be broken separately. These windows also will produce dangerous glass shards.

Plate Glass

Commercial plate glass is a stronger, thicker glass used in large window openings. Although it is being replaced by tempered glass for safety reasons, commercial plate glass can still be found frequently in older large buildings, storefronts, and residential sliding doors. It can easily be broken with a sharp object such as a Halligan tool or a pike pole. When broken, commercial plate glass will create many large, sharp pieces.

Laminated Glass

Laminated glass, also known as safety glass, is used to prevent windows from shattering and causing injury. Laminated glass is molded with a sheet of plastic between two sheets of glass. Laminated glass is most commonly used in vehicle windshields, but may also be found in other applications such as doors or building windows.

Tempered Glass

Tempered glass is specially heat-treated, making it four times stronger than regular glass. It is commonly found in side and rear windows in vehicles, in commercial doors, in newer sliding glass doors, and in other locations where a person might accidentally walk into the glass and break it. Tempered glass breaks into small pellets without sharp edges to help prevent injury during accidents. The best way to break tempered glass is by using a sharp, pointed object in the corner of the frame. During a vehicle extrication, a center punch is often used to break a vehicle window.

Wired Glass

Wired glass is tempered glass with wire reinforcing. It may be clear or frosted, and is often used in fire-rated doors that require a window or sight-line from one side of the door to the other. This glass is difficult to break and force.

Frame Designs

Window frames come in many styles. This chapter covers the most common ones, but fire fighters must be familiar with both the types in their response area and forced entry techniques for each type.

Double-Hung Windows

Design

Double-hung windows have two movable sashes, usually of wood or vinyl, that move up and down (► **Figure 11-15**). They are common in residences and have wood, plastic, or metal runners. Newer double-hung window sashes may be removed or swung in for cleaning. They may have one locking mechanism in the center of the window, or two locks on each side of the lower sash that prevent the sashes from moving up or down.

Forcing Entry

Forcing entry through double-hung windows involves opening or breaking the locking mechanism. Place a prying tool under the lower sash and force it up to break the lock or remove it from the track. This technique must be done carefully because it may cause the glass to shatter. Because forcing the lock(s) causes extensive damage to the window, breaking the glass and opening the lock(s) may be a less expensive method of gaining entry. To force entry through a wooden double-hung window, follow the steps in (► **Skill Drill 11-4**).

1. Size-up the window for any safety hazards and find the locking mechanism(s). **(Step 1)**
2. Place the pry end of the Halligan tool under the bottom sash in line with the locking mechanism. **(Step 2)**
3. Pry the bottom sash upward to displace the locking mechanism.
4. Secure the window so that it does not close. **(Step 3)**

Single-Hung Windows

Design

Single-hung windows are similar to double-hung windows, except that the upper sash is fixed, and only the lower sash moves (► **Figure 11-16**). The locking mechanism is the same as on double-hung windows as well. It may be difficult to distinguish between single-hung and double-hung windows from the exterior of a building.

Forcing Entry

Forcing entry through a single-hung window uses the same technique as forced entry through a double-hung window. Place a prying tool under the lower sash and force it up, which will usually break the lock or remove the sash from

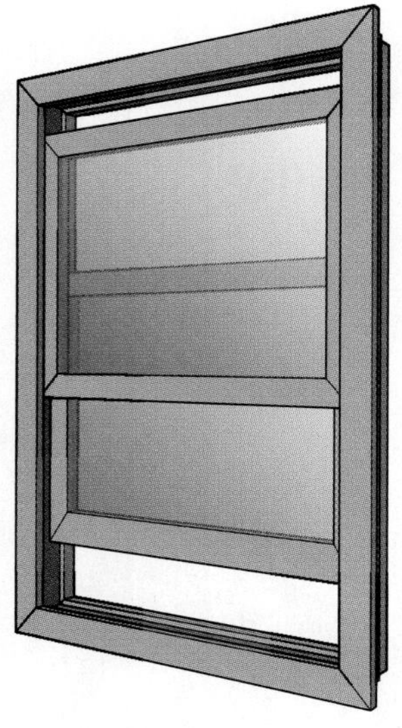

(Figure 11-15) Double-hung windows allow the inner and outer sashes to move freely up and down.

(Figure 11-16) A single-hung window has only one moveable sash. Used with permission of Andersen Corporation

Fire Fighter Tips

When forcing entry through older, double-hung windows, try sliding a hacksaw blade or similar blade up between the sashes. The teeth on the blade will grip the locking mechanism and can often move it into the unlocked position. This will not work on newer double-hung windows, because of the newer latch design.

the track. Again, be careful because the glass will probably shatter. Breaking the glass and opening the window is generally easier with a single-hung window.

Jalousie Windows

Design

A jalousie window is made of adjustable sections of tempered glass in a metal frame that overlap each other when closed (▼ Figure 11-17). This type of window is often found in mobile homes and is operated by a small hand-wheel or crank located in the corner of the window.

Forcing Entry

Forcing entry through jalousie windows can be difficult because they are made of tempered glass and have several panels. Forcing open the panels or breaking a lower one to operate the crank is possible, but doesn't leave a big enough opening to enter. Removing or breaking the panels one at a time is time-consuming and does not always leave a clean entry point with the framing system still in place. The best strategy is to avoid these windows.

(Figure 11-17) Jalousie windows are opened and closed with a small hand-crank.

11-4 Skill Drill

Forcing Entry Through a Wooden Double-Hung Window

1 Size-up the window for any safety hazards, and find the locking mechanism(s).

2 Place the pry end of the Halligan tool under the bottom sash in line with the locking mechanism.

3 Pry the bottom sash upward to displace the locking mechanism. Secure the window so that it does not close.

Awning Windows

Design

Awning windows are similar in operation to jalousie windows, except that they usually have one large or two medium-sized glass panels instead of many small ones (► Figure 11-18). Awning windows are operated by a hand crank located in the corner or in the center of the window. Awning windows can be found in residential, commercial, or industrial structures. Residential awning windows may be framed in wood, vinyl, or metal, while commercial and industrial windows will usually be metal framed.

Commercial and industrial awning windows often use a lock and a notched bar to hold the window open, rather than a crank. These are called projected windows and will be discussed later.

Forcing Entry

Forcing entry through an awning window is the same as for jalousie windows. Break or force the lower panel and operate the crank, or break out all the panels. Depending on the size of the panels and the window frame, it may be easier to access an awning window, and the larger opening makes entry easier as well.

Figure 11-18 An awning window has larger panels than a jalousie window.

Figure 11-19 Horizontal-sliding windows work like sliding doors. Used with permission of Andersen Corporation.

Horizontal-Sliding Windows

Design

Horizontal-sliding windows are similar to sliding doors (► Figure 11-19). The latch is similar as well and attaches to the window frame. People often place a rod or pole in the track to prevent break-ins. Newer sliding windows have latches between the windows, similar to those on double-hung windows.

Forcing Entry

Forcing entry through sliding windows is just like forcing entry through sliding doors. Place a pry bar near the latch to break the latch or the plate. If there is a rod in the track, look for another entry point or break the glass, although this should be the last resort.

Casement Windows

Design

Casement windows have a steel or wood frame and open away from the building with a crank mechanism (► Figure 11-20). Although they are similar to jalousie or awning windows, casement windows have a side hinge, rather than a top hinge. Several types of locking mechanisms can be used. Like jalousie windows, these windows should be avoided because they are difficult to force open.

Forcing Entry

The best way to force entry through a casement window is to break out the glass from one or more of the panes, and operate the locking mechanisms and crank manually. To force entry through a casement window, follow the steps in (Skill Drill 11-5).

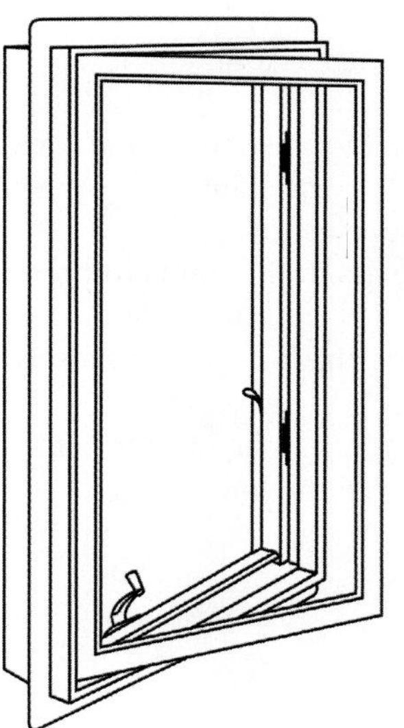

Figure 11-20 Casement windows open to the side with a crank mechanism. Used with permission of Andersen Corporation.

1. Size-up the window to check for any safety hazards and locate the locking mechanism.
2. Select an appropriate tool to break out a windowpane.
3. Stand to the windward side of the window and break out the pane closest to the locking mechanism, keeping the tool lower than your hands.
4. Remove all of the broken glass in the pane to prevent injuries.
5. Reach in and manually operate the window crank to open the window.

Projected Windows

Design

Projected windows, also called factory windows, are usually found in older warehouse or commercial buildings (► Figure 11-21). They can project inward or outward on an upper hinge. Screens are rarely used with these windows, but forcing entry may not be easy, depending on the integrity of the frame, the type of locking mechanism used, and their distance off the ground. These windows may have fixed, metal-framed wire glass panes above them.

Forcing Entry

Avoid forcing entry through a projected window when possible. They are often difficult to force open and may be difficult to enter. To force entry, access one pane, unlock the mechanism, and open the window by hand. If the opening created is not large enough, break out the entire window assembly. To force entry through a projected or factory window, follow the steps in (Skill Drill 11-6).

1. Size-up the window to check for any safety hazards and locate the locking mechanism.
2. Select an appropriate tool, such as a pike pole, to break a pane.
3. Stand to the windward side of the window and break the pane closest to the locking mechanism, keeping the tool head lower than your hands.
4. Remove all of the broken glass in the frame to prevent injuries.
5. Reach in and manually open the locking mechanism and the window.
6. A cutting tool, such as a torch, can be used to remove the window frame and enlarge the hole.

(Figure 11-21) Projected windows may open inward or outward.

Locks

Locks have been used for hundreds of years. They range from very simple push button locks found in most homes to complex computer-operated locks found in banks and high security areas.

Parts of a Door Lock

All door locks have the same basic parts. Gaining entry will be easier with an understanding of these parts and how they operate (► Figure 11-22). The major parts of a door lock include:

- Latch—The part of the lock that "catches" and holds the door frame
- Operator lever—The handle, doorknob, or keyway that turns the latch to lock it or unlock it
- Deadbolt—A second, separate latch that locks and reinforces the regular latch

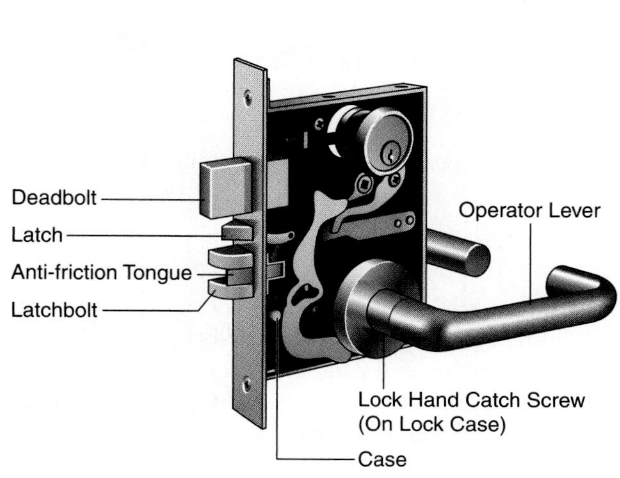

Figure 11-22 All door locks have the same basic parts.

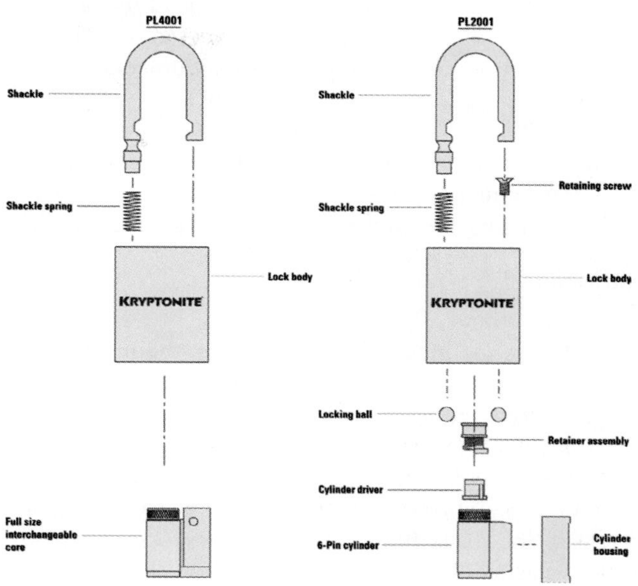

Figure 11-23 All padlocks have three basic parts.

Parts of a Padlock

All padlocks, like all door locks, have similar parts (▶ Figure 11-23). The basic parts of a padlock include:

- Shackles—The U-shaped top of the lock that slides through a hasp and locks in the padlock itself
- Unlocking device—The key way, combination wheels, or combination dial used to open the padlock
- Lock body—The main part of the padlock that houses the locking mechanisms and the retention part of the lock

Safety

Tool safety is important when executing forcible entry through a door or padlock. The tools used for cutting the cylinders must be kept sharp, and fire fighters must be careful to avoid being cut. Care also must be taken when using the striking tools to avoid pinching or crushing fingers or hands. Fire fighters must wear proper protective equipment such as gloves and use tools carefully to minimize the risk of injury.

Types of Locks

The four major lock categories are: cylindrical locks, padlocks, mortise locks, and rim locks.

Cylindrical Locks

Design

Cylindrical locks are the most common fixed lock in use today (▶ Figure 11-24). They are relatively inexpensive and

Figure 11-24 A cylindrical lock.

are standard in most doors. The locks and handles are set into predrilled holes in the doors. One side of the door usually has a key-in-the-knob lock; the other side will have a keyway, a button, or some type of locking/unlocking device.

Forcing Entry

Forcing entry into a cylindrical lock is usually very easy because most mechanisms are not very strong. Place a pry bar near the locking mechanism and lever it to force the lock.

Padlocks

Design

Padlocks are the most common locks on the market today and come in many designs and strengths. Both regular-duty and heavy-duty padlocks are available.

Padlocks come with various unlocking devices, including keyways, combination wheels, or combination dials. Operating the unlocking device opens one side of the lock to release the shackle and allow entry. Shackles for regular padlocks generally have a diameter of 1/4" or less and are not made of case-hardened metal. Shackles for heavy-duty padlocks are 1/4" or larger in diameter and made of case-hardened steel or some other case-hardened metal. The design of some padlocks hides the latch, making forced entry difficult.

Forcing Entry

Several ways can be used to force entry through padlocks without causing extensive property damage. Before breaking the padlock, consider cutting the shackle or hasp first. This saves the lock and makes securing the building easier. Breaking the shackle is the best method for forcing entry through padlocked doors. If the padlock is made of case-hardened steel, many conventional methods of breaking the lock will be ineffective. The most common tools used to force entry through a padlock are bolt cutters, duck-billed lock breakers, a bam-bam tool, locking pliers and chain, and a torch.

- Bolt cutters—Bolt cutters can quickly and easily break regular-duty padlocks, but cannot be used on heavy-duty, case-hardened steel padlocks. To use bolt cutters on a padlock, open the jaws as wide as possible. Close the jaws around one side of the lock shackles to cut through the shackle. Once one shackle of a regular-duty lock is cut, the other side will spin freely and allow access.
- Duck-billed lock breakers —The duck-billed lock breaker has a large metal wedge attached to a handle. Place the narrow end of the wedge into the center of the shackle and force it through with another striking tool. The wedge will spread the shackle until it breaks, freeing the padlock and allowing access.

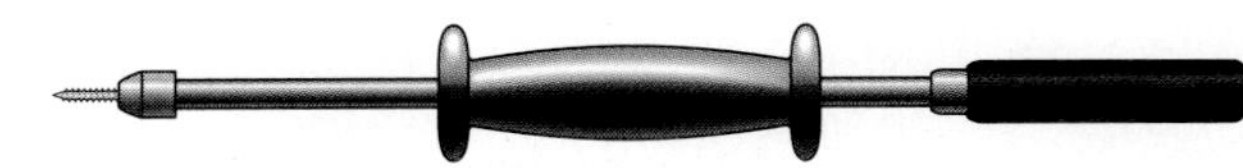

Figure 11-25 A bam-bam tool removes the lock cylinder from a padlock.

- Bam-bam tool—The bam-bam tool can pull the lock cylinder out of a regular-duty padlock, but will not work on higher-end padlocks that add security rings to hold the cylinder in place (▲ Figure 11-25). This tool has a cased-hardened screw that is placed in the keyway. Once the screw is set, the sliding hammer will pull the tumblers out of the padlock so that the trip-lever mechanism inside the lock can be opened manually.
- Locking pliers and chain—The locking pliers and chain are attached to a padlock to secure it in a fixed position so that it can be cut. Use a rotary saw with a metal-cutting blade or a torch to cut the padlock after it has been secured. Securing the padlock is necessary because it is too dangerous to hold the lock with a gloved hand or to cut into it while it is loose.

Mortise Locks

Design

Mortise locks, like cylindrical locks, are designed to fit in predrilled openings inside a door, and are commonly found in hotel rooms (► Figure 11-26). They have both a latch and a bolt built into the same mechanism, which operate independently of each other. The latch will lock the door, but the bolt can also be deployed for added security.

Forcing Entry

The design and construction of mortise locks makes them difficult to force. A door with a mortise lock may be forced with conventional means, but will probably require the use of a through-the-lock technique.

Rim Locks/Dead Bolts

Design

Rim locks and dead bolts are different types of locks that can be surface mounted on the interior of the door frame (► Figure 11-27). They are commonly found in residences as secondary locks to support the through-the-handle locks. They can be identified from the outside by the keyway that has been bored into the door. These locks have a bolt that extends at least 1″ into the door frame, making the door more difficult to force.

Forcing Entry

Rim locks and dead bolts are the hardest of the locks to break. They can be very difficult to force with conventional methods, so that through-the-lock methods may be the only option. Skill Drills 11-7 through 11-9 present various through-the-lock techniques. To force entry using a K tool, follow the steps in ► **Skill Drill 11-7**.

1. Size-up the lock area to check for any safety hazards. Determine what type of lock is used, whether it is regular or heavy-duty construction, and whether the cylinder has a case-hardened collar, which may hamper proper cutting.
2. Place the K tool over the face of the cylinder, noting the location of the keyway. **(Step 1)**
3. Using a Halligan or similar pry tool, tap the K tool into the cylinder. **(Step 2)**
4. Place the adz end of the Halligan tool into the slot on the K tool, and strike the end of the Halligan tool with a flat-head axe to drive the K tool further into the lock cylinder. **(Step 3)**
5. Pry up on the Halligan to pull out the lock and expose the locking mechanism. **(Step 4)**
6. Using the small tools that come with the K tool, turn the mechanism to open the lock. **(Step 5)**

To force entry using an A tool, follow the steps in ► **Skill Drill 11-8**.

1. Size-up the door and lock area to check for any safety hazards.
2. Place the cutting edges of the A tool at the top of the cylinder, between the cylinder and the door frame. **(Step 1)**
3. Using a flat-head axe or similar tool, drive the A tool into the cylinder. **(Step 2)**
4. Pry up on the A tool to remove the cylinder from the door. **(Step 3)**
5. Insert a key tool into the hole to manipulate the locking mechanism and open the door. **(Step 4)**

To force entry by unscrewing the lock, follow the steps in ► **Skill Drill 11-9**.

1. Size-up the door and lock area to check for any safety hazards.
2. Using a set of vise grips, lock the pliers onto the outer housing of the lock with a good grip. **(Step 1)**
3. Unscrew the housing until it can be fully removed. **(Step 2)**
4. Manipulate the locking mechanism with lock tools to disengage the arm, then open the door. **(Step 3)**

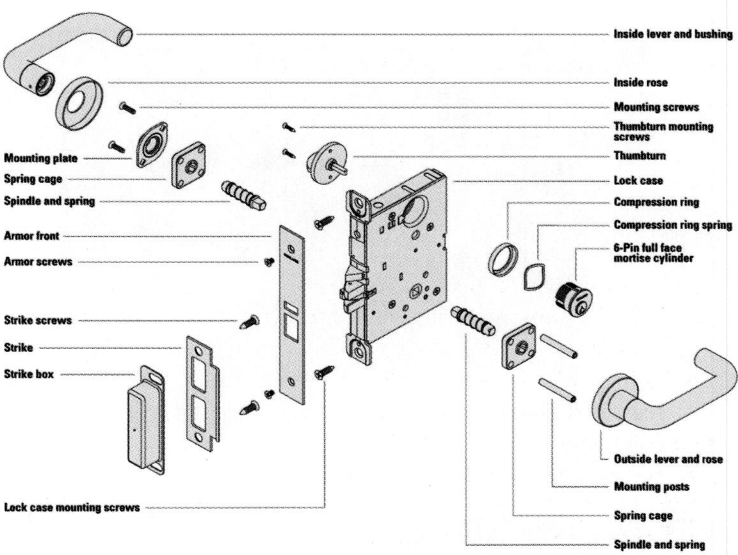

Figure 11-26 Mortise locks have a latch and a bolt, which operate independently of each other.

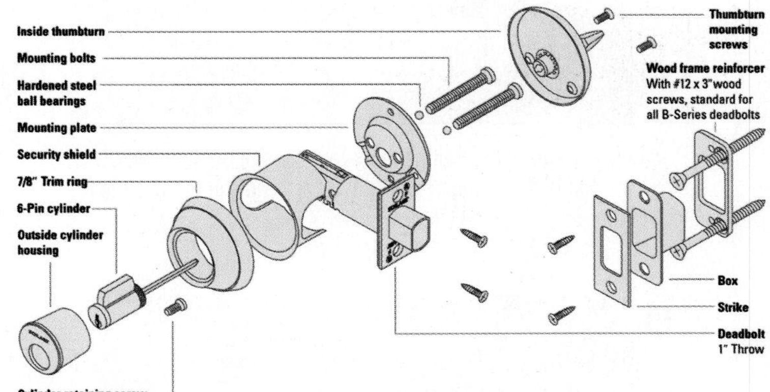

Figure 11-27 Rim locks and dead bolts are mounted on the inside of a residence door.

Skill Drill

Forcing Entry Using a Through-the-Lock Technique (K Tool)

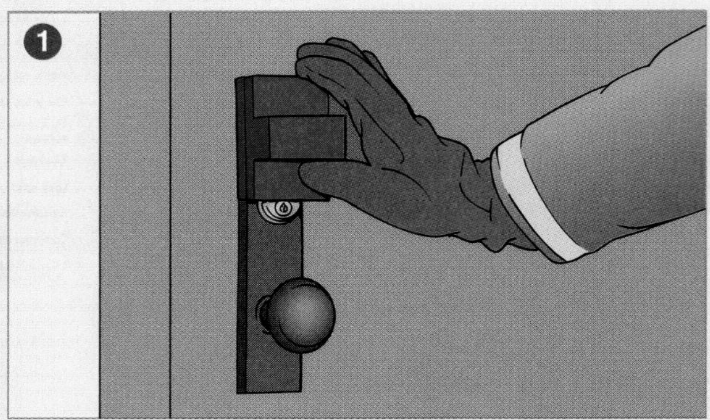

Place the K tool over the face of the cylinder, noting the location of the keyway.

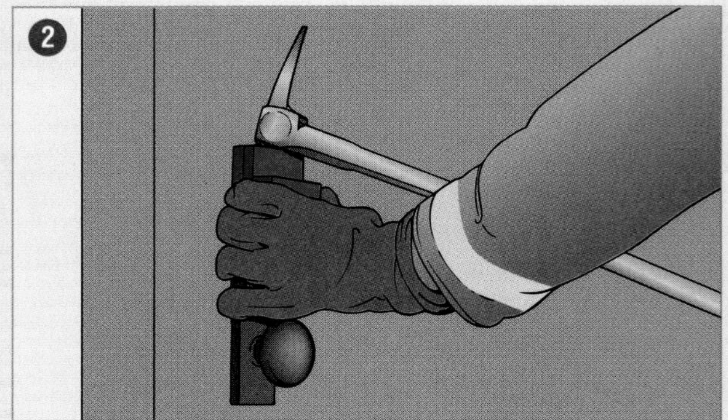

Using a Halligan or similar pry tool, tap the K tool down into the cylinder.

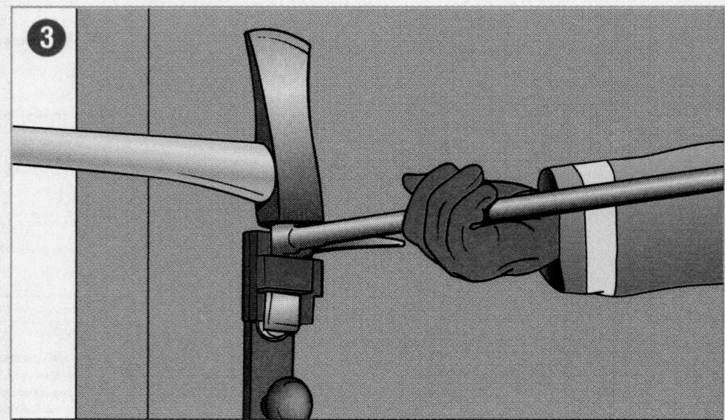

Place the adz end of the Halligan tool into the slot on the K tool, and strike the end of the Halligan tool with a flat-head axe to drive the K tool further into the lock cylinder.

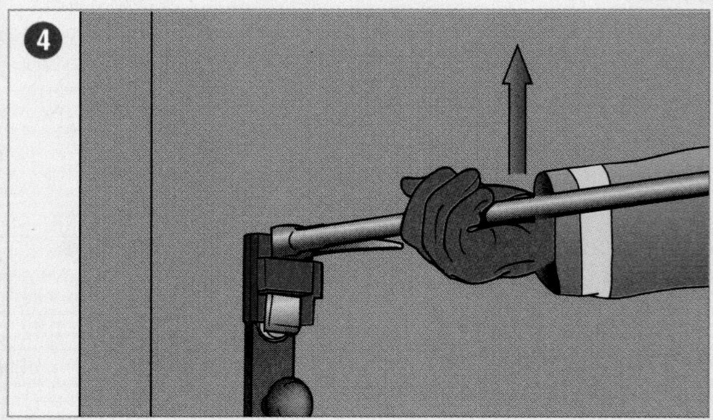

Pry up on the Halligan tool to pull out the lock and expose the locking mechanism.

Using the small tools that come with the K tool, turn the mechanism to open the lock.

11-8 Skill Drill

Forcing Entry Using an A Tool

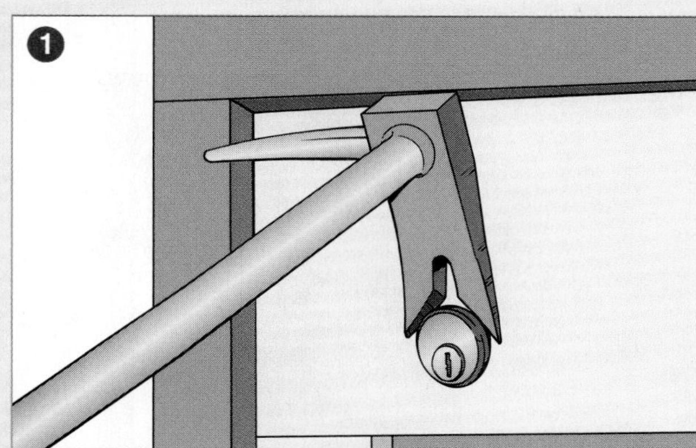

1 Place the cutting edges of the A tool at the top of the cylinder, between the cylinder and the door frame.

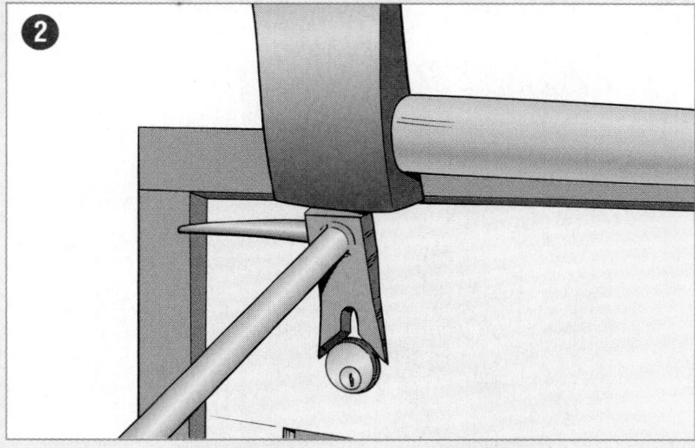

2 Using a flat-head axe or similar tool, drive the A tool into the cylinder.

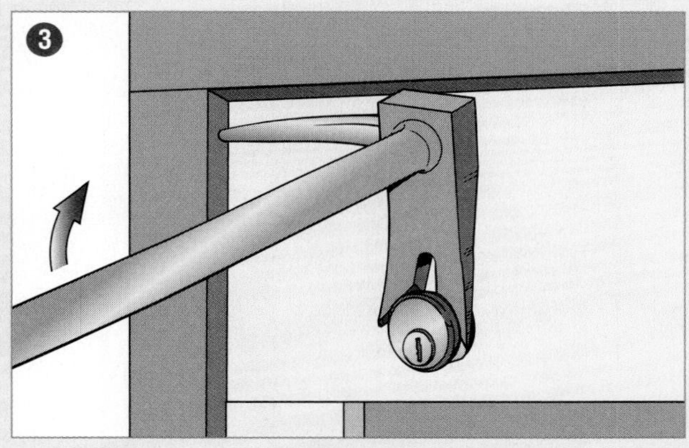

3 Pry up on the A tool to remove the cylinder from the door.

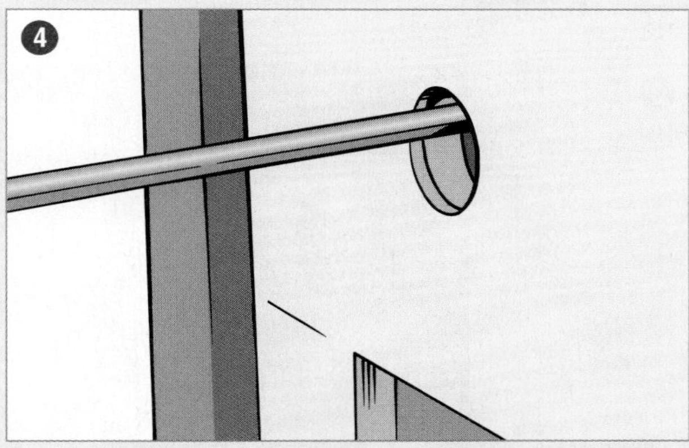

4 Insert a key tool into the hole to manipulate the locking mechanism and open the door.

Skill Drill

Forcing Entry Using a Through-the-Lock Technique (Unscrewing the Lock)

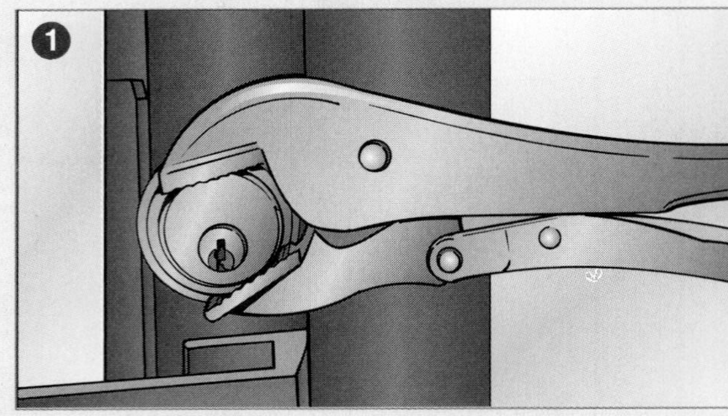

Using a set of vise grips, lock the pliers onto the outer housing of the lock with a good grip.

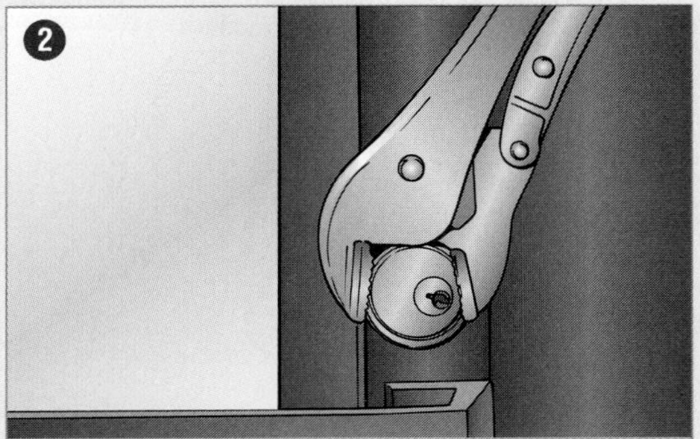

Unscrew the housing until it can be removed.

Manipulate the locking mechanism with lock tools to disengage the arm, then open the door.

Load-Bearing Wall

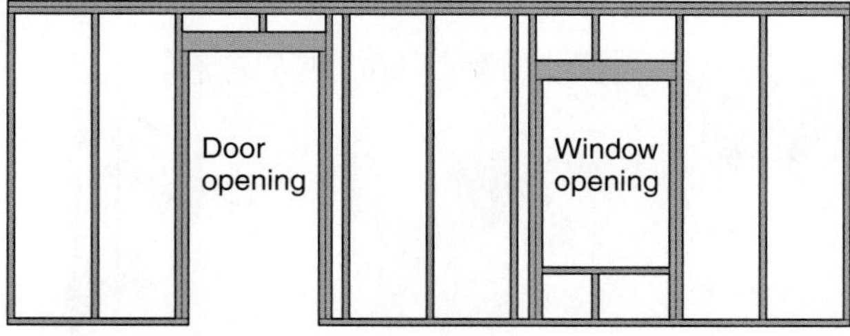

Nonbearing Wall

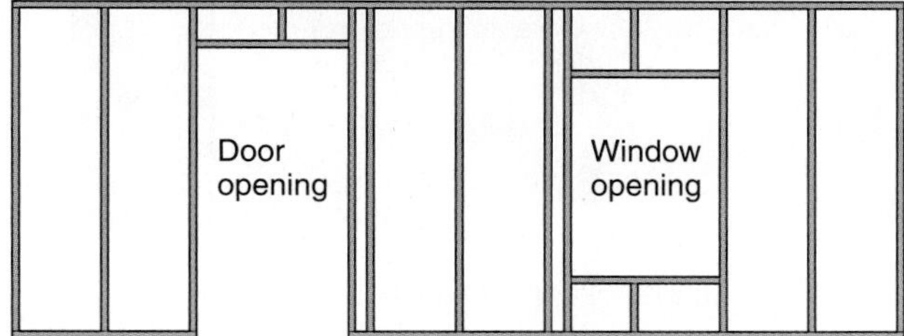

Figure 11-28 A load-bearing wall supports the building's ceiling and/or rafters; a nonbearing wall simply partitions the space.

Figure 11-29 Residences may have exterior walls of multiple materials, such as wood and masonry block.

Breaching Walls and Floors

On occasion, forced entry through a window, lock, or door may not be possible or may take too long. In these cases, consider the option of breaching a wall or floor. An understanding of basic construction concepts is required to safely and quickly execute this technique.

Load-bearing/Nonbearing Walls

Before breaching a wall, first consider whether the wall is load-bearing. A load-bearing wall supports the building's ceiling and/or rafters. Removing or damaging a load-bearing wall could cause the building or wall to collapse. A nonbearing wall can be removed safely and without danger (▲ Figure 11-28). Nonbearing walls are also called partition walls or simply partitions. Refer to the chapter on building construction for a review of wall construction.

Exterior Walls

Exterior walls can be constructed of one or more materials. Many residences have both wood and brick, aluminum siding, or masonry block construction (► Figure 11-29). Commercial buildings usually have concrete, masonry, or metal exterior walls. Commercial buildings also may have steel I-beam construction; older commercial structures may have heavy timber construction. Because there are exceptions to every rule, it is important to know the buildings in the specific response area.

Whether to attempt to breach an exterior wall is a difficult decision. Masonry, metal, and brick are formidable materials, and breaking through them can be very difficult. The best tools to use in breaching a concrete or masonry exterior wall are a battering ram, a sledgehammer, or a rotary saw with a concrete blade.

Interior Walls

Interior walls in residences are usually constructed of wood or metal studs covered by plaster, gypsum, or sheetrock. Commercial buildings may have concrete block interior walls. Breaching an interior wall can be dangerous to fire fighters for several reasons. Many interior walls contain electrical wiring, plumbing, cable wires, and phone wires, which present various hazards. Interior walls may also be load-bearing. Although load-bearing walls may be breached, extreme care should be taken, particularly if studs are removed.

After determining whether the wall is load-bearing or not, sound it to locate a stud away from any electrical outlets or switches. Tap on the wall; the area between studs will make a hollow sound compared to the solid sound directly over the stud.

After locating an appropriate site, make a small hole to check for any obstructions. If the area is clear, expand the opening to reveal the studs. Walls should be breached as close as possible to the studs because this makes a large opening, and cutting is easier. If possible, enlarge the opening by removing at least one stud to enable quick escapes. To breach a wall frame, follow the steps in (Skill Drill 11-10).

1. Size-up the wall looking for safety hazards such as electrical outlets, wall switches, or any signs of plumbing. Inspect the overall scene to ensure that the wall is not load-bearing.
2. Using a striking or cutting tool, sound the wall to locate any studs, and make a hole between the studs.

3. Cut as close to the studs as possible.
4. Enlarge the hole by extending it from stud to stud and as high as necessary.

To breach a masonry wall, you will create an upside-down V in the wall. Follow the steps in (Skill Drill 11-11).

1. Size-up the wall for any safety hazards such as electrical outlets, wall switches, and plumbing. Inspect the overall area to ensure that the wall is not load-bearing.
2. Select a row of masonry blocks that is at or near floor level.
3. Using a sledgehammer, knock two holes each in five masonry blocks along the selected row. Each hole should pierce into the hollow core of the masonry blocks.
4. Repeat the process on four masonry blocks above the first row.
5. Repeat the process on three masonry blocks above the second row.
6. Repeat the process on two masonry blocks above the third row.
7. Repeat the process on one masonry block above the fourth row. An upside down V has now been created.
8. Begin knocking out the remaining portion of the masonry blocks by hitting the masonry blocks parallel to the wall rather than perpendicular to it.
9. Clear any reinforcing wire using bolt cutters.
10. Enlarge the hole as needed.

To breach a metal wall, follow the steps in (Skill Drill 11-12).

1. Size-up the wall for any safety hazards such as electrical outlets, wall switches, or plumbing. Inspect the overall area to ensure that the wall is not load-bearing.
2. Select an appropriate tool such as a rotary saw with a metal-cutting blade or a torch.
3. Open the wall using a diamond-shaped or V-cut.
4. Protect or pad the metal edges of the cut to ensure that no one who goes through the opening is cut.

Floors

The two most popular floor materials found in residences and commercial buildings are wood and poured concrete. Both are very resilient and may be difficult to breach. Conduct a thorough size-up; breaching a floor should be a last resort measure (► Figure 11-30). A rotary saw with an appropriate blade is the best tool to breach a floor. To breach a floor, follow the steps in (Skill Drill 11-13).

1. Size-up the floor area to ensure there are no hazards in the area to be cut.
2. Sound the floor with an axe or similar tool to locate the floor joists.
3. Use an appropriate cutting tool to cut one side of the hole first, then cut the opposite side.

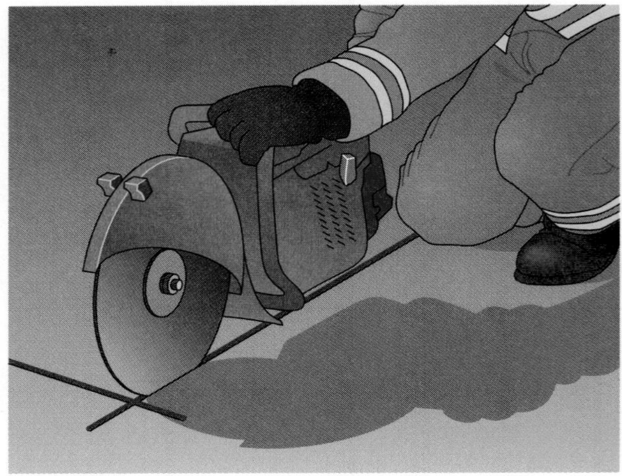

(Figure 11-30) Breaching a floor should be a last resort.

4. Remove any flooring such as carpet, tile, or floorboards that loosens.
5. Cut a similar opening into the subfloor, until the proper size hole is achieved.
6. Secure the area around the hole to prevent others working in the area from falling through the hole.

Forcible Entry and Salvage

Before Entry

The decision of which type of forcible entry to use will have a direct impact on salvage operations. In a nonemergency, through-the-lock methods will keep damage to a minimum, thus salvaging the property from excessive repairs. In an emergency situation, more urgent methods may cause massive amounts of destruction, requiring expensive repairs to a door, window, or wall.

After Entry

When the operation is over, and normal salvage and overhaul have begun, the structure must be secured before fire fighters leave the scene. The use of conventional forced-entry techniques and specialized tools requires that doors and windows be secured to prevent break-ins or theft after the call. If through-the-lock methods are used, a hasp and padlock may be all that is needed to secure a site. Broken windows or doors can be boarded with sheets of plywood. Police will often increase patrols around a damaged building when needed to deter potential thieves.

Remember that the fire department is charged not only with saving lives but also with protecting property. Do everything possible to reduce the amount of damage caused by both the fire and your own actions. Homeowners will be especially appreciative of support from the fire department after the fire is extinguished.

Wrap-Up

Ready for Review

This chapter examined the major principles behind forcible entry. An understanding of the tools used, building construction (including door, window, lock, wall, and floor construction), and specific techniques is needed to conduct forcible entry assignments successfully. This chapter also reviewed the proper precautions needed to handle tools safely, proper carrying techniques, and basic maintenance procedures. It examined specific skills required for forcible entry and discussed the impact of forcible entry on salvage and overhaul operations.

Chief Concepts

- Forcing entry into a structure can be performed with a basic understanding of forcible entry tools as well as window, door, wall, and lock construction.
- Using proper safety techniques is the most important part of using forcible entry tools.
- Always use full personal protective equipment when performing forcible entry.
- Striking tools are used for driving other tools or for hitting doors, windows, walls, or other objects.
- Cutting tools are used for cutting into doors, windows, or locks to gain entry.
- Prying tools are used to spread doors and windows away from their frames.
- Lock tools are specialized tools used for attacking the lock itself, rather than a door or window.
- An understanding of door construction, composition, and opening techniques will facilitate successful forcible entry.
- An understanding of window construction, composition, and glazing will facilitate successful forcible entry.
- An understanding of how a lock works, what it is made of, and how it secures a door or window will facilitate successful forcible entry.
- Floors and walls can be used for forcible entry if precautions are taken and all other methods have failed. Due to potential safety problems, this should be the last type of entry considered.

Hot Terms

A tool A cutting tool with a pry bar built into the cutting part of the tool.

Adz The prying part of the Halligan tool.

Annealed The process of forming standard glass.

Awning windows Windows that have one large or two medium panels operated by a hand crank from the corner of the window.

Bam-bam tool A tool with a case-hardened screw, which is secured in the keyway of a lock, to remove the keyway from the lock.

Battering ram A tool made of hardened steel with handles on the sides used to force doors and to breach walls. Larger versions are used by up to four people; smaller versions are made for one or two people.

Bite A small opening made to enable better tool access in forcible entry.

Bolt cutter A cutting tool used to cut through thick metal objects, such as bolts, locks, and wire fences.

Case-hardened steel A process that uses carbon and nitrogen to harden the outer core of a steel component, while the inner core remains soft. Case-hardened steel can be cut only with specialized tools.

Casement windows Windows in a steel or wood frame that open away from the building via a crank mechanism.

Claw The forked end of a tool.

Cutting tools Tools that are designed to cut into metal or wood.

Cylindrical locks The most common fixed lock in use today. The locks and handles are placed into predrilled holes in the doors. One side of the door will usually have a key-in-the-knob lock, and the other will have a keyway, a button, or some other type of locking/unlocking device.

Deadbolt A secondary locking device used to secure a door in its frame.

Dead bolts Surface or interior mounted locks on or in a door with a bolt that provide additional security.

Door An entryway; the primary choice for forcing entry into a vehicle or structure.

Double-hung windows Windows that have two movable sashes that can go up and down.

Double-pane glass Window design that traps air or inert gas between two pieces of glass to help insulate a house.

Wrap-Up

Duck-billed lock breaker A tool with a point that can be inserted in the shackles of a padlock. As the point is driven further into the lock, it gets larger and forces the shackles apart until they break.

Exterior wall A wall—often made of wood, brick, metal, or masonry—that makes up the outer perimeter of a building. Exterior walls are often load-bearing.

Forcible entry Gaining access to a structure or vehicle when normal means of entry have been blocked or cannot be used.

Glazed Transparent glass.

Hardware The parts of a door or window that enable it to be locked or opened.

Hollow-core A door made of panels that are honeycombed inside, creating an inexpensive and lightweight design.

Horizontal-sliding windows Windows that slide open horizontally.

Interior wall A wall inside a building that divides a large space into smaller areas. Also known as a partition.

Irons A combination tool, normally the Halligan tool and the flat-head axe.

Jalousie windows Windows made of small slats of tempered glass that overlap each other when closed. Often found in trailers and mobile homes, jalousie windows are held by a metal frame and operated by a small hand wheel or crank found in the corner of the window.

Jamb The part of a doorway that secures the door to the studs in a building.

J tool A tool that is designed to fit between double doors equipped with panic bars.

K tool A tool designed to remove the face plate of a lock cylinder.

Laminated glass Also known as safety glass; the lamination process places a thin layer of plastic between two layers of glass, so that the glass does not shatter and fall apart when broken.

Latch The part of the door lock that secures into the jamb.

Load-bearing wall A wall that supports structural members or upper floors of a building.

Lock body The part of a padlock that holds the main locking mechanisms and secures the shackles.

Mortise locks Door locks with both a latch and a bolt built into the same mechanism; the two locking devices operate independently of each other. This is a common lock found in places such as hotel rooms.

Nonbearing wall A wall that does not support a ceiling or structural member, but simply divides a space. See interior wall and partition.

Operating lever The handle of a door that turns the latch to open it.

Padlocks The most common lock on the market today, built to provide regular-duty or heavy-duty service. Several types of locking mechanism are available, including a keyway, combination wheels, or combination dials.

Partition A wall that does not support a ceiling or structural member, but simply divides a space. See interior or nonbearing wall.

Pick The pointed end of a pick axe that can be used to make a hole or bite in a door, floor, or wall.

Plate glass A type of glass with additional strength so it can be formed in larger sheets, but will still shatter upon impact.

Projected windows Also called factory windows. Usually found in older warehouse or commercial buildings. They project inward or outward on an upper hinge.

Pry axe A specially designed hand axe that serves multiple purposes. Similar to a Halligan bar, it can be used to pry, cut, and force doors, windows, and many other types of objects. Also called a multipurpose axe.

Rabbet A type of door frame that has the stop for the door cut into the frame.

Rim locks Surface or interior mounted locks on or in a door with a bolt that provide additional security.

Shackles The U-shaped part of a padlock that runs through a hasp and secures back into the lock body.

Solid-core A door design that consists of wood filler pieces inside the door. This creates a stronger door that may be fire-rated.

Striking tools Tools designed to strike other tools or objects such as walls, doors, or floors.

Tempered glass A type of safety glass that is heat-treated so that it will break into small pieces that are not as dangerous.

Unlocking device A key way, combination wheel, or combination dial.

Fire Fighter in Action

It is 11:30 a.m. when your ladder truck company is dispatched to a commercial structure fire in a strip mall. The fire is reported to be located deep within the building in a dry cleaning business. You are the first ladder truck to arrive on the scene. The incident commander instructs your lieutenant to gain access to the rear of the fire building for the suppression crew. Lieutenant Christiansen directs you and your partner to force entry and coordinate entry with the suppression crew. As you walk to the rear of the building you observe Engine 1 stretching an attack line and back up line and preparing for entry.

1. Basic safety rules that must be followed when using striking tools include:
   A. wear bunker gear and gloves.
   B. wear eye protection.
   C. ensure there is a clear area prior to swinging a striking tool.
   D. swing the striking tool with short and controlled swings.
   E. all of the above

2. Lieutenant Christiansen instructs you to take the K tool and irons. The irons consists of a(n):
   A. pick-head axe and bolt cutters.
   B. flat-head axe and a Halligan tool.
   C. pry bar and a circular saw.
   D. A tool and a bam-bam tool.

You meet the suppression crew at the rear of the building. The only opening to the rear of the fire building was a metal outward opening door. As you size up the door you notice that the hinges are sealed and the door has two locks: a deadbolt and a cylindrical lock in the door handle. The metal door is encased in a wood frame.

3. Dead bolts are the most difficult type of lock to break into. What forcible entry method should you use?
   A. Through the lock
   B. Conventional
   C. Specialty
   D. None of the above

4. Once the dead bolt has been pulled and unlocked you are now ready to force the cylindrical lock that is in the door handle. Based on this scenario, what would be the best forcible entry method?
   A. Pull the hinge pins
   B. Use the battering ram
   C. Use the irons to force the door
   D. Use a cutting torch to cut the hinges

www.FireFighter.jbpub.com

Ladders

Technology Resources
www.FireFighter.jbpub.com
Chapter Pretests
Interactivities
Hot Term Explorer
Web Links
Review Manual
FireLearn
Chapter Features
Skill Drills
Voices of Experience
Fire Marks
Teamwork Tips
Fire Fighter Safety Tips
Fire Fighter Tips
Canadian Perspectives
Hot Terms
Wrap-Up

Chapter 12

NFPA 1001 Standard

Fire Fighter I

5.3.6 Set up ground ladders, given single and extension ladders, an assignment, and team members if needed, so that hazards are assessed, the ladder is stable, the angle is correct for climbing, extension ladders are extended to the necessary height with the fly locked, the top is placed against a reliable structural component, and the assignment is accomplished.

5.3.6 (A) *Requisite Knowledge.* Parts of a ladder, hazards associated with setting up ladders, what constitutes a stable foundation for ladder placement, different angles for various tasks, safety limits to the degree of angulations, and what constitutes a reliable structural component for top placement.

5.3.6 (B) *Requisite Skills.* The ability to carry ladders, raise ladders, extend ladders and lock flies, determine that a wall and roof will support the ladder, judge extension ladder height requirements, and place the ladder to avoid obvious hazards.

5.5.3 Clean and check ladders, ventilation equipment, self-contained breathing apparatus (SCBA), ropes, salvage equipment, and hand tools, given cleaning tools, cleaning supplies, and an assignment, so that equipment is clean and maintained according to manufacturer's or departmental guidelines, maintenance is recorded, and equipment is placed in a ready state or reported otherwise.

Fire Fighter II

NFPA 1001 contains no Fire Fighter II Job Performance Requirements for this chapter.

Additional NFPA Standards

NFPA 1901, *Standard for Automotive Fire Apparatus*

NFPA 1931, *Standard on Design of and Design Verification Tests for Fire Department Ground Ladders*

NFPA 1932, *Standard on Use, Maintenance, and Service Testing of Fire Department Ground Ladders*

Knowledge Objectives

After studying this chapter, you will be able to:

- List and describe the parts of a ladder.
- Describe the different types of ladders.
- Describe how to clean and inspect ladders.
- Describe the hazards with ladders.
- Describe how to deploy a ladder.
- Describe how to work on a ladder.

Skills Objectives

After completing this chapter, you will be able to perform the following tasks:

- One-fire fighter carry
- Two-fire fighter shoulder carry
- Three-fire fighter shoulder carry
- Two-fire fighter suitcase carry
- Three-fire fighter suitcase carry
- Three-fire fighter flat carry
- Four-fire fighter flat carry
- Three-fire fire flat-shoulder carry
- Four-fire fighter flat-shoulder carry
- One-fire fighter rung raise for ladders under 14'
- One-fire fighter rung raise for ladders over 14'
- Two-fire fighter beam raise
- Two-fire fighter rung raise
- Three-fire fighter rung raise
- Four-fire fighter rung raise
- Tie the halyard
- Climb a ladder
- Work from a ladder
- Deploy a roof ladder

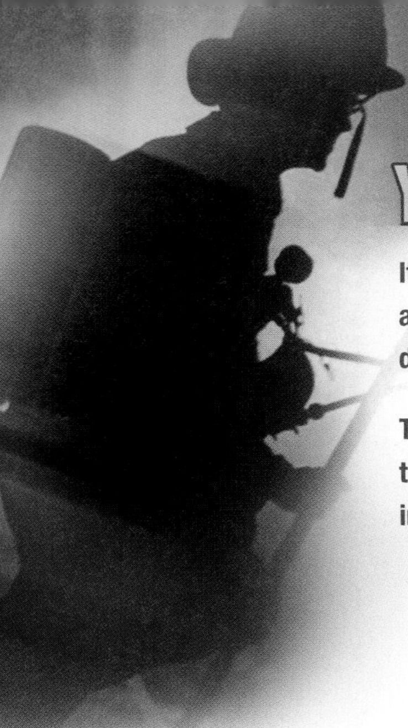

You Are the Fire Fighter

It is 11:30 p.m. and you have been dispatched to a residential structure fire. Your engine company arrives just seconds behind the first-due engine. A quick size-up reveals a two-story single family dwelling with a walkout basement. Smoke and flames are visible at the top of the chimney flue.

The first-due engine company stretches an attack line to the front door and enters the structure to search for indications of fire outside the flue. You are assigned to assist two other fire fighters in discharging dry chemical agent down the chimney flue.

1. ***What size ladder will you need to access the chimney flue?***
2. ***What type of carry and raise will your crew use to place the ladder?***
3. ***Where can you safely place the ladder?***
4. ***Does your engine have a ladder long enough to do the job?***

Introduction

Despite many technological advances in the fire service, one fundamental piece of equipment has not changed much. The ladder remains one of the fire fighter's basic tools, carried on nearly every piece of structural firefighting apparatus.

The portable ladder is one of the most functional, versatile, durable, inexpensive, easy-to-use, and rapidly deployable tools used by fire fighters. No advanced technology can substitute for a ladder as a means of rapid, safe vertical access for fire suppression and rescue operations. Every fire fighter must be proficient in the basic skills of working with ladders.

Functions of a Ladder

Ladders provide a vertical path, either up or down, from one level to another. Ladders can provide access to an area or egress (a method to exit or escape) from an area (► Figure 12-1). At times, a ladder can be used as a work platform so that a fire fighter can perform various functions in locations that cannot be reached otherwise.

Fire Marks

On March 16, 1993, the Chicago Fire Department responded to a fire in a four-story 1930s vintage hotel. The fire rapidly made the interior stairways and hallways impassable. Because the building was occupied by approximately 160 people, fire fighters were faced with a massive rescue situation.

Although 19 people lost their lives, fire fighters made over 100 successful rescues. Many of those rescues resulted from the quick and effective ladder work by the first-arriving ladder companies.

Ladders are most often used to provide access to and egress from areas above grade (the level at which the ground intersects the foundation of a structure). Used outside a building, ladders can enable fire fighters to reach the roof for ventilation operations, to enter a window for an interior search, or to rescue a victim. Exterior ladders also provide an emergency exit for crews working inside a structure. Within the building, ladders can be used to access the attic or cockloft or to provide a safe path between floors, avoiding a damaged stairway. A response team can use a ladder to access and exit from the top of a rail car during a hazardous materials incident.

Fire fighters also use ladders to access and exit areas below grade. Lowering a ladder into a trench or manhole allows fire fighters to reach the level of an incident or to escape safely. A ladder can be used to reach an injured person in a ravine or to access a highway accident from an overpass.

Figure 12-1 Ladders can provide access to or exit from a structure.

Ladders are useful even at grade level. A ladder can be used to create a bridge across a small opening or to enable fire fighters to climb over a fence or obstruction to reach an emergency scene.

Ladders also can be used as work platforms. For example, a fire fighter who has climbed a ladder to a window can stand on the ladder to open the window for ventilation or to direct a hose stream into the building. Ladders also can be used during overhaul operations, providing a platform for a fire fighter to remove exterior trim in a wood-frame building.

Secondary Functions of Ladders

Ladders serve several other functions in the fire service. Roof ladders provide stable footing and distribute the weight of fire fighters during pitched roof operations. A two-section ladder, rigged with rope and webbing, forms a makeshift lift called a **ladder gin**. Ladder gins are used to lower a rescuer into a trench or manhole or to raise a victim safely from a below-grade site (► Figure 12-2).

Ladders provide elevated platforms for equipment as well as for fire fighters. They are especially useful when elevation would extend the reach or enhance the operations of the equipment. For example, a hose line attached to a ladder section can protect exposures or confine a fire. A smoke ejector may be mounted on a ladder to provide ventilation. Portable lighting also may be secured to ladders to illuminate working areas.

Ladders can be used as ramps for moving equipment, or for hoisting or lowering victims. They can be used to help shore up a damaged wall, to support a hose line over an opening, or to guide a rope up, down, or over an obstacle. With a tarp or salvage cover, fire fighters can turn a ladder into a channel or chute for water and debris.

Figure 12-2 Rescuers can use a ladder gin to remove a fall victim from an underground vault.

Ladder Construction

In its most basic design, a ladder consists of two beams connected by a series of parallel rungs. Fire service ladders, however, are specialized tools with several different parts. To use and maintain ladders properly, fire fighters must be familiar with the different types of ladders, as well as with the different parts and terms used to describe them.

Basic Ladder Components

The basic components of a straight ladder are used in most other types of ladders as well (► Figure 12-3).

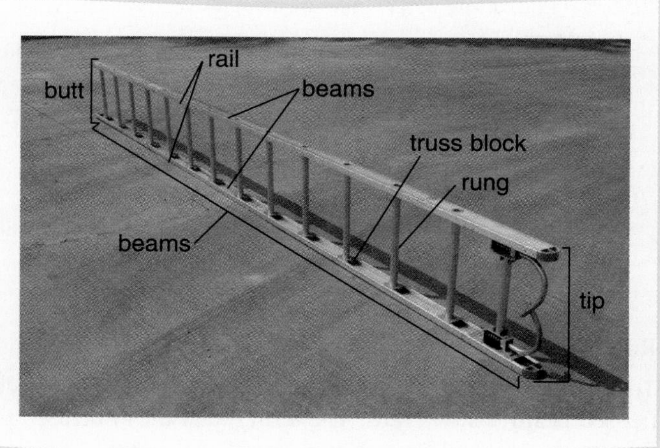

Figure 12-3 Basic components of a straight ladder.

Beams

A **beam** is one of the two main structural components that run the entire length of most ladders or ladder sections. The beams support the rungs and carry the load of a person from the rungs down to the ground.

There are three basic types of beam construction: trussed beam, I-beam, and solid beam.

- **Trussed beam**: A trussed beam has a top and a bottom rail, joined by a series of smaller pieces called truss blocks. The rungs are attached to the truss blocks. Trussed beams are usually constructed of aluminum or wood.
- **I-beam**: An I-beam has thick sections at the top and bottom, connected by a thinner section. The rungs are

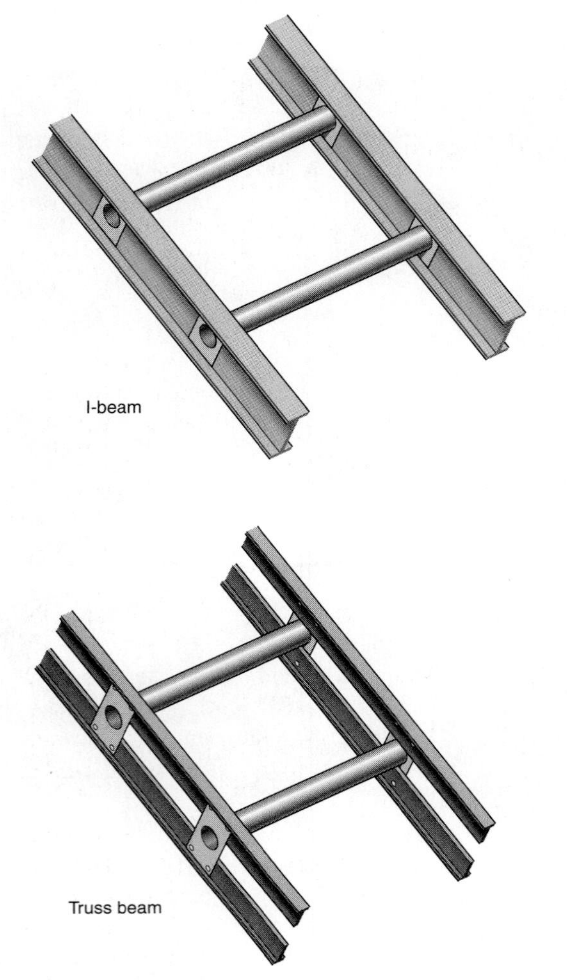

Figure 12-4 Two types of beam construction are truss beam and I-beam.

attached to the thinner section of the beam. This type of beam is usually made from fiberglass (▲ Figure 12-4).

- Solid beam: A solid beam has a simple rectangular cross section. Many wooden ladders have solid beams. Rectangular aluminum beams, which are usually hollow or C-shaped, are also classified as solid beams.

Rail

The rail is the top or bottom section of a trussed beam. Each trussed beam has two rails. The term rail also can be used to refer to the top and bottom surfaces of an I-beam.

Truss Block

A truss block is a piece or assembly that connects the two rails of a trussed beam. The rungs are attached to the truss blocks. Truss blocks can be made from metal or wood.

Rung

A rung is a crosspiece that spans the two beams of a ladder. The rungs serve as steps and transfer the weight of the user to the beams. Most portable ladders used by fire departments have aluminum rungs, but wooden ladders are still constructed with wood rungs.

Tie Rod

A tie rod is a metal bar that runs from one beam of the ladder to the other to keep the beams from separating. Tie rods are typically found in wood ladders.

Tip

The tip is the very top of the ladder.

Butt

The butt is the end of the ladder that is placed against the ground when the ladder is raised. It is sometimes called the heel or base.

Butt Spurs

Butt spurs are metal spikes attached to the butt of a ladder. The spurs prevent the butt from slipping out of position.

Butt Plate or Footpad

A butt plate or footpad is an alternative to a simple butt spur. It is a swiveling plate attached to the butt of the ladder and incorporates both a spur and a cleat or pad.

Roof Hooks

Roof hooks are spring-loaded, retractable, curved metal pieces attached to the tip of a roof ladder. The hooks are used to secure the tip of the ladder to the peak of a pitched roof.

Heat Sensor Label

A heat sensor label identifies when the ladder has been exposed to specific heat conditions that could damage the structural integrity of the ladder. The label changes colors when exposed to a particular temperature. Ladders that have been exposed to excessive heat must be removed from service and tested before being used again. A ladder that fails the structural stability test should be removed from service permanently.

Protection Plates

Protection plates are reinforcing pieces placed on a ladder at chaffing and contact points to prevent damage from friction or contact with other surfaces.

Extension Ladder Components

An extension ladder is an assembly of two or more ladder sections that fit together and can be extended or retracted to adjust the length. Extension ladders have additional parts (► Figure 12-5).

Bed Section

The bed section is the widest section of an extension ladder. It serves as the base; all other sections are raised from the bed section. The bottom of the bed section rests on the supporting surface.

Fly Section

A fly section is the part of an extension ladder that is raised or extended from the bed section. Extension ladders often have more than one fly section, each of which extends from the previous section.

Dogs

Dogs are mechanical locking devices used to secure the extended fly section(s). Dogs are also referred to as pawls, ladder locks, or rung locks.

Guides

Guides are strips of metal or wood that guide a fly section as it is being extended. Channels or slots in the bed or fly section may also serve as guides.

Halyard

The halyard is the rope or cable used to extend or hoist the fly section(s) of an extension ladder. The halyard runs through the pulley.

Pulley

The pulley is a small grooved wheel used to change the direction of the halyard pull. A downward pull on the halyard creates an upward force on the fly section(s), extending the ladder.

Stops

Stops are pieces of wood or metal that prevent the fly section(s) of a ladder from overextending and collapsing the ladder. Stops are also referred to as stop blocks.

Staypole

Staypoles (or tormentors) are long metal poles attached to the top of the bed section. They are used to help stabilize the ladder as it is being raised and lowered. One pole is attached to each beam of a long (40' or longer) extension ladder with a swivel joint. Each pole has a spur on the other end. Ladders with staypoles are typically referred to as Bangor ladders.

Types of Ladders

Ladders used in the fire service can be classed into two broad categories. Aerial ladders are permanently mounted and operated from a piece of fire apparatus. Portable ladders are carried on fire apparatus, but are designed to be removed and used in other locations.

Aerial Apparatus

Aerial apparatus vary greatly in design and function. This discussion is limited to a brief overview of the basic styles of aerial apparatus. Detailed requirements for aerial apparatus are documented in NFPA 1901 *Standard for Automotive Fire Apparatus*. This standard sets minimum performance requirements for apparatus referred to as either "aerial ladders" or "elevating platforms."

Aerial ladders are permanently mounted, power-operated ladders with a working length of at least 50'. They have at least two sections (▼ Figure 12-6). Some aerial ladders have a permanently mounted waterway and monitor nozzle. Aerial ladders are often referred to as "straight-stick" aerials.

An elevating platform apparatus includes a passenger-carrying platform (bucket) attached to the tip of a ladder or boom (► Figure 12-7). The ladder or boom must have at least two sections, which may be telescoping or articulating (jointed). An elevating platform apparatus that is 110' or less

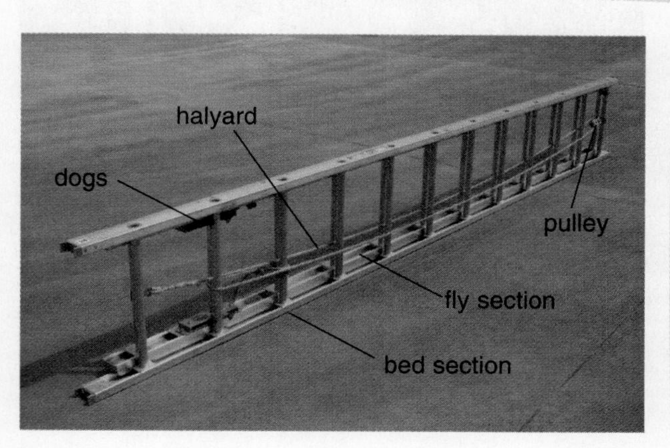

Figure 12-5 Extension ladders have additional parts not found in simple ladders.

Figure 12-6 An aerial ladder can be used to access the roof of a commercial structure.

Figure 12-7 An elevated platform serves as a secure working platform for fire fighters.

in length must also have a prepiped waterway and a permanently mounted monitor nozzle. An elevated platform supported by a boom may have an aerial ladder for continuous access to the platform.

Portable Ladders

Portable ladders, often called ground ladders, are carried on most fire apparatus. Portable ladders are designed to be removed from the apparatus and used in different locations. They may be general purpose ladders, such as straight or extension ladders, or specialized ladders. Specialized ladders have names that indicate their particular function.

The number and lengths of portable ladders used by a fire department depend on the maximum height of buildings in the response area. Most fire department engines carry 24′ or 28′ ladders, which can reach the roof of a typical two-story building. Ladder trucks usually carry 35′ or 40′ ladders, which can reach the roofs of most three-story buildings.

Generally, fire service portable ladders are limited to a maximum length of 50′. If building heights exceed the capabilities of portable ladders, aerial apparatus must be used. (► **Table 12-1**) shows the various types of portable ladders used by fire departments, their construction, and their weights.

Fire Fighter Safety Tips

Fire service portable ladders should be constructed and certified as compliant with the most recent edition of the NFPA 1931, *Standard on Design of and Design Verification Tests for Fire Department Ground Ladders.*

Straight Ladder

A straight ladder is a single-section, fixed-length ladder. Straight ladders may also be called wall ladders or single ladders. They are lightweight, can be raised quickly, and can reach windows and roofs of one- and two-story structures. Straight ladders are commonly 12′ to 20′ long, but can be up to 30′ and longer. Longer straight ladders are rarely used by the fire service because they are difficult to store and to handle.

Roof Ladder

A roof ladder (sometimes called a hook ladder) is a straight ladder equipped with retractable hooks at one end. The hooks secure the tip of the ladder to the peak of a pitched roof, so that the ladder lies flat on the roof. A roof ladder provides stable footing and distributes the weight of fire fighters and their equipment. This helps reduce the risk of structural failure in the roof assembly. Although roof ladders are available in lengths from 10′ to 30′, they are usually 12′ to 18′ long (► **Figure 12-8**).

Figure 12-8 Roof ladders are commonly 12′ to 18′ long.

Extension Ladder

An extension ladder is an adjustable-length ladder with multiple sections.

Figure 12-9 An extension ladder can be used at any length, from fully retracted to fully extended.

Table 12-1 Lengths and Weights of Various Ladders

Length	Type	No. of Sections	Construction	Material	Weight
20′	Straight	1	Solid beam	Aluminum	45 lb
20′	Straight	1	Solid beam	Fiberglass	50 lb
20′	Extension	2	Solid beam	Aluminum	60 lb
20′	Extension	2	Truss beam	Aluminum	76 lb
20′	Extension	2	Solid beam	Fiberglass	95 lb
24′	Extension	2	Solid beam	Aluminum	72 lb
24′	Extension	2	Truss beam	Aluminum	90 lb
24′	Extension	2	Solid beam	Fiberglass	110 lb
30′	Extension	2	Solid beam	Aluminum	107 lb
30′	Extension	3	Solid beam	Aluminum	108 lb
30′	Extension	2	Truss beam	Aluminum	131 lb
30′	Extension	2	Solid beam	Fiberglass	140 lb
30′	Extension	3	Truss beam	Aluminum	164 lb
30′	Extension	3	Solid beam	Fiberglass	170 lb
35′	Extension	2	Solid beam	Aluminum	114 lb
35′	Extension	2	Solid beam	Aluminum	129 lb
35′	Bangor	2	Truss beam	Aluminum	153 lb
35′	Extension	2	Solid beam	Fiberglass	160 lb
35′	Bangor	3	Truss beam	Aluminum	179 lb
35′	Extension	3	Solid beam	Fiberglass	195 lb
40′	Bangor	2	Solid beam	Aluminum	171 lb
40′	Bangor	3	Solid beam	Aluminum	193 lb
40′	Bangor	2	Truss beam	Aluminum	210 lb
40′	Bangor	3	Truss beam	Aluminum	245 lb
45′	Bangor	2	Solid beam	Aluminum	182 lb
45′	Bangor	2	Truss beam	Aluminum	255 lb
45′	Bangor	3	Truss beam	Aluminum	265 lb
45′	Bangor	3	Solid beam	Aluminum	280 lb
50′	Bangor	2	Truss beam	Aluminum	229 lb
50′	Bangor	3	Truss beam	Aluminum	297 lb

The bed or base section supports one or more fly sections. Pulling on the halyard extends the fly sections along a system of brackets or grooves (◄ **Figure 12-9**). The fly sections lock in place at set intervals so that the rungs are aligned to facilitate climbing.

An extension ladder is usually heavier than a straight ladder of the same length and requires more than one person to set up. Because the length is adjustable, an extension ladder

Fire Fighter Safety Tips

Roof ladders are not free-hanging ladders. The roof hooks will not support the full weight of the ladder and anyone on it when the ladder is in a vertical position.

Figure 12-10 Staypoles are used to stabilize a Bangor ladder while it is being raised or lowered.

Figure 12-11 Combination ladders can be used in several different configurations.

can replace several straight ladders, and can be stored in places where a longer straight ladder would not fit.

Bangor Ladder

Bangor ladders are extension ladders with staypoles for added stability during raising and lowering operations. Staypoles are required on ladders of 40′ or greater length and can be found on 35′ ladders. The poles help keep these heavy ladders under control during maneuvers (▲ Figure 12-10). When the ladder is positioned, the staypoles are planted in the ground on either side for additional stability.

Combination Ladder

A combination ladder can be converted from a straight ladder to a stepladder configuration (A-frame), or from an extension ladder to a stepladder configuration. Combination ladders are convenient for indoor use and for maneuvering in tight spaces. Combination ladders are generally 6′ to 10′ in the A-frame configuration and 10′ to 15′ in the extension configuration (► Figure 12-11).

Folding Ladder

A folding ladder (or attic ladder) is a narrow, collapsible ladder designed to allow access to attic scuttle holes and confined areas (► Figure 12-12). The two beams fold in for portability. Folding ladders are commonly available in 8′ to 14′ lengths.

Fresno Ladder

A Fresno ladder is a narrow, two-section extension ladder, designed to provide attic access. The Fresno ladder is generally short, just 10′ to 14′, so it has no halyard; it is extended manually (► Figure 12-13). A Fresno ladder can be used in tight space applications, such as bridging over a damaged section of an interior stairway.

Pompier Ladder

A pompier ladder, also known as a scaling ladder, is a lightweight, single-beam ladder. Fire fighters once used pompier ladders to climb up the outside of a building, one

Fire Fighter Safety Tips

It is very easy to pinch your hands when opening and closing a folding ladder. Always wear gloves when working with folding ladders.

Figure 12-12 Folding ladders are designed to be used in tight spaces.

Fire Marks

Chris Hoell, a St. Louis Fire Department Lieutenant, is credited with developing the pompier or scaling ladder in the late 1800s. The device, originally called the Hoell Life Saving Appliance, caught the attention of the Fire Department of New York City (FDNY), which hired Hoell to train its members to use the device. The pompier ladder was used by FDNY companies from the early 1880s until 1996.

or two floors at a time, to rescue a trapped person. Today, pompier ladders are used only in situations where no other option is available.

The beam of a pompier ladder has a large hook at the tip. The rungs extend through the beam, creating steps and handholds on both sides. The fire fighter would hook the ladder over a windowsill or solid object one or two floors above ground level, and climb up (► Figure 12-14). The fire fighter would then dismount, pull the ladder up, and secure the hook at a higher level, repeating the process up to the desired level. The fire fighter would then reverse direction and carry the person down to safety.

Inspection, Maintenance, and Service Testing of Portable Ladders

The portable ladders used by the fire service must be able to withstand extreme conditions. NFPA 1931, *Standard on Design of and Design Verification Tests for Fire Department Ground Ladders* establishes requirements for the construction of new ladders. The requirements are based on probable conditions during emergency operations.

Ladders can be dropped, overloaded, or exposed to temperature extremes during use. NFPA 1932, *Standard on Use, Maintenance, and Service Testing of Fire Department Ground Ladders* provides general guidance for ground ladder users. Portable ladders must be regularly inspected, maintained, and service-tested, following the NFPA Standard. In addition, ladders should always be inspected and maintained in accordance with the manufacturer's recommendations.

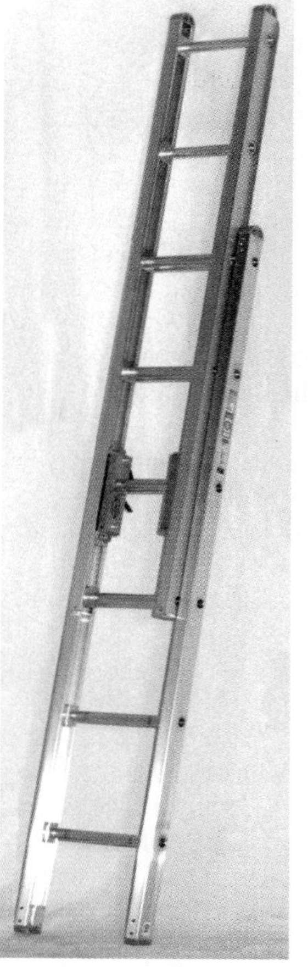
Figure 12-13 A Fresno ladder is extended manually.

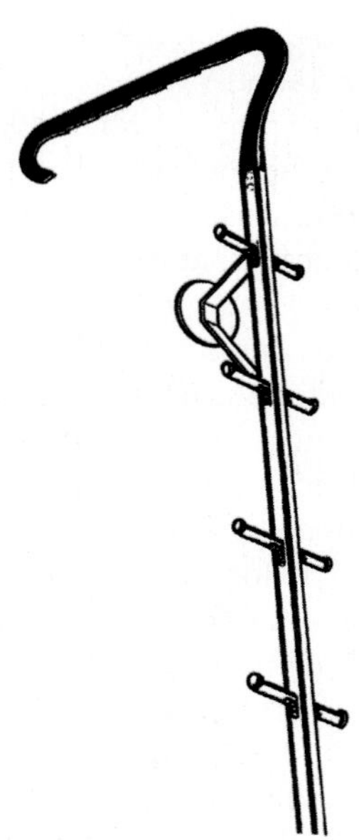
Figure 12-14 The pompier ladder enables a fire fighter to scale a vertical wall.

Inspection

Ground ladders must be visually inspected at least monthly. A ladder should also be inspected after each use. A general visual inspection should, at a minimum, note the following items:

- Beams: Check the beams for cracks, splintering, breaks, gouges, checks, wavy conditions, or deformations.
- Rungs: Check all rungs for snugness, tightness, punctures, wavy conditions, worn serrations, splintering, breaks, gouges, checks, or deformations.
- Halyard: Check the halyard for fraying or kinking; ensure that it moves smoothly through the pulleys.
- Wire-Rope Halyard Extensions: Check the wire-rope halyard extensions on three- and four-section ladders for snugness. This check should be performed when the ladder is in the bedded position, to ensure that upper sections will align properly during operation.

- Ladder Slides: Check the ladder slide areas for chaffing. Also check for adequate wax, if the manufacturer requires wax.
- Dogs: Check the dog (pawl) assemblies for proper operation.
- Butt Spurs: Check the butt spurs for excessive wear or other defects.
- Heat Sensors: Check the labels to see if the sensors indicate that the ladder has been exposed to excessive heat.
- Bolts and Rivets: Check all bolts and rivets for tightness; bolts on wood ladders should be snug and tight without crushing the wood.
- Welds: Check all welds on metal ladders for cracks or apparent defects.
- Roof Hooks: Check the roof hooks for sharpness and proper operation.
- Metal Surfaces: Check metal surfaces for signs of surface corrosion.
- Fiberglass and Wood Surfaces: Check fiberglass ladders for loss of gloss on the beams. Check for damage to the varnish finish on wooden ground ladders.

If the inspection reveals any deficiencies, the ladder must be removed from service until repairs are made. Minor repairs that require simple maintenance can often be performed by properly trained fire fighters at the fire station. Repairs involving the structural or mechanical components of a ladder must be performed only by qualified personnel at a properly equipped repair facility.

Maintenance

All fire fighters should be able to perform routine ladder maintenance. Maintenance is simply the regular process of keeping the ladder in proper operating condition. Fundamental maintenance tasks include:

- Clean and lubricate the dogs, following the manufacturer's instructions (► Figure 12-15).
- Clean and lubricate the slides on extension ladders in accordance with the manufacturer's recommendations.
- Replace worn halyards and wire rope on extension ladders when they fray or kink (► Figure 12-16).
- Clean and lubricate hooks. Remove rust and other contaminants and lubricate the folding roof hook assemblies on roof ladders to keep them operational (► Figure 12-17).

Fire Fighter Safety Tips

Fire fighters should perform routine ladder maintenance. Repairs to portable ladders should be performed ONLY by trained repair personnel.

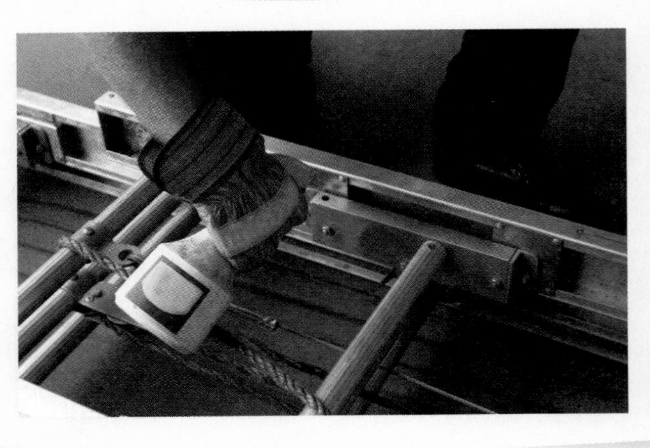

Figure 12-15 Extension ladder dogs must operate smoothly.

Figure 12-16 Replace the halyard if it is worn or damaged.

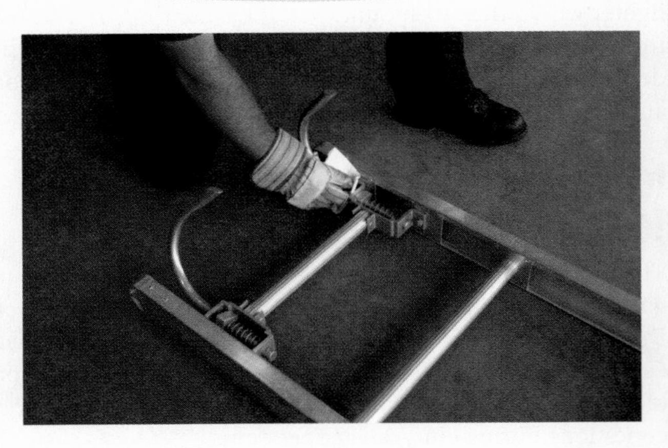

Figure 12-17 Roof hooks must operate smoothly.

- Check the heat sensor labels. Replace heat sensors when they reach their expiration date. Remove a ladder that has been exposed to high temperatures from service for testing.
- Maintain the finish on fiberglass and wooden ladders in accordance with the manufacturer's recommendations.
- Portable ladders should NOT be painted except for the top and bottom 18" of each section. Paint can hide structural defects in the ladder. The tip and butt are painted for purposes of identification and visibility.

Ladders that are in storage should be placed on racks or in brackets and protected from the weather. Fiberglass ladders can be damaged by prolonged exposure to direct sunlight (► Figure 12-18).

Cleaning

Ladders must be regularly cleaned to remove road grime and dirt that build up during storage on the apparatus. Ladders should be cleaned before each inspection to ensure that any hidden faults can be observed. Ladders should also be cleaned after each use to remove dirt and debris.

Use a soft-bristle brush and water to clean ladders. A mild, diluted detergent may be used, if allowed by the manufacturer's recommendations (► Figure 12-19). Remove any tar, oil, or grease deposits with a safety solvent as recommended by the manufacturer.

Rinse and dry the cleaned ladder before replacing it on the apparatus. Wipe dry any ladders exposed to water or used in the rain.

Service Testing

Service testing is performed periodically to evaluate the continued usefulness of a ladder during its life. Service tests should be performed on new ladders as well as on ladders that have been in use for some time. Service testing is different from design verification testing. Design verification tests are conducted by testing laboratories to ensure that new ladders are constructed in compliance with manufacturing specifications. Service tests measure the structural integrity of a portable ladder.

Service testing of portable ladders must follow NFPA 1932, *Standard on Use, Maintenance, and Service Testing of Fire Department Ground Ladders*.

Service tests should be conducted before a new ladder is used and annually while it is in service. A ladder that has been exposed to extreme heat, overloaded, impact- or shock-loaded, visibly damaged, or is suspected of being unsafe for any other reason must be removed from service until it has passed a service test. A repaired ladder must also pass a service test before it can be returned to service, unless a halyard replacement was the only repair.

The horizontal-bending test evaluates the structural strength of a ladder. The ladder is placed in a horizontal position across a set of supports. A weight is then put on the ladder, and the amount of deflection or bending caused by the weight is measured to evaluate the strength of the ladder (► Figure 12-20). Additional tests are performed on the extension hardware of extension ladders. The hooks on roof ladders are also tested.

Service testing of ground ladders requires special training and equipment and must be conducted only by qualified

Fire Fighter Safety Tips

Do not get any solvent on the halyard of an extension ladder. Contact with solvents can damage halyard ropes.

Figure 12-18 Portable ladders should be stored on racks or in brackets, out of the weather or direct sunlight.

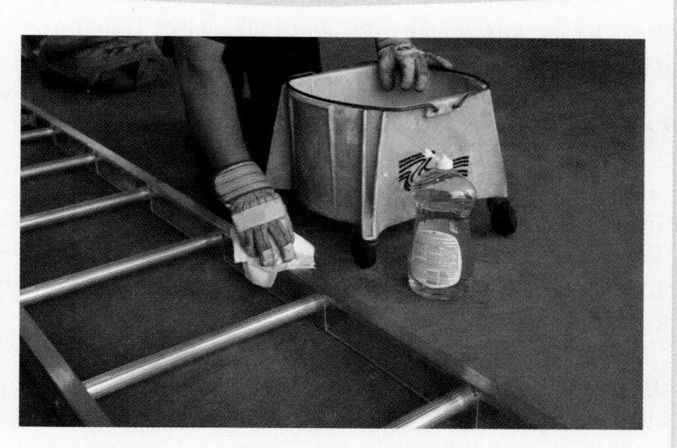

Figure 12-19 Ladders should be cleaned regularly, as well as after each use, with a mild detergent.

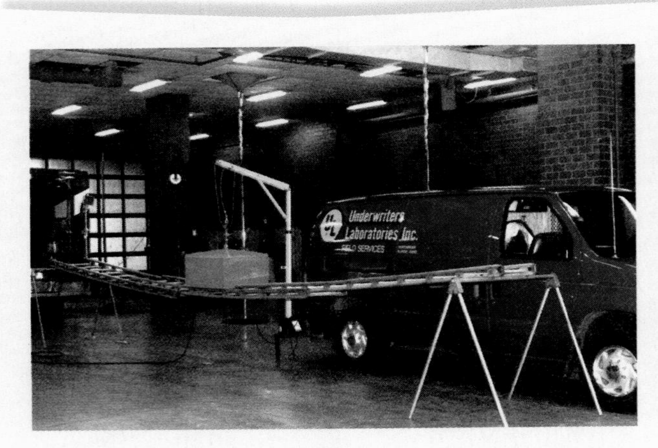

Figure 12-20 Horizontal bend test evaluates the structural strength of a ladder.

personnel. Many fire departments use outside contractors to perform these tests. The results of all service tests must be recorded and kept for future reference.

Ladder Safety

Several potential hazards are associated with ladder use. These risks are easily overlooked during emergency operations. Many fire fighters have been seriously injured or killed in ladder accidents, both on the fireground and during training sessions.

Ladders must be used with caution; follow standard procedures and regularly reinforce your skills through training. Many safety precautions should be followed from the time a ladder is removed from the apparatus until it is returned to the apparatus. Basic safety issues include:

- general safety
- lifting and moving ladders
- placement of ground ladders
- working on the ladder
- rescue operations
- ladder damage

These issues are explained in general here and will be revisited under the specifics of using ladders.

General Safety Requirements

In any firefighting operation, full turnout clothing and protective equipment are essential for working with ground ladders. Turnout gear provides protection from mechanical injuries, as well as from fire and heat injuries. Helmets, coats, pants, gloves, and protective footwear provide protection from falling debris, impact injuries, or pinch injuries. Fire fighters must be able to work with and on ladders while wearing self-contained breathing apparatus. You should practice working with ladders and tying knots while wearing all of your protective equipment, including gloves and eye protection.

Lifting and Moving Ladders

Teamwork is essential when working with ladders. Some ladders are heavy and can be very awkward to maneuver, particularly when they are extended. Crew members must use proper lifting techniques and coordinate all movements.

Never attempt to lift or move a ladder that weighs more than you are capable of lifting safely. Because of their shape, ladders may be awkward to carry, especially over uneven terrain, through gates, or over snow and ice. It is better to ask for help to move a ladder, rather than risk injury or delay the placement of the ladder.

Placement of Ground Ladders

Fire fighters working on the fireground should survey the area where a portable ladder is going to be used before placing the ladder or beginning to raise it. If possible, inspect the area before retrieving the ladder from the apparatus. If you note any hazards, consider changing the position of the ladder. Work in the safest available location.

Sometimes a ladder has to be placed in a potentially hazardous location. If you are aware of the hazard, you can take corrective action and special precautions to avoid accidents.

The most important check is the location of overhead utility lines. If a ladder comes in contact with an electrical wire, the fire fighters who are handling it can be electrocuted. If the line falls, it can electrocute other fire fighters as well. Avoid placing a ladder against a surface that has been energized by a damaged or fallen power line.

Do not assume that it is safe to use wood or fiberglass ladders around power lines. Metal ladders will conduct electricity, but a wet or dirty wood or fiberglass ladder can also conduct electricity.

If possible, do not raise ladders anywhere near overhead wires. At a minimum, keep ladders at least 10' away from any power lines while using or raising them. If a power line is nearby, make sure that there are enough fire fighters to keep the ladder under control as it is raised. One fire fighter should watch to make sure that the ladder does not come too close to the line. If the ladder falls into a power line, the results can be deadly.

Other types of overhead obstructions can also be hazards during a ladder raise. If the ladder hits something as it is being raised, the weight and momentum of the ladder will shift. This can cause the fire fighters to lose control and drop the ladder.

Ladders must be placed on stable and relatively level surfaces (► Figure 12-21). The shifting weight of a climber can easily tip a ladder placed on a slope or unstable surface. Always evaluate the stability of the surface before placing the ladder, and check it again during the course of the firefighting operations. A ladder placed on snow or mud may become unstable as the snow melts or as the base of the ladder sinks into the mud. Water from the firefighting operations can make soft ground even more unstable.

Figure 12-21 Ladders must be placed on stable and relatively level surfaces.

Figure 12-22 To prevent a ladder from slipping, fire fighters should secure the base by heeling the ladder.

Ladders should not be exposed to direct flames or extreme temperatures. Excessive heat can cause permanent damage or catastrophic failure. Check the heat sensor labels immediately after a ladder has been exposed to high temperatures.

Working on a Ladder

Before climbing a ladder, be sure it is set at the proper climbing angle (approximately 75°) for maximum load capacity and safety. Dogs must be locked and the halyard tied before anyone climbs an extension ladder.

To prevent slipping, another fire fighter should secure the base by heeling (footing) the ladder (► Figure 12-22). This technique uses the fire fighter's weight to keep the base of the ladder from slipping. The base of the ladder also can be mechanically secured to a solid object. An alternative to securing the base is to use a rope or strap to secure the tip of the ladder to the building.

All ladders have weight limits. Most portable ladders are designed to support a weight of 750 lbs. In a rescue situation, this weight limit is equivalent to one person and two fire fighters with full protective clothing and equipment. This weight limit would also accommodate two fire fighters climbing or working on the ladder along with their equipment.

The weight should be distributed along the length of the ladder. Only one fire fighter should be on each section of an extension ladder at any time.

While climbing the ladder, be prepared for falling debris, misguided hose streams, or people jumping from the building that could knock you off the ladder. This is another reason why fire fighters should wear full turnout gear when working on ladders. It will help protect a fire fighter who is knocked to the ground. It is even more important that fire fighters working from ladders near the fire wear protective clothing and equipment.

If fire conditions should change rapidly while you are working on a ladder, you must be prepared to climb down quickly. For example, if the fire suddenly flashes over or if flames break out through a window near the ladder, you must be prepared to move quickly out of danger. Turnout gear will not protect you from direct exposure to the flames for more than a few seconds.

Fire Fighter Safety Tips

Always maintain a minimum 10′ clearance between a ladder and utility lines to prevent electrocution.

Fire Fighter Tips

Operators of aerial equipment are trained to extend the bucket or tip of the aerial ladder above a trapped individual during a rescue. They then lower it down to the person. This generally prevents the individual from jumping onto the ladder.

A fire fighter working from a ladder is in a less stable position than one working on the ground. You must constantly adjust your balance, especially when swinging a hand tool or reaching for a trapped occupant. There is the danger of falling as well as the risk to people below if something falls or drops. Fire fighters who are working from a ladder should use a safety belt or a leg lock to secure themselves to the ladder.

Rescue

Rescue is a fire fighter's most important and unpredictable duty. Portable ladders are often used to reach and remove trapped occupants from the upper stories of a building. However, fire fighters must address several important safety concerns before going up a ladder to rescue someone.

A person who is in extreme danger may not wait to be rescued. Jumpers risk their own lives and may endanger the fire fighters trying to rescue them. Several fire fighters have been seriously injured by persons who jumped before a rescue could be completed.

A trapped person might try to jump onto the tip of an approaching ladder, or to reach out for anything or anyone nearby. You might be pulled or pushed off the ladder by the person you are trying to rescue. If several people are trapped, they might all try to climb down the ladder at the same time.

It is important to make verbal contact with any person you are trying to rescue. You must remain in charge of the situation and not let the individual panic. Tell the person to remain calm and wait to be rescued. If there are enough fire fighters, one could maintain contact with the person while others raise the ladder.

After you reach the person, you still have to get him or her down to the ground safely. Try to make the person realize that any erratic movements could tip over the ladder. It is often helpful to have one fire fighter assist the person, while a second fire fighter acts as a guide and back-up to the rescuer. Modern fire ladders are designed to support this load, but a ladder that is overloaded during a rescue operation must be removed from service for inspection and service testing. See Chapter 13 for specific ladder rescue techniques.

Ladder Damage

Ladders may easily be damaged while in use. A ladder might be overloaded or used at a low angle during a rescue. An unexpected shift in fire conditions could bring the ladder in direct contact with flames. Whenever a ladder is used outside of its recommended limits, it should be taken out of service for inspection and testing, even if there is no visible damage.

Using Portable Ladders

Portable ladders are often urgently needed during emergency incidents. Because an accident or error in handling or using a ladder can result in death or serious injury, all fire fighters must know how to work with ladders.

Using a ladder requires that fire fighters complete a series of consecutive tasks. The first step is to select the best ladder for the job from those available. Fire fighters must then remove the ladder from the apparatus and carry it to the location where it will be used. The next step is to raise and secure the ladder. At the end of the operation, the ladder must be lowered and returned to the apparatus. Each of these tasks is important to the safe and successful completion of the overall objective.

Ladder Selection

The first step in using a portable ladder is to select the appropriate ladder from those available. Fire fighters must be familiar with all of the ladders carried on their apparatus. Engine and ladder company apparatus usually carry several portable ladders of various lengths. Many other types of apparatus, such as tankers (water tenders) and rescue units, often carry additional portable ladders. NFPA 1901 *Standard for Automotive Fire Apparatus* provides minimum requirements for portable ladders for each type of apparatus (▶ Table 12-2).

Selecting an appropriate ladder requires that fire fighters estimate the heights of windows and rooflines. (▶ Table 12-3) provides general guidelines for estimating heights for residential and commercial buildings.

Ladder placement will also affect the size and length of the ladder needed. When the ladder is used to access a roof, the tip of the ladder should extend several feet above the roofline. This provides a handhold and footing for fire fighters as they mount and dismount. The extra length also makes the tip of the ladder visible to fire fighters working on the roof (▶ Figure 12-23). A common rule of thumb is to be sure at least five ladder rungs show above the roofline.

Ladders used to provide access to windows must be longer than those used in rescue operations. During access operations, the ladder is placed next to the window, with the ladder tip even with the top of the window opening (▶ Figure 12-24). However, during rescue operations, the tip of the ladder should be immediately below the windowsill. This prevents the ladder from obstructing the window opening while a trapped occupant is removed (▶ Figure 12-25).

The final factor in determining the correct length to use is the angle formed by the ladder and the placement surface (ground). A portable ladder should be placed at an angle of

Table 12-2 Minimum Ladder Complement for Apparatus as Specified in NFPA 1901, *Standard for Automotive Fire Apparatus* (1999 ed.)

Pumper
1 Attic Ladder
1 Roof Ladder
1 Extension Ladder

Quick Attack
1 Extension Ladder must be 12′ or longer

Aerial/Ladder
Portable ladders that have a total length of 115′ or more and contain a minimum of:

- 1 Attic Ladder
- 2 Roof Ladders
- 2 Extension Ladders

Quint
Portable ladders that have a total length of 85′ or more and contain a minimum of:

- 1 Attic Ladder
- 1 Roof Ladder
- 1 Extension Ladder

Table 12-3 Approximate Distances for Residential and Commercial Construction

Residential Floor-to-Floor Height	8′ to 10′
Residential Floor-to-Windowsill Height	3′
Commercial Floor-to-Floor Height	12′
Commercial Floor-to-Windowsill Height	4′

approximately 75° for maximum strength and stability. This means that the ladder will have to be slightly longer than the vertical distance between the ground and the target point. Generally, a ladder requires an additional 1′ in length for every 15′ of vertical height.

To reach a window 30′ feet above grade level, the ladder would have to be at least 32′ long. Because ladders used in roof operations need to extend at least above the roofline, accessing a roof that is 30′ above grade level requires a ladder at least 35′ long.

Removing the Ladder from Apparatus

Ladders are mounted on apparatus in various ways. Ladders should be mounted in locations where they will not be exposed to excessive heat, engine exhaust, or mechanical damage. Fire fighters must be familiar with how ladders are

Figure 12-23 The tip of the ladder should extend above the roof line during roof operations.

Figure 12-24 The tip of the ladder should be even with the top of the window opening to provide access through the window.

Figure 12-25 For rescue operations, the tip of the ladder should be just below the windowsill.

Voices of Experience

"Everything was going smoothly when suddenly the halyard broke before the locks could be secured."

Several years ago I was instructing a ground ladders class at our annual state fire academy. During the practical exercise, a group of students were raising a 35′ extension ladder. Everything was going smoothly when suddenly the halyard broke before the locks could be secured. "Look out!" shouted one of the students as the fly section of the ladder retracted rapidly into the bed.

Fortunately, the students had been instructed to always support the ladder by the rails. None of the students had their hands on the rungs. Had they been holding the rungs, their fingers would have been severed immediately.

It is essential to remember that all of the equipment you work with can be potentially dangerous at all times. Never be lax about following safety precautions. Always practice the correct procedure while performing each and every task. It only takes one mistake to end a career.

Tommy Fuqua
Roanoke County Fire and Rescue
Roanoke, Virginia

mounted on their apparatus and practice removing them safely and quickly.

Ladders are often nested one inside another and mounted on brackets on the side of a pumper. Fire fighters should note the nesting order and location of the ladders relative to the brackets when removing or replacing them. Ladders that are not needed should not be placed on the ground in front of an exhaust pipe because they could be damaged by the hot exhaust.

Ladders can also be stored on overhead hydraulic lifts (► Figure 12-26). The hydraulic mechanism keeps them out of the way until they are needed, when they can be lowered to a convenient height.

On some vehicles, portable ladders are stored in compartments under the hose bed or aerial device. The ladders may lie flat or vertically on one beam (► Figure 12-27). Fire fighters slide the ladders out the rear of the vehicle to remove or replace them.

Lifting Ladders

Many ladders are heavy and awkward to handle. As noted in Table 12-1, some ladders weigh more than 200 lb. To prevent lifting injuries while handling ladders, fire fighters must work together to lift and carry long or heavy ladders.

When fire fighters are working as a team to lift or carry a ladder, one fire fighter must act as the leader, providing direction and coordinating the actions of all team members. The lead fire fighter needs to call out the intended movements clearly, using standard commands and terminology. For example, when lowering a ladder, the lead should say "Prepare to lower," followed by the command "Lower." Team members have to communicate in a clear and concise fashion, using the same terminology for commands. There should be no confusion about the specific meaning of each command. Because terminology may differ among jurisdictions, fire fighters newly or temporarily assigned to an apparatus should verify what commands are used and what they mean.

A pre-arranged method should exist for determining the lead fire fighter for ladder lifts. In some fire departments, the fire fighter on the right side at the butt end of the ladder is the standard leader. In other departments, the fire fighter at the tip of the ladder on the right side may be the leader. The departmental policy should be consistent for all ladder operations.

Additionally, fire fighters must use good lifting techniques when handling ladders. When bending to pick up a ladder, bend at the knees and keep the back straight (► Figure 12-28). Lift and lower the load with the legs rather than the back. Take care to avoid twisting motions during lifting and lowering, because these motions often lead to back strains.

Carrying

Once the ladder has been removed from the apparatus or lifted from the ground, it must be carried to the placement site. Ladders can be carried at shoulder height or at arm's

Figure 12-26 This hydraulic mechanism lowers the ladders to a convenient height when they are needed.

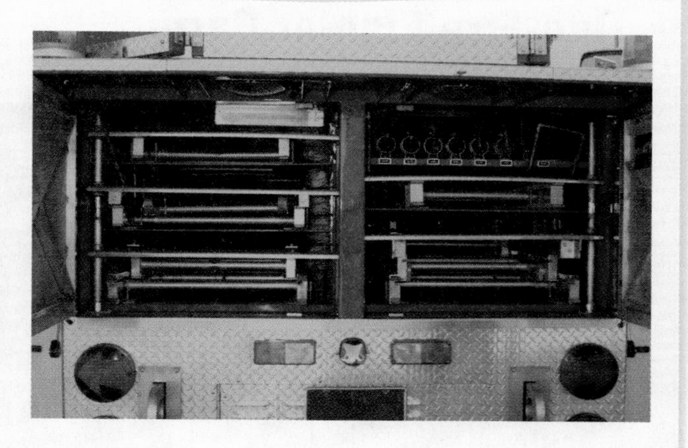

Figure 12-27 Ladders in rear compartments can be stored flat or vertically.

Figure 12-28 Bend the knees and keep the back straight when lifting or lowering a ladder.

length, and on edge or flat. The most common carries are explained in the following sections.

One-Fire Fighter Carry

Most straight ladders and roof ladders less than 18′ long can be safely carried by a single fire fighter. The steps involved in the one-fire fighter carry are outlined in ▼ Skill Drill 12-1.

1. Start with the ladder mounted in a bracket or standing on edge (on one beam). Locate the center of the ladder. **(Step 1)**
2. Place an arm between two rungs of the ladder just to one side of the middle rung. **(Step 2)**
3. The top beam of the ladder rests on the fire fighter's shoulder as it is carried. **(Step 3)**

A straight ladder is carried with the butt end first and pointed slightly downward. A roof ladder should be carried tip first. Most ladders are carried with the butt end forward, because the placement of the butt determines the positioning of the ladder. Roof ladders are carried with the hooks toward the front. A roof ladder usually must be carried up another ladder to reach the roof. Then, the roof ladder is usually pushed up the pitched roof slope from the butt end until the hooks engage the peak.

Two-Fire Fighter Shoulder Carry

The two-fire fighter shoulder carry is generally used with extension ladders up to 35′ long. It can also be used to carry straight ladders or roof ladders that are too long for one per-

Skill Drill

One-Fire Fighter Carry

Locate the center of the ladder.

Place one arm through two rungs, just to one side of the middle rung.

Bring the top beam to rest on the fire fighter's shoulder.

son to handle. To perform the two-fire fighter shoulder carry, follow the steps in ▼ Skill Drill 12-2.

1. Start with the ladder mounted in a bracket or standing on edge. Both fire fighters stand on the same side of the ladder, facing the butt, one near the butt and one near the tip. **(Step 1)**
2. Facing the butt end of the ladder, each fire fighter places an arm between two rungs and lifts the ladder onto the shoulder. The ladder is carried butt end first. **(Step 2)**
3. The fire fighter closest to the butt covers the butt spur with a gloved hand to prevent injury to other fire fighters in the event of a collision. **(Step 3)**

Three-Fire Fighter Shoulder Carry

Three fire fighters may be needed to carry a heavy ladder. This carry is similar to the two-fire fighter shoulder carry, with the additional fire fighter at the middle of the ladder. All three fire fighters stand on the same side of the ladder. To perform the three-fire fighter shoulder carry, follow the steps in ► Skill Drill 12-3.

1. Start with the ladder mounted in a bracket or standing on edge. All three fire fighters stand on the same side of the ladder, facing the butt, one near each end and one at the middle. **(Step 1)**
2. Each fire fighter places an arm between two rungs and hoists the ladder onto the shoulder. **(Step 2)**

12-2 Skill Drill

Two-Fire Fighter Shoulder Carry

1 Both fire fighters approach the ladder from the same side, facing the butt. One fire fighter stands near the butt and the other near the tip.

2 Each fire fighter places an arm between two rungs and lifts the ladder onto the shoulder.

3 The butt spurs are covered with a gloved hand while the ladder is transported.

3. The fire fighter closest to the butt covers the butt spur with a gloved hand to prevent injury to other fire fighters in the event of a collision. **(Step 3)**

Two-Fire Fighter Suitcase Carry

The two-fire fighter suitcase carry is commonly used with straight and extension ladders. The ladder is carried at arm's length. To perform the two-fire fighter suitcase carry, follow the steps in (► **Skill Drill 12-4**).

1. Begin with the ladder resting on the ground on one beam. Both fire fighters stand on the same side of the ladder, at opposite ends, and face the butt. **(Step 1)**
2. The fire fighters reach down and grasp the upper beam of the ladder. **(Step 2)**
3. Pick the ladder up from the ground and carry it with the butt end forward. **(Step 3)**

Three-Fire Fighter Suitcase Carry

A three-fire fighter suitcase carry can be used for heavier ladders. This carry is similar to the two-fire fighter suitcase carry, with the addition of a third fire fighter at the center of the ladder. All three fire fighters remain on the same side of the ladder. To perform the three-fire fighter suitcase carry, follow the steps in (► **Skill Drill 12-5**).

Skill Drill

Three-Fire Fighter Shoulder Carry

All three fire fighters approach the ladder from the same side and face the butt end. Two fire fighters stand at each end and one in the middle.

Each fire fighter places an arm between two rungs and lifts the ladder onto the shoulder.

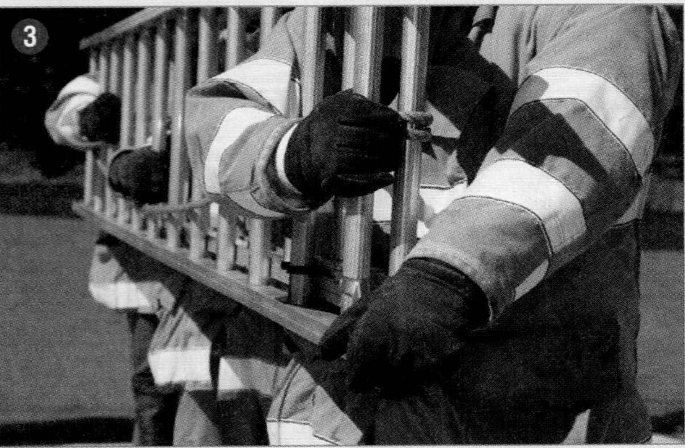

The butt spurs are covered with a gloved hand while the ladder is transported.

1. The carry begins with the ladder resting on the ground on one beam. One fire fighter stands near the butt of the ladder, one near the center, and the third near the tip. All three fire fighters stand on the same side and face the butt of the ladder. All three fire fighters reach down and grasp the upper beam of the ladder. **(Step 1)**
2. Pick the ladder up from the ground and carry it at arm's length with the butt forward. **(Step 2)**

12-4 **Skill Drill**

Two-Fire Fighter Suitcase Carry

1 Both fire fighters face the butt of the ladder, at opposite ends.

2 The upper beam of the ladder is grasped.

3 Pick up the ladder using good lifting techniques.

12-5 **Skill Drill**

Three-Fire Fighter Suitcase Carry

1 Three fire fighters stand on one side of the ladder, facing the butt. All three fire fighters grasp the upper beam.

2 Pick up the ladder using good lifting techniques.

Three-Fire Fighter Flat Carry

The three-fire fighter flat carry is typically used with extension ladders up to 35′ long. To perform the three-fire fighter flat carry, follow the steps in ▼ Skill Drill 12-6.

1. The carry begins with the bed section of the ladder flat on the ground. Two fire fighters stand on one side of the ladder, one at the butt and one at the tip. The third fire fighter stands on the opposite side of the ladder near the center. **(Step 1)**
2. All three fire fighters kneel down, facing the butt end, and grasp the closer beam at arm's length. **(Step 2)**
3. The fire fighters rise to a standing position lifting the ladder to arm's length. **(Step 3)**

Four-Fire Fighter Flat Carry

This carry is similar to the three-fire fighter flat carry described above except that two fire fighters are positioned on each side of the ladder. On each side, one fire fighter is at the tip and the other is at the butt end of the ladder. To perform the four-fire fighter flat carry, follow the steps in ► Skill Drill 12-7.

1. The carry begins with the bed section of the ladder flat on the ground. Two fire fighters stand on each side of the ladder, one at the butt and one at the tip. **(Step 1)**
2. All four fire fighters kneel down, facing the butt end of the ladder and grasp the closer beam at arm's length. **(Step 2)**
3. The fire fighters rise to a standing position lifting the ladder at arm's length and walk with the butt end forward. **(Step 3)**

Three-Fire Fighter Flat Shoulder Carry

The three-fire fighter flat shoulder carry is similar to the three-fire fighter flat carry. The difference is that the ladder is carried on the shoulders instead of at arm's length. The fire fighters face the tip of the ladder as they lift it, then pivot into the ladder as they raise it to shoulder height. This carry is useful when fire fighters must carry the ladder over short obstacles. Because raising the ladder to shoulder height increases the potential for back strain, fire fighters must follow proper lifting techniques. To perform the three-fire fighter flat shoulder carry, follow the steps in ► Skill Drill 12-8.

1. The carry begins with the bed section of the ladder flat on the ground. On one side of the ladder a fire fighter is located at the butt and another is located at the tip. On the other side of the ladder, a fire fighter is positioned near the center of the ladder. All three fire fighters face the tip of the ladder. The fire fighters kneel and grasp the closer beam. **(Step 1)**
2. The fire fighters stand, raising the ladder. As the ladder approaches chest height, the fire fighters all pivot into the ladder. **(Step 2)**
3. The ladder rests on the shoulders of the fire fighters, with all three facing the butt of the ladder. The ladder is carried in this position with the butt moving forward. **(Step 3)**

Skill Drill

Three-Fire Fighter Flat Carry

Two fire fighters stand at the ends of the ladder on one side and the third stands in the middle on the opposite side.

All three face the butt of the ladder and kneel and grasp the closer beam at arm's length.

The fire fighters rise and carry the ladder at arm's length.

12-7 Skill Drill

Four-Fire Fighter Flat Carry

Two fire fighters stand on each side of the ladder facing the butt, one at the butt and one at the tip.

All four kneel down, facing the butt end, and grasp the closer beam at arm's length.

The ladder is raised to arm's length.

12-8 Skill Drill

Three-Fire Fighter Flat Shoulder Carry

Two fire fighters are positioned on the same side at the ends of the ladder, facing the tip. The third is on the opposite side at the middle. All three fire fighters kneel and grasp the closer beam.

As the ladder approaches chest height, the fire fighters all pivot into the ladder.

The ladder is carried on the shoulders facing the butt.

12-9 Skill Drill

Four-Fire Fighter Flat Shoulder Carry

1 Two fire fighters are positioned on each side of the ladder at the ends. All four face the tip.

2 All four fire fighters kneel and grasp the closer beam.

3 As the ladder approaches chest height, the fire fighters all pivot toward the ladder, bringing it to rest on the shoulder. The ladder is carried on the shoulders facing the butt.

Four-Fire Fighter Flat Shoulder Carry

This carry is similar to the three-fire fighter flat shoulder carry described above, except that there are two fire fighters on each side of the ladder. On each side, one fire fighter is at the tip and the other at the butt end of the ladder. To perform the four-fire fighter flat shoulder carry, follow the steps in ▲ **Skill Drill 12-9**.

1. The carry begins with the bed section of the ladder flat on the ground. On each side of the ladder a fire fighter is located at the butt and another is located at the tip. All four fire fighters face the tip of the ladder. **(Step 1)**
2. The fire fighters kneel and grasp the closer beam. **(Step 2)**
3. The fire fighters stand, raising the ladder. As the ladder approaches chest height, the fire fighters all pivot toward the ladder. The ladder ends at rest on the shoulders of the fire fighters, with all four facing the butt of the ladder. The ladder is carried at this position with the butt moving forward. **(Step 3)**

Placing a Ladder

The first step in raising a ladder is selecting the proper location for the ladder. Generally, an officer or senior fire fighter will select the general area for ladder placement, and the fire fighter at the butt of the ladder will determine the exact site. Both the officer ordering the ladder and the fire fighter placing the ladder need to consider several factors in their decisions.

A raised ladder should be at an angle of approximately 75° to provide the best combination of strength, stability, and vertical reach. This angle also creates a comfortable climbing angle. When a fire fighter is standing on a rung, the rung at shoulder height will be about an arm's length away. The ratio of ladder height (vertical reach) to distance from the structure should be 4:1. For example, if the vertical reach of a ladder is 20′, the butt of the ladder should be placed 5′ out from the wall. Most new ladders have an inclination guide on the beams so that the ladder will be set at the proper angle.

When calculating vertical reach, remember to add additional footage for rooftop operations. If the butt of the ladder is placed too close to the building, the climbing angle will be too steep. The tip of the ladder could pull away from the building as the fire fighter climbs, making the ladder unstable. If the butt is too far from the structure, the climbing angle will be too shallow, reducing the load capacity of

Fire Fighter Safety Tips

Do not place chocks or wood blocks under one beam of a ladder to position it on sloping ground. The ladder could slip off the blocks and overturn.

Figure 12-29 A ladder placed on an uneven surface can easily tip over.

the ladder and increasing the risk that the butt could slip out from under the ladder.

The ladder should be placed on a stable and relatively level surface. Ladders placed on uneven surfaces are prone to tipping (▲ Figure 12-29). As a fire fighter climbs the ladder, the moving weight will cause the load to shift back and forth between the two beams. Unless both butt ends are firmly placed on solid ground, this can start a rocking motion that will tip the ladder over.

Ladders should not be placed on top of manholes or trap doors. The weight of the ladder, fire fighters, and equipment could cause the cover or door to fail, injuring the fire fighter(s).

Ladders should only be placed where there are no overhead obstructions. If a ladder comes into contact with overhead utility lines, particularly electric power lines, the fire fighters working on it, as well as those stabilizing it, could be injured or killed. This is true for all ladders, regardless of their composition (fiberglass, wood, or metal).

A ladder can be energized even if it does not actually touch an electric line. A ladder that enters the electromagnetic field surrounding a power line can become energized. Ladders should stay at least 10' away from energized power lines.

Finally, portable ladders should not be placed in high traffic areas unless no other alternative is available. For example, because the main entrance to a structure is heavily used, a ladder should not be placed where it would obstruct the door.

Raising a Ladder

Once the position has been selected, the ladder must be raised. Two common techniques for raising portable ladders are the beam raise and the rung raise. A beam raise is usually used when the ladder must be raised parallel to the target surface. A rung raise is often used when the ladder can be raised from a position perpendicular to the target surface.

The number of fire fighters required to raise a ladder depends on the length and weight of the ladder, as well as on the available clearance from obstructions. A single fire fighter can safely raise many straight ladders and lighter extension ladders. Two or more fire fighters are required for longer and heavier ladders.

One-Fire Fighter Rung Raises

There are two variations of a one-fire fighter rung raise. One is used by a single fire fighter to raise a small, straight ladder, typically 14' or less in length. To perform this raise, follow the steps in (► Skill Drill 12-10).

1. Start with the fire fighter carrying a ladder using the one-fire fighter carry described earlier. Check for overhead hazards. **(Step 1)**
2. The fire fighter places the ladder flat on the ground. The heel of the ladder should be positioned approximately where it will be when the ladder is in the raised position.
3. Standing at the tip, the fire fighter raises the ladder to hip level.
4. The fire fighter walks hand-over-hand down the rungs until the ladder is vertical. **(Step 2)**
5. The fire fighter places one foot against the beam of the ladder and leans it into place. **(Step 3)**

The other variation of the one-fire fighter rung raise is generally used with straight ladders longer than 14' and with extension ladders that can be safely handled by the fire fighter. Each fire fighter will have a different safety limit, depending on his or her strength and the weight of the ladder. To perform this variation of the one-fire fighter rung raise, follow the steps in (► Skill Drill 12-11).

1. Start with the fire fighter carrying a ladder using the one-fire fighter carry described earlier. Check for overhead hazards. **(Step 1)**
2. The fire fighter places the butt of the ladder on the ground directly against the structure and rotates the ladder so both spurs contact the ground and structure. Then the fire fighter lays the ladder on the ground. If the ladder is an

Fire Fighter Safety Tips

Fire fighters must recognize their individual limits. If you need assistance, do not try to raise the ladder alone. The fireground is no place to prove your strength or daring. If you overexert and injure yourself, your team will be a member short, making it more difficult to perform necessary rescues.

12-10 Skill Drill

One-Fire Fighter Rung Raise for Ladders Under 14'

1 Carry the ladder to the structure. Check for overhead hazards before raising the ladder.

2 By the tip, raise the ladder to hip level. Walk hand-over-hand down the rungs until the ladder is vertical.

3 Heel the ladder and lean it into place.

extension ladder, the base section should be against the ground (fly section up). **(Step 2)**

3. The fire fighter takes hold of a rung near the tip, brings that end of the ladder to chest height and then steps beneath the ladder and pushes upward on the rungs. **(Step 3)**
4. The ladder is raised using a hand-over-hand motion as the fire fighter walks toward the structure until the ladder is vertical and against the structure. **(Step 4)**
5. If an extension ladder is being used, the fire fighter holds the ladder vertical against the structure and extends the fly section by pulling the halyard smoothly, with a hand-over-hand motion, until desired height is reached and the dogs are locked.
6. The butt of the ladder is then pulled out from the structure to create the proper climbing angle. To move the butt away from the structure, the fire fighter grips a lower rung and lifts slightly while pulling outward. At the same time, pressure should be applied to an upper rung to keep the tip of the ladder against the structure. **(Step 5)**
7. If the ladder is an extension ladder, it will be necessary to rotate the ladder so the fly section is out. The halyard should be tied as described in Skill Drill 12-12.

12-11 Skill Drill

One-Fire Fighter Rung Raise for Ladders Over 14'

1 Carry the ladder to the structure. Check for overhead hazards before raising the ladder.

2 Place the butt against the base of the structure and rotate ladder so both spurs contact the ground and structure. Lay the ladder flat on the ground.

3 Grasp a rung near the tip, bring that end of the ladder to chest height, step beneath the ladder, and push upward on the rungs.

4 Walk toward the structure, lifting the rungs hand-over-hand until the ladder is vertical against the structure.

5 Pull the butt away from the structure.

12-12 Skill Drill

Tying the Halyard

Wrap the excess halyard around two rungs and pull it tight over the upper rung.

Tie a clove hitch around the upper rung.

Pull the knot tight and add a safety knot.

Tying the Halyard

The halyard of an extension ladder should always be tied after the ladder has been extended and lowered into place. A tied halyard stays out of the way and provides a safety backup to the dogs for securing the fly section. To tie the halyard, follow the steps in ▲ Skill Drill 12-12.

1. The fire fighter wraps the excess halyard rope around two rungs of the ladder and pulls the rope tight across the upper of the two rungs. **(Step 1)**
2. The fire fighter ties a clove hitch around the upper rung and the vertical section of the halyard. Refer to Chapter 9, Ropes and Knots, to review how to tie a clove hitch. **(Step 2)**
3. Pull the clove hitch tight and place an overhand safety knot as close to the clove hitch as possible to prevent slipping. **(Step 3)**

Two-Fire Fighter Beam Raise

The two-fire fighter beam raise is used with midsized extension ladders up to 35′ long. To perform the two-fire fighter beam raise, follow the steps in ► Skill Drill 12-13.

1. The two-fire fighter beam raise begins with a shoulder or suitcase carry. One fire fighter is near the butt of the ladder and one near the tip. The fire fighters check for overhead hazards. **(Step 1)**
2. The fire fighter at the butt of the ladder places the butt of the lower beam on the ground, while the fire fighter at the tip of the ladder holds the other end. The fire fighter at the butt of the ladder places a foot on the butt of the beam that is in contact with the ground and grasps the upper beam. **(Step 2)**
3. The fire fighter at the tip of the ladder begins to walk toward the butt, while raising the lower beam in a hand-over-hand fashion until the ladder is vertical. **(Step 3)**
4. The two fire fighters pivot the ladder into position as necessary. **(Step 4)**
5. The fire fighters face each other, one on each side of the ladder, and heel the ladder by each placing the toe of one boot against the opposing beams of the ladder.
6. One fire fighter extends the fly section by pulling the halyard smoothly with a hand-over-hand motion until the fly section is the height desired and the dogs are locked. The other fire fighter stabilizes the ladder by holding the outside of the two beams—so that if the fly comes down suddenly it will not strike the fire fighter's hands. **(Step 5)**
7. The fire fighter facing the structure places one foot against one beam of the ladder and then both fire fighters lean the ladder into place. The halyard is tied as described previously. **(Step 6)**

Fire Fighter Safety Tips

Pull the halyard of the extension ladder straight down in line with the ladder to avoid pulling the ladder over.

Two-Fire Fighter Beam Raise

Begin with one fire fighter at the butt of the ladder and one at the tip. Check for overhead hazards.

The fire fighter at the butt lowers the ladder until one beam is on the ground. The fire fighter at the butt of the ladder places a foot on the butt of the beam that is in contact with the ground and grasps the upper beam.

The fire fighter at the tip walks toward the butt of the ladder, raising the lower beam hand-over-hand until it is vertical.

The two fire fighters stand on opposite sides and pivot the ladder into position, as necessary.

Each fire fighter places one foot against the butt of a beam to brace the ladder. The fire fighter with his back to the building extends and locks the fly section.

The fire fighter on the outside heels the ladder while both lean it into place. Secure the halyard before climbing the ladder.

Two-Fire Fighter Rung Raise

The two-fire fighter rung raise is also commonly used with midsized extension ladders up to 35′ long. To perform the two-fire fighter rung raise, follow the steps in ▼ Skill Drill 12-14.

1. The two-fire fighter rung raise begins from a shoulder carry or suitcase carry, with one fire fighter near the butt of the ladder and one near the tip. Check for overhead hazards. The fire fighter at the butt of the ladder places the butt of the lower beam on the ground, while the fire fighter at the tip holds the other end.
2. The fire fighter at the tip rotates the ladder so that both beams are in contact with the ground.
3. The fire fighter at the butt of the ladder stands on the bottom rung, grasps a higher rung with both hands, crouches down and leans backward.
4. The fire fighter at the tip of the ladder swings under the ladder and walks toward the butt, advancing down the ladder and lifting the rungs in a hand-over-hand fashion until the ladder is vertical. The two fire fighters stand on opposite sides of the ladder and pivot it into position as necessary. **(Step 1)**
5. The fire fighters face each other, one on each side of the ladder, and heel the ladder by each placing the toe of one boot against the opposite beams of the ladder. If using an extension ladder, one fire fighter extends the

12-14 **Skill Drill**

Two-Fire Fighter Rung Raise

1 Begin with one fire fighter near the butt of the ladder and one at the tip. Check for overhead hazards. The fire fighter at the butt places the butt of the lower beam on the ground. The fire fighter at the tip rotates the ladder until both beams are in contact with the ground. The fire fighter at the butt places both feet on the bottom rung, grasps a higher rung, crouches, and leans backward. The fire fighter at the tip swings under the ladder and walks toward the butt, raising the rungs hand-over-hand until the ladder is vertical. On opposite sides, the fire fighters pivot the ladder into position, if necessary.

2 Each fire fighter places one foot against the butt of a beam to brace the ladder. The fire fighter with his back to the structure extends and locks the fly section.

3 The fire fighter on the outside heels the ladder while both lean it into place. Secure the halyard before climbing the ladder.

Fire Fighter Safety Tips

When raising the fly section of an extension ladder, do not wrap the halyard around your hand. If the ladder falls or the fly section unexpectedly retracts, your hand could be caught in the halyard, and you could be seriously injured. Also, do not place your foot under the fly sections. If the rope slips, your foot will be crushed.

fly section by pulling the halyard smoothly with a hand-over-hand motion until the tip is at the desired height and the dogs are locked. The other fire fighter stabilizes the ladder by holding the outside of the base section's beams—so that if the fly comes down suddenly it will not strike the fire fighter's hands. **(Step 2)** The fire fighter facing the structure places one foot against one beam of the ladder and then both fire fighters lean the ladder into place. The halyard is tied as described in Skill Drill 12-12. **(Step 3)**

Three- and Four-Fire Fighter Rung Raises

The three- and four-fire fighter rung raises are used for very heavy ladders. The basic steps in these raises are similar to the two-fire fighter rung raise. In a three-fire fighter rung raise, the third fire fighter assists with the hand-over-hand raising of the ladder.

When four fire fighters are raising the ladder, two anchor the butt of the ladder while the other two raise the ladder. The two fire fighters at the butt each place their inside foot on the bottom rung of the ladder and bend over, grasping a rung in front of them. To perform the three-fire fighter rung raise, follow the steps in **► Skill Drill 12-15**.

1. The three-fire fighter rung raise begins from a shoulder carry or suitcase carry, with one fire fighter near the butt of the ladder, one in the middle, and one near the tip. **(Step 1)**
2. The fire fighters check for overhead hazards. **(Step 2)**
3. The fire fighter at the butt of the ladder places the butt of the lower beam on the ground, while the fire fighter at the tip holds the other end. The fire fighter in the middle moves to the tip. The fire fighters at the tip rotate the ladder so that both butts are in contact with the ground. **(Step 3)**
4. The fire fighter at the butt of the ladder stands on the bottom rung, grasps a higher rung with both hands, crouches down and leans backward. **(Step 4)**
5. The fire fighters at the tip of the ladder begin to walk toward the butt, advancing down the ladder and lifting the rungs in a hand-over-hand fashion until the ladder is vertical. **(Step 5)**
6. The three fire fighters pivot the ladder into position as necessary. **(Step 6)**
7. Two fire fighters face each other, one on each side of the ladder, heel the ladder by each placing the toe of one boot against the opposite beams of the ladder, and grasp the outsides of the beams. **(Step 7)**
8. One fire fighter extends the fly section by pulling the halyard smoothly with a hand-over-hand motion until the tip is at the desired height and the dogs are locked. **(Step 8)**
9. One fire fighter heels the ladder while the other two lean the ladder into place.
10. The halyard is tied as described previously. **(Step 9)**

To perform the four-fire fighter rung raise, follow the steps in **Skill Drill 12-16**.

1. The four-fire fighter rung raise begins with a four-fire fighter flat carry. Two fire fighters are at the butt of the ladder and two fire fighters are at the tip. Check for overhead hazards.
2. The fire fighters at the butt of the ladder place both butts on the ground while the fire fighters at the tip of the ladder hold the other end.
3. The fire fighters at the butt of the ladder stand side-by-side, facing the ladder. Each fire fighter places the inside foot on the bottom rung and the other foot on the ground outside the beam. Both then crouch down, grab a rung and the beam, and lean backward.
4. The two fire fighters at the tip of the ladder begin to walk toward the butt of the ladder advancing down the rungs in a hand-over-hand fashion until the ladder is vertical.
5. The fire fighters pivot the ladder into position as necessary.
6. Two fire fighters heel the ladder by placing a boot against each beam. Each fire fighter places the toe of one boot against one of the beams. The third fire fighter stabilizes the ladder by holding it on the outside of the rails.
7. The fourth fire fighter extends the fly section by pulling the halyard smoothly with a hand-over-hand motion until the tip reaches the desired height and the dogs are locked.
8. The two fire fighters facing the structure each place one foot against one beam of the ladder while the other two fire fighters lower the ladder into place.
9. The halyard is tied as described previously.

Fly Section Orientation

When raising extension ladders, fire fighters must know whether the fly section should be placed toward the building (fly in) or away from the building (fly out). The ladder manufacturer and department SOPs will specify whether the fly should be facing in or out. In general, manufacturers of fiberglass and metal ladders recommend that the fly sections be placed away from the structure. Wood fire-service ladders

12-15 Skill Drill

Three-Fire Fighter Rung Raise

1. Begin with one fire fighter at the butt of the ladder, one in the middle, and one at the tip.

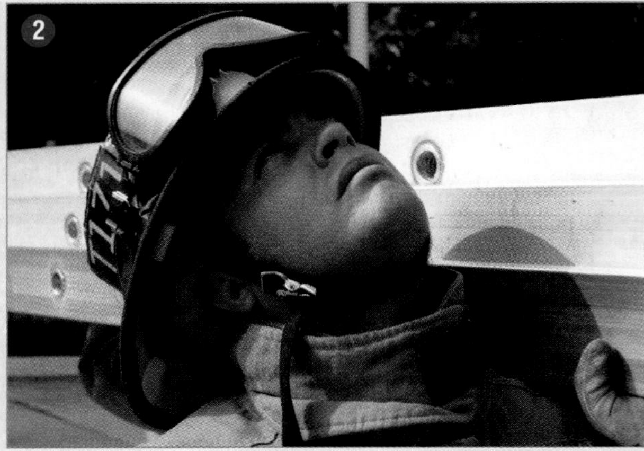

2. Check for overhead hazards.

3. The fire fighter at the butt lowers the ladder until the butt of one beam is on the ground. The fire fighter in the middle moves to the tip. The fire fighters at the tip rotate the ladder until both butts are on the ground.

4. The fire fighter at the butt of the ladder places both feet on the bottom rung, grasps a higher rung, crouches, and leans backward.

5. The fire fighters at the tip walk toward the butt of the ladder, raising the rungs hand-over-hand until the ladder is vertical.

Skill Drill

The three fire fighters pivot the ladder into position, as necessary.

Two fire fighters stabilize the ladder, each placing one foot against the butt of a beam and grasping the outsides of the beams with both hands.

One fire fighter extends and locks the fly section.

The fire fighter on the outside heels the ladder while two fire fighters lean it into place. The halyard is secured before climbing the ladder.

are often designed to be used with the fly in.

Securing the Ladder

There are several ways to prevent a ladder from moving once it is in place. One option is to have a fire fighter stand between the ladder and the structure, grasp the beams, and lean back to pull the ladder into the structure (► **Figure 12-30**). This is called heeling the ladder. A fire fighter(s) on the outside of the ladder, facing the structure, can also heel the ladder by placing a boot against the beam(s) of the ladder (► **Figure 12-31**).

A rope, a rope-hose tool, or webbing also can be used to secure a ladder in place. The lower part of the ladder can be tied to any solid object to keep the base from kicking out. The base should always be secured if the ladder is used at a low angle. The tip of the ladder can be tied to a secure object near the top to keep it from pulling away from the building. The best method is to secure both the tip and the base.

Climbing the Ladder

Climbing a ladder on the fire or emergency scene should be done in a deliberate and controlled manner. Always make sure the ladder is secure (tied or heeled). Before climbing an extension ladder, verify that the dogs are locked and the halyard is secured.

The proper climbing angle should be checked as well. Some ladder beams have level guide stickers placed there by the manufacturer. The bottom line of the indicator will be parallel to the ground when the ladder is positioned properly. Standing at the base of the ladder and extending the arms straight out is another way to check the climbing angle (► **Figure 12-32**). The hands should comfortably reach the beams or rungs if the angle is appropriate. Finally, the angle can be measured by ensuring that the butt of the ladder is one-fourth of the working height out from the base of the structure.

Keep bounce and shifting to a minimum while climbing. Eyes should be focused forward with only occasional glances upward. This will prevent debris such as falling glass from injuring the face or eyes. Lower the protective face shield for additional protection.

Use a hand-over-hand motion on the rungs of the ladder, or slide both hands along the underside of the beams while climbing. Sliding the hands along the beams is more secure because it maintains three contact points with the ladder at all times (both hands and one foot).

If tools must be moved up or down, it is better to hoist them by ropes than to carry them up a ladder. Carrying tools on a ladder reduces the fire fighter's grip and increases the potential for injury if a tool slips or falls. To climb a ladder while holding a hand tool, the fire fighter should hold the tool against one beam with one hand and maintain contact with the opposite beam with the other hand. Tools such as pike poles can be hooked onto the ladder and moved up every few rungs. To climb a ladder while carrying a tool, follow the steps in (► **Skill Drill 12-17**).

Figure 12-30 The fire fighter's weight pulls the ladder against the structure to secure the ladder.

1. The fire fighter prepares to climb the ladder by placing the tool in one hand and holding it against the beam of the ladder. **(Step 1)**
2. The fire fighter wraps the other hand around the opposite beam and begins climbing. Contact is maintained between the free hand and the beam by sliding the tool along the beam while climbing. **(Step 2)**

Be sure not to overload the ladder. There should be no more than two fire fighters on a ladder at one time. A properly placed ladder should be able to support two rescuers (with their protective clothing and equipment) and one victim.

Dismounting the Ladder

When a ladder is used to reach a roof or a window entry, the fire fighter will have to dismount the ladder. Fire fighters can minimize the risk of slipping and falling by making sure that the surface is stable before dismounting. Test the stability with a tool before dismounting. For example, fire fighters who dismount from a ladder onto a roof typically sound the roof with an axe before stepping onto the surface (► **Figure 12-33**).

Figure 12-31 A fire fighter can push against the ladder to heel it.

Figure 12-32 The climbing angle can be checked by standing on the bottom rung and holding the arms straight out.

Skill Drill

Climbing the Ladder while Carrying a Tool

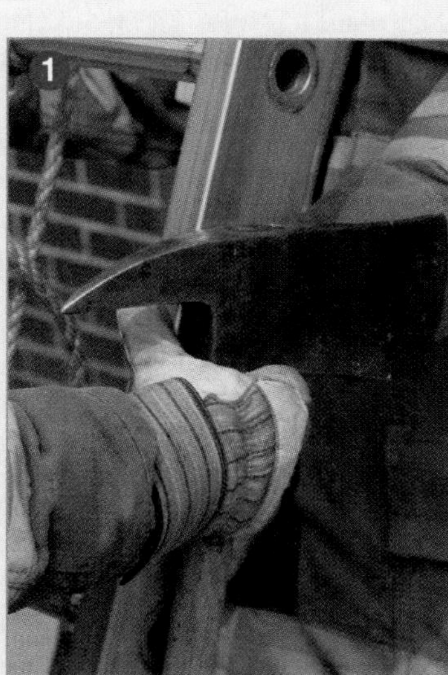

1 Hold the tool in one hand and place it against the beam.

2 Slide the tool up the beam, while sliding the opposite hand up the other beam.

Figure 12-33 Check the stability of the roof before dismounting from the ladder.

Figure 12-34 A ladder belt has a large hook that is designed to secure a fire fighter to a ladder.

Try to maintain contact with the ladder at three points when dismounting. For example, a fire fighter who steps onto a roof should keep two hands and one foot on the ladder while checking the footing. This is particularly important if the surface slopes or is covered with rain, snow, or ice. Do not shift your weight onto the roof until you have tested the footing.

It is also important to check the footing when entering a window from a ladder positioned to the side of the window. Before stepping into the window, be sure that the interior surface is structurally sound and offers secure footing. The three points of contact rule also applies in this situation.

Under heavy fire and smoke conditions, fire fighters sometimes dismount by climbing over the ladder tip and sliding over the windowsill into the building. Sound the floor inside the window before entering to be sure it is solid and stable. To remount a ladder under these conditions, back out the window feet first and rest your abdomen on the sill until you can feel the ladder under your feet. Under better conditions, sit on the windowsill with legs out and roll onto the ladder.

Working from a Ladder

Fire fighters must often work from a ladder. To avoid falling while working from a ladder, the fire fighter must be secured to the ladder. Two different methods are used by fire fighters to secure themselves to a ladder.

The first method is a **ladder belt**. A ladder belt is specifically designed to secure a fire fighter to a ladder or elevated surface (▲ Figure 12-34). Fire fighters must be sure to use only ladder belts designed and certified to NFPA 1983, *Standard on Fire Service Life Safety Rope and System Components*. Utility belts designed to carry tools should never be used as ladder belts. The alternate method is to apply the leg lock. The leg lock is simple, secure, and requires no special equipment. To work from a ladder, follow the steps of (► Skill Drill 12-18).

1. The fire fighter climbs to the desired work height and steps up to the next higher rung. **(Step 1)**
2. The fire fighter notes the side of the ladder where the work will be performed. The leg on the opposite side is extended between the rungs. **(Step 2)**
3. Once the leg is between the rungs, the fire fighter bends the knee and brings that foot back under the rung and through to the climbing side of the ladder. **(Step 3)**
4. The foot is secured against the next lower rung or the beam of the ladder. The fire fighter uses the thigh for support and steps down one rung with the opposite foot. **(Step 4)**

Working from a Ladder

The fire fighter climbs the ladder to the desired work height and then one rung higher.

The fire fighter notes the side where the work will be performed. The opposite leg is extended between the rungs.

The knee is bent around the rung and the foot is brought back under the rung.

The foot is secured around the lower rung or the beam. The fire fighter moves the other leg down one rung.

The fire fighter is now free to work with both hands.

Fire Fighter Safety Tips

Working with ladders requires attention to basic safety precautions. Fire fighters should always check these important safety items:

- Choose the proper ladder for the job.
- Wear protective gear, including gloves, when working with ladders.
- Use the proper number of fire fighters for each raise.
- Use leg muscles, not back or arm muscles, when lifting ladders.
- Never raise ladders into or near electrical wires.
- Check the ladder for proper angle.
- Check the ladder locks to be sure they are seated over the rungs.
- Check staypoles to be sure they are set properly.
- Make sure the ladder is secure at the top or the bottom (preferably both) before climbing.
- Climb smoothly and rhythmically.
- Always tie in with a leg lock or a ladder belt when working from a ladder.
- Do not overload the ladder.
- Inspect ladders for damage and wear after each use.

5. The fire fighter is now free to lean out to the side of the ladder and work with two hands on the tool. **(Step 5)**

Mastering the leg lock and using a ladder belt enable fire fighters to accomplish several advanced tasks while on a portable ladder. Among these tasks are placing a roof ladder for ventilation or using a hose stream to apply water to a hard-to-reach location.

Placing a Roof Ladder

Several methods can be used to properly position a roof ladder on a sloping roof. The most common method is described in **Skill Drill 12-19**.

1. The fire fighter carries the roof ladder to the base of a ladder that is already in place to provide access to the roofline.
2. The fire fighter places the roof ladder on the ground and rotates the hooks of the roof ladder to the open position.
3. The fire fighter uses a one-fire fighter beam or rung raise to lean the roof ladder against one beam of the other ladder with the hooks oriented outward.
4. The fire fighter climbs the lower ladder until reaching the mid point of the roof ladder. The fire fighter then

slips one shoulder between two rungs of the roof ladder and shoulders the ladder.

5. The fire fighter climbs to the roofline of the structure carrying the roof ladder on one shoulder.
6. The fire fighter then uses a ladder belt to secure to the ladder or applies a ladder leg lock (described in Skill Drill 12-18).
7. The fire fighter places the roof ladder on the roof surface with hooks down. The ladder is pushed up toward the peak of the roof with a hand over hand motion.
8. Once the hooks have passed the peak, the fire fighter pulls back on the roof ladder to set the hooks and checks to ensure they are secure.
9. A roof ladder is removed from the roof by reversing the process described above. After releasing the hooks from the peak, it may be necessary to turn the ladder on one of its beams so that it can slide down the roof without catching the hooks on the roofing material.

Wrap-Up

Ready for Review

This chapter covered the basic knowledge and skills necessary to use portable ladders safely. It also provided guidance on selecting, inspecting, and maintaining portable ladders. Fire fighters must be able to identify the proper situations for using a ladder, which ladder to use in each situation, and how to properly place a ladder so that it provides both access to and egress from a building. Fire fighters also work from ladders while making ventilation openings or discharging water into a building.

The chapter began by identifying the functions ladders serve on the emergency scene. A ladder may be the only feasible method of rescuing a trapped occupant. A ladder can provide safe entry to a building when fire blocks the primary entry. A ladder also can provide fire fighters with access to the roof of a building to make ventilation openings, or to other structures, such as chimneys.

The basic terminology used to describe the construction and parts of a ladder was presented. The fire fighter must understand the different components of a ladder and how they work together.

This chapter also covered the types of portable ladders and the selection of the appropriate ladder for a task. Different ladders are used on the fireground, depending on the situation. An aerial ladder is effective for gaining access to a roof. A folding ladder is needed to get into a small opening such as an attic scuttle.

The chapter also outlined the steps to be followed in carrying, placing, and raising a portable ladder. Properly placing the ladder is critical to safe ladder operations on the fireground. Climbing ladders safely and working from ladders were also covered.

Chief Concepts

- The primary function of portable ladders is to provide safe access to and egress from otherwise inaccessible areas.
- Portable ladders can be used for several auxiliary purposes including channeling debris, serving as a lift point, and holding other firefighting equipment.
- To inspect and maintain portable ladders, the fire fighter must become familiar with ladder construction and the terminology used to describe ladders.
- Portable ladders must be regularly inspected, cleaned, and maintained.
- To select the appropriate portable ladder for a job, the fire fighter must know what types of portable ladders are available as well as their uses and limitations.
- Communication is key to coordinating efforts when working with ladders.
- To deploy a portable ladder, fire fighters must be able to carry, place, raise, and climb the ladder safely, using common techniques.
- Fire fighters must be able to work safely from ladders.

Hot Terms

Aerial ladder A power-operated ladder permanently mounted on a piece of apparatus.

Bangor ladder A ladder equipped with tormentor poles or staypoles that stabilize the ladder during raising and lowering operations.

Beam One of the two main structural pieces running the entire length of each ladder or ladder section. The beams support the rungs.

Bed section The lowest and widest section of an extension ladder. The fly sections of the ladder extend from the bed section.

Butt Often called the heel or base, the butt is the end of the ladder that is placed against the ground when the ladder is raised.

Butt plate (footpad) An alternative to a simple butt spur; a swiveling plate with both a spur and a cleat or pad that is attached to the butt of the ladder.

Butt spurs The metal spikes attached to the butt of a ladder. The spurs help prevent the butt from slipping out of position.

Combination ladder A ladder that converts from a straight ladder to a step ladder configuration (A-frame) or from an extension ladder to a step ladder configuration.

Dogs (also referred to as pawls, ladder locks, and rung locks) A mechanical locking device used to secure the fly section(s) of a ladder after they have been extended.

Egress A method of exiting from an area or a building.

Extension ladder An adjustable-length, multiple-section ladder.

Fly section A section of an extension ladder that is raised or extended from the base section or from another fly section. Some extension ladders have more than one fly section.

Folding ladder A ladder that collapses by bringing the two beams together for portability. Unfolded, the folding ladder is narrow and used for access to attic scuttle holes and confined areas.

Fresno ladder A narrow, two-section extension ladder that has no halyard. Because of its limited length, it can be extended manually.

Grade The level at which the ground intersects the foundation of a structure.

Guides Strips of metal or wood that serve to guide a fly section during extension. Channels or slots in the bed or fly section may also serve as guides.

Halyard The rope or cable used to extend or hoist the fly section(s) of an extension ladder.

Heat sensor label A piece of heat-sensitive material on each section of a ladder that identifies when the ladder has been exposed to high heat conditions.

I-beam A ladder beam constructed of one continuous piece of I-shaped metal or fiberglass to which the rungs are attached.

Ladder belt A belt specifically designed to secure a fire fighter to a ladder or elevated surface.

Ladder gin An A-shaped structure formed with two ladder sections. It can be used as a makeshift lift when raising a trapped person. One form of the device is called an A-frame hoist.

Pompier ladder (scaling ladder) A lightweight, single beam ladder.

Portable ladder Ladder carried on fire apparatus, but designed to be removed from the apparatus and deployed by fire fighters where needed.

Protection plates Reinforcing material placed on a ladder at chaffing and contact points to prevent damage from friction and contact with other surfaces.

Pulley A small, grooved wheel through which the halyard runs. The pulley is used to change the direction of the halyard pull, so that a downward pull on the halyard creates an upward force on the fly section(s).

Rail The top or bottom piece of a trussed-beam assembly used in the construction of a trussed ladder. The term rail is also sometimes used to describe the top and bottom surfaces of an I-beam ladder. Each beam will have two rails.

Roof hooks The spring-loaded, retractable, curved metal pieces that allow the tip of a roof ladder to be secured to the peak of a pitched roof. The hooks fold outward from each beam at the top of a roof ladder.

Roof ladder (hook ladder) A straight ladder equipped with retractable hooks so that the ladder can be secured to the peak of a pitched roof. Once secured, the ladder lies flat against the surface of the roof, providing secure footing for fire fighters.

Rung A ladder crosspiece that provides a climbing step for the user. The rung transfers the weight of the user out to the beams of the ladder or back to a center beam on an I-beam ladder.

Solid beam A ladder beam constructed of a solid rectangular piece of material, typically wood, to which the ladder rungs are attached.

Staypole (also known as a tormentor) A long piece of metal attached to the top of the bed section of an extension ladder and used to help stabilize the ladder during raising and lowering. The pole attaches to a swivel point and has a spur on the other end. One pole is attached to each beam of long (40′ or longer) extension ladders.

Stop A piece of material that prevents the fly section(s) of a ladder from overextending and collapsing the ladder.

Tie rod A metal rod that runs from one beam of the ladder to the other to keep the beams from separating. Tie rods are typically found in wood ladders.

Tip The very top of the ladder.

Truss block A piece of wood or metal that ties the two rails of a trussed beam ladder together and serves as the attachment point for the rungs.

Trussed beam A ladder beam constructed of top and bottom rails joined by truss blocks that tie the rails together and support the rungs.

Fire Fighter in Action

It is 1:00 a.m. on a rainy November morning when your engine company is dispatched on a second alarm for a commercial structure fire in a 2-story warehouse. Your engine company arrives 15 minutes after the initial operations had begun. The suppression crews are reporting that smoke is to the floor with no visibility, and they are having difficulty locating the fire. A ventilation crew is preparing to go to the roof and begin vertical ventilation. Your engine is assigned to the roof division. The roof division supervisor tells your lieutenant to ladder side B for a second means of egress for the ventilation team. Lieutenant Johnson tells you and the other crew members to remove the ladder from the engine and ladder the building. Your engine carries the following ground ladders: one 12' attic, one 16' roof, and one 35' extension ladder.

1. Which ladder would be the most appropriate ladder for a second means of egress?
   - **A.** Roof ladder
   - **B.** Extension ladder
   - **C.** Aluminum ladder
   - **D.** Straight ladder

2. Which of the following is not a safety rule for raising a ladder?
   - **A.** Utilize the least amount of personnel needed to save resources.
   - **B.** Wear full bunker gear.
   - **C.** Raise the ladder at least 10' from overhead utility lines.
   - **D.** Place the ladder on a relatively level surface.

You arrive back at the station. Lieutenant Johnson tells you to clean and inspect the ladder prior to placing it back into service.

3. Ladders should be cleaned using a soft bristled brush and a mild detergent:
   - **A.** monthly and after each use.
   - **B.** before inspection, regularly, and after each use.
   - **C.** biannually.
   - **D.** prior to use and monthly.

4. Once the ladder has been cleaned, it needs to be inspected. Which of the following is not a recommended inspection procedure?
   - **A.** Check the heat labels to see if the sensors indicate that the ladder has been exposed to excessive heat.
   - **B.** Check the beams for cracks, splintering, breaks, gouges, or deformations
   - **C.** Check the rungs for snugness, punctures, breaks, gouges, or deformations
   - **D.** Check the paint job for chips, scrapes, and scratches. Sand down and repaint all surfaces to keep the ladder looking good.

www.FireFighter.jbpub.com

www.FireFighter.jbpub.com
Chapter Pretests
Interactivities
Hot Term Explorer
Web Links
Review Manual
FireLearn

Search and Rescue

Technology Resources

www.FireFighter.jbpub.com

- Chapter Pretests
- Interactivities
- Hot Term Explorer
- Web Links
- Review Manual
- FireLearn

Chapter Features

- Skill Drills
- Voices of Experience
- Fire Marks
- Teamwork Tips
- Fire Fighter Safety Tips
- Fire Fighter Tips
- Canadian Perspectives
- Hot Terms
- Wrap-Up

Chapter 13

NFPA 1001 Standard

Fire Fighter I

5.3.9 Conduct a search and rescue in a structure operating as a member of a team, given an assignment, obscured vision conditions, personal protective equipment, a flashlight, forcible entry tools, hose lines, and ladders when necessary, so that ladders are correctly placed when used, all assigned areas are searched, all victims are located and removed, team integrity is maintained, and team members' safety—including respiratory protection—is not compromised.

5.3.9 (A) *Requisite Knowledge.* Use of forcible entry tools during rescue operations, ladder operations for rescue, psychological effects of operating in obscured conditions and ways to manage them, methods to determine if an area is tenable, primary and secondary search techniques, team members' roles and goals, methods to use and indicators of finding victims, victim removal methods (including various carries), and considerations related to respiratory protection.

5.3.9 (B) *Requisite Skills.* The ability to use SCBA to exit through restricted passages, set up and use different types of ladders for various types of rescue operations, rescue a fire fighter with functioning respiratory protection, rescue a fire fighter whose respiratory protection is not functioning, rescue a person who has no respiratory protection, and assess areas to determine tenability.

Fire Fighter II

NFPA 1001 contains no Fire Fighter II Job Performance Requirements for this chapter.

Knowledge Objectives

After studying this chapter, you will be able to:

- Define search and rescue.
- Describe the importance of scene size-up in search and rescue.
- Describe search techniques.
- Describe the primary search.
- Describe search patterns.
- Describe the secondary search.
- Describe how to ensure fire fighter safety during a search.
- Describe ladder rescue techniques.

Skills Objectives

After completing this chapter, you will be able to:

- Demonstrate the one-person walking assist.
- Demonstrate the two-person walking assist.
- Demonstrate the two-person extremity carry.
- Demonstrate the two-person seat carry.
- Demonstrate the two-person chair carry.
- Demonstrate the cradle-in-arms carry.
- Demonstrate the clothes drag.
- Demonstrate the blanket drag.
- Demonstrate the webbing sling drag.
- Demonstrate the fire fighter drag.
- Demonstrate the one-person emergency drag from a vehicle.
- Demonstrate the long backboard rescue.
- Demonstrate rescuing a conscious person from a window.
- Demonstrate rescuing an unconscious person from a window.
- Demonstrate rescuing an unconscious child or small adult from a window.
- Demonstrate rescuing a large adult from a window.

You Are the Fire Fighter

A discussion about a recent rescue situation is in progress when you arrive at the station for your shift. The incident was a house fire. When the fire department arrived at the scene, the homeowner reported that her seven-year-old daughter, who slept in the back bedroom, was missing. The engine company was assigned search and rescue. The search team could not find the young girl until they searched the closet in her room.

1. ***What other places might be easily overlooked when searching a house?***
2. ***What other groups of people are most likely to require rescue from a fire?***

Search and Rescue

The mission of the fire department is to save lives and protect property. Saving lives is the highest priority at a fire scene. That is why search and rescue are so important at any incident. The first fire fighters to arrive must always consider the possibility that lives could be in danger and act accordingly.

Saving lives remains the highest priority until it is determined that everyone who was in danger has been found and moved to a safe location, or until it is no longer possible to rescue anyone successfully. The first situation occurs when a thorough search is done and there is no one remaining to be rescued. The second situation might occur if fire conditions or other factors make it likely that no one could still be alive to be rescued.

Search and rescue are almost always performed in tandem. A **search** is done to look for victims who need assistance to leave a dangerous area. **Rescue** is the physical removal of a person from confinement or danger, such as when a fire fighter leads an occupant to an exit or carries an unconscious victim out of a burning building and down a ladder. Both are examples of rescues because the victim is physically removed from imminent danger through the actions of a rescuer.

Although any fire department company or unit could be assigned to search-and-rescue operations, many fire departments routinely assign this responsibility to ladder (or truck) companies and rescue companies. All fire fighters must be trained and prepared to perform search-and-rescue functions.

Search-and-rescue operations must be conducted quickly and efficiently. A systematic approach will ensure that everyone who can possibly be saved is successfully located and removed from danger. Fire fighters should practice the specific search and rescue procedures used by their departments to sharpen their skills and increase their efficiency.

Often, the first-arriving fire fighters will not know if anyone is inside a burning building or how many people could be inside. Fire fighters should never assume that a building is unoccupied. The only way to be sure that everyone has safely evacuated a building is to conduct a thorough search. Every building where a fire occurs should be searched for potential victims. There are times, however, when a search cannot be conducted immediately. Depending on the circumstances, the search could be delayed or limited to a specific area.

Coordinating Search and Rescue with Fire Suppression

Although search and rescue are always the first priorities at a fire, they are never the only actions taken by first-arriving fire fighters. Fire fighters must plan and coordinate all other activities to support the search-and-rescue priority. When search-and-rescue operations are finished, fire fighters can focus totally on fire suppression.

Often, fire fighters must take action to confine or control the fire before search-and-rescue operations can begin. It might be necessary to position hose lines to keep the fire away from potential victims or to protect the entry and exit paths, so that the victims can be found and safely removed (► **Figure 13-1**). In some cases, the best way to save lives is to control the fire and eliminate the danger quickly.

Other fire scene activities also must be coordinated with search and rescue. Forcible entry might be needed to provide entrances and exits for search-and-rescue teams. Well-placed ventilation can reduce interior temperatures and improve visibility, enabling search teams to locate victims more rapidly. Portable lighting can provide valuable assistance to interior search crews.

Often, the search for potential victims also provides valuable information about the location and extent of the fire within a building. The searchers act as a reconnaissance team to deter-

Figure 13-1 To support search-and-rescue operations, fire fighters may first have to position hose lines to protect the means of egress.

Figure 13-2 The IC must weigh the risk to fire fighters against the possibility of saving anyone inside before authorizing an interior search.

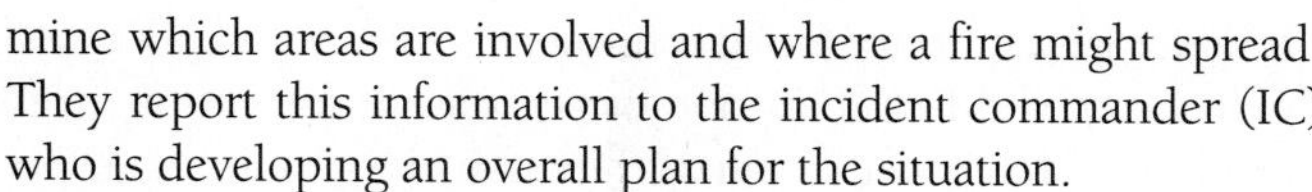

mine which areas are involved and where a fire might spread. They report this information to the incident commander (IC) who is developing an overall plan for the situation.

Search-and-Rescue Size-Up

The size-up process at every fire should include a specific evaluation of the critical factors for search and rescue. These include the number of occupants in the building, their location, the degree of risk to their lives, and their ability to evacuate by themselves. Usually, this information is not immediately available, so actions have to be based on a combination of observations and expectations. The type of occupancy; the size, construction, and condition of the building; and the apparent smoke and fire conditions, as well as the time of day and the day of the week, are important observations that could indicate whether and how many people may need to be rescued.

Fire fighters can then develop a search-and-rescue plan based on their conclusions. This plan identifies the areas to be searched, the priorities for searching different areas, the number of search teams required, and any additional actions needed to support the search-and-rescue activities. One search-and-rescue team might be sufficient for a small building, but multiple teams might be necessary to search a large building.

Risk-Benefit Analysis

Plans for search and rescue must take into consideration the risks and benefits of the operation. In some situations, search-and-rescue operations must be limited or cannot be performed. For example, if a building is exhibiting potential backdraft conditions, the risk to fire fighters might be too high and the possibility of saving anyone inside might be too low to justify sending a search-and-rescue team into the building (► Figure 13-2). A similar decision might be made if the fire is in an abandoned building or a building in danger of structural collapse. In other situations, it might not be possible to conduct an interior search until the fire has been extinguished.

Occupancy Factors

When a building is known or believed to be occupied, fire fighters should first rescue the occupants who are in the most immediate danger, followed by those who are in less danger. Search teams should be assigned on the basis of these priorities.

Several factors determine the level of risk faced by the occupants of a burning building. These include the location of the fire within the building, the direction of spread, the volume and intensity of the fire, and smoke conditions in different areas.

In establishing search priorities, fire fighters should also consider where occupants are most likely to be located. Occupants who are close to the fire, above the fire, or in the path of spread are usually at greater risk than those who are farther away from the fire. A person in the apartment where the fire started is probably in immediate danger, while the occupants of a lower floor are relatively safe.

Building occupants who are at the windows or on balconies calling for help obviously realize that they are in danger and want to be rescued. But there may be other occupants in the building who cannot be seen from the exterior. These people could be asleep, unconscious, incapacitated, or trapped. A search has to be conducted to locate them before they can be rescued.

Occupants who are asleep are at greater risk than those who are awake. Young children and elderly people, who probably need assistance to evacuate, are at greater risk than adolescents and adults. Patients confined to a bed or wheelchair are at greater risk than people who can move freely.

Fire Fighter Safety Tips

Search-and-Recue Size-Up Considerations

- Occupancy
- Size of building
- Construction of building
- Time of day and day of week
- Number of occupants
- Degree of risk to the occupants presented by the fire
- Ability of occupants to exit on their own

More fire fighters will be needed for search and rescue for a fire in a nursing home than for one in an office building, where the occupants can usually make their own way out.

A fire in a single-family dwelling generally places the members of one family at risk, while a fire in a large apartment building endangers several families. Generally, the life risk in residential occupancies is higher at night and on weekends, when more people are at home. Risk is greatest late at night, when the occupants are probably asleep. A fire that occurs in a residential occupancy on a weekday afternoon would present a significantly different rescue problem from a fire in the same location at two o'clock in the morning.

An office building that is fully occupied on a weekday afternoon is generally unoccupied at midnight. A warehouse fire probably presents a direct risk to few occupants. But a nightclub fire on a Friday or Saturday night could endanger hundreds of lives. The risk factor increases depending on the building construction; occupants of an unprotected, wood-frame building are in more danger than those in a building with fire-resistive construction and automatic sprinklers.

Observations

Fire fighters should never assume that a building is completely unoccupied. An observant fire fighter will notice clues that indicate whether or not a building is occupied and how many people are likely to be present. The initial size-up at an incident can provide valuable information on the probability of finding occupants, the number of occupants, and the most likely location of any occupants.

For example, the first-responding unit at a house fire might see two cars in the driveway and numerous toys on the front lawn. These signs indicate that the house is probably occupied by a family with children. In such a situation, the IC would probably assign several search-and-rescue teams and emphasize the importance of quickly searching every room where victims might be located. An absence of cars, locked doors, and an overflowing mailbox suggest that the house is vacant. The IC would still order a search, just in case someone is inside, but the assignment would go to a single search-and-rescue team (► Figures 13-3 A & B).

Similar observations can be made for other structures. Although an empty parking lot does not guarantee that a building is empty, the presence of a car probably indicates that at least one maintenance worker or security guard is inside. Boarded windows and doors on a building surrounded by a chain link fence are indications that the building is probably unoccupied.

Occupant Information

Occupants who have already escaped and neighbors who know the occupants often can provide valuable information about how many people might still be inside and where they might be. Although this information can be valuable, it is not always accurate. Parents and children standing outside a burning house is a good sign, but does not necessarily mean that everyone is safe. Fire fighters must make sure that everyone is accounted for. An important question to ask is, "Are you sure that no one else was home?"

It is often difficult to obtain accurate information from people who have just escaped from a burning building, especially if they believe that family members, close friends, or even a beloved pet might still be inside. People may be too emotional to speak or even to think clearly. If someone says that they think occupants are still inside, fire fighters should obtain as much information as possible. Ask specific questions, such as "Where is your son's room located?" and "Where was he when you last saw him?" It is always easier to find someone if you know whom you are looking for and where to look. Many fire fighters can tell stories about searching for a missing baby, only to discover that "Baby" is the family dog.

Building Size and Arrangement

The size and arrangement of the building are other important factors to consider in planning and conducting a search. A large building, with many rooms, must be searched on a systematic basis. Access to an interior layout or floor plan is often helpful when planning and assigning teams to search a building. To ensure that each area is searched promptly and that no areas are omitted, there must be a systematic way of assigning specific areas to different teams. Teams must report when they have completed searching each area. Assignments are often based on the stairway locations, corridor arrangements, and the apartment or room numbering system. Because it is difficult to determine this information in the midst of an emergency, fire fighters often conduct preincident surveys of buildings such as apartment complexes and offices. Fire fighters assemble information about the building during the survey and use it to prepare preincident plans, so they are ready when an emergency occurs.

Figures 13-3 A & B Exterior observations can often provide a good indication if a building is occupied. **A.** Parked cars indicate the residence is occupied. **B.** Boarded windows indicate the residence is unoccupied.

Preincident plans can include valuable information such as:

- Corridor layouts
- Exit locations
- Stairway locations
- Apartment layouts
- Number of bedrooms in apartments
- Locations of handicap apartments
- Special function rooms or areas

Fire fighters should note how the floors of a building are numbered. Buildings constructed on a slope may appear to have a different number of levels when viewed from different sides. For example, a fire may appear to be on the third floor to a fire fighter standing in front of the building, while a fire fighter at the rear of the building might report that it is on the fifth floor. Without knowing how floors are numbered, fire fighters can be sent to work above a fire instead of below it.

Search Coordination

The overall plan for the incident must focus on the life-safety priority as long as search-and-rescue operations are still underway. As soon as all searches are complete, the priority can shift to controlling and extinguishing the fire.

The IC makes search assignments and serves as the search coordinator. As the teams complete their search of each area, they must notify the IC of the results. An "all clear" report indicates that an area has been searched and all victims have been removed.

If fire fighters who have been assigned to perform some other task discover a victim or come upon a critical rescue situation, they must notify the IC immediately. Because life safety is always the highest priority, the IC will adjust the overall plan to support the rescue situation. For example, another company might be assigned to perform the first team's task and to assist with the rescue.

Another aspect of search coordination is keeping track of everyone who was rescued or escaped without assistance. This information should be tracked at the command post so that reports of missing persons can be matched to reports of rescued persons. Fire fighters should also conduct an exterior search for any missing people. A person who escaped from the fire may be lying unconscious on the ground nearby or in the care of neighbors. Someone who jumped from a window could be injured and unable to move.

Search Priorities

A search begins with the areas where victims are at the greatest risk. One or two search teams can usually go through all of the rooms in a single-family dwelling in less than five minutes. Multiple teams and a systematic division of the building are needed in larger structures such as apartment buildings. Area search assignments should be based on a system of priorities:

- The first priority is to search the area immediately around the fire, then the rest of the fire floor.
- The second priority is to search the area directly above the fire and the rest of that floor.
- The next priority is higher level floors, working from the top floor down, because smoke and heat are likely to accumulate in these areas.
- Generally, areas below the fire floor are a lower priority.

At a high-rise building fire, the IC might assign two or more search-and-rescue teams to each floor. The teams must work together closely and coordinate their searches to ensure that all areas are covered. For example, one team might search all of the apartments on the right side of a corridor, while the other searches apartments on the left side.

Search Techniques

Fire fighters should employ standard techniques to search assigned areas quickly, efficiently, and safely. Searchers should always operate in teams of two and should always stay together (► Figure 13-4). The partners must remain in direct visual, voice, or physical contact with each other.

At least one member of each search team must also have a radio to maintain contact with the command post or someone outside the building. The team uses the radio to call for help if they become disoriented, trapped by fire, or need assistance. If the search team finds a victim, they must notify the IC so that help is available to remove the victim from the building and provide medical treatment. The team must also notify the IC when the search of each area has been completed so the IC can make informed decisions.

Two types of searches are performed in buildings. A **primary search** is a quick attempt to locate any potential victims who are in danger. The primary search should be as thorough as time permits and should cover any place where victims are likely to be found. Fire conditions might make it impossible to conduct a primary search in some areas, or may limit the time available for an exhaustive search.

A **secondary search** is conducted after the situation is under control. During this follow-up search, fire fighters should take the time to look everywhere and ensure that no one is missing. If possible, the secondary search should be conducted by a different team, so that each area of the building is examined with a fresh set of eyes.

Primary Search

During the primary search, fire fighters rapidly search the accessible areas of a burning structure to locate any potential victims. The objective of the primary search is to find any potential victims as quickly as possible and remove them from danger. When fire fighters complete the primary search, they have gone as far as they could and have removed anyone that could be rescued. The phrase "primary all clear" is used to report that the primary search has been completed.

By necessity, the primary search is conducted quickly and gives priority to the areas where victims are most likely to be located. Time is always a critical concern, because fire fighters must reach potential victims before they are burned, overcome by smoke and toxic gases, or trapped by a structural collapse. Search teams may have only a few minutes to conduct a primary search. In that limited time, fire fighters must try to find anyone who could be in danger and remove them to a safe area. Often, active fire conditions may limit the areas that can be searched quickly as well as the time that can be spent in each area.

Fire fighters should try to check all of the areas where victims might be, such as beds, cribs, chairs, and sofas. Adults who tried to escape on their own are often found near doors or windows. Some people, particularly children, may try to hide in a closet, in a bathtub or shower enclosure, or under a bed or another piece of furniture. Young children may also be frightened by the appearance of a fire fighter wearing full personal protective gear.

Figure 13-4 Search teams consist of at least two members.

The primary search is frequently conducted in conditions that expose both fire fighters and victims to the risks presented by heavy smoke, heat, structural collapse, and entrapment by the fire. Search teams must often work in conditions of zero visibility and may have to crawl along the floor to stay below layers of hot gases. Because the beam from a powerful hand light might be visible for only a few inches inside a smoke-filled building, search-and-rescue teams must practice searching for victims in total darkness. They must know how to keep track of their location and how to get back to their entry point. Practicing these skills in a controlled environment will enable fire fighters to perform confidently under similar conditions at an emergency incident.

Fire fighters must rely on their senses when they search a building. The three most important senses during a search are:

- Sight—Can you see anything?
- Sound—Can you hear someone calling for help, moaning, or groaning?
- Touch—Do you feel a victim's body?

If smoke and fire restrict visibility, searchers must use sound and touch to find victims. Every few seconds a member of the search team should yell out, "Is anyone in here? Can anyone hear me?" Then listen. Hold your breath to quiet your breathing regulator and stop moving. People who have suffered burns or are semi-conscious may not be able to speak intelligibly, so you will need to listen for guttural sounds, cries, faint voices, groans, and moans. Focus on the direction of any sounds you hear.

As you search, feel around for hands, legs, arms, and torsos of potential victims. Use hand tools to extend your reach, sweeping ahead and to the sides with the handle as you crawl. Use the tool to reach toward the center of the room as

Fire Fighter Tips

Primary search teams are usually the first fire fighters to enter a burning building. As they search for possible victims, team members will often obtain valuable information about the location and spread of the fire. This reconnaissance information should be communicated to the IC to help in managing the overall operation.

you move along the walls (► Figure 13-5). Practice until you can tell the difference between a piece of furniture and a human body. The type of furniture you encounter may provide clues to finding victims. Spindle-type legs in a bedroom may indicate a baby crib that needs to be searched. An unusually low bed may be the bottom berth of a children's bunk bed.

Zero-visibility conditions can be very disorienting. Searchers must follow walls and note turns and doorways to avoid becoming lost. After locating a victim or completing a search of an area, the search team must be able to retrace its path and return to its entry point. Search teams must also have a secondary escape route in case fire conditions change and block their planned exit. During the search, fire fighters should note the locations of stairways, doors, and windows. Searchers should always be aware of the nearest exit and an alternate exit.

If the situation deteriorates rapidly, a window could become the emergency exit from a room. Searchers should keep track of the exterior walls and window locations, even reaching up to feel for the windows if conditions force them to crawl.

Searchers must remain in contact with someone on the outside and must be able to state their location at all times, particularly if they need assistance. The search team must be able to give their location, including the building section and floor, to the IC. If the team needs a ladder for the rescue, the IC will need to know where the ladder should be placed (which side of the building) and how long a ladder will be needed (which floor). Searchers should use standard incident management system terminology to describe their location over the radio.

Search Patterns

Each room should be searched using a standard pattern. If the rooms are relatively small, searchers should follow the walls around the perimeter of each room and reach toward the middle to feel for victims. In larger rooms, one team member can maintain contact with the wall, while the other maintains contact with the first, and the two work their way around the room in tandem.

Some fire departments use a clockwise pattern to search a room while others use a counterclockwise pattern. In a clockwise search (also known as a left-hand search) fire fighters turn left at the entry point, keep the left hand in contact with the wall, and use the right arm to sweep the room. At each corner the searchers make a right turn, eventually returning to the entry door (▲ Figure 13-6). A counterclockwise search moves around the room in the opposite direction (► Figure 13-7). Fire fighters should regularly practice and use the standard system adopted by their department.

Figure 13-5 Use your tools to extend your reach and sweep the area in front of you.

Figure 13-6 To use a clockwise search pattern, turn left when you enter, then right at each corner around the room.

Fire fighters can use the handle of a hand tool to extend their reach while sweeping across the floor and under furniture. Searchers should also check on top of beds, in chairs, and in other locations where victims may hide, such as closets, bathrooms, bathtubs, or shower enclosures.

The same clockwise/counterclockwise search pattern also applies to searches of areas that are divided up into small spaces, such as an apartment or apartment building. Searchers

Figure 13-7 In a counterclockwise search pattern, turn right when you enter, then left at each corner around the room.

should work their way around the apartment from room to room in the standard direction, searching each individual room in the same manner. At the end of the search, they will be back at the entry point.

If a door is closed, fire fighters should check the temperature of the door to determine if there is an active fire on the other side. Fire fighters should follow their department's standard procedures for performing such a check. Some departments permit a fire fighter to remove a glove just long enough to feel the temperature of a door. Other departments consider this to be a risky procedure. A **thermal imaging device**, which shows heat images instead of visual images, can be used to determine if there is a fire behind the door. A hot door should not be opened unless there is a hose line ready to douse the fire. It is usually better to leave a hot door closed and move on to search adjacent rooms.

Searchers must keep track of their position in relation to the entry door so they can find their way out of the room. To keep the door open during the search of a small room, one team member should remain at the door while the second team member performs the search. A chock can be used to keep the door from closing or a flashlight can be placed in the doorway to serve as a beacon and to help searchers maintain their orientation. Searchers should always exit through the same door that they entered.

The searched room or apartment should be marked so other personnel will know that it has been searched. Chalk, crayons, felt tip markers, or masking tape can be used to mark the door. Some fire departments use a two-part marking system to indicate when a search is in progress and when it has been completed. A "/" indicates a search in progress, and an "X" indicates that the search has been completed (▶ Figure 13-8). Other departments place an object in the doorway or attach a tag or latch strap to the doorknob to indicate that the room has been searched.

Figure 13-8 In large buildings, doors should be marked after the rooms are searched.

Thermal Imaging Devices

A thermal imaging device is a valuable tool for conducting a primary search in a smoke-filled building. A thermal imaging device is similar to a television camera, except that it captures heat images instead of visible light images. The images appear on a display screen and show the relative temperatures of different objects. The device can be set to distinguish small temperature differences, enabling fire fighters to conduct a search quickly and thoroughly.

The major benefit of using a thermal imaging device is that it can "see" the image of a person in conditions of total darkness or through smoke that totally obscures normal vision (▶ Figure 13-9). Fire fighters will be able to identify the shape of a human body with a thermal image scan because the body will be either warmer or cooler than its surroundings.

Temperature differences also enable the thermal imaging

Fire Fighter Tips

Learn the features of your thermal imager and practice with it regularly to get the maximum advantage from it.

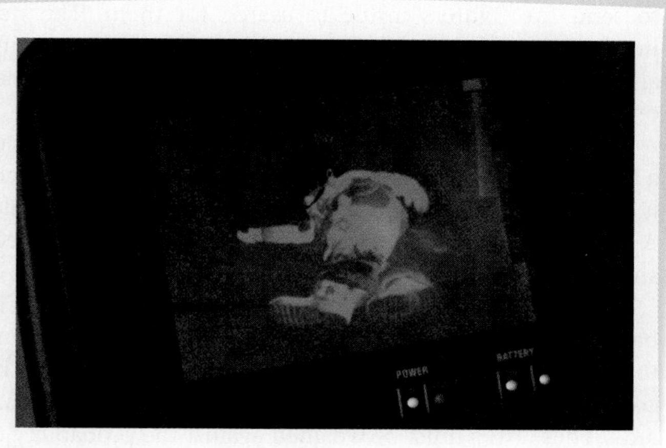

Figure 13-9 The thermal imaging device can capture an image of a person through thick smoke.

Figure 13-10 A thermal imaging device can be used to check for hidden fire.

device to show furniture, walls, doorways, and windows. This enables a fire fighter to navigate through the interior of a smoke-filled building. The thermal imaging device can even be used to locate a fire in a smoke-filled building or behind walls or ceilings (► Figure 13-10). A thermal image scan of the exterior of a building can be used to locate the fire source and the direction of fire spread. Scanning a door before opening it can indicate whether the room is safe to enter.

There are several different types of thermal imaging devices. Some thermal imagers are hand-held devices, similar to a hand-held computer with a built-in monitor. Others are helmet-mounted and produce an image in front of the user's SCBA facepiece. Fire fighters need training and practice to become proficient in using these devices, interpreting the images, and maneuvering confidently through a building while wearing a thermal imager.

Search Ropes

Special techniques must be used to search large, open areas, such as warehouses or gymnasiums, when visibility is limited. Search ropes should be used when it is impossible to cover the interior by following the walls. They also should be used in areas with interconnected rooms or spaces, or with multiple aisles created by display counters or storage racks. Without a search rope, the search teams might not be able to find their way out of the area.

Search ropes should be anchored at the entry point to provide a direct path to this location for each searcher. One fire fighter should remain at the anchor point to tend the ropes and maintain accountability for the fire fighters who enter. One fire fighter stretches a large diameter rope down the center of a room. The other searchers then attach their individual search ropes to the main line and branch off to cover the area. The search ropes always provide a reliable return path to the entry point.

Figure 13-11 A secondary search is conducted after the fire is controlled.

Secondary Search

A secondary search is conducted after the fire is under control or fully extinguished. During the secondary search, fire fighters locate any victims that might have been missed in the primary search and search any areas not included in the primary search. Conditions in the building are usually better during the secondary search because the fire is under control and there is adequate ventilation. Fire fighters will be able to see better and move around without having to crawl under layers of hot gases (▲ Figure 13-11). After a secondary search is completed, fire fighters should notify the IC by reporting, "Secondary all clear."

Even though the fire is under control, safety remains a primary consideration during a secondary search. SCBA must be used until the air is tested and pronounced safe to breathe. The levels of carbon monoxide and other poisonous gases often remain high for a long time after a fire. In burned areas, fire fighters must have a hose line available in case the fire rekindles or starts up again. The structural stability of the building also must be evaluated before beginning a secondary search. Portable lighting should be used, and holes in floors and other possible hazards should be marked.

The secondary search should be started as soon as the fire is under control and sufficient resources are available. It is conducted slowly and methodically to ensure that no areas are overlooked. Whenever possible, a different team of fire fighters should perform the secondary search. They should check all places where someone, especially a child, could hide, including closets, bathtubs, shower enclosures, behind doors, under windows, under beds and furniture, and inside toy chests.

The secondary search must include areas that were involved in the fire. In these areas, the secondary search should be conducted as a body recovery process. Fire fighters should move carefully and look closely at the debris to identify the remains of a deceased victim.

Search Safety

Search-and-rescue operations often present the highest risk to fire fighters during emergency situations. During these operations, fire fighters are exposed to the same risks that endanger the lives of potential victims. Even though fire fighters have the advantages of protective clothing, protective equipment, training, teamwork, and standard operating procedures, they can still be seriously or fatally injured. Safety must be an essential consideration in all search-and-rescue operations.

Risk Management

Search-and-rescue situations require a very special type of risk management. Every emergency operation involves a degree of unavoidable, inherent risk to fire fighters. During search-and-rescue operations, fire fighters will probably encounter situations that involve a significantly higher degree of personal risk. This level of risk is acceptable only if there is a reasonable probability of saving a life.

The IC is responsible for managing the level of risk during emergency operations. The IC must determine which actions are taken in each situation. Actions that present a high level of risk to the safety of fire fighters are justified only if lives can be saved. Only a limited risk level is acceptable to save property. When there is no possibility of saving either lives or property, no risk is acceptable.

The acceptance of unusual risk can be justified only when victims are known or believed to be in immediate danger, and there is a reasonable probability that they can be rescued. It is not acceptable to risk the lives of fire fighters when there is no possibility of rescuing any victims. For example, if a building is fully involved in flames, there is no possibility that any occupants would still be alive, so there is no reason to send fire fighters inside to search for them. The risk involved in conducting search-and-rescue operations must always be weighed against the probability of finding someone who is still alive to be rescued.

A primary search is conducted during the active stages of a fire, when time is a critical factor. During a primary search, it may be necessary to accept a high level of risk to save a life. A secondary search, conducted after the fire is extinguished, should not expose fire fighters to any avoidable risks.

The IC must decide whether to conduct a primary search, based on a risk–benefit evaluation of the situation. In making this decision, the IC must consider the stage of the fire, the condition of the building, and the presence of any other hazards. These risks must be weighed against the probability of finding any occupants who could be rescued. This on-scene evaluation is often guided by established policies. For example, many fire departments have policies that prohibit fire fighters from entering abandoned structures known to be in poor structural condition. This policy is based on the probability that such a building has a high risk of collapse and the low probability that anyone would be inside.

The IC may decide not to conduct a primary search because the risk to fire fighters is too great or the possibility of making a successful rescue is too remote. Such a decision might be based on advanced fire conditions, potential backdraft conditions, imminent structural collapse, or other circumstances (► Figure 13-12). The IC may be able to identify these conditions from the exterior, or may learn of them from a team assigned to conduct a search. A search team that encounters conditions that make entry impossible should report their findings back to the IC. In these situations, a secondary search is conducted when it is feasible to enter.

Search-and-Rescue Equipment

To perform search and rescue properly, fire fighters must have the appropriate equipment. Search-and-rescue equipment includes:

- Personal protective equipment
- Portable radio
- Hand light or flashlight
- Forcible entry tools
- Hose lines
- Thermal imaging devices
- Ladders
- Long rope(s)
- A piece of tubular webbing or short rope (16′ to 24′)

Full personal protective equipment, which is always required for structural fire fighting, is essential for interior search and rescue. The proper attire includes helmet, protective hood, bunker coat, turnout pants, boots, and gloves. Each fire fighter must use SCBA and carry a flashlight or hand light. Before entering the building, fire fighters must activate their PASS devices.

Fire Fighter Safety Tips

During search-and-rescue operations, fire fighters should remember to:

- Work from a single plan.
- Maintain radio contact with the IC, through the chain of command and portable radios.
- Monitor fire conditions during the search.
- Coordinate ventilation with search-and-rescue activities.
- Maintain your accountability system.
- Stay with a partner.

Fire Fighter Tips

Risk-Benefit Analysis

The IC must always balance the risks involved in an emergency operation with the potential benefits.

- Actions that present a high level of risk to the safety of fire fighters are justified only if there is a potential to save lives.
- Only a limited level of risk is acceptable to save valuable property.
- It is not acceptable to risk the safety of fire fighters when there is no possibility to save lives or property.

Figure 13-12 A primary search should not be conducted in a building that is fully involved in fire.

NFPA 1500 Requirements

Specific safety requirements for search-and-rescue operations are defined in NFPA 1500, *Standard on Fire Department Occupational Safety and Health Program,* and in regulations enforced by the Occupational Safety and Health Administration. The NFPA requirements state that a team of at least two fire fighters must enter together, and at least two other fire fighters must remain outside the danger area, ready to rescue the fire fighters who are inside the building. This is sometimes called the **two-in/two-out rule**.

An exception to this rule is permitted in an imminent life-threatening situation, where immediate action can prevent the loss of life or serious injury. Only under such specific circumstances are fire fighters allowed to take actions at a higher level of risk. The initial IC must evaluate the situation and determine if the risk is justified. The IC also must be prepared to explain his or her decision.

At least one member of each search team should be equipped with a portable radio. If possible, each individual should have a radio. If a fire fighter gets into trouble, the radio is the best means to obtain assistance.

Each fire fighter assigned to search and rescue should carry at least one forcible entry hand tool, such as an axe, Halligan tool, or short ceiling hook. These tools can be used both to open an area for a search and, if necessary, to open an emergency exit path. A hand tool can also be used to extend the fire fighter's reach during a sweep for unconscious victims.

A search-and-rescue team working close to the fire should have a hose line or be accompanied by another team with a hose line. A hose line can protect the fire fighters and enable them to search a structure more efficiently. A hose line is essential when a search team is working close to or directly above the fire. The hose line can be used to knock down the fire, to protect a means of egress (stairway or corridor), or to protect the victims as they are escaping.

A hose line also can be used to guide searchers out of a structure. Searchers can follow the hose line to a coupling, feel the coupling to determine the male and female ends, and then follow the hose in the direction of the male coupling. If the team is working without a hose line, it should have search ropes, particularly if it would be difficult for fire fighters to find their way out of a search area. More information on emergency exit procedures is presented in Chapter 17, Fire Fighter Survival.

Fire fighters also must pay attention to their air supplies during search-and-rescue operations. Fire fighters must have adequate air to make a safe exit. This limits the time that can be spent searching and the distance that fire fighters can penetrate into a building.

Methods to Determine if an Area is Tenable

Fire fighters must make rapid, accurate, and on-going assessments about the safety of the building while working at a structure fire. When they arrive, fire fighters must quickly determine the type of structure involved, the possibility of collapse, and the life safety risk involved. They also must evaluate the stability of the structure and the potential for backdraft or flashover. Sagging walls, chipped or cracked mortar or cement, warped or failing structural steel, and any partial collapse are significant indicators of an impending collapse. Dark black smoke, blackened windows, the appearance of a "breathing" building, and intense heat are signs of possible flashover or backdraft.

Even after the decision has been made to enter a burning structure, fire fighters must continue to reevaluate the safety of the operation. Fire fighters working inside the building must rely on other fire fighters and the IC outside the structure to notify them of changing fire conditions. The IC may know or see things that fire fighters working inside may not know, and may call for an evacuation as the fire situation changes. Conditions inside the structure must constantly be evaluated. Fire fighters should check and recheck the surface they are working on. Floors that have been burnt through or feel soft or spongy should be avoided. A rapid rise in the amount of heat or flame "rollover" may indicate a potential flashover.

Rescue Techniques

Rescue is the removal of a person who is unable to escape from a dangerous situation. Fire fighters rescue people not only from fires but also from a wide variety of accidents and mishaps. Although this section refers primarily to rescuing occupants from burning buildings, some of the techniques can be used for other types of situations. As a fire fighter, you must learn and practice the assists, carries, drags, and other techniques used to rescue people from fires. After mastering these techniques, you will be able to rescue victims from a life-threatening situation.

Rescue is the second component of search and rescue. When you locate a victim during a search, you must direct, assist, or carry the person to a safe area. Rescue can be as basic and simple as verbally directing an occupant toward an exit. Or it can be as demanding as extricating a trapped, unconscious victim and physically carrying that person out of the building. The term rescue is generally applied to situations where the rescuer physically assists or removes the victim from the dangerous area.

Most people who realize that they are in a dangerous situation will attempt to escape on their own. Children and elderly persons, physically or developmentally handicapped persons, and ill or injured persons, however, may be unable to escape and will need to be rescued. People who are sleeping or under the influence of alcohol or drugs may not become aware of the danger in time to escape on their own. Toxic gases may incapacitate even healthy individuals before they can reach an exit. Victims also may be trapped when a rapidly spreading fire, an explosion, or a structural collapse cuts off potential escape routes.

Because fires are life-threatening to both victims and rescuers, the first priority is to remove the victim from the fire building or dangerous area as quickly as possible. It is usually better to move the victim to a safe area first, and then provide any necessary medical treatment. The assists, lifts, and carries described in this chapter should not be used if you suspect that the victim has a spinal injury, unless there is no other way to remove him or her from the life-threatening situation.

Always use the safest and most practical means of egress when removing a victim from a dangerous area. A building's normal exit system, such as interior corridors and stairways, should be used if it is open and safe. If the regular exits cannot be used, an outside fire escape, a ladder, or some other method of egress must be found. Ladder rescues, which are covered in this chapter, can be both difficult and dangerous, whether the victim is conscious and physically fit or unconscious and injured (▼ Figure 13-13).

Shelter-in-Place

In some situations, the best option is to shelter the occupants in place instead of trying to remove them from a fire building. This option should be considered when the occupants are conscious and in a part of the building that is adequately protected from the fire by fire-resistive construction and/or fire suppression systems. If smoke and fire conditions block the exits, they might be safer staying in the sheltered location than attempting to evacuate through a hazardous environment.

Such a situation could occur in a high-rise apartment building, with a fire confined to one apartment. The stair-

Figure 13-13 Bringing an unconscious victim or a person who needs assistance down a ladder can be difficult and dangerous.

Fire Fighter Safety Tips

To exit from a fire, follow the hose line in the direction of the male coupling.

ways and corridors could be filled with smoke, but the occupants who are remote from the fire would be very safe on their balconies or in their apartments with the windows open to provide fresh air. They would be exposed to more risk if they attempted to exit than if they remained in their apartments until the fire is extinguished. This decision must be made by the IC.

Exit Assist

The simplest rescue is the exit assist. The victim is responsive and able to walk without assistance or with very little assistance. The fire fighter may only need to guide the person to safety or to provide a minimal level of physical support. Even if the victim can walk without assistance, the fire fighter should take the person's arm or use the one-person walking assist (see below) to make sure that the victim does not fall or become separated from the rescuer.

The following assists can be used to help responsive victims exit a fire situation:

- One-person walking assist
- Two-person walking assist

One-Person Walking Assist

The one-person walking assist can be used if the person is capable of walking.

To perform a one-person walking assist, follow the steps in ▼ Skill Drill 13-1.

1. Help the person to stand next to you, facing the same direction. **(Step 1)**
2. Have the person place his or her arm behind your back and around your neck. Hold the person's wrist as it drapes over your shoulder. **(Step 2)**
3. Put your free arm around the person's waist and help the victim to walk. **(Step 3)**

13-1 Skill Drill

One-Person Walking Assist

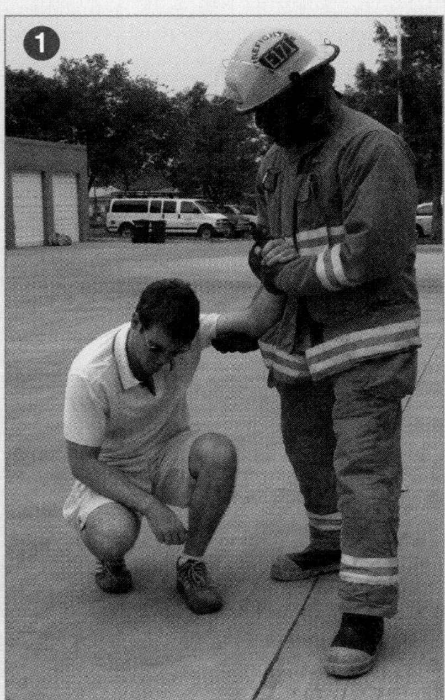

Help the person to stand.

Have the person place his or her arm around your neck, and hold on to the person's wrist, which should be draped over your shoulder.

Put your free arm around the person's waist and help the victim to walk.

Two-Person Walking Assist

The two-person walking assist is useful if the victim cannot stand and bear weight without assistance. The two rescuers completely support the victim's weight. It may be difficult to walk through doorways or narrow passages using this type of assist. To perform a two-person walking assist, follow the steps in ► Skill Drill 13-2.

1. Two fire fighters stand facing the victim, one on each side of the victim. **(Step 1)**
2. Both fire fighters assist the victim to a standing position. **(Step 2)**
3. Once the victim is fully upright, place the victim's right arm around the neck of the fire fighter on the right side. Place the victim's left arm around the neck of the fire fighter on the left side. The victim's arms should drape over the fire fighters' shoulders. The fire fighters hold the person's wrist in one hand. **(Step 3)**
4. Both fire fighters put their free arms around the person's waist, grasping each other's wrists for support and locking arms together behind the victim. **(Step 4)**
5. Both fire fighters slowly assist the victim to walk. Fire fighters must coordinate their movements and move slowly. **(Step 5)**

Simple Victim Carries

Four simple carries can be used to move a victim who is conscious and responsive, but incapable of standing or walking:

- Two-person extremity carry
- Two-person seat carry
- Two-person chair carry
- Cradle-in-arms carry

Two-Person Extremity Carry

The two-person extremity carry requires no equipment and can be performed in tight or narrow spaces, such as mobile home corridors, small hallways, and narrow spaces between buildings. The focus of this carry is on the victim's extremities. To perform a two-person extremity carry, follow the steps in ► Skill Drill 13-3.

1. Two fire fighters help the victim to sit up. **(Step 1)**
2. One fire fighter kneels behind the victim, reaches under the victim's arms, and grasps the victim's wrists. **(Step 2)**
3. The second fire fighter backs in-between the victim's legs, reaches around, and grasps the victim behind the knees. **(Step 3)**
4. At the command of the first fire fighter, both fire fighters stand up and carry the victim away, walking straight ahead. Fire fighters must coordinate their movements. **(Step 4)**

Two-Person Seat Carry

The two-person seat carry is used with victims who are disabled or paralyzed. This carry requires two fire fighters, and moving through doors and down stairs may be difficult. To perform a two-person seat carry, follow the steps in ► Skill Drill 13-4.

1. Two fire fighters kneel near the victim's hips, one on each side of the victim. **(Step 1)**
2. Both fire fighters raise the victim to a sitting position and link arms behind the victim's back. **(Step 2)**
3. The fire fighters' remaining free arms are then placed under the victim's knees and linked. **(Step 3)**
4. If possible, the victim puts his or her arms around the fire fighters' necks for additional support. **(Step 4)**

Two-Person Chair Carry

The two-person chair carry is particularly suitable when a victim must be carried through doorways, along narrow corridors, or up or down stairs. Two rescuers use a chair to transport the victim. A folding chair cannot be used, and the chair must be strong enough to support the weight of the victim while being carried. The victim should feel much more secure with this carry than with the two-person seat carry. The victim should be encouraged to hold on to the chair. To perform a two-person chair carry, follow ► Skill Drill 13-5.

1. Tie the victim's hands together or have the victim grasp hands together.
2. One fire fighter stands behind the seated victim, reaches down, and grasps the back of the chair. **(Step 1)**
3. The fire fighter tilts the chair slightly backward on its rear legs so that the second fire fighter can step back between the legs of the chair and grasp the tips of the chair's front legs. The victim's legs should be between the legs of the chair. **(Step 2)**
4. When both fire fighters are correctly positioned, the fire fighter behind the chair gives the command to lift and walk away.
5. Because the chair carry may force the victim's head forward, watch the victim for airway problems. **(Step 3)**

Fire Fighter Safety Tips

Keep your back as straight as possible and use the large muscles in your legs to do the lifting!

13-2 Skill Drill

Two-Person Walking Assist

Two fire fighters stand facing the victim, one on each side of the victim.

The fire fighters assist the victim to a standing position.

Once the victim is fully upright, drape the victim's arms around the necks and over the shoulders of the fire fighters, who each hold one of the victim's wrists.

Both fire fighters put their free arm around the person's waist, grasping each other's wrists for support and locking their arms together behind the victim.

Assist walking at the victim's speed.

13-3 Skill Drill

Two-Person Extremity Carry

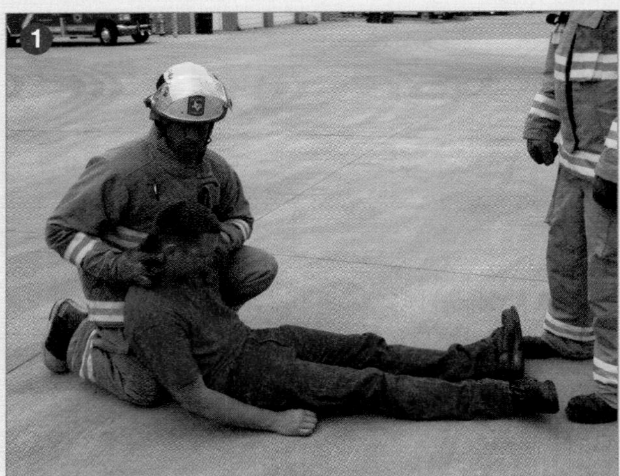

1 Two fire fighters help the victim to sit up.

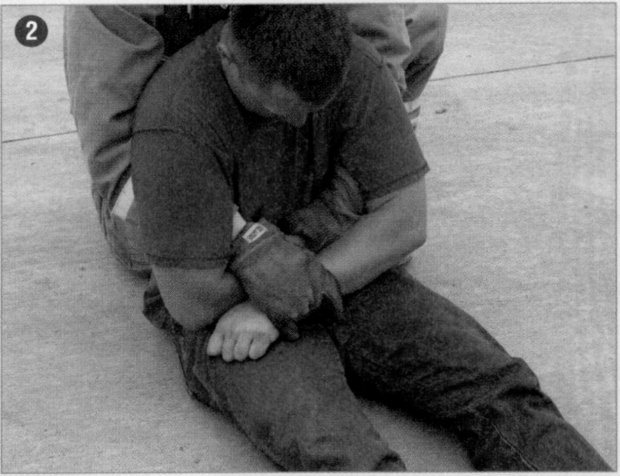

2 The first fire fighter kneels behind the victim, reaches under the victim's arms, and grasps the victim's wrists.

3 The second fire fighter backs in-between the victim's legs, reaches around, and grasps the victim behind the knees.

4 The first fire fighter gives command to stand and carry the victim away, walking straight ahead. Fire fighters must coordinate their movements.

Two-Person Seat Carry

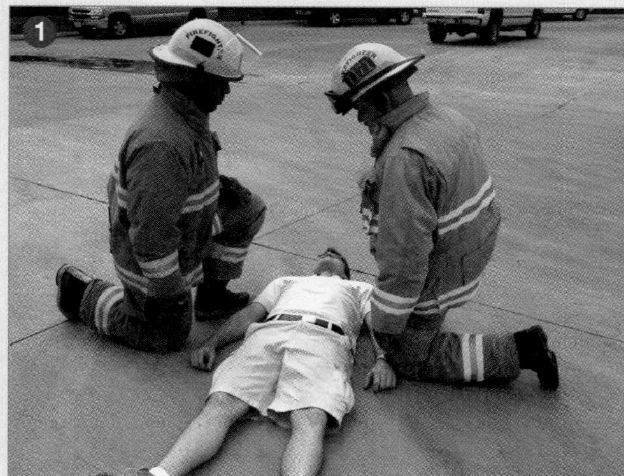

Kneel beside the victim near the victim's hips.

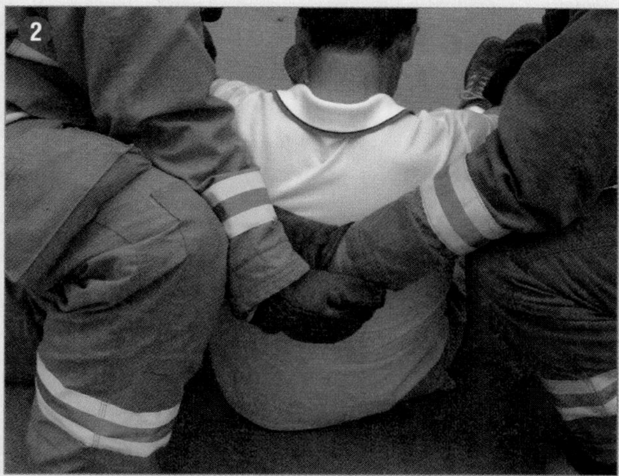

Raise the victim to a sitting position and link arms behind the victim's back.

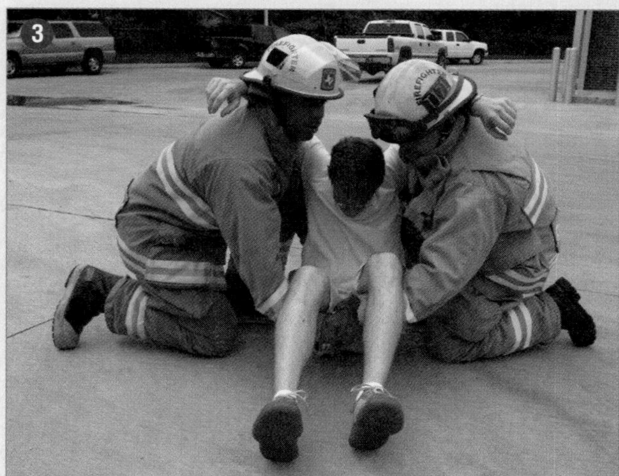

Place your free arms under the victim's knees and link arms.

If possible, the victim puts his or her arms around your necks for additional support.

13-5 Skill Drill

Two-Person Chair Carry

One fire fighter stands behind the seated victim, reaches down, and grasps the back of the chair.

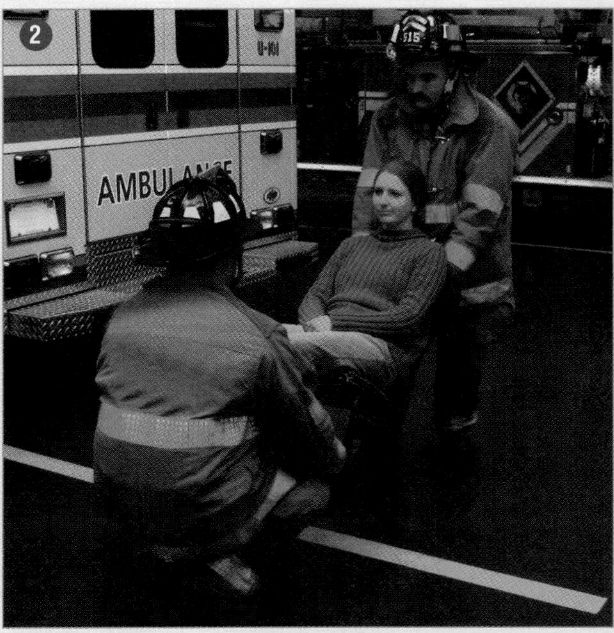

The fire fighter tilts the chair slightly backward on its rear legs so that the second fire fighter can step back between the legs of the chair and grasp the tips of the chair's front legs. The victim's legs should be between the legs of the chair.

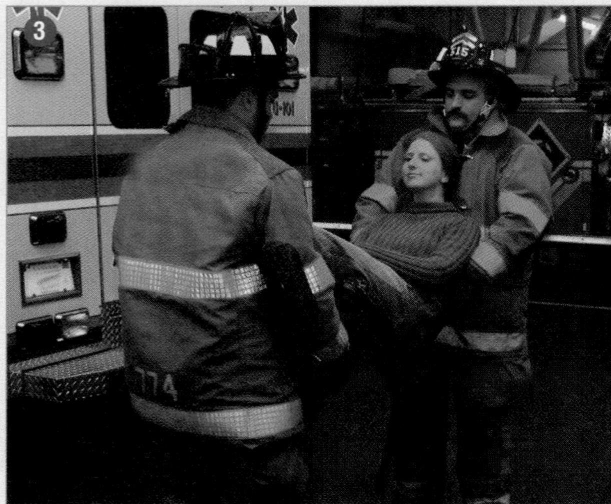

When both fire fighters are correctly positioned, the fire fighter behind the chair gives the command to lift and walk away. Because the chair carry may force the victim's head forward, watch the victim for airway problems.

Figure 13-14 Cradle-in-arms carry.

Cradle-in-Arms Carry

The cradle-in-arms carry can be used by one fire fighter to carry a child or a small adult (▲ Figure 13-14). The fire fighter should be careful of the victim's head when moving through doorways or down stairs. To perform the cradle-in-arms carry, follow the steps in (Skill Drill 13-6).

1. Kneel beside the child and place one arm around the child's back and the other arm under the thighs.
2. Lift slightly and roll the child into the hollow formed by your arms and chest.
3. Be sure to use your leg muscles to stand.

Emergency Drags

The most efficient method to remove an unconscious or unresponsive victim from a dangerous location is a drag. Five emergency drags can be used to remove unresponsive victims from a fire situation:

- Clothes drag
- Blanket drag
- Webbing sling drag
- Fire fighter drag
- Emergency drag from a vehicle

When using an emergency drag, the rescuer should make every effort to pull the victim in line with the long axis of the body to provide as much spinal protection as possible. The victim should be moved head first to protect the head.

Clothes Drag

The clothes drag is used to move a victim who is on the floor or ground and is too heavy for one rescuer to lift and carry. The rescuer drags the person by pulling on the clothing in the neck and shoulder area. The rescuer should grasp the clothes just behind the collar, use the arms to support the victim's head, and drag the victim away from danger. To perform the clothes drag, follow the steps in (▼ Skill Drill 13-7).

1. Crouch behind the victim's head, grab the shirt or jacket around the collar and shoulder area, and support the head with your arms. **(Step 1)**
2. Lift with your legs until you are fully upright. Walk backwards, dragging the victim to safety. **(Step 2)**

Skill Drill

Clothes Drag

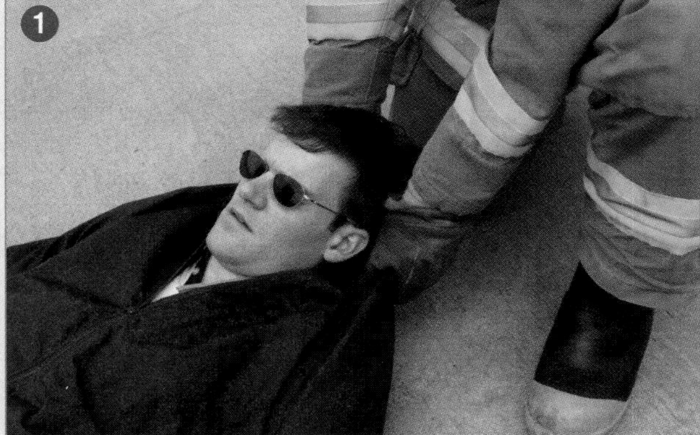

Crouch behind the victim's head and grab the shirt or jacket around the collar and shoulder area.

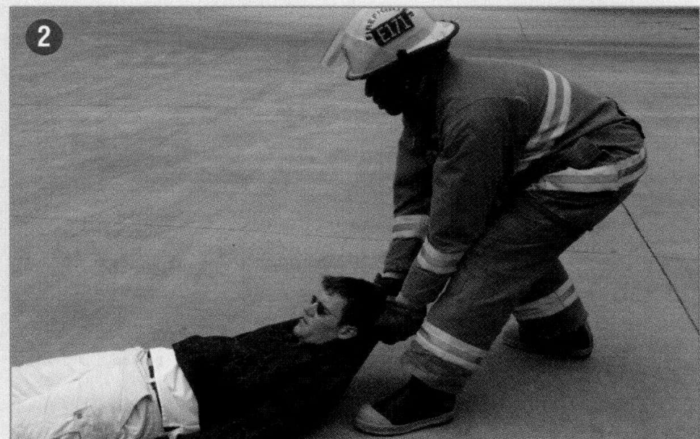

Lift with your legs until you are fully upright. Walk backwards, dragging the victim to safety.

13-8 Skill Drill

Blanket Drag

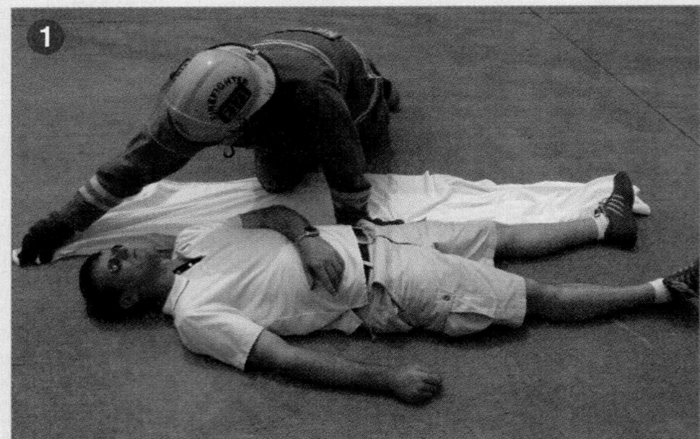

Stretch out the material you are using next to the victim.

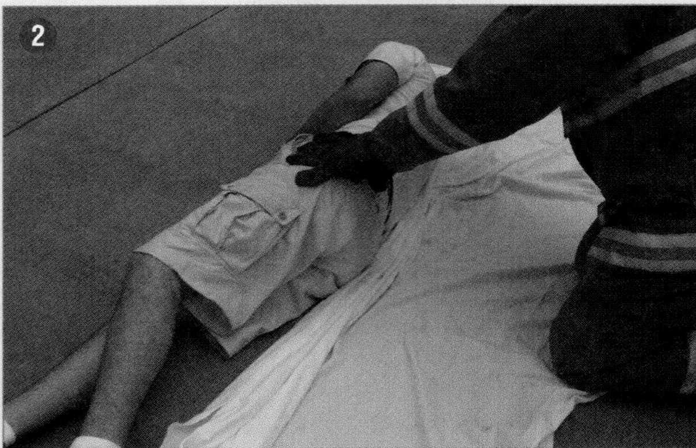

Roll the victim onto one side. Neatly bunch one third of the material against the victim's body.

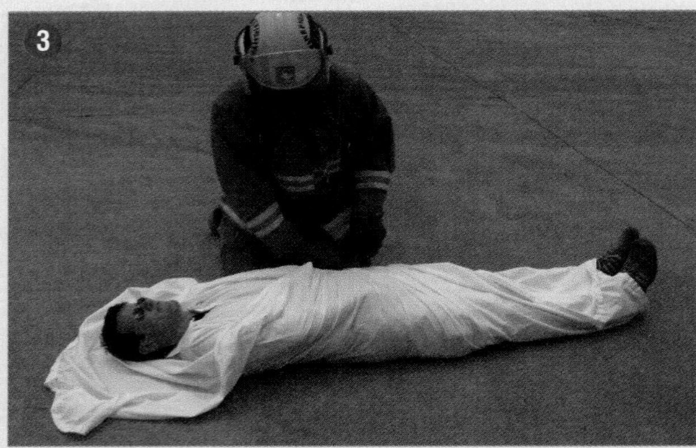

Lay the victim back down (supine). Pull the bunched material out from underneath the victim and wrap it around the victim.

Grab the material at the head and drag backwards to safety.

Blanket Drag

The blanket drag can be used to move a victim who is not dressed or is dressed in clothing that is too flimsy for the clothes drag (for example, a nightgown). This procedure requires the use of a large sheet, blanket, curtain, or rug. Place the item on the floor and roll the victim onto it, then pull the victim to safety by dragging the sheet or blanket.

To perform a blanket drag, follow the steps in ▲ **Skill Drill 13-8**.

1. Lay the victim supine (face up) on the ground. Stretch out the material for dragging next to the victim. **(Step 1)**
2. Roll the victim onto the right or left side. Neatly bunch one third of the material against the victim so the victim will lie approximately in the middle of the material. **(Step 2)**
3. Lay the victim back down (supine) on the material. Pull the bunched material out from underneath the victim and wrap it around the victim. **(Step 3)**
4. Grab the material at the head and drag backwards to safety. **(Step 4)**

The standing drag is performed by following the steps in **Skill Drill 13-9**:

1. Stand at the head of the supine victim. Then kneel at the victim's head.
2. Raise the victim's head and torso 90°. The victim is leaning against you.
3. Reach under the victim's arms, wrap your arms around the victim's chest, and lock your arms.
4. Stand straight up using your legs.
5. Drag the victim out.

13-10 Skill Drill

Webbing Sling Drag

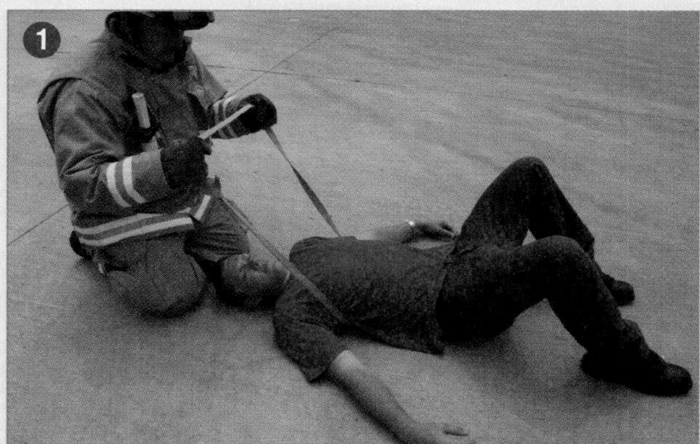

Place the victim in the center of the loop so the webbing is behind the patient's back.

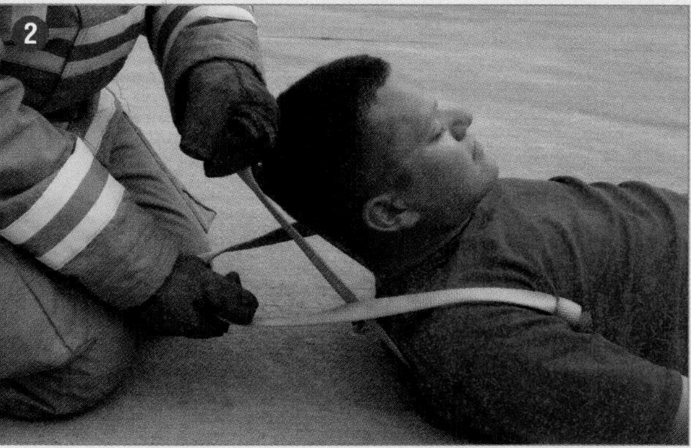

Take the large loop over the victim and place it above the victim's head. Reach through, grab the webbing behind the victim's back, and pull through all the excess webbing. This creates a loop at the top of the victim's head and two loops around the victim's arms.

Adjust hand placement to protect the victim's head while dragging.

Webbing Sling Drag

The webbing sling drag provides a secure grip around the upper part of a victim's body, for a faster removal from the dangerous area. In this drag, a sling is placed around the victim's chest and under the armpits, and used to drag the victim. The webbing sling helps support the victim's head and neck. A webbing sling can be rolled and kept in a turnout coat pocket.

A carabiner can be attached to the sling to secure the straps under the victim's arms and provide additional protection for the head and neck.

To perform the webbing sling drag, follow the steps in **▲ Skill Drill 13-10**.

1. Using a prepared webbing sling, place the victim in the center of the loop so the webbing is behind the victim's back in the area just below the armpits. **(Step 1)**
2. Take the large loop over the victim and place it above the victim's head. Reach through, grab the webbing behind the victim's back, and pull through all the excess webbing. This creates a loop at the top of the victim's head and two loops around the victim's arms. **(Step 2)**
3. Adjust hand placement to protect the victim's head while dragging. **(Step 3)**

Skill Drill

Fire Fighter Drag

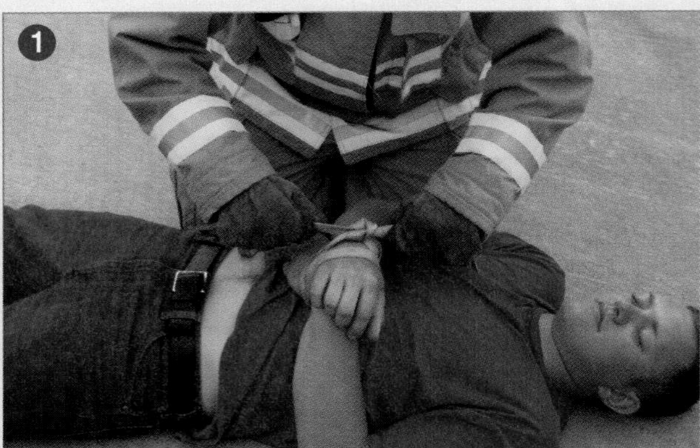

Tie the victim's wrists together with anything that is handy.

Get down on hands and knees and straddle the victim.

Pass the victim's tied hands around your neck, straighten your arms, and drag the victim across the floor by crawling on your hands and knees.

Fire Fighter Drag

The fire fighter drag can be used if the victim is heavier than the rescuer because it does not require lifting or carrying the victim. To perform the fire fighter drag, follow the steps in (▲ **Skill Drill 13-11**).

1. Tie the victim's wrists together with anything that is handy: a cravat (a folded triangular bandage), gauze, belt, or necktie. **(Step 1)**
2. Get down on hands and knees and straddle the victim. **(Step 2)**
3. Pass the victim's tied hands around your neck, straighten your arms, and drag the victim across the floor by crawling on your hands and knees. **(Step 3)**

13-12 Skill Drill

One-Person Emergency Drag from a Vehicle

Grasp the victim under the arms and cradle the head between your arms.

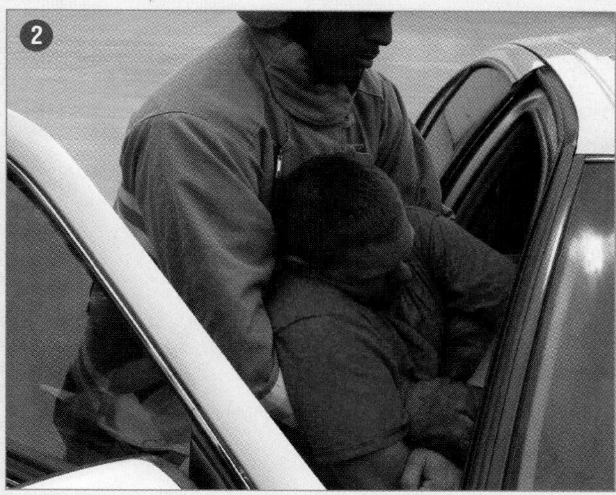

Gently pull the victim out of the vehicle.

Lower the victim down into a horizontal position in a safe place.

Emergency Drag from a Vehicle

An emergency drag from a vehicle is performed when the victim must be quickly removed from a vehicle to save his or her life. The drags described below might be used if the vehicle is on fire or if the victim requires CPR.

One Rescuer

There is no effective way for one person to remove a victim from a vehicle without some movement of the neck and spine. Preventing excess movement of the victim's neck, however, is important. To perform an emergency drag from a vehicle with only one rescuer, follow the steps in (▲ **Skill Drill 13-12**).

1. Grasp the victim under the arms and cradle the head between your arms. **(Step 1)**
2. Gently pull the victim out of the vehicle. **(Step 2)**
3. Lower the victim down into a horizontal position in a safe place. **(Step 3)**

13-13 Skill Drill

Long Backboard Rescue

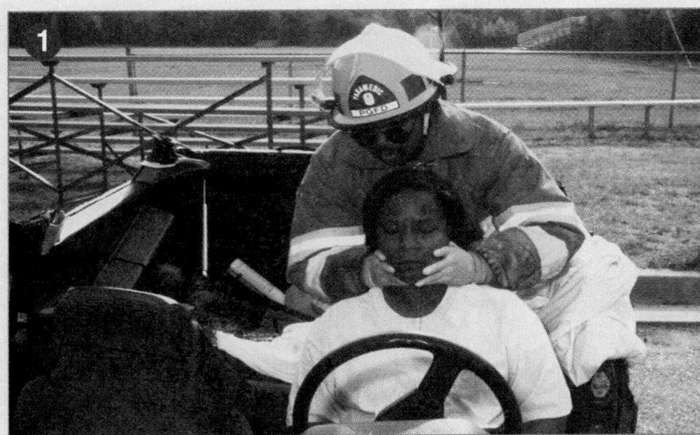

The first fire fighter provides in-line manual support of the head and cervical spine.

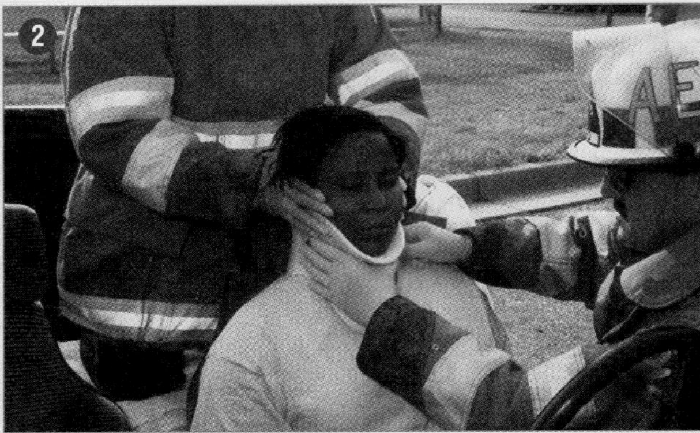

The second fire fighter gives commands and applies a cervical collar.

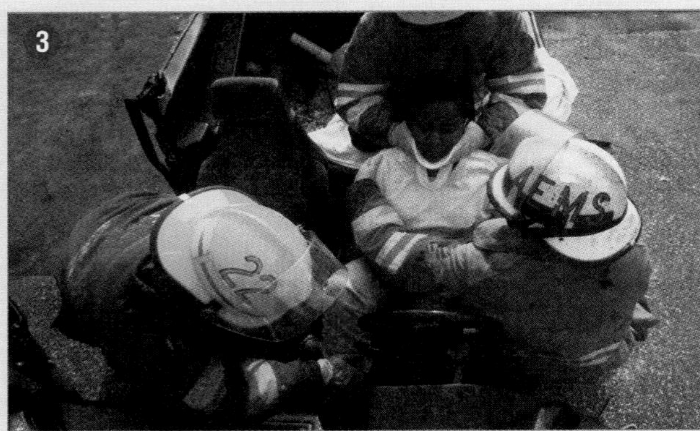

The third fire fighter frees the victim's legs from the pedals and moves the legs together without moving the pelvis or spine.

The second and third fire fighters rotate the victim as a unit in several short, coordinated moves. The first fire fighter (relieved by the fourth fire fighter as needed) supports the head and neck during rotation (and later steps).

Long Backboard Rescue

If four or more fire fighters are present, one fire fighter can support the victim's head and neck, while the second and third fire fighters move the victim by lifting under the arms. The victim can then be moved in line with the long axis of the body, with the head and neck stabilized in a neutral position. Whenever possible, a long backboard should be used to remove a victim from the vehicle. Follow the steps in ▲ **Skill Drill 13-13** to perform this skill.

1. The first fire fighter supports the victim's head and cervical spine from behind. Support may be applied from the side, if necessary, by reaching through the driver's side doorway. **(Step 1)**
2. The second fire fighter serves as team leader and, as such, gives the commands until the patient is supine on the backboard. Because the second fire fighter lifts and turns the victim's torso, he or she must be physically capable of moving the patient. The second fire fighter works from the driver's side doorway. If the first fire fighter is also working from that doorway, the second fire fighter should stand closer to the door hinges toward the front of the vehicle. The second fire fighter applies a cervical collar. **(Step 2)**
3. The second fire fighter provides continuous support of the victim's torso until the victim is supine on the backboard. Once the second fire fighter takes control

13-13 Skill Drill

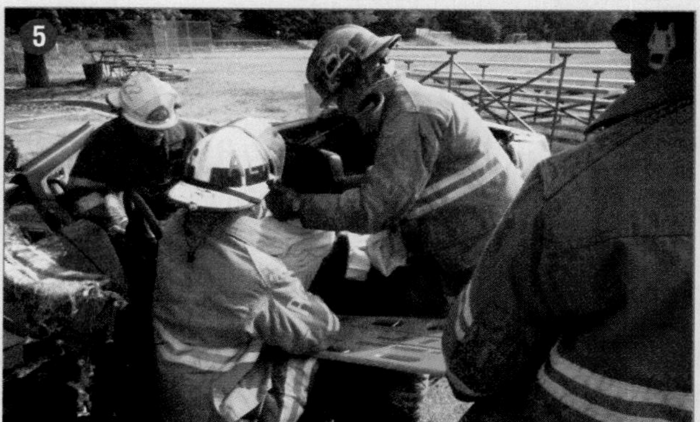

The first (or fourth) fire fighter places the backboard on the seat against the victim's buttocks. The second and third fire fighters lower the victim onto the long backboard.

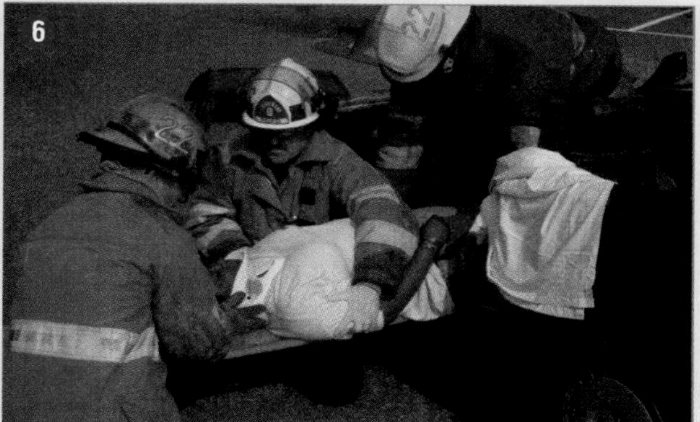

The third fire fighter moves to an effective position for sliding the victim. The second and third fire fighters slide the victim along the backboard in coordinated, 8" to 12" moves until the hips rest on the backboard.

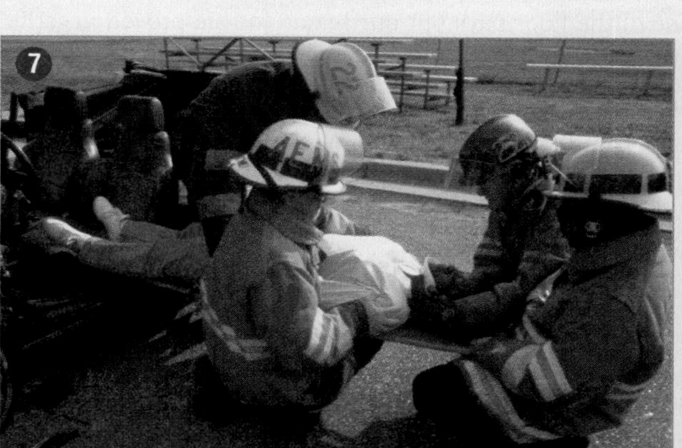

The third fire fighter exits the vehicle, moves to the backboard opposite the second fire fighter, and together they continue to slide the victim until the victim is fully on the backboard.

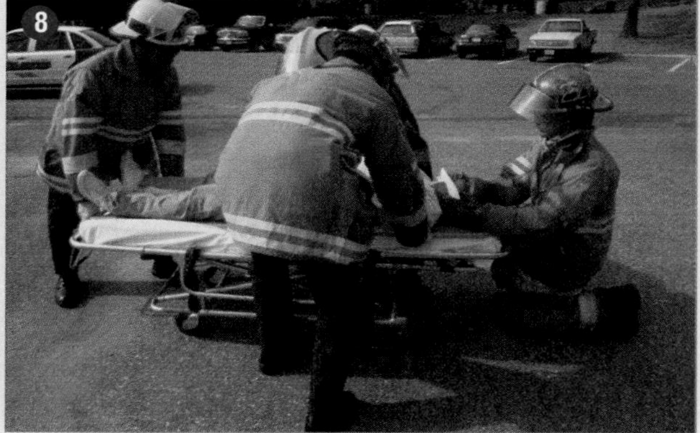

The first (or fourth) fire fighter continues to stabilize the victim's head and neck while the second, third, and fourth fire fighters carry the victim away from the vehicle.

of the torso, usually in the form of a body hug, he or she should not let go of the victim for any reason. Some type of cross-chest shoulder hug usually works well, but you will have to decide what method works best for you on any given victim. You must remember that you cannot simply reach into the car and grab the victim; this will only twist the victim's torso. You must rotate the victim as a unit.

4. The third fire fighter works from the front passenger's seat and is responsible for rotating the victim's legs and feet as the torso is turned, ensuring that they are free of the pedals and any other obstruction. With care, the third fire fighter should first move the victim's nearer leg laterally without rotating the victim's pelvis and lower spine. The pelvis and lower spine rotate only as the third fire fighter moves the second leg during the next step. Moving the nearer leg early makes it much easier to move the second leg in concert with the rest of the body. After the third fire fighter moves the legs together, they should be moved as a unit. **(Step 3)**
5. The victim is rotated 90° so that the back is facing out the driver's door and the feet are on the front passenger's seat. This coordinated movement is done in three or four short, quick "eighth turns." The second fire fighter directs each quick turn by saying, "Ready,

Voices of Experience

“We later found out that the mother was actually found first but could not be removed due to her weight and position in the bathtub. The child was found later in her arms, hidden from the initial search team's view.”

It was a rainy winter afternoon when we responded to a fire in a four-story building with sixteen apartments. The fire was located in the second floor rear apartment and was radiating to the third floor. Additionally, the high humidity would not allow the heavy smoke condition to lift.

I was assigned to a six-person rescue company. Upon arrival, we were assigned to search the floors above the fire. My partner and I were sent to the third floor while the other members went to the fourth floor. The smoke condition on the third floor was heavy. We found fire in one of the floor bays but our search for life proved negative. We called the IC and requested a handline to fight the fire in the floor bay.

As we were completing our operations, the team on the fourth floor contacted us via radio. They had found victims and needed assistance. We quickly worked our way up to the fourth floor.

The visibility on the fourth floor was zero. Upon reaching the stairway landing, we were met by a company member removing a young semi-conscious child who was soaking wet. He advised us that there was another victim in the bathroom two doors down.

We found a woman in her thirties lying in a bathtub full of water. We had seen this before; it is a common response for confused occupants to climb into a bathtub full of water, believing that the water will protect them from the fire. Smoke incapacitates the occupants who lose consciousness in a bathtub full of water.

We later found out that the mother was actually found first but could not be removed due to her weight and position in the bathtub. The child was found later in her arms, hidden from the initial search team's view.

We attempted to pull the woman out of the tub, but her weight plus the additional weight of her waterlogged clothes made it quite difficult. We decided to use my 24′ length of tubular webbing to sling the victim. This made it possible to remove the victim from the tub and drag her down the long narrow hallway to the apartment threshold where a Stokes basket and the remainder of the company were waiting. Both mother and child survived due to teamwork, equipment, and practical training in victim search and rescue.

Mark Gregory
New York City Fire Department
New York, New York

turn" or "Ready, move." Hand position changes should be made between moves.

6. In most cases, the first fire fighter will be working from the back seat. At some point, either because the door-post is in the way or because he or she cannot reach farther from the back seat, the first fire fighter will be unable to follow the torso rotation. At that time, the third fire fighter should assume temporary support of the head and neck until the first fire fighter can regain control of the head from outside the vehicle. If a fourth fire fighter is present, he or she stands next to the second fire fighter. The fourth fire fighter takes control of the head and neck from outside the vehicle without involving the third fire fighter. As soon as the change has been made, the rotation can continue. **(Step 4)**
7. Once the victim has been fully rotated, the backboard should be placed against the victim's buttocks on the seat. Do not try to wedge the backboard under the victim. If only three fire fighters are present, be sure to place the backboard within arm's reach of the driver's door before the move so that the board can be pulled into place when needed. In such cases, the far end of the board can be left on the ground. When a fourth fire fighter is available, the first fire fighter exits the rear seat of the car, places the backboard against the patient's buttocks, and maintains pressure in toward the vehicle from the far end of the board. (Note: When the door opening allows, some fire fighters prefer to insert the backboard onto the car seat before the victim is rotated.)
8. As soon as the victim has been rotated and the backboard is in place, the second fire fighter and third fire fighter lower the patient onto the board while supporting the head and torso so that neutral alignment is maintained. The first fire fighter holds the backboard until the victim is secured. **(Step 5)**
9. Next, the third fire fighter must move across the front seat to be in position at the victim's hips. If the third fire fighter stays at the victim's knees or feet, he or she will be ineffective in helping to move the body's weight. The knees and feet follow the hips.
10. The fourth fire fighter maintains support of the head and now takes over giving the commands. The second fire fighter maintains direction of the extrication. The second fire fighter stands with his or her back to the door, facing the rear of the vehicle. The backboard should be immediately in front of the third fire fighter. The second fire fighter grasps the patient's shoulders or armpits. Then, on command, the second fire fighter and the third fire fighter slide the victim 8" to 12" along the backboard, repeating this slide until the victim's hips are firmly on the backboard. **(Step 6)**
11. At that time, the third fire fighter gets out of the vehicle and moves to the opposite side of the backboard, across from the second fire fighter. The third fire fighter now takes control at the shoulders, and the second fire fighter moves back to take control of the hips. On command, these two fire fighters move the victim along the board in 8" to 12" slides until the victim is placed fully on the board. **(Step 7)**
12. The first (or fourth) fire fighter continues to maintain support of the head. The second fire fighter and third fire fighter now grasp their side of the board, and then carry it and the victim away from the vehicle onto the prepared cot nearby. **(Step 8)**

These steps must be considered a general procedure to be adapted as needed. Two-door cars differ from four-door models. Larger cars differ from smaller compact models, pickup trucks, and full-size sedans and four-wheel-drive vehicles. You will handle a large, heavy adult differently from a small adult or child. Every situation will be different—a different car, a different patient, and a different crew. Your resourcefulness and ability to adapt are necessary elements to successfully perform this technique.

Assisting a Person Down a Ground Ladder

Using a ground ladder to rescue a trapped occupant is one of the most critical, stressful, and demanding tasks performed by fire fighters. Assisting someone down a ladder involves a considerable risk of injury to both fire fighters and occupants. The fire fighter must use the proper technique to safely accomplish a ladder rescue. In addition, the fire fighter must have the physical strength and stamina to accomplish the rescue without injury to anyone involved. Although circumstances could require an individual fire fighter to work alone, at least two fire fighters should work as a team to rescue a person whenever possible.

Time is a critical factor in many rescue situations. Someone waiting to be rescued is often in immediate danger and may be preparing to jump. Fire fighters may have only a limited time to work. They must quickly and efficiently raise a ladder and assist the occupant to safety.

Ladder rescue begins with proper placement of the ladder. A ladder used to rescue a person from a window should have its tip placed just below the windowsill. This positioning makes it easier for the person to mount the ladder. If possible, one or more fire fighters in the interior should help the person onto the ladder, and one fire fighter should stay on the ladder to assist the individual down.

A ladder used for rescue must be heeled or tied in. The weight of an occupant and one or two fire fighters, all moving on the ladder at the same time, can easily destabilize a ladder that is not adequately secured.

Rescuing a Conscious Person From a Window

When a rescue involves a conscious person, fire fighters should establish verbal contact as quickly as possible to reassure the individual that help is on the way. Many people have jumped to their deaths seconds before a ladder could be raised to a window. All ladder rescues should be performed with two fire fighters whenever possible; Skill Drill 13-14 presents a technique that could, if necessary, be performed by a single fire fighter.

1. The rescue team will place the ladder into the rescue position with the tip of the ladder just below the windowsill, and secure the ladder in place.
2. The first fire fighter should climb the ladder, make contact with the victim, and climb inside the window to assist the victim. Contact should be made as soon as possible to calm the victim. The victim should be encouraged to stay at the window until the rescue can be performed.
3. The second fire fighter should climb up to the window, standing one rung below the windowsill. This leaves at least one rung available for the victim. When ready, the fire fighter should advise the victim to slowly come out onto the ladder, feet first and facing the ladder.
4. The fire fighter should form a semi-circle around the victim, with both hands on the beams of the ladder.
5. The fire fighter and victim should then proceed slowly down the ladder, one rung at a time, with the fire fighter one rung below the victim.
6. If the victim slips or loses footing, the fire fighter's legs should keep the victim from falling.
7. The fire fighter can take control of the victim at any time by leaning in toward the ladder and squeezing the victim against the ladder.

Rescuing an Unconscious Person from a Window

If the trapped person is unconscious, one or more fire fighters will have to climb inside the building and pass the person out of the window to a fire fighter on the ladder. Caution should be used when lowering an unconscious person down a ladder, because it is very easy for the person's arms or legs to get caught in the ladder. To rescue an unconscious person via a ladder, follow the steps in ► Skill Drill 13-15.

1. The rescue team will place and secure the ladder in the rescue position with the tip of the ladder just below the windowsill. **(Step 1)**
2. One fire fighter should climb up the ladder and into the window to assist from the inside. The second fire fighter climbs up to the window opening and waits for the victim. **(Step 2)**
3. The fire fighter on the ladder should have a firm grip on the ladder with both hands on the rungs. One leg should be straight and the other should be bent so that the thigh is horizontal to the ground, with the knee at a 90° angle. The foot of the straight leg should be one rung below the foot of the bent leg.
4. When both fire fighters are ready, the interior fire fighter will pass the victim out through the window and onto the ladder. The victim's back should be toward the ladder, so the victim is face-to-face with the fire fighter on the ladder. **(Step 3)**
5. The victim should be lowered so that the groin of the victim will rest on the horizontal leg of the fire fighter. The fire fighter's arms should be under the arms of the victim and holding onto the rungs. It is important to keep the balls of both feet on the rungs of the ladder. It is much more difficult to move your feet in this position if the heels are close to the rungs. **(Step 4)**
6. The fire fighter can now climb down the ladder slowly one rung at a time. The victim is always supported at the groin by one of the fire fighter's legs. The fire fighter's arms are under the victim's arms to support the upper torso. As an option the interior fire fighter may tie the victim's hands together and place them over the neck of the fire fighter on the ladder. **(Step 5)**

Rescuing an Unconscious Child or Small Adult from a Window

Small adults and children can be cradled across a fire fighter's arms during a rescue. The child must be light enough that the fire fighter can descend safely, using only arm strength to support the victim. To carry an unconscious child down a ladder, follow the steps in ► Skill Drill 13-16.

1. The rescue team will set up and secure the ladder in the rescue position, with the tip of the ladder just below the windowsill. **(Step 1)**
2. The first fire fighter should climb the ladder and enter the window to assist from the interior. The second fire fighter will climb the ladder to the window opening and wait for the victim. Both arms should be level with hands on the beams. **(Step 2)**
3. When ready, the interior fire fighter should pass the victim to the fire fighter on the ladder so the victim is cradled across the fire fighter's arms. **(Step 3)**
4. The fire fighter can now climb down the ladder slowly with the victim being held in his or her arms. The fire fighter's hands should slide down the beams. **(Step 4)**

Rescuing a Large Adult

Three fire fighters using two ladders may be needed to rescue very tall or heavy adults. To rescue a large adult using a ladder, follow the steps in Skill Drill 13-17.

1. The rescue team will place and secure two ladders, side-by-side, in the rescue position. The tips of the two ladders should be just below the windowsill.
2. Multiple fire fighters may be required to enter the window to assist from the inside.
3. Two fire fighters, one on each ladder, should climb up to the window opening and wait for the victim.
4. When ready, the victim should be lowered down across the arms of the fire fighters, with one supporting the victim's legs and the other supporting the victim's arms. Once in place, the fire fighters can slowly descend the ladder, using both hands to hold onto the ladder rungs.

Rescuing an Unconscious Victim from a Window

1 The tip of the ladder is placed just below the windowsill.

2 One fire fighter enters to assist the victim. The second fire fighter climbs to the window.

3 The fire fighter waiting on the ladder places both hands on the rungs with one leg straight and the other horizontal to the ground with the knee at 90°. The interior fire fighter will then pass the victim through the window onto the ladder with the victim's back toward the ladder.

4 Lower the victim to straddle the fire fighter's leg. The fire fighter's arms should be under the victim's arms holding onto the rungs.

5 Step down one rung at a time, transferring the victim's weight from one leg to the other. The victim's arms can be secured around the fire fighter's neck.

13-16 Skill Drill

Rescuing an Unconscious Child or Small Adult from a Window

1 The ladder is placed in rescue position, with the tip below the windowsill.

2 One fire fighter enters the window to assist the victim. The second fire fighter stands on the ladder to receive the victim, with both arms level and hands on beams.

3 The victim is placed in the fire fighter's arms.

4 The fire fighter descends, keeping arms level and sliding the hands down the beams.

Fire Fighter Tips

If you are removing a deceased person, first place the victim in a body bag, then place the body bag on a backboard or basket stretcher. Carrying a body bag without a rigid support is difficult and awkward.

Fire Fighter Tips

In some situations, it may be necessary to raise or lower a victim using a rope and a Stokes basket or harness. These techniques, which are not included in this text, require additional training and regular practice.

Removal of Victims by Ladders

Ladders should be used to remove victims only when it is not possible to use interior stairways or fire escapes. A ladder rescue is often frightening to conscious victims. Rescuing an unconscious victim by ladder is dangerous and difficult, but may be the best way to save a life.

Aerial ladders and platforms: An aerial ladder or platform often is used for rescue operations. The same basic rescue techniques are used with both aerial and ground ladders, but aerial ladders have several advantages over ground ladders. Aerial ladders are much stronger and have a longer reach. They also are wider and more stable, with side rails for additional security (► Figure 13-15).

Aerial platforms are even more suitable for rescue operations. These devices reduce the risk of slipping and falling because the victim is lowered to the ground mechanically. An aerial platform is usually preferred for rescue work if one is available.

Ground ladders: Before a ground ladder can be used in a rescue, it must be properly positioned and secured. Positioning and securing ground ladders are covered in Chapter 12, Ladders. Additional personnel will be needed to secure the ladder and to assist in bringing the victim down the ladder.

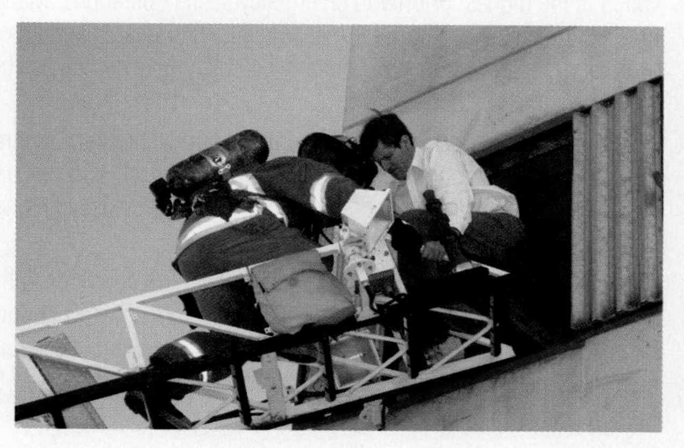

Figure 13-15 An aerial ladder is stronger and more stable than a ground ladder.

Wrap-Up

Ready for Review

This chapter discussed the activities involved with search and rescue. Search-and-rescue activities are directly related to life safety, which is the highest priority in emergency scene operations. Many fire departments assign primary responsibility for search and rescue to truck or rescue companies, but any company may get the assignment when the need arises.

Search-and-rescue activities must be integrated with other firefighting activities. Ventilation, fire suppression, and search-and-rescue activities must be carefully coordinated because these activities have a direct impact on each other.

This chapter discussed the importance of using fire conditions and safety considerations to determine whether search activities should be started. Because search activities are time-dependent, the areas closest to the fire should be searched first. Top priorities include the fire floor and the floor above the fire.

The importance of wearing full personal protective clothing and carrying appropriate hand tools, lights, and radios was covered. Searching requires fire fighters to use their senses of sight, touch, and hearing. The use of thermal imaging devices increases the effectiveness and efficiency of a search. Fire fighters must be aware of possible escape routes at all times.

This chapter emphasized the function of the primary search and the importance of conducting it safely and efficiently. The role of the secondary search was discussed.

The techniques used to rescue a victim from a fire also were discussed. Different assists, lifts, moves, and carries were outlined. The steps in bringing responsive and unresponsive victims down ladders were presented.

Chief Concepts

- Search and rescue are the highest priorities at a fire scene and may be assigned to any type of fire company.
- Search and rescue must be integrated with other firefighting functions.
- The initial overview and observations of the fire scene provide valuable information for making decisions about search-and-rescue functions.
- The IC must be able to make an informed decision about whether it is safe to begin a search.
- Search-and-rescue priorities start with the fire floor and then move to the floor above the fire.
- Fire fighters must be properly dressed and equipped for search and rescue.
- Fire fighters must use the senses of sight, touch, and hearing when searching.
- Thermal imaging equipment can improve the effectiveness and efficiency of a search in a smoke-filled building.
- Fire fighters must always be aware of possible escape routes.
- The primary search should be as thorough as possible in the time available.
- The secondary search is made after the fire is under control.
- Rescue techniques include assists, drags, and carries.

Hot Terms

Primary search An initial search conducted to determine if there are victims who must be rescued.

Rekindle A situation where a fire, which was thought to be completely extinguished, reignites.

Rescue Those activities directed at locating endangered persons at an emergency incident, removing those persons from danger, treating the injured, and providing for transport to an appropriate health care facility.

Search The process of looking for victims who are in danger.

Search rope A guide rope used by fire fighters that allows them to maintain contact with a fixed point.

Secondary search A more thorough search undertaken after the fire is under control. This search is done to ensure that there are no victims still trapped inside of the building.

Thermal imaging devices Electronic devices that detect differences in temperature based on infrared energy and then generate images based on that data. Commonly used in obscured environments to locate victims.

Two-in/two-out rule A safety procedure that requires a minimum of two personnel to enter a hazardous area and a minimum of two back-up personnel to remain outside the hazardous area during the initial stages of an incident.

Fire Fighter in Action

It is a rainy Thursday morning when your engine is dispatched to a residential structure fire. Dispatch reports that calls indicate that fire is coming out a window on the second floor. You arrive on the scene and your lieutenant gives a size up and establishes command. You are met at the driveway by a hysterical man. He states that his elderly mother is still upstairs. Lieutenant Cox tells you and your crew to prepare for a primary search and rescue. You are ascending the stairs and are met with heavy smoke and heat conditions.

1. The following statements are true about search and rescue risk-benefit except:
   - **A.** It is not acceptable to risk the safety of fire fighters when there is no possibility of saving lives or property
   - **B.** Pets are considered family members and should have the same risk consideration as human loss of life.
   - **C.** Only a limited level of risk is acceptable to save valuable property
   - **D.** Actions that present a high level of risk to the safety of fire fighters are justified only if there is a potential to save lives

2. You use the thermal imaging device to scan the hallway. The thermal imaging device captures:
   - **A.** infrared light rather than visible light images.
   - **B.** light using a photocathode to convert photon energy to electrons and to strike a phosphor screen that emits an image you can see.
   - **C.** heat images rather than visible light images.
   - **D.** none of the above

With the aid of the thermal imaging device, you and your partner find the victim quickly. She is unconscious but breathing. You need to remove the victim fast. The smoke is heavy and visibility is poor.

3. What would be the most appropriate rescue technique?
   - **A.** A webbing sling drag
   - **B.** A two-person walking assist
   - **C.** An exit assist
   - **D.** A cradle-in-arms carry

4. When removing a victim from a dangerous area you should:
   - **A.** use the safest and most practical means of egress.
   - **B.** use the buildings normal exit system if clear and safe.
   - **C.** use a ladder rescue whenever possible.
   - **D.** both A and B

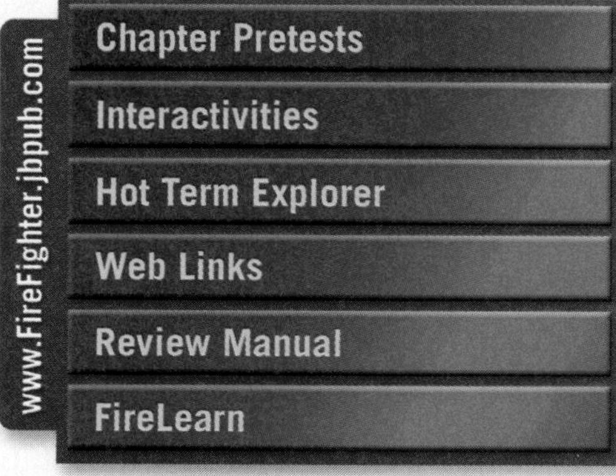

Ventilation

Chapter 14

NFPA 1001 Standard

Fire Fighter I

5.3.11 Perform horizontal ventilation on a structure operating as part of a team, given an assignment, personal protective equipment, ventilation tools, equipment, and ladders, so that the ventilation openings are free of obstructions, tools are safely used as designed, ladders are correctly placed, ventilation devices are correctly placed, and the structure is cleared of smoke.

5.3.11 (A) *Requisite Knowledge.* The principles, advantages, limitations, and effects of horizontal, mechanical, and hydraulic ventilation; safety considerations when venting a structure; fire behavior within a structure; the products of combustion found within a structure fire; the signs, causes, and effects, and prevention of backdrafts; and the relationship of oxygen concentration to life safety and fire growth.

5.3.11 (B) *Requisite Skills.* The ability to transport and operate ventilation tools and equipment and ladders and to use safe procedures for breaking window and door glass and removing obstructions.

5.3.12 Perform vertical ventilation on a structure as part of a team, given an assignment, personal protective equipment, ground and roof ladders, and tools, so that ladders are positioned for ventilation, a specified opening is created, all ventilation barriers are removed, structural integrity is not compromised, products of combustion are released from the structure, and the team retreats from the area when ventilation is accomplished.

5.3.12 (A) *Requisite Knowledge.* The methods of heat transfer; the principles of thermal layering within a structure on fire; the techniques and safety precautions for venting flat roofs, pitched roofs, and basements; basic indicators of potential collapse or roof failure; the effects of construction type and elapsed time under fire conditions on structural integrity; and the advantages and disadvantages of vertical and trench/strip ventilation.

5.3.12 (B) *Requisite Skills.* The ability to transport and operate ventilation tools and equipment; hoist ventilation tools to a roof; cut roofing and flooring materials to vent flat roofs, pitched roofs, and basements; sound a roof for integrity; clear an opening with hand tools; select, carry, deploy, and secure ground ladders for ventilation activities; deploy roof ladders on pitched roofs while secured to a ground ladder; and carry ventilation-related tools and equipment while ascending and descending ladders.

Fire Fighter II

6.3.2 Coordinate an interior attack line team's accomplishment of an assignment in a structure fire, given attack lines, personnel, personal protective equipment, and tools, so that crew integrity is established; attack techniques are selected for the given level of the fire (for example, attic, grade level, upper levels, or basement); attack techniques are communicated to the attack teams; constant team coordination is maintained; fire growth and development is continuously evaluated; search, rescue and ventilation requirements are communicated or managed; hazards are reported to the attack teams, and Incident Command is apprised of changing conditions.

6.3.2 (A) *Requisite Knowledge.* Selection of the nozzle and hose for fire attack given different fire situations; selection of adapters and appliances to be used for specific fire ground situations; dangerous building conditions created by fire and fire suppression activities; indicators of building collapse; the effects of fire and fire suppression activities on wood, masonry (brick, block, stone), cast iron, steel, reinforced concrete, gypsum wall board, glass, and plaster on lath; search and rescue and ventilation procedures; indicators of structural instability; suppression approaches and practices for various types of structural fires and the association between specific tools and special forcible entry needs.

6.3.2 (B) *Requisite Skills:* The ability to assemble a team, choose attack techniques for various levels of fire (e.g., attic, grade level, upper levels, or basement), evaluate and forecast a fire's growth and development, select tools for forcible entry, incorporate search and rescue procedures and ventilation procedures in the completion of the attack team efforts, and determine developing hazardous building or fire conditions.

Knowledge Objectives

After studying this chapter, you will be able to:

- Define ventilation as it relates to fire suppression activities.
- List the effects of properly performed ventilation on fire and fire suppression activities.
- Describe how fire behavior principles affect ventilation.
- Describe how building construction features within a structure affect ventilation.
- List the principles, advantages, limitations, and effects of horizontal ventilation.
- List the principles, advantages, limitations, and effects of natural ventilation.
- List the principles, advantages, limitations, and effects of mechanical ventilation.
- List the principles, advantages, limitations, and effects of negative-pressure and positive-pressure ventilation.
- List the principles, advantages, limitations, and effects of hydraulic ventilation.
- List the principles, advantages, limitations, and effects of vertical ventilation.
- List safety precautions for ventilating roofs.
- List the basic indicators of roof collapse.
- Explain the role of ventilation in the prevention of backdraft and flashover.

Skills Objectives

After completing this chapter, you will be able to:

- Break glass with a hand tool.
- Break a window with a ladder.
- Break windows on upper floors using the Halligan toss.
- Establish negative-pressure ventilation.
- Establish positive-pressure ventilation.
- Sound a roof.
- Operate a power saw.
- Perform a rectangular cut.
- Perform a louver cut.
- Perform a triangular cut.
- Perform a peak cut.
- Perform a trench cut.

You Are the Fire Fighter

Your engine company is responding to an early morning report of a dwelling fire. When you arrive, the street is obscured by dense brown smoke. Visibility is so restricted that it is difficult to determine which dwelling is on fire. The engine advances slowly and stops near a three-story wood-frame dwelling. Smoke is seeping from around the closed doors and windows, and then lingering and hanging near the ground level in the front yard. You begin to stretch the line with your officer and company.

The volume of smoke and the heat levels inside the house will make entry difficult. Proper and aggressive ventilation will be essential and must be closely coordinated with search-and-rescue and suppression efforts. Simultaneous ventilation will be needed so that the engine company can enter, locate and attack the fire, and rescue the occupants, if they have not already succumbed.

1. ***If conditions are this severe upon your arrival, how long has the fire been burning, and what fire conditions can you expect upon entry?***
2. ***If the volume of smoke and heat is so high, are you making a rescue or a recovery?***
3. ***What are the priorities?***

Introduction

Fire service **ventilation** is the process of removing smoke, heat, and toxic gases from a burning building and replacing them with cooler, cleaner, more oxygen-rich air. When ventilation is coordinated with fire attack and water, it can save lives and reduce property damage. Proper ventilation assists in the location and rescue of victims, enables hose teams to advance and locate the source of the fire, and prevents fire spread. The lack of ventilation or improper ventilation techniques can spread the fire, injure both fire fighters and civilians, and increase property damage.

During its normal progression and growth, a fire gives off smoke, heat, and toxic gases. As long as there is fuel and oxygen, the fire will continue to burn and produce these three **products of combustion**. When the fire is inside a building, the structure acts as a container or box, trapping the products of combustion. As the fire grows and develops, the smoke, heat, and toxic gases spread throughout the structure, presenting a direct risk to the lives of occupants and fire fighters, reducing visibility, and increasing property damage. In some cases, the trapped products of combustion create a potential for explosion. Ventilation removes these products from the interior atmosphere or allows them to escape in a controlled manner.

The primary principle that controls the spread of smoke, heat, and toxic gases within a room or a building is **convection**. See Chapter 5, Fire Behavior, for more information. Heated gases expand, becoming less dense than cooler gases. As a result, the hot gases produced by a fire in a closed room will rise to the ceiling and spread outward, displacing cooler air and pushing it toward the floor (▲ **Figure 14-1**). As the fire continues to burn, the hot layer of gas banks (curves) down closer to the floor.

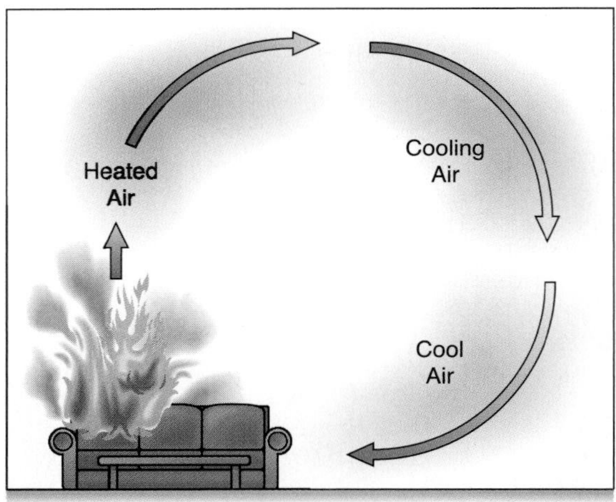

Figure 14-1 Convection currents cause heated products of combustion to rise within a room and spread along the ceiling.

If the heated products of combustion escape from the room, the same principles will apply as they spread throughout a building. Smoke, heat, and toxic gases will spread horizontally, along the ceiling, until they find a path such as a stairway, elevator shaft, or pipe **chase** (an open space within a wall where wires and pipes can run) that allows them to reach a higher level. They will then flow upward through the vertical opening until they reach another horizontal obstruc-

Figure 14-2 The heated products of combustion spread horizontally along the ceiling until they find a path that allows them to flow upward.

Figure 14-4 Self-contained breathing apparatus is essential when entering an area contaminated by products of combustion.

Figure 14-3 Mushrooming occurs as the heated products of combustion flow outward and downward.

Figure 14-5 Fire spreads where there is fuel and wherever the hot products of combustion flow or accumulate.

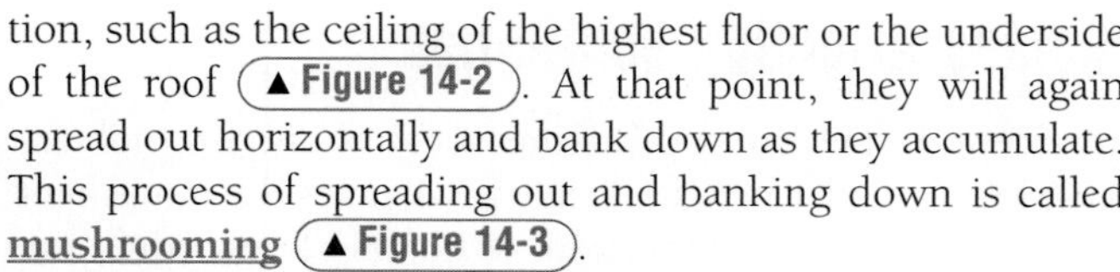

tion, such as the ceiling of the highest floor or the underside of the roof (▲ Figure 14-2). At that point, they will again spread out horizontally and bank down as they accumulate. This process of spreading out and banking down is called **mushrooming** (▲ Figure 14-3).

As long as the products of combustion are trapped within the structure, they present a series of risks and dangers to the occupants and to fire fighters. Most of the gases produced by a fire, most notably carbon monoxide, are toxic. The contaminated atmosphere they create poses a life-threatening condition to occupants as well as to anyone who enters without self-contained breathing apparatus (SCBA) (► Figure 14-4). The gases may be so hot that simply breathing them can cause fatal respiratory burns. As the fire continues to consume oxygen from the atmosphere, an additional respiratory hazard may be created. The particulate matter in smoke can severely obscure visibility, making it impossible for occupants to find their way to an exit or for rescuers to locate trapped occupants. Smoke is also irritating to the eyes and the mucous membranes of the respiratory system.

Convection, the flow of heated gases produced by the fire, is one of the primary mechanisms of fire spread. The gases may be hot enough to ignite combustible materials along their path. The fire spreads along the paths taken by the heated gases and in the areas where they accumulate. If the hot gases spread into additional areas, through vertical or horizontal openings, there is a high probability that those areas will also become involved in the fire (▲ Figure 14-5).

In addition to igniting combustible materials, the hot gases, smoke, and other products of combustion can themselves explode. In many cases, they may include a rich supply of partially burned fuels that are hot enough to ignite, but lack sufficient oxygen to support combustion. If these products are mixed with clean air, the atmosphere itself can ignite or, in an extreme situation, explode in a **backdraft** (sudden explosive ignition of fire gases when oxygen is introduced into a superheated space). Even if they do not ignite, the products of combustion are often hot enough to cause thermal burns to exposed skin.

Yet another concern is the property damage caused by exposure to smoke and heat. Soot and other residues of smoke often cause as much property damage as flames. Significant heat damage can occur in areas that were exposed only to products of combustion, not to the fire itself. Even when a fire is contained and under control, prompt and effective ventilation of the accumulated heat and smoke can be an important factor in limiting property losses.

Benefits of Proper Ventilation

Ventilation is the process of controlling the flow of smoke, heat, and toxic gases so that they are released safely and effectively from a building. Ventilation must be closely coordinated with all other activities being carried out at a fire scene.

The life safety benefits of ventilation are of primary importance. Ventilation should allow the smoke to lift slightly so that fire fighters can locate trapped occupants more rapidly. In addition, ventilation can provide clean air to occupants who may be overcome by toxic products of combustion.

Ventilation's role in removing heat is important for fire fighters advancing an initial attack hose line. As the fire fighters advance, ventilation can relieve the pressure and intense heat of the fire and create a much safer environment. A vent in front of the attack hose line pulls away the steam that is created when water cools the flames. It also cools the atmosphere and prevents the steam from banking down on top of the attack line. The ventilation opening also causes the smoke to lift and allows fire fighters to aim the hose stream directly at the **seat of the fire** (the main area of the fire). By reducing the potential for a backdraft or a **flashover** (the sudden ignition of all combustible objects in a room), effective ventilation also increases the safety of the attack team.

Proper ventilation also helps to limit fire spread within the structure. Ventilating near the source of the fire can limit the area of involvement and reduce the spread of heat and toxic gases throughout the structure (► Figure 14-6). Initially, this may allow more oxygen to reach the fire, causing a flare-up. A hose line should be in place and ready to hit the seat of the fire prior to ventilation. Careful timing of ventilation will enable fire fighters to confine and extinguish a fire quickly.

Releasing trapped heat and smoke from the upper level of a building will often prevent the fire from spreading horizontally into adjoining areas or neighboring buildings. A well-placed opening will release heat and smoke to the outside so that fire fighters can operate effectively inside the building.

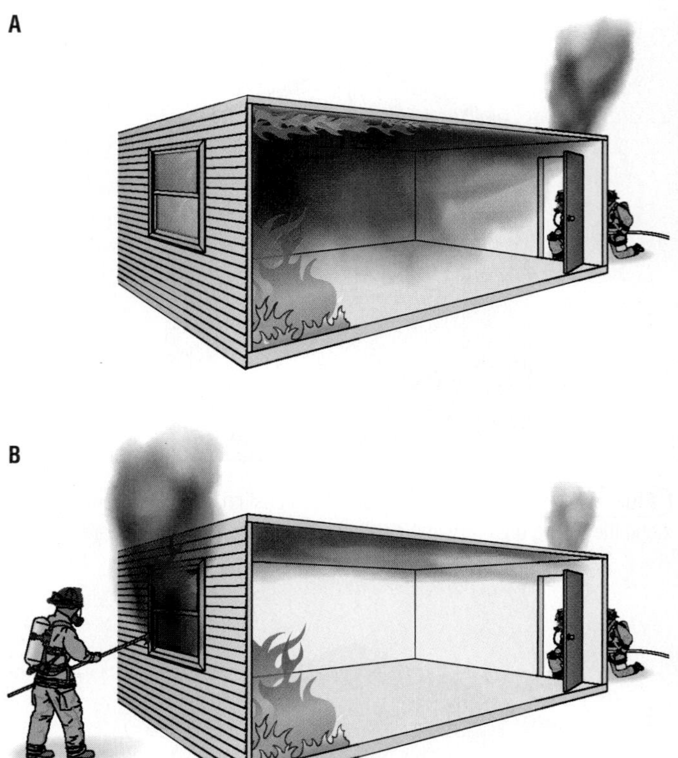

Figure 14-6 Proper ventilation enables fire fighters to control a fire rapidly. **A.** Unvented structure. **B.** Vented structure.

After the fire is under control, ventilation should be continued until all of the heat, smoke, and toxic products of combustion have been removed from the building. Complete ventilation creates a clear atmosphere for overhauling the fire and limits property losses caused by smoke damage.

Factors Affecting Ventilation

In order to ventilate a structure successfully, fire fighters must first consider how fire behavior dictates the movement of the products of combustion. With this knowledge, fire fighters can develop a plan of where and when to create openings to release these products and limit smoke and fire spread. Fire behavior is directly related to the principles of heat transfer, which involve convection, conduction, and radiation. The movement of smoke and heat within a structure is primarily related to convection.

Convection

Convection refers to the transfer of heat through a circulating medium of liquid or gas. Heated air naturally expands and rises, carrying smoke and gases as it flows up and out from a fire. This heated mixture flows across the underside of ceilings, through doorways, and up stairways, elevator shafts, and vertical concealed spaces such as pipe chases. Unless it is released to the exterior, the heat will accumulate in any accessible space in the upper levels of a building.

As the heated air rises, it displaces cooler air, which is drawn toward the seat of the fire. The cooler air brings oxygen to support combustion and causes the fire to intensify. As the fire continues to burn, the products of combustion spread throughout the building, eventually banking down to lower levels. Within any room or space, the hottest gases are at the highest level and cooler gases are stratified below.

Convection currents can carry smoke and superheated gases to uninvolved areas within the structure or into adjoining spaces, if there is an opening available. The flow of heated gases will always follow the path of least resistance and can be altered by making an opening that offers less resistance.

Fire fighters use this basic principle by making ventilation openings that will cause the convective flow to draw the heated products out of the building. Doorways, windows, or roof openings can be used for ventilation. To accomplish this, fire fighters must identify the most appropriate type of openings as well as the most feasible locations for openings in the structure.

Mechanical Ventilation

In addition to making or controlling openings to influence convective flow, fire fighters can also use mechanical ventilation to direct the flow of combustion gases. Fans can be used to draw or pull smoke through openings (negative-pressure ventilation) or to introduce fresh air to displace smoke and other products of combustion (positive-pressure ventilation). Hose streams can be used to create air currents to ventilate an area. Some buildings have ventilation systems designed to remove smoke or prevent smoke from entering certain areas.

Wind and Atmospheric Forces

Wind and other atmospheric forces can play a significant role in ventilation. The wind should always be considered when determining where and how to ventilate a building. Even a slight breeze can have an impact on the effectiveness of a ventilation opening.

For example, a wind blowing against one side of a building will prevent smoke, heat, or other products of combustion from escaping through an opening on that side. If a door or window on that side of the building is opened, a strong current of oxygen-rich air could enter the structure and accelerate the fire. The wind would push the heat and other products of combustion throughout the structure, creating serious life safety risks for those inside the building.

Conversely, opening a ventilation outlet on the leeward (the side the wind is blowing on) side of a building could be especially effective on a windy day. As the wind blows around the building, it creates a negative-pressure zone that literally removes smoke and heat out of the building. After the fire has been extinguished, opening additional doors and windows on the windward side will rapidly clear residual smoke (▲ Figure 14-7).

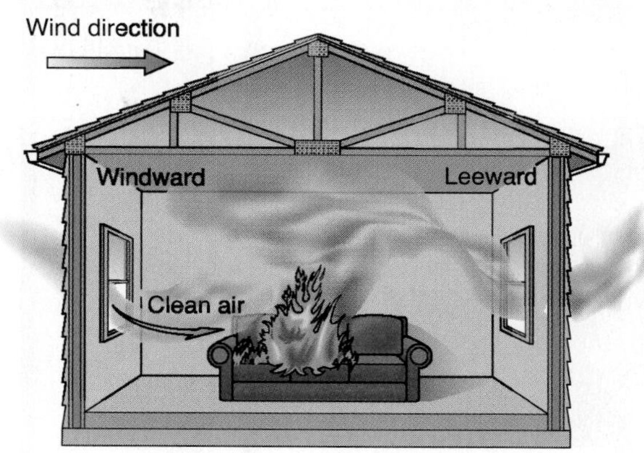

Figure 14-7 On a windy day, proper ventilation will remove smoke and heat from a structure.

Temperature and humidity also can affect ventilation, particularly in tall buildings and large-area structures. On a cold day, there will probably be a strong updraft through a heated high-rise building. A downdraft is more likely on a hot day in an air-conditioned building. Because humidity makes the atmosphere denser, removing cool smoke from a building could be difficult on a humid day.

Wind, temperature, and humidity are powerful, persistent forces of nature. It is more efficient and more effective to work with them than against them.

Building Construction Considerations

The way a building is constructed will affect ventilation operations. Each construction type presents its own set of problems. More information on building construction can be found in Chapter 6, Building Construction.

Fire-Resistive Construction

Fire-resistive construction refers to a building in which all of the structural components are made of noncombustible or limited-combustible materials. This type of construction generally has spaces divided into compartments, which limit potential fire spread. A fire is most likely to involve the combustible contents of an interior space within the building.

Although fire-resistive construction is intended to confine a fire, there are potential avenues for fire spread. Openings for mechanical systems, such as heating and cooling ducts, plumbing and electrical chases, elevator shafts, and stairwells all provide paths for fire spread. A fire can also spread from one floor to another through exterior windows,

Figure 14-8 Leap-frogging occurs when the fire jumps from floor to floor through exterior windows.

Figure 14-9 Vertical ventilation is essential when a fire is burning in an attic or cockloft.

a phenomenon called **leap-frogging** (or auto-exposure) ▲ Figure 14-8.

Smoke can travel through a fire-resistive building using the same routes. Heating, ventilation, and air conditioning (HVAC) systems, stairways, and elevator shafts are common pathways. Because the windows in these buildings are often sealed and are not broken easily, there may be limited opportunities for creating ventilation openings. In some cases, one stairway can be used as an exhaust shaft, while other stairways are cleared for rescue and access. Ventilation fans are often required to direct the flow of smoke.

The roof on a fire-resistive building is usually supported by either a steel or a concrete roof deck. It may be difficult or impossible to make vertical ventilation holes in such a roof. Similar roofing materials are used in buildings with noncombustible construction.

Ordinary Construction

Ordinary construction buildings have exterior walls made of noncombustible or limited-combustible materials, which support the roof and floor assemblies. The interior walls and floors are usually wood construction; the roof usually has a wood deck and a wood structural support system. The roof can be cut with power saws or axes to open vertical ventilation holes. Ordinary construction buildings usually have windows and doors that can be used for horizontal ventilation.

This type of construction often includes numerous openings within the walls and floors for plumbing and electrical chases. If a fire breaks through the interior finish materials (plaster or drywall), it can spread undetected within the void spaces to other portions of the structure. Interior stairwells generally allow a fire to extend vertically within the building.

The vertical openings in ordinary construction often provide a path for fire to extend into an attic or **cockloft** (open space between the ceiling of the top floor and the underside of the roof). Heat and smoke can accumulate and spread laterally within these spaces between sections of a large building or into attached buildings. Vertical ventilation is essential when fire is burning in an attic or cockloft ▲ Figure 14-9.

Wood-Frame Construction

Wood-frame construction has many of the same features as ordinary construction. The primary difference is that the exterior walls in a wood-frame building are not required to be constructed of masonry or noncombustible materials.

Wood-frame buildings often have many void spaces where fire can spread, including attics or cocklofts. Wood truss roofs and floors, which can fail quickly under fire conditions, are common in these buildings.

Older buildings of wood-frame construction were often assembled with **balloon-frame construction**. This type of construction has direct vertical channels within the exterior walls, so a fire can spread very quickly from a lower level to the attic or cockloft. Heated smoke and gases can accumulate under the roof, requiring rapid vertical roof ventilation.

Modern wood-frame construction uses **platform-frame techniques**. In these buildings, the structural frame is built one floor at a time. Between each floor, there is a plate at the floor and the ceiling that acts as a fire stop. This prevents the fire from spreading up and contains it on a single floor.

Tactical Priorities

Ventilation is directly related to the three major tactical priorities in structural firefighting operations: life safety, fire containment, and property conservation.

Venting for Life Safety

Life safety is the primary goal of the fire service, whether it is rescuing civilians or protecting fire department personnel. Ventilation helps to clear smoke, heat, and toxic gases from the structure, which gives occupants a better chance to survive. It provides firefighting crews increased visibility and makes the structure more tenable, enabling more rapid searches for victims.

Ventilation also limits fire spread and allows fire fighters to advance hose lines more safely and rapidly to attack the fire. Controlling the fire reduces the risk to fire fighters as well as occupants.

A dedicated ventilation team can often have a significant impact on life safety. Interior search and fire attack teams should be prepared to create ventilation openings as needed.

Venting for Fire Containment

A fire fighter's second priority is to contain the fire and gain control of the situation. Fire spread can often be controlled or limited through effective ventilation. Releasing smoke and superheated gases to the exterior prevents them from spreading throughout the interior of a building or into adjoining spaces.

Attack teams can be more effective when there is less smoke and heat. This enables attack lines to be advanced more easily to extinguish the fire. A coordinated fire attack and ventilation effort should consider the timing and the direction of the attack as well as the type and location of possible ventilation openings.

Venting for Property Conservation

The third priority of the fire service is property conservation, which involves reducing losses caused by means other than direct involvement in the fire. Ventilation can play a significant role in limiting property damage. If a structure is ventilated rapidly and correctly, the damage caused by smoke, heat, water, and overhaul operations can all be reduced.

Location and Extent of Smoke and Fire Conditions

A fire fighter must be able to recognize when ventilation is needed and where it should be provided, based on the circumstances of each fire situation. There are many factors that must be considered, including the size of the fire, the stage of combustion, the location of the fire within a building, and the available ventilation options. An experienced fire fighter learns to recognize these significant factors and quickly interpret each situation.

Figure 14-10 The color, location, and amount of smoke can provide valuable clues.

When feasible, ventilation should be as close to the fire as possible. The most desirable options are to provide a ventilation opening directly over the seat of the fire or to open a door or window that will release heat and smoke from the fire directly to the exterior. If it is not possible to create a ventilation opening in the immediate area of the fire, fire fighters must be able to predict how ventilation from another location will affect the fire. Usually, fire will travel toward a ventilation opening, following the flow of heat and smoke. To make the fire travel in the opposite direction, fire fighters could use the wind or a fan to push fresh air through a ventilation opening.

The color, location, movement, and amount of smoke can provide valuable clues about the fire's size, intensity, and fuel (▲ Figure 14-10). Thin, light-colored smoke, moving lazily out of the building, usually indicates a small fire involving ordinary combustibles. Thick, dark gray smoke "pushing" out of a structure suggests a larger, more intense fire. A fire involving petroleum products will produce large quantities of black, rolling smoke that rises in a vertical column.

Smoke movement is a good indicator of the fire's temperature. A very hot fire will produce smoke that moves quickly, rolling and forcing its way out through an opening. The hotter the fire is, the faster the smoke will move. Cooler smoke moves more slowly and gently. On a cool, damp day with very little wind, this type of smoke might hang low to the ground (a phenomenon known as **smoke inversion**).

An unusual phenomenon occurs in buildings with automatic sprinkler systems. The water discharged from the sprinklers cools the smoke, so that it hardly moves within the building. This type of smoke can fill a large warehouse space from floor to ceiling with an opaque mixture that behaves like fog on a damp day. Mechanical ventilation is needed to clear this cold smoke from the building. A similar situation occurs when smoke is trapped within a building long enough to cool to ambient temperature.

Figure 14-11 Windows are frequently used in horizontal ventilation.

Types of Ventilation

To remove the products of combustion and other airborne contaminants from a structure, fire fighters use two basic types of ventilation: horizontal and vertical. **Horizontal ventilation** utilizes the doors and windows on the same level as the fire, as well as any other horizontal openings that are available. In some cases, fire fighters might make additional openings in a wall to provide horizontal ventilation.

Vertical ventilation involves openings in roofs or floors so that heat, smoke, and toxic gases escape from the structure in a vertical direction. Pathways for vertical ventilation can include stairwells, exhaust vents, and roof openings such as skylights, scuttles, or monitors. Additional openings can be created by cutting a hole in the roof or the floor.

Ventilation can be either natural or mechanical. Natural ventilation depends on convection currents and other natural forces, such as the wind, to move heat and smoke out of a building and allow clean air to enter. Mechanical ventilation uses fans or other powered equipment to exhaust heat and smoke and/or to introduce clean air.

When referring to ventilation techniques, fire fighters use the term contaminated atmosphere to describe the products of combustion that must be removed from a building and the term clean air to refer to the outside air that replaces them. The contaminated atmosphere can include any combination of heat, smoke, and gases produced by combustion, including toxic gases. A contaminated atmosphere also may be any dangerous or undesirable atmosphere not caused by a fire. For example, natural gas, carbon monoxide, ammonia, and many other products can contaminate the atmosphere as a result of a leak, spill, or similar event. The same basic techniques of ventilation can be applied to many of these situations as well.

Horizontal Ventilation

Horizontal ventilation uses horizontal openings in a structure, such as windows and doors, and can be employed in many situations, particularly in small fires (▲ Figure 14-11). Horizontal ventilation is commonly used in residential fires, room-and-contents fires, and fires that can be controlled quickly by the attack team.

Horizontal ventilation can be a rapid, generally easy way to clear a contaminated atmosphere. Often, outlets for horizontal ventilation can be made simply by opening a door or window. In other cases fire fighters may need to break a window or to use forcible entry techniques. Search-and-attack teams operating inside a structure, as well as fire fighters operating outside the building, use horizontal ventilation to reduce heat and smoke conditions.

Horizontal ventilation is most effective when the opening goes directly into the room or space where the fire is located. Opening an exterior door or window allows heat and smoke to flow directly outside and eliminates the problems that can occur when these products travel through the interior spaces of a building.

Horizontal ventilation is more difficult if there are no direct openings to the outside or if the openings are inaccessible. In these situations, it may be necessary to direct the air flow through other interior spaces or to use vertical ventilation. Horizontal ventilation may have to be used in situations where an inaccessible or damaged roof makes vertical ventilation impossible.

Horizontal ventilation also can be used in less urgent situations if the building can be ventilated without additional structural damage. Low urgency situations might include residual smoke caused by a stovetop fire or a small natural gas leak.

Horizontal ventilation tactics include both natural and mechanical methods. Mechanical ventilation involves the use of fans or other powered equipment.

Natural Ventilation

Natural ventilation depends on convection currents, wind, and other natural air movements to allow a contaminated atmosphere to flow out of a structure. The heat of a fire creates convection currents that move smoke and gases up toward the roof or ceiling and out away from the fire source. Opening or breaking a window or door will allow these products of combustion to escape through natural ventilation.

Natural ventilation can only be used when the natural air currents are adequate to blow the contaminated atmosphere out of the building and replace it with fresh air. Mechanical devices can be used if natural forces do not provide adequate ventilation.

Natural ventilation is often used when quick ventilation is needed, such as when attacking a room-and-contents fire or a first-floor residential fire with people trapped on the second floor. For search-and-attack teams to enter and act quickly, smoke and heat must be ventilated immediately.

Wind speed and direction play an important role in natural ventilation. If possible, windows on the leeward side of a building should be opened first so the contaminated atmosphere flows out. Openings on the windward side can then be used for cross ventilation, bringing in clean air. Opening a window on the windward side first could push the fire into uninvolved areas of the structure, particularly on a windy day.

Breaking Glass

Natural ventilation uses pre-existing or created openings in a building. The methods used to create horizontal openings employ a variety of tools and techniques. For example, some windows can simply be opened by hand. But if the window cannot be opened and the need for ventilation is urgent, fire fighters should not hesitate to break the glass. Breaking a window is a fast and simple way to create a ventilation opening.

When breaking glass, the fire fighter should always use a hand tool (Halligan tool, axe, or pike pole) and keep hands above or to the side of the falling glass. This will prevent pieces of glass from sliding down the tool and potentially injuring the fire fighter. Then the tool should be used to clear the entire opening of all remaining pieces of glass. This creates the largest opening possible and provides a way for fire fighters to enter or exit through the window in the event of an emergency.

Clearing the broken glass also reduces the risk of injury to anyone who may be in the area and eliminates the danger of sharp pieces falling out later. Before breaking glass, particularly if the window is above the ground floor, fire fighters must look out to ensure that no one will be struck by the falling glass. To break glass on the ground floors with a hand tool, follow the steps in (► **Skill Drill 14-1**).

> **Fire Fighter Tips**
>
> Opening windows on the windward side of a building before opening leeward side windows may force the fire into uninvolved areas of the structure.

1. The fire fighter should wear full personal protective equipment, including eye protection.
2. Select a hand tool and position yourself to the side of the window. **(Step 1)**
3. With back against the wall, swing backward forcefully with the tip of the tool striking the top $^1/_3$ of the glass. **(Step 2)**
4. Clear remaining glass from the opening with the hand tool. **(Step 3)**

Breaking a Window from a Ladder

A similar technique can be used to break a window on an upper floor, with the fire fighter working from a ladder. The ladder should be positioned to the side of the window. The fire fighter should climb to a position level with the window and lock in to the ladder for safety. Using a hand tool, the fire fighter strikes and breaks the window, then clears the opening completely. The fire fighter must not be positioned below the window, where glass could slide down the handle of the tool (▼ **Figure 14-12**).

Breaking a Window with a Ladder

Fire fighters on the ground can use the tip of a ladder to break and clear a window when immediate ventilation is needed on an upper floor. Dropping the tip of the ladder into the upper half of the glass will break it. Moving the tip of the ladder back and forth inside the opening will usually clear the remaining glass.

This technique requires proper ladder selection. Usually, second-floor windows can be reached by 16′ or 20′ roof ladders, as well as by bedded 24′ or 28′ extension ladders. Extension ladders are needed to break windows on the third or higher floors.

The ladder can be raised directly into the top half of the window or it can be raised next to the window to determine the proper height for the tip. The ladder is then rolled

Figure 14-12 When breaking an upper-story window, the fire fighter should place the ladder to the side of the window.

into the window, drawn back at the tip, and forcibly dropped into the top third of the window. The objective is to push the broken glass into the window opening, but there is always a risk that some of the glass will fall outward. Fire fighters working below the window must wear full personal protective equipment to shield themselves from falling glass. A word of caution—the higher the window, the greater the danger of falling glass.

When a ladder is used to break a window, there may be shards of glass left hanging from the edges of the opening. If the opening will be used for access, it will need to be completely cleared with a tool before anyone enters through it. To break a window with a ladder, follow the steps in Skill Drill 14-2.

1. Wear full personal protective equipment, including eye protection.
2. Select the proper size ladder for the job.
3. Check for overhead lines. Use standard procedures for performing a ladder raise utilizing one, two, or three fire fighters, depending on the size of the ladder.
4. If a roof ladder is used, extend the hooks toward the window.
5. Raise the ladder next to the window. Extend the tip so it is even with the top third of the window.
6. Position the ladder in front of the window.
7. The ladder is forcibly dropped into the window.
8. Exercise caution; falling glass can cause serious injury. Glass may slide down the beams of the ladder.

14-1 Skill Drill

Breaking Glass with a Hand Tool

1. Position yourself to the side of the window.

2. With back against the wall, swing backward forcefully with the tip of the tool striking the top 1/3 of the glass.

3. Clear remaining glass from the opening with the hand tool.

9. Raise the ladder from the window and move it to the next window to be ventilated. Either carry the ladder vertically or pivot the ladder on its feet.

Breaking a Window from Above

In some instances, fire fighters are above the window that must be broken. They may be on the roof, performing roof ventilation. They may be on a floor above the fire. Or, there may not be a ladder available that can reach the window.

One way to break a window from above is to attach a Halligan tool to a rope or chain, toss it over the roof edge, and allow it to swing into the window. An alternative method is to use a long-handled tool, such as a pike pole, and to reach down from the edge of the roof, from a balcony, or from a window on the floor above the fire. Remember that smoke and heat will be released when the window is broken so be sure to stand to the side to keep from being enveloped in them. To break a window from above with a Halligan tool, follow the steps in (▼ **Skill Drill 14-3**).

1. Look over the roof edge to locate the window to be broken. Always be sure to check if crews are operating below you when you do this maneuver as glass and debris may fall down.
2. Attach the Halligan tool securely to the rope or chain using a clip or hook.
3. Lower the Halligan tool over the edge of the roof until the tool is located in the middle of the window. Mark this spot on the rope or chain; this is the pivot point for the toss. Raise the tool back to the roof. **(Step 1)**
4. Toss the Halligan tool over the roofline and out approximately the distance down to the window. This action allows the tool to swing into the window with enough force to break the glass. **(Step 2)**
5. Be sure to be in a safe position as this technique will put you above and in the direct path of the products of combustion that are being vented. Always be aware of the location of the roof edge, and your safety zone on the roof.

14-3 **Skill Drill**

Breaking Windows on Upper Floors—Halligan Toss

1. Lower the Halligan tool over the edge of the roof to measure the distance.
2. Toss the Halligan tool over the roofline so that it swings back to break the glass.

6. The tool may become lodged in the window. If it does, give it a few shakes and pulls, it should free itself, breaking whatever it is caught on. Always be sure to check if crews are operating below you when you do this maneuver as glass and debris may fall down.

Opening Doors

Door openings also can be used for natural ventilation operations. Doorways have certain advantages when used for ventilation. They are large openings, often twice the size of an average window. Some doorways may be large enough to drive a truck through. A locked door can be opened using forcible entry techniques.

The problem with using doorways as ventilation openings is that this compromises entry and exit for interior fire attack teams, as well as search-and-rescue teams. If heat and smoke are pouring through a doorway, fire fighters cannot enter or exit through that doorway. Because protecting exit

Fire Fighter Tips

Insulated Thermopane™ Windows

Concerns for energy savings have increased the use of insulated thermal glass both in new construction and as a replacement for conventional glass in older windows. Thermopane windows have multiple layers of glass sealed together, with a small airspace between the layers. These windows are much more difficult to break than conventional glass windows. A fire fighter may have to use multiple strikes and additional force to break a thermopane window.

Thermopane windows are used to increase the energy efficiency of a structure. The windows fit tightly so there is little air leakage, and the glass resists heat transfer through it. The tight fitting of the window is not a feature unique to thermopane windows, but rather generally to the age of the installation. Older window installations leak more air. Thermopane windows may have a more substantial frame, but if they do not fit properly, they will leak air as will any window. This can cause potentially dangerous fire conditions to develop in some situations. Any tight fitting window can cause potentially dangerous fire conditions to develop.

Conventional windows often break from the heat of a fire, providing immediate ventilation. Thermopane windows are more likely to stay intact as the fire develops to an advanced stage because the fire would have to break multiple layers of glass. By this time, the fire may consume most of the available oxygen. When the window is finally broken, the sudden rush of fresh air can cause instantaneous burning of these gases with explosive forces being created (▼ Figure 14-13 A & B).

Fire fighters should evaluate the structure and fire conditions for signs of backdraft before creating any type of ventilation opening. Refer to Chapter 5, Fire Behavior, for a list of indications.

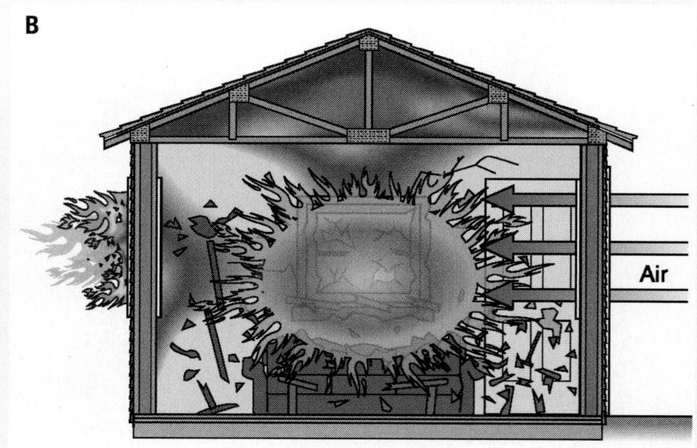

Figure 14-13 A & B **A.** Building structure and fire conditions should be evaluated for signs of backdraft before any type of ventilation opening is made. **B.** Improperly ventilating a fire in an oxygen deficient atmosphere can cause a backdraft.

paths is a priority, doorways are more suitable as entries for clean air. The contaminated atmosphere should be discharged through a different opening.

Fire fighters will approach a fire through the interior of a building to attack from the unburned side. Opening an exterior door that leads directly to the fire creates a vent and pushes the heat and smoke out of the building. This plan also creates a natural clean air entrance, as clear, cool air flows toward the fire with the attack team.

Doorways are also good places to put mechanical devices when natural ventilation needs to be augmented. Fans can be used to push clean air in (positive-pressure ventilation) or to pull contaminated air out (negative-pressure ventilation).

Mechanical Ventilation

Mechanical ventilation uses mechanical means to augment natural ventilation. There are three different methods of mechanical ventilation. Negative-pressure ventilation uses fans called smoke ejectors to exhaust smoke and heat from a structure. Positive-pressure ventilation uses fans to introduce clean air into a structure and push the contaminated atmosphere out. Hydraulic ventilation moves air by using fog or broken-pattern fire streams to create a pressure differential behind and in front of the nozzle.

In some buildings, HVAC systems can be used to provide mechanical ventilation. The system needs to be configured so that the zones can be controlled correctly and effectively. The building engineer or maintenance staff may have the specific knowledge required to assist in this effort. Some HVAC systems can be used to exhaust smoke or supply fresh air if a fire occurs, but others should be shut down immediately. Information about HVAC systems should be obtained during pre-incident planning surveys.

Negative-Pressure Ventilation

The basic principles of air flow are used in negative-pressure ventilation. Fire fighters locate the source of the fire, and with a fan or blower, exhaust the products of combustion out through a window or door. The fan draws the heat, smoke, and fire gases out by creating a slightly negative pressure inside the building. In turn, fresh air is drawn into the structure, replacing the contaminated air.

Smoke ejectors are usually 16" to 24" in diameter, although larger models are available. Ejectors can be powered by electricity, gasoline, or water (▶ **Figure 14-14**).

Negative-pressure ventilation can be used to clear smoke out of a structure after a fire, particularly if natural ventilation would be too slow or if no natural cross ventilation exists. In large buildings, negative-pressure ventilation may be used to pull heat and smoke out while fire fighters are working, although it is not generally used before or during fire attack. Negative-pressure ventilation is usually associated with horizontal ventilation, but an ejector can be used in a roof-vent operation to pull heated gases out of an attic.

Figure 14-14 A smoke ejector pulls products of combustion out of a structure.

Negative-pressure ventilation has several limitations, including positioning, power source, maintenance, and air flow control. Often, fire fighters must enter the heated and smoke-filled environment to set up the ejector, and they may need to use braces and hangers to position it properly. Because most ejectors run on electricity, a power source and a cord are required. The ejectors also get very dirty during use and must be taken apart and cleaned each time.

Air flow control is also difficult. The fan must be completely sealed so that the exhausted air is not immediately drawn back into the building. This effect is called churning and reduces the effectiveness of the ejector. Churning can be eliminated by completely blocking the opening around the fan.

Smoke ejectors usually have explosion-proof motors, which makes them excellent choices for venting flammable or combustible gases and other hazardous environments. To perform negative-pressure ventilation, follow the steps in (▶ **Skill Drill 14-4**).

1. Determine the area to be ventilated and the outside wind direction.
2. If possible, place the smoke ejector to exhaust on the leeward side of the building.
3. The fan should be hung as high in the selected opening as possible. **(Step 1)**
4. Remove any obstructions from the area used to ventilate smoke, including curtains, screens, and other debris.
5. Reduce churning by covering the area around the fan with salvage covers or plastic sheeting. **(Step 2)**
6. Provide an opening on the windward side of the structure to provide cross ventilation. **(Step 3)**

Positive-Pressure Ventilation

Positive-pressure ventilation uses large, powerful fans to force fresh air into a structure. The fans create a positive pressure

14-4 Skill Drill

Negative-Pressure Ventilation

1 Hang the fan in the upper part of the opening.

2 Use salvage covers to prevent churning.

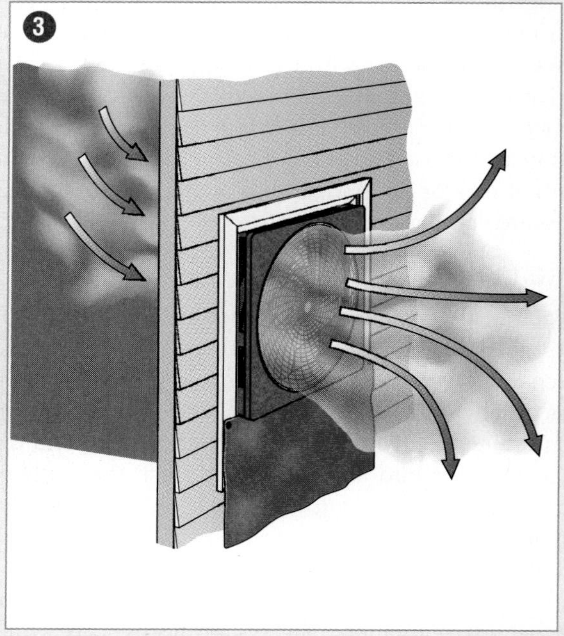

3 Provide openings on the windward side for cross ventilation.

inside the structure, displace the contaminated atmosphere, and push the heat and products of combustion out. Positive-pressure ventilation can be used to reduce interior temperatures and smoke conditions in coordination with an initial attack, or to clear a contaminated atmosphere after a fire.

Positive-pressure fans are usually set up at exterior doorways, often at the same opening used by the attack team (▶ Figure 14-15). If the fan is placed several feet away from the entry point, it will produce a cone of air that completely covers the opening. This pushes clean air in the same direction as the advancing hose line, so fire fighters move toward the fire in a stream of clean, cool air. Heat and smoke are pushed back and out of the building through an opening on the opposite side of the building.

For this system to work effectively, the integrity of the building must be intact and ventilation openings kept to a minimum. Positive-pressure ventilation will not work properly if there are too many openings. There must be an opening near the seat of the fire for the heat and smoke to exhaust (▶ Figure 14-16). The entry and exit openings should be approximately the same size to create the desired positive pressure within the structure. If there is no exhaust opening, or if it is too small, the heat and smoke can flow into other areas of the building or back toward the attack team. If the exhaust opening is too large, it will not create a buildup of pressure inside the structure that creates the velocity to effectively remove the smoke.

To increase the efficiency of positive-pressure ventilation, fire fighters should close doors to unaffected areas of the structure. Because smaller areas can be ventilated rapidly, venting one room at a time or one section of the building at a time can quickly clear trapped smoke out of a building after a fire. In a high-rise building, positive-pressure fans can blow fresh air up through a stair shaft. By opening the doors, one floor at a time can be cleared. Very large structures can be ventilated by placing multiple fans side-by-side or one behind the other.

Positive-pressure ventilation has several advantages over negative-pressure ventilation. A single fire fighter can set up a fan very quickly. Because the fan is positioned outside the structure, the fire fighter does not have to enter a hazardous environment, and the fan does not interfere with interior operations. Positive pressure is both quick and efficient, because the entire space within the structure is under pressure. When used properly, positive-pressure ventilation can help confine a fire to a smaller area and increase safety for interior operating

Figure 14-15 Positive-pressure fans are usually set up at exterior doorways, often at the same opening used by the attack team.

Figure 14-16 There must be an opening near the seat of the fire to allow the heat and products of combustion to exhaust.

crews, as well as any building occupants, by reducing interior temperatures. Positive-pressure fans do not require as much cleaning and maintenance as negative-pressure fans, because the products of combustion never move through the fan.

Positive-pressure ventilation does have its disadvantages, however. The most important concern is that improper use of positive pressure can spread a fire. If the fire is burning in a structural space such as in a pipe chase or above a ceiling, positive pressure can push it into unaffected areas of the structure. If the fire is in structural void spaces, positive-pressure ventilation should not be used until access to these spaces is available and attack crews are in place. Similar

Fire Fighter Tips

Doors used for ventilation must be kept open to ensure that ventilation is sustained. Depending on the time available and the type of door and lock, fire fighters may remove the door, disable the closure device, or place a wedge in the hinge side or under the door.

Overhead doors must be blocked in the open position, so they cannot accidentally roll down and close.

problems can occur if positive pressure is used without an adequate exhaust opening for the heat and smoke.

Positive-pressure fans operate at high velocity and can be very noisy. Most are powered by internal combustion engines and can increase carbon monoxide levels if they are run for significant periods of time after the fire is extinguished. Natural ventilation should be used after the structure is cleared to prevent carbon monoxide build-up. Because positive-pressure fan motors can get hot, they are unsafe to use if combustible vapors are present. **► Skill Drill 14-5** outlines the steps in performing positive-pressure ventilation.

1. Determine the location of the fire within the building and the direction of attack.
2. Place the fan 4′ to 10′ in front of the opening to be used for attack. **(Step 1)**
3. Provide an exhaust opening at or near the fire. This opening can be made before starting the fan or when the fan is started. **(Step 2)**
4. Check for interior openings that could allow the products of combustion to be pushed into unwanted areas.
5. Start the fan and check the cone of air produced. It should completely cover the opening. This can be checked by running a hand around the door frame to feel the direction of air currents.
6. Allow smoke to clear—usually 30 seconds to 1 minute depending on the size of the area to be ventilated and smoke conditions. **(Step 3)**

Hydraulic Ventilation

Hydraulic ventilation uses the water stream from the hose line to exhaust smoke and heated gases from a structure. The fire fighter working the hose line directs a narrow fog stream or a broken stream from a smooth bore nozzle out of the building through an opening, such as a window or doorway. The contaminated atmosphere is drawn into a low-pressure area behind the nozzle. An induced draft created by the high-pressure stream of water pulls the smoke and gas out through the opening **► Figure 14-17**. A well-placed fog stream can move a tremendous volume of air through an opening.

Hydraulic ventilation is most useful in clearing smoke and heat out of a room after the fire is under control. To perform hydraulic ventilation, a fire fighter must enter the room and remain close to the ventilation opening. The operator

14-5 Skill Drill

Positive-Pressure Ventilation

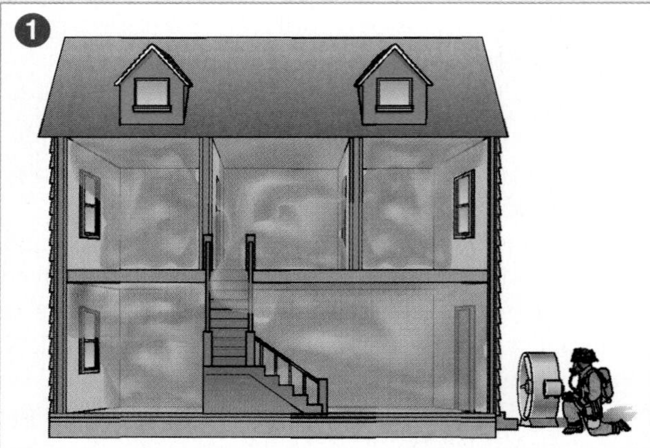

1 Place the fan in front of the opening to be used for attack.

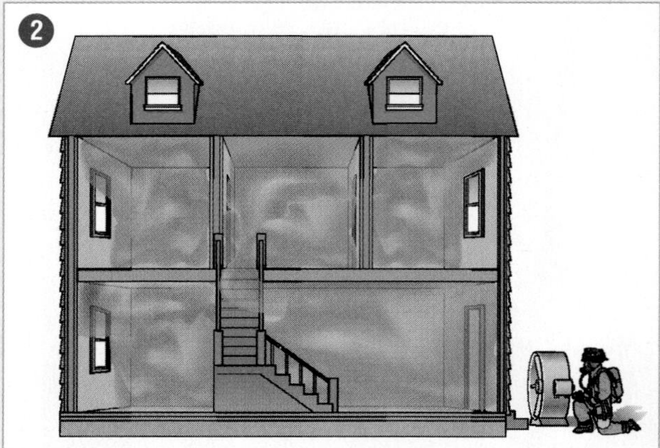

2 Provide an exhaust opening at or near the fire.

3 Start fan and allow smoke to clear.

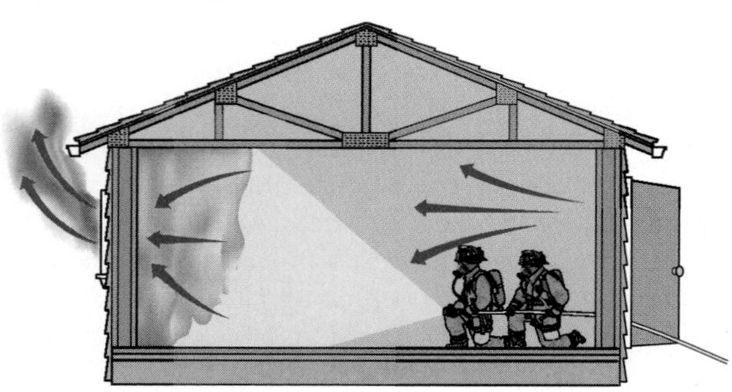

Figure 14-17 Hydraulic ventilation can be used to draw smoke out of a building after the fire is controlled.

places the nozzle through the opening and opens the nozzle to a narrow fog or broken spray pattern. The operator keeps directing the stream outside and backs into the room until the fog pattern almost fills the opening. The nozzle should be 2′ to 4′ inside the opening.

The fire fighter operating the nozzle must stay low, out of the heat and smoke, or to one side to keep from partially obstructing the opening. This technique is covered more fully in Chapter 16, Fire Hose, Nozzles, Streams, and Foam.

Hydraulic ventilation can move several thousand cubic feet of air per minute and is effective in clearing heat and smoke from a fire room. Because it does not require any spe-

Fire Fighter Safety Tips

If the fire is in structural spaces, do not use positive-pressure ventilation until access to these spaces is available and attack crews are in place.

cialized equipment, it can be performed by the attack team, using the same hose line that was used to control the fire.

There are some disadvantages to using hydraulic ventilation. Fire fighters must enter the heated, toxic environment and remain in the path of the products of combustion as they are being exhausted. If the water supply is limited, using water for ventilation must be balanced with using water for fire attack. This technique can cause excessive water damage if it is used improperly or for long periods of time. Also, in cold climates, ice build-up can occur on the ground, creating a safety hazard.

Ideally, the ventilation opening should be created before the hose line advances into the fire area. The hose stream also can be used to break a window before it is directed on the fire. The steam created when the water hits the fire will push the heat and smoke out through the vent. Hydraulic ventilation can then be used to clear the remaining heat and smoke from the room.

Vertical Ventilation

Vertical ventilation refers to the release of smoke, heat, and other products of combustion into the atmosphere in a vertical direction. Vertical ventilation will occur naturally, due to convection currents, if there is an opening available above a fire. Convection causes the products of combustion to rise and flow through the opening. In some situations, vertical ventilation can be assisted by mechanical means such as fans or hose streams.

Although vertical ventilation refers to any opening that allows the products of combustion to travel up and out, it is most often applied to operations on the roof of a structure. The roof openings can be existing features, such as skylights or bulkheads, or created by fire fighters who cut through the roof covering. The choice of roof openings depends primarily on the building's roof construction.

Fire Fighter Tips

Remember, ventilation openings allow products of combustion, as well as steam from the attack lines, to escape from the fire site. If ventilation is not coordinated with attack, the smoke, heat, steam, and embers will be forced back at the attack team.

Figure 14-18 The vertical-ventilation opening should be as close as possible to the seat of the fire.

A vertical-ventilation opening should be made as close as possible to the seat of the fire (▲ Figure 14-18). Smoke issuing from the roof area, melted asphalt shingles, or steam coming from the roof surface are all signs that fire fighters can use to identify the hottest point.

Safety Considerations in Vertical Ventilation

Vertical ventilation should be performed only when it is necessary and can be done safely. Rooftop operations involve significant inherent risks. If horizontal ventilation can sufficiently control the situation, vertical ventilation is unnecessary. Before performing vertical ventilation, fire fighters must evaluate all the pertinent safety issues and avoid unnecessary risks.

The most obvious risk in a rooftop operation is the possibility that the roof will collapse. Many fire fighters have died during vertical ventilation attempts because the fire had compromised the structural integrity of the roof support system. If the structural system supporting the roof fails, the roof and the fire fighters standing on it will fall into the fire. Fire fighters also can fall through an area of the roof deck that has been weakened by the fire.

Falling from a roof—either off of the building or through a ventilation opening, open shaft, or skylight into the building—is another possibility. Smoke can cause poor visibility and disorientation, increasing the risk of falling off the edge of a roof. Factors such as darkness, snow, ice, or rain can create hazardous situations. A sloping roof presents even greater risks than a flat roof.

Effective vertical ventilation often results in rapid fire control and reduces the risks to fire fighters operating inside the building. In some situations, prompt vertical ventilation can save the lives of building occupants by relieving interior conditions, opening exit paths, and allowing fire fighters to enter and perform search-and-rescue operations.

Vertical ventilation should always be performed as quickly and efficiently as possible. If they provide an ade-

Fire Fighter Safety Tips

Whenever possible, fire fighters conducting vertical ventilation should operate from the safety of roof ladders or aerial apparatus (▼ Figure 14-19). This reduces the danger of falling from the roof or into the building if the roof collapses.

Figure 14-19 Fire fighters operating from the safety of aerial apparatus.

quate opening, skylights, ventilators, bulkheads, and other existing openings should be opened first. If fire fighters must cut through the roof covering to create a hole, the location should be identified and the operation should be performed promptly and efficiently.

Fire fighters working on the roof should always have two safe exit routes. A second ground ladder or aerial device should be positioned to provide a quick alternative exit, in case conditions on the roof deteriorate. The two escape routes should be separate from each other, preferably in opposite directions from the operation site. The team working on the roof should always know where these exit routes are located, and plan ventilation operations to work toward these locations.

The ventilation opening should never be between the fire fighters and their escape route. A ventilation opening will release smoke and hot gases from the fire. If this column of smoke (and possibly fire) gets between the fire fighters and their exit from the roof, they could be in serious danger. A charged hose line should be ready for use on the roof to protect fire fighters and exposures, but a hose stream should never be directed into a ventilation opening (► Figure 14-20).

Once the ventilation opening has been made, the fire fighters should withdraw to a safe location. There is no good reason to stand around the vent hole and look at it; this simply exposes the fire fighters to an unnecessary risk.

Before fire fighters climb or walk onto any roof, they should test the roof to ensure that it will support their weight. If the roof is spongy or soft, it may not support a crew of fire fighters and their equipment.

Figure 14-20 Be prepared for smoke and flames that may be released when the roof is opened.

Sounding is a way to test the stability of a roof with a tool such as an axe or pike pole. Simply strike the surface of the roof with the blunt end of the tool. If the roof is in good condition, the impact should produce a firm rebound and a reassuring sound. If the impact does not produce this sound or if the tool penetrates the roof, the structure is not safe. During ventilation operations, fire fighters should continually sound the roof, probing ahead and around themselves with the handle of an axe or a pike pole to ensure that the area is solid.

Modern lightweight construction frequently uses thin plywood for roof decking, with a variety of waterproof and insulating coverings. A relatively minor fire under this type of roof can cause the plywood to delaminate, without any visible indication of damage or weakness from above. The weight of a fire fighter on a damaged area could be enough to open a hole and cause the fire fighter to fall through the roof.

Sounding is not a foolproof method to test the condition of a roof and may give fire fighters a false sense of security on some types of roofs. For example, slate or tile roofs can appear solid even if the supporting structure is ready to fail, due to the rigid surface of the tiles. Always look for other visible indicators of possible collapse and take the roof construction into account. To sound a roof for safety, follow the steps in (► Skill Drill 14-6).

1. Before stepping onto a roof, use a hand tool (such as an axe, pike pole, sledge, etc.), to strike the roof decking with considerable force. Listen and feel for rebound. **(Step 1)**
2. Sound the area ahead and to each side of your path. The material should sound solid and remain intact. If the tool penetrates, the roof is not safe.
3. Locate support members by listening to the sound and watching the bounce of the tool. Support members create a solid sound and make the tool bounce from the roof. The spaces between supports will sound hollow, and the tool will not bounce.

14-6 Skill Drill

Sounding a Roof

Use a hand tool to check the roof before stepping onto it.

Use the tool to sound ahead and to both sides as you walk. Locate support members by sound and rebound. Check conditions around your work area periodically.

Sound the roof along your exit path.

4. Continue to sound the area around you periodically to monitor conditions. **(Step 2)**
5. Remember to sound the roof as you leave the area as well. You are not completely safe until your feet are back on solid ground. **(Step 3)**

A fire fighter's path to a proposed vent hole site should follow the areas of greatest support and strength. These include the roof edges, which are supported by bearing walls, and the hips and valleys where structural materials are doubled for strength. These areas should be stronger than interior sections of the roof.

Fire fighters should always be aware of their surroundings when working on a roof. Knowing where the roof's edge is located, for example, can help avoid accidentally walking or falling off the roof. This is especially important, although difficult, in darkness, adverse weather, or heavy smoke conditions. It is equally important to watch for openings in the roof, such as skylights and ventilation shafts.

The order of the cuts in making a ventilation opening should be carefully planned. Fire fighters should be upwind, have a clear exit path, and be standing on a firm section of the roof or using a roof ladder. If fire fighters are upwind from the open hole, the wind will push the heat and smoke away from them. The escaping heat and smoke should not block their exit path. To prevent an accidental fall into the building, fire fighters should stand on a portion of the roof that is firmly supported, particularly when making the final cuts.

Basic Indicators of Roof Collapse

Roof collapse is the greatest risk to fire fighters performing vertical ventilation. Some roofs, particularly truss roofs, give little warning that they are about to collapse.

Fire fighters assigned to vertical ventilation tasks should always be aware of the condition of the roof. They should immediately retreat from the roof if they notice any of the following signs:

- Any spongy feeling or indication that the roof is not as solid as when the venting operation began
- Any visible indication of sagging roof supports
- Any indication that the roof assembly is separating from the walls, such as the appearance of fire or smoke near the roof edges
- Any structural failure of any portion of the building, even if it is some distance from the ventilation operation
- Any sudden increase in the intensity of the fire from the roof opening

Roof Construction

As we learned in Chapter 6, Building Construction, roofs can be constructed of many different types of materials in several configurations. All roofs have two major components—a support structure and a roof covering. The roof support system provides the structural strength to hold the roof in place. It must also be able to bear the weight of any rain or snow accumulation, any rooftop machinery or equipment, and any other loads placed on the roof, such as fire fighters or other people walking on the roof.

The support system can be constructed of solid beams of wood, steel, or concrete, or a system of trusses. Trusses are produced in several configurations and are made of wood, steel, or a combination of wood and steel.

The **roof covering** is the weather-resistant surface of the roof and may have several layers. Roof covering materials include shingles and composite materials, tar and gravel, rubber, foam plastics, and metal panels. The **roof decking** is a rigid layer made of wooden boards, plywood sheets, or metal panels. In addition to the decking, the roof covering generally includes a waterproof membrane and insulation to retain heat in winter and limit solar heating in the summer. Fire will very quickly burn through some roofs, but others will retain their integrity for long periods.

Vertical ventilation operations often involve cutting a hole through the roof covering. The selection of tools, the technique used, the time required to make an opening, and the personnel requirements all depend on the roofing materials and the layering configurations.

Each type of material used in roof construction is affected differently by fire. These reactions will either increase or decrease the time available to perform roof operations. For example, solid beams are generally more fire resistant than truss systems, because they are larger and heavier. Wooden beams are affected by fire more quickly than concrete beams. Steel, unlike wood, does not burn but elongates and loses strength when heated.

Most roofs will eventually fail as a result of fire exposure, some very quickly and others more gradually. In some situations, the fire will weaken or burn through the supporting structure so that it collapses while the roof covering remains intact. Structural failure usually results in sudden and total collapse of the roof. A structural failure involving other building components could also cause the roof to collapse.

In other cases, the fire may burn through the roof covering, while the structure is still sound. This type of roof failure usually begins with a "burn through" close to the seat of the fire or at a high point above the fire. As combustible products in the roof covering become involved in the fire, the opening spreads, eventually causing the roof to collapse.

The inherent strength and fire resistance of a roof is often determined by local climate conditions. In areas that receive winter snow loads, roof structures can usually support a substantial weight and may include multiple layers of insulating material. A serious fire could burn under this type of roof with very little visible evidence from above. Heavily constructed roofs, supported by solid structural elements, can burn for hours without collapsing.

In warmer climates, roof construction is often very light. These roofs may simply function as umbrellas. The supporting structure may be just strong enough to support a thin deck with a waterproof, weather-resistant membrane. This type of roof may burn through very quickly, and the supporting structure may collapse after a short fire exposure.

Solid-Beam Versus Truss Construction

The two major structural support systems for roofs use either solid-beam or truss construction. It may be impossible to determine which system is used simply by looking at the roof from the exterior of the building, particularly when the building is on fire. This information should be obtained during preincident planning surveys.

The basic difference between solid-beam and truss construction is in the way individual load-bearing components are made. Solid-beam construction uses solid components, such as girders, beams, and rafters. Trusses are assembled from smaller, individual components. In most cases, it makes no difference whether solid-beam or truss construction is used. For fire fighters, however, there is an important difference: A truss system can collapse quickly and suddenly when it is exposed to a fire.

Trusses are constructed by assembling relatively small and lightweight components in a series of triangles. The resulting system can efficiently span long distances while supporting a load. In most cases, a solid beam that spanned an equal distance and carried an equivalent load would be much larger and heavier than a truss (► Figure 14-21). The disadvantage of truss construction is that some trusses can fail completely when only one of the smaller components is weakened or when one of the connections between components fails.

A truss is not necessarily bad or inherently weak construction. Some trusses are very strong and are assembled with substantial components that can resist fire as well as solid beams. However, many trusses are made of lightweight materials and are designed to carry as much load as possible

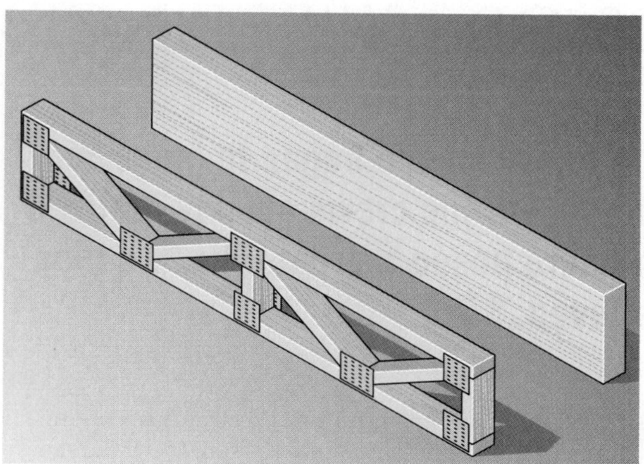

Figure 14-21 A lightweight truss can carry as much weight as a much heavier solid beam.

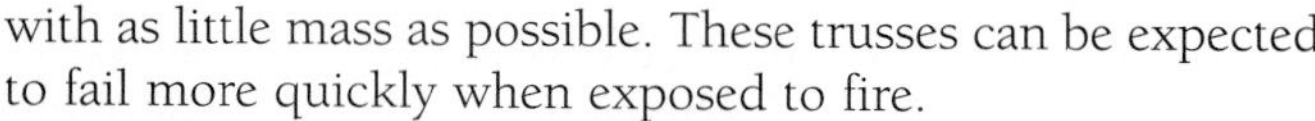

Figure 14-22 Gusset plates can fail in a fire, weakening the truss.

with as little mass as possible. These trusses can be expected to fail more quickly when exposed to fire.

Truss construction can be used with almost any type of roof or floor. Trusses can support flat roofs, pitched roofs, arched roofs, or overhanging roofs. Fire fighters should assume that any modern construction uses truss construction for the roof support system until proven otherwise.

The individual components of a lightweight truss used for roof support are often wood 2′ × 4′ sections or a combination of wood and lightweight steel bars or tubes. Because these components are small, they can be weakened quickly by a fire. Even more critical, however, are the points where the individual components join together. These connection points are often the weakest portion of the truss and the most common point of failure.

For example, the components of a truss constructed with 2″ × 4″ wood pieces can be connected by heavy-duty staples or by **gusset plates** (connecting plates made of wood or lightweight metal) ► Figure 14-22. These connections can fail quickly in a fire. When one connection fails, the truss loses its ability to support a load.

If only one truss fails, other trusses in the system may be able to absorb the additional load. However, the overall strength of the system will be compromised. A fire that causes one truss to fail will probably weaken additional trusses in the same area. These trusses also can fail, sometimes as soon as the additional load is transferred. In most cases, a series of trusses will fail in rapid succession, resulting in a total collapse of the roof. Sometimes the weight of a fire fighter walking on the roof will be enough to trigger a collapse.

Trusses also can be made of metal, usually individual steel bars or angle sections that are welded together. Lightweight steel trusses, known as bar joists, often support flat roofs on commercial or industrial buildings. When exposed to the heat of a fire, these metal trusses will expand and lose strength. A roof supported by bar joists will probably sag before it collapses, a warning that fire fighters should heed. The expansion can stretch the trusses, which may cause the supporting walls to collapse suddenly. Horizontal cracks in the upper part of a wall indicate that the steel roof supports are pushing outward.

Roof Designs

Fire fighters must be able to identify the types of roofs and the materials used in their construction. Creating ventilation openings may require the use of assorted tools and techniques, depending on the type of roof decking.

Flat Roof

Flat roofs can be constructed with many different support systems, roof decking systems, and materials. Although they are classified as flat, most roofs have some slope so water can flow to roof drains or scuppers.

Flat roof construction is generally very similar to floor construction. The roof structure can be supported by solid components, such as beams and **rafters** (the elements that support the roof), or by trusses. The horizontal beams or trusses often run from one exterior **bearing wall** (wall that supports the weight of a floor or roof) to another bearing wall. In some buildings, the roof is supported by a system of vertical columns and/or interior bearing walls.

The roof deck is usually constructed of multiple layers, beginning with wooden boards, plywood sheets, or metal decking. Then come one or more layers of roofing paper, insulation, tar and gravel, rubber, gypsum, lightweight concrete, or foam plastic. Some roof decks have only a single layer, such as metal sheets or precast concrete sections.

Flat roofs often have vents, skylights, scuttles (small openings or hatches with movable lids), or other features that penetrate the roof deck. Removing the covers from these openings will provide vertical ventilation without cuts through the roof deck.

Voices of Experience

“Once the heavy smoke began to clear and visibility improved, the crews quickly realized that the structure posed a greater hazard than initially believed.”

One morning my crew and I were called to a working structure fire at a light industrial complex. When we arrived on scene, heavy smoke was pouring out of the building's air-conditioning vents. To gain access to the building, we immediately made a triangle cut with a K-12 tool in a metal garage door.

The initial attack team headed into the structure with a 1 3/4" line. The second crew on scene established a water supply for the engine. After the water supply had been established, the ventilation crew began positive-pressure ventilation from the exterior of the structure, using a fan to ventilate through the triangle cut. The smoke quickly began to rise through the building, eventually escaping through preexisting vents near the top of the structure.

Once the heavy smoke began to clear and visibility improved, the crews quickly realized that the structure posed a greater hazard than initially believed. Inside the warehouse we discovered three large, open-topped vats of hydrofluoric acid with fire burning underneath.

If ventilation had not been performed quickly and properly, a number of disasters could have occurred. The crews could have aimed their hoses blindly and inadvertently hit the hydrofluoric acid, which would have caused a violent and hazardous reaction. Also, the vats could have melted or failed, spilling the hydrofluoric acid onto the fire fighters. Because of the proper use of ventilation techniques, my crew and I were able to quickly, effectively, and safely extinguish the fire.

Todd Gilgren
Arvada Fire Protection District
Arvada, Colorado

Flat roofs also may have **parapet walls**, freestanding walls that extend above the normal roofline. A parapet can be an extension of a firewall, a division wall between two buildings, or a decorative addition to the exterior wall.

Pitched Roofs

Pitched roofs have a visible slope for rain, ice, and snow runoff. The pitch or angle of the roof usually depends on both local weather conditions and the aesthetics of construction. A pitched roof can be supported either by trusses or by a system of rafters and beams. The rafters usually run from one bearing wall up to a center ridge pole and back to another bearing wall.

Most pitched roofs have a layer of solid sheeting, which can be metal, plywood, or wooden boards. This layer is often covered by a weather-resistant membrane and an outer covering such as shingles, slate, or tiles. Some pitched roofs have a system of **laths** (thin, parallel strips of wood) instead of solid sheeting to support the outer covering.

As with flat roofs, the type of roof construction material will dictate how to ventilate a pitched roof. For example, a slate or tile pitched roof may be opened by breaking the tiles and pushing them through the supporting laths. A tin roof can be cut and "peeled" back like the lid on a can. Opening a wooden roof usually requires cutting, chopping, or sawing.

Pitched roofs may have steep or gradual slopes. Roof ladders should be used to provide a stable support while working on a pitched roof. A ground or aerial ladder is used to access the lower part of the roof; the roof ladder is placed on the sloping surface and hooked over the center ridge pole.

Arched Roofs

Arched roofs are generally found in commercial structures because they create large open spans without columns (▼ Figure 14-23). Arched roofs are common in warehouses, supermarkets, bowling alleys, and similar buildings. Arched roofs can be supported by trusses or by other types of construction, including large wood, steel, or concrete arches.

Figure 14-23 An arched roof is commonly used for warehouses, supermarkets, and bowling alleys.

Arched roofs are often supported by bowstring trusses, which give the roof its distinctive curved shape. Bowstring trusses are usually constructed of wood and spaced 6′ to 20′ apart. They support a roof deck of wooden boards or plywood sheeting and a covering of waterproof membrane. Although these roofs are quite distinctive when seen from above or outside, they may not be evident from inside the building because a flat ceiling is attached to the bottom chords of the trusses. This creates a huge attic space that is often used for storage. A hidden fire within this space can severely and quickly weaken the bowstring trusses.

The collapse of a bowstring truss roof is usually very sudden. A large area may collapse due to the long spans and wide spaces between trusses. Fire fighters should be wary of any fire involving the truss space in these buildings. Many fire fighters have been killed when a bowstring truss roof collapsed.

Buildings with this type of roof construction should be identified and documented during preincident planning surveys. Parapet walls, smoke, and other obstructions may make it difficult to recognize the distinctive arched shape of a bowstring truss roof during an incident.

Vertical Ventilation Techniques

Roof Ventilation

Among the different roof openings that can provide vertical ventilation are built-in roof openings, inspection openings to locate the optimal place to vent, primary expandable openings directly over the fire, and defensive secondary openings to prevent fire spread (▼ Table 14-1).

The objective of any roof ventilation operation is simple: to provide the largest opening in the appropriate location, using the least amount of time and the safest technique.

Before starting any vertical ventilation operation, fire fighters must make an initial assessment. Construction features and indications of possible fire damage should be noted, safety zones and exit paths established, and built-in roof openings that can be used immediately should be identified. The sooner a building is ventilated, the safer it will become for hose teams working inside.

Ventilation operations must not be conducted in unsafe locations. A less efficient opening in a safe location is better

Table 14-1 Types of Vertical-Ventilation Openings

- Built-in roof openings
- Inspection openings
- Primary (expandable) openings
- Secondary (defensive) openings

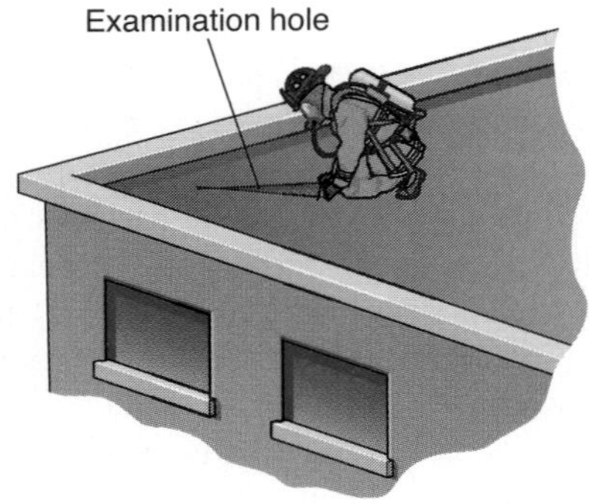

Figure 14-24 A triangular cut can be used as an examination hole.

Figure 14-25 Built-in rooftop openings can often provide vertical ventilation very quickly.

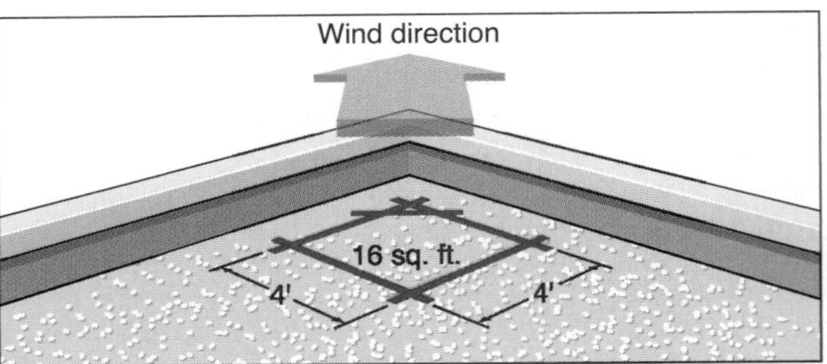

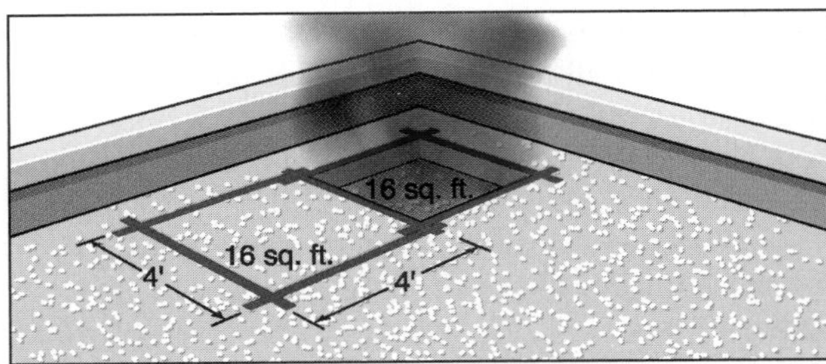

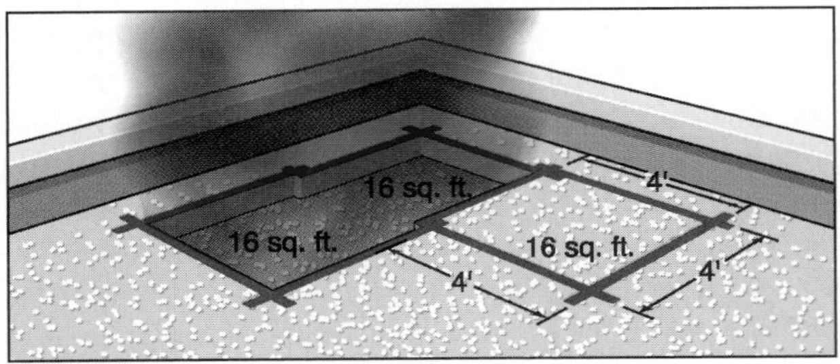

Figure 14-26 A ventilation hole can be quickly expanded by extending the cuts.

than an optimal opening in a location that jeopardizes the lives of the ventilation team. In some cases, it may not be safe to conduct any rooftop operations and fire fighters will have to rely on horizontal ventilation techniques.

Vertical ventilation is most effective when the opening is at the highest point, directly over the fire. Information on the fire's location, as well as visible clues from the roof, can be used to pinpoint the best location to vent. The incident commander or the interior crews may be able to provide some direction, and the ventilation team should look for signs such as smoke issuing through the roof, tar bubbling up, or other indications of heat. If the roof is wet, rising steam could indicate the hottest area.

Examination holes can be made to evaluate conditions under the roof and to verify the proper location for a ventilation opening. A triangular examination hole can be created very quickly by making three small cuts in the roof. A power saw can also be used to make a **kerf cut**, which is only as wide as the saw blade. Examination holes also are used to determine how large an area is involved, whether a fire is spreading, and which direction it is moving (▲ Figure 14-24).

Once the spot directly over the fire is located, the ventilation team should determine the most appropriate type of opening to make. Built-in rooftop openings provide readily available ventilation openings (◄ Figure 14-25). Skylights, rooftop stairway exit doors, louvers, and ventilators can quickly become ventilation openings simply by removing a cover or an obstruction. This also results in less property damage to the building.

If the roof must be cut to provide a ventilation opening, cutting one large hole is better than making several small ones. The original hole should be rectangular and at least 4′ long by 4′ wide. If necessary, this hole can be expanded by continuing the cuts to make a larger opening (▲ Figure 14-26). If the crews inside the building report that smoke conditions are lifting and temperature levels are dropping, the vent is effective and probably large enough. If interior crews do not see a difference, the opening might be obstructed or need to be expanded.

Once the roof opening is made, a hole of the same size should be made in the ceiling material below to allow heat and smoke to escape from the interior of the building. If the ceiling is not opened, only the heat and smoke from the space between the ceiling and the roof will escape through the roof opening. The blunt end of a pike pole or hook can be used to push down as much of the ceiling material as possible. A roof opening will be much less effective if there is no ceiling hole or if the ceiling hole is too small.

Tools Used in Vertical Ventilation

Several tools can be used in roof ventilation. Power saws are commonly used to cut the vent openings, but many hand tools are also used. Axes, Halligan tools, pry bars, tin cutters, pike poles, and other hooks can be used to remove coverings from existing openings, cut through the roof decking, remove sections of the roof, and punch holes in the interior ceiling. The specific tools that will be needed can usually be determined by looking at the building and the roof construction features.

Ventilation team members should always carry a standard set of tools, as well as a utility rope for hauling additional equipment if needed. All personnel involved in roof operations should use breathing apparatus and wear full protective clothing. Ground ladders or an aerial apparatus can provide access to the roof.

Power saws will effectively cut through most roof coverings. A rotary saw with a wood- or metal-cutting blade or a chain saw can be used. Special carbide-tipped saw blades can cut through typical roof construction materials. To operate a power saw, follow the steps in (Skill Drill 14-7).

1. The saw should always be checked during daily apparatus inspection.
2. Be sure the cutting device or blade is appropriate for the material anticipated. If it is not, put the correct blade on before going to the roof.
3. Briefly inspect the blade or chain for obvious damage.
4. Ensure that your proper protective gear is in place, including eye protection, SCBA, and full personal protective equipment.
5. The saw should be started to ensure that it runs properly before going to the roof. Set the choke to halfway position, or as recommended by the manufacturer.
6. Stay clear of any moving parts of the saw, use a foot or knee to anchor the saw to the ground, and pull the starter cord as recommended to start the saw.
7. Run the saw briefly at full throttle to verify proper operation.
8. Shut the saw down, wait for the blade to stop completely, then carry the saw to the roof.
9. If possible, always work off of a roof ladder or aerial platform for added safety.
10. Start the saw in an area slightly away from where you intend to cut.
11. Always run a saw at maximum throttle when cutting. The saw should be running at full speed before the blade touches the roof decking. Keep the throttle fully open while cutting and removing the blade from the cut to reduce the tendency for the blade to bind.

Types of Roof Cuts

Roof construction is a major consideration in determining the type of cut to use. Some roofs are thin and easy to cut with an axe or power saw; others have multiple layers that are difficult to cut. Familiarity with local roof types and the best methods for opening them will increase your efficiency and effectiveness in ventilation operations.

Rectangular Cut

A 4′ × 4′ rectangular cut is the most common vertical ventilation opening. It requires four cuts completely through the roof decking, using an axe or power saw. When using a power saw, the fire fighter must carefully avoid cutting through the structural supports. After the four cuts are completed, a section of the roof deck can be removed.

The fire fighter should stand upwind of the opening, with an unobstructed exit path. The first and last cuts should be made parallel to and just inside the roof supports. The fire fighter making the cuts must always stand on a solid portion of the roof. A triangular cut in one corner of the planned opening can be used as a starting point for prying the deck up with a hand tool.

Depending on the roof construction, it may be possible to lift out the entire section at once. If there are several layers of roofing material, fire fighters may have to peel them off in layers. The decking itself could be plywood sheets or individual boards that have to be removed one at a time. To perform the rectangular cut, follow the steps in (► Skill Drill 14-8).

1. Sound the roof with a tool to locate the roof supports.
2. Make the first cut parallel to a roof support. The cut should be made so that the support will be outside the area that will be opened. **(Step 1)**
3. Make a small triangular cut at the first corner of the opening. **(Step 2)**
4. Follow with two cuts perpendicular to the roof supports. Do not cut through roof supports. Rock the saw over them to avoid damaging the integrity of the roof structure. Then make the final cut parallel to another roof support. **(Step 3)**
5. Make the final cut parallel to and slightly inside a roof support. Be sure to stand on the solid portion of the roof when making this cut.
6. Use a hand tool to pull out the corner triangle or push it through to create a small starter hole. **(Step 4)**
7. If possible, use the hand tool to pull the entire roof section free and flip it over onto the solid roof. It may be necessary to pull the decking out, one board at a time.
8. After opening the roof deck, use a pike pole to punch out ceiling below. This hole should be the same size as the opening in the roof decking. Be careful for a sudden updraft of hot gases or flames. **(Step 5)**

Fire Fighter Safety Tips

Always carry and handle tools and equipment in a safe manner. When carrying tools up a ladder, hold the beam of the ladder instead of the rungs and run the hand tool up the beam. This technique will work for most hand tools, including axes, pike poles and hooks, and tin roof cutters. Hoist tools using a rope with an additional tagline to keep them away from the building. Carry power saws and equipment in a sling across your back. Always TURN OFF power equipment before you carry it.

Louver Cut

Another common cut is the louver cut, which is particularly suitable for flat or sloping roofs with plywood decking.

Using power saws or axes, fire fighters make two parallel cuts, approximately 4′ apart, perpendicular to the roof supports. Then the fire fighters make cuts parallel to the roof supports, approximately half way between each pair of supports. Using each roof support as a fulcrum, the cut sections are tilted to create a series of louvered openings.

Louver cuts can quickly create a large opening. Continuing the same cutting pattern in any direction creates additional louver sections. To make a louver cut, follow the steps in (▶ **Skill Drill 14-9**).

1. Sound the roof with a tool to locate the roof supports. **(Step 1)**
2. Make two parallel cuts, perpendicular to the roof supports. Do not cut through the roof supports. Rock the saw over them to avoid damaging the integrity of the roof structure. **(Step 2)**
3. Make cuts parallel to the supports and between pairs of supports in a rectangular pattern. **(Step 3)**
4. Strike the nearest side of each section of the roofing material with an axe or maul, pushing it down on one side; use the support at the center of each panel as a fulcrum. Tilt the panel over the middle support. **(Step 4)**
5. Moving horizontally along the roof, make additional louver openings.
6. Open the interior ceiling area below the opening by using the butt end of a pike pole. This hole should be the same size as the opening made in the roof decking.

Triangular Cut

The triangular cut works well on metal roof decking because it prevents the decking from rolling away as it is cut. Using saws or axes, the fire fighter removes a triangle-shaped section of decking. Smaller triangular cuts can be made between supports, so that the decking falls into the opening, or over supports to create a louver effect. Because triangular cuts are generally smaller than other types of roof ventilation openings, several may be needed to create an adequately sized vent. To make a triangular cut, follow the steps in (▶ **Skill Drill 14-10**).

1. Locate the roof supports. **(Step 1)**
2. The first cut is made from just inside a support member in a diagonal direction toward the next support member. **(Step 2)**
3. The second cut begins at the same location as the first, and is made in the opposite diagonal direction, forming a "V" shape. **(Step 3)**
4. The final cut is made along the support member and connects the first two cuts. Cutting from this location allows fire fighters the full support of the member directly below them while performing ventilation.**(Step 4)**

Peak Cut

Peak cuts are limited to peaked roofs sheeted with plywood. The 4′ × 8′ sheets of plywood are applied starting at the bottom edge of the roof and working toward the top. A partial sheet is usually placed at the top, along the roof peak. A small gap is left between the two sides of the roof to allow the plywood sheeting to expand and contract with temperature changes.

To make a peak cut, the fire fighter first uses a hand tool to clear the outer covering along the peak (roof cap). This reveals the roof supports through the gap in the sheeting. Using a power saw or axe, the fire fighter makes a series of vertical cuts between the supports from the top to the bottom of the plywood sheet. The individual panels are then struck with an axe and louvered or pried up with a hook.

Repeating this technique on both sides of the roof essentially removes the entire peak. Unless there are multiple layers of roof decking, there is no need to make horizontal cuts. The opening can be expanded either vertically or horizontally with a minimum of cutting. To make a peak cut, follow the steps in (**Skill Drill 14-11**).

1. Sound the roof with a tool to locate the roof supports.
2. Clear the roofing materials away from the roof cap.
3. Make the first cut vertically, at the furthest point away. Start at the roof peak in the area between the support members and cut down to the bottom of the first plywood panel.
4. Make parallel downward cuts between supports, moving horizontally along the roofline to make additional ventilation openings.
5. Strike the nearest side of the roofing material with an axe or maul, pushing it in, using the support located at the center as a fulcrum.
6. This causes one end of the roofing material to go downward into the opening and the other to rise up.
7. This process can be repeated on both sides of the peak, horizontally across the peak, or vertically toward the roof edge.

Skill Drill

Rectangular Cut

Locate the roof supports by sounding. Make the first cut parallel to the roof support.

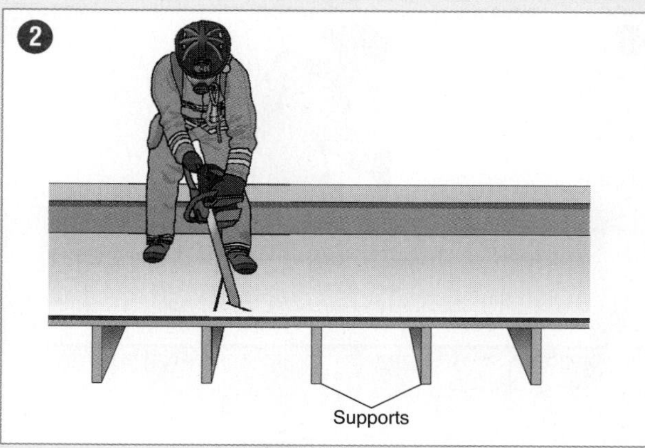

Make a triangle cut at the first corner.

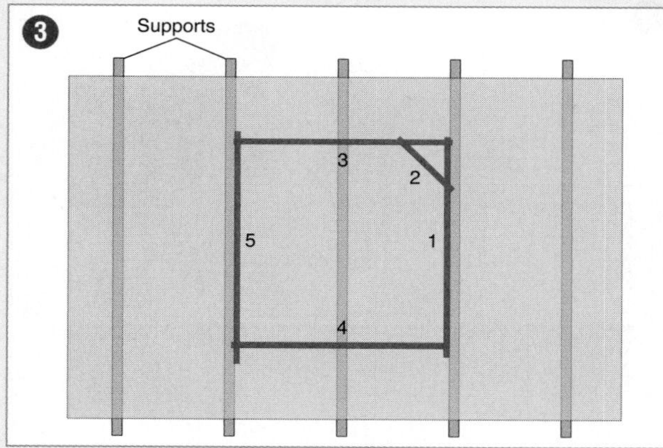

Make two cuts perpendicular to the roof supports (3 and 4). Then make the final cut parallel to another roof support (5).

Pull out or push in the triangle cut.

Punch out the ceiling below. Be careful for a sudden updraft of hot gases or flames.

14-9 Skill Drill

Louver Cut

Locate the roof supports by sounding.

Make two parallel cuts perpendicular to the roof supports.

Cut parallel to the supports and between pairs of supports in a rectangular pattern.

Tilt the panel to a vertical position.

8. Open the interior ceiling area below the opening by using the butt end of a pike pole. This hole should be the same size as the vent opening made in the roof decking.

Trench Cut

Trench cut ventilation is used to stop fire spread in long narrow buildings, such as strip malls and small storage complexes. The trench cut creates a large opening ahead of the fire, by removing a section of fuel and letting heated smoke and gases flow out of the building. Essentially, it is a firebreak in the roof.

Trench cuts are a defensive ventilation tactic to stop the progress of a large fire, particularly one that is advancing through an attic or cockloft. The incident commander who chooses this tactic is "writing off" part of the building and identifying a point where crews will be able to stop the fire.

A trench cut is made from one exterior wall across to the other. It begins with two parallel cuts, spaced 2′ to 4′ apart.

Skill Drill

Triangular Cut

Locate the roof supports.

The first cut is made from just inside a support member in a diagonal direction toward the next support member.

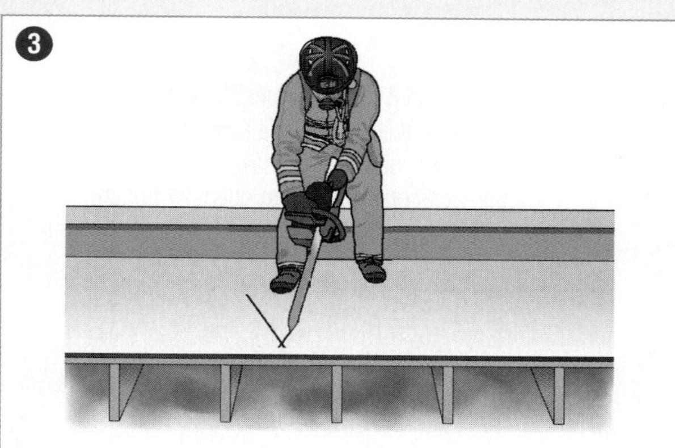

The second cut begins at the same location as the first, and is made in the opposite diagonal direction, forming a "V" shape.

The final cut is made along the support member and connects the first two cuts. Cutting from this location allows fire fighters the full support of the member directly below them while performing ventilation.

About every 4′, fire fighters make short perpendicular cuts between the two parallel cuts. They can then lift the roof covering out in sections, completely opening a section of the roof. On a pitched roof, the trench should run from the peak down. On a sloping roof, it should start at the higher end and work toward the lower end.

A trench cut is a **secondary cut**, used to limit the fire spread, rather than a **primary cut** located over the seat of the fire. A primary vent should still be made before crews start working on the trench cut.

A trench cut must be made far in advance of the fire. The ventilation crew must be able to complete the cut and the interior crew needs time to get into position before the fire passes the trench. Inspection holes should be made to ensure that the fire has not already passed the chosen site before the trench cut is completed. Crew lives will be in danger and the tactic is useless if the fire advances beyond the trench before it is opened.

Although trench cuts are effective, they require both time and manpower. As with all types of vertical ventilation, careful coordination between the ventilation crew and the interior attack crew is essential. While the trench cut is being made, hose teams should be deployed inside the building to defend the area in front of the cut (► Figure 14-27). To make a trench cut, follow the steps in (▼ Skill Drill 14-12).

1. After a primary cut has been made over the seat of the fire, the crew cuts a number of small inspection holes to identify a point sufficiently ahead of the fire travel.
2. Make two parallel cuts, 2′ to 4′ apart, across the entire roof, starting at the ridge pole (for peaked roofs) or a bearing wall (for flat roofs).
3. Cut between the two long cuts to make a row of rectangular sections.
4. Remove the rectangular panels to open the trench. **(Step 1)**

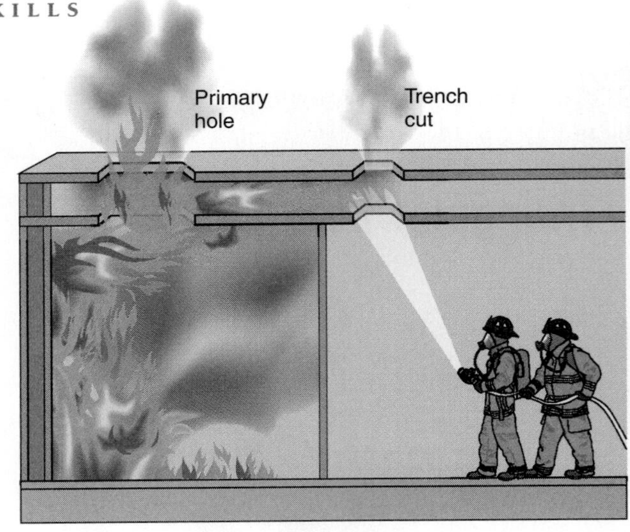

Figure 14-27 Ventilation and attack teams work together to use a trench cut effectively.

14-12 Skill Drill

Trench Cut

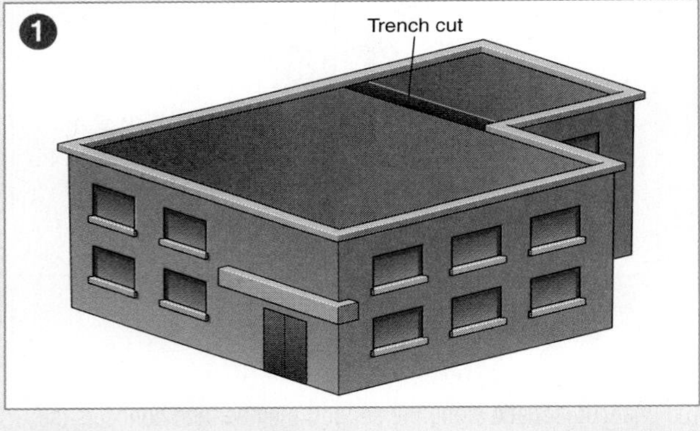

Make two parallel cuts, 2′ to 4′ apart, across the entire roof, starting at the ridge pole (for peaked roofs) or a bearing wall (for flat roofs). Cut between the two long cuts to make a row of rectangular sections. Remove the rectangular panels to open the trench.

Special Considerations

Many obstacles can be encountered during ventilation operations. Fire fighters must be creative and remember the basic objectives of ventilation—to create high openings as rapidly as possible so that hose lines can be advanced into the building.

Poor access or obstructions such as trees, fences, or tight exposures can prevent fire fighters from getting close enough to place ladders. Many residential and commercial roofs have multiple layers; some may even have a new roof built on top of an older one. Window openings in abandoned buildings may be boarded or sealed.

Steel bars, shutters, and other security features can hamper ventilation efforts. A building that requires forcible entry is also likely to present ventilation challenges. Some commercial buildings are roofed with steel plating for security purposes. In these buildings, fire fighters should not attempt vertical ventilation through the roof but should use alternative ventilation techniques.

Ventilating a Concrete Roof

Some commercial or industrial structures have concrete roofs. Concrete roofs can be constructed with "poured-in-place" concrete, with precast concrete sections of roof decking placed on a steel or concrete supporting structure, or with T-beams. Concrete roofs are generally flat and difficult to breach. The roof decking is usually very stable, but fire conditions underneath could weaken the supporting structural components or bearing walls, leading to failure and collapse.

There are few options for ventilating concrete roofs. Even special concrete-cutting saws are generally ineffective. Fire fighters should use alternative ventilation openings such as vents, skylights, and other roof penetrations or horizontal ventilation.

Ventilating a Metal Roof

Metal roofs and metal roof decks also present many challenges. Because metal conducts heat more quickly than other roofing materials, discoloration and warping may indicate the seat of the fire. Tin-cutter hand tools can be used to slice through thin metal coverings; special saw blades may be needed to cut through metal roof decking. In many cases, the metal is on the bottom and supports a built-up or composite roof covering.

Metal roof decking is often supported by lightweight steel bar joists, which can sag or collapse when exposed to a fire. Because the metal decking is lightweight, the supporting structure may be relatively weak, with widely spaced bar joists. The resulting assembly may fail quickly with only limited fire exposure.

As the fire heats the metal deck, the tar roof covering can melt and leak through the joints into the building, where it can release flammable vapors. This can quickly spread the fire over a wide area under the roof decking. Fire fighters should look for indications of dripping or melting tar, and begin rapid ventilation to dissipate the flammable vapors before they can ignite. Hose streams should be used to cool the roof decking from below to stop the tar from melting and producing vapors.

When a metal roof deck is cut, the metal can roll down and create a dangerous slide directly into the opening. The triangular cut prevents the decking from rolling away as easily, so it is the preferred option, even though several cuts may be needed to create an adequately sized vent.

Venting a Basement

Basement fires are especially difficult to ventilate. Basements generally have just a few small windows, if they have any at all. Basement stairways may lead only to the ground floor interior, with no exterior exit.

If a basement fire occurs, windows or exterior doorways into the basement should be opened or broken to provide as much ventilation as possible. If the basement has few or no exterior windows or doors, the interior stairways and other vertical openings will act as chimneys. Heat, smoke, and gases will rise up them into the rest of the structure. In buildings with balloon-frame construction, a basement fire can travel through unprotected wall spaces directly up to the underside of the roof.

A combination of vertical and horizontal ventilation can sometimes be used in attacking a basement fire. A direct path is created from the basement to the first floor, where the heat and smoke are pushed out through a door or window. A stairway or some other opening can be used for vertical ventilation, or a hole can be cut in the floor directly over the fire.

The opening from the basement to the ground floor should be near a window or door that can be used for horizontal ventilation. The heat and smoke can be exhausted using a hose line (hydraulic ventilation) or a negative-pressure smoke ejector. A hose line must be ready in case the fire spreads into the ground floor.

A basement fire presents problems to fire fighters working in the basement and above it. To attack the fire, fire fighters must enter the basement. Smoke and hot gases moving up the stairway as the fire fighters descend can make entry difficult or impossible. Fire fighters operating above the fire are in danger if the floor or stairway collapses.

The preferred method of attacking a basement fire is to create as many ventilation openings as possible on one side of the basement. This draws the heat and smoke in that direction. Fire fighters can then enter the basement from the opposite side, along with clean air. The direction of the attack must be determined and the entry and ventilation points must be clearly identified before this operation begins.

Ventilation openings in basement fires help push heat and smoke away from the attacking fire fighters. As the products of combustion rise, fire fighters can sometimes crawl beneath them to reach the basement. Because the basement floor is usually much cooler than the stairway, fire fighters can attack the fire while the heat and smoke exit over their heads. However, if the entrance stairway is the only ventilation opening, when the water hits the fire, the steam that is produced will push the heat and smoke back toward the attack team. A ventilation opening through the floor into the basement provides a second exit for the products of combustion.

High-Rise Buildings

Ventilating a high-rise building can be challenging. A high-rise building resembles a stack of individual floor compartments connected by stairways, elevator shafts, and other vertical passages. Many high-rise buildings have sealed windows that are difficult to break. Additionally, high-rise buildings have unique patterns of smoke movement. Smoke can be trapped on individual floors or it can move up or down within the vertical shafts.

Many newer high-rise buildings have incorporated smoke management capabilities in their HVAC systems. This enables different areas to be pressurized with fresh air or exhausted directly to the outside. If the HVAC system does not have this capacity, it may complicate problems by circulating smoke to different areas of the building.

A phenomenon called the **stack effect** can occur in high-rise structures. The stack effect is a response to the differences in temperature inside and outside a building. A cold

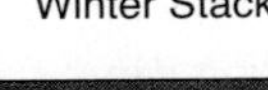

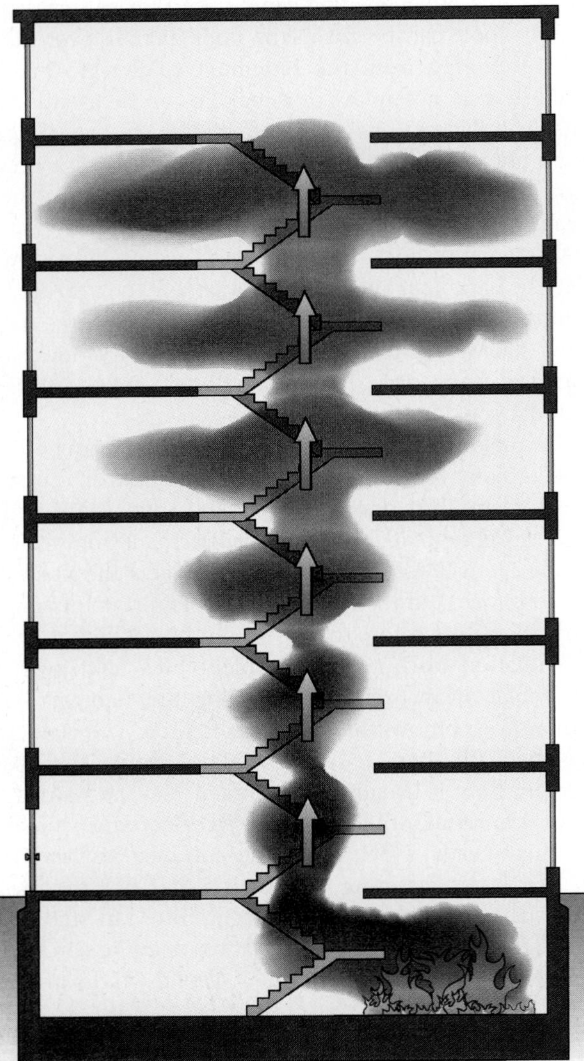

Figure 14-28 A cold outer atmosphere and a heated interior create a winter stack effect.

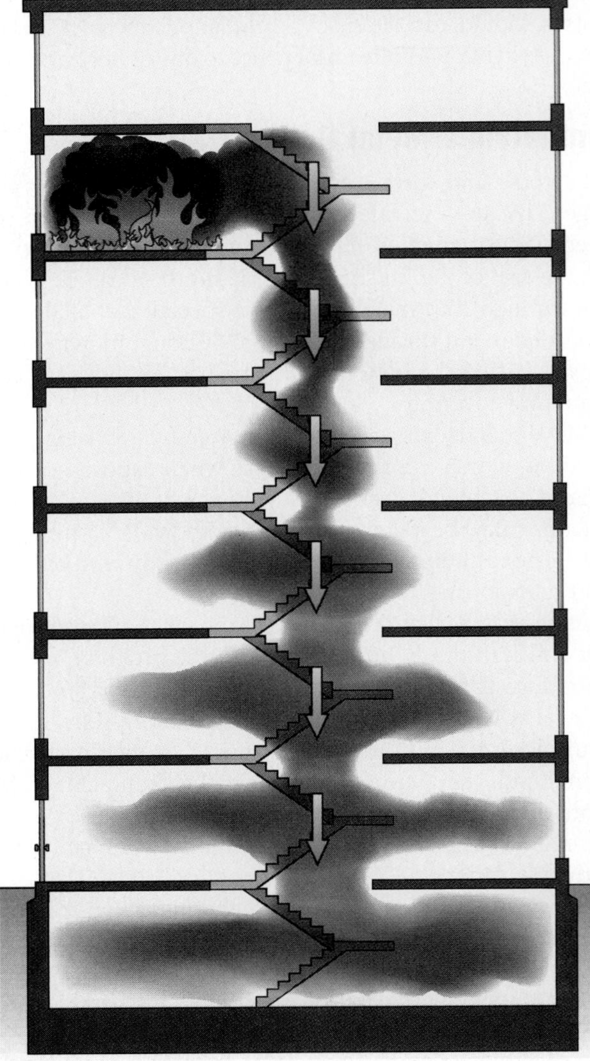

Figure 14-29 A summer stack effect occurs when the interior temperature is much cooler than the outside air.

outer atmosphere and a heated interior will cause smoke to rise quickly through stairways, elevator shafts, and other vertical openings, filling the upper levels of the building ▲ Figure 14-28.

The opposite situation can occur on a hot summer day, when the interior temperature is much cooler than the outside atmosphere. The heavy cooler air pushes the smoke down the vertical openings, toward a lower level exit ► Figure 14-29.

The situation can change as the fire produces enough heat to alter the temperature profile within the building. The air currents within a tall building may be stronger than the convection currents that normally govern the flow of smoke and other products of combustion. A strong wind through an open or broken window may change the direction of smoke. Contaminated air may suddenly move toward the opening or be pushed back as a strong draft of fresh air enters the building.

After smoke mixes with fresh air or is hit by water from sprinklers or hose streams, it cools and can "sit" in one location. This cold smoke can fill several floors and will usually need to be cleared by mechanical means.

A key objective in ventilating a high-rise building is to manage the air movement in stairways and elevator shafts. Positive-pressure ventilation can help keep stairways clear of smoke and ventilate individual floors in high-rise structures.

At least one stairwell should be kept well-ventilated and designated as an occupant and rescue route. These stairs

Figure 14-30 Structures without windows pose significant problems in ventilation.

must be used by rescuers and escaping victims ONLY. Some buildings have smokeproof stair towers or pressurized stair shafts designed to keep smoke out of the stairway. Otherwise, positive-pressure fans can be used in a stairway to keep smoke out. Place the positive-pressure fans at a ground floor doorway, so that fresh air blows into the stairway.

A pressurized stairway also can be used to clear smoke from a floor. Opening a door from the stairway allows fresh air to enter the floor. The contaminated atmosphere is vented out through a window or another stairway.

Windowless Buildings

Many structures do not have windows (▲ Figure 14-30). Some buildings are designed without windows; others have bricked-up or covered windows. These buildings pose two significant risks to fire fighters: Heat and products of combustion are trapped, and fire fighters have no secondary exit route.

Windowless buildings are similar to basements. Any ventilation will need to be as high as possible and will probably require mechanical assistance. Using existing rooftop openings, cutting openings in the roof, reopening boarded-up windows or doors, and making new openings in exterior walls are all possible ways to ventilate windowless structures.

Large Buildings

Providing adequate ventilation is more difficult in large buildings than in smaller ones. A ventilation hole in the wrong location can draw the fire toward the opening, spreading the fire to an area that had not been involved. This underscores the importance of coordinating ventilation operations with the overall fire attack strategy.

Smoke will cool as it travels into unaffected portions of a large building. A sprinkler suppression system will also cool the smoke, causing it to stratify. As the cold smoke fills the area, it becomes more difficult to clear.

If possible, fire fighters should use interior walls and doors to create several smaller compartments in the building. This can limit the spread of heat and smoke. The smaller areas can be cleared one at a time with positive-pressure fans. Several fans can be used in a series or in parallel lines to clear smoke from a large area.

Backdraft and Flashover Considerations

Ventilation is a major consideration in two significant fireground phenomena: backdraft and flashover. Both can be deadly situations and fire fighters should exercise great caution when conditions indicate that either is possible.

Backdraft

A backdraft can occur when a building is charged with hot gases, and most of the available oxygen has been consumed. There may be few flames, but the hot gases contain rich amounts of unburned or partially burned fuel. If clean air is introduced to the mixture, the fuel can ignite and explode. To reduce the danger, fire fighters must release as much heat and unburned products of combustion as possible, without allowing clean air to enter.

A ventilation opening as high as possible within the building or area can help to eliminate potential backdraft conditions. A roof opening will draw the hot mixture up and relieve the interior pressure. As the mixture rises into the open atmosphere, it may ignite. The ventilation crew should have a charged line ready to protect themselves and nearby exposures.

The attack crews should charge their hose lines outside the building. They should not force entry or begin to spray water until the structure is ventilated. Once they see flaming combustion inside the structure, they can open their hose streams to cool the interior atmosphere as quickly as possible.

Flashover

Both ventilation and cooling are needed to relieve potential flashover conditions. Flashover can occur when the air in a room is very hot, and exposed combustibles in the space are near their ignition point. Applying water will cool the upper atmosphere, while ventilation draws the heated smoke and gases out of the space. Ventilation openings should be placed to draw the heat and flames away from the hose crew. Simultaneously, the water stream should be used to push the products of combustion toward the opening. Although flame conditions may intensify briefly, the heat will be directed away from the hose crew.

Wrap-Up

Ready for Review

Ventilation is a critical function that must be addressed in the first few minutes after fire fighters arrive at a burning building. Proper ventilation can save lives, improve operating conditions for fire fighters advancing hose lines, and limit property damage to the structure.

Several factors affect ventilation decisions, including building construction, built-in openings, and the fire itself. Fire fighters must be able to identify factors that might inhibit ventilation. Ventilation crews must use every option to release superheated gases and internal pressure so that interior crews can enter quickly to find victims and suppress the fire. Proper, expedient ventilation enables fire fighters to find victims, to advance hose lines quickly and safely, and to limit fire spread.

Fire fighters should always be safety-conscious when performing ventilation. Ventilation operations are generally above the fire and in the path of superheated smoke and gases. Conditions can deteriorate quickly, so safety is a paramount concern.

Chief Concepts

- Ventilation is a process that helps remove heat, smoke, and toxic gases from a burning building.
- Ventilation is a critical component of fire attack and must be coordinated with the advancement of attack hose lines.
- Ventilation saves lives and enhances fire fighter safety.
- Vertical ventilation should be provided at the highest point directly over the source of the fire.
- Horizontal ventilation should be provided as close to the seat of the fire as possible.
- Fire fighters should know and recognize the warning signs of flashover and backdraft.
- Fire fighters should be able to recognize differences in building construction that will enhance or inhibit ventilation efforts.
- Fire fighters should be able to identify the hazards associated with ventilation operations.
- Fire fighters should be able to identify when ventilation is necessary and when it may be an unacceptable risk.
- Fire fighters should understand natural and mechanical ventilation techniques and when each should be used.

Hot Terms

Arched roof A rounded roof usually associated with a bow-truss.

Backdraft The sudden explosive ignition of fire gases when oxygen is introduced into a superheated space previously deprived of oxygen.

Balloon-frame construction An older type of wood frame construction in which the wall studs extend vertically from the basement of a structure to the roof without any fire stops.

Bearing wall A wall that is designed to support the weight of a floor or roof.

Chase Open space within walls for wires and pipes.

Churning Recirculation of exhausted air that is drawn back into a negative-pressure fan in a circular motion.

Cockloft The concealed space between the top floor ceiling and the roof of a building.

Convection Heat transfer by circulation within a medium such as a gas or a liquid.

Ejectors Electrical fans used in negative-pressure ventilation.

Fire-resistive construction Buildings that have structural components of noncombustible materials with a specified fire resistance. Materials can include concrete, steel beams, and masonry block walls. Type I building construction, as defined in NFPA 220, *Types of Building Construction.*

Flashover The condition where all combustibles in a room or confined space have been heated to the point at which they release vapors that will support combustion, causing all combustibles to ignite simultaneously.

Flat roofs Horizontal roofs often found on commercial or industrial occupancies.

Gusset plates The connecting plate made of wood or lightweight metal used in trusses.

Horizontal ventilation The process of making openings so that smoke, heat, and gases can escape horizontally from a building through openings such as doors and windows.

Hydraulic ventilation Ventilation that relies upon the movement of air caused by a fog stream.

Kerf cut A cut that is only the width and depth of the saw blade. It is used to inspect cockloft spaces from the roof.

Laths Thin strips of wood used to make the supporting structure for roof tiles.

Leap-frogging A fire spread from one floor to the other through exterior windows (auto-exposure).

Louver cut A cut that is made using power saws and axes to cut along and between roof supports so that the sections created can be tilted into the opening.

Mechanical ventilation Ventilation that uses mechanical devices to move air.

Mushrooming The process that occurs when rising smoke, heat, and gases encounter a horizontal barrier such as a ceiling and begin to move out and back down.

Natural ventilation Ventilation that relies upon the natural movement of heated smoke and wind currents.

Negative-pressure ventilation Ventilation that relies upon electric fans to pull or draw the air from a structure or area.

Ordinary construction Buildings where the exterior walls are noncombustible or limited-combustible, but the interior floors and walls are made of combustible materials. Also known as Type III building construction, as defined in NFPA 220, *Types of Building Construction.*

Parapet walls Walls on a flat roof that extend above the roof line.

Peak cut A ventilation opening that runs along the top of a peaked roof.

Pitched roof A roof with sloping or inclined surfaces.

Positive-pressure ventilation Ventilation that relies upon fans to push or force clean air into a structure.

Primary cut The main ventilation opening made in a roof to allow smoke, heat, and gases to escape.

Products of combustion Heat, smoke, and toxic gases.

Rafters Solid structural components that support a roof.

Roof covering The material or assembly that makes up the weather-resistant surface of a roof.

Roof decking The rigid component of a roof covering.

Seat of the fire The main area of the fire.

Secondary cut An additional ventilation opening to create a larger opening, or to limit fire spread.

Smoke inversion Smoke hanging low to the ground due to the cold air.

Sounding The process of striking a roof with a tool to determine if the roof is solid enough to support the weight of a fire fighter.

Stack effect The vertical movement of smoke and products of combustion within a high-rise building caused by temperature differentials between the interior and exterior.

Trench cut A cut that is made from bearing wall to bearing wall to prevent horizontal fire spread in a building.

Triangular cut A triangle-shaped ventilation cut in the roof decking that is made using saws or axes.

Truss A collection of lightweight structural components joined in a triangular configuration that can be used to support either floors or roofs.

Ventilation The process of removing smoke, heat, and toxic gases from a burning structure and replacing them with clean air.

Vertical ventilation The process of making openings so that the smoke, heat, and gases can escape vertically from a structure.

Wood frame construction Buildings with exterior walls, interior walls, floors, and roof made of combustible wood material. Type V building construction, as defined in NFPA 220, *Types of Building Construction.*

Fire Fighter in Action

You have just joined Ladder Company 3 for a 90-day orientation period. Captain Brass assigns you to work with Fire Fighter Hess on the ventilation team. Your first job is to check the ladder truck for the equipment you will be using.

1. There are four gasoline-powered saws in one compartment on the ladder truck: one chain saw and three circular saws with different types of blades. Why does one truck require so many different power saws?
   - **A.** Different types of saws and blades are used to cut through different materials.
   - **B.** It may be necessary to cut in multiple locations at the same time.
   - **C.** Changing blades takes valuable time when a saw is needed.
   - **D.** All of the above are true.
2. Fire Fighter Hess tells you that you must always start a power saw and run it at full throttle for 30 seconds before taking it to the roof of a building to create a ventilation opening. Why do you think he tells you this?
   - **A.** It is easier to start a saw on the ground than on a roof.
   - **B.** The saw will run more smoothly after it has been warmed up.
   - **C.** You want to ensure that the saw runs properly before taking it to the roof.
   - **D.** If there is heavy smoke on the roof, the saw may not be able to run properly.
3. Your company is called for a room-and-contents fire in a single-family dwelling. The fire is in a room at the rear of the house. While the engine company is preparing to enter with an attack line, Captain Brass calls for positive-pressure ventilation. Fire Fighter Hess tells you to set up the positive-pressure fan while he checks for an exhaust opening. What should you do?
   - **A.** Start the fan immediately to clear smoke out ahead of the attack crew.
   - **B.** Wait until your partner tells you that an opening is ready at the back of the house.
   - **C.** Wait until the attack crew has entered and then start the fan.
   - **D.** Follow your partner to the rear of the house to set up the fan.
4. You and your partner are on the flat roof of a single-story shopping center. Two engine companies are inside trying to locate the fire. They report over the radio that conditions are "zero visibility and very hot." What should you do?
   - **A.** Find the highest point on the roof to make a peak cut.
   - **B.** Find a point ahead of the fire to make a trench cut.
   - **C.** Locate the hottest area and make a rectangular opening directly above the fire.
   - **D.** Make a series of kerf cuts to see if the fire is in the cockloft.
5. You have just arrived at a working fire in a three-story garden apartment building. Gray smoke is coming from an apartment on the second floor at the front of the building. The engine company is advancing a hose line up the stairway toward the apartment. The most appropriate ventilation technique to use is:
   - **A.** Vertical ventilation
   - **B.** Horizontal ventilation
   - **C.** Positive-pressure ventilation
   - **D.** Negative-pressure ventilation
6. The search-and-rescue team reports that the interior stairway is full of smoke and visibility is zero. They ask for ventilation to improve visibility. Captain Brass tells you and your partner to take care of this problem. What should you do?
   - **A.** Go to the roof and look for a bulkhead or hatch that can be opened at the top of the stairs.
   - **B.** Set up a positive-pressure fan to blow fresh air into the stairway.
   - **C.** Set up a smoke ejector to exhaust the smoke through an apartment on the third floor.
   - **D.** Obtain a hose line to provide hydraulic ventilation in the stairway.

7. You have been dispatched to a fire in an auto parts store in the predawn hours. As you arrive, you observe heavy dark smoke coming from the building and hanging close to the ground. The front windows are dark and a dull red glow can be seen inside. Based on these observations, what would you anticipate?
   A. Potential structural collapse
   B. Potential rollover conditions
   C. Potential flashover conditions
   D. Potential backdraft conditions

8. The preincident plan indicates that the auto parts store has a flat roof supported by lightweight wood trusses. Based on this information, what orders should you expect?
   A. Go to the roof as quickly as possible for vertical ventilation.
   B. Set up a positive-pressure fan in front of the building.
   C. Set up a negative-pressure smoke ejector at the rear of the building.
   D. Prepare for defensive operations.

Water Supply

Chapter 15

NFPA 1001 Standard

Fire Fighter I

5.3.15 Connect a fire department pumper to a water supply as a member of a team, given supply or intake hose, hose tools, and a fire hydrant or static water source, so that connections are tight and water flow is unobstructed.

5.3.15 (A) *Requisite Knowledge.* Loading and off-loading procedures for mobile water supply apparatus; fire hydrant operation; and suitable static water supply sources, procedures, and protocol for connecting to various water sources.

5.3.15 (B) *Requisite Skills.* The ability to hand lay a supply hose, connect and place hard suction hose for drafting operations, deploy portable water tanks as well as the equipment necessary to transfer water between and draft from them, make hydrant-to-pumper hose connections for forward and reverse lays, connect supply hose to a hydrant, and fully open and close the hydrant.

Fire Fighter II

6.5.4 Test the operability of and flow from a fire hydrant, given a Pitot tube, pressure gauge, and other necessary tools, so that the readiness of the hydrant is assured and the flow of the water from the hydrant can be calculated and recorded.

6.5.4 (A) *Requisite Knowledge.* How water flow is reduced by hydrant obstructions; direction of hydrant outlets to suitability of use; the effect of mechanical damage, rust, corrosion, failure to open the hydrant fully, susceptibility to freezing; and the meaning of the terms static, residual, and flow pressure.

6.5.4 (B) *Requisite Skills.* The ability to operate a pressurized hydrant, use a Pitot tube and pressure gauges, detect damage, and record results of test.

Additional NFPA Standards

NFPA 24, *Standard for the Installation of Private Fire Service Mains and Their Appurtenances*

NFPA 1142, *Standard on Water Supplies for Suburban and Rural Fire Fighting*

Knowledge Objectives

After studying this chapter, you will be able to:

- Describe the sources of water for a municipal water supply system.
- Explain the purpose of a water treatment facility.
- Describe the major features of a municipal water distribution system.
- Describe dry-barrel fire hydrants and wet-barrel fire hydrants.
- Discuss maintaining and testing a fire hydrant.
- Define static pressure, residual pressure, and flow pressure.
- Discuss rural water supplies.
- Describe how portable tanks are used to supply water for firefighting.

Skills Objectives

After completing this chapter, you will be able to perform the following tasks:

- Operate a fire hydrant.
- Shut down a fire hydrant.
- Unload and assemble a portable water tank.

You Are the Fire Fighter

You are a member of the attack crew at a house fire. A couch fire has spread to the other contents of the living room, and flames are blowing out through the front window. You advance a preconnected hose line into the building through the front door and begin to douse the fire with water as your partner advances the hose line.

Another crew advances a hose line to the second floor while a third crew starts search-and-rescue operations. You realize that everyone in the building is depending on you to control the fire and stop it from spreading.

On your portable radio, you can hear the pump operator arranging for a supply line to the engine. You imagine the chaos and tragedy if you ran out of water and are grateful for the steady supply.

1. ***If you use too much water too quickly, can everyone get out safely?***
2. ***What might happen if the supply line is compromised?***
3. ***How many hose lines can the engine supply from the on-board water tank?***

Introduction

The importance of a dependable and adequate **water supply** for fire-suppression operations is self-evident. The hose line is not only the primary weapon for fighting fire, it is also the fire fighter's primary defense against being burned or driven out of a burning building. The basic plan for fighting most fires depends on having an adequate supply of water to confine, control, and extinguish the fire.

If the water supply is interrupted while crews are working inside a building, the fire fighters can be trapped, injured, or killed. Fire fighters entering a burning building need to be confident that their water supply is both reliable and adequate to operate hose lines for their protection and to extinguish the fire.

Ensuring a dependable water supply is a critical fireground operation that must be accomplished as soon as possible. A water supply should be established at the same time as other initial fireground operations. At many fire scenes, size-up, forcible entry, raising ladders, search and rescue, ventilation, and establishing a water supply all occur concurrently.

Fire fighters obtain water from one of two sources. **Municipal** and **private water systems** furnish water under pressure through fire hydrants (► Figure 15-1). Rural areas may depend on **static water sources** such as lakes and streams. They serve as drafting sites for fire department apparatus to obtain and deliver water to the fire scene.

Figure 15-1 The water that comes from a hydrant is provided by a municipal or private water system.

Often, the water that is carried in a tank on one of the first-arriving vehicles is used in the initial attack. Although many fires are successfully controlled using tank water, this

Figure 15-2 Mobile water supply apparatus can deliver large quantities of water to the scene of a fire.

Figure 15-3 Impurities are removed at the water treatment facility.

tactic does not ensure the adequacy and reliability of the water supply. The establishment of an adequate, continuous water supply then becomes the primary objective to support the fire attack (▲ Figure 15-2). The operational plan must ensure that an adequate and reliable water supply is available before the tank is empty.

Municipal Water Systems

Municipal water systems make clean water available to people in populated areas and provide water for fire protection. As the name suggests, most municipal water systems are owned and operated by a local government agency, such as a city, county, or special water district. Some municipal water systems are privately owned; however, the basic design and operation of both private and government systems are very similar.

Municipal water is supplied to homes, commercial establishments, and industries. Hydrants make the same water supply available to the fire department. In addition, most automatic sprinkler systems and many standpipe systems are connected to a municipal water source. There are three parts to a municipal water system: the water source, the treatment plant, and the distribution system.

Water Sources

Municipal water systems can draw water from wells, rivers, streams, lakes, or man-made storage facilities called **reservoirs**. The source will depend on the geographic and hydrologic features of the area. Many municipal water systems draw water from several sources to ensure a sufficient supply. Underground pipelines or open canals supply some cities with water from sources that are many miles away.

The water source for a municipal water system needs to be large enough to meet the total demands of the service area. Most municipal water systems include large storage facilities, so they will be able to meet the community's water supply demands if the primary water source is interrupted. The back-up supply for some systems can provide water for several months or years, while in other systems, the supply may last only a few days.

Water Treatment Facilities

Municipal water systems also include a water treatment facility, where impurities are removed (▲ Figure 15-3). The nature of the treatment system depends on the quality of the untreated source water. Water that is clean and clear from the source requires little treatment. Other systems must use extensive filtration to remove impurities and foreign substances. Some treatment facilities use chemicals to remove impurities and improve the water's taste. All of the water in the system must be suitable for drinking.

Chemicals also are used to kill bacteria and harmful organisms and to keep the water pure as it moves through the distribution system to individual homes or businesses. After the water has been treated, it enters the distribution system.

Water Distribution System

The distribution system delivers water from the treatment facility to the end users and fire hydrants through a complex network of underground pipes, known as **water mains**. In most cases the distribution system also includes pumps, storage tanks, reservoirs, and other necessary components to ensure that the required volume of water can be delivered, where and when it is needed, at the required pressure.

Water pressure requirements differ, depending on how the water will be used. Generally, water pressure ranges from 20 pounds per square inch (psi) to 80 psi at the delivery point. The recommended minimum pressure for water coming from a fire hydrant is 20 psi, but it is possible to operate with lower hydrant pressures under some circumstances.

Most water distribution systems rely on an arrangement of pumps to provide the required pressure, either directly or indirectly. Some water distribution systems use pumps to supply direct pressure. If the pumps stop operating, the pressure is lost, and the system will be unable to deliver adequate water to the end users or to hydrants. Most municipal systems have multiple pumps and back-up power supplies to reduce the risk of a service interruption due to a pump failure. The extra pumps can sometimes be used to boost the flow for a major fire or a high demand period.

In a pure **gravity-feed system**, the water source, treatment plant, and storage facilities are located on high ground while the end users live in lower-lying areas, such as a community in a valley (► Figure 15-4). This type of system may not require any pumps, because gravity, through the elevation differentials, provides all the necessary pressure to deliver the water. In some systems, the elevation pressure is so high that pressure-control devices are needed to keep from over-pressurizing parts of the system.

Most municipal water supply systems use both pumps and gravity to deliver water. Pumps may be used to deliver the water from the treatment plant to **elevated water storage towers** or to reservoirs located on hills or high ground areas. The elevated storage facilities maintain the desired water pressure in the distribution system, so that water can be delivered under pressure, even if the pumps are not operating (► Figure 15-5). When the elevated storage facilities need refilling, large supply pumps are used. Additional pumps may be installed to increase the pressure in particular areas, such as a booster pump that provides additional pressure for a hilltop neighborhood.

A combination pump-and-gravity-feed system must maintain enough water in the elevated storage tanks and reservoirs to meet anticipated demands. If more water is being used than the pumps can supply, or if the pumps are out of service, some systems will be able to operate for several days using only their elevated storage reserves, while others will be able to function only for a few hours.

The underground water mains that deliver water to the end users come in several different sizes. Large mains, known as **primary feeders**, carry large quantities of water to a section of the town or city. Smaller mains, called **secondary feeders**, distribute the water to a smaller area. The smallest pipes, called **distributors**, carry water to the users and hydrants along individual streets.

The size of the water mains required depends on the amount of water needed both for normal consumption and for fire protection in that location. Most jurisdictions specify the minimum size main that can be installed in a new municipal water system to ensure an adequate flow. Some municipal water systems, however, may still have undersized water mains in older areas of the community. Fire fighters must know the arrangement and capacity of the water systems in their response areas.

Water mains in a well-designed system will follow a grid pattern. A grid arrangement provides water flow to a fire hydrant from two or more directions and establishes multiple paths from the source to each area. This helps to ensure an adequate flow of water for firefighting. The grid design also helps to minimize downtime for the other portions of the system if a water main breaks or needs maintenance work. With a grid, the water flow can be diverted around the affected section.

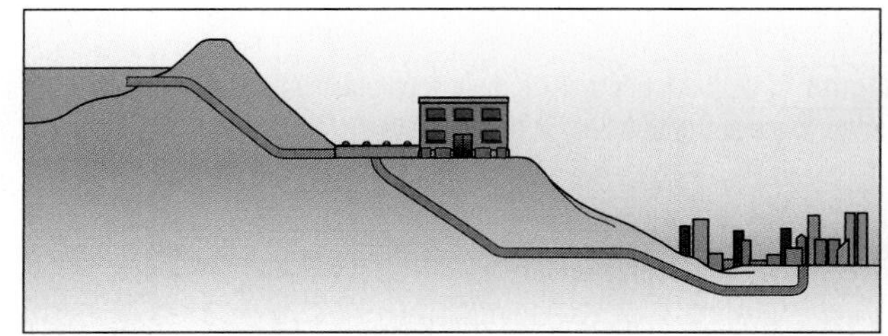

Figure 15-4 A gravity-feed system can deliver water to a low-lying community without the need for pumps.

Figure 15-5 Water that is stored in an elevated tank can be delivered to the end users under pressure.

Older water distribution systems may have dead-end water mains, which supply water from only one direction. Dead-end water mains may still be found in the outer reaches of a municipal system. Hydrants on a dead-end water main will have a limited water supply. If two or more hydrants on the same dead-end main are used to fight a fire, the upstream hydrant will have more water and water pressure than the downstream hydrants.

Control valves installed at intervals throughout a water distribution system allow different sections to be turned off or isolated. These valves are used when a water main breaks or when work must be performed on a section of the system.

Shut-off valves are located at the connection points where the underground mains meet the distributor pipes. Shut-off valves control the flow of water to individual customers or to individual fire hydrants (▼ Figure 15-6). If the water system in a building or to a fire hydrant is damaged, the shut-off valves can be closed to prevent further water flow.

The fire department should notify the water department when fire operations will require prolonged use of large quantities of water. The water department may be able to increase the normal volume and/or pressure by starting additional pumps. In some systems, the water department can open valves to increase the flow to a certain area in response to fire department operations at major fires.

Fire Hydrants

Fire hydrants provide water for firefighting purposes. Public hydrants are part of the municipal water distribution system and draw water directly from the public water mains. Hydrants also can be installed on private water systems supplied by the municipal water system or from a separate source. The water source as well as the adequacy and reliability of the supply to private hydrants must be identified to ensure that they will be sufficient in fighting fires.

Most fire hydrants consist of an upright steel casing (barrel) attached to the underground water distribution system. The two main types of hydrants are the dry-barrel hydrant and the wet-barrel hydrant. Hydrants are equipped with one or more valves to control the flow of water through the hydrant. One or more outlets are provided to connect fire department hoses to the hydrant. These outlets are sized to fit the $2^1/_2$" or larger fire hoses used by the local fire department.

Dry-Barrel Hydrants

Dry-barrel hydrants are used in climates where temperatures can be expected to fall below freezing. The valve that controls the flow of water into the barrel of the hydrant is located at the base, below the frost line, to keep the hydrant from freezing (▼ Figure 15-7 A & B). The length of the barrel depends on the climate and the depth of the valve. Water only enters the barrel of the hydrant when it is going to be used. Turning the nut on the top of the hydrant rotates the operating stem, which opens the valve so that water flows up into the barrel of the hydrant.

Whenever the hydrant is not in use, the barrel must remain dry. If the barrel contains standing water, it will

Figure 15-6 A shut-off valve controls the water supply to an individual user or fire hydrant.

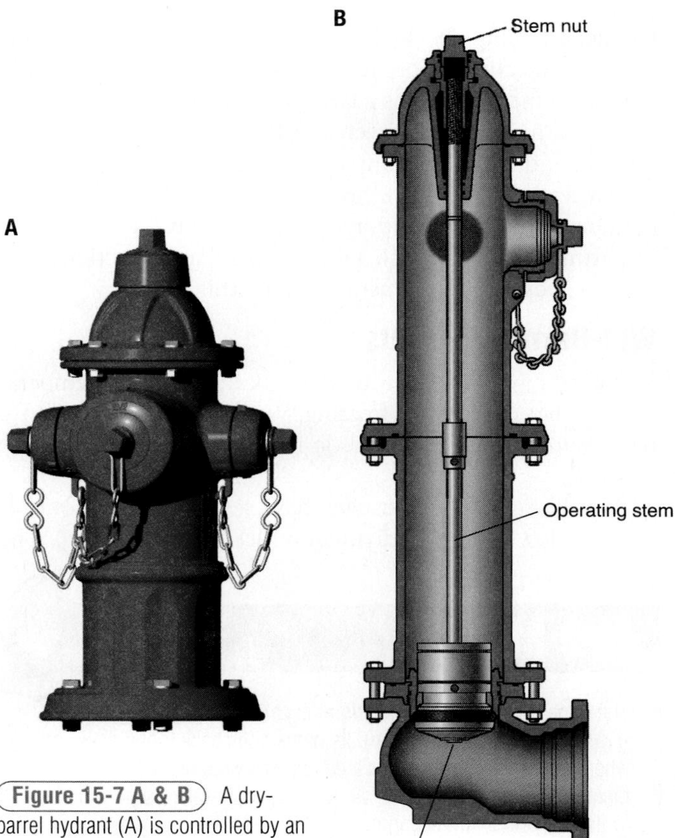

Figure 15-7 A & B A dry-barrel hydrant (A) is controlled by an underground valve (B).

freeze in cold weather and render the hydrant inoperable. After each use, the water drains out through an opening at the bottom of the barrel. This drain is fully open when the hydrant valve is fully shut. When the hydrant valve is opened, the drain closes. This prevents water from being forced out of the drain when the hydrant is under pressure.

A partially opened valve means that the drain is also partially open, and pressurized water can flow out. This can erode (undermine) the soil around the base of the hydrant and may damage the hydrant. For this reason a hydrant should always be either fully opened or fully closed. A fully opened hydrant also makes the maximum flow available to fight a fire.

Most dry-barrel hydrants have only one large valve controlling the flow of water (► **Figure 15-8**). Each outlet must be connected to a hose or an outlet valve, or have a hydrant cap firmly in place before the valve is turned on. Many fire departments use special hydrant valves so additional connections can be made after water is flowing through the first hose. If additional outlets are needed later, separate outlet valves can be connected before the hydrant is opened. For more information on hoses, see Chapter 16, Fire Hose, Nozzles, Streams, and Foam.

In some areas, vandals will dispose of trash or foreign objects in the empty barrels of dry-barrel hydrants. These materials can obstruct the water flow or damage a fire department pumper if they are drawn into the pump. Many fire departments check the operation and flush out debris before connecting a hose to a hydrant. The fire fighter making the connection opens a large outlet cap and then releases the valve just enough to ensure that water flows into the hydrant and flushes out any foreign matter. This should only take a few seconds. The fire fighter then closes the valve, connects the hose, and reopens the valve all the way. Departments should also perform regular inspections and tests to keep hydrants operating smoothly.

Wet-Barrel Hydrants

Wet-barrel hydrants are used in locations where temperatures do not drop below freezing. Wet-barrel hydrants always have water in the barrel and do not have to be drained after each use.

Wet-barrel hydrants usually have separate valves controlling the flow to each individual outlet (► **Figure 15-9**). The fire fighter can hook up one hose line and begin flowing water, and later attach a second hose line and open the valve for that outlet, without shutting down the hydrant.

Fire Marks

The large opening on a fire hydrant is called the steamer port (or steamer connection). Its name goes back to the days when horse-drawn steam-powered engines were used. Large-diameter outlets on hydrants allowed as much water as possible to flow directly into the engine.

Operation of Fire Hydrants

Fire fighters must be proficient in operating a fire hydrant. (► **Skill Drill 15-1**) outlines the steps in getting water from a dry-barrel hydrant efficiently and safely. These same steps, with the modifications noted, apply to wet-barrel hydrants as well.

1. Remove the cap from the outlet you will be using. **(Step 1)**
2. Quickly look inside the hydrant opening for any objects that may have been thrown into the hydrant. **(Step 2)** (Omit this step for a wet-barrel hydrant.)
3. Check to ensure that the remaining hydrant caps are snugly attached. **(Step 3)** (Omit this step for a wet-barrel hydrant.)

Figure 15-8 Most dry-barrel hydrants only have the one large valve at the bottom of the barrel controlling the flow of water.

Figure 15-9 A wet-barrel hydrant has a separate valve for each outlet.

4. Place the hydrant wrench on the stem nut. Check the top of the hydrant for the arrow indicating which direction to turn the nut to open the hydrant valve. **(Step 4)**
5. Open the hydrant just enough to determine that there is a good flow of water and to flush out any objects that may have been put into the hydrant. **(Step 5)** (Omit this step for a wet-barrel hydrant.)
6. Shut off the flow of water. **(Step 6)** (Omit this step for a wet-barrel hydrant.)
7. Attach the hose or valve to the hydrant outlet. **(Step 7)**
8. When instructed by your officer or the pump operator, start the flow of water. Turn the hydrant wrench to fully open the valve. This may take 12 to 14 turns, depending on the type of hydrant. **(Step 8)**
9. Open the hydrant slowly to avoid a pressure surge. Once the flow of water has begun, you can open the hydrant valve more quickly. Make sure that you open the hydrant valve completely. If the valve is not fully opened, the drain hole will remain open. **(Step 9)**

NOTE: This skill drill applies to a dry-barrel hydrant. For a wet-barrel hydrants, omit steps 2, 3, 5, and 6. Open the valve for the particular outlet that will be used.

Individual fire departments may vary these procedures. For example, some departments specify that the wrench be left on the hydrant. Other departments require that it be removed and returned to the fire apparatus so that an unauthorized person cannot interfere with the operation. Always follow the standard operating procedures for your department.

Shutting Down a Hydrant

Shutting a hydrant down properly is just as important as opening a hydrant properly. If the hydrant is damaged during shutdown, it cannot be used until it has been repaired. Following the steps in (► **Skill Drill 15-2**) will enable you to shut down a hydrant efficiently and safely.

1. Turn the hydrant wrench slowly until the valve is closed. **(Step 1)**
2. Allow the hose to drain by opening a drain valve or disconnecting a hose connection downstream. Slowly disconnect the hose from the hydrant outlet, allowing any remaining pressure to escape. **(Step 2)**
3. On dry-barrel hydrants, leave one outlet open until the water drains from the hydrant. **(Step 3)**
4. Replace the hydrant cap. **(Step 4)**

NOTE: Do not leave or replace the caps on a dry-barrel hydrant until you are sure that the water has completely drained from the barrel. If you feel suction on your hand when you place it over the opening, the hydrant is still draining. In very cold weather, you may have to use a hydrant pump to remove all of the water and prevent freezing.

Locations of Fire Hydrants

Fire hydrants are located according to local standards and nationally recommended practices. Fire hydrants may be placed a certain distance apart, perhaps every 500′ in residential areas and every 300′ in high-value commercial and industrial areas. In many communities, hydrants are located at every street intersection, with mid-block hydrants if the distance between intersections exceeds a specified limit.

In some cases, the requirements for locating hydrants are based on the occupancy, construction, and size of a building. A builder may be required to install additional hydrants when a new building is constructed so that no part of the building will be more than a specified distance from the closest hydrant.

Knowing the plan for installing fire hydrants makes them easier to find in emergency situations. Fire fighters who perform fire inspections or develop preincident plans should identify the locations of nearby fire hydrants for each building or group of buildings.

Inspecting and Maintaining Fire Hydrants

Because hydrants are essential to fire suppression efforts, fire fighters must understand how to inspect and maintain them. Hydrants should be checked on a regular schedule, no less than once a year, to ensure that they are in proper operating condition. During inspections, fire fighters may encounter some common problems and should know how to correct them.

The first factors to check when inspecting hydrants are visibility and accessibility. Hydrants should always be visible from every direction, so they can be easily spotted. A hydrant should not be hidden by tall grass, brush, fences, debris, dumpsters, or any other obstructions (▼ **Figure 15-10**). In winter, hydrants must be clear of snow. No vehicles should be allowed to park in front of a hydrant.

Figure 15-10 Hydrants should not be hidden or obstructed.

Skill Drill

Operating a Fire Hydrant

Remove the cap from the outlet you will be using.

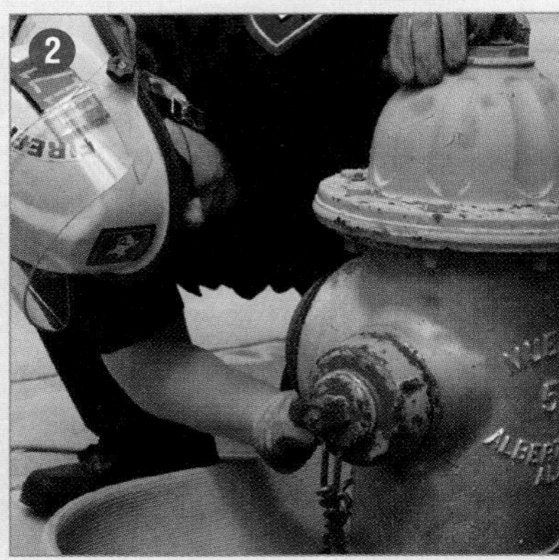

Quickly look inside the hydrant opening for foreign objects. (Dry-barrel hydrant only.)

Check to ensure that the remaining caps are snugly attached. (Dry-barrel hydrant only.)

Attach the hydrant wrench to the stem nut. Check for an arrow indicating the direction to turn to open.

Open the hydrant enough to verify flow and flush hydrant. (Dry-barrel hydrant only.)

Shut off the flow of water. (Dry-barrel hydrant only.)

Attach hose or valve to the hydrant outlet(s).

When instructed, turn the hydrant wrench to fully open the valve.

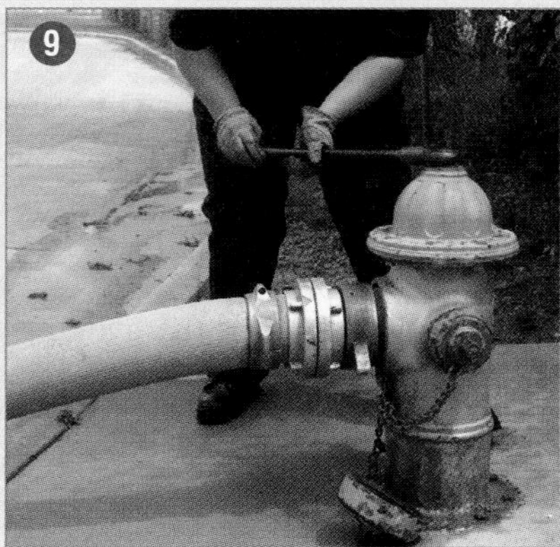

Open slowly to avoid pressure surge.

15-2 Skill Drill

Shutting Down a Hydrant

Turn the wrench to slowly close the hydrant valve.

Drain the hose line. Slowly disconnect the hose from the hydrant outlet.

Leave one hydrant outlet open until the hydrant is fully drained.

Replace the hydrant cap.

In many communities, hydrants are painted in bright reflective colors for increased visibility. The bonnet (the top of the hydrant) may also be color-coded to indicate the available flow rate of a hydrant. Colored reflectors are sometimes mounted next to hydrants or placed in the pavement in front of them to make them more visible at night.

Hydrants should be installed at an appropriate height above the ground. The outlets should not be so high or so low that fire fighters have difficulty connecting hose lines to them. NFPA 24, *Standard for the Installation of Private Fire Service Mains and Their Appurtenances* requires a minimum of 18" from the center of a hose outlet to the finished grade.

Fire Fighter Tips

Before you leave the fire scene, make sure that dry-barrel hydrants are completely drained, even if the weather is warm. By winter, any water left in the hydrant could freeze. Fire fighters will lose valuable time connecting a hose to a frozen hydrant only to discover that it will not operate. If this happens, fire fighters will be without water until they can locate a working hydrant and can reposition and reconnect the hose lines.

Hydrants should be positioned so that the connections, especially the large steamer connection, are facing the street.

During a hydrant inspection, check the exterior for signs of damage. Open the steamer port of dry-barrel hydrants to ensure that the barrel is dry and free of debris. Make sure that all caps are present and that the outlet hose threads are in good working order (► Figure 15-11).

The second part of the inspection ensures that the hydrant works properly. Open the hydrant valve just enough to ensure that water flows out and flushes any debris out of the barrel. After flushing, shut down the hydrant. Leave the cap off dry-barrel hydrants to ensure that they drain properly. A properly draining hydrant will create suction against a hand placed over the outlet opening (► Figure 15-12). When the hydrant is fully drained, replace the cap.

If the threads on the discharge ports need cleaning, use a steel brush and a small triangular file to remove any burrs in the threads. Also check the gaskets in the caps to make sure they are not cracked, broken, or missing. Replace worn gaskets with new ones, which should be carried on each apparatus. Follow the manufacturer's recommendations for any parts that require lubrication.

Testing Fire Hydrants

The amount of water available to fight a fire at a given location is a crucial factor in planning an attack. Will the hydrants deliver enough water at the needed pressure to enable fire fighters to control a fire? If not, what can be done to improve the water supply? How can fire fighters obtain additional water if a fire does occur?

Fire-suppression companies are often assigned to test the flow from hydrants in their districts. The procedures for testing hydrants are relatively simple, but a basic understanding of the concepts of hydraulics and careful attention to detail are required. This section explains some of the basic theory and terminology of hydraulics, and describes how the tests are conducted and the results are recorded.

Flow and Pressure

To understand the procedures for testing fire hydrants, fire fighters must understand the terminology that is used. The flow or quantity of water moving through a pipe, hose, or nozzle is measured by its volume, usually in gallons (or liters) per minute. Water pressure refers to an energy level and is measured in pounds per square inch (psi) (or kilopascals). Volume and pressure are two different but mathematically related measurements.

Water that is not moving has potential energy. When the water is moving, it has a combination of potential energy and kinetic energy. Both the quantity of water flowing and the pressure under a specific set of conditions must be measured as part of testing any water system, including hydrants.

Static pressure is the pressure in a system when the water is not moving. Static pressure is potential energy, because it would cause the water to move if there were some place the water could go. Static pressure causes the water to flow out of an opened fire hydrant. If there were no static pressure, nothing would happen when fire fighters opened a hydrant.

Static pressure is generally created by **elevation pressure** and/or pump pressure. An elevated storage tank creates ele-

Figure 15-11 All hydrants should be checked at least annually.

Figure 15-12 You should feel suction to indicate the hydrant is draining.

vation pressure in the water mains. Gravity also creates elevation pressure in a water system as the water flows from a hilltop reservoir to the water mains in the valley below. Pumps create pressure by bringing the energy from an external source into the system.

Static pressure in a water distribution system can be measured by placing a pressure gauge on a hydrant port and opening the hydrant valve. There cannot be any water flowing out of the hydrant when static pressure is measured.

Static pressure measured in this way assumes that there is no flow in the system. Because municipal water systems deliver water to hundreds or thousands of users, there is almost always water flowing within the system. In most cases, a static pressure reading is actually measuring the normal operating pressure of the system.

Normal operating pressure refers to the amount of pressure in a water distribution system during a period of normal consumption. In a residential neighborhood, for example, people are constantly using water to care for lawns, wash clothes, bathe, and do other normal household activities. In an industrial or commercial area, normal consumption occurs during a normal business day as water is used for various purposes. The system uses some of the static pressure to deliver this water to residents and businesses. A pressure gauge connected to a hydrant during a period of normal consumption will indicate the normal operating pressure of the system.

Fire fighters need to know how much pressure will be in the system when a fire occurs. Because the regular users of the system will be drawing off a normal amount of water even during a fire fight, the normal operating pressure is sufficient for measuring available water.

Residual pressure is the amount of pressure that remains in the system when water is flowing. When fire fighters open a hydrant and start to draw large quantities of water out of the system, some of the potential energy of still water is converted to the kinetic energy of moving water. However, not all of the potential energy turns into kinetic energy; some of it is used to overcome friction in the pipes. The pressure remaining while the water is flowing is residual pressure.

Residual pressure is important because it provides the best indication of how much more water is available in the system. The more water that is flowing, the less residual pressure there is. In theory, when the maximum amount of water is flowing, the residual pressure is zero, and there is no more potential energy to push more water through the system. In reality, 20 psi is considered the minimum usable residual pressure, to reduce the risk of damage to underground water mains or pumps.

At the scene of a fire, a pump operator uses the difference between static pressure and residual pressure to determine how many more attack lines or appliances can be operated from the available water supply. The operator can refer to a set of tables to calculate the maximum amount of available water. The tables are based on the static and residual pressure readings taken during hydrant testing. **Flow pressure** measures the quantity of water flowing through an opening during a hydrant test. When a stream of water flows out through an opening (known as an orifice), all of the pressure is converted to kinetic energy. To calculate the volume of water flowing, fire fighters measure the pressure at the center of the water stream as it passes through the opening, and factor in the size and flow characteristics of the orifice. A **Pitot gauge** is used to measure flow pressure in pounds per square inch (or kilopascals), and to calculate the flow in gallons per minute (or liters per minute).

Knowing the static pressure, the flow in gallons (liters) per minute, and the residual pressure enables fire fighters to calculate the amount of water that can be obtained from a hydrant or a group of hydrants on the same water main. The test procedure is described in the following section.

Hydrant Testing Procedure

The procedure for testing hydrant flows requires two adjacent hydrants, a Pitot gauge, and an outlet cap with a pressure gauge. Fire fighters will measure static pressure and residual pressure at one hydrant, and will open the other to let water flow out. The two hydrants should be connected to the same water main and preferably at about the same elevation (▼ Figure 15-13).

The cap gauge is placed on one of the outlets of the first hydrant. The hydrant valve is then opened to allow water to fill the hydrant barrel. The initial pressure reading on this gauge is recorded as the static pressure.

At the second hydrant, fire fighters remove one of the discharge caps and open the hydrant. They put the Pitot gauge into the middle of the stream and take a reading. This is recorded as the Pitot pressure. At the same time, fire fighters at the first hydrant record the residual pressure reading.

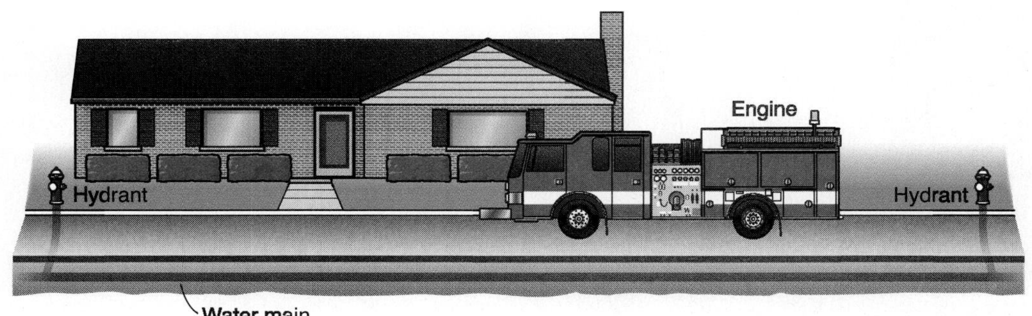

Figure 15-13 Testing hydrant flow requires two hydrants on the same water main.

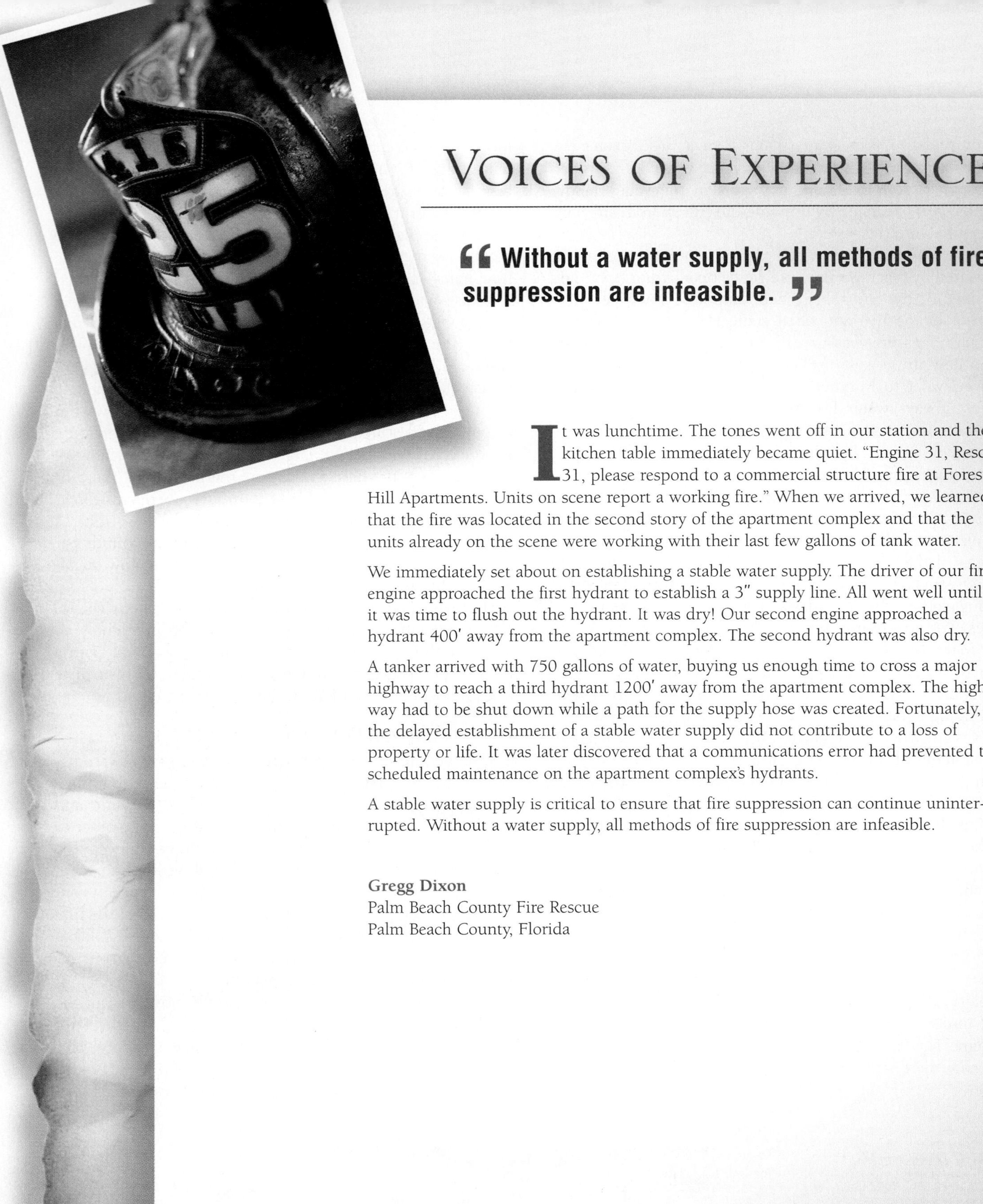

Voices of Experience

"Without a water supply, all methods of fire suppression are infeasible."

It was lunchtime. The tones went off in our station and the kitchen table immediately became quiet. "Engine 31, Rescue 31, please respond to a commercial structure fire at Forest Hill Apartments. Units on scene report a working fire." When we arrived, we learned that the fire was located in the second story of the apartment complex and that the units already on the scene were working with their last few gallons of tank water.

We immediately set about on establishing a stable water supply. The driver of our first engine approached the first hydrant to establish a 3″ supply line. All went well until it was time to flush out the hydrant. It was dry! Our second engine approached a hydrant 400′ away from the apartment complex. The second hydrant was also dry.

A tanker arrived with 750 gallons of water, buying us enough time to cross a major highway to reach a third hydrant 1200′ away from the apartment complex. The highway had to be shut down while a path for the supply hose was created. Fortunately, the delayed establishment of a stable water supply did not contribute to a loss of property or life. It was later discovered that a communications error had prevented the scheduled maintenance on the apartment complex's hydrants.

A stable water supply is critical to ensure that fire suppression can continue uninterrupted. Without a water supply, all methods of fire suppression are infeasible.

Gregg Dixon
Palm Beach County Fire Rescue
Palm Beach County, Florida

Using the size of the discharge opening (usually $2^1/_2$") and the Pitot pressure, fire fighters can calculate the flow in gallons per minute or look it up in a table. The table usually incorporates factors to adjust for the shape of the discharge opening. Fire fighters can use special graph paper or computer software to plot the static pressure and the residual pressure at the test flow rate. The line defined by these two points shows the number of gallons (liters) per minute that are available at any residual pressure. The flow available for fire suppression is usually defined as the number of gallons (liters) per minute available at 20 psi residual pressure.

Several special devices are available to simplify taking accurate Pitot readings. Some outlet attachments have smooth tips and brackets that hold the Pitot gauge in the exact required position (► Figure 15-14). The flow can also be measured with an electronic flow meter instead of a Pitot gauge.

Figure 15-14 The Pitot gauge.

Rural Water Supplies

Many fire departments protect areas that are not serviced by municipal water systems. In these areas, residents usually depend on individual wells or cisterns to supply the water for domestic uses. Because there are no fire hydrants in these areas, fire fighters must depend on water from other sources. Fire fighters in rural areas must know how to get water from the sources that are available.

Static Sources of Water

Several potential static water sources can be used for fighting fires in rural areas. Both natural and man-made bodies of water such as rivers, streams, lakes, ponds, oceans, reservoirs, swimming pools, and cisterns can be used to supply water for fire suppression (► Figure 15-15). Some areas have many different static sources, while others have few or none at all.

Water from a static source can be used to fight a fire directly, if it is close enough to the fire scene. Otherwise, it must be transported to the fire using long hose lines, engine relays, or mobile water supply tankers.

Figure 15-15 Any accessible body of water can be used as a static source.

Static water sources must be accessible to a fire engine or portable pump. If there is a road or hard surface within 20' of the water source, a fire engine can drive close enough to draft water directly into the pump through a hard suction hose. Rural fire departments should identify these areas and practice establishing drafting operations at all of these locations. Some fire departments construct special access points so engines can approach the water source.

Dry hydrants also provide quick and reliable access to static water sources. A dry hydrant is a pipe with a strainer on one end and a connection for a hard suction hose on the other end. The strainer end should be below the water's surface and away from any silt or potential obstructions. The other end of the pipe should be accessible to fire apparatus, with the connection at a convenient height for an engine hook-up (► Figure 15-16). When a hard suction hose is connected to the dry hydrant, the engine can draft water from the static source.

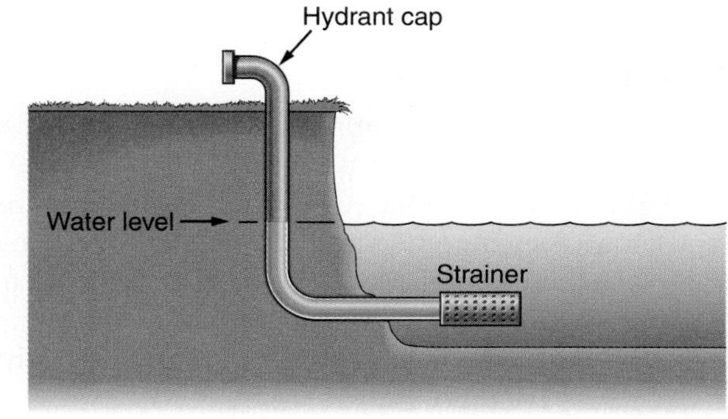

Figure 15-16 A dry hydrant can be placed at an accessible location near a static water source.

Dry hydrants are often installed in lakes and rivers and close to clusters of buildings where there is a recognized need for fire protection. Dry hydrants also may be installed in farm cisterns or connected to swimming pools on private property to make water available for the local fire department. In some areas, dry hydrants are used to enable fire fighters to reach water under the frozen surface of a lake or river. NFPA 1142, *Standard on Water Supplies for Suburban and Rural Fire Fighting* has more information about dry hydrants.

The portable pump is another alternative for areas that are inaccessible to fire apparatus (► Figure 15-17). The portable pump can be hand-carried or transported by an off-road vehicle to the water source. Portable pumps can deliver up to 500 gallons of water per minute.

Figure 15-17 A portable pump can be used if the water source is inaccessible to a fire department engine.

Mobile Water Supply Apparatus

Mobile water supply apparatus (often called tankers or water tenders) also can deliver water to fight fires in rural areas. These trucks are designed specifically to carry water to a fire. Although most fire department engines carry at least 500 gallons of water in the booster tank, fire department tankers generally carry 1,000 to 3,500 gallons of water (► Figure 15-18). Some tankers can transport as much as 5,000 gallons of water.

Figure 15-18 A fire department tanker.

If mobile water supply apparatus is the only source of water for fighting a structural fire, the attack must be carefully planned. There must be enough water on the scene to supply the required hose lines. If the water supply is exhausted before the fire is extinguished, the attack team will be in serious danger and the building will probably be lost. If water use is too conservative, fire-suppression efforts will probably be unsuccessful. To be both safe and successful, the fire department must be able to deliver enough water and to maintain a reliable supply of water to the fire scene.

Some fire departments will begin the attack using water from the tank of the first-arriving unit. Tankers will pump additional water directly into the attack pumper to keep it full. In rural areas, several tankers will be dispatched for a structural fire. These tankers deliver water to the attack engine in sequence, refilling at a remote location and returning to the fire scene as needed.

Portable Tanks

Portable tanks carried on fire apparatus can be quickly set up at a fire scene. Portable tanks typically hold between 600 and 5,000 gallons of water. One engine drafts water from the tank, using a hard suction hose just as it would if the water was from a static source (► Figure 15-19). A tanker is used to fill the portable tank. The pump operator primes the pump and begins drafting water out of the portable tank, while the tanker leaves to get another load of water.

Figure 15-19 An engine is set up to draft water from the portable tank.

Speed is a primary advantage in using a portable tank system. Tankers do not have to hook up to the attack pumper

Figure 15-20 A dump valve allows a tanker to discharge water into the portable tank quickly.

and slowly transfer the water. A **dump valve** enables the tankers to off-load up to 3,000 gallons of water in one minute into a portable tank (▲ Figure 15-20). The faster the tanker can off-load its water, the quicker it can return to the fill site for another load.

Another advantage of the portable tank system is its ability to expand rapidly. Additional portable tanks can be set up and linked to increase water storage capacity, additional pumpers can be used to draft water from the portable tanks, and additional tankers can be added to deliver more water at a faster rate. Several tankers can be used as shuttles, dumping water simultaneously or in sequence.

Follow the steps in (Skill Drill 15-3) to set up a portable tank.

1. Two fire fighters lift the portable tank off the apparatus. The tank may be mounted on a side rack, or on a hydraulic rack that lowers it to the ground.
2. Place the portable tank on level ground beside the engine. The pump operator will indicate the best location.
3. Expand the tank (metal-frame type) or lay it flat (self-expanding type).
4. One fire fighter helps the pump operator place the strainer on the end of the suction hose, put the suction hose into the tank, and connect it to the engine.
5. The other fire fighter helps the tanker driver discharge water into the portable tank. If the tank is self-expanding, the fire fighters may need to hold the collar until the water level is high enough for the tank to support itself.

Fire Fighter Tips

You have learned the steps needed to establish a water supply. Now you can understand how important this task is for the whole fire-fighting team, and how identifying the location of fire hydrants and static water sources during preincident planning can save time and lives.

Tanker Shuttles

When a large volume of water is needed for an extended period at a fire, a **tanker shuttle** can be used to deliver water from a fill site to the scene. The number of tankers needed will depend on the distance between the fill site and the fire scene, the time it takes the tanker to dump and refill, and the flow rate required at the fire scene.

All of the components of a tanker shuttle must be set up so water moves efficiently from the fill site to the fire scene. The fill site must refill the tankers without delay. The routes in both directions must be planned so tankers make efficient round trips, without having to back up or make U-turns. At the fire scene, the tankers should be able to drive up, dump their water into the portable tanks, and immediately return to the fill site. The portable tanks must be large enough to receive the full load of each tanker as it arrives.

An effective tanker shuttle can deliver several hundred gallons of water per minute without interruption. A separate component of the incident command structure is usually established to coordinate a tanker-shuttle operation. Departments must practice these operations to ensure proper coordination and effective water delivery.

Wrap-Up

Ready for Review

This chapter covered the steps needed to establish a water supply to fight a fire. It covered the design and components of a municipal water supply system, water sources, water treatment facilities, and water distribution systems. It described the construction and operation of dry-barrel and wet-barrel hydrants. The importance of maintaining and testing fire hydrants was also reviewed.

This chapter also described static water sources and mobile water supplies that are available in areas without municipal water supplies. The use of portable tanks and tanker shuttles for supplying water were illustrated.

Chief Concepts

- Municipal water systems consist of a water supply, water treatment facilities, and a water distribution system.
- Fire hydrants enable fire fighters to obtain water from the municipal water supply system. Fire fighters must know how to operate and hook up hoses to a fire hydrant.
- In areas without municipal water supply systems, water can be obtained from various natural or man-made sources. Fire department tankers, portable tanks, and tanker shuttles are all means of delivering water to a fire scene.

Hot Terms

Distributors Relatively small-diameter underground pipes that deliver water to local users within a neighborhood.

Dry-barrel hydrant A type of hydrant used in areas subject to freezing weather. The valve that allows water to flow into the hydrant is located underground and the barrel of the hydrant is normally dry.

Dry hydrant A permanent piping system that provides access to draft water from a static water source.

Dump valve A large valve installed on a mobile water supply apparatus to allow the contents of the tank to be discharged quickly.

Elevated water storage tower An above-ground water storage tank that is designed to maintain pressure on a water distribution system.

Elevation pressure The amount of pressure created by gravity.

Flow pressure The amount of pressure created by moving water.

Gravity feed system A water distribution system that depends on gravity to provide the required pressure. The system storage is usually located at a higher elevation than the end users.

Mobile water supply apparatus A fire department vehicle that is designed to transport water to the scene of a fire; often called a tanker.

Municipal water system A water distribution system that is designed to deliver potable water to end-users for domestic, industrial, and fire protection purposes.

Normal operating pressure The observed static pressure in a water distribution system during a period of normal demand.

Pitot gauge A type of gauge that is used to measure the velocity pressure of water that is being discharged from an opening. Used to determine the flow of water from a hydrant.

Portable tanks Folding or collapsible tanks that are used at the fire scene to hold water for drafting.

Primary feeder The largest diameter pipes in a water distribution system, carrying the greatest amounts of water.

Private water system A privately owned water system operated separately from the municipal water system.

Residual pressure The pressure remaining in a water distribution system while water is flowing. The residual pressure indicates how much more water is potentially avaiable.

Reservoir A water storage facility.

Secondary feeder Smaller diameter pipes that connect the primary feeders to the distributors.

Shut-off valve Any valve that can be used to shut down water flow to a water user or system.

Static pressure The pressure in a water pipe when there is no water flowing.

Static water source A water source such as a pond, river, stream, or other body of water that is not under pressure.

Steamer port The large diameter port on a hydrant.

Tanker shuttle A method of transporting water from a source to a fire scene using a number of mobile water supply apparatus.

Water main The generic term for any underground water pipe.

Water supply A source of water.

Wet-barrel hydrant A hydrant used in areas that are not susceptible to freezing. The barrel of the hydrant is normally filled with water.

Fire Fighter in Action

It is 9:00 a.m. on a sunny July morning when your engine company is dispatched to a second alarm for a commercial structure fire in a lumberyard. As you are responding you look toward the general location of the fire and observe a large thermal column of smoke. The first arriving engine reports that four large lumber racks are on fire. The first engine is setting up a defensive operation. The incident commander instructs your engine to hook up the hydrant and supply the first in engine. Lieutenant Phillips tells you to prepare for a forward hose lay and hook up to the hydrant. The hydrants in your area are dry barrel type.

1. You should open and flow water out of the hydrant prior to hooking up the supply hose to ensure that:
   - **A.** the hydrant works.
   - **B.** all foreign matter is flushed out.
   - **C.** you are turning the hydrant stem the right direction.
   - **D.** the drain works.

2. Opening the hydrant partially will:
   - **A.** cause the valve to prematurely fail.
   - **B.** undermine the soil around the hydrant.
   - **C.** allow the maximum flow of water.
   - **D.** both b and c

It took over 4 hours to extinguish the fire using master streams. Lieutenant Phillips tells you to shut down the hydrant and begin reloading the hose.

3. You can check to see if a dry-barrel hydrant is draining water by:
   - **A.** using a flashlight and looking through an open port.
   - **B.** attaching a static gauge and checking for a vacuum.
   - **C.** holding your hand over the open port to feel for suction.
   - **D.** attaching a hydrant pump and creating a vacuum.

4. The purpose of draining a dry-barrel hydrant is to prevent:
   - **A.** freezing in cold temperatures.
   - **B.** corrosion.
   - **C.** potential contaminating algae bloom.
   - **D.** water hammer that could damage the municipal water system.

Fire Hose, Nozzles, Streams, and Foam

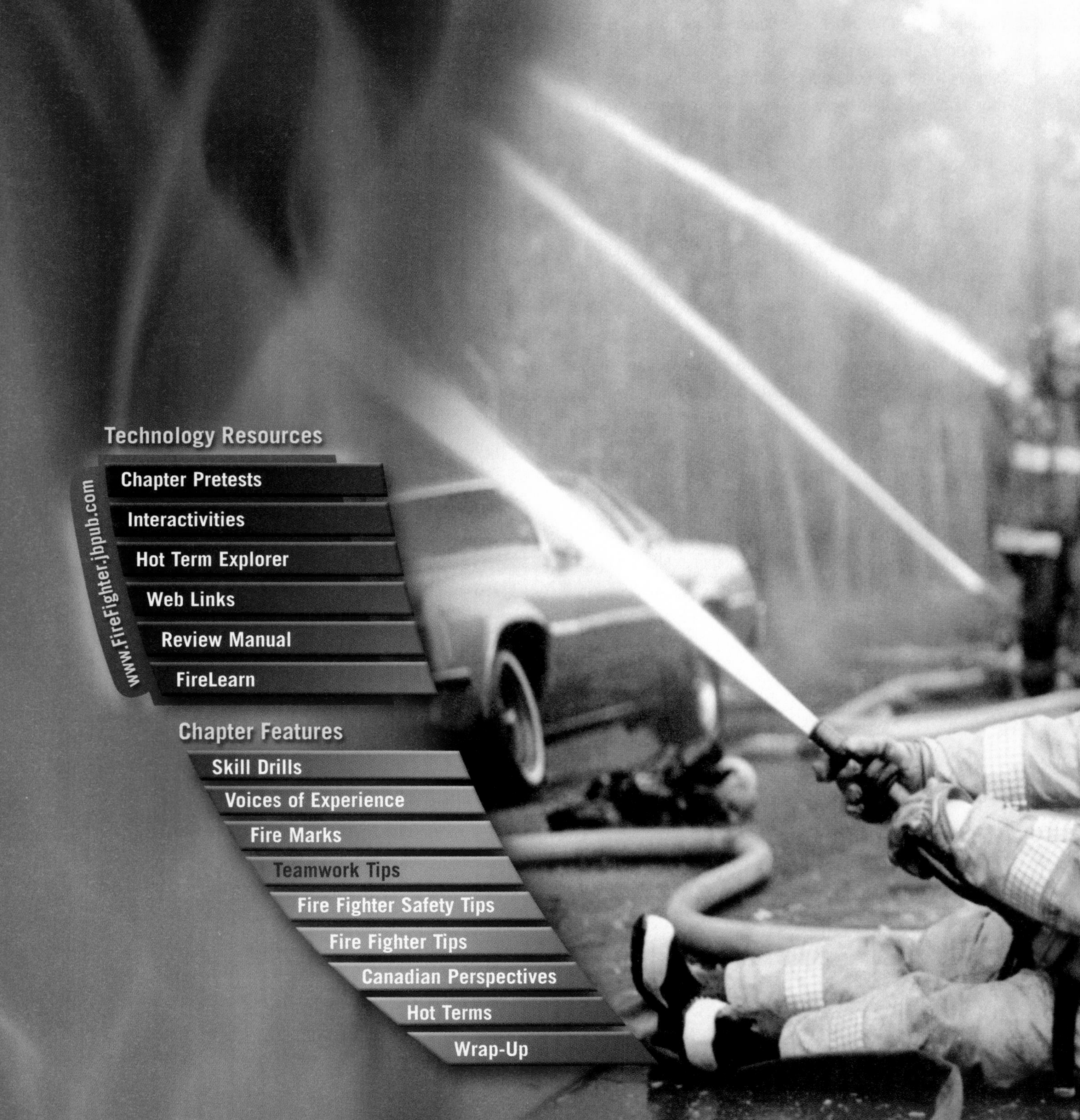

Chapter 16

Fire Hose, Nozzles, Streams, and Foam

NFPA 1001 Standard

Fire Fighter I

5.3.10* (A) *Requisite Knowledge.* Principles of fire streams; types, design, operation, nozzle pressure effects, and flow capabilities of nozzles; precautions to be followed when advanced hose lines to a fire; observable results that a fire stream has been properly applied; dangerous building conditions created by fire; principles of exposure protection; potential long-term consequences of exposure to products of combustion; physical states of matter in which fuels are found; common types of accidents or injuries and their causes; and the application of each size and type of attack line, the role of the backup team in fire attack situations, attack and control techniques for grade level and above and below grade levels, and exposing hidden fires.

5.3.10* (B) *Requisite Skills.* The ability to prevent water hammers when shutting down nozzles; open, close, and adjust nozzle flow and patterns; apply water using direct, indirect, and combination attacks; advance charged and uncharged 1½" (38 mm) diameter or larger hose lines up ladders and up and down interior and exterior stairways; extend hose lines; replace burst hose sections; operate charged hose lines of 1½" (38 mm) diameter or larger while secured to a ground ladder; couple and uncouple various hand-line connections; carry hose; attack fires at grade level and above and below grade levels; and locate and suppress interior wall and subfloor fires.

5.3.15 Connect a fire department pumper to a water supply as a member of a team, given supply or intake hose, hose tools, and a fire hydrant or static water source, so that connections are tight and water flow is unobstructed.

5.3.15 (A) *Requisite Knowledge.* Loading and off-loading procedures for mobile water supply apparatus; fire hydrant operation; and suitable static water supply sources, procedures, and protocol for connecting to various water sources.

5.3.15 (B) *Requisite Skills:* The ability to hand lay a supply hose, connect and place hard suction hose for drafting operations, deploy water tanks as well as the equipment necessary to transfer water between and draft from them, make hydrant-to-pumper hose connections for forward and reverse lays, connect supply hose to a hydrant, and fully open and close the hydrant.

5.5.4 Clean, inspect, and return fire hose to service; given washing equipment, water, detergent, tools, and replacement gaskets, so that damage is noted and corrected, the hose is clean, and the equipment is placed in a ready state for service.

5.5.4 (A) *Requisite Knowledge.* Departmental procedures for noting defective hose and removing it from service, cleaning methods, and hose rolls and loads.

5.5.4 (B) *Requisite Skills.* The ability to clean different types of hose; operate hose washing and drying equipment; mark defective hose; and replace coupling gaskets, role hose, and reload hose.

Fire Fighter II

6.3.1 Extinguish an ignitable liquid fire, operating as a member of a team, given an assignment, an attack line, personal protective equipment, a foam proportioning device, a nozzle, foam concentrates, and a water supply, so that the correct type of foam concentrate is selected for the given fuel and conditions, a properly proportioned foam stream is applied to the surface of the fuel to create and maintain a foam blanket, fire is extinguished, re-ignition is prevented, team protection is maintained with a foam stream, and the hazard is faced until retreat to safe haven is reached.

6.3.1 (A) *Requisite Knowledge.* Methods by which foam prevents or controls a hazard; principles by which foam is generated; causes for poor foam generation and corrective measures; difference between hydrocarbon and polar solvent fuels and concentrates that work on each; the characteristics, uses, and limitations of fire-fighting foams; the advantages and disadvantages of using fog nozzles versus foam nozzles for foam application; foam stream application techniques; hazards associated with foam stream usage; and methods to reduce or avoid hazards.

6.3.1 (B) *Requisite Skills.* The ability to prepare a foam concentrate supply for use, assemble foam stream components, master various foam application techniques, and approach and retreat from spills as part of a coordinated team.

6.5.3 Perform an annual service test on fire hose, given a pump, a marketing device, pressure gauges, a timer, record sheets, and related equipment, so that procedures are followed, the condition of the hose is evaluated, any damaged hose is removed from service, and the results are recorded.

6.5.3 (A)* *Requisite Knowledge.* Procedures for safely conducting hose service testing, indicators that dictate any hose be removed from service, and recording procedures for hose test results.

6.5.3 (B) *Requisite Skills.* The ability to operate hose testing equipment and nozzles and to record results.

Additional NFPA Standards

NFPA 1145 *Use of Class A Foams in Manual Structural Fire Fighting*

Knowledge Objectives

After studying this chapter, you will be able to:

- Describe how to prevent water hammers.
- Describe how a hose is constructed.
- Describe the types of hoses used in the fire service.
- Describe how to clean and maintain a hose.
- Describe how to inspect a hose.
- Describe how to note a defective hose.
- Describe how to roll a hose.
- Describe how to lay a supply line.
- Describe how to load a hose.
- Describe how to connect a hose to a water supply.
- Describe how to carry and advance a hose.
- Describe the types and designs of nozzles.
- Describe pressure effects and flow capabilities of nozzles.
- Describe how foam works.
- List the types of foam.
- Describe how to make foam.
- Describe how to apply foam.

Skills Objectives

After studying this chapter, you will be able to:

- Replace the swivel gasket.
- Perform the one-fire fighter foot-tilt method of coupling a fire hose.
- Perform the two-fire fighter method for coupling a fire hose.
- Perform the one-fire fighter knee-press method of uncoupling a fire hose.
- Perform the two-fire fighter stiff-arm method.
- Uncouple a hose with spanners.
- Connect two lines with damaged coupling.
- Clean hoses.
- Mark a defective hose.
- Perform the straight hose roll.
- Perform the single donut hose roll.
- Perform the twin donut hose roll.
- Perform the self-locking twin donut roll.
- Perform the forward lay.
- Perform the four-way hydrant valve.
- Perform the reverse lay.
- Perform the split hose lay.
- Perform a flat hose load.
- Perform a horseshoe load.
- Perform an accordion hose load.
- Attach a soft suction hose to a fire hydrant.
- Attach a hard suction hose to a fire hydrant.
- Load the minuteman hose load.
- Advance the minuteman hose load.
- Loading the preconnected flat load.
- Advance the preconnected flat hose load.
- Load the triple layer hose load.
- Advance the triple layer hose load.
- Unload and advance the wyed lines.
- Perform a working hose drag.
- Perform a shoulder carry
- Advance an accordion load.
- Advance a hose line up a stairway.
- Advance a hose line down a stairway.
- Advance an uncharged hose line up a ladder.
- Use a hose stream from a ladder.
- Connect to a standpipe system.
- Advance from a standpipe.
- Replace a hose section.
- Drain a hose and carry.
- Operate a smooth bore nozzle.
- Operate a fog nozzle.
- Place a foam line in service.
- Perform the sweep method of applying foam.
- Perform the bankshot method of applying foam.
- Perform the rain-down method of applying foam.

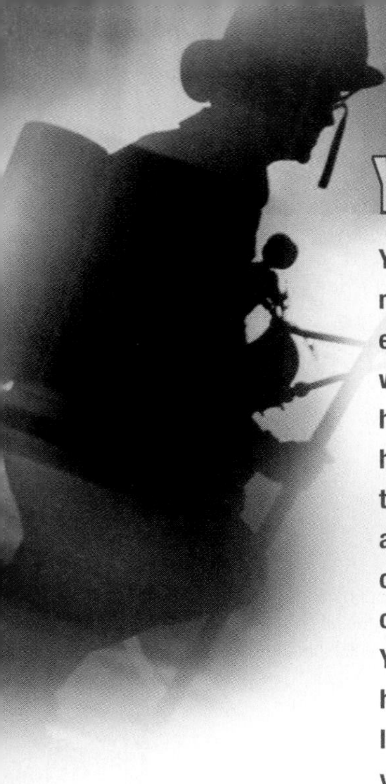

You Are the Fire Fighter

You are responding as the second due engine company on a first-alarm assignment for an apartment fire. As you approach the scene, the incident commander (IC) orders your company to establish a supply line for the first due engine. Your lieutenant orders you to catch the hydrant with the 5″ supply line. When the engine stops, you exit the apparatus and pull the end of the 5″ hose with attached adaptor and hydrant wrench from the rear hose bed. You drag the end of the hose to the hydrant, wrap it around the hydrant, and signal the engine to proceed. The engine then drives slowly toward the fire, laying a supply line for the first engine. You flush the hydrant and hook the hose to the large steamer outlet. When you get the signal that the line has been connected to the first engine, you open the hydrant to deliver the water. When you rejoin your company, your lieutenant and the other fire fighter are getting ready to advance a 1¾″ attack line. You and your partner pull the attack line and advance it toward the fire building until all of the hose is out of the bed. Your crew stops at the front door to don the SCBA facemasks, and the lieutenant signals the engine operator to charge the line. After bleeding the air from the hose line, you advance inside the building, following the first due company's attack line.

1. ***What types of hose are used in the fire service?***
2. ***What are types of nozzles used in the fire service?***
3. ***What are the different methods of advancing a hose line?***

Fire Hydraulics

Fire hydraulics deal with the properties of energy, pressure, and water flow as related to fire suppression. When operating hose lines at a fire, it is important to understand some basic principles of hydraulics. You need to understand the basic concepts of friction loss in different size hose lines, changes in pressure due to elevation, and water hammer. Fire fighters who advance to the position of pump operator will learn more about fire service hydraulics.

Flow

Flow refers to the volume of water that is being moved through a pipe or hose. In fire hydraulics, flow is measured in gallons per minute (gpm). Canadian fire departments use the metric system and measure flow in liters per minute (lpm). The conversion factor is: 1 gallon (U.S.) = 3.785 liters.

Pressure

The amount of energy in a body or stream of water is measured as pressure. In fire hydraulics, pressure is measured in pounds per square inch (psi). Pressure is required to push water through a hose, to expel water through a nozzle, or to lift water up to a higher level. A pump adds energy to a water stream, causing an increase in pressure. In the metric system, pressures are measured in kilopascals (kPa). The conversion factor is: 1 psi = 6.894 kPa.

Friction Loss

Friction loss is a loss of pressure as water moves through a pipe or hose. This loss of pressure represents the energy required to push the water through the hose. Friction loss is influenced by the diameter of the hose, the volume of water traveling through the hose, and the distance the water travels. In a given size hose, a higher flow rate produces more friction loss. In a 2½″ hose, a 300 gpm flow causes much more friction loss than a 200 gpm flow. At a given flow rate, the smaller the diameter of the hose, the greater the friction loss. At a flow of 500 gpm, the friction loss in a 2½″ hose is much greater than the friction loss in a 4″ hose. With any combination of flow and diameter, the friction loss is directly proportional to the distance. At a flow of 250 gpm in a 2½″ hose, the friction loss in 200′ is double the friction loss in 100′.

Elevation Pressure

Elevation affects water pressure. An elevated water tank supplies pressure to a municipal water system because of the difference in height between the water in the water tank and the underground delivery pipes. If a fire hose is laid down a

Fire Fighter Tips

Open and close hydrants and nozzles slowly to prevent water hammer.

hill, the water at the bottom will have additional pressure due to the change in elevation. If a fire hose is advanced upstairs to the third floor of a building, it will lose pressure due to the energy that required to lift the water. The fire pump operator has to take elevation changes into account when setting the discharge pressure.

Water Hammer

Water hammer is a surge in pressure caused by suddenly stopping the flow of a stream of water. A fast-moving stream of water has a large amount of kinetic energy. If the water suddenly stops moving when a valve is closed, all of the kinetic energy is converted to an instantaneous increase in pressure. Because water cannot compress, the additional pressure is transmitted along the hose or pipe as a shock wave. Water hammer can rupture a hose, cause a coupling to separate, or damage the plumbing on a piece of fire apparatus. Severe water hammer can even damage an underground piping system. Fire fighters have been injured by equipment that was damaged by a water hammer.

A similar situation can occur if a valve is opened too quickly and a surge of pressurized water suddenly fills a hose. The surge in pressure can damage the hose or cause the fire fighter at the nozzle to lose control of the stream.

To prevent water hammer, always open and close fire hydrants valves slowly. Pump operators also need to open and close the valves on fire engines slowly. When you are operating the nozzle on an attack line, open the nozzle slowly. Most importantly, when you close the shut-off valve on an attack line, do it slowly.

Fire Hoses

Functions of Fire Hoses

Fire hoses are used for two main purposes. The hoses used to discharge water from an attack engine onto the fire are called **attack hoses**, or attack lines. Most attack hoses carry water directly from the attack engine to a nozzle that is used to direct the water onto the fire. In some cases, an attack line is attached to a deck gun, aerial device, or some other type of master stream appliance. Attack lines can also be used to deliver water to a fire department connection that supplies a standpipe or sprinkler system inside a building.

Hoses used to deliver water to an attack engine are called **supply hoses**, or **supply lines**. The water can come directly from a hydrant or it can come from another engine that is being used to provide a water supply for the attack engine. Supply line sizes are 2½", 3", 4", 5", and 6". Attack hoses usually operate at higher pressures than supply lines. Supply hoses are designed to carry larger volumes of water at lower pressures.

Fire Marks

A fire department engine that is being used to deliver water directly to one or more attack hose lines is called an attack engine. An engine that is being used to deliver water to an attack engine is called a supply engine. Most fire department engines are designed to function as either an attack engine or a supply engine, depending on the situation.

Sizes of Hose

Fire hoses range in size from 1" to 6" in diameter (▼ Figure 16-1). The nominal hose size refers to the inside diameter of the hose when it is filled with water. The smaller diameter hoses are used as attack lines and the larger diameter hoses are almost always used as supply lines. Medium diameter hose sizes can be used as either attack lines or supply lines.

Small diameter hose (SDH) lines range in size from 1" to 2" in diameter. Many fire department vehicles are equipped with a reel of ¾" or 1" hard rubber hose called a **Booster hose**, or **booster line**, which is used for small outdoor fires. A lightweight collapsible 1" hose, known as forestry hose, is often used to fight brush fires.

The hoses that are most commonly used to attack interior fires are either 1½" or 1¾" in diameter. These hoses are usu-

Figure 16-1 Fire hose comes in a wide range of sizes for different uses and situations.

Canadian Perspectives

Fire departments in Canada use metric measurements to refer to hose sizes and pressures. The following tables provide the equivalent measures for standard fire hose sizes.

Hose Diameter

Inches (in)	Millimeters (mm)
1	25
1½	38
1¾	45
2	52
2½	65
3	75
3½	90
4	100
5	125
6	150

Pressure Ratings

In Canada, attack hose is service tested to a pressure of at least 2100 kPa and is intended to be used at pressures up to 1900 kPa. Supply hose is service tested to a pressure of at least 1400 kPa and is intended to be used at pressures up to 1300 kPa.

Note: The nominal hose diameters in millimeters and pressure ratings in kilopascals are not exact mathematical equivalents to the hose diameters in inches and pressures in psi. The actual hose sizes and construction are the same. The numbers are rounded to allow for easier field calculations. The standard lengths for hose sections in Canada are 15 meters and 30 meters, which are equivalent to 49.2 feet and 98.4 feet respectively.

ally connected directly to a handline nozzle. Some fire departments also use 2″ attack lines. Each section of attack hose is usually 50″ long.

Hoses of 2½″ or 3″ in diameter are called **medium diameter hose (MDH)**. Hoses in this size range can be used as either supply lines or attack lines. Large handline nozzles are often used with 2½″ hose to attack larger fires. When used as an attack hose, the 3″ size is more often used to deliver water to a master stream device or a fire department connection. These hose sizes also come in 50′ lengths.

Hoses 3½″ or more in diameter are called **large diameter hoses (LDH)**. Standard LDH sizes include 4″ and 5″ diameters, which are used as supply lines by many fire departments. The largest LDH size is 6″ in diameter. Standard lengths of either 50′ or 100′ feet are available for LDH.

Fire hose is designed to be used as either attack hose or supply hose. Attack hose must withstand higher pressures and is designed to be used in a fire environment where it can be subjected to high temperatures, sharp surfaces, abrasion and other potentially damaging conditions. LDH supply hose is constructed to operate at lower pressures than attack hose and in less severe operating conditions; however, it must still be durable and resistant to external damage. Attack hose can be used as supply hose, but LDH supply hose must never be used as attack hose.

Attack hose must be tested annually at a pressure of at least 300 psi and is intended to be used at pressures up to 275 psi. Supply hose must be tested annually at a pressure of at least 200 psi and is intended to be used at pressures up to 185 psi. Most large diameter hose is constructed as supply line; however, some fire departments use special large diameter hose that can withstand higher pressures.

Hose Construction

Most fire hose is constructed with an inner waterproof liner surrounded by either one or two outer layers. The outer layers provide the strength to withstand the high pressures that are exerted by the water inside the hose. The strength is provided by a woven mesh made from high strength synthetic fibers such as nylon that are resistant to high temperatures, mildew, and many chemicals. These fibers can also withstand some mechanical abrasion.

Double jacket hose is constructed with two layers of woven fibers. The outer layer serves as a protective covering while the inner layer provides most of the strength. The tightly woven outer jacket can resist abrasion, cutting, hot embers, and other external damage. The woven fibers are treated to resist water and provide added protection from many common hazards that are likely to be encountered at the scene of a fire.

Instead of a double jacket, some fire hoses are constructed with a durable rubber-like compound as the outer covering. This material is bonded to a single layer of strong woven fibers that provides the strength to keep the hose from rupturing under pressure. This type of construction is called **rubber–covered hose**, or **rubber-jacket hose** (▶ Figure 16-2).

Both types of hose are designed to be stored flat and to fold easily when there is no water inside the hose. This allows a much greater length of hose to be stored in the hose compartments on a fire apparatus.

The **hose liner**, or **hose inner jacket** is the inner part of the hose (▶ Figure 16-3). This liner prevents the water from leaking out of the hose and provides a smooth inside surface for the water to move against. Without this smooth surface, there would be excessive friction between the moving water and inside of the hose, reducing the amount of pressure that could reach the nozzle. The inner liner is usually made of a synthetic rubber compound or a thin flexible membrane material that can be flexed and folded without developing leaks. In double jacket hose, the liner is bonded to the inner woven jacket. In a rubber-covered hose, the inner and outer layers are usually bonded together and the woven fibers are contained within.

Figure 16-2 Rubber-covered hose.

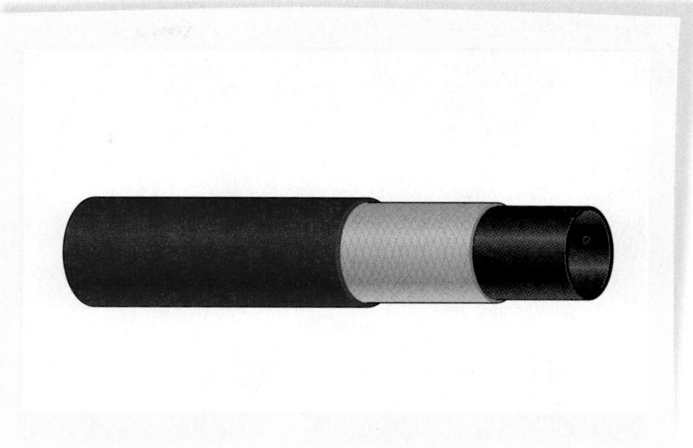

Figure 16-3 The liner inside a fire hose can be made from synthetic rubber or a variety of membrane materials.

Hose Couplings

Couplings are used to connect individual lengths of fire hose together. Couplings are also used to connect a hose line to a hydrant; to an intake or discharge valve on a engine; or to a variety of nozzles, fittings, and appliances. A coupling is permanently attached to each end of a section of fire hose. The two most common types of fire hose couplings are **threaded hose couplings** or nonthreaded **(Storz-type couplings)**.

Threaded Couplings

Threaded couplings are used on most hoses up to 3" in diameter and on soft suction hose and hard suction hose. A set of threaded couplings consists of a male coupling, which has the threads on the outside, and a female coupling, which has matching threads on the inside (► Figure 16-4). The female coupling has a swivel, so the male and female ends can be attached together without twisting the hose. A length of fire hose has a male coupling on one end and a female coupling on the other end.

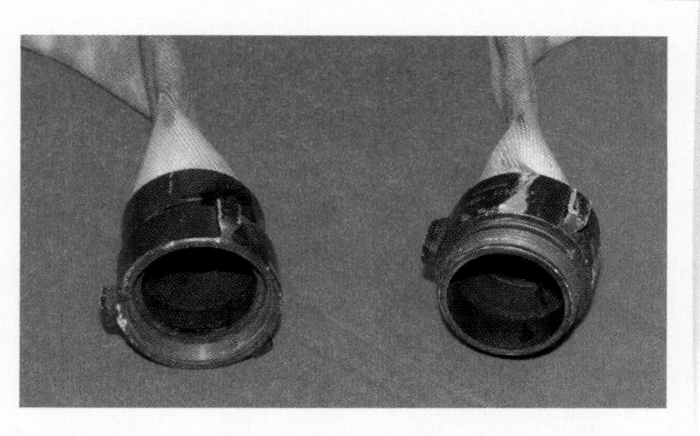

Figure 16-4 A set of threaded couplings includes one male and one female coupling. The male coupling has exposed threads, while the threads on the female coupling are inside the swivel.

Standardized hose threads are used by most fire departments so that fire hose from different departments can be connected together. The use of standardized hose threads is not universal among all jurisdictions, so some fire departments have to carry special adapters on their apparatus for use during mutual aid firefighting operations.

When connecting fire hoses with threaded couplings, you have to make sure the threads are properly aligned so the male and female couplings will engage fully. When the couplings are properly aligned, the two ends should attach together with minimal resistance. The swivel on the female coupling should be turned until the connection is snug, but only hand tight, so the couplings can be easily disconnected.

If there is any leakage after the hose is filled with water, further tightening may be needed. You can use a **spanner wrench** to gently tighten the couplings until the leakage is stopped. Spanner wrenches are used to connect and disconnect hose couplings (► Figure 16-5). Normally two spanners are used together to rotate the two couplings in opposing directions.

The coupling are constructed with either **rocker lugs**, or **rocker pins**, to engage a spanner wrench. Using a wrench to tighten couplings on an empty hose or overtighten couplings on a filled hose can damage the gaskets and cause them to leak. You may need to use a spanner wrench to uncouple the hose after it has been pressurized with water.

Higbee indicators (sometimes called Highbee notch) show the position where the ends of the threads on a pair of couplings are properly aligned with each other. Using the Higbee indicators will help you to couple hose more quickly. When the indicators on the male and female couplings are aligned, the two couplings should connect quickly and easily (► Figure 16-6).

An important part of a threaded coupling is the rubber

Figure 16-5 A spanner wrench is used to tighten or loosen a hose coupling.

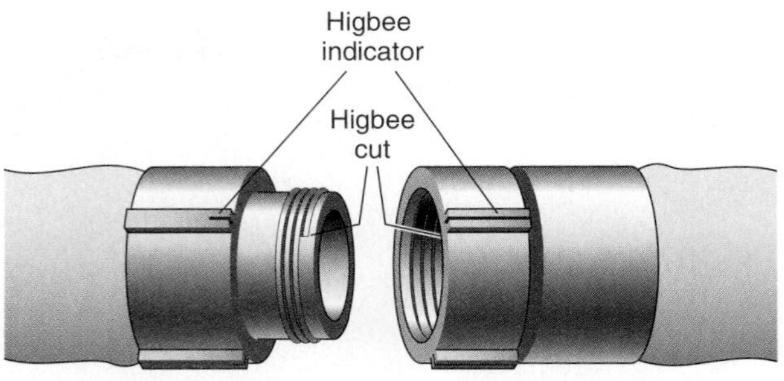

Figure 16-6 Higbee indicators show the position where the threads on a pair of couplings are properly aligned with each other.

Figure 16-7 Storz-type couplings are designed so that the couplings on both ends of a length of hose are the same.

gasket. The gasket is an O-shaped piece of rubber that sits inside the swivel section of the female coupling. When the male coupling is tightened down against it, a seal is formed that stops water from leaking. If the gasket is damaged or missing, the coupling will leak. These gaskets can deteriorate with time and can also be damaged by overtightening the coupling. Periodically, the gaskets have to be changed as part of the maintenance on the hose.

While a leaking coupling is not a critical problem during most firefighting operations, it can result in unnecessary water damage. During cold weather, a leaking coupling can cause ice to form and create a significant safety hazard. The best way to prevent leaks is to make sure the gaskets are in good condition and replace any gaskets that are missing or damaged. To replace the swivel gasket, follow the steps in ► Skill Drill 16-1.

1. Fold the new gasket, bringing the thumb and forefinger together, creating two loops. **(Step 1)**
2. Place either of the two loops into the coupling and against the gasket seat. **(Step 2)**
3. Using the thumb, push the remaining unseated portions into the coupling until the entire gasket is properly positioned against the coupling seat. **(Step 3)**

Storz-Type Couplings

Storz-type couplings are designed so that the couplings on both ends of a length of hose are the same. There is no male or female end to the hose. When this system is used, each coupling can be attached to any other coupling of the same diameter ◄ Figure 16-7. Storz-type couplings are made for all hose sizes; however, in North America they are most often used on a large diameter hose (LDH).

Storz-type couplings are connected by mating the two couplings face-to-face and then turning clockwise one third of a turn. To disconnect a set of couplings, the two parts are rotated counterclockwise one third of a turn. A spanner wrench can be used to tighten a leaking coupling or to release a connection that cannot be loosed by hand.

Adaptors are used to connect Storz-type couplings to threaded couplings or to connect couplings of different sizes together. Many fire departments use large diameter hose (LDH) with Storz-type couplings as a supply line between a hydrant and an engine.

Coupling and Uncoupling Hose

There are several techniques for attaching and releasing hose couplings. Depending upon the circumstances, each is effective. A fire fighter should learn how to perform each. To perform the one-fire-fighter foot-tilt method of coupling fire hose, follow the steps in ► Skill Drill 16-2.

1. Place one foot on the hose behind the male coupling.
2. Push down with your foot to tilt the male coupling upward. **(Step 1)**
3. Place one hand behind the female coupling and grasp the hose. **(Step 2)**

16-1 Skill Drill

Replacing the Swivel Gasket

Fold the new gasket, bringing the thumb and forefinger together, creating two loops.

Place either of the two loops into the coupling and against the gasket seat.

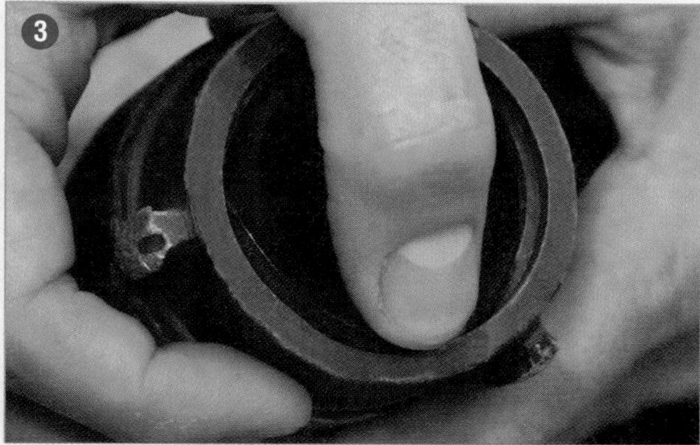

Using the thumb, push the remaining unseated portions into the coupling until the entire gasket is properly positioned against the coupling seat.

4. Place the other hand on the coupling swivel.
5. Bring the two couplings together and align the Higbee indicators. Rotate the swivel in a clockwise direction to connect the hoses. **(Step 3)**

To perform the two-fire-fighter method for coupling a fire hose, follow the steps in (► Skill Drill 16-3).

1. Pick up the male end of the coupling. Grasp it directly behind the coupling and hold it tightly against the body. **(Step 1)**
2. The second fire fighter holds the female coupling firmly with both hands. **(Step 2)**
3. The second fire fighter brings the female coupling to the male coupling. **(Step 3)**
4. The second fire fighter aligns the female coupling with the male coupling. Use the Higbee indicators for easy alignment. **(Step 4)**
5. The second fire fighter turns the female coupling counterclockwise until it clicks. This indicates the threads are aligned. **(Step 5)**
6. Turn the female coupling clockwise to couple the hoses. **(Step 6)**

Charged hose lines should never be disconnected while the water inside the hose is under pressure. The loosened couplings can flail around wildly and cause serious injury to personnel or bystanders in the vicinity. Always shut off the water supply and bleed off the pressure before uncoupling.

Skill Drill

Performing the One-Fire Fighter Foot-Tilt Method of Coupling a Fire Hose

1. Place one foot on the hose behind the male coupling. Push down with your foot to tilt the male coupling upward.

2. Place one hand behind the female coupling and grasp the hose.

3. Place the other hand on the coupling swivel. Bring the two couplings together and align the Higbee indicators. Rotate the swivel in a clockwise direction to connect the hoses.

Fire Fighter Safety Tips

NEVER attempt to uncouple charged hose lines.

The water pressure will make it difficult to uncouple a charged hose line. If the coupling resists an attempt to uncouple, check to make sure the pressure is relieved before using spanner wrenches to loosen the coupling.

To perform the one-fire fighter knee-press method of uncoupling a fire hose, follow the steps in ▶ **Skill Drill 16-4**.

1. Pick up the connection by the female coupling end. **(Step 1)**
2. Turn the connection upright, resting the male coupling on a firm surface. **(Step 2)**
3. Place a knee on the female coupling and with body weight press down (this compresses the gasket). **(Step 3)**
4. Turn the female swivel counterclockwise and loosen the coupling. **(Step 4)**

16-3 Skill Drill

Performing the Two-Fire Fighter Method for Coupling a Fire Hose

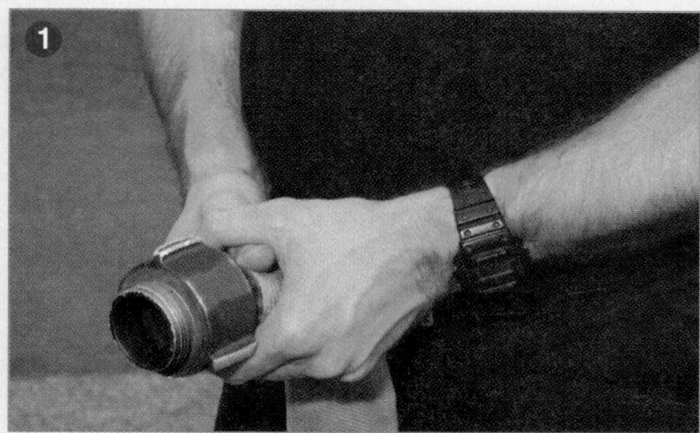

Pick up the male end of the coupling. Grasp it directly behind the coupling and hold it tightly against the body.

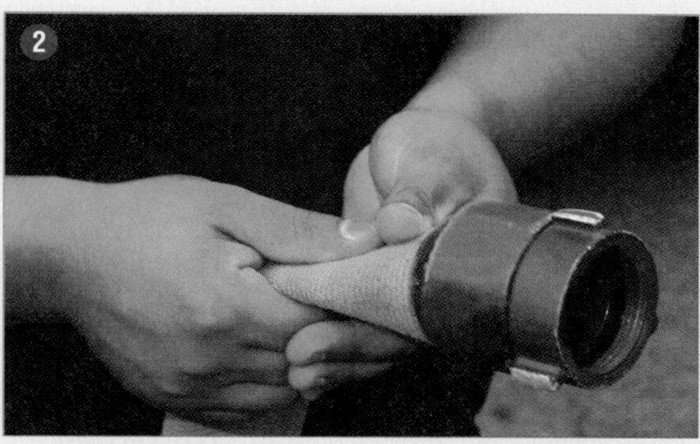

The second fire fighter holds the female coupling firmly with both hands.

The second fire fighter brings the female coupling to the male coupling.

The second fire fighter aligns the female coupling with the male coupling. Use the Higbee indicators for easy alignment.

The second fire fighter turns the female coupling counterclockwise until it clicks. This indicates the threads are aligned.

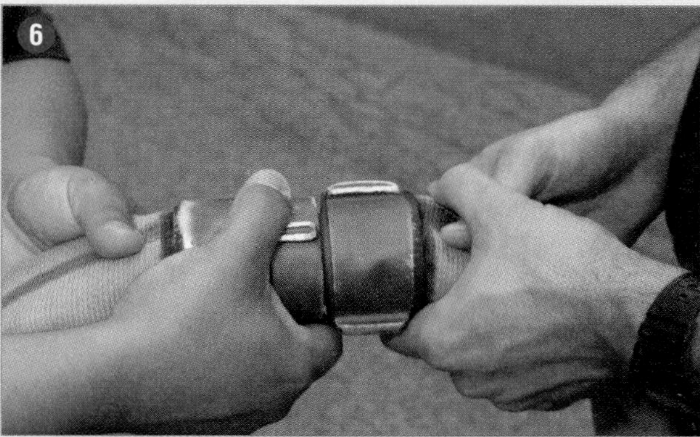

Turn the female coupling clockwise to couple the hoses.

16-4 Skill Drill

Performing the One-Fire Fighter Knee-Press Method of Uncoupling a Fire Hose

1 Pick up the connection by the female coupling end.

2 Turn the connection upright, resting the male coupling on a firm surface.

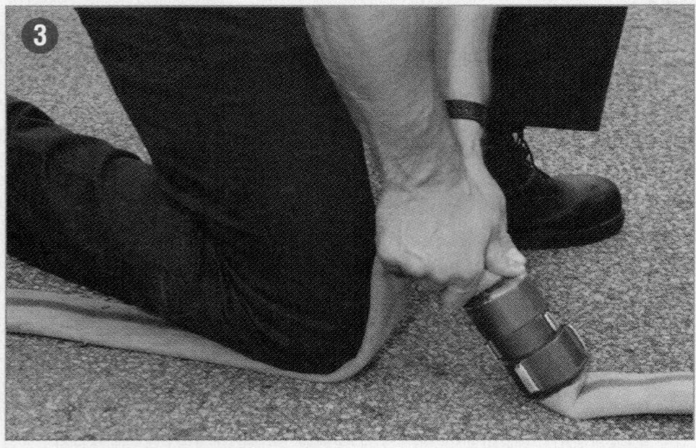

3 Place a knee on the female coupling and with body weight press down.

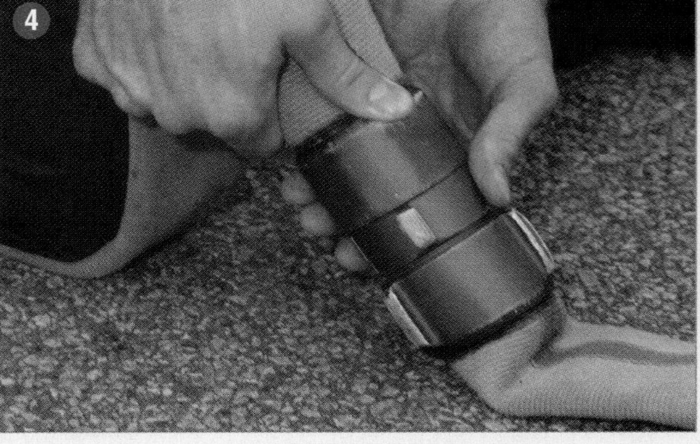

4 Turn the female swivel counterclockwise and loosen the coupling.

To perform the two-fire fighter stiff-arm method of coupling a hose, follow the steps of ► **Skill Drill 16-5**.

1. Two fire fighters face each other and firmly grasp their respective coupling. **(Step 1)**
2. With elbows locked straight, they push towards each other. **(Step 2)**
3. While pushing towards each other, the fire fighters turn the coupling counterclockwise, loosening the coupling. **(Step 3)**

To uncouple a hose with spanners, follow the steps in ► **Skill Drill 16-6**.

1. With the connection on the ground, straddle connection above the female coupling. **(Step 1)**
2. Place one spanner wrench on the female coupling with handle to the left. **(Step 2)**
3. Place the second spanner wrench on the male coupling with the handle to the right. **(Step 3)**
4. Push both spanner handles down toward the ground, loosening the connection. **(Step 4)**

16-5 Skill Drill

Performing the Two-Fire Fighter Stiff-Arm Method

Two fire fighters face each other and firmly grasp their respective coupling.

With elbows locked straight, the fire fighters push towards each other.

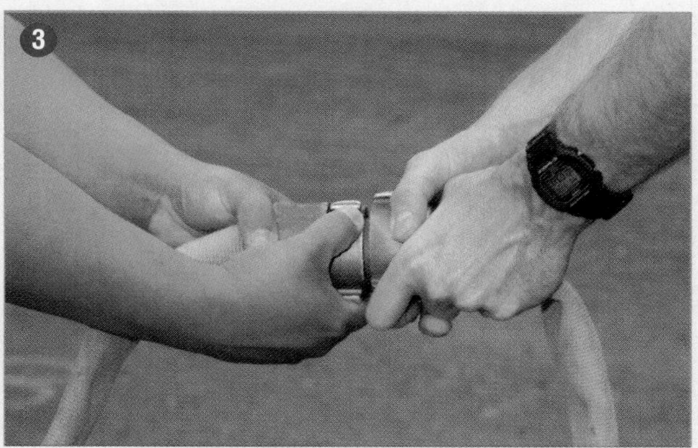

While pushing towards each other, the fire fighters turn the coupling counterclockwise, loosening the coupling.

To connect two lines with damaged coupling, follow the steps of Skill Drill 16-7.

1. Using a hose jacket, open the hose jacket and place the damaged coupling in one end.
2. Place the second coupling in the other end of the jacket.
3. Close the hose jacket, insuring that the latch is secure.

Slowly bring the hose line up to pressure, allowing the gaskets to seal around the hose ends.

Attack Hose

Attack hose is designed to be used for fire suppression where it can be exposed to heat and flames, hot embers, broken glass, sharp objects, and many other potentially damaging conditions. It must be tough, but flexible and light in weight.

Most fire departments use two sizes of hose as attack lines for fire suppression. The smaller size is usually either 1½" or 1¾" in diameter, while 2½" hose is most often used for heavy interior attack lines. These lines can be either double jacket or rubber covered construction.

1½" and 1¾" Attack Hose

Most fire departments use either 1½" or 1¾" hose as the primary attack line for most fires. Both sizes of hose use the

16-6 Skill Drill

Uncoupling Hose with Spanners

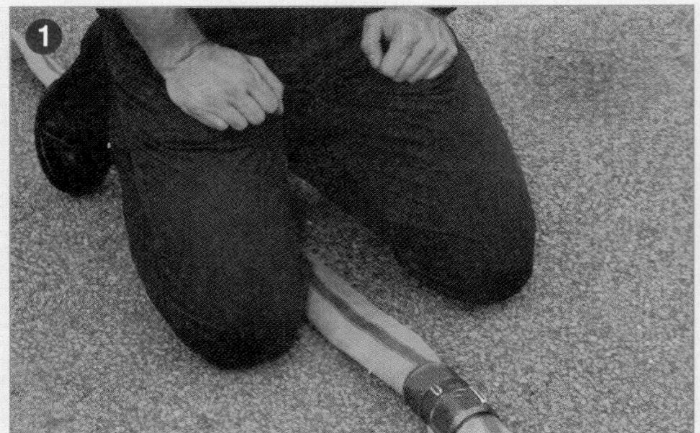

1. With the connection on the ground, straddle connection above the female coupling.

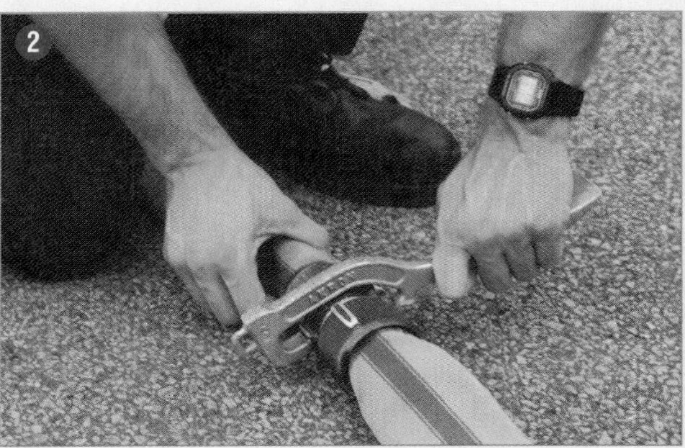

2. Place one spanner wrench on the female coupling with handle to the left.

3. Place the second spanner wrench on the male coupling with the handle to the right.

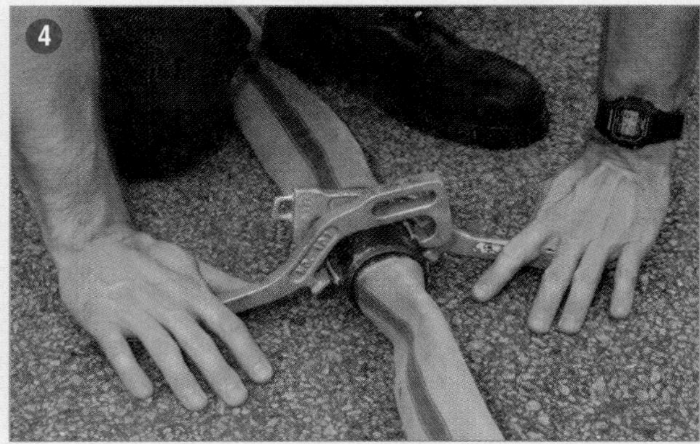

4. Push both spanner handles down toward the ground, loosening the connection.

same 1½" couplings. This is the attack hose that is used most often during basic fire training. Handlines of this size can usually be operated by one fire fighter, although a second person on the line makes it much easier to advance and control. This hose is often stored on fire apparatus as a preconnected attack line in lengths of 150′ to 350′, ready for immediate use.

The primary difference between 1½" and 1¾" hose is the amount of water that can flow though the hose. Depending on the pressure in the hose and the type of nozzle that is used, a 1½" hose can generally flow between 60 and 125 gallons of water per minute. An equivalent 1¾" hose can flow between 120 and 180 gpm. This is important because the amount of fire that can be extinguished is directly related to the amount of water that is applied to it. A 1¾" hose can deliver much more water and is only slightly heavier and more difficult to advance than a 1½" hose line.

2½" Attack Hose

A 2½" hose is used as an attack line for fires that are too large to be controlled by a 1½" or 1¾" hose line. A 2½" handline hose is generally considered to flow about 250 gallons of water per minute. It takes at least two fire fighters to safely control a 2½" handline hose due to the weight of the hose and the water and the nozzle reaction force. A 50′ length of dry 2½" hose weighs about 30 pounds. When the hose is charged and filled with water it weighs as much as 200 pounds per length.

Higher flows, up to approximately 350 gpm, can be achieved with higher pressures and larger nozzles; however, it is difficult to operate a handline hose at these high flow rates. These flows are more likely to be used to supply a master stream device.

Booster Hose

A booster hose is usually carried on a hose reel that holds 150′ or 200′ of rubber hose. Booster hose contains a steel wire that gives it a rigid shape. The rigid shape of this hose allows it to flow water without pulling all the hose off the reel. It is light in weight and can be advanced quickly by one person.

The disadvantage of booster hose is its limited flow. The normal flow from a 1″ booster hose is 40 to 50 gpm, which is not an adequate flow for structure fires. The use of booster hose is usually limited to small outdoor fires and trash dumpsters. Booster hose should not be used for structural firefighting.

Forestry Lines

Small diameter hose, typically 1″ or 1½″ in diameter, is often used for fighting wildland and ground fires. Since large volumes of water are not often needed for these fires, the small diameter provides extreme maneuverability through brush and trees. These hose lines are sometimes extended for hundreds of feet.

Supply Hose

Supply hose is used to deliver water to an attack engine from a pressurized source, which could be a hydrant or another engine working in a relay operation. Supply lines range from 2½″ all the way up to 6″ in diameter. The choice is made based on the preferences and operating requirements of each fire department. This depends on the amount of water needed to supply the attack engine, the distance from the source to the attack engine, and the pressure that is available at the source.

Fire department engines are normally loaded with at least one bed of hose that can be laid out as a supply line. When threaded couplings are used, this hose can be loaded to lay out from the hydrant to the fire (known as a straight lay) or from the fire to the hydrant (known as a reverse lay). Sometimes engines are loaded with two beds of hose so they can easily drop a supply line in either direction. If Storz-type couplings are used or the necessary adaptors are provided, hose from the same bed can be laid in either direction.

When 2½″ hose is used as a supply line, it is usually the same type of hose used for attack lines. This size hose has a limited flow capacity, but it can be effective at low to moderate flow rates and over short distances. Sometimes two parallel lines of 2½″ hose are used to provide a more effective water supply.

Figure 16-8 Soft suction hose.

Large diameter supply lines are much more efficient than 2½″ hose for moving larger volumes of water over longer distances. Many fire departments use 4″ or 5″ hose as their standard supply line. A single 5″ supply line can deliver flows exceeding 1,500 gpm under some conditions. Large diameter hose is heavy and difficult to move after it has been charged with water. This hose comes in 50′ and 100′ lengths. A typical fire engine may have anywhere from 750′ to 1250′ of supply hose on it.

Soft Suction

A **soft suction hose** is a short section of large diameter hose used to connect a fire department engine directly to the large steamer outlet on a hydrant (▲ Figure 16-8). The soft suction hose is used to allow as much water as possible to flow from the hydrant to the pump through a single hose. A soft suction hose has a female connection on each end, with one end matching the local hydrant threads and the other end matching the threads on a large diameter inlet to the engine. The couplings have large handles to allow for quick tightening by hand. The hose can be from 4″ to 6″ in diameter and is usually between 10′ and 25′ in length.

Hard Suction

A **hard suction hose** is a special type of supply hose used to draft water from a static source such as a river, lake, or portable drafting basin (► Figure 16-9). The water is drawn through this hose into the pump on a fire department engine or into a portable pump. It is called a hard suction hose because it is designed to remain rigid and will not collapse when a vacuum is created in the hose to draft the water into the pump.

Hard suction hose normally comes in 10′ or 20′ sections. The diameter is based on the capacity of the pump and can be as large as 6″. The hose can be made from either rubber

Figure 16-9 A hard suction hose.

or plastic; however, the newer plastic versions are much lighter and more flexible.

Long handles are provided on the female couplings of hard suction hose to assist in tightening the hose. In order to draft water, it is essential to have an airtight connection at each coupling. Sometimes it may be necessary to gently tap these handles with a rubber mallet to tighten the hose or to disconnect it. Tapping these handles with anything metal could cause damage to the handles or the coupling.

Hose Care, Maintenance, and Inspection

Fire hose should be regularly inspected and tested following the procedures in NFPA 1962, *Inspection, Care and Use of Fire Hose, Couplings, and Nozzles and Testing of Fire Hose.* Hoses that are not properly maintained can deteriorate over time and eventually burst. The gaskets in female couplings need to be checked regularly and replaced when they are worn or damaged.

Causes and Prevention of Hose Damage

The fire hose is a lifeline for fire fighters. Every time fire fighters fight a fire, they have to rely on fire hose to deliver the water needed to attack the fire and protect themselves from the fire. Fire hose is a highly engineered product designed to perform well under adverse conditions. We must be careful to prevent damage to the hose that could result in premature or unexpected failure. The most common factors that can cause damage to fire hose include mechanical causes, chemicals, heat, cold, and mildew.

Mechanical Damage

Mechanical damage can occur from many sources. Hose that is dragged over rough objects or along a roadway can be damaged by abrasion. Broken glass and sharp objects can cut through the hose. Be especially careful if you need to place a hose line through a broken window; remove any protruding sharp edges of glass first. Particles of grit caught in the fibers can damage the jacket or puncture holes in the liner. Reloading dirty hose can cause damage to the fibers in the hose jacket.

Fire hose is likely to be damaged if it is run over by a vehicle. Hose ramps should be used if traffic has to drive over a hose that is in the roadway. Hose coupling can also be damaged by other mechanical forces, such as dropping them on the ground. The exposed threads on male couplings are easily damaged if they are dropped. Avoid dragging hose couplings, since this can cause damage to the threads and to the swivels.

Heat and Cold

Hoses can be damaged by heat and cold and prolonged exposure to sunlight. Heat is an obvious concern when fighting a fire. A hose that is directly exposed to a fire can burn through and burst quickly. Burning embers and hot coals can also damage the hose, causing small leaks or weakening the hose so that it is likely to burst under pressure. Always visually inspect any hose that has been in direct contact with a fire.

Avoid storing a hose in places where it will be in contact with hot surfaces, such as a heating unit or the exhaust pipe on a vehicle. If the apparatus is parked outside, use a hose cover to protect the hose from sunlight.

In cold weather, freezing is a threat to hoses. Freezing can rupture the inner liner and break fibers in the hose jacket. When working in below freezing temperatures, water should be kept flowing through the hose to prevent freezing. If a line has be shut down temporarily, the nozzle should be left partly open to keep the water moving and the stream should be directed to a location where it will not cause additional water damage. When a line is no longer needed, the hose should be drained and rolled before it freezes.

Hose that is frozen or encased in ice can often be thawed out with a steam generator. Another option is to carefully chop the hose out using an axe, being careful not to cut the hose itself. The hose can then be transported back to the fire station to thaw. Do not attempt to bend a section of frozen hose. In situations where the hose is frozen solid, it may be necessary to transport the hose back to the fire station on a flatbed truck.

Chemicals

Many chemicals can cause damage to fire hoses. These chemicals can be encountered at incidents in facilities where chemicals are manufactured, stored, or used and in locations where their presence is not anticipated. Most vehicles contain a wide variety of chemicals that can damage fire hose, including battery acid, gasoline, diesel fuel, antifreeze, motor oil, and transmission fluid. The hose can come in contact with these chemicals at vehicle fires or at the scene

Fire Fighter Safety Tips

Any time a fire hose has suffered possible damage, it should be thoroughly inspected and tested according to NFPA 1962, *Inspection, Care and Use of Fire Hose, Couplings, and Nozzles and Testing of Fire Hose,* before it is returned to service.

of a collision where chemicals are spilled on the roadway. Supply hoses often come in contact with residue from these chemicals when lines are laid in the roadway. It is important to remove chemicals from the hose as soon as possible and to wash the hose with an approved detergent, thoroughly rinse it, and let it dry thoroughly.

Mildew

Mildew is a type of fungus that can grow on fabrics and materials in warm, moist conditions. A fire hose that has been packed away while it is still wet and dirty is a natural breeding ground for mildew. Mildew feeds on nutrients found in many natural fibers and can cause them to rot and deteriorate. In the days when cotton fibers were used in fire hose jackets, mildew was a large problem. Hose had to be washed and completely dried after every use before it could be placed back on the apparatus.

Modern fire hose is made from synthetic fibers that are resistant to mildew, and most types can be repacked without drying. Mildew can still grow on exposed fibers if they are soiled with contaminants that will provide mildew with the necessary nutrients. The fibers in rubber-covered hose are protected from mildew.

Cleaning and Maintaining Hoses

Hose that is dirty or contaminated should be cleaned, following the steps in (Skill Drill 16-8).

1. Lay the hose out flat.
2. Rinse the hose with water.
3. Gently scrub the hose with mild detergent, paying attention to soiled areas.
4. Turn over the hose and repeat steps two and three.
5. Give a final rinse to the hose with water.
6. Hang the hose and allow it to dry before properly storing it.

Hose Inspections

Each length of hose should be tested at least annually, according to the procedures listed in NFPA 1962, *Inspection, Care and Use of Fire Hose, Couplings, and Nozzles and Testing of Fire Hose*. The hose testing procedures are complicated and require special equipment. The hose testing equipment must be operated according to the manufacturer's instructions.

Visual hose inspections should be performed at least quarterly. A visual inspection should also be performed after each use, either while the hose is being cleaned and dried or when it is reloaded onto the apparatus. If any defects are found, that length of hose should be immediately removed from service and tagged with a description of the problem. The appropriate notifications must be made to have the hose repaired.

To clearly mark a defective hose, follow the steps in (Skill Drill 16-9).

1. Inspect the hose for defects.
2. Upon finding a defect, mark the area on the hose and remove the hose from service.
3. Tag the hose as defective with a description of the defect, take out of service and notify your superiors.

Hose Records

Hose records are important documents. A hose record is a written history of each individual length of fire hose. Each length of hose should be identified with a unique number stenciled or painted on it. A hose record will contain information such as:

- Hose size, type, and manufacturer
- Date the hose was manufactured
- Date the hose was purchased
- Dates when the hose was tested
- Any repairs that have been made to the hose

Some fire departments keep hose records on cards or paper files, while others keep the records in a database in a computer system.

Hose Appliances

A hose appliance is any device used in conjunction with a fire hose for the purpose of delivering water. You should be familiar with a wye, water thief, Siamese connection, double-male and double-female adaptors, reducers, hose clamps, hose jackets, and hose rollers. It is important for you to learn how to use the hose appliances and tools required by your fire department. You should understand the purpose of each device and be able to utilize each appliance.

Wyes

A wye is a device that splits one hose stream into two hose streams. The word wye refers to a Y-shaped part or object. When threaded couplings are used, a wye has one female connection and two male connections.

The wye that is most commonly used in the fire service splits one $2^1/_2''$ hose line into two $1^1/_2''$ hose lines. A gated wye is equipped with two-quarter turn ball valves so that the flow of water to each of the split lines can be controlled independently (► Figure 16-10). A gated wye enables you to initially attach and operate one hose line and then add a second hose later. The use of a gated wye avoids the need to

Voices of Experience

"It was clear to me that the industrial fire was a prime situation for dual $2^1/_2$" attack streams."

As a young fire fighter, I responded to a large industrial fire involving a number of mutual aid engines and trucks due to the intensity, size, and type of the incident. I drove the department squad car behind the two engines from headquarters station. The need for good communication, quick planning, and precise action was paramount.

When my department's engine stopped in front of the fire, the captain issued commands, and the crew pulled our fire department's dual hose lays. It was clear to me that the industrial fire was a prime situation for dual $2^1/_2$" attack streams. I pulled the squad car to the side of the road and ran directly to the wye and loosened it to replace it with a $2^1/_2$" nozzle. I couldn't find the nozzle so I sped to another department's engine, but nothing looked familiar. I then realized that the captain had not given the order for the dual $2^1/_2$" attack streams. There I was, with an open hose lying in the street. I had to quickly replace the wye before water arrived.

Fire fighters should remember that the incident commander makes appropriate decisions and directs the crew based on his or her assessment of the scene. No matter how insightful and skilled fire fighters may be, only the incident commander can be expected to make the call for action. Then, and only then, should fire fighters react. Any action other than that which is called for can add chaos to an already chaotic situation. An open hose, under pressure, in the wake of the need for attack lines, only slows down a process that is designed for speed and efficiency. In this case, both would have been sacrificed and a scramble to recoup would have been precious time wasted.

Roger Land
Lake Superior State University
School of Criminal Justice and Fire Science
Sault Ste. Marie, Michigan

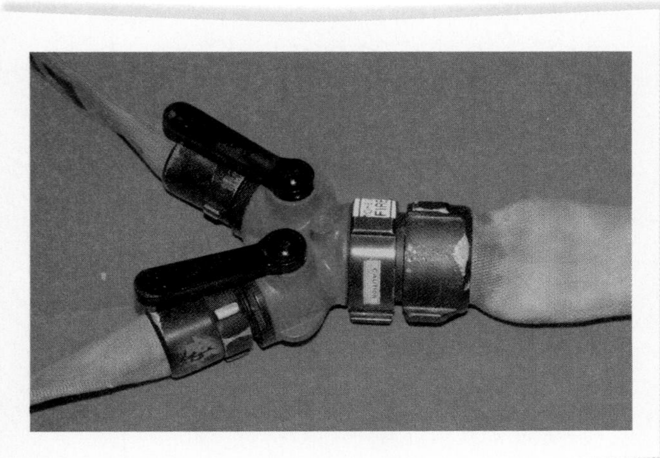

Figure 16-10 A gated wye is used to split one 2 1/2" hose line into two 1 1/2" lines.

Figure 16-11 A water thief.

shut down the hose line supplying the wye in order to attach the second line.

Water Thief

A water thief is similar to a gated wye, with an additional 2 1/2" outlet (► Figure 16-11). The water that comes from a single 2 1/2" inlet can be directed to two 1 1/2" outlets and one 2 1/2" outlet. Under most conditions, it will not be possible to supply all three outlets at the same time because the capacity of the supply hose is limited.

A water thief can be placed near the entrance to a building to provide the water for interior attack lines. One or two 1 1/2" attack lines can be used and, if necessary, can be shut down and a 2 1/2" line can be substituted. Sometimes the 2 1/2" line is used to knock down a fire and the two 1 1/2" lines are used for overhaul.

Siamese

A Siamese is a hose appliance that combines two hose lines into one. The most commonly used type of Siamese combines two 2 1/2" hose lines into a single line (► Figure 16-12). This increases the flow of water on the outlet side of the Siamese. A Siamese that is used with threaded couplings has two female connections on the inlets and one male connection on the outlet.

A Siamese connection is sometimes used on a engine inlet to allow water to be received from two different supply lines. Siamese connections are also used to supply master stream devices and ladder pipes. Siamese connections are commonly installed on the fire department connections that are used to supply water to standpipe and sprinkler systems in buildings.

Adaptors

Adaptors are used for connecting hose couplings of the same diameter that have dissimilar threads. Dissimilar threads could be encountered when different fire departments are working together or in industrial settings where the hose threads do not match the threads of the municipal fire department. Adaptors are also used to connect threaded couplings to Storz-type couplings.

Figure 16-12 A typical Siamese connection has two female inlets and a single outlet.

Adaptors can also be used when it is necessary to connect two female couplings or two male couplings. A double-female adaptor is used to join two male hose couplings. A double-male adaptor is used to join two female hose couplings. Double male and double female adaptors are commonly used when performing a reverse hose lay (► Figure 16-13).

Reducers

A reducer is used to attach a smaller hose to a larger hose (► Figure 16-14). Usually the larger end has a female connection and the smaller end has a male connection. A common type of reducer is used to reduce a 2 1/2" hose line

Figure 16-13 Double male and double female adaptors are used to join two couplings of the same sex.

Figure 16-14 A reducer is used to connect a smaller hose line to the end of a larger line.

Figure 16-15 A hose jacket is used to repair a leaking hose line.

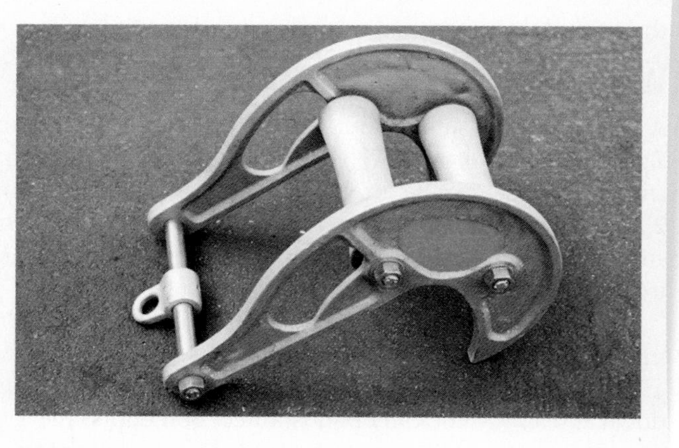

Figure 16-16 A hose roller is used to protect a hose when it is hoisted over a sharp edge of a roof or a windowsill.

to a 1½" line. Many 2½" nozzles are constructed with a built-in reducer, so that a 1½" line can be attached for overhaul.

Hose Jacket

A **hose jacket** is a device that is placed over a leaking section of hose to stop a leak (► Figure 16-15). The best way to handle a leak in a section of hose is to replace the defective section of hose. A hose jacket can provide a temporary fix until the section of hose can be replaced. A hose jacket should be used only in cases where it is not possible to quickly replace the leaking section of hose.

The hose jacket consists of a split metal cylinder that fits tightly over the outside of a hose line. The cylinder is hinged on one side to allow it to be placed over the leak, then a fastener is used to clamp the cylinder tightly around the hose. Gaskets on each end of the hose jacket prevent water from leaking out the ends of the hose jacket.

Hose Roller

A **hose roller** is used to protect a hose line that is being hoisted over the edge of a roof or over a windowsill (▲ Figure 16-16). The hose roller keeps the hose from chafing or kinking at the sharp edge. A hose roller is sometimes called a hose hoist because it makes it easier to raise or hoist a hose over the edge of the building. Hose rollers are also used to protect ropes when hoisting an object over the edge of a building and during rope rescue operations.

Hose Clamp

A **hose clamp** is used to temporarily stop the flow of water in a hose line. Hose clamps are often used on supply lines,

Figure 16-17 A hose clamp is used to temporarily interrupt the flow of water in a hose line.

Figure 16-18 A deck gun is mounted on top of an engine or other apparatus.

so that the hydrant can be opened before the line is hooked up to the intake of the attack engine. The fire fighter at the hydrant does not have to wait for the pump operator to connect the line to the pump intake before opening the hydrant. As soon as the line is connected, the clamp is released. A hose clamp can also be used to stop the flow in a line if a hose ruptures or it has to be connected to a different appliance (▲ Figure 16-17).

Master Stream Devices

A master stream device is a large capacity nozzle that can be supplied by two or more hose lines through a Siamese connection. Master stream devices include deck guns and portable ground monitors. A deck gun is usually attached to the top of an engine and may be supplied by a direct pipe connection from the pump (► Figure 16-18). A ground monitor can be removed from the apparatus and placed on the ground. When it is placed on the ground, the water can be supplied by 2½″ or larger hose lines (► Figure 16-19). Some devices can be used as either a deck gun or removed from the apparatus and used as a ground monitor. Master stream devices are used during defensive fire fighting operations.

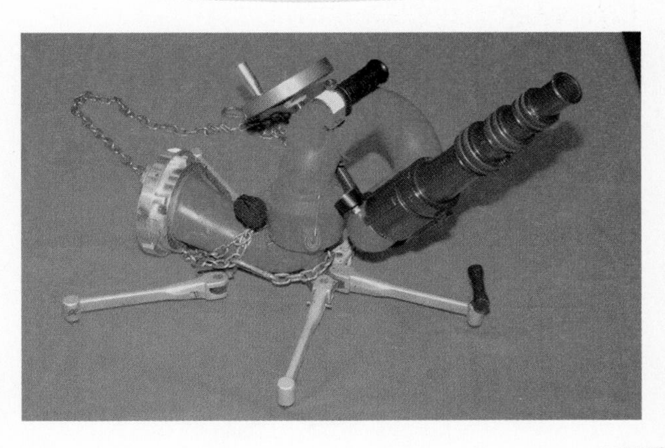

Figure 16-19 A ground monitor can be removed from the apparatus and placed on the ground.

Valves

Valves are used to control the flow of water in a pipe or hose line. Several different types of valves are used on fire hydrants, fire apparatus, standpipe and sprinkler systems, and hose lines. The important thing to remember when opening and closing any valve or nozzle is to do it S-L-O-W-L-Y to prevent water hammer.

Some of the common valves that you will encounter include:

- Ball Valves: These types of valves are used on nozzles, gated wyes, and engine discharge gates. Ball valves are made up of a ball with a hole in the middle. When the hole is in line with inlet and outlet, water flows through it. As the ball is rotated, the flow of water is gradually reduced until it is shut off completely (► Figure 16-20A).
- Gate Valves: These valves are found on hydrants and on sprinkler systems. Rotating a spindle causes a gate to move slowly across the opening. The spindle is rotated by turning it with a wrench or a wheel-type handle (► Figure 16-20B).
- Butterfly valves: These valves are often found on the large pump intake connections where a hard suction hose or soft suction hose is connected. They are opened or closed by rotating a handle one-quarter turn (► Figure 16-20C).

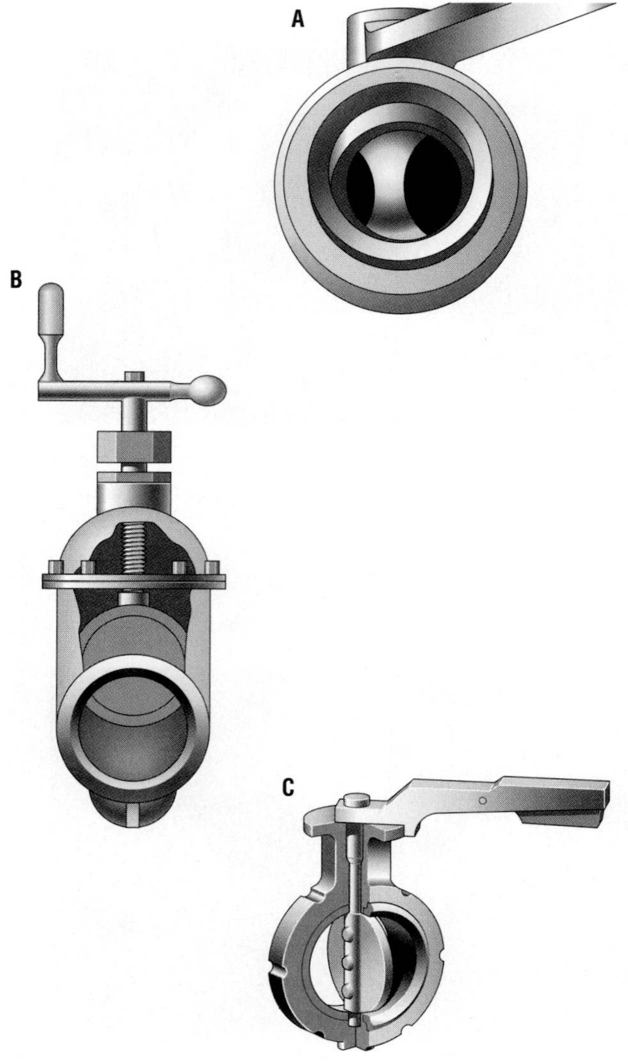

Figure 16-20 A, B, C A. Ball valve. B. Gate valve. C. Butterfly valve.

Hose Rolls

Overview of Hose Rolls

An efficient way to transport a single section of fire hose is in the form of a roll. Rolled hose is compact and easy to manage. A fire hose can be rolled many different ways, depending on how it will be used. Follow the SOPs of your department.

Straight or Storage Hose Roll

The straight roll is a simple and frequently used hose roll. It is used for general handling and transporting of hose. This roll is also used for rack storage of hose. To roll a straight roll, follow the steps of ► Skill Drill 16-10.

1. Lay the length of hose to be rolled flat and straight. **(Step 1)**
2. Begin by folding the male coupling over on top of the hose. **(Step 2)**
3. Roll the hose to the female coupling. **(Step 3)**
4. Set the hose roll on its side and tap any protruding hose flat with a foot. **(Step 4)**

Single Donut Hose Roll

The single donut roll is used when the hose will be put into use directly from the rolled state. The hose has both couplings on the outside of the roll. The hose can be connected and extended by one fire fighter. As the hose is extended, it unrolls. To perform a single donut roll, follow the steps in ► Skill Drill 16-11.

1. Place the hose flat and in a straight line. **(Step 1)**
2. Locate the mid-point of the hose. **(Step 2)**
3. From the mid-point, move 5′ toward the male coupling end.
4. Start rolling the hose toward the female coupling. **(Step 3)**
5. At the end of the roll, wrap the excess hose of the female end over the male coupling to protect the threads. **(Step 4)**

Twin Donut Hose Roll

The twin donut roll is used primarily to make a small compact roll that can be carried. To perform the twin donut hose roll, follow the steps in ► Skill Drill 16-12.

1. Lay the hose flat and in a straight line. **(Step 1)**
2. Bring the male coupling alongside the female coupling. **(Step 2)**
3. Fold the far end over and roll toward the couplings, creating a double roll. **(Step 3)**
4. The roll can be carried by hand, by a rope, or strap. **(Step 4)**

Self-Locking Twin Donut Hose Roll

The self-locking twin donut is similar to the twin donut with the exception that it forms its own carry loop. To perform the self-locking twin donut roll, follow the steps in ► Skill Drill 16-13.

1. Lay the hose flat and bring the coupling alongside each other. **(Step 1)**
2. Move one side of the hose over the other creating a loop. This loop creates the carrying shoulder loop. **(Step 2)**
3. Bring the loop back toward the couplings to the point where the hose crosses. **(Step 3)**
4. From the point where the hose crosses, begin to roll the hose toward the couplings with the loop as its center. This creates a loop on each side of the roll. **(Step 4)**
5. On completion of the rolling, position the couplings on the top of the rolls. **(Step 5)**
6. Position the loops so one is larger than the other. Then pass the larger loop over the couplings and through the smaller loop. This secures the rolls together and forms the shoulder loop. **(Step 6)**

16-10 Skill Drill

Rolling a Straight Hose Roll

Lay the length of hose to be rolled flat and straight.

Begin by rolling the male coupling over on top of the hose.

Roll the hose to the female coupling.

Set the hose roll on its side and tap any protruding hose flat with a foot.

16-11 Skill Drill

Performing a Single Donut Roll

Place the hose flat and in a straight line.

Locate the mid-point of the hose.

From the mid-point, move 5′ toward the male coupling end. Start rolling the hose toward the female coupling.

At the end of the roll, wrap the excess hose of the female end over the male coupling to protect the threads.

16-12 Skill Drill

Twin Donut Roll

Lay the hose flat and in a straight line.

Bring the male coupling alongside the female coupling.

Fold the far end over and roll toward the couplings, creating a double roll.

The roll can be carried by hand, rope, or strap.

Performing a Self-Locking Twin Donut

Lay the hose flat and bring the coupling alongside each other.

Move one side of the hose over the other creating a loop. This loop creates the carrying shoulder loop.

Bring the loop back toward the couplings to the point where the hose crosses.

From the point where the hose crosses, begin to roll the hose toward the couplings with the loop as its center. This creates a loop on each side of the roll.

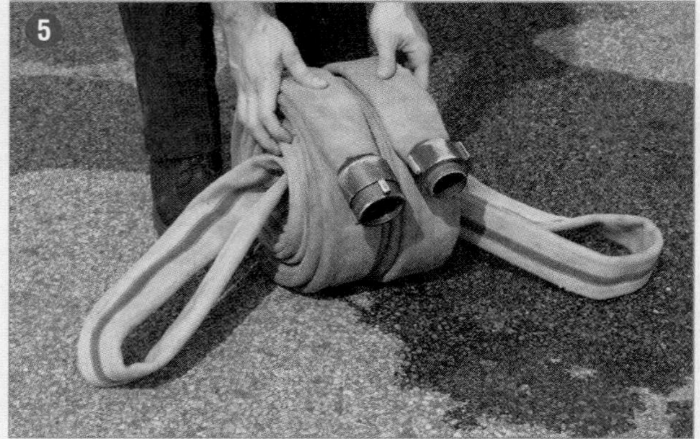

On completion of the rolling, position the couplings on the top of the rolls.

Position the loops so one is larger than the other. Then pass the larger loop over the couplings and through the smaller loop.

Fire Hose Evolutions

Fire hose evolutions are standard methods of working with fire hose to accomplish different objectives in a variety of situations. Most fire departments set up their equipment and conduct regular training in order to be prepared to perform a set of standard hose evolutions. Hose evolutions involve specific actions that are assigned to each member of a crew, depending on their riding positions on the apparatus. Every fire fighter should know how to perform all of the standard evolutions quickly and proficiently. When an officer calls for a particular evolution to be performed, each crew member should know exactly what to do.

Hose evolutions are divided into supply line operations and attack line operations. Supply line operations involve laying hose lines and making connections between a water supply source and an attack engine. Attack line operations involve advancing hose lines from an attack engine to apply water onto the fire.

Supply Line Operations

The objective of laying a supply line is to deliver water from a hydrant or an alternative water source to an attack engine. In most cases, this involves laying a hose line with a moving vehicle or dropping a continuous line hose out of a bed as the vehicle moves forward. This can be done using either a forward lay or reverse lay. A forward lay starts at the hydrant and proceeds toward the fire; the hose is laid in the same direction as the water flows, from the hydrant to the fire. A reverse lay involves laying the hose from the fire to the hydrant; the hose is laid in the opposite direction to the water flow. Each fire department determines their own preferred methods and procedures based on apparatus, water supply, and regional considerations.

Forward Hose Lay

The forward hose lay is most often used by the first arriving engine company at the scene of a fire (▼ Figure 16-21). This method allows the engine company to establish a water supply without assistance from an additional company. A forward lay also places the attack engine close to the fire, allowing access to additional hoses, tools, and equipment that are carried on the apparatus.

Fire Fighter Safety Tips

When performing a forward lay, the fire fighter who is connecting the supply line to the hydrant must not stand between the hose and the hydrant. When the apparatus starts to move off, the hose could become tangled and suddenly be pulled taut. Anyone standing between the hose and the hydrant could be seriously injured.

A forward hose lay can be performed using medium diameter hose lines (2½" or 3" hose) or with large diameter hose (3½" and larger). The larger the diameter of the hose, the more water can be delivered to the attack engine through a single supply line. When medium diameter hose is used and the beds are arranged to lay dual lines, a company can lay two parallel lines from the hydrant to the fire.

If the fire hydrant is close to the fire, it may supply a sufficient quantity of water from the hydrant alone. A 5" hose can supply 700 gallons of water per minute over a distance of 500' and lose only about 40 pounds of pressure due to friction loss. To perform a forward hose lay, follow the steps in (Skill Drill 16-14).

1. The fire apparatus should stop within about 10' of the hydrant.
2. Grasp enough hose to reach to the hydrant and to loop around the hydrant. Step off the apparatus with the hydrant wrench and all necessary tools.
3. Loop the end of the hose around the hydrant or secure the hose as specified in the SOP. Ensure that you are not between the hose and the hydrant. Never stand on the hose.
4. Signal the driver/operator to proceed to the fire once the hose is secured.
5. Once the apparatus has moved off and a length of supply line has been removed from the apparatus and is lying on the ground, the fire fighter should remove

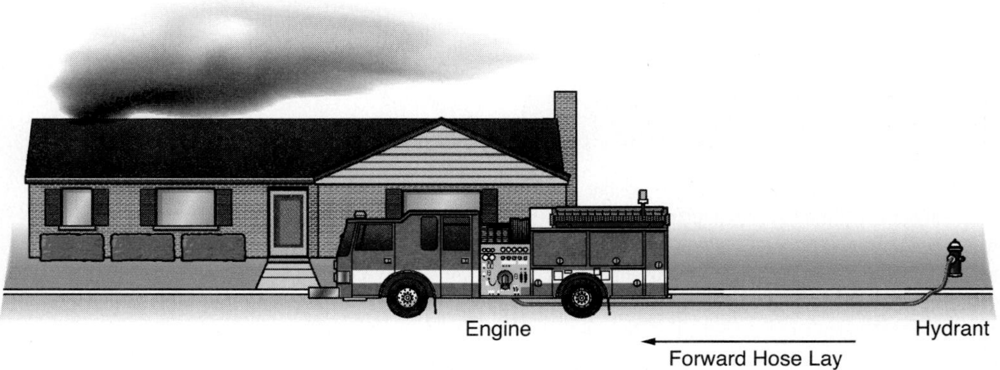

Figure 16-21 Forward hose lay.

the appropriate size hydrant cap nearest to the fire. Follow local SOP for checking the operating condition of the hydrant.

6. Attach the supply hose to the outlet. An adaptor may be needed if a large diameter hose with Storz-type couplings is used.
7. Attach the hydrant wrench to the hydrant.
8. The driver/operator should uncouple the hose and attach the end of the supply line to the pump inlet or clamp the hose close to the pump, depending on local SOPs.
9. The driver/operator should signal to charge the hose by prearranged hand signal, radio, or air horn.
10. Slowly open the hydrant completely.
11. Follow the hose back to the engine and remove any kinks from the supply line.

Four-Way Hydrant Valve

In cases where long supply lines are needed or when medium diameter hose is used, it is often necessary to place a supply engine at the hydrant. The supply engine pumps water through the supply line in order to increase the flow to the attack engine. Some departments use a **four-way hydrant valve** to connect the supply line to the hydrant, so that the supply line can be charged with water immediately and still allow for a supply engine to connect into the line later.

When the four-way valve is placed on the hydrant, the water flows initially from the hydrant through the valve to the supply line, which delivers the water to the attack engine. The second engine can hook up to the four-way valve and redirect the flow by changing the position of the valve. The water then flows from the hydrant to the supply engine. The supply engine boosts the pressure and discharges the water into the supply line, boosting the flow of water to the attack engine. This can be accomplished without uncoupling any lines or interrupting the flow. To use a four-way valve, follow the steps in **Skill Drill 16-15**.

1. The attack engine should stop about 10′ after the hydrant to be used.
2. Grasp the four-way valve, the attached hose, and enough hose to reach to and loop around the hydrant. Remove the four-way valve off the apparatus along with the hydrant wrench and any other needed tools.

Fire Fighter Tips

When connecting a supply line to a hydrant, wait until you get the signal from the driver/operator to charge the line. If the hydrant is opened prematurely, the hose bed could become charged with water or a loose hose line could discharge water at the fire scene. Either situation will disrupt the operation and could cause serious injuries. Be sure that you know your department's signal to charge a hose line and do not become so excited or rushed that you make a mistake.

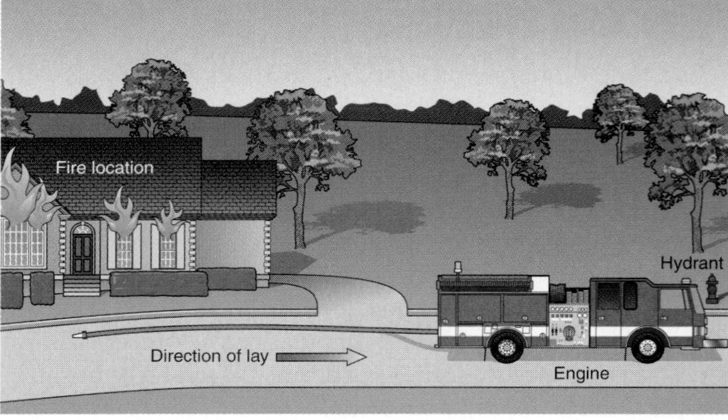

Figure 16-22 Reverse hose lay.

3. Loop the end of the hose around the hydrant or secure the hose with a rope as specified in the local SOPs. DO NOT stand between the hydrant and the hose.
4. Signal the driver/operator to proceed to the fire.
5. Once enough hose has been removed from the apparatus and is lying in the street, the hydrant person should remove the steamer port from the fire hydrant. Follow local SOPs for checking the operating condition of the hydrant.
6. Attach the four-way valve to the hydrant outlet (an adaptor may be needed).
7. Attach the hydrant wrench to the hydrant.
8. The driver/operator uncouples the hose and attaches the end of the supply line to the pump inlet.
9. The driver/operator signals you by prearranged hand signal, radio, or air horn to charge the supply line.
10. Slowly open the hydrant completely.
11. Initially, the attack engine is supplied with water from the hydrant. When the supply engine arrives at the fire scene, the driver/operator should stop at the hydrant with the four-way valve.
12. The driver/operator should attach a hose from the four-way valve outlet to the intake side of the engine.
13. Attach a second hose to the inlet side of the four-way valve and connect the other end to the pump discharge.
14. Change the position of the four-way valve to direct the flow of water from the hydrant through the supply engine and into the supply line.

Reverse Hose Lay

The reverse hose lay is the opposite of the forward lay **▲ Figure 16-22**. In the reverse lay, the hose is laid out from the fire to the hydrant, in the direction opposite to the flow of the water. This evolution can be used when the attack engine arrives at the fire scene without a supply line. This could be a standard tactic in areas where there are sufficient hydrants available and additional companies that can assist in establishing a water supply will be arriving quickly. One

Fire Fighter Tips

When laying out supply hose with threaded couplings, you may find that the wrong end of the hose is on top of the hose bed. **Double male connectors** and **double female connectors** will allow you to attach a male coupling to a male discharge or to attach a female coupling to a female coupling. A set of adaptors (one double male and one double female) should be easily accessible for these situations. Some departments place a set of adaptors on the end of the supply hose for this purpose.

Fire Fighter Tips

Split Hose Beds

A **split hose bed** is a hose bed that is divided into two or more sections. This is done for several purposes.

- One compartment in a split hose bed can be loaded for forward lay (female coupling out) and the other side can be loaded for a reverse lay (male coupling out). This allows a line to be laid in either direction without adaptors.
- Two parallel hose lines can be laid at the same time. (This is sometimes called laying dual lines.) Dual lines are beneficial if the situation requires more water than one hose line can supply.
- The split beds can be used to store different sized hoses. For example, one side of the hose bed could be loaded with 2½" hose that can be used as a supply line or as an attack line. The other side of the hose bed could be loaded with 5" hose for use as a supply line. This set-up enables the use of the most appropriate sized hose for a given situation.
- All of the hose from both sides of the hose bed can be laid out as a single hose line. This is done by coupling the end of the hose in one bed to the beginning of the hose in the other bed.

A variation of the split hose bed is known as a combination load, where the last coupling in one bed is normally connected to the first coupling in the other bed. When one long line is needed, all of the hose plays out of one bed first, then the hose continues to play out from the second bed. To lay dual lines, the connection between the two hose beds is uncoupled and the hose can play out of both beds simultaneously. When the two sides of a split bed are loaded with the hose in opposite directions, either a double female or a double male adaptor is used to make the connection between the two hose beds.

of these companies will be assigned to lay a supply line from the attack engine to a hydrant.

In this case, the attack engine will focus on immediately attacking the fire using water from the on-board tank. The supply engine stops close to the attack engine and hose is pulled from the bed of the supply engine to an intake on the attack engine. The supply engine then drives to the hydrant (or alternative water source) and pumps water back to the attack engine. Usually the supply engine parks so that hose can be pulled from the supply engine to the inlet to the

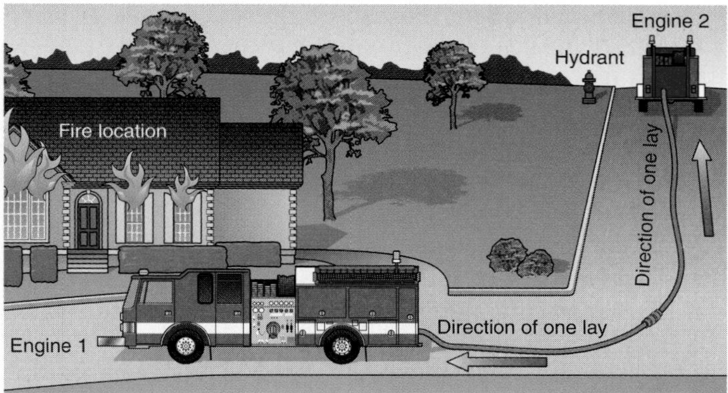

Figure 16-23 Split hose lay.

attack engine. To perform a reverse lay, follow the steps in Skill Drill 16-16.

1. Pull sufficient hose to reach from the supply engine to the inlet of the attack engine.
2. Anchor the hose to a stationary object if possible. Wrapping a portion of the hose around the front wheels of the attack engine is one possible method and a tree is another. DO NOT stand between the hose and the stationary object.
3. The driver/operator of the supply engine drives away, laying out hose from the attack engine to the fire hydrant or other static water source.
4. Connect the supply line to the inlet of the attack engine.
5. The driver/operator of the supply engine should uncouple the supply hose from the hose remaining in the hose bed and attach the supply hose to the discharge side of the supply engine.
6. The driver/operator will then connect the supply engine to the hydrant or water source. The four-way hydrant valve can be used here if needed to boost the supply pressure.
7. Upon signal from the attack engine, the supply engine driver/operator should charge the supply line.

Split Hose Lay

A split hose lay is performed by two engine companies in situations where hose must be laid in two different directions to establish a water supply (▲ Figure 16-23). This evolution could be used when the attack engine has to approach a fire along a dead-end street with no hydrant or down a long driveway. To perform a split hose lay, the attack engine drops the end of its supply hose at the corner of the street and performs a forward lay toward the fire. The supply engine stops at the intersection, pulls off enough hose to connect to the end of the supply line that is already there, and then performs a reverse lay to the hydrant or water source. When the two lines are connected together, the supply engine can provide water to the attack engine.

A split lay often requires coordination by two-way radio, because the attack engine must advise the supply engine of the plan and where the end of the supply line is being dropped. In many cases, the attack engine is out of sight when the supply engine arrives at the split point. **NOTE:** A split hose lay does not necessarily require split hose beds. It can be performed with or without split beds if the necessary adaptors are used. To perform a split hose lay, follow the steps in Skill Drill 16-17.

1. The driver/operator of the first arriving engine company will stop at the intersection or driveway entrance.
2. Remove the supply line from the hose bed and anchor it in a kneeling position.
3. The driver/operator should proceed slowly towards the structure fire.
4. Either proceed by foot to the structure fire or wait at the intersection for the supply engine. Check with local SOPs.
5. If the supply engine returns to the structure fire, the supply engine needs to stop and connect the supply hose to the hose end laid in the street by the attack engine. If threaded couplings are used, a double male adaptor may be required.
6. If the supply engine remained at the intersection, anchor the supply line from the second engine company. After the hose is laid to a hydrant, connect the two lines to form one supply line.
7. The driver/operator of the attack engine will start pumping water from the booster tank and also connect the supply hose to the pump intake.
8. The driver/operator of the supply engine should position the apparatus at the hydrant according to local SOPs.
9. Then, the driver/operator of the supply engine should pull off hose from the hose bed until the next coupling. The hose is broken at this connection and is connected to the pump discharge or the four-way hydrant valve. Check with local SOPs.
10. Upon signal from the attack engine, the supply engine driver/operator will charge the supply line.

Loading Supply Hose

This section describes the basic procedures for loading supply line hose into the hose beds on a fire apparatus. Hose can be loaded in several different manners, depending on the way the hose is planned to be laid out. The hose must be easily removable from the hose bed, without kinks or twists and without the possibility of becoming caught or tangled. The ideal hose load would be easy to load, avoid wear and tear on the hose, have few sharp bends, and allow the hose to play out of the hose bed smoothly and easily.

You must learn the specific hose loads used by your fire department. When loading hose, always remember that the time and attention that goes into loading the hose properly will become worthwhile when it is time to use the hose at a fire. You should wear appropriate safety gear when loading hose—at the least your helmet and boots.

There are three basic hose loads that are commonly used for supply line hose: the flat load, the horseshoe load, and the accordion load. Any one of these methods can be used to load hose for either a straight lay or for a reverse lay.

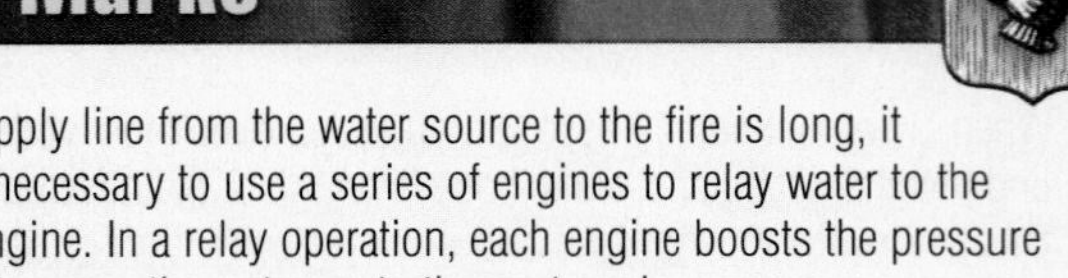

If the supply line from the water source to the fire is long, it may be necessary to use a series of engines to relay water to the attack engine. In a relay operation, each engine boosts the pressure enough to move the water on to the next engine.

Fire Fighter Tips

When refering to the hosebed, the end closest to the cab is called the front of the hosebed. The end closet to the tailboard is called the rear of the hosebed.

Flat Hose Load

The flat hose load is the easiest to load and can be used for any size of hose, including large diameter hose. Because the hose is placed flat in the hose bed, it should lay out flat without twists or kinks. Wear and tear on the edges of the hose from the movement and vibration of the vehicle during travel are limited. The flat hose load can be used with a single hose bed or a split bed. There are many variations of the flat hose load. Follow the standard operating procedures of your department. To load hose in a flat load, follow the steps in ► Skill Drill 16-18.

1. If you are loading hose with threaded couplings, determine whether the hose will be used for a forward or reverse lay.
2. To set up the hose for a forward lay, place the male hose coupling in the hose bed first. To set up the hose for a reverse lay, place the female hose coupling in the hose bed first. **(Step 1)**
3. Start the hose load with the coupling at the front end of the hose compartment. **(Step 2)**
4. Fold the hose back on itself at the rear of the hose bed. **(Step 3)**
5. Run the hose back to the front end on top of the previous length of hose.
6. Fold the hose back on itself so the top of the hose is on the previous length. **(Step 4)**
7. While laying the hose back to the rear of the hose bed, angle the hose to the side of the previous fold. **(Step 5)**
8. Continue to lay the hose in neat folds until the whole hose bed is covered with a layer of hose.
9. To make this hose load neat, make every other layer of hose slightly shorter or alternating the folds. This keeps the ends from getting too high at the folds.
10. Continue to load the layers of hose until the required amount of hose is loaded. **(Step 6)**

16-18 Skill Drill

Performing a Flat Hose Load

To set up the hose for a forward lay, place the male hose coupling in the hose bed first. To set up the hose for a reverse lay, place the female hose coupling in the hose bed first.

Start the hose lay with the coupling at the front end of the hose compartment.

Fold the hose back on itself at the rear of the hose bed.

Run the hose back to the front end on top of the previous length of hose. Fold the hose back on itself so the top of the hose is on the previous length.

While laying the hose back to the rear of the hose bed, angle the hose to the side of the previous fold.

Continue to lay the hose in neat folds until the whole hose bed is covered with a layer of hose. Continue to load the layers of hose until the required amount of hose is loaded.

Horseshoe Hose Load

The **horseshoe hose load** is accomplished by standing the hose on its edge and placing it around the perimeter of the hose bed in a U-shape. At the completion of the first U-shape, the hose is folded inward to form another U in the opposite direction. This continues until a complete layer is filled; then another layer is started above the first. When the hose load is completed, the hose in each layer is in the shape of a horseshoe. A major advantage of the horseshoe load is that it contains fewer sharp bends than the other hose loads.

The horseshoe load cannot be used for large diameter hose because the hose tends to fall over when it stands on edge. The horseshoe load causes more wear on the hose because the weight of the hose is supported by the edges. Also, when laying out a horseshoe load, the hose tends to lay out in a wave-like manner from one side of the street to the other. There are many variations of the horseshoe hose load. Follow the SOPs of your department. To perform a horseshoe hose load, follow the steps in (► Skill Drill 16-19).

1. If you are loading threaded hose, determine whether you want to load the hose for a forward lay or a reverse lay. For a forward lay, start with the male coupling in the rear corner of the hose bed. For a reverse lay, start with the female coupling in the rear corner of the hose bed. **(Step 1)**
2. Lay the first length of hose on its edge against the right or left wall of the hose bed. **(Step 2)**
3. At the front of the hose bed, lay the hose across the width of the bed and continue down the opposite side toward the rear. **(Step 3)**
4. When the hose reaches the rear of the hose bed, fold the hose back on itself and continue laying it back toward the front of the hose bed. Keep the hose tight to the previous row of hose around the hose bed until it is back to the rear on the starting side. Fold the hose back on itself again and continue packing the hose tight to the previous row. **(Step 4)**
5. Continue to pack the hose on the first layer. Each fold of hose will decrease the amount of space available inside of the horseshoe. Once the center of the horseshoe is filled in, begin a second layer by bringing the hose from the rear of the hose bed and laying it around the perimeter of the hose bed. Complete additional layers using the same pattern as you did for the first layer. Finish the hose load with any adaptors or appliances used by your department. **(Step 5)**

Accordion Hose Load

The **accordion hose load** is performed with the hose placed on its edge. The hose is laid side-to-side in the hose bed. One advantage of the accordion load is that it is easy to load in the hose bed. One layer is loaded from left to right and then the next layer is loaded above it from left to right.

Fire Fighter Tips

When a split hose bed is loaded for a combination load, the end of the last length of hose in one bed is coupled to the beginning of the first length in the opposite bed. Begin loading the first bed with the initial coupling hanging out. This coupling will be the last to be deployed from the first bed when the hose is laid out. Leave enough of the end of the hose hanging down to reach the other hose bed. Load the second bed in the normal manner. When both beds have been loaded, connect the hanging coupling from the bottom of the first hose bed to the end coupling on the top of the adjoining bed.

Fire Fighter Tips

The following tips will help you to do a better job when loading hose:

- Drain all of the water out of the hose before loading.
- Rolling the hose first will result in a flatter hose load, because there will be no air in the hose.
- Do not load hose too tightly. Leave enough room so that you can slide a hand between the folds of hose. If hose is loaded too tightly, it may not lay out properly.
- Load hose so that couplings do not have to turn around as the hose is pulled out of the hose bed. Make a short fold in the hose close to the coupling to keep the hose properly oriented. This short fold is called a **Dutchman** (▼ Figure 16-24).
- Couple sections of hose with the flat sides oriented in the same direction.
- Check gaskets before coupling hose.
- Tighten couplings hand tight only. With a good gasket, the hose should not leak.

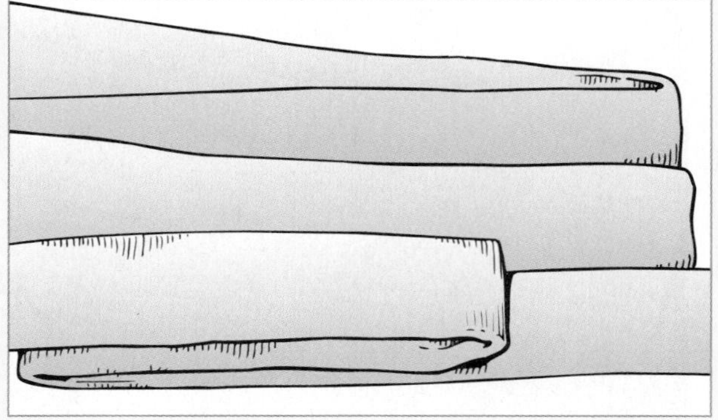

Figure 16-24 A Dutchman is used so that a coupling will not have to turn around and possibly become stuck as the hose is laid out.

16-19 Skill Drill

Performing a Horseshoe Hose Load

For a forward lay, start with the male coupling in the rear corner of the hose bed. For a reverse lay, start with the female coupling in the rear corner of the hose bed.

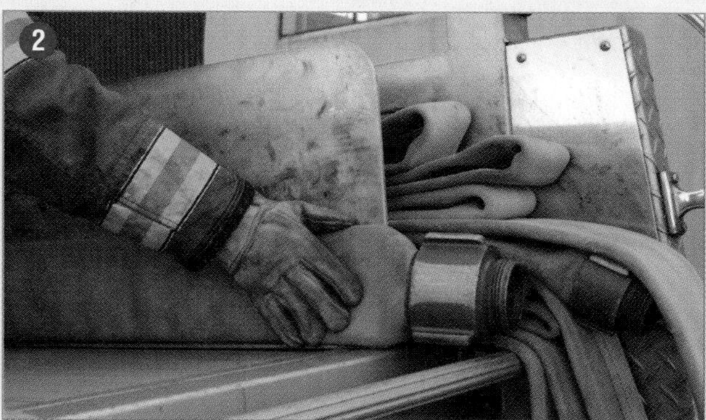

Lay the first length of hose on its edge against the right or left wall of the hose bed.

At the front of the hose bed, lay the hose across the width of the bed and continue down the opposite side toward the rear.

When the hose reaches the rear of the hose bed, fold the hose back on itself and continue laying it back toward the front of the hose bed. Keep the hose tight to the previous row of hose around the hose bed until it is back to the rear on the starting side. Fold the hose back on itself again and continue packing the hose tight to the previous row.

Continue to pack the hose on the first layer. Each fold of hose will decrease the amount of space available inside of the horseshoe. Once the center of the horseshoe is filled in, begin a second layer by bringing the hose from the rear of the hose bed and laying it around the perimeter of the hose bed. Complete additional layers using the same pattern as you did for the first layer. Finish the hose load with any adaptors or appliances used by your department.

16-20 Skill Drill

Performing an Accordion Hose Load

Lay the first length of hose in the hose bed on its edge against the side of the hose bed.

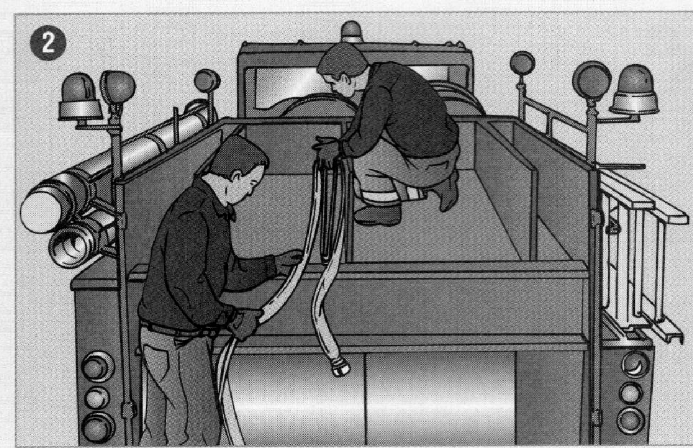

Double the hose back on itself at the rear of the hose bed. Leave the female end extended so that the two hose beds can be cross-connected.

Lay the hose next to the first length and bring it to the front of the hose bed.

Fold the hose at the front of the hose bed so the bend is even to the edge of the hose bed. Continue to lay folds of hose across the hose bed.

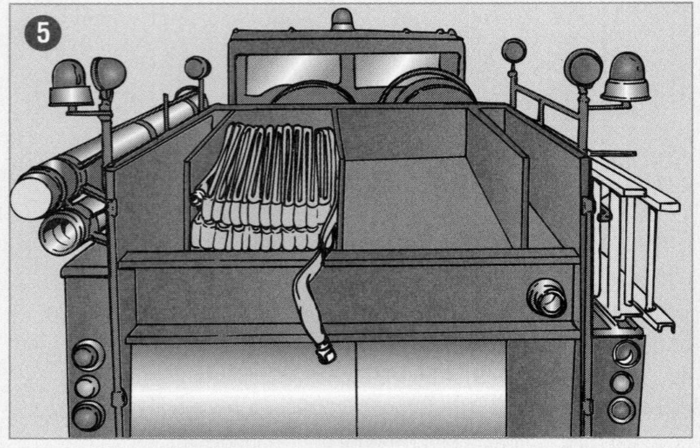

Alternate the length of the hose folds at each end to allow more room for the folded ends. When the bottom layer is completed, angle the hose upward to begin the second tier. Continue the second layer by repeating the steps you used to complete the first layer.

There are disadvantages to the accordion hose load. Because the hose is stacked on its side, there is more wear on the hose than with a flat load. The accordion hose load is not recommended for large diameter hose because large diameter hose tends to collapse when placed on its side. There are many variations of the accordion hose load. Follow the SOPs of your department. To perform the accordion hose load, follow the steps in ◄ Skill Drill 16-20.

1. Determine whether the hose will be used for a forward or reverse lay.
2. To set up the hose for a forward lay, place the male hose coupling in the hose bed first. To set up the hose for a reverse lay, place the female hose coupling in the hose bed first.
3. Start the hose lay with the coupling at the front end of the hose compartment.
4. Lay the first length of hose in the hose bed on its edge against the side of the hose bed. **(Step 1)**
5. Double the hose back on itself at the rear of the hose bed. Leave the female end extended so that the two hose beds can be cross-connected. **(Step 2)**
6. Lay the hose next to the first length and bring it to the front of the hose bed. **(Step 3)**
7. Fold the hose at the front of the hose bed so the bend is even to the edge of the hose bed. Continue to lay folds of hose across the hose bed. **(Step 4)**
8. Alternate the length of the hose folds at each end to allow more room for the folded ends.
9. When the bottom layer is completed, angle the hose upward to begin the second tier.
10. Continue the second layer by repeating the steps you used to complete the first layer. **(Step 5)**

Connecting a Fire Department Engine to a Water Supply

When an engine is located at a hydrant, a supply hose must be used to deliver the water from the hydrant to the engine. This is a special type of supply line that is intended to deliver as much water as possible over a short distance. In most cases, a soft suction hose is used to connect directly to a hydrant. This connection can also be made with a hard suction hose or with a short length of large diameter supply hose.

Attaching a Soft Suction Hose to a Hydrant

The large soft suction hose is normally used to connect an engine directly to a hydrant. To attach a soft suction hose to a hydrant, follow the steps in ► Skill Drill 16-21.

1. The driver/operator positions the apparatus so that its inlet is the correct distance from the hydrant. **(Step 1)**
2. Remove the hose, any needed adaptors, and the hydrant wrench. **(Step 2)**

16-21 Skill Drill

Attaching a Soft Suction Hose to a Fire Hydrant

The driver/operator positions the apparatus so that its inlet is the correct distance from the hydrant.

Remove the hose, any needed adaptors, and the hydrant wrench.

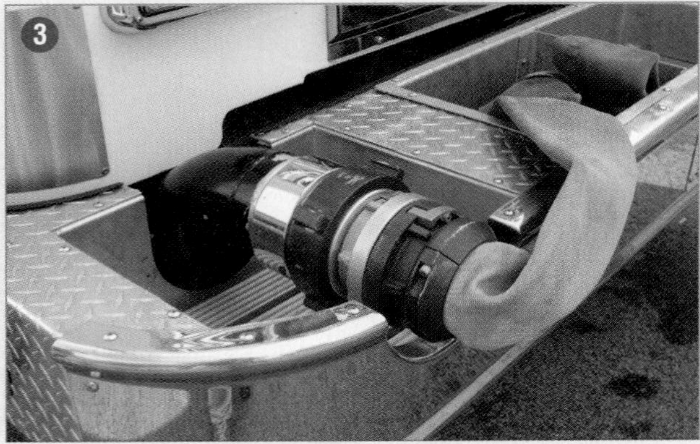

Attach the soft suction hose to the inlet of the engine.

Unroll the hose.

Remove the large hydrant cap.

Attach the soft suction hose to the hydrant.

16-21 Skill Drill

Ensure that there are no kinks or sharp bends in the hose that will restrict the flow of water.

Open the hydrant slowly when so indicated by the pump operator. Check all connections for leaks. Tighten if necessary.

Place chafing blocks under the hose where it contacts the ground to prevent mechanical abrasion.

3. Attach the soft suction hose to the inlet of the engine. In some departments this end of the hose is preconnected. **(Step 3)**
4. Unroll the hose. **(Step 4)**
5. Remove the large hydrant cap. **(Step 5)**
6. Attach the soft suction hose to the hydrant. Sometimes it will be necessary to use an adaptor. **(Step 6)**
7. Ensure that there are no kinks or sharp bends in the hose that will restrict the flow of water. **(Step 7)**
8. Open the hydrant slowly when so indicated by the pump operator. Check all connections for leaks. Tighten if necessary. **(Step 8)**
9. Place chafing blocks under the hose where it contacts the ground to prevent mechanical abrasion. **(Step 9)**

Attaching a Hard Suction Hose to a Hydrant

Although it is not commonly done, a hard suction hose is sometimes used to connect an engine to a hydrant. This can be a difficult task because the hard suction is heavy and has limited flexibility, so additional personnel will often be

16-22 Skill Drill

Attaching a Hard Suction Hose to a Fire Hydrant

1 The driver/operator will position the apparatus so that the intake is located the correct distance from the hydrant.

2 Remove the pump inlet cap. Remove the hydrant steamer outlet cap. Place an adaptor on the hydrant if needed. With your partner's help, remove a section of hard suction hose from the engine.

3 Connect the hard suction hose to the large intake on the engine.

4 Connect the opposite end to the hydrant. Slowly open the hydrant when instructed to do so by the driver/operator.

needed to lift and attach the hose. The apparatus must be carefully positioned to make this connection properly. To attach a hard suction hose to a hydrant, follow the steps in ▲ **Skill Drill 16-22**.

1. The driver/operator will position the apparatus so that the intake on the apparatus is located the correct distance from the hydrant. **(Step 1)**
2. Remove the pump inlet cap on the apparatus.
3. Remove the hydrant steamer outlet cap, which is the largest one.
4. Place an adaptor on the hydrant if needed.
5. With your partner's help, remove a section of hard suction hose from the engine. **(Step 2)**
6. Connect the hard suction hose to the large intake on the engine. **(Step 3)**
7. Connect the opposite end to the hydrant, using a double female adaptor with the hydrant thread on one side and the suction hose thread on the other side.
8. The driver/operator may need to move the apparatus slightly to position the apparatus to make the final connection.
9. Slowly open the hydrant when instructed by the driver/operator. **(Step 4)**

Attack Line Evolutions

Attack lines are the hose lines used to deliver water from an attack engine to a nozzle, which discharges the water onto the fire. Any hose line that is used to discharge water onto the fire without going through another pump is defined as an attack line. Attack lines can use several different hose sizes and any length of hose.

Attack lines are usually stretched from an engine or an apparatus that is functioning as an attack engine to the fire. The attack engine is usually located close to the fire and attack lines are stretched manually by fire fighters. In some situations, an engine will drop an attack line at the fire and drive from the fire to a hydrant or water source. This is similar to a reverse lay evolution as described in the supply line section, except that the hose will be used as an attack line.

Most engines are equipped with preconnected attack lines, which provide a predetermined length of attack hose that is already equipped with a nozzle and connected to a pump discharge outlet. An additional supply of attack hose is usually carried in a hose bed or compartment that is not preconnected. To create an attack line with this hose, the desired length of hose is removed from the bed; then a coupling is disconnected and attached to a pump discharge outlet. This hose can also be used to extend a preconnected attack line or to attach to a wye or a water thief.

The attack hose is loaded so that it can be quickly and easily deployed. There are many ways to load attack lines into a hose bed. This section presents some of the most common hose loads. Your department may use a variation of one of these loads. It is important for you to master the hose loads used by your department.

Preconnected Attack Lines

Preconnected hose lines are intended for immediate use as attack lines. A preconnected hose line has a predetermined length of hose with the nozzle already attached and is connected to a discharge outlet on the fire engine. The most commonly used attack lines are $1^{3}/_{4}''$ hose, generally from 150′ to 250′ in length. Many engines are also equipped with a preconnected $2^{1}/_{2}''$ hose line for quick attack on larger fires.

Attack lines should be loaded in the hose bed so they can be quickly stretched from the attack engine to the fire. It should be possible for one or two fire fighters to quickly remove the hose from the hose bed and advance the hose to the fire. In many cases, an attack line is stretched in two stages. First, the hose from the attack engine to the building entrance or to a location close to the fire will be laid out. From that point, the hose is advanced into the building to reach the fire. Extra hose should be deposited at the entrance to the fire building.

Fire Marks

A transverse preconnected hose located over the pump is called a "Mattydale Hose Bed" or a "Mattydale lay" because this set-up was first used by the fire department in Mattydale, New York.

Several different hose loads can be used. The hose should not get tangled as it is being removed from the bed and advanced. Laying out the hose should not require multiple trips between the engine and the fire building and it should be easy to lay the hose around obstacles and corners. It should also be possible to repack the hose quickly and with minimal personnel. There is no perfect hose load that works well for every situation. The three most common hose loads for preconnected attack lines are the minuteman load, the flat load, and the triple layer load. Variations in circumstances will make one type of hose load preferable for your community.

Because of different types of situations, most departments load attack lines of different lengths. An engine could be equipped with a 150′ preconnected line and a 250′ preconnected line. Fire fighters should pull an attack line that is long enough to reach the fire, but not so long that there will be an excess of hose to slow down the operation and become tangled.

Preconnected hose lines can be placed in several different locations on a fire engine. A section of a divided hose bed at the rear of the apparatus can be loaded with a preconnected attack line. Transverse hose beds are installed above the pump on many engines and loaded so the hose can be pulled off from either side of the apparatus. Preconnected lines can be loaded into special trays that are mounted on the side of fire apparatus.

Many engines have a special compartment in the front bumper that can store a short preconnected hose line. This shorter preconnected hose line often is used for vehicle fires and dumpster fires, where the apparatus can drive up close to the incident and a longer hose line is not needed. Booster

16-23 Skill Drill

Loading the Minuteman Hose Load

1. Connect the female end of the first length of hose to the discharge outlet.

2. Lay the hose flat to the edge of the hose bed and leave the remaining hose with the male coupling out of the front of the bed to be connected later. Do not connect to the rest of the load.

3. Assemble the remaining hose sections and attach the nozzle.

4. Place the nozzle on the first length already in the hose bed.

hose is another type of preconnected attack line. Booster reels holding ¾" or 1" diameter hose can be mounted in a variety of locations on fire apparatus.

The Minuteman Load

To prepare a 200′ minuteman hose load, connect the female coupling to the preconnect discharge. Lay the first two sections in the hose bed using a flat load. After the first loop of hose, make short pulling handles at the end of the hose bed. Leave the male end of the second section of hose to the side of the hose compartment. Attach the remaining two sections of hose together, and place the nozzle on the male end of the last section of hose. Place these sections of hose in the hose bed, starting with the nozzle end. Leave long pulling loops in the first loops attached to the nozzle. When you reach the female end of the last section of hose, attach it to the male end of the second section of hose placed into the hose bed. To prepare the minuteman load, follow the steps in ▲ Skill Drill 16-23.

Fire Fighter Safety Tips

When loading hose on an apparatus, you should always use caution in climbing up and down on the apparatus. If you are loading hose at a fire scene, watch out for wet, slippery surfaces, ice, or other hazards. Also, wet hose can be heavy and you may need to reach, stretch, or lift the hose to get it into the hose bed. Use caution!

16-23 Skill Drill

Load the remaining hose flat into the bed, alternating the folds from the left to right sides of the bed.

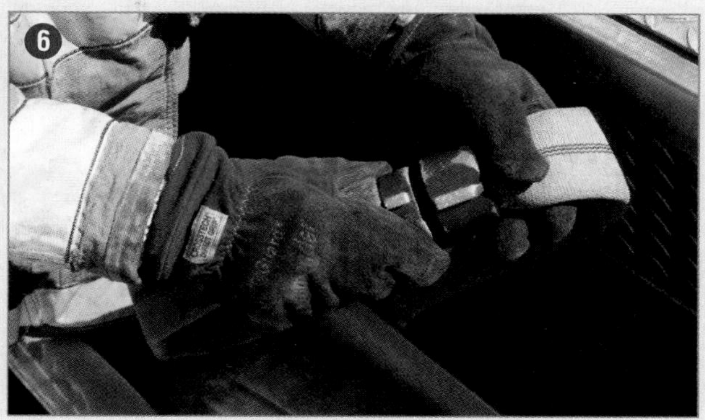

Connect the last section loaded to the first section placed in the bed.

Lay the remaining loose hose on top of the load.

1. Connect the female end of the first length of hose to the discharge outlet. **(Step 1)**
2. Lay the hose flat to the edge of the hose bed and leave the remaining hose with the male coupling out of the front of the bed to be connected later. Do not connect to the rest of the load. **(Step 2)**
3. Assemble the remaining hose sections and attach the nozzle. **(Step 3)**
4. Place the nozzle on the first length already in the hose bed. **(Step 4)**
5. Load the remaining hose flat into the bed, alternating the folds from the left to right sides of the bed. **(Step 5)**
6. Connect the last section loaded to the first section placed in the bed. **(Step 6)**
7. Lay the remaining loose hose on top the load. **(Step 7)**

To advance the minuteman hose load, follow the steps in ▶ **Skill Drill 16-24**.

1. Grasp the nozzle and the folds next to it. **(Step 1)**
2. Pull the load approximately one-third out of the bed. **(Step 2)**
3. Turn away from the hose bed and place the load on the shoulder. Walk away from the apparatus until all hose is clear of the hose bed. **(Step 3)**
4. Continue walking away, pulling the remaining hose from the hose bed and then allow the hose to deploy from the top of the load on the shoulder. **(Step 4)**

16-24 Skill Drill

Advancing the Minuteman Hose Load

Grasp the nozzle and the folds next to it.

Pull the load approximately one-third out of the bed.

Turn away from the hose bed and place the load on the shoulder. Walk away from the apparatus until all hose is clear from the hose bed.

Continue walking away, pulling the remaining hose from the hose bed and then allow the hose to deploy from the top of the load on the shoulder.

The Preconnected Flat Load

To prepare the preconnected flat load, attach the female end of the hose to the preconnect discharge. Begin placing the hose flat in the hose bed. When about one-third of the hose is in the bed, make an 8″ loop at the end of the hose bed. This loop will be used as a pulling handle. When two-thirds of the hose is loaded, make a second pulling loop about twice the size of the first loop. Finish loading the hose, attach the nozzle, and place it on top of the hose bed. The preconnected flat load is now ready for use. To prepare the preconnected flat hose load, follow the steps in ► Skill Drill 16-25.

1. Attach the female end of the hose to the preconnect discharge. **(Step 1)**
2. Begin laying the hose flat in the hose bed. **(Step 2)**
3. When about one-third of the hose is in the bed, make an 8″ loop at the end of the hose bed. This loop will be used as a pulling handle.
4. When two-thirds of the hose is loaded, make a second pulling loop about twice the size of the first loop.
5. Finish loading the hose, attach the nozzle, and place it on top of the hose bed. The preconnected flat load is now ready for use. **(Step 3)**

To advance the preconnected flat hose load, follow the steps in ► Skill Drill 16-26.

1. Place arm through the larger lower loop.
2. Grasp the smaller loop with the same hand. **(Step 1)**
3. Grasp the nozzle with the opposite hand. **(Step 2)**

16-25 Skill Drill

Loading the Preconnected Flat Load

Attach the female end of the hose to the preconnect discharge.

Begin laying the hose flat in the hose bed.

When about one-third of the hose is in the bed, make an 8" loop at the end of the hose bed. This loop will be used as a pulling handle. When two-thirds of the hose is loaded, make a second pulling loop that is about twice the size of the first loop. Finish loading the hose, attach the nozzle, and place it on top of the hose bed. The preconnected flat load is now ready for use.

4. Pull the load from the bed. **(Step 3)**
5. Walk away from the vehicle. **(Step 4)**
6. As the load deploys, drop the small loop.
7. Extend the remaining hose to length. **(Step 5)**

The Triple Layer Load

To prepare the triple layer load, attach the female end of the hose to the preconnect discharge. Connect the sections of hose together. Extend the hose directly from the hose bed. Pick up the hose two-thirds of the distance from the discharge to the hose nozzle. Carry the hose back to the apparatus, forming a three-layer loop. Pick up the entire length of folded hose (this will take several people.) Lay the tripled folded hose in the hose bed in an S-shape with the nozzle on top. To prepare the triple layer hose load, follow the steps in ► Skill Drill 16-27.

1. To make the triple layer load, attach the female end of the hose to the preconnect discharge. **(Step 1)**
2. Connect the sections of hose together. **(Step 2)**
3. Extend the hose directly from the hose bed. Pick up the hose two-thirds of the distance from the discharge to the hose nozzle. **(Step 3)**
4. Carry the hose back to the apparatus, forming a three-layer loop. **(Step 4)**
5. Pick up the entire length of folded hose. (This will take several people.) **(Step 5)**
6. Lay the tripled folded hose in the hose bed in an S-shape with the nozzle on top. **(Step 6)**

16-26 Skill Drill

Advancing the Preconnected Flat Hose Load

Place the arm through the larger lower loop. Grasp the smaller loop with the same hand.

Grasp the nozzle with the opposite hand.

Pull the load from the bed.

Walk away from the vehicle.

As the load deploys, drop the small loop. Extend the remaining hose to length.

16-27 Skill Drill

Loading the Triple Layer Hose Load

1 Attach the female end of the hose to the preconnect discharge.

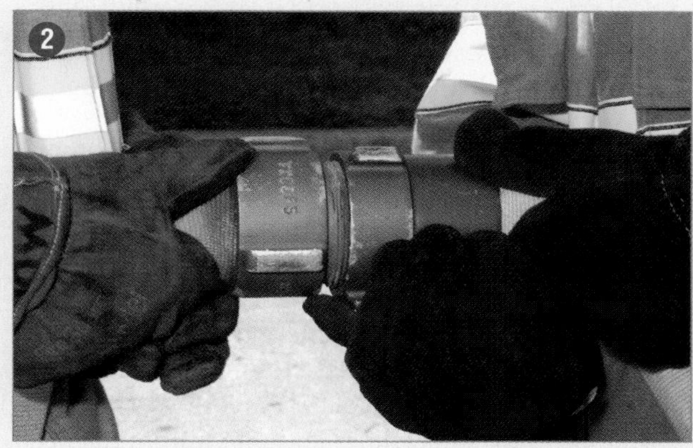
2 Connect the sections of hose together.

3 Extend the hose directly from the hose bed. Pick up the hose two-thirds of the distance from the discharge to the hose nozzle.

4 Carry the hose back to the apparatus, forming a three-layer loop.

5 Pick up the entire length of folded hose.

6 Lay the tripled folded hose in the hose bed in an S-shape with the nozzle on top.

To advance the triple layer hose load, follow the steps in ► Skill Drill 16-28.

1. Grasp the nozzle and the top fold. **(Step 1)**
2. Turn away from the hose bed and place the hose on the shoulder. **(Step 2)**
3. Walk away from vehicle until the entire load is out of the bed. **(Step 3)**
4. When the load is out of the bed, drop the fold. **(Step 4)**
5. Extend the nozzle the remaining distance. **(Step 5)**

Wyed Lines

In order to reach a fire that may be some distance from the engine, it may be necessary to first advance a larger diameter line, such as 2½″ hose line, and then split it into two 1¾″ attack lines. This is accomplished by attaching a gated wye or a water thief to the end of the 2½″ line and then attaching the two attack lines to the gated outlets. To unload and advance the wyed lines, follow the steps of ► Skill Drill 16-29.

1. Grasp one of the attack lines and pull it from the bed. **(Step 1)**
2. Pull the second attack line from the bed and place it far enough from the first line so that you can walk between the hose lines. This will keep the lines from becoming entangled. **(Step 2)**
3. Grasp the gated wye and pull it from the bed.
4. The apparatus will deploy the remaining hose from the wye back to a water source. Place the gated wye so that one attack line is on the other side. **(Step 3)**
5. The individual attack lines can now be extended to desired positions. **(Step 4)**

Hose Carries and Advances

Several different techniques are used to carry and advance fire hose. The best technique for a particular situation will depend on the size of the hose, the distance it must be moved, and the number of fire fighters available to perform the task. The same techniques can be used for supply lines or attack lines.

Whenever possible, a hose line should be laid out and positioned as close as possible to the location where it will be operated before it is charged with water. A charged line is much heavier and more difficult to maneuver than a dry hose line. A suitable amount of extra hose should be available to allow for maneuvering and advancement after the line is charged.

Working Hose Drag

The working hose drag technique is used to deploy hose from a hose bed and advance the line a relatively short distance to the desired location. Depending on the size and length of the hose, several fire fighters may be required to perform this task. To perform a working hose drag, follow the steps of ► Skill Drill 16-30.

1. Place the end of the hose over your shoulder. **(Step 1)**
2. Hold onto the coupling with your hand. **(Step 2)**
3. Walk in the direction you want to advance the hose. **(Step 3)**
4. As the next hose coupling is ready to come off the hose bed, have a second fire fighter grasp the coupling and place the hose over the shoulder. **(Step 4)**
5. Continue this process until you have enough hose out of the hose bed. **(Step 5)**

Shoulder Carry

The shoulder carry is used to transport full lengths of hose over a longer distance than it is practical to drag the hose. The shoulder carry is also useful when a hose line has to be advanced around obstructions. For example, this technique could be used to stretch an attack line from the front of a building, around to the rear entrance, and up to the second floor. This technique could also be employed to stretch an additional supply line to an attack engine in a location where the hose cannot be laid out by another engine.

This technique requires practice and good teamwork in order to be successful. By working together to complete tasks efficiently, we achieve our goal of extinguishing the fire in the shortest period of time. To perform a shoulder carry, follow the steps in ► Skill Drill 16-31.

1. Stand at the tailboard of the engine. Grasp the end of the hose and place it over your shoulder so the coupling is at chest height. **(Step 1)**
2. Have a second fire fighter place additional hose on your shoulder so that the ends of the folds reach to about knee level. Continue to place enough folds on your shoulder that you can safely carry. **(Step 2)**
3. Hold the hose to prevent it from falling off your shoulder. Continue to hold the hose and move forward about 15′.
4. A second fire fighter should stand at the tailboard to receive a load of hose. **(Step 3)**
5. When enough fire fighters have received hose loads, the hose can be uncoupled from the hose bed. **(Step 4)**
6. To move the hose, all fire fighters must coordinate their movements. The driver/operator should connect this coupling to the engine.
7. All of the fire fighters should start walking towards the fire. The last fire fighter should not start off-loading hose from his or her shoulder until it has all been laid out. The next fire fighter in line then starts laying out hose from his or her shoulder.
8. Each fire fighter lays out their supply of hose until the entire length is laid out. **(Step 5)**

To advance an accordion load using a shoulder carry, follow the steps in ► Skill Drill 16-32.

1. Find the end of the accordion load, whether it is a nozzle or coupling.
2. Using two hands, grasp the end of the load and the number of folds it will take to make an adequate shoulder load. **(Step 1)**
3. Pull the accordion load about one-third of the way off the apparatus. **(Step 2)**

16-28 Skill Drill

Advancing the Triple Layer Hose Load

Grasp the nozzle and the top fold.

Turn away from the hose bed and place the hose on the shoulder.

Walk away from the vehicle until the entire load is out of the bed.

When the load is out of the bed, drop the fold.

Extend the nozzle the remaining distance.

16-29 Skill Drill

Unload and Advance the Wyed Lines

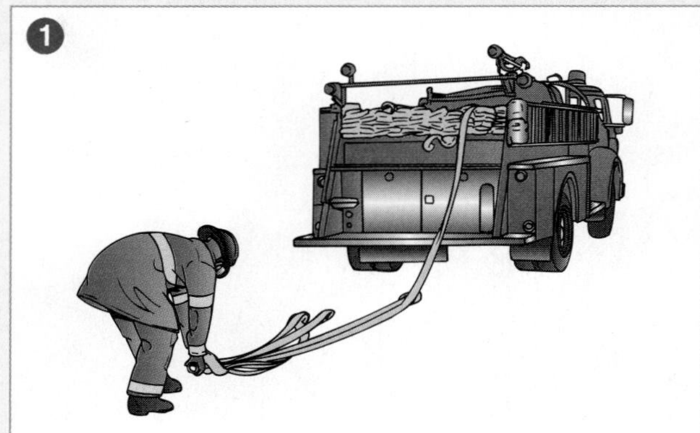

1 Grasp one of the attack lines and pull it from the bed.

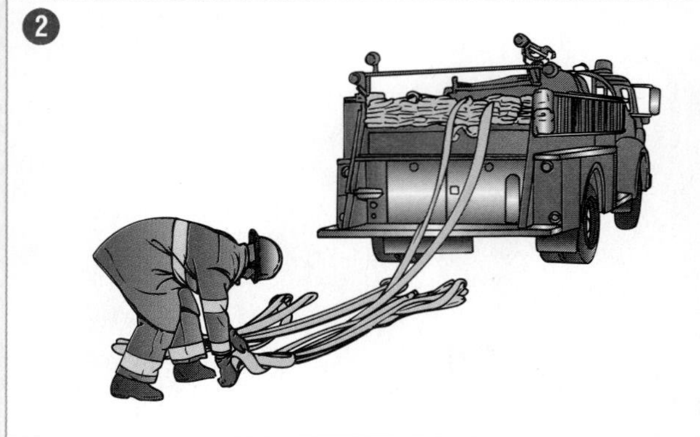

2 Pull the second attack line from the bed and place it far enough from the first line so that you can walk between the hose lines. This will keep the lines from becoming entangled.

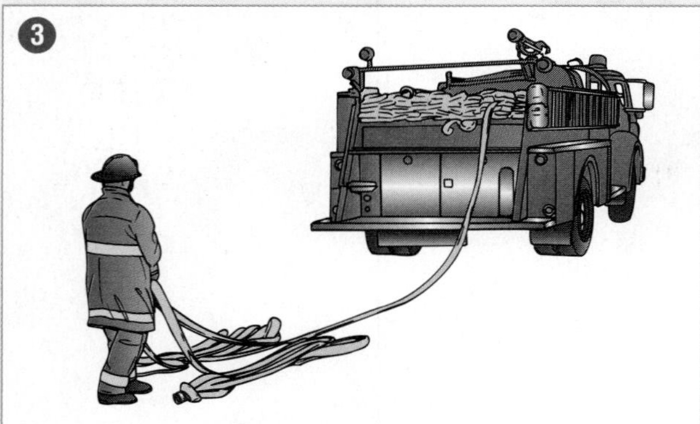

3 Grasp the gated wye and pull from the bed. The apparatus deploys the remaining hose from the wye back to a water source. Place the gated wye so that one attack line is on the other side.

4 The individual attack lines can now be extended.

16-30 Skill Drill

Performing a Working Hose Drag

1 Place the end of the hose over your shoulder.

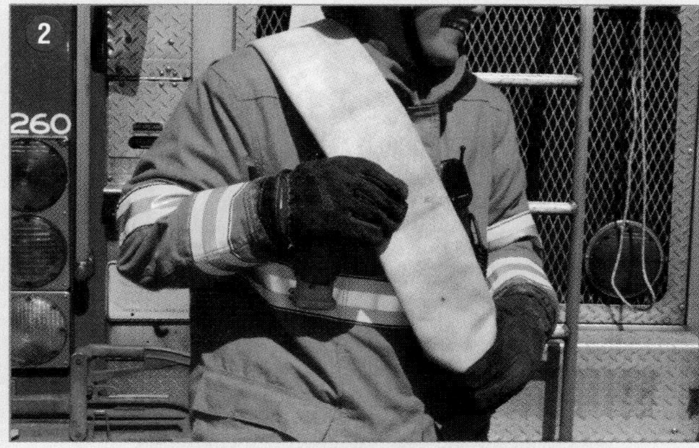

2 Hold onto the coupling with your hand.

3 Walk in the direction you want to advance the hose.

4 As the next hose coupling is ready to come off the hose bed, a second fire fighter grasps the coupling and places the hose over the shoulder.

5 Continue this process until you have enough hose out of the hose bed.

16-31 Skill Drill

Performing a Shoulder Carry

1 Stand at the tailboard of the engine. Grasp the end of the hose and place it over your shoulder so the coupling is at chest height.

2 Have a second fire fighter place additional hose on your shoulder so that the ends of the folds reach to about knee level. Continue to place enough folds on your shoulder that you can safely carry.

3 Hold the hose to prevent it from falling off your shoulder. Continue to hold the hose and move forward about 15′. A second fire fighter should stand at the tailboard to receive a load of hose.

4 When enough fire fighters have received hose loads, the hose can be uncoupled from the hose bed.

5 All of the fire fighters start walking towards the fire. The last fire fighter should not start off-loading hose from his or her shoulder until it is all laid out. The next fire fighter then starts laying out hose from his or her shoulder. Each fire fighter lays out their supply of hose until the entire length is laid out.

16-32 Skill Drill

Advancing an Accordion Load

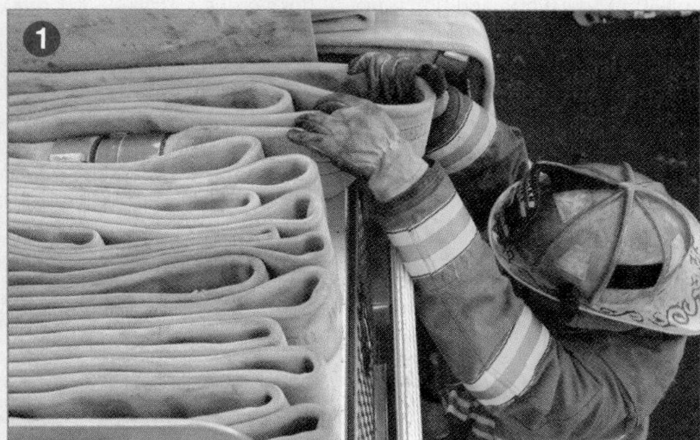

Using two hands, grasp the end of the load and the number of folds it will take to make an adequate shoulder load.

Pull the accordion load about one-third of the way off the apparatus.

The folds must then be twisted so they become flat with the end of the accordion load (nozzle or coupling) on the bottom of the now flat shoulder load.

Transfer the hose to the opposite shoulder while turning so you face in the direction you will walk.

Place the shoulder load over your shoulder and grasp tightly with both hands.

Walk away from the apparatus, pulling the shoulder load out of the hose bed.

4. The folds must then be twisted so they become flat with the end of the accordion load (nozzle or coupling) on the bottom of the now flat shoulder load. **(Step 3)**
5. Transfer the hose to the opposite shoulder while turning so you face in the direction you will walk. **(Step 4)**
6. Place the shoulder load over your shoulder and grasp tightly with both hands. **(Step 5)**
7. Walk away from the apparatus, pulling the shoulder load out of the hose bed. **(Step 6)**
8. If additional fire fighters are needed, they may follow steps 1-7 to assist in removing the required amount of hose.

Advancing an Attack Line

In order to attack an interior fire, an attack line is usually advanced in two stages. The first stage involves laying out the hose to the building entrance. The second stage is to advance the line into the building to the location where it will be operated.

When the attack line has been laid out to the entry point, the extra hose that will be advanced into the building should be flaked out in a serpentine pattern so that it will not become tangled when it is charged. The hose should be flaked out with the lengths of hose running parallel to the front of the fire building so that it can be easily advanced into the building (▼ **Figure 16-25**). The flaked hose should be set back from the doorway so that it does not obstruct the entry and exit path.

You should make sure you flake out the hose BEFORE it is charged with water. Once the line is charged, the hose becomes much more difficult to maneuver and advance. It is also important to be sure that there is enough hose to reach the location where it will be needed inside the building.

While you are flaking out the hose and preparing to enter the building, your officer will be completing the size-up. Other members of your team should be carrying out other tasks to support the operation. They may be forcing entry, getting into position for ventilation, establishing a water supply, and performing search and rescue. All of these tasks must be performed in sequence to maintain a safe environment and to efficiently extinguish the fire.

Figure 16-25 Hose should be flaked out in a serpentine pattern outside the building entrance.

Fire Fighter Tips

To be prepared to advance the hose, form an upright loop about 4' in diameter outside the door to the fire building. If more hose is needed, one person can roll this loop toward the nozzle. This will yield several additional feet of hose at the nozzle.

Once the hose is flaked out, signal the driver/operator to charge the line. Open the nozzle slowly to bleed out any trapped air and to make sure the hose is operating properly. If you are using an adjustable nozzle, make sure the nozzle is set to deliver the appropriate stream. This will usually be a straight stream pattern for an interior attack. Once this is done, slowly close the nozzle.

This is the time to don your SCBA mask and protective hood and get ready for action. Place your helmet on your head and secure the chinstrap. Quickly recheck all parts of your PPE. Make sure your coat is fastened and your collar is turned up and fastened in front. Check your gloves and be sure you have a hand light. Check your partner's equipment and have your partner check your equipment. If you have time, catch your breath while you are breathing ambient air. Be ready to start breathing air from your SCBA and to advance the charged hose line as soon as your officer directs you.

When you are given the command to advance the hose, keep safety as your number one priority. Make sure the other members of the nozzle team are ready. Do not stand in front of the door as it is opened. You do not know what may happen when the door is opened.

As you move inside, stay low to avoid the greatest amount of heat and smoke. If you cannot see because of the dense smoke, use your hands to feel the pathway in front of you. Feel in front of you so you do not fall into a hole or opening. Look for the glow of fire, and check for the sensation of heat coming through your face piece. Communicate with the other members of the nozzle team as you advance.

As you advance the hose line, you need to have enough hose to enable you to move forward. A good hose line crew consists of at least two members at the nozzle and a third member outside the door. As resistance is encountered in advancing the line, the second fire fighter at the nozzle can help to pull more hose, while the fire fighter at the door is responsible for feeding more hose into the building. If necessary, the second fire fighter can retrace the hose back to relieve an obstruction that prevents the hose from advancing. Charged hose lines are not easy to advance through a house or other building. It is only with good teamwork that efficient hose line advancement can occur.

16-33 Skill Drill

Advancing a Hose Line Up a Stairway

1 Use a shoulder carry to advance up the stairs.

2 When ascending the stairway, lay the hose against the outside of the stairs to reduce tripping hazards. Avoid sharp bends.

3 Arrange excess hose so that it is available to fire fighters entering the fire floor.

Advancing a Hose Line up a Stairway

When advancing a hose line up stairs, arrange an adequate amount of extra hose close to the bottom of the stairs. Make sure all members of the team are ready to move on command. It is hard to move a charged hose line up a set of stairs while flowing water through the nozzle. Shutting down the hose line while moving up the stairs will often allow you to get to the top of the stairs more quickly and safely. Follow the directions of your officer. It is easiest to position an uncharged hose line. To advance an uncharged hose line up a stairway, follow the steps in **▲ Skill Drill 16-33**.

1. Use a shoulder carry to advance up the stairs. **(Step 1)**
2. When ascending the stairway, lay the hose against the outside of the stairs to avoid sharp bends and kinks and to reduce tripping hazards. **(Step 2)**
3. Arrange excess hose so that it is available to fire fighters entering the fire floor. **(Step 3)**

Advancing a Hose Line Down a Stairway

Advancing a charged hose line down a stairway is also difficult. The smoke and flames from the fire tend to travel up the stairway. You want to get down the stairway and position yourself below the heat and smoke as quickly as possible. Keep as low as possible to avoid the worst of the heat and smoke. The one thing you have going for you as you advance a hose line down a stairway is that gravity is working with you to bring the hose line down the stairs. You should never advance towards a fire unless your hose line is charged and ready to flow water.

Wearing PPE and SCBA changes your center of gravity. If you try to crawl down a stairway headfirst, you are likely to find yourself tumbling head over heels. Move down the stairway feet first, using your feet to feel for the next step. Move carefully, but as quickly as possible to get below the worst of the heat and smoke. To advance a hose line down a stairway, follow the steps in (Skill Drill 16-34).

1. Advance a charged hose line.
2. Descend stairs feet first.
3. Position fire fighters at areas where hose lines could snag.

Advancing a Hose Line up a Ladder

If a hose line has to be advanced up a ladder, this should be done before the line is charged. To do this, place the hose on the left side of the ladder. Place the hose across your chest with the nozzle draped over your right shoulder. Climb up the ladder with the uncharged hose line in this position. If the line is mistakenly charged while you are climbing, it will not push you away from the ladder.

Additional fire fighters should pick up the hose about every 25′ and help to advance it up the ladder. The nozzle is passed over the top rung of the ladder and into the fire building. Additional hose should be fed up the ladder until sufficient hose is inside the building to reach the fire. The hose should be secured to the ladder with a hose strap to keep it from becoming dislodged. To advance a hose line up a ladder, follow the steps in (► Skill Drill 16-35).

1. If a hose line needs to be advanced up a ladder, it should be advanced before it is charged.
2. Advance the hose line to the ladder. **(Step 1)**
3. Pick up the nozzle; place the hose across your chest with the nozzle draped over your shoulder. **(Step 2)**
4. Climb up the ladder with the uncharged hose line. **(Step 3)**
5. Once the first fire fighter reaches the first fly section of the ladder, a second fire fighter should shoulder the hose to assist advancing the hose line up the ladder. To avoid overloading of the ladder, limit one fire fighter per fly section. **(Step 4)**
6. The nozzle is placed over the top rung of the ladder and advanced into the fire area. **(Step 5)**
7. Additional hose can be fed up the ladder until sufficient hose is in position.
8. The hose can be secured to the ladder with a hose strap to support its weight and keep it from becoming dislodged. **(Step 6)**

Operating a Hose Stream From a Ladder

A hose stream can be operated from a ladder and directed into a building through a window or other opening. To operate a fire hose from a ladder, follow the steps in (Skill Drill 16-36).

1. Climb the ladder with a hose line to the height at which the line will be operated.
2. Apply a leg lock or use a ladder belt.
3. Place the hose between two rungs and secure the hose to the ladder with a rope hose tool, rope, or piece of webbing.
4. Carefully operate the hose stream from the ladder. Be careful when opening and closing nozzles and redirecting the stream because of the nozzle back-pressure. This force could destabilize the ladder.

Extending an Attack Line

When choosing a preconnected hose line or assembling an attack line, it is better to have too much hose than a line that is too short. With a hose that is longer than necessary, you can flake out the excess hose. With a hose line that is too short, you cannot advance it to the seat of the fire without shutting it down and taking the time to extend it.

There may be circumstances where it is necessary to extend a hose line by adding additional lengths of hose. This could occur if the fire is further from the apparatus than initially estimated. More hose line could be needed to reach the burning area or to complete extinguishment.

There are two basic ways to extend a hose line. The first way is to disconnect the hose from the discharge gate on the attack engine and add the extra hose at that location. This requires advancing the full length of the attack line to take advantage of the extra hose, which could take time and considerable effort. The alternative is to add the hose to the discharge end of the hose. This can be done easily if the nozzle is a breakaway type that can be separated between the shut-off and the tip. With this type of nozzle, the valve is shut down, the nozzle tip is removed, and the extra hose is attached to the shut-off valve. The nozzle tip can then be installed on the male end of the added hose. This evolution allows you to lengthen the attack hose without shutting it down at the pump. A standpipe kit can be used to supply the added section of hose.

Connecting Hose Lines to Standpipe and Sprinkler Systems

Fire department connections on buildings are provided so that the fire department can pump water into a standpipe or sprinkler system. A hose line that is used to deliver water to a fire department connection is an attack line by definition, because it is connected to the discharge side of the attack engine. The function of the hose line is to provide either a primary or secondary water supply for the sprinkler or standpipe system. The same basic techniques are used to connect the hose lines to either type of system.

Standpipe systems are used to provide a water supply for attack lines that will be operated inside the building. Outlets are provided inside the building where fire fighters can connect attack lines. The fire fighters inside the building have to depend on fire fighters outside to supply the water to the fire department connection (► Figure 16-26).

There are two types of standpipe systems. A dry standpipe system depends on the fire department to provide all of the water. A wet standpipe system has a built-in water supply, but the fire department connection is provided to deliver

16-35 Skill Drill

Advancing an Uncharged Hose Line Up a Ladder

Advance the hose line to the ladder.

Pick up the nozzle; place the hose across your chest with the nozzle draped over your shoulder.

Climb up the ladder with the uncharged hose line.

Once the first fire fighter reaches the first fly section of the ladder, a second fire fighter should shoulder the hose to assist advancing the hose line up the ladder. To avoid overloading of the ladder, limit one fire fighter per fly section.

The nozzle is placed over the top rung of the ladder and advanced into the fire area.

Additional hose can be fed up the ladder until sufficient hose is in position. The hose can be secured to the ladder with a hose strap to support its weight and keep it from becoming dislodged.

Figure 16-26 A standpipe connection.

Fire Fighter Safety Tips

Some departments use large diameter hose with Storz-type couplings to connect to fire departments connections. Hose that is rated for use as attack hose should be used to connect to a standpipe system.

Large diameter supply line hose is rated for a safe working pressure of only 185 psi. Standpipe systems generally require at least 150 psi to be provided at the fire department connection and a water hammer can cause a spike of much higher pressure. The excess pressure could burst the hose and place fire fighters working inside the building in danger.

Many fire department SOPs require two hose lines to be connected to a fire department connection. If one line breaks or the flow is interrupted, water will still be delivered to the system through the other line.

a higher flow or to boost the pressure. The pressure requirements for standpipe systems depend on the height where the water will be used inside the building.

The fire department connection for a sprinkler system is also used to supplement the normal water supply. The required pressures and flows for different types of sprinkler systems can vary significantly. As a guideline, sprinkler systems should be fed at 150 psi unless there is more specific information available. To connect a hose line to supply a fire department connection, follow the steps in Skill Drill 16-37.

1. Locate the fire department connection to the standpipe or sprinkler system.
2. Extend a hose line from the engine discharge to the fire department connection using the size hose required by the fire department's SOPs. Some fire departments use a single hose line, while others call for two or more lines to be connected.
3. Remove the caps on the standpipe inlet. Some caps are threaded into the connections and have to be unscrewed. Other caps are designed to break away when struck with a tool such as a hydrant wrench or spanner.
4. Visually inspect the interior of the connection to ensure that there is no debris that could obstruct the water flow. Never stick your hand or fingers inside the connections; fire fighters have been injured from broken glass or needles inside these connections.
5. Attach the hose line(s) to the connection(s).
6. Notify the driver/operator when the connection has been completed.

Advancing an Attack Line from a Standpipe Outlet

The standpipe outlets inside a building are provided for fire fighters to connect attack hose lines. This eliminates the need to advance hose lines all the way from the attack engine on the street outside to an upper floor or a fire deep inside a large area building. In tall buildings, it would be impossible to advance hose lines up the stairways in a reasonable time and to supply sufficient pressure to fight a fire on an upper floor.

The standpipe outlets are often located in stairways and standard operating procedures generally require attack lines to be connected to an outlet one floor below the fire. The working space in the stairway and around the outlet valve is usually limited. Before opening the door, it is important to properly flake out the hose line so it will be ready to advance into the fire floor. Before charging the hose line, the hose should be flaked out on the stairs going up from the fire floor. When the hose line is charged and advanced into the fire floor, gravity will help to move the line forward. This is much easier than having to pull the charged hose line up the stairs.

To connect and advance an attack line from a standpipe outlet, follow the steps in Skill Drill 16-38.

1. Carry a standpipe hose bundle to the standpipe connection below the fire. Remove the cap from standpipe.
2. Attach the proper adaptor or appliance such as a gated wye.
3. Flake the hose up the stairs to the floor above the fire.
4. Extend the hose to the fire floor and prepare for your fire attack.

Replacing a Defective Section of Hose

With proper maintenance and testing, the risk of fire hose failure should be low, but it is always possible. You need to know what to do if a section of hose bursts or develops a major leak while it is being used. Every fire fighter should know how to quickly replace a length of defective hose and restore the flow.

A burst hose line should be shut down as soon as possible. If the line cannot be shut down at the pump or at a control valve, a hose clamp can be used to stop the flow in an undamaged section of hose upstream from the problem. After the water flow has been shut off, quickly remove the damaged section of hose and replace it with two sections of hose.

16-40 Skill Drill

Drain Hose and Carry

Lay the section of hose straight on a flat surface.

Start at one end of the section and lift the hose to shoulder level.

Move down the length of hose, flaking it back and forth over your shoulder.

Continue down the length until the entire section is on your shoulder.

Using two sections of hose will ensure that the replacement hose is long enough to replace the damaged section. To replace a hose section, follow the steps in Skill Drill 16-39.

1. Shut down or clamp off damaged line.
2. Remove damaged section of hose.
3. Replace with two sections to insure length will be adequate.
4. Restore water flow.

Draining and Picking Up Hose

In order to put the hose back into service, the hose must be drained of water. That is accomplished by laying the hose straight on a flat surface. Then lift one end of the hose to shoulder level. Gravity will allow the water to flow to the lower portion of the hose and eventually out of the hose. As you proceed down the length of hose, fold the hose back and forth over your shoulder. When you reach the end of the section, you will have the whole section of hose on your shoulder. To drain the hose, follow the steps in ▲ Skill Drill 16-40.

1. Lay the section of hose straight on a flat surface. **(Step 1)**
2. Start at one end of the section and lift the hose to shoulder level. **(Step 2)**
3. Moving down the length of hose, fold it back and forth over your shoulder. **(Step 3)**
4. Continue down the length until the entire section is on your shoulder. **(Step 4)**

Unloading Hose

There are times other than a fire where you will need to unload the hose from an engine. Hose should be unloaded and reloaded on a regular basis to place the bends in different portions of the hose. Leaving bends in the same locations for long periods of time is likely to cause weakened areas. Hose might have to be unloaded to change out apparatus. Sometimes all of the equipment from one engine must be transferred to another vehicle. It could also be necessary to offload the hose for annual testing to be conducted.

The following procedure should be used to unload a hose bed:

- A large area such as a parking lot should be used for this procedure.
- Disconnect any gate valves or nozzles from the hose before you begin.
- Grasp the end of the hose, and pull it off the engine in a straight line.
- When a coupling comes off the engine, disconnect the hose and pull off the next section of hose.
- When all of the hose has been removed from the hose bed, use a broom to brush off any dirt or debris on both sides of the hose jacket.
- Sweep out any debris or dirt from the hose bed.
- Roll all of the hose into doughnut rolls. Place the male end out for supply lines, and female end out for hand lines.
- Store hose rolls off the floor on a rack, in a cool dry area.

Nozzles

Nozzles are attached to the discharge end of attack lines to give fire streams shape and direction. Without a nozzle, the water discharged from the end of a hose would only reach a short distance. Nozzles are used on all sizes of handlines as well as on master stream devices. Nozzles can be classified into three groups. **Low volume nozzles** flow 40 gpm or less. These are primarily used for booster hoses and their use is limited to small outside fires. **Handline nozzles** are used on hose lines ranging from $1^1/_2''$ to $2^1/_2''$ in diameter. Handline steams usually flow between 60 and 350 gpm. **Master stream nozzles** are used on deck guns, portable monitors, and ladder pipes that flow more than 350 gpm.

Low volume and handline nozzles incorporate a shut-off valve that is used to control the flow of water. The control valve for a master stream is usually separate from the nozzle itself. All nozzles have some type of device or mechanism to direct the water stream into a certain shape. Some nozzles also incorporate a mechanism to automatically adjust the flow based on the water volume and pressure that are available.

Fire Fighter Safety Tips

Always open and close nozzles slowly to prevent water hammer.

Fire Fighter Tips

Some nozzles are made so that the tip of the nozzle can be separated from the shut-off valve. This is a **breakaway nozzle**. This allows you to shut-off the flow, unscrew the nozzle tip, and then add additional lengths of hose to extend the hose line without shutting off the valve at the engine.

Nozzle Shut Offs

The **nozzle shut off** enables a person at the nozzle to start or stop the flow of water. The most common nozzle shut off mechanism is a quarter-turn valve. The handle that controls this valve is called a bale. Some nozzles incorporate a rotary control valve operated by rotating the nozzle in one direction to open and the opposite direction to shut off the flow of water.

Two different types of nozzles are manufactured for the fire service. These are the **smooth bore nozzles** and **fog stream nozzles**. Smooth bore nozzles produce a solid column of water. Fog stream nozzles separate the water into droplets. The size of the water droplets and the discharge pattern can be varied by adjusting the nozzle setting. Nozzles must have an adequate volume of water and an adequate pressure in order to produce a good fire stream. The volume and pressure requirements vary according to the type and size of nozzle.

Smooth Bore Nozzles

The simplest smooth bore nozzle consists of a shut off valve and a **smooth bore tip** that gradually decreases the diameter of the stream to a size smaller than the hose diameter (► Figure 16-27). Smooth bore nozzles are manufactured to fit both handlines and master stream devices. Smooth bore nozzles that are used for master stream and ladder pipes often consist of a set of stacked tips. Each successive tip in the stack has a smaller diameter opening. Tips can be quickly added or removed to provide the desired stream size. This allows different sizes of streams to be produced under different conditions.

There are several advantages of using a smooth bore nozzle. A good smooth bore has a longer reach than a combination fog nozzle operating at a straight stream setting. A smooth bore is capable of deeper penetration into burning materials, resulting in quicker knock-down and extinguishment. Smooth bore nozzles also operate at lower pressures than adjustable stream nozzles. Most smooth bore nozzles are designed to operate at 50 psi, while adjustable stream nozzles generally require 75 to 100 psi. Lower nozzle pressure makes it easier for a fire fighter to handle the nozzle.

Figure 16-27 Smooth bore nozzle.

Figure 16-28 Fog stream nozzle.

A straight stream extinguishes a fire with less air movement and less disturbance of the thermal layering than a fog stream. This makes the heat conditions less intense for fire fighters during an interior attack. It is also easier for the operator to see the pathway of a solid stream than a fog stream.

There are also disadvantages with smooth bore nozzles. Smooth bore nozzles do not absorb heat as readily as fog streams and are not as effective for hydraulic ventilation. You cannot change the setting of a smooth bore nozzle to produce a fog pattern; however, a fog nozzle can be set to produce a straight stream. To operate a smooth bore nozzle, follow the steps in (► Skill Drill 16-41).

1. Select the desired tip size and attach to nozzle shut-off valve. **(Step 1)**
2. Obtain a stable stance (if standing). **(Step 2)**
3. Slowly open the valve, allowing water to flow. **(Step 3)**
4. Open the valve completely to achieve maximum effectiveness. (Failure to fully open the valve will result in reduced flow depriving you of the necessary gallons per minute.) **(Step 4)**
5. Direct the stream to the desired location. **(Step 5)**

Fog Stream Nozzles

Fog stream nozzles produce fine droplets of water (► Figure 16-28). The advantage of creating these droplets of water is that they absorb heat much more quickly and efficiently than a solid column of water. This is an important characteristic when immediate reduction of room temperature is needed to avoid a flashover. Fog nozzles can produce a variety of stream patterns from a straight stream to narrow a fog cone of less than 45° to a wide-angle fog pattern that is close to 90°.

The straight streams produced by fog nozzles have openings in the center. Therefore, a fog nozzle can produce a straight stream, but not a solid stream. The straight stream from a fog stream nozzle will break up faster and will not have the reach of a solid stream. A straight stream will be affected more by wind than a solid stream.

There are several advantages of using fog stream nozzles. First, fog stream nozzles can be used to produce a variety of stream patterns by rotating the tip of the nozzle. Fog streams are effective at absorbing heat and can be used to create a water curtain to protect fire fighters from extreme heat.

Fog nozzles move large volumes of air along with the water. This can be an advantage or a disadvantage, depending on the situation. A fog stream can be used to exhaust smoke and gases through hydraulic ventilation. This air movement can also result in sudden heat inversion in a room that pushes hot steam and gases down onto the fire fighters. If used incorrectly, a fog pattern can push the fire into unaffected areas of a building.

In order to produce an effective stream, nozzles must be operated at the pressure recommended by the manufacturer.

16-41 Skill Drill

Operating a Smooth Bore Nozzle

1. Select the desired tip size and attach to nozzle shut-off valve.

2. Obtain a stable stance (if standing).

3. Slowly open the valve, allowing water to flow.

4. Open the valve completely to achieve maximum effectiveness.

5. Direct the stream to the desired location.

For many years, the standard operating pressure for fog stream nozzles was 100 psi. In recent years, some manufacturers have produced low-pressure nozzles that are designed to operate at 50 psi or 75 psi. The advantage of low-pressure nozzles is that they produce less reaction force, which makes them easier to control and advance. Lower nozzle pressure also decreases the risk that the nozzle will get out of control. To operate a fog nozzle, follow the steps in ► **Skill Drill 16-42**.

1. Select the desired nozzle.
2. Obtain a stable stance (if standing). **(Step 1)**
3. Slowly open the valve allowing water flow. **(Step 2)**
4. Open the valve completely. (Failure to open the valve fully will restrict water flow reducing the necessary gpm). **(Step 3)**
5. Select the desired water pattern by rotating the bezel of the nozzle.
6. Apply water where needed. **(Step 4)**

Types of Fog Stream Nozzles

There are three types of fog stream nozzles. The difference between the types is in the water delivery capability. A fixed gallonage fog nozzle will deliver a preset flow in gpm at the rated discharge pressure. The nozzle could be designed to flow 30 gpm, 60 gpm, or 100 gpm.

An adjustable gallonage fog nozzle allows the operator to select a desired flow from several settings. This is done by rotating a selector bezel to adjust the size of the opening. For example, a nozzle could have the options of flowing 60 gpm, 95 gpm, or 125 gpm. Once the setting is chosen the nozzle will only deliver the rated flow as long as the rated pressure is provided at the nozzle.

An automatic adjusting fog nozzle can deliver a wide range of flows. The nozzle has an internal spring-loaded piston. As the pressure at the nozzle increases or decreases, its piston moves in or out to adjust the size of the opening. The amount of water flowing through the nozzle is adjusted to maintain the rated pressure and produce a good stream. A typical automatic nozzle could have an operating range of 90 to 225 gpm while maintaining 100 psi discharge pressure.

Other Types of Nozzles

There are other types of nozzles that are used for special purposes. Piercing nozzles are used to make a hole in automobile sheet metal, aircraft, or building walls, in order to extinguish fires behind these surfaces ► **Figure 16-29**.

Cellar nozzles and Bresnan distributor nozzles are used to fight fires in cellars and other inaccessible places ► **Figure 16-30**. These nozzles discharge water in a wide circular pattern as the nozzle is lowered vertically through a hole into the cellar. They work like a large sprinkler head.

Water curtain nozzles are used to deliver a flat screen of water to form protective sheet of water on the surface of an exposed building ► **Figure 16-31**. The water curtains must be directed onto the exposed building because radiant heat

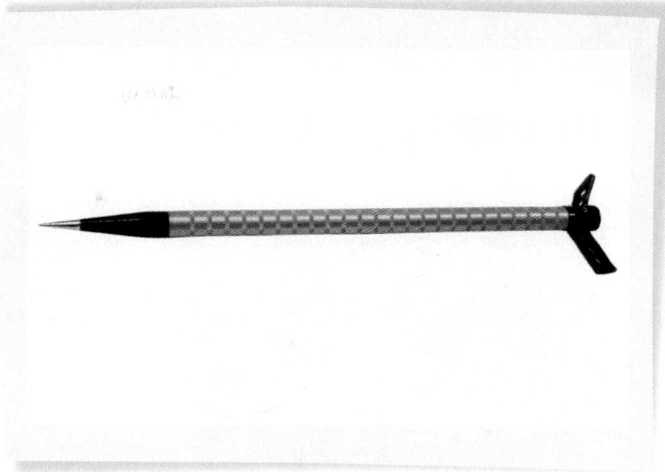

Figure 16-29 Piercing nozzle.

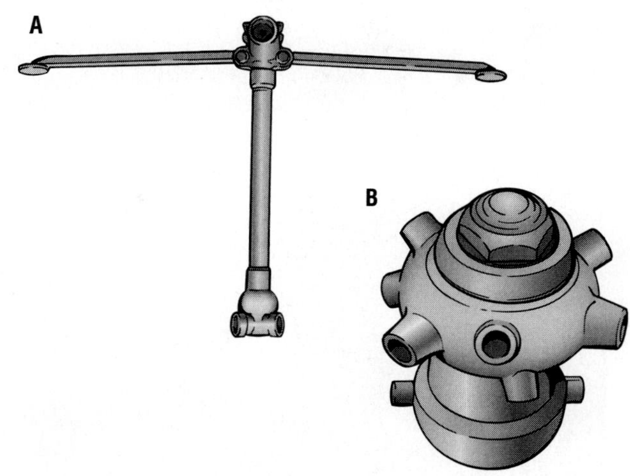

Figure 16-30 Cellar nozzles (A) and Bresnan distributor nozzle (B).

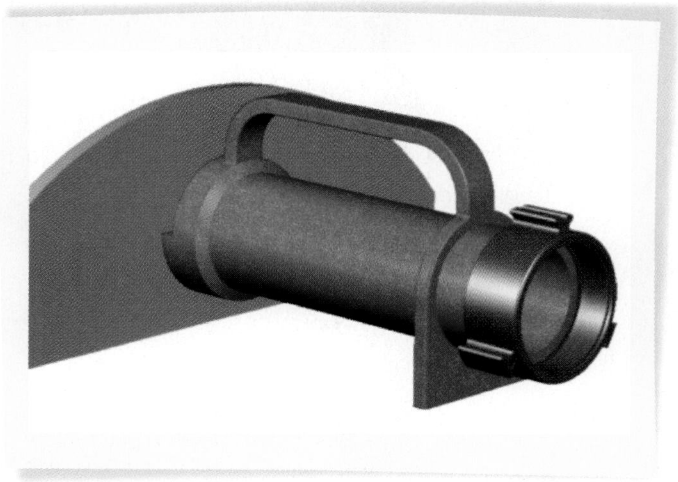

Figure 16-31 One type of a water curtain nozzle.

Skill Drill

Operating a Fog Nozzle

Obtain a stable stance (if standing).

Slowly open the valve allowing water flow.

Open the valve completely.

Select the desired water pattern, by rotating the bezel of the nozzle. Apply water where needed.

can pass through the water curtain. If your fire department has other types of specialty nozzles, you need to become proficient in the use and operation of them.

Nozzle Maintenance and Inspection

Nozzles should be inspected on a regular basis, along with all of the equipment on a fire department vehicle. A nozzle should be checked after each use before being placed back on the apparatus. They should be kept clean and clear of debris. Debris inside the nozzle will affect the performance of the nozzle, possibly reducing flow. Dirt and grit can interfere with the valve operation and prevent opening and closing fully. A light grease on the valve ball will keep it operating smoothly.

On fog nozzles, inspect the fingers on the face of the nozzle. Make sure all fingers are present and the finger ring can spin freely. Any missing fingers or failure of the ring to spin will drastically affect the fog pattern. Any problems noted should be referred to a competent technician for repair.

Foam

Firefighting foam can be used to fight several different types of fires and also to prevent the ignition of materials that could become involved in a fire. The use of foam is increasing as many new types of foam have become available and efficient systems for applying them have been developed. Foams also have been developed for use in neutralizing hazardous materials and decontamination. Many fire departments are using several different types of foam for a variety of situations.

Firefighting foam is produced by mixing **foam concentrate** with water and air to produce a solution that can be used as an effective extinguishing agent. There are several different types of foam used for fires involving different types of fuels. Each type of foam requires the appropriate type of concentrate, the proper equipment to mix the concentrate with water in the required proportions, and the proper application equipment and techniques. As a fire fighter, you need to become familiar with the specific types of foam used by your fire department and the proper techniques for using them. It is particularly important to learn where and when to use each type of foam that is available in your department.

Foam Classifications

The basic classifications of fire fighting foams are either Class A or Class B. There are many types of foam within each classification. **Class A foam** is used to fight fires involving ordinary combustible materials, such as wood, paper, and textiles. It is effective on organic materials such as hay and straw. Class A foam is particularly useful for protecting buildings in rural areas during forest and brush fires when the supply of water is limited.

Fire Marks

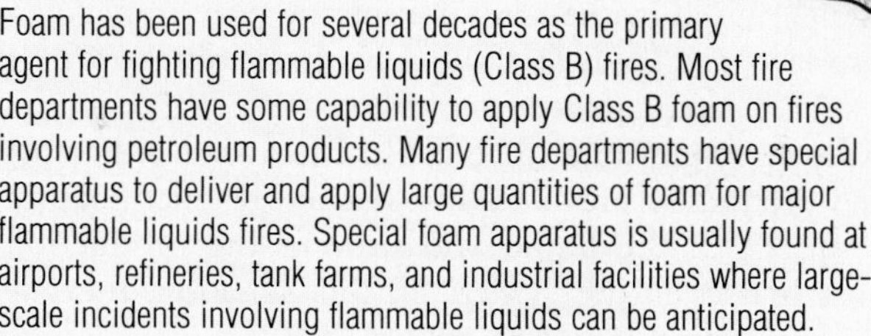

Foam has been used for several decades as the primary agent for fighting flammable liquids (Class B) fires. Most fire departments have some capability to apply Class B foam on fires involving petroleum products. Many fire departments have special apparatus to deliver and apply large quantities of foam for major flammable liquids fires. Special foam apparatus is usually found at airports, refineries, tank farms, and industrial facilities where large-scale incidents involving flammable liquids can be anticipated.

Class A foam increases the effectiveness of water as an extinguishing agent by reducing the surface tension of water. This allows the water to penetrate dense materials instead of running off the surface and allows more heat to be absorbed. The foam also keeps water in contact with unburned fuel to prevent ignition. Class A foam can be added to water streams and applied with several different types of nozzles.

Class B foam is used to fight Class B fires—flammable and combustible liquids. There are several different types of Class B foam that are formulated to be effective on different types of flammable liquids. Some liquids are incompatible with different foam formulations and will destroy the foam before the foam can control the fire.

In a flammable liquid fire, the liquid itself does not burn. Only the flammable vapors that are evaporating from the surface of the liquid and mixing with air can burn. Depending upon the temperature and the physical properties of the liquid, the amount of vapors that are being released from the surface of a liquid will vary. For example, gasoline produces flammable vapors down to a temperature of minus 45°F. Some vapors are lighter than air and rise up into the atmosphere. Other vapors are heavier than air and flow across the surface and along the ground, collecting in low spots.

Foam extinguishes flammable liquid fires by separating the fuel from the fire. When a blanket of foam completely covers the surface of the liquid, the release of flammable vapors is stopped. Preventing the production of additional vapors eliminates the fuel source for the fire, so the fire is extinguished.

Once a foam blanket has been applied, it must NOT be disturbed. The fuel under the foam blanket is still hot and capable of producing flammable vapors. If the foam blanket is disturbed by wind, by someone walking through the liquid, or by hose streams breaking up the foam blanket, flammable vapors will be released and could be easily reignited.

When using foam to extinguish a flammable liquid fire, it is critically important to apply enough foam to fully cover the liquid surface. If there is not enough foam available to completely cover the surface at one time, the fire will have the ability to continue burning and the suppression efforts will be ineffective. The rate of foam application must also be high enough to cover the surface and maintain a blanket on top of the liquid. If the application rate is too low, thermal updrafts caused by the fire will keep the foam from covering the surface, and the heat of the fire will destroy the foam that has already been applied.

Class B foam can also be applied to a spill of flammable liquid product to prevent a fire. A foam blanket floating on the surface of the liquid will prevent the production of vapors that could be ignited. In this situation, the rate of foam application is not as critical, because there is no fire working to destroy the foam while it is being applied.

It is important to use the proper foam for the situation that is encountered. For example, an alcohol-resistant foam must be used for any incident involving a polar solvent. An ordinary foam would be broken down quickly if it came in contact with this type of product. Make sure to use the proper foam for the product involved.

Class A foams are not designed to resist hydrocarbons or to self-seal on a liquid surface. If Class A foam is used on a flammable liquid, it would be difficult to extinguish a fire and the foam blanket would not mend itself if it became disrupted. The risk of reignition could create a dangerous situation.

Foam Concentrates

Foam concentrate is the product that is mixed with water in different ratios to produce **foam solution**. The product that is actually applied to extinguish a fire or onto a spill is foam solution.

Class A foams are usually formulated to be mixed with water in ratios from 0.1% (1 gallon of concentrate to 999 gallons of water) to 1.0% (1 gallon of concentrate to 99 gallons of water). The end product can be produced with different properties by varying the percentage of foam concentrate in the mixture and the application method. It is possible to produce "wet" foam that will have good penetrating properties or "stiff" foam that is more effective for applying a protective layer of foam onto a building.

Fire Fighter Safety Tips

Most foam concentrates are somewhat corrosive in nature; however, there is little health risk to fire fighters from normal use if normal precautions are taken. Some fire fighters have experienced minor skin irritations from coming into contact with foam concentrate. Rubber gloves and eye protection should be used when handling foam concentrate. As always, before using any equipment, the manufacturer's directions should be read and followed.

Standard personal protective clothing provides appropriate protection when working with foam agents. It is advised to rinse skin after coming into contact with foam and to flush all equipment with clear water after foam use. Protective clothing should be rinsed with plain water.

Most Class B foam concentrates are designed to be used in strengths of either 3% or 6%. A 3% foam mixes 3 gallons of foam concentrate with 97 gallons of water to produce 100 gallons of foam solution. A 6% foam mixes 6 gallons of foam concentrate with 94 gallons of water to produce 100 gallons of foam solution. Some foams are designed to be used at 3% for ordinary hydrocarbons and at 6% for polar solvents. It is important to know what type of fuel you are working with in order to select the correct proportioning rate.

The compatibility of foam agents with other extinguishing agents needs to be considered. For example, some combinations of dry chemical extinguishing agents and foam agents can cause an adverse reaction. The compatibility data is available from the manufacturers of the agents.

It is important not to mix different types of foam concentrate or even different brands of the same type unless they are known to be compatible. Some concentrates have been known to react with other types of foams, causing a congealing of the concentrate in the storage containers or in the proportioning system. This type of reaction can plug a foam system and render it useless.

Some constituents in Class B foam concentrates that have been widely used in the past are being phased out due to environmental concerns. Newer concentrates have been developed that are equally effective without the undesirable properties. The major categories of Class B foam concentrate include:

- Protein foam
- Fluoroprotein foam
- Aqueous film-forming foam (AFFF)
- Alcohol-resistant foams

Protein Foam

Protein foams are made from animal byproducts. They are effective on Class B hydrocarbon fires and are applied in 3% or 6% delivery rates.

Fluoroprotein Foam

Fluoroprotein foams are made from the same base materials as protein foam along with flurochemical surfactant additives. The additives allow this foam to produce a fast-spreading membrane across the surface of a flammable liquid and provide a greater ability to seal against the edges of a tank or objects that penetrate the surface.

Aqueous Film-Forming Foam (AFFF)

Aqueous film-forming foam (AFFF) is a synthetic-based foam that is particularly suitable for spill fires involving gasoline and light hydrocarbon fuels. It can form a seal across a surface quickly and has excellent vapor suppression capabilities.

Alcohol-Resistant Foam

Alcohol-resistant foam has properties similar to AFFF; however, it is formulated so that alcohols and other polar solvents will not dissolve the foam. Regular foams cannot be used on these products.

Foam Equipment

Foam equipment includes the proportioning equipment used to mix foam concentrate and water to produce foam solution, as well as the nozzles and other devices that are used to apply the foam. There are many different types of proportioning and application systems. Most engine companies carry the necessary equipment to place at least one foam attack line into operation. Structural firefighting apparatus can also be designed with built-in foam proportioning systems and on-board tanks of foam concentrate to provide greater capabilities. Many fire departments specify both Class A and Class B integrated foam systems on new apparatus or have special foam apparatus available for situations when large quantities of foam are needed.

Foam Proportioning Systems

A foam proportioner is the device that mixes the foam concentrate into the fire stream in the proper percentage. There are two types of proportioners, eductors, or injectors, in a wide range of sizes and capacities. Foam solution can also be produced by batch mixing or premixing.

Foam Eductors

A foam eductor uses a venturi effect to draw foam concentrate from a container or storage tank into a moving stream of water. An eductor can be built into the plumbing of an engine or a portable eductor can be inserted in an attack hose line. A foam eductor is usually designed to work at a predetermined pressure and flow rate. A metering valve can be adjusted to set the percentage of foam concentrate that is educted into the stream.

The most common type of portable in-line eductor used by fire departments is sized to work with a 1½" attack line (► Figure 16-32). This type of eductor requires 200 psi of water pressure to draw foam concentrate from a portable container into the stream. An attack line with of 150′ of 1½" hose can be connected on the discharge side of the eductor to deliver the foam to a nozzle

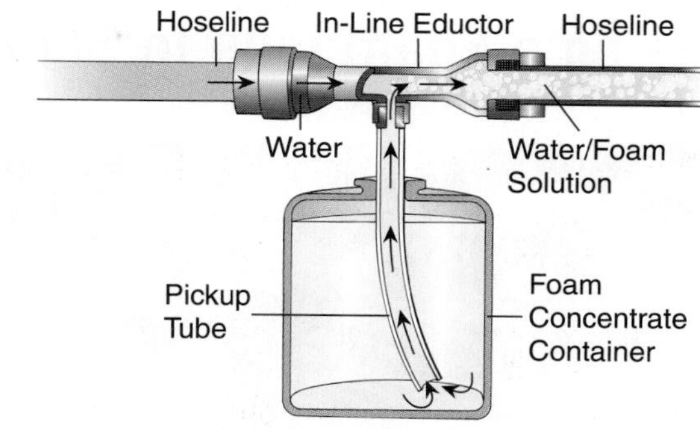

Figure 16-32 A typical portable in-line eductor will operate a single foam handline.

Injectors

Foam injectors add the foam concentrate to the water stream under pressure. Most injector proportioning systems will work across a range of flow rates and pressures. A metering system measures the flow rate and pressure of the water and adjusts the injector to add the proper amount of foam concentrate. This type of system is often installed on special foam apparatus.

To place a foam line in service, follow the steps in (► Skill Drill 16-43).

1. Make sure all necessary equipment is available. This includes an in-line foam eductor and an air aspirating nozzle. Ensure that there is enough foam concentrate available to suppress the fire.
2. Fire fighters should be wearing all of their PPE.
3. An attack line needs to be procured and the nozzle removed and replaced with the air aspirating nozzle.
4. The in-line eductor is placed in the hose line no more than 150′ from the nozzle.
5. Place the foam concentrate container(s) next to the eductor, check the percentage that the foam concentrate should be used at (found on container label), and set the metering device on the eductor accordingly.
6. Place the pick-up tube from eductor into the foam concentrate making sure to keep both items at similar elevations to ensure sufficient induction of foam concentrate.
7. Charge the hose line with water, ensuring there is a minimum of 200 psi at the eductor. Flow water through the hose line until foam starts to come out of the nozzle. The hose line is now ready to be advanced onto burning flammable liquids.
8. Apply foam using one of the three application methods (sweep technique, bankshot technique, rain-down technique) depending upon the situation. **(Step 1)**

Batch Mixing

Foam concentrate can be poured directly into an apparatus booster tank to produce foam solution. This technique is called batch mixing. If the booster tank has a capacity of 500 gallons, 15 gallons of 3% foam concentrate should be added. If 6% foam concentrate is used, 30 gallons of foam concentrate should be added to the booster tank. It may be necessary to drain sufficient water from the tank first to make room for the foam concentrate. After the concentrate has been added, the solution should be mixed by circulating the water through the pump before it is discharged.

Placing a Foam Line in Service

Make sure all necessary equipment is available. This includes an in-line foam eductor and an air aspirating nozzle. Ensure that there is enough foam concentrate available to suppress the fire.

Fire fighters should be wearing all of their PPE.

An attack line needs to be procured and the nozzle removed and replaced with the air aspirating nozzle.

The in-line eductor is placed in the hose line no more than 150' from the nozzle.

Place the foam concentrate container(s) next to the eductor, check the percentage that the foam concentrate should be used at (found on container label), and set the metering device on the eductor accordingly.

Place the pick-up tube from eductor into the foam concentrate making sure to keep both items at similar elevations to ensure sufficient induction of foam concentrate.

Charge the hose line with water, ensuring there is a minimum of 200 psi at the eductor. Flow water through the hose line until foam starts to come out of the nozzle. The hose line is now ready to be advanced onto burning flammable liquids.

Apply foam using one of the three application methods (sweep technique, bankshot technique, rain-down technique) depending upon the situation.

Premixing

Premixed foam is commonly used in $2^1/_2$-gallon portable fire extinguishers. Foam fire extinguishers are filled with premixed foam solution and pressurized with compressed air or nitrogen. Some vehicles are equipped with a large tank holding 50 or 100 gallons of premixed foam, which operate in the same manner.

Foam Application Systems

Foam can be applied to a fire or spill through portable extinguishers, handlines, or master stream devices, or a variety of fixed systems for special applications. Most fire departments apply foam through either handlines or master stream devices on fire apparatus. Several types of nozzles are used to produce different types of foam. The major difference between these nozzles is the manner of introducing air into the stream of foam solution to produce the desired consistency of foam and air bubbles.

Foam can be applied with a wide range of expansion rates, depending on the amount of air that is mixed into stream and the size of the bubbles that are produced. Low expansion foam has little entrained air and a small bubble structure. This type of foam is often produced with standard adjustable fog nozzles. The air is entrained by the flowing stream and mixed into the foam solution. This type of nozzle is often used to apply AFFF or Class A foam.

Medium expansion foam is produced with special aerating nozzles that are designed to introduce more air into the stream and produce a consistent bubble structure (aeration). This type of nozzle is generally used with protein and fluoroprotein foams to produce a thicker blanket of foam. This type of nozzle is also recommended for use with alcohol resistant foams to produce a thicker foam blanket.

High expansion foam has a much greater proportion of air and large bubbles. This type of system uses a high-expansion foam generator to introduce large quantities of air into the discharge stream. High expansion foam is sometimes used in automatic systems that are designed to completely fill a large space with foam. These systems are most likely to be found in aircraft hangars or other large storage areas.

Compressed air foam systems (CAFS) inject the air into the stream of foam solution under pressure and discharge a mixture of foam solution and compressed air through the nozzle. This mixture expands quickly as it is released from the nozzle to produce a bubbly product. It is usually applied through a smooth bore nozzle.

Foam Application Techniques

When applying foam from a handline, the correct application techniques must be used to produce the desired quality of foam and successfully blanket the surface of a burning liquid or spill. These techniques include the sweep, bankshot, and rain-down methods of application.

Sweep Method

The sweep (or roll-on) method should only be used on a pool of flammable product that is on open ground. The fire fighter sweeps the stream along the ground just in front of the target to produce a quantity of foam and then uses the energy of the stream to push the foam blanket across the surface. The stream should be moved back and forth in a slow steady horizontal motion to gently push the foam forward until the area is covered. It is important to push the foam gently so that the blanket is not broken. It may be necessary to move to different positions to be sure that the entire surface of the product is covered by the foam blanket.

To perform the sweep method of applying foam, follow the steps in Skill Drill 16-44.

1. Open the nozzle and test to ensure that foam is being produced.
2. Move into a safe range of the product and open the nozzle.
3. Direct the stream of foam onto the ground just in front of the pool of product.
4. Allow the foam to roll across the top of the pool of product until it is completely covered.
5. Be aware that the fire fighters may have to change positions along the spill in order to adequately cover the entire pool.

Bankshot Method

The bankshot (or bank down) method is used at fires where there is an object that can be used to deflect the foam stream and let it flow down onto the burning surface. This method could be used to apply foam to an open-top storage tank or a rolled-over transport vehicle. The operator should sweep the foam back and forth against the object while the foam flows down and spreads back across the surface. As with all foam application methods, it is important to let the foam blanket flow gently on the surface of the flammable liquid to form a blanket.

To perform the bankshot method, follow the steps in Skill Drill 16-45.

1. Open the nozzle and test to ensure that foam is being produced.
2. Move into a safe range of the product and open the nozzle.
3. Direct the stream of foam onto a solid structure such as a wall or metal tank so that the foam is directed off the object and onto the pool of product.
4. Allow the foam to flow across the top of the pool of product until it is completely covered.
5. Be aware that the fire fighters may have to bank the foam off several areas of the solid object in order to extinguish the burning product.

Rain-down Method

The rain-down application method is performed by lofting the foam stream into the air above the fire and letting it fall down gently onto the surface. The stream should be broken so that the amount of foam falling in the same area does not cause splashing or break-up the blanket that has already been applied. Carefully observe how the foam blanket is building it up and direct your stream so that the entire surface is covered.

To perform the rain-down method, follow the steps in Skill Drill 16-46.

1. Open the nozzle and test to ensure that foam is being produced.
2. Move into a safe range of the product and open the nozzle.
3. Direct the stream of foam into the air so that the foam breaks apart in the air and falls onto the pool of product.
4. Allow the foam to flow across the top of the pool of product until it is completely covered.
5. Be aware that the fire fighters may have to move to several locations and shoot the foam into the air in order to extinguish the burning product.

Back Up Resources

When attempting to use foam to extinguish a flammable liquid fire, it is critically important to be sure that enough foam concentrate is available to complete the job. If the flow of foam has to be interrupted while additional foam supplies are obtained, the fire will destroy the foam that has already been applied. It is best to wait until an adequate supply of foam concentrate is on hand than to waste the limited supply that is immediately available.

There are specific formulas provided by the foam manufacturers to calculate how much foam is required to extinguish fires of a certain size. Most fire departments have contingency plans to deliver quantities of foam to the scene of a major incident. This foam could be on designated vehicles or in storage at fire stations. Back up sources often include airport crash vehicles and petroleum facilities that have their own foam apparatus and often keep large quantities of foam in storage. The manufacturers of foam products also have emergency programs to deliver large quantities of foam to the scene of exceptionally large scale incidents.

Foam Apparatus

Some fire departments operate apparatus that is specifically designed to produce and apply foam. The most common examples are used at airports for aircraft rescue and fire fighting (ARFF). These are large vehicles that carry the foam concentrate and water on board and are designed to quickly apply large quantities of foam to a flammable liquids fire. Remote control monitors can be used to apply foam while the vehicle is in motion.

Wrap-Up

Ready for Review

Hose, nozzles, and appliances are the most basic of firefighting tools. They have been in service since the first fire brigades were formed. Technology has vastly improved their performance, but the basic principles of their use remain the same. Hose comes in various sizes and types. Each has different uses based on that fact. Hose is used to supply pumps, and it is used to deliver water to the fire. That hose, if it is properly maintained, can serve for decades. Appliances, adaptors, reducers, and fittings allow hose to be configured and used in countless ways. This enables fire fighters endless combinations and possibilities for just about any situation that presents itself. There are also many methods on how hose is loaded and deployed from the fire apparatus. Again, each method has its special attributes. The main objective still remains—to quickly and efficiently deploy and extinguish fire.

The use of firefighting foam has emerged as one of the newer and more effective tools available to the fire fighter today. As technology advances, more and more applications for foam are being discovered.

Chief Concepts

- Fire hose is the most fundamental firefighting tool.
- Fire hose comes in different sizes for a variety of uses.
- Hose delivers water to fire pumps and from pumps to the fire.
- Through the use of hose appliances, fire fighters have many options to choose from for configuring fire hose to meet the requirements of the situation.
- There are many different ways to carry hose and transport it on fire apparatus. The method chosen is usually determined by the advantages of the method and the needs of the individual fire department.
- Fire hose needs proper care and maintenance just as any other firefighting equipment. Properly cared for hose will have a long service life.
- Fire hydraulics influences hose selection and configurations.
- Nozzle selection is determined by the effects desired. Examples are fog stream and straight stream.
- Firefighting foam is becoming one of the most valuable tools in modern firefighting.
- Foam has more applications today than exclusively for the flammable liquid fires of years past.

Hot Terms

Accordion hose load A method of loading hose on a vehicle whose appearance resembles accordion sections. It is achieved by standing the hose on its edge, then placing the next fold on its edge and so on.

Adaptor A device that joins hose couplings of the same type, such as male to male or female to female.

Adjustable gallonage fog nozzle A nozzle that allows the operator to select a desired flow from several settings.

Aeration Inducting air into the foam solution, which expands and finishes the foam.

Alcohol resistant foam A synthetic foam that resists mixing with polar solvents (alcohol type liquids).

Aqueous film-forming foam (AFFF) A water-based extinguishing agent used on Class B fires that forms a foam layer over the liquid and stops the production of flammable vapors.

Attack engine The engine from which the attack lines have been pulled.

Attack hose (attack line) The hose that delivers water from a fire pump to the fire. Attack hoses range in size from 1″ to 2½″.

Automatic adjusting fog nozzle A nozzle that can deliver a wide range of water stream flows. It operates by an internal spring loaded piston.

Ball valves Valves used on nozzles, gated wyes, and engine discharge gates. Made up of a ball with a hole in the middle of the ball.

Bankshot (bank down) method A method that applies the stream onto a nearby object, such as a wall, instead of directly aiming at the fire.

Batch mixing Pouring foam concentrate directly into the fire apparatus water tank, mixing a large amount at one time.

Booster hose (booster lines) A rigid hose that is ¾″ or 1″ in diameter. This hose only delivers 30 to 60 gpm, but can do so at high pressures. It is used for small outdoor fires.

Breakaway nozzle A nozzle with a tip that can be separated from the shut-off valve.

Bresnan distributor nozzle A device that can be placed in confined spaces. The nozzle spins, spreading water over a large area.

Butterfly valves Valves that are found on the large pump intake valve where the hard or soft suction hose connects.

Cellar nozzles Nozzles used to fight fires in cellars and other inaccessible places. These devices work by spreading water in a wide pattern as the nozzle is lowered by a hole into the cellar.

Wrap-Up

Hot Terms

Class A foam A firefighting foam for use on a particular type of fire. Class A refers to fires involving ordinary combustible materials such as wood, paper, and textiles.

Class B foam A firefighting foam for a particular type of fire. Class B refers to fires involving flammable liquids.

Double-female adaptor Used to join two male couplings.

Double female connection A hose adaptor that is equipped with two female connectors. It allows two hoses with male couplings to be connected together.

Double jacket hose A hose constructed with two layers of woven fibers.

Double-male adaptor Used to join two female hose couplings.

Double male connection A hose adaptor that is equipped with two male connectors. It allows two hoses with female couplings to be connected together.

Dutchman A term used for a short fold placed in a hose when loading it into the bed. This fold prevents the coupling from turning in the hose bed.

Fire hydraulics The physical science of how water flows through a pipe or hose.

Fixed gallonage fog nozzle A nozzle delivers a set number of gallons per minute that the nozzle was designed for, no matter what pressure is applied to the nozzle.

Flat hose load A method of putting a hose on a vehicle in which the hose is laid flat and stacked on of top the previous section.

Fluoroprotein foam A blended organic and synthetic foam that is made from animal byproducts and synthetic surfactants.

Foam concentrate Foam that is in its raw state.

Foam eductor A device placed in the hose line that draws foam concentrate from a container and introduces it into the fire stream.

Foam injector A device installed on a fire pump that meters foam by pumping or injecting it into the fire stream.

Foam proportioner A device that meters and introduces the foam concentrate into the fire stream at a proper rate.

Foam solution Foam that has been mixed with water.

Fog stream nozzle Device placed at the end of a fire hose that separates water into fine droplets to aid in heat absorption.

Forward lay A method of laying a supply line where the line starts at the water source and ends at the attack engine.

Four-way hydrant valve A specialized type of valve that can be placed on a hydrant that allows another engine to increase the supply pressure without interrupting flow.

Friction loss The reduction in pressure due to the water being in contact with the side of the hose. This contact requires force to overcome the drag the wall of the hose creates.

Gate valves Valves found on hydrants and sprinkler systems.

Gated wye A valved device that splits a single hose into two separate hoses, allowing each hose to be turned on and off independently.

Handline nozzles Used on hoses ranging from 1½" to 2½" hose lines, usually flow between 90 and 350 gallons per minute.

Hard suction hose A hose designed to prevent collapse under vacuum conditions so that it can be used for drafting water from below the pump (lakes, rivers, wells, or sea water, etc.).

Higbee indicators An indicator on the male and female threaded couplings that indicate where the threads start. These indicators should be aligned before starting to thread the couplings together.

Horseshoe hose load A method of loading hose where the hose is laid into the bed along the three walls of the bed, resembling a horseshoe.

Hose appliance Any device used in conjunction with fire hose for the purpose of delivering water.

Hose clamp A device used to compress a fire hose to stop water flow.

Hose jacket A device used to stop a leak in a fire hose or to join hoses that have damaged couplings.

Hose liner (hose inner jacket) The inside portion of a hose that is in contact with the flowing water.

Hose roller A device that is placed on the edge of a roof and is used to protect hose as it is hoisted up and over the roof edge.

Large Diameter Hose (LDH) Hose in the 4", 5", and 6" range.

Low volume nozzles Nozzles that flow 40 gallons per minute or less.

Master stream device A large capacity nozzle that can be supplied by two or more hose lines of fixed piping. Can flow 300 gallons per minute. Includes deck guns and portable ground monitors.

Master stream nozzles A nozzle used on deck guns, portable monitors, and ladder pipes that flow more than 350 gallons per minute.

Medium diameter hose (MDH) Hose of 2½" or 3" size.

Mildew A condition that can occur on hose if it is stored wet. Mildew can damage the jacket of a hose.

Nozzle shut off Device that enables the person at the nozzle to start or stop the flow of water.

Nozzles Attachments to the discharge end of attack hoses to give fire streams shape and direction.

Piercing nozzle A nozzle that can be driven through sheet metal or other material to deliver a water stream to that area.

Premixed foam Commonly used in fire extinguishers, fire foam extinguishers are charged with compressed air or nitrogen, which discharges the foam solution through an attached aerating nozzle.

Protein foam An organic foam that is made from animal byproducts.

Reducer A device that can join two hoses of different sizes.

Reverse lay A method of laying a supply line where the line starts at the attack engine and ends at the water source.

Rocker lug (rocker pins) Fittings on threaded couplings that aid in coupling the hoses.

Rubber-covered hose (rubber-jacket hose) Hose whose outside covering is made of rubber, said to be more resistant to damage.

Siamese A device that allows two hoses to be connected together and flow into a single hose.

Small Diameter Hose (SDH) Hose in the 1″ to 2″ range.

Smooth bore nozzle Nozzles that produce a straight stream that is a solid column of water.

Smooth bore tip A nozzle device that is a smooth tube, used to deliver a solid column of water.

Soft suction hose A large diameter hose that is designed to be connected to the large port on a hydrant (steamer connection) and into the engine.

Spanner wrench A type of wrench used in coupling or uncoupling hoses by turning the rocker lugs on the connections.

Split hose bed A hose bed arranged such that a 1″-long supply line can be laid out, or two supply lines can be laid out.

Split hose lay A scenario where the attack engine will lay a supply line from an intersection to the fire, and the supply engine will lay a supply line from the hose left by the attack engine to the water source.

Storz-type coupling A hose coupling that has the property of being both the male and female coupling. It is connected by engaging the lugs and turning the coupling a one-third turn.

Supply hose (supply line) The hose used to deliver water from a source to a fire pump.

Sweep (roll-on) method Foam applying method that is done by sweeping the stream just in front of the target.

Threaded hose couplings A type of coupling that requires a male and female fitting to be screwed together.

Triple layer load A hose loading method that utilizes folding the hose back onto itself to reduce the overall length to one-third before loading in the bed. This load method reduces deployment distances.

Water curtain nozzles Nozzles used to deliver a flat screen of water to form a protective sheet of water.

Water hammer An event that occurs when flowing water is suddenly stopped; the velocity force of the moving water is transferred to everything it is in contact with. These can be tremendous forces that can damage equipment and cause injury.

Water Thief A device that has a 2½″ inlet and a 2½″ outlet in addition to two 1½″ outlets. The device is used to supply many hoses from one source.

Wye A device used to split a single hose into two separate lines.

Fire Fighter in Action

Your engine company is dispatched to a 3-alarm fire in a large church. When you arrive on the scene, the church is 60% involved and fire is coming out of the windows. The incident comander has decided on a defensive strategy and has instructed your lieutenant to set up a water supply and ground monitor attack on side C of the structure.

1. Which of the following is a method of laying a supply line from the hydrant to the fire?

A. Forward hose lay

B. Reverse hose lay

C. Split hose lay

D. Flat hose lay

2. A ground monitor can be supplied by what type of hose?

A. Multiple booster hose lines

B. Multiple $1^1/_2$" or $1^3/_4$" hose lines

C. Multiple $2^1/_2$" or larger hose lines

D. Multiple forestry hose lines

The fire is successfully knocked down without spreading to any exposures. However, there are hot spots that need to be extinguished. Lieutenant Cox tells you to disconnect the ground monitor and attach a gated wye and two 150′ $1^1/_2$" hose lines with nozzles for overhaul. Once the overhaul is completed, Lieutenant Cox tells you to reload the hose.

3. A gated wye:

A. combines two hose lines into one.

B. splits one hose stream into two.

C. joins two female hose couplings.

D. joins two hose lines of similar diameter but differing threads.

4. Large diameter hose should be loaded using a(n):

A. horseshoe load

B. accordion load.

C. Dutchman load.

D. flat load.

www.FireFighter.jbpub.com

www.FireFighter.jbpub.com

Chapter Pretests
Interactivities
Hot Term Explorer
Web Links
Review Manual
FireLearn

Fire Fighter Survival

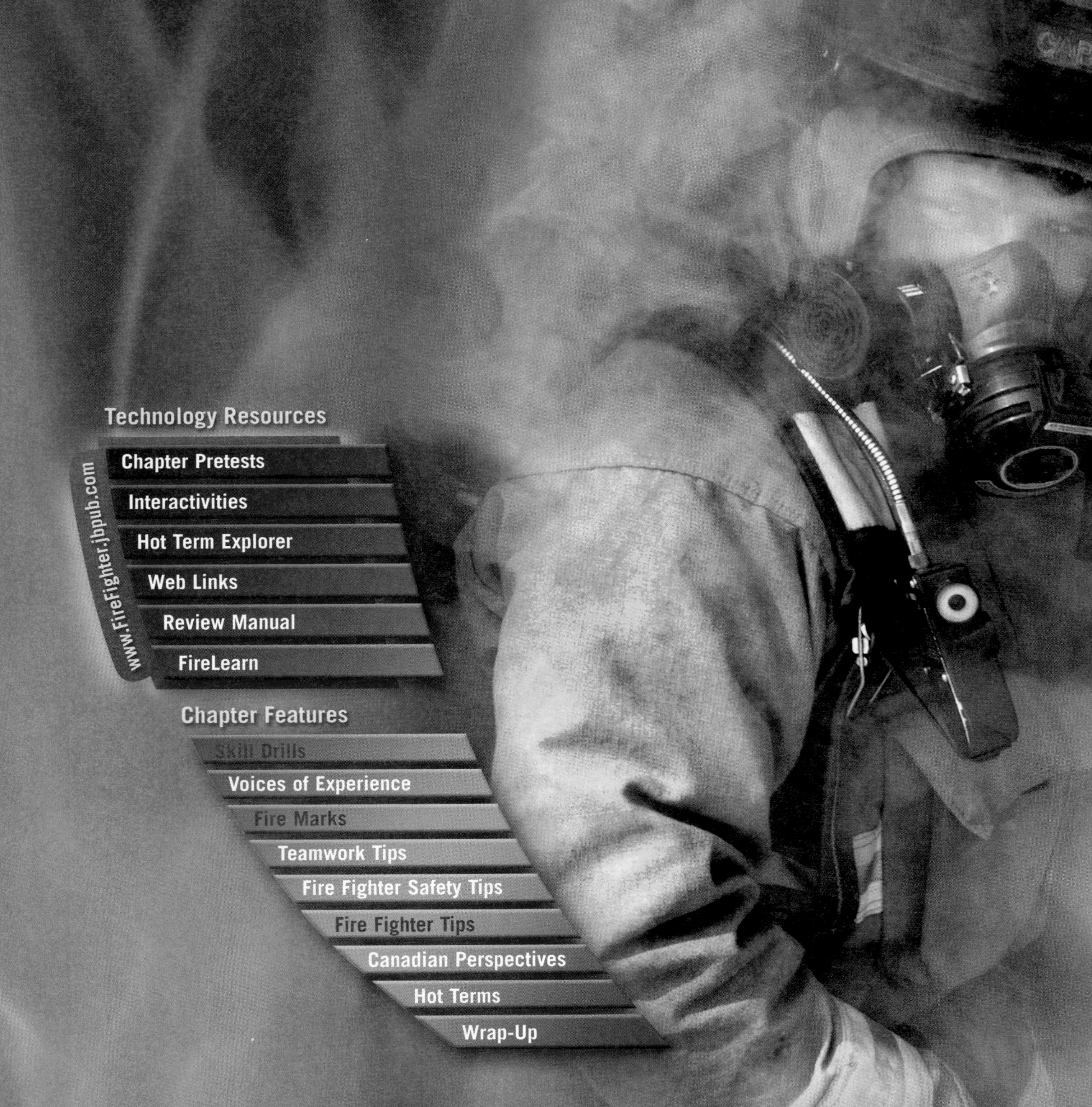

Chapter 17

NFPA 1001 Standard

Fire Fighter I

5.3.1 (B) *Requisite Skills.* The ability to control breathing, replace SCBA air cylinders, use SCBA to exit through restricted passages, initiate and complete emergency procedures in the event of SCBA failure or air depletion, and complete donning procedures.

5.3.5 Exit a hazardous area as a team, given vision-obscured conditions, so that a safe haven is found before exhausting the air supply, others are not endangered, and the team integrity is maintained.

5.3.5 (A) *Requisite Knowledge:* Personal accountability systems, communications procedures, emergency evacuation methods, what constitutes a safe haven, elements that create or indicate a hazard, and emergency procedures for loss of air supply.

5.3.5 (B) *Requisite Skills:* The ability to operate as a team member in vision-obscured conditions, locate and follow a guideline, conserve air supply, and evaluate areas for hazards and identify a safe haven.

Fire Fighter II

NFPA 1001 contains no Fire Fighter II job performance requirements for this chapter.

Additional NFPA Standards

NFPA 1500, *Standard on Fire Department Occupational Safety and Health Program.*

Knowledge Objectives

After studying this chapter, you will be able to:

- Describe the procedure for making an appropriate risk-benefit analysis.
- Describe the procedures for the personnel accountability system.
- Describe the role of the rapid intervention crew (RIC).
- Define self-rescue techniques.
- Describe how to conserve SCBA air supply.
- Describe the critical incident stress management process.

Skills Objectives

There are no skills objectives for this chapter.

You Are the Fire Fighter

At 3 am, your engine company is dispatched to a first alarm for "smoke in the building." The building is a three-story wood frame structure containing seven dwelling units. Upon arrival, no smoke is visible from the outside and a few tenants are standing on the front porch. You walk up to the porch with your lieutenant, carrying a pressurized water extinguisher.

The tenants inform you that there is light smoke in the second floor corridor. Your first inclination is to go inside and investigate, but your lieutenant stops you. He tells you to wait until the rest of the team members are ready to go in with a hose line. He tells the pump operator to be ready to charge the line and to collect the accountability tags from the other arriving companies. He transmits an initial report on the radio and tells the second engine company due to lay a supply line and then act as the rapid intervention crew. The four team members from the ladder company split into a search team and a ventilation team.

1. ***Why does your lieutenant stop you from going inside immediately?***
2. ***What is the purpose of the accountability tags?***
3. ***What does a company assigned to RIC do in this situation?***

Introduction

Fire fighter survival is the primary desired outcome in any successful fire department operation. Fire fighters often perform their duties in environments that are inherently dangerous. Survival is a positive outcome that can only be achieved by working in a manner that recognizes risks and hazards and consistently applies safe operating policies and procedures. "Everybody goes home," should be the goal of every member of the team. This chapter will introduce you to the actions, attitudes, and systems that are important in achieving that goal.

Fire fighters often encounter situations where survival depends on making the right decisions and taking the appropriate actions. Many of the factors that cause fire fighter deaths and injuries appear again and again in post-incident studies. These findings identify risks that are often present at fires and other emergency incidents. Fire fighters must learn to recognize dangerous situations and to take actions appropriate for the situation.

Risk-Benefit Analysis

A risk-benefit analysis approach to emergency operations can limit the risk of fire fighter deaths and injuries. Risk-benefit analysis is based upon comparing the positive results that can be achieved with the probability and severity of potential negative consequences. A standard approach to risk-benefit analysis should be incorporated into the standard operating procedures for all fire departments. NFPA 1500, *Standard on Fire Department Occupational Safety and Health Program* incorporates this approach. This standard is an excellent reference guide to fire department operations.

In the fire service, risk-benefit analysis has to be practiced at several different levels. The incident commander (IC) is responsible for the high-level risk-benefit analysis that determines the strategy and tactics to be implemented at every incident scene. A company officer processes risk and benefits to ensure the safety of a group of fire fighters working as a team. As a fire fighter, you should perform a similar risk-benefit analysis of situations you encounter and any actions you are involved in performing.

The IC must always assess the risks and benefits before committing crews to the interior of a burning structure. If the potential benefits are too low and the risks are too high, fire fighters should not be committed to conduct an interior attack. The IC must also reassess the risks and benefits periodically during an operation. There should be no hesitation to terminate an interior attack and move to an exterior defensive operation at any time if the risks outweigh the benefits.

A simply stated risk-benefit philosophy for a fire department can be expressed as follows:

- We will not risk our lives at all for persons or property that are already lost.
- We will accept a limited level of risk, under measured and controlled conditions, to save property of value.
- We will accept a higher level of risk only where there is a reasonable and realistic possibility of saving lives.

Figure 17-1 If a fire has already reached the stage no occupants could survive and no property of value can be saved, there is no justification for risking the lives of fire fighters.

This method attempts to predict and recognize the potential risks that fire fighters would face when entering a burning structure and weighs them against the results that could be achieved. If a building is known to be unoccupied and has no value, there is no justification for risking the lives of fire fighters to save it. Similarly, if a fire has already reached the stage where no occupants could survive and no property of value can be saved, there is no justification for risking the lives of fire fighters (▲ Figure 17-1).

In a situation where no lives are at stake, but there is a reasonable possibility that property could be saved, this policy allows for fire fighters to be committed to an interior attack. This policy recognizes that a standard set of hazards is anticipated in an interior fire attack situation, but the combination of personal protective clothing and equipment, training, and standard operating procedures (SOP) is designed to allow fire fighters to work safely in this type of environment. All standard approaches to operational safety must be followed without compromise. No property is worth the life of a fire fighter.

It is only permissible to risk the life of a fire fighter in a situation where there is a reasonable and realistic possibility of saving a life. Even in these situations, the actions must be conducted in a manner that is as safe as possible. The determination that a risk is acceptable in a particular situation does not justify taking unsafe actions. It only justifies taking actions that involve a higher level of risk.

Company officers and safety officers are involved in risk-benefit analysis on a continuous basis. If there is a change in the balance of risk to benefit, from their perspective, it is their responsibility to report their observations and recommendations to the IC. The IC reevaluates the risk-benefit balance whenever conditions change and decides if the strategy must be changed.

In addition, each individual fire fighter involved in the operation should make a risk-benefit analysis from his or her perspective. A report from an observant fire fighter to his or her company officer can be crucial to the safety of an operation.

Hazard Indicators

As you progress as a fire fighter, it will become evident that fire suppression and other types of emergency operations involve many different types of hazards. Fire fighters must be capable of working safely in an environment that includes a wide range of inherent hazards. While the danger of fire fighting should never be taken for granted or thought of as routine, a fire fighter has to learn to routinely follow safe SOPs. Some of these hazards are encountered at almost every incident, while others are rare occurrences. A fire fighter should recognize many different types of hazardous conditions and react to them appropriately.

An example of a common hazard would be the presence of smoke inside a structure. All fires produce smoke and it has long been known that breathing smoke is dangerous. The proper response to this hazard is to always wear a breathing apparatus where smoke is in the atmosphere. This is a situation when an obvious hazard is recognized and a standard solution is applied.

Hazardous conditions may or may not be evident by observation. For example, smoke and flames are usually visible, while hazards related to the construction details of a building might not be easily detectable. You should be aware of the indicators of hazards that present themselves in a less obvious manner. This requires knowledge of what can be hazardous, given a particular set of circumstances. Hazard recognition becomes easier through study and experience. Some examples are:

- **Building construction:** Knowledge of building construction can help in anticipating fire behavior in different types of buildings and recognizing the potential for structural collapse. Is this building designed to withstand a fire for some period of time or is it likely to be weakened quickly? Has the building been modified from its original use? Does the structure incorporate void spaces that would allow a fire to spread quickly?
- **Weather conditions:** Inclement weather might make an otherwise routine incident hazardous. Rain and snow make operations on a roof extra dangerous. Raising a ladder close to power lines during windy weather involves an increased level of risk.
- **Occupancy:** A name such as "Central Plating Company" on a building should cause fire fighters to anticipate that hazardous materials could be present. A warning placard outside would be a more specific hazardous materials indicator (► Figure 17-2).

These are just a few examples of observable factors that might indicate a hazard. At each and every incident, you

Figure 17-2 The NFPA 704 diamond indicates that hazardous materials are present.

Figure 17-3 A team attempts to achieve a common goal.

should observe and learn what could be a critical indication of a hazard.

Safe Operating Procedures

SOPs define the manner in which a fire department conducts operations at an emergency incident. Many of the SOPs are based on fire fighter health and safety. The most important component of fire fighter survival is to consistently follow safe operating procedures.

Safe operating procedures must be learned and practiced before they can be implemented. Training is the only way for individuals to become proficient at performing any set of skills. In order to be effective, the same skills must be applied routinely and consistently at every training session and at every emergency incident. It has been shown that when under great pressure, most people will revert to habit. When self-survival is at stake, well-practiced habits can be a lifesaver.

Team Integrity

Teamwork is an essential requirement in fire fighting (► Figure 17-3). A team can be defined as two or more individuals attempting to achieve a common goal. Due to the intense physical labor necessary to pull hose or raise ladders, team efforts are imperative. The most important reason for team integrity in emergency operations is for safety; team members keep track of each other and provide immediate assistance to each other when needed. The standard operational team in firefighting operations is a company, usually consisting of three to five fire fighters working under the direct supervision of a company officer. Career fire fighters may arrive as a company on the same apparatus and on-call volunteer fire fighters should assemble in companies upon arrival.

Individual companies are assigned as a group to perform tasks at the scene of an emergency incident. The IC keeps track of companies and communicates with company officers. Every company officer must have a portable radio to maintain contact with the IC. The company officer is responsible for knowing the location of every company member at all times. The company officer should also know exactly what each individual fire fighter is doing at all times.

Team integrity means that a company arrives at a fire together, works together, and leaves together. The members of a company should always be oriented to each other's location, activities, and condition. In a hazardous atmosphere, such as inside a structure, you should always be able to contact your company members by voice, sight, or touch. When air cylinders have to be replaced, all of the company members leave the building together, change cylinders together, and return to work together. If the members require rehabilitation, the full company goes to rehabilitation together and returns to action together.

In some cases a company will be divided into smaller teams to perform specific tasks. For example, a ladder company could be split into a roof team and an interior search team. The initial attack on a room and contents fire in a single-family dwelling might be made by two fire fighters, while two other fire fighters remain outside, following the two-in/two-out rule. For safety and survival reasons, fire fighters should always use a buddy system, working in teams of at least two. If you should become incapacitated, your partner can provide help immediately and call for assistance. If you are alone, no one will know that you need help. The two team members must remain in direct contact with each other. At least one member of every team must have a portable radio to maintain contact with the IC; in many fire departments, every member carries a portable radio.

Personnel Accountability System

A personnel accountability system can be defined as a systematic way to keep track of the location and function

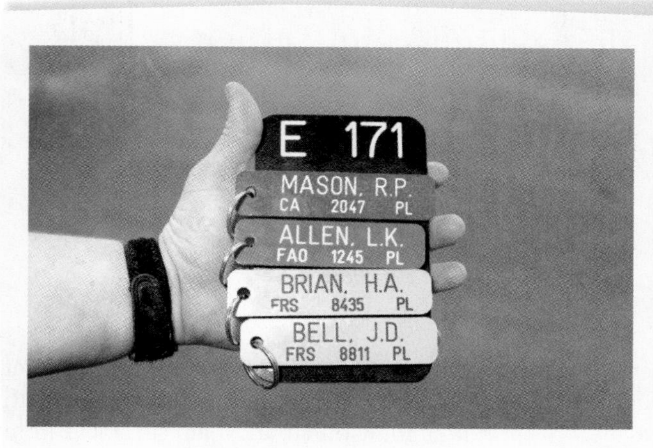

Figure 17-4 Personnel accountability system.

of all personnel operating at the scene of an incident (▲ Figure 17-4). The IC has the responsibility to account for each fire fighter involved in an emergency incident. A personnel accountability system is designed to establish standard procedures for performing this function and to manage the necessary information.

An accountability system has to keep track of every company or team of fire fighters working at the scene of the incident from the time they arrive until the time they are released. In addition, the system must be able to identify each individual fire fighter who is assigned to that company or team. The system assumes that all of the company members will be together, working as a unit under the supervision of the company officer, unless the IC has been advised of a change. At any point in the operation, the IC should be able to identify the location and function of each company and, within seconds, know the individual fire fighters assigned to that company.

The personnel accountability systems used by fire departments can take different forms. Whatever type of system is used, the only way for it to be effective is if every member takes responsibility for its use at all incidents. This includes even the smallest of incidents that appear to pose no immediate hazards. Some fire departments start with a written roster of the team members assigned to each company, while others use a computer database tied to a daily staffing program. Lists, roll calls, or electronic tracking databases should be used to account for all personnel at the beginning of every tour of duty or when changes are made to assignments throughout the tour.

A list of assigned company members is carried on each piece of apparatus. One common system uses a Velcro™ or magnetic nametag for each company member. The tags for all of the members assigned to the apparatus are affixed to a special board carried in the cab. This board is sometimes called a passport. These boards should be kept up to date throughout the tour of duty and denote any change in the assigned team. In the case of a volunteer company, the tags can be affixed to the board at the time of an alarm.

At the scene of an incident, these boards are physically left with a designated person at the command post or at the entry point to a hazardous area. Depositing the board and the tags indicates that the company and those individual fire fighters have entered the hazardous area. When a company leaves the hazardous area or is released from the scene, the company officer must retrieve the board. Retrieving the board indicates that the company has safely left the hazardous area.

A **Personnel Accountability Report (PAR)** is a roll call taken by each supervisor at an emergency incident. When the IC requests a PAR, each company officer physically verifies that all assigned members are present and confirms this information to the IC. This should occur at regular, predetermined intervals throughout the incident, generally every 10 minutes. A PAR should also be requested at tactical benchmarks, such as a change from offensive strategy to defensive strategy.

The company officer must be in visual or physical contact with all company members in order to verify their status. The company officer then communicates with the IC by radio to report a PAR. Any time a fire fighter cannot be accounted for, he or she is considered missing until proven otherwise. A report of a missing fire fighter always becomes the highest priority at the incident scene.

If unusual or unplanned events occur at an incident, a PAR should always be performed. For example, if there is a report of an explosion, a structural collapse, or a fire fighter missing or in need of assistance, a PAR should immediately take place so it can be determined how many personnel might be missing, identify the missing individuals, and establish their last known location and assignment. This last location would be the starting point for any search and rescue crews.

Emergency Communications Procedures

Communication is a critical part of any emergency operation (► Figure 17-5). This is particularly important in relation to fire fighter safety and survival. Breakdown in the communication process is a major contributing cause of injury and death to fire fighters. Good communications can save the lives of fire fighters in a dangerous situation.

Chapter 3, Fire Service Communications, discussed the proper radio communications procedures. The following is a model of good radio communication:

"Ladder 3 to Central IC."

"This is Central IC. Go ahead, Ladder 3."

"Ladder 3 has completed roof ventilation and is available for another assignment."

"IC copies. Roof ventilation completed and Ladder 3 available for reassignment."

Figure 17-5 Communication is a critical part of any emergency operation.

These standard communications procedures ensure that the message is clearly stated and then repeated back as confirmation that it has been received and understood.

When a fire fighter is missing or in trouble, or an imminent safety hazard is detected, it is even more important to communicate clearly and ensure that the message is understood. There are standard methods to transmit emergency information, standard information items to convey, and standard responses to emergency messages. Reserved phrases, sounds, and signals for emergency messages should be a part of your department's SOP. A separate set of standard phrases should be used for emergency situations and hazardous conditions. These phrases should be known and practiced by everyone in the department. The same procedures and terminology should be used by all of the fire departments that can be expected to work together at emergency incidents. In many areas, these procedures are coordinated regionally.

The word 'Mayday' is used to indicate that a fire fighter is in trouble and requires immediate assistance. Mayday could be used if a fire fighter is lost, trapped, has a self-contained breathing apparatus (SCBA) failure, or is running out of air. A fire fighter can transmit a Mayday to request assistance or another fire fighter could transmit a Mayday to report that a fire fighter is missing or in trouble. A Mayday takes precedence over all other radio communications.

The term 'Emergency Traffic' is used to indicate an imminent fire ground hazard, such as a potential explosion or structural collapse. An Emergency Traffic message would also be used to order fire fighters to immediately withdraw from interior offensive attack positions and switch to defensive strategy. An Emergency Traffic message takes precedence over all other radio communications with the exception of a Mayday.

Any use of the term Mayday should bring an immediate response. The IC will interrupt any other communications and direct the person reporting the Mayday to proceed. That person will then state the emergency, including all pertinent information; for instance, "This is Fire Fighter Jones with Engine 12. I'm running low on air and I am separated from my crew. I'm on the third floor and I believe that I'm in the northwest section of the building." The IC will repeat this information back and then initiate procedures to rescue Fire Fighter Jones. This could include committing the rapid intervention crew (RIC), calling for additional resources, and redirecting other teams to support a search and rescue operation.

In many fire departments, the communications center will sound a special tone over the radio to alert all members that an emergency situation is in progress and then repeat the information to be certain that everyone at the incident scene heard the message correctly.

Similar procedures are followed for emergency traffic situations. All imminent hazards and emergency instructions should be announced in a manner that captures the attention of everyone at the incident scene and ensures that the message is clearly understood.

Fire Fighter Safety Tips

Standard Radio Terminology

MAYDAY is used to report a fire fighter in trouble and requiring immediate assistance.
EMERGENCY TRAFFIC is used to report a hazardous condition or situation.

Initiating a Mayday

A Mayday message, as described previously, is used to indicate that a fire fighter is in trouble and needs immediate assistance. The analysis of fire fighter fatalities and serious injuries has shown that fire fighters often wait until it is too late to call for assistance. Instead of initiating a Mayday when first getting into trouble, too often a fire fighter will wait until the situation is absolutely critical before requesting assistance. This could be due to a fear of embarrassment if the situation turns out to be less severe than anticipated, combined with a hope that the fire fighter will be able to resolve the problem without assistance.

The failure to act promptly can be fatal in many situations. Systems have been designed to do everything possible to protect fire fighters and to rescue fire fighters from dangerous situations. These systems can only be effective if you act appropriately when you find yourself in trouble. Do not hesitate to call for help when you think you need it.

To initiate a Mayday follow the SOPs for your fire department. In most cases, this will involve transmitting "Mayday-Mayday-Mayday" over the radio to initiate the process. When acknowledged by the IC, clearly state your identity, the nature of the problem, and your location or approximate location.

Manually activate your PASS device so that other fire fighters can hear the signal and come to your location. Also activate the emergency alert button on your portable radio, if it has this feature. All of these actions will increase the probability that the RIC or other fire fighters will be able to locate you.

Rapid Intervention Company/Crew (RIC)

Rapid intervention companies/crews (RIC) are established for the sole purpose of rescuing fire fighters who are operating at emergency incidents. In some fire departments, the terms rapid intervention team (RIT) or fire fighter assistance and safety team (FAST) are used to describe this assignment. A RIC is a company or crew that is assigned to stand by at the incident scene, fully dressed and equipped for action, and be ready to deploy immediately when assigned by the IC.

A RIC is an extension of the two-in/two-out rule. During the early stages of an interior fire attack, a minimum of two fire fighters are required to establish an entry team and a minimum of two additional fire fighters are required to remain outside the hazardous area. The two fire fighters who remain outside the hazardous area can perform other functions, but they must have full PPE clothing and SCBA, so that they can immediately enter and assist the entry team if the interior fire fighters need to be rescued. The two fire fighters who remain outside the hazardous area are the first stage of a RIC. The second stage is a dedicated RIC, specifically assigned to stand by for fire fighter rescue assignments.

NFPA 1500, *Standard for a Fire Department Occupational Safety and Health Program* includes a complete description of RIC procedures and examples of their use at typical scenes. A number of options are available for establishing and organizing RIC crews or teams, including dispatching one or more additional companies to incidents with this specific assignment. Some small fire departments depend on an automatic response from an adjoining community to provide a dedicated RIC at working incidents.

SOPs should dictate when a RIC is required, how it is assigned, and where it should be positioned at an incident scene. A RIC should be in place at any incident where fire fighters are operating in conditions that are immediately dangerous to life and health (IDLH), which includes any structure fire where fire fighters are operating inside a building and using SCBA. Many fire departments specify a special set of rescue equipment that will be brought to the standby location. All the RIC members must wear full PPE and SCBA, and be ready for use.

The IC should immediately deploy the RIC to any situation where a fire fighter needs immediate assistance. A situation that calls for the use of the RIC would include a lost or missing fire fighter, an injured fire fighter who has to be removed from a hazardous location, or a fire fighter trapped by a structural collapse. NOTE: The U.S. Federal regulation that establishes requirements for using respiratory protection, OSHA 29 CFR 1910.134, mandates the use of back-up teams at incidents where operations are conducted in an unsafe atmosphere (potentially IDLH).

Canadian Perspectives

Within Canada, the provinces set minimum standards for the protection of fire fighters. Individual provinces establish regulations or guidelines that include RIC requirements. The Department of National Defence (DND), which operates fire departments on military bases, requires a minimum of five fire fighters to be on the scene of a fire before initiating an interior attack.

Fire Fighter Survival Procedures

Some of the most important procedures you need to learn are directly related to your personal safety and the safety of the fire fighters who will be working with you. These procedures are designed to keep you from getting into dangerous situations and to guide you in situations where you could be in immediate danger. Your personal safety will depend on learning, practicing, and consistently following these procedures.

Maintaining Orientation

As you practice working with SCBA, you will find out that it is very easy to become disoriented in a dark, smoke-filled building. Sometimes the visibility inside a burning building can be measured in inches. In these conditions, it is extremely important to stay oriented. If you get lost, you could run out of air before you find your way out. The atmosphere outside your facemask could be fatal in one or two breaths. Even if you call for assistance, rescuers might not be able to find you. Safety and survival inside a fire building can be directly related to remaining oriented within the building.

Several methods can be used to stay oriented inside a smoke-filled building. Before entering a building, look at it from the outside to get an idea of the size, shape, arrangement, and number of stories. After entering, follow walls and pay attention to where you go, whether you are moving toward the front or the rear, toward the left side or right side, and upstairs or downstairs.

One of the most basic methods to remain oriented is to always stay in contact with a hose line. This can be accomplished by always keeping a hand or foot on the hose line to remain oriented. To find your way out, follow the hose line back from the nozzle to the entry point.

Team integrity is an important factor in maintaining orientation. Everyone works as a team to stay oriented. When the team members cannot see each other, they have to stay in direct physical contact or within the limits of verbal contact. Standard procedures are employed to work efficiently as a team, such as having one member remain at the door-

Voices of Experience

"The captain was able to reach a window and push the other fire fighter through, but he was unable to get through the window himself."

On a Saturday morning, my engine company was dispatched to a reported structure fire in a hotel near the Naval Submarine Base. Upon arrival at the scene, the incident commander reported heavy smoke showing from the basement and the first floor, with the possibility of people trapped inside on the second floor.

My two person crew was instructed to take a $1\frac{1}{2}''$ hose line to the second floor and assist another two person crew with search and rescue operations. As we opened the doors to the hotel rooms to search for victims, conditions began to change dramatically. Air horns sounded to evacuate the building. We joined the engine company operating at the other end of the corridor and, before evacuating, made a quick count of all personnel believed to be on the second story of the hotel. Our count showed that the other two person crew that was supposed to be doing search and rescue was missing. We immediately radioed the incident commander to report the missing fire fighters.

Through radio communication, the incident commander pinpointed the location of the missing crew, consisting of a captain and a fire fighter, to the basement. It appeared they had become disoriented in their attempts to find a route of egress, and ended up trapped in a room in the basement. The two were able to reach a window. The captain pushed the other fire fighter through, but he was unable to get through the window himself. The fire in the corridor outside the room was growing rapidly, and the heat inside the room was getting intense. Then the captain's SCBA cylinder ran out of air. The captain remained at the window and we passed a mask from another SCBA to him to allow him to breathe. An attack crew entered the hotel and knocked down the fire in the basement corridor, providing an exit for the trapped captain. He spent the night in the hospital for observation and treatment for minor burns. Had we not taken a head count, we may not have known that the crew was missing, and the incident could have ended tragically.

John F. McDonald III
Submarine Base Fire Department
Groton, Connecticut

Teamwork Tips

At a minimum, one member of every company must remain oriented to their location in the building.

way into an apartment while the other members perform a systematic search inside. The searchers work their way around the apartment in one direction until they get back to the fire fighter at the door.

A **guideline** can be used for orientation when inside a structure. This is a rope attached to an object on the exterior or a known fixed location. The guideline is stretched out as a team enters the structure. They can follow the guideline back out or another team can follow the guideline in to find them. Some fire departments use a series of knots to mark distances along the rope to indicate how far it is stretched. The guideline technique requires intense practice.

Training sessions in conditions with no visibility will build confidence. This type of training can be conducted inside any building, while using an SCBA facemask with the lens covered. More realistic conditions can be simulated by using smoke machines inside training facilities. Fire fighters should practice navigating around furniture, following charged hose lines, climbing and descending stairs, and fitting through narrow spaces. Distracting noises, such as operating fans, can add a realistic feel. Team integrity must always be maintained in these practice sessions.

Self-Rescue

In the event that you become separated from your team, or lost, disoriented, or trapped inside a structure, several techniques can be used for **self-rescue**. The first thing to do is to call for assistance. Do not wait until it is too late for anyone to find you before making it known that you are in trouble. As described in the section "Initiating a Mayday," time is critical. You need to initiate the process as soon as you think you are in trouble, not when you are absolutely sure that you are in grave danger.

If you are simply separated from your team, your best option is to take a normal route of egress to the exterior. Following a hose line back to an open doorway, descending a ladder, or climbing out through a ground floor window can all provide a direct exit and require no special tools or techniques. You must immediately communicate with your company officer or the IC to inform them that you have exited safely.

There are many more complicated techniques that fire fighters can use to escape from dangerous predicaments. These include some standard methods, such as breaching a wall or using a rescue line and harness to rappel down to the ground. These methods must be learned and practiced regularly so that you will be ready to perform them without hesitation if you are ever in a situation that requires them.

Disentanglement is an important skill that needs to be learned and practiced. When crawling in a smoke filled building, a fire fighter can easily become entangled in wires, cables, debris, or whatever may be hidden in the smoke. Many fire fighters carry small tools in the pockets of their PPE that can be used to cut through wires or small cables. This can be very difficult if visibility does not allow the entangling material to be seen and identified.

Some additional self-rescue methods involve using tools and equipment in manners that they were not designed for. These are considered last resort methods. They should only be taught by experienced and qualified instructors and practiced with strict safety measures in place. There is controversy over several of these methods. Some fire departments consider them unacceptable due to the high risk of injury when they are used or even practiced. Other fire departments train in using them because they could save the life of a fire fighter in an extreme situation. This is a policy decision that must be made by each fire department.

Safe Havens

A **safe haven** is a temporary location that provides refuge while awaiting rescue or finding a method of self-rescue

Fire Fighter Safety Tips

Some proven actions to assist in self-rescue are:

- If you have a portable radio, use it. Follow the procedures for initiating a Mayday (or whatever terminology your department uses to relay critical information). Activate the emergency button on your radio, if it has one.
- Manually activate the alarm mode of your personal alert safety system (PASS).
- Turn on your flashlight. In darkness, this could help lead rescuers to your location.
- Conserve air. Remain as calm as possible, regulate your rate of breathing, and take shallow breaths or skip breaths when possible.
- Locate a window or a door. Place yourself as close as possible to locations where rescuers are likely to enter. Try to make it easy for them to find you.
- If possible, close the door to the room you are in before opening or breaking the window.
- Break a window to make yourself visible and to reach fresh air.
- Look for a safe haven where you can wait for help.

Remember: The only way for other fire fighters to know you are in trouble is to tell them. Call for assistance as soon as you realize you are in trouble. Do not wait until it is too late!

from an extremely hazardous situation. The term safe in this case is relative; it may not be particularly safe, but it is less dangerous than the alternatives. Wildland fire fighters carry lightweight reflective emergency shelters as a form of safe haven. When overrun by a fire, taking refuge in the shelter might be the only option for survival.

Safe havens are important when situations become critical and few alternatives are available. When a critical situation occurs, fire fighters should know where to look for a safe haven and recognize one when it presents itself.

If fire fighters are trapped inside a burning building, a room with a door and a window could be a safe haven. Closing the door could keep the fire out of the room for at least a few minutes and breaking the window could provide fresh air and an escape route. The safe haven provides time for a rescue team to reach the fire fighters, a ladder to be raised to the window, or another group of fire fighters to control the fire.

A roof or floor collapse often leaves a void adjacent to an exterior wall. If trapped fire fighters can reach this void, it could be the best place to go. Breaching the wall might be the only way out of the void or the only way for a rescue team to reach them.

Maintaining team integrity is important when escaping to a safe haven. Fire fighters have survived in safe havens by buddy breathing with SCBA and by working together to create an opening. The ability to use a radio to call for help and to describe the location where fire fighters are trapped can be critical.

As with all of the emergency procedures described in this chapter, these activities require good instruction and practice and must follow your department's operating guidelines.

Air Management

Air management is important to all fire fighters when using an SCBA. Air management relates to the basic fact that air equals time. The time that you can survive in a dangerous atmosphere depends on how much air is in your SCBA and how quickly you use it. You have to manage your air supply in order to manage your time in the hazardous area.

Your time in the hazardous atmosphere has to include the time it takes to enter and get to the location where you plan to operate and the time it will take to exit safely. Your work time is limited by the amount of air that you have in your SCBA cylinder, after allowing for entry and exit time. If it will take six minutes for you to reach a safe atmosphere and you only have enough air left for three minutes, the results could be disastrous.

Remember that the time rating for an SCBA is based on a standard rate of consumption for a typical adult under low exertion conditions in a test laboratory. Fire fighters typically consume air much more rapidly than the standard test. A fire fighter can often consume a 30-minute rated air supply in 10 to 12 minutes.

The actual rate of air consumption varies significantly among individual fire fighters and depends on the activities that are being performed. Chopping holes with an axe or pulling ceilings with a pike pole will cause you to consume air more rapidly than conducting a survey with a thermal image camera. Even given the same workload, no two individuals will use exactly the same amount of air from their SCBA cylinder. Differences in body size, fitness levels, and even heredity factors cause air consumption rates to be faster or slower than the average.

Air management has to be a team effort as well as an individual responsibility. You will always be working as part of a team when using SCBA. The team members will use air at different rates. The member who uses the air supply most rapidly determines the working time for the team.

A good way to determine your personal air usage rate is to participate in an SCBA consumption exercise. This drill puts fire fighters through a series of exercises and obstacles to learn how quickly each fire fighter uses air when performing different types of activities. Repeated practice can help you learn to use air more efficiently by building your confidence and learning to control your breathing rate. Keeping physically fit results in a more efficient use of oxygen and allows you to work longer with a limited air supply.

Practice sessions should be done in a nonhazardous atmosphere. A typical skill-building exercise might include the following:

- While wearing full PPE and breathing air from your SCBA, perform typical activities such as carrying hose packs up stairs, hitting an object with and axe or sledgehammer, or crawling along a hallway.
- Record the following: How much air is in the cylinder when you start; how long you work; whether the physical labor is light or heavy; the time when the low air alarm sounds; and the time when the cylinder is completely empty.

This type of exercise will allow you to determine your personal rate of consumption and work on improving your times with practice. With practice you will be able to predict how long your air supply should last under a given set of conditions.

Knowledge of your team members' physical conditions and their workloads can help you look out for their safety as well. Since many tasks are divided up in any given operation, one member may be working to his or her maximum level while another may be acting only in a supportive role. This could cause your team member to use up his or her air supply much faster without realizing it.

When using SCBA at an emergency incident, be aware of the limitations of the device. Do not enter a hazardous area unless your air cylinder is full. (Some fire departments allow a cylinder to be used if it is at least 90% full.) Keep track of the time you are using air and check the gauge regularly to know how much air you have left at all times. Do not wait until the low-pressure alarm sounds to start thinking about leaving the hazardous area; it may not provide enough time to make a safe exit.

Even using the best of air management practices, emergency situations can occur. An SCBA can malfunction or fire fighters can be trapped inside a building by a structural collapse or an unexpected situation. These are the times when remaining calm and knowing how to use all of the emergency features of an SCBA can be critically important. Remaining calm, controlling your breathing rate, and taking shallow breaths can slow air consumption. All of these techniques require practice to build confidence.

Rescuing a Downed Fire Fighter

One of the most critical and demanding situations in the fire service is when fire fighters have to rescue another fire fighter who is in trouble. Air management has particular implications in a fire fighter rescue situation. Air management has to be considered for the rescuers as well as the fire fighter who is in trouble.

If you reached a downed fire fighter, the first thing to do is assess the fire fighter's condition. Is the fire fighter conscious and breathing? Does the fire fighter have a pulse? Is the fire fighter trapped or injured? Make a rapid assessment and notify the IC of your situation and location. Have the RIC deployed to your location and quickly determine the additional resources you will need.

The most critical decision at this point is whether the fire fighter can be moved out of the hazardous area quickly and easily or if considerable time and effort will be required. If it will take more than a minute or two to remove the fire fighter from the building, air supply will be an important consideration. You need to provide enough air to keep the fire fighter breathing for the duration of the rescue operation.

Fire fighters must know and understand all aspects of their SCBA in this type of situation. If the fire fighter has a working SCBA, determine how much air remains in the cylinder. A fire fighter who is breathing and has an adequate air supply is not in immediate, life-threatening danger. If there is very little air or no air in the SCBA, this is a critical priority. You need to move the fire fighter out of the hazardous area immediately or provide an additional air supply. Can a spare air cylinder be brought to your location and connected to the downed fire fighter's SCBA? Can the RIC bring in a whole replacement SCBA?

If the fire fighter's SCBA is not working, check to see if a simple fix can make it operational. Did a hose or regulator come off? Was the cylinder valve accidentally closed? If regulator failure is a possibility, open the bypass valve.

Many newer SCBA units are designed with an additional hose or hose connections exclusively for buddy breathing. Buddy breathing is a method for two fire fighters to breathe air from the same SCBA cylinder. If you have the proper equipment, you need to know the policies and procedures for their use in your fire department.

Rehabilitation

Chapter 19 covers Fire Fighter Rehabilitation. The purpose of **rehabilitation** is to reduce the effects of fatigue during an emergency operation (▼ Figure 17-6). Firefighting involves very demanding physical labor. When combined with the extremes of weather and the mental stresses of emergency incidents, it can challenge the strength and endurance of fire fighters who are physically fit and well prepared. Rehabilitation helps fire fighters retain the ability to perform at the current incident and restores their capacity to work at later incidents.

At the simplest of incidents, rehabilitation can be set up on the tailboard of an apparatus with a water cooler. At larger incidents, a complete rehabilitation operation should be established, including personnel to monitor the vital signs of fire fighters and provide first aid. In whatever size or form it takes, rehabilitation is integral to fire fighter safety and survival. The personnel accountability system must con-

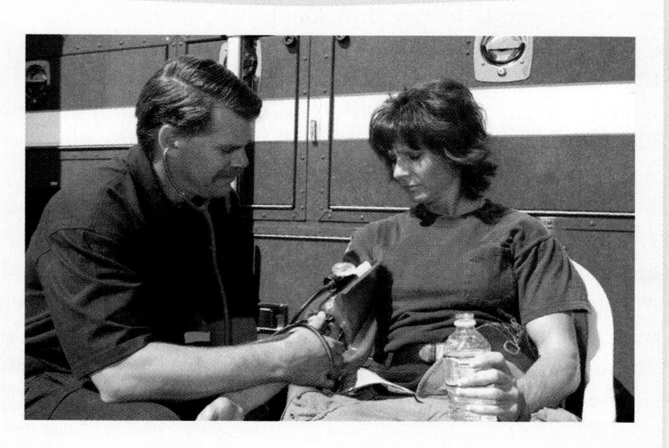

Figure 17-6 Rehabilitation reduces the effects of fatigue caused by an operation.

tinue to track fire fighters who are assigned to report to rehabilitation and when they are released from rehabilitation for another assignment.

Critical Incident Stress

Fire fighters are often exposed to stressful situations. There are times when fire fighters are involved in situations that are described as critical incidents. These incidents challenge the capacity of most individuals to deal with stress. Examples of critical incidents include:

1. Line-of-duty deaths (police, fire/rescue, EMS)
2. Suicide of a colleague
3. Serious injury to a colleague
4. Situations that involve a high level of personal risk to fire fighters
5. Events in which the victim is known to the fire fighters
6. Multicasualty/disaster/terrorism incidents
7. Events involving death or life-threatening injury/illness to a victim, especially a child
8. Events that are prolonged or end with a negative or unexpected outcome

This list is not complete, nor is it a fact that any of these situations will seriously trouble every individual. Normal coping mechanisms help many fire fighters to handle many situations. Some individuals have a high capacity to deal with stressful situations through exercise, talking to friends and family, or turning to their religious beliefs. These are healthy, nondestructive ways to deal with the pressures of being exposed to a critical incident.

Sometimes individuals react to critical incidents in ways that are not positive—such as alcohol abuse, depression, the inability to function normally, or a negative attitude towards life and work. These symptoms can occur in anyone, even individuals who normally have healthy coping skills. Reactions will vary from one individual to the next, both in type and severity. Many times fire fighters do not realize they are affected in a deeply negative manner. A somewhat routine incident can trigger negative reactions from a critical incident that occurred in the past. Critical incident stress can also be cumulative. Sometimes it is called burnout and cannot be traced to any one incident.

Figure 17-7 Critical incident stress debriefings are used to alleviate the stress reactions caused by high-stress emergency situations.

The aim of **Critical Incident Stress Management (CISM)** is to prevent these reactions from having a negative impact on the fire fighter's work and life, both short-term and long-term (▲ **Figure 17-7**). Fire fighters should understand the CISM system and how they can access it. The concept of CISM is a simple but important part of fire fighter survival.

The recognized stages of emotional reaction experienced by fire fighters and other rescue personnel after a critical incident can include:

1. Anxiety
2. Denial/disbelief
3. Frustration/anger
4. Inability to function logically
5. Remorse
6. Grief
7. Reconciliation/acceptance

These stages can occur within minutes or hours, or they can take several days or months. Not all of the steps will occur for every event and they do not necessarily occur in order.

CISM helps fire fighters recognize and deal with these reactions in the most positive manner possible. There are some variations in the way this is accomplished, but the purpose remains the same. In-house, peer-driven programs guided and assisted by mental health care professionals are a proven method for CISM. Team members who have faced the same situations and experienced similar feelings can be understanding and nonjudgmental when providing assistance with stress management.

Most CISM programs operate in a similar manner. To begin with, there is an informal process where a trained CISM team member will have a conversation with an individual or small group that has been involved in a critical incident. This is a simple and quick approach to see if any further action is required for the group or for any individual fire fighter. This might take place on-scene, at the fire station, or another location away from the critical incident scene.

If an incident is recognized as requiring further intervention, a series of steps might follow. These steps range from a semi-formal group discussion of the incident to the inclusion of health care professionals. The most common form of CISM is peer defusing, when a group informally discusses events that they experienced together. Critical incident stress debriefings are meetings between fire fighters and specially trained leaders. The purpose of a debriefing is to allow an open discussion of feelings, fears, and reactions to the situation that occurred. A debriefing is not an investigation or an interrogation. Debriefings are usually held within 24 to 72 hours after a major incident. The leaders offer suggestions and information on overcoming stress.

Your fire department should have some type of CISM program in place. All department members should know how it is activated and be supportive of its intent and function. Fire fighters should realize that emotional and mental health must be protected, just as much as physical health and safety.

Wrap-Up

Ready for Review

Fire fighter survival is dependent on many factors. While it is mandated that the safety of fire fighters is a primary responsibility of the incident commander, every member has a personal responsibility to keep safe.

As each situation presents itself, fire fighters should take a risk-management approach to identify safe and unsafe situations. Knowledge, training, and experience all help to reduce exposure to potential risks. Attitude is a starting point. A personal dedication to operating safely is extremely important. Training sessions should be conducted with the idea that how we practice will determine how we operate under real conditions. The ability to use equipment under a variety of conditions can lead to safety and survival. Survival is mental and emotional as well as physical.

Chief Concepts

- Risk-benefit analysis should be done continuously by all members when operating at hazardous incidents.
- Hazards of some kind are always present. They should be identified and used to make a risk-benefit analysis.
- The personnel accountability system is designed to keep track of all personnel operating at an incident.
- Communication procedures should be known and used in a consistent manner at all incidents. This process is especially important when fire fighters are in trouble.
- Rapid Intervention Crews are in place on the fireground for one central purpose: to rescue fire fighters.
- Fire fighters should learn, by whatever means possible, to recognize and analyze hazards when operating at any incident.
- Air management is a critical skill that takes training and practice.
- Emergency evacuation methods are used when normal means of egress are blocked, and include the use of normal tools and techniques in nontraditional ways.
- Critical incident stress is a known hazard that emergency personnel can learn to deal with. Organizational help is usually available in the form of in-house programs or referral to professionals. This hazard deals with long-term health, safety, and survival.

Hot Terms

Air management The way in which an individual utilizes a limited air supply to be sure it will last long enough to enter a hazardous area, accomplish needed tasks, and return safely.

Critical Incident Stress Management (CISM) A system to help personnel deal with critical incident stress in a positive manner; its aim is to promote long-term mental and emotional health after a critical incident.

Guideline A rope used for orientation when inside a structure when there is low or no visibility; it is attached to a fixed object outside the hazardous area.

Personnel Accountability Report (PAR) A report confirming that all members of a company are present.

Personnel accountability system A method of tracking the identity, assignment, and location of fire fighters operating at an incident scene.

Rapid intervention company/crew (RIC) A minimum of two fully equipped personnel on site, in a ready state, for immediate rescue of injured or trapped fire fighters. In some departments, this is also known as a Rapid Intervention Team.

Rehabilitation A systematic process to provide periods of rest and recovery for emergency workers during an incident; usually conducted in a designated area away from the hazardous area.

Risk-benefit analysis The process of weighing predicted risks against potential benefits and making decisions based on the outcome of that analysis.

Safe haven A temporary place of refuge to await rescue.

Self-rescue The activity of a fire fighter using techniques and tools to remove himself from a hazardous situation.

Fire Fighter in Action

It is 4:00 p.m. on a summer day, and you are preparing for your afternoon workout when your engine is dispatched to a residential structure fire. While responding to the fire, the dispatcher radios an update and states that a 3-year-old child is missing and was last seen inside the residence. Your engine is first on the scene and your lieutenant tells you to start a search of the second floor.

1. What are the risk-benefit philosophies that should be considered in this incident?
   - **A.** Accept a higher level of risk when there is a higher possibility to save lives.
   - **B.** Accept limited risk to save property of value.
   - **C.** Deny risk for persons or property that are already lost.
   - **D.** All of the above

2. What risk factors should be considered in this scenario?
   - **A.** Smoke, heat, building construction, length of burn time, location of fire, and utilities
   - **B.** Radioactive materials, snow, and smoke
   - **C.** Industrial equipment, weather, and large HVAC systems on the roof
   - **D.** High fire load, hazardous materials, and high voltage lines

You and your partner enter the structure and begin to do a left wall search from the top of the stairs. While performing your search you become disoriented and lose your partner and your hose line.

3. As soon as you realize that you are lost, you should:
   - **A.** try to find your way out of the structure.
   - **B.** wait until the situation becomes critical to save yourself some embarrassment.
   - **C.** immediately seek assistance by calling for a Mayday.
   - **D.** Both A and B

4. The incident commander should call for a PAR after a fire fighter is missing. A PAR is a(n):
   - **A.** personal accountability safety system.
   - **B.** roll call taken by each supervisor at an emergency incident.
   - **C.** team of fire fighters designated for fire fighter rescue.
   - **D.** emergency signal that is signaled through your portable radio.

www.FireFighter.jbpub.com

Chapter Pretests
Interactivities
Hot Term Explorer
Web Links
Review Manual
FireLearn

Salvage and Overhaul

Chapter 18

NFPA 1001 Standard

Fire Fighter I

5.3.10 (A) *Requisite Knowledge.* Principles of fire streams; types, design, operation, nozzle pressure effects, and flow capabilities of nozzles; precautions to be followed when advancing hose lines to a fire; observable results that a fire stream has been properly applied; dangerous building conditions caused by a fire; principles of exposure protection; potential long-term consequences of exposure to products of combustion; physical states of matter in which fuels are found; common types of accidents or injuries and their causes; and the application of each size and type of attack line, the role of the backup team in fire attack situations, attack and control techniques for grade level and above and below grade levels, and exposing hidden fires.

5.3.13 Overhaul a fire scene, given personal protective equipment, attack line, hand tools, a flashlight, and an assignment, so that structural integrity is not compromised, all hidden fires are discovered, fire cause evidence is preserved, and the fire is extinguished.

5.3.13 (A) *Requisite Knowledge.* Types of fire attack lines and water application devices most effective for overhaul, water application methods for extinguishment that limit water damage, types of tools and methods used to expose hidden fire, dangers associated with overhaul, obvious signs of area of origin or signs of arson, and reasons for protection of fire scene.

5.3.13 (B) *Requisite Skills.* The ability to deploy and operate an attack line; remove flooring, ceiling, and wall components to expose void spaces without compromising structural integrity; apply water for maximum effectiveness; expose and extinguish hidden fires in walls, ceilings, and subfloor spaces; recognize and preserve obvious signs of area of origin and arson; and evaluate for complete extinguishment.

5.3.14 Conserve property as a member of a team, given salvage tools and equipment and an assignment, so that the building and its contents are protected from further damage.

5.3.14 (A) *Requisite Knowledge.* The purpose of property conservation and its value to the public, methods used to protect property, types of and uses for salvage covers, operations at properties protected with automatic sprinklers, how to stop the flow of water from an automatic sprinkler head, identification of the main control valve on an automatic sprinkler system, and forcible entry issues related to salvage.

5.3.14 (B) *Requisite Skills.* The ability to cluster furniture; deploy covering materials; roll and fold salvage covers for reuse; construct water chutes and catch-alls; remove water; cover building openings, including doors, windows, floor openings, and roof openings; separate, remove, and relocate charred material to a safe location while protecting the area of origin for cause determination; stop the flow of water from a sprinkler with sprinkler wedges or stoppers; and operate a main control valve on an automatic sprinkler system.

5.3.17 Illuminate the emergency scene, given fire service electrical equipment and an assignment, so that designated areas are illuminated and all equipment is operated within the manufacturer's listed safety precautions.

5.3.17 (A) *Requisite Knowledge.* Safety principles and practices, power supply capacity and limitations, and light deployment methods.

5.3.17 (B) *Requisite Skills.* The ability to operate department power supply and lighting equipment, deploy cords and connectors, reset ground-fault interrupter (GFI) devices, and locate lights for best effect.

Fire Fighter II

6.5.2 Maintain power plants, power tools, and lighting equipment, given tools and manufacturers' instructions, so that equipment is clean and maintained according to manufacturer and departmental guidelines, maintenance is recorded, and equipment is placed in a ready state or reported otherwise.

6.5.2 (A) *Requisite Knowledge.* Types of cleaning methods, correct use of cleaning solvents, manufacturer and departmental guidelines for maintaining equipment and its documentation, and problem-reporting practices.

6.5.2 (B) *Requisite Skills.* The ability to select correct tools; follow guidelines; complete recording and reporting procedures; and operate power plants, power tools, and lighting equipment.

Knowledge Objectives

After studying this chapter, you will be able to:

- Describe the safety precautions that need to be considered when performing salvage.
- List the tools that are used for salvage.
- Describe how fire fighters can limit losses from smoke and heat.
- Describe the steps needed to protect building contents using a salvage cover.
- Describe some general steps that can be taken to limit water damage.
- Describe the steps needed to stop the flow of water from activated sprinkler heads.
- Describe overhaul.
- List the safety concerns that must be addressed to ensure safety for fire fighters performing overhaul.
- Describe how to preserve structural integrity during overhaul.
- List the tools that are used for overhaul.
- Describe the importance of adequate lighting at the fire scene and in the fire building.

Skills Objectives

After studying this chapter, you will be able to:

- Perform the one fire fighter salvage cover fold.
- Perform the two fire fighter salvage cover fold.
- Fold and roll a salvage cover.
- Perform a one-person salvage cover roll.
- Perform a shoulder toss.
- Perform a balloon toss.
- Stop water from a sprinkler head using a sprinkler stop.
- Stop water from a sprinkler head using sprinkler wedges.
- Close and open main control valve (outside screw-and-yoke).
- Close and open main control valve (post-indicator).
- Construct a water chute.
- Construct a water catch-all.
- Open a ceiling to check for fire with a pike pole.
- Open an interior wall.
- Inspect and run a generator.

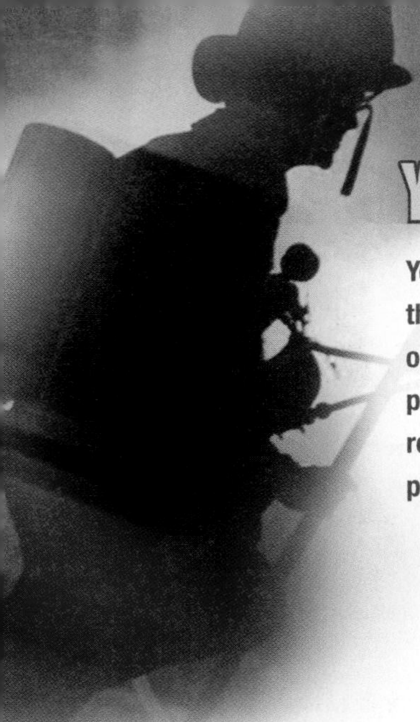

You Are the Fire Fighter

Your ladder company is dispatched to assist units at a residential fire. When you arrive, you find that search-and-rescue operations have been completed, and fire suppression crews are working on the second floor and in the attic. Your crew is assigned to salvage operations. You and your partner take salvage covers into the dining room, pull the furniture to the middle of the room, remove hanging photographs, and add them to the pile. You top the pile with a salvage cover to protect everything from water damage.

1. ***What are the most important objects to try to save in a house?***
2. ***Should you concentrate your salvage operations on certain rooms?***

Introduction

The basic goals of firefighting are first to save lives, second to control the fire, and third to protect property. Salvage and overhaul operations help meet that third goal by limiting and reducing property losses resulting from a fire. **Salvage** efforts protect property and belongings from damage, particularly from the effects of smoke and water. **Overhaul** ensures that a fire is completely extinguished by finding and exposing any smoldering or hidden pockets of fire in an area that has been burned.

Salvage and overhaul are usually conducted in close coordination with each other and have a lower priority than search and rescue or controlling the fire. If a department has enough personnel on scene to address more than one priority, salvage can be done concurrent with fire suppression. Overhaul, however, can only begin when the fire is under control.

Throughout the salvage and overhaul process, fire fighters must attempt to preserve evidence related to the cause of a fire, particularly when working near the suspected area of fire origin and when the fire cause is not obvious. In general, disturb as little as possible in or near the area of origin until a fire investigator determines the cause of the fire and gives permission to clear the area. Knowing what caused a fire can help prevent future fires and enable victims to recoup losses under their insurance coverage. This subject is covered in greater depth in Chapter 37, Fire Cause Determination.

Fire fighters performing salvage and overhaul must be able to see where they are going, what they are doing, and what potential hazards are present. The smoke may have cleared, but electrical power may still be unavailable. Portable lights should be set up to fully illuminate areas during salvage and overhaul.

Salvage

Salvage operations are conducted to save property from a fire and to reduce the damage that results from a fire. Often, the damage caused by smoke and water can be more extensive and costly to repair or replace than the property that is burned. Salvage operations are also conducted in other situations that endanger property, such as floods or water leaking from burst pipes.

Salvage efforts usually are aimed at preventing or limiting **secondary losses** that result from smoke and water damage, fire suppression efforts, and other causes. They must be performed promptly to preserve the building and protect the contents from avoidable damage. Salvage operations include expelling smoke, removing heat, controlling water runoff, removing water from the building, securing a building after a fire, covering broken windows and doors, and temporarily patching ventilation openings in the roof to protect the structure and contents.

Fire fighters must help protect property from avoidable loss or damage as part of their professional responsibilities. At the scene of a fire or any other type of emergency, the victims' property is entrusted to the care of fire fighters who are expected to protect that property to the best of their ability.

Salvage efforts at residential fires are often focused on protecting personal property. These items often have a personal impact on the victims' lives. Fire fighters can often identify and protect critical personal belongings, such as wallets, purses, car keys, or medications. The value of some items, such as photographs and family heirlooms, cannot be measured in dollars. These items are treasured by victims and are often irreplaceable.

Salvage priorities at commercial or industrial fires are quite different. The preincident plan should identify critical items and enable fire fighters to focus their salvage efforts appropriately. For example, a law firm might identify its files as the most valuable property, while a library places more importance on books and microfiche files than on furniture. A preincident plan should note whether the contents of a warehouse are valuable electronic components or industrial steel pipe. Factory machinery could be a higher priority than the finished products stored on premises.

Fire Fighter Safety Tips

At the emergency scene, the Incident Commander (IC) should know where every crew is working and what they are doing at all times. The IC relies on your company officer to know your location and activities and to relay this information to the command post.

Safety Considerations During Salvage Operations

Safety is a primary concern during salvage operations. Personal protective equipment (PPE) required will depend on the site of the salvage operations, the stage of the fire, and any hazards present. Salvage operations often begin while the fire is burning and continue for several hours after it has been extinguished. During firefighting operations and immediately afterward, fire fighters should wear a full set of protective clothing and equipment, including self-contained breathing apparatus (SCBA). After the fire is out, the area must be ventilated, and the atmosphere tested and determined to be safe by the Safety Officer before fire fighters can work without SCBA.

If salvage is conducted in a different part of the building, several floors below the fire, or in a neighboring building, fire fighters may have other protective clothing requirements. Hazards in these areas may be quite different from those in the fire area. Fire fighters conducting salvage operations in remote areas can wear partial PPE (boots, turnout pants, helmet, and gloves) with the approval of the Safety Officer.

Structural collapse is always a possibility during salvage operations because of fire damage to the building's structural components. Lightweight trusses can collapse quickly when damaged by a fire. Heavy objects, such as air conditioning units and appliances, might crash through roofs or ceilings weakened by fire damage.

Even the water used to douse the fire can contribute to structural collapse. Each gallon of water weighs 8.3 lb. Extinguishing a major fire could load the structure with tons of extra water weight. If the structure is already weakened by the fire, the extra water weight could cause a catastrophic structural collapse, with the potential to injure or kill fire fighters.

At times, this danger can be masked by absorbent materials such as bolts of cloth or bales of rags stored in the building. Uneven absorption also presents a danger. Stacked rolls of newsprint in a printing factory could collapse if the bottom layer becomes waterlogged and unable to support the stack.

Water can also create potential hazards for fire fighters conducting salvage operations on floors below the fire. Water used to extinguish a fire leaks down through the building and can cause major damage. Water trapped in the void space above a ceiling can cause sections of the ceiling to collapse. Fire fighters should watch for warning signs such as water leaking through ceiling joints and sagging areas of ceilings. Water that accumulates in a basement can hide a stairway and present additional hazards if utility connections are also located in the basement.

Damage to the building's utility systems presents significant risks to fire fighters conducting salvage operations. Gas and electrical services should always be shut off to eliminate the potential hazards of electrocution or explosion stemming from gas leaks.

Salvage Tools

Fire fighters conducting salvage operations use special tools and equipment as listed in ▼ Table 18-1. Salvage covers and other protective items shield and cover building contents. Other salvage tools remove smoke, water, and other products that can damage property ▼ Figure 18-1.

Table 18-1 Salvage Tools

- Salvage covers; treated canvas or plastic
- Box cutter for cutting plastic
- Floor runners
- Wet/dry vacuums
- Squeegees
- Submersible pumps and hose
- Sprinkler shut-off kit
- Ventilation fans, power blowers
- Small tool kit

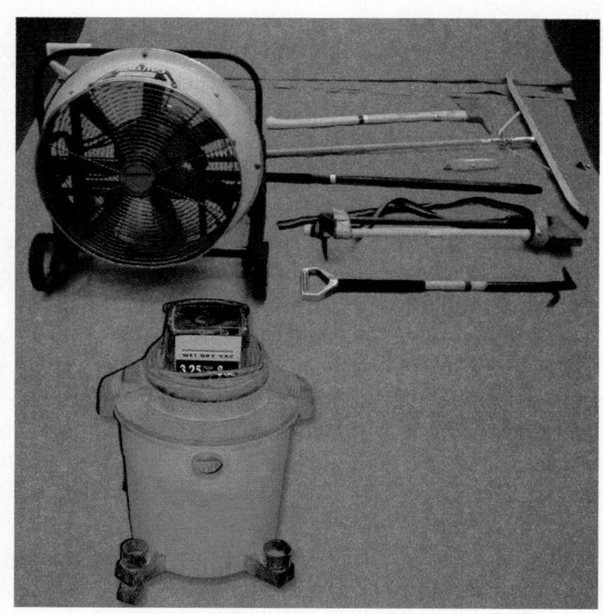

Figure 18-1 Salvage tools.

Limiting Smoke and Heat Damage

One way to reduce property loss is to keep heat and smoke out of areas that are not involved in the fire. A closed door can effectively keep smoke and heat out of a room or area. Fire fighters doing search-and-rescue operations should remember to close doors after a room is searched.

Properly timed and effective ventilation practices can limit smoke and heat damage. Rapid ventilation will often reduce smoke damage in an area already filled with smoke. The visible components of smoke are mostly soot particles and other products of combustion, including corrosive chemicals that will settle on any horizontal surface when the smoke cools. Blowing the smoke out of the building is a better option than allowing contaminants to settle on the contents. All members of the fire suppression team—not just the salvage crews—should recognize opportunities to prevent smoke and heat damage. Ventilation is discussed in detail in Chapter 14, Ventilation.

Salvage Covers

The most common method of protecting building contents is to cover them with salvage covers. **Salvage covers** are large square or rectangular sheets of heavy canvas or plastic material used to protect furniture and other items from water runoff, falling debris, soot, and particulate matter in smoke residue.

Salvage crews usually begin on the floor immediately below the fire to prevent water damage to furniture and other contents on lower floors. If the fire is in an attic, fire fighters may have enough time to spread covers over the furniture in the rooms directly below the fire before pulling the ceilings to attack the flames.

The most efficient way to protect a room's contents is to move all the furniture to the center of the room, away from the walls, where water could damage the backs of the furniture. This reduces the total area that must be covered, enabling one or two fire fighters to cover the pile quickly and move on to the next room (► Figure 18-2).

Remove any pictures from the walls and place them with the furniture. Put smaller pictures and valuable objects in drawers or wherever they will be protected from breakage. If there is time, roll any rugs and place them on the pile.

Some departments use rolls of construction-grade polyethylene film instead of salvage covers. This material comes in rolls as long as 120′ so it can be unrolled over the building contents and cut to the correct length with a box cutter. It is useful for covering long surfaces such as retail display counters. This material is disposable and can be left behind after the fire, unlike traditional salvage covers, which must be picked up, washed, dried, and properly folded for storage in fire apparatus compartments after each use. Occasionally, salvage covers may be left behind, as protection for the building contents, but should be picked up when they are no longer needed.

Figure 18-2 Move furniture to the center of the room so that a single salvage cover can protect all of the contents.

Special folding and rolling techniques are used to store salvage covers so that they can be deployed quickly by one or two fire fighters. Be certain to follow your fire department's Standard Operating Procedures (SOPs) when folding or rolling a salvage cover.

To fold a salvage cover to prepare it for one fire fighter to deploy, follow the steps in (► Skill Drill 18-1).

1. Spread the salvage cover flat on the ground. Stand at one end, facing your partner standing at the other end. **(Step 1)**
2. You and your partner each place one hand on the outer edge of the cover and the other hand one quarter of the way in from the edge. **(Step 2)**
3. Together, flip the outside edge in 3″ from the middle of the cover, creating a fold at the quarter point. **(Step 3)**
4. Flip the outside fold in to the same point of the cover, creating a second fold. Repeat steps 2, 3, and 4, from the opposite side of the cover. The folded edges should meet at the middle of the cover, with the folds 6″ apart. **(Step 4)**
5. Fold the two halves of the salvage cover together. **(Step 5)**
6. Starting from the middle of the cover, use a broom to brush the air out of the cover. **(Step 6)**
7. Move to the newly created narrow end of the salvage cover. **(Step 7)**
8. Fold the narrow end of the salvage cover 3″ from the middle of the cover, creating a fold at the quarter point.
9. Flip the outside fold of the narrow end in to the same point of the cover, creating a second fold. The folded edge should meet at the middle of the cover, with the folds 6″ apart. **(Step 8)**
10. Fold the two halves of the salvage cover together. **(Step 9)**

Skill Drill

One Fire Fighter Salvage Cover Fold

Spread the salvage cover flat on the ground. Stand at one end, facing your partner standing at the other end.

You and your partner each place one hand on the outer edge of the cover and the other hand one quarter of the way in from the edge.

Together, flip the outside edge in 3" from the middle of the cover, creating a fold at the quarter point.

Flip the outside fold in to the same point of the cover, creating a second fold. Repeat steps 2, 3, and 4, from the opposite side of the cover. The folded edges should meet at the middle of the cover, with the folds 6" apart.

Fold the two halves of the salvage cover together.

Starting from the middle of the cover, use a broom to brush the air out of the cover.

Continued

18-1 Skill Drill

One Fire Fighter Salvage Cover Fold—continued

Move to the newly created narrow end of the salvage cover.

Fold the narrow end of the salvage cover 3" from the middle of the cover, creating a fold at the quarter point. Flip the outside fold of the narrow end in to the same point of the cover, creating a second fold. The folded edge should meet at the middle of the cover, with the folds 6" apart.

Fold the two halves of the salvage cover together.

To fold a salvage cover to prepare it for two fire fighters to deploy, follow the steps in (► **Skill Drill 18-2**).

1. Spread the salvage cover flat on the ground. Stand at one end, facing your partner standing at the other end.
2. You and your partner each place one hand on the outer edge of the cover and the other hand one quarter of the way in from the edge. **(Step 1)**
3. Together, fold the cover in half. **(Step 2)**
4. Together, grasp the unfolded edge and fold the cover in half again. **(Step 3)**
5. Starting from the middle of the cover, use a broom to brush the air out of the cover. **(Step 4)**
6. Move to the newly created narrow end of the salvage cover.
7. Fold the salvage cover in half length-wise. **(Step 5)**
8. Fold the salvage cover in half length-wise again. Make certain that the open end is on top.
9. Fold the cover in half a third and final time. **(Step 6)**

Two Fire Fighter Salvage Cover Fold

Spread the salvage cover flat on the ground. Stand at one end, facing your partner standing at the other end. You and your partner each place one hand on the outer edge of the cover and the other hand one quarter of the way in from the edge.

Together, fold the cover in half.

Together, grasp the unfolded edge and fold the cover in half again.

Starting from the middle of the cover, use a broom to brush the air out of the cover.

Move to the newly created narrow end of the salvage cover. Fold the salvage cover in half length-wise.

Fold the salvage cover in half length-wise again. Make certain that the open end is on top. Fold the cover in half a third and final time.

To fold and roll a salvage cover, follow the steps in ► Skill Drill 18-3.

1. Spread the salvage cover flat on the ground. Stand at one end, facing your partner standing at the other end. **(Step 1)**
2. You and your partner each place one hand on the outer edge of the cover and the other hand one quarter of the way in from the edge. **(Step 2)**
3. Together, flip the outside edge in to the middle of the cover, creating a fold at the quarter point. **(Step 3)**
4. Flip the outside fold in to the middle of the cover, creating a second fold. Repeat steps 2, 3, and 4, from the opposite side of the cover. **(Step 4)**
5. The folded edges should meet at the middle of the cover, with the folds touching, but not overlapping. **(Step 5)**
6. Tightly roll up the folded salvage cover from the end. **(Step 6)**

Although it takes two people to fold and roll a salvage cover for storage, this technique enables one person to unroll the cover easily and quickly. To perform the one-person salvage cover roll, follow the steps in ► Skill Drill 18-4.

1. Stand in front of the end of the object that you are going to cover. **(Step 1)**
2. Start to unroll the cover over one end of the object. **(Step 2)**
3. Continue unrolling the cover over the top of the object. Allow the remainder of the roll to settle at the other side of the object. **(Step 3)**
4. Spread the cover, unfolding each side outward over the object to the first fold. **(Step 4)**
5. Unfold the second fold on each side to drape the cover completely over the object. **(Step 5)**
6. Tuck in all loose edges of the cover around the object. **(Step 6)**

A single fire fighter can also use a shoulder toss to spread a salvage cover. Follow the steps in ► Skill Drill 18-5 to do a shoulder toss.

1. Place a folded salvage cover over one arm. **(Step 1)**
2. Toss the cover over the salvaged object with a straight-arm movement. **(Step 2)**
3. Unfold the cover until it completely drapes the object. **(Step 3)**

Two fire fighters can use a balloon toss to cover a pile of building contents quickly. The balloon toss is shown in ► Skill Drill 18-6.

1. Place the cover on the ground beside the object. **(Step 1)**
2. Unfold the cover so that it runs along the entire base of the object. Each fire fighter grabs one edge of the cover and brings it up to waist height. **(Step 2)**
3. Together, lift the cover quickly so that it fills with air like a balloon. **(Step 3)**
4. Move quickly to the other side of the item, using the air to help support the cover, and spread the entire cover over the object. **(Step 4)**

Skill Drill

Fold and Roll a Salvage Cover

Spread the salvage cover flat on the ground. Stand at one corner, facing your partner standing at the other end.

You and your partner each place one hand on the outer edge of the cover and the other hand one quarter of the way in from the edge.

Together, flip the outside edge in to the middle of the cover, creating a fold at the quarter point.

Flip the outside fold in to the middle of the cover, creating a second fold. Repeat steps 2, 3, and 4 from the opposite side of the cover.

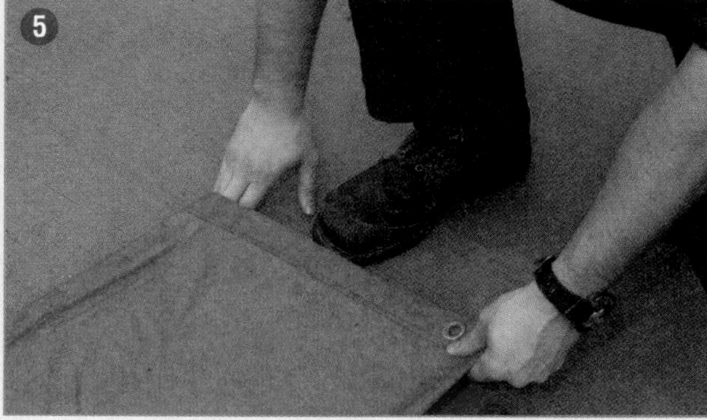

The folded edges should meet at the middle of the cover, with the folds touching, but not overlapping.

Tightly roll up the folded salvage cover from the end.

18-4 Skill Drill

One-Person Salvage Cover Roll

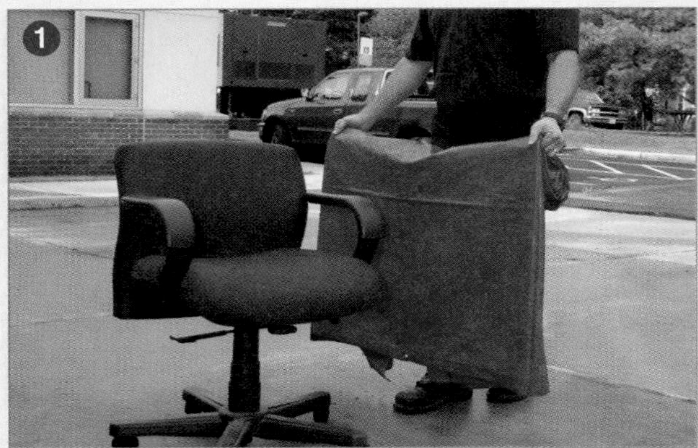
1. Stand in front of the end of the object that you are going to cover.

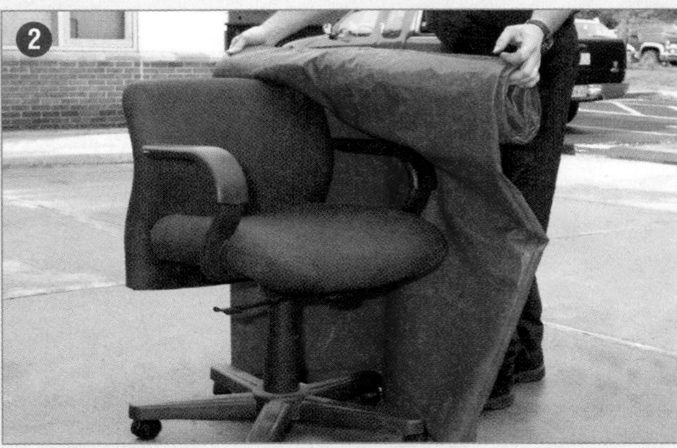
2. Start to unroll the cover over one end of the object.

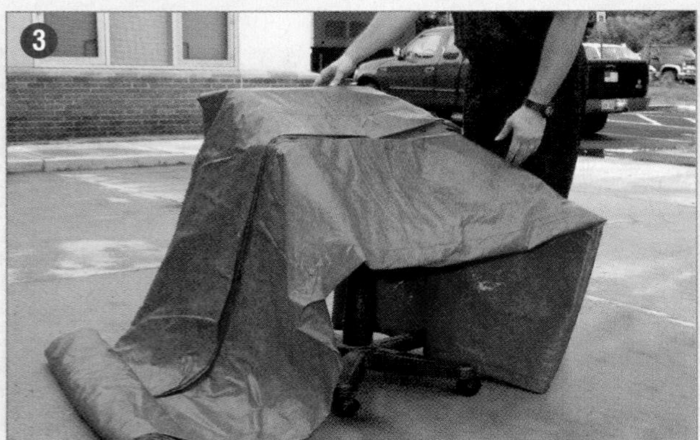
3. Continue unrolling the cover until you reach the top of the object. Allow the remainder of the cover to unroll and settle at the end of the object.

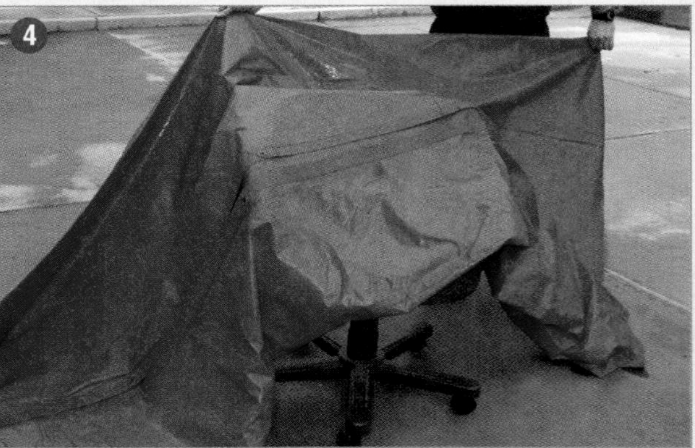
4. Spread the cover, unfolding each side outward over the object to the first fold.

5. Unfold the second fold on each side and drape the cover completely over the object.

6. Tuck in all loose edges of the cover around the object.

18-5 Skill Drill

Shoulder Toss

Place a folded salvage cover over one arm.

Toss the cover over the salvaged object with a straight-arm movement.

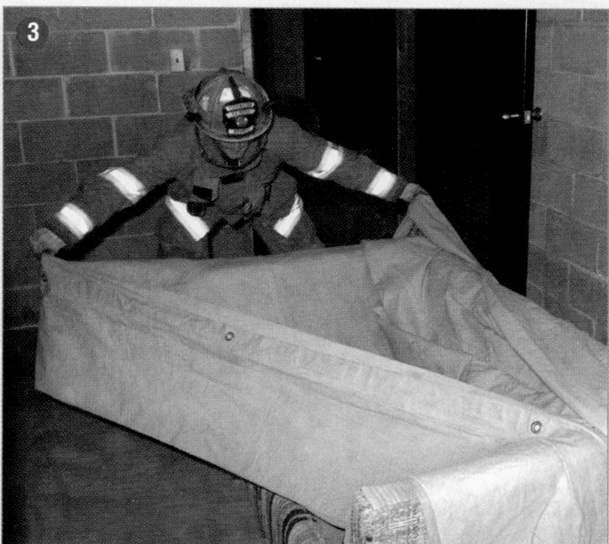

Unfold the cover until it completely drapes the object.

18-6 Skill Drill

Balloon Toss

Place the cover on the ground beside the object.

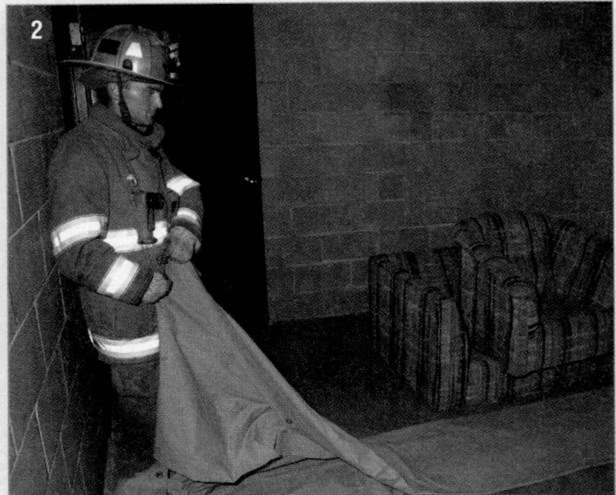

Unfold the cover so that it runs along the entire base of the object. Each fire fighter grabs one edge of the cover and brings it up to waist height.

Together, lift the cover quickly so that it fills with air like a balloon.

Move quickly to the other side of the item, using the air to help support the cover, and spread the entire cover over the object.

Floor Runners

A floor runner is a long section of protective material used to cover a section of flooring or carpet. Floor runners protect carpets or hardwood floors from water, debris, fire fighters' boots, and firefighting equipment. Fire fighters entering an area for salvage operations should unroll the floor runner ahead of themselves and stay on the floor runner while working in the area.

Preventing Water Damage

The best way to prevent water damage at a fire scene is to limit the amount of water used to fight the fire. Use only enough water to knock down the fire quickly. Do not use more water than is needed. Do not spray water blindly into smoke or continue to douse a fire that is already out; this only creates unnecessary water damage.

Turn off hose nozzles when they are not in use. If a nozzle must be left partly open to prevent freezing during cold weather, direct the flow through a window. Take time to tighten any leaky hose couplings after the fire has been knocked down. This prevents additional water damage during overhaul. Look for leaking water pipes in the building; if you find any, shut off the building's water supply.

Deactivating Sprinklers

Buildings equipped with sprinkler systems require special steps to limit water damage. Sprinklers can control a fire efficiently, but will keep flowing until the activated sprinklers are shut off or the entire system is shut down. This can cause water damage.

Sprinklers should be shut down as soon as the IC declares that the fire is under control. Do not shut down sprinklers prematurely; the fire may rekindle, causing major damage. Between the time that sprinklers are shut down and overhaul is completed, a fire fighter with a portable radio should stand by and be ready to reactivate the sprinklers if necessary.

Sprinkler systems are usually designed so that only those sprinkler heads directly over or close to the fire will activate. Most fires can be contained or fully extinguished by the release of water from only one or two sprinkler heads (▶ Figure 18-3). Although movies may show all sprinkler heads activating at once, this does not occur in real life. More information on sprinkler systems is presented in Chapter 36, Fire Protection, Detection, and Suppression Systems.

If only one or two sprinkler heads have been activated, inserting a sprinkler wedge or a sprinkler stop can quickly stop the flow. This also keeps the rest of the system operational during overhaul.

A sprinkler wedge is a simple triangular piece of wood. Inserting two wedges on opposite sides, between the orifice and the deflector of the sprinkler, and pushing them together effectively plugs the opening, stopping the flow from standard upright and pendant sprinkler heads.

Figure 18-3 Most fires are controlled by the activation of only one or two sprinkler heads.

A sprinkler stop is a more sophisticated mechanical device with a rubber stopper that can be inserted into a sprinkler head. Several types of sprinkler stops are available, including some that work only with specific sprinkler heads. Not all sprinkler heads, however, can be shut off with a wedge or a sprinkler stop. Recessed sprinklers, which are often installed in buildings with finished ceilings, are usually difficult to shut off. If the individual heads cannot be shut off, stop the water flow by closing the sprinkler control valve.

To stop the flow of water from a sprinkler head using a sprinkler stop, follow the steps in (Skill Drill 18-7).

1. Have a sprinkler stop in hand.
2. Place the flat-coated part of the sprinkler stop over the sprinkler head orifice and between the frame of the sprinkler head.
3. Push the lever to expand the sprinkler stop until it snaps into position.

To stop the flow of water from a sprinkler head using a pair of sprinkler wedges, follow the steps in (▶ Skill Drill 18-8).

1. Hold one wedge in each hand. **(Step 1)**
2. Insert the two wedges, one from each side, between the discharge orifice and the sprinkler head deflector. **(Step 2)**
3. Bump the wedges securely into place to stop the water flow. **(Step 3)**

18-8 Skill Drill

Using Sprinkler Wedges

1 Hold one wedge in each hand.

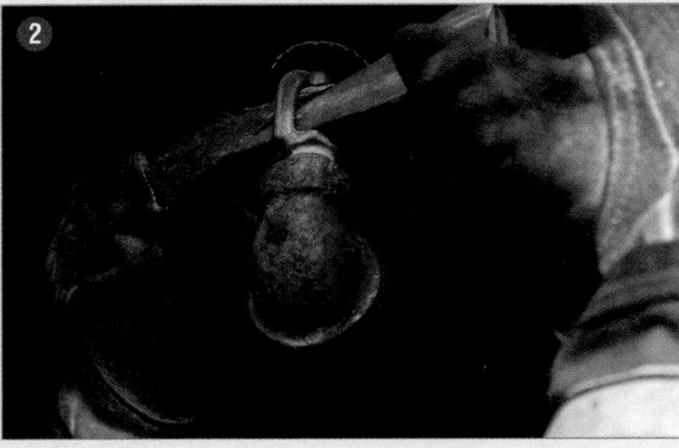
2 Insert the two wedges, one from each side, between the discharge orifice and the sprinkler head deflector.

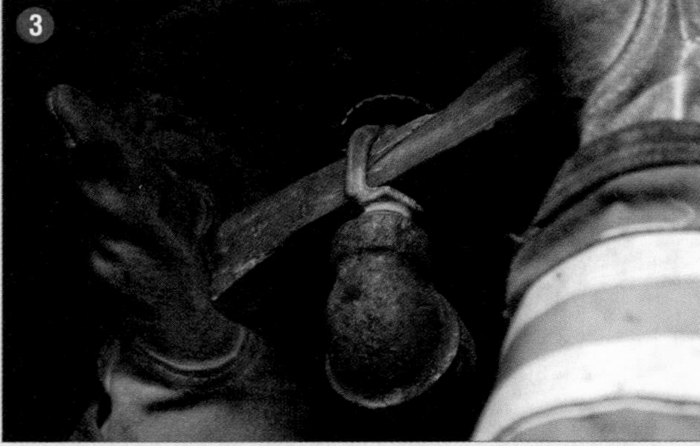
3 Bump the wedges securely into place to stop the water flow.

If individual heads cannot be plugged or if several sprinklers have been activated, stop the flow by closing the sprinkler control valve. After the valve is closed, the water trapped in the piping system will drain out through the activated sprinkler heads for several minutes. If there is a drain valve near the control valve, open it to drain the system quickly. This also directs the water flow to a location where it will not cause additional damage.

The main control valve for a sprinkler system is usually an **outside stem and yoke (OS&Y) valve** or a **post indicator valve (PIV)**. An OS&Y valve is typically found in a mechanical room in the basement or on the ground floor level of a building. Post indicator valves are located outside the building or on an exterior wall. Some sprinkler systems also have zone valves that control the flow of water to sprinklers in different areas of the building. High-rise buildings usually have a zone valve for each floor. Closing a zone valve stops the flow to sprinklers in the zone; in the rest of the building, sprinklers remain operational. The locations of the main control valves and zone valves should be identified during preincident planning visits to a building.

Sprinkler control valves should always be locked in the open position, usually with a chain and padlock, and equipped with a tamper alarm tied into the fire alarm system. Fire fighters sent to shut off a sprinkler control valve should take a pair of bolt cutters to remove the lock if the key is not available.

To close and re-open the main control valve, follow the steps in (► **Skill Drill 18-9**).

1. Locate the OS&Y valve as indicated on the preincident plan. It may be inside the building or close by outside. The stem of an OS&Y valve protrudes from the valve handle in the open position.
2. Identify the valve that controls sprinklers in the fire area. Smaller systems will have only one OS&Y valve. Large buildings may have multiple sprinkler systems or one main control valve and additional zone valves controlling the flow to different areas.
3. If the valve is locked in the open position with a chain and padlock, and the key is readily available, unlock and remove the chain. If no key is available, cut the lock or the chain with a pair of bolt cutters. Cut a link close to the padlock so that the chain can be reused. **(Step 1)**
4. Turn the valve handle clockwise to close the valve. Keep turning until resistance is strong and little of the stem is visible. **(Step 2)**
5. To open the OS&Y valve, turn the handle counterclockwise until resistance is strong and the stem is visible again. Lock the valve in the open position. **(Step 3)**

To close and open a post indicator valve, follow the steps in (► **Skill Drill 18-10**).

1. Locate the PIV (usually located outside the building). The location of the valve should be indicated on a preincident plan.
2. Most PIVs will be locked in the open position by a padlock. Unlock the padlock if the key is readily available. If no key is available, cut the lock with a pair of bolt cutters. **(Step 1)**
3. Remove the handle from its storage position on the PIV and place it on top of the valve, similar to a hydrant wrench.
4. Turn the valve stem in the direction indicated on top of the valve to close the valve. Keep turning until resistance is strong and the visual indicator changes from "OPEN" to "SHUT." **(Step 2)**
5. To reopen the PIV, turn in the opposite direction until resistance is strong and the indicator changes back to "OPEN." The valve should then be locked in the open position. **(Step 3)**

Before a sprinkler system can be restored to normal operation, the sprinkler heads that have been activated must be replaced. Every sprinkler system should have spare heads stored somewhere, usually near the main control valves. Some fire departments may carry a selection of different heads in a sprinkler kit on the apparatus. An activated sprinkler head must be replaced with another head of the same design, size, and temperature rating.

The main sprinkler control valve or the appropriate zone valve must be closed, and the system must be drained before a sprinkler head can be changed. Special wrenches must be used when replacing sprinkler heads to prevent damage to the operating mechanism.

After the activated heads are replaced, the sprinkler system can be restored to service. Restoring a sprinkler system to service takes special training and must be performed only by qualified individuals.

18-9 Skill Drill

Close and Re-Open Main Control Valve (OS&Y)

1 Locate the OS&Y valve as indicated on the preincident plan. It may be inside the building or close by outside. The stem of an OS&Y valve protrudes from the valve handle in the open position. Identify the valve that controls sprinklers in the fire area. Smaller systems will have only one OS&Y valve. Large buildings may have multiple sprinkler systems or one main control valve and additional zone valves controlling the flow to different areas. If the valve is locked in the open position with a chain and padlock, and the key is readily available, unlock and remove the chain. If no key is available, cut the lock or the chain with a pair of bolt cutters. Cut a link close to the padlock so that the chain can be reused.

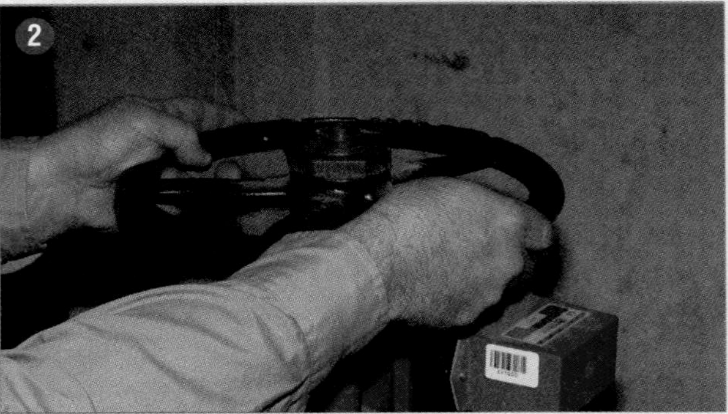

2 Turn the valve handle clockwise to close the valve. Keep turning until resistance is strong and little of the stem is visible.

3 To open the OS&Y valve, turn the handle counterclockwise until resistance is strong and the stem is visible again. Lock the valve in the open position.

Removing Water

Water that accumulates within a building or drips down from higher levels should be channeled to a drain or to the outside of the building to prevent or limit water damage. A pump may be needed to help remove the water in some cases.

Some buildings have floor drains that will funnel water into the below-ground sewer system. Floor drains should be kept free from debris so that the water can drain freely. Fire fighters can use squeegees to direct the water into the drain.

In buildings or houses without floor drains, it may be possible to shut off the water supply to a toilet, remove the nuts that hold the toilet bowl to the floor, and remove the toilet from the floor flange. This creates a large drain capable of handling large quantities of water, as long as someone keeps the opening from becoming clogged with debris.

Skill Drill

Close and Open Main Control Valve (PIV)

Locate the PIV. The PIV is usually located outside the building as indicated on a preincident plan. Most PIVs will be locked in the open position with a padlock. If the key is readily available, unlock the padlock. If no key is available, cut the lock with a pair of bolt cutters.

Remove the handle from its storage position on the PIV and place it on top of the valve. Turn the valve stem in the direction indicated on top of the valve to close the valve. Keep turning until resistance is strong and the visual indicator changes from "OPEN" to "SHUT."

To reopen the PIV, turn in the opposite direction until resistance is strong and the indicator changes back to "OPEN." The valve should then be locked in the open position.

Water on a floor at grade level can often be channeled to flow outside through a doorway or other opening. Water on a floor above ground level can sometimes be drained to the outside by making an opening at floor level in an exterior wall of the building. This type of opening is called a **scupper**.

Water chutes or **water catch-alls** are often used to collect water leaking down from firefighting operations on higher floor levels and to help protect property on calls involving burst pipes or leaking roofs. Water chutes and catch-alls can be constructed quickly using salvage covers.

A water chute catches dripping water and directs it toward a drain or to the outside through a window or doorway. A water catch-all is a temporary pond that holds dripping water in one location. The accumulated water must then be drained to the outside of the building.

Voices of Experience

"When done properly, salvage and overhaul enable you to show care and compassion for the occupants, and can help the property owner save thousands of dollars."

Throughout your fire service career, you will respond to a variety of incidents and serve people from all walks of life. You will respond to incidents at homes and businesses valued at more than a million dollars. You will marvel at the beautiful furnishings, the expensive computer equipment and electronics, and the precious family pictures that you see, and understand the need for careful salvage and overhaul operations to protect all the memories and valuables in this property.

Do not forget that it is also your duty to protect smaller homes and business that do not have such items of luxury and expense. The salvage work that we do at these properties is every bit as necessary as at any other occupancy. Sometimes the occupants do not have insurance, and salvage and overhaul operations are even more critical at these incidents.

Salvage work is probably one of the best public relations activities that fire departments can perform. When done properly, salvage and overhaul enable you to show care and compassion for the occupants, and can help the property owner save thousands of dollars. You and your crew can save the property and its contents from needless destruction due to smoke and water damage. Irreplaceable and valuable possessions should be saved, regardless of the type of occupancy you are working in.

Mark Cogswell
Massachusetts Firefighting Academy
Stow, Massachusetts

To construct a water chute, follow the steps in ► Skill Drill 18-11. A water chute can be constructed with a single salvage cover, or with a cover and two pike poles for support.

1. Fully open a large salvage cover flat on the ground. **(Step 1)**
2. Roll the cover tightly from one edge toward the middle. If using pike poles, lay one pole on the edge and roll the cover around the handle. Roll the opposite edge tightly toward the middle in the same manner. Stop when the rolls are 1′ to 3′ apart. **(Step 2)**
3. Turn the cover upside down. Position the chute so it collects the dripping water and channels it toward a drain or outside opening. The chute can be placed on the floor, with one end propped-up by a chair or other object.
4. Use a stepladder or other tall object to support chutes constructed with pike poles. **(Step 3)**

To construct a water catch-all, follow the steps in ► Skill Drill 18-12.

1. Fully open a large salvage cover flat on the ground. **(Step 1)**
2. Roll two edges inward from the opposite sides, approximately 3′ on each side. **(Step 2)**
3. Fold each of the four corners over at a 90° angle, starting each fold approximately 3′ in from the edge. **(Step 3)**
4. Roll the remaining two edges inward approximately 2′. **(Step 4)**
5. Lift the rolled edge over the corner flaps, and tuck it in under the flaps to lock the corners in place. **(Step 5)**

Water Vacuum

Special vacuum cleaners that suck up water also can be used during salvage operations. There are two types of **water vacuums**: a small capacity backpack-type and a larger wheeled unit. The backpack vacuum cannot be used by someone who is wearing SCBA. Additionally, wet/dry shop vacs may also be used as a low cost alternative.

Fire Fighter Safety Tips

You cannot operate a gasoline-powered pump safely inside a structure without wearing SCBA because it generates poisonous carbon monoxide gas.

Drainage Pumps

Drainage pumps remove water that has accumulated in basements or below ground level. Portable electric submersible pumps can be lowered directly into the water to pump it out of a building. Gasoline-powered portable pumps are placed outside because they exhaust poisonous carbon monoxide gas. These pumps use a hard-suction hose lowered into the basement through a window to draft water out of the building.

Other Salvage Operations

The best way to protect the contents of a building may be to move them to a safe location. The IC will make this determination. Any items removed from the building should be placed in a dry, secure area, preferably a single location. In some cases, salvaged items can be moved to a suitable location within the same building. If items are moved outside, they should be protected from further potential damage caused by firefighting operations or the weather. Valuable items should be placed in the care of a law enforcement officer if the property owner is not present.

There are times when the building contents can provide clues to the cause or spread of the fire. In these situations, fire investigators should be consulted and supervise the removal process.

Salvage operations sometimes extend outside the building to include valuable property such as vehicles or machinery. Use a salvage cover to protect outside property or move vehicles if this can be done without compromising the fire suppression efforts.

18-11 Skill Drill

Construct a Water Chute

Fully open a large salvage cover flat on the ground.

Roll the cover tightly from one edge toward the middle. If using pike poles, lay one pole on the edge and roll the cover around the handle. Roll the opposite edge tightly toward the middle until the two rolls are 1′ to 3′ apart.

Turn the cover upside down. Position the chute so it collects the dripping water and channels it toward a drain or outside opening. The chute can be placed on the floor, with one end propped-up by a chair or other object. Use a stepladder or other tall object to support chutes constructed with pike poles.

Skill Drill

Construct a Water Catch-All

Fully open a large salvage cover flat on the ground.

Roll two edges inward from the opposite sides, approximately 3′ on each side.

Fold each of the four corners at a 90° angle, starting each fold approximately 3′ in from the edge.

Roll the remaining two edges inward approximately 2′.

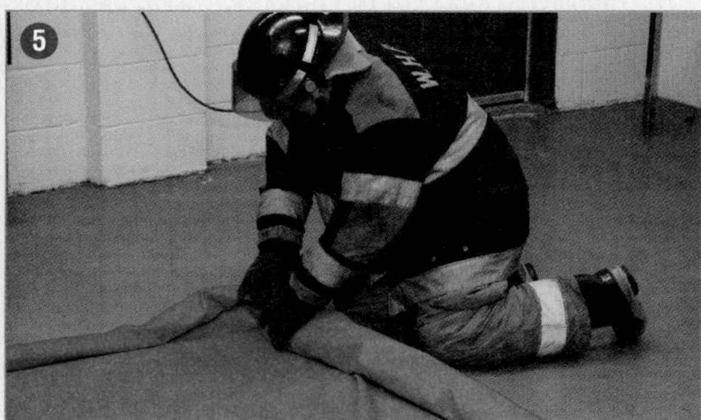

Lift the rolled edge over the corner flaps, and tuck it in under the flaps, to lock the corners in place.

Overhaul

Overhaul is the process of searching for and extinguishing any pockets of fire that remain after a fire has been brought under control. The risk of a new fire remains even if 99% of a fire is out and just 1% is smoldering. A single pocket of embers can **rekindle** after fire fighters leave the scene and cause even more damage and destruction than the original fire. A fire cannot be considered fully extinguished until the overhaul process is complete.

The process of overhaul begins after the fire is brought under control. Overhaul can be a time-consuming, physically demanding process. The greatest challenge during overhaul is to identify and open any void spaces in a building where the fire might be burning undetected. If the fire extends into any void spaces, fire fighters must open the walls and ceilings to expose the burned area. Any materials that are still burning must be soaked with water or physically removed from the building. This process must continue until all of the burned material is located and unburned areas are exposed. Overhaul is also required for non-structure fires such as automobiles, bulk piles (tires or woodchips), vegetation, and even garbage.

Safety Considerations During Overhaul

Many injuries can occur during overhaul. Overhaul is strenuous work conducted in an area already damaged by fire. Fire fighters who were involved in the fire suppression efforts may be physically fatigued and overlook hazards. Many fire departments summon fresh crews to conduct overhaul, giving the crews that fought the fire time to recover.

Several hazards may be present in the overhaul area. The structural safety of the building could be compromised. Heavy objects could collapse roofs or ceilings, debris could litter the area, and there could be holes in the floor. Visibility is often limited, so fire fighters may have to depend on portable lighting. Wet or icy surfaces make falls more likely. Smoldering areas may burst into flames, and the air is probably not safe to breath. Dangerous equipment, including axes, pike poles, and power tools, are used in close quarters.

Fire fighters must be aware of these hazards and proceed with caution. When necessary, take extra time to evaluate the hazards and determine the safest way to proceed. There is no need to rush during overhaul, and there is no excuse for risking injury during this phase of operations.

A charged hose line must always be ready for use during overhaul operations because flare-ups and explosions may occur. A smoldering fire in a void space could flare up suddenly when fire fighters open the space and allow fresh air to enter. Such a fire could quickly become intense and dangerous. Finely grained combustible materials such as sawdust can smolder for a long time and ignite explosively when they are disturbed and oxygenated.

Fire fighters must wear full PPE during overhaul. SCBA use is mandatory until the air is tested and found safe to breathe. Because working for extended periods in full PPE creates heat stress and dehydration, overhaul crews should work for short periods and take frequent breaks. Fresh crews or crews that have been properly rehabbed can replace fatigued crews.

> **Fire Fighter Safety Tips**
>
> A charged hose line must always be available during overhaul.

A Safety Officer should always be present during overhaul operations to note any hazards and ensure that operations are conducted safely. Company officers should supervise operations, look for hazards, and make sure that all crew members work carefully. The work should proceed at a cautious pace. Too many people working in a small area creates chaos.

Always evaluate the structural condition of a building before beginning overhaul. Look for the following indicators of possible structural collapse:

- Lightweight and/or truss construction
- Cracked walls, walls out of alignment, sagging floors
- Heavy mechanical equipment on the roof
- Overhanging cornices or heavy signs
- Accumulations of water

Do not compromise the structural integrity of the building. When opening walls and ceilings, remove only the outer coverings and leave the structural members in place. If more invasive overhaul is required, be careful not to compromise the structure's load-bearing members. Avoid cutting lightweight wood tresses, load-bearing wall studs, and floor or ceiling joists, especially when using power tools.

If fire damages the structural integrity of a building, the IC can call for a "hydraulic overhaul" rather than a standard overhaul. In this situation, large-caliber hose streams are used to completely extinguish a fire from the exterior. This strategy is appropriate if there are excessive risks to fire fighters and such extensive damage to the property that it has no salvage value. Heavy mechanical equipment may be used to demolish unsafe buildings and expose any remaining hot spots.

If a complete overhaul cannot be conducted, the IC can establish a fire watch. The fire watch team remains at the fire scene and watches for signs of rekindling. The team can request additional help if needed.

Coordinating Overhaul with Fire Investigators

Overhaul crews must work with fire investigators to ensure that important evidence is not lost or destroyed. Ideally, a fire investigator should examine the area before overhaul operations begin, identifying evidence and photographing the scene before it is disturbed.

If an investigator is not immediately available, the overhaul crews should make careful observations and report to the investigator later. When performing overhaul in or near the suspected area of fire origin, note burn patterns on the

Fire Fighter Safety Tips

Before overhaul operations begin, the Safety Officer should confirm that electric power and gas service in the overhaul area are shut off. Beware—even though the gas supply may be turned off, there could still be gas under pressure in the lines that must be bled off.

walls or ceilings that could indicate the exact site of origin or the path of fire travel. When moving appliances or other electrical items, note whether they were plugged in or turned on. Always look for evidence that the fire investigator can use to determine the cause of the fire.

If you observe anything suspicious, particularly indications of arson, delay overhaul until an investigator can examine the scene. Ensure that the fire will not rekindle, but do not go any further until the investigator arrives. Chapter 37, Fire Cause Determination has additional information on evidence preservation.

Where to Overhaul

Determining when, where, and how much property needs to be overhauled requires good judgment. Overhaul must be thorough and extensive enough to ensure that the fire is completely out. At the same time, fire fighters should try not to destroy any more property than necessary.

Generally, it is better to make sure that the fire is definitely out than to be too careful about damaging property. If the fire rekindles, there will be more property damage, and the fire department could be held responsible for the additional losses.

The area that must be overhauled depends on the building's construction, its contents, and the size of the fire. All areas directly involved in the fire must be overhauled. If the fire was confined to a single room, all of the furniture in that room must be checked for smoldering fire. The overhaul process also must ensure that the fire did not extend into any void spaces in the walls, above the ceiling, or into the floor. If there is any indication that the fire spread into the structure itself or into the void spaces, all suspect areas must be opened to expose any hidden fire.

If the fire involved more than one room, overhaul must include all possible paths of fire extension. The paths available for a fire to spread within a building are directly related to the type of construction. Learn to anticipate where and how a fire is likely to spread in different types of buildings to find any hidden pockets of fire.

Fire-resistive construction can help contain a fire and keep it from spreading within a building, although this is not guaranteed. Look for any openings that would allow the fire to spread, including utility shafts, pipe chases, and fire doors or dampers that failed to close tightly.

Wood-frame and ordinary-construction buildings may have several areas where a hidden fire could be burning. These structures often have void spaces under the wall covering materials. When a serious fire strikes a building of ordinary or wood-frame construction, fire fighters may need to open every wall, ceiling, and void space to check for fire extension. Neighboring buildings may also need to be overhauled, if the fire could have spread into them.

In **balloon-frame** construction, a fire can extend directly from the basement to the attic, without obvious signs of fire on any other floor (▼ Figure 18-4). These buildings require a thorough floor-by-floor overhaul. The problems could be compounded if the attic insulation is blown-in cellulose materials because these materials can smolder for a long time. Additional information on the effect that building construction can have on fire spread can be found in Chapter 6, Building Construction.

If a building has been extensively remodeled, overhaul presents special challenges. Fire can hide in the space between a dropped ceiling and the original ceiling or extend into a different section of the building through doors and windows covered by new construction. Some buildings have two roofs, one original and one added later, with a void space between them. A fire in this void space presents very difficult overhaul problems.

The cause of the fire also can indicate the extent of necessary overhaul. A kitchen stove fire will probably extend into the kitchen exhaust duct and could ignite combustible materials in the immediate vicinity. Follow the path of the duct and open around it to locate any residual fire. A lightning strike releases enormous energy through the wiring and piping systems, which can start multiple fires in different parts of the building. Overhaul after this incident must be quite extensive.

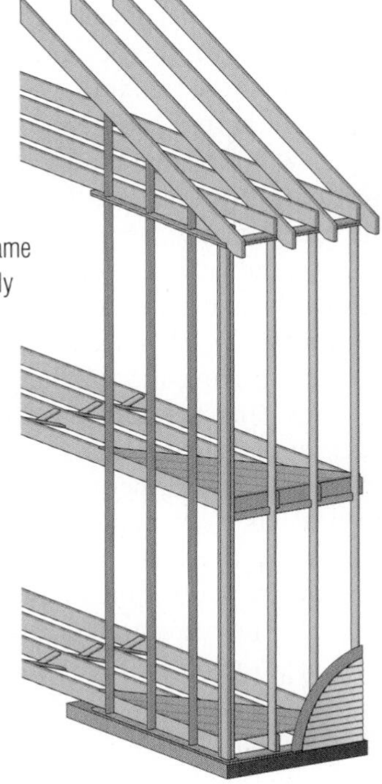

Figure 18-4 A fire in a balloon-frame construction building can spread directly from the basement to the attic.

Using Your Senses

Efficient and effective overhaul requires the use of all your senses. Look, listen, and feel to detect signs of potential burning.

Look for:

- Smoke seeping from cracks or from around doors and windows
- Fresh or new smoke
- Red, glowing embers in dark areas
- Burnt areas
- Discolored material
- Peeling paint or cracked plaster

Listen for:

- Crackling sounds that indicate active fire
- Hissing sounds that indicate water has touched hot objects

Feel for:

- Heat, using the back of the hand (only if it is safe to remove your glove)

Experience is a valuable trait during overhaul situations. By using the senses in a systematic manner, an experienced fire fighter can often determine which parts of the structure need to be opened and which areas remain untouched by the fire.

Thermal Imager

The **thermal image device** is a valuable, high-tech tool used during overhaul. The same type of thermal imager used for search and rescue can also be used to locate hidden hot spots or residual pockets of fire. It can quickly differentiate between unaffected areas and areas that need to be opened. Using a thermal imager can decrease the time needed to overhaul a fire scene and reduce the amount of physical damage to the building.

The thermal imager distinguishes between objects or areas with different temperatures and displays hot areas and cold areas as different colors on the video screen. Because it is very sensitive, the thermal imager can "see" a hot spot, even if the heat source is behind a wall. The imager also can show the pattern of fire within a wall, even if there is only a few degrees' difference in the wall's surface temperature.

Interpreting the readings from a thermal image device requires practice and training (► Figure 18-5). The imager displays relative temperature differences, so an object will appear to be warmer or cooler than its surroundings. If the room has been superheated by fire, all of the contents, including the walls, floor, and ceiling, will appear hot. An extra-hot spot behind a wall could be difficult to distinguish. It might be better to look at the wall from the side that was not exposed to the fire to identify any hot spots.

Figure 18-5 Thermal imaging devices can help locate hot spots.

The thermal imager, however, is only a tool; it cannot completely rule out the possibility of concealed fire. The only way to ensure that no hidden fires exist is to open a ceiling or wall and do a direct inspection.

Overhaul Techniques

The objective of overhaul is to find and extinguish any fires that could still be burning after a fire is brought under control. Any fire uncovered during overhaul must be thoroughly extinguished. Overhaul operations should continue until the IC is satisfied that no smoldering fires are left.

During overhaul operations, a charged hose line should be available to douse any sudden flare-ups or exposed pockets of fire. If necessary, use a direct stream from the hose line but avoid unnecessary water damage. Extinguish small pockets of fire or smoldering materials with the least possible amount of water by using a short burst from the hose line or simply drizzling water from the nozzle directly onto the fire.

Extinguish smoldering objects that can be safely picked up by dropping them into a bathtub or bucket filled with water. Remove materials prone to smoldering, such as mattresses and cushioned furniture, from the building and thoroughly soak them outside. Roll mattresses and secure them with rope or webbing to move them out of the building and decrease the possibility of rekindling.

Place debris that is moved outside far enough away from the building to prevent any additional damage if it reignites. Do not allow debris to block entrances or exits. In some cases, a window opening can be enlarged so that debris can be removed easily. Heavy machinery, such as a front-end loader, may be used to remove large quantities of debris from commercial buildings.

Fire Fighter Tips

Overhaul is physically demanding work. Pace yourself and use proper technique to increase your efficiency. Ask experienced members of your team for tips on improving your technique.

Overhaul Tools

Tools used during overhaul are designed for cutting, prying, and pulling so fire fighters can access spaces that might contain hidden fires. Many of the tools used for overhaul are also used for ventilation and forcible entry. The tools used for overhaul operations include:

- Pike poles and ceiling hooks for pulling ceilings and removing gypsum wallboard
- Crowbars and Halligan-type tools for removing baseboards and window or door casings
- Axes for chopping through wood, such as floor boards and roofing materials
- Power tools such as battery-powered saws for opening up walls and ceilings
- Pitchforks and shovels for removing debris
- Rubbish hooks and rakes for pulling things apart

Because overhaul situations usually do not require high pressures or large volumes of water, a 1½" or 1¾" hose line is usually sufficient to extinguish hot spots. Follow your department's SOPs.

Buckets, tubs, wheelbarrows, and carryalls (rubbish carriers) are used to remove debris from a building (▶ Figure 18-6). A carryall is a 6'-square piece of heavy canvas material with rope handles in each corner.

Opening Walls and Ceilings

Pike poles are used to open ceilings and walls to expose hidden fire. To pull down a ceiling with a pike pole, follow the steps in (▶ Skill Drill 18-13).

1. Select the appropriate length pike pole based on the height of the ceiling. For most residential applications, a 6' pole is sufficient; longer poles are needed for higher ceilings. **(Step 1)**
2. Determine what area of the ceiling will be opened. In most cases, the officer in charge makes this determination. Typically, the most heavily damaged areas will be opened first, followed by the surrounding areas. **(Step 2)**
3. Position yourself to begin work with your back toward a door, so the debris you pull down will not block your access to the exit. **(Step 3)**
4. Using a strong, upward-thrusting motion, penetrate the ceiling with the tip of the pike pole. Face the hook side of the tip away from you. **(Step 4)**
5. Pull down and away from your body, so the ceiling material falls away from you. **(Step 5)**

Figure 18-6 A carryall is used to remove debris during overhaul.

6. Continue pulling down sections of the ceiling until the desired area is open. Pull down any insulation, such as rolled fiberglass, found in the ceiling. **(Step 6)**

Use the pike pole to break through and pull down large sections of ceilings made with gypsum board. Pulling down laths and breaking through plaster ceilings requires more force. Power saws may be required to cut through ceilings made with plywood or solid boards.

Pike poles, axes, power saws, and handsaws can be used to open a hole in a wall. Use the same technique to open a wall with a pike pole as you do to open a ceiling. When using an axe, make vertical cuts with the blade of the axe, then pull the wallboard away from the studs by hand or with the pick end of the axe. A power saw also can be used to make vertical cuts. Pull the wall section away with another tool or by hand.

To open an interior wall with a pick-head axe, follow the steps in (▶ Skill Drill 18-14).

1. Determine what area of the wall will be opened. The officer in charge will usually make this determination. Typically, the areas most heavily damaged by the fire are opened first, followed by the surrounding areas, working outward. **(Step 1)**
2. Use the axe blade to begin cutting near the top of the wall. Cut downward between wall studs. Survey the wall for electrical switches or receptacles as they are evidence of electric wires behind the wall. **(Step 2)**
3. After making two vertical cuts, use the pick end of the axe to pull the wall material away from the studs and open the wall. Work from top to bottom. **(Step 3)**
4. If necessary, remove items such as baseboards or window and door trim with a Halligan tool or axe. **(Step 4)**
5. Continue opening additional sections of the wall until the desired area is open. Pull out any insulation, such as rolled fiberglass, found behind the wall. **(Step 5)**

Skill Drill

Pull a Ceiling Using a Pike Pole

Select the appropriate length pike pole based on the height of the ceiling. For most residential applications, a 6′ pole is sufficient; longer poles are needed for higher ceilings.

Determine what area of the ceiling will be opened. Typically, the most heavily damaged areas are opened first, followed by the surrounding areas.

Position yourself to begin work with your back toward a door, so the debris you pull down will not block your access to the exit.

Using a strong, upward-thrusting motion, penetrate the ceiling with the tip of the pike pole. Face the hook side of the tip away from you.

Pull down and away from your body, so the ceiling material falls away from you.

Continue pulling down sections of the ceiling until the desired area is opened. Pull down any insulation, such as rolled fiberglass, found in the ceiling.

NOTE: Use the pike pole to break through and pull down large sections of ceilings made with gypsum board. Pulling down laths and breaking through plaster ceilings will require more force. Power saws may be required to cut through ceilings made of plywood or solid boards.

Skill Drill

Open an Interior Wall

Determine what area of the wall will be opened up. The officer in charge usually makes this determination. Typically, the areas most heavily damaged by the fire are opened first, followed by the surrounding areas, working outward.

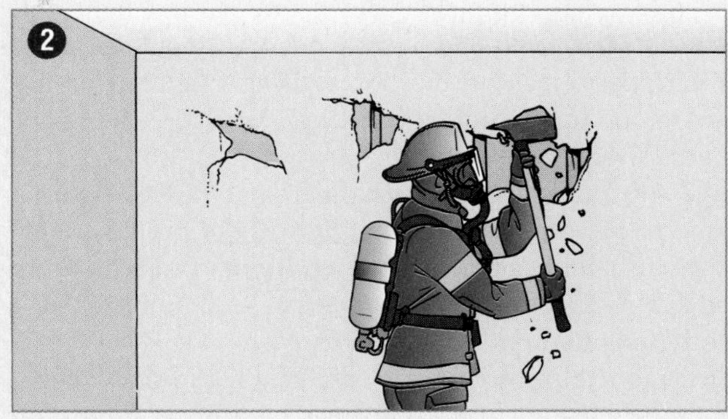

Use the axe blade to begin cutting near the top of the wall. Cut downward between wall studs. Survey the wall for electrical switches or receptacles as they are evidence of electric wires behind the wall.

After making two vertical cuts, use the pick end of the axe to pull the wall material away from the studs and open the wall. Work from top to bottom.

If necessary, remove items such as baseboards or window and door trim with a Halligan tool or axe.

Continue opening sections of the wall until the desired area is open. Pull out any insulation, such as rolled fiberglass, found behind the wall.

Lighting

Lighting is an important concern at many fires and other emergency incidents. Because many emergency incidents occur at night, lighting is required to illuminate the scene and enable safe, efficient operations. Inside fire buildings, heavy smoke can completely block out natural light and electrical service is often interrupted or disconnected for safety purposes, making interior lighting essential for fire fighters to conduct search and rescue, ventilation, fire suppression, salvage, and overhaul.

Most fire departments have different types of lighting equipment for various situations. **Spotlights** project a narrow concentrated beam of light. **Floodlights** project a more diffuse light over a wide area. Lights can be portable or permanently mounted on fire apparatus.

Although the primary responsibility for fire-scene lighting often belongs to truck companies or support units, any company may be assigned this duty. Lighting equipment can be mounted on any apparatus, and some fire departments equip special vehicles with high-powered lights to illuminate incident scenes.

Battery-Powered Lights

Battery-powered lights are generally used by individual fire fighters to find their way in dark areas or to illuminate their immediate work area. Battery-powered lights are most often used during the first few critical minutes of an incident because they are lightweight, easily transported, do not require power cords, and can be used immediately (▼ Figure 18-7). These lights are powered by either disposable or rechargeable batteries and have a limited operating time before the batteries need to be recharged or replaced.

Powerful lights are needed to pierce the smoke inside a fire building. Large hand lights that project a powerful beam of light are preferred for search-and-rescue and interior fire suppression operations when fire fighters must penetrate smoke-filled areas quickly. These lights can be equipped with a shoulder strap for easy transport. Every crew member entering a fire building should be equipped with a high-powered hand light.

Figure 18-7 Hand lights manufactured for use by fire fighters have light outputs ranging from 3,500 to 75,000 candlepower.

Fire Fighter Safety Tips

Never run a portable generator in a confined space or area without adequate ventilation or wearing SCBA. The exhaust contains carbon monoxide, which can incapacitate you.

A personal flashlight is another type of battery-operated light used by fire fighters. Always carry a flashlight with your personal protective equipment; it could be a lifesaver if your primary light source fails. Flashlights are not as powerful as the larger hand lights, and they will not operate for as long on one set of batteries. A fire fighter's flashlight should be rugged and project a strong light beam.

Electrical Generators

More powerful lighting equipment requires a separate source of electricity, generally a 110-volt AC (alternating current) delivered through power cords. The electricity can be supplied by a **generator**, an **inverter**, or a building's electrical system, if the power has not been interrupted.

Power inverters convert 12-volt DC (direct current) from a vehicle's electrical system to 110-volt AC power. An inverter can provide a limited amount of AC current and is usually used to power one or two lights mounted directly on the apparatus. Some vehicles have additional outlets for operating a portable light or a small electrically powered tool. Most power inverters do not produce enough power to operate high intensity lighting equipment, large power tools, or ventilation fans. Connecting devices that draw too much current to an inverter can seriously damage it.

Electrical generators are powered by either gasoline or diesel engines and can be portable or permanently mounted on fire apparatus. Portable generators, which are small enough to be removed from the apparatus and carried to the fire scene, come in various sizes and produce up to 6,000 watts (6 kilowatts) of power. A portable generator has a small gasoline- or diesel-powered motor with its own fuel tank.

Apparatus-mounted generators have much larger capacities, sometimes exceeding 20 kilowatts. The smaller units, which are similar to portable generators, can be permanently mounted in a compartment on the apparatus. Larger generators are usually mounted directly on the vehicle, often have diesel engines, and draw fuel directly from the vehicle's fuel tank. Permanently mounted generators can also be powered by the apparatus engine through a power take-off and a hydraulic pump.

Another possible power source for lighting is a building's normal power supply. This is usually not an option if there is a serious fire in the building because the current will be disconnected for safety reasons. If the fire is relatively minor, the power might not have to be interrupted and can be used for lighting. Sometimes power can be obtained from a nearby building.

Safety Principles and Practices

The lighting and power equipment used at a fire scene generally operates on 110-volt AC, which is the same as standard household current. Some systems require higher voltage. All cords, junction boxes, lights, and power tools must be maintained properly and handled carefully to avoid electrical shocks. All electrical equipment must be properly grounded. Electrical cords must be well insulated, without cuts or defects, and properly sized to handle the required amperage.

Generators should have ground fault interrupters (GFI) to prevent a fire fighter from receiving a potentially fatal electric shock. A GFI senses when there is a problem with an electrical ground and interrupts the current, shutting down both the power source and the equipment it is feeding.

Some portable generators are equipped with a grounding rod that must be inserted into the ground. Always use a grounding device if one is provided. Avoid areas of standing or flowing water when placing power cords and junction boxes at a fire scene, and put electrical equipment on higher ground whenever possible.

Lighting Equipment

Portable lights can be taken into buildings to illuminate the interior. They can also be set up outside to illuminate the fire or emergency incident scene (► Figure 18-8). Portable lights usually range from 300 watts to 1500 watts and can use several types of bulbs, including quartz and halogen bulbs.

Portable light fixtures are connected to the generator with electrical cords. Electrical cords should be stored neatly coiled or on permanently mounted reels attached to the fire apparatus. The cord is pulled from the reel to where the power is needed. Portable reels can be taken from the apparatus into the fire scene.

Junction boxes are used as mobile power outlets. They are placed in convenient locations so that cords for individual lights and electrical equipment can be attached. Junction boxes used by fire departments are protected by waterproof covers and are often equipped with small lights so that they can be easily located.

The connectors and plugs used for fire department lighting have special connectors that attach with a slight clockwise twist. This keeps the power cords from becoming unplugged during fire department operations.

Lights also can be permanently mounted on fire apparatus to illuminate incident scenes (► Figure 18-9). The vehicle operator can immediately illuminate the emergency scene simply by pressing the generator starter button. Some vehicle-mounted lights can be manually raised to illuminate a larger area. Mechanically operated light towers, which can be raised and rotated by remote control, can create nearly daylight conditions at an incident scene.

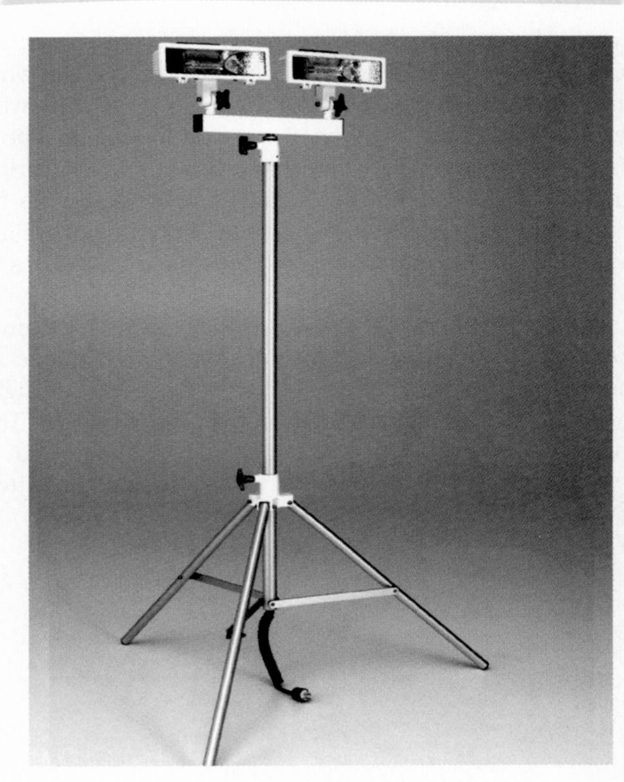

Figure 18-8 Portable emergency lights come in various sizes and levels of brightness.

Figure 18-9 An apparatus mounted light tower can provide near daylight conditions at an incident scene.

Lighting Methods

Effective lighting improves the efficiency and safety of all crews at an emergency incident, but requires practice to set up quickly and efficiently. Departmental SOPs or the IC will dictate responsibility for setting up lighting.

Exterior Lighting

Exterior lighting should be provided at all incident scenes during hours of darkness so that fire fighters can see what they are doing, recognize hazards, and find victims who need rescuing. Scene lighting also makes fire fighters more visible to drivers approaching the scene or maneuvering emergency vehicles. Powerful exterior lights can often be seen through the windows and doors of a dark, smoke-filled building and can guide disoriented fire fighters or victims to safety.

Apparatus operators should turn on their apparatus-mounted floodlights and position them to illuminate as much of the area as possible. Position vehicles with light towers to use their lights for maximum effectiveness. The entire area around a fire building should be illuminated. If some areas cannot be covered with apparatus-mounted lights, use portable lights.

Interior Lighting

Portable lights can provide interior lighting at a fire scene. During interior operations, quickly extend a power cord and set up a portable light at the entry point. Fire fighters looking for the building exit can use the light at the door as a beacon. Extend additional lights into the building to illuminate interior areas as needed and as time permits.

Interior operations often progress through the search-and-rescue and fire suppression stages before effective interior lighting can be established. If interior operations are lengthy, portable lighting should be used to provide as much light as possible in the areas where fire fighters are working.

When operations reach the salvage and overhaul phase, take extra time to provide adequate interior lighting in all areas so that crews can work safely. Proper lighting enables fire fighters to see what they are doing and to observe any dangerous conditions that need to be addressed.

Cleaning and Maintenance

Portable electrical equipment should be cleaned and properly maintained to ensure that it will work when it is needed. Test and run generators on a weekly or monthly basis to ensure they start, run smoothly, and produce power. Run gasoline-powered generators for 15 to 30 minutes to reduce any deposit build-up that could foul the spark plugs and make the generator hard to start.

At the same time, inspect and test other electrical equipment on the apparatus. Take junction boxes out and check them for cracked covers or broken outlets. Plug them in to make sure the GFI works properly. Check power cords for tears in the protective covering, mechanical damage, fraying, heat damage, or burns. Inspect plugs for loose or bent prongs. Inspect and test run all power tools and equipment on the apparatus.

After shutting down the generator, refill the fuel tank. Follow the manufacturer's recommendations for generator engine maintenance, including regular oil changes. To conduct a weekly/monthly generator test, follow the steps in Skill Drill 18-15.

1. Remove the generator from the apparatus compartment (if required) or open all doors as needed and install grounding rod (as needed).
2. Check oil and fuel levels and start generator.
3. Connect the power cord or junction box to the generator.
4. Connect a load (fan or lights) and listen as the engine revs up to proper speed. Check the voltage and amperage gauges to confirm efficient operation.
5. Run the generator under load for 15 to 30 minutes.
6. Turn off the load and listen as the generator slows down to idle speed.
7. Allow generator to idle for approximately 2 minutes, then turn it off.
8. Disconnect all power cords and junction boxes, and remove grounding rod (as needed).
9. Clean all power cords and tools and replace them in proper storage areas.
10. Allow the generator to cool for 5 minutes before refueling and checking oil.
11. Refill with fuel and oil as needed.
12. Return the generator to its compartment.

Fire Fighter Safety Tips

A good rule to remember is: Light early, light often, and light safely.

Wrap-Up

Ready for Review

This chapter covered two major loss control operations: salvage and overhaul. Salvage and overhaul are part of property conservation and are a lower priority than life safety, but still require fire fighters to follow good safety practices. The objectives of salvage are to protect property and contents by expelling smoke, removing heat, and preventing water damage. Prompt, effective ventilation and targeted, efficient water use during fire suppression support salvage efforts. Damage from smoke can be limited with simple measures such as closing doors; water damage can be limited by placing contents in the middle of a room and covering them with a salvage cover. Water can be channeled into a drain or out of the building. Activated sprinkler heads must be turned off and replaced to limit damage in buildings with sprinkler systems. All or part of the system may need to be shut down. Sprinkler systems should not be shut down until the officer in charge has determined that the fire is under control.

Overhaul is the process of searching for and extinguishing any remaining pockets of fire after the main fire is under control. Building construction and the extent of the fire determine how much overhaul is necessary. An officer must be in charge of developing an overhaul plan and supervising the overhaul process. Fire fighters must look, listen, and feel to determine where overhaul is needed. Thermal imaging devices can help find hidden pockets of fire and prevent unnecessary destruction of property. Proper safety procedures, including appropriate PPE, must be followed during overhaul because of the hazardous environment and fire fighter fatigue.

Appropriate salvage measures—including saving family pictures, heirlooms, identification, and medication—reduce fire loss and are greatly appreciated by homeowners. Careful attention to overhaul reduces property loss from rekindled fires and prevents return runs to the same location.

Lighting equipment and procedures also were covered in this chapter. Effective lighting enables the safe, efficient performance of emergency scene operations. Lighting equipment includes battery-powered lights used by individual fire fighters, exterior lights to illuminate the incident scene, and portable lights that can be taken inside a fire building.

Chief Concepts

- Loss control seeks to minimize damage during and after the incident.
- Primary loss refers to the damage caused by the fire. Secondary losses are caused by smoke, heat, and water damage as well as the damage from the fire fighting operations.
- Salvage and overhaul are important fireground activities.
- Salvage is a property conservation issue. It should be started as soon as life safety and incident stabilization are addressed.
- The objective of salvage is to reduce secondary property losses.
- Good ventilation helps to reduce the damage caused by smoke.
- Objects can be removed or covered.
- It may be necessary to pump water out of a building or to rechannel water to remove it from a building.
- Proper safety practices must be followed when engaging in salvage and overhaul.
- The objective of overhaul is to locate and extinguish any remaining fire.
- Use your senses of sight, hearing, and touch to determine where overhaul is needed.
- Overhaul may be quickly completed for a minor fire or may be extensive and time-consuming for a major fire.
- Lighting is needed at emergency scenes to improve safety and efficiency.

Hot Terms

Balloon-frame construction An older type of wood frame construction in which the wall studs extend vertically from the basement of a structure to the roof without any fire stops.

Carryall A piece of heavy canvas with handles, which can be used to tote debris, ash, embers, and burning materials out of a structure.

Floodlight A light that can illuminate a broad area.

Floor runner A piece of canvas or plastic material, usually 3' to 4' wide and in various lengths, used to protect flooring from dropped debris and/or dirt from shoes and boots.

Generator An engine-powered device that provides electricity.

Inverter A device that converts the direct current from an apparatus electrical system into alternating current.

Junction box A device that attaches to an electrical cord to provide additional outlets.

Outside stem and yoke (OS&Y) valve A sprinkler control valve with a valve stem that moves in and out as the valve is opened or closed.

Overhaul Examination of all areas of the building and contents involved in a fire to ensure that the fire is completely extinguished.

Post indicator valve (PIV) A sprinkler control valve with an indicator that reads either open or shut depending on its position.

Rekindle A situation where a fire thought to be out reignites.

Salvage Removing or protecting property that could be damaged during firefighting or overhaul operations.

Salvage cover Large square or rectangular sheets made of heavy canvas or plastic material spread over furniture and other items to protect them from water run-off and falling debris.

Scupper An opening through which water can be removed from a building.

Secondary loss Property damage that occurs due to smoke, water, or other measures taken to extinguish the fire.

Spotlight A light designed to project a narrow, concentrated beam of light.

Sprinkler stop A mechanical device inserted between the deflector and the orifice of a sprinkler head to stop the flow of water.

Sprinkler wedge A piece of wedge-shaped wood placed between the deflector and the orifice of a sprinkler head to stop the flow of water.

Thermal imaging devices Electronic devices that detect differences in temperature based on infrared energy and then generate images based on that data. Commonly used in obscured environments to locate victims.

Water catch-all A salvage cover folded to form a container to hold water until it can be removed.

Water chute A salvage cover folded to direct water flow out of a building or away from sensitive items or areas.

Water vacuum A device similar to a household vacuum cleaner, but with the ability to pick up liquids. Used to remove water from buildings.

Fire Fighter in Action

It is 5:30 a.m. when your ladder company is dispatched to assist with salvage and overhaul operations following a structure fire in a row of 3-story town houses. Two units had been involved in the upper floors. Upon arrival at the scene you are assigned to Division A. The Division A supervisor tells your lieutenant to salvage the second floor. Lieutenant Lawrence instructs you and the crew to retrieve the salvage equipment and begin salvaging the second floor. Another crew is on the floor above you overhauling the fire.

1. What salvage equipment will you potentially need?
   - **A.** Submersible pump, chainsaw, pike pole, and shovel.
   - **B.** Box knife, ladder, circular saw, and Halligan bar.
   - **C.** Salvage cover, small tool kit, floor runners, and a wet/dry vacuum.
   - **D.** Floor runner, sprinkler shut-off kit, rope, and clothespins.

2. The purpose of salvage is to:
   - **A.** find and extinguish hidden fires.
   - **B.** prevent or limit secondary losses and preserve the building and protect the contents.
   - **C.** remove potential fuel for the fire so that it will slow the spread of fire.
   - **D.** remove smoke from furniture and rugs.

You reach the second floor, which has a formal living room, family room, kitchen, dining room, and a bathroom. This home is nicely decorated and appears to have expensive furniture.

3. What items should be the highest priority to salvage?
   - **A.** Photographs and family heirlooms
   - **B.** The furniture
   - **C.** The carpets and throw rugs
   - **D.** The fine china

4. What are some safety concerns associated with salvage?
   - **A.** Breathing in products of combustion
   - **B.** Structural collapse
   - **C.** Electrocution
   - **D.** All of the above

www.FireFighter.jbpub.com

www.FireFighter.jbpub.com

Chapter Pretests
Interactivities
Hot Term Explorer
Web Links
Review Manual
FireLearn

Fire Fighter Rehabilitation

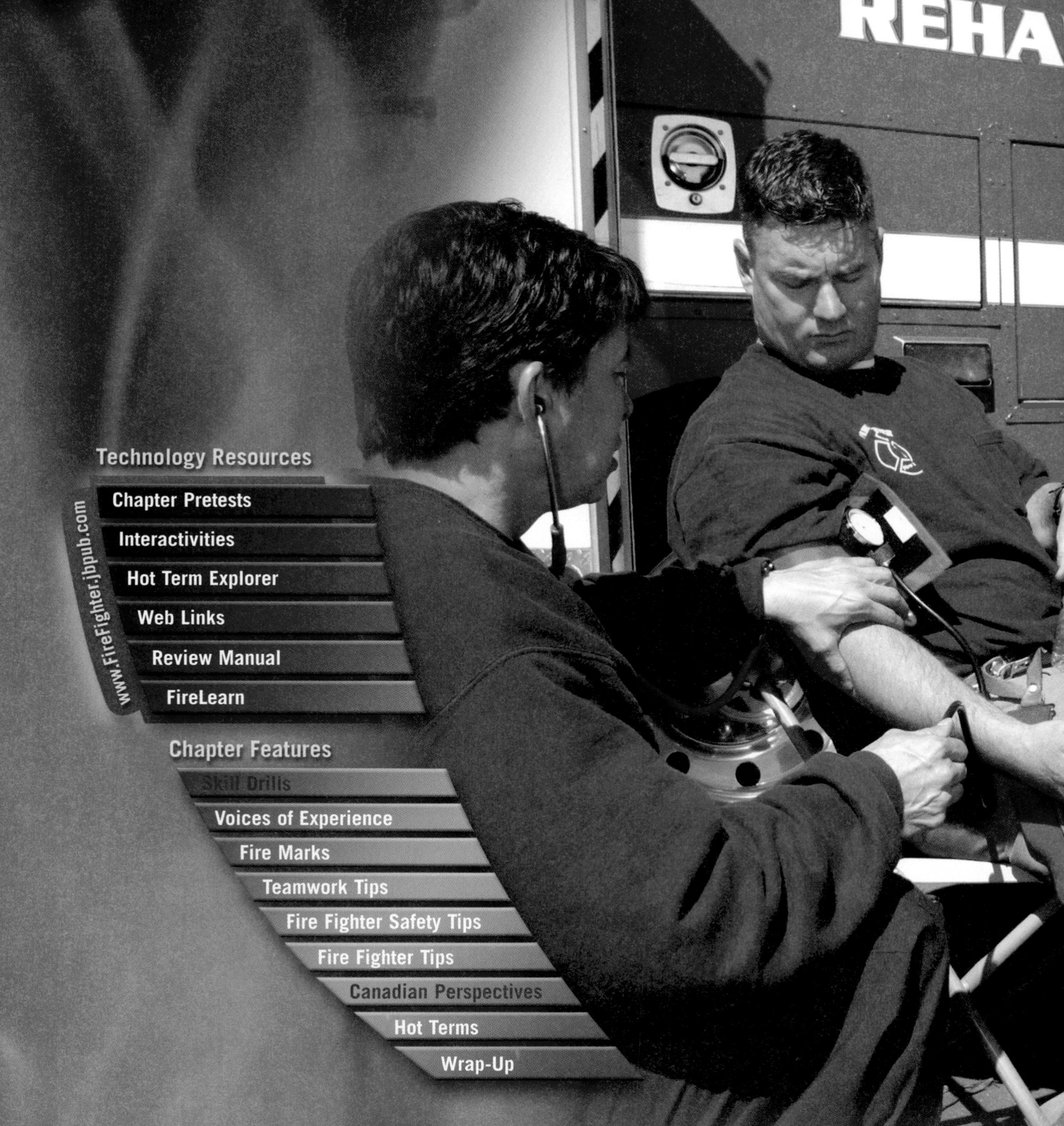

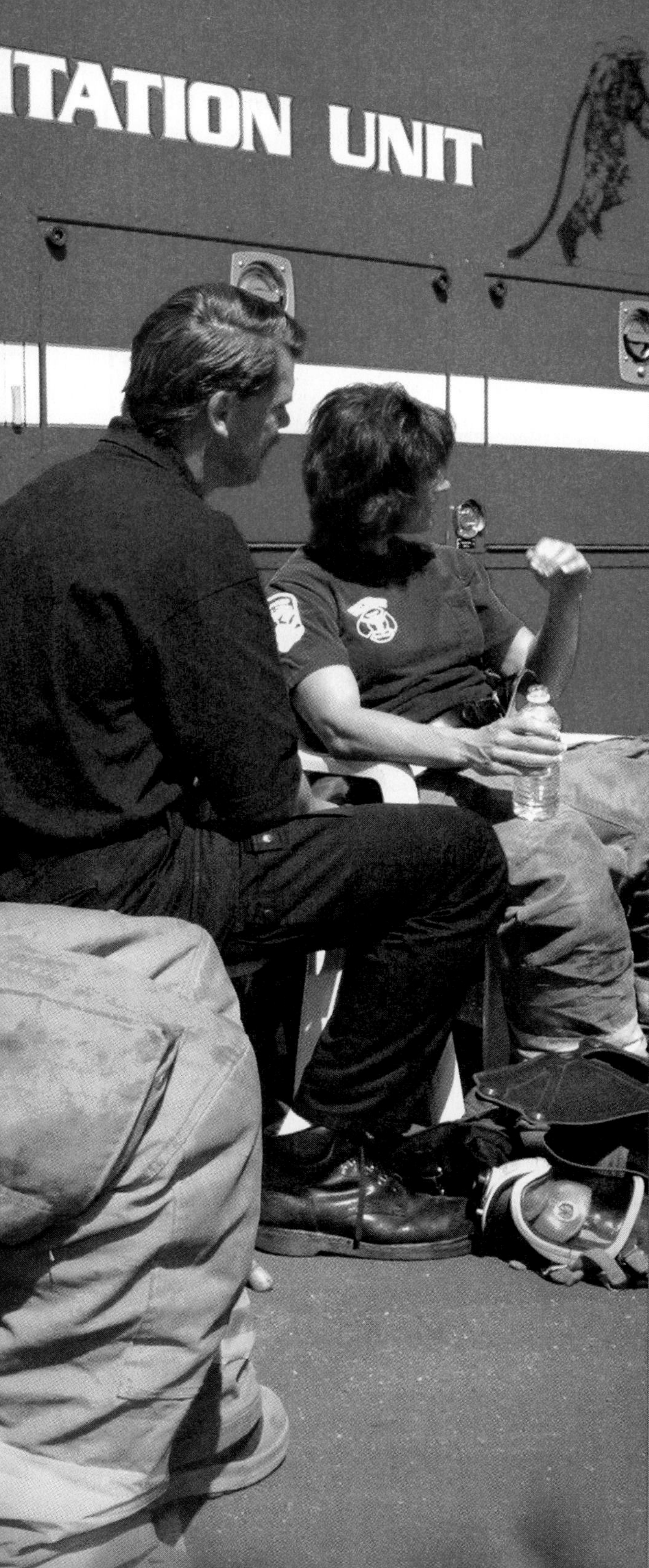

Chapter 19

NFPA 1001 Standard

NFPA 1001 contains no Fire Fighter I Job Performance Requirements for this chapter.

NFPA 1001 contains no Fire Fighter II Job Performance Requirements for this chapter.

Knowledge Objectives

After studying this chapter, you will be able to:

- Define emergency incident rehabilitation.
- Describe why fire fighters need emergency incident rehabilitation.
- List and describe the types of extended fire incidents where fire fighters need emergency incident rehabilitation.
- Describe four other types of incidents where fire fighters would benefit from emergency incident rehabilitation.
- Describe the seven functions of a rehabilitation center.
- List four parts of revitalization.
- Describe the types of fluids that are well suited for fire fighters to drink during emergency incident rehabilitation.
- Describe the types of food that are well suited for fire fighters to eat during emergency incident rehabilitation.
- Describe the personal responsibilities related to emergency incident rehabilitation.

Skills Objectives

There are no skill objectives for this chapter.

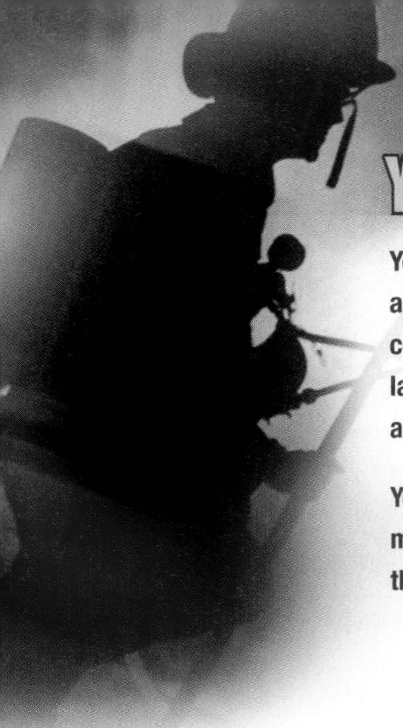

You Are the Fire Fighter

You have completed your initial fire training and are now assigned to a truck company. An hour ago, you were dispatched as the second-due truck company on a working fire in an apartment complex. Your company is assigned to perform search and rescue. As you finish searching the last apartment, your second bottle of air runs low. As your company leaves the building, the alarm on your captain's SCBA starts to sound.

Your captain gives her report to the incident commander. The incident commander says the primary search is now complete and the hose crew has control of the fire. The incident commander then assigns your company to rehabilitation.

1. ***What is rehabilitation, and why is it so important for active fire fighters?***
2. ***Why is the incident commander assigning you to rehabilitation?***

Introduction

A common saying within the fire service is: "Take care of yourself first, take care of the rest of your team second, and take care of the people involved in the incident third." At first glance, this statement may seem to be self-centered. Fire fighters have a deep commitment to help others. If you were to put your personal comfort first, you probably wouldn't want to get up in the middle of a cold winter night to answer a call. But this statement has a deeper meaning. You must take care of yourself physically and mentally so you can continue to help others. You must place a high priority on the health and well-being of your teammates for the good of the department as a whole. Only if you and your teammates are physically fit and mentally alert will you be able to perform the tasks necessary to save lives and protect property.

To **rehabilitate** means to restore to a condition of health or to a state of useful and constructive activity. Fighting fires is a job that requires excellent physical conditioning to combat the rigors of heat, cold, smoke, flames, physical exertion, and emotional stress. Even a seasoned, well-conditioned fire fighter can quickly become fatigued when battling a tough fire (► **Figure 19-1**). New fire fighters are frequently amazed at the amount of energy expended during fire suppression activities. This exertion takes its toll on the body, leaving it tired and dehydrated. The goal of rehabilitation, or rehab, is to take time in order to "recharge the body's batteries" so you can continue to be a productive member of the team. Rehabilitation is a critical factor in maintaining health and well-being.

Emergency Incident Rehabilitation is part of the overall emergency effort. Fire fighters and other emergency workers who are exhausted, thirsty, hungry, ill, injured, or emotionally upset can take a break for rest, fluids, food, medical evaluation, and treatment of illnesses and/or injuries (► **Figure 19-2**). Without the opportunity to rest and

Figure 19-1 Firefighting often involves extreme physical exertion and results in rapid fatigue.

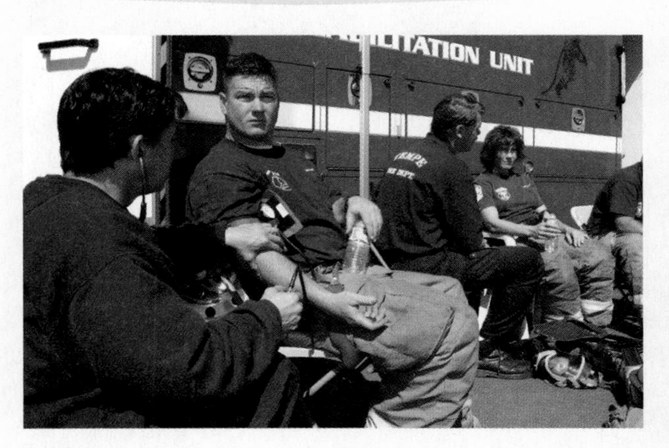

Figure 19-2 Rehabilitation provides fire fighters an opportunity to take a break for rest, fluids, food, medical evaluation, and treatment of illnesses and injuries.

Fire Fighter Tips

The amount of rest needed to recover from physical exertion is directly related to the intensity of the work performed. Fire fighters who have exerted tremendous efforts will require a longer recovery period than those who have performed moderate work.

Fire Fighter Tips

The SAID Principle (Specific Adaptation to Imposed Demands) identifies a need for training that mimics the type of work to be performed. In other words, the best way to prepare for physical work is to do activities that mimic the type, intensity, and duration of the work. The optimal training program for the fire fighter will include both high- and low-intensity cardiovascular training and muscular work that involves strength, endurance, and power.

recover, you may develop physical symptoms such as fatigue, headaches, or gastrointestinal problems. Severe or prolonged stress during a major urban or wildland fire may lead to depression, flashbacks, or amnesia. Rehabilitation gives you the opportunity to address these issues.

Rehabilitation enables fire fighters to perform more safely and effectively at an emergency scene. A tired or dehydrated fire fighter is more likely to be injured and runs the risk of collapsing. The effort that is required to rescue a collapsed or injured fire fighter takes time and resources away from fire suppression activities. Rehabilitation is essential to fight fires safely and effectively.

Factors, Cause, and Need for Rehabilitation

Many conditions come together during a fire fight to produce a stressful environment. Consider the stresses involved in a typical, middle-of-the-night call for a working fire. The loud, jarring sound of the alarm jolts your sleeping body awake. Without hesitation, you must immediately get up, get dressed, and put on personal protective equipment. You may not have time to eat or get something to drink. You may have to drive an emergency vehicle, haul hoses, position a ladder, and climb to the roof to cut a ventilation hole. All these tasks require a significant amount of energy and concentration. You must be able to move into action quickly with no time to warm up your muscles as athletes do before an event.

You may be called to a fire on the hottest day of the year, the coldest day of the year, and under all types of adverse circumstances. Because you know that lives and property are at risk, you may feel an added emotional stress, which affects the body. Fire fighters often work in unfamiliar, smoke-filled environments, which makes the job more difficult and stressful.

Fire Fighter Safety Tips

A tired or dehydrated fire fighter is more likely to be injured and runs the risk of collapsing. Rehabilitation is essential to correct imbalances in the body that, if left untreated, could endanger the fire fighter, coworkers, and others.

Personal Protective Equipment

Personal protective equipment (PPE), the protective clothing and breathing apparatus used by fire fighters to reduce and prevent injuries, adds heat stress on the body (▼ Figure 19-3). PPE can weigh 40 pounds or more, and the extra weight increases the amount of energy needed simply to move around. PPE creates a protective envelope around a fire fighter that protects the body from the smoke, flames, heat, and steam of a fire. But at the same time, it traps almost all body heat inside the protective envelope.

Figure 19-3 The personal protective equipment that is worn to protect a fire fighter also contributes to heat stress.

Normally, evaporating perspiration helps cool the body so that it does not overheat. PPE acts as a vapor barrier that keeps harmful liquids and vapors out, but it also prevents most of a fire fighter's perspiration from evaporating. When a fire fighter is wearing PPE, perspiration soaks the inner clothing, but the evaporative cooling does not occur.

Dehydration

Dehydration is a state in which fluid losses are greater than fluid intake into the body, leading to shock and even death if untreated. Fighting fires is a very strenuous activity, and the large amounts of muscular energy required can produce a significant amount of heat. During this exertion, the body loses a substantial amount of water through perspiration. Fire fighters in action can lose up to 2 quarts of fluid in less than 1 hour (► Figure 19-4). Dehydration reduces strength, endurance, and mental judgment. Replenishing fluids during rehabilitation is essential to correct this imbalance in the body.

Energy Consumption

Food provides the fuel the body needs to do muscular work. In times of strenuous activity, the body burns carbohydrates and fats for energy. These energy sources then need to be replenished. Without a sufficient supply of the right food for energy, the body cannot continue to perform at peak levels for extended periods. During rehabilitation, it is essential to refuel the body with nutritious food.

Tolerance for Stress

Each individual has a different tolerance level for the stresses encountered when fighting fires. For example, younger individuals tend to have greater endurance and can tolerate higher levels of stress. A person who is well-rested and well-conditioned will have more endurance than one who is tired and in poor condition. Carrying extra weight and performing strenuous tasks will strain the heart and increase the risk of heart attack.

Conditioning plays a significant role in a fire fighter's level of endurance. A well-conditioned person with good

Figure 19-4 When perspiring, fire fighters can lose up to 2 quarts of fluid in less than 1 hour.

Figure 19-5 A well-conditioned fire fighter will have a greater tolerance for the stresses encountered when fighting fires.

Figure 19-6 Taking short breaks to rehabilitate reduces the risks of injury and illness.

cardiovascular capacity, good flexibility, and well-developed muscles will be better able to tolerate the stresses of fighting fires than a person who is out of shape (◄ Figure 19-5). However, even the most impressive conditioning will not keep a fire fighter from becoming exhausted under physically stressful situations.

The Body's Need for Rehabilitation

Rehabilitation provides periods of rest and time to recover from the fatigue and stresses of fighting fires and participating in emergency operations. Studies have shown that proper rehabilitation is one way to prevent fire fighters from collapsing or suffering injuries during fire suppression activities. Taking short breaks, replacing fluids, and obtaining energy from food reduces the risks of injury and illness (◄ Figure 19-6). Rehabilitation even helps to improve the quality of decision-making, because people who are tired tend to make poor decisions.

Types of Incidents Affecting Fire Fighter Rehabilitation

The concept of rehabilitation needs to be addressed at all incidents, but it will not be necessary to implement all the components of a rehabilitation center for every incident. For example, fire departments should always have fluids and high-quality energy foods available for rehabilitation. Fire fighters putting out a small fire in a single room might require only water for rehydration while those involved in extinguishing a major wildland fire may need a full rehabilitation station.

Extended Fire Incidents

Structure Fires

Large emergency incidents require full-scale rehabilitation efforts. Major structure fires that involve extended time on the scene will be hard on crews (► Figure 19-7). Crews working on the interior will become dehydrated and fatigued quickly due to the intense heat and stressful conditions. They will require rehabilitation so they can continue working at a healthy and safe level. Rotating crews off the fire ground and bringing in fresh crews also promotes health and safety and helps get the job done in an efficient manner.

High-Rise Fires

High-rise fires place increased stress on fire fighters at the same time that they drain energy resources. It takes considerable energy to walk up several flights of stairs while wearing personal protective clothing and self-contained breathing apparatus. In addition, other equipment such as high-rise hose packs, hand tools, and extra air bottles must be carried up staircases that are already crowded with people exiting the building. Just getting everyone and everything up to the location of the fire may require a truly marathon effort, even before the attack on the fire begins. It is no wonder that fatigue and dehydration can soon occur.

During a high-rise fire, a department may assign three companies to do the work that is normally done by one company. This enables the companies to rotate between attacking the fire, replenishing their air supplies, and meeting their needs for rest, fluids, and nutrition. At high-rise

Fire Marks

The introduction of structured rehabilitation procedures has had a very positive impact on reducing injuries due to heat stress and exhaustion. The old philosophy was to "fight hard until you drop," and it was not unusual to see exhausted fire fighters being carried out of burning buildings to waiting ambulances. However, this disrupted effective fire fighting operations because the fire fighters who could still function were busy rescuing their injured and exhausted coworkers.

Fire Fighter Tips

Various activities can help develop optimal stair-climbing endurance. At the health club, the fire fighter may participate in step aerobics classes or use the stair climbing devices (with or without gear) to develop leg muscles and lung capacity. Outside of the health club, a fire fighter can use step boxes or stairwells to enhance climbing capacity.

Figure 19-7 Major fires often require a strong effort for an extended duration. Rehabilitation and crew rotation are important to limit the risk of exhaustion and injuries to fire fighters.

Teamwork Tips

During active operations, fire fighters may be very reluctant to admit that they need a rest. Asking an obviously exhausted fire fighter if he or she needs to go to rehabilitation almost invariably brings the response, "No—I'm OK." All company members need to watch out for one another, and company officers must monitor their crews for indications of fatigue. It is better to go to rehabilitation a few minutes early than to wait too long and risk the consequences of injury or exhaustion. The incident commander should always plan ahead so that a fresh or rested crew is ready to rotate with a crew that needs rehabilitation.

fires, the rehabilitation center will often be located two or three floors below the level of the fire to reduce the time and energy expended walking up and down the stairs.

Figure 19-8 Wildland fires often involve hundreds of fire fighters working for days or weeks. Rehabilitation is conducted on a large scale at these incidents.

Wildland Fires

Emergency incident rehabilitation was first used extensively for fighting fires in wilderness areas such as grasslands and forests. Given the size and intensity of major wildland fires, crews need to work in shifts so their bodies can recover from the heat, smoke, and stresses of the fire. Minimal rehabilitation efforts may be all that are needed for a ground fire over a small area that can be easily extinguished. However, a large forest fire may require hundreds of fire fighters and several weeks to control and extinguish; such a fire requires a significant and ongoing rehabilitation effort for all crews involved (► Figure 19-8). Proper nutrition and rehydration are essential in keeping the crews healthy and ready to resume duty.

Other Types of Incidents Requiring Rehabilitation

Other incidents may also require extensive rehabilitation efforts to maintain the health and safety of fire fighters. Hazardous materials incidents that require fire fighters to wear **fully encapsulated suits** (a protective suit that fully covers the responder, including the breathing apparatus) are especially draining (► Figure 19-9). Hazardous materials incidents can expose fire fighters to strenuous conditions for extended periods of time. These incidents may also require that the staging area be located at a distance from the hazardous materials, so fire fighters will have to walk long distances while wearing the heavy PPE. As a result, hazardous materials incidents require adequate rehabilitation for fire fighters.

Figure 19-9 Fire fighters wearing fully encapsulated suits must be carefully monitored for symptoms of heat stress.

At times, the fire department is involved in long-duration search-and-rescue activities or other incidents that require the presence of public safety agencies for extended periods of time. These situations can be both mentally and physically stressful. The establishment of a rehabilitation center that fire fighters can use during these incidents is essential.

The need for rehabilitation is not limited to emergency situations. Training exercises, athletic events, and even

Figure 19-10 The same type of rehabilitation procedures should be implemented for training activities as those used for actual emergency incidents.

stand-by assignments may also require rehabilitation. Whenever fire fighters are required to be ready for action for an extended period of time, some provision should be made for providing nourishing foods and replenishing fluids.

Large-scale training activities, including live burn exercises, involve the same concerns as major fire incidents, and rehabilitation should be incorporated in the planning for these activities (▲ Figure 19-10). Training exercises may be conducted over a full day and involve a series of activities, as well as time to set-up each exercise. As part of the process, time should be set aside for the participants to go to rehabilitation between strenuous activities.

Weather conditions can have a significant impact on the need for rehabilitation. Fire fighters should always dress appropriately for the weather, and plans for rehabilitation procedures should take into account the anticipated environmental conditions. Whether it is hot or cold, returning the body's temperature back to normal is one of the primary goals of rehabilitation.

Emergencies that occur when the temperature is very hot will increase the need to rotate crews and allow extensive rehabilitation. Crews working on the interior fire attack may not notice much difference, because their environment during the attack will be hot regardless of the temperature outside. However, hot weather will affect crews working on the outside. These fire fighters will tend to become dehydrated and fatigued much faster than they would if temperatures were in a more comfortable range. Another factor that must be considered is the humidity of the air. Humidity plays an important role in evaporative cooling. High humidity reduces evaporative cooling, making it more difficult for the body to regulate its internal temperature.

Cold weather also increases the need for rehabilitation and crew rotation. The cold can be just as dangerous as the heat.

Figure 19-11 Winter presents a different set of problems for fire fighters. Rehabilitation procedures must be conducted in a warm space, and fire fighters should be given hot liquids and food as well as dry clothing.

Hypothermia, a condition in which the internal body temperature falls below 95°F (35°C), can lead to loss of coordination, muscle stiffness, coma, and death (▲ Figure 19-11). Even in cold weather, the weight of PPE and the physical exertion of fighting a fire will cause the body to sweat inside the protective clothing. Damp clothing and cold temperatures can quickly lead to hypothermia.

How Does Rehabilitation Work?

One way to understand an emergency incident rehabilitation center is to look at the functions it is designed to perform. The most common model has seven different functions (► Table 19-1):

1. Physical Assessment
2. Revitalization
3. Medical Evaluation and Treatment
4. Regular Monitoring of Vital Signs
5. Transportation
6. Critical Incident Stress Management
7. Reassignment

Table 19-1 Seven Functions of Rehabilitation
1. Physical Assessment
2. Revitalization
3. Medical Evaluation and Treatment
4. Regular Monitoring of Vital Signs
5. Transportation
6. Critical Incident Stress Management
7. Reassignment

Physical Assessment

Whenever possible, whole companies should be assigned to rehabilitation and stay together in the rehabilitation center. All fire fighters should be identified as they enter and leave the rehabilitation center to maintain accountability. When fire fighters are ready to be released, the company should be released as a unit.

The first function of emergency incident rehabilitation is physical assessment. A fire fighter's vital signs, including pulse, respiration, blood pressure, and temperature, should be taken and recorded by assigned medical personnel when the fire fighter arrives at the rehabilitation center (▼ Figure 19-12). Many signs and symptoms of fatigue will indicate the need for rehabilitation and are a function of many factors, including poor nutrition and mental, physical, and emotional exhaustion. Each crew member should be questioned and observed for indications of emotional stress. From here, the fire fighter may be referred for medical treatment or sent for rest, rehydration, and food before being reassigned to operational activities.

Table 19-2 Four Components of Revitalization
1. Rest
2. Fluid replacement
3. Nutrition
4. Temperature stabilization

Revitalization

Revitalization is the main part of the rehabilitation process. The four components of revitalization are (▲ Table 19-2):

1. Rest
2. Fluid replacement
3. Nutrition
4. Temperature stabilization

Rest

Rest begins as soon as the fire fighter arrives for rehabilitation. Rehabilitation should be located away from the central activity of the emergency, so the fire fighter can disengage from all other stressful activities and remove personal protective clothing (▼ Figure 19-13). Rest continues as the fire fighter goes through the other parts of revitalization. Rehabilitation centers are equipped with chairs or cots so fire fighters can sit or lie down and relax.

Fluid Replacement

When the body becomes overheated, it sweats so that evaporative cooling can reduce body temperature. Perspiration is composed of water and other dissolved substances such as salt. During fire suppression activities, a fire

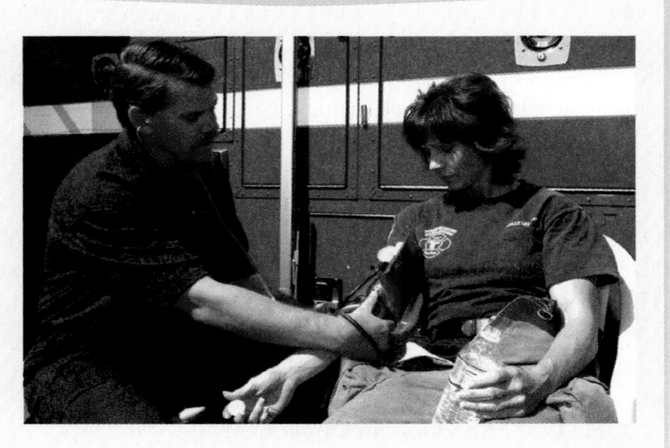

Figure 19-12 The physical condition of each fire fighter who arrives at the rehabilitation center should be evaluated and vital signs should be recorded by assigned medical personnel.

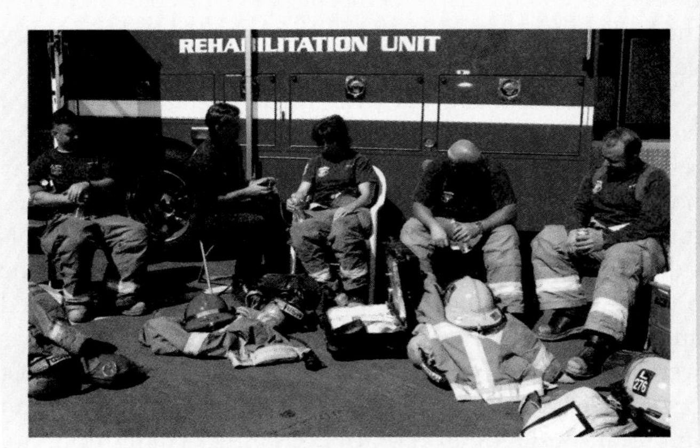

Figure 19-13 In the rehabilitation center, fire fighters should be able to remove personal protective clothing and get some rest before returning to action.

fighter can perspire enough to lose 2 quarts of water in the time it takes to go through 2 bottles of air. Because 1 pint of water weighs 1 pound, losing 2 quarts of water is equivalent to a 2% loss in body weight for a 200-pound fire fighter. The loss of that much body fluid can result in impaired body temperature regulation.

Further dehydration can result in reduced muscular endurance, reduced strength, and heat cramps. Severe cases of dehydration can contribute to heat stroke. Because a fire fighter can lose so much water during fire suppression activities, it is important to replace fluids before severe dehydration takes place (▼ Figure 19-14).

When the body perspires, it loses **electrolytes** (certain salts and other chemicals that are dissolved in body fluids and cells) as well as water. If a fire fighter loses large amounts of water through perspiration, the electrolytes must be replenished as well as the water. This is why sport-type drinks are often used in rehabilitation. Sport drinks supply water, balanced electrolytes, and some sugars. These drinks can be mixed with twice the recommended amount of water to dilute the sugars.

Rehydrating your body is not as easy as just drinking lots of fluids quickly. Drinks like colas, coffee, and tea should be avoided because they contain caffeine. Caffeine acts as a diuretic that causes the body to excrete more water. Sugar-rich carbonated beverages are not tolerated or absorbed as well as straight water or diluted sports drinks. Too much sugar is difficult to digest and causes swings in the body's energy levels. Drinks that are too cold or too hot may be hard to consume and may prevent you from ingesting enough liquids.

Figure 19-14 The rehabilitation center should have plenty of fluids to rehydrate fire fighters. Plain water or diluted sports drinks are preferred.

An additional concern is the rate at which fluids can be absorbed from the stomach. Drinking too much too quickly can cause bloating, a condition in which air fills the stomach. This causes a feeling of fullness and can lead to discomfort, nausea, and even vomiting.

Studies demonstrate that the stomach can absorb 1 to 1.5 quarts of fluid per hour. However, the body can lose up to 2 quarts of fluid per hour. Thus, the body loses fluids much more rapidly than they can be replaced. Once the initial 2 quarts of water are lost, the body will require 1 to 2 hours to recover.

Nutrition

A fire fighter performs more physical tasks and exerts more energy than the average worker. Like an engine that runs on diesel fuel, the body runs on **glucose**. Glucose, also known as blood sugar, is carried throughout the body by the blood stream and is needed to burn fat efficiently and release energy.

In order for the body to work properly, the glucose (blood sugar) levels need to be in balance. If blood sugar drops too low, the body becomes weak and shaky. If blood sugar is too high, the body becomes sluggish. Blood sugar levels can be balanced by eating a proper diet of carbohydrates, proteins, and fats.

Fire Fighter Tips

The sensation of thirst is not a reliable indicator of the amount of water the body has lost. Thirst develops only after the body is already dehydrated. Do not wait until you feel thirsty to be concerned with your hydration. Start to drink before you get thirsty. Remember to drink early and often.

Try to drink enough fluids to keep the body properly hydrated at all times, particularly during hot weather. This will decrease the risk of dehydration when an emergency incident occurs. Adequate hydration also ensures peak physical and mental performance.

The best indicator of proper hydration is the color of urine. If urine is dark or amber in color, the body is dehydrated. If urine is light or clear, the body is properly hydrated. Fatigue and heat sensitivity are also signs of dehydration. Fire fighters should strive to drink a quart of water an hour during periods of work.

Voices of Experience

"If this kind of breakdown can take place in a controlled environment with physically fit fire fighters, imagine the potential when dealing with fire fighters in a chaotic atmosphere full of unknowns."

On a hot August day in 1996, the Salt Lake City Fire Department recruits spent the morning donning and doffing Class A hazardous materials suits (heavy, encapsulating suits that envelop both the wearer and SCBA totally) and conducting decontamination drills. After a short lunch break, the afternoon was filled with ladder drills. In the late afternoon, one of the recruits began having difficulty completing the evolutions. As her condition worsened, she was unable to think clearly and demonstrate basic skills. She became confused, and one of the training officers asked if she was all right. When she was unable to give a clear answer, she was told to take a break and rehydrate. A fire fighter approached her to see how she was feeling and quickly realized that something was not right. The fire fighter took the recruit's vital signs and discovered that her temperature, pulse, and blood pressure were all elevated. Her skin was now bright red and she showed no sign of sweating. The decision was made immediately to start an IV and transport her to the hospital. The assessment in the emergency room revealed that the recruit was dehydrated and had low blood sugar levels. She was treated for her condition and released when her vital signs returned to normal.

This incident involved a recruit, in training, who was in above average physical condition. If this kind of breakdown can take place in a controlled environment with physically fit fire fighters, imagine the potential when dealing with fire fighters in a chaotic atmosphere full of unknowns. Rehabilitation on the fire ground, or any other physically demanding scene, should be as much a part of actual fire ground activities as putting water on the fire. Fire fighter fitness levels and events leading up to the incident, such as recent workouts, heavy meals, or illness, create a level of unpredictability when it comes to crew capabilities. Salt Lake City Fire Department has made many advances in ensuring proper rehabilitation for all members in training or on the emergency scene.

It is interesting to note that this fire fighter did not stop working without being approached. Fire fighters need to take responsibility for themselves and their crewmates. There are many distractions on the fire ground. A heightened awareness of your own body's signals, and paying close attention to your crewmates' actions, can allow for early intervention. A simple 10- to 15-minute break could make the difference between being a functional, contributing fire fighter, or becoming part of the problem.

Martha Ellis
Salt Lake City Fire Department
Salt Lake City, Utah

Figure 19-15 Carbohydrates are a major source of fuel for the body.

Figure 19-16 Proteins perform many vital functions in the body.

Figure 19-17 Plans for rehabilitation at extended duration incidents should include complete balanced meals.

Carbohydrates are a major source of fuel for the body and can be found in grains, vegetables, and fruits (▲ Figure 19-15). The body converts carbohydrates into glucose, making "carbs" an excellent energy source. A common dietary myth is that carbohydrates are fattening and should be avoided. In fact, carbohydrates should make up 55% to 65% of calories in a balanced diet. Carbohydrates have the same number of calories per gram as proteins and fewer calories than fat. Additionally, carbohydrates are the only fuel that the body can readily use during high-intensity physical activities such as fighting fires.

Proteins perform many vital functions within the body (► Figure 19-16). Most protein comes from meats and dairy products. Smaller amounts of protein are found in grains, nuts, legumes, and vegetables. Proteins (amino acids) are used by the body to grow and repair tissues and are only used as a primary fuel source in extreme conditions such as starvation. Like other nutrients, excess proteins are converted and stored in the body as fat. A balanced diet has 10% to 12% of calories from proteins.

Fats are also essential for life. Fats are used for energy, for insulating and protecting organs, and for breaking down certain vitamins. Some fats, such as those found in fish and nuts, are healthier and more beneficial to the body than others, such as the fats found in margarine. However, no more than 25% to 30% of the diet should come from fats. Excess fat consumption, particularly of saturated fats (which come mostly from animal products), is linked to high cholesterol, high blood pressure, and cardiovascular disease.

Candy and soft drinks contain sugar. The body can quickly absorb and convert these foods to fuel. But simple sugars also stimulate the production of insulin, which reduces blood glucose levels. That's why eating a lot of sugar can actually result in lower energy levels.

For a fire fighter to sustain peak performance levels, it is necessary to refuel during rehabilitation. During short incidents, low-sugar, high-protein sports bars can be used to keep the glucose balance steady. During extended incidents, a fire fighter should eat a more complete meal (▲ Figure 19-17). The proper balance of carbohydrates, pro-

Fire Fighter Safety Tips

Personal protective equipment (PPE) is designed to provide protection from hazards encountered in emergency operations. Fire fighters who are overheated should remove their PPE as soon as possible to permit evaporative cooling. PPE must only be removed in a safe place, outside the hazard area. The incident commander or a safety officer may approve working without full PPE in situations where the risks have been fully evaluated.

teins, and fats will maintain energy levels throughout the emergency. To ensure peak performance, the meal should include complex carbohydrates such as whole grain breads, whole grain pasta, rice, and vegetables. It is also better to eat smaller meals; larger meals can increase glucose levels and slow down the body.

Healthy, balanced eating should be a lifestyle. Proper nutrition reduces stress, improves health, and provides more energy. It is just as important to keep blood sugar levels balanced throughout the day as it is to remain hydrated. Then, you will be ready to react to an emergency at any time.

Temperature Stabilization

The fourth part of revitalization is stabilizing body temperature. Body temperature must return to a normal range before a fire fighter resumes strenuous activities. Fire fighters exposed to hot or cold temperatures need to have a place where their body temperature can return to normal levels before resuming further activities.

As mentioned previously, fighting fires in a complete ensemble of personal protective equipment, sometimes known as **turnout gear**, can result in some degree of heat stress (► Figure 19-18). The heat generated by the body during the intense physical exertion, coupled with the increased temperature from the fire, can increase the body's internal temperature and produce profuse sweating. In most situations, turnout gear should be removed as soon as possible to allow the body to cool. If necessary, additional steps such as the use of cold compresses should be taken to reduce body temperature. When ambient temperatures are low, damp clothes should be removed and blankets should be used to stabilize body temperature.

Stabilizing body temperature is further complicated during periods of hot weather or high humidity. At these times, fire fighters cannot cool off even when they take a break and remove their turnout gear. For this reason, rehabilitation centers need to be climate-controlled so that fire fighters can achieve a normal body temperature before resuming active duties. Some fire departments have air-conditioned vehicles for rehabilitation. Others use buses or establish rehabilitation centers in an already cooled area such as a shopping center.

Figure 19-18 Fighting fires in complete turnout gear can result in some degree of heat stress.

Cold temperatures cause problems as well. Fire fighters responding to incidents during cold weather are subject to hypothermia and **frostbite** (damage to tissues resulting from prolonged exposure to cold) (► Figure 19-19). In these cases, the rehabilitation center needs to be warm enough so they can get warmed up before returning to the chilly environment. Fire fighters who are wet or severely chilled should be wrapped in warming blankets and moved into a well-heated area before they remove their turnout gear. As soon as possible, all wet clothing should be removed and replaced with warm, dry clothing.

Medical Evaluation and Treatment

A fire fighter who has abnormal vital signs, is suffering pain, or is injured needs to have further medical evaluation. A medical evaluation and treatment team should be assigned to the rehabilitation center (► Figure 19-20). Signs of illness or injury should be checked in the rehabilitation center

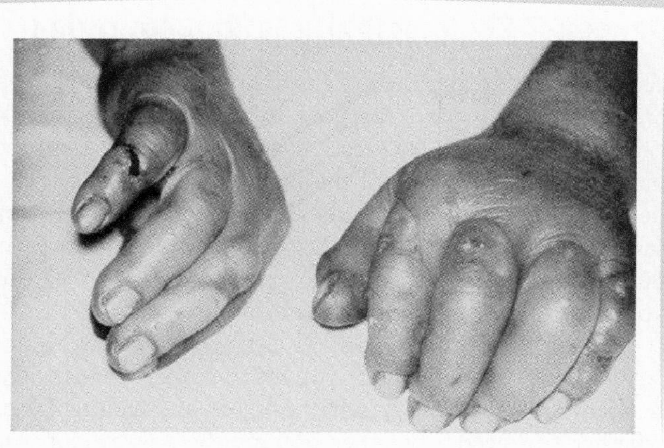

Figure 19-19 Prolonged exposure to freezing weather can result in severe frostbite.

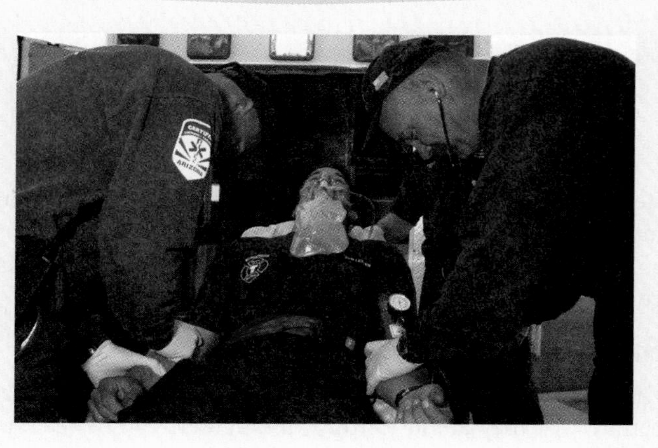

Figure 19-20 A medical team should be part of rehabilitation to check vital signs and treat any injuries that occur.

before the fire fighter resumes activities. Identifying problems in the rehabilitation center can prevent later disruption on the fire scene. A fire fighter who collapses or becomes incapacitated while in action endangers the entire team.

Monitoring of Vital Signs

Pulse, respiration, blood pressure, and temperature should be monitored at regular intervals during rehabilitation. These readings should be compared with readings taken when the fire fighter first enters the rehabilitation center. Vital signs must be taken repeatedly to ensure that they return to normal limits before the fire fighter is reassigned. If vital signs do not return to normal levels, a fire fighter may need to spend more time in the rehabilitation center or be further evaluated by medical personnel.

Transportation to a Hospital

Another function of the rehabilitation center is to transport ill or injured fire fighters to the hospital (► Figure 19-21). An ambulance should be available to ensure that ill or injured fire fighters receive the best possible care.

Critical Incident Stress Management

When emergencies involve stressful situations, particularly mass casualty incidents or fire fighter fatalities, rehabilitation should include staff trained in critical incident stress management (CISM). CISM is a process that confronts the responses to critical incidents, defuses them, and directs the fire fighter toward physical and emotional balance. The CISM team members should talk to each company in rehabilitation to assess the emotional state of fire fighters. Talking with a CISM member in the rehabilitation center is the first

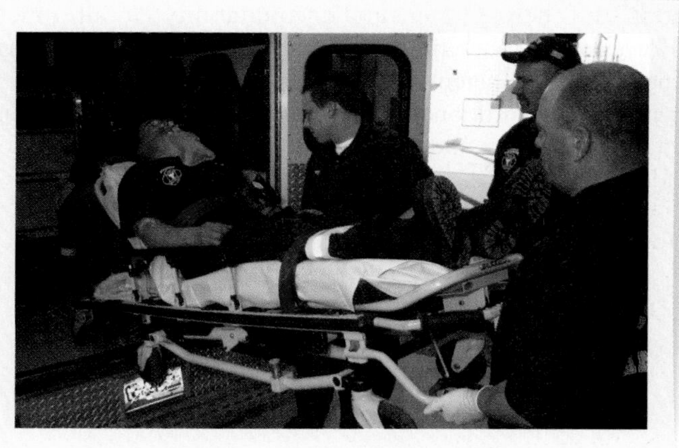

Figure 19-21 An ambulance should always be ready near the rehabilitation center to transport an ill or injured fire fighter to a hospital.

Fire Fighter Safety Tips

Stress management skills are crucial for high-quality performance, regardless of the type of stressor involved or what role the person plays in emergency assistance. A fire fighter needs to know basic stress management skills to deliver high-quality performance. Exercise, proper nutrition, and emotional support are all helpful in managing stress.

Fire Fighter Tips

You must take care of yourself first to be able to help others.

step in the CISM system. After the incident, CISM team members may meet with companies or individual fire fighters to discuss the event and work through it. CISM is discussed further in Chapter 17, Fire Fighter Survival.

Reassignment

Once rested, rehydrated, refueled, and rechecked to make certain that they are fit for duty, fire fighters can be released from rehabilitation and reassigned to active duty. Reassignments may be to the same job performed before or to a different task, depending on the decision of the officer in charge. Generally, companies stay together during the rehabilitation process because this creates a more effective work unit and helps to ensure accountability at all times. However, it is sometimes necessary to temporarily reassign members to different companies. This allows fire fighters who require additional rest or medical treatment to remain in rehabilitation and releases fully-staffed companies for additional duty.

Personal Responsibility in Rehabilitation

The goal of a fire fighting team is to save lives first and property second. To achieve this goal, however, you need to take care of yourself first, take care of the rest of your team second, and then take care of the people involved in the incident. Another way of looking at this is to remember that safety begins with you. The other members of your company depend on you to bear your share of the load. To do this, you need to maintain your body in peak condition.

Part of your responsibility is to know your own limits. No one else can know what you ate, whether you are lightheaded or dehydrated, whether you are feeling ill, or whether you need a breather. You are the only person who knows these things. Therefore, you may be the only person who knows when you need to request rehabilitation. It may be difficult to say "I need a break" while your team members are still hard at work. But it is better to break when you need it than to push yourself too far and have to be rescued by other members of your department.

Regular rehabilitation enables fire fighters to accomplish more work during a major incident. It decreases the risk of stress-related injuries and illness. Remember, safety begins and ends with you.

Wrap-Up

Ready for Review

This chapter covered the information about emergency incident rehabilitation that you should know as a beginning fire fighter. Rehabilitation is a specially designated area where emergency personnel can rest, receive fluids and nourishment, and be evaluated for medical and CISM problems. Rehabilitation centers help prevent injuries and illnesses, enable fire fighters to accomplish more during an emergency incident, and reduce the need for rescuing fire fighters from the fire scene.

Structure fires that are large or continue for a long period of time, high-rise fires, and wildland fires frequently require the use of rehabilitation centers. Other emergency situations that may require the establishment of a rehabilitation center include hazardous material operations, search-and-rescue operations, and other extended-duration incidents. Training activities and athletic events are also situations that may require a rehabilitation center. The need for rehabilitation increases when the weather conditions are very cold, very hot, or very humid.

The seven parts to rehabilitation are: physical assessment, revitalization, medical evaluation and treatment, monitoring of vital signs, transportation to a hospital, critical incident stress support, and reassignment. Revitalization is the step that concerns you most as a new fire fighter. Revitalization consists of rest, fluid replacement, nutrition, and temperature stabilization.

Fluid replacement should occur before the signs of dehydration become obvious. Water or diluted sports drinks are recommended. Carbonated drinks, drinks containing caffeine, and highly sugared drinks should be avoided. Fluids are better tolerated if they are not too hot or too cold.

Nutritional needs during minor incidents can be met with low-sugar, high-protein sports bars. During longer incidents, small frequent meals that contain a balance of carbohydrates, proteins, and fats will help to maintain energy levels.

While you are replacing fluids and meeting your nutritional needs, you can also rest and allow your body temperature to stabilize. It is important to rest, physically and mentally, from the rigors of the incident. You need to allow your body to return to its normal temperature range so it can work efficiently.

Your responsibilities in rehabilitation are to know your limits, to listen to your body, and to use rehabilitation facilities prudently. The objective of rehabilitation is to keep you operating in an efficient manner for the duration of the incident or as long as you are needed. You must avoid overextending yourself to the point where you collapse or become ill or injured. Safety begins with you.

Chief Concepts

- Rehabilitation is a specially designated area where emergency personnel can rest, receive fluids and nourishment, and be evaluated for medical and CISM problems.
- Rehabilitation centers help prevent injuries and illnesses, enable fire fighters to accomplish more during an emergency incident, and reduce the need for rescuing exhausted or collapsed fire fighters from the fire scene.
- Types of incidents that may necessitate rehabilitation centers include major structure fires, high-rise fires, and wildland fires. Rehabilitation is also needed for hazardous materials incidents, search-and-rescue operations, training activities, and athletic events.
- The seven parts of rehabilitation are: physical assessment, revitalization, medical evaluation and treatment, monitoring of vital signs, transport, critical incident stress management, and reassignment. Revitalization consists of rest, fluid replacement, nutrition, and temperature stabilization.
- Your responsibilities in rehabilitation are to know your limits, to listen to your body, and to use the rehabilitation facilities when needed.

Wrap-Up

Hot Terms

Critical Incident Stress Management (CISM) A system to help personnel deal with critical incident stress in a positive manner; its aim is to promote long-term mental and emotional health after a critical incident.

Dehydration A state in which fluid losses are greater than fluid intake into the body, leading to shock and even death if untreated.

Electrolytes Certain salts and other chemicals that are dissolved in body fluids and cells.

Emergency incident rehabilitation A function on the emergency scene that cares for the well-being of the fire fighters. It includes physical assessment, revitalization, medical evaluation and treatment, and regular monitoring of vital signs.

Frostbite Damage to tissues as the result of exposure to cold; frozen or partially frozen body parts.

Fully-encapsulated suits A protective suit that completely covers the fire fighter, including the breathing apparatus, and does not let any vapor or fluids enter the suit.

Glucose The source of energy for the body. One of the basic sugars, it is the primary fuel, along with oxygen.

Hypothermia A condition in which the internal body temperature falls below 95°F, usually a result of prolonged exposure to cold or freezing temperatures.

Personal protective equipment (PPE) Gear worn by fire fighters that includes helmet, gloves, hood, coat, pants, SCBA, and boots. The personal protective equipment provides a thermal barrier for fire fighters against intense heat.

Rehabilitate To restore to a condition of health or to a state of useful and constructive activity.

Turnout gear Defined as helmet, hood, coat, pants, boots, gloves, personal alert safety system, and self-contained breathing apparatus. Sometimes referred to as personal protective equipment.

www.FireFighter.jbpub.com

- Chapter Pretests
- Interactivities
- Hot Term Explorer
- Web Links
- Review Manual
- FireLearn

Fire Fighter in Action

It is a hot, humid August midafternoon (temperature: 94°F/35°C; relative humidity: 95%). Your engine company is dispatched on the third alarm for a fire in an apartment complex where companies have already been working on the scene for almost 30 minutes. When you arrive, the staging officer tells your lieutenant that your company is to establish a rehabilitation center. Lieutenant Jones points to a grassy area in the park across the street from the fire and tells you and the other crew members to set-up the rehabilitation center.

1. What equipment would you NOT bring with you?
   - **A.** Medical equipment bag
   - **B.** Drinking water and ice
   - **C.** Self-Contained Breathing Apparatus
   - **D.** Clipboards and pens
2. Lieutenant Jones tells you and your partner to check in the crews as they report to rehabilitation. This means that you will:
   - **A.** Write down the identity of each company reporting to rehabilitation.
   - **B.** Write down the time each company reports to rehabilitation.
   - **C.** Record the name of each individual fire fighter reporting to rehabilitation.
   - **D.** Determine if anyone reporting to rehabilitation is injured.
   - **E.** All of the above.
3. Should this company go back into the building?
   - **A.** Yes—the fire is still burning and they are needed urgently.
   - **B.** Yes—they should be fine for another work period after a quick drink.
   - **C.** No—they cannot leave rehabilitation until you have written down all of their names.
   - **D.** No—they need more time to cool down and rehydrate.
4. What should this company do to revitalize?
   - **A.** Remove protective clothing, drink a carbonated soda, and eat a donut.
   - **B.** Keep protective clothing on and drink cold water or sports beverage.
   - **C.** Remove protective clothing, eat a protein bar, and drink cold water or diluted sports beverage.
   - **D.** Remove protective clothing, drink coffee, and eat a donut.

Wildland and Ground Fires

Technology Resources

www.FireFighter.jbpub.com

- Chapter Pretests
- Interactivities
- Hot Term Explorer
- Web Links
- Review Manual
- FireLearn

Chapter Features

- Skill Drills
- Voices of Experience
- Fire Marks
- Teamwork Tips
- Fire Fighter Safety Tips
- Fire Fighter Tips
- Canadian Perspectives
- Hot Terms
- Wrap-Up

Chapter 20

NFPA 1001 Standard

Fire Fighter I

5.3.19 Combat a ground cover fire operating as a member of a team, given protective clothing, SCBA if needed, hose lines, extinguishers or hand tools, and an assignment, so that threats to property are reported, threats to personal safety are recognized, retreat is quickly accomplished when warranted, and the assignment is completed.

5.3.19 (A) *Requisite Knowledge.* Types of ground cover fires, parts of ground cover fires, methods to contain or suppress ground cover fires, and safety principles and practices.

5.3.19 (B) *Requisite Skills.* The ability to determine exposure threats based on fire spread potential, protect exposures, construct a fire line or extinguish with hand tools, maintain integrity of established fire lines, and suppress ground cover fires using water.

Fire Fighter II

NFPA 1001 contains no Fire Fighter II Job Performance Requirements for this chapter.

Knowledge Objectives

After studying this chapter, you will be able to:

- Define the terms wildland and ground fires.
- Define light fuels, heavy fuels, subsurface fuels, surface fuels, and aerial fuels.
- Describe how weather factors and topography influence the growth of wildland fires.
- Define the parts of a wildland and ground fire.
- Describe how wildland and ground fires can be suppressed.
- List the hazards associated with wildland and ground firefighting.
- Describe the personal protective equipment needed for wildland firefighting.
- Explain the problems created by the wildland urban interface.

Skills Objectives

There are no skills objectives for this chapter.

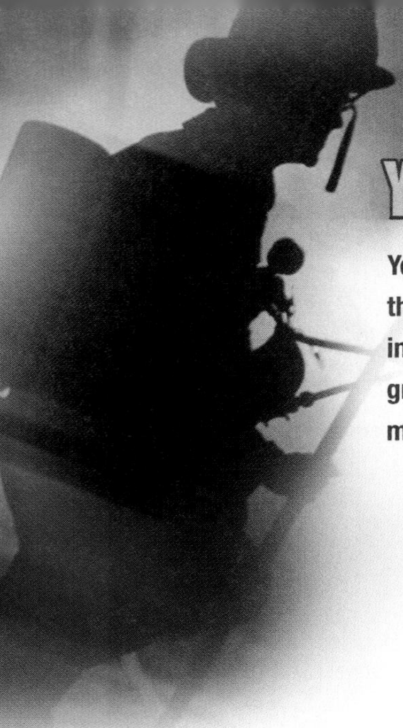

You Are the Fire Fighter

You have just finished lunch when your engine company is dispatched for a ground cover fire on the edge of your town. It is a bright summer day with low humidity. There is a light breeze coming in from the south. When you arrive at the scene, you find a small brush fire approaching a grove of trees separating a housing development from the main road. It has not rained in a month, so the ground is quite dry. Your incident commander begins to call for assistance.

1. ***What weather conditions could have an impact on this fire?***
2. ***What special hazards are associated with fighting wildland fires?***
3. ***Why is an effective incident command system so important at wildland fires?***

What Are Wildland Fires?

Wildland fires are defined by the NFPA as unplanned and uncontrolled fires burning in vegetative fuel (grass, leaves, and trees) that sometimes includes structures. Wildland fires can consume grasslands, brush, and trees of all sizes. Some wildland fires burn vegetation that is located many feet above the ground. The incidence of these fires varies from season to season and from one year to the next. The threat and intensity of wildland fires is closely tied to the moisture content of the vegetative fuels, and to weather conditions and topography.

Wildland fires are sometimes referred to by different terminology, including brush fires, forest fires, grass fires, ground cover fires, ground fires, natural cover fires, and wildfires. While these names provide a description of specific types of wildland fires, all are properly classified as wildland fires.

Ground cover fires burn loose debris on the surface of the ground. This debris includes vegetation such as grass. It also includes dead leaves, needles, and branches that have fallen from shrubs and trees. In most fire departments, ground cover fires are the most common type of wildland fire encountered.

Some fire departments respond to more wildland and ground fires than to structural fires. While many calls for wildland and ground fires are for small incidents, some calls will evolve into major incidents. Most fire departments are responsible for some areas that can be classified as wildland. Wildland is land in an uncultivated natural state that is covered by timber, woodland, brush, or grass. Most cities have parklands that are covered with natural vegetation.

Many structural fire fighters are called on to extinguish wildland and ground fires at some point. The information in this chapter is intended to teach you how to extinguish wildland and ground fires safely. This chapter will not train you to be a wildland fire fighter. Fire fighters whose primary responsibility is suppressing wildland fires need to complete wildland firefighting training.

Large wildland fires are handled by agencies that specialize in this type of firefighting. Each state has an agency designated to coordinate wildland firefighting. Federal government agencies such as the U.S. Forest Service, Bureau of Land Management, Bureau of Indian Affairs, National Park Service, and the U.S. Fish and Wildlife Service are also responsible for coordinating firefighting activities at large incidents and incidents that occur on federal lands. The Incident Management System is also activated at these incidents. See Chapter 4, Incident Management System, for more information.

Wildland and Ground Fires and the Fire Triangle

In order to safely and effectively extinguish wildland and ground fires, you need to understand the factors that cause fire ignition and affect the growth and spread of wildland fires. The fire triangle provides a good model for explaining the behavior of wildland and ground fires (► Figure 20-1).

The fire triangle, as you learned in Chapter 5, Fire Behavior, consists of three elements: fuel, oxygen, and heat. Wildland and ground fires require the same three elements as structural fires. The difference in wildland and ground fires is the conditions under which the fuel, oxygen, and heat come together to produce an uncontrolled fire. Since weather conditions have a great impact on the behavior of wildland fires, we will also consider how each side of the fire triangle is influenced by certain weather conditions.

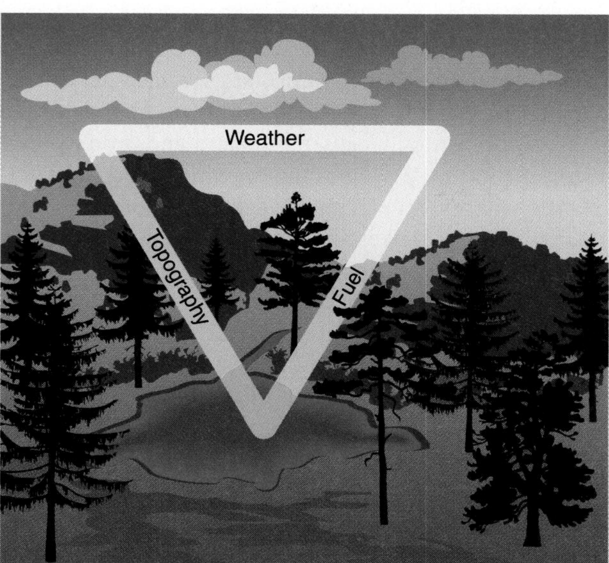

Figure 20-1 The wildland fire triangle.

Fire Fighter Safety Tips

All fires of fine or heavy fuels should be treated with the same caution and safety. Many fire fighters have died because they underestimated fire behavior. Treat all wildland and ground fires with respect. Never be caught in the trap of underestimating what a wildland and ground fire may do.

Fire Fighter Safety Tips

Heavy fuels such as trees and stumps require close attention. Spraying water on the outside will not extinguish the fire. Heavy fuels have a tendency to continue to burn on the inside. Heavy fuels need to be cut out to be sure that you have extinguished all the fire and to prevent rekindles.

Fuel

The primary fuel for wildland and ground fires is the vegetation (grasslands, brush, and trees) in the area. The amount of fuel in an area will range from sparse grass to heavy underbrush and large trees. Some fuels ignite readily and burn rapidly when dry. Others are harder to ignite and burn more slowly. Vegetative fuels can be located under the ground (roots), on the surface (grass and fallen leaves), or above the ground (tree branches). Fuels are classified as fine fuels or heavy fuels.

Fine Fuels

Fine fuels include dried vegetation such as twigs, leaves, needles, grass, moss, and light brush. Ground duff, the partly decomposed organic material on a forest floor, is another type of fine fuel. Fine fuels have a large surface area relative to their volume. This causes them to ignite easily and burn quickly. Fine fuels are usually the main type of fuel present in ground cover fires. Fine fuels aid the ignition of heavier fuels and cause fire to burn and spread with great intensity. They burn rapidly and unpredictably. Fires spread more quickly in fine fuels than in heavy timber and brush.

Heavy Fuels

Heavy fuels are of a larger diameter than fine fuels. They include large brush, heavy timber, stumps, branches, and dead timber on the ground. Another type of heavy fuel is slash. Slash consists of the leftovers of a logging or land clearing operation. Slash includes large and small pieces of logs, branches, bark, stumps, and other vegetative debris. Heavy fuels do not spread a fire as rapidly as fine fuels do, but heavy fuels can burn with a high intensity.

Subsurface, Surface, and Aerial Fuels

Subsurface fuels are those fuels located under the ground. Roots, moss, duff, and decomposed stumps are examples of subsurface fuels. Fires involving subsurface fuels are hard to locate and hard to extinguish.

Surface fuels are located close to the surface of the ground. Surface fuels include grass, leaves, twigs, needles, small trees, and logging slash. Brush less than six feet above the ground is also considered to be a surface fuel. Surface fuels are sometimes called ground fuels. This type of fuel is involved in ground cover fires.

Aerial fuels (or canopy fuels) are located more than six feet above the ground. Aerial fuels are usually trees. This includes tree limbs, leaves and needles on limbs, and moss attached to the tree limbs.

Other Fuel Characteristics

Other characteristics of wildland fuels may determine how quickly the fuel ignites, how rapidly it burns, and how readily it spreads to other areas. These characteristics include the size and shape of the fuel, the compactness of the fuel, its continuity, volume, and moisture level.

The size and shape of a fuel influences how it burns. Fine fuels burn more quickly than heavy fuels and require less heat to reach their ignition temperature. Fine fuels have a greater surface area relative to the total volume of the fuel.

Fuel compactness influences the rate at which a fuel will burn. Air cannot circulate in and around fuels that are tightly compacted. For this reason, subsurface fuels burn more slowly than aerial fuels. **Fuel continuity** refers to the relative closeness of wildland fuels. Fuels that have continuity are close together or touch each other. Fuel continuity allows fire to spread from one area of fuel to the next. Fuels that are continuous have a sufficient supply of air to support rapid combustion. For this reason, fuels that are continuous burn much more rapidly than fuels that are compact. **Fuel volume** refers to the quantity of fuel available in a specific area. The amount of fuel in a given area influences the growth and intensity of the fire.

Fuel moisture refers to the amount of moisture contained in a fuel. The amount of moisture in a fuel influences the speed of ignition, the rate of spread, and the intensity of the fire. Fuels with high moisture content will not ignite and burn as readily as fuels that have low moisture content. Some types of vegetation naturally contain more moisture than others. The amount of moisture in a fuel is related to the season of the year. Fuel moisture also varies with the amount of rain that has fallen. In a year with decreased rainfall, the moisture contained in vegetative fuels will be lower than the moisture the same vegetation contains in a rainy year.

Oxygen

The second side of the fire triangle is oxygen. Oxygen is needed to initiate and support the process of combustion. In a structure fire, a limited supply of oxygen results in a slow burning fire. If additional oxygen becomes available to a fire, the fire will begin to burn more rapidly.

In wildland and ground fires, which are usually free burning, oxygen is not usually an important variable in the ignition or spread of the fire. Unlike structure fires, oxygen is available in unlimited quantities. However, air movement around wildland and ground fires will influence the speed with which a fire moves. Wind blowing on a wildland and ground fire brings more oxygen to the fire, speeds the process of combustion, and influences the direction the fire travels.

Heat

The third side of the fire triangle is heat. Sufficient heat must be applied to fuel in the presence of adequate oxygen to produce a fire. Three categories of factors may ignite wildland and ground fires. These are natural causes, accidental causes, and intentional causes. Lightning sparks are the source of almost all of the naturally caused wildland and ground fires. Some storms produce dry lightning, in which there are lightning strikes without rain. A single lightning storm can produce many lightning strikes and start many fires. A variety of accidental causes result in wildland and ground fires, such as discarded smoking materials and improperly extinguished campfires. Downed electric wires provide a heat source that can ignite vegetative fuels. Unfortunately, some wildfires and ground fires are the result of intentional acts of arson.

Other Factors That Affect Wildland Fires

Once a wildland and ground fire has started, the growth of the fire is influenced by weather factors and by the topography of the land involved in the fire. Weather conditions and topography have a greater effect on wildland and ground fires than on structure fires.

Weather

Weather conditions have a large impact on the course of a wildland fire. The two weather conditions that influence a wildland fire the most are moisture and wind.

Moisture

Moisture can be present in the form of relative humidity or precipitation. **Relative humidity** is the ratio of the amount of water vapor present in the air compared to the maximum amount the air can hold at a given temperature. (Warm air has a higher capacity for moisture content than cool air.) If the air contains the maximum amount of water vapor possible, the relative humidity is said to be 100%. On a rainy day, the relative humidity is usually 100%. In dry climates, the relative humidity may be 20% or less. This means the water vapor contained in the air is only 20% of the maximum that the air is capable of holding.

Relative humidity is a major factor in the behavior of wildland and ground fires. When relative humidity is low, the vegetative fuels dry out, making them more susceptible to ignition. When relative humidity is high, the moisture from the air is absorbed by the vegetative fuels, making them less susceptible to ignition. Changes in relative humidity affect light fuels more than heavy fuels.

Relative humidity varies with the time of day. As the temperature increases throughout the day, the relative humidity drops. As the temperature drops during the evening and night hours, the relative humidity rises. A rise in the relative humidity can help to slow the spread of a wildland fire and give an added advantage to fire crews trying to control the fire. In some parts of the country, the average relative humidity varies with the season of the year. These changes contribute to variations in the moisture content of vegetative fuels. This helps to explain why many parts of the country have seasonal patterns of wildland and ground fires. The seasons with the most wildland and ground fires are often the seasons with the lowest relative humidity.

The second type of moisture that influences the behavior of wildland and ground fires is precipitation. Moisture falling from the sky helps to increase the relative humidity, and much precipitation is absorbed by plants, which increases the moisture of the plants and makes them less susceptible to combustion. In seasons with adequate precipitation, the incidence of wildland and ground fires is much lower than in years with below average precipitation.

Wind

Air movement also has an effect on the spread and direction of wildland and ground fires. On a still day, a wildland and ground fire can burn and grow in size. However, a fire with similar conditions will spread more rapidly if there is adequate wind to push the fire into additional fuels. Wind has the ability to move a fire at great speed. The effect of wind on a wildland and ground fire is similar to fanning a fire to help it burn more rapidly.

Topography

The movement of wildland and ground fires is also related to the topography of the land. Topography refers to the changes of elevation in the land, as well as the position of natural and manmade features. Topography has a great impact on the behavior of wildland and ground fires. Heat from these fires rises just like the heat in a structure fire. When wildland and ground fires burn on flat land, much of the heat from the fire will rise into the air. However, where the elevation rises in the direction the fire is traveling, the heat from the fire will ignite a greater quantity of fuel and will increase the speed at which the fire spreads.

Other features of the topography also influence the spread of a fire. Natural barriers, such as streams and lakes, may help contain fires. Built barriers such as highways also make it easier to contain a fire.

Extinguishing Wildland Fires

In order to safely attack and effectively extinguish wildland and ground fires, consider how the factors of weather and topography influence the behavior of the fire. You also need to know the common terms that are used to describe different parts of the fire in order to understand directions and to report information back to fire officers.

Anatomy of a Wildland Fire

The location where wildland and ground fires begin is called the area of origin. As the fire grows and moves into new fuel, the most rapidly moving area—the traveling edge of the fire—is called the head of the fire. As the fire gets bigger, the area close to the area of origin is often referred to as the heel of the fire or the rear of the fire (▲ Figure 20-2).

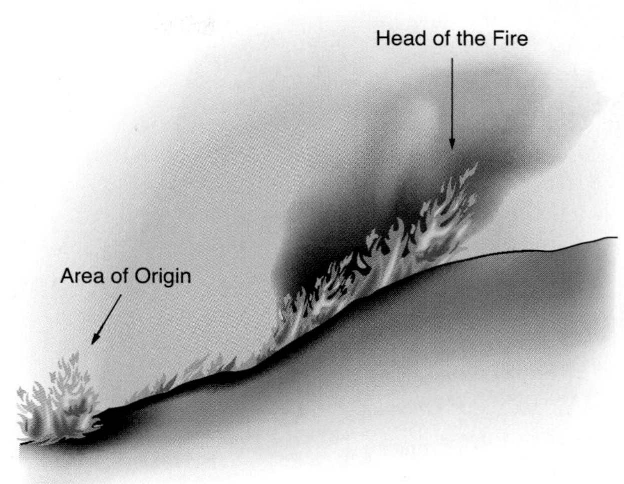

Figure 20-2 The anatomy of a wildland fire.

As the fire grows, a change in wind or topography can cause it to move into a long narrow extension of fire that projects out from the head of the fire. This is called a finger. A finger can grow and produce a secondary direction of travel for the fire. An unburned area between a finger and the main head of the fire is called a pocket. A pocket is a dangerous place for fire fighters because it is an area of unburned fuel surrounded on three sides by fire. An area of land that is left untouched by the fire, but is surrounded by burned land, is called an island.

A spot fire is a new fire that starts outside the perimeter of the main fire. Spot fires usually start from flaming vegetation such as an ember or spark that is picked up by the convection currents generated by the fire and dropped some distance away.

Green and black are terms often used by wildland fire fighters. Green describes the area of unburned fuels. Because this area contains unburned fuels, it is more likely to become involved in the fire than the black area. Black is used to describe areas that have already been burned. Because black areas contain few unburned fuels, these areas are often relatively safe for fire fighters.

Methods of Extinguishment

The methods used to extinguish wildland and ground fires involve disrupting the conditions represented by the fire triangle. Wildland and ground fires can be controlled and extinguished by cooling the fuel, removing fuel from the fire, or by smothering the fire.

Cooling the Fuel

Water is used to cool the fuel in wildland and ground fires. The principles involved in cooling wildland and ground fires are the same as those used for structural fires, but the methods used to apply the water may be different.

Voices of Experience

"Although it seemed excessive, using the deck gun was the only strategy that worked and worked safely."

It was an early summer evening when a ground fire erupted on the side of a highway embankment. The brush along the highway was dry and had caught on fire. The street running beside the highway provided an access route to the fire, but an 8′ chain link fence stopped us from gaining total access to the area of origin.

The first engine company began shooting water through the fence in an attempt to suppress the ground fire. This did not work and the ground fire began to build. A call for a second engine company went out.

The first engine company then placed ladders against the fence and prepared to go over the top with handlines. This scene looked more like a keystone cops film than a training video—hoses snagged, the ladders swayed, and the ground fire grew in intensity.

The call went out for a third engine company. Arriving at the scene, the third company officer ordered the deck gun on top of the apparatus to be set-up. The driver/operator immediately pulled the levers to supply water to the deck gun. A huge stream of water rushed out and in a matter of seconds the ground fire was completely out.

All three companies stood silent in amazement. Previously, we had only used deck guns on major structural fires. Although it seemed excessive, using the deck gun was the only strategy that worked and worked safely.

"In tribute to Captain Paul Leitch (retired)."

Bill Kehoe
Alexandria Fire Department
Alexandria, Virginia

For small fires with a light fuel load, **backpack pump extinguishers** may be effective. These can be transported to the fire and quickly put into operation before the fire has an opportunity to grow. Portable backpack pumps hold between 4 and 8 gallons of water and expel the water using a hand pump.

Water from booster tanks carried on structural fire apparatus or on special wildland fire trucks is another means of cooling a wildland fire. The hoses used for wildland fires may be smaller than the hose required for structural fire fighting. It is not always possible to get trucks into a position where they can attack a wildland fire. The water supply is limited to the amount of water carried by the apparatus, unless static water sources can be tapped by fire pumpers or portable pumps. Backpack pumps and small hose lines are used to extinguish most ground cover fires.

A third means of applying water to a wildland fire is from aircraft. Fixed wing aircrafts can take on a load of water from a lake and apply it to the fire. Rotary wing aircrafts (helicopters) can carry a container of water and drop it on the fire. The use of aircrafts can be an effective means to control wildland fires; however, it requires close communication between different agencies to coordinate this activity. Aircrafts are most commonly used for fires involving heavy fuels.

Removing Fuel

The second method of controlling and extinguishing wildland and ground fires is to remove the fuel from the fire. This can be done by fire fighters using hand tools or mechanized equipment.

Removal of fine fuels can be accomplished with a fire broom to sweep away light grass or weeds. Steel fire rakes can be used to create a fire line in light brush. A combination hoe and rake called a **McLeod tool** is also used to create a fire line. For heavier brush, an **adze** can be used to grub out brush to create a fire break. The **Pulaski axe** combines an adze and an axe for brush removal. These are some of the hand tools most commonly used to create a fire line by removing fuel from the fire.

In some situations, saws are used to remove heavy brush and trees from the fire. These saws range from hand saws to gasoline powered chain saws. Other powered equipment includes tractors, plows, and bulldozers.

Backfiring is another technique used to remove the fuel from a wildland fire. When properly set, a backfire can burn an area of vegetation in front of the fire, thereby creating a wide area devoid of vegetation. Backfires must be set at the right time and at the proper place to work correctly. When improperly set, a backfire can worsen a wildland fire or create a new one. Backfires should be set only by people who have had special training in the control of wildland and ground fires.

Removing Oxygen

The third method that can be used to control and extinguish wildland fires is to remove oxygen from the fire. Some vegetative fires in grasses can be smothered using a special fire beater sheet constructed of heavy rubber. Smothering is most commonly used when overhauling the last remnants of a wildland and ground fire. Earth is often thrown on smoldering vegetation to prevent flare-ups. Smothering is not as useful during the more active phases of a fire. **Compressed air foam systems (CAFS)** combine foam concentrate, water, and compressed air to produce a foam that will stick to vegetation and structures that are in a wildland and ground fire's path. When the heat of the fire reaches the foam, it absorbs the fire's heat and breaks down the foam, thereby cooling the fuel. CAFS can also extinguish a fire by coating the fuel and preventing oxygen from combining with the fuel. CAFS can be carried in a backpack extinguisher or mounted on apparatus.

Types of Attack

Wildland and ground fires can advance quickly and can change directions quickly. They present a very hazardous environment to fire fighters. For these reasons, it is important for the incident commander to match the type of attack to the conditions present at a fire. There are two types of attacks used to contain and extinguish wildland fires: direct attacks and indirect attacks.

Direct Attack

A **direct attack** on a wildland and ground fire is mounted by containing and extinguishing the fire at its burning edge. A direct attack can be made with a fire crew and hand tools, a structural engine company, a wildland engine company, a plow, or aircraft. Fire fighters might smother the fire with dirt, use hoses to apply water to cool the fire, or remove fuel. A direct attack is more often used on a small wildland fire before it has grown to a large size.

A direct attack can be dangerous if not properly done. It needs to be coordinated through the use of the incident command system. Before a direct attack is started, the incident commander will assess the resources present and determine the best way to attack the fire.

A direct attack is dangerous to fire fighters because they must work in smoke and heat close to the fire. A direct attack has the advantage of accomplishing quick contain-

Fire Fighter Tips

Two primary methods of attack are used for attacking wildland and ground fires: the pincer attack and the flanking attack. The pincer attack uses two crews of fire fighters to mount a two-pronged attack with one crew suppressing one side of the fire and the other crew suppressing the other side of the fire. The two crews work together to "pinch out" the fire. The flanking attack requires only one crew. The incident commander poisitons the crew on one side of the fire and the crew suppresses it.

ment of a fire. By extinguishing the fire while it is small, the risk the fire poses is reduced. Most wildland and ground fires are kept small by the initial crews using an aggressive direct attack on the fire.

Indirect Attack

An **indirect attack** is most often used for large wildland and ground fires that are too dangerous to approach through a direct attack. An indirect attack is mounted by building a fire line along natural fuel breaks, favorable breaks in the topography, or at considerable distance from the fire and burning out the intervening fuel. An indirect attack is similar to using a defensive attack on a structure fire. The fire line may be constructed fairly close to the fire or it can be constructed several miles away. An indirect attack can be mounted using hand tools or by using mechanized machinery.

Indirect attacks are used when there is not enough equipment or personnel to mount a direct attack. An indirect attack is also most appropriate when the topography is so rough that a direct attack is dangerous or impossible.

Priorities of Attack

Before determining how to attack a wildland fire, it is necessary for the incident commander to assess and evaluate the priorities for preserving lives and property. The top priority is to assure the safety of both fire fighters and civilians. The second priority is to preserve any structures threatened by the blaze.

Safety in Wildland Firefighting

Fighting wildland and ground fires is hazardous duty. The wildland and ground fire environment shares many of the hazards of structural firefighting as well as additional hazards associated with driving, falls, smoke, fire, and falling trees.

Hazards of Wildland Firefighting

Responses for wildland and ground fires involve all the hazards encountered by personnel responding to structural fires. In addition, responses to wildland fires may involve hazards associated with driving fire apparatus on poorly maintained roads or trails, and driving apparatus off-road. Driving on unimproved roads and steep terrain greatly increases the chance of fire apparatus rollovers.

It is important for drivers to understand the operating characteristics of their fire apparatus and to operate the apparatus within the safe limits for which it was designed. Always keep your seat belt fastened whenever the apparatus is moving.

Working in rough terrain places, wildland and ground fire fighters are at risk for falls. Rough ground often contains holes that are hard to see in smoky conditions. Steep terrain also increases the likelihood of falls. Since wildland and ground firefighting involves working with sharp tools, it is important to prevent injuries caused by these tools.

The hazards of fighting wildland and ground fires include burns and smoke inhalation. Because the personal protective equipment (PPE) worn by wildland and ground fire fighters provides less protection than the PPE worn by structural fire fighters, it is important to keep far enough from the heat of the fire to prevent burns. And since much wildland and ground firefighting is done without SCBA, fire fighters need to avoid inhaling poisonous gases and suspended smoke particles. Use your SCBA in conditions where needed.

When engaged in wildland and ground firefighting, be alert for the hazards of falling trees. During a fire, the lower parts of trees may burn away and weaken the support of the rest of the tree. Trees of all sizes can fall with little warning.

Also be alert for the presence of electrical hazards. Electrical transmission lines and other electrical wires may be present in the location of a wildland fire. Wires that drop on vegetation may ignite a wildland and ground fire and continue to pose an electrical hazard to fire fighters. Many of these safety hazards can be difficult to see at night and in smoky conditions.

Personal Protective Equipment

It is important for wildland fire fighters to wear appropriate PPE. They should be equipped with a one-piece jumpsuit, or a coat, shirt, and trousers that meet the requirements of NFPA 1977, *Standard on Protective Clothing and Equipment for Wildland Fire Fighting*. These garments should be constructed of a fire-resistant material like Nomex. This assures that the clothing will stop burning as soon as the heat and flame are removed. This type of clothing reduces the chance of fire fighters being burned. Wildland fire fighters should also wear an approved helmet with a protective shroud, eye protection, gloves, and protective footwear. All this equipment should meet the provisions of NFPA 1977. (► **Figure 20-3**) shows the PPE for wildland and ground firefighting.

Respiratory protection for wildland and ground firefighting is usually limited to a filter mask that filters small particles of smoke. These masks provide a very limited degree of protection and should not be relied upon beyond the limits for which they are approved. SCBA should be used in conditions of heavy smoke and poisonous gases.

The PPE that is provided for wildland and ground firefighting varies from one location to another, depending on the types of wildland fires encountered and on the climate in a given location. Follow the standards of your department regarding PPE.

Fire Shelters

One of the most important pieces of PPE for wildland and ground fire fighters is the **fire shelter**. This lifesaving piece of equipment should be issued to all wildland and ground fire fighters. Fire shelters are made of a thin reflective foil

Figure 20-3 PPE for a wildland fire fighter.

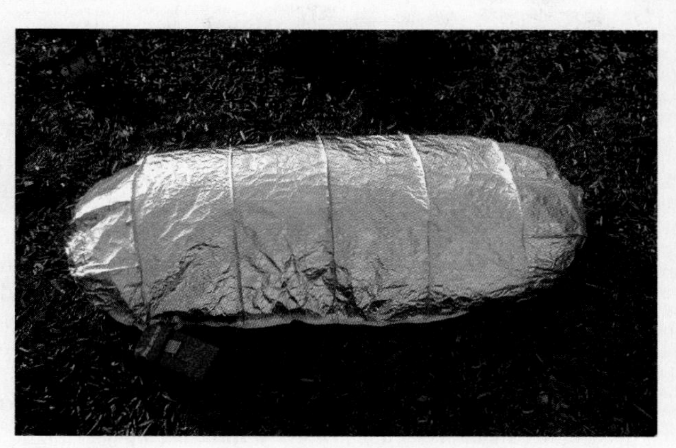

Figure 20-4 A fire shelter.

layer that is attached to a layer of fiberglass. These shelters are designed to reflect about 95% of a fire's radiant heat for a short period of time. This allows a rapidly moving fire to pass over a fire fighter who has deployed a fire shelter.

Fires shelters are carefully folded and carried in a protective pouch on the fire fighter's belt. When fire fighters are in danger of being overrun by rapidly moving fire, they should try to get to a safe location. Only when they cannot escape should they use their fire shelters. To use a fire shelter, the fire fighter opens the shelter, lies face down on the ground, and covers himself or herself with the shelter. When properly used by well-trained personnel, these shelters can save lives. As with all equipment, it is important to receive proper training in order to use this device safely. ► Figure 20-4 shows how the shelter is deployed.

The Challenge of the Wildland Urban Interface

When we think of wildland fires, we think of burning grass, brush, and forests. However, wildland fires as defined by the NFPA are "unplanned and uncontrolled fires burning in vegetation, including structures and other improvements." This means that many wildland fires threaten not only vegetative fuels, but also improvements that have been made on the land. Today many people want a house that is surrounded by the beauty of nature. They want a house that has modern conveniences, but they want it to be set in the middle of the woods. The mixing of wildlands and developed areas creates a massive problem for fire departments in many parts of the country. Wildland fires regularly ignite buildings and become structure fires. Structure fires regularly ignite vegetative fuels and become wildland fires. In many cases, it is not possible to clearly separate one type of fire from the other. This phenomenon explains why most fire departments are involved in fighting some types of wildland fires.

Fire Fighter Safety Tips

The use of structural PPE for wildland firefighting should be avoided because wearing structural PPE for long periods of time will result in the retention of body heat and will greatly increase the chance of dehydration, heat exhaustion, and heat stroke.

The term **wildland urban interface** is used to explain the mixing of wildland with developed areas. The wildland urban interface is defined as an area where undeveloped land with vegetative fuels is mixed with manmade structures. The wildland urban interface presents a great challenge for fire fighters. Since many people live in the wildland urban interface, fires in this zone present a significant life safety hazard. And since there are so many structures built in this zone, there is a huge potential for property loss. Many parts of the wildland urban interface do not have adequate municipal water systems.

The wildland urban interface presents a unique challenge for fire fighters and emphasizes the importance of having structural fire fighters learn the basic principles of fighting wildland fires. The material presented in this chapter will help you to understand the basic principles on fighting wildland fires. It is not intended to provide all the skills and knowledge needed by wildland fire fighters. If your department is regularly involved in fighting wildland fires, you should complete a separate course on wildland fire fighting based on NFPA 1051, *Standard for Wildland Fire Fighter Professional Qualifications.*

Wrap-Up

Ready for Review

This chapter has covered the basics of wildland and ground firefighting. It covered the role structural fire fighters have in battling wildland and ground fires. Wildland fire is defined as an unplanned and uncontrolled fire burning natural vegetative fuels.

Different types of fuel are encountered in wildland and ground fires. These fuels may be located below the ground, on the ground, or above the ground. The weather can influence the fuel and oxygen supply of wildland fires, and topography has an important impact on the growth of wildland and ground fires.

The terminology of wildland and ground fires relates to the factors of the fire triangle. Accordingly, there are three approaches to extinguishing wildland and ground fires; cooling the fuel, removing the fuel, and removing the oxygen supply. There are numerous examples of each method, and specific hand and power tools that can be used. The advantages and disadvantages of direct and indirect attacks on fires depend on a fire's size and stage of development, and the resources available.

An understanding of the hazards associated with wildland and ground firefighting leads to the steps that can be taken to improve safety. The importance of using proper personal protective equipment is a key safety point. The wildland urban interface presents great challenges for fire fighters. For the many departments that will fight wildland fires as well as structure fires, specialized training is essential.

Chief Concepts

- Most fire departments have a role in fighting wildland and ground fires.
- Wildland and ground fires consume vegetative fuels and structures that have been built in wildland areas.
- Wildland and ground fires feed on a variety of fuels, combined with oxygen and a source of heat.
- Moisture, wind, and topography have significant impacts on wildland and ground fires.
- Wildland and ground fires can be extinguished by cooling, removing the fuel, or smothering the fire using direct or indirect attacks.
- Safety is the primary priority when fighting wildland and ground fires.
- The wildland urban interface creates a challenge for fire fighters.

Hot Terms

Adze A hand tool constructed of a thin arched blade set at right angles to the handle. It is used to chop brush for clearing a fire line, or to mop up a wildland fire.

Aerial fuel Fuels more than six feet off the ground, usually part of or attached to trees.

Area of origin The room or area where a fire began.

Backfiring A planned operation to remove fuel from a wildland fire by burning out large selected areas. This should be done only under the close supervision of experienced, authorized fire officers.

Backpack pump extinguisher A portable fire extinguisher consisting of a four-gallon to eight-gallon water tank worn on the user's back and features a hand-powered piston pump for discharging the water.

Black An area that has already been burned.

Compressed air foam systems (CAFS) Combines foam concentrate, water, and compressed air to produce foam that can stick to both vegetation and structures.

Direct attack (wildland and ground fire) A method of fire attack mounted by containing and extinguishing the fire at its burning edge.

Fine fuel Fuel that ignites and burns easily, such as dried twigs, leaves, needles, grass, moss, and light brush.

Finger A narrow point of fire caused by a shift in wind or a change in topography.

Fire shelter A special tent-like device that protects fire fighters from the radiant heat of fire for a short period of time.

Fuel compactness How tightly packed fuels are.

Fuel continuity The relative closeness of wildland fuels, a factor in a fire's ability to spread from one area of fuel to another.

Fuel moisture The amount of moisture present in a fuel, a factor in how readily the fuel will ignite and burn.

Fuel volume The amount of fuel present in a given area.

Green An area of unburned fuels.

Ground cover fire A fire that burns loose debris on the surface of the ground.

Ground duff Partly decomposed organic material on a forest floor; a type of light fuel.

Head of the fire The main or running edge of a fire. The part of the fire that spreads with the greatest speed.

Heavy fuels Fuels of a large diameter, such as large brush, heavy timber, snags, stumps, branches, and dead timber on the ground; these ignite and are consumed more slowly than light fuels.

Heel of the fire The side opposite the head of the fire, often close to the area of origin.

Indirect attack (wildland and ground fire) A method of attack in which the control line is located along natural fuel breaks, favorable breaks in the topography, or at considerable distance from the fire, and the intervening fuel is burned out.

Island An unburned area surrounded by fire.

McLeod tool A hand tool used for constructing fire lines and overhauling wildland fires. One side of the head consists of a five-toothed to seven-toothed fire rake and the other side is a hoe.

Pocket A deep indentation of unburned fuel along the fire's perimeter, often between a finger and the head of the fire.

Pulaski axe A hand tool that combines an adze and an axe for brush removal.

Rear of the fire The side opposite the head of the fire, also called the heel of the fire.

Relative humidity The ratio of the amount of water vapor present in the air compared to the maximum amount the air can hold at a given temperature.

Slash The leftovers of a logging operation; includes large and small pieces of logs, branches, bark, stumps, and other vegetative debris.

Spot fire A new fire that starts outside areas of the main fire; usually caused by flying embers and sparks.

Subsurface fuels Partially decomposed matter that lies beneath the ground, such as roots, moss, duff, and decomposed stumps.

Surface fuels Fuels that are close to the surface of ground, such as grass, leaves, twigs, needles, small trees, logging slash, and low brush; also called ground fuels.

Topography The features of the earth's surface; changes in land elevation and the position of natural and manmade features.

Wildland Land in an uncultivated natural state that is covered by timber, woodland, brush, or grass.

Wildland fire An unplanned and uncontrolled fire burning in vegetative fuels.

Wildland urban interface An area where undeveloped land with vegetative fuels is mixed with manmade structures.

Fire Fighter in Action

It is 3:00 p.m. on a hot and humid summer afternoon. Your engine is dispatched to a grass fire threatening multiple structures. Upon arrival you see a 100' by 50' running grass field fire that is being fanned by light wind. It is approximately 100' from a barn

1. What type of PPE should you wear for this incident?
   - **A.** Structural firefighting boots, gloves, coat, pants, and helmet
   - **B.** Nomex coveralls or shirt and pants, helmet, gloves, eye protection, and footwear that meet NFPA 1997.
   - **C.** Coveralls or shirt and pants, gloves, eye protection, and mask that meet NFPA 1977.
   - **D.** Nomex shorts, shirt, boots, helmet, gloves, and eye protection.

2. What type of fuel would grass be classified as?
   - **A.** Light fuel
   - **B.** Heavy fuel
   - **C.** Subsurface fuel
   - **D.** Aerial fuel

Lieutenant Wigal tells you and the crew to prepare for a direct attack on the head of the fire using forestry hose lines attacking from the black. You don the appropriate PPE and cautiously head towards the head of the fire.

3. What are some of the hazards that should be considered in this scenario?
   - **A.** Smoke inhalation
   - **B.** Burns
   - **C.** Electrical wires
   - **D.** Obscured vision while walking
   - **E.** All of the above

4. "Attacking from the black" means to:
   - **A.** fight the fire from the unburned fuels.
   - **B.** fight the fire from the burned area.
   - **C.** fight the fire from its origin.
   - **D.** fight the fire from an island of unburned fuel.

www.FireFighter.jbpub.com

Chapter Pretests
Interactivities
Hot Term Explorer
Web Links
Review Manual
FireLearn

www.FireFighter.jbpub.com

Fire Suppression

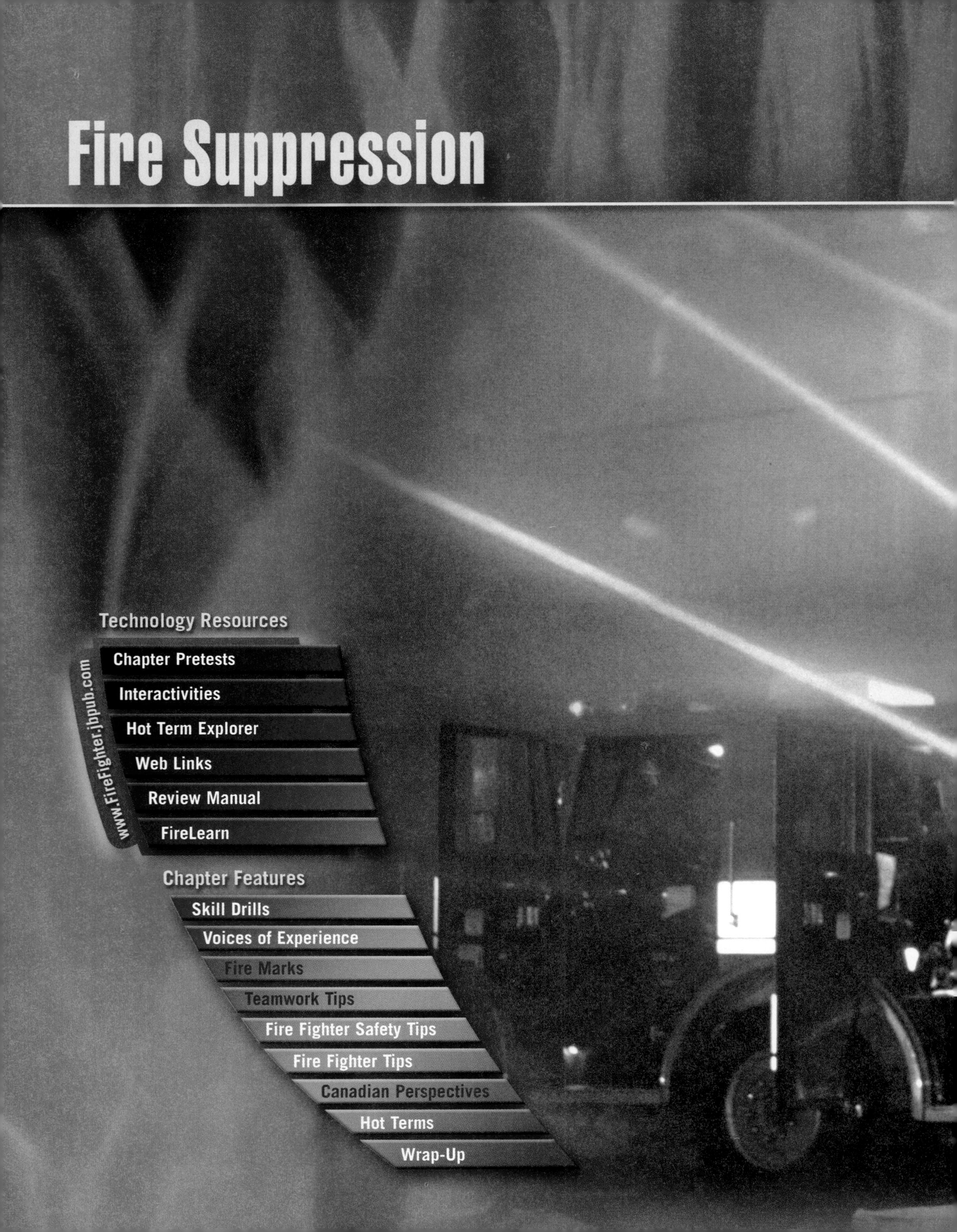

Technology Resources

www.FireFighter.jbpub.com

- Chapter Pretests
- Interactivities
- Hot Term Explorer
- Web Links
- Review Manual
- FireLearn

Chapter Features

- Skill Drills
- Voices of Experience
- Fire Marks
- Teamwork Tips
- Fire Fighter Safety Tips
- Fire Fighter Tips
- Canadian Perspectives
- Hot Terms
- Wrap-Up

Chapter 21

Fire Suppression

NFPA 1001 Standard

Fire Fighter I

5.3.7 Attack a passenger vehicle fire operating as a member of a team, given personal protective equipment, attack line, and hand tools, so that hazards are avoided, leaking flammable liquids are identified and controlled, protection from flash fires is maintained, all vehicle compartments are over-hauled and the fire is extinguished.

5.3.7 (A) *Requisite* Knowledge. Principles of fire streams as they relate to fighting automobile fires; precautions to be followed when advancing hose lines toward an automobile; observable results that a fire stream has been properly applied; identify alternative fuels and the hazards associated with them; dangerous conditions created during an automobile fire; common types of accidents or injuries related to firefighting automobile fires and how to avoid them; how to access locked passenger, trunk, and engine compartments; and methods for overhauling an automobile.

5.3.7 (B) *Requisite Skills.* The ability to identify automobile fuel type; assess and control fuel leaks; open, close, and adjust the flow and pattern on nozzles; apply water for maximum effectiveness while maintaining flash fire protection; advance $1^1/_2$-in. (38-mm) or large diameter attack lines; and expose hidden fires by opening all automobile compartments.

5.3.8* Extinguish fires in exterior Class A materials, given fires in stacked or piled and small unattached structures or storage containers that can be fought from the exterior, attack lines, hand tools and master stream devices, and an assignment, so that exposures are protected, the spread of the fire is stopped, collapse hazards are avoided, water application is effective, the fire is extinguished, and signs of the origin area(s) and arson are preserved.

5.3.8* (A) *Requisite Knowledge.* Types of attack lines and water streams appropriate for attacking stacked, piled materials and outdoor fires; dangers—such as collapse—associated with stacked and piled materials; various extinguishing agents and their effect on different material configurations; tools and methods to use in breaking up various types of materials; the difficulties related to complete extinguishment of stacked and piled materials; water application methods for exposure protection and fire extinguishment; dangers such as exposure to toxic or hazardous materials associated with storage building and container fires; obvious signs of origin and cause; and techniques for the preservation of fire cause evidence.

5.3.8* (B) *Requisite Skills.* The ability to recognize inherent hazards related to the material's configuration, operate handlines or master streams, break up material using hand tools and water streams, evaluate for complete extinguishment, operate hose lines and other water application devices, evaluate and modify water application for maximum penetration, search for and expose hidden fires, assess patterns for origin determination, and evaluate for complete extinguishment.

5.3.10* Attack an interior structure fire operating as a member of a team, given an attack line, ladders when needed, personal protective equipment, tools, and an assignment, so that team integrity is maintained, the attack line is deployed for advancement, ladders are correctly placed when used, access is gained into the fire area, effective water application practices are used, the fire is approached correctly, attack techniques facilitate suppression given the level of the fire, hidden fires are located and controlled, the correct body posture is maintained, hazards are recognized and managed, and the fire is brought under control.

5.3.10* (A) *Requisite Knowledge.* Principles of fire streams; types, design, operation, nozzle pressure effects, and flow capabilities of nozzles; precautions to be followed when advanced hose lines to a fire; observable results that a fire stream has been properly applied; dangerous building conditions created by fire; principles of exposure protection; potential long-term consequences of exposure to products of combustion; physical states of matter in which fuels are found; common types of accidents or injuries and their causes; and the application of each size and type of attack line, the role of the backup team in fire attack situations, attack and control techniques for grade level and above and below grade levels, and exposing hidden fires.

5.3.10* (B) *Requisite Skills.* The ability to prevent water hammers when shutting down nozzles; open, close, and adjust nozzle flow and patterns; apply water using direct, indirect, and combination attacks; advance charged and uncharged $1^1/_2$-in. (38-mm) diameter or larger hose lines up ladders and up and down interior and exterior stairways; extend hose lines; replace burst hose sections; operate charged hose lines of $1^1/_2$-in. (38-mm) diameter or larger while secured to a ground ladder; couple and uncouple various handline connections; carry hose; attack fires at grade level and above and below grade levels; and locate and suppress interior wall and subfloor fires.

Fire Fighter II

6.3 *Fireground Operations.* This duty involves performing activities necessary to insure life safety, fire control, and property conservation, according to the following job performance requirements.

6.3.2* Coordinate an interior attack line for team's accomplishment of an assignment in a structure fire, given attack lines, personnel, personal protective equipment, and tools, so that crew integrity is established; attack techniques are selected for the given level of the fire (for example, attic, grade level, upper levels, or basement); attack techniques are communicated to the attack teams; constant team coordination is maintained; fire growth and development is continuously evaluated; search, rescue, and ventilation requirements are communicated or managed; hazards are reported to the attack teams; and incident command is apprised of changing conditions.

6.3.2* (A) *Requisite Knowledge.* Selection of the nozzle and hose for fire attack given different fire situations; selection of adapters and appliances to be used for specific fire ground situations; dangerous building conditions created by fire and fire suppression activities; indicators of building collapse; the effects of fire and fire suppression activities on wood, masonry (brick, block, stone), cast iron, steel, reinforced concrete, gypsum wall board, glass, and plaster on lath; search and rescue and ventilation procedures; indicators of structural instability; suppression approaches and practices for various types of structural fires, and the association between specific tools and special forcible entry tools.

6.3.2* (B) *Requisite Skills.* The ability to assemble a team, choose attack techniques for various levels of a fire (e.g., attic, grade level, upper levels, or basement), evaluate and forecast a fire's growth and development, select tools for forcible entry, incorporate search and rescue procedures and ventilation procedures in the completion of the attack team efforts, and determine developing hazardous building or fire conditions.

6.3.3 Control a flammable gas cylinder fire operating as a member of a team, given an assignment, a cylinder outside of a structure, an attack line, personal protective equipment, and tools, so that crew integrity is maintained, contents are identified, safe havens are identified prior to advancing, open valves are closed, flames are not extinguished unless the leaking gas is eliminated, the cylinder is cooled, cylinder integrity is evaluated, hazardous conditions are recognized and acted upon, and the cylinder is faced during approach and retreat.

6.3.3 (A) *Requisite Knowledge.* Characteristics of pressurized flammable gases, elements of a gas cylinder, effects of heat and pressure on closed cylinders, boiling liquid expanding vapor explosion (BLEVE) signs and effects, methods for identifying contents, how to identify safe havens before approaching flammable gas cylinder fires, water stream usage and demands for pressurized cylinder fires, what to do if the fire is prematurely extinguished, valve types and their operation, alternative actions related to various hazards and when to retreat.

(B) 6.3.3 (B) *Requisite Skills.* The ability to execute effective advances and retreats, apply various techniques for water application, assess cylinder integrity and changing cylinder conditions, operate control valves, choose effective procedures when conditions change.

Knowledge Objectives

After studying this chapter, you will be able to:

- Describe offensive versus defensive operations.
- Describe how to operate hose lines.
- Describe how to attack an interior structure fire.
- Describe exposure protection.
- Describe how to attack a vehicle fire.
- Describe how to extinguish a flammable gas cylinder fire.
- Describe a BLEVE.
- Describe how to attack fires involving electricity.

Skills Objectives

After completing this chapter you will be able to:

- Apply water using the direct attack.
- Apply water using an indirect attack.
- Apply water using the combination attack.
- Use a large handline using the one-fire fighter method.
- Use a large handline line using the two-fire fighter method.
- Operate deck guns.
- Operate portable monitors.
- Locate and suppress interior wall and subfloor fires.
- Extinguish a fire in a flammable gas cylinder.

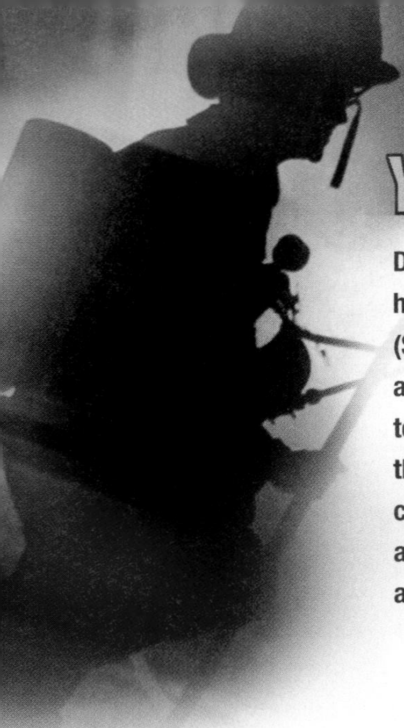

You Are the Fire Fighter

During your shift at your fire station, you are dispatched to a reported car fire. You know the hazards involved with fighting a vehicle fire, so you don your self-contained breathing apparatus (SCBA) and personal protective equipment (PPE) before boarding the fire engine. As your crew arrives on the scene, you see thick black smoke coming from the rear of a detached garage next to a large two-story house. You pull the appropriate hose line and make your way to the seat of the fire. As your firefighting team turns the corner, the situation changes. You see a fully involved car fire. Next to the burning vehicle, you notice a welders' cylinder of compressed gas. The radiant heat from the car fire is warping and melting the siding on the garage, and it may ignite at any time.

1. ***What are the hazards at this fire?***
2. ***How should you attack this fire?***

Introduction

The term fire suppression refers to all of the tactics and tasks that are performed on the fire scene to achieve the final goal of extinguishing the fire. Fire suppression can be accomplished through a variety of methods that will stop the combustion process. All of these methods involve removal of one of the four components of the fire tetrahedron. A fire can be extinguished by removing the oxygen, the fuel, or the heat from the combustion process. Interrupting the chemical chain reactions will also stop the combustion process and extinguish the fire.

This chapter presents the methods that are most frequently used by fire fighters for extinguishing fires. The apparatus and equipment used by most fire departments is designed to apply large volumes of water to a fire in order to cool the fuel below its ignition temperature. Although fire departments typically use a variety of extinguishing agents for different situations, water is used more often than any other agent.

Water is a very effective extinguishing agent for many different types of fires, because tremendous quantities of heat energy are required to convert water into steam. When water is applied to a fire, all of the heat that is used to create steam is removed from the combustion process. If a sufficient quantity of water is applied, the fuel is cooled below its ignition temperature and the fire is extinguished.

Offensive versus Defensive Operations

All fire suppression operations are classified as either offensive or defensive. When fire fighters advance hose lines into a building to attack a fire, the strategy is offensive. Defensive operations are conducted from the exterior, by directing water streams from a safe distance.

Offensive operations expose fire fighters to the heat and smoke of the fire inside the building, as well as several other risk factors, such as the possibility of being trapped by a structural collapse. The objective in an offensive operation is for fire fighters to get close enough to the fire to apply extinguishing agents at close range. This allows the extinguishing agent to be applied directly where it is needed to overpower the fire. When an offensive attack is successful, the fire can be controlled with the least amount of property damage.

Offensive strategy is used in situations where the fire is not too large to be extinguished by applying water from handlines or other extinguishing agents. Small handlines are used for most interior fire attack operations. Large handlines are often used to conduct an interior attack on a large fire, although they are much heavier and more difficult to maneuver inside a building, the volume of fire may require their use.

Large handlines and master streams are more often used in defensive operations. Defensive strategy is used in situations where the fire is too large to be controlled by an offensive attack and in situations where the level of risk to fire fighters conducting an interior attack would be unacceptable. The primary objective in a defensive operation is to prevent the fire from spreading. Water is directed into the building through doorways, windows, and other openings, or onto exposures to keep the fire from spreading, while fire fighters remain outside the building and operate from safe positions.

The decision whether to conduct offensive or defensive operations must be made by the incident commander (IC) at the beginning of each fire suppression operation and is periodically reevaluated throughout the incident. This decision must be made before operations begin and must be clearly communicated and understood by everyone involved in the operation. There is no room for confusion between offensive and defensive operations. It would be an extremely dangerous situation if one group of fire fighters initiated defensive operations while another group of fire fighters was inside a building, conducting an offensive attack.

If the decision is made to switch from offensive to defensive or from defensive to offensive strategy at any point dur-

Fire Fighter Safety Tips

Offensive (interior attack) and defensive (exterior attack) operations must NEVER be performed simultaneously.

Fire Fighter Safety Tips

An incident commander should never risk the lives of fire fighters when there are no lives to save.

ing an operation, the change must also be clearly communicated and understood. The change from offensive to defensive operations could occur if an interior attack is unsuccessful or the risk factors are determined to be too great to justify having fire fighters inside the building. Sometimes the strategy is switched from defensive to offensive after an exterior attack has reduced the volume of fire inside a building to the point at which fire fighters can enter and complete extinguishments with handlines. An offensive fire attack requires well-planned coordination among crews performing different tasks, such as ventilating, operating hose lines, and conducting aggressive search and rescue.

A defensive strategy should be implemented when the IC determines that it would be impossible to enter the burning building and control the fire with handlines, as well as in situations where the risk of injury or death of a fire fighter is excessive. A defensive operation involves the use of large hose lines and master stream devices from the exterior of the structure to confine the fire. Exposure protection should be a high priority during a defensive operation.

Command Considerations

The IC has to evaluate a whole range of factors to decide if offensive strategy (interior attack) or defensive strategy (exterior attack) should be used at a particular fire. If the risk factors are too great, an exterior attack is the only acceptable option. If the decision is made to launch an interior attack, the IC has to determine where and how to attack, considering both safety and the effectiveness of the operation. The factors to be evaluated when considering whether to enter the structure to mount an attack include the following:

- What are the risks versus the potential benefits?
- Is it safe to send fire fighters into the building?
- Are there any structural concerns?
- Are there any lives at risk?
- Does the size of the fire prohibit entry?
- Are there enough fire fighters on the scene to mount an interior attack? (Remember the two-in/two-out rule.)
- Is there an adequate water supply?
- Can proper ventilation be performed to support offensive operations?

After sizing up the situation, the IC must determine which type of attack is appropriate. As a new fire fighter, you are not responsible for determining the type of fire attack that will be used, but you should understand the factors that go into making these decisions and why the IC orders different types of fire attacks for different types of fires.

Operating Hose Lines

Some of the most basic skills that must be mastered by every fire fighter involve the use of hose lines to apply water onto a fire. You must be able to advance and operate a hose line effectively to extinguish a fire. The proper operation of a hose line is also essential to protect yourself, your crew, and any trapped victims from the fire. Fire attack operations are often conducted under extremely stressful conditions, with high heat conditions and limited or zero visibility in unfamiliar surroundings. Care should be taken not to have opposing hoselines, where two crews are operating at the same time "against" one another.

Fire fighters have to learn how to operate both large and small handlines, as well as master stream appliances. Small handlines can be up to 2" in diameter. The most frequently used sizes for interior fire attack are $1\frac{1}{2}$" or $1\frac{3}{4}$" in diameter. One fire fighter can usually operate the nozzle on a small hose line. A second fire fighter can provide valuable assistance when a small hose line has to be advanced and maneuvered.

Large handlines are defined as hoses that are at least $2\frac{1}{2}$" in diameter. Water can flow through these hoses at more than 250 gallons per minute. Large handlines are heavier and less maneuverable than smaller lines. At least two fire fighters are required to advance and control a large handline. One fire fighter can control a large handline if it is well-anchored.

Master streams are used when even larger quantities of water are needed to control a very large fire. A master stream can deliver at least 350 gallons of water per minute onto a fire, and some master stream devices can flow more than 2000 gallons per minute. The most commonly used master stream devices flow between 500 and 1000 gallons per minute. Master stream devices are operated from a fixed position, either on the ground, on top of a piece of fire apparatus, or on an elevating device. Master streams are only used for defensive operations.

Fire Streams

Several different types of fire streams can be produced by different types of nozzles. The nozzle defines the pattern and form of the water that is discharged onto the fire. A fire stream can be produced with either a smooth bore nozzle or an adjustable nozzle. Fire department policies and standard operating procedures (SOPs) will usually determine the type of nozzle that is used with different types of hose lines. The nozzle operator must know which type of nozzle should be used in different situations. When an adjustable nozzle is

Fire Fighter Tips

Although every effort should be made to minimize water damage, remember, property can be dried out but it cannot be unburned.

used, the nozzle operator must know how to set the discharge pattern to produce different kinds of streams.

The first major distinction in nozzle discharge patterns is between a fog stream and a straight stream (► Figure 21-1). A fog stream divides water into droplets, which have a very large surface area and can absorb heat efficiently. When heat levels in a building need to be lowered quickly, a fog stream is usually the pattern of choice. A fog stream can also be used to protect fire fighters from the heat of a large fire. Most adjustable nozzles can be adjusted from a straight stream, to a narrow fog pattern, to a wide fog pattern, depending on the reach that is required and how the stream will be used.

A **straight stream** provides more reach than a fog stream, so it can hit the fire from a greater distance. A straight stream also keeps the water concentrated in a small area, so it can penetrate through a hot atmosphere to reach and cool the burning materials. A straight stream is produced by setting an adjustable nozzle to the narrowest pattern it can discharge. This type of stream is made up of a highly concentrated pattern of droplets that are all discharged in the same direction.

A **solid stream** is produced by a smooth bore nozzle (► Figure 21-2). A solid stream has more reach and penetrating power than a straight stream, because it is discharged as a continuous column of water.

One important point that must be remembered when selecting and operating nozzles is the amount of air that is moved along with the water. A fog stream naturally moves a large quantity of air along with the mass of water droplets. This air flows into the fire area along with the water. When this air movement is combined with steam production as the water droplets encounter a heated atmosphere, the thermal balance is likely to be disrupted quickly. The hot fire gases and steam can be displaced back toward the nozzle operator or pushed into a different area of the building. Straight and solid streams move little air in comparison with a fog stream, so there are fewer concerns with displacement and disruption of the thermal balance.

The air movement created by a fog stream can be used for ventilation. Discharging a fog stream out through a window or doorway will draw smoke and heat out in the same manner as an exhaust fan. This operation must be performed carefully to prevent accidentally drawing hidden fire toward the nozzle operator.

Interior Fire Attack

An interior fire attack is an offensive operation that requires fire fighters to enter a building and discharge an extinguishing agent (usually water) onto the fire. An interior structure fire is a fire that occurs inside a building or structure. The fuel could be the contents of the building, or the structure itself could be burning. The larger the fire, the greater the challenge and the greater the risks that are involved in interior fire suppression.

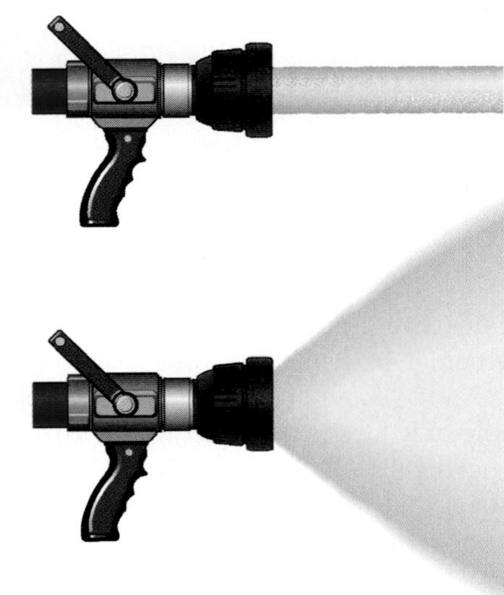

Figure 21-1 A straight stream and a fog stream are both produced using a fog nozzle.

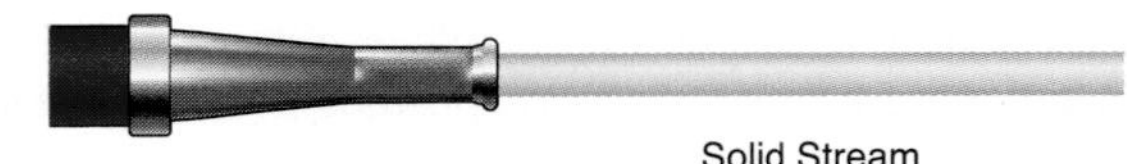

Figure 21-2 A solid stream is produced by a smooth bore nozzle.

Interior fire attack can be conducted on many different scales. In many cases an interior attack is conducted on a fire that is only burning in one room, and the fire can be controlled quickly by one attack hose line. Larger fires require more water, which could be provided by two or more small handlines working together or one or more larger handlines. Fires that involve multiple rooms, large spaces, or concealed spaces are more complicated and require much more coordination; however, the basic techniques for attacking the fire are similar.

Direct attack and indirect attack are two different methods of discharging water onto a fire. A combination attack is performed in two stages, beginning with an indirect attack and then continuing with a direct attack.

Fire Fighter Safety Tips

It is imperative to extinguish all fire as you proceed to its seat. Failure to do so may allow the fire to grow behind you, entrapping you by cutting off your escape route.

Figure 21-3 In a direct attack, a straight or solid hose stream is used to deliver water directly onto the base of the fire.

Figure 21-4 In an indirect attack, fire fighters direct a fog stream at the ceiling of the intensely heated area to create steam.

Direct Attack

The most effective means of fire suppression in most situations is the **direct attack**. The direct attack uses a straight or solid hose stream to deliver water directly onto the base of the fire (▲ Figure 21-3). The water cools the fuel until it is below its ignition temperature. The water is directed into the fire in short bursts and in a controlled method. Fire fighters should not apply more water than necessary to extinguish the fire, in order to keep damage from excess water to a minimum. To perform a direct attack, follow the steps in (Skill Drill 21-1).

1. Exit the fire apparatus wearing full PPE, including SCBA.
2. Select the proper hose line used to fight the fire depending on the fire's size, location, and type.
3. Advance the hose line from the apparatus to the entry point of the structure.
4. Don face piece, activate SCBA and PASS device prior to entering the building.
5. Signal the pump operator that you are ready for water.
6. Open the nozzle to purge air from the system and make sure water is flowing.
7. Make sure ventilation is completed or in progress.
8. Enter into the structure and locate the seat of the fire.
9. Apply water in either a straight or solid stream onto the base of the fire in short bursts.
10. Watch for changes in fire condition; use only enough water to extinguish the fire.
11. Locate and extinguish hot spots.

Indirect Attack

Indirect application of water is used in situations where the temperature is increasing and it appears that the room or space is ready to flashover. A short burst of water is aimed at the ceiling to cool the superheated gases in the upper levels of the room or space. This action can prevent or delay flashover long enough for fire fighters to apply water directly to the seat of the fire or to make a safe exit. Follow the SOP of your department regarding the application of water.

The objective of an **indirect attack** is to quickly remove as much heat as possible from the fire atmosphere. An indirect attack is particularly effective at preventing flashover from occurring. This method should be used when a fire has produced a layer of hot gases at the ceiling level. When the water is injected into the hot fire gases, it is converted to steam, absorbing tremendous quantities of heat in the process. The atmosphere is cooled quickly down to 212°F (100°C), the boiling point of water. The heat that is used to convert the water to steam is removed from the combustion process.

An indirect attack can be performed by fire fighters using a straight stream, a solid stream, or a narrow fog stream. The fire fighters direct their water at the ceiling of the intensely heated area where the hot gases are layered, in order to create steam (▲ Figure 21-4). This is often referred to as "painting the ceiling" with the water stream. The water is distributed over a large surface area to absorb heat as quickly as possible. Once the temperature has been reduced and the area has been properly ventilated, the fire fighters can switch to a direct attack to complete extinguishment.

As soon as enough steam has been produced to reduce the fire, the fire stream should be shut down so that the thermal layering of the superheated gases is disturbed as little as possible. When water is converted to steam, it expands to occupy a volume 1700 times greater than the volume of the water. This expansion tends to displace the hot gases that were near the ceiling and push them down toward the floor. This mixture of steam and hot gases is capable of causing serious steam burns to fire fighters through their PPE. Serious injuries can occur if fire fighters put too much water into the upper atmosphere and the hot gases are forced down on top of them.

To perform an indirect attack, follow the steps in Skill Drill 21-2.

1. Exit the fire apparatus wearing full PPE, including SCBA.
2. Select the correct hose line to be used to attack the fire depending on the type of fire, its location, and size.
3. Advance the hose line from the apparatus to the opening in the structure where the indirect attack will be made.
4. Don face piece, activate SCBA and PASS device.
5. Notify the pump operator that you are ready for water.
6. Open the nozzle to be sure that air is purged from the hose line and that water is flowing. If using a fog nozzle, ensure proper nozzle pattern for entry. Shut down the nozzle until you are in a position to apply water.
7. Advance with a charged hose line to the location where you are going to apply water.
8. Direct the water stream toward the upper levels of the room and ceiling into the heated area overhead and move the stream back and forth. Flow water until the room begins to darken. Then shut the nozzle off and assess conditions.
9. Watch for changes and reduction in the amount of fire. Once the fire is reduced, shut down the nozzle.
10. Check to make sure ventilation has been completed.
11. Attack remaining fire and hot spots until extinguished.

Combination Attack

The combination attack employs both the indirect and direct attack methods sequentially. This method should be used when a room's interior has been heated to the point where it is nearing a flashover condition. The fire fighters should first use an indirect attack method to cool the fire gases down to safer temperatures and prevent flashover from occurring. This is followed with a direct attack on the main body of fire.

When using this method, the fire fighter who is operating the nozzle should be given plenty of space to maneuver it. Only enough water as needed to control the fire should be used, thus avoiding unnecessary water damage.

To perform a combination attack, follow the steps in Skill Drill 21-3.

1. Wearing full PPE and SCBA, select the correct hose line to accomplish the suppression task at hand.
2. Stretch the hose line to the entry point of the structure and signal the pump operator that you are ready to receive water.
3. Open the nozzle to get the air out and be sure that water is flowing.
4. Enter the structure and locate the room or area of origin of the fire.
5. Aim your nozzle at the upper left corner of the fire and make either a "T," "O," or "Z" pattern with the nozzle. Remember to start high and work the pattern down to the fire level.
6. Use only enough water to darken down the fire without upsetting the thermal layering.

Fire Fighter Safety Tips

Indicators of possible building collapse

- Leaning walls
- Smoke emitting from cracks
- Creaking sounds
- Sagging roofs or floors

Considerations for possible building collapse

- Exposure to serious fire
- Water loading due to master streams
- Age of building

Fire Fighter Tips

When using the indirect or combination attack, watch for droplets of water raining down. This indicates lowered ceiling temperatures. If no droplets fall, the ceiling is still too hot.

7. Once the fire has been reduced, find the remaining hot spots, and complete extinguishments using a direct attack.

Large Handlines

Large handlines can be used for either offensive fire attack or for defensive operations. In an offensive attack situation, a $2^1/_2''$ attack line can be advanced into a building to apply a heavy stream of water onto a large volume of fire. The same direct and indirect attack techniques as were described for small hose lines can also be used with large handlines. A $2^1/_2''$ handline can overwhelm a substantial interior fire if it can be discharged directly into the involved area. The extra reach of the stream can also be valuable for an interior attack in a large building.

It is more difficult for fire fighters to advance and maneuver a large handline inside a building, particularly in tight quarters or around corners. At least three fire fighters are usually needed to advance and maneuver a $2^1/_2''$ handline inside a building. The fire fighters have to contend with the nozzle reaction force, as well as the combined weight of the hose and the water. In situations where the hose line has to be advanced a considerable distance into a building, additional fire fighters will be required to move the line. The extra effort is balanced by the powerful fire suppression capabilities of a large handline.

Large handlines are often used in defensive situations to direct a heavy stream of water onto a fire from an exterior position. In these cases the nozzle is usually positioned to be operated from a single location by one or two fire fighters. The stream can be used to attack a large exterior fire or to protect exposures. The stream can also be directed into a building through a doorway or window opening to knock

Skill Drill 21-4

One-Fire Fighter Method for Operating a Large Handline

Close the nozzle and then make a loop with the hose, assuring that the nozzle is UNDER the hose line that is coming from the fire apparatus.

Lash the hose together at the section of hose where they cross, or use your body weight to kneel or sit on the hose line at the point where the hose crosses itself.

Be sure to allow enough hose to extend past the section where the line crosses itself for maneuverability.

Open the nozzle and direct water onto the designated area.

down a large volume of fire inside. If the exterior attack is successful in reducing the volume of fire, the incident commander might make the decision to switch to an offensive (interior) attack to complete extinguishment.

One-Fire Fighter Method

One fire fighter can control a large attack hose by forming a large loop of hose about 2′ behind the nozzle. By placing the loop over the top of the nozzle, the weight of the hose stabilizes the nozzle and reduces the nozzle reaction. To add more stability, lash the hose loop to the hose behind the nozzle where they cross. This reduces the energy needed to control the line if it is necessary to maintain the water stream for a long period of time. This method does not allow the hose to be moved while water is flowing, but it is good for protecting exposures when operating in a defensive attack mode. To perform the one-fire fighter method, follow the steps in (▲ Skill Drill 21-4).

1. Select the correct size fire hose for the task to be performed.
2. In full PPE and SCBA, advance the hose into the position from which you plan to attack the fire.
3. Signal the pump operator that you are ready for water.
4. Open the nozzle to allow air to escape the hose and to assure that water is flowing.
5. Close the nozzle and then make a loop with the hose, assuring that the nozzle is UNDER the hose line that is coming from the fire apparatus. **(Step 1)**

6. Lash the hose together at the section of hose where they cross or use your body weight to kneel or sit on the hose line at the point where the hose crosses itself. **(Step 2)**
7. Be sure to allow enough hose to extend past the section where the line crosses itself for maneuverability. **(Step 3)**
8. Open the nozzle and direct water onto the designated area. **(Step 4)**

Two-Fire Fighter Method

When two fire fighters are available to operate a large handline, one should be the nozzle operator, while the other provides a back up. The nozzle operator grasps the nozzle with one hand and holds the hose behind the nozzle with the other hand. The hose should be cradled across the fire fighter's hip for added stability. The backup fire fighter should be positioned about 3′ behind the nozzle operator. This person grasps the hose with both hands and holds the hose against a leg or hip. A hose strap can also be used to provide a better hand grip on a large hose line. When the line is operated from a fixed position, the second fire fighter can kneel on the hose with one knee to stabilize it against the ground. **(▶ Skill Drill 21-5)** demonstrates this skill.

1. Don all PPE and SCBA.
2. Select the correct hose line for the task at hand.
3. Stretch the hose line from the fire apparatus into position. **(Step 1)**
4. Signal the pump operator that you are ready for water.
5. Open the nozzle a small amount to allow air to escape and to assure that water is flowing.
6. Advance the hose line as needed. **(Step 2)**
7. Before attacking the fire, the fire fighter on the nozzle should cradle the hose on his or her hip while grasping the nozzle with one hand and supporting the hose with the other hand. **(Step 3)**
8. The second fire fighter should stay approximately 3′ behind the fire fighter on the nozzle. The second fire fighter should grasp the hose with two hands and, if necessary, use a knee to stabilize the hose against the ground.
9. A hose strap can be used to provide a better grip on the hose.
10. Open the nozzle in a controlled fashion and direct water onto the fire or designated exposure. **(Step 4)**

If you need to advance a flowing $2\frac{1}{2}''$ handline a short distance and only two fire fighters are available, be aware of the large reaction force exerted by the flowing water. It is much easier to shut down the nozzle momentarily and move it to the new position than to relocate a flowing line. If the line must be moved while water is flowing, both fire fighters must brace the hose against their bodies to keep it under control. Three fire fighters can stabilize and advance a large handline more comfortably and safely than two.

Fire Fighter Safety Tips

Master streams should NEVER be directed into a building while fire fighters are operating inside the structure. The force of the stream can push the heat, smoke, and fire onto the fire fighters.

Master Stream Devices

Master stream devices are used to produce high-volume water streams for large fires. There are several different types of master stream devices, including portable monitors, deck guns, ladder pipes, and other elevated stream devices. Most master streams discharge between 500 and 1000 gallons of water per minute, although much larger capacities are available for special applications. In addition, the stream that is discharged from a master stream device has a greater range than a handline, so it can be effective from a greater distance.

A master stream device can be either manually operated or directed by remote control. Many master stream devices can be set up and then left to operate unattended. This can be extremely valuable in a high-risk situation, because there is no need to leave a fire fighter in an unsafe location or hazardous environment to operate the device.

Master streams are only used during defensive operations. A master stream should NEVER be directed into a building while fire fighters are operating inside the structure. The master stream can push heat, smoke, or fire onto the fire fighters. The force and impact of the stream can also dislodge loose materials or cause a structural collapse.

Deck Guns

A deck gun is permanently mounted on a vehicle and equipped with a piping system that delivers water to the device. These devices are sometimes called turret pipes or wagon pipes **(▼ Figure 21-5)**. If the vehicle is equipped with

Figure 21-5 A deck gun is permanently mounted on a vehicle and equipped with a piping system that delivers water to the device.

Skill Drill

Two-Fire Fighter Method for Operating a Large Handline

Stretch the hose line from the fire apparatus into position.

Advance the hose line as needed.

Before attacking the fire, the fire fighter on the nozzle should cradle the hose on his or her hip while grasping the nozzle with one hand and supporting the hose with the other hand.

The second fire fighter should stay approximately 3′ behind the fire fighter on the nozzle. The second fire fighter should grasp the hose with two hands and may use a knee to stabilize the hose against the ground if necessary. Open the nozzle in a controlled fashion and direct water onto the fire or designated exposure.

Voices of Experience

"The pressurized gas fire he was fighting kept pushing through the fog pattern."

The nozzle you select will not be of service to you if you do not use it properly. I have seen that, in many cases, selecting the proper nozzle or adjusting its pattern or flow setting can be just as important to an incident's outcome as the incident commander's strategy. I remember one incident in particular when this was especially true.

While fighting a propane fire during a training evolution, I noticed that one of the other fire fighters was unable to control the fire with his fire stream. He kept the fog pattern open to the wide fog position because he believed that that was the best protection pattern available. The pressurized gas fire he was fighting kept pushing through the fog pattern. I told him to narrow his pattern to more of a "power-cone" pattern, and he was able to "capture" the fire jetting out and control it until the fuel source could be valved out. It was necessary for the water pattern to exert more pressure on the fire than the fire was exerting on the water for the fire stream to be effective and control the fire.

Sometimes, a very small adjustment is all that is necessary. Knowing the proper use and limits of your equipment is indeed important, and will enable you to suppress the fire safely and effectively. Only through familiarization and training can fire fighters learn how their equipment can best serve them when they need it most.

Robert Havens
Lamar Regional Fire Academy
Lamar Institute of Technology
Beaumont, Texas

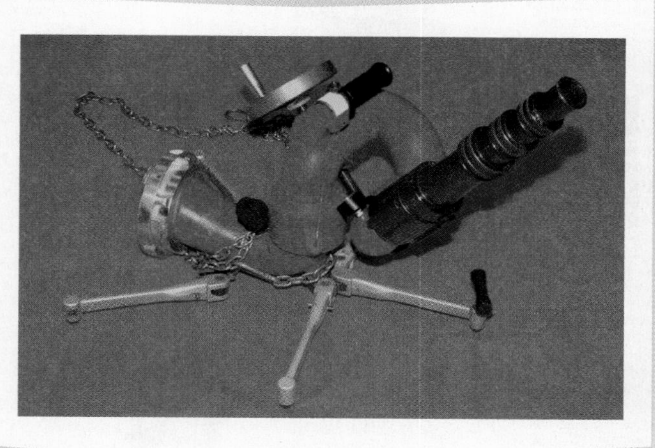

Figure 21-6 A portable monitor is placed on the ground and supplied with water from one or more hose lines.

Figure 21-7 Elevated master stream devices can be mounted on aerial apparatus.

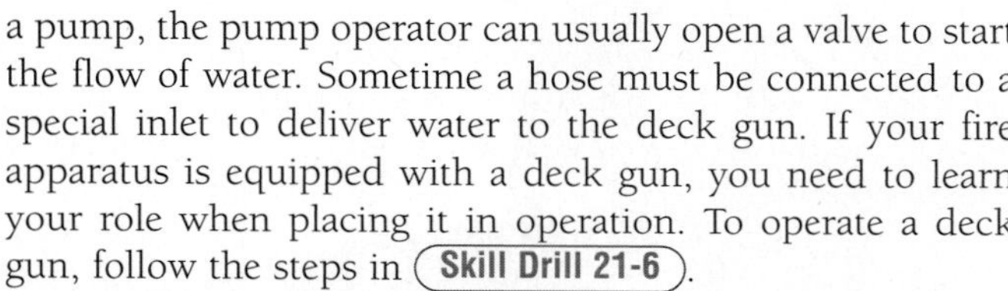

a pump, the pump operator can usually open a valve to start the flow of water. Sometime a hose must be connected to a special inlet to deliver water to the deck gun. If your fire apparatus is equipped with a deck gun, you need to learn your role when placing it in operation. To operate a deck gun, follow the steps in Skill Drill 21-6.

1. Make sure that all firefighting personnel are out of a structure before using a deck gun.
2. Place the deck gun in position.
3. Aim the deck gun at the fire or at the intended exposure.
4. Signal the pump operator that you are ready for water.
5. Once water is flowing, adjust the angle, aim, or water flow as necessary.

Portable Monitor

A **portable monitor** is a master stream device that can be positioned wherever a master stream is needed and placed on the ground (▲ Figure 21-6). Hose lines are connected to the portable monitor to supply the water. Most portable monitors are supplied with either two or three 2½" inlets or with one large-diameter hose inlet. Some monitors can be used as a deck gun or taken off the fire apparatus and used as a portable monitor.

To deploy a portable monitor, remove it from the apparatus and carry it to the location where it will be used. Advance an adequate number of hose lines from the engine to the monitor. The number of hose lines needed depends on the volume of water needed and the size of the hose lines. Form a large loop in the end of each hose line in front of the monitor and then attach the male coupling to the inlets of the monitor. The loops serve to counteract the force created by the flow of the water through the nozzle.

To set up and operate a portable monitor, follow the steps in Skill Drill 21-7.

1. Remove the portable monitor from the fire apparatus and move it into position.
2. Attach the necessary hose lines to the monitor as per SOPs or manufacturer's instructions.
3. Be sure to loop the hose lines in front of the monitor to counteract the force created by water flowing out of the nozzle.
4. Signal the pump operator that you are ready for water.
5. Aim the water stream at the fire or onto the designated exposure and adjust as necessary.

The nozzle reaction force can cause a portable monitor to move from the position where it was placed if it is not adequately secured. A moving portable monitor can be extremely dangerous to anyone in its path. Many portable monitors are equipped with a strap or chain that must be secured to a fixed object to prevent the monitor from moving. Pointed feet on the base also help to keep a portable monitor from moving. If the stream is operated at a low angle, the reaction force will tend to make the monitor unstable. A safety lock is usually provided to prevent the monitor from being lowered beyond a safe limit of 35°. When setting up any portable monitor, always follow the manufacturer's instructions and the SOPs of your department for safe and effective operation.

Elevated Master Streams

Elevated master stream devices can be mounted on aerial ladders, aerial platforms, or special hydraulically operated booms (▲ Figure 21-7). A **ladder pipe** is an elevated stream device that is mounted at the tip of an aerial ladder. On many aerial ladders the ladder pipe is only attached to the top of the ladder when it is needed, and a hose is run up the ladder to deliver water to the device. Most newer aerial ladders and tower ladders are equipped with a fixed piping system to deliver water to a permanently mounted master stream device at the top. This saves valuable set-up time at a fire scene. If your apparatus is equipped with a ladder pipe, you need to learn how to assist in its set-up.

Protecting Exposures

Protecting exposures refers to actions that are taken to prevent the spread of a fire to areas that are not already burning. Exposure protection is a consideration at every fire; however, it is a much more important consideration when the fire is large. If the fire is relatively small and contained within a limited area, the best way to protect exposures is usually to extinguish the fire; when the fire is extinguished, the exposure problem ceases to exist. In cases where the fire is too large to be controlled by an initial attack, exposure protection becomes a priority. In some cases the best outcome you can hope to accomplish is to stop the fire from spreading.

The IC has to consider the size of the fire and the risk to exposures in relation to the amount of firefighting capability that is available and how quickly it can be assembled. In some cases the IC will direct the first companies to protect exposures while a second group of companies prepares to attack the fire. Sometimes the IC has to identify a point where the progress of the fire can be stopped and direct all firefighting efforts toward that objective.

Protecting exposures involves very different tactics from offensive fire attack. At a large fire, the first priority is to protect exposed buildings and property from a combination of radiant heat, convective heat, and burning embers (► Figure 21-8). The best option is usually to direct the first hose streams at the exposures rather than the fire itself. Wetting the exposures will keep the fuel from reaching its ignition temperature. Because radiant heat can travel through a water stream, directing water onto the exposed surface is more effective than aiming a stream between the fire and the exposure. Master stream devices such as deck guns, portable monitors, and elevated master streams are excellent tools for protecting exposures. Large volumes of water can be directed onto the exposures from a safe distance.

Figure 21-8 Protecting an exposure from radiant heat.

Figure 21-9 Fires may be hidden behind walls.

Ventilation

Before any interior attack is initiated, it is important that the structure be ventilated. Ventilation must be coordinated with the suppression efforts to ensure that both events occur simultaneously and in a manner that supports the attack plan. Proper ventilation allows for the hot gases and smoke to be removed from the building, improving visibility and tenability in the building for any trapped victims and fire fighters. Improper ventilation can create conditions that allow a fire to burn more aggressively and make it more difficult for fire fighters to enter and attack the fire. Coordination is essential to ensure that the hose lines will be ready to attack when the ventilation openings are made. The ventilation openings must be located so that the hot smoke and gases will be drawn away from the attack crews.

Concealed-Space Fires

Fires in ordinary and wood frame construction can burn in combustible void spaces behind walls and under subfloors. In order to prevent the fire from spreading, these fires must be found and suppressed (▲ Figure 21-9).

To locate and suppress fires behind walls and under subfloors, follow the steps in (Skill Drill 21-8).

1. Locate the area of the fire building where a hidden fire is believed to be.
2. Look for signs of fire such as smoke coming from cracks or openings in walls, charred areas with no outward evidence of fire, and peeling or bubbled paint or wallpaper. Listen for cracks and pops or hissing steam.
3. If available, use a thermal imager to look for areas of heat that may indicate a hidden fire.
4. Using the back of your hand, feel for heat coming from a wall or floor.
5. If a hidden fire is suspected, use a tool such as an axe or haligan to remove the building material over the area.

6. If fire is located, expose the area as well as possible and extinguish the fire using conventional firefighting methods. Be sure to expose as much area as needed without causing unnecessary damage.

Basement Fires

Fires in basements or below grade level present several different challenges. Basements are difficult and dangerous to enter, and they have limited routes of egress. Basements are usually difficult to ventilate, which means that an interior attack often has to be made in conditions of high heat and low visibility. This also means that it will be difficult to remove the fire gases and steam produced by the attack lines. It can be difficult to see even after ventilation has been performed. Basements are often used for storage, and it may be hard to keep your sense of orientation in the narrow disorganized spaces.

Fire fighters should identify the safest means of entry and exit into the area where firefighting operations will be conducted. An exterior access point allows fire fighters to enter a basement without passing through the hot gas layers at the basement ceiling level. If the only point of entry is an interior stairway, fire fighters must protect that stairway opening to keep the fire from extending to the upper floors (► Figure 21-10). Ventilation must be planned and conducted early. If ventilation is not managed properly, the interior stairwell will act as a chimney and bring heat and smoke up from the basement.

Always consider the possibility of a fire burning below the ground floor level when you enter a fire building. A fire that appears to be on the ground floor could have originated in the cellar and weakened the floor. Cellar fires can also spread to upper floors in houses with balloon construction.

Fires above Ground Level

Advancing charged hose lines up stairs and along narrow hallways requires much more physical effort than fighting a fire on the ground level in an open area. It is always important to protect stairways and other vertical openings between floors when fighting a fire in a multiple-level structure. Hose lines must be placed to keep the fire from extending vertically and to ensure that exit paths are always available.

When advancing a hose above the ground floor, advance them uncharged until you reach the fire floor and have extra hose available. This allows for easier advancement of attack lines.

Fire Fighter Safety Tips

Any time a floor feels hot through your PPE, consider the possibility of a fire burning in the level below you. Beware of a weakened floor caused by a hidden basement fire!

Figure 21-10 Fire below grade.

Figure 21-11 With fires above grade, stairways in structures must be protected from fire.

Interior fire crews must always look for a secondary exit path in case their entry route is blocked by the fire or by a structural collapse. The secondary exit could be a second interior stairway, an outside fire escape, a ground ladder placed to a window, or an aerial device (▲ Figure 21-11).

Be aware of the risk of structural instability and collapse. Check the floor that you are working on and listen for the sounds associated with a failing ceiling or roof. Do not use more water than is needed to extinguish the fire. Additional water adds weight to the structure and could lead to structural collapse. If the structure is designed to be residential, then a thorough primary search must be accomplished early in the firefighting efforts.

In high-rise buildings, the standpipe system will be used to supply water for hose lines. Fire fighters must practice connecting hose lines to standpipe outlets and extending lines from stairways into remote floor areas. Additional hose, tools, air cylinders, and Emergency Medical Service (EMS) equipment should be staged one or two floors below the fire.

Fires in Large Buildings

Many large buildings have floor plans that can cause fire fighters to become lost or disoriented while working inside, particularly in low-visibility or zero-visibility conditions. The use of guide lines may be necessary to keep fire fighters from becoming lost and running out of air. A well-organized preincident plan of the structure can be essential when fighting this type of fire. Knowing the occupancy and the other hazards beforehand will help in determining the best strategy and tactics.

Fires in Buildings during Construction, Renovation, or Demolition

Buildings that are under construction, renovation, or demolition are all at increased risk for destruction by fire. These buildings often have large quantities of combustible materials exposed, without the fire-resistant features of a finished building. If the building lacks windows and doors, an almost unlimited supply of oxygen could be available to fuel a fire. Fire detection, fire alarm, and automatic fire suppression systems might not be operational. Construction workers may be using torches and other flame-producing devices. These buildings are often unoccupied and could be targets for arson.

Under these conditions a fire in a large building could be impossible to extinguish. If there are no life safety hazards involved, fire fighters should use a defensive strategy for this type of fire. No fire fighters should enter the building, and a collapse zone should be established. A defensive exterior operation should be conducted using master streams and large handlines to protect exposures.

Fires in Lumberyards

Lumberyard fires are often prime candidates for a defensive firefighting strategy. A typical lumberyard contains large quantities of highly combustible material. This fuel is stored in the open or in sheds where plenty of air is available to support combustion. A lumberyard fire will usually produce tremendous quantities of radiant heat and release burning embers that will cause the fire to spread quickly from stack to stack. Nearby buildings and other structures may become involved in fire.

Exposure protection is often the primary objective at a lumberyard fire, since there is little life safety risk. Exposure protection should be dealt with early in the operation by placing large handlines and master stream devices where they can be most effective. A collapse zone must be established around any stacks of burning material to keep fire fighters out of dangerous positions.

Fires in Stacked or Piled Materials

Fires occurring in stacked or piled materials can present a variety of hazards. The greatest danger to fire fighters is the possibility that a stack of heavy material such as rolled paper or baled rags will collapse without warning. This can occur if a fire has damaged the stacked materials or if water has soaked into them. Absorbed water can greatly increase the weight of many materials and also weaken cardboard and paper products. The water discharged by automatic sprinklers alone can be sufficient to make some stacked materials unstable. A tall stack of material that falls on top of a team of fire fighters can cause injury or death.

Fires in stacked materials should be approached cautiously. All fire fighters must remain outside potential collapse zones. Mechanical equipment should be used to move material that has been partially burned or water-soaked.

Conventional methods of fire attack can often be used to gain control of the fire; however, water must penetrate into the stacked material in order to fully extinguish the residual combustion. Class A foam can often be used to extinguish smoldering fire in tightly packed combustible materials. Overhaul will require the materials to be separated to expose any remaining deep-seated fire. This can be a labor intensive process unless mechanical equipment can be used to dig through the material.

Trash Container and Rubbish Fires

Trash container (dumpster) fires usually occur outside of any structure and appear to present fewer challenges than a fire inside a building. Fire fighters still must be vigilant in wearing full PPE and using SCBA when fighting trash container fires. There is no way of knowing what might be inside a trash container.

If the fire is deep seated, the trash will have to be overhauled. Manual overhaul involves pulling the contents of a trash container apart with pike poles and other hand tools so that water can reach the burning material. This can be labor intensive and involves considerable risk to fire fighters. The fire fighters are exposed to whatever contaminants are in the container, as well as the risks of injury from burns, smoke, or other causes. Considering the value of the contents of a trash container, it is difficult to justify any risk to fire fighters.

Class A foam is useful for many trash container fires, because it allows water to soak into the materials. This can eliminate the need for manual overhaul. Some fire departments use the deck gun on the top of an engine to extinguish large trash container fires and then complete extinguishments by filling the container with water. This is done by pointing the deck gun at the dumpster and slowly opening the discharge gate to lob water into the container (▼ Figure 21-12).

Trash containers are often placed behind large buildings and businesses. If the container is close to the structure, be sure to check for extension. Also look above the container for telephone, cable, and power lines that might have been damaged by the fire.

Confined Spaces

Fires and other emergencies can occur in confined spaces. Fires in underground vaults and utility rooms such as transformer vaults are too dangerous to enter. Fire fighters should summon the utility company and keep the area around manhole covers and other openings clear while awaiting the arrival of the utility personnel. For emergencies in these areas, the Occupational Safety and Health Administration (OSHA) requires specially trained entry teams. Learn what your fire department's operational procedures are for these incidents. It is also important to know about confined spaces that exist in industries within individual response areas. These areas should be visited with plant personnel so that preincident plans can be developed.

Other hazards exist in confined spaces, including oxygen deficiencies, toxic gases, and standing water. Conditions might appear to be safe, but there might not be enough oxygen in the confined space to sustain life. It is common for a victim to enter these spaces and pass out, usually followed by another victim who comes in to assist and also passes out. If entering these spaces to attempt rescue, full SCBA cylinder should be worn, and fire fighters should be attached to a lifeline. An additional lifeline should be lowered or brought with the fire fighter to tie the victim in a bowline on a bight (rescue knot), which allows surface personnel to raise the victim out of the hole. Due to the lack of ventilation in most confined spaces, fire fighters may notice an intense amount of heat once in the space. Fire fighters will tire quickly and must recognize the signs of heat exhaustion and heat stroke.

Figure 21-12 Some fire departments use a deck gun on top of an engine to extinguish large trash container fires.

Confined spaces commonly have low oxygen levels and high levels of combustible gases, such as methane. Fire fighters without breathing apparatus will quickly be overcome in confined spaces with these conditions. Fire fighters who must enter a confined space should take a monitoring device. Air quality must be checked constantly, looking for the buildup of explosive gases as well as oxygen levels.

A strict accountability system must be adhered to when fire fighters enter into a confined space. This ensures that only those with proper training and equipment enter the confined space. It is important for a safety officer to track the movement of personnel and the time that they are in the confined space.

Fire suppression in confined spaces must not begin until all utilities and industrial processes have been turned off. Suppression agents include hose streams, high and low expansion foams, carbon dioxide flooding systems, and built-in sprinkler systems.

Vehicle Fires

Automobile fires are common occurrences in most communities. Modern automobiles contain hundreds of pounds of plastics, which give off toxic smoke when they burn. For this reason it is important to wear SCBA when fighting a vehicle fire. A 1½" or 1¾" hose line should always be used to attack a car fire, in order to provide sufficient protection from a sudden flare-up. Many modern vehicles have shock absorbers, bumpers, and trunk or hatchback components that are gas-filled. Be aware that when these cylinders are heated they can burst and send the object hurdling from the vehicle at high velocity.

If the owner or driver is present, ask about any specific hazards that may be present in the vehicle, such as a barbecue grill gas cylinder, cans of spray paint, or any other hazardous materials. If no driver or occupant is around, do not assume that the vehicle is safe; always be cautious as you approach a vehicle fire.

Attacking Vehicle Fires

Fires under the Hood

For fires under the hood or engine area, approach from the uphill and upwind side at a 45° angle (► Figure 21-13). Using the reach of the stream, direct the water into the wheel wells and through the front grill. This technique gets some water onto the fire quickly to cool the shock absorbers and the pistons of shock-absorbing bumpers. These are enclosed cylinders capable of exploding when heated. An exploding cylinder can travel up to 20′ and pack enough force to break bones and cause severe burns. A front bumper assembly can travel about 3′ if the cylinders explode. The small pistons used in hatchbacks and sometimes hoods to keep the hood

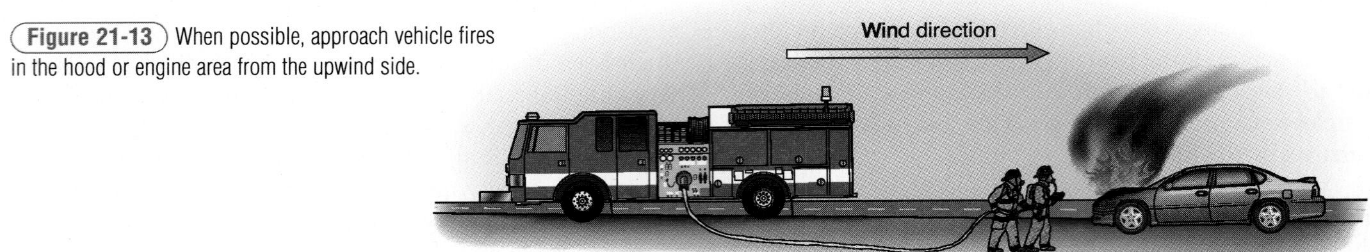

Figure 21-13 When possible, approach vehicle fires in the hood or engine area from the upwind side.

raised are also a safety concern. These are also sealed cylinders that can explode and shoot a spear-like rod through the front or rear of a car.

Have another fire fighter stabilize the car to keep it from rolling, and check the interior of the car for occupants. The fire fighter should open the driver's door and pull the hood release latch, which is usually found at the bottom of the "A" post. This could be ineffective if the fire has melted the release cable, but it is a worthwhile first step. Make sure that the shock absorbers are cool before attempting this.

Once the car has been cooled and some of the fire has been darkened, try to open the hood. If the hood release operated, a gloved hand can trip the secondary latch, which is usually located in the middle of the front of the hood. If the release cable has melted, a Halligan bar can be used to force the corner of the hood up enough to allow a nozzle to be inserted into the engine compartment. Another method is to break the front grill, cut the release cable close to the latch mechanism, and grasp the cable with vise grip pliers to release the hood. Once the hood is raised, extinguish the fire, being careful not to splash battery acid. Usually the plastic case of the battery will have melted during the fire and exposed the acid.

It is also important to be aware of any leaking fluids on the ground. These could possibly ignite, spreading the fire beyond the vehicle. In addition, these fluids could enter nearby storm drains or waterways, creating an environmental problem.

After the main body of the fire has been extinguished, it is important to overhaul a vehicle, just like a structure fire. All of the seating should be removed or exposed to ensure that fire has not spread into this combustible material. Other concealed areas of the vehicle should be checked for fire extension. Since the vehicle has probably already been extensively damaged from the fire and firefighting operations, the liberal use of water will not cause any undue damage beyond what has already occurred.

Fires in the Passenger Area

Fires in the passenger area should be approached from the upwind side at a 90° angle from the car. Using the reach of the 1¾" hose line, start approximately 50′ from the car, with the nozzle set on straight stream, and darken the fire down by slowly sweeping the stream back and forth in a horizontal motion. Extinguish all visible fire while walking toward the car. Observe the area under the car during the approach for any sign of leaking flammable liquids. If burning flammable liquids are present, widen the spray pattern on the nozzle. Foam can be used to extinguish the burning liquid and provide a vapor barrier to prevent re-ignition.

After all visible fire has been knocked down, allow a few minutes for the steam and smoke to dissipate before starting overhaul. This will allow visibility to improve so that overhaul can be completed safely. During overhaul of interior fires, remember that air bags can deploy without warning in a burning automobile. Never place any part of your body in the path of a front or side air bag.

Fire in the Trunk

A fire in the trunk area of an automobile can be accessed by using the pike of the Halligan bar to force the lock into the trunk, then using a screwdriver or key tool (K-tool lock puller) to turn the lock cylinder in a clockwise direction. A charged hose line must be ready when the trunk lid is raised.

Fires in the rear of light trucks and vans must always be approached cautiously. Vans are often used by couriers and could contain medical waste, lab specimens, and radioactive material.

Alternative Fuel Vehicles

Always be alert for signs that a burning vehicle could be powered by an alternate fuel, such as compressed natural gas (CNG) or liquefied propane gas (LPG). Fully involved fires in vehicles powered by either type of fuel should be fought with an unmanned master stream to prevent injuries from exploding gas cylinders.

Automobiles powered by CNG contain storage cylinders that are very similar to SCBA cylinders. These cylinders are usually located in the trunk and contain CNG at high pressures. They must be cooled and protected like any gas cylinder. CNG is a nontoxic lighter-than-air gas that will rise and dissipate if it is released into the atmosphere.

Propane is stored in the same types of cylinders as are used for heating or cooking purposes. Propane is heavier than air, so the vapors will pool or collect in low areas.

Hybrid automobiles have small gasoline-powered engines and large battery banks. The batteries power electric motors that drive the wheels, much like a train locomotive. There are two noteworthy hazards in these vehicles:

- The nickel metal hydride batteries are very hazardous when burning and may explode. The runoff from water used in firefighting will also be hazardous and should be avoided.
- High-voltage direct current cables connect the batteries to the electric motors that power the wheels. Cutting these orange cables can cause serious injury or death. The cables usually run from the battery bank to the front of the car via the undercarriage. They are usually placed so that they will pass directly under the center of the driver's chair. Extra care should be used when using hydraulic metal cutters or spreaders on these types of autos.

Flammable Liquids Fires

Flammable liquids fires can be encountered in almost any type of occupancy. Most fires involving a vehicle (plane, train, ship, car, truck) are also likely to involve a combustible or flammable liquid. Special tactics must be used when attempting to extinguish a flammable liquids fire. Special extinguishing agents such as foam or dry chemicals may be needed.

Hazards

Fires involving flammable liquids such as gasoline require special extinguishing agents. Most flammable liquids can be extinguished using either foam or dry chemicals. Class B extinguishing agents are approved for use on Class B (flammable liquids) fires.

Flammable liquids fires can be classified as either two-dimensional or three-dimensional. A two-dimensional fire refers to a spill, pool, or open container of liquid that is burning only on the top surface. A three-dimensional fire refers to a situation where the burning liquid is dripping, spraying, or flowing over the edges of a container.

A two-dimensional flammable liquid fire can usually be controlled by applying the appropriate Class B foam onto the burning surface. There are several different formulations of Class B foams that are suitable for different liquids and situations. The foam will flow across the surface and create a seal that stops the fuel from vaporizing. This separates the fuel from the oxygen and extinguishes the fire. Foam will also cool the liquid and reduce the possibility of re-ignition.

Fire Fighter Safety Tips

Some vehicle components (engines or body) may be made out of magnesium or metal alloys that can react violently when water is applied during suppression. A Class D extinguishing agent should be used instead of water.

Fire fighters should look out for hot surfaces or open flames that could cause the vapors to re-ignite after a fire has been extinguished. It is important to determine the identity of the liquid that is involved in order to select the appropriate extinguishing agent and to determine whether the vapors are lighter or heavier than air.

A three-dimensional fire is much more difficult to extinguish with foam, because the foam cannot establish an effective seal between the fuel and the oxygen. Either dry chemical or gaseous extinguishing agents are usually more effective than foam in controlling three-dimensional fires. These agents can also be used to extinguish two-dimensional fires; however, they do not provide a long-lasting seal between the fuel and the oxygen. In some cases a fire can be extinguished with a dry chemical, and then the surface can be covered with foam to prevent re-ignition.

Fire fighters should avoid standing in pools or contaminated runoff from flammable liquids. Fire fighters' PPE could absorb the product and become contaminated. If it is seriously contaminated, the PPE itself could become flammable.

Suppression

The skills for suppressing flammable liquids fires are presented in the portable extinguishers chapter, Chapter 7.

Flammable Gas Cylinders

Flammable gas cylinders can be found anywhere. There are many types of flammable gases that are stored in different types and sizes of containers. Many different types of flammable gases can be found in industrial occupancies.

Fire Fighter Safety Tips

Do not attempt to extinguish a propane gas fire until the fuel flow is shut off.

Propane Gas

The popularity of propane gas for heating and cooking has caused these cylinders to become commonplace in residential areas and many other locations. In addition, propane is used as an alternative fuel for vehicles and is often stored to power emergency electrical generators. Fire fighters should be familiar with the basic hazards and characteristics of propane and procedures for fighting propane fires.

Propane or LPG exists as gas in its natural state at temperatures above –44°F (–42.2°C). When it is placed into a storage cylinder under pressure, it is changed into a liquid. Storing propane as a liquid is very efficient, because it has an expansion ratio of 270 to 1. (One cubic foot of liquid propane will convert into 270 cubic feet of gas when it is released into the atmosphere.) A large quantity of fuel can be stored in a small container.

Inside a propane container, there is a space filled with propane gas above the level of the liquid propane. As the contents of the cylinder are used, the liquid level becomes lower and the vapor space increases. The internal piping is arranged to draw product from the vapor space.

Propane gas containers come in a variety of sizes and shapes, with capacities ranging from a few ounces to thousands of gallons. The cylinder itself is usually made of steel or aluminum. A discharge valve keeps the gas inside the cylinder from escaping into the atmosphere and controls the flow of gas into the system where it is used. This valve should be easily visible and accessible. In the event of a fire, closing this valve should stop the flow of the product and extinguish the fire. The valve should be clearly marked to indicate the direction it should be turned or moved to reach the closed position.

A connection to a hose, tubing, or piping allows the propane gas to flow from the cylinder to its destination. In the case of portable tanks, this connection is often the most likely place for a leak to occur. If the gas is ignited, this area could become involved in fire.

A propane cylinder is always equipped with a relief valve to allow excess pressure to escape in order to prevent an explosion if the tank is overheated. Propane cylinders must be stored in an upright position so that the relief valve remains within the vapor space. If a propane cylinder is placed on its side, the relief valve could be below the liquid level. If a fire were to heat the tank and cause an increase in pressure, the relief valve would release liquid propane, which would then expand by the 270 to 1 ratio. This would create a huge cloud of potentially explosive propane gas.

Propane Hazards

Propane is highly flammable. It is nontoxic, but it can displace oxygen and cause asphyxiation. By itself propane is odorless, and leaks could not be detected by a human sense of smell. Mercaptan is added to propane to create a distinctive odor. Propane gas is heavier than air, so it will flow along the ground and accumulate in low areas.

When responding to a reported LPG leak, fire fighters and their apparatus should stage uphill and upwind of the scene. Fire fighters should be aware that an explosion could happen at any time, so full PPE and SCBA must be worn. When using meters to check for LPG, be sure to check storm drains, basements, and other low-lying areas for concentrations of the gas. Life safety should be the highest priority; depending on the type and size of the leak, an evacuation might be necessary.

The greatest danger with propane and similar products is a BLEVE (boiling-liquid, expanding-vapor explosion). If an LPG tank is exposed to heat from a fire, the temperature of the liquid inside the container will increase. The fire could be fueled by propane escaping from the tank or from an external source. As the temperature of the product increases, the vapor pressure will also increase. The increasing pressure creates added stress on the container. If the pressure exceeds the strength of the cylinder, it could rupture catastrophically.

To protect the container from rupture, the relief valve will open to release some of the pressure. The relief valve should exhaust vapor until the pressure drops to a preset level, and then the valve should close or reseat. As the heating and relieving cycles continue, the liquid can begin to boil within the container. If the flame is impinging directly on the tank, the container can weaken and fail somewhere above the liquid line. When this happens, the container will rupture and release its contents with explosive speed. The boiling liquid will expand, vaporize, and ignite in a giant fireball, accom-

Fire Fighter Safety Tips

A propane tank with the relief valve operating should not be approached. Remotely directing water onto the tank allows it to cool until the relief valve resets. Then the tank may be approached and the valve turned off.

panied by flying fragments of the ruptured container. Many fire fighters have been killed in these explosions.

The best method to prevent a BLEVE is to direct heavy streams of water onto the tank from a safe distance. The water should be directed at the area where the tank is being heated. The fire fighters operating these streams should work from shielded positions or utilize remote control or unmanned monitors. Horizontal tanks are designed to fail at the ends if there should be a catastrophic failure, so fire fighters should only operate from the sides of the tank.

Propane Fire Suppression

Fighting fires involving LPG or other flammable gas cylinders requires careful analysis and logical procedures. If the gas itself is burning because of a pipe or regulator failure, the best way to extinguish the fire is to shut off the main discharge valve at the cylinder. If the fire is extinguished and the fuel continues to leak, there is a high probability that it will re-ignite explosively. Do not attempt to extinguish the flames unless the source of the fuel has been shut off or all of the fuel has been consumed. If the fire is heating the storage tank, hose streams should be used to cool the cylinder, being careful not to extinguish the fire.

Unless there is a remote shutoff valve, the flow can only be stopped if it is safe to approach the cylinder. The integrity of the cylinder should be inspected from a distance before any attempt is made to approach and shut off the valve. If the container is damaged or the valve is missing, the fuel should be allowed to burn off, while hose streams continue to cool the tank from a safe distance.

Approach the fire with two 1¾" hose lines working together. When approaching a horizontal LPG tank, always approach from the sides. The nozzles should be set on a wide fog pattern, with the discharge streams interlocked to create a protective curtain. The team leader should be located between the two nozzle operators. On the commands of the leader, the crew should move forward, remaining together and never turning their backs to the burning product. Upon reaching the valve, the fire fighter in the center turns off the valve, stopping the flow of gas. Any remaining fire is then extinguished by normal means. Continue the flow of water as a protective curtain and to reduce sources of ignition.

Should the fire be extinguished prematurely, the valve should still be turned off as soon as the team reaches it. Always approach and retreat from these types of fires facing the objective with water flowing, in case of re-ignition.

Unmanned master streams should be used to protect flammable gas containers that are exposed to a severe fire. Direct the water to cool the vapor space or upper part of the container (► Figure 21-14). If the container is next to a fully involved building or fire that is too large to control, evacuate the area and do not fight the fire. If there is nothing to save, risk nothing.

Figure 21-14 Using a master stream to protect a flammable gas container exposed to fire.

Keep in mind that if the relief valve is open, the container is under stress. Exercise extreme caution when this is occurring. As the gas pressure is relieved, it will sound like the whistle on a teakettle; if the sound is rising in frequency, an explosion could be imminent and evacuation should be ordered.

To suppress a flammable gas cylinder fire, follow the steps in **Skill Drill 21-9**.

1. Cool the tank from a distance until the relief valve resets.
2. Wearing full PPE, two teams of fire fighters using a minimum of two 1¾". hose lines advance using an interlocking 90°-wide fog pattern for protection. Do not approach the cylinder from the ends. The team leader is located between the two nozzle persons and coordinates and directs the advance upon the cylinder.
3. When the cylinder is reached, the two nozzle teams isolate the shutoff valve from the fire with their fog streams while the leader closes the tank valve, eliminating the fuel source.
4. After the burning gas is extinguished, the fire fighters continue to apply water to the cylinder to cool the metal so as to prevent tank failure resulting in a BLEVE.
5. As cooling continues, fire fighters slowly back away from the cylinder, never turning their backs to it.

Shutting Off Gas Service

Many structures use either natural gas or propane gas for heat or cooking. There are also many industrial applications for these two energy sources. If a gas line inside a structure is compromised during a fire, the escaping gas can add fuel to the fire. The method in which the gas is supplied to the structure must be located to stop the flow.

Most residential gas supplies are delivered through a gas meter connected to an underground utility network or from a storage tank located outside the building. If the gas is supplied by an underground distribution system, the flow can be stopped by closing a quarter-turn valve on the gas meter. If there is an outside LPG storage cylinder, closing the cylinder valve will stop the flow. After the gas service has been shut off, use a lockout tag to be sure that it is not turned back on. Only a professional can re-establish the flow of gas to a structure.

Fires Involving Electricity

The greatest danger with most fires involving electrical equipment is the possibility of electrocution. Only Class C extinguishing agents should be used when energized equipment is involved in a fire. All electrical equipment should be considered as potentially energized until the power company or a qualified electrical professional confirms that the power is off. Once the electrical service has been disconnected, most fires in electrical equipment can be controlled using the same tactics and procedures as a Class A fire.

When a fire occurs in a building, the electrical service should be turned off as soon as possible to reduce the risks of injury or death to fire fighters, even if there is no involvement of electrical equipment in the fire. If possible, this should be done at the main circuit breaker box. A lockout tag should be used to make sure someone does not accidentally turn the electrical current back on.

In cases where it is not possible to turn the electricity off at the breaker box, it will be necessary to notify the electrical utility company to send a representative to the fire scene to disconnect the service from a location outside the building. In many cases, the local utility company is automatically notified of any working fire.

Suppression

Fire suppression methods for fires involving electrical equipment depend on the type of equipment and the power supply. In many cases the best approach is to wait until the power is disconnected and then use the appropriate extinguishing agents to control the fire. If the power cannot be disconnected or the situation requires immediate action, only Class C extinguishing agents, such as halon, carbon dioxide (CO_2), or dry chemical, should be used.

When delicate electronic equipment is involved in a fire, either halon or CO_2 should be used to limit the damage as much as possible. These agents will cause less damage to computers and sensitive equipment than water or dry chemical agents.

When power distribution lines or transformers are involved in a fire, special care must be taken for the safety of the emergency personnel as well as the public. No attempt should be made to attack these fires until the power has been disconnected. In some cases it will be necessary to protect exposures or extinguish fire that has spread to other combustible materials, if this can be accomplished without coming in contact with the electrically energized equipment. If a hose stream comes in contact with the energized equipment, the current can flow back through the water to the nozzle and electrocute fire fighters who are in contact with the hose line.

Many electrical transformers contain a cooling liquid that includes PCBs, a cancer-causing material. DO NOT apply water to a burning transformer. Water can cause the liquid to spill or splash, contaminating fire fighters as well as the environment. If the transformer is on a pole, it should be allowed to burn until power company professionals arrive and disconnect the power. Dry chemical extinguishers can then be used to control the fire. Fires in ground-mounted transformers can also be extinguished with dry chemical agents after the power has been disconnected. Fire fighters should stay out of the smoke and away from any liquids that are discharged from a transformer, and full PPE and SCBA must be worn to attack the fire.

Some very large transformers contain large quantities of cooling oil and require foam for extinguishment. The foam can only be applied after the power has been disconnected. Until the power is off, fire fighters can only protect exposures while taking care to avoid contamination.

Underground power lines and transformers are often located in vaults beneath the surface. Explosive gases can build up within these vaults. A spark can ignite the gases and cause an explosion that can lift a manhole cover from a vault and hurl it a considerable distance. Products of combustion can also leak into buildings through the underground conduits. Fire fighters should never enter an underground electrical vault while the equipment is energized. Either CO_2 or dry chemical extinguishers can be discharged into the vault from above. Replacing the lid will help to maintain the concentration of extinguishing agent in the vault. Even after the power has been disconnected, these vaults should be considered as confined spaces containing potentially toxic gases, explosive atmospheres, or oxygen-deficient atmospheres. Special precautions are required to enter this type of confined space.

Large commercial and residential structures often have high-voltage electrical service connections and interior rooms containing transformers and distribution equipment. These areas should be well marked with electrical hazard signs and should not be entered by fire fighters unless there is a rescue to be made. Until the power has been disconnected, fire suppression efforts should be limited to protecting exposures. Fire fighters must wear full PPE and SCBA due to the inhalation hazard represented by the burning of the plastics and cooling liquids that are often used with equipment of this size.

Wrap-Up

Ready for Review

Fire fighters will be faced with fires of different magnitude and different priorities. The first goal at any fire is life safety of both emergency personnel and the public. Once the issue of life safety is addressed, a decision will be made on the best and most effective way to attack the fire. By successfully removing one part of the fire tetrahedron, fire fighters can suppress any fire. By knowing the hazards of fighting different types of fires with different fuels and different extinguishing agents, fire fighters can reduce the risk of injury or death through proper knowledge and training.

Chief Concepts

- Wear appropriate PPE to all fires.
- Understand the fire tetrahedron and how eliminating one part of it will suppress any fire.
- Know and understand your department's SOPs on dealing with different types of fires and extinguishing agents.
- Never turn your back on any fire.
- There are three primary fire streams: fog, straight, and solid.
- There are three types of interior attack: direct, indirect, and combination.
- Vehicle fires are common in most communities. Given their explosive potential, proceed with caution.

Hot Terms

Boiling-liquid, expanding-vapor explosion (BLEVE) An explosion that can occur when a tank containing a volatile liquid is heated.

Combination attack A type of attack employing both the direct and indirect attack methods.

Deck gun Apparatus-mounted master stream device intended to flow large amounts of water directly onto a fire or exposed building.

Direct attack Firefighting operations involving the application of extinguishing agents directly onto the burning fuel.

Elevated master stream Nozzle mounted on the end of an aerial device capable of delivering large amounts of water onto a fire or exposed building from an elevated position.

Indirect application of water Using a solid object such as a wall or ceiling to break apart a stream of water, causing it to create more surface area on the water droplets and thereby absorb more heat.

Indirect attack Firefighting operations involving the application of extinguishing agents to reduce the buildup of heat released from a fire without applying the agent directly onto the burning fuel.

Ladder pipe Nozzle attached to the end of a straight ladder truck designed to provide large volumes of water from an elevated position.

Master stream device A large capacity nozzle that can be supplied by two or more hose lines or fixed piping. Can flow in excess of 300 gallons per minute. Includes deck guns and portable ground monitors.

Portable monitor A master stream appliance designed to be set up and then left to operate unattended. This device is designed to flow large amounts of water onto a fire or exposed building from a ground-level position.

Solid stream A stream made by using a smooth bore nozzle to produce a penetrating stream of water.

Straight stream A stream made by using an adjustable nozzle that is set to provide a straight stream of water.

Fire Fighter in Action

It is 10:30 a.m. when your engine is dispatched to a large motor home fire on a major interstate in your response area. Upon arrival you find a fully involved motor home. Traffic is stopped in both directions. Your lieutenant tells you to prepare for an attack from outside of the vehicle. Your don your SCBA and prepare to attack.

1. What operational strategy has your lieutenant chosen?
   - A. Offensive
   - B. Defensive
   - C. Transitional
   - D. Aggressive

2. What size hose line should be used in this incident?
   - A. Booster hose line
   - B. Forestry hose line.
   - C. 1 $^1/_2$" or 1 $^3/_4$" hose line.
   - D. Supply hose line

You and your partner are extinguishing the fire by directing your streams through the windows and door of the motor home. Suddenly you hear a roar coming from the other side of the vehicle. You see that the source of the roar is a propane tank that is venting through its pressure relief valve.

3. What action should you take to appropriately handle the venting propane tank?
   - A. Try to evacuate the area and wait for it to BLEVE.
   - B. Direct your streams directly on the propane tank to cool it.
   - C. Extinguish the fire that is coming out of the relief valve to protect exposures.
   - D. Evacuate everyone because propane is toxic and poses a health hazard.

4. What does BLEVE stand for?
   - A. Boiling Liquid Expanding Vapor Explosion
   - B. Boiling Liquid Explosion Vapor Event
   - C. Butane Liquid Expanding Vapor Event
   - D. Better Leave and Evacuate Very Expediently

www.FireFighter.jbpub.com

Preincident Planning

Chapter 22

NFPA 1001 Standard

Fire Fighter I

NFPA 1001 contains no Fire Fighter I Job Performance Requirements for this chapter.

Fire Fighter II

6.5 *Prevention, Preparedness, and Maintenance.* This duty involves performing activities related to reducing the loss of life and property due to fire through hazard identification, inspection, and response readiness, according to the following job performance requirements.

6.5.1 Prepare a preincident survey, given forms, necessary tools, and an assignment, so that all required occupancy information is recorded, items of concern are noted, and accurate sketches or diagrams are prepared.

6.5.1 (A) Requisite Knowledge. The sources of water supply for fire protection; the fundamentals of fire suppression and detection system; common symbols used in diagramming construction features, utilities, hazards, and fire protection systems; departmental requirements for a preincident survey and form completion; and the importance of accurate diagrams.

6.5.1 (B) *Requisite Skills.* The ability to identify components of fire suppression and detection systems; sketch the site, buildings, and special features; detect hazards and special considerations to include in the preincident sketch; and complete all related departmental forms.

Additional NFPA Standards

NFPA 220, *Standard on Types of Building Construction*

NFPA 704, *Standard for Identification of the Hazards of Materials for Emergency Response*

NFPA 1620, *Standard for Recommended Practice for Preincident Planning*

Knowledge Objectives

After studying this chapter, you should be able to:

- Conduct a preincident survey.
- Prepare an accurate sketch or diagram.
- Obtain the required occupancy information.
- Note any items of concern regarding occupancy.

Skills Objectives

There are no skills objectives for this chapter.

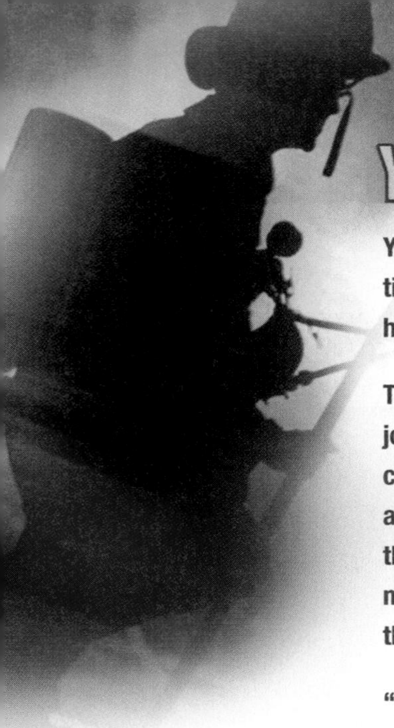

You Are the Fire Fighter

You've been a probationary fire fighter for three months. Today, your captain announces that it is time to review the preincident plans prepared by the "A" shift. You yawn and wish your company had something more exciting to do.

The captain begins by saying, "Let me tell you a story. Two months before you new fire fighters joined us, we were dispatched to a fire at a fast food restaurant on Main Street long after it had closed. When we arrived, there was heavy smoke showing. My first reaction was to begin an aggressive interior attack. However, the preincident plan contained two red flags. It noted that the building was constructed with lightweight trusses and that the air conditioning units were mounted on the roof. Based on this information, I ordered an exterior attack. Two minutes later, the roof collapsed."

"If there had not been a preincident plan for this building, we could have lost three fire fighters that day. I strongly believe that preincident planning is one of the most important things we can do. There is no doubt in my mind that preplanning saves lives." Suddenly you realize that you are no longer tired!

1. ***Why is preincident planning so important to a fire fighter's survival?***
2. ***What types of buildings require preincident planning?***

Introduction

Preincident planning gives you the tools and knowledge to become a much more effective fire fighter. Without a preincident plan, you are going into an emergency situation "blind." You may not be familiar with the structure, the location of hydrants, or the potential hazards. With a preincident plan, you not only know where the hydrants and exits are, but also what hazards to anticipate. A preincident plan puts all that information at your fingertips, either on paper or in a computer file, when you respond to a fire.

Preincident planning helps your department to make better command decisions because important information is assembled before the emergency occurs. At the emergency scene, the incident commander (IC) can use the preincident information to direct the emergency operations much more effectively (◄ **Figure 22-1**). Because a preincident plan identifies potentially hazardous situations before the emergency occurs, fire fighters can be made aware of hidden dangers and prepare for them.

Figure 22-1 The preincident information is supplied to the IC at the emergency scene.

Preincident Plan

Preincident planning is the process of obtaining information about a building or a property and storing the information in a system so that it can be retrieved quickly for future reference. Preincident planning is performed under the direction of an officer. The completed **preincident plan** should be available to all units that would respond to an incident at that location.

The preincident plan is intended to help the IC make informed decisions when an emergency incident occurs at the location (► **Figure 22-2**). Some of the information would be useful to any fire company or unit responding to an incident at that location. A preincident plan can also be used in training activities to help fire fighters become familiar with properties within their jurisdiction.

Tactical Priorities

Address: 1500, 1510, 1520		
Occupancy Name:		
Preplan #: 02-N-01	Number Drawings: 1	Revised Date: 12/2002
District: E275A	Subzone: 60208	By: ACEVEDO
Rescue Considerations: Yes ( ) No (X)		
Occupancy Load Day:	Occupancy Load Night:	
Building Size:	Best Access:	
Knox Box: NONE	Knox Switch: NONE	Opticom: NONE
Roof Type: X	Attic Space: Yes ( ) No ( )	Attic Height: X
Ventilation Horizontal:	Ventilation Vertical:	
Sprinklers: Yes (X) No ( )	Full (X) Partial ( )	
Standpipes: Yes (X) No ( )	Wet ( ) Dry ( )	
Gas: Yes ( ) No ( )	Lpg ( )	
Hazardous Materials: Yes (X) No ( ) DIESEL GENERATORS 1,000 GALLON TANKS BATTERY ROOM		
Firefighter Safety Considerations: ELEVATOR PIT		
Property Conservation And Special Considerations: VENTILATION: AUTOMATIC SMOKE REMOVAL SYSTEM 3 OFFICE BLDGS; 2 PARKING STRUCTURES 6 FLRS - 1230 W. WASHINGTON ST. 4 FLRS - 1500 N. PRIEST DR. 4 FLRS - PARKING GARAGE		

Figure 22-2 An example of a preincident plan.

The objective of a preincident plan is to make valuable information available during an emergency incident that otherwise would not be readily evident or easily determined. The amount and the nature of the information provided for different properties will depend on the size and complexity of the property, the types of risks that are present, and the particular hazards or challenges that are likely to be encountered.

The use of computers has greatly increased the ability of fire departments to capture, store, organize, update, and quickly retrieve preincident planning information. Before this technology became widely available, preincident information was often limited by the need to carry hard copy information in fire apparatus and command vehicles. With computers and mobile data terminals, information such as drawings, maps, aerial photographs, descriptive text, lists of hazardous materials, and material safety data sheets is easily available.

A preincident plan makes information immediately available to the fire fighters who need it during an emergency incident. It is particularly important to inform fire fighters of potential safety hazards. The most critical information, like hydrant locations and life hazards, should be instantly available, with additional data accessible as needed. All of the information must be presented in an understandable format.

A preincident plan usually includes one or more diagrams to show details such as the building location and arrangement, access routes, entry points, exposures, and hydrant locations or alternative water supplies. The location and nature of any special hazards to the public or to fire fighters should be particularly noted on the diagrams. Information about the actual building should include the

Fire Fighter Tips

Preincident planning is different from fire prevention. The goal of fire prevention is to identify hazards and minimize or correct them so that fires do not occur or have limited consequences. Preincident planning assumes that a fire will occur and compiles information that responding fire fighters would need. In many cases, a preincident survey identifies the need for a preincident plan or for a fire prevention inspection. Many departments conduct preincident surveys simultaneously with fire safety inspections. If this method is used, different members of the department should be assigned to conduct the inspection and to complete the preincident survey. This allows each fire fighter to concentrate on the objectives of one specific assignment.

Figure 22-3 Copies of preincident plans can be kept in three-ring binders.

Table 22-1 Typical Target Hazard Properties

- Bulk oil facilities and refineries
- High-rise buildings
- Hospitals
- Hotels and rooming houses
- Large apartment buildings
- Lumberyards
- Manufacturing plants
- Nursing homes and assisted-living facilities
- Public assembly occupancies
- Schools
- Shopping centers
- Storage structures for hazardous materials
- Warehouses

height and overall dimensions, type of construction, nature of the occupancy, and the types of contents in different areas. Additional information should include interior floor plans, stairway and elevator locations, utility shutoff locations, and information about built-in fire protection systems. For more information, see NFPA 1620, *Standard for Recommended Practice for Preincident Planning*.

Target Hazards

Most fire departments are not able to create a preincident plan for every individual property in their jurisdiction. Instead, they identify properties that are particularly large and/or where unusual risks are present. These properties are identified as **target hazards** (▲ Table 22-1). Target hazard properties pose an increased risk to fire fighters.

A preincident plan should be prepared for every property that involves a high life safety hazard to the occupants or presents safety risks for responding fire fighters. Preincident plans should also be prepared for properties that have the potential to create a large fire or **conflagration** (a large fire involving multiple structures).

Properties that have an increased life safety hazard include:

- Hospitals
- Nursing facilities
- Assisted-living facilities
- Large apartment buildings
- Hotels and rooming houses
- Schools
- Public assembly occupancies

Developing a Preincident Plan

The information that goes into a preincident plan is gathered during a **preincident survey**. The survey is usually performed by one of the crews that would respond to an emergency incident at the location. This enables the crew to visit the property and become familiar with the location as they collect the preincident information.

The survey data is compiled in a standard preincident plan format and filed by property address in an information management system. The actual preincident plan is usually prepared at the fire station, where the fire officer takes the time to organize the information properly and to create drawings that can be used effectively during an emergency.

The information can be stored on paper or entered into a computer system. If the preincident plan is stored on paper, copies should be made and distributed to all of the companies and command officers who would respond to the location on an initial alarm. The plans should be kept in binders or in a filing system on each vehicle (▲ Figure 22-3).

The communications center should also have a copy of the completed plan so that the dispatcher has access to the information. In some jurisdictions, the communications center can fax a copy to a portable fax machine in a command unit at the scene of an incident.

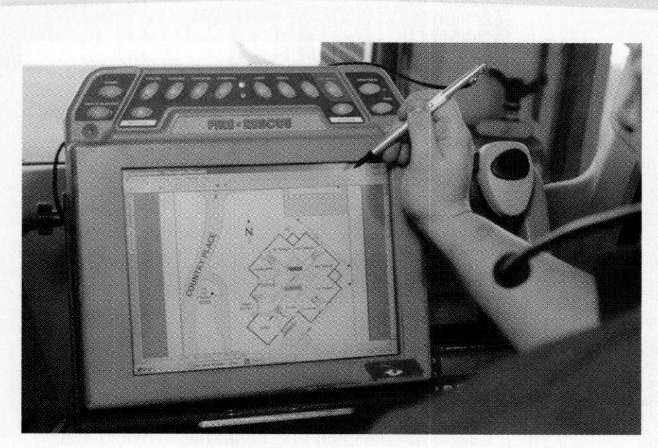

Figure 22-4 Computers allow access to preincident plans from any location.

Using a computer system to develop and store preincident plans has many advantages. Plan data can be accessed directly, through computer terminals at fire stations, and through mobile data terminals in vehicles (▲ Figure 22-4). The preincident plan for a specific property can be automatically transmitted to a dispatch terminal and printed out at the fire station when a call to that address is made. Updating preincident plans is easier when they are stored on computers. As soon as the master copy is updated, everyone has access to the new information. There is no need to make and distribute new copies to everyone.

Conducting a Preincident Survey

A preincident survey should be conducted with the knowledge and cooperation of the property owner or occupant. The property owner or a representative should be contacted before the preincident survey is conducted. This enables the fire department to schedule an acceptable time, to explain the purpose and the importance of the preincident survey, and to clarify that the information is needed to prepare fire fighters in the event that an emergency occurs at the location.

The team members who conduct the survey should dress and conduct themselves in a manner appropriate to the department's mission. A representative of the property should accompany the survey team to answer questions and provide access to different areas. Every effort should be made to obtain accurate, useful information.

The preincident survey is conducted in a systematic fashion, following a uniform format. Begin with the outside of the building. Gather all of the information about the building's geographic location, external features, and access points. Then survey the inside to collect information about every interior area. A good, systematic approach starts at the roof and works down through the building, covering every level of the structure, including the basement. If the property is large and complicated, it may be necessary to make more than one visit to ensure that all the required information is obtained and accurately recorded.

Table 22-2 Information Gathered During the Preincident Survey

- Building location
- Apparatus access to exterior of the building
- Access points to the interior of the building
- Hydrant locations and/or alternative water supply
- Size of the building (height, number of stories, length, width)
- Exposures to the fire building and separation distances
- Type of building construction
- Building use
- Type of occupancy (assembly, institutional, residential, commercial, industrial)
- Floor plan
- Life hazards
- Building exit plan and exit locations
- Stairway locations (note if enclosed or unenclosed)
- Elevator locations and emergency controls
- Built-in fire protection systems (sprinklers, sprinkler control valves, standpipes, standpipe connections)
- Fire alarm systems and **fire alarm annunciator panel** (part of the fire alarm system that indicates the location of an alarm within the building) location
- Utility shut off locations
- Ventilation locations
- Presence of hazardous materials
- Presence of unusual contents or hazards
- Type of incident expected
- Sources of potential damage
- Special resources required
- General firefighting concerns

The same set of basic information must be collected for each property that is surveyed (▲ Table 22-2). Additional information should be gathered for properties that are unusually large, complicated, or where particularly hazardous situations are likely to occur. Most fire departments use standard forms to record the survey information. After the team returns to the fire station, they can use the information to develop the preincident plan.

The fire fighters conducting the survey should prepare sketches or drawings to show the building layout and the location of important features such as exits. It takes practice and experience to learn how to sketch the required information while doing the survey, and then to convert the infor-

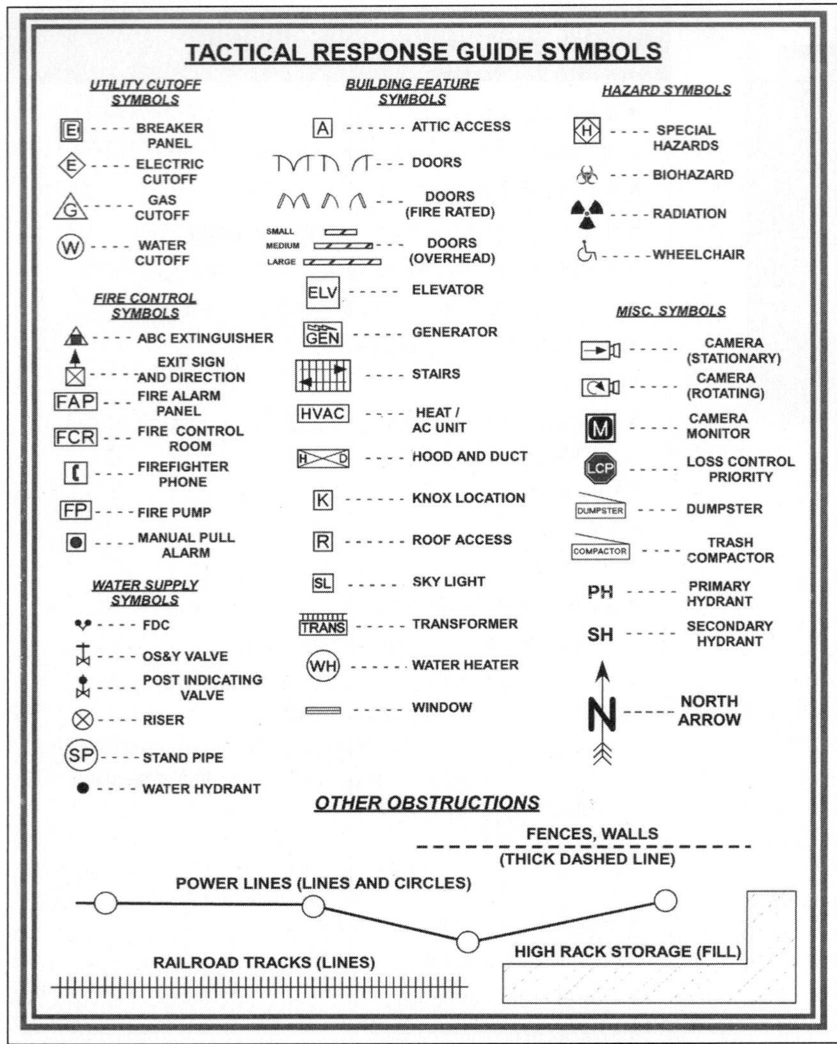

Figure 22-5 Preincident plan symbols.

Canadian Perspectives

In Canadian communities, the Fire Underwriters' Survey, a division of the Insurers' Advisory Organization, determines the Public Fire Protection Classification (PFPC). The PFPC rating is then used by insurance companies in setting premiums. Forty percent of the PFPC grading is based on the fire department's effectiveness. A key element of this rating is the thoroughness and completeness of preincident planning within the community. The Insurance Services Office (ISO) rating in the United States is very similar, although pre-planning is not heavily weighted.

mation to a final drawing. In some cases, the building owner can provide the survey team with a copy of a plot plan or a floor plan. Many fire departments also use digital cameras to record information during the survey.

The completed drawing should use standard, easily understood map symbols (▲ Figure 22-5 and ► Figure 22-6). Many fire departments use computer-assisted drawing software to create and store these diagrams.

Preincident Planning for Response and Access

Building layout and access information is particularly important during the response phase of an emergency incident. A preincident plan should provide information that would be valuable to units en route to an incident. For instance, the plan could identify the most efficient route to the fire building and also note an alternate route if the time of the day and local traffic patterns would affect the primary route. Alternate routes should also be established if primary routes require crossing railroad tracks, drawbridges, or other potentially blocked routes.

When conducting a preincident survey, fire fighters should ensure that the building address is easily visible to save time in locating an emergency incident. If the building is part of a complex, the best route to each individual building, section, or apartment should be clearly indicated on a map. The locations of hydrants and fire department connections for sprinkler and standpipe systems should always be indicated.

Points of access into the building must also be noted, particularly if there are multiple entrances or if access points are not easily seen from the street. In some cases, different entrances should be used, depending on the location of the emergency within the building. In other cases, the first unit should always respond to a designated entrance where the fire alarm annunciator panel is located or to meet a security guard who can act as a guide.

Diagrams should clearly indicate where gates, fences, or other barriers block access to parts of a building. The preincident survey should also note whether the topography makes one or more sides of a building inaccessible to fire apparatus or if an underground parking area will not support the weight of apparatus.

Access to the Exterior of the Building

The preincident plan should address access to the exterior of the building. For example, you should ask the following questions during your survey:

- Are there several roads that lead to the building, or just a few?
- Where are hydrants located?
- Where are the fire department connections for automatic sprinkler and standpipe systems located?

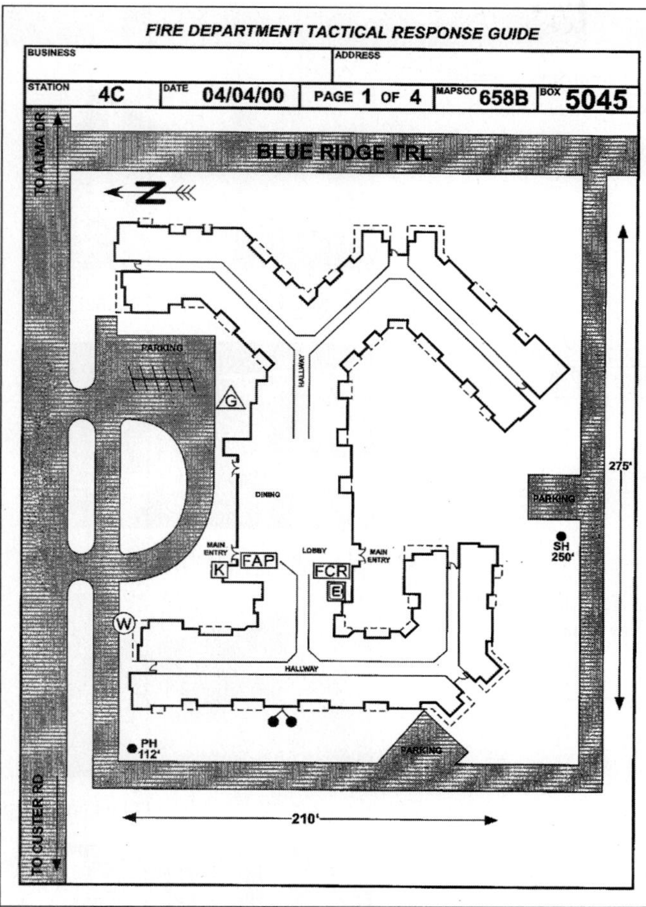

Figure 22-6 The completed drawing should use standard, easily understood map symbols.

- Are there security barriers that will limit access to the site?
- Are there fire lanes to provide access to specific areas?
- Are there barricades, gates, or other obstructions that are not wide enough or high enough to allow passage of apparatus?
- Are there bridges or underground structures that will not support the weight of apparatus?
- Are there gates that require keys or a code to gain entry?
- Will it be necessary to cut fences?
- Are there natural barriers such as streams, lakes, or rivers that limit access?
- Does the topography limit access to any parts of the building?
- Does the landscaping or snow prevent access to certain parts of the building?

Access to the Interior of the Building

The preincident survey should also consider access to the interior of the building. The following questions are helpful:

- Is there a lockbox containing keys to the building? Do the keys work?
- Where is the lockbox located? Is it easily visible?
- Is the lockbox within easy reach? Sometimes a ladder is required to reach the lockbox for security reasons.
- Are key codes needed to gain access to the building? Who has them?
- Does the building have security guards? Is a guard always on duty? Does the guard have access to all areas of the building?
- Is a key holder available to respond to the alarm within a reasonable amount of time? How can this person be reached?
- Where is the fire alarm annunciator panel located?
- Is the fire alarm annunciator panel properly programmed so you can quickly determine the exact location of an alarm?

Preincident Planning for Scene Size-Up

The preincident survey must also obtain essential information about the building that is important for size-up (the ongoing observation and evaluation of factors that are used to develop objectives, strategies, and tactics for fire suppression). This information should include the building construction, height, area, use, and occupancy, as well as the presence of hazardous materials or other risk factors. The location of other buildings or structures that may be jeopardized by a fire in the building should also be recorded.

Fire protection system information is also important for size-up. The preincident survey should identify areas that are protected by automatic sprinklers or other types of fire extinguishing systems, the locations of standpipes, and the locations of firewalls and other features designed to limit the spread of a fire. The survey should also note any areas where fire protection is lacking, such as an area without sprinklers in a building that has a sprinkler system.

Features inside a building that would allow a fire to spread but are not readily visible from the outside should also be noted. A common attic space over several occupancies, unprotected openings between floors, or buildings connected by overhead passages or conveyor systems are examples of these features.

Construction

Building construction is an important factor to identify in the preincident survey. Two similar-looking buildings may be constructed differently and thus will behave differently during a fire. ▶ Table 22-3 lists five different types of construction, as defined in NFPA 220, *Standard on Types of Building Construction*. Building construction is discussed in detail in Chapter 6, Building Construction. Preincident planning enables you to take the time that is needed to determine the exact type of building construction and iden-

Table 22-3 Types of Building Construction

Type	Description	Figure
Type I: **Fire Resistive**	Buildings where the structural members are of noncombustible materials that have a specified fire resistance (► Figure 22-7). Materials include concrete, steel beams, masonry block walls, etc.	 Figure 22-7 A Type I building.
Type II: **Noncombustible**	Buildings where the structure members are of noncombustible materials, but may not have fire resistance protection (► Figure 22-8). Includes unprotected steel beams, etc.	 Figure 22-8 A Type II building.
Type III: **Ordinary**	Buildings where the exterior walls are noncombustible or limited-combustible, but the interior floors and walls are made of combustible materials (► Figure 22-9)	 Figure 22-9 A Type III building.
Type IV: **Heavy Timber**	Buildings where the exterior walls are noncombustible or limited-combustible, but the interior walls and floors are made of combustible materials (► Figure 22-10). The dimensions of the interior materials are greater than that of ordinary construction (typically with minimum dimensions of 8′× 8′).	 Figure 22-10 A Type IV building.
Type V: **Wood Frame**	Buildings where the exterior wall, interior walls, floors, and roof are made of combustible wood material (► Figure 22-11)	 Figure 22-11 A Type V building.

tify any problem areas. Then, if an emergency occurs, the IC can check the preincident plan for information on construction, instead of hazarding a guess.

Lightweight Construction

It is important to identify whether or not the building contains lightweight construction. Lightweight construction uses assemblies of small components, such as trusses or fab-

Fire Fighter Safety Tips

Lightweight construction is a relative term. A structure may be "lightweight" in comparison to other methods of construction, but its components are still heavy enough to cause serious injury or death if they collapse on a fire fighter.

Fire Marks

Because lightweight construction uses truss supports, a floor or roof may appear sturdier than it actually is. Two fire fighters in Houston, Texas died while battling an early-morning blaze in a fast food restaurant. The fire burned through the lightweight wood truss roof supports, and the roof-mounted air conditioner dropped into the building, trapping and killing the fire fighters. The fire was later determined to be arson.

Fire Fighter Safety Tips

Buildings that are being remodeled or demolished are at higher risk for major fires. You should consider doing a special preincident survey for buildings undergoing substantial remodeling during the construction phase. This will enable you to gather more thorough information for a preincident plan of the completed building.

ricated beams, as structural support materials. This type of construction can be found in newer buildings as well as in older buildings with extensive remodeling. In many cases, the lightweight components are located in void spaces or concealed above ceilings and are not readily visible.

A wood truss, constructed from 2′ × 4′ or 2′ × 6′ pieces of wood, can be used to span a wide area and support a floor or roof. A truss uses the principle of a triangle to build a structure that can support a great deal of weight with much less supporting material than conventional construction methods (► Figure 22-12 and Figure 22-13). Lightweight construction is used in houses, fast food restaurants, auto dealerships, and many other types of buildings.

Remodeled Buildings

Buildings that have been remodeled or renovated present special situations. Remodeling can remove some of the original, built-in fire protection and create new hazards. For example, a remodeled building may contain multiple ceilings with void spaces between them. The new construction may have put a concrete topping over a wooden floor assembly. There may be new openings between floors or through walls that originally were fire resistant.

If you can conduct the preincident survey during construction or remodeling, you will be able to see the construction from the inside out, before everything is covered over. It is also important to realize that buildings under construction, whether they are being remodeled or demolished, are especially vulnerable to fire. Unfinished construction is open, without many of the fire resistant features and fire detection/suppression systems that will be part of the finished structure.

Figure 22-12 A lightweight wood truss roof assembly.

Figure 22-13 A lightweight wood truss floor assembly.

Building Use

The second major consideration during size-up is the building's use and occupancy, so it is important to identify these during a preincident survey. Buildings are used for many different purposes such as residences, offices, stores, restaurants, warehouses, factories, schools, churches, and many other purposes. For each use, different types of problems, concerns, and hazards may be present.

For example, building use can help to determine the number of occupants and their ability to escape if a fire occurs. Building use is also a key factor in determining the

Table 22-4 Classifications of Buildings

Major Use Classification	Occupancy Sub-Categories
Public Assembly	Theatres, auditoriums, and churches
	Arenas and stadiums
	Convention centers and meeting halls
	Bars and restaurants
Institutional	Hospitals and nursing homes
	Schools
Commercial	Retail stores
	Industrial factories
	Warehouses
	Parking garages
	Offices

probable contents of the building. These factors can have a major impact on the problems and hazards encountered by fire fighters responding to an emergency incident at the location.

A building is usually classified by major use. This identifies the basic characteristics of the building (▲ Table 22-4). Within the major use classification, there are occupancy sub-classifications that provide a more specific description of possible uses and their associated characteristics.

Many large building complexes contain multiple occupancy sub-categories under one roof. For example, a shopping mall usually has both retail stores and restaurants. The mall also could be part of a larger complex that includes offices, residential condominiums, a hotel, a movie theatre, an underground parking garage, and a subway station. Each area may have very different characteristics and risk factors.

Occupancy Changes

Building use may change over time. An outdated factory may be transformed into a residential building, or an unused school may be converted into an office building. A warehouse that once stored concrete blocks may now be filled with foam-plastic insulation or swimming pool chemicals. This is an important reason why preincident plans should be checked and updated on a regular schedule. A building's current occupancy information must always be determined during a preincident survey.

Exposures

An **exposure** is any other building or item that may be in danger if an incident occurs in another building or area. An exposure could be an attached building, separated by a common wall, or it could be a building across an alley or street. Exposures can include other buildings, vehicles, outside storage, or anything else that could be damaged by or involved in a fire. For example, heat from a burning warehouse could ignite adjacent buildings and spread the fire.

A preincident survey should identify any potential exposures to the property that is being evaluated. This should take into account the size, construction, and **fire load** (the amount of combustible material and the rate of heat release) of the property being evaluated, the distance to the exposure, and the ease of ignition by radiation, convection, or conduction of heat.

Fire Fighter Tips

Remember that a building can have multiple uses during its lifetime or can incorporate more than one use within the same structure.

Built-In Fire Protection Systems

The preincident survey should identify built-in fire protection and fire suppression systems on the property. These systems include automatic sprinklers, standpipes, fire alarms, and fire detection systems, as well as systems designed to control or extinguish particular types of fires. Some buildings have automatic smoke control or exhaust systems. Most high-rise buildings have systems that control elevator function during an emergency. Each are covered in detail in Chapter 36, Fire Protection, Suppression, and Detection Systems.

Automatic Sprinkler Systems

A properly designed and maintained automatic **sprinkler system** can help control or extinguish a fire before the arrival of the fire department. When properly designed and maintained, these systems are extremely effective and can play a major role in reducing the loss of life and property at an incident. They also create a safer situation for fire fighters.

The preincident survey should determine if the building has a sprinkler system and what parts of the building the system covers. Note the location of valves that control water flow to different sections of the system. These valves should always be open.

In addition to the control valves, sprinkler systems should have a fire department connection outside the building. The fire department connection is used to supplement the water supply to the sprinklers by pumping water from a hydrant or other water supply through fire hoses into this connection. The location of the fire department connection for the sprinkler system must be marked on the preincident plan.

Standpipe Systems

Standpipe systems are installed to deliver water to fire hose outlets on each floor of a building. This eliminates the need to extend hoselines from an engine at the street level up to fire level. Standpipes may also be used in low-rise buildings, such as convention centers. Fire fighters can bring attack hoselines inside the building and connect them to a standpipe outlet close to the fire, while the engine delivers water to a fire department connection outside the building.

The location of the fire department connection, as well as the locations of the outlets on each floor, should be marked on the preincident survey. The locations of nearby hydrants that will be used to supply water to the fire department connection must also be recorded. There may be multiple fire department connections to deliver water to different parts of a large building. The area or floor levels served by each connection should be carefully noted on the preincident plan and labeled at the connection.

Fire Alarm and Fire Detection Systems

The primary role of a fire alarm system is to alert the occupants of a building so that they can evacuate or take action when an incident occurs. Some fire alarm systems are connected directly to the fire department; others are monitored by a service that calls the fire department when the system is activated.

In some cases, the fire alarm system must be manually activated. In other cases, the alarm system is activated automatically. A smoke or heat detection system can trigger the alarm or it may respond to a device that indicates when water is being discharged from the sprinkler system.

The annunciator panel, which is usually located close to a building entrance, indicates the location and type of device that activated the alarm system. In most cases, fire fighters who respond to the alarm must check the annunciator panel to determine the actual source within the building or complex. The preincident survey should identify what type of system is installed, where the annunciator panel is located, and whether the system is remotely monitored.

Special Fire Extinguishing Systems

There are several different types of fixed fire extinguishing systems that can be installed to protect areas where automatic sprinklers are not suitable. Most commercial kitchens and computer rooms are required to have installed fire suppression systems (► Figure 22-14). Special extinguishing systems are also found in many industrial buildings. The type of system and the area that is protected should be identified in the preincident survey.

Areas where flammable liquids are stored or used can have sophisticated foam or dry chemical fire suppression systems. The location of these systems should be noted on the preincident plan. Details about the method of operation, location of equipment, and foam supplies should also be recorded during the preincident survey. More information about built-in systems is presented in Chapter 36, Fire Protection, Suppression, and Detection Systems.

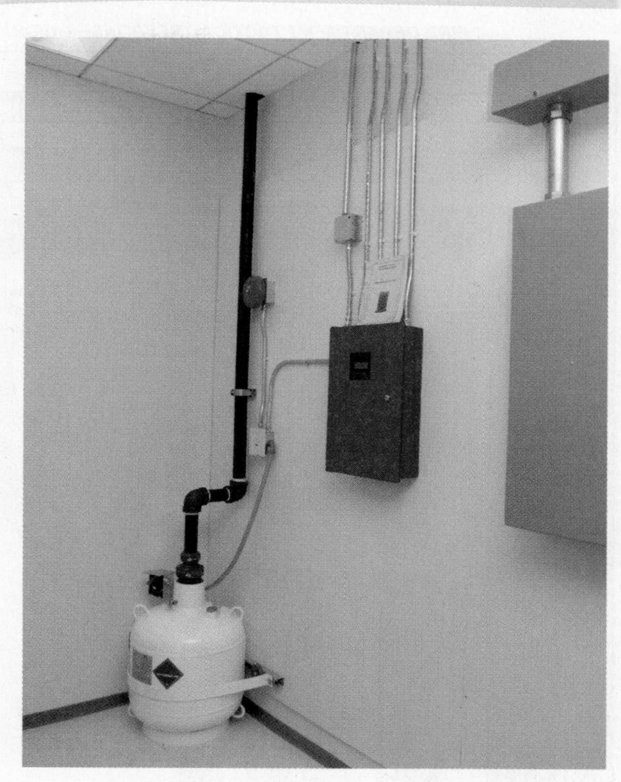

Figure 22-14 Most computer rooms have installed fire suppression systems.

Tactical Information

There are many different categories of tactical information that can be obtained during a preincident survey and documented in a preincident plan. This is information that would be of particular value to an IC in directing operations during an emergency incident or to fire fighters who are assigned to perform specific functions. When conducting a survey, all of the factors that could be significant when conducting emergency operations at that location should be considered.

Considerations for Water Supply

During the preincident survey, the amount of water needed to fight a fire in the building should be determined. The water supply source should be identified as well. The required flow rate, measured in gallons per minute or liters per minute, can be calculated, based on the building's size, construction, contents, and exposures.

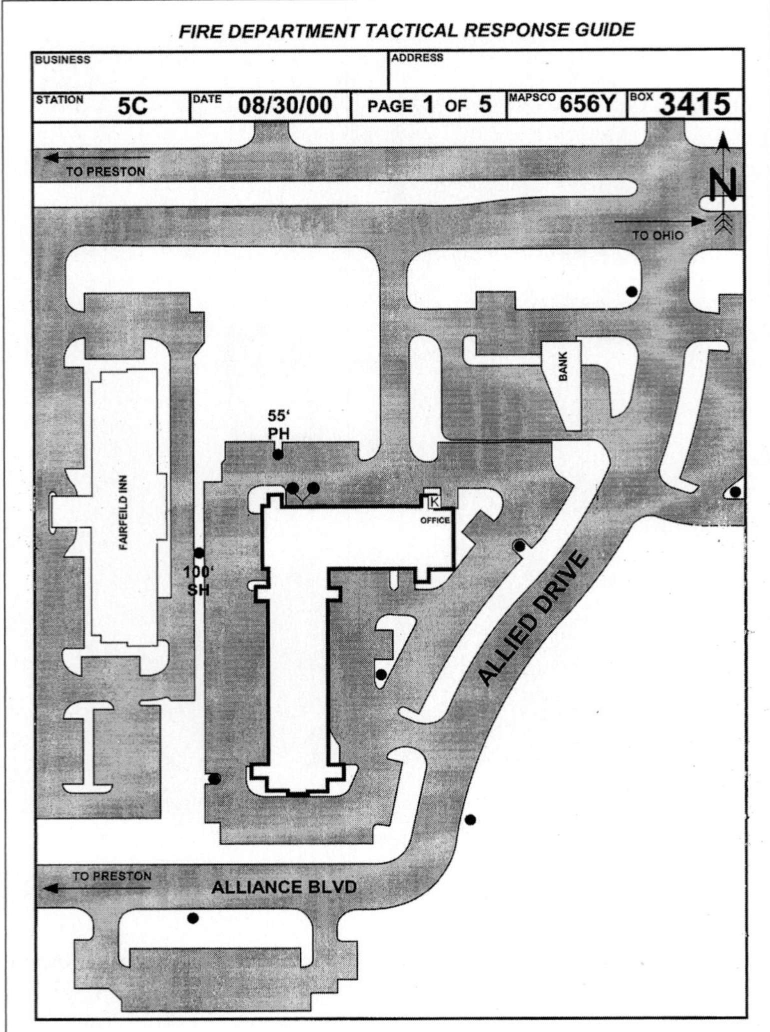

Figure 22-15 A diagram showing hydrant locations around a building.

In most urban areas, the water supply will come from municipal hydrants. It is important to locate the hydrants closest to the building. For large buildings, it will be necessary to locate enough hydrants to supply the volume of water required to control a fire. The ability of the municipal water system to provide the required flow must also be determined. It may be necessary to use hydrants that are supplied by different water mains of sufficient size to achieve the needed water flow (▲ Figure 22-15).

In areas without municipal water systems, the water may have to be obtained from a **static water supply**, such as a lake or stream, or delivered by fire department tankers. When static water sources are used, the preincident plan must identify **drafting sites** (locations where an engine can draft water directly from the static source) or the locations of the nearest **dry hydrants** (an arrangement of pipes that is permanently connected to a static water supply). It is also important to measure the distance from the water source to the fire building to determine whether a large-diameter hose can be used and whether additional engines will be needed. The preincident plan should also outline the operation that would be required to deliver water to the fire.

If a **tanker shuttle** (a system of tankers transporting water from a water supply to the fire scene) will be used to deliver water, several additional details must be included in the preincident plan. Sites for filling the tankers and for discharging their loads must be identified (► Figure 22-16). The preincident plan should also identify how many tankers are needed based on the distance they must travel, the quantity of water each vehicle can transport, and the time it takes to empty and refill the vehicle.

A large industrial or commercial complex may have its own water supply system that provides water for automatic sprinkler systems, standpipes, and private hydrants. These systems usually include storage tanks or reservoirs as well as fixed fire pumps to deliver the water under pressure. The details and arrangement of the private water supply system must be determined during the preincident survey.

The survey should indicate how much water is stored on the property and where the tanks are located. If there is a fixed fire pump, its location, its capacity in gallons per minute, and its power source (electricity or a diesel engine) should be noted. It is also important to make sure that these private water systems are maintained in good operating condition.

The private water supply system on an abandoned property may be useless because it is no longer properly maintained.

In many cases, the same private water main provides water for both the sprinkler system and the private hydrants. If the fire department uses the private hydrants, the water supply to

Fire Marks

A private water system may not be able to supply both sprinklers and hoses effectively. A commercial property in Bluffton, Indiana was destroyed when the responding fire department hooked its pumpers to private fire hydrants. The reduction in water flow that resulted rendered the building's sprinklers ineffective.

Figure 22-16 The points where tankers can be filled and where they can discharge their loads must be identified.

Fire Fighter Tips

Identifying the available water supply for a particular building is important. Consider the following:

- How much water will be needed to fight a fire in the structure?
- Where are the nearest hydrants located?
- How much water is available and at what pressure?
- Where are hydrants that are on different water mains located?
- How much hose would be needed to deliver the water to the fire?

If there are no hydrants, considerations should include:

- What is the nearest water supply?
- Is the water readily available year-round? Does it freeze during the winter or dry up during the summer?
- How many tanker trucks would be needed to deliver the water?
- How long will it take to establish a tanker shuttle?

the sprinklers may be compromised. The preincident plan for these sites should note whether public, off-site hydrants should be used instead of the private hydrants.

Utilities

During an emergency incident, it may be necessary to turn off utilities such as electricity or natural gas as a safety measure. The preincident survey should note the locations of shut-offs for electricity, natural gas, propane gas, fuel oil, and other energy sources. These sites, as well as contact information for the appropriate utility company, should be included in the preincident plan. In some cases, special knowledge or equipment is required to disconnect utilities. This must be noted in the preincident plan, along with the procedure for contacting the appropriate individual or organization.

Fire Fighter Tips

The location of utility shut-offs is as varied as the layout of structures. While some structures may have their utility shut-offs outside on the top of a utility pole, other structures may have their utility shut-offs inside in a basement closet. It is important to note the location of utility shut-offs on the preincident plan in order to prevent lengthy delays in shutting utilities down during an emergency.

Fire Fighter Safety Tips

Only qualified personnel should shut down utilities.

Electrical wires pose particular problems. Electricity may be supplied by overhead wires or through underground utilities. Overhead wires can be deadly if a ground ladder or aerial ladder comes in contact with them. The preincident plan should show the locations of high-voltage electrical lines and equipment that could be dangerous. If electricity is supplied by underground cables, the shut-off may be located inside the basement of a building or in an underground vault. These locations should also be noted on the preincident plan.

If propane gas or fuel oil is stored on the property, the preincident plan should show the locations and note the capacity of each tank. The presence of an emergency generator should also be noted, along with the fuel source and a list of equipment powered by the generator.

Preincident Planning for Search and Rescue

Fire fighters conducting search-and-rescue operations will need to know where the occupants of a building are located as well as the location of exits. The preincident survey should identify all entrances and exits to the building, including fire escapes and roof exits.

(Figure 22-17) Fire fighters should plan to assist occupants who are trying to use the exits.

The preincident plan should include information on how to contact utility companies to shut down gas and electrical service.

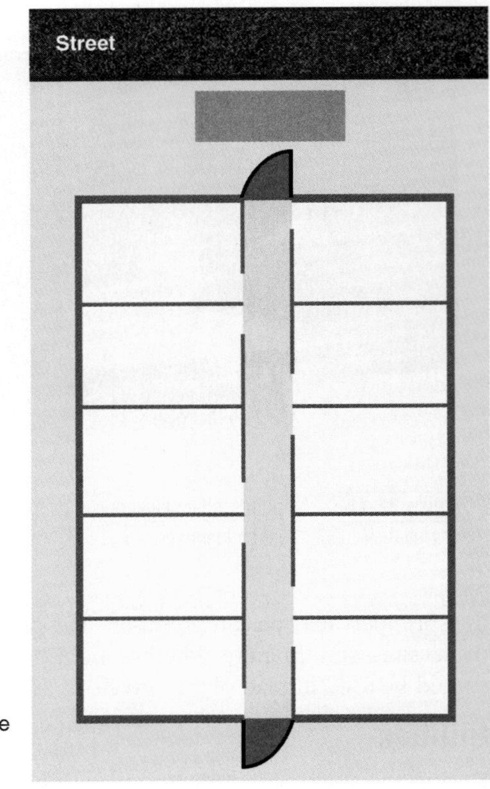

Exit Access
Exit
Exit Discharge

(Figure 22-18) An interior floor plan showing the exits should be obtained during a preincident survey.

In addition to conducting a search for occupants unable to escape on their own, search-and-rescue teams may need to assist occupants who are trying to use the exits (▲ Figure 22-17). Fire fighters may need to assist occupants descending fire escapes or to place ladders near windows or other locations for occupants to use.

An interior floor plan should be obtained during a preincident survey (► Figure 22-18). Knowing the floor plan of a building can be a lifesaver when the building is filled with smoke. It is much easier to understand a floor plan when you can tour a building under non-emergency conditions. In large buildings, it may be necessary to plan for the use of ropes during search and rescue to prevent disorientation in conditions of limited visibility.

Preincident Planning for Forcible Entry

As previously noted, the preincident survey should consider both exterior and interior access problems. Locations where forcible entry may be required should be identified and marked on the site diagrams and building floor plan. Making a note of what tools would be needed to gain entry can save time during the actual emergency. The location of a lock box and instructions on obtaining keys also should be noted.

Preincident Planning for Ladder Placement

The preincident survey is an excellent time to identify the best locations for placing ground ladders or aerial apparatus (► Figure 22-19). The length of ladder needed to reach a roof or entry point should be noted. When planning ladder placement, pay careful attention to electrical wires and other obstructions that might not be visible at night or in a smoky atmosphere (► Figure 22-20).

Preincident Planning for Ventilation

While performing a preincident survey, fire fighters should consider what information would be valuable to the members of a ventilation team during a fire. For example, what would be the best means to provide ventilation? How useful are the existing openings for ventilation? Are there windows and doors that would be suitable for **horizontal ventilation**? Where could fans be placed? Can the roof be opened to provide **vertical ventilation**? What is the best way to reach the roof? Is the roof safe? Are there ventilators or skylights that can be easily removed or bulkhead doors that can be opened easily? Will saws and axes be needed to cut through the roof? Are there multiple ceilings

Figure 22-19 Considerations for use of ladders should include identifying the best locations to place ground ladders or aerial apparatus.

Figure 22-20 The survey should note overhead obstructions, particularly electrical wires that might not be visible at night or in a smoky atmosphere.

Fire Marks

A church fire in Lake Worth, Texas claimed the lives of three fire fighters. The fire spread into the roof, which was supported by lightweight wood trusses. Fire fighters entered the building to initiate an interior attack while other crews went to the roof to start ventilation operations. The fire burned through the trusses, and the roof collapsed, trapping three fire fighters. An NFPA Fire Investigation Report is available on this fire.

Fire Fighter Tips

Determine where and how ventilation can be accomplished for different parts of the building.

that will have to be punctured to allow smoke and heat to escape?

It is also important to know if the **HVAC (heating, ventilation, and air-conditioning) system** can be used to remove smoke without circulating it throughout the building. Many buildings with sealed windows have controls that enable the fire department to set the HVAC system to deliver outside air to some areas and exhaust smoke from other areas. The instructions for controlling the HVAC system should be included in the preincident plan.

Roof construction must also be evaluated to determine whether it would be safe to work on the roof when there is a fire below. If the roof is constructed with lightweight trusses, the risk of collapse may be too great to send fire fighters above the fire. The existence of an attic that will allow a fire to spread quickly under the roof should also be noted.

Occupancy Considerations

Each type of occupancy involves particular considerations that should be taken into account when preparing a preincident plan. Fire fighters should keep these factors in mind when conducting a preincident survey.

High-Rise Buildings

A high-rise building is generally defined as a structure that is more than 75 feet high, which is usually 6 or 7 stories. High-rise buildings present problems during an emergency because of the difficulty in gaining access and because of the large numbers of occupants. Major fires in high-rise buildings often result in injuries, fatalities, and millions of dollars in property losses.

A preincident survey should identify both the building construction and any special features that have been installed. It should note all of the systems that are present in a particular building and how they are designed to function. This information should become part of the preincident plan.

The fire protection features of a high-rise building will vary depending on the age of the building and the specific building and fire code requirements in different jurisdictions. Older high-rise buildings are generally constructed of noncombustible materials and designed to confine a fire within a limited area. Newer buildings are generally con-

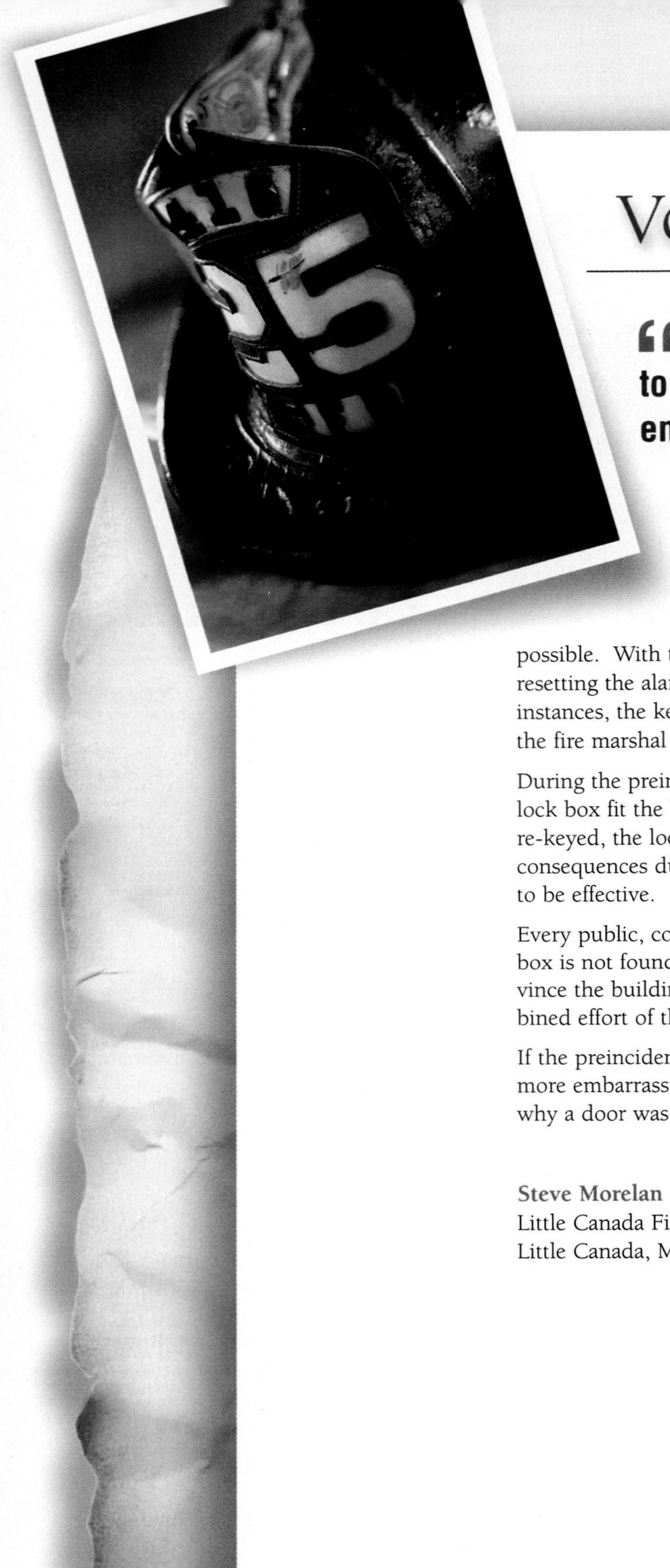

Voices of Experience

"Having the wrong set of keys can lead to disastrous consequences during an emergency."

A lock box is the best forcible entry tool a fire fighter can have. Lock boxes allow you to gain access to secured buildings in the safest, quickest, and least damaging way possible. With the aid of a lock box, fire fighters have responded to many false alarms, resetting the alarm and securing the premises, all before a keyholder arrived. In some instances, the keyholder did not even know fire fighters had been at the property until the fire marshal visited the property the next day!

During the preincident survey, the fire department should verify that the keys in the lock box fit the building's locks. Many times when a building changes ownership or is re-keyed, the lock box is forgotten. Having the wrong set of keys can lead to disastrous consequences during an emergency. Like any other tool, a lock box must be maintained to be effective.

Every public, commercial, and industrial property should have a lock box. If a lock box is not found during the preincident survey, every effort should be made to convince the building owner to install one. This can usually be accomplished with a combined effort of the fire chief, fire marshal, and building inspector.

If the preincident plan indicates that a lock box is present, use it! There are few things more embarrassing for the incident commander to explain to the property owner than why a door was forced open with a pry bar when the keys were in the lock box.

Steve Morelan
Little Canada Fire Department
Little Canada, Minnesota

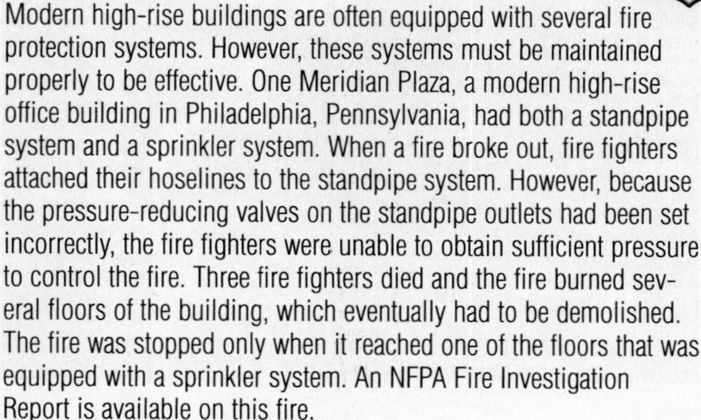

Fire Marks

Modern high-rise buildings are often equipped with several fire protection systems. However, these systems must be maintained properly to be effective. One Meridian Plaza, a modern high-rise office building in Philadelphia, Pennsylvania, had both a standpipe system and a sprinkler system. When a fire broke out, fire fighters attached their hoselines to the standpipe system. However, because the pressure-reducing valves on the standpipe outlets had been set incorrectly, the fire fighters were unable to obtain sufficient pressure to control the fire. Three fire fighters died and the fire burned several floors of the building, which eventually had to be demolished. The fire was stopped only when it reached one of the floors that was equipped with a sprinkler system. An NFPA Fire Investigation Report is available on this fire.

structed with automatic sprinklers, smoke detection and alarm systems, emergency generators, elevator control systems, smoke control systems, and building control stations where all these systems are monitored and controlled. The details of each system must be documented in the preincident plan. In many cases, expert consultants will be used to develop an emergency plan for the building. These plans are intended to be used by the building tenants and management as well as the fire department.

Assembly Occupancies

Large public assembly venues, such as stadiums and arenas, present the possibility that large numbers of people could become involved in an emergency incident. These structures are often very large and complicated and may have complex systems to manage emergency situations. Gaining access to the location of the fire or emergency situation may be difficult when all of the occupants are trying to evacuate at the same time.

Health Care Facilities

Health care facilities such as hospitals, nursing homes, and ambulatory health care centers require special preincident planning. Hospitals are often very large and include different areas, ranging from operating rooms to walk-in clinics. The most challenging problem during an emergency incident at a health care facility is protecting nonambulatory patients.

A **defend-in-place** philosophy is used when designing fire protection for these facilities. This approach presumes that patients will not be able to escape from a fire without assistance and that there may not be enough staff present to move all of the patients. Therefore, the facility itself is designed to protect the patients from the fire. Most health care facilities are built of fire resistant construction and have fire sprinkler systems.

Fire Marks

Several fires have occurred in home improvement stores. In one case, pool chemicals ignited, creating a special challenge for the fire department because of the hazardous materials involved. In another case, the fire started in a seat cushion display and spread rapidly through several aisles. The responding fire fighters were faced with a large fire, no visibility, and product dropping onto them from the upper shelves. Reports on these fires are available from NFPA Fire Investigations.

If it is necessary to move patients from a dangerous area, **horizontal evacuation** (moving patients from a dangerous area to a safe area on the same floor) is often used. It is much easier to wheel a bed to another area on the same floor level than to carry patients down stairways.

Detention and Correctional Facilities

By their nature, detention and correctional facilities are designed to ensure that the occupants cannot leave the building, even under normal conditions. The preincident plan must consider the problems involved in removing the inmates from a dangerous situation while protecting the fire fighters from angry or frightened inmates. Additionally, security concerns can make it difficult for fire fighters to gain rapid access into the building.

Residential Occupancies

Preincident plans are usually prepared for multifamily residential properties such as apartment complexes and condominiums. Most fire departments do not prepare individual preincident plans for one- and two-family residential buildings, unless there are unusual risk factors, such as a dwelling with handicapped occupants. In addition to the standard preincident plan information, the following additional information will be helpful when developing a preincident plan for a private residence:

- Detailed floor plan
- Location of sleeping areas
- Information about any handicapped occupants
- Escape routes and an outside rendezvous area
- Fire hydrant or water source location

Even if the individual homes are not surveyed, many fire departments will survey different neighborhoods making special note of addresses, access routes, and hydrant locations. Knowing what styles and types of private residences are in your jurisdiction will help you prepare for emergency incidents.

A homeowner may request an individual fire safety survey of the property. These surveys are intended to identify fire hazards and to educate the homeowner about steps that can be taken to reduce the risk of fire.

Locations Requiring Special Considerations

Preincident planning should extend beyond planning for fires and other types of emergency situations that could occur in a building. Preincident planning should also anticipate the types of incidents that could occur at other locations, such as airports, bridges, and tunnels, as well as incidents along highways or railroad lines, or at construction sites ▶ Figure 22-21.

Detailed preincident plans are prepared for airports and should note the various types of aircraft that use the airport. In addition, the airport facilities themselves often require extensive preincident planning because of their size, the numbers of people who may be present, and security issues. Similar planning should be done for bridges, tunnels, and many other locations where complicated situations could occur.

Other special locations that would require preincident planning include:

- Gas or liquid fuel transmission pipelines
- Electrical transmission lines ▶ Figure 22-22
- Ships and waterways ▶ Figure 22-23
- Subways ▶ Figure 22-24
- Railroads ▶ Figure 22-25

Figure 22-21 Preincident planning should also anticipate the types of incidents that are likely to occur at other locations, like a tunnel.

Special Hazards

One of the most important reasons for developing a preincident plan is to identify any special hazards and to provide information that would be valuable during an emergency incident. This includes chemicals or hazardous materials that are stored or used on the premises, structural conditions that could result in a building collapse, industrial processes that pose hazards, high-voltage electrical equipment and confined spaces. Preincident plans warn fire fighters of potentially dangerous situations and could include detailed instructions about what to do. This information could save the lives of fire fighters.

Fire fighters who are conducting a preincident survey should always look for special hazards and obtain as much information as possible at that time ▶ Figure 22-26. If the information is not readily available, it may be necessary to contact specialists for advice or conduct further research before completing the preincident plan.

Fire Marks

The presence of hazardous materials presents special challenges to fire fighters. Pool chemicals stored in a warehouse in Phoenix, Arizona ignited and the resulting fire overwhelmed the sprinkler system. Fire fighters were not able to apply enough water to control the fire before it destroyed the warehouse. An NFPA Fire Investigation Report is available on this fire.

Hazardous Materials

A preincident survey should obtain a complete description of all hazardous materials that are stored, used, or produced at the property. This includes an inventory of the types and quantities of hazardous materials that are on the premises, as well as information about where they are located, how they are used, and how they are stored. The appropriate actions and precautions for fire fighters in the event of a spill, leak, fire, or other emergency incident should also be listed.

Many jurisdictions require a business that stores or uses hazardous materials to obtain a fire department permit. They may also require that hazardous materials specialists conduct special inspections. If the quantity of hazardous materials on hand exceeds a specified limit, federal and state regulations also require a property owner to provide the local fire department with current inventories and **Material Safety Data Sheet** (MSDS) documents ▶ Figure 22-27. Fire fighters conducting preincident surveys should ensure that the required information has been provided and is up-to-date.

Fire fighters should expect to encounter hazardous materials at certain types of occupancies, such as chemical companies, garden centers, swimming pool supply stores, hardware stores, and laboratories. Hazardous materials can also be found in unexpected locations. Fire fighters should always be alert for hazardous materials.

Some communities put placards on the outside of any building that contains hazardous materials. The placards should use the marking system specified in NFPA 704, *Standard for Identification of the Hazards of Materials for Emergency Response*.

Some hazardous materials require special suppression techniques or extinguishing agents. This information should be noted in the preincident plan. The plan should also contain contact information for individuals or organizations that can provide advice if an incident occurs. Hazardous materials are discussed in detail in Chapters 27–33.

Figure 22-22 Electrical transmission lines.

Figure 22-23 Ships and waterways.

Figure 22-24 Subway station.

Figure 22-25 Railroad.

Figure 22-26 During the preincident survey, look for special hazards and obtain as much information as possible.

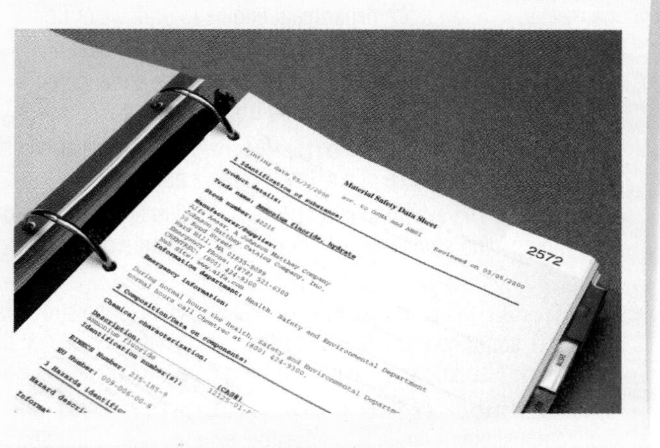

Figure 22-27 MSDS document.

Wrap-Up

Ready for Review

This chapter discussed preincident planning. Preincident planning involves both a preincident survey of the property by fire fighters to gather information and the development of a plan by fire officers. A preincident plan is a systematic gathering and recording of information so that it will be available to other members of the department and other public safety agencies when an incident occurs.

During the preincident survey, the following information should be obtained and noted: the building location and site plan, apparatus access, entry points, building size and construction, building use, floor plan, exits, built-in fire protection and alarm systems, water supply and fire-fighting resources required, life hazards, presence of hazardous materials, exposures to the fire building, location of utility shut-offs, and locations for ventilation.

Preincident surveys must also consider the type of occupancy being surveyed. The fire fighter will have different areas of concern when surveying a high-rise than surveying a hospital. Preincident surveys are also not limited to buildings; fire fighters must plan for emergencies at other locations, from airports to subway stations.

Hot Terms

Conflagration A large fire, often involving multiple structures.

Defend-in-place A strategy where the victims are protected from the fire without relocation.

Drafting sites Location where an engine can draft water directly from a static source.

Dry hydrant An arrangement of pipe that is permanently connected to allow a fire department engine to draft water from a static source.

Exposure Any person or property that may be endangered by flames, smoke, gases, heat, or runoff from a fire.

Fire alarm annunciator panel Part of the fire alarm system that indicates the source of an alarm within a building.

Fire load The weight of combustibles in a fire area or on a floor in buildings and structures including either contents or building parts, or both.

Fire resistive construction Buildings where the structural members are of noncombustible materials that have a specified fire resistance. Also known as Type I building construction.

Heavy timber construction Buildings constructed with noncombustible or limited-combustible exterior walls and interior walls and floors made of large dimension combustible materials. Also known as Type IV building construction.

Horizontal evacuation Moving occupants from a dangerous area to a safe area on the same floor level.

Horizontal ventilation The process of making openings so that the smoke, heat, and gases can escape horizontally from a building through openings such as doors, windows, etc.

HVAC system Heating, ventilation, and air-conditioning system in large buildings.

Lightweight construction The use of small dimension members such as as 2′ × 4′ or 2′ × 6′ wood assemblies as structural supports in buildings.

Material Safety Data Sheet (MSDS) A form, provided by manufacturers and compounders (blenders) of chemicals, containing information about chemical composition, physical and chemical properties, health and safety hazards, emergency response, and waste disposal of the material.

Noncombustible construction Buildings where the structural members are of non-combustible materials without fire resistance. Also know as Type II building construction.

Ordinary construction Buildings where the exterior walls are noncombustible or limited-combustible, but the interior floors and walls are made of combustible materials. Also know as Type III building construction.

Preincident plan A written document resulting from the gathering of general and detailed information to be used by public emergency response agencies and private industry for determining the response to reasonable anticipated emergency incidents at a specific facility.

Preincident survey The process used to gather information to develop a preincident plan.

Size-up The ongoing observation and evaluation of factors that are used to develop objectives, strategy, and tactics for fire suppression.

Sprinkler system An automatic fire protection system designed to turn on sprinklers if a fire occurs.

Standpipe system An arrangement of piping, valves, and hose connections installed in a structure to deliver water for fire hoses.

Static water supply A water supply that is not under pressure, such as a pond, lake, or stream.

Tanker shuttle A method of transporting water from a source to a fire scene using a number of mobile water supply apparatus.

Target hazard Any occupancy type or facility that presents a high potential for loss of life or serious impact to the community resulting from fire, explosion, or chemical release.

Truss A collection of lightweight structural components joined in a triangular configuration that can be used to support either floors or roofs.

Vertical ventilation The process of making openings so that the smoke, heat, and gases can escape vertically from a structure.

Wood frame construction Buildings where the exterior walls, interior walls, floors, and roof structure are made of wood. Also known as Type V building construction.

Fire Fighter in Action

Your captain has directed your engine company to conduct a preincident survey on a new building that was built in your response area. The building that you are going to be surveying is the local school district's performing arts center. It has a rated occupancy capacity of 800 people. Your lieutenant has contacted the school district and has scheduled a time to meet with a school representative and walk through. Up to this point, you have studied preincident surveys and this is your first opportunity to participate in the development process.

1. Which of the following pieces of information should be gathered during the preincident survey?
   A. Access points to the exterior and interior of the building
   B. Hydrant locations
   C. Floor plans
   D. Building location
   E. All of the above

2. Which of the following is not information obtained about the fire protection system for preincident planning for scene size-up?
   A. Areas that are protected by automatic sprinklers
   B. Location of standpipes
   C. Location of firewalls
   D. Location of pull stations

During your survey, you note that the structural members are made of steel and are not protected by a fire resistive coating. Talking with the school representative, you find that the building will mostly be used for school productions, plays, and special presentations.

3. What type of construction is the performing arts center?
   A. Fire Resistive
   B. Noncombustible
   C. Ordinary
   D. Heavy Timber

4. What would be the classification of this building?
   A. Institutional
   B. Commercial
   C. Public Assembly
   D. Industrial

Chief Concepts

- Preincident planning enables a fire department to evaluate the conditions of target hazards before an emergency.
- Preincident planning must be systematically gathered, recorded, updated, and it must be made available to members of the fire department who might respond to that location.
- The fire officer takes the information collected from the preincident survey and creates a preincident plan.
- The information that goes into the preincident plan is gathered by conducting a preincident survey.
- The fire fighter conducting the preincident survey should prepare sketches or drawings to note the layout and location information.
- The preincident plan provides tactical information to the IC during an emergency.
- Fire fighters should take into account the specific considerations of each type of occupancy when creating a preincident survey.
- Preincident plans should also prepare for incidents that could occur in locations like airports and subways.

www.FireFighter.jbpub.com
Chapter Pretests
Interactivities
Hot Term Explorer
Web Links
Review Manual
FireLearn

www.FireFighter.jbpub.com

Fire and Emergency Medical Care

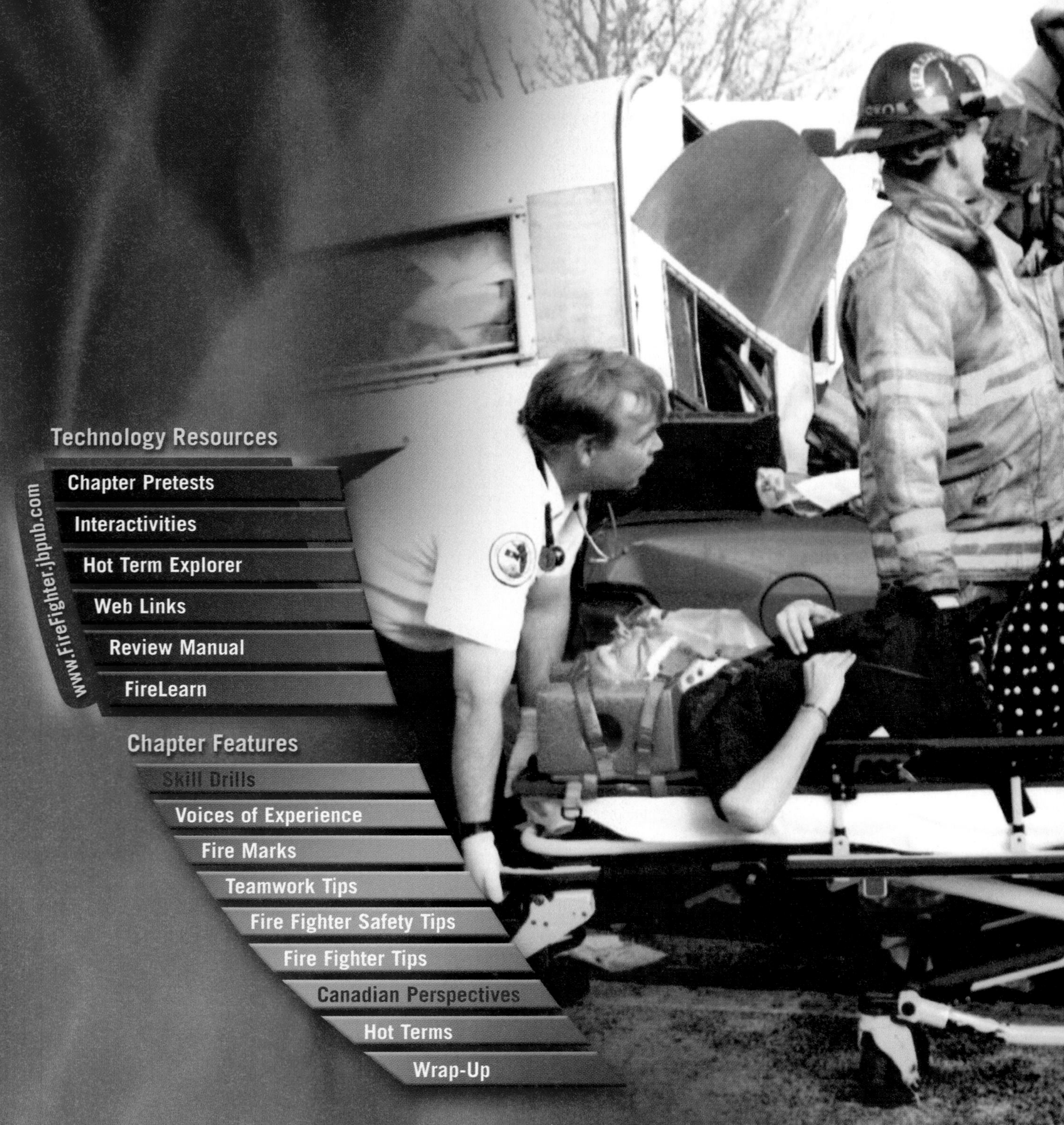

Technology Resources

www.FireFighter.jbpub.com

- Chapter Pretests
- Interactivities
- Hot Term Explorer
- Web Links
- Review Manual
- FireLearn

Chapter Features

- Skill Drills
- Voices of Experience
- Fire Marks
- Teamwork Tips
- Fire Fighter Safety Tips
- Fire Fighter Tips
- Canadian Perspectives
- Hot Terms
- Wrap-Up

Chapter 23

NFPA 1001 Standard

NFPA 1001 contains no Fire Fighter I Job Performance Requirements for this chapter.

NFPA 1001 contains no Fire Fighter II Job Performance Requirements for this chapter.

Knowledge Objectives

After studying this chapter, you will be able to:

- Describe how the delivery of Emergency Medical Services (EMS) fits into the mission of the fire department.
- Describe how EMS delivery can be a customer service issue.
- Distinguish between basic life support and advanced life support.
- Identify the skills performed by advanced life support EMS that are not performed by basic life support EMS.
- Define the two levels of EMS training within basic life support.
- Define the two levels of EMS training within advanced life support.
- Explain why EMS can work well within a fire department.
- Differentiate a combination EMS system from a fire department EMS system.
- List three groups that interact with an EMS provider.

Skills Objectives

There are no skill objectives for this chapter.

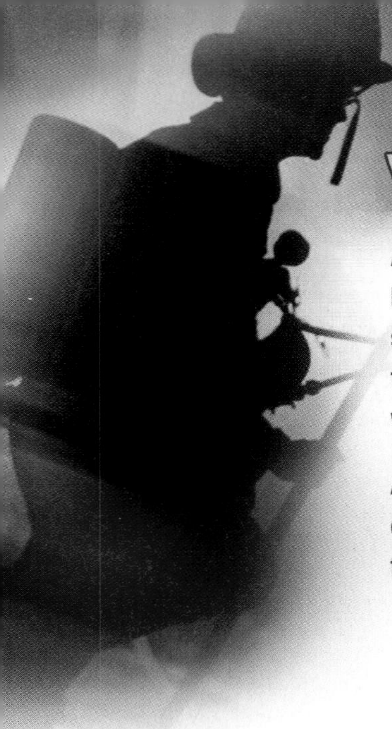

You Are the Fire Fighter

As part of your training as a new fire fighter, you have been assigned to a ride along with Captain Lewis at Station #3. After greeting you and introducing you to the rest of the crew, Captain Lewis shows you the equipment on the engine. Just as you are walking back to the kitchen, you hear the tones go off. "Engine 403 and Medic 433, respond to 11285 Main Street for a 63-year-old male with chest pain. The patient is reporting shortness of breath."

As you strap yourself into your assigned seat, you wonder why fire fighters are responding to a call for medical assistance for someone having a heart attack. Then you start thinking of questions you could ask the Captain later.

1. ***Why does the fire department respond to emergency medical calls?***
2. ***Will you have a role in delivering emergency medical services?***
3. ***Will you need additional training?***
4. ***How does the delivery of emergency medical services relate to your duties as a fire fighter?***

Introduction

This chapter will help you understand the role played by the fire department in providing emergency medical services (EMS). This role varies from one department to another. By learning the components of the EMS system, you will have a better understanding of the relationship between fire departments and EMS, including:

- The importance of EMS as part of the fire department mission of saving lives and protecting property
- The difference between basic life support (BLS) and advanced life support (ALS) levels of emergency medical care
- The different types of EMS training and certification
- The importance of EMS training for fire fighters
- The different types of EMS systems provided by fire departments
- The variety of interactions between EMS providers and other members of the health care team

The material presented in this chapter is not designed to make you an EMS provider. To achieve that goal, you will need to take an approved course leading to certification or registration as a First Responder, an Emergency Medical Technician-Basic (EMT-Basic), an Emergency Medical Technician-Intermediate (EMT-Intermediate), or an Emergency Medical Technician-Paramedic (EMT-Paramedic) (► Table 23-1). Instead, this material introduces you to the different levels of care that EMS systems provide, the types of training available in EMS, the types of EMS systems found in fire departments, and the other health care providers who are also part of the EMS system.

Table 23-1

EMS Certification Levels

- First Responder
- Emergency Medical Technician-Basic (EMT-Basic)
- Emergency Medical Technician-Intermediate (EMT-Intermediate)
- Emergency Medical Technician-Paramedic (EMT-Paramedic)

The Importance of EMS to the Fire Service and the Community

The mission of the fire service is to save lives and protect property. When many people think of their fire department, they think of fire suppression first. However, in some fire departments, more than 80% of the emergency calls are requests for emergency medical services (► Figure 23-1). Many fire departments have added "EMS" or "rescue" to their names to reflect this change. As a new fire fighter, you will have an important role in providing emergency medical services to the citizens of your community.

Most fire departments provide some level of emergency medical services, although the degree of involvement varies. Some departments provide only extrication and medical assistance for vehicle collisions. Others are the first responders for emergency medical calls, or provide basic or advanced life support until another agency arrives to transport the patient. Still other fire departments provide both medical first response and patient transportation services.

Figure 23-1 In some fire departments, more than 80% of emergency calls are for emergency medical services.

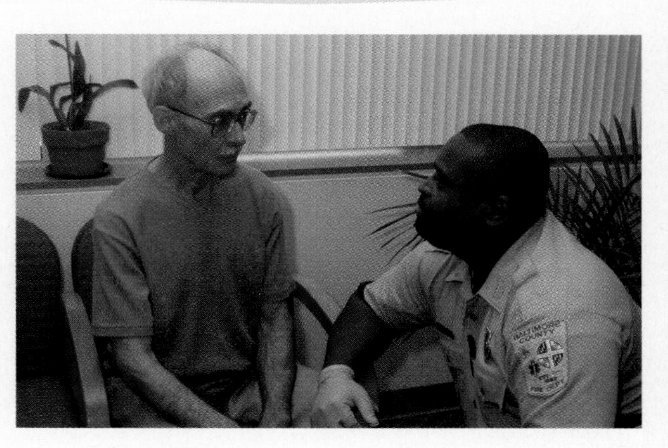

Figure 23-2 EMS providers need to be supportive of the patient in all types of EMS calls.

Fire Marks

According to a survey of the International Association of Fire Chiefs (IAFC), 94% of their member's agencies provide EMS to their community. Of that, 56% provide patient transportation services.

Delivering emergency medical service fits directly into the fire department mission of protecting lives and property. Strict fire codes and effective public education efforts have been successful in reducing the numbers of fires. This has enabled the fire service to take a greater role in providing emergency medical care.

The same fire stations, emergency vehicles, and fire fighters that are available to respond quickly to fires can also be used to deliver assistance to people who need urgent medical care. Relatively small investments in EMS training and equipment can increase the fire department's ability to deliver life-saving services to the public.

In most communities, much of the funding for the fire department's operations comes from tax dollars paid by the citizens of that community. In this sense, the citizens of your community are your customers, and the services you provide should be customer-oriented. Their continued support is vital for the department's fire suppression and EMS activities. EMS functions deliver both an essential service to the citizens and positive public relations for the department. The community responds positively, and its support enables your department to maintain its effectiveness in all areas of operation.

When you respond to an emergency, you should realize that the citizen who called 9-1-1 probably had little or no training in identifying a medical emergency or knowing what to do to help a person in distress. Sometimes, what the caller perceives as an emergency is actually a minor incident. At other times, the call may not be made until the patient's condition is literally life-threatening. A competent and caring EMS provider will always remain supportive of the patient and the caller, even when the call was unnecessary or delayed. Remember, from the caller's perspective, this is an emergency (▲ Figure 23-2).

Levels of Service

Emergency medical services are provided at the basic life support (BLS) level and at the advanced life support (ALS) level.

Basic Life Support

Basic life support is the level of medical care that can be provided by persons trained to perform a limited set of emergency medical skills. These skills include:

- Scene control
- Evaluating conditions for responder safety
- Patient assessment
- Basic airway management techniques
- Cardiopulmonary resuscitation (CPR)
- Administering oxygen
- Splinting
- Controlling external bleeding and bandaging
- Treating for shock
- Lifting and moving patients (► Figure 23-3)
- Transporting patients to an appropriate medical facility

Figure 23-3 Basic life support services include administering oxygen, controlling bleeding, and lifting and moving patients.

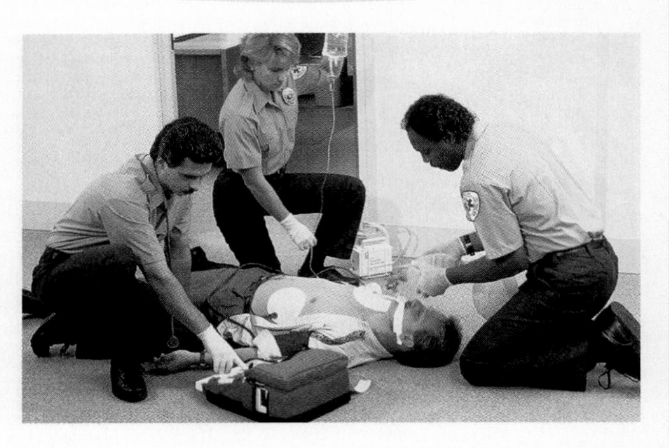

Figure 23-4 Advanced life support services include airway techniques such as endotracheal intubation and administration of intravenous fluids or medications.

BLS providers can perform cardiac defibrillation to return an irregular heartbeat to a normal rhythm by using an automated external defibrillator (AED). But basic life support services do not include the administration of medications beyond assisting the patient with his or her prescribed medications, or the administration of intravenous fluids. People qualified to perform BLS skills are trained as medical first responders or as EMT-Basics.

Advanced Life Support

People who provide **advanced life support** (ALS) services have extensive training in advanced lifesaving procedures. Advanced life support systems require a greater investment in equipment and personnel training than BLS systems. Advanced life support skills include:

- Advanced airway management techniques including endotracheal intubation to keep the airway open and help the patient breathe
- Administering intravenous fluids to treat shock ► Figure 23-4
- Administering medications for various conditions
- Monitoring and interpreting heart rhythms
- Electrically pacing the heart by using medical intervention
- Defibrillating the heart
- Removing trapped air from the chest

Advanced life support personnel operate as the extension of a physician and use standing orders, protocols, and/or radio direction from the physician to assure uniform and correct treatment. At times, providing advanced life support requires more personnel at an emergency because the care the patient receives is more complex. People who provide advanced life support are trained as EMT-Paramedics, although there are also EMT-Intermediate levels of training that include limited advanced life support skills.

Training

Most fire fighters will need to learn some emergency medical care techniques. Many departments require that all personnel be trained in EMS as well as in fire suppression. The required training level will depend on the design of the local emergency medical service delivery system and on the functions performed by the fire department.

Basic Life Support Training

EMS providers may be trained to provide basic or advanced life support. EMS courses in the United States follow a national standard curriculum, established by the U.S. Department of Transportation, that assures uniformity of knowledge and skills. Basic life support courses are offered at the first responder level and at the EMT-Basic level.

First Responder

The **first responder** course is designed for people such as police officers, teachers, and day care providers who encounter medical emergencies as part of their jobs. The first responder is expected to assess and provide basic care for the patient until EMT-Basics or EMT-Paramedics arrive to provide further care and transportation to a medical facility. The first responder course typically exceeds 40 hours in length and is designed to train a person to stabilize a patient at the scene by providing immediate, lifesaving care such as controlling bleeding, establishing an airway, initiating CPR, and using an AED ► Figure 23-5.

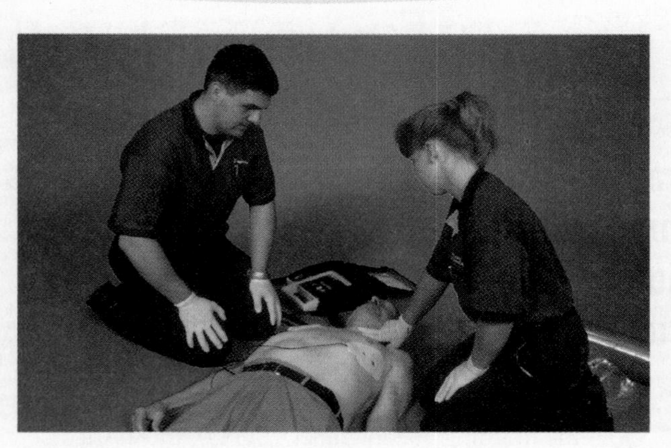

Figure 23-5 A typical EMS training course will teach you how to use an AED.

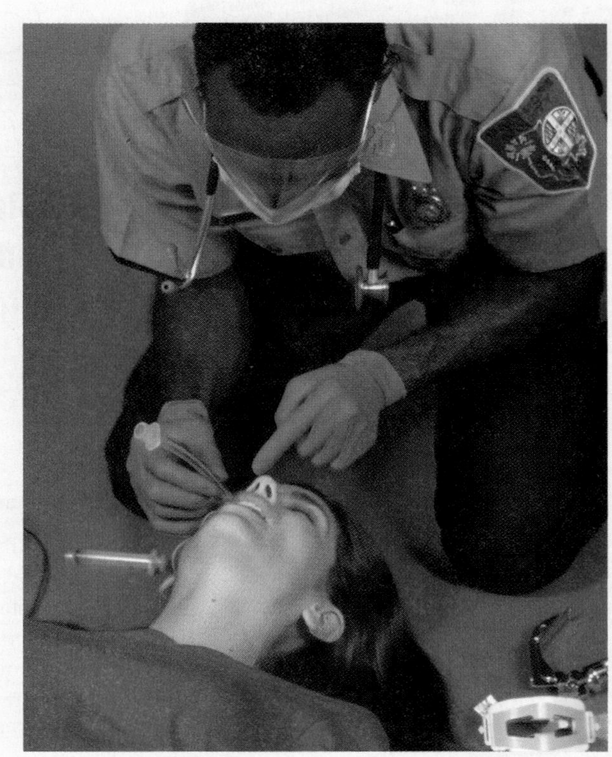

Figure 23-6 EMT-Paramedics learn how to insert endotracheal tubes into the airway.

A first responder course covers many of the same topics as the EMT-Basic course but without as much detail. For example, skills like lifting and moving patients are only touched upon because the first responder is not usually responsible for moving and transporting the patient. A first responder course will not prepare a person to work as an emergency responder on an ambulance. Instead, it is designed for people who will treat patients at emergency scenes either under the direction of an EMT or prior to the arrival of EMT-Basics or EMT-Paramedics. First responders can be certified in some states and/or nationally registered through the National Registry of Emergency Medical Technicians.

Emergency Medical Technician-Basic

The Emergency Medical Technician-Basic (EMT-Basic) course involves more than 110 hours of training. The EMT-Basic course teaches:

- Scene size-up
- Securing the scene and scene safety
- Patient assessment
- Simple airway techniques
- CPR
- Splinting
- Bandaging
- Administering oxygen
- Lifting and moving ill and injured patients
- Controlling external bleeding
- Treating for shock
- Ambulance operations
- Performing cardiac defibrillation using an automated external defibrillator (AED)

The course consists of lectures, skills practice, and observation time in an emergency department and on an ambulance. EMT-Basic certification is the minimal level of training necessary to provide care to a patient in an ambulance. After completing EMT-Basic training, a student may become a state-certified EMT-Basic, a nationally registered EMT-Basic, or both, depending on the state in which he or she was trained.

Advanced Life Support

Emergency Medical Technician-Paramedic

The Emergency Medical Technician-Paramedic (EMT-Paramedic) program is designed to teach advanced life support skills and techniques. Most EMT-Paramedic courses require at least one year to complete and build on the EMT-Basic course. Covered areas include:

- Using electrocardiograms to evaluate the patient
- Administering medications
- Inserting endotracheal tubes into the airway ▲ Figure 23-6
- Electrically pacing the heart

The EMT-Paramedic course also covers the causes and treatments of diseases. The course includes both clinical time in the hospital and time working on an advanced life support

Voices of Experience

"He was instrumental in maintaining cervical spine immobilization on the combative victim and initially opening the airway of the unconscious victim."

On the morning of September 2, 2000, three Arizona State college students were driving to San Diego for the Labor Day weekend. Their sports utility vehicle left the roadway and rolled, end over end, ejecting the two women in the back seat.

As a full-time paramedic working for AMR in the Rural East County of San Diego, I arrived to find two major trauma victims and Rick Williams, an off-duty Santee Fire Department Captain/Paramedic, attending to the injured victims. My initial triage indicated two "immediate" victims and one "delayed." Both "immediate" victims had airway problems that required aggressive treatment. A closer look at the most critical victim's chief complaint revealed that she had to be intubated. She was intubated and ventilated while en route to the trauma center, arriving approximately 47 minutes after our initial dispatch.

I have been a licensed full-time paramedic for over 25 years and have run on thousands of calls. There are occasions in EMS when everything seems to go right. When this happens, the victim benefits with a positive outcome. Such was the case on this call. Using his cell phone, Rick had called for an air ambulance after calling 9-1-1. He was instrumental in maintaining cervical spine immobilization on the combative victim and initially opening the airway of the unconscious victim.

It is most rewarding when victims call to personally thank you for saving their lives, as this young woman did. She could have very easily died from her injuries, had not everything fallen into place in just the right way. That day the EMS system performed perfectly, thanks to the heroic action of one off-duty Fire Captain/Paramedic.

Rick Foehr
American Medical Response
San Diego, California
Grossmont Rural EMS
Alpine, California

Teamwork Tips

All of the EMS providers in a system must be trained to work together and must carefully coordinate their activities. Several providers may need to work together to treat a patient with a life-threatening condition, with some providers performing ALS functions and others performing BLS functions. When a patient's life depends on the team's combined skills, there is no room for confusion or errors.

Fire Marks

The popular 1970s television program "Emergency" first depicted the delivery of advanced life support skills to the public. This dramatic series, based on the activities of fire fighters in Los Angeles County, is partly responsible for the recognition of fire department EMS programs during that period.

unit. A student who completes the EMT-Paramedic program is eligible to become a state-certified EMT-Paramedic, a nationally registered EMT-Paramedic, or both, depending on the state in which he or she was trained.

Emergency Medical Technician-Intermediate

Most states have established standards for an intermediate level of care. The Emergency Medical Technician-Intermediate (EMT-Intermediate) can perform a limited number of advanced life support skills. The skills that an EMT-Intermediate can perform vary from one state to another. Like EMT-Basics and EMT-Paramedics, EMT-Intermediates can be state-certified, nationally registered, or both, depending on the state in which he or she was trained. Both EMT-Paramedics and EMT-Intermediates work under the direction of physician medical directors.

Training Agencies

EMS training may be offered through a fire-training academy, a community college, a vocational training center, or a hospital. In some areas, fire fighters receive this training before beginning work with the department; in other jurisdictions, the training is provided by the fire department to employees.

Figure 23-7 Typical continuing medical education classes cover such topics as pediatric emergency care, geriatric emergency care, and initial assessment skills.

Continuing Medical Education

To keep up with changes in the EMS field and to refresh current skills and knowledge, EMS providers must attend continuing medical education (CME) classes (also commonly referred to as continuing education units or CEUs) (◄ Figure 23-7). These classes are important to ensure that EMS providers are performing at a satisfactory level and are aware of changes in medical protocols or procedures. They also provide an opportunity for EMS crews to review some of their more interesting calls, which can be an invaluable learning experience. Because state and national requirements for recertification include on-going education, these sessions can count toward recertification. CME is also an important part of a fire department's efforts to improve the quality of its services.

EMS Delivery Systems

Emergency medical services can be delivered in several ways. Various organizations, including fire departments, EMS departments, law enforcement agencies, hospitals, volunteer organizations, and for-profit companies operate EMS delivery systems. Some EMS systems operate with a combination of these organizations performing different roles.

Local and state government officials usually decide which agency will handle EMS, and it is important to realize that one type of system is not inherently better or worse than another. Quality EMS can be provided by any of these agencies. The quality depends on the training and dedication of the EMS providers and the quality assurance measures that are implemented within the system, not on the type of system. Regardless of who provides emergency medical care, all emergency service providers must cooperate to deliver quality service to the customer/patient.

There are several advantages to having EMS systems located in the fire department. For example, the entry requirements for fire fighters help assure quality employees. Most fire departments are required to regularly monitor the health and fitness of their employees and have mechanisms in place for continuing education. Fire departments also have the infrastructure in place to support EMS operations, including radio systems, dispatch services, and fire stations located strategically throughout the community. This enables the fire department to respond quickly to EMS calls throughout the jurisdiction. Furthermore, the fire department is familiar with the community and its people, which can be invaluable when responding to and treating patients. In short, many communities find that when they need to find a home for their EMS system, the fire department is a good fit.

Types of Systems

There are two primary types of EMS delivery systems within fire departments. In a combination EMS system, the fire department provides only medical first response, while EMS transport is provided by a separate agency (▼ Figure 23-8). In fire department EMS systems, the fire department provides both medical first response and EMS transport.

Combination EMS Systems

In a combination EMS system, the fire department provides medical first response while another agency operates the ambulances that transport the patients. The EMS first response services may be part of an engine or truck company's responsibility, or provided by a special EMS unit. First response may be provided at the basic or advanced life support levels.

Although this system limits the number of fire department personnel dedicated to EMS and does not require them to transport patients, it also has disadvantages. For example, a fire company may be out of service for a long time if it must wait for a transport unit from another agency to arrive. Patient care can be adversely affected if there are not enough EMS transport units available.

A combination EMS system requires that fire department personnel and transport agency personnel have a good working relationship and effective communications. The responsibility for patient care must be transferred from the fire department to the agency that operates the ambulance.

Fire Fighter Tips

Fire fighters should have a fitness assessment and participate in a regular physical fitness program. If your department does not have this type of program, consider starting one for your fire fighter community. There are many organizations that help train and certify qualified fitness leaders. Contact your local college or health club for information about fitness education organizations in your region.

Fire Department EMS Systems

Fire department EMS systems provide the medical first response and also transport patients. Both medical and transport services may be provided at either basic life support or advanced life support levels (▼ Figure 23-9). As in a combination system, engine companies, truck companies, or special EMS units may be used for medical first response. Because all of the personnel work for the same agency, training is easily accomplished and efforts are coordinated under a united control.

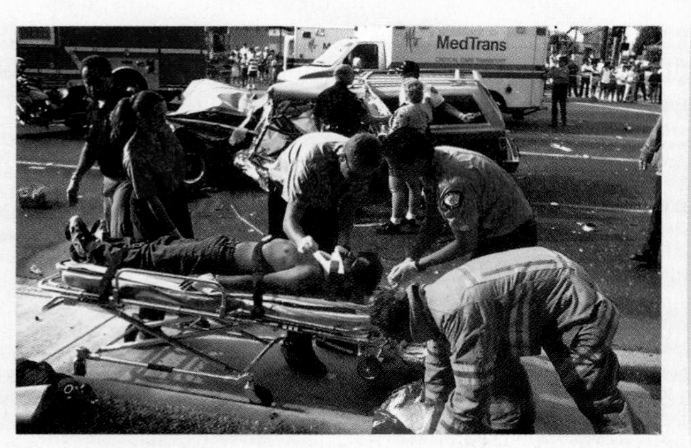

Figure 23-8 A combination EMS system in action.

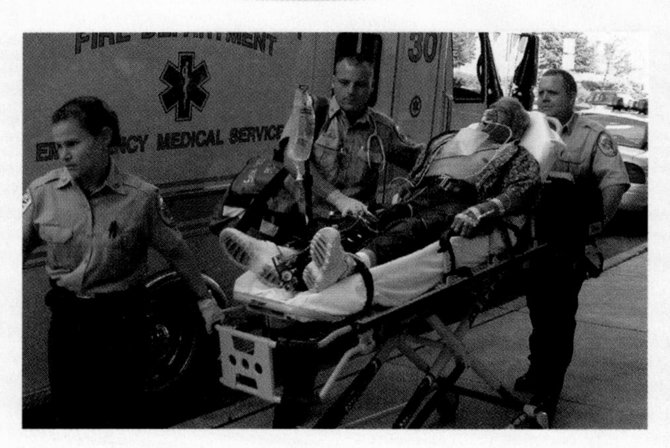

Figure 23-9 A fire department EMS system delivers medical services and transports patients.

Fire Fighter Safety Tips

One advantage of having the medical first response part of the EMS system in the fire department is that injured or ill fire fighters can receive immediate care.

Operational Considerations

Some fire departments that supply EMS service cross-train personnel in both fire suppression and emergency medical services. Cross-training enables the department to rotate personnel between fire suppression duty and EMS duty. Other departments train the personnel in either fire suppression or EMS, resulting in a separate staff of EMS providers within the department.

Interactions

Patients

If you become an EMS provider, you will come into contact with many people in your community. Do your best to provide prompt, efficient, competent care for all members of your community, regardless of their age, socioeconomic background, or nature of injury. Your attitude will affect the way patients, their family members, and bystanders respond to you. Your attitude may even affect the way people judge the care you deliver. Remember that an emergency situation is stressful for everyone, but especially for the patient and the patient's family members. Your calm, caring attitude will help calm them as well. Your patient deserves the best care you can give. Treat your patient as you would like to be treated if you were the patient (► Figure 23-10).

Figure 23-10 Treat your patient as you would like to be treated if you were the patient.

Medical Director

Each EMS system has a physician **medical director** who authorizes the providers in the service to deliver medical care in the field (► Figure 23-11). The appropriate care for each injury, condition, or illness that EMS providers will encounter is determined by the medical director and is described in a set of written standing orders, or protocols. **Medical control** is either offline (indirect) or online (direct), as authorized by the medical director. **Standing orders** are a type of indirect medical control. The standing orders direct the EMS providers to take specific action when they encounter different situations. Each EMS provider must know and follow the protocols developed by the system's medical director.

Figure 23-11 A physician medical director authorizes the providers in the service to deliver medical care in the field.

Direct or online medical control is provided by an EMS physician who can be reached by radio or telephone during a call. Local protocols define when EMS providers should give a radio report or obtain online medical direction. Once the crew has initiated any immediate, urgent care according to standing orders, the online medical control physician is contacted. The crew provides a report that describes the patient's condition and any treatment provided. The physician will either confirm or modify the proposed treatment plan; he or she may also prescribe special orders that the EMS providers are to follow for that patient.

The medical director is the ongoing working liaison for the medical community, hospitals, and the EMS providers. If treatment problems arise or different procedures should be considered, the medical director must be consulted for a decision and directions. The medical director also determines and approves the continuing education and training courses that are required to ensure that proper training standards are met.

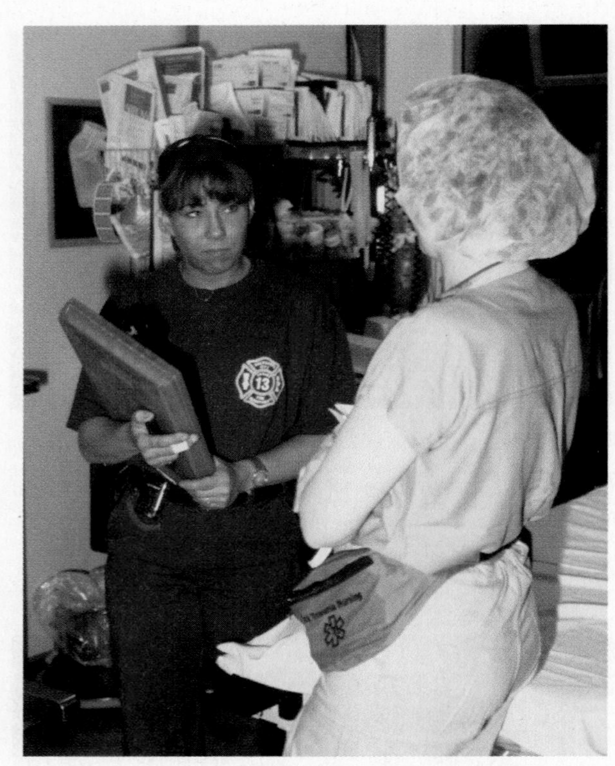

Figure 23-12 Communication is a key component of quality patient care.

Hospital Personnel

When EMS providers transport a patient to a hospital for additional care, they must transfer the care of the patient to hospital personnel. A good working relationship between EMS providers and other members of the health care team helps ensure good patient care (◄ Figure 23-12). For example, patient care reports must be clear and complete so hospital personnel know what has been done and what the patient's reaction to treatment has been. An understanding of the hospital procedures and policies will also help EMS providers who may be required to assist hospital personnel during the transfer.

Suggested Contents of a First Responder Life Support Kit

Patient Examination Equipment

1 flashlight

Personal Safety Equipment

5 pairs of gloves

5 face masks

Resuscitation Equipment

1 mouth-to-mask resuscitation device

1 portable hand-powered suction device

1 set oral airways

1 set nasal airways

Bandaging and Dressing Equipment

10 gauze-adhesive strips 1″

10 gauze pads 4″ x 4″

5 gauze pads 5″ x 9″

2 universal trauma dressings 10″ x 30″

1 occlusive dressing for sealing chest wounds

4 conforming gauze rolls 3″ x 5 yd

4 rolls 4½″ x 5 yd

6 triangular bandages

1 adhesive tape 2″

1 burn sheet

Patient Immobilization Equipment

2 (each) cervical collars: small, medium, large or 2 adjustable cervical collars

3 rigid conforming splints (SAM™ splints) **or**

1 set air splints for arm and leg **or**

2 (each) cardboard splints 18″ and 24″

Extrication Equipment

1 spring-loaded center punch

1 pair heavy leather gloves

Miscellaneous Equipment

2 blankets (disposable)

2 cold packs

1 bandage scissors

Other Equipment:

1 set personal protective clothing (helmet, eye protection, EMS jacket)

1 reflective vest

1 fire extinguisher (5 lb. ABC dry chemical)

1 *Emergency Response Guidebook*

6 fusees

1 binoculars

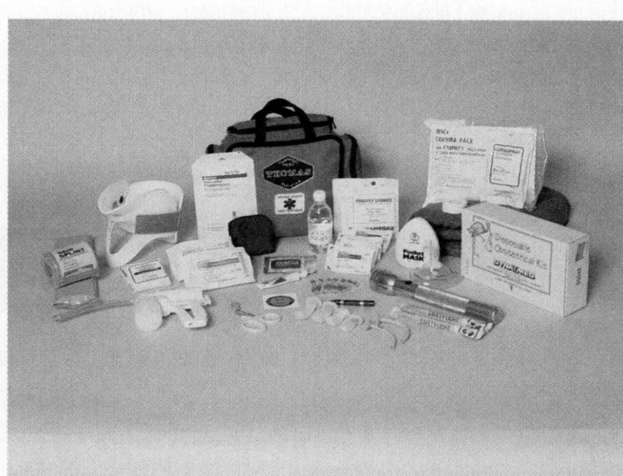

Wrap-Up

Ready for Review

This chapter explored some reasons for a fire department to offer EMS. Most fire departments are responsible for delivering some emergency medical services. EMS fits in with the fire department's mission of saving lives and protecting property and builds positive public relations with the community. EMS care is offered at both basic and advanced life support levels. First responders and EMT-Basics complete basic life support training programs, while EMT-Intermediates and EMT-Paramedics must complete advanced life support training programs. In combination EMS delivery systems, the fire department provides medical first response and another agency provides EMS transport. The fire department may also supply both the medical first response and the EMS transport. The interactions an EMS provider has with patients, the medical director, and hospital personnel are very important. By working hard to maintain good relations with these groups, an EMS provider can help ensure good care for all patients.

Chief Concepts

- EMS fits the fire department's mission to save lives and protect property.
- Most fire departments are involved in providing some emergency medical services.
- EMS delivers a positive customer service.
- The two levels of emergency medical services are basic life support and advanced life support.
- The levels of training for basic life support providers are First Responder and EMT-Basic (EMT-B).
- The levels of training for advanced life support providers are EMT-Intermediate (EMT-I) and EMT-Paramedic (EMT-P).
- Local government decides which agency (or agencies) will provide EMS.
- Multiple factors, including entry requirements, communications, and equipment maintenance, make a fire department suitable for providing EMS.
- In combination EMS systems, the fire department provides medical first response and another agency provides transport.
- In fire department EMS systems, both medical first response and transport are provided by the fire department.
- EMS providers interact with a wide range of citizens and health care providers.

Hot Terms

Advanced life support (ALS) Advanced lifesaving procedures, such as cardiac monitoring, administration of IV fluids and medications, and use of advanced airway adjuncts.

Basic life support (BLS) Noninvasive emergency lifesaving care that is used to treat airway obstruction, respiratory arrest, or cardiac arrest.

Combination EMS system In this system, the fire department provides medical first response, while another agency transports the patient to the hospital emergency room.

Emergency Medical Technician (EMT)-Basic EMT-Basics account for most of the EMS providers in the country. EMT-Basics have training in basic emergency care skills, including oxygen therapy, bleeding control, cardiopulmonary resuscitation (CPR), automated external defibrillation, use of basic airway devices, and assisting patients with certain medications.

Emergency Medical Technician (EMT)-Intermediate EMT-Intermediate personnel can perform limited procedures that usually fall between those provided by an EMT-Basic and those provided by an EMT-Paramedic, including intravenous (IV) therapy, interpretation of cardiac rhythms, defibrillation, and airway intubation.

Emergency Medical Technician (EMT)-Paramedic An EMT-Paramedic has the highest level of training in EMS, including cardiac monitoring, administering drugs, inserting advanced airways, manual defibrillation, and other advanced assessment and treatment skills.

Fire department EMS system In this system, the fire department provides both the medical first response and transportation of the patient to the hospital emergency room.

First responder The first trained individual to arrive at the scene of an emergency to provide initial medical assistance.

Medical control Physician instructions that are given directly by radio or indirectly by protocol/guidelines as authorized by the medical director of the service.

Medical director The physician providing direction for patient care activities in the prehospital setting.

Standing orders Written documents, signed by the system's medical director, that outline specific directions, permissions, and sometimes prohibitions regarding patient care; also called protocols.

Fire Fighter in Action

Your department operates engine and ladder companies, as well as medic units. All fire fighters are required to be trained and certified at the EMT-Basic level when hired and must maintain this certification as a condition of employment. Fire fighters trained as paramedics are assigned to the medic units. Patients who require ALS treatment are transported in the medic units, while BLS patients are transported by a private ambulance company.

1. You are assigned to an engine company. What level of emergency medical care can this unit provide?
   - A. ALS
   - B. BLS
   - C. First responder
   - D. ALS transport
2. You respond to a medical call for a patient who requires advanced life support treatment and arrive at the same time as the medic unit. Your role is to:
   - A. do nothing—only paramedics can provide advanced life support.
   - B. assist the paramedics by performing BLS functions.
   - C. start an IV and administer drugs as instructed by the medical director.
   - D. provide ALS treatment under the direction of the paramedics.
3. Your engine company is first to arrive at the scene of a one-car motor crash. The driver of the car is unconscious and appears to be seriously injured. What do you do?
   - A. Evaluate the driver's condition and provide BLS treatment until the medic unit arrives.
   - B. Work on removing the driver from the car, but do not treat the patient because you cannot provide ALS treatment.
   - C. Wait until the paramedics arrive before doing anything.
   - D. Extricate the driver and call for a private ambulance to transport the driver to the hospital.
4. Two passengers who were in the car are out of the vehicle and walking around. They appear to have minor cuts and bruises. The medic unit arrives and the paramedics begin treating the driver. What do you do?
   - A. Assist the paramedics who are treating the driver.
   - B. Begin treating the passengers who only require BLS care.
   - C. Call for a private ambulance to treat the passengers.
   - D. Call the medical director and ask for instructions.

www.FireFighter.jbpub.com

Chapter Pretests
Interactivities
Hot Term Explorer
Web Links
Review Manual
FireLearn

www.FireFighter.jbpub.com

Emergency Medical Care

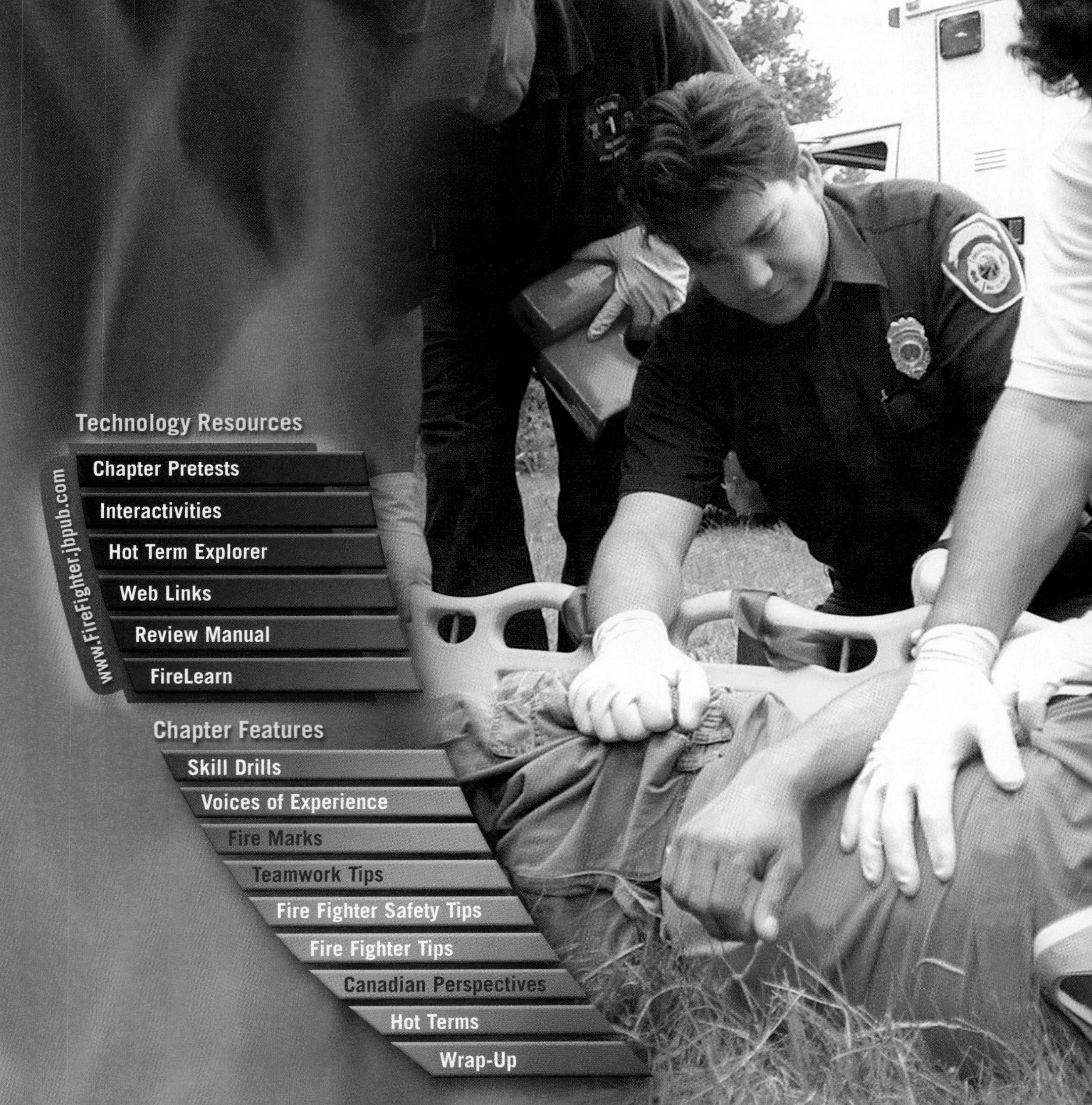

Chapter 24

NFPA 1001 Standard

4.3 Emergency Medical Care. Minimum emergency medical care performance capabilities for entry level personnel shall be developed and validated by the authority having jurisdiction to include infection control, CPR, bleeding control, and shock management.

Knowledge Objectives

After studying this chapter, you will be able to:

- Describe the steps needed to provide infection control for victims and for fire fighters.
- Describe the steps needed to perform cardiopulmonary resuscitation (CPR) on adult, child, and infant victims.
- Describe the steps used to manage shock.
- Explain the steps needed to control external bleeding.
- Discuss triage at a mass casualty incident.
- Describe how to assure safety at emergency medical services incidents.

Skills Objectives

After studying this chapter, you will be able to:

- Perform the head tilt-chin lift technique.
- Perform jaw-thrust technique.
- Clear the airway using finger sweeps.
- Perform the recovery position.
- Perform mouth-to-mask rescue breathing.
- Perform mouth-to-barrier rescue breathing.
- Perform mouth-to-mouth rescue breathing.
- Perform the Heimlich Maneuver.
- Treat airway obstruction in an unconscious adult.
- Treat airway obstruction in a conscious infant.
- Treat airway obstruction in an unconscious infant.
- Perform external chest compressions on an adult.
- Perform one-rescuer adult CPR.
- Perform two-rescuer adult CPR.
- Perform one-rescuer infant CPR.
- Perform one-rescuer child CPR.
- Treat shock.

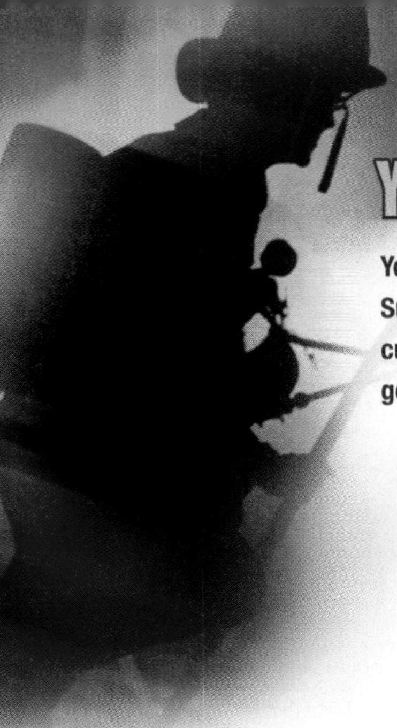

You Are the Fire Fighter

You have just completed your fire fighter training and are relaxing at home after a busy shift. Suddenly Mr. Barnes, your retired neighbor, runs over to your house. He tells you that his wife has cut herself while preparing dinner. "Please come and help her," he pleads. You grab your emergency kit and hurry next door.

1. ***Explain what is meant by the statement, "Your training as a fire fighter is something that you always carry with you."***
2. ***Why is it important for fire fighters to be trained in basic emergency medical care?***

Introduction

This chapter covers the basic emergency medical care skills included in NFPA 1001, *Standard for Fire Fighter Professional Qualifications*. Because fire fighters encounter ill or injured people on the job, a basic knowledge of these skills is important. This chapter covers infection control, cardiopulmonary resuscitation, controlling bleeding, and shock. This chapter emphasizes the importance of assuring your safety, the safety of other rescuers, the safety of victims, and safety of bystanders at the scene of emergency medical incidents. Finally, this chapter describes the concept of triage used at the scene of mass casualty incidents to provide the best treatment to large numbers of victims.

The knowledge and skills covered in this chapter are not meant to replace or substitute formal courses required for CPR, fire fighter, or emergency medical technician certification. Most states have training and certification requirements for those people who respond to medical emergencies. It is important for you to follow the requirements imposed by your state regulations and local ordinances.

Infection Control

In recent years, the **acquired immunodeficiency syndrome (AIDS)** epidemic and the growing concern about tuberculosis and hepatitis have increased awareness of infectious diseases. Some understanding of the most common infectious diseases is important so you can protect yourself from unnecessary exposure to these diseases and so you do not become unduly alarmed about them.

Federal regulations require all health care workers, including fire fighters, to assume that all victims in all settings are potentially infected with the viruses that can lead to AIDS (**human immunodeficiency virus [HIV]**) and hepatitis B (**hepatitis B virus [HBV]**), hepatitis C (**hepatitis C virus [HCV]**), or other bloodborne **pathogens**. Regulations require that all health care workers use protective equipment to prevent possible exposure to blood and bodily fluids of victims. This concept is known as **body substance isolation (BSI)**.

HIV is transmitted by direct contact with infected blood, semen, or vaginal secretions. There is no scientific documentation that the virus is transmitted by contact with sweat, saliva, tears, sputum, urine, feces, vomitus, or nasal secretions, unless these fluids visibly contain blood.

HBV is also spread by direct contact with infected blood. Fire fighters should follow the **universal precautions** described

Fire Fighter Safety Tips

The CDC recommends that all health care workers use the following universal precautions:

- Always wear gloves when handling victims, and change gloves after contact with each victim. Wash your hands immediately after removing gloves. (Note that leather gloves are not considered to be safe—leather is porous and traps fluids.)
- Always wear protective eyewear or a face shield when you anticipate that blood or other bodily fluids may splatter. Wear a gown or apron if you anticipate splashes of blood or other bodily fluids, such as those that occur with childbirth and major trauma.
- Wash your hands and other skin surfaces immediately and thoroughly if they become contaminated with blood and other bodily fluids. Change contaminated clothes and wash exposed skin thoroughly.
- Place used needles directly in a puncture resistant container designed for "sharps."
- Even though saliva has not been proven to transmit HIV, you should use a face shield, pocket mask, or other airway adjunct if the victim needs resuscitation.

A 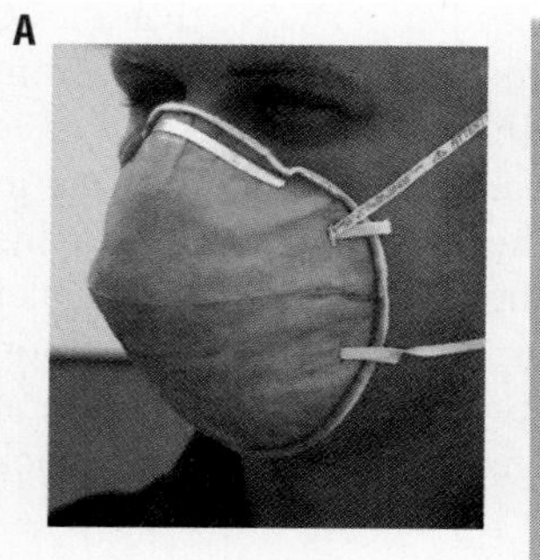

B 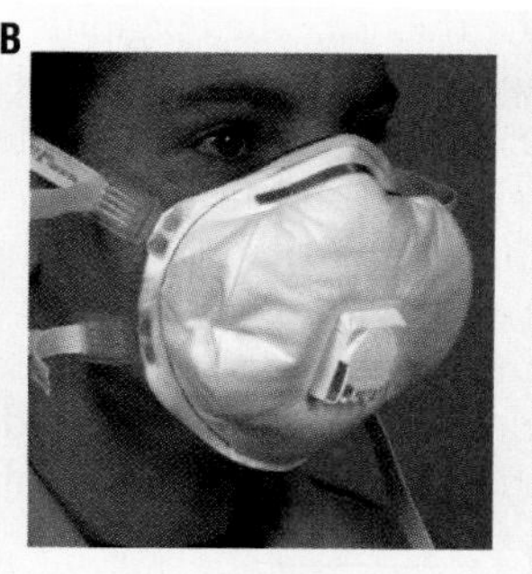

Figure 24-1 Two types of HEPA respirators. HEPA respirators reduce the transmission of airborne diseases.

in the next section to reduce their chance of contracting hepatitis B and C. Check with your medical director about receiving injections of the hepatitis B vaccine to protect you against this infection. This vaccine should be made available to you at no cost.

Tuberculosis is also becoming more common, and the presence of drug-resistant strains makes this disease dangerous. Tuberculosis is spread through the air when an infected person coughs or sneezes. Wear a face mask or a high-efficiency particulate air (HEPA) respirator, and place an oxygen mask on the victim to minimize your exposure (▲ Figure 24-1). If no oxygen mask is available, place a face mask on the victim. Fire fighters should have a skin test for tuberculosis every year.

Severe Acute Respiratory Syndrome (SARS) is a recently identified pneumonia-like illness that is fatal in 4% of victims. SARS is spread through water droplets in the air and by direct contact. It is important for you to keep up to date and follow the recommendations of the Centers for Disease Control (CDC) and your medical director to minimize the chance of spreading or contracting this disease.

Universal Precautions

You will not always be able to tell whether a victim's bodily fluids contain blood. Therefore, the CDC recommends that all health care workers use universal precautions, based on the assumptions that all victims are potential carriers of bloodborne pathogens.

Federal agencies, such as Occupational Safety and Health Administration (OSHA), and state agencies, such as state public health departments, have regulations about body substances isolation. Because these regulations are constantly changing, it is important for your department to keep up to date on these regulations.

Immunizations

Certain immunizations are recommended for emergency medical care providers, including tetanus and hepatitis B vaccine. Tuberculin testing is also recommended. Your medical director can determine the immunizations and tests needed for members of your department.

Fire Fighter Safety Tips

Simple, portable safety equipment can help prevent injuries and illnesses.

- Medical gloves, masks, and eye protection prevent the spread of infectious diseases.
- Brightly colored clothing or vests make you more visible to traffic in the daytime; reflective striping or vests make you more visible in the dark.
- Heavy gloves can help prevent cuts at a motor vehicle accident scene.
- A hard hat or helmet is needed when you are at an industrial or motor vehicle accident scene.

Some situations may require additional safety equipment. Do not hesitate to call for additional equipment as needed. See NFPA 1581, *Standard on Fire Department Infection Control Program,* for further information.

Airway

This section introduces two important lifesaving skills: airway care and rescue breathing. Victims must have an open airway passage and must maintain adequate breathing to survive. By learning and practicing the simple skills in this section, you can often be the difference between life and death for a victim.

A review of the major structures of the respiratory system is needed before you practice airway and rescue breathing skills. Once you learn the functions of these structures, you will be a long way down the road to becoming proficient in performing these skills.

The skills of CPR are as easy as A, B, C—where "A" is airway, "B" is breathing, and "C" is circulation. Because you must assess and correct the airway before you turn your attention to the victim's breathing status, it is helpful to remember the ABC sequence. A mnemonic that will be used throughout this section is "check and correct." By using this two-step sequence for each of the ABCs, you will be able to remember the steps needed to check and correct problems involving the victim's airway, breathing, and circulation.

The "A," or airway section, presents airway skills, including how to check the level of consciousness and manually correct a blocked airway by using the head tilt-chin lift and jaw-thrust techniques. You must check the victim's airway for foreign objects. If you find foreign objects, you must correct the problem by removing the objects.

The "B," or breathing section, describes how to check victims to determine whether or not they are breathing adequately. You will learn how to correct breathing problems by using three rescue breathing techniques: mouth-to-mask, mouth-to-barrier device, and mouth-to-mouth.

Anatomy and Function of the Respiratory System

To maintain life, all organisms must receive a constant supply of certain substances. In human beings, these basic life-sustaining substances are food, water, and oxygen. A person can live several weeks without food because the body can use nutrients it has stored. Although the body does not store as much water, it is possible to live several days without fluid intake. But lack of oxygen, even for a few minutes, can result in irreversible damage and death.

The most sensitive cells in the human body are in the brain. If brain cells are deprived of oxygen and nutrients for 4 to 6 minutes, they begin to die. Brain death is followed by the death of the entire body. Once brain cells have been destroyed, they cannot be replaced. This is why it is so important to understand the anatomy and function of the respiratory system.

The main purpose of the respiratory system is to provide oxygen and to remove carbon dioxide from the red blood cells as they pass through the **lungs**. This action forms the basis for your study of the lifesaving skill of CPR.

The parts of the body used in breathing are shown in ► Figure 24-2, and include the mouth, nose, throat, **trachea** (windpipe), lungs, **diaphragm** (the dome-shaped muscle between the chest and the abdomen), and numerous chest muscles. Air enters the body through the nose and mouth. In an unconscious victim lying on his or her back, the passage of air through both nose and mouth may be blocked by the tongue ▼ Figure 24-3. The tongue is attached to the lower jaw. When a person loses consciousness, the jaw relaxes and the tongue falls backward into the rear of the mouth, effectively blocking the passage of air from both nose and mouth to the lungs. A partially blocked airway often produces a snoring sound. At the bottom of the throat are two passages, the **esophagus** (the food tube) and the trachea. The epiglottis is a thin flapper valve that allows air to enter the trachea but prevents food or water from doing so. Air passes from the throat to the **larynx** (voice box), which can be seen externally as the Adam's apple in the neck. Below the trachea, the **airway** divides into the **bronchi** (two large tubes). The bronchi branch into smaller and smaller airways in the lungs. The lungs are located on either side of the heart and are protected by the sternum at the front and by the rib cage at the sides and back ► Figure 24-4.

The airways branch into smaller and smaller passages, which end as tiny air sacs called **alveoli**. The alveoli are surrounded by very small blood vessels, called **capillaries**. The actual exchange of gases takes place across a thin membrane

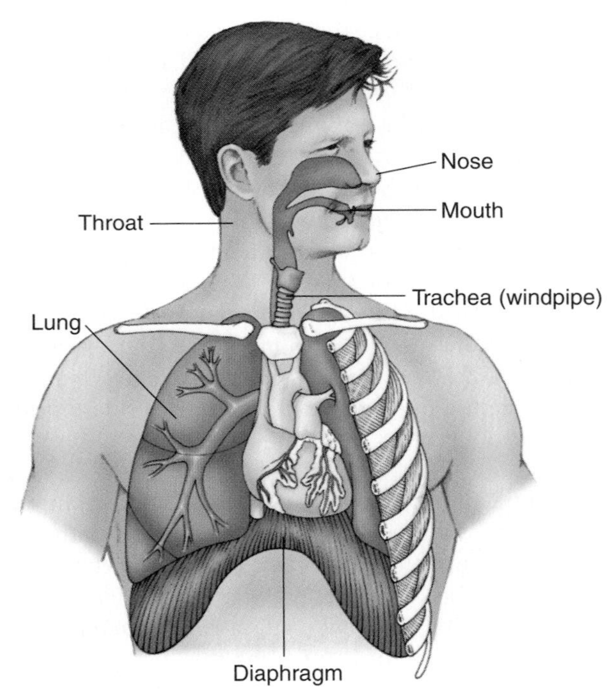

Figure 24-2 The respiratory system.

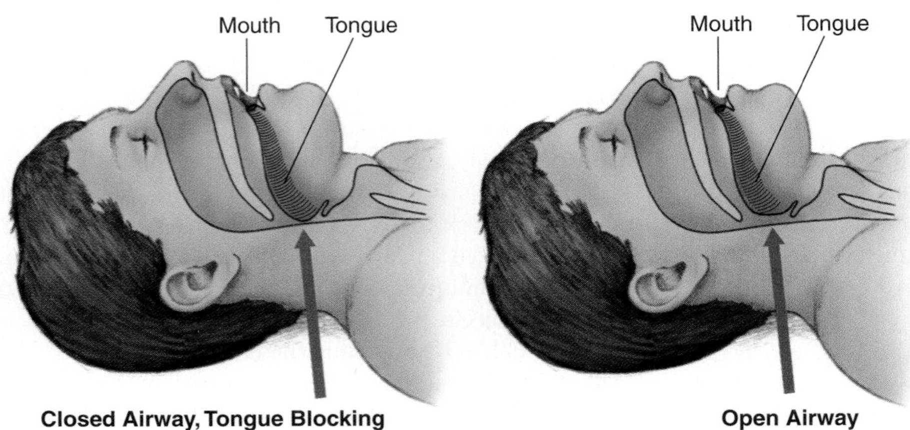

Figure 24-3 In an unconscious patient, the airway may be blocked by the tongue (left). An open airway is shown on the right.

that separates the capillaries of the circulatory system from the alveoli of the lungs (▶ Figure 24-5). The incoming oxygen passes from the alveoli into the blood, and the outgoing carbon dioxide passes from the blood into the alveoli.

The lungs consist of soft, spongy tissue with no muscles. Therefore, movement of air into the lungs depends on movement of the rib cage and the diaphragm. As the rib cage expands, air is drawn into the lungs through the trachea. The diaphragm, a muscle that separates the abdominal cavity from the chest, is dome-shaped when it is relaxed. When the diaphragm contracts, it flattens and moves downward. This action increases the size of the chest cavity and draws air into the lungs through the trachea. In normal breathing, the combined actions of the diaphragm and the rib cage automatically produce adequate inhalation and exhalation (▼ Figure 24-6).

Fire Fighter Tips

Infants and Children

- The respiratory system structures in children and infants are smaller than in adults. Thus, the air passages of children and infants may be more easily blocked by secretions or by foreign objects.
- In children and infants, the tongue is proportionally larger than in adults. Thus, the tongue of these smaller victims is more likely to block the airway than it would in an adult victim.
- Because the trachea of an infant or child is more flexible than an adult's, it is more likely to become narrowed or blocked.
- The head of a child or an infant is proportionally larger than an adult's. You will have to learn slightly different techniques for opening a child's airway.
- Children and infants have smaller lungs than adults. You need to give them smaller breaths when you perform rescue breathing.
- Most children and infants have healthy hearts. When a child or infant suffers cardiac arrest (stoppage of the heart), it is usually because the victim has a blocked airway or has stopped breathing, not because of a problem with the heart.

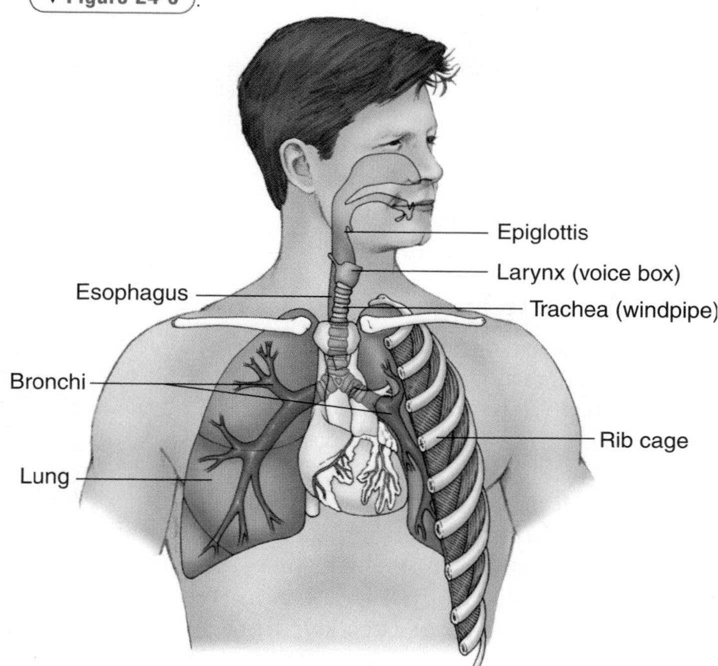

Figure 24-4 Anatomy of the respiratory system.

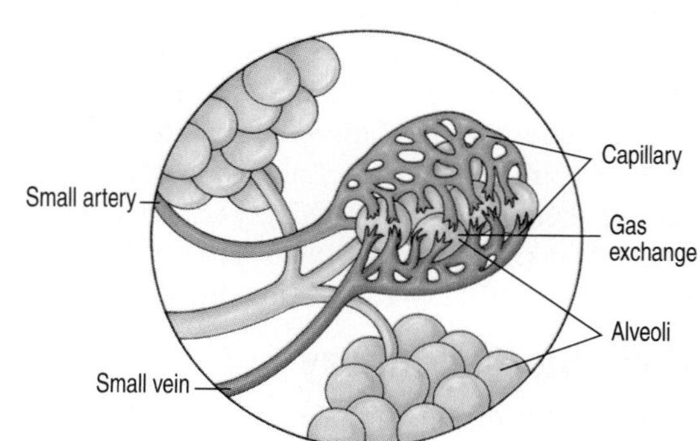

Figure 24-5 The exchange of gases occurs in the alveoli of the lungs.

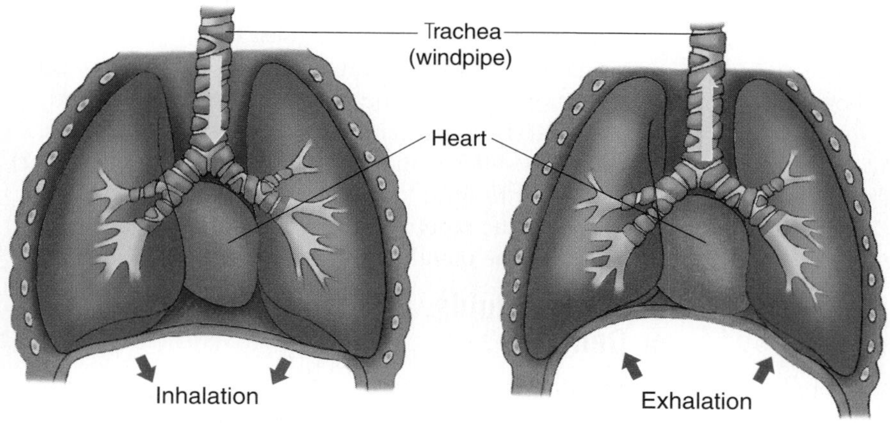

Figure 24-6 Normal mechanical act of breathing.

Fire Fighter Safety Tips

Use BSI techniques whenever you may be in contact with body secretions that might contain blood.

"A" Is for Airway

The victim's airway is the pipeline that transports life-giving oxygen from the air to the lungs and transports the waste product, carbon dioxide, from the lungs to the air. In healthy individuals, the airway automatically stays open. An injured or seriously ill person, however, may not be able to protect the airway, and it may become blocked. If a victim cannot protect his or her airway, you must take certain steps to check the condition of the victim's airway and correct the problem to keep the victim alive.

Check for Responsiveness

The first step in assessing a victim's airway is to check the victim's level of responsiveness. When you first approach a victim, you can immediately find out whether the victim is conscious or unconscious by asking, "Are you okay? Can you hear me?" If you get a response, you can assume that the victim is conscious and has an open airway. If there is no response, grasp the victim's shoulder and gently shake the victim (► Figure 24-7). Then, repeat your question. If the victim still does not respond, you can assume the victim is unconscious and that you will need more help. Before doing anything for the victim, call 9-1-1 ("phone first") if the EMS system has not already been activated, especially if you are the only rescuer. Position the victim by supporting the victim's head and neck and placing the victim on his or her back.

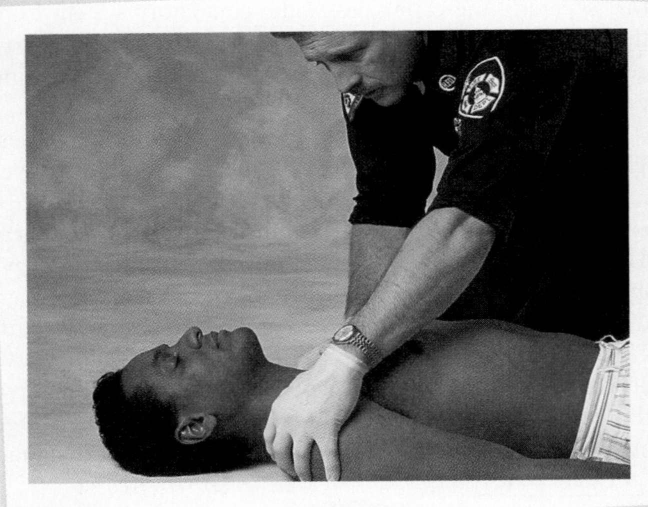

Figure 24-7 Establish the level of consciousness.

Correct the Blocked Airway

An unconscious victim's airway may be blocked because the tongue has dropped back and is obstructing it. In this case, simply opening the airway may enable the victim to breathe spontaneously.

Head Tilt-Chin Lift Technique

To open the airway, place one hand on the forehead and place the fingers of the other hand under the bony part of the lower jaw near the chin. Push down on the forehead and lift up and forward on the chin. Be certain you are not merely pushing the mouth closed when you use this technique. This method of opening the airway is called the **head tilt-chin lift technique**. Do not use this method if a neck injury is suspected.

To perform the head tilt-chin lift, follow the steps in (► Skill Drill 24-1).

1. Place one hand on the victim's forehead and apply firm pressure. Move the victim's head back as far as possible. **(Step 1)**
2. Place the tips of your fingers under the lower jaw. Move the victim's head back as far as possible. **(Step 2)**
3. Lift the chin forward. **(Step 3)**

Jaw-Thrust Technique

The **jaw-thrust technique** or maneuver is another way to open a victim's airway. If the victim was injured in a fall, diving mishap, or automobile accident, do not tilt the head to open the airway. If the victim has a neck injury, using the head tilt method may cause permanent paralysis. If you suspect a neck injury, use the jaw-thrust technique. Open the airway by placing your fingers under the angles of the jaw and pushing upward. At the same time, use your thumbs to open the mouth slightly. The jaw-thrust technique should open the airway without extending the neck.

To perform the jaw-thrust technique, follow the steps in (► Skill Drill 24-2).

1. Place the victim on his or her back and kneel at the top of the victim's head. Place your fingers behind the angles of the victim's lower jaw and move the jaw forward with firm pressure. **(Step 1)**
2. Tilt the head backward to a neutral or slight **sniffing position**, where the head and chin are thrust slightly forward. Do not extend the cervical spine in a victim who has suffered an injury to the head or neck. **(Step 2)**
3. Use your thumbs to pull the victim's lower jaw down, opening the mouth enough to allow breathing through the mouth and nose. **(Step 3)**

Check for Fluids, Foreign Bodies, or Dentures

After you have opened the victim's airway by using either the head tilt-chin lift or the jaw-thrust technique, look in the

24-1 Skill Drill

Head Tilt-Chin Lift Technique

Place one hand on the victim's forehead and apply firm pressure. Move the victim's head back as far as possible.

Place the tips of your fingers under the lower jaw. Move the victim's head back as far as possible.

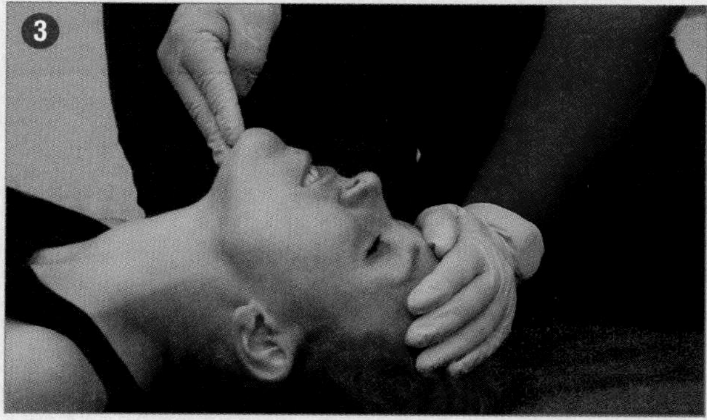

Lift the chin forward.

victim's mouth to see if anything is blocking the victim's airway. Potential blocks include secretions, such as vomitus, mucus, or blood; foreign objects, such as candy, food, or dirt; and dentures or false teeth that may have become dislodged and are blocking the victim's airway (► Figure 24-8). If you find anything in the victim's mouth, remove it.

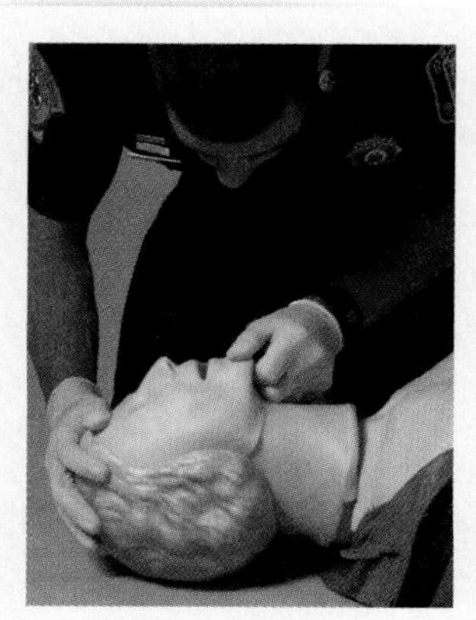

Figure 24-8 Check for fluids, foreign bodies, and dentures.

Correct the Airway Using Finger Sweeps or Suction

Vomitus, mucus, blood, and foreign objects must be cleared from the victim's airway. This can be done by using finger sweeps or moving the victim into the recovery position.

24-2 Skill Drill

Jaw-Thrust Technique

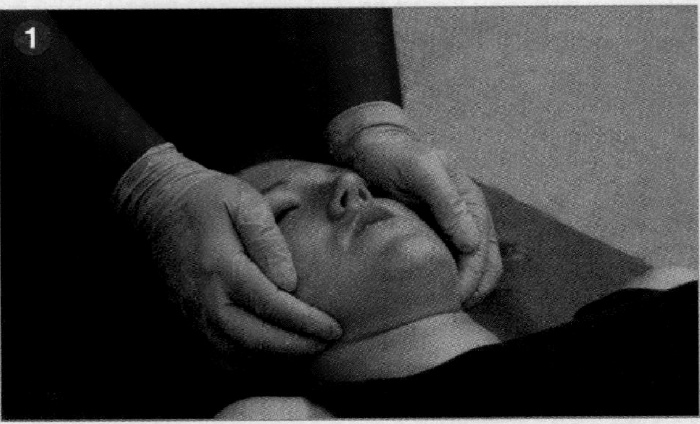

Place your fingers behind the angles of the victim's lower jaw and move the jaw forward.

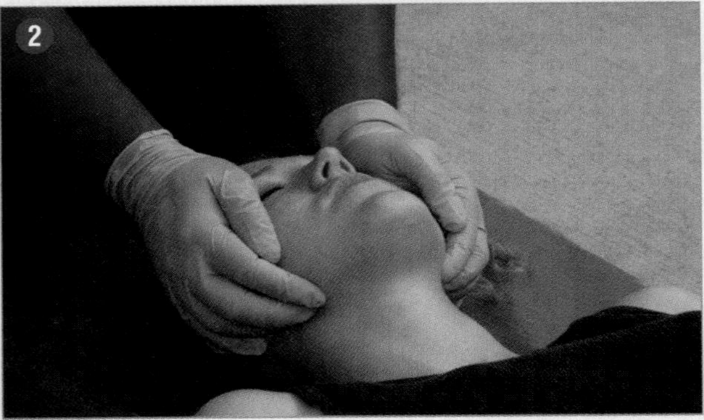

Tilt the head backward to a neutral or slight sniffing position.

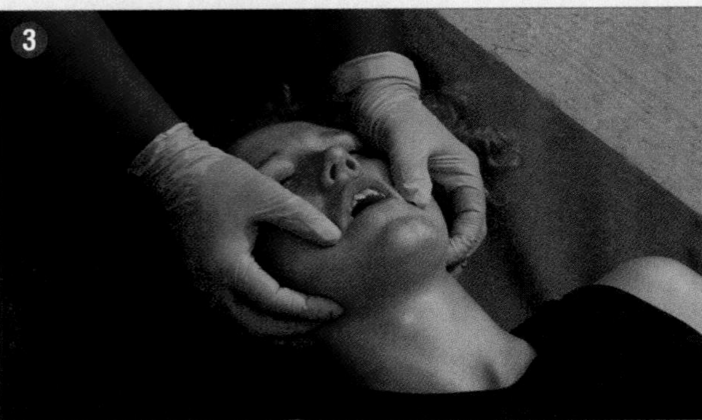

Open the mouth by using your thumbs to pull the victim's lower jaw down.

Finger Sweeps

Finger sweeps can be done quickly and require no special equipment except a set of medical gloves.

To perform a finger sweep, follow the steps in **► Skill Drill 24-3**.

1. Turn the victim's head to one side. **(Step 1)**
2. Insert your gloved index and middle fingers into the victim's mouth. **(Step 2)**
3. Curve your finger into a C-shape and sweep it from one side of the back of the mouth to the other. **(Step 3)**
4. Repeat the finger sweeps until you have removed all the foreign material in the victim's mouth.

Maintain the Airway

If a victim is unable to keep the airway open, you must open the airway manually. Unconscious victims will not be able to keep their airway open. You can continue to keep their airway open by using the head tilt-chin lift or jaw-thrust technique. If the victim is breathing adequately, you can keep the airway open by placing the victim in the recovery position.

Recovery Position

If an unconscious victim is breathing and has not suffered trauma, one way to keep the airway open is to place the vic-

24-3 Skill Drill

Clearing the Airway Using Finger Sweeps

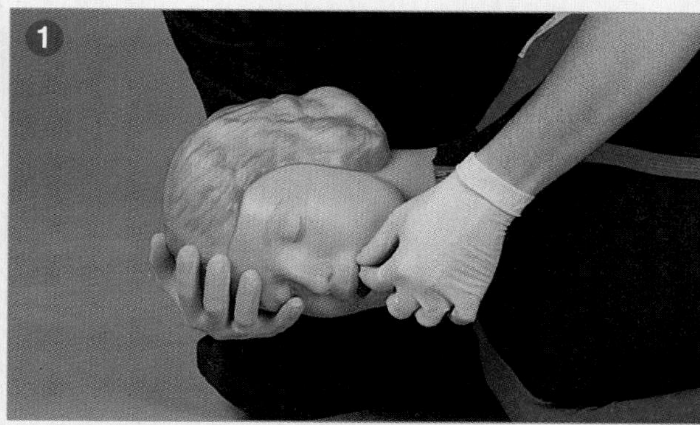

Turn the victim onto his or her side.

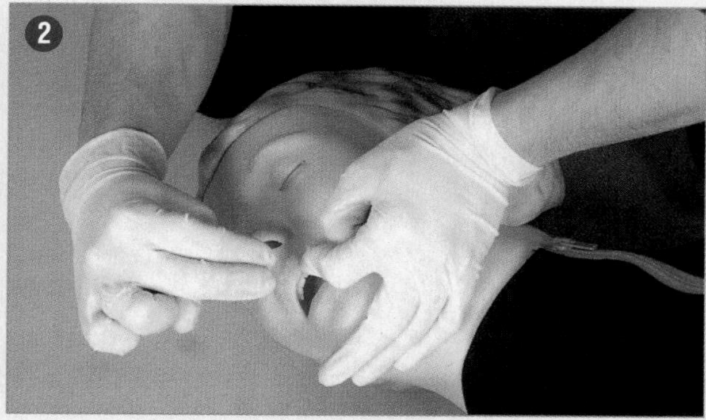

Insert your fingers into the victim's mouth.

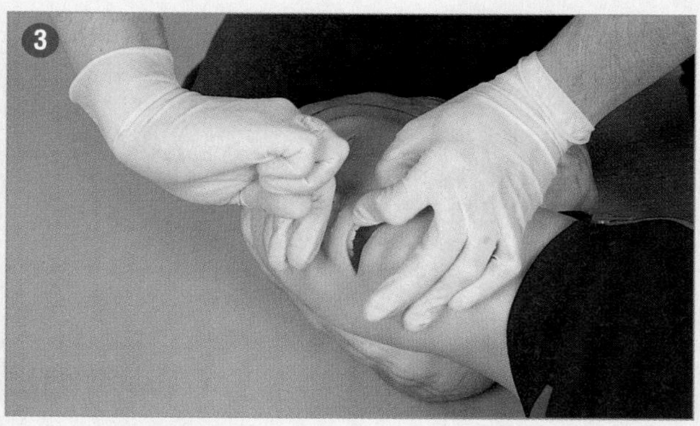

Curve your fingers into a C-shape and sweep it from one side of the back of the mouth to the other.

tim in the recovery position. The recovery position helps keep the victim's airway open by allowing secretions to drain out of the mouth instead of draining into the trachea. It also uses gravity to help keep the victim's tongue and lower jaw from blocking the airway.

To place a victim in the recovery position, follow the steps in (► Skill Drill 24-4).

1. Carefully roll the victim onto one side as you support the victim's head. Roll the victim as a unit without twisting the body. You can use the victim's hand to help hold his or her head in a good position. **(Step 1)**
2. Place the victim's face on its side so any secretions drain out of the mouth. **(Step 2)**
3. The head should be in a position similar to the tilted back position of the head tilt-chin lift technique. **(Step 3)**

"B" Is for Breathing

After you have checked and corrected the victim's airway, you should move on to check and correct the victim's breathing. To do this, you must understand the signs of adequate breathing, the signs of inadequate breathing, and the signs and causes of respiratory arrest.

24-4 Skill Drill

Recovery Position

Roll the victim as a unit without twisting the body.

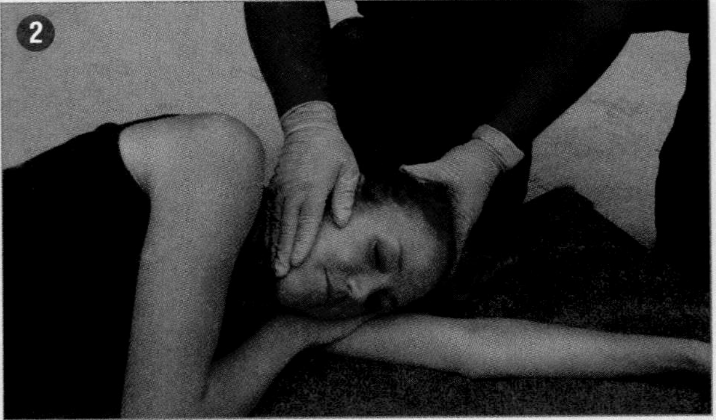

Place the victim's face on its side so any secretions drain out of the mouth.

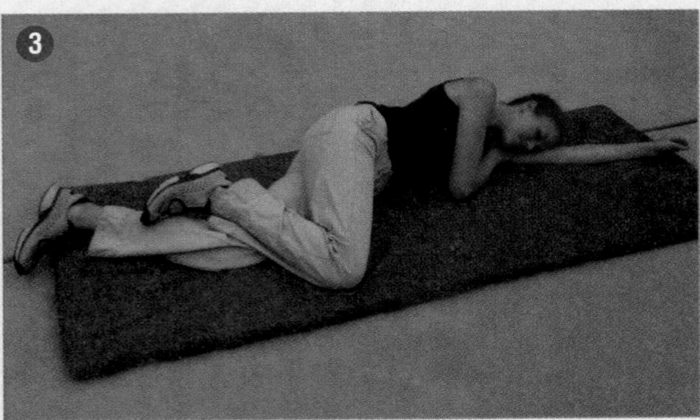

The head should be in a position similar to the tilted back position of the head tilt-chin lift technique.

Signs of Adequate Breathing

To check for adequate breathing, you must look, listen, and feel at the same time. If a victim is breathing adequately, you can look for the rise and fall of the victim's chest, listen for the sounds of air passing into or out of the victim's nose and mouth, and feel the air moving on the side of your face. Place the side of your face close to the victim's nose and mouth and watch the victim's chest. In this way, you can look for chest movements, listen for the sounds of air moving, and feel the air as it moves in and out of the victim's nose and mouth. A normal adult has a resting breathing rate of approximately 12 to 20 breaths per minute. Remember that one breath includes both an inhalation and an exhalation.

Signs of Inadequate Breathing

If a victim is breathing inadequately, you will detect signs of abnormal respirations. Noisy respirations, wheezing, or gurgling indicate partial blockage or constriction somewhere along the respiratory tract. Rapid or gasping respirations may indicate that the victim is not receiving adequate oxygen due to illness or injury. The victim's skin may be pale or even blue, especially around the lips or fingernail beds.

The most critical sign of inadequate breathing is **respiratory arrest** (total lack of respirations). This critical state is characterized by no chest movements, no breath sounds, and no air against the side of your face. In victims with severe **hypothermia** (when the internal body temperature falls below 95°F), respirations can be slowed (and/or shallow) to the point that the victim appears clinically dead.

There are many causes of respiratory arrest. By far the most common cause is heart attack, which claims more than 500,000 lives each year. Other major causes of respiratory arrest include:

- Mechanical blockage or obstruction caused by the tongue
- Vomitus, particularly in a victim weakened by an illness such as a stroke
- Foreign objects: teeth, dentures, balloons, marbles, pieces of food, or pieces of hard candy (especially in small children)
- Illness or disease such as heart attack or severe stroke
- Drug overdose
- Poisoning
- Severe loss of blood
- Electrocution by electrical current or lightning

Check for the Presence of Breathing

After establishing the loss of consciousness and opening the airway of the unconscious victim, check for breathing by looking, listening, and feeling (► Figure 24-9):

- Look for the rising and falling of the victim's chest.
- Listen for the sound of air moving in and out of the victim's nose and mouth.
- Feel for the movement of air on the side of your face and ear.

Continue to look, listen, and feel for at least 3 to 5 seconds; if you do not, you risk checking the victim between breaths and missing any signs of breathing that may be present. Your breathing check should take no more than 10 seconds. If there are no signs of breathing, proceed to the next step and correct the lack of breathing by beginning rescue breathing. If the victim is breathing adequately (about 12 to 20 times a minute), you can continue to maintain the airway and monitor the rate and depth of respirations to ensure adequate breathing continues.

Correct Breathing

You must breathe for any victim who is not breathing. As you perform rescue breathing, keep the victim's airway open by using the head tilt-chin lift method (or the jaw-thrust method for victims with head or neck injuries). To perform rescue breathing, blow your air into the victim's mouth. Pinch the victim's nose with your thumb and forefinger, take a deep breath, and blow slowly for 2 seconds (► Figure 24-10).

Use slow, gentle, sustained breathing and just enough breath to make the victim's chest rise. This minimizes the amount of air blown into the stomach. Remove your mouth, and allow the lungs to deflate. Breathe for the victim a second time. After these first two breaths, breathe once into the victim's mouth every 4 to 5 seconds. The rate of breaths should be 10 to 12 per minute for an adult.

Rescue breathing can be done by using a mouth-to-mask device, a barrier device, or just your mouth. The mouth-to-mask and barrier devices prevent you from putting your mouth directly on the victim's mouth. These devices should be available to you as a fire fighter.

If a rescue breathing device is not available, you must weigh the potential good to the victim against the limited chance that you will contract an infectious disease if you perform mouth-to-mouth rescue breathing.

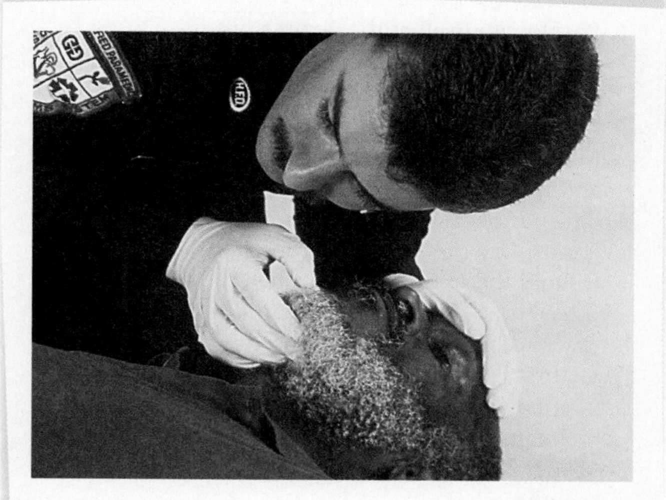

Figure 24-9 Check for breathing by looking, listening, and feeling.

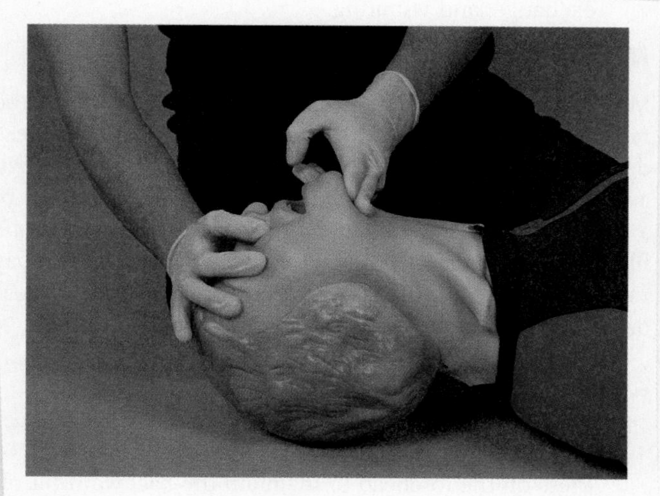

Figure 24-10 To perform rescue breathing, pinch the victim's nose with your thumb and forefinger.

Mouth-to-Mask Rescue Breathing

Your fire fighter life support kit should contain an artificial ventilation device that enables you to perform rescue breathing without mouth-to-mouth contact with the victim. This simple piece of equipment is called a mouth-to-mask ventilation device. A mouth-to-mask ventilation device consists of a mask that fits over the victim's face, a one-way valve, and a mouthpiece through which the rescuer breathes. It may also have an inlet port for supplemental oxygen and a tube between the mouthpiece and the mask. These devices are shown in ► Figure 24-11. Because mouth-to-mask devices prevent direct contact between you and the victim, they reduce the risk of transmitting infectious diseases.

To use a mouth-to-mask ventilation device for rescue breathing, follow the steps in ► Skill Drill 24-5.

1. Position yourself at the victim's head.
2. Use the head tilt-chin lift or jaw-thrust technique to open the victim's airway. **(Steps 1 and 2)**
3. Place the mask over the victim's mouth and nose. Make sure that the mask's nose notch is on the nose and not the chin. **(Step 3)**
4. Grasp the mask and the victim's jaw, using both hands. Use the thumb and forefinger of each hand to hold the mask tightly against the face. Hook the other three fingers of each hand under the victim's jaw and lift up to seal the mask tightly against the victim's face.
5. Maintain an airtight seal as you pull up on the jaw to maintain the proper head position.
6. Take a deep breath, then seal your mouth over the mouthpiece.
7. Breathe slowly into the mouthpiece for 2 seconds. Breathe until the victim's chest rises. **(Step 4)**
8. Monitor the victim for proper head position, air exchange, and vomiting.

Figure 24-11 Types of mouth-to-mask ventilation devices.

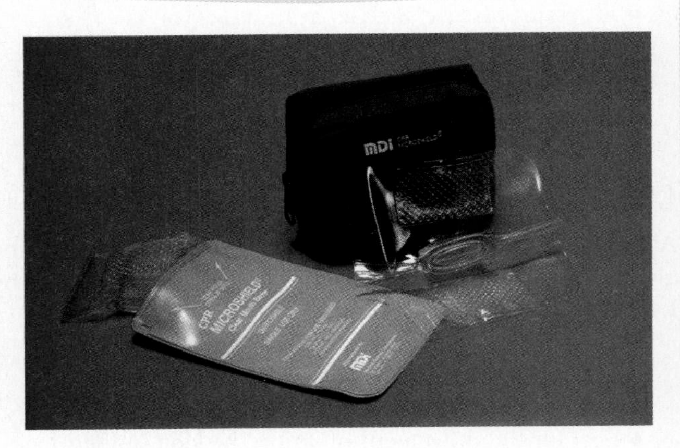

Figure 24-12 Barrier devices.

Mouth-to-Barrier Rescue Breathing

Mouth-to-barrier devices also provide a barrier between the rescuer and the victim ► Figure 24-12. Some of these devices are small enough to carry in your pocket. Although a wide variety of devices is available, most of them consist of a port or hole that you breathe into and a mask or plastic film that covers the victim's face. Some also have a one-way valve that prevents backflow of secretions and gases. These devices provide variable degrees of infection control.

To perform mouth-to-mouth rescue breathing with a barrier device, follow the steps in ► Skill Drill 24-6.

1. Open the airway with the head tilt-chin lift technique. **(Step 1)**
2. Press on the forehead to maintain the backward tilt of the head.
3. Place the barrier device over the victim's mouth.
4. Pinch the victim's nostrils together with your thumb and forefinger. **(Step 2)**
5. Keep the victim's mouth open with the thumb of whichever hand you are using to lift the victim's chin.
6. Take a deep breath, then make a tight seal by placing your mouth on the barrier device around the victim's mouth.
7. Breathe slowly into the victim's mouth for 2 seconds. Breathe until the victim's chest rises. **(Step 3)**
8. Remove your mouth and allow the victim to exhale passively. Check to see that the victim's chest falls after each exhalation.
9. Repeat this rescue breathing sequence 10 to 12 times per minute (one breath every 4 to 5 seconds) for an adult.

Mouth-to-Mouth Rescue Breathing

Mouth-to-mouth rescue breathing is an effective way of providing artificial ventilation for nonbreathing victims. It requires no equipment. However, because there is a some-

24-5 Skill Drill

Mouth-to-Mask Rescue Breathing

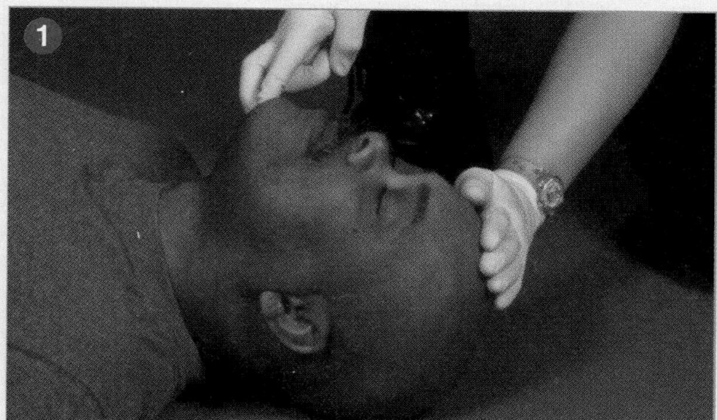
Open the airway using the head tilt-chin lift technique.

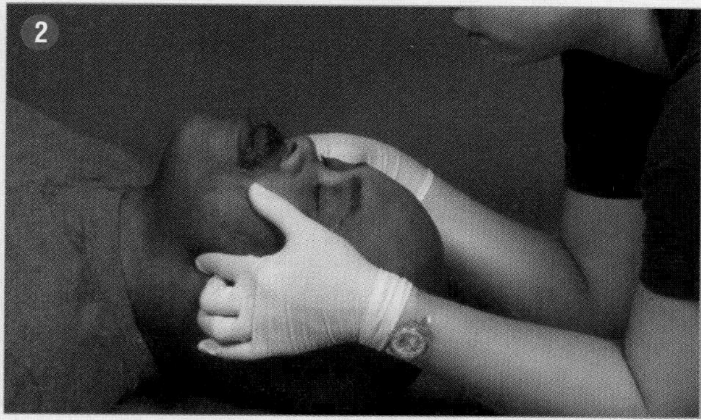
If there is a suspected head injury, open the airway using the jaw-thrust technique.

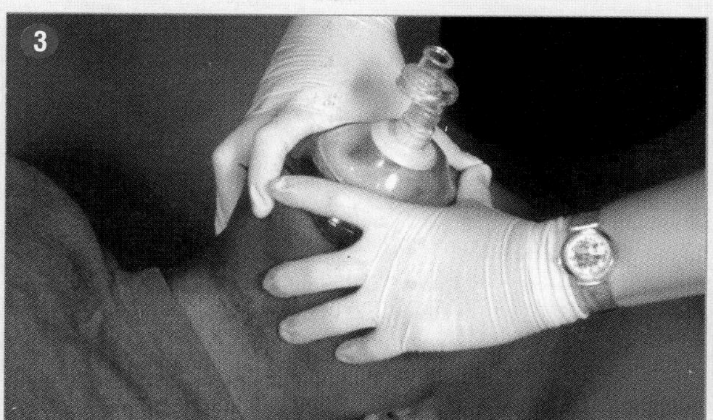
Seal the mask against the victim's face.

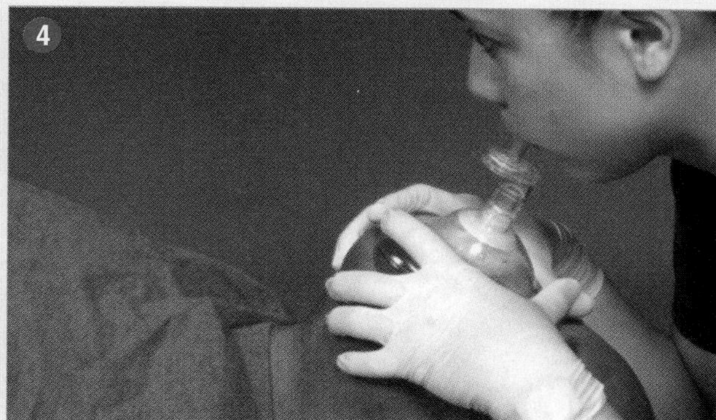
Breathe through the mouthpiece.

what higher risk of contracting a disease when using this method, you should use a mask or barrier breathing device if available. If a rescue breathing device is not available, you must weigh the potential good to the victim against the limited chance that you will contract an infectious disease from mouth-to-mouth rescue breathing.

To perform mouth-to-mouth rescue breathing, follow the steps in Skill Drill 24-7.

1. Open the airway with the head tilt-chin lift technique.
2. Press on the forehead to maintain the backward tilt of the head.
3. Pinch the victim's nostrils together with your thumb and forefinger.
4. Keep the victim's mouth open with the thumb of whichever hand you are using to lift the victim's chin.
5. Take a deep breath, and then make a tight seal by placing your mouth over the victim's mouth.
6. Breathe slowly into the victim's mouth for 2 seconds. Breathe until the victim's chest rises.
7. Remove your mouth and allow the victim to exhale passively. Check to see that the victim's chest falls after each exhalation.
8. Repeat this rescue breathing sequence 10 to 12 times per minute for adult victims and about 20 times per minute for children and infants.

24-6 Skill Drill

Mouth-to-Barrier Rescue Breathing

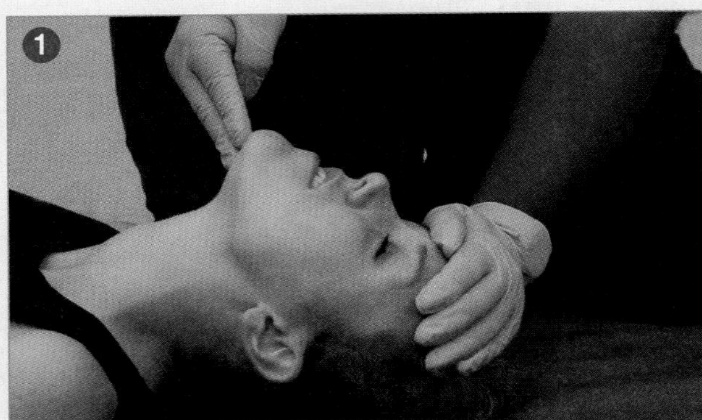
1 Open the airway using the head tilt-chin lift technique.

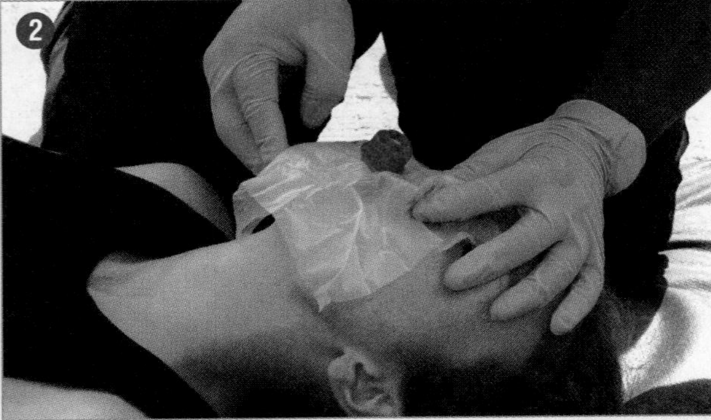
2 Place the barrier device over the victim's mouth. Pinch the victim's nostrils together.

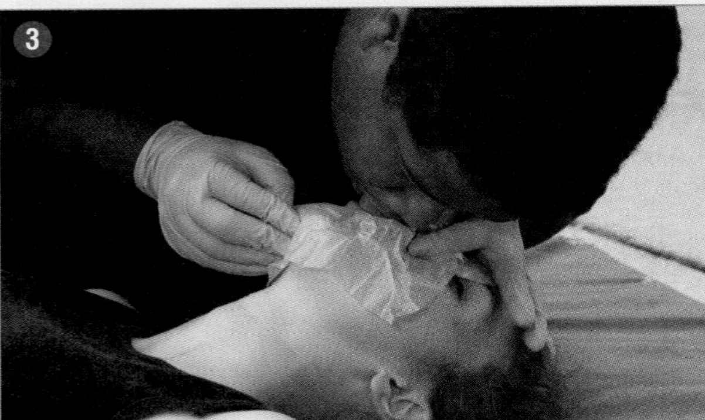
3 Perform rescue breathing.

Performing Rescue Breathing on Children and Infants

The "A" steps required to check and correct the victim's airway and the "B" steps needed to check and correct the victim's breathing are similar for adults, children, and infants. However, there are some differences. You must learn and practice the different airway and breathing sequences for children and infants.

Rescue Breathing for Children

For purposes of performing rescue breathing, a child is a person between one and eight years of age. The steps for determining responsiveness, checking and correcting airways, and checking and correcting a child's breathing are essentially the same as for an adult victim, but you should keep the following differences in mind:

1. Children are smaller, and you will not have to use as much force to open their airways and tilt their heads.
2. You should perform rescue breathing for 1 minute before activating the EMS system ("phone fast") if you are on the scene by yourself and EMS has not been notified.
3. Each rescue breath should be slightly shorter for a child, given over a period of 1 to $1^1/_2$ seconds instead of over a period of 2 seconds for an adult victim.

4. The rate of rescue breathing is slightly faster for children. Give 1 rescue breath every 3 seconds (about 20 rescue breaths per minute), instead of the adult rate of one rescue breaths every 4 to 5 seconds (about 12 rescue breaths per minute).

Rescue Breathing for Infants

If the victim is an infant (under 1 year of age), you must vary rescue breathing techniques slightly. Keep in mind that an infant is tiny and must be treated very gently. The steps in rescue breathing for an infant are as follows:

Airway

1. *Check* for responsiveness by gently shaking the infant's shoulder or tapping the bottom of the foot. If the infant is unresponsive, place the infant on his or her back and proceed to step 2.
2. *Correct* the airway if it is closed, by using the head tilt-chin lift technique. Do not tip the infant's head back too far because this may block the infant's airway. Tilt it only enough to open the airway.
3. *Check* for any visible secretions or foreign objects.
4. *Correct* any airway problems. Use finger sweeps only if there is a foreign object visible. If suction is needed, use it gently.
5. *Maintain* the airway by continuing to use the head tilt-chin lift technique.

Breathing

1. *Check* for the presence of breathing:
   - *Look* for the rising and falling of the infant's chest.
   - *Listen* for the sound of air moving in and out of the infant's mouth and nose.
   - *Feel* for the movement of air on the side of your face and ear.
   - Continue to look, listen, and feel for at least 3 to 5 seconds. If there is adequate breathing, place the victim in the recovery position. If there is no breathing, go to step 2.
2. *Correct* the lack of breathing by performing rescue breathing. Cover the infant's mouth and nose with your mouth. Blow gently into the infant's mouth and nose for 1 to $1^1/_2$ seconds. Watch the chest rise with each breath. Remove your mouth and allow the lungs to deflate. Breathe for the infant a second time. After these first two breaths, breathe into the infant's mouth and nose every 3 seconds (20 rescue breaths per minute).

Foreign Body Airway Obstruction

Causes of Airway Obstruction

Your attempt to perform rescue breathing may not be effective if there is an airway obstruction. The most common airway obstruction is the tongue. If the tongue is blocking the airway, the head tilt-chin lift technique or the jaw-thrust technique should move it sufficiently to open the airway. However, if a foreign body is lodged in the air passages, you must use other techniques.

Food is the most common foreign object that causes an airway obstruction. An adult may choke on a large piece of meat; a child may inhale candy, a peanut, or a piece of a hot dog. Children may put small objects in their mouths and inhale such things as toys or balloons. Vomitus may obstruct the airway of a child or an adult (▼ Figure 24-13).

Types of Airway Obstruction

Airway obstruction may be partial or complete. The first step in caring for a conscious person who may have an obstructed airway is to ask, "Are you choking?" If the victim can reply to your question, the airway is not completely blocked. If the victim is unable to speak or cough, the airway is completely blocked.

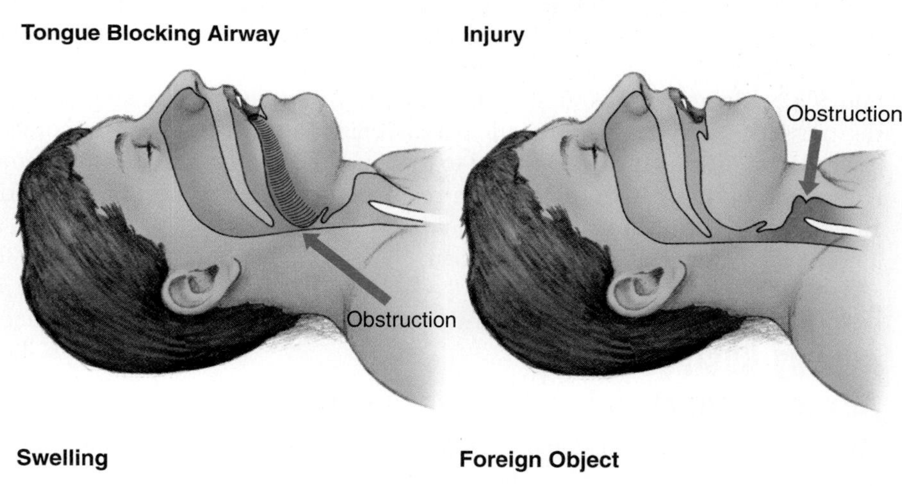

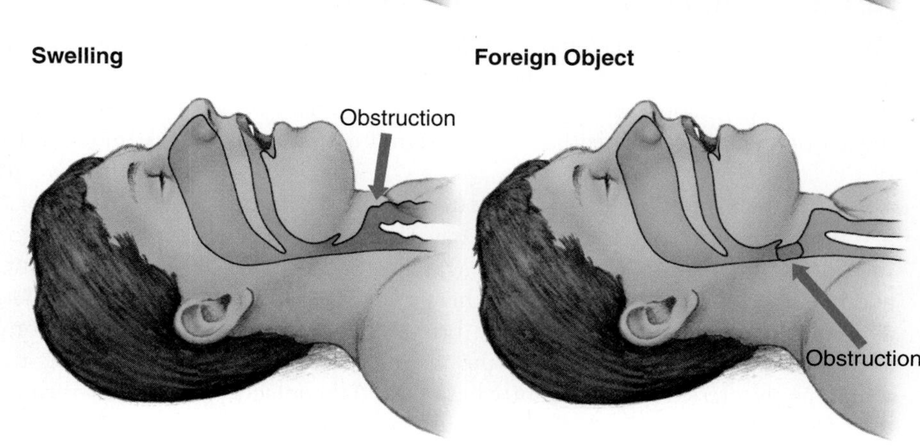

Figure 24-13 Common causes of airway obstruction.

Partial Airway Obstruction

In partial airway obstruction, the victim coughs and gags. This indicates that some air is passing around the obstruction. The victim may even be able to speak, although with difficulty.

To treat a partially obstructed airway, encourage the victim to cough. Coughing is the most effective way of expelling a foreign object. If the victim is unable to expel the object by coughing (if, for example, a bone is stuck in the throat), you should arrange for the victim's prompt transport to an appropriate medical facility. Such a victim must be monitored carefully while awaiting transport and during transport because a partial obstruction can become a complete obstruction at any moment.

Complete Airway Obstruction

A victim with a complete airway obstruction will have different signs and symptoms. The body quickly uses all the oxygen breathed in with the last breath. The victim is unable to breathe in or out, and, because he or she cannot exhale air, speech is impossible. If the airway is completely obstructed, the victim will lose consciousness in 3 to 4 minutes.

The currently accepted treatment for a completely obstructed airway in an adult or child is the **Heimlich maneuver**. Abdominal thrusts compress the air that remains in the lungs, pushing it upward through the airway so that it exerts pressure against the foreign object. The pressure pops the object out, in much the same way that a cork pops out of a bottle after the bottle has been shaken to increase the pressure. Many rescuers report that abdominal thrusts can cause an obstructing piece of food to fly across the room. A person who has had an obstruction removed from their airway by the Heimlich maneuver should be transported to a hospital for examination by a physician.

Management of Foreign Body Airway Obstructions

Airway Obstruction in a Conscious Adult

The steps to treat complete airway obstruction vary, depending on whether the victim is conscious or unconscious. If the victim is conscious, stand behind the victim and perform the abdominal thrusts while the victim is standing or seated in a chair.

Locate the **xiphoid process** (the bottom of the sternum) and the navel. Place one fist above the navel and well below the xiphoid process, thumb side against the victim's abdomen. Grasp your fist with your other hand. Then apply abdominal thrusts sharply and firmly, bringing your fist in and slightly upward. Do not give the victim a bear hug; rather, apply pressure at the point where your fist contacts the victim's abdomen. Each thrust should be distinct and forceful. Repeat these abdominal thrusts until the foreign object is expelled or until the victim becomes unconscious.

Review the following sequence until you can carry it out automatically. To assist a conscious victim with a complete airway obstruction by using the Heimlich maneuver, follow the steps in (► Skill Drill 24-8).

1. Ask, "Are you choking? Can you speak?" If there is no response, assume that the airway obstruction is complete.
2. Stand behind the victim and deliver an abdominal thrust (position the thumb side of your fist above the victim's navel and well below the xiphoid process). Press into the victim's abdomen with a quick upward thrust. **(Step 1)**
3. Repeat the abdominal thrusts until either the foreign body is expelled or the victim becomes unconscious. **(Step 2)**
4. If the victim is obese or in the late stages of pregnancy, use chest thrusts instead of abdominal thrusts. Chest thrusts are done by standing behind the victim and placing your arms under the victim's armpits to encircle the victim's chest. **(Step 3)**

Airway Obstruction in an Unconscious Adult

You can perform abdominal thrusts on victims who are unconscious or so large that you cannot encircle them with your arms by placing the victim flat on his or her back. Straddle the victim. Locate the xiphoid process and the navel. Place the heel of one hand slightly above the navel and well below the xiphoid process. Place your other hand directly on top of the first hand. Lean your shoulders forward and quickly press in and slightly upward. The force of the thrust should come from your shoulders and be delivered in the midline of the abdomen, not to either side.

Review the following sequence until you can carry it out automatically. To assist an unconscious victim with a complete airway obstruction, follow the steps in (► Skill Drill 24-9):

1. Verify unconsciousness. Ask the victim, "Are you okay?" Tap or gently shake the victim's shoulder. If the victim is unresponsive, call 9-1-1 immediately ("phone first") to activate the EMS system if this has not been done. Support the victim's head and neck as you place the victim on his or her back.
2. Open the airway by using the head tilt-chin lift technique. If trauma to the head or neck is suspected, use the jaw-thrust technique. **(Step 1)**
3. Check for breathing for at least 3 to 5 seconds. If breathing is absent, attempt rescue breathing. If there is an obstruction, air will not move into or out of the victim. **(Step 2)** In this case, proceed to the next step.
4. Reposition the head to improve the likelihood of opening the airway. Repeat the head tilt-chin lift or jaw-thrust technique to open the airway. Attempt rescue breathing again. **(Step 3)** If there is still no air movement, assume the presence of complete obstruction and proceed to the next three steps.

24-8 Skill Drill

Heimlich Maneuver

Stand behind the victim and deliver an abdominal thrust. Press into the victim's abdomen with a quick upward thrust.

Repeat the abdominal thrusts until either the foreign body is expelled or the victim becomes unconscious.

If the victim is obese or in the late stages of pregnancy, use chest thrusts instead of abdominal thrusts.

5. Deliver an abdominal thrust. Straddle the victim and place the heel of your hand slightly above the navel and well below the xiphoid process. Place your other hand on top of the first and press into the abdomen with a quick thrust inward and slightly upward. If the foreign body is not expelled, repeat the abdominal thrust four times. If this does not work, proceed to the next step. **(Step 4)**
6. Use the tongue-jaw lift to open the victim's mouth so you can check for obstructions. Grasp the lower jaw, placing your thumb on the tongue, and wrap your gloved fingers around the chin. Lift the jaw to open the mouth. Holding the victim's tongue down against the lower jaw with your thumb may offer better access to sweep an object from the throat. Use your finger to sweep the mouth clear. Curve your finger into a C shape and sweep it from one side of the back of the mouth to the other. **(Step 5)**
7. Attempt rescue breathing again. Repeat abdominal thrusts, finger sweeps, and breathing attempts until the airway is cleared. **(Step 6)**

24-9 Skill Drill

Airway Obstruction in an Unconscious Adult

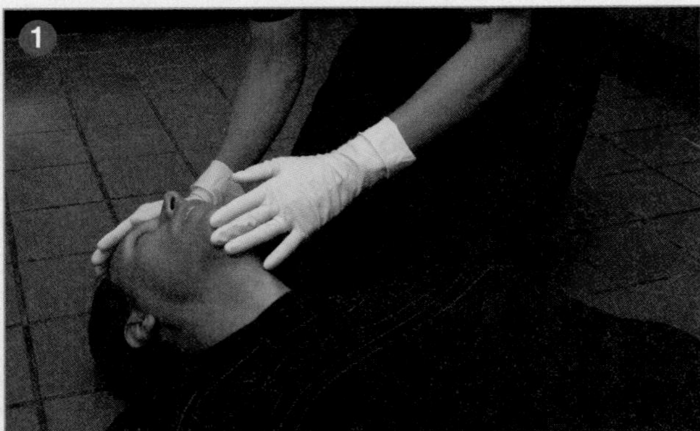

1 Open the airway by using the head tilt-chin lift technique. If trauma to the head or neck is suspected, use the jaw-thrust technique.

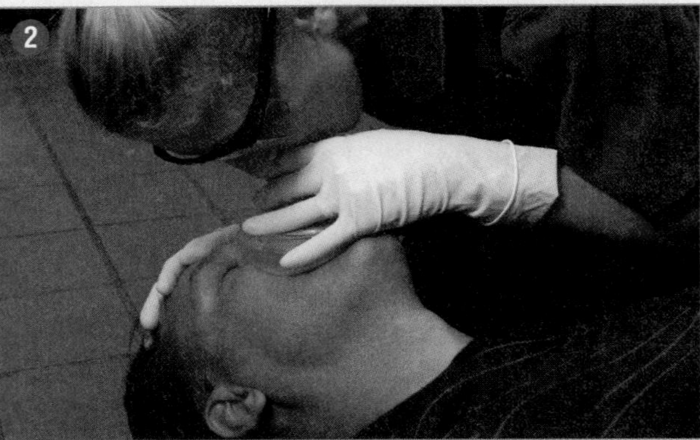

2 Check for breathing for at least 3 to 5 seconds. Attempt rescue breathing. If there is an obstruction, air will not move into or out of the victim.

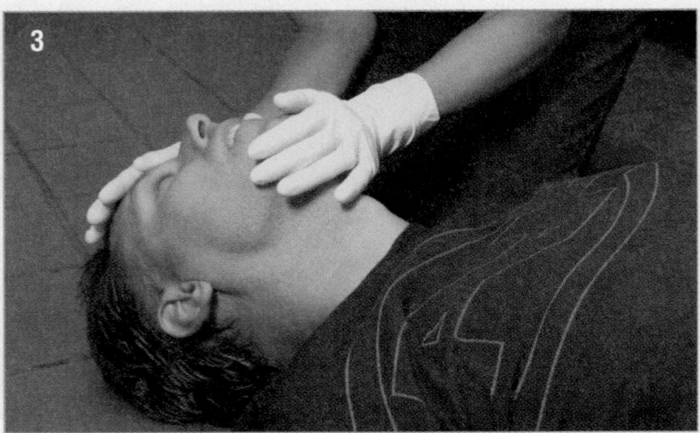

3 Reposition the head to improve the likelihood of opening the airway. Attempt rescue breathing again.

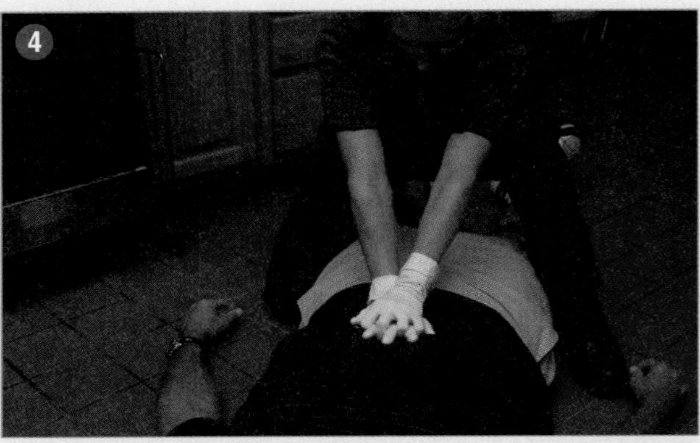

4 Deliver an abdominal thrust.

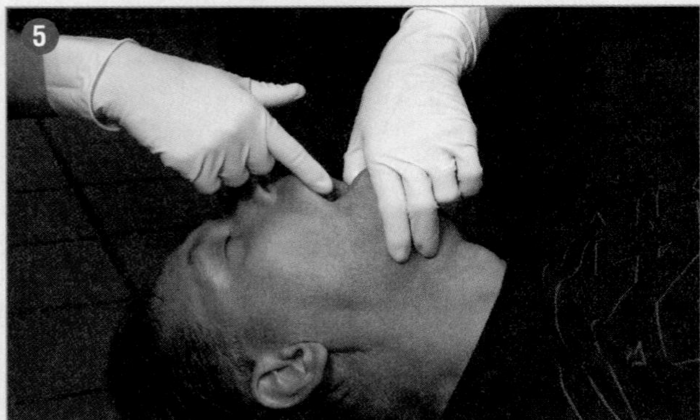

5 Use the tongue-jaw lift to open the victim's mouth so you can check for obstructions. Use your finger to sweep the mouth clear.

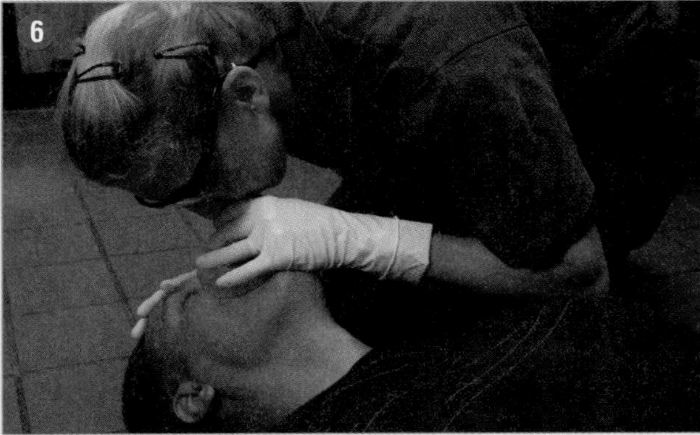

6 Attempt rescue breathing again. Repeat abdominal thrusts, finger sweeps, and breathing attempts until the airway is cleared.

Airway Obstruction in a Conscious Child

The steps for relieving an airway obstruction in a conscious child (1 to 8 years old) are the same as for an adult victim. The anatomic differences between adults and children/infants require that you make some adjustments in your technique. When opening the airway of a child or infant, tilt the head back just past the neutral position. Tilting the head too far back (hyper extending the neck) can actually obstruct the airway of a child or infant.

Airway Obstruction in an Unconscious Child

The steps for relieving a complete airway obstruction in an unconscious child (1 to 8 years old) are almost the same as those for an unconscious adult. The only difference is that, for a child, you should perform a finger sweep only if you can see the object obstructing the airway. Because the child's airway is so much smaller than an adult's, it is important that you do not push the foreign object farther down into the child's airway. If you are alone, you should attempt to remove the foreign body obstruction for 1 minute before you stop to activate the EMS system.

Airway Obstruction in a Conscious Infant

The steps for relieving an airway obstruction in a conscious infant (less than 1 year old) must take into consideration that an infant is very fragile. An infant's airway structures are very small, and they are more easily injured than those of an adult. If you suspect an airway obstruction, assess the infant to determine if there is any air exchange. If the infant is crying, the airway is not completely obstructed. Ask the person who was with the infant what was happening when the infant stopped breathing. This person may have seen the infant put a foreign body into his or her mouth.

If there is no movement of air from the infant's mouth and nose, suspect an airway obstruction. To relieve an airway obstruction in an infant, use a combination of back blows and chest thrusts. You must have a good grasp of the infant in order to alternate the back blows and the chest thrusts.

Review the following sequence until you can carry it out automatically. To assist a conscious infant with a complete airway obstruction, follow the steps in Skill Drill 24-10.

1. Assess the infant's airway and breathing status. Determine that there is no air exchange.
2. Place the infant in a face-down position over one arm so that you can deliver five back blows.
3. Support the infant's head and neck with one hand, and place the infant face down, with the head lower than the trunk. Rest the infant on your forearm and support your forearm on your thigh. Using the heel of your hand, deliver up to five back blows forcefully between the infant's shoulder blades.
4. Supporting the head, turn the infant face up by sandwiching the infant between your hands and arms. Rest the infant on his or her back with the head lower than the trunk.
5. Deliver up to five chest thrusts in the middle of the sternum. Use two fingers to deliver the chest thrusts in a firm manner.
6. Repeat the series of back blows and chest thrusts until the foreign object is expelled or until the infant becomes unconscious.

Airway Obstruction in an Unconscious Infant

Use the same sequence of back blows and chest thrusts steps for relieving an airway obstruction in an unconscious infant (less than 1 year old). However, the steps taken to determine unconsciousness and the presence of an airway obstruction are somewhat different.

Review the following sequence until you can carry it out automatically. To assist an unconscious infant with a complete airway obstruction, follow the steps in Skill Drill 24-11.

1. Determine unresponsiveness by gently shaking the shoulder or by gently tapping the bottom of the foot.
2. Position the infant on a firm, hard surface and support the head and neck.
3. Open the airway using the head tilt-chin lift technique. Be careful not to tilt the infant's head back too far.
4. Check for breathing by placing your ear close to the infant's mouth and nose. Listen and feel for the infant's breathing.
5. If the infant is not breathing, begin rescue breathing. If rescue breathing is unsuccessful, proceed to the next step.
6. Reposition the airway and reattempt rescue breathing. Note: Up to this point, the steps are similar to those used for infant rescue breathing.
7. Deliver up to five back blows using the same technique as for a conscious infant.
8. Deliver up to five chest thrusts using the same technique as for a conscious infant.
9. Perform a tongue-jaw lift and remove any foreign object that you can see.
10. Repeat the sequence of back blows and chest thrusts until the foreign body is expelled.

Gastric Distention

Gastric distention occurs when air is forced into the stomach instead of the lungs. This makes it harder to get an adequate amount of air into the victim's lungs, and it increases the chance that the victim will vomit. Breathe slowly into the victim's mouth, just enough to make the chest rise. Remember that the lungs of children and infants are smaller, and require smaller breaths during rescue breathing. The excess air may enter the stomach and cause gastric distention. Preventing gastric distention is much better than trying to cure the results of it.

Dental Appliances

Do not remove firmly attached dental appliances. They may help keep the mouth fuller so you can make a better seal between the victim's mouth and your mouth or a breathing device. Loose dental appliances, however, may cause problems. Partial dentures may become dislodged during trauma or while you are performing airway care and rescue breathing. If you discover loose dental appliances during your examination of the victim's airway, remove the dentures to prevent them from occluding the airway. Try to put them in a safe place so they will not get damaged or lost.

Airway Management in a Vehicle

If you arrive on the scene of an automobile accident and find that the victim has airway problems, how can you best assist the victim and maintain an open airway?

- If the victim is lying on the seat or floor of the car, you can apply the standard jaw-thrust technique. Use the jaw-thrust technique if there is any possibility that the accident could have caused a head or spinal injury.
- When the victim is in a sitting or semireclining position, approach him or her from the side by leaning in through the window or across the front seat.
- Grasp the victim's head with both hands. Put one hand under the victim's chin and the other hand on the back of the victim's head just above the neck.
- Maintain slight upward pressure to support the head and cervical spine and to ensure that the airway remains open.

This technique will often enable you to maintain an open airway without moving the victim. This technique has several advantages:

- You do not have to enter the automobile.
- You can easily monitor the victim's carotid pulse and breathing patterns by using your fingers.
- It stabilizes the victim's cervical spine.
- It opens the victim's airway.

Anatomy and Function of the Circulatory System

The **circulatory system** is similar to a city water system because both consist of a pump (the heart), a network of pipes (the blood vessels), and fluid (blood). After blood picks up oxygen in the lungs, it goes to the heart, which pumps the oxygenated blood to the rest of the body. The heart functions as a pump. It is about the size of your fist, and is located in the chest between the lungs. The cells of the body absorb oxygen and nutrients from the blood and produce waste products (including carbon dioxide) that the blood carries back to the lungs. In the lungs, the blood exchanges the carbon dioxide for more oxygen. Blood then returns to the heart to be pumped out again.

The human heart consists of four chambers, two on the right side and two on the left side. Each upper chamber is called an **atrium**. The right atrium receives blood from the veins of the body; the left atrium receives blood from the lungs. The bottom chambers are the **ventricles**. The right ventricle pumps blood to the lungs; the left ventricle pumps blood throughout the body. The most muscular chamber of the heart is the left ventricle, which needs the most power because it must force blood to all parts of the body. Together the four chambers of the heart work in a well-ordered sequence to pump blood to the lungs and to the rest of the body (▼ Figure 24-14).

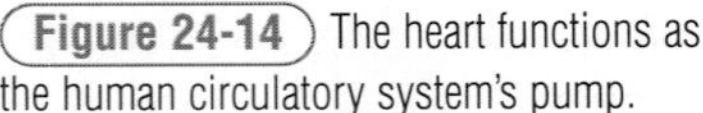
Figure 24-14 The heart functions as the human circulatory system's pump.

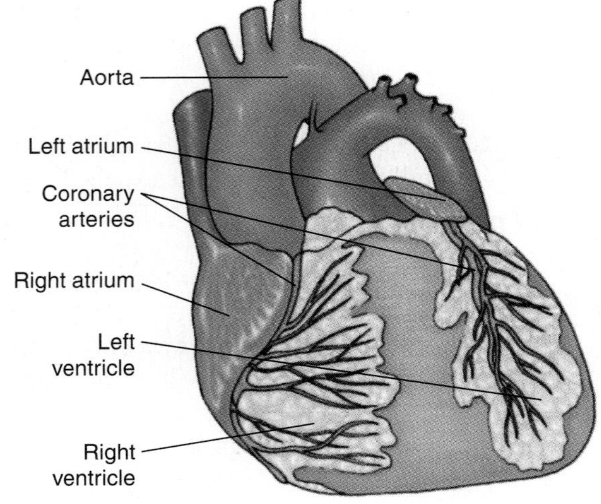

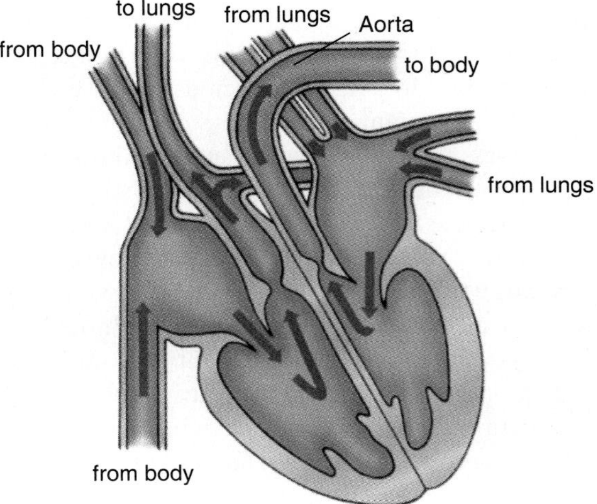

One-way check valves in the heart and veins allow the blood to flow in only one direction through the circulatory system. The arteries carry blood away from the heart at high pressure and are therefore thick-walled. The main artery carrying blood away from the heart is quite large (about 1" in diameter) but arteries become smaller farther away from the heart.

The four major arteries are the carotid artery in the neck, radial artery in the wrist, brachial artery in the arm, and the femoral artery in the groin. The locations of these arteries are shown in ▼ Figure 24-15. Because these arteries lie between a bony structure and the skin, they can be used to measure the victim's pulse.

A pulse is the pressure wave generated by the pumping action of the heart. The beat you feel is caused by the surge of blood from the left ventricle as it contracts and pushes blood out into the main arteries of the body. In counting the number of pulsations per minute, you also count heartbeats per minute. In other words, the pulse rate reflects the heart rate.

The capillaries are the smallest pipes in the system. Some capillaries are so small that only one blood cell at a time can go through them. At the capillary level, oxygen passes from the blood cells into the cells of body tissues, and carbon dioxide and other waste products pass from the tissue cells to the blood cells, which then return to the heart, and then to the lungs. Veins are the thin-walled pipes of the circulatory system that carry blood back to the heart.

Blood has several components: plasma (a clear, strawcolored fluid), red blood cells, white blood cells, and platelets. The red blood cells give blood its red coloring. Red blood cells carry oxygen from the lungs to the body and bring carbon dioxide back to the lungs. The white blood cells are called "infection fighters" because they devour bacteria and other disease-causing organisms ► Figure 24-16. Platelets start the blood-clotting process.

A

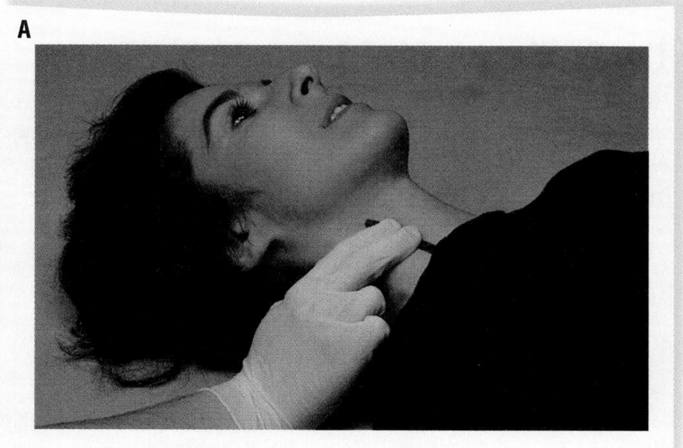

B

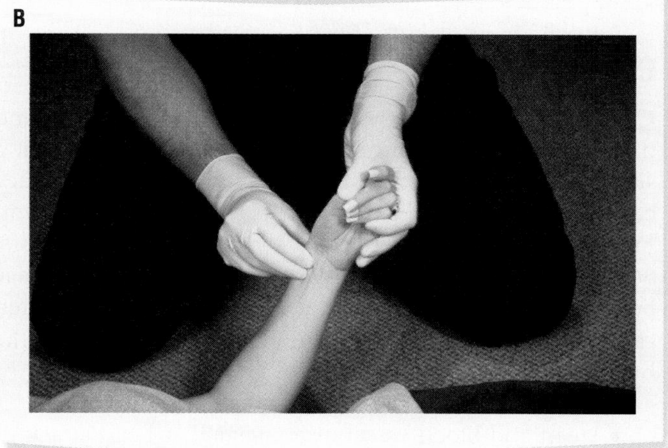

C

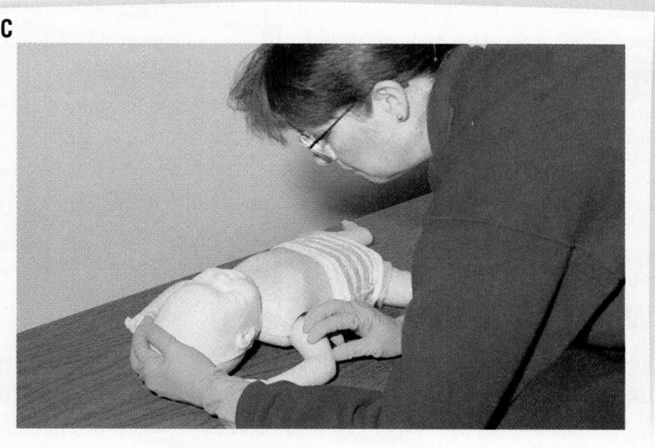

D

Figure 24-15 Locations for assessing the victim's pulse. **A.** Neck or carotid pulse. **B.** Wrist or radial pulse. **C.** Arm or brachial pulse. **D.** Groin or femoral pulse.

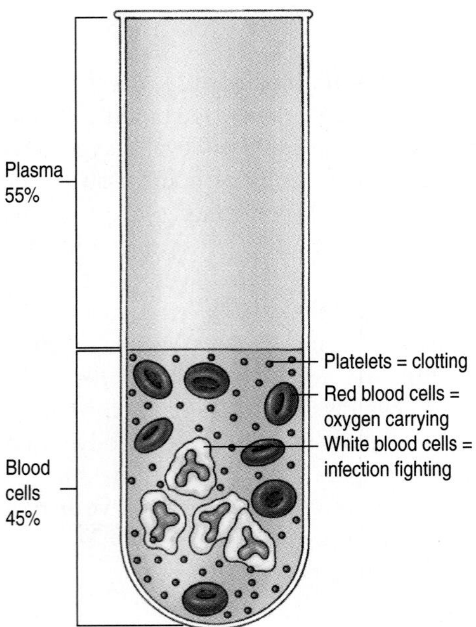

Figure 24-16 The components of blood.

Cardiac Arrest

Cardiac arrest occurs when the heart stops contracting, and no blood is pumped through the blood vessels. Without a supply of blood, the cells of the body will die because they cannot get any oxygen and nutrients, and they cannot eliminate waste products. As the cells die, organ damage occurs. Some organs are more sensitive than others. Brain damage begins within 4 to 6 minutes after the victim has suffered cardiac arrest. Within 8 to 10 minutes, the damage to the brain may become irreversible.

Cardiac arrest may have many different causes:

- Heart and blood vessel diseases such as heart attack and stroke
- Respiratory arrest, if untreated
- Medical emergencies such as epilepsy, diabetes, allergic reactions, electrical shock, and poisoning
- Drowning
- Suffocation
- Trauma and shock caused by massive blood loss

A victim who has suffered cardiac arrest is unconscious and is not breathing. You cannot feel a pulse, and the victim looks dead. Regardless of the cause of cardiac arrest, the initial treatment is the same: CPR.

Components of CPR

The technique of cardiopulmonary resuscitation (CPR) requires three types of skills: the A (airway) skills, B (breathing) skills, and C (circulation) skills. In the previous section, you learned the airway and breathing skills. You learned how to check victims to determine if the airway is open and to correct a blocked airway by using the head tilt-chin lift or jaw-thrust technique. You learned how to check victims to determine if they are breathing by using the look, listen, and feel technique. You learned to correct the absence of breathing by performing rescue breathing.

To perform CPR, you must combine the airway and breathing skills with circulation skills. You begin by checking the victim for a pulse. If there is no pulse, you must correct the victim's circulation by performing external chest compressions. The airway and breathing skills you know will push oxygen into the victim's lungs. External chest compressions move the oxygenated blood throughout the body. By depressing the victim's sternum (breastbone), you change the pressure in the victim's chest and force enough blood through the system to sustain life for a short period of time.

CPR by itself cannot sustain life indefinitely. However, it should be started as soon as possible to give the victim the best chance of survival. By performing all three parts of the CPR sequence, you can keep the victim alive until more advanced medical care can be administered. In many cases, the victim will need defibrillation and medication in order to recover from cardiac arrest.

The Cardiac Chain of Survival

In most cases of cardiac arrest, CPR alone is not sufficient to save lives. But it is the first treatment in the American Heart Association's "Chain Of Survival." The links in the chain include:

1. Early access to the emergency medical services (EMS) system
2. Early CPR
3. Early defibrillation
4. Early advanced care by paramedics and hospital personnel

As a fire fighter, you can help the victim by providing early CPR and by making sure that the EMS system has been activated. Some fire fighters are trained in the use of automated defibrillators. By keeping these links of the chain strong, you will help keep the victim alive until early advanced care can be administered by paramedics and hospital personnel.

Just as an actual chain is only as strong as its weakest link, this CPR chain of survival is only as good as its weakest link. Your actions in performing early CPR are vital to giving cardiac arrest victims their best chance for survival (► Figure 24-17).

When to Start and Stop CPR

When to Start CPR

CPR should be started on all nonbreathing, pulseless victims, unless they are obviously dead or unless they have a Do Not Resuscitate (DNR) order (also known as a "Do Not Attempt Resuscitation" order) that is valid in your jurisdiction. Few reliable criteria exist to determine death immediately.

The following criteria are reliable signs of death and indicate that CPR should not be started:

Figure 24-17 The CPR chain of survival.

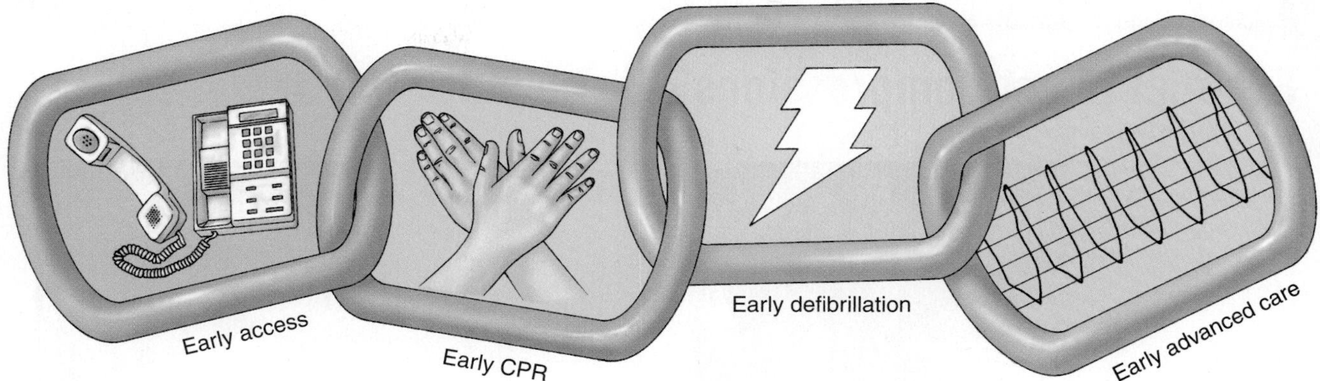

- Decapitation. Decapitation occurs when the head is separated from the rest of the body. When this occurs, there is obviously no chance of saving the victim.
- Rigor mortis. This is the temporary stiffening of muscles that occurs several hours after death. The presence of this stiffening indicates the victim is dead and cannot be resuscitated.
- Evidence of tissue decomposition. Tissue decomposition or actual flesh decay occurs only after a person has been dead for more than a day.
- Dependent lividity. Dependent lividity is the red or purple color that appears on the parts of the victim's body closest to the ground. It is caused by blood seeping into the tissues on the dependent, or lower, part of the person's body. Dependent lividity occurs after a person has been dead for several hours.

If any of the preceding signs is present in a pulseless, nonbreathing person, you should not begin CPR. If none of these signs is present, you should activate the EMS system and then begin CPR. It is far better to start CPR on a person who is later declared dead by a physician than to withhold CPR from a victim whose life might have been saved.

When to Stop CPR

You should discontinue CPR only when:

- Effective, spontaneous circulation and ventilation are restored.
- Resuscitation efforts are transferred to another trained person who continues CPR.
- A physician assumes responsibility for the victim.
- The victim is transferred to properly trained EMS personnel.
- Reliable criteria for death (as previously listed) are recognized.
- You are too exhausted to continue resuscitation, environmental hazards endanger your safety, or continued resuscitation would place the lives of others at risk.

The Technique of External Cardiac Compression in an Adult

A victim in cardiac arrest is unconscious and is not breathing. You cannot feel a pulse and the victim looks dead. If you suspect that the victim has suffered cardiac arrest, first check and correct the airway, then check and correct the breathing, and finally check for circulation. Check for circulation by feeling the carotid pulse and looking for signs of coughing or movement. To check the carotid pulse, place your index and middle fingers on the larynx (Adam's apple). Now slide your fingers into the groove between the larynx and the muscles at the side of the neck (▼ Figure 24-18). Keep your fingers there for 5 to 10 seconds to be sure the pulse is absent and not just slow. As you check the pulse, look for signs of coughing or movement that may indicate circulation is present.

If there is no carotid pulse in an unresponsive victim, you must begin chest compressions. For chest compressions to be effective, the victim must be lying on a firm, horizontal

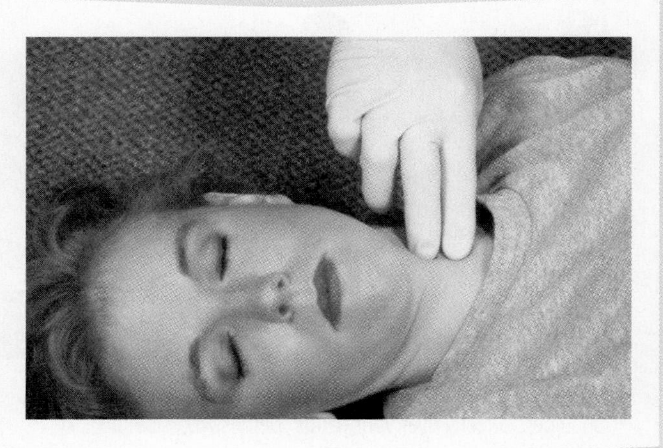

Figure 24-18 Check the victim's carotid pulse.

24-12 Skill Drill

External Chest Compressions on an Adult

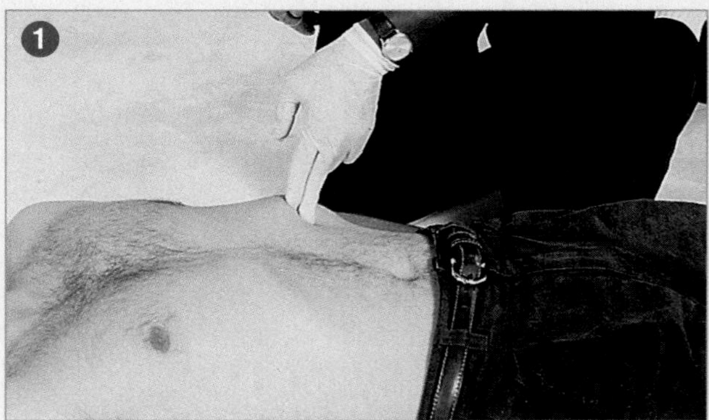

1 Slide your index fingers along the rib cage to the notch in the center of the chest.

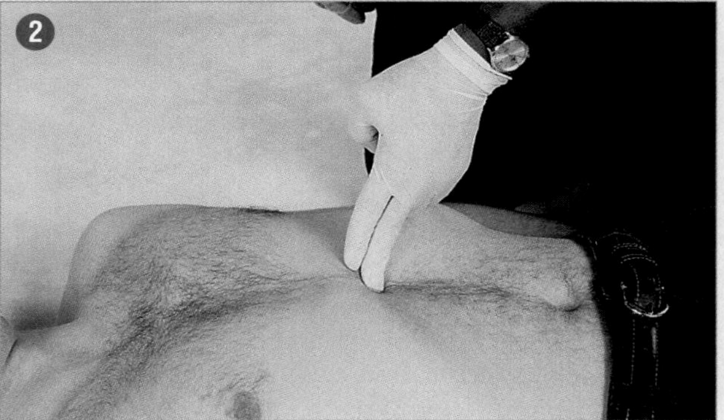

2 Push the middle finger high into the notch, and lay the index finger on the lower portion of the sternum.

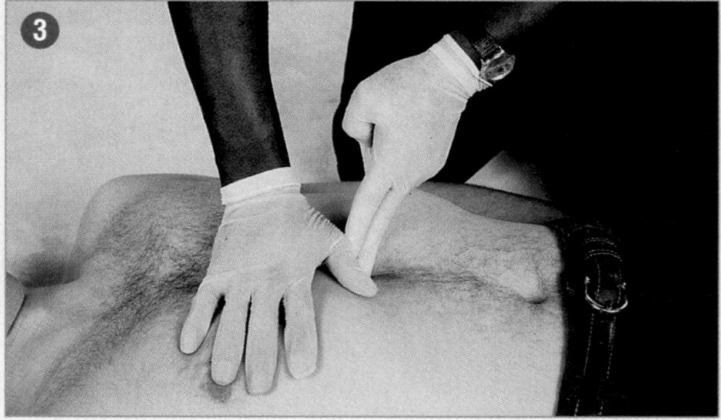

3 Place the heel of the second hand on the lower half of the sternum, touching the index finger of your first hand.

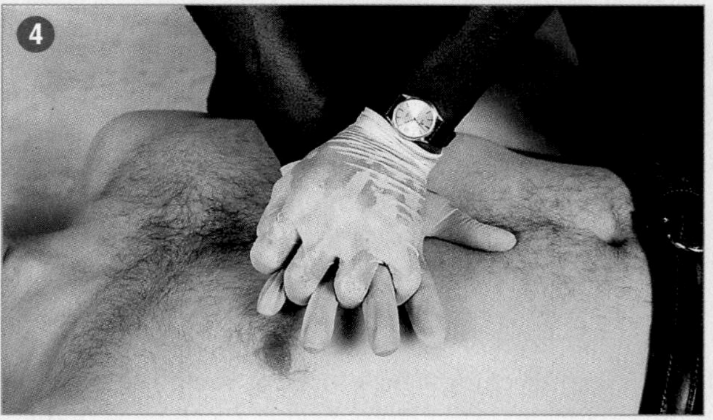

4 Remove your first hand from the notch, and place it over the hand on the sternum.

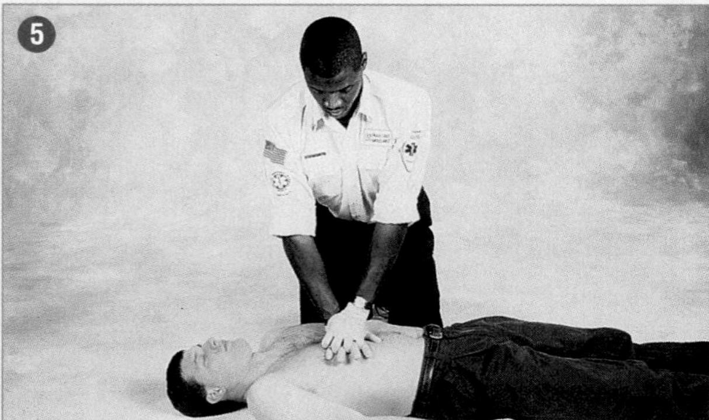

5 With your arms straight, lock your elbows, and position your shoulders directly over your hands. Depress the sternum $1\frac{1}{2}$" to 2", using a rhythmic motion.

surface. If the victim is on a soft surface, such as a bed, it is impossible to compress the chest. Immediately place all victims needing CPR on a firm, level surface.

To perform chest compressions effectively, follow the steps in ◄ **Skill Drill 24-12**.

1. Slide your index fingers along the rib cage to the notch in the center of the chest. **(Step 1)**
2. Push the middle finger high into the notch, and lay the index finger on the lower portion of the sternum. **(Step 2)**
3. Place the heel of the second hand on the lower half of the sternum, touching the index finger of your first hand. **(Step 3)**
4. Remove your first hand from the notch, and place it over the hand on the sternum. **(Step 4)**
5. With your arms straight, lock your elbows, and position your shoulders directly over your hands. Depress the sternum 1½" to 2", using a rhythmic motion. **(Step 5)**

It is important to locate and maintain the proper hand position while applying chest compressions. If your hands are too high, the force you apply will not produce adequate chest compressions. If your hands are too low, the force you apply may damage the liver. If your hands slip sideways off the sternum and onto the ribs, the compressions will not be effective, and you may damage the ribs and lungs.

After you have both hands in the proper position, compress the chest of an adult 1½" to 2" straight down. For effective compressions, kneel close to the victim's side, and lean forward so that your arms are directly over the victim. Keep your back straight and your elbows stiff so you can apply the force of your whole body to each compression, not just your arm muscles. Between compressions, keep the heel of your hand on the victim's chest but completely release the pressure.

Compressions must be rhythmic and continuous. Each compression cycle consists of one downward push followed by a rest so that the heart can refill with blood. Compressions should be at the rate of 100 compressions per minute in an adult victim. After every 15 chest compressions, give 2 rescue breaths. Practice on a manikin until you can compress the chest smoothly and rhythmically.

External Chest Compressions on an Infant

Infants (under 1 year of age) who have suffered cardiac arrest will be unconscious and not breathing. They will have no pulse. To check for cardiac arrest, you first need to check and correct the airway. Remember not to tilt the head back too far because this may occlude the infant's airway. Next, check for breathing by using the look, listen, and feel technique. Correct the absence of breathing by giving mouth-to mouth-and-nose rescue breathing.

To check an infant's circulation, feel for the brachial pulse on the inside of the upper arm ◄ **Figure 24-19**. Use two fingers of one hand to feel for the pulse, and use the other hand to maintain the head tilt. If there is no pulse, begin chest compressions. Draw an imaginary horizontal line between the two nipples, and place your index finger below the imaginary line in the middle of the chest. Place your middle and ring fingers next to your index finger. Use your middle and ring fingers to compress the sternum. Make sure you compress above the xiphoid process. Compress the sternum about ½" to 1" (approximately one-third to one-half the depth of the chest). Compress the sternum at a rate of at least 100 times per minute. Give one rescue breath after every five chest compressions.

Place the infant on a solid surface such as a table, or cradle the infant in your arm, as shown in, when doing chest compressions ▼ **Figure 24-20**. You will not need to use much force to achieve adequate compressions on infants because they are so small and their chests are so pliable.

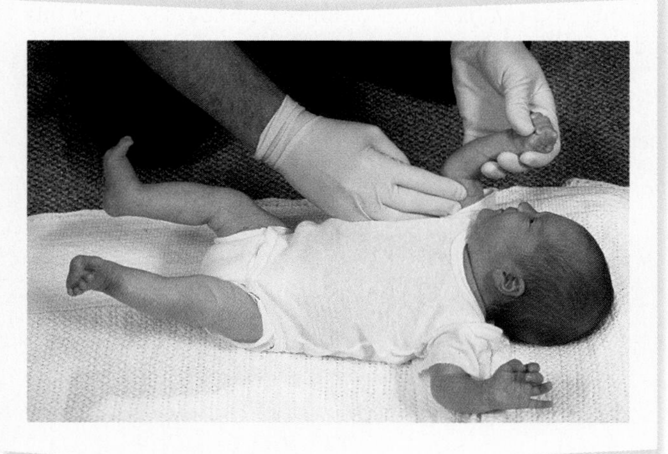

Figure 24-19 Check the brachial pulse on the inside of the infant's arm.

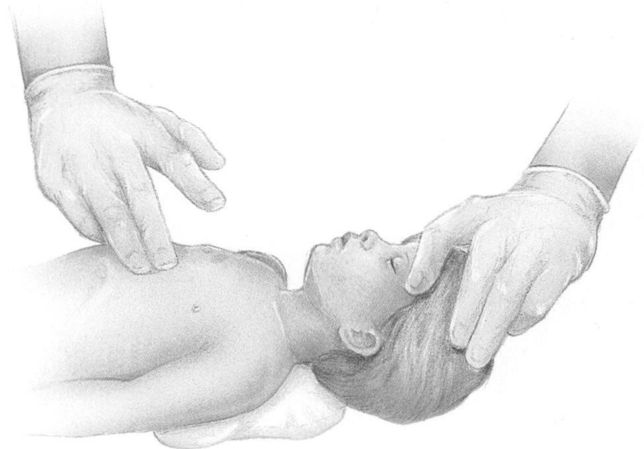

Figure 24-20 Positioning the infant victim for proper CPR.

Voices of Experience

“The first responder was holding c-spine and talking to the victim, telling him not to move and assuring him that everything was being done to help him.”

On November 22, 1996, my partner and I were called to a motor vehicle crash in a rural area of our county. There was a lengthy drive, about fifteen minutes or so, from our location to the crash. While en route we received a report called in by a nearby neighbor who said that the victim of the crash had been ejected and was lying in the field in very bad shape.

The first responder fire department arrived on the scene five minutes before we did, and was treating the young male victim who indeed had been ejected and thrown several feet from the vehicle. Upon our arrival, we made our way to the victim and took a glimpse of the car. It was destroyed by the crash. I asked the first responder at the head, “What do we have?” He responded with, “I am not sure, but I am doing what I was taught.” This was indeed true—he was holding c-spine and talking to the victim, telling him not to move and assuring him that everything was being done to help him.

The situation we had was a male in his early twenties who had left school where he was a coach. He was on his way home when he lost control of his car and rolled over several times before being ejected. Upon assessment we found that the victim had no feeling or movement from the nipple line down. Immediately we realized our situation was critical. We rapidly loaded the victim into the ambulance and en route to the hospital started a large bore IV, high-flow oxygen, and checked for other signs of trauma. This was all being done while the victim was still conscious and asking why he couldn't move his lower body. Once we arrived at the local emergency department, we turned care over to an orthopaedic physician.

We later found out that he had received an injury to his thoracic spinal cord and was going to need several weeks of rehabilitation. This is an excellent example of how fire fighters fit into EMS operations, and can have a major impact on how a scene of this magnitude can be resolved in an effective manner.

Kevin Sargent
Mt. Vernon Fire Department
Mt. Vernon, Illinois
Rend Lake College
Ina, Illinois

External Chest Compressions on a Child

The signs of cardiac arrest in a child (from 1 year to 8 years of age) are the same as those for an adult and for an infant. If you suspect that a child is in cardiac arrest, first check and correct the child's airway, then check and correct the child's breathing, and, finally, check for circulation. Check the carotid pulse by placing two or three fingers on the larynx. Slide your fingers into the groove between the Adam's apple and the muscle. Feel for the carotid pulse with one hand and maintain the head-tilt position with the other hand.

To perform chest compressions on a child, locate the proper position for your hands by placing the fingers of one hand on the upper end of the sternum and using the fingers of the other hand to locate the xiphoid process at the bottom of the sternum. Place the heel of one hand on the lower half of the sternum. Do not put the heel of your other hand on top; the force of the heel of one hand is sufficient to perform chest compressions on a child. Compress the sternum approximately one-third to one-half the depth of the chest, or approximately 1" to 1½". Compress the chest at a rate of 100 times per minute. Give one rescue breath after every five compressions.

Place the infant on a solid surface such as a table, or cradle the infant in your arm, when doing chest compressions. You will not need to use much force to achieve adequate compressions on infants because they are so small and their chests are so pliable.

One-Rescuer Adult CPR

CPR consists of three skill sets: checking and correcting the airway, checking and correcting the breathing, and checking and correcting the circulation. Now that you have learned these separate skills, you are ready to put these three skills together to perform CPR. If you are the only trained person at the scene, you must perform one-rescuer CPR.

To perform one-rescuer CPR, follow the steps in ▶ Skill Drill 24-13.

1. Establish the victim's level of consciousness. Ask the victim, "Are you okay?" Shake the victim's shoulder. If there is no response, call for additional help by activating the EMS system. (If you are alone, phone 9-1-1 before you begin CPR. "Phone first.") **(Step 1)**
2. Turn the victim on his or her back, supporting the head and neck as you do.
3. Open the airway. Use the head tilt-chin lift technique or, if the victim is injured, use the jaw-thrust technique. Maintain the open airway. **(Step 2)**
4. Check for breathing. Place the side of your face and your ear close to the nose and mouth of the victim. Look, listen, and feel for the movement of air: Look for movement of the chest, listen for sounds of air exchange, and feel for air movement on the side of your face. Your breathing check should last at least 3 to 5 seconds. If there are no signs of breathing, begin rescue breathing. Use a mouth-to-mask ventilation device, if one is available, or place your mouth over the victim's mouth, seal the victim's nose with your thumb and index finger, and begin mouth-to-mouth rescue breathing. **(Step 3)**
5. Give two breaths. Blow slowly for 2 seconds using just enough force to make the chest rise. Allow the lungs to deflate between breaths. **(Step 4)**
6. Look for signs of circulation by checking the carotid pulse and looking for signs of coughing or movement. Find the carotid pulse by locating the victim's larynx with your index and middle fingers, then sliding your fingers into the groove between the larynx and the muscles at the side of the neck. Check for 5 to 10 seconds. If the pulse is absent, proceed to the next step. (If the pulse is present, continue rescue breathing.) **(Step 5)**
7. Begin chest compressions. Place the heel of one hand on the lower half of the sternum, two finger widths above the xiphoid process. Place the other hand on top of the first, so the hands are parallel. Now press down to compress the chest about 1½" to 2". Apply 15 compressions at the rate of 100 compressions per minute. Count the compressions out loud: "One and two and three and..." **(Step 6)**
8. After 15 chest compressions, give two full-size rescue breathes.
9. Continue alternating compressions and ventilations. Deliver a sequence of 15 compressions followed by two ventilations.
10. Check for a pulse. After 1 minute and every few minutes thereafter, check for a carotid pulse.

When performing one-rescuer CPR, you must deliver chest compressions and rescue breathing at a ratio of 15 compressions to two breaths. Immediately give two lung inflations after each set of 15 chest compressions. Because you must interrupt chest compressions to ventilate, you should perform each series of 15 chest compressions in 10 seconds (a rate of 100 compressions per minute). At this rate, the victim will actually receive about 60 compressions per minute.

Although one-rescuer CPR can keep the victim alive, two-rescuer CPR is preferable for an adult victim because it is less exhausting for the rescuers. Whenever possible, CPR for an adult should be performed by two rescuers.

Two-Rescuer Adult CPR

In many cases, a second trained person will be on the scene to help you perform CPR. Two-rescuer CPR is more effective than one-rescuer CPR. One rescuer can deliver chest compressions while the other performs rescue breathing. Chest compressions and ventilations can be given more regularly

Skill Drill

One-Rescuer Adult CPR

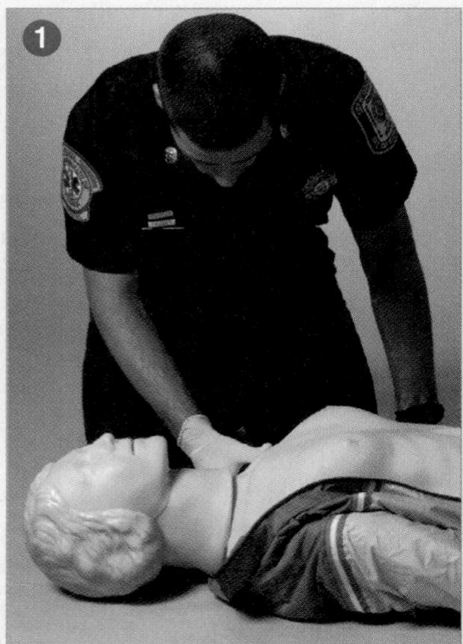

Establish responsiveness.

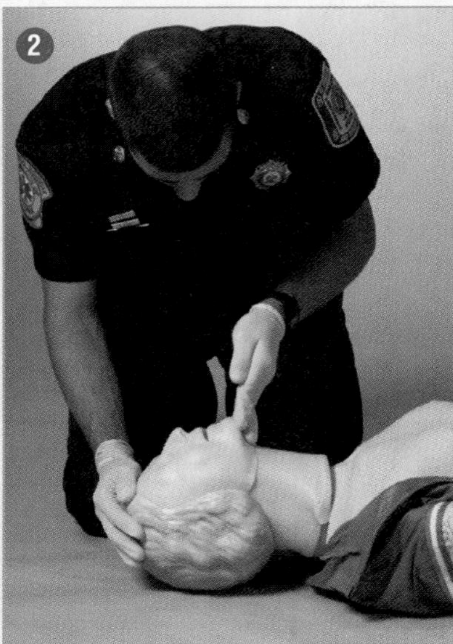

Open the airway.

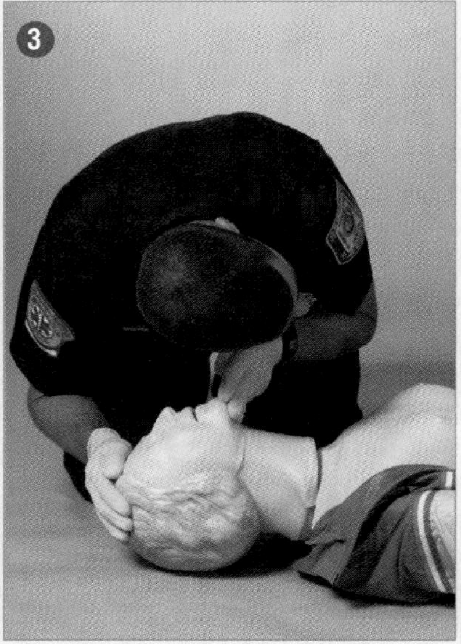

Check for breathing.

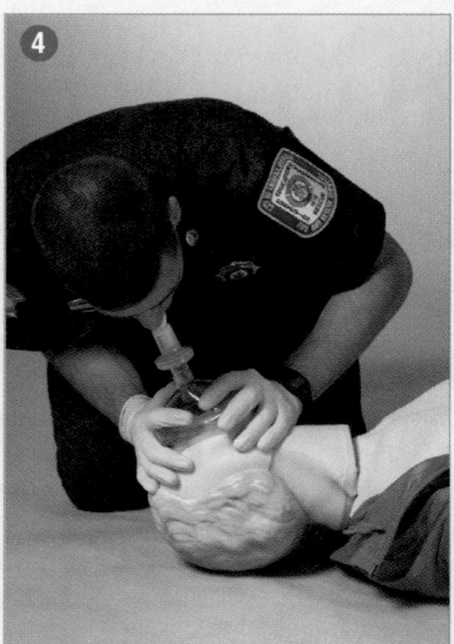

Perform rescue breathing.

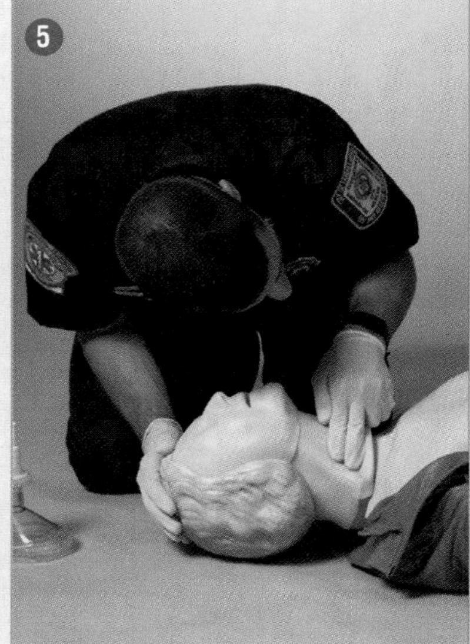

Check for signs of circulation.

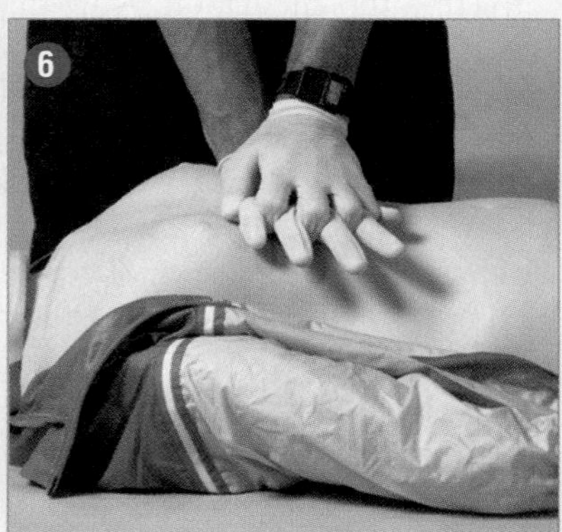

Perform chest compressions.

and without interruption. Two-rescuer CPR is also less tiring for the rescuers. You can continue to perform effective CPR for a longer period of time with a partner than you could alone.

In two-rescuer CPR, one rescuer delivers ventilations (mouth-to-mouth or mouth-to-mask breathing) and the other gives chest compressions. Position yourselves on either side of the victim, one near the head and the other near the chest. The sequence of steps is the same as for one-rescuer CPR, but the tasks are divided.

To perform two-rescuer CPR, follow the steps in ► **Skill Drill 24-14**.

1. Rescuer One (at the victim's head) determines the victim's state of consciousness by asking, "Are you okay?" If there is no response, the rescuer shakes the person's shoulder. **(Step 1)**
2. Call 9-1-1 to activate the EMS system if the victim is unconscious. If other people are present, ask them to call for EMS. If two rescuers are present, Rescuer Two should make the call.
3. Turn the victim on his or her back. Turn the victim as a unit, supporting the head and neck to protect the spine.
4. Rescuer One opens the airway using the head tilt-chin lift or jaw-thrust technique. **(Step 2)**
5. Rescuer One checks for breathing by placing the side of his or her face close to the mouth and nose of the victim to look for chest movements, listen for breathing sounds, and feel for air movement for at least 3 to 5 seconds. **(Step 3)** If there are no signs of breathing, proceed to step 6.
6. Rescuer One gives two 2-second breaths. Allow time for complete deflation between breaths. **(Step 4)**
7. Rescuer One checks for signs of circulation by checking for the presence of a carotid pulse and looking for signs of coughing and movement. If there is no carotid pulse, proceed to step 8.
8. Rescuer Two applies 15 chest compressions at a rate of 100 compressions per minute. Count "one and two and three" and so forth to maintain the proper rate between compressions and to let Rescuer One know when to ventilate. **(Step 5)**
9. After Rescuer Two completes 15 chest compressions, Rescuer One gives two ventilations. Rescuer Two should pause just long enough for Rescuer One to ventilate twice.
10. Periodically, Rescuer One should place his or her fingers on the carotid pulse of the victim as Rescuer Two continues the compressions. If the compressions are being done correctly, Rescuer One should feel a pulse with each compression. This confirms that the CPR is adequate.
11. Every few minutes, Rescuer One should feel for the carotid pulse and ask Rescuer Two to stop compressions. If the victim's heart starts beating on its own, Rescuer One will continue to feel a pulse. In this case, Rescuer Two can stop doing compressions. Rescue breathing should be continued until spontaneous breathing resumes. If there is no pulse, CPR should be resumed. **(Step 6)**

Compressions and ventilations should remain rhythmic and uninterrupted. By counting out loud, Rescuer Two can continue to deliver compressions at the rate of 100 per minute and also let Rescuer One know when to ventilate the victim's lungs. Once you and your partner establish a smooth pattern of CPR, do not stop, except for 5 seconds to check for a return of the victim's pulse or to move the victim.

One-Rescuer Infant

An infant is defined as anyone under 1 year of age. The principles of CPR are the same for adults and infants. In actual practice, however, you must use slightly different techniques for an infant.

To perform one-rescuer infant CPR, follow the steps in **Skill Drill 24-15**.

1. Position the infant on a firm surface.
2. Establish the infant's level of responsiveness. An unresponsive infant is limp. Gently shake or tap the infant to determine whether he or she is unconscious. Call for additional help if the victim is unconscious. Activate the EMS system.
3. Open the airway. This is best done by the head tilt-chin lift method. Be careful as you tilt the infant's head back, because tilting too far can obstruct the airway. Tilt the head back to a sniffing position, but do not tilt it back as far as it will go. Continue holding the head with one hand.
4. Check for breathing. Place the side of your face close to the mouth and nose of the infant as you would for an adult. Look, listen, and feel for at least 3 to 5 seconds.
5. Give two slow breaths, each lasting 1 to $1\frac{1}{2}$ seconds. To breathe for an infant, place your mouth over the infant's mouth and nose. Because an infant has very small lungs, you should give only very small puffs of air, just enough to make the chest rise. Do not use large or forceful breaths.
6. Check for circulation. Check the brachial pulse rather than the carotid pulse. The brachial pulse is on the inside of the arm. You can feel it by placing your index and middle fingers on the inside of the infant's arm halfway between the shoulder and the elbow. Check for 5 to 10 seconds.
7. Begin chest compressions. An infant's heart is located relatively higher in the chest than an adult's heart. Therefore, you must deliver chest compressions by pressing on the middle (rather than lower) portion of the sternum. Imagine a horizontal line drawn between

Skill Drill

Two-Rescuer Adult CPR

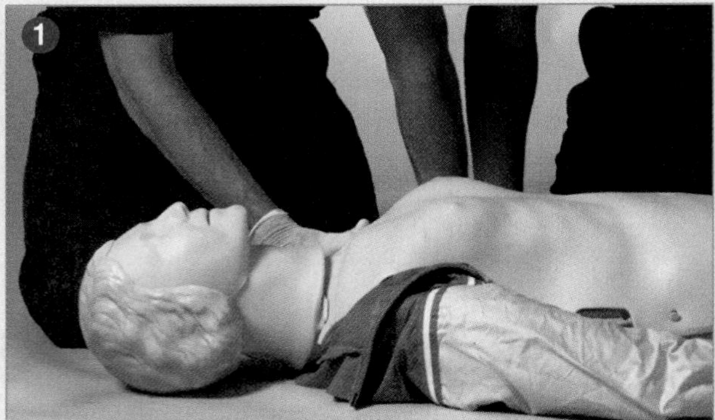

Establish responsiveness.

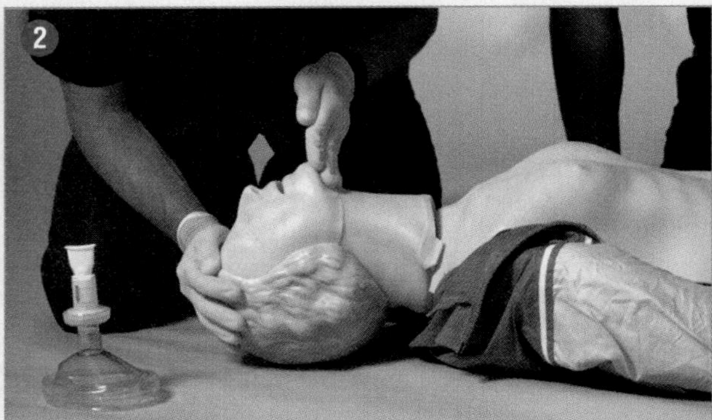

Open the airway using the head tilt-chin lift or jaw-thrust technique.

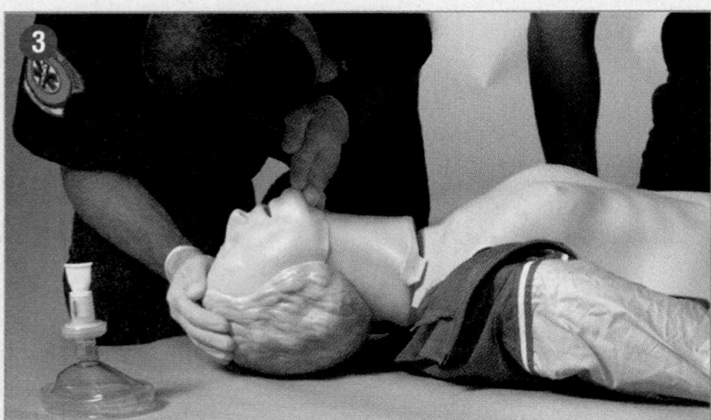

Check for breathing—look, listen, and feel.

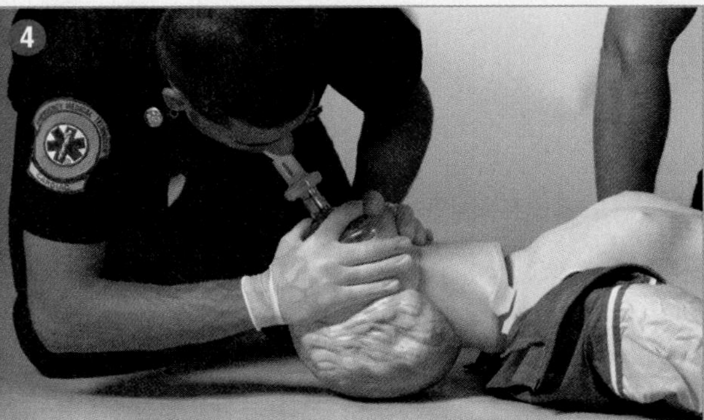

Perform rescue breathing; give two breaths.

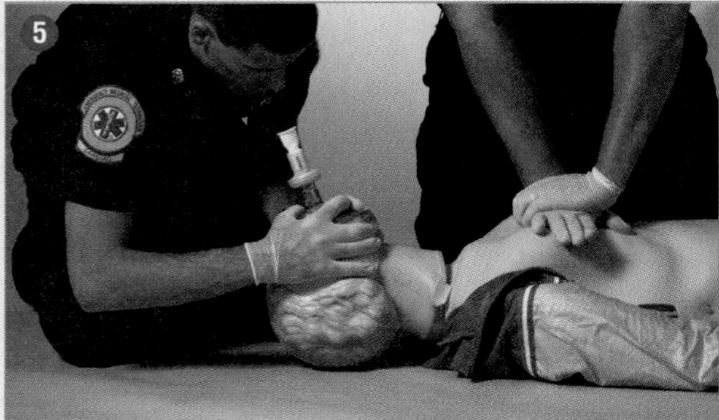

Check for signs of circulation using the carotid pulse. If there is no pulse, begin chest compressions.

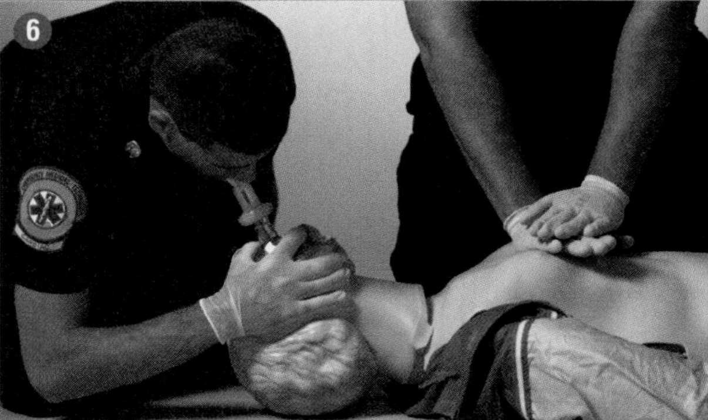

Perform rescue breathing; give two breaths. Check for signs of circulation every few minutes.

the infant's nipples. Place your index finger below that line in the middle of the chest. Place your middle and ring fingers next to your index finger. Use the middle and ring fingers to compress the sternum. Because the chest of an infant is smaller and more pliable than the chest of an adult, use only two fingers to compress the chest. Compress the sternum $^1/_2$" to 1". To achieve good results, the infant must be lying on a firm surface. Because the heart rate of an infant is faster than an adult's, you must deliver compressions at the rate of at least 100 per minute. The ratio of cardiac compressions to ventilations in an infant is 5 to 1 instead of the 15-to-2 ratio used for adults. The size of the infant makes it harder to perform two-rescuer CPR.
8. Continue compressions and ventilations. Give one slow ventilation after each set of five compressions during a 1-second to $1^1/_2$-second pause.
9. Reassess the victim after 20 sets of compressions and ventilations (about 1 minute) and every few minutes thereafter.

One-Rescuer Child CPR

A child is defined as a person between the ages of 1 and 8 years. The steps for child CPR are almost the same as for an adult; however, some steps must be modified for a child. These variations include:

- Use less force to ventilate the child. Ventilate only until the child's chest rises.
- Use only one hand to depress the sternum 1" to $1^1/_2$".
- Use less force to compress the child's chest.
- Give one rescue breath after every five chest compressions for a child, instead of two rescue breaths after 15 chest compressions for an adult.

To administer CPR to a child, follow the steps in Skill Drill 24-16.

1. Establish the child's level of responsiveness. Tap and gently shake the shoulder and shout, "Are you okay?" If a second rescuer is available, have him or her activate the EMS system.
2. Turn the child on his or her back as you support the head and neck.
3. Open the airway. Use the head tilt-chin lift technique or, if the child is injured, use the jaw-thrust technique. Maintain the open airway.
4. Check for breathing. Place the side of your face and your ear close to the nose and mouth of the child. Look, listen, and feel for the movement of air: Look for movement of the chest, listen for sounds of air exchange, and feel for air movement on the side of your face. Check for breathing for at least 3 to 5 seconds. If signs of breathing are absent, place your mouth over the child's mouth, seal the child's nose with your thumb and index finger, and begin mouth-to-mouth rescue breathing. A mouth-to-mask ventilation device may be used.
5. Give two effective breaths. Blow slowly for 1 to $1^1/_2$ seconds, using just enough force to make the chest rise. Allow the lungs to deflate between breaths.
6. Check for circulation. Locate the larynx with your index and middle fingers. Slide your fingers into the groove between the larynx and the muscles at the side of the neck to feel for the carotid pulse. Check for at least 5 to 10 seconds. If the pulse is absent, proceed to the next step. (If the pulse is present, continue rescue breathing.)
7. Begin chest compressions. Place the heel of one hand on the lower half of the sternum, two finger widths above the xiphoid process. Apply five compressions, using the heel of one hand. Each compression should be about 1" to $1^1/_2$" and at the rate of 100 compressions per minute. Count the compressions out loud: "One and two and three and. . . ."
8. After five chest compressions, ventilate the victim's lungs. Deliver one effective breath.
9. Continue compressions and ventilations. Continue a sequence of five compressions followed by one ventilation.
10. Check for a pulse. After 1 minute and every few minutes thereafter, check for a carotid pulse.

In large children, you may need to use two hands to achieve an adequate depth of compression.

Signs of Effective CPR

It is important to know the signs of effective CPR so you can assess your efforts to resuscitate the victim. The signs of effective CPR are:

1. A second rescuer feels a carotid pulse while you are compressing the chest.
2. The victim's pupils constrict when they are exposed to light.
3. The victim's skin color improves (from blue to pink).
4. Independent breathing begins or the victim gasps.
5. An independent heartbeat, which is the goal of CPR, begins. This does not occur often without defibrillation and other advanced life support procedures.

If some of these signs are not present, evaluate your technique to see if it can be improved.

Complications of CPR

A discussion of CPR would not be complete without mention of its complications: broken bones, gastric distention, and vomiting. They can be minimized by using proper technique.

Broken Bones

If your hands slip to the side of the sternum during chest compressions, or if your fingers rest on the ribs, you may break ribs while delivering a compression. To prevent this, use proper hand positioning and do not let your fingers come in contact with the ribs. If you hear a cracking sound while performing CPR, check and correct your hand position but continue CPR. Sometimes you may break bones or cartilage even with proper CPR technique.

Vomiting

Vomiting is common during CPR, so you must be prepared to deal with it. There is not much you can do to prevent vomiting, except to keep air out of the victim's stomach. Vomiting is likely to occur if the patient has suffered cardiac arrest. When cardiac arrest occurs, the muscle that keeps food in the stomach relaxes. If there is any food in the stomach, it backs up, causing the patient to vomit.

If the victim vomits as you are administering CPR, immediately turn the victim onto his or her side to allow the vomitus to spill out of the mouth. Then clear the victim's mouth of remaining vomitus, first with your fingers and then with a clean cloth (if one is handy).

The victim may experience several rounds of vomiting, so you must be prepared to take these actions repeatedly. EMS units carry a suction machine that can clear the victim's mouth. As a fire fighter, however, you cannot wait until the suction machine arrives before beginning or resuming CPR. You must simply deal with the vomiting as it occurs.

Do your best to clear any vomitus from the victim's airway. If the airway is not cleared, three problems may arise:

- The victim may breathe in (aspirate) the vomitus.
- You may force vomited material into the lungs with the next artificial ventilation.
- It takes a strong stomach and the realization that you are trying to save the victim's life to continue with resuscitation after the victim has vomited. But you must continue. Remove the vomitus with a towel, the victim's shirt, your fingers, or any other available object. As soon as you have cleared away the vomitus, take a deep breath and continue rescue breathing.

Creating Sufficient Space for CPR

As a fire fighter, you may find yourself alone with a victim experiencing cardiac arrest. One of the first things you must do is to create or find a space where you can perform CPR. Ask yourself, "Is there enough room in this location to perform CPR?" To perform CPR effectively, you need 3′ to 4′ of space on all sides of the victim. This will give enough space so that rescuers can change places, advanced life support procedures can be implemented, and an ambulance stretcher can be brought in.

If there is not enough space around the victim, you have two options:

- Quickly rearrange the furniture in the room to make space.

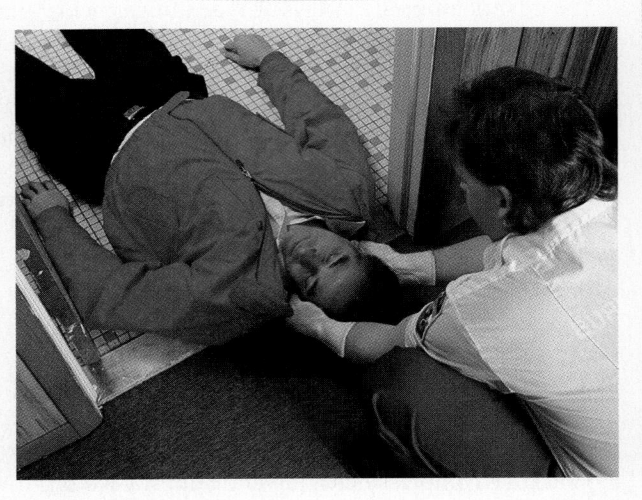

Figure 24-21 Create sufficient space for CPR.

- Quickly drag the victim into an area that has more room, for instance, out of the bathroom and into the living room—not the hallway (▲ Figure 24-21).

Space is essential to a smooth rescue operation for a cardiac arrest victim. It only takes 15 to 30 seconds either to clear a space around the victim or to move the victim into a larger area.

CPR Training

As a fire fighter, you should successfully complete a CPR course through a recognized agency such the American Heart Association (AHA), the National Safety Council (NSC), or the American Red Cross (ARC). You should also regularly update your skills by successfully completing a recognized recertification course.

You cannot achieve proficiency in CPR unless you have adequate practice on adult and infant manikins. Your department should schedule periodic reviews of CPR theory and practice for all people who are trained as fire fighters.

Legal Implications of CPR

Living wills, advance directives, and DNR orders are legal documents that specify the patient's wishes regarding specified medical procedures. Fire fighters sometimes wonder if they should start CPR on a person who has a living will or an advance directive. Because you are not in a position to determine if the living will or advance directive is valid, CPR should be started on all patients unless they are obviously dead. If a patient has a living will or advance directive, the physician at the hospital will determine whether or not to stop CPR. You should check your department's protocols and state regulations on this matter.

Do not hesitate to begin CPR on a pulseless, nonbreathing victim. Without your help, the victim will certainly die.

You may have legal problems if you begin CPR on a victim who does not need it and this action harms the victim. However, the chances of this happening are minimal if you assess the victim carefully before beginning CPR.

Another potential legal pitfall is abandonment—the discontinuing of CPR without the order of a licensed physician or without turning the victim over to someone who is at least as qualified as you are.

If you avoid these pitfalls, you need not be overly concerned about the legal implications of performing CPR. Your most important protection against a possible legal suit is to become thoroughly proficient in the theory and practice of CPR.

Bleeding and Shock

This section presents the skills you need to recognize and care for victims who are suffering from shock, bleeding, or soft-tissue injuries. Because most soft-tissue injuries result in bleeding, maintaining good body substance isolation is important when you are caring for these injuries. The section describes four types of wounds: abrasions, lacerations, punctures, and avulsions. Techniques for controlling external bleeding are stressed. It is important that you learn the techniques for dressing and bandaging wounds presented here.

Damage to internal soft tissues and organs can cause life-threatening problems. Internal bleeding causes the victim to lose blood in the circulatory system and results in shock. More trauma victims die from shock than any other reason. Your ability to recognize the signs and symptoms of shock and to take simple measures to aid shock victims will give them the best chance for survival.

Shock

Shock is defined as failure of the circulatory system. Circulatory failure has many possible causes, but the three primary causes are discussed here.

Pump Failure

Cardiogenic shock occurs if the heart cannot pump enough blood to supply the needs of the body. Pump failure can result if the heart has been weakened by a heart attack. Inadequate pumping of the heart can cause blood to back up in the vessels of the lungs, resulting in a condition known as **congestive heart failure (CHF)**.

> **Fire Fighter Safety Tips**
>
> Most soft-tissue injuries involve some degree of bleeding. Any time you approach a victim with a potential soft-tissue injury, you need to consider your BSI strategy.

Pipe Failure

Pipe failure is caused by the expansion (dilation) of the capillaries to as much as three or four times their normal size. This causes blood to pool in the capillaries instead of circulating throughout the system. When blood pools in the capillaries, the rest of the body, including the heart and other vital organs, is deprived of blood. **Blood pressure** falls and shock results.

In shock caused by sudden expansion of the capillaries, blood pressure may drop so rapidly that you are unable to feel either a radial or a carotid pulse. Three types of shock caused by capillary expansion are:

- Shock induced by fainting
- Anaphylactic shock
- Spinal shock

The least serious type of shock caused by pipe failure is fainting. Fainting is the body's response to a major psychological or emotional stress. The capillaries suddenly expand to three or four times their normal size. Fainting is a short-term condition that corrects itself once the victim is placed in a horizontal position.

Anaphylactic shock is caused by an extreme allergic reaction to a foreign substance, such as venom from bee or insect stings, penicillin, or certain foods. Shock develops very quickly following exposure. The victim may suddenly start to itch, a rash or hives may appear, the face and tongue may swell very quickly, and a blue color may appear around the mouth. The victim appears flushed (reddish), and breathing may quickly become difficult, with wheezing sounds coming from the chest.

Blood pressure drops rapidly as the blood pools in the expanded capillaries. The pulse may be so weak that you cannot feel it. Pipe failure has occurred, and death will result if prompt action to counteract the toxin is not taken.

Spinal shock may occur in victims who have suffered a spinal cord injury. The injury to the spinal cord allows the capillaries to expand, and blood pools in the lower extremities. The brain, heart, lungs, and other vital organs are then deprived of blood, resulting in shock.

Fluid loss caused by excessive bleeding (hemorrhage) is the most common cause of shock. Blood escapes from the normally closed circulatory system through an internal or external wound, and the system's total fluid level (blood volume) drops until the pump cannot operate efficiently. To compensate for fluid loss, the heart begins to pump faster to maintain pressure in the pipes. But as the fluid continues to drain out, the pump eventually loses its prime and stops pumping altogether.

External bleeding is not difficult to detect because you can see blood escaping from the circulatory system to the outside. With internal bleeding, blood escapes from the system, but you cannot see it. Even though the escaped blood remains inside the body, it cannot reenter the circulatory system and is not available to be pumped by the heart. Whether

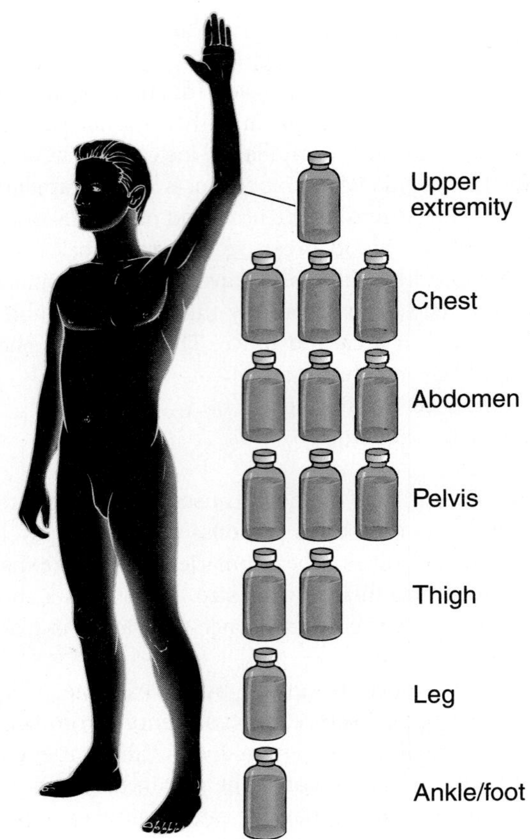

Figure 24-22 Potential blood loss from injuries in various parts of the body. Each bottle equals one pint.

external or internal, unchecked bleeding causes shock, eventual pump failure, and death.

An average adult has about 12 pints of blood circulating in the system. Losing a single pint of blood will not produce shock in a healthy adult. In fact, one pint per donor is the amount that blood banks collect. However, the loss of two or more pints of blood can produce shock. This amount of blood can be lost as a result of injuries such as a fractured femur. ▲ Figure 24-22 shows the amount of blood that can be lost as a result of various injuries.

Signs and Symptoms of Shock

Shock deprives the body of sufficient blood to function normally. As shock progresses, the body alters some of its functions in an attempt to maintain sufficient blood supply to its vital parts. A victim who is suffering from shock may exhibit some or all of the signs and symptoms shown in the Fire Fighter Tip. Initially, the victim's breathing may be rapid and deep, but as shock progresses in severity, breathing becomes rapid and shallow.

Changes in mental status may be the first signs of shock, so monitoring the overall mental status of a victim can help you detect shock. Any change in mental status may be significant. In severe cases, the victim loses consciousness. If a trauma victim who has been quiet suddenly becomes agitated, restless, and vocal, you should suspect shock. If a trauma victim who has been loud, vocal, and belligerent becomes quiet, you should also suspect shock and begin treatment. If the victim has dark skin, you will not be able to use skin color changes to help you detect shock. Therefore, you must be especially alert for other signs of shock. The capillary refill test and the condition of the skin (cool and clammy) will help you recognize shock in victims who have dark skin.

> **Fire Fighter Tips**
>
> **Signs and Symptoms of Shock**
>
> - Confusion, restlessness, or anxiety
> - Cold, clammy, sweaty, pale skin
> - Rapid breathing
> - Rapid, weak pulse
> - Increased capillary refill time
> - Nausea and vomiting
> - Weakness or fainting
> - Thirst

General Treatment for Shock

As a fire fighter, you can combat shock from any cause and keep it from getting worse by taking several simple but important steps.

To treat shock, follow these general steps (Skill Drill 24-17):

1. Position the victim correctly.
2. Maintain the victim's ABCs.
3. Treat the cause of shock, if possible.
4. Maintain the victim's body temperature by placing blankets under and over the victim.
5. Make sure the victim does not eat or drink anything.
6. Assist with other treatments (such as administering oxygen, if available).
7. Arrange for immediate and prompt transportation to an appropriate medical facility.

Position the Victim Correctly

If there is no head injury, extreme discomfort, or difficulty breathing, lay the victim flat on the back on a horizontal surface. Place the victim on a blanket, if available. Elevate the victim's legs 12″ to 18″ off the floor or ground ► Figure 24-23. This enables some blood to drain from the legs back into the circulatory system. If the victim has a head injury, do not elevate the legs.

If the victim is having chest pain or difficulty breathing (which is likely to occur in cases of heart attack and

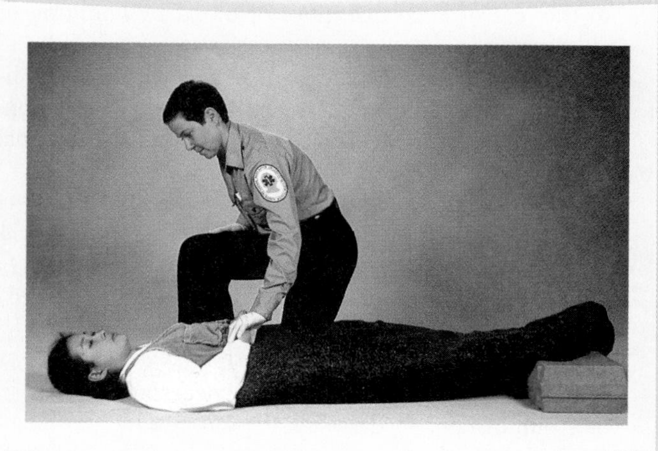

Figure 24-23 Position for treatment of shock if there is no head injury. Note the elevated legs. If the victim is experiencing difficulty breathing, place the victim in a sitting or semireclining position.

emphysema), place the victim in a sitting or semireclining position.

Maintain the Victim's ABCs

Check the victim's airway, breathing, and circulation at least every five minutes. If necessary, open the airway, perform rescue breathing, or begin CPR.

Treat the Cause of Shock if Possible

Most causes of shock must be treated in the hospital. Often this treatment consists of surgery by specially trained physicians. However, you will be able to treat one common cause of shock—external bleeding. By controlling external bleeding with direct pressure, elevation, or pressure points, you will be able to temporarily treat this cause of shock until the victim can be transported to an appropriate medical facility for more advanced treatment.

Maintain the Victim's Body Temperature

Attempt to keep the victim comfortably warm. A victim with cold, clammy skin should be covered. It is as important to place blankets under the victim to keep body heat from escaping into the ground as it is to cover the victim with blankets.

Make Sure the Victim Does Not Eat or Drink Anything

Even if a victim in shock is very thirsty, do not give liquids by mouth. There are two reasons for this:

- A victim in shock may be nauseated, and eating or drinking may cause vomiting.
- A victim in shock may need emergency surgery. Victims should not have anything in their stomachs before surgery.

If you are working in an area where ambulance response time is more than 20 minutes, you may give victims a clean cloth or gauze pad that has been soaked in water to suck. This relieves dryness of the mouth but does not quench thirst. No matter how thirsty victims are, do not permit them to drink anything.

Arrange for Prompt Ambulance Transport to an Appropriate Medical Facility

As soon as you realize that you have a victim who is suffering from shock, you should make sure that an ambulance has been dispatched. When the ambulance arrives, give the EMS personnel a concise hand-off report emphasizing the signs and symptoms of shock that you noted. The EMTs or paramedics should make sure that the victim is quickly prepared for immediate and prompt transport to a medical facility that can handle a victim with these severe problems. The cure for shock is usually surgical repair, and the sooner the victim gets to the hospital, the better the chance of survival will be.

Treatment for Shock Caused by Pump Failure

Victims suffering from pump failure may be confused, restless, anxious, or unconscious. Their pulse is usually rapid and weak. Their skin is cold and clammy, sweaty, and pale. Their respirations are often rapid and shallow.

Pump failure is a serious condition. Your proper treatment and prompt transport by ambulance to a medical center will give these victims their best chance for survival.

Follow these steps to treat shock caused by pump failure:

1. Keep the victim lying down unless breathing is easier in a sitting position.
2. Maintain the victim's ABCs. Be prepared to do CPR if necessary.
3. Conserve the victim's normal body temperature.
4. Make sure the victim does not eat or drink anything.
5. Keep the patient quiet and do any necessary moving for him or her.
6. Provide reassurance.
7. Arrange for prompt ambulance transport to an appropriate medical facility.
8. Provide high-flow oxygen as soon as it is available.

Treatment for Shock Caused by Pipe Failure

Victims who have fainted, who are suffering from anaphylactic shock, or who have suffered a severe spinal cord injury will have pipe failure. The size of their capillaries may increase three or four times, causing signs and symptoms of shock.

Follow these steps to treat shock caused by pipe failure:

1. Examine the victim to make sure there is no injury.
2. Keep the victim lying down with legs elevated 12″ to 18″ off the floor or ground. This enables the blood to drain from the legs back into the central circulatory system.
3. Maintain the ABCs.

4. Maintain the victim's normal body temperature.
5. Provide reassurance.

Treatment for Anaphylactic Shock

The initial treatment for anaphylactic shock (shock caused by an allergic reaction) is similar to the treatment for any other type of shock. Anaphylactic shock is an extreme emergency and the victim must be transported as soon as possible. Paramedics, nurses, and doctors can give medications that may reverse the allergic reaction.

Follow these steps to treat anaphylactic shock:

1. Keep the victim lying down. Elevate the legs 12" to 18" off the floor or ground.
2. Maintain the victim's ABCs. Anaphylactic shock may cause airway swelling. In severe reactions, the victim may require mouth-to-mask breathing or full CPR.
3. Maintain the victim's normal body temperature.
4. Provide reassurance.
5. Arrange for rapid transport by ambulance to an appropriate medical facility.

Treatment for Shock Caused by Fluid Loss

Shock may be caused by internal blood loss (blood that escapes from damaged blood vessels and stays inside the body) or by external blood loss (blood that escapes from the body). Excessive bleeding is the most common cause of shock.

Follow these steps to treat shock caused by external blood loss:

1. Control bleeding by applying direct pressure on the wound, elevating the injured part, and using pressure points (compressing a major artery against the bone). This is the most important step.
2. Make sure the victim is lying down. Elevate the legs 12" to 18" off the floor or ground.
3. Maintain the victim's ABCs.
4. Maintain the victim's body temperature.
5. Make sure the victim does not eat or drink anything.
6. Provide reassurance.
7. Provide high-flow oxygen as soon as it is available.
8. Arrange for prompt transportation by ambulance to an appropriate medical facility.

Shock Caused by Internal Blood Loss

Victims die quickly and quietly from internal bleeding following abdominal injuries that rupture the spleen, liver, or large blood vessels. You must be alert to catch the earliest signs and symptoms of internal bleeding and to begin treatment for shock. If you are treating several injured victims, those with internal bleeding should be transported promptly to the medical facility first, as immediate surgery may be needed.

Bleeding from stomach ulcers, ruptured blood vessels, or tumors can all cause internal bleeding and shock. This bleeding can be spontaneous, massive, and rapid, often leading to the loss of large quantities of blood by vomiting or bloody diarrhea.

It is important that you recognize the signs and symptoms of internal bleeding and take prompt corrective action.

Follow these steps to treat shock caused by internal blood loss:

1. Keep the victim lying down. Elevate the legs 12" to 18" off the floor or ground.
2. Maintain the victim's ABCs.
3. Maintain the victim's normal body temperature.
4. Make sure the victim does not eat or drink anything.
5. Provide reassurance.
6. Keep the victim quiet and do any necessary moving for him or her.
7. Provide high-flow oxygen as soon as it is available.
8. Monitor the victim's vital signs at least every five minutes.
9. Arrange for prompt transportation by ambulance to an appropriate medical facility.

Bleeding

Controlling External Blood Loss

There are three types of external blood loss: capillary, venous, and arterial (► Figure 24-24). The most common type of external blood loss is capillary bleeding. In **capillary bleeding** (such as from a cut finger), the blood oozes out. You can control capillary bleeding simply by applying direct pressure to the site.

The next most common type of bleeding is **venous bleeding**. This bleeding has a steady flow. Bleeding from a large vein may be profuse and life-threatening. To control venous bleeding, apply direct pressure.

The most serious type of bleeding is **arterial bleeding**. Arterial blood spurts or surges from the laceration or wound with each heartbeat. Blood pressure in arteries is higher than in capillaries or veins, and unchecked arterial bleeding can result in death from loss of blood in a short time. To control arterial bleeding, exert direct pressure and, if necessary, apply pressure to a pressure point. This is done by compressing a major artery against the underlying bone, as explained below. Because many injured victims actually die from shock caused by blood loss, it is vitally important that you control external bleeding quickly.

Direct Pressure

Most external bleeding can be controlled by applying direct pressure to the wound. Place a dry, sterile dressing directly on the wound and press on it with your gloved hand (► Figure 24-25). Wear the gloves from your fire fighter life support kit. If you do not have a sterile dressing or gauze bandage, use the cleanest cloth available.

To maintain direct pressure on the wound, wrap the dressing and wound snugly with a roller gauze bandage. Do

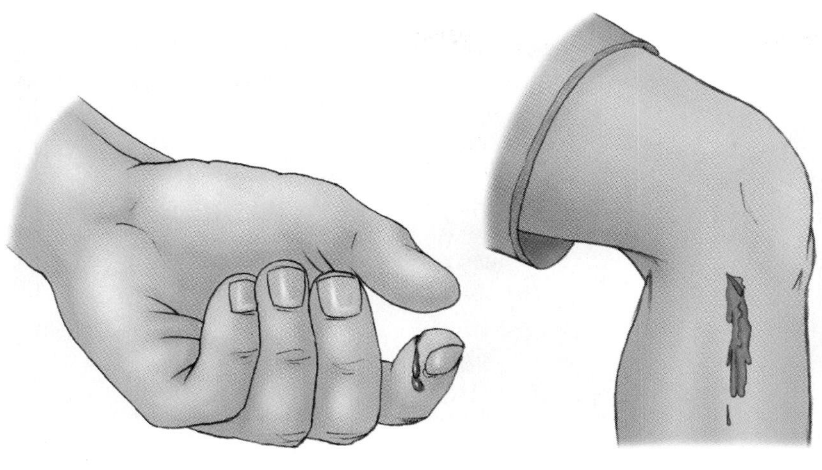
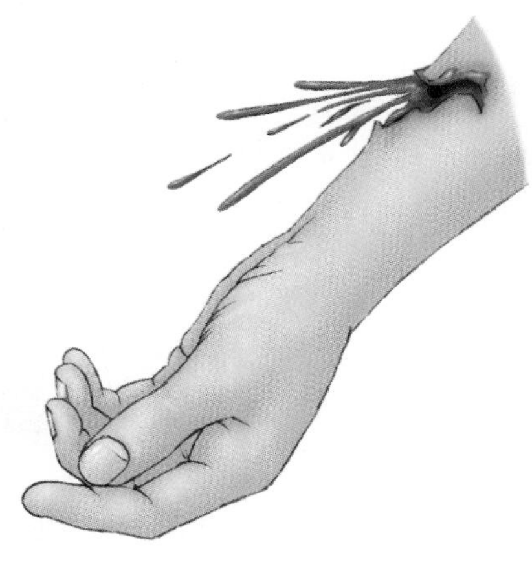

Figure 24-24 Recognizing the three types of external bleeding. **A.** Capillary **B.** Venous **C.** Arterial.

not remove a dressing after you have applied it, even if it becomes blood-soaked. Place another dressing on top of the first and keep them both in place.

According to the American Medical Association, it is extremely unlikely that you—as a fire fighter providing emergency care—will contract AIDS from a victim who is bleeding. For more information, call the Centers for Disease Control National AIDS Hotline at 1-800-342-AIDS. Direct exposure to blood must be reported to the emergency physician.

Elevation

If direct pressure does not stop external bleeding from an extremity, elevate the injured arm or leg as you maintain direct pressure. Elevation, in conjunction with direct pressure, will usually stop severe bleeding (► Figure 24-26).

Pressure Points

If the combination of direct pressure and elevation does not control bleeding from an arm or leg wound, you must try to control it indirectly by preventing blood from flowing into the limb. This is accomplished by compressing a major artery against the bone at a pressure point.

Fire Fighter Tips

A word about tourniquets: No! Tourniquets are mentioned only to discourage their use. A tourniquet is made of a band of material placed around an arm or leg above a wound and tightened to stop severe bleeding. Tourniquets are difficult to make from improvised materials. They are also difficult to apply properly. If a tourniquet is applied improperly, it can actually increase bleeding.

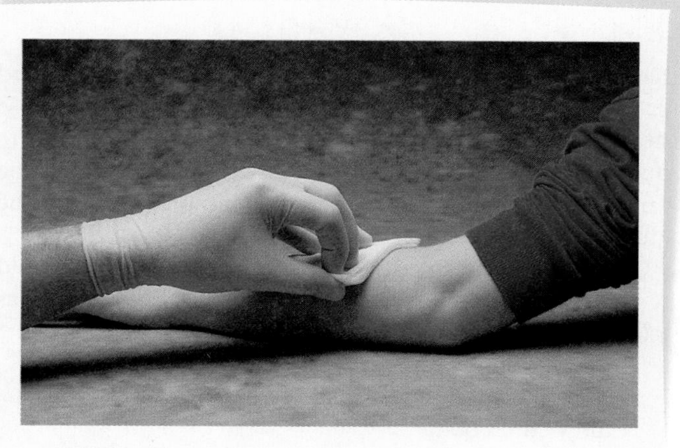

Figure 24-25 Applying direct pressure to a wound.

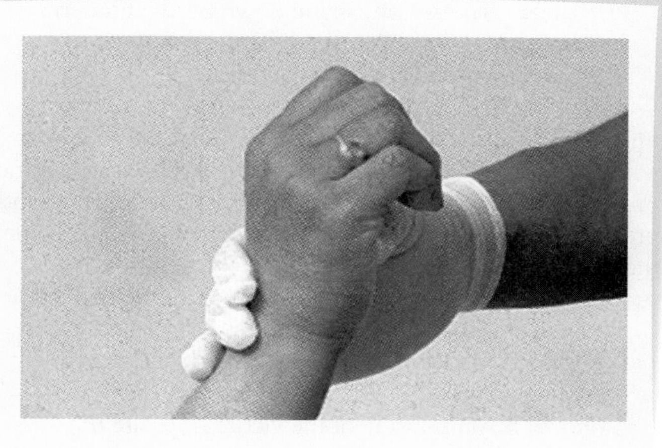

Figure 24-26 Elevate an arm while maintaining direct pressure to control external bleeding.

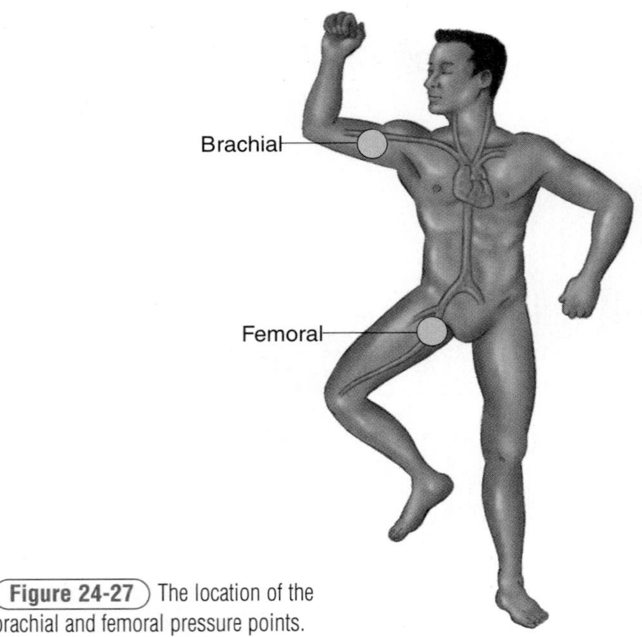

Figure 24-27 The location of the brachial and femoral pressure points.

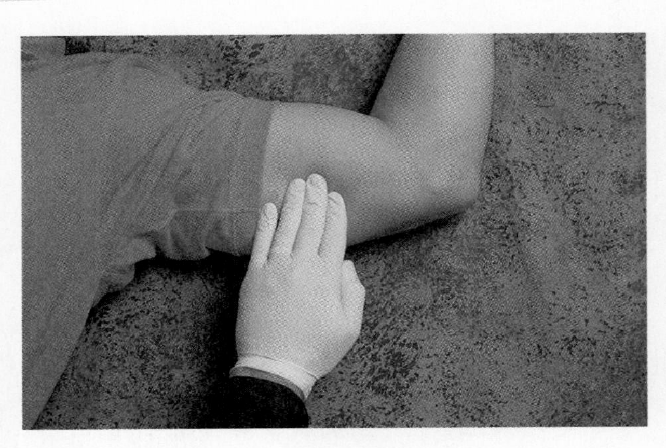

Figure 24-28 Applying pressure to the brachial artery.

Compressing the artery at a pressure point stops blood flow in much the same way that stepping on a garden hose stops the flow of water. Although there are several **pressure points** in the body, the brachial artery pressure point (in the upper arm) and the femoral artery pressure point (in the groin) are the most important (▲ Figure 24-27).

In applying pressure to the brachial artery, you should remember the words "slap, slide, and squeeze." Proceed as follows:

1. Position the victim's arm so the elbow is bent at a right angle (90°), and hold the upper arm away from the victim's body.
2. Gently "slap" the inside of the biceps with your fingers halfway between the shoulder and the elbow to push the biceps out of the way.
3. "Slide" your fingers up to push the biceps away.
4. "Squeeze" (press) your hand down on the humerus (upper arm bone). You should be able to feel the pulse as you press down.

If the victim is sitting down, "squeeze" by placing your fingers halfway between the shoulder and the elbow and your thumb on the opposite side of the victim's arm. If done properly, this technique (in combination with direct wound pressure and elevation) will quickly stop any bleeding below the point of application (► Figure 24-28).

The femoral artery pressure point is more difficult to locate and squeeze. Proceed as follows:

1. Position the victim on his or her back and kneel next to the victim's hips, facing the victim's head. You should be on the side of the victim opposite the extremity that is bleeding.
2. Find the pelvis and place the little finger of your hand closest to the injured leg along the front crest on the injured side (► Figure 24-29A).
3. Rotate your hand down firmly into the groin area between the genitals and the pelvic bone. This compresses the femoral artery and usually stops the bleeding, when combined with elevation and direct pressure over the bleeding site (► Figure 24-29B).
4. If the bleeding does not slow immediately, reposition your hand and try again.

Wounds

A wound is an injury caused by any physical means that leads to damage of a body part. Wounds are classified as closed or open. In a **closed wound**, the skin remains intact; in an **open wound**, the skin is disrupted.

Closed Wounds

The only closed wound is a **bruise** (contusion). A bruise is an injury of the soft tissue beneath the skin. Because small blood vessels are broken, the injured area becomes discolored and swells. The severity of these closed soft-tissue injuries varies greatly. A simple bruise heals quickly. In contrast, bruising and swelling following an injury may also be a sign of an underlying fracture that could take months to heal. Whenever you encounter a significant amount of swelling or bruising, suspect a possible underlying fracture.

Open Wounds

Abrasion

Abrasions are commonly called scrapes, road rashes, or rug burns. They occur when the skin is rubbed across a rough surface (► Figure 24-30).

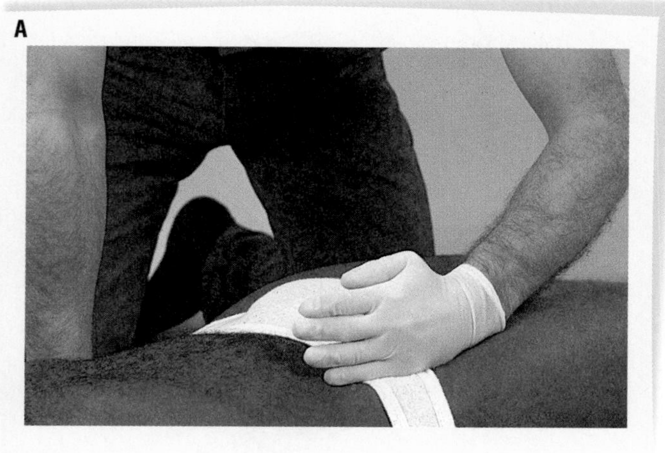

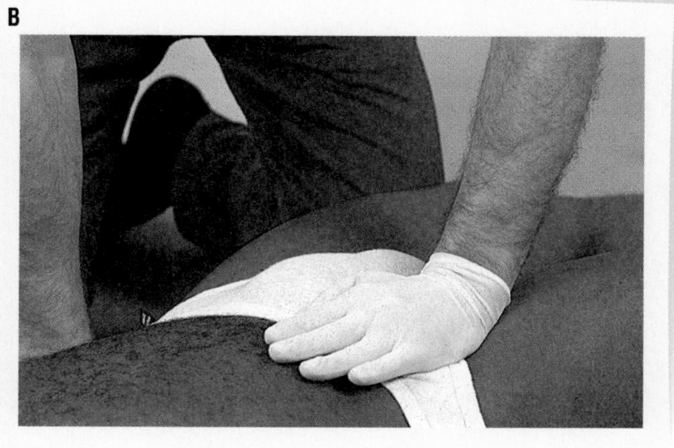

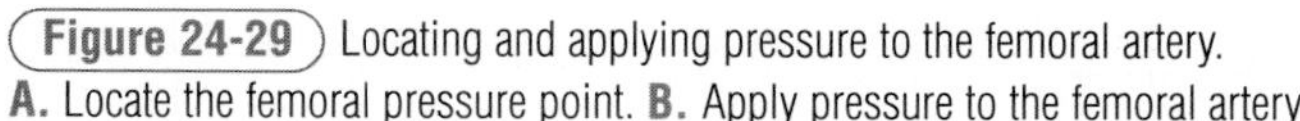

Figure 24-29 Locating and applying pressure to the femoral artery. **A.** Locate the femoral pressure point. **B.** Apply pressure to the femoral artery.

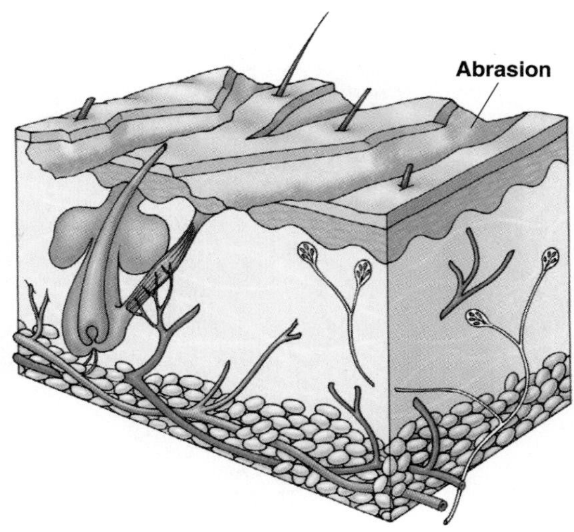

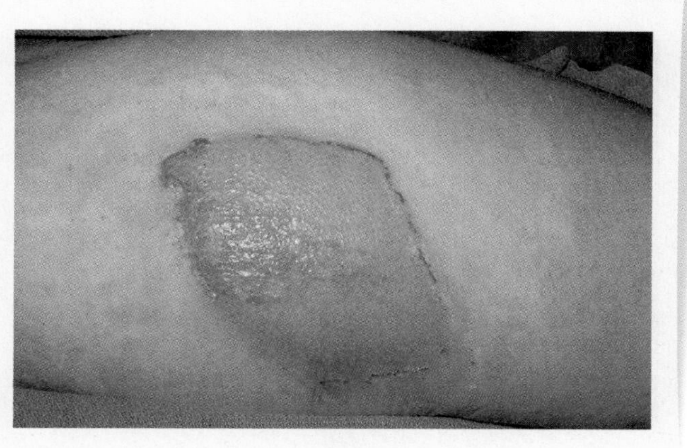

Figure 24-30 Abrasions involve variable depths of the skin; they are often called scrapes or road rashes.

Puncture

Puncture wounds are caused by a sharp object that penetrates the skin (► Figure 24-31). These wounds may cause significant deep injury that is not immediately recognized. Puncture wounds do not bleed freely. If the object that caused the puncture wound remains sticking out of the skin, it is called an impaled object.

A gunshot wound is a special type of puncture wound. The amount of damage done by a gunshot depends on the type of gun used, and the distance between the gun and the victim. A gunshot wound may appear as an insignificant hole but can do massive damage to internal organs. Some gunshot wounds are smaller than a dime, and some are large enough to destroy significant amounts of tissue. Gunshot wounds usually have both an entrance wound and an exit wound. The entrance wound is usually smaller than the exit wound. Most deaths from gunshot wounds result from internal blood loss caused by damage to internal organs and major blood vessels as the bullet passes through the body. There is often more than one gunshot wound. A thorough victim examination is important to be sure that you have discovered all of the entrance and exit wounds.

Laceration

The most common type of open wound is a laceration (► Figure 24-32). Lacerations are commonly called cuts. Minor lacerations may require little care, but large lacerations can cause extensive bleeding and even be life-threatening.

Avulsion

An avulsion is a tearing away of body tissue (► Figure 24-33). The avulsed part may be totally severed from the body or it may be attached by a flap of skin. Avulsions may involve small or large amounts of tissue. If an entire body part is torn away,

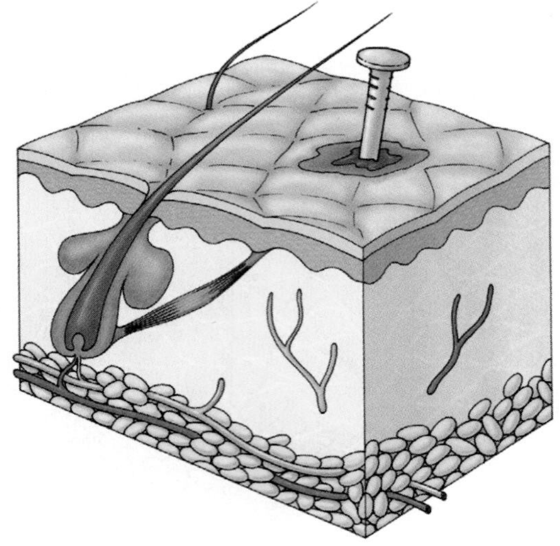

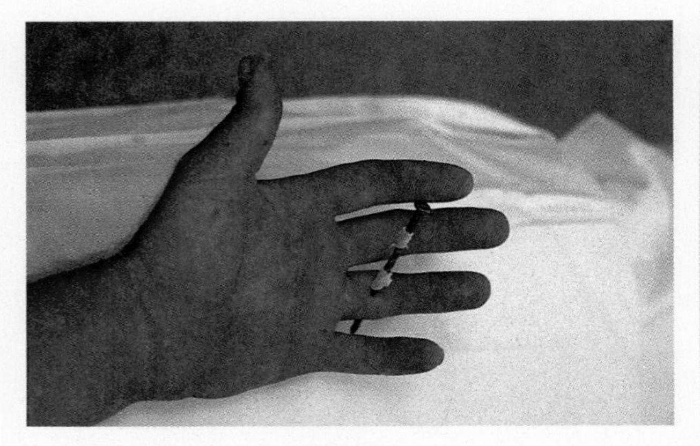

Figure 24-31 Puncture wounds may penetrate the skin to any depth.

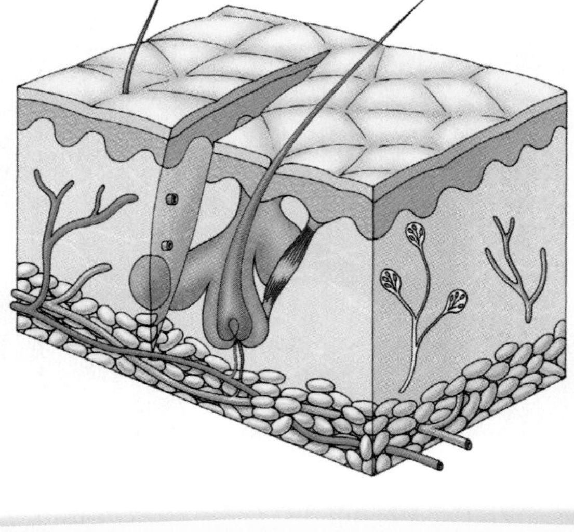

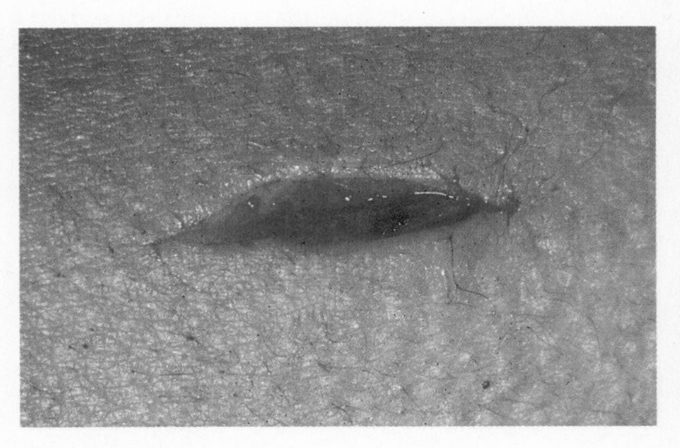

Figure 24-32 Lacerations are cuts produced by sharp objects.

the wound is called a traumatic amputation (► Figure 24-34). Any amputated body part should be located, placed in a clean plastic bag, kept cool, and taken with the victim to the hospital for possible reattachment (reimplantation). If a clean plastic bag is not available, you can use a surgical glove turned inside out. Cold pacs or iced water can be used to keep the detached body parts cold.

Principles of Wound Treatment

Very minor bruises need no treatment. Other closed wounds should be treated by applying ice and gentle compression, and by elevating the injured part. Because extensive bruising may indicate an underlying fracture, all contusions should be splinted.

The major principles of open-wound treatment are:

- Control bleeding.
- Prevent further contamination of the wound.
- Immobilize the injured part.
- Stabilize any impaled object.

It is important to stop bleeding as quickly as possible using the cleanest dressing available. You can usually control bleeding by covering an open wound with a dry, clean, or sterile dressing and applying pressure to the dressing with your hand. If the first dressing does not control the bleeding, reinforce it with a second layer. Additional ways to control bleeding include elevating an extremity and using pressure points.

A dressing should cover the entire wound to prevent further contamination. Do not attempt to clean the contaminated wound in the field, because cleaning will only cause more bleeding. A thorough cleaning will be done at the hospital. All dressings should be secured in place by a compression bandage. Learning to dress and bandage wounds requires practice. As a trained fire fighter, you should be able to bandage all parts of the body quickly and competently (► Figure 24-35).

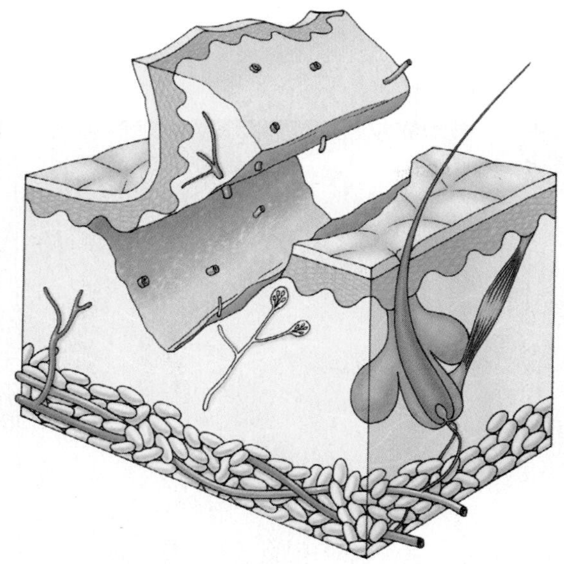

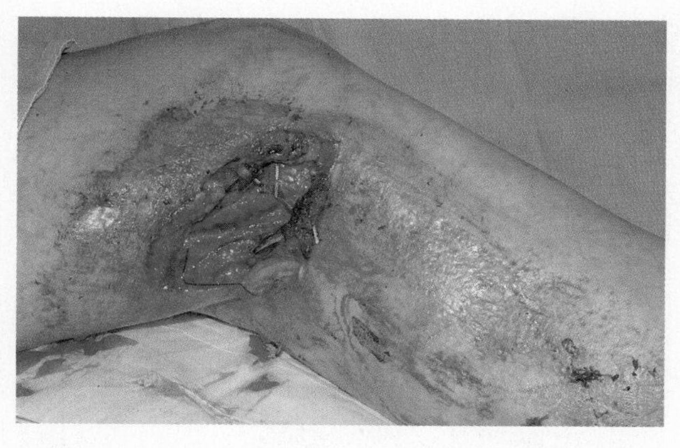

Figure 24-33 Avulsions raise flaps of tissue; significant bleeding is common.

Dressings

A dressing is an object placed directly on a wound to control bleeding and prevent further contamination. Once a dressing is in place, apply firm, direct, manual pressure on it to stop the bleeding. It is important to stop severe bleeding as quickly as possible using the cleanest dressing available. If no equipment is available, you may have to apply direct pressure with your hand to a wound that is bleeding extensively, even though you should normally wear gloves.

Sterile dressings come packaged in many different sizes. Three of the most common sizes are 4″ by 4″ gauze squares (commonly known as 4 x 4s). Heavier pads measure 5″ by 9″ (5 x 9s). A trauma dressing is a thick sterile dressing that measures 10″ by 30″ (► Figure 24-36). Use a trauma dressing to cover a large wound on the abdomen, neck, thigh, or scalp—or as padding for splints.

When you open a package containing a sterile dressing, touch only one corner of the dressing (► Figure 24-37). Place it on the wound without touching the side of the dressing that will be next to the wound. If bleeding continues after you have applied a compression dressing to the wound, put additional gauze pads over the original dressing. Do not remove the original dressing because the blood-clotting process will have already started and should not be disrupted. When you are satisfied that the wound is sufficiently dressed, proceed to bandage.

Bandaging

A bandage is used to hold the dressing in place. Two types of bandages commonly used in the field are roller gauze and triangular bandages. The first type, conforming roller gauze, stretches slightly and is easy to wrap around the body part

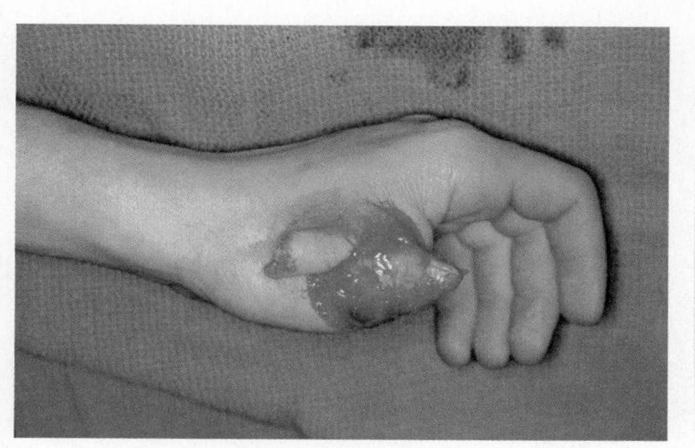

Figure 24-34 An amputated thumb. Amputated parts can often be reattached. You should attempt to locate the part and transport it to the hospital with the victim.

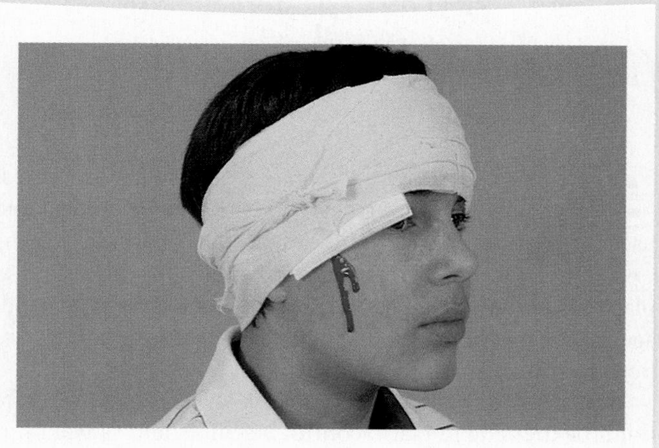

Figure 24-35 Head bandage.

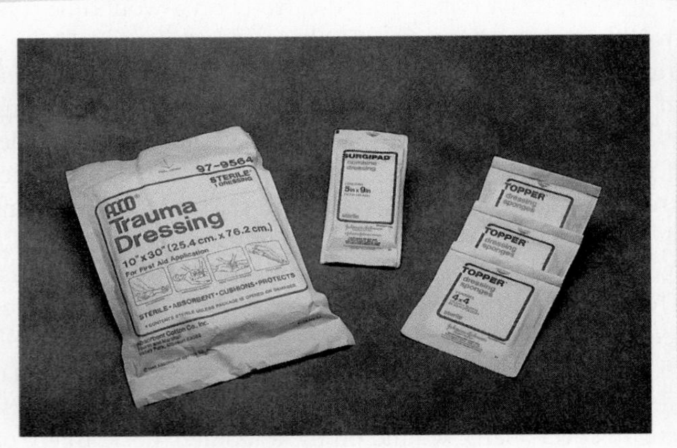

Figure 24-36 Common sizes of wound dressings are 10" x 30", 5" x 9", and 4" x 4".

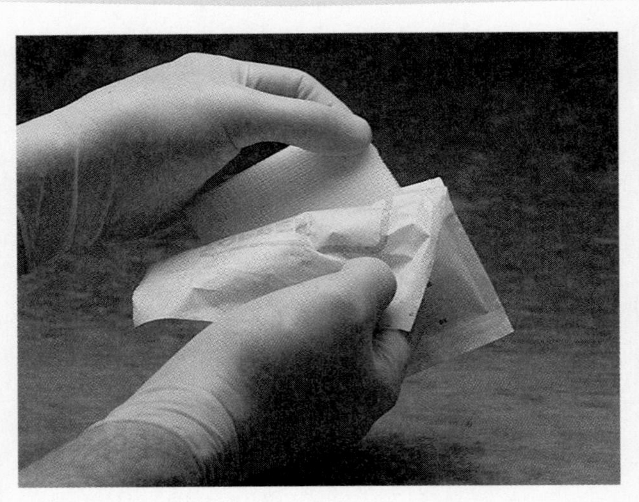

Figure 24-37 Open the package containing a sterile dressing carefully.

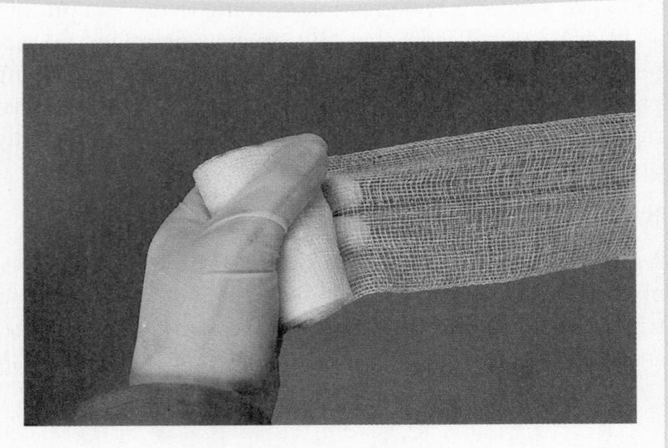

Figure 24-38 Roller gauze bandage.

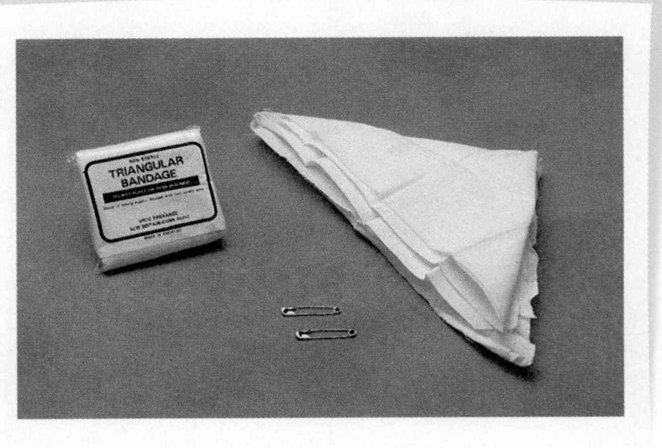

Figure 24-39 Triangular bandage.

▲ Figure 24-38. Triangular bandages are usually 36" across ► Figure 24-39. A triangular bandage can be folded and used as a wide cravat, or it can be used without folding ► Figure 24-40. Roller gauze is easier to apply and stays in place better than a triangular bandage, but a triangular bandage is very useful for bandaging scalp lacerations and lacerations of the chest, back, or thigh.

You must follow certain principles if the bandage is to hold the dressing in place, control bleeding, and prevent further contamination. Before you apply a bandage, check to ensure that the dressing completely covers the wound and extends beyond all sides of the wound ► Figure 24-41. Wrap the bandage just tightly enough to control bleeding.

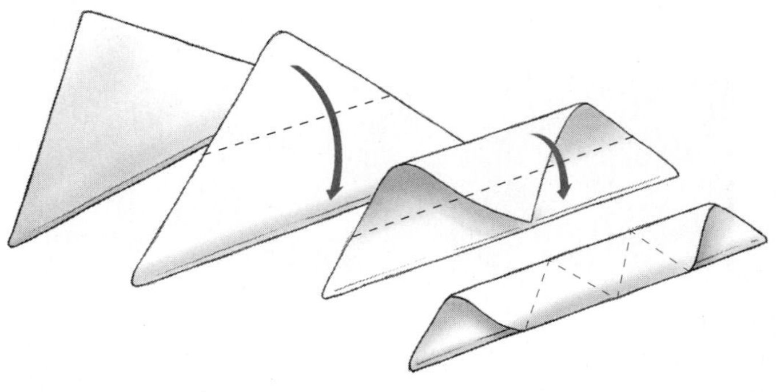

Figure 24-40 Folding a triangular bandage to make a cravat.

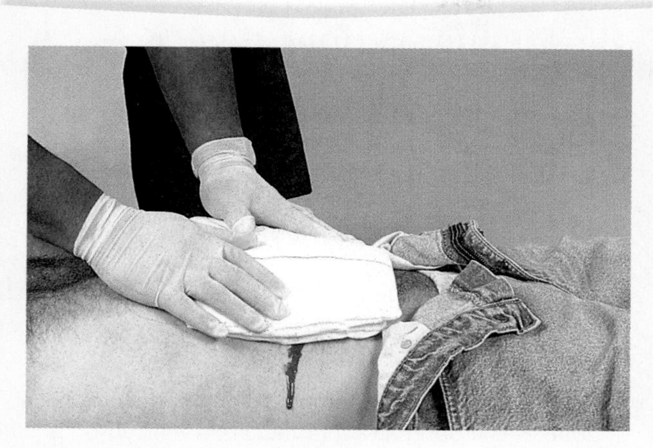

Figure 24-41 Check to make sure the dressing completely covers the wound.

Do not apply it too tightly because it may cut off all circulation. It is important to regularly check circulation at a point further away from the heart than the injury itself, because swelling may make the bandage too tight. If this happens while the victim is under your care, remove the roller gauze or triangular bandage and reapply it, making sure that you do not disturb the dressing beneath.

Once bandaging is complete, secure the bandage so it cannot slip. Tape, tie, or tuck in any loose ends. Practice these bandaging techniques for several types of wounds using both roller gauze and triangular bandages. Although the principles of bandaging are simple, some parts of the body are difficult to bandage. It is important to practice bandaging different parts of the body to assure competency in emergency situations.

Mass-Casualty Incidents

A mass-casualty incident refers to any call that involves multiple victims, as well as any situation that places such a great demand on available equipment or personnel that the system is stretched to its limit or beyond. Airplane, bus, or railroad crashes, and earthquakes are obvious examples of mass-casualty incidents.

Triage

Triage is essential at all mass-casualty incidents. Triage is the sorting of two or more victims based on the severity of their conditions to establish priorities for care based on available resources.

In a smaller scale mass-casualty incident, the first provider on scene with the highest level of training usually begins the triage process. When back-up ambulances and crews are readily available, victims are ranked in order of severity. The victim with the most severe injuries is given priority attention. After counting the number of victims and notifying the dispatcher of the additional help that is needed, initial assessment of all victims begins. As personnel arrive, the fire fighter should assign crews and equipment to priority victims first.

Triage at a large-scale mass-casualty incident should be done in several steps. The following triage steps are accepted by the majority of larger scale mass-casualty operations:

- Life-saving care rapidly administered to those in need.
- Color coding to indicate priority for treatment and transportation at the scene. Red-tagged victims are the first priority, yellow-tagged victims are the second priority, and those tagged green or black are the lowest priority.
- Rapid removal of red-tagged victims for field treatment and transportation as ambulances are available.
- Use of a separate treatment area to care for red-tagged victims if transport is not immediately available. Yellow-tagged victims can also be monitored and cared for in the treatment area while waiting for transportation.
- When there are more victims waiting for transport than there are ambulances, the Transportation Sector Officer decides which victim is the next to be loaded.
- Specialized transportation resources (air ambulance, paramedic ambulances, etc.) require separate decisions when these resources are available but limited.

Triage Priorities

Victims should be color coded early to visually identify the severity of the condition and to eliminate the need for individual victim assessments to be performed by each fire fighter who comes along later.

Victims with red tags should later be reassessed in the treatment area to determine who should receive limited resources such as paramedic assessment and care. The sorting of multiple red-tagged victims in the treatment area who need to be seen immediately by paramedics will depend on the number of paramedics available at that time. The order in which the victims will be transported is determined after the initial triage and treatment is completed.

If victims are entrapped, extrication is required. If circumstances such as heavy smoke or a potential hazardous materials exposure exist, triage will be difficult if not impossible. The immediate concern will be the removal of the victims to a safe area for triage to be done later. The triage area is the name usually given to such a collecting area for victims to be initially triaged and color coded. For victims located in nonhazardous areas, this initial triage can begin right away.

Violent Situations

The safety of you and your team is of primary concern. Civil disturbances, domestic disputes, and crime scenes, especially those involving gangs, can create many hazards for firefighting personnel. Large gatherings of hostile or potentially hostile people are also dangerous. Several agencies will respond to large civil disturbances. In these instances, it is important for you to know who is in command and will be issuing orders. However, you may be on your own when a group of people seems to grow larger and become increasingly hostile. In these cases, call law enforcement immediately if they are not already there. You may need to wait for law enforcement to arrive before you can begin treatment.

Remember that you and your crew must be protected from the dangers at the scene before you can provide victim care. Law enforcement must make sure the scene is safe before you and your partner enter. A crime scene often poses potential problems for fire fighters. If the perpetrator is still somewhere on the scene, this person could reappear and threaten you or attempt to further injure the victim you are treating. Bystanders who are trying to be helpful may interfere with you. Family members may be very distraught and not understand what you are doing when you attempt to splint an injured extremity and the victim cries out with pain. Be sure that you have adequate assistance from the appropriate public safety agency in these cases.

Behavioral Emergencies

Many emergency situations include people who are suffering from mental illness. While most people with mental problems are not violent, look for the following indicators that may be associated with violence in some people:

- Past history—has this victim previously exhibited hostile, overly aggressive, or violent behavior? This information should be solicited by personnel at the scene or requested from law enforcement personnel, family, previous records, or hospital information.
- Posture—how is this person sitting or standing? Does the victim appear to be tense, rigid, or sitting on the edge of the bed, chair, or wherever he or she is positioned? The observation of increased tension by physical posture is often a warning signal for hostility.
- Vocal activity—what is the nature of the speech the victim is using? Loud, obscene, erratic, and bizarre speech patterns usually indicate emotional distress. The victim who is conversing in quiet, ordered speech is not as likely to strike out against others as is the victim who is yelling and screaming.
- Physical activity—perhaps one of the most demonstrative factors to look for is the motor activity of a person who is undergoing a behavioral crisis. The victim who is pacing, cannot sit still, or is displaying protection of his or her boundaries of personal space needs careful watching. Agitation is a prognostic sign to be observed with great care and scrutiny.
- Other factors to take into consideration for potential violence include the following:
  - Poor impulse control
  - The combination of truancy, fighting, and uncontrollable temper
  - Instability of family structure, inability to keep a steady job
  - Substance abuse
  - Functional disorder (If the victim says that he or she is hearing voices that say to kill, believe it!)
  - Depression, which accounts for 20% of violent attacks

Fire Fighter Safety Tips

Remember that your personal safety is of the utmost importance. As part of your scene size-up, evaluate the scene for the potential for violence. If violence is a possibility, call for additional resources. Rely on the advice of law enforcement personnel, as they have more experience and expertise in handling these situations.

Wrap-Up

Ready for Review

This chapter covers the concepts of universal precautions and body substance isolation, which are needed to protect you and your victims from infectious diseases. It describes the steps for performing CPR by determining responsiveness, checking and correcting the airway, checking and correcting breathing, and checking and correcting circulation. The steps for performing CPR on adults, children, and infants are covered. The steps needed to relieve an airway obstruction are also detailed.

This chapter describes the signs and symptoms of arterial, venous, and capillary bleeding. It describes how to control bleeding by using direct pressure, elevation, and pressure points. It describes the signs and symptoms of shock and details the steps needed to treat a victim suffering from shock.

The importance of assuring your safety, the safety of other rescuers, the safety of victims, and safety of bystanders at emergency medical incidents is emphasized. And this chapter describes the concept of triage and how it is used at mass casualty incidents to provide the best treatment to large numbers of victims.

The knowledge and skills covered in this chapter are meant to give you limited knowledge and skills in dealing with some medical emergencies. The knowledge presented in this chapter is not intended to replace or substitute formal training courses in CPR or courses leading to certification as an emergency medical technician or paramedic. Most states have training and certification requirements for people who regularly respond to medical emergencies. It is important for you to follow the requirements imposed by your state regulations and local ordinances.

Chief Concepts

- Body substance isolation prevents the spread of disease organisms.
- The respiratory system is made up of many different parts that all work together.
- The airway must be open in order to perform CPR.
- Check and correct the victim's breathing.
- You can remove an obstruction in the victim's upper airway.
- The circulatory system is made up of interrelated parts.
- When the heart stops, the victim is in cardiac arrest.
- Airway, breathing, and circulation make up the components of CPR.
- The cardiac chain of survival is only as strong as its weakest link.
- External chest compressions are used to provide circulation.
- Body substance isolation is important when dealing with soft tissue injuries.
- The circulatory system consists of a pump, pipe, and fluids.
- Shock can be caused by pump failure, pipe failure, or fluid failure.
- Bleeding can be internal or external.
- Wounds require treatment to control bleeding and infection.
- Mass casualty incidents require triage to provide the best care to the greatest number of victims.

Hot Terms

Abrasion Loss of skin as a result of a body part being rubbed or scraped across a rough or hard surface.

Acquired immunodeficiency syndrome (AIDS) An immune disorder caused by HIV, resulting in an increased vulnerability to infections and to certain rare cancers.

Airway The passages from the openings of the mouth and nose to the air sacs in the lungs through which air enters and leaves the lungs.

Airway obstruction Partial or complete obstruction of the respiratory passages resulting from blockage by food, small objects, or vomitus.

Alveoli The air sacs of the lungs where the exchange of oxygen and carbon dioxide takes place.

Anaphylactic shock Severe shock caused by an allergic reaction to food, medicine, or insect stings.

Arterial bleeding Serious bleeding from an artery in which blood frequently pulses or spurts from an open wound.

Atrium Either of the two upper chambers of the heart.

Avulsion An injury in which a piece of skin is either torn completely loose from all of its attachments or is left hanging by a flap.

Blood pressure The pressure of the circulating blood against the walls of the arteries.

Body substance isolation (BSI) An infection control concept that treats all bodily fluids as potentially infectious.

Brachial artery The major vessel in the upper extremity that supplies blood to the arm.

Bronchi The two main branches of the windpipe that lead into the right and left lungs. Within the lungs, they branch into smaller airways.

Bruise Injury caused by a blunt object striking the body and crushing the tissue beneath the skin; also called a contusion.

Capillaries The smallest blood vessels that connect small arteries and small veins. Capillary walls serve as the membrane to exchange oxygen and carbon dioxide.

Capillary bleeding Bleeding from the capillaries in which blood oozes from the open wound.

Cardiac arrest A sudden ceasing of heart function.

Cardiogenic shock Shock resulting from inadequate functioning of the heart.

Cardiopulmonary resuscitation (CPR) The artificial circulation of the blood and movement of air into and out of the lungs in a pulseless, nonbreathing victim.

Carotid artery The major artery that supplies blood to the head and brain.

Circulatory system The heart and blood vessels, which together are responsible for the continuous flow of blood throughout the body.

Closed wound Injury in which soft-tissue damage occurs beneath the skin even though there is no break in the surface of the skin.

Congestive heart failure (CHF) Heart disease characterized by breathlessness, fluid retention in the lungs, and generalized swelling of the body.

Decapitation Occurs when the head is separated from the rest of the body.

Dependent lividity The red or purple color that appears on the parts of the victim's body closest to the ground. It is caused by blood seeping into the tissues on the dependent, or lower, part of the person's body.

Diaphragm A muscular dome that separates the chest from the abdominal cavity. Contraction of the diaphragm and the chest wall muscles brings air into the lungs; relaxation expels air from the lungs.

Do Not Resuscitate (DNR) Written documentation giving orders to medical personnel to not attempt resuscitation in the event of cardiac arrest.

Dressing A bandage.

Esophagus The tube through which food passes. It starts at the throat and ends at the stomach.

External chest compressions A means of applying artificial circulation by applying rhythmic pressure and relaxation on the lower half of the sternum.

Femoral artery The principal artery of the thigh.

Gastric distention Inflation of the stomach caused when excessive pressures are used during artificial ventilation and air is directed into the stomach rather than the lungs.

Head tilt-chin lift technique Opening the airway by tilting the victim's head backward and lifting the chin forward, bringing the entire lower jaw with it.

Wrap-Up

Heimlich maneuver A series of manual thrusts to the abdomen to relieve upper airway obstruction.

Hepatitis B virus (HBV) A virus that causes inflammation of the liver.

Hepatitis C virus (HCV) A virus that causes inflammation of the liver. It is transmitted through blood.

Human immunodeficiency virus (HIV) The virus that causes AIDS.

Hypothermia A condition in which the internal body temperature falls below 95°F after prolonged exposure to cool or freezing temperatures.

Jaw-thrust technique Opening the airway by bringing the victim's jaw forward without extending the neck.

Laceration An irregular cut or tear through the skin.

Larynx A structure composed of cartilage in the neck that guards the entrance to the windpipe and functions as the organ of voice; also called the voice box.

Lungs The organs that supply the body with oxygen and eliminate carbon dioxide from the blood.

Mass-casualty incident An emergency situation involving more than one victim, and which can place such great demand on equipment or personnel that the system is stretched to its limit or beyond.

Mouth-to-mask ventilation device A piece of equipment that consists of a mask, a one-way valve, and a mouthpiece. Rescue breathing is performed by breathing into the mouthpiece after placing the mask over the victim's mouth and nose.

Open wound Injury that breaks the skin or mucous membrane.

Pathogens Microorganisms capable of causing disease.

Plasma The fluid part of the blood that carries blood cells, transports nutrients, and removes cellular waste materials.

Platelets Microscopic disc-shaped elements in the blood that are essential to the process of blood clot formation, the mechanism that stops bleeding.

Pressure points Points where a blood vessel lies near a bone; pressure can be applied to these points to help control bleeding.

Pulse The wave of pressure created by the heart as it contracts and forces blood out into the major arteries.

Puncture wounds A wound resulting from a bullet, knife, ice pick, splinter, or any other pointed object.

Radial artery The major artery in the forearm; it is palpable at the wrist on the thumb side.

Red blood cells Cells that carry oxygen to the body's tissues.

Rescue breathing Artificial means of breathing for a victim.

Respiratory arrest Sudden stoppage of breathing.

Rigor mortis The temporary stiffening of muscles that occurs several hours after death.

Severe Acute Respiratory Syndrome (SARS) A viral respiratory illness caused by a coronavirus.

Shock A state of collapse of the cardiovascular system; the state of inadequate delivery of blood to the organs of the body.

Sniffing position Position where the head and chin are thrust slightly forward.

Sternum The breastbone.

Trachea The windpipe.

Triage The process of sorting victims based on the severity of injury and medical needs to establish treatment and transportation priorities.

Universal precautions Procedures for infection control that treat blood and certain bodily fluids as capable of transmitting bloodborne diseases.

Venous bleeding External bleeding from a vein, characterized by steady flow; the bleeding may be profuse and life-threatening.

Ventricle Either of the two lower chambers of the heart.

White blood cells Blood cells that play a role in the body's immune defense mechanisms against infection.

Xiphoid process The flexible cartilage at the lower end of the sternum (breastbone), a key landmark in the administration of CPR and the Heimlich maneuver.

Fire Fighter in Action

It is 10:00 a.m. on a sunny July morning when your engine company is dispatched to a second alarm for a commercial structure fire in a warehouse. Upon arrival, your engine crew is assigned to a rapid intervention team in Division A. The first suppression crews were headed to the rehabilitation area when a 45-year-old male fire fighter collapsed to the ground. Your rapid intervention team is directed to provide care. Advanced life support is not available on the scene but is en route. You approach the collapsed fire fighter and find that he is unresponsive and does not appear to be breathing.

1. The best way to check that the victim is breathing is to:
   A. look, listen, and feel for at least 3 to 5 seconds.
   B. give one rescue breath to see if there is resistance.
   C. shake the victim and shout at him to see if you can get a response.
   D. hold a small mirror to the victim's nose to see if it will fog up.

2. On an unresponsive adult, which artery should you check for a pulse?
   A. Radial
   B. Brachial
   C. Carotid
   D. Femoral

You have adjusted the airway using the head tilt chin lift. After assessing for breathing, you give two rescue breaths. You check for a pulse and it is absent. You remove the victim's PPE and begin two-rescuer CPR.

3. What is the proper technique for performing chest compressions?
   A. Technique does not matter as long as the chest gets depressed.
   B. Bend your elbows and press down in a rhythmic motion.
   C. Straddle the victim and press downward with your shoulders directly over your hands.
   D. Put your knees close to the side of the victim, keep your arms straight, elbows locked, and your shoulders directly over your hands.

4. How far should you depress an adult sternum during chest compressions?
   A. 1" to 2"
   B. 1 1/2" to 2"
   C. 2" to 2 1/2"
   D. 1" to 1 1/2"

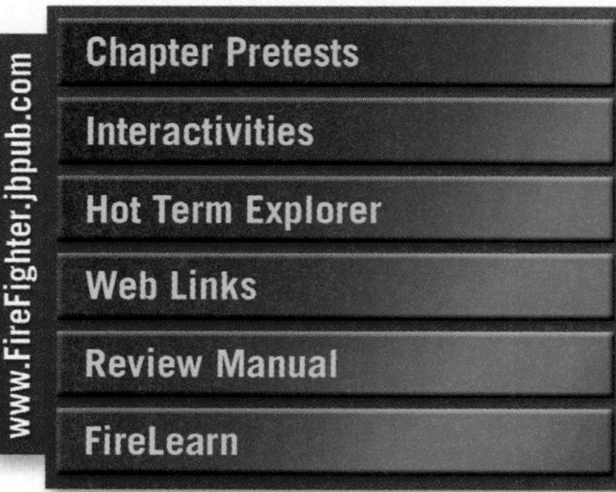

Vehicle Rescue and Extrication

Chapter 25

NFPA 1001 Standard

Fire Fighter I

5.3.7 (A) *Requisite Knowledge.* Principles of fire streams as they relate to fighting automobile fires; precautions to be followed when advancing hose lines toward an automobile; observable results that a fire stream has been properly applied; identify alternative fuels and the hazards associated with them; dangerous conditions created during an automobile fire; common types of accidents or injuries related to firefighting automobile fires and how to avoid them; how to access locked passenger, trunk, and engine compartments; and methods for overhauling an automobile.

Fire Fighter II

6.4 *Rescue Operations.* This duty involves performing activities related to accessing and disentangling victims from motor vehicle accidents and helping special rescue teams, according to the following job performance requirements.

6.4.1 Extricate a victim entrapped in a motor vehicle as part of a team, given stabilization and extrication tools, so that the vehicle is stabilized, the victim can be disentangled without undue further injury, and hazards are managed.

6.4.1 (A) *Requisite Knowledge.* The fire department's role at a vehicle accident, points of strength and weakness in auto body construction, dangers associated with vehicle components and systems, the uses and limitations of hand and power extrication equipment, and safety procedures when using various types of extrication equipment.

6.4.1 (B) *Requisite Skills.* The ability to operate hand and power tools used for forcible entry and rescue in a safe and efficient manner; use cribbing and shoring material; and choose and apply appropriate techniques for moving or removing vehicle roofs, doors, windshields, windows, steering wheels or columns, and the dashboard.

Knowledge Objectives

After studying this chapter, you will be able to:

- Describe the vehicle anatomy.
- List the hazards involved in responding to an emergency scene.
- List the hazards to look for when arriving on the scene of a vehicle extrication situation.
- Describe cribbing.
- Describe the extrication tools that are used for stabilizing, bending, cutting, and disassembling.
- Describe how to gain access to the victim.

Skills Objectives

After completing this chapter you will be able to:

- Break tempered glass with a spring-loaded punch.

You Are the Fire Fighter

You are assigned to an engine-rescue company in a suburban city just outside of a busy high-tech metropolitan area. Your company is dispatched to a two-car motor vehicle crash at the intersection of two crossroads, with possible personal injuries. While responding, you mentally review the department's guidelines for vehicle crashes.

After arriving, the driver of your vehicle positions it to protect emergency workers from oncoming traffic and creates a safe working area. The impact to the first vehicle deployed both frontal airbags, and the battery is leaking fluid. The driver's front-side quarter panel on the second vehicle is damaged from the area just behind the wheel well extending back into and jamming the driver-side door. The second vehicle's bumper is resting against a Jersey barrier on the passenger-side doors, making entry on that side inaccessible. No airbags were deployed.

The senior fire fighter completes his safety survey (parameter check) of the crash scene then indicates to the other fire fighters that both vehicles are safe to approach to access the victims.

1. ***How would you gain access to the victims in each of these vehicles?***
2. ***What concerns are raised due to the airbag system in the second vehicle?***

Introduction

Beginning fire fighters must understand the process of extrication and have some proficiency in extrication skills. Most fire departments are involved in some part of the extrication process. This chapter presents the knowledge and skills needed to assist rescuers in extricating victims from motor vehicles involved in crashes.

While this chapter provides the knowledge and skills a beginning fire fighter needs, it does not cover all the knowledge and skills needed by members of special extrication teams. Fire fighters who are members of special rescue teams should complete a course in rescue techniques that meets the requirements NFPA 1006, *Standard for Rescue Technician Professional Qualifications,* and NFPA 1670, *Standard on Operations and Training for Technical Rescue Incidents.*

Types of Vehicles

Most of the vehicles on the road are **conventional vehicles**, which use internal combustion engines for power. Internal combustion engines burn gasoline or diesel fuel to produce power. Fuel creates a hazard if it leaks after a crash. Other hazards of these vehicles include short circuits and battery acid leaks after a collision. The 12-volt electrical system poses a minimal threat to rescuers.

Alternative-powered vehicles are those that are powered by compressed natural gas, which is carried in cylinders that may be mounted in the trunk of a vehicle or in another convenient location. City buses, delivery vans, and fleet sedans may be powered by compressed natural gas. Compressed natural gas-powered vehicles should be identified by a "CNG" sticker mounted on the front and back of the vehicle (▼ Figure 25-1). After a crash, there could be damage to the natural gas cylinders or fuel lines, resulting in the escape of flammable natural gas. In addition, a fire in a natural gas-powered vehicle poses the threat of a BLEVE (a boiling liquid, expanding vapor explosion). Any sign of fire in a vehicle powered by compressed natural gas should be treated with extreme caution.

Electric-powered vehicles are propelled by an electric motor that is powered by batteries. These vehicles contain a large number of batteries to provide the needed power. Hazards posed by these vehicles include the large amount of energy stored in the batteries, the potential for electrical

Figure 25-1 The CNG sticker identifies a vehicle that is powered by compressed natural gas.

Fire Fighter Safety Tips

Some vehicles like fork lifts, mass transit buses, and school buses run on propane (liquefied petroleum gas). These vehicles can produce flammable vapors.

shorts, and leakage from damaged batteries. You should be aware that these vehicles often use a higher voltage than the 12-volt system found in conventional vehicles.

Hybrid vehicles use battery-powered electric motors and a gasoline-powered engine. These cars use an electric motor at low speeds. The gasoline-powered engine is used at higher speeds. When maximum power is needed, the electric motor and the engine operate at the same time. When sitting at a traffic light, both sources of power are shut off, and the vehicle is said to be hibernating. Hybrid vehicles contain the hazards associated with conventional gasoline-powered vehicles. In addition, they pose hazards from the electrical system. Electrical hazards come from the large number of batteries used. Some hybrid vehicles use a much higher voltage than the voltage found in conventional vehicles. It is important to become familiar with the features of these vehicles. Vehicle manufacturers maintain web sites that you can access to learn more about the operation and hazards of hybrid vehicles.

Vehicle Anatomy

In order to reduce confusion and mistakes at extrication scenes, it is important to use standardized terminology when referring to specific parts of vehicles. The front of a vehicle normally travels down the road first. The hood is located on the front of the vehicle. The rear of a vehicle is where the trunk is usually located.

The left side of a vehicle is on your left as you sit in the vehicle. In the United States and Canada, the driver's seat is on the left side of the vehicle. The right side of a vehicle is where the passenger's seat is located. Always refer to left and right as they relate to the vehicle. Do not refer to left and right from where you might be standing.

Vehicles contain "A," "B," and "C" **posts**, which are the vertical support members of a vehicle that hold up the roof and form the upright columns of the passenger compartment. The **"A" posts** are located closest to the front of the vehicle. The "A" posts form the sides of the windshield. In four-door vehicles, the **"B" posts** are located between the front and rear doors of a vehicle. In some vehicles, these posts do not reach all of the way to the roof of the vehicle. In four-door vehicles, the **"C" posts** are located behind the rear doors. They are located behind the rear passenger windows.

The hood covers the engine compartment. The structure that divides the engine compartment from the passenger compartment is called the **bulkhead**. The passenger compartment includes the front and back seats. This part of a vehicle is sometimes called the occupant cage or the occupant compartment (▲ Figure 25-2).

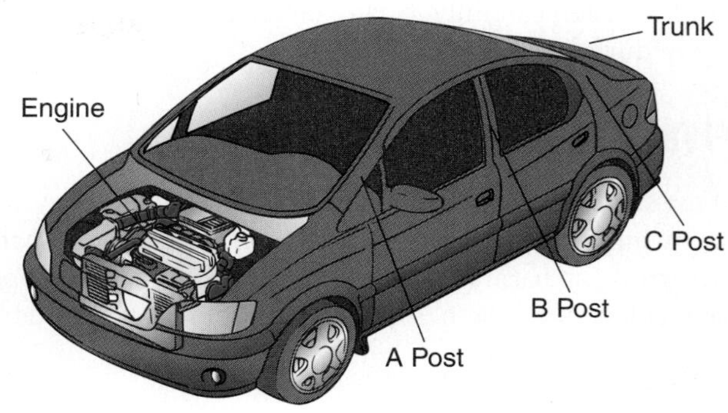

Figure 25-2 The anatomy of a vehicle.

There are two common types of vehicle frames: the **platform frame** construction and the **unibody (Unit body)** construction. Platform frame construction uses beams to fabricate the load-bearing frame of a vehicle. The engine, transmission, and body components are then attached to this basic platform frame. This type of frame construction is found primarily in trucks and sport utility vehicles. It is not commonly found in smaller passenger cars. Platform frame construction provides a structurally sound place for stabilizing the vehicle and an anchor point for attaching cables or extrication tools. Unibody construction is used for most modern cars. This type of construction combines the vehicle body and the frame into a single component. By folding multiple thicknesses of metal together, a column can be formed that is strong enough to serve as the frame for a lightweight vehicle. Unibody construction has the advantage of enabling auto manufacturers to produce lighter weight vehicles. When extricating a person from a unibody vehicle, remember that unibody vehicles do not have the frame rails that are present with platform frame-constructed vehicles.

Respond to the Scene

The first step in the extrication process is response. In order to be able to perform extrication, you need to safely and efficiently arrive on the scene. Safe response includes picking the best route for the time of the day, driving in a safe manner, and keeping in mind the limitations of your vehicle and its warning devices. Always fasten your seatbelt. You should follow the same safety practices while responding to a motor vehicle crash that you practice when responding to a structural fire call.

When responding to an emergency, evaluate the dispatch information that you have received, as it may help you to anticipate the necessary types of equipment and the procedures that you may need to perform. Use this time to mentally consider the possible scenarios that you might encounter. Listen carefully for any direction that your officer may give during the response phase.

Arrival and Size-Up the Scene

The next step in the extrication process is arrival and size-up. After arriving at the scene of a motor vehicle crash, it is important to assess the hazards present and to determine the scope of the incident and whether additional resources are needed.

Traffic

Determine where to locate your emergency vehicle. This decision needs to take into account the safety of emergency workers, the victims, and the motorists traveling along the road. Whenever possible, place emergency vehicles in a manner that will assure safety and not disrupt traffic any more than necessary. However, do not hesitate to request that the road be closed if necessary.

Use large emergency vehicles to provide a barrier for motorists who fail to heed emergency warning lights. Many departments place apparatus at an angle to the crash (▼ Figure 25-3). This position helps to push the apparatus to the side of a crash in the event that the emergency apparatus is struck from behind. Traffic cones or flares can be placed to direct motorists away from the crash (► Figure 25-4).

Fire fighters need to be visible at a crash scene. Personnel protective equipment (PPE) should be bright to help assure your visibility during daylight hours. PPE that is used at night needs to be equipped with reflective material to increase your visibility in the darkness. PPE needs to be worn at all motor vehicle crashes. Before exiting fire apparatus at an emergency scene, be alert for any vehicles that might cause you injury. Do not assume that motorists will always heed your warning lights. Let law enforcement personnel coordinate traffic control.

> **Fire Fighter Safety Tips**
>
> Hybrid vehicles and electric-powered vehicles contain powerful batteries and may contain an electrical system that uses a higher voltage than encountered in a conventionally powered vehicle. Be alert for special electrical hazards in these vehicles.

The incident commander (IC) will usually perform a size-up of the scene by conducting a 360° walk around the scene. During this size-up, the IC evaluates the hazards present and the number of crash victims. Using the information obtained from the scene size-up, the IC can create an action plan and call of additional resources, if needed.

Fire Hazards

Look for spilled fuel and other flammable substances. Motor vehicles carry a variety of fuels and lubricants that pose a fire hazard. Look for the presence of fire. A short in the electrical system or a damaged battery may cause a post-crash fire. Other post-crash fires can be started by sparks created during the crash, igniting spilled fuel. These fires may trap the occupants of the vehicle and require fire suppression.

Electrical Hazards

Downed power lines create an electrical hazard. Look closely to determine whether the crash has damaged any electrical power poles. Downed power lines may be clearly visible or difficult to see. Downed power lines create a deadly hazard for rescuers, the victims of the crash, and bystanders.

Other Hazards

Environmental conditions can present hazards at a crash scene. Crashes that occur in rain, sleet, or snow present an added hazard for rescuers and the victims of the crash.

Figure 25-3 Many departments place an apparatus at an angle to the crash.

Figure 25-4 Traffic cones or fuses can be placed to direct motorists away from the crash.

Crashes that occur on hills are harder to handle than those that occur on level ground.

Look for signs of hazardous materials. Assess the scene for hazardous material placards, unusual odors, or leaking liquids. More information about hazardous materials is presented Chapters 27 through 33.

Be especially alert for the presence of infectious bodily substances. Be prepared for the presence of blood and exercise universal precautions. Do not let blood or other bodily fluids come in contact with your skin. Be prepared by wearing gloves that will protect you from contaminated fluids and also from sharp objects that are present at a crash site.

Some crash scenes may present threats of violence. Intoxicated people or those who are upset with other motorists may present a threat to you or to other people present at the scene. Be alert for weapons that are carried in civilian vehicles.

Occasionally animals present a hazard at crash scenes. Dogs and other family pets may be protective of their owners and create a threat for rescuers. Vehicles that are transporting farm animals or horses that have been involved in a crash may need care.

Stabilize the Scene

The next step of the extrication process is scene stabilization. This step consists of reducing, removing, or mitigating the hazards at the scene. These hazards should have been identified during arrival and scene size-up. The order in which these hazards are stabilized will depend on the specific conditions at a scene and the amount of risk each hazard poses.

Traffic Hazards

The hazards posed by traffic need to be handled quickly before they create additional crashes. In some cases, the most critical issue to be addressed is getting traffic slowed or stopped before additional damage and injuries can occur. As mentioned previously, emergency vehicles can be used to block traffic from the crash scene. Traffic cones and flares can slow motorists and direct them in a safe pattern around the crash scene.

Traffic hazards are best handled by the appropriate law enforcement agency. This is their area of expertise, and it leaves fire department personnel free to handle other extrication tasks. You need to work in concert with law enforcement officials to control traffic in a manner that is safe for emergency workers, victims of the crash, and motorists. If law enforcement officials are not on the scene when you arrive, verify that they are aware of the incident and that they have been dispatched.

Fire Hazards

Because there is a significant risk of spilled fuel in many motor vehicle crashes, it should be a standard operating procedure (SOP) to advance a charged hose line close to the damaged vehicles to protect them from the threat of fire. This should be at least a 1.5" hose and should be staffed by a trained fire fighter in full turnout gear. This hose line can serve to provide protection for rescuers and the victims of the crash. Crashes that pose large fire hazards or actual fires may require additional fire suppression resources. These should be requested as soon as possible. Small fuel spills can be handled by using an absorbent material to remove the fuel from the area around the damaged vehicle.

Electrical Hazards

Disconnecting the electrical system of the vehicle can mitigate electrical hazards posed by damaged vehicles. Follow your department's SOP for disconnecting the electrical system. You must weigh the need to disconnect the electrical system with the advantage of being able to operate electrically powered components of the vehicle.

Stabilizing electrical hazards posed by downed electrical wires is essential before approaching the crash site. At times it may be necessary to instruct victims of a crash to remain in their vehicle until the power can be turned off. Disconnecting the power to damaged electrical lines is a job for properly trained members of the power company. Do not attempt to approach downed electrical lines until the power company has turned off the electricity. Although it is frustrating to delay extrication and treatment because of downed power lines, it is essential to avoid injuries and death to rescuers and victims of the crash.

Other Hazards

Stabilize the environmental hazards present at a crash scene. At scenes where it is hot, it may be necessary to provide some shade for a victim during a prolonged extrication. Rescue crews require rehabilitation more quickly in a hot environment. At scenes where it is cold, provide covering to help the victim maintain body warmth.

Many crashes occur at night. It is important to provide adequate lighting at crash scenes so that rescuers can work quickly and safely. Be careful when it is wet or icy. It is important to prevent slips and falls. Sprinkling sand or a kitty litter type of material may help to give solid footing to rescuers.

Vehicle crashes produce a variety of sharp objects. Mangled pieces of sheet metal, plastic, and glass have sharp edges that pose significant hazards for rescuers and victims of the crash. Proper personnel protective clothing for rescuers will reduce the chance of injury. A second way to reduce the threat of injury from these edges is to cover them with some type of padding. Short sections of old fire hose will fit over some sharp edges. Be especially careful when moving victims to be sure that they are protected from any sharp objects.

In incidents involving animals, it may be necessary to deal with the animals before proceeding with other activities. Dogs can be very possessive of their owners. In cases in which the dog's owner is injured, remove the dog from the crash site and place it with an uninjured family member or other responsible person. If a pet is injured, call the agency that is responsible for transporting them to a veterinary clinic. You should learn who is responsible in your community for caring for large animals at rescue scenes.

Cribbing

Unstable objects pose a threat to rescuers and the victims of the crash. These objects need to be stabilized before your approach. The objects that need to be stabilized most often are the damaged vehicles.

Vehicles that end up on their wheels need to be stabilized with **cribbing** in the front and back of the wheels. Cribbing is short lengths of robust timber (4″ × 4″, 4″ × 18″, and 4″ × 24″). It is used to stabilize a vehicle. Cribbing prevents the vehicle from rolling backward or forward. ▶ Figure 25-5 shows the application of cribbing in the front and back of the vehicle wheels.

Figure 25-5 The application of cribbing in the front and back of the vehicle wheels stabilizes the vehicle.

After the cribbing has been placed, a vehicle can still move because of the give and motion that the suspension system causes. This motion occurs as rescuers get into the vehicle and the victims are extricated from the vehicle. This type of instability can cause further injuries to the victims of the crash.

The suspension system of most vehicles can be stabilized with **step blocks**, which are shaped like stair steps and are placed under the side of the vehicle. Place one step block toward the front of the vehicle and a second step block toward the rear of the vehicle. ▶ Figure 25-6 shows step blocks being used to stabilize a vehicle. Once the step blocks are in place, the tires can be deflated by pulling out the valve stem or by puncturing the tires. This creates a stable vehicle.

Figure 25-6 Step blocks are used to stabilize a vehicle.

In cases in which step blocks are not available or the right size, you can build a box crib by placing cribbing at right angles to the preceding layer of cribbing. ▶ Figure 25-7 shows a box crib being constructed to stabilize a vehicle.

Figure 25-7 A box crib is constructed to stabilize a vehicle.

After a crash, some vehicles will come to rest on their roofs or sides. These vehicles are very unstable, and the slightest weight on them can cause them to move. These vehicles can be stabilized using box cribs or step blocks on each end of the vehicle.

Besides the cribbing, **wedges** are used to snug loose cribbing under the load, or when using lift airbags, to fill the void between the crib and the object as it is raised ▶ Figure 25-8. Wedges should be the same size width as the cribbing, with the tapered end no less than 0.25″ thick. When the ends are feathered to less than 0.25″, the end will commonly fracture under a load.

Rescue Lift Airbags

Rescue lift airbags are pneumatic-filled bladders made out of rubber or other synthetic material ▶ Figure 25-9. They are used to lift an object or spread one or more objects away from each other to assist in freeing a victim. Rescue lift airbags are NEVER used to shore or stabilize a vehicle by themselves. Cribbing is always required in conjunction with lift airbags.

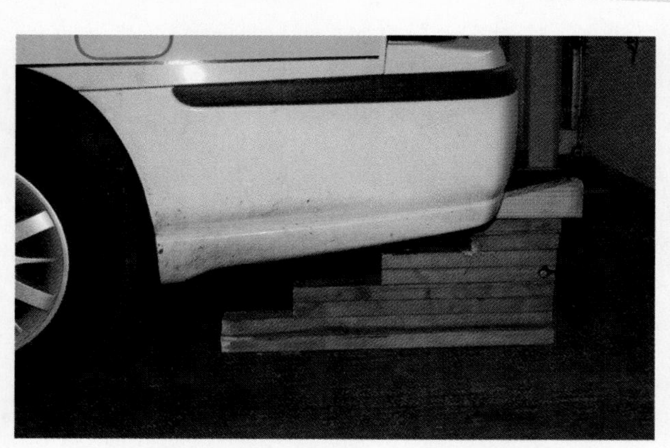

Figure 25-8 Wedges are used to snug loose cribbing under the load, or when using lift airbags, to fill the void between the crib and the object as it is raised.

Figure 25-9 A rescue lift airbag.

Rescue lift airbags are often used to lift a vehicle or object off a victim. Fire fighters should use extreme caution when using rescue lift airbags and should adhere to all the safety precautions as outlined in the rescue lift airbag owner's manual. Cribbing must be used whenever lifting a load. Instability can occur from weight shifts, or the lift airbags can fail under a load. It is therefore a must, not an option, that cribbing be used in conjunction with rescue lift airbags. The top layer of cribbing should be solid underneath the rescue lift airbag, otherwise the cribbing may spread apart, causing failure.

There are currently three types of commonly used pneumatic rescue lift airbags: low, medium, and high pressure. No rescue lift airbag should be used without first properly blocking the wheels opposite of the lift airbag and cribbing the vehicle or object as it is lifted. This is a safety precaution in case of a catastrophic failure of a rescue lift airbag. Rescue lift airbags can fail as they age. They should be tested according to the manufacturer's recommendations. Rescue lift airbags reaching 10 years of age have generally reached their useful life.

Fire Fighter Safety Tips

Rescue lift airbags are NEVER used to stabilize a vehicle by themselves; always stabilize the vehicle as you go with cribbing. This is a safety issue in case there is a rescue lift airbag failure.

Fire Fighter Safety Tips

- Never stack **high-pressure lift airbags** more than two high.
- Do not use a rescue lift airbag to pull a steering column.
- Do not use a rescue lift airbag to stabilize a vehicle solely; cribbing must be your primary stabilizer.
- Never operate your rescue lift airbag system without having been properly trained and fully knowing and understanding how the system works.
- Do not use boards or plywood between or above rescue lift airbags.
- Clean rescue lift airbags using the recommendations of the individual manufacturer.
- Rescue lift airbags should be tested.
- Never store a lift airbag with gasoline.

Low-Pressure Lift Airbags

The low-pressure lift airbags are commonly used for recovery operations and are sometimes used by departments for vehicle rescue operations. Low-pressure airbags come in many sizes and shapes; however, square airbags offer greater stability. Because of light-weight airbag construction, low-pressure lift airbags can be less stable until fully inflated as compared with the high-pressure type of airbags, which have a lower height lift and stronger construction.

Medium-Pressure Lift Airbags

The medium-pressure lift airbags have either two or three cells inside their design. For general vehicle rescue, the medium pressure is not needed. It is more suitable for aircraft, medium or heavy truck, or bus rescue, and recovery work.

High-Pressure Lift Airbags

The most common type rescue lift airbags used by fire fighters are the high-pressure lift airbags. The high-pressure airbags are heavily constructed and generally made by vulcanized rubber mats that are reinforced by steel or other material woven into a fiber mat and then covered with rubber.

Principles of Gaining Access and Disentangling the Victim

Before discussing the steps of gaining access and disentanglement, it is important to understand the principles involved with these steps of extrication. Although there are many different techniques and methods used to gain access and disentangle victims from crashed vehicles, the objective of these techniques is accomplished by performing four different functions: stabilize, bend, cut, and disassemble. The first function is to stabilize or hold an object or vehicle. Vehicles are stabilized when cribbing is used to keep them from moving. The second function is to bend, distort, or displace. An example of this is bending a vehicle door back to get it out of the way. The third function is to cut or sever. An example of this is cutting a roof. The fourth function is to dissemble. An example of this is removing a vehicle door by unbolting the door hinges.

Tools Used for Extrication

There are many different tools that can be used for extrication. It is important to understand the purpose of each tool that you will be using for extrication. Different situations mean that a certain tool will be more appropriate. You need to know which tools are available in your department and how to operate each of these tools.

Cribbing, rescue lift airbags, and step blocks are stabilization devices. When using these tools, be certain that they are placed firmly under the vehicle and that the vehicle is stable before attempting to enter it.

Hand tools and powered tools can be used to displace, bend, or distort metal parts of a vehicle in order to gain access to a victim and to disentangle a victim from the vehicle. Hand tools include pry axe, short pry bars, long pry bars, and Halligan tools. These tools can be used to bend sheet metal so that it can be pulled away from a victim. Manual hydraulic rams and manual hydraulic spreaders can exert a large amount of force onto the vehicle. Powered hydraulic rams and powered hydraulic spreaders use a hydraulic pump powered by an electric motor, a gasoline engine, or the engine on a rescue truck.

Hand tools that can be used for cutting include axes, bolt cutters, cable cutters, hacksaws, and manual hydraulic cutters. Each of these cutting tools is designed to cut certain materials. If you try to use a tool on an inappropriate material, you will not only fail to cut the material, but you also may damage the tool or injure the victim or yourself.

A variety of power tools for cutting or severing are available. Some of these are air powered. Some are electrically powered, and some are hydraulically powered. Air chisels are powered by pressurized air from portable SCBA air cylinders or from a cascade system mounted on a rescue truck. Air chisels are efficient at cutting sheet metal and some types of plastic. They can be used to cut metal posts. (► Figure 25-10) shows one type of air chisel.

Figure 25-10 An air chisel is useful for cutting sheet metal.

A variety of powered saws are used for extrication. Some electric saws require power cords, whereas other electric saws operate on battery packs. Powered saws produce different types of motion. Reciprocating saws move back and forth very quickly. Rotary saws move in a circular motion. One of the most common extrication tools is the powered hydraulic cutter, which can quickly cut through the roof posts of a vehicle.

Specialized equipment such as the hand-operated **come along** is also used during extraction. You can learn more about the proper use of this and other specialized equipment through special rescue courses or in-service training.

Sometimes the most effective way to remove a part of a vehicle is to disassemble it. This method of extrication has the advantage of being controlled. It may also provide an added degree of safety to the victim and rescuers. The types of tools needed for disassembly are the same tools that are used for construction and repair of vehicles. A mechanic's tool kit that contains an assortment of wrenches and screwdrivers will enable a qualified rescuer to disassemble some parts of a vehicle efficiently and safely.

Gain Access to the Victim

Open the Door

After stabilizing the vehicle, the simplest way to access a victim of a crash is to open a door. Try all of the doors first—even if they appear to be badly damaged. It is an embarrassing waste of time and energy to open a jammed door with heavy rescue equipment when another door can be opened easily and without any special equipment.

Attempt to unlock and open the least damaged door first. Make sure the locking mechanism is released. Then try the outside and inside handles at the same time if possible.

Break Tempered Glass

If a victim's condition is serious enough to require immediate care and you cannot enter through a door, consider

Fire Fighter Tips

To help reduce the tempered glass pieces from falling on victims, cover the window with duct tape or shelf paper prior to breaking. The majority of the glass can be removed as a unit.

breaking a window. Do not try to break and enter through the windshield because it is made of **laminated windshield glass**, which is difficult to break. The side and rear windows are made of **tempered glass**, and will break easily into small pieces when hit with a sharp, pointed object such as a spring-loaded center punch or the point of an axe. Because these windows do not pose as great of a safety threat, they can be your primary access route.

If you must break a window to unlock a door or gain access, try to break a window that is away from the victim. If the victim's condition warrants your immediate entry, do not hesitate to break the closest window. Small pieces of tempered glass do not usually pose a danger to victims trapped in cars. Advise EMS personnel if a victim is covered with broken glass so that they can notify the hospital emergency department.

After breaking the window, use your gloved hands to pull the remaining glass out of the window frame so it does not fall onto any victims or injure any rescuers. If you are using something other than a spring-loaded center punch to break the window, always aim for a low corner. The window frame will help prevent the tool (such as an axe or screw driver) from sailing into the vehicle and hitting the victim inside. To break tempered glass, follow the steps in (Skill Drill 25-1).

1. All fire fighters should be in proper PPE, including dust mask, safety glasses, or goggles.
2. Attempt to lower windows as far as possible before breaking glass, and then select a spring-loaded center punch or axe.
3. Ensure that the victim and other fire fighters are properly protected.
4. Personnel are warned by the verbal command "Breaking Glass," unless a stop/freeze call is made, while you continue with glass removal starting with a window furthest from the victim.
5. Place the center punch in the lower corner of the window and apply pressure until the spring is activated, or strike the lower corner with the axe.
6. Remove any loose glass around the window opening with a tool.
7. Follow this procedure until all glass has been removed.

Once you have broken the glass and removed the pieces from the frame, try to unlock the door again. Release the locking mechanism, and then use both the inside and outside door handles at the same time. This will often enable you to force a jammed locking mechanism, even in a door that appears to be badly damaged.

(Figure 25-11) The most common technique for gaining access is door displacement.

Breaking the rear window will sometimes provide an opening large enough to enable a rescuer to gain access to the victim if there is no other rapid means for gaining access. Using the simple techniques of opening a door or breaking the rear window will enable you to gain access to most vehicle crash victims, even those in an upside down vehicle.

Force the Door

If you cannot gain access by the previously mentioned methods, you will need heavier extrication tools to gain access to the victim. The most common technique for gaining access is door displacement (▲ Figure 25-11). The opening and displacement of vehicle doors may be difficult and somewhat unpredictable.

When it is necessary to force a door to gain access to the victim, choose a door that will not endanger the safety of the victim. For example, do not try to force a door open if the victim is leaning against it.

This is usually done by using a prying tool to bend the sheet metal at the edge of the door near the locking mechanism. Once this metal has been exposed, a hydraulic spreading tool can be inserted into the space between the door and the pillar of the vehicle and can pull the locking mechanism apart. Once the door has been removed, it should be placed away from the vehicle where it will not be a safety hazard to rescuers.

Powered hydraulic tools are the most efficient and widely used tools for opening jammed doors, which can be opened by releasing the door from the latch side or the hinge side of the door. The decision regarding which method to use will depend on the structure of the vehicle and the type of damage the door has sustained.

The first step in opening a vehicle door is to use a pry tool to bend the sheet metal away from the edge of the door where the hydraulic tool is to be inserted. This gives the spreader of the hydraulic tool a place to be inserted.

Place the spreader in a position so that it is not in the pathway that the door will take when the hinges or latch break. Do not stand in a position that might put you in danger. Activate the hydraulic tool to push apart the outer sheet metal skin of the vehicle to expose the hinges or the door latch. Once the outer sheet metal has been exposed, close the tips of the spreader and remove them.

Insert the closed tips onto the inner skin of the door and the doorjamb just above the latch or above the hinges. Activate the spreader to spread the tips until the latch or the hinge separates. When separating a door at the latch side, you can place a 4″ × 4″ cribbing under the bottom of the door to hold it up and then start to separate the hinges of the door. Some hydraulic tools are capable of cutting door hinges. Check the manufacturer's recommendations for the proper operation of your hydraulic tool.

When separating a door from the hinge side, place the spreader on the top of the bottom hinge. Use the hydraulic spreader to separate the door from the hinge. Once the hinges have separated, place 4″ × 4″ cribbing underneath the door to hold it in place while you work on the latch side of the door.

Provide Initial Medical Care

As soon as you have secured access to the victim, begin to provide emergency medical care. A qualified emergency medical provider should perform this function. Being trapped in a damaged vehicle is a frightening experience. It is important to have one caregiver remain with the victim and to provide emotional comfort as well as physical care. Victim care should be provided at the same time that extrication is being done. Although it may be necessary to delay for a short period one part of the these processes, it is important to work toward the goal of getting the victim stabilized and removed from the vehicle as quickly and safely as possible.

Disentangle The Victim

The next step in the extrication sequence is disentangling the victim. The purpose of this step is to remove the parts of the vehicle that are trapping the victim. It is important to note that the goal of this step is to remove the sheet metal and plastic from around the victim. It is not, as you may have read in the popular press, to "cut the victim out of the vehicle."

Before beginning disentanglement, study the situation. What is trapping the victim in the vehicle? Perform only those disentanglement procedures that are necessary to remove the victim safely from the vehicle. The order in which these procedures are performed will be dictated by the conditions at that scene. Many times it will be necessary to perform one procedure before you can access the parts of the vehicle in order to perform another procedure.

As you work to disentangle the victim from the vehicle, be sure to protect the victims by covering them with a blanket or by using a backboard. Be sure that the victim understands what is being done. The sounds made by extrication procedures are frightening for victims.

Figure 25-12 In frontal crashes, the vehicle may become compressed.

In order to learn all of the methods of disentangling a victim, you need to take an approved extrication course. The five procedures presented here are the ones that are most commonly performed.

Displace the Seat

In frontal and rear end crashes, the vehicle may become compressed (▲ **Figure 25-12**). As the front of the vehicle is pushed back, the space between the steering wheel and the seat becomes smaller. In some cases, the driver may be trapped between the steering wheel and the back of the front seat. Displacing the seat can relieve pressure on the driver and give rescuers more space for removal.

If it is necessary to displace a seat backward, start with the simplest steps. Many times you can gain some room by moving the seat backward on its track. This works with short drivers who have their seat forward. To move a seat back, first be sure that the victim is supported. With manually operated seats, you need to release the seat-adjusting lever and carefully slide the seat back as far as it will go. If the seat is electrically operated, check to be certain that the car still has power, and engage the lever to electrically move the seat backward. Attempt to use these simple methods first.

For seats that have adjustable backs and adjustable heights, you can use these features to move the seat back to give the victim more room or lower the seat to help disentangle the victim. Be certain that the victim is adequately supported before attempting these maneuvers.

If this is unsuccessful, perform a dash displacement. As a last resort, you can use a manual hydraulic spreader or a powered hydraulic tool to move the seat back. Place one tip of the hydraulic tool on the bottom of the seat. Avoid pushing on the seat channel that is attached to the floor of the vehicle. Place the other tip of the spreader at the bottom of the "A" post doorjamb. Support the victim carefully. Engage the seat adjustment lever on manually operated seats, and open the spreader in a careful and controlled fashion. The seat should move backward in a controlled fashion.

In some cases, it will be helpful to remove the back of the seat. To accomplish this, cut the upholstery away from the bottom of the seat back where it joins the main part of the seat. A reciprocating saw or a hydraulic cutter can be used to cut the supports for the seat back. Be certain that the victim is supported and protected during this procedure.

Displacing the seat or removing the seatback will give you more room to administer treatment and remove the victim from the vehicle.

Remove the Windshield

A second technique that is often part of disentanglement is the removal of glass. Removing the rear window, the side glass, or the windshield improves communication between rescue personnel inside the vehicle and personnel outside the vehicle. Sometimes all that you need to do is to roll a window down to gain this advantage. Open windows provide a good route for passing medical care supplies to the inside caregiver. When the roof of a vehicle needs to be removed, it is necessary to first remove all of the glass from the vehicle.

Remember that the side windows and the rear window on most vehicles are tempered glass that can be removed by striking the glass in a lower corner with a sharp object, such as a spring-loaded center punch. Whenever it is necessary to do this, you should protect the victim whenever possible by covering them to prevent the victim from getting cut from the small pieces of glass.

Windshields cannot be broken with a spring-loaded center punch. Windshields consist of plastic laminated glass, which is actually a sandwich. The two pieces of bread are a thin sheet of specially constructed glass. The filling of the sandwich is a thin layer of a special flexible plastic. When laminated glass is struck by a sharp stone or by a spring-loaded center punch, a small mark is formed, but the structure of the glass remains intact. Because of this special construction, it is necessary to remove the windshield in one large piece.

Fire Fighter Safety Tips

Some newer vehicles may soon be equipped with side and rear windows made out of a new type of glass that cannot be broken with a sharp object. This new type of glass will provide more protection during a crash and also provide an increased level of protection against theft. New methods of breaking this glass will need to be developed.

Figure 25-13 Removing the windshield with an axe by cutting close to the "A" post.

The windshields of most passenger vehicles are glued in place with a plastic type of glue that is very strong. There are several ways to remove a windshield after a collision. One method is to cut the windshield with an axe.

To remove a windshield using an axe, first be certain that the victim is protected from flying glass. One rescuer begins to cut the top of the windshield at the middle of the windshield, using short strokes of the axe. Continue cutting along the top of the windshield. Next cut down the side of the windshield close to the "A" post (▲ **Figure 25-13**). Finally, the rescuer should cut along the bottom of the windshield. At this point, the first rescuer stabilizes the half of the windshield that has been cut free.

Next, a second rescuer starts at the top of the windshield where the initial cuts were made by the first rescuer and cuts the second half of the windshield following the same sequence of cuts used by the first rescuer.

When the second rescuer has completed cutting the windshield, it is lifted out of its frame and placed under the vehicle or in another safe place, where it will not present a safety hazard.

One variation of this technique is to use a saw to remove the glass. When using a saw, the windshield is removed by making the cuts in the same order as those made with an axe. Removing a windshield is an essential step to removing the roof of a vehicle. It will also provide added space when administering emergency medical care to an injured victim.

Remove the Steering Wheel

During normal driving, the steering wheel must be located close to the driver. During a crash, any compression of the front part of the vehicle may cause the steering wheel to be pushed back into the victim's abdomen or chest. Removing the steering wheel can be a major help in disentangling a victim from their vehicle.

VOICES OF EXPERIENCE

"Upon entering the auto, I found the victim with an 18″ long, 2″ by 2″ piece of wood impaled in his right eye socket."

"The man inside has a stick in his head!" That's what a bystander told the first arriving units when they reached the scene of a pre-dawn accident on a frigid morning.

A car had crashed through a board fence lined with barbed wire, and overturned in a field. The sole occupant of the car, a 21-year-old male, was conscious but pinned in the automobile. Upon entering the auto, I found the victim with an 18″ long, 2″ by 2″ piece of wood impaled in his right eye socket. In addition, a 1″ by 4″ fence board has lacerated his scalp and was, in effect, supporting his head and upper body. The victim's only complaint was that he was cold (the outside temperature was about 15° F).

I was a fire fighter/EMT at the time, and I assessed the victim while other fire fighters began stabilizing the scene. Access to the automobile was made and the auto was stabilized using ropes, shoring, and chocks.

A paramedic fire fighter joined me and began advanced life support (ALS) care while another fire fighter and I worked on stabilizing the victim and the 1″ by 4″ board that was supporting him. Other fire fighters used helmets and blankets to "shore up" the victim so that the board entrapping the victim could be removed. Fire fighters were finally able to safely remove a rear window and slide the board out, freeing the victim.

As this was going on, we carefully stabilized the impaled wood. Fire fighters slid a spine board under the victim as the three of us inside the car lifted him. The victim was loaded into the ALS ambulance. He was further stabilized and transported to a waiting helicopter for airlift to a trauma center. Evaluation at the trauma center showed no other injuries, and the victim was taken into surgery. The impaled board was removed, and the wound examined. There was no damage beyond the eye socket, and the surgical team was able to replant the eye.

Within a few days, the victim had regained sight in his right eye. The ophthalmologic surgeon felt a full recovery was likely. The surgeon credited the rapid and conscientious actions of the fire fighters—who had the victim evaluated, stabilized, extricated, treated, transported, and into surgery in less than two hours—with saving the young man's sight. Of particular importance, he said, was the stabilization of the impaled object: one of the "basics" taught in EMS training.

Gordon M. Sachs
Fairfield Fire and EMS (retired)
Fairfield, Pennsylvania

Fire Fighter Safety Tips

Airbag safety

- Steering wheels on most recently manufactured vehicles contain the driver's side airbag, which is a life-saving safety feature for the occupants of the vehicle. Front passenger airbags may also be present.
- If the airbag has deployed during the crash, it does not present a safety hazard for rescuers.
- If not deployed during the crash, an airbag presents a hazard for the occupant of the vehicle and for rescue personnel. An undeployed airbag could deploy if wires are cut or if it became activated during the rescue operation.
- If the airbag did not deploy, disconnect the battery and allow the airbag capacitor to discharge. The time required to discharge the capacitor varies from one model of airbag to another.
- Do not place a hard object such as a backboard between the victim and an undeployed airbag.
- Do not attempt to cut a steering wheel if the airbag has not deployed.
- For your safety, never get in front of an undeployed airbag. You could suffer serious injury if it is unexpectedly activated.
- Some vehicles contain side-mounted airbags or curtains that provide lateral protection for occupants. Check vehicles for the presence of these devices.

Steering wheels can be cut with hand tools such as a hacksaw or bolt cutters. Hydraulic cutters and reciprocating saws will also cut steering wheel spokes and the steering wheel hoop. One method of removing a steering wheel is to cut the spokes as close to the hub of the steering wheel as possible. A second method is to cut the hoop or ring of the steering wheel. The steering wheel ring can be removed completely, or one section can be cut and removed. Cutting the steering wheel hoop or spokes leaves sharp edges that present a safety hazard. These sharp edges need to be covered to prevent injury to the victim and to rescuers. Cutting the steering wheel is one more technique that can be used in certain situations to aid disentangling the vehicle from the victim.

Displace the Dash

Often during frontal crashes the dash will get pushed down or backward. When a victim is trapped by the dash, it is necessary to remove it (► Figure 25-14). The technique that is used to remove the dash is called the dash displacement or a dash roll-up. The objective of the dash displacement is to lift the dash up and move it forward.

Dash displacement requires a cutting tool such as a hack saw, a reciprocating saw, an air chisel, or a hydraulic cutter. A cutting tool is needed to make a cut on the "A" post. A mechanical high lift jack or a hydraulic ram is needed to

Figure 25-14 When a victim is trapped by the dash, it is necessary to remove it.

push the dash forward, and cribbing is needed to maintain the opening made with these tools.

The first step of the dash displacement is to open both front doors. Tie them in the open position so that they will not move as the dash is being displaced. Alternatively, the doors can be removed, as discussed in the section on gaining access.

Place a backboard or other protective device between the victim and the bottom part of the "A" post where the relief cut will be made. Cut the bottom of the "A" post where it meets the sill or floor of the vehicle. This cut can be done with a hacksaw, a reciprocating saw, a hydraulic cutter, or an air chisel. It is critical to make the cut perpendicular to the "A" post. Failure to do so may cut the fuel or power line located in the rocker panels. Place the base of a high lift mechanical jack or a hydraulically powered ram at the base of the "B" post where the sill and the "B" post meet. Place the tip of the jack or ram at the bend in the "A" post. This is located toward the top of the "A" post.

In a controlled manner, extend the jack or ram to push the dash up and off of the victim. Once the dash has been removed from the victim, build a crib to hold the sill in the proper position and to prevent the dash from moving. Once this is done, the jack or ram can be removed.

Performing the dash displacement requires careful monitoring of the movement of the dash to be certain that it is moving away from the victim and that it is not causing any additional harm to the victim. Be sure that the victim is protected and that someone is telling the victim what is being done.

Displace the Roof

Removing the roof of a vehicle has several advantages. It allows equipment to be passed to the emergency medical provider. It increases the amount of space available to perform medical care and increases the visibility and space for performing disentanglement. Visibility and fresh air supply are improved. The increased space helps to reduce the feeling of panic caused by the confined space of the crashed vehicle. Removing the roof also provides a large exit route for the victim (► Figure 25-15).

One method of displacing the roof is to cut the "A" posts and fold the roof back toward the rear of the vehicle. This method provides limited space and takes about the same time as removing the entire roof. Therefore, in most cases, it is preferable to remove the entire roof.

Roof displacement can be accomplished with hand tools such as hacksaws, air chisels, or manual hydraulic cutters. Appropriate power tools include reciprocating saws and powered hydraulic cutters. A hydraulic cutter is an efficient tool for this task.

The first step in displacing the roof is to assure the safety of rescuers and the victim inside of the vehicle. Remember, as you cut the posts that support the roof, rescuers must support the roof to keep it from falling on the victim.

To remove the roof, first remove the glass to prevent it from falling on the victims. Cut the vehicle posts farthest away from the victim. Cut the posts at a level to ensure that the least amount of post will remain after roof removal. When cutting the wider rear posts, cut them at the narrowest point of the post. As each post is cut, a rescuer needs to support that post. Cut the post closest to the victim last. This provides some safety for the victim.

Working together, remove the roof and place it away from the vehicle, where it does not pose a safety hazard. Cover the sharp ends of the cut posts with a protective device.

Figure 25-15 Removing the roof provides a large exit route for the victim.

Remove and Transport the Victim

The final phase of the extrication process consists of removing the victim from the crashed vehicle. During this phase, the victim needs to be stabilized and packaged in preparation for removal. The definitive treatment of trauma victims needs to be done at a hospital; therefore, the amount of stabilization that should be done while the victim is in the vehicle should be limited to those steps that are needed to prevent further injury to the victim. See Chapter 13, Search and Rescue, for more information on using a long backboard to remove a victim from a vehicle.

Develop a plan for removal and be sure that a clear exit pathway is available. The actual victim removal needs to be directed by a designated person and needs to be conducted with clear commands. Be sure that there are an adequate number of rescuers who are able to assist with the victim removal. Be sure that everyone involved in this process understands the commands that will be given.

Try to make the removal process as seamless as possible. When possible, locate the ambulance cot close by so that the victim can be placed on the cot without any delay. Follow the direction of the medical care provider in charge of victim care. The victim should be moved to an ambulance as soon as possible.

The victim should be transported to an appropriate medical facility as soon as the initial stabilization has been completed. If your department uses a helicopter service to transport victims, you need to be familiar with the procedures used to load victims.

Secure the Scene and Prepare for the Next Call

The final step in the extrication process is to secure the scene and prepare for the next call. The objective of this phase of extrication is to gather all of the equipment used, inspect it, clean it, sharpen it, refuel it, and place it back in the proper location. Sometimes this can be accomplished at the scene so that the unit is ready for another call. At other times after a major incident, it may be necessary to return to the station to clean and restock the apparatus before it can be returned to service.

Personnel need to prepare for the next call. After long or draining extrications, rehabilitation may be necessary. At times, personnel may need to take a shower, change clothes, and get something to eat before they are ready for another call.

During this phase of the extrication, it is important to maintain safe practices. It is easy to remove protective clothing and then get cut on the sharp edges of sheet metal that you have cut. Remember that even though the victims have been transported from the scene, safety hazards remain that can cause injuries for rescue personnel.

Wrap-Up

Ready for Review

This chapter has covered the knowledge and skills that you need as a beginning fire fighter to assist in extricating a victim from a vehicle after a motor vehicle crash. It covered the factors that you need to address to prepare for a response to a motor vehicle crash. It stressed the importance of a safe and efficient response to the scene. It addressed the hazards that you need to identify and stabilize at a motor vehicle crash.

The tools that are commonly used for vehicle extrication include tools for stabilizing, bending, cutting, and disassembling. The hand tools and power tools that perform each of these functions were discussed. It is important to use standard terminology when talking about vehicles.

Methods to gain access to a victim by opening the door, breaking tempered glass, and forcing a door were explained. Disentangling victims requires a combination of cutting, distorting, and disassembling. Techniques of using these principles to displace a seat, remove a windshield, remove a steering wheel, displace a dash, or displace a roof were explained.

The importance of properly stabilizing victims before moving them was stressed. The importance of securing the scene and preparing for the next call was addressed.

Although this chapter provides the knowledge and skills that you need as a fire fighter I and II, it does not cover all of the knowledge and skills that are needed by members of special extrication teams. Fire fighters who are members of special rescue teams should complete a course in rescue techniques that meets the requirements of NFPA 1006, *Standard for Rescue Technician Professional Qualifications,* and NFPA 1670, *Standard on Operations and Training for Technical Rescue Incidents.*

Chief Concepts

- Most fire departments are involved in extricating victims from vehicles.
- Extrication should follow logical steps.
- It is important to prepare equipment and personnel for extrication emergencies.
- Respond safely and efficiently to an extrication incident.
- Size-up needs to evaluate all of the hazards that are present.
- Stabilizing the hazards is essential to assuring safety for all people at an emergency scene.
- Tools used for gaining access and disentanglement work by stabilizing, bending, cutting, or disassembling.
- You can gain access to victims by opening doors, breaking tempered glass, or forcing doors.
- Techniques for disentangling victims include displacing a seat, removing the windshield, removing the steering wheel, displacing the dash, or displacing the roof.
- Victims need to be properly packaged and carefully removed to prevent further pain and injuries.
- Equipment needs to be cleaned and returned to apparatus to be ready for the next call.

Wrap-Up

Hot Terms

A posts Vertical support members that form the sides of the windshield.

Alternative-powered vehicles Vehicles that are powered by compressed natural gas.

B posts Vertical support members located between the front and rear doors.

Bulkhead The wall that separates the engine compartment from the passenger compartment, formally known as a firewall.

C posts Vertical support members located behind the rear doors.

Come along A hand-operated tool used for dragging or lifting heavy objects. Sometimes known as lever blocks.

Conventional vehicle A vehicle that uses an internal combustion engine and is fueled with either gasoline or diesel fuel.

Cribbing Short lengths of robust and usually hardwood timber (4″ × 4″ and 18″ to 24″ long) that are used in a variety of ways, usually in the stabilization of vehicles.

Electric-powered vehicles A vehicle propelled by an electric motor that is powered by batteries.

High-pressure lift airbags An extrication tool, consisting of airbags, filler hoses, air regulators, control valves, and a supply of compressed air.

Hybrid vehicle A vehicle that uses battery-powered electric motors and a gasoline-powered engine.

Laminated windshield glass Type of window glazing that incorporates a sheeting material that stops the glass from breaking into individual shards. Commonly used in windshields.

Platform frame Type of vehicle frame resembling a ladder type made up of two parallel rails joined by a series of cross members. Typically used for luxury vehicles, sport utility vehicles, and all types of trucks.

Posts One of the vertical support members of a vehicle that holds up the roof and forms the upright columns of the passenger compartment.

Rescue lift airbags An extrication tool, consisting of airbags, filler hoses, air regulators, control valves, and a supply of compressed air. Used to lift vehicles involved in crashes.

Step blocks Specialized cribbing assemblies made out of wood or plastic blocks in a step configuration. These are usually used to stabilize vehicles.

Tempered glass A type of safety glass that is heat-treated so that it will break into small pieces that are not as dangerous.

Unibody (Unit body) Most commonly used frame construction in vehicles. The base unit is made of formed sheet metal and structural components then added to the base to form the passenger compartment. Subframes are attached to each end. This type of construction does away with rail-beams used in platform framed vehicles.

Wedges Used to snug loose cribbing under the load or when using lift airbags to fill the void between the crib and the object as it is raised.

Fire Fighter in Action

It is 6:30 a.m. on a cold and icy morning when your rescue company is dispatched to a vehicle accident on a two-lane highway. While responding to the incident, dispatch confirms that the accident is blocking traffic, and that the injured victim is trapped inside the vehicle. Your lieutenant reminds you and the crew to look out for traffic.

1. How should fire fighters protect the scene from traffic?
   - A. Use large emergency vehicles to provide a barrier from traffic
   - B. Wear reflective personal protective equipment
   - C. Strategically place fuses and cones to direct traffic away from the crash
   - D. Close the roadway if necessary
   - E. All of the above

2. What steps need to be taken to stabilize the scene?
   - A. Reduce, remove, or mitigate the hazards at the incident scene.
   - B. Crib the vehicle and ensure the rescue vehicle has wheel chocks
   - C. Deploy airbags, break window glass, and remove the steering wheel.
   - D. Perform size-up, determine type of vehicle, and orient yourself to the anatomy of the vehicle.

When you arrive on the scene, you find a vehicle that has crashed head-on into a utility pole. The driver is barely conscious and is complaining of head, chest, and bilateral leg pain. Lieutenant Christiansen tells you and your partner to stabilize the car with airbags and cribbing and break the tempered glass to allow emergency medical personnel access to the victim.

3. Safety considerations for working with airbags include:
   - A. never stack high-pressure bags more than two high.
   - B. do not use a rescue lift airbag to stabilize a vehicle by itself; use cribbing.
   - C. place plywood above lift airbags.
   - D. Both a and b

4. When breaking tempered glass you should:
   - A. strike the window with a blunt object such as a hammer.
   - B. choose a window close to the victim.
   - C. use a sharp pointed object and strike the window in the lower corner.
   - D. Both a and b

www.FireFighter.jbpub.com

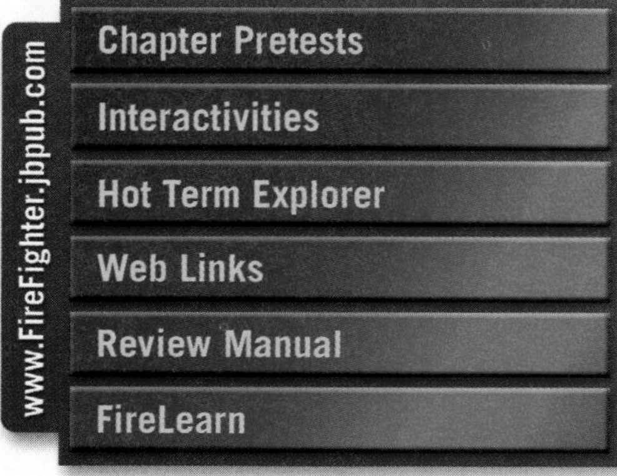

Assisting Special Rescue Teams

Chapter 26

NFPA 1001 Standard

Fire Fighter I

NFPA 1001 contains no Fire Fighter I Job Performance Requirements for this chapter.

Fire Fighter II

6.4 *Rescue Operations.* This duty involves performing activities related to accessing and disentangling victims from motor vehicle accidents and helping special rescue teams, according to the following job performance requirements.

6.4.2 Assist rescue operation teams, given standard operating procedures, necessary rescue equipment, and an assignment, so that procedures are followed, rescue items are recognized and retrieved in the time as prescribed by the AHJ, and the assignment is completed.

6.4.2 (A) *Requisite Knowledge.* The fire fighter's role at a special rescue operation, the hazards associated with special rescue operations, types and uses for rescue tools, and rescue practices and goals.

6.4.2 (B) *Requisite Skills.* The ability to identify and retrieve various types of rescue tools, establish public barriers, and assist rescue teams as a member of the team when assigned.

A.4.4.2 The Fire Fighter II is not expected to be proficient in special rescue skills. The Fire Fighter II should be able to help special rescue teams in their efforts to safely manage structural collapses, trench collapses, cave and tunnel emergencies, water and ice emergencies, elevator and escalator emergencies, energized electrical line emergencies, and industrial accidents. (***Annex Material***)

Additional NFPA Standards

NFPA 1670, *Operations and Training For Technical Rescue Incidents*

Knowledge Objectives

After studying this chapter, you will be able to:

- Define the types of special rescues encountered by fire fighters.
- Describe the steps of a special rescue.
- Describe the general procedures at a special rescue scene.
- Describe how to safely approach and assist at a vehicle or machinery rescue incident.
- Describe how to safely approach and assist at a confined space rescue incident.
- Describe how to safely approach and assist at a rope rescue incident.
- Describe how to safely approach and assist at a trench and excavation rescue incident.
- Describe how to safely approach and assist at a structural collapse rescue incident.
- Describe how to safely approach and assist at a water or ice rescue incident.
- Describe how to safely approach and assist at a wilderness search and rescue incident.
- Describe how to safely approach and assist at a hazardous materials rescue incident.
- Describe how to safely respond to an elevator or escalator rescue.

Skills Objectives

There are no skill objectives for this chapter.

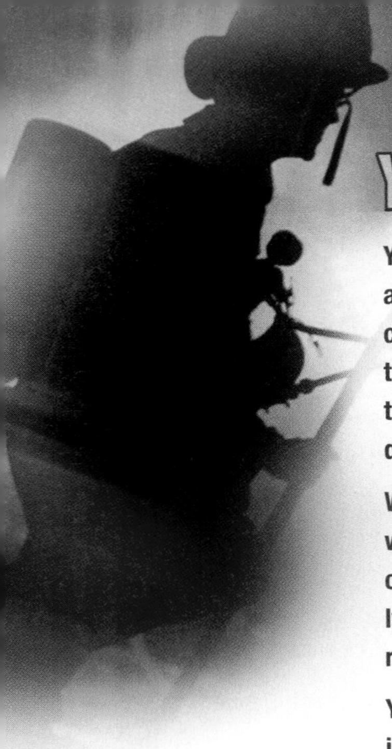

You Are the Fire Fighter

You are a probationary fire fighter working in an engine company. Your company is dispatched to a reported trench collapse, along with another engine company, a ladder company, and a rescue company. Since it is in another district, other companies have arrived on the scene first, and from the initial report given by the first arriving company, you know that there is a worker trapped by the collapse. You hear the incident commander (IC) make a request for additional units to be dispatched because a fire fighter is also trapped.

When your company arrives on the scene, your officer orders the crew to stay with the engine, while she reports to the command post for an assignment. You can see that there is a large pile of dirt next to a trench, and that part of it has slid down into the trench. There are hand tools lying next to the trench, and a backhoe is nearby, but no one appears to be doing anything to rescue the two trapped men.

You realize how dangerous a trench can be. A technical rescue incident (TRI) is not an everyday incident and requires specialized knowledge. As you are helping to unload the equipment off of the rescue squad, you understand the complexity of technical rescues.

1. ***What should you be looking for when responding to a structural collapse rescue?***
2. ***What concepts about trench rescue safety does this scenario illustrate?***
3. ***What should have been done with the soil removed from the trench to help prevent the collapse of the trench in this situation?***

Introduction

In recent years the number of fires and associated human and property loss from fire in the United States have generally decreased. This decrease is due to better fire prevention, better fire safety education, better codes, and better fire safety design. At the same time, the mission of fire departments has expanded. As fire departments take on added roles, such as emergency medical services, hazardous materials response, and technical rescue responses, they are generally seen as the agency that can handle any and all types of emergencies.

A technical rescue incident (TRI) is a complex rescue incident involving vehicles or machinery, water or ice, rope techniques, trench or excavation collapse, confined spaces, structural collapse, wilderness search and rescue, or hazardous materials that requires specially trained personnel and special equipment.

This chapter shows you how to assist specially trained rescue personnel in carrying out the tasks for which they have been trained. The chapter will not make you an expert in the skills that require specialized training to handle the eight types of rescue situations discussed. The more training you receive in these areas, the better you will understand how to conduct yourself at these scenes.

Training in technical rescue areas is conducted at three levels: awareness, operations, and technician.

- Awareness level (TRI): This training is an introduction to the topic with an emphasis on recognizing the hazards, securing the scene, and calling for appropriate assistance. There is no actual use of rescue skills at the awareness level.
- Operations level (TRI): This level is geared toward working in the "warm zone" of an incident (the area directly around the hazard area). This will allow you to directly assist those conducting the rescue operation and use certain skills and procedures to conduct the rescue.
- Technician level (TRI): This is the level that allows you to be directly involved in the rescue operation itself. Training includes use of specialized equipment, care of victims during the rescue, and management of the incident and of all personnel at the scene.

Most of the training you receive as a beginning fire fighter is aimed toward the awareness level so that you can identify the hazards and secure the scene to prevent further people from becoming victims.

Types of Rescues Encountered by Fire Fighters

Most fire departments respond to a variety of special rescue situations (► Figure 26-1). These include the following:

- Machinery and vehicle rescue
- Confined space rescue
- Rope rescue
- Trench and excavation rescue
- Structural collapse rescue
- Water and ice rescue
- Wilderness rescue
- Hazardous materials incidents

In order to become proficient in handling these situations, you must take a formal course to gain specialized knowledge and skills. NFPA 1670, *Operations and Training for Technical Rescue Incidents* covers these knowledge and skills areas.

It is important for awareness-level responders to have an understanding of these special types of rescues. Often, the first emergency unit to arrive at a rescue incident is a fire department engine or truck company. The initial actions taken by that company may determine the safety of the victims and the safety of the rescuers. Initial actions will also determine how efficiently the rescue is completed.

It is also important for you to understand and recognize many of the tools needed for rescue situations. You may not have to know how to use all of them, but you may be called upon to assist with a tool at any time.

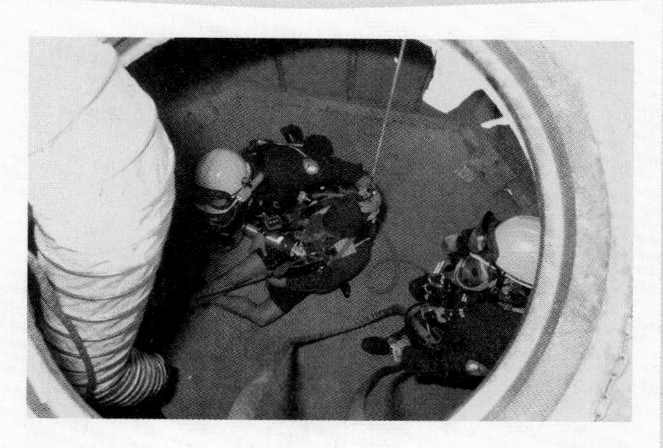

Figure 26-1 Most fire departments respond to a variety of special rescue situations.

Guidelines for Operations

When assisting rescue team members, keep in mind the five guidelines that you follow during other firefighting operations:

- Be safe.
- Follow orders.
- Work as a team.
- Think.
- Follow the Golden Rule of public service.

Be Safe

Rescue situations have many hidden hazards. These can include oxygen-deficient atmospheres, weakened floors, and strong water currents. Knowledge and training are required to recognize the signs that indicate a hazardous rescue situation exists. Once the hazards are recognized, determine what actions are necessary to ensure your own safety, as well as the safety of your team members, the victims of the incident, and bystanders. It requires experience and skill to determine that a rescue scene is not safe to enter, and that determination could save lives.

Follow Orders

As you begin as a fire fighter, you will have limited training and experience. Your officers and the rescue teams with whom you will work on special rescue incidents have received extensive specialized training. They have been chosen for their duties because they have experience and skills in a particular area of rescue. It is critical to follow the orders of those who understand exactly what needs to be done to ensure safety and to mitigate the dangers involved in the rescue situation. Orders should be followed exactly as given. If you do not understand what is expected of you, *ask*. Have the orders clarified so you will be able to complete your assigned task safely.

To do well as a fire fighter, you must have a strong appreciation for the basic philosophy of many firefighting organizations. Most fire departments are run like a military organization. Grasping the paramilitary structure will enable you to understand the command and control concept of fire departments.

- The fire officer's knowledge base and experience is greater than yours.
- Orders come from superiors. Legitimate orders are only those given by a fire officer or other designated person.
- Follow rules and procedures. A fire fighter is required to follow rules, procedures, and guidelines regardless of personal opinions.
- You must do your own job.
- Get the job done. In emergency situations, time is critical. But you must not act beyond your own skill and training level, and you must not violate any rules, procedures, standard operating guidelines, or orders of a superior in order to get the job done quickly.

Figure 26-2 There are personnel that are trained in specific tasks, such as rope rescue.

Fire Fighter Tips

F-A-I-L-U-R-E

In any TRI, it is important to know why rescuers fail. The reasons for rescue failures are most commonly referred to by the acronym "FAILURE":

F—Failure to understand the environment, or underestimating it

A—Additional medical problems not considered

I—Inadequate rescue skills

L—Lack of teamwork or experience

U—Underestimating the logistics of the incident

R—Rescue versus recovery mode not considered

E—Equipment not mastered

If you as a fire fighter and your team as a whole can learn to avoid these traps, you will have a greater chance of succeeding in any TRI.

You will find that acting with these ideas clear in your head will enable you to deal with the orders of your commanding fire officers quickly and confidently.

Work as a Team

Rescue efforts often require many people to complete a wide variety of tasks. There are personnel that are trained in specific tasks, such as rope rescue or swift water rescue (▲ Figure 26-2). However, they cannot do their jobs without the support and assistance of other fire fighters.

Firefighting requires members of the team to work together to complete the goal of fire suppression. Rescue is a different goal, but it requires the same kind of team effort. Your role in the rescue effort is an essential part of this team effort.

Think

As you are working on a rescue situation, you must constantly assess and reassess the scene. You may see something that your company officer does not see. If you think your assigned task may be unsafe, bring this to the attention of your superior officer. Do not try to reorganize the total rescue effort that is being directed by people who are highly trained and experienced, but do not ignore what is going on around you either.

Observations that a fire fighter should bring to the attention of a superior fire officer include changing weather conditions that might affect the rescue scene operations, suspicious packages or items on the scene, and broken equipment. In a hazardous materials rescue incident, wind direction and wind speed are important to know. An increase in wind speed would cause hazardous vapors to be spread more quickly. This knowledge would affect the response plan.

Follow the Golden Rule of Public Service

As you get involved in carrying out a rescue effort, it is easy to concentrate on the technical parts of the rescue and forget that there is a scared person involved who needs your emotional support and encouragement. It is helpful to have a rescuer stay with the victim whenever possible. This rescuer can tell the victim what actions will be performed during the rescue process. This rescuer is often the one the victim remembers and the one who helps to fulfill the golden rule.

Steps of Special Rescue

Though special rescue situations may take many different forms, there are basic steps that all rescuers take in order to perform special rescues in a safe, effective, and efficient manner. The ten steps of the special rescue sequence are as follows:

- Preparation
- Response
- Arrival and size-up
- Stabilization
- Access
- Disentanglement
- Removal
- Transport
- Security of the scene and preparation for the next call
- Postincident analysis

Preparation

You can prepare for responses to emergency rescue incidents by training with other fire departments in your area. Doing this will better enable you to respond to a mutual aid call. It will also inform you of the type of rescue equipment other departments have access to, as well as the training levels of their personnel.

Know the terminology used in the field. This will make communicating with other rescuers easier and more effective. Know also the different types of rescue situations that you could encounter. Prior to a technical rescue call, your department must consider these issues:

- Does the department have the personnel and equipment to handle a TRI from start to finish?
- Does the department meet NFPA and OSHA standards for technical rescue calls?
- What will the department send on a technical rescue call?
- Do members of the department know the hazard areas of the department's response area; have they visited hazard areas with local representatives?

Response

A TRI should have a dispatch protocol. If your agency has its own technical rescue team, it will usually respond with a rescue squad, medic unit, engine company, and chief. This initial response will satisfy all the basic requirements of any TRI. Departments that do not have their own team will often respond with an EMS unit, engine company, and chief. The rescue squad will come from an outside agency. Often, it is necessary to notify power and utility companies during a TRI for possible assistance. Some technical rescues involve electricity, sewer pipes, or factors that may otherwise create the need for additional heavy equipment, to which utility companies have ready access.

Arrival and Size-Up

Immediately upon arrival, the first company officer will assume command. A rapid and accurate size-up is needed to avoid placing rescuers in danger and to determine what additional resources may be needed. What is the extent of injuries and how many victims are involved? This will help to determine how many medic units and other resources are needed.

When responding to a worksite or industrial facility, the officer should make contact with the foreman or supervisor, also known as the responsible party (RP). These individuals can often provide valuable information about the worksite and the emergency situation. The most important part of any rescue is the identification of any hazards and the decision of recovery versus rescue. With this information, a decision is made to call for any additional resources, and actions are taken to stabilize the incident.

Do NOT rush into the incident scene until an assessment can be made of the situation. A fire fighter approaching a trench collapse may cause further collapse. A fire fighter entering a swiftly flowing river might be quickly knocked down and carried downstream. A fire fighter climbing down into a well to evaluate an unconscious victim may be overcome by an oxygen-deficient atmosphere. Stop and think about the dangers that may be present. Do not make yourself part of the problem.

Stabilization

Once the resources are on the way and the scene is safe to enter, it is time to stabilize the incident. An outer perimeter is established to keep the public and media out of the staging area (cold zone (TRI)), and a smaller perimeter is established directly around the rescue (warm and hot zones). A rescue area is an area that surrounds the incident site (collapsed structure, or collapsed trench, or hazardous spill area, etc.) and whose size is proportional to the hazards that exist. The zones should be established by identifying and evaluating the hazards that are discovered at the scene, observing the geographical area, noting the routes of access and exit, observing weather and wind conditions, and considering evacuation problems and transport distances. Three controlled zones should be established:

- Hot zone: This area is for entry teams and rescue teams only. This zone immediately surrounds the dangers of the site (e.g., hazardous materials releases) in order to protect personnel outside the zone.
- Warm zone: This area is for properly trained and equipped personnel only. This zone is where personnel and equipment decontamination and hot zone support take place.
- Cold zone: This area is for staging vehicles and equipment. This zone contains the command post. The public and the media should be clear of the cold zone at all times.

The most common method of establishing the control zones for an emergency incident site is the use of police or fire line tape. Police or fire line tape is available in a variety of colors. The most commonly used colors are red, for the hot zone; orange, for the warm zone; and yellow, for the cold zone. For other methods of securing the perimeters of the controlled zones, see Chapter 31, Hazardous Materials: Scene Safety.

Fire Fighter Safety Tips

Often at fire and emergency scenes, fire line tape is set up to keep bystanders back. This fire line tape is routinely ignored by emergency responders operating at such a scene who simply duck under or climb over the tape. This complacency toward fire line tape as a warning device creates a potential hazard for responders at TRIs and other emergency scenes where the tape is placed to warn not just bystanders but responders as well. All fire line tape should be considered a hazard warning to fire fighters unless and until it is proven to be for bystanders only.

Figure 26-3 Lockout and tagout systems are methods of ensuring that the electricity has been shut down.

Once the controlled zones have been established by the fire department, responders should ensure that the zones of the emergency scene are enforced. Due to lack of firefighting personnel, scene control activities are sometimes assigned to law enforcement personnel.

Lockout and tagout systems should be used at this time to secure a safe environment (▲ **Figure 26-3**). Lockout and tagout systems are methods of ensuring that the electricity has been shut down and that electrical switches are "locked" so that they cannot be switched on.

In an emergency situation, all power must be turned off and locked, and switches or valves must be tagged with labels to protect personnel and emergency response workers from accidental machine start-up. Lockout and tagout procedures are used to warn personnel and ensure that the electrical power is disconnected. Only qualified, authorized, and trained personnel can disconnect the source of power, lock it out, and tag it. Locks and tags are used for everyone's protection against electrical dangers. For your safety and that of others, *never* remove or ignore a lock or tag.

During stabilization, atmospheric monitoring should also be started to identify any **immediately dangerous to life and health (IDLH)** environments for rescuers and victims. The next steps involve looking at the type of incident and planning on how to safely rescue victims. In a trench rescue, this would include setting up ventilation fans for airflow, setting up lights for visibility, and protecting the trench from further collapse.

Access

Once the scene is stabilized, access to the victim must be gained. How is the victim trapped? In a trench situation, it may be a dirt pile. In a rope rescue, it may be scaffolding that has collapsed. In a confined space, it may be a hazardous atmosphere that has caused someone to collapse. In order to reach a victim who is buried or trapped beneath debris, it is sometimes necessary to dig a tunnel as a means of rescue and escape. Identify the actual reason for the rescue, and work towards freeing the victim safely.

Communicate with the victims at all times during the rescue to make sure they are not injured further by the rescue operation. Even if they are not injured, they need to be reassured that the team is working as quickly as possible to free them.

Emergency medical care should be initiated as soon as access is made to the victim. Technical rescue paramedics are vital resources at TRIs; not only can these responders start IVs and treat medical conditions, but they also know how to deal with the equipment and procedures that are going on around them. It is vitally important that EMS personnel are effectively coordinated into ongoing operations during rescue incidents. Their main functions are to treat victims and to stand by in case a rescue team member needs medical assistance. As soon as a rescue area or scene is secured and stabilized, the EMS personnel must be allowed access to the victims for medical assessment and stabilization. Some fire departments have trained paramedics to the technical rescue level so that they can enter hazardous areas and provide direct assistance to the victim. Throughout the course of the rescue operation, which can sometimes span many hours, EMS personnel must continually monitor and ensure the stability of the victim and, therefore, must be allowed access to the victim.

Gaining access to the victim depends on the type of incident. For example, in an incident involving a vehicle, the location and position of the vehicle, the damage to the vehicle, and the position of the victim are important considerations. The means of gaining access to the victim must take into account the victim's injuries and their severity. The chosen means of access may have to be changed during the course of the rescue as the nature or severity of the victim's injuries becomes apparent.

Disentanglement

Once precautions have been taken and the reason for entrapment has been identified, the victim needs to be freed as safely as possible. A team member should remain with the victim to direct the rescuers who are performing the disen-

Figure 26-4 Victim removal.

tanglement. In a trench incident, this would include digging either with a shovel or by hand to free the victim.

In a vehicle accident, the most important point to remember is that the vehicle is to be removed from around the victim rather than trying to remove the victim through the wreckage. Various parts of the vehicle may trap the occupants, such as the steering wheel, seats, pedals, and dashboard. Disentanglement is the cutting of a vehicle (and or machinery) away from trapped or injured victims. In a vehicle accident, this is accomplished by using extrication and rescue tools with various extrication methods.

Removal

Once the victim as been disentangled, efforts will be redirected to removing the victim (▲ Figure 26-4). In some instances this may simply amount to having someone assist the victim up a ladder. Usually it will require removal with spinal immobilization due to possible injuries. There are a wide variety of choices such as Stokes baskets, backboards, stretchers, and other immobilization devices that are commonly used to remove an injured victim from a trench, confined space, or an elevated point.

Preparing the victim for removal involves maintaining continued control of all life-threatening problems, dressing wounds, and immobilizing suspected fractures and spinal injuries. The use of standard splints in confined areas is difficult and frequently impossible, but stabilization of the arms to the victim's trunk and of the legs to each other will often be adequate until the victim is positioned on a long spine board, which may serve as a splint for the whole body. The short spine board is used most frequently for stabilization of the sitting victim.

Sometimes a victim may have to be removed quickly (rapid extrication) because the victim's general condition is deteriorating, and time will not permit meticulous splinting and dressing procedures. Quick removal also occurs if there are hazards present, such as spilled gas or hazardous materials that could endanger the victim or the rescue personnel. The only time the victim should be moved prior to completion of initial care, assessment, stabilization, and treatment is when the victim's life or the emergency responder's life is in immediate danger.

Packaging is preparing the victim for movement as a unit and is often accomplished by means of a long spine board or similar device. The boards are essential for moving victims with potential or actual spine injuries.

Rough-terrain rescues may require passing a multiple-person stretcher up and over rough terrain, fording streams, or rock climbing. Rescues in rough terrain often require ingenuity to suspend or pad a stretcher so that the victim is provided a reasonably comfortable ride. Padding, such as inflated inner tubes, 4″ to 5″ foam padding, or loosely rolled blankets may be used; this is superior to "slinging" the stretcher on straps, which may allow excessive swaying and bouncing.

The overriding objective for each rescue, transfer, and removal is to complete the process as safely and efficiently as possible. It is important that the rescuer utilize good body mechanics, victim-packaging, removal, and transportation skills.

Transport

Once the victim has been removed from the hazard area, transport to an appropriate medical facility will be done by EMS. Depending upon the severity of the injuries and the distance to the medical facility, the type of transport will vary. For example, if a victim is critically injured or if the rescue is taking place some distance from the hospital, air transportation may be more appropriate than a ground ambulance.

In rough-terrain rescues, four-wheel drive, high-clearance vehicles may be required to transport victims on stretchers to an awaiting ambulance. Snowmobiles with attached sleds can be used to transport victims down a snowy mountain. Helicopters are increasingly used for quick evacuation from remote areas, but they have their limitations due to weather conditions.

Transporting victims who have been injured in a hazardous materials incident may pose many problems. Prior to transport, it is extremely important that adequate decontamination takes place. It is dangerous to the victim and the rescue personnel to place a poorly decontaminated victim inside an ambulance or helicopter and then close the doors. Any toxic fumes given off by the victim or by the victim's clothing can contaminate the inside of the transport vehicle, causing injury to the EMS personnel. Bags of contaminated clothing or other personal effects must be properly sealed if they are going with the victim. Receiving hospitals should be notified of the impending arrival of victims who have been involved in a hazardous materials exposure incident. This will alert the hospital triage teams so that the appropriate precautions can be taken once the victims begin to arrive.

VOICES OF EXPERIENCE

" Technical rescues can be very complex. "

"We have a man down and he is up there!" the construction superintendent shouted when I arrived on the scene of a reported person down at a construction site. "Up there" was atop a sky crane standing 167' in the air. The crane operator appeared to have suffered a heart attack. He had climbed outside of the control booth and fell alongside it onto an 18" catwalk. He had remained motionless since calling and complaining of chest pain on his radio.

Our rescue unit shortly arrived along with an engine company. I quickly discussed our strategy, which was to climb the crane utilizing an exterior ladder system and assess the victim. As I looked over the available rescue personnel, I realized that only myself and one other fire fighter were trained in high angle rescue techniques. I immediately transferred command to the engine company officer after explaining what rope and rescue systems we would need to rig in order to lower the victim to the ground.

After climbing the crane, we evaluated the operator and saw encouraging signs of a savable victim. We quickly began life support measures. During this time our rescue team was establishing a lowering system for transferring the victim to the ground. However, we were confronted with a serious problem. The limited amount of rescue equipment on the scene was not adequate for this type of structure, and the ropes were not long enough due to the tremendous height of the crane. Also, due to the soft ground and other obstacles at the base of the crane, an aerial ladder would not work either.

As we continued to provide life support measures, we evaluated the resources needed to successfully complete the rescue. By including the construction superintendent in our rescue planning, we discovered there was another crane with an extra jib (long crane tip) on the construction site. As life-saving efforts continued with time running out for our victim, I requested that the second crane be rigged with the extra jib so that it could be used to transfer the victim to the ground. After a very expedient setup and placement of the second crane at our position by construction personnel, we were able to effectively and safely transfer the victim to the ground. He was immediately transported to the hospital.

This rescue incident illustrates the need for rescue services personnel to always expect the unexpected. Technical rescues can be very complex. Rescue services personnel should be prepared to identify and utilize all available resources to meet the challenges posed by these incidents.

Sam Hansen
Vestavia Hills Fire Department
Vestavia Hills, Alabama

Postincident Duties

Security of the Scene and Preparation for the Next Call

Once the rescue is complete, the scene must be stabilized by the rescue crew to ensure that no one else becomes injured. In a trench incident, this includes filling the trench with dirt and roping off the surrounding area.

In a hazardous materials incident, the cleanup of equipment and personnel takes place after the hazardous materials incident has been completely controlled and all victims have been treated and transported. Trained disposal crews should be available to clean the site. All equipment, protective gear, and clothing, as well as the rescue personnel must be decontaminated.

In a vehicle rescue incident, the vehicle is removed, usually transported by a tow truck to a storage lot. The street or roadway is cleaned of all debris from the vehicle. This includes the cleanup of any spilled fluids, including gasoline, from the vehicle that was involved in the incident.

In an industrial setting, the supervisor of the facility is responsible for securing the scene, but the technical rescue team still must follow up with the supervisor or the safety coordinator to ensure that further problems are prevented.

Once you have secured the scene and packed up your equipment, it is important to return to the station and fully inventory, clean, service, and maintain all of the equipment to prepare it for the next call. Some items will need repair, but most will need simple maintenance before being placed back on the apparatus and considered in service.

Back at the station, as you prepare for the next call, complete paperwork and document the rescue incident. Record keeping serves several important purposes. Adequate reporting and the keeping of accurate records ensure the continuity of quality care, guarantees proper transfer of responsibility, and fulfills the administrative needs of the fire department for local, state, and federal reporting requirements. In addition, the reports can be used to evaluate response times, equipment usage, and other areas of administrative interest.

Postincident Analysis

As with any type of call, the best way to prepare for the next rescue call is to review the last one and identify any strengths and weaknesses. What could have been done better? What equipment would have made the rescue safer or easier? If a death or serious injury occurred during the call, a critical incident stress management (CISM) session may occur to assist fire fighters. Reviewing a TRI with everyone involved will allow everyone to learn from the call and make the next call even more successful.

General Rescue Scene Procedures

At any scene you respond to—whether it is a fire, EMS, or technical rescue call—the safety of you, your company, and the public is paramount. At a TRI there are many things that need to be considered. While the temptation may be to approach the victim or the accident area, it is critically important to slow down and properly evaluate the situation.

In confined space rescue incidents, the potential hazards can be deep or isolated spaces, multiple complicating hazards (such as water or chemicals), failure of essential equipment or service, and environmental conditions (such as snow or rain). In all rescue incidents, fire fighters should consider the potential general hazards and risks of utilities, hazardous materials, confined spaces, and environmental conditions, as well as hazards that are immediately dangerous to life and health.

Approaching the Scene

As you approach the scene of a TRI you will not always know what kind of scene you are going into. Is it a construction scene? Are there piles of dirt that would indicate a trench? Has a structural collapse occurred in a building? What actions are the civilians taking? Are they attempting to rescue trapped people, possibly placing themselves at great danger?

From the initial dispatch of the rescue call, the fire fighter should be compiling facts and factors about the call. Size-up begins with the information gained from the person reporting the incident and then from the bystanders at the scene upon arrival.

The information received when an emergency call is received is important to the success of the rescue operation. The information should include the following:

- Location of the incident
- Nature of the incident (kinds and number of vehicles)
- Condition and position of victims
- Condition and position of vehicles
- Number of people trapped or injured, and types of injuries
- Any specific or special hazard information
- Name of person calling and a number where the person can be reached

Once on the scene, life-threatening hazards can be identified and corrective measures can be taken. If additional resources are needed, they should be ordered by the incident commander.

A size-up should include the initial and continuous evaluation of the following:

- Scope and magnitude of the incident
- Risk and benefit analysis
- Number of known and potential victims
- Hazards
- Access to the scene
- Environmental factors

- Available and necessary resources
- Establishment of control perimeter

Careful size-up and coordination of rescue efforts are a must in order to avoid further injuries to the victims and to provide for the safety of the fire fighters.

Utility Hazards

Are there any downed electrical wires that are near the scene (► Figure 26-5)? Is the equipment or machinery present electrically charged so as to present a hazard to the victim or the rescuers? The IC should ensure that the proper procedures have been taken to shut off the utilities in the area where the rescuers will be working.

Utility hazards require the assistance of trained personnel. For electrical hazards, such as downed lines, park at least one apparatus span away. Watch for falling utility poles; a damaged pole may bring other poles down with it. Do not touch any wires, power lines, or other electrical sources until they have been deactivated by a power company representative. It is not just the wires that are hazardous; any metal that they touch is also energized. Metal fences that become energized are energized for their entire unbroken length. Be careful around running or standing water, since it is an excellent conductor of electricity.

Both natural gas and liquefied petroleum gas (LPG) are nontoxic but are classified as asphyxiants because they displace breathing air. In addition, both types of gas are explosive. If a call involves leaking gas, call the gas company immediately. If a victim has been overcome by leaking gas, wear positive-pressure self-contained breathing apparatus. Remove the victim from the hazardous atmosphere before beginning treatment.

Scene Security

Has the area been secured to prevent people from entering the area? Often coworkers, family members, and sometimes other rescuers will enter an unsafe scene and become additional victims. The IC should coordinate with law enforcement to help secure the scene and control access. A strict accountability system should be used by the fire department to control access to the rescue scene.

Protective Equipment

Firefighting gear is designed to protect the body from the high temperatures of fire. It does, however, restrict movement. Technical rescues generally require the ability to move around freely. Firefighting gear does not work well in a TRI. Most specialized teams also carry items such as harnesses; smaller, lighter helmets; and jumpsuits that are easier to move in than turnout gear (► Figure 26-6).

For personnel operating in a water rescue hazard zone, the minimum personal protective equipment includes a **personal flotation device (PFD)**, thermal protection, a helmet appropriate for water rescue, a cutting device, a whistle, and contamination protection (if necessary).

Figure 26-5 Downed electrical wires present a hazard.

Figure 26-6 Technical rescue technicians need to be able to move in their gear.

A handheld strobe light may help fire fighters keep track of each other in a crowd or in rural or wilderness locations. When working along highways, fire fighters can hook these lights onto their belts or attach them to their upper arm to provide additional visibility to oncoming vehicles. Strobe lights are lightweight, quite durable, and readily visible at night at a distance of approximately one mile. A durable, lightweight strobe light provides increased visibility of the rescuer in many situations.

Other equipment items that are easily carried by fire fighters are binoculars, chalk or spray paint for marking searched areas, a compass for wilderness rescues, first aid kits, a whistle, a handheld global positioning system, and cyalume-type light sticks. The IC and the technical rescue team will help to determine what protective equipment you will need to wear while assisting. Also see NFPA 1951, *Protective Ensemble for USAR Operations.*

Incident Management System (IMS)

The first arriving officer immediately assumes command and starts using the incident management system or IMS. This is critically important because many TRIs will eventually become very complex and require a large number of assisting units. Without the IMS in place, it will be difficult, if not impossible, to ensure the safety of the rescuers.

Accountability

Accountability should be practiced at all emergencies, no matter how small. The **accountability system** is the single most important process that any rescuer needs in order to ensure safety. An accountability system tracks the personnel on the scene, including their identities, assignments, and locations. This system ensures that only rescuers who have been given specific assignments are operating within the area where the rescue is taking place. By using an accountability system and working within IMS, an IC can track the resources at the scene, task out assignments, and ensure that every person at the scene operates safely.

Making Victim Contact

At any rescue scene, try to communicate with the victim if at all possible. Technical rescue situations often last for hours, with the victim left alone for long periods of time. Sometimes it is just not possible, but try to communicate via a radio, cell phone, or by yelling. Reassure the victim that everything is being done to ensure the victim's safety.

It is important for the fire fighter to stay in communication with the victim. If possible, it is advisable to have someone assigned to talk to the victim, while others are involved with the rescue. Realize that the victim could be sick or injured and is probably frightened. If the fire fighter is calm, this will calm the victim. To help keep a victim calm do the following:

- Make and keep eye contact with the victim.
- Tell the truth. Lying destroys trust and confidence. You may not always tell the victim everything, but if the victim asks a specific question, answer truthfully.
- Communicate at a level that the victim can understand.
- Be aware of your own body language.
- Always speak slowly, clearly, and distinctly.
- Use the victim's name.
- If a victim is hard of hearing, speak clearly and directly at the person, so that the person can read your lips.
- Allow time for the victim to answer or respond to your questions.
- Try to make the victim comfortable and relaxed whenever possible.

Many of the victims at TRIs require medical care, but this care should only be given if it can be done so *safely.* Do not become a victim during a rescue attempt. Many fire fighters have been killed entering a TRI trying to help a dead victim.

Assisting Rescue Crews

Every TRI will be different. If you have the role of assisting a technical rescue team, training with the team is probably the most important thing you can do. By training with the team you will respond with, you will get a feel for how they operate and they will get an idea of what they can trust you with. The more knowledge you have, the more you will be able to do.

At any TRI, follow the orders of the company officer who receives direction from the IC. Many of the tasks will involve moving equipment and objects from one place to another. Other tasks will involve protecting the team and victims. Do not take these tasks lightly; they may not seem important but they are essential to the effort. Without scene security, traffic control, protective hose lines, and people to maintain equipment, the rescue cannot and will not happen.

As you read the information that follows about assisting at specific types of rescue scenes, bear in mind the three factors about safely approaching the scene that apply to them all:

- Approach the scene cautiously.
- Position apparatus properly.
- Assist specialized team members as needed.

Vehicles and Machinery

There are a wide variety of rescue situations involving vehicles and machinery. By far, the most common type that we encounter as fire fighters is motor vehicle crashes. The actions commonly taken at the scene of motor vehicle crashes are covered in Chapter 25, Vehicle Rescue and Extrication. This section includes other types of rescues involving vehicles and machinery. These include motorcycles, trucks, buses, trains, mass transit vehicles, aircraft, watercraft, farm machinery, agriculture implements, construction machinery, and industrial machinery. Rescues involving mass transit often involve many victims. Most other types of rescues involve single victims.

Safe Approach

Vehicle and machinery rescues present a wide variety of problems to rescuers and victims. There are hazards such as flammable liquids, electrical hazards, and unstable machines or vehicles. Electricity is an invisible hazard. Any machine that is encountered should be considered electrically charged until proven otherwise. It is important to lockout and tagout any electrical source before approaching. The operator of the machine or a maintenance person who is familiar with the machine can be a valuable resource. For vehicle calls it is extremely important to make sure that traffic has been controlled. Fire fighters and EMS workers are struck and killed every year at accident scenes by passing drivers who are distracted by the accident. Placement of the apparatus is key to the protection of you and those around you.

Stabilization of the rescue scene includes making sure that the machine cannot move. Access to some victims will be easy. In other situations access to the victim can occur only after extensive stabilization of the scene.

Unusual and special rescue situations include extrication or disentanglement operations at incidents involving vehicles on their tops or sides, trucks and large commercial vehicles, and vehicles on top of other vehicles. To ensure responder safety, beware of fuel spilled at these scenes.

How You Can Assist

At a vehicle or machinery rescue call, you may be called upon to assist in the extrication and treatment of the victim. Protection of the victim (C-Spine immobilization, holding of blankets, etc) will allow the technical rescue team to operate hydraulic tools and cut the vehicle or machine apart without further hurting the victim. Many of the hydraulic tools are heavy and may require you to help support them while technicians operate them.

You may also be called upon to do the following:

- Assist with controlling site security and the perimeter of the rescue incident
- Obtain information from witnesses
- Retrieve more rescue tools and equipment from the fire apparatus or rescue trucks
- Console a family member of a victim
- Keep bystanders out of the way
- Assist in moving items that are in the way of the rescue team
- Service equipment
- Set up power tools
- Stand by with a hose line
- Extinguish a fire

Tools Used

There are many tools required for a successful vehicle or machinery rescue situation, and some are inexpensive items. Simple tools such as a spring-loaded punch can safely break out a vehicle window to access a victim. Tools and items that will be required for a vehicle or machinery rescue include the following:

- Personal protective equipment (PPE)
- Hydraulic tools (spreaders, cutters, rams)
- Halligan tool
- Cutting torch
- Air chisels
- Cribbing
- Saber saw
- Windshield cutter
- Spring-loaded punch
- Chains
- Airbags (high and low pressure)
- Basic hand tools
- Come-along
- Portable generator
- Seat belt cutters
- Hand lights
- Hose lines (protection)
- Blanket

Confined Space

A **confined space** is a location surrounded by a structure that is not designed for people to occupy. Confined spaces have limited openings for entrance and exit and may have limited ventilation to provide air circulation and exchange. Confined spaces can occur in farm, commercial, and industrial settings. Grain silos, industrial pits, tanks, and below-ground structures are all confined spaces. Cisterns, well casings, and septic tanks are confined spaces that are found in residential settings.

Confined spaces present a special hazard, because they may be oxygen deficient or contain poisonous gases. Entering a confined space without testing the atmosphere for safety and without the proper breathing apparatus can result in death. A confined space call is sometimes dispatched as a heart attack or medical illness call, because the caller assumes the person who entered a confined space and became unresponsive suffered a heart attack or medical illness.

Safe Approach

As you approach a rescue scene, look for a bystander who might have witnessed the emergency. Information gathered prior to the technical rescue team's arrival will save valuable time during the actual rescue. Do not believe a person in a pit has simply suffered a heart attack; always assume that there is an IDLH atmosphere at any confined space call. An IDLH atmosphere can immediately incapacitate anyone who enters the confined space without breathing protection. There can be toxic gases present, or there may not be enough oxygen to support life. When a rescue involves a confined space, remember that it will take some time for

Fire Fighter Safety Tips

Remember, the greatest hazards in a confined space are the lack of oxygen and the presence of poisonous gases. If rescuers fail to identify a confined space and to ensure that it contains a safe atmosphere, injury and death can occur to rescuers as well as to the original victim. There have been cases where well-meaning rescuers have gone into a confined space and become incapacitated because they did not properly evaluate the situation.

qualified rescuers to arrive on the scene and prepare for a safe entry into the confined space. The victim of the original incident may have died before your arrival. Do not put your life in danger for a dead victim.

How You Can Assist

Your main role in confined space rescues is to secure the scene, preventing other people from entering the confined space until additional rescue resources arrive. As additional highly trained personnel arrive, your company may provide help by giving the rescuers a situation report.

The first responding company must share whatever information is discovered at the rescue scene with the arriving crew. Anything that may be important to the response should be noted by the first arriving unit. Observed conditions should be compared to reported conditions and a determination should be made as to the relative change over the time period. Whether an incident appears to be stable or has changed greatly since the first report will affect the operation strategy for the rescue. A size-up should be quickly completed immediately upon arrival, and this information should be relayed to the special rescue team members upon their arrival at the scene. Other items of importance that should be included in a situation report are a description of any rescue attempts that have been made, exposures, hazards, extinguishment of fires, the facts and probabilities of the scene, the situation and resources of the fire company, the identity of any hazardous materials present, and a progress evaluation so far.

Many engine or ladder companies carry atmospheric or gas detection devices; if you have this capability, obtain readings to determine if there is a hazard. If you do encounter an oxygen-deficient atmosphere, set up a ventilation fan to help remove toxic gases and improve airflow to the victim. Sometimes you can help a victim without entering the confined space by passing down an SCBA, oxygen, first aid supplies, or even a ladder to climb out.

Confined space rescues can be complex and can take a long time to complete. You may be asked to assist by bringing rescue equipment to the scene, maintaining a charged hose line, or assisting with crowd control. By understanding the hazards of confined spaces, you will be better prepared to assist a specialized team that is dealing with an emergency involving a confined space.

Tools Used

One of the key components of any confined space rescue is a **supplied air breathing apparatus system (SABA)** (► Figure 26-7). These are similar to SCBA, except that instead of carrying the air supply in a cylinder on your back, you are connected by a hose line to an air supply located outside of the confined space. This provides the rescuer with a continuous supply of air that is not limited by the capacity of a back-mounted air cylinder.

Figure 26-7 SABA.

There are a number of tools and pieces of equipment that can be used during a confined space rescue operation:

- SABA
- Tripod for raising and lowering rescuers and victims
- Pry bar or manhole cover remover
- Rescue rope
- Personnel harnesses
- Gas monitors
- Ventilation fans
- Explosion-proof lights
- Radios
- Protective gear
- Lockout, tagout, and blankout kits
- Accountability boards
- Victim removal devices such as Stokes baskets, backboards, and other commercially made devices
- Medical equipment
- Carabiners and locking "D" rings
- Descenders with ears
- Safety goggles
- Hand light (battery-operated)
- Hearing protection headsets

Rope Rescue

Rope rescue skills are used in a variety of different TRIs. It may be necessary to lift out an injured victim from inside of a tank using a raising system. Rescuers may need to lower themselves down to the area where victims are trapped in a structural collapse. Rope rescue skills are the most versatile and widely used technical rescue skills.

Rope rescue incidents are divided into low-angle and high-angle operations. **Low-angle operations** are situations

Figure 26-8 Low-angle operations.

where the slope of the ground over which the fire fighters are working is less than 45°, fire fighters are dependent on the ground for their primary support, and the rope system becomes the secondary means of support (▲ Figure 26-8). An example of a low-angle system is a rope stretched from the top of an embankment and used for support by rescuers who are carrying a victim up an incline. The rope is the secondary means of support for the rescuers. The primary support is the rescuer's contact with the ground. There are multiple examples of low-angle rescue operations, which help to make the job easier and safer.

Low-angle operations are used when the scene only requires ropes to be used as assistance to pull or haul up a victim or rescuer. This is usually necessary when adequate footing is not present in areas such as a dirt or rock embankment. In such an incident, a rope will be tied to the rescuer's harness and the rescuer will climb the embankment on his or her own, only using the rope to make sure he or she does not fall. Low-angle operations also include lifelines during ice or water rescues.

Ropes can also be used to assist in carrying a Stokes basket. This is done to aid the rescuers and to free them from having to carry all of the weight over rough terrain. Rescuers at the top of the embankment can help to pull up or lower the basket using a rope system.

High-angle operations are situations where the slope of the ground is greater than 45°, and rescuers or victims are dependent on life safety rope and not a fixed surface of support such as the ground. High-angle rescue techniques are used to raise or lower a person when other means of raising or lowering are not readily available. Sometimes a rope rescue is performed to remove a person from a position of peril. At other times rope rescues are needed to remove ill or injured persons.

In Chapter 8, Ropes and Knots, you learned how to tie some basic knots and use them to lift and lower selected tools. Rope rescue training courses build on the skills you have learned. There is a lot more to learn before you are ready to perform high-angle rescues. You need to be able to perform complex tasks quickly and safely. Rope rescue operations require a cohesive team effort.

Safe Approach

If you respond to an incident that may require a rope rescue operation, consider your safety and the safety of those around you. Rope rescues are among the most time consuming calls that you will encounter. There is extensive setup and equipment that needs to be assembled prior to initiating any rescue. Protect your safety by remaining away from the area under the victim and away from any loose materials that may fall. Work to control the scene so that the bystanders and friends of the victim move to an area where they will not be injured. You can do a lot to stabilize the scene and prevent further injuries by remaining calm and putting your skills to work. A rope can be used to secure a rescuer when making a rescue attempt, haul a stretcher or litter up an embankment, or lower a rescuer into a trench or cave.

How You Can Assist

Rope rescues are labor-intensive operations. By remembering the introductory rope skills you have learned, you may be able to assist with many phases of the rescue operation. You may be assigned to a technical rescue team member to tie knots and get anchors ready. Do not be offended if the team members go back and check your work. It is protocol to check a system two or three times to ensure safety. Remember to avoid stepping on ropes. Any damage or friction to the outer layers of a rope can decrease its tensile strength and possibly cause a catastrophic failure. Keep everyone clear of areas where a falling object could cause injuries. Keep in mind the importance of following the IMS. Work to complete your assigned tasks.

Tools Used

Rope rescues rely almost completely on the equipment used. When you are rappelling down to rescue a victim 300' off of the ground, you need to have equipment that has been properly designed and maintained to keep you alive. Some of the tools and equipment that you will need to recognize and use are as follows:

- PPE
- Personnel harnesses
- Stokes basket
- Harness for Stokes basket
- Rescue ropes
- Carabiners
- Webbing

- Miscellaneous hardware (racks, pulleys)
- Electrical cords (with plug and adapters)
- Electrical lights and hand lights (battery-operated)
- Ladders
- Tools (hacksaw, rubber mallet, pipe wrench, bolt cutter, crow bar, pry bar)

Trench and Excavation Collapse

Trench and excavation rescues occur when the earth has been removed for a utility line or for other construction and the sides of the excavation collapse, trapping a worker (▶ Figure 26-9). **Entrapments** can occur when children play around a pile of sand or earth that collapses. Many entrapments occur because the required safety precautions were not taken.

Any time there has been collapse, you need to understand that the collapsed product is unstable and prone to further collapse. Earth and sand are heavy, and a person partly entrapped cannot be pulled out. They must be carefully dug out. This can be done only after **shoring** has stabilized the sides of the excavation.

Vibration or additional weight on top of displaced earth will increase the probability of a **secondary collapse**. A secondary collapse is one that occurs after the initial collapse. It can be caused by equipment vibration, personnel standing at the edge of the trench, or water eroding away the soil. Safe removal of trapped persons requires a special rescue team that is trained and equipped to erect shoring to protect the rescuers and the entrapped person from secondary collapse.

Safe Approach

Safety is of paramount importance when approaching a trench or excavation collapse. Walking close to the edge of a collapse can trigger a secondary collapse. Stay away from the edge of the collapse and keep all workers and bystanders away. Vibration from equipment and machinery can cause secondary collapses, so shut off all heavy equipment. Vibrations caused by nearby traffic can also cause collapse, so it may be necessary to have traffic stopped or diverted.

Soil that has been removed from the excavation and placed in a pile is called the **spoil pile**. This material is unstable and may collapse if placed too close to the excavation. Avoid disturbing the spoil pile.

Make verbal contact with the trapped person if possible, but do NOT place yourself in danger while doing this. If you are going to approach the trench at all, approach it from the narrow end where the soil will be more stable. However, it is best *not* to approach the trench unless absolutely necessary. Stay out of a trench unless properly trained.

Provide reassurance by letting the trapped person know that a trained rescue team is on the way. By removing people from the edges of the excavation, shutting down machinery, and establishing contact with the victim, you start the rescue process.

Figure 26-9 Trench rescue.

You can also size-up the scene by looking for evidence that would indicate where the trapped victims may be located. Hand tools are an indicator of where the victims may have been working. Hard hats are another indicator. By questioning the bystanders you can also determine where the victims were last seen.

How You Can Assist

As the rescue team starts to work, your company will be assigned certain tasks. These may range from unloading lumber for shoring to assisting with cutting timbers a safe distance away from the entrapment. This type of rescue can take a long time. If it is hot or cold, a rehabilitation sector may need to be set up. Early implementation of the IMS will help make this type of rescue go smoothly. If someone on your company has a specialty such as carpentry, this should be made known, since this skill is valuable at this type of rescue. Cutting and measuring timber and shores for a trench operation is an important aspect of the rescue. Just as in a confined space, trenches may have an IDLH atmosphere inside due to poisonous gases like methane or other sewer gases. Setting up ventilation fans can often make a difference in victim and rescuer survivability.

It may also be necessary to pump water out of a trench. If the collapse occurred because of a ruptured pipe or during a rainstorm, rising water in the trench can endanger the trapped victim and cause additional soil to collapse into the trench. By removing this water, the situation can be stabilized.

When extricating victims from the trench, it may be necessary to lift them out using a raising system, just like a high-angle rope rescue operation. This will require that the rescuers have rescue ropes, harnesses, pulleys, carabiners, and all of the other associated equipment available. Personnel must be trained in how to use and apply these systems in this type of rescue scenario.

Tools Used

Trench and excavation rescue shares equipment, rescue techniques, and skills with both confined space rescue and structural collapse rescue. Tools and equipment used in trench and excavation rescue include the following:

- PPE: helmet, gloves, personal protective clothing, harness, flashlight, work boots, knee pads, elbow pads, eye protection, SCBA, SABA
- Hydraulic, pneumatic, and wood shores
- Lumber and plywood for shoring
- Cribbing
- Power cutting tools, saws
- Carpenter hand tools
- Shovels
- Buckets for moving soil
- Rescue rope, harnesses, webbing, associated hardware
- Utility rope
- Ventilation fans
- Pumps
- Lighting
- Ladders
- Folding shovel
- Extrication equipment such as Stokes baskets or backboards
- Harness set for Stokes baskets
- Medical equipment

Structural Collapse

Structural collapse is the sudden and unplanned fall of part or all of a building (► Figure 26-10). Collapses occur because of fires, removal of supports during construction or renovation, vehicle crashes, explosions, rain, wind or snowstorms, earthquakes, and tornados. Chapter 6, Building Construction, emphasizes the importance of all building components working together to provide a stable building. Consider the type of building construction when determining the potential for collapse. When any part of a building is compromised, the dynamics of the building change; fire fighters should always be alert for signs of a possible building collapse. A partial building collapse may be hazardous to rescuers because of the potential for secondary or further collapse.

Figure 26-10 Structural collapse is the sudden and unplanned fall of part or all of a building.

Safe Approach

Because of the variety of factors that can cause building collapses, you must approach the scene carefully. The cause of some collapses, such as a vehicle crashing through the wall of a house, will be evident. Others, like a natural gas explosion, may require extensive investigation before the cause is determined. As you approach a building collapse, consider the need to shut off utilities. Entering a structure with escaping natural gas or propane is extremely hazardous. Electricity from damaged wiring can also present a deadly hazard.

A prime safety consideration is the stability of the building. Even a well-trained engineer cannot always determine the stability of a building by looking only at its exterior. Therefore, you must operate as though the building may experience a secondary collapse at any time. The IC must make the decision regarding whether the building is safe to enter. This is a difficult decision, especially since the bystanders cannot always tell you whether there were people in the building before the collapse. Be aware of the potential for secondary collapse.

How You Can Assist

Rescue operations at a structure collapse vary depending on the size of the building and the amount of damage to the building. In cases of large building collapses, the rescue operation will be sizeable. Urban search and rescue teams or structural collapse teams have received special training in dealing with these types of situations. They are trained in shoring and specialized techniques for gaining access and extricating victims that enable them to systematically search the affected building. These types of rescue situations take a lot of time and require a lot of personnel.

Personnel without special training will usually be used in support operations. You may be assigned to be part of a

Fire Fighter Tips

New fire fighters must learn when to offer helpful suggestions and when to remain quiet. There may be times when you see something that the company officer does not see, or have an idea that the company officer has not considered. However, the need to provide helpful information must be balanced against the risk of overwhelming a company or chief officer with "helpful" suggestions. There can be no hard and fast rule about this, but stay mindful of the fact that remaining quiet is at times the most helpful thing you can do.

bucket brigade that works to remove debris from the building. There may be a substantial amount of manual digging and searching that is physically taxing. You may be able to assist with this if the work is not in an unstable location.

In order for a rescue effort of this type to work, there must be teamwork and a well-organized and well-implemented incident command system. Without a solid command structure, most large-scale rescue efforts are doomed to fail.

Tools Used

Fire fighters should know the tools and equipment designed for structural collapse emergency rescue incidents:

- PPE: helmet, gloves, work boots, harness, elbow pads, knee pads, eye protection, SCBA, SABA
- Shoring equipment
- Lumber for shoring and cribbing
- Power tools
- Hand tools
- Lighting
- Rescue ropes, harnesses, webbing, and associated hardware
- Utility ropes
- Buckets
- Folding shovels

Water and Ice Rescue

Almost all fire departments have the potential for being called to perform a water rescue (► **Figure 26-11**). Water is present in small streams and large rivers. It fills lakes, oceans, reservoirs, and swimming pools. A static source such as a lake may have no current, while a white water stream or flooded river may have a swift current. During a flood, a dry wash in the desert can become a raging monster.

In North America, the most common swift water rescue scenario is a car that has tried to drive through a pool of water created by a flooded stream. The vehicle stalls because of the depth of the water, leaving the vehicle occupants stranded in a rising stream with a swift current. If the water is high enough, the vehicle can be swept away. These incidents are especially dangerous for rescue personnel, because it is hard to determine the depth of the water around the vehicle. Fire departments should be prepared for the types of water rescue incidents that may occur in their communities. Fixed bodies of water in your community are easy to identify. Identifying the areas in your community where flooding is likely to occur will require preplanning.

Figure 26-11 Water rescue.

Safe Approach

When responding to water rescue incidents, your safety and the safety of other teammates is your primary concern. Your turnout gear is designed for structural firefighting. It provides amazing protection for that purpose. It is not designed for water rescue activities. Fire fighters who fall into the water while wearing turnout gear quickly find that their movements are severely limited by the bulky gear. When working at a water rescue scene, you should use personal protective equipment designed for water rescue, not structural firefighting. Anytime you are within ten feet of the water, you should be wearing an approved personal floatation device (PFD). Shoes that provide solid traction are preferred to fire boots.

If you are part of the first arriving engine or truck company, and the endangered people are in a vehicle or holding on to a tree or other solid object, try to communicate with them. Let them know that additional help is on the way.

Do not exceed your level of training. If you cannot swim, operating around or near water is not recommended. A person who is trained as a lifeguard for still water is not prepared to enter flowing water with a strong current, such as in a river, stream, or ocean. Trained rescuers may consider using a throw rope, a pike pole, or a ladder to reach a person. Many people who are in trouble in the water are close to shore. Do not use a boat unless you are trained in its use. Ensure that bystanders do not try to rescue the victim and place themselves in a situation where they need to be rescued too.

Ice rescues are common in colder climates. Many departments in colder regions have developed specialized equipment to assist in ice rescues. Throwing a rope or floatation device may be helpful initially. Ladders can be used to distribute the weight of the rescuer on ice-covered water. Specially designed hose lines have been used with end-caps and an air line to create a floatation buoy. There are many other possibilities in regard to things that can be used. Special rescue suits are available for ice rescues to prevent hypothermia and provide floatation for the rescuer. If your department is involved in ice rescues, you should receive training in these specialized procedures.

How You Can Assist

At water and ice rescues, dangers to rescuers are always present. PFDs should be used by everyone operating at the scene. Tying basic knots to attach the ropes being used and just acting as another set of eyes for the incident are important roles you will fill.

Be alert for changing weather conditions, such as an increase in wind speed, that could affect the rescue operations. By providing strict scene and site control, you can prevent other people from becoming victims at the scene. You can assist the special rescue teams with changing their air supplies and with other equipment needs. You can also relay communications from rescuers to the IC and keep EMS personnel advised of rescue progress.

Tools Used

Fire fighters should become familiar with the tools used in water rescue operations:

- PFD
- SCUBA equipment
- Helmet
- Whistle
- Throw bag with rope
- Boat
- Rescue rope, webbing, and associated hardware
- Lighting, flashlight
- Gloves
- Goggles or other eye protection
- Suitable footwear for the environmental and rescue conditions

Wilderness Search and Rescue

Wilderness search and rescue (SAR) is an activity that is conducted by a limited number of fire departments. It is included in NFPA 1670, *Standard on Operations and Training for Technical Rescue Incidents,* as part of technical rescue training. SAR missions consist of two parts, search and rescue. Search is defined as looking for a lost or overdue person. Rescue, in the SAR context, is defined as removing a victim from a hostile environment.

There are several types of situations that result in the initiation of SAR missions. Small children wander off and are unable to find their way back to a known place. Older adults who are suffering from disorders such as Alzheimer's disease fail to remember where they are going and become lost. People who are hiking, hunting or participating in other wilderness activities become lost because they do not have the proper training or equipment for that activity, because the weather changes unexpectedly, or because they become sick or injured.

Safe Approach

As a beginning fire fighter, you may respond to a possible SAR mission as part of your fire company. In cases of a lost person, some small fire departments will request all available personnel to respond to assist with a search. It is important to respond as part of your department and to work together as a team. The IC for such a mission should be a person who is well trained in directing SAR missions. In some communities, SAR is the responsibility of law enforcement personnel. In others, it becomes the responsibility of volunteer search and rescue groups.

"Wilderness" can include varied environments, such as forests, mountains, deserts, parks, and animal refuges. Depending on the terrain and environmental factors, the wilderness can be as little as a few minutes into the backcountry or a few feet off the roadway. Incidents with a short access time could require an extended evacuation and thus qualify as a wilderness incident. Examples of terrain include cliffs, steep slopes, rivers, streams, valleys, mountainsides, and beaches. Terrain hazards include cliffs, caves, wells, mines, avalanches, and rock slides.

When you participate in SAR missions, prepare for the weather conditions by bringing suitable clothing. Make sure that you do not exceed your physical limitations, and do not get in situations that are beyond your ability to handle yourself in the wilderness. Call for a special wilderness rescue team, depending on the needs of the situation and your local protocols.

How You Can Assist

The fire service teaches working in a buddy system; this is another case in which you never go out alone. By working in teams of at least two and having a radio, teams can be methodically deployed and assigned to a search. The more knowledge you have about the search area, the more vital your role is during the search. Knowing your response area and the places where a child or lost person might hide is always beneficial to the search. A well-coordinated team can be an effective SAR force. An unorganized group of people will create increasing chaos. Do not enter the search area before the search team arrives. If the team will be using dogs, your scent will distract them.

Tools Used

Each specialized rescue team will bring equipment to the scene. Many of the same tools, equipment, and personal protective clothing that you would use for other specialized rescue situations will be used in SAR emergencies:

- Personal clothing appropriate for the environment
- Water
- Food
- Lighting, flashlights
- Communications equipment, including radios
- Medical equipment
- Maps, compass, global positioning system (GPS)
- Shelter
- Rescue ropes, harnesses, webbing, and associated hardware
- Extrication equipment such as Stokes basket and harness
- Flare gun and flares, whistles, or other signaling devices

Hazardous Materials Incidents

Hazardous materials are defined as any materials or substances that pose a significant risk to the health and safety of persons or to the environment if they are not properly handled during manufacture, processing, packaging, transportation, storage, use, or disposal (► Figure 26-12).

While hazardous materials incidents often involve a petroleum product, there are many other chemicals in our society that have toxic effects when not handled properly. We tend to think of hazardous materials incidents as occurring during transportation or at a large industrial setting, but it is important to consider that many retail businesses contain significant quantities of hazardous materials. Home and garden stores and pool supply stores are two places where hazardous materials can be found in significant quantities.

With the current threat of domestic terrorism, many of the agents that could be used as weapons also fall into this category. Most fire departments are trained and equipped to recognize these incidents, contain the hazards, and evacuate people if necessary. See Chapters 27–34 for complete coverage of hazardous materials incidents and terrorism.

Safe Approach

Hazardous materials incidents do not always get dispatched as hazardous materials incidents. You must be able to recognize the signs that indicate that there may be hazardous materials present as you approach the scene. You may see an escaping chemical or smell a suspicious odor. Warning **placards** are required for most hazardous materials either for storage or in transit.

Once you have recognized the presence of a hazardous material, you must protect yourself by staying out of the area exposed to the hazardous material. If you have happened upon a hazardous materials incident, it is important to have the hazardous materials team dispatched as soon as possible. In addition to standard action and precautions, call for a special hazardous materials rescue team immediately upon arrival at such an incident, implement site control and scene management, and assist specialized personnel after their arrival according to your training level.

Figure 26-12 Hazardous materials incident.

How You Can Assist

In order to assist at a hazardous materials incident you must have formal training. Training at the awareness level will provide you with the knowledge and skills you need to be able to recognize the presence of a hazardous material, protect yourself, call for appropriate assistance, and evacuate or secure the affected area. As part of your training, you will learn how to assist other hazardous materials responders.

The four major objectives of training at the operational level are to analyze the magnitude of the hazardous materials incident, plan an initial response, implement the planned response, and evaluate the progress of the actions taken to mitigate the incident.

Tools Used

Tools used in a hazardous materials incident include the following:

- Personal protective equipment appropriate to the level of the hazard
- Two-way radios
- Lighting
- Gas monitors
- Various sensing and monitoring tools to identify the materials involved
- Extensive research material
- Decontamination equipment
- Wide variety of hand tools (hammers, screwdrivers, axe, wrenches, and pliers)
- Hand tools
- Devices for sealing breached containers
- Leak-control devices
- Binoculars
- Fire line tape
- Control agents

Elevator and Escalator Rescue

Elevator and escalator rescues are technical responses that require technical knowledge. Rescue crews should review the American Society of Mechanical Engineers' (ASME) A17.4 Guide for Emergency Personnel, and the Occupational Health and Safety Administration's (OSHA) lockout and tagout procedures. All elevator and escalator responses begin with knowledge of the machinery located within your department's jurisdiction. This can only be accomplished through a preincident analysis of these structures. Specialized training classes are also an essential component of safe and successful operations.

Figure 26-13 Elevator emergencies are increasing in frequency.

Elevator Rescue

As metropolitan areas continue to grow, building heights are also increasing. Elevator emergencies are increasing in frequency (► Figure 26-13). The growing number of elevator emergencies is a factor that fire fighters must contend with.

Fire fighters should never attempt to move or relocate an elevator under any circumstances. Only professional elevator technicians who are thoroughly trained and authorized to do so should consider this. When responding to elevator incidents, consider these recommendations:

1. **Always cut power to a malfunctioning elevator,** and secure the power supply through the lockout and tagout procedures. Remember, elevator technicians and building maintenance personnel have been summoned to the same elevator incident. They may try to turn on the elevator equipment. To eliminate any possibility of this happening, use the lockout and tagout procedures at every incident. The elevator machine room is where the power supply should be turned off. Identify the location of the stalled elevator within the hoist-way or elevator shaft. Are there victims in the stalled elevator? If so, communicate with them, and reassure them that the situation is under control, they are not in danger, and the situation will soon be corrected.
2. **Incident risk management and risk assessment.** Can the victims in this disabled elevator be removed without increasing the risk of injury or death to fire fighters as well as the victims? Elevator cabs are passenger compartments that are securely locked to ensure that the victims are contained within the cab for their own safety. Fire fighters must (a) carefully evaluate the situation and the inherent risks involved in a rescue and extrication operation, and (b) understand that the victims are in a much safer environment inside than outside. The best option may be to reassure the victims and wait for the elevator technician to arrive and correct the situation.

3. **Are there enough trained personnel and the proper equipment on scene** to perform the rescue and extrication operation safely? Elevator hall door release keys, safety harnesses, rope bags, ladders, forcible entry tools, portable lighting, and an assortment of hand tools will be necessary. Always have additional trained personnel on hand, and anticipate obstacles and problems.
4. **Once the incident has been resolved, leave the power supply off.** Always leave the building with the elevator power supply turned off.

Escalator Rescue

Escalator emergencies can be challenging and complex. Fire fighters should never attempt to move an escalator under any circumstances. Only professional escalator technicians who are thoroughly trained and authorized to do so should consider this. This is particularly true if a victim (usually a child) is caught or entangled in the machinery. Good incident risk management and risk assessment skills are crucial with this type of incident; an error in judgment or a poorly executed reaction on the part of rescue personnel can easily turn an abrasion, laceration, or fractured bone into an amputated limb. Victims who are caught in machinery must be reassured, calmed, and stabilized medically by the rescuers.

Consider these recommendations:

1. **The escalator machinery must be stopped.** Stop switches are located at the top and bottom of most escalators. One of your company must maintain constant pressure on the stop switch until the power supply can be shut down. If the escalator machinery is stopped upon your arrival, which is generally the case, secure the power source. Escalator power supplies are usually located under the top landing or step plate of the escalator. *Always* turn off the power supply to the escalator and secure the de-energized (power off) power source with the lockout and tagout procedures.
2. **Are personnel competently trained for this type of incident?** Is the proper equipment on the scene to perform the rescue and extrication operation safely? These are reasonable questions, and the answer may be that you do not have the experienced, adequately trained personnel or the equipment to handle the situation. Again, exercise good judgment through risk management and risk assessment; simply reacting to the situation could have a very negative outcome. Unnecessary or excessive damage to the escalator equipment is also undesirable.
3. **The best action may be no action,** until a trained professional arrives on the scene to apply special expertise regarding this type of incident. The machinery may have to be dismantled in order to reduce the possibility of causing more damage to the victim entangled (usually the victim's hands or feet), or the equipment involved.
4. **Leave the power supply to the escalator turned off** after the escalator incident has been resolved. Assume nothing. Remember, safety is the number one priority.

Wrap-Up

Ready for Review

Since you will often be the first fire department company to arrive at a rescue scene, you must know the steps you can take to safely approach and stabilize the incident before specialty rescue personnel arrive at the scene. You will best be able to assist specially trained rescue personnel by following the five general guidelines for fire-fighting operations: be safe, follow orders, work as a team, think, and follow the golden rule of public service.

A full rescue response involves advanced preparation, thoughtful and safety-conscious decisions when first at the scene, cooperation with specialized rescue teams, and postincident analysis. In making decisions about approaching dangerous scenes, responder safety must take top priority.

The eight types of rescue incidents each have their own distinct characteristics that demand specialized techniques and equipment and, therefore, specialized training. Responders need personal protective equipment that is appropriate to the type of rescue, and they must operate within the IMS.

Chief Concepts

- Fire fighters encounter many different kinds of rescues. The basic principles of rescue apply to all types.
- Vehicle and machinery rescues occur in many settings. These situations require responders to stabilize the machinery and ensure that the electricity is off.
- It is important to recognize a confined space. Rescuers must ensure there is a pure air supply before entering confined spaces.
- Trench and excavation rescues are hazardous and require responders to minimize the chance for secondary collapses.
- A damaged building is prone to structural collapse. Any time a building has been damaged, assume it may collapse.
- Water rescue training is needed in almost all communities. Proper PFDs and appropriate clothing are important to ensure rescuer safety.
- Wilderness rescue may be called for even when initial access to a lost or stranded individual is quick.
- Hazardous materials training at the awareness level is required for fire fighter certification at Level I. This includes recognizing hazardous materials, protecting responders, calling for appropriate assistance, and evacuating and securing the affected area. Training at the operational level is required for fire fighter certification at Level II.

Hot Terms

Accountability system A method of accounting for all personnel at an emergency incident and ensuring that only personnel with specific assignments are permitted to work within the various zones.

Awareness level (TRI) Provides training that gives first responders the ability to recognize a potential hazardous materials emergency, to isolate the area, and to call for assistance; awareness-level responders take protective actions.

Cold zone (TRI) A safe area at a hazardous materials incident for those agencies involved in the operations; the incident commander, command post, EMS providers, and other support functions necessary to control the incident should be located in the cold zone. This zone may also be referred to as the clean zone or the support zone.

Confined space A space with limited or restricted access that is not meant for continuous occupancy, such as a manhole, well, or tank.

Entrapment A condition in which a victim is trapped by debris, soil, or other material and is unable to extricate himself or herself.

Hazardous materials Any materials or substances that pose an unreasonable risk of damage or injury to persons, property, or the environment if not properly controlled during handling, storage, manufacture, processing, packaging, use and disposal, or transportation.

High-angle operations A rope rescue operation where the angle of the slope is greater than 45°; rescuers depend on life safety rope rather than a fixed support surface such as the ground.

Hot zone The area immediately surrounding a hazardous materials spill/incident site that is directly dangerous to life and health. All personnel working in the hot zone must wear complete, appropriate protective clothing and equipment. Entry requires approval by the IC or a designated sector officer. Complete back-up, rescue, and decontamination teams must be in place at the perimeter before operations begin.

Immediate Danger to Life and Health (IDLH) An atmospheric concentration of any toxic, corrosive, or asphyxiant substance that poses an immediate threat to life or could cause irreversible or delayed adverse health effects. There are three general IDLH atmospheres: toxic, flammable, and oxygen-deficient.

Lockout/tagout systems Lockout and tagout are methods of ensuring that electricity and other utilities have been shut down and switches are "locked" so that they cannot be switched on, in order to prevent flow of power or gases into the area where rescue is being conducted.

Low-angle operations A rope rescue operation on a mildly sloping surface (less than 45°) or flat land where fire fighters are dependent on the ground for their primary support, and the rope system is a secondary means of support.

Operations level (TRI) Responders trained to the operations level should be able to recognize a potential hazardous materials incident, isolate the incident, deny entry to other responders and the public, and take defensive actions such as shutting off valves and protecting drains without having contact with the product. Operations-level responders act in a defensive fashion.

Packaging Preparing the victim for movement as a unit, often accomplished with a long spine board or similar device.

Personal flotation device (PFD) Also commonly known as a life vest; it allows the body to float in water.

Placards Signage required to be placed on all four sides of highway transport vehicles, railroad tank cars, and other forms of hazardous materials transportation that identifies the hazardous contents of the vehicle, using a standardized system: $10\ ^3/_4 \times 10\ ^3/_4$ in. diamond-shaped indicators.

Secondary collapse A collapse that occurs following the primary collapse. This can occur in trench, excavation, and structural collapses.

Shoring A method of supporting a trench wall or building components such as walls, floors, or ceilings using either hydraulic, pneumatic, or wood shoring systems; it is used to prevent collapse.

Spoil pile The pile of dirt that has been removed from an excavation, which may be unstable and prone to collapse.

Supplied air breathing apparatus (SABA) Emergency breathing systems similar to SCBA, but which utilize an airline running from the rescuers to a fixed air supply located outside of the confined space.

Technical rescue incident (TRI) A complex rescue incident involving vehicles or machinery, water or ice, rope techniques, a trench or excavation collapse, confined spaces, a structural collapse, an SAR, or hazardous materials, and which requires specially trained personnel and special equipment.

Technical rescue team A group of rescuers specially trained in the various disciplines of technical rescue.

Technician level (TRI) At this level, responders are allowed to enter heavily contaminated areas using the highest levels of protection. This is the most aggressive level of training as identified in NFPA 472. Hazardous materials technicians take offensive actions.

Warm zone The area located between the hot zone and the cold zone at an incident. The decontamination corridor is located in the warm zone.

Wilderness search and rescue (SAR) The process of locating and removing a victim from the wilderness.

Fire Fighter in Action

Your engine is dispatched to assist with the rescue of a victim who has slipped down an embankment. Your unit is the first to arrive on the scene. Your lieutenant speaks to the victim's spouse. She states that the victim was mowing the lawn when he stepped backwards and slipped approximately 35' down an embankment to a small ledge. Lieutenant Brown asks you to try to contact the victim while she gathers more information from the spouse. You yell over the side of the embankment and make contact with the victim. The high-angle rescue team is enroute to your location.

1. What are some tips for communicating with the victim to ensure that he remains calm?
   - **A.** Talk in technical jargon so the victim feels comfortable that you know what you are doing.
   - **B.** Speak clearly and slowly so the victim can understand you, use the victim's name, and always tell the truth.
   - **C.** Ignore the victim.
   - **D.** Lie to the victim so that he thinks that he will be rescued soon.

2. What can you do while you wait for the high-angle rescue team to arrive?
   - **A.** Throw a rope down to the victim so he can try to self extricate.
   - **B.** Try to peer over the embankment to get a visual.
   - **C.** Control the area around the embankment by marking it with fire line tape.
   - **D.** Both a and b

The high-angle rescue team arrives and begins working to access, remove, and transport the victim.

3. How can you assist on a rope rescue?
   - **A.** Tie knots, secure anchors, and help to haul the rope.
   - **B.** Put together pulley systems and tie-up the Stokes basket.
   - **C.** Put on a harness and repel down the embankment.
   - **D.** Be the safety officer.

4. Once you have secured the scene you should:
   - **A.** fully inventory, clean, service, and maintain all equipment.
   - **B.** complete all paperwork and document the incident.
   - **C.** leave all the work for the oncoming crew.
   - **D.** Both A and B

www.FireFighter.jbpub.com

www.FireFighter.jbpub.com

Chapter Pretests

Interactivities

Hot Term Explorer

Web Links

Review Manual

FireLearn

Hazardous Materials: Overview

Chapter 27

NFPA 472 Standard

4.1.1 Introduction

4.1.1.1 First responders at the awareness level shall be trained to meet all competencies of this chapter.

4.1.1.2 They also shall receive any additional training to meet applicable United States Department of Transportation (DOT), United States Environmental Protection Agency (EPA), Occupational Safety and Health Administration (OSHA), and other state, local, or provincial occupational health and safety regulatory requirements.

4.1.2 Goal

4.1.2.1 The goal of the competencies at the awareness level shall be to provide first responders with the knowledge and skills to perform the tasks in 4.1.2.2 safely.

4.1.2.2 When first on the scene of an emergency involving hazardous materials, the first responder at the awareness level shall be able to perform the following tasks:

(1) Analyze the incident to determine both the hazardous materials present and the basic hazard and response information for each hazardous material by completing the following tasks:
- (a) Detect the presence of hazardous materials.
- (b) Survey a hazardous materials incident from a safe location to identify the name, UN/NA identification number, or type placard applied for any hazardous materials involved.
- (c) Collect hazard information from the current edition of the *Emergency Response Guidebook*.

(2) Implement actions consistent with the local emergency response plan, the organization's standard operating procedures, and the current edition of the *Emergency Response Guidebook* by initiating and completing the following tasks:
- (a) Protective actions
- (b) Notification process

5.1.1 Introduction

5.1.1.1 First responders at the operational level shall be trained to meet all competencies at the first responder awareness levels and the competencies of this chapter.

5.1.1.2 First responders at the operational level also shall receive any additional training to meet applicable DOT, EPA, OSHA, and other state, local, or provincial occupational health and safety regulatory requirements.

5.1.2 Goal The goal of the competencies at the operational level shall be to provide first responders with the knowledge and skills to perform the tasks in 5.1.2.1 safely.

5.1.2.1 The first responder at the operational level shall be able to perform the following tasks:

(1) Analyze a hazardous materials incident to determine the magnitude of the problem in terms of outcomes by completing the following tasks:
- (a) Survey the hazardous materials incident to identify the containers and materials involved, determine whether hazardous materials have been released, and evaluate the surrounding conditions.
- (b) Collect hazard and response information from MSDS; CHEMTREC/CANUTEC/SETIQ; local, state, and federal authorities; and shipper/manufacturer contacts.
- (c) Predict the likely behavior of a material as well as its container.
- (d) Estimate the potential harm at a hazardous materials incident.

(2) Plan an initial response within the capabilities and competencies of available personnel, personal protective equipment, and control equipment by completing the following tasks:
- (a) Describe the response objectives for hazardous materials incidents.
- (b) Describe the defensive options available for a given response objective.
- (c) Determine whether the personal protective equipment provided is appropriate for implementing each defensive option.
- (d) Identify the emergency decontamination procedures.

(3) Implement the planned response to favorably change the outcomes consistent with the local emergency response plan and the organization's standard operating procedures by completing the following tasks:
- (a) Evaluate the status of the defensive actions taken in accomplishing the response objectives.
- (b) Communicate the status of the planned response.
- (c) Don, work in, and doff personal protective equipment provided by the authority having jurisdiction.
- (d) Perform defensive control functions identified in the plan of action.

(4) Evaluate the progress of the actions taken to ensure that the response objectives are being met safely, effectively, and efficiently by completing the following tasks:
- (a) Evaluate the status of the defensive actions taken in accomplishing the response objectives.
- (b) Communicate the status of the planned response.

Knowledge Objectives

After studying this chapter, you will be able to:
- Define a hazardous material.
- Describe the different levels of hazardous materials training: Awareness, Operations, and Technician.
- Understand the laws that govern hazardous material response activities.
- Explain the difference between hazardous materials incidents and other emergencies.
- Explain the need for a planned response to a hazardous materials incident.

Skills Objectives

There are no skills objectives for this chapter.

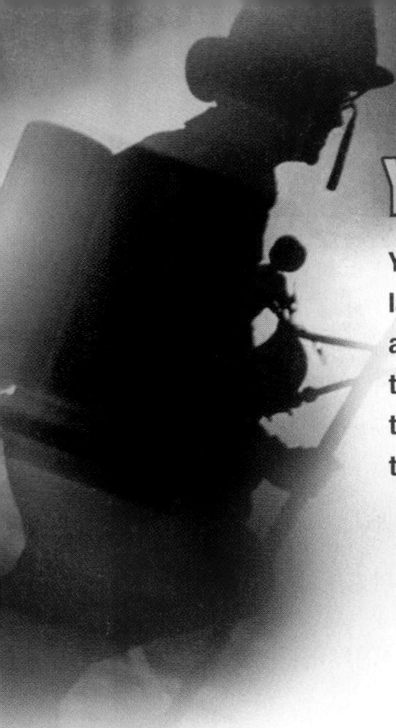

You Are the Fire Fighter

Your engine company has been called to investigate an odor coming from the cargo area of a large delivery truck. The truck is parked by the loading ramp at the rear of a building. As you approach, you can see that it has a red placard with a number "3" on the bottom. The driver says that he started feeling dizzy as he was unloading some packages. Other site employees say that they can smell something like "paint thinner" in the air. The driver is unable to provide any further identification of the contents of the truck.

1. ***What clues do you have for handling this incident in a safe, effective manner?***
2. ***What actions will you be able to take at the scene?***

Hazardous Materials Overview

The modern fire service must respond to both existing and emerging threats to lives and property. In addition to fighting fires, today's fire fighters may be called to incidents involving chemical spills, gas leaks, emergencies at industrial plants, or railroad or truck crashes with resultant material leaks or spills. These incidents frequently involve hazardous materials that threaten lives, property, and the environment.

This chapter is designed to give you the basic information required to recognize, respond to, and take initial action at a hazardous materials incident. The information is required for fire fighter I and II certification and is designed to build a foundation for safe and efficient response procedures.

The following material describes the competency levels for first responders at the awareness and operations level as found in NFPA 472, *Standard for Professional Competence of Responders to Hazardous Materials Incidents.*

Figure 27-1 The ability to recognize a potential hazardous material incident is critical to ensure your safety.

What Is a Hazardous Material?

A hazardous material, as defined by the United States Department of Transportation (DOT), is a material that poses an unreasonable risk to the health and safety of operating emergency personnel, the public, and/or the environment if it is not properly controlled during handling, storage, manufacture, processing, packaging, use and disposal, or transportation.

Fire fighters must be able to recognize the presence of a hazardous material, analyze the available information, and take appropriate action. The ability to recognize a potential hazardous material incident is critical to ensure your own safety (► **Figure 27-1**). To avoid entering a dangerous area, you must be alert to the possibility that hazardous materials are present at the scene. Next, you must attempt to identify and isolate the released material. Then prevent and address any injuries or exposures encountered. After life hazards are addressed, you must decide what level of action, along with appropriate personal protective equipment, is required to alleviate or resolve the incident.

Based on the DOT definition, a hazardous material can be almost anything. Milk, for example, is not regarded as a dangerous chemical substance, but 5,000 gallons of milk leaking into a creek does, in fact, pose an unreasonable risk to the environment. A large chlorine cloud fits the definition, as does a sustained leak from a 55-gallon drum of sulfuric

Figure 27-2 A hazardous material can be found anywhere.

acid. A fire fighter must be able to identify the chemicals involved and place them in the proper context. In other words, ask yourself this question: What is the chemical and what are the conditions under which it was released?

Hazardous materials can be found anywhere. From hospitals to petrochemical plants, pure chemicals and chemical mixtures are used to create millions of consumer products (▲ Figure 27-2). Currently, more than 80,000 chemicals are registered for use in commerce in the United States, and an estimated 2,000 new ones are introduced annually. The bulk of the new chemical substances are industrial chemicals, household cleaners, and lawn care products.

In addition, manufacturing processes sometimes generate hazardous waste. **Hazardous waste** is the material that remains after a process or manufacturing plant has used some of the material and it is no longer pure. Hazardous waste can be just as dangerous as pure chemicals. Hazardous waste can be mixtures of several chemicals, resulting in a hybrid substance. It may be difficult to determine how such a substance will react when it is released or comes in contact with other chemicals.

Fire Fighter Safety Tips

An awareness and an understanding of the surroundings is your first defense against danger. Be aware of the hazards in your response area.

Canadian Perspectives

Transport Canada is the focal point for the Canadian national program to promote public safety during the transportation of dangerous goods. The department's Transport Dangerous Goods Directorate serves as the major source of regulatory development, information, and guidance on dangerous goods transport for the public, industry, and government employees. This Directorate also works closely with other federal and provincial agencies to implement safety programs. CANUTEC, the Canadian Transport Emergency Centre, is operated by Transport Canada to assist emergency response personnel in handling hazardous materials emergencies. Established in 1979, CANUTEC responds to more than 30,000 calls a year, ranging from information requests to emergency notifications.

The Canadian Centre for Occupational Health and Safety (CCOHS) promotes a safe and healthy working environment by providing information and advice about occupational health and safety.

The federal *Hazardous Products Act* defines which materials (i.e., controlled products) are included in the Workplace Hazardous Materials Information System (WHMIS) and what information suppliers must provide to employers for controlled products used in the workplace through material safety data sheets (MSDS).

The good news is that, compared to the volume of chemicals used and the waste generated, there are very few hazardous material accidents that require fire fighter intervention. When accidents do happen, the fire department must usually deal with the problem.

Levels of Training: Regulations and Standards

To understand where your training fits in, you must first understand some of the regulatory process that govern hazardous materials response, beginning with the difference between a regulation and a standard. Regulations are issued and enforced by governmental bodies such as the U.S. **Occupational Safety and Health Administration (OSHA)** and the U.S. **Environmental Protection Agency (EPA)**. Standards are issued by nongovernmental entities and are generally consensus-based. Essentially, consensus organizations like the **National Fire Protection Association (NFPA)** issue standards that the public can comment on before committee members agree to adopt them.

NFPA standards governing hazardous materials response come from the Technical Committee on Hazardous Materials Response Personnel. This group includes members from private industry, the fire service, and state and federal agencies.

Canadian Perspectives

Those jurisdictions that have adopted a standard for training personnel in Hazardous Materials response have selected NFPA 472, *Standard for Professional Competence of Responders to Hazardous Materials Incidents.*

Fire Fighter Tips

Each state has the right to adopt and supercede safety and health regulations put forth by the federal OSHA. States that have exercised this right are called state-plan states. California, for example, is a state-plan state; its regulatory body is called CAL-OSHA. About half of the states in the U.S. are state-plan states.

The NFPA established three standards: NFPA 471, *Recommended Practice for Responding to Hazardous Materials Incidents;* NFPA 472, *Standard for Professional Competence of Responders to Hazardous Materials Incidents;* and NFPA 473, *Standard for Competencies for EMS Personnel Responding to Hazardous Materials Incidents.* The NFPA levels of training are awareness, operations, and technician.

In the United States, OSHA regulations are also important. The federal document containing the hazardous materials response competencies is known as **HAZWOPER** (Hazardous Waste Operations and Emergency Response). This set of regulations, issued in the late 1980s, standardized training for hazardous materials response and for hazardous waste site operations. The regulations can be found in the Code of Federal Regulations (CFR), book number 29, part 1910.120. Fire departments are primarily concerned with subsection (q) Emergency Response. The training levels found in HAZWOPER, much like the NFPA training levels, are identified as awareness, operations, technician, specialist, and incident commander. Fire fighters must be trained at the operations level.

Fire departments follow both the OSHA and the NFPA requirements for hazardous materials response. NFPA 472 clearly states: "First responders at the operational level also shall receive any additional training to meet applicable DOT, EPA, OSHA, and other state, local, or provincial occupational health and safety regulatory requirements."

The following descriptions provide a general overview (as found in NFPA 472 and the HAZWOPER regulations) of the different levels of hazardous materials training and competencies.

- **Awareness level**: This level of training enables first responders to recognize a potential hazardous materials emergency, isolate the area, and call for assistance. Awareness-level responders take *protective* actions.
- **Operations level**: Fire fighters trained to the operations level should be able to recognize a potential hazardous materials incident, isolate and deny entry to other responders and the public, evacuate persons in danger, and take defensive actions such as shutting off valves and protecting drains without having contact with the product. Operations-level responders act in a *defensive* fashion (► Figure 27-3).

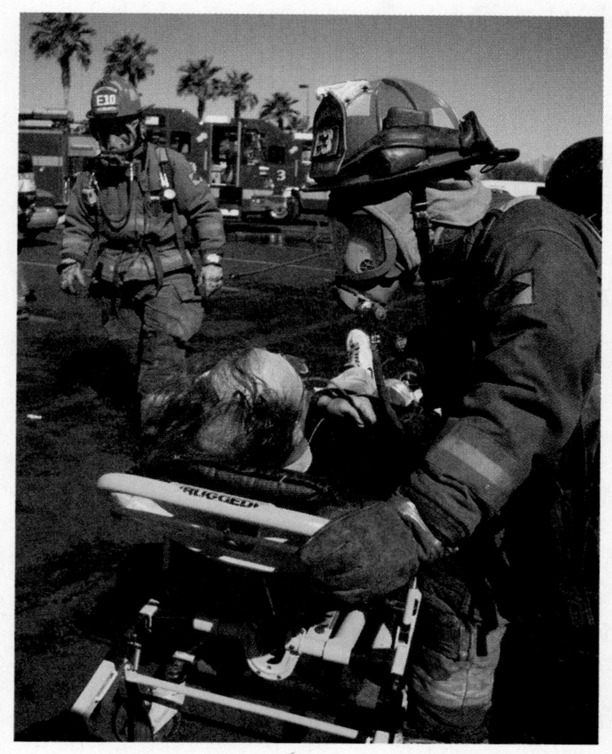

Figure 27-3 Operations-level responders.

- Technician level: At this level, fire fighters are trained to enter heavily contaminated areas using the highest levels of personal protection. This is the highest level of training identified in NFPA 472. Hazardous materials technicians take *offensive* actions (▶ Figure 27-4).
- Specialist level: Identified only in the HAZWOPER standard, this level receives more specialized training than a hazardous materials technician. Practically speaking, however, the two levels are not very different.
- Hazardous materials incident commander (IC Hazardous Materials): This level of training is intended for those assuming command of a hazardous materials incident beyond the operations level. Individuals trained as incident commanders should have at least operations-level training as well as additional training specific to commanding a hazardous materials incident.

Figure 27-4 Hazardous materials technicians.

NFPA 472 also spells out competencies for hazardous materials branch officers, hazardous materials branch safety officers, and other specialist employees. In addition to the initial training requirements listed above, OSHA regulations require annual refresher training. Consult your local agency for more specific information on refresher training.

Other Hazardous Materials Laws, Regulations, and Regulatory Agencies

Several government agencies are concerned with hazardous materials. The DOT, for example, enforces and publicizes laws and regulations that govern the transportation of goods by highways, rail, air, and, in some cases, marine transport.

The EPA regulates and governs issues relating to hazardous materials in the environment. The EPA's version of HAZWOPER can be found in Title 40, Protection of the Environment, Part 311: Worker Safety. As previously mentioned, OSHA is very concerned about hazardous materials response. In addition to the HAZWOPER regulations, OSHA issues guidance on respiratory protection, personal protective equipment, and a multitude of other topics regarding worker safety.

The Superfund Amendments and Reauthorization Act (SARA) was one of the first laws to affect how fire departments respond in a hazardous material emergency. Finalized in 1986, SARA was the original driver for the HAZWOPER regulations. Indirectly, it indicated that workers handling

VOICES OF EXPERIENCE

“No longer can we count on hazardous materials incidents to be few and far between, or only at industrial or motor vehicle accidents.”

In 1977 when I was a new fire fighter, "hazardous materials" was the latest buzz word in the fire service. Hazardous materials was the new specialized area that the fire service was undertaking. Fire departments across the United States were looking to fund these specialized units within their departments. Hazardous materials incidents were few and far between, and usually were the result of an industrial or motor vehicle accident.

Today, the buzz word in the fire service is "weapons of mass destruction." A fire service colleague of mine has defined weapons of mass destruction as "nothing more than a hazardous materials incident with an attitude." I could not agree more with this definition. No longer can we count on hazardous materials incidents to be few and far between, or only at industrial or motor vehicle accidents. Hazardous materials can be found at any routine emergency we respond to.

Fire fighters at all levels must possess the knowledge and understanding of hazardous materials as part of their regular duties. Advances in science, technology, and industry have led to the development of more and more chemical-based products. As such, we can no longer rely on specialized fire department units to handle hazardous materials incidents; all fire fighters must have a solid background in and understanding of hazardous materials. Having the ability to identify when hazardous materials are involved and what actions are needed to secure the scene will enable fire fighters to protect themselves, the public, the environment, and their fellow fire fighters.

Paul R. Gerardi
Fairview Fire Department
White Plains, New York

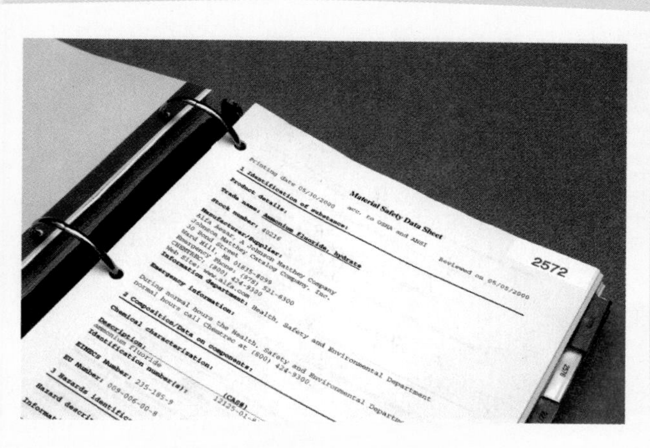

Figure 27-5 A material safety data sheet (MSDS) details the properties of a specific chemical.

hazardous wastes should have a minimum amount of training. Additionally, this law laid the foundation that ultimately allowed local fire departments and the community at large to obtain information on how and where hazardous materials were stored in their community. The **Emergency Planning and Community Right to Know Act (EPCRA)** requires a business that handles chemicals to report storage type, quantity, and storage methods to the fire department and the local emergency planning committee.

Local Emergency Planning Committees (LEPC) gather and disseminate information about hazardous materials to the public. These committees are comprised of members of industry, transportation, media, fire and police agencies, and the public at large. Essentially, LEPCs ensure that local resources are adequate to respond to a chemical event in the community. Fire departments should be familiar with their LEPC and know how their department works with this committee.

One piece of data collected by the LEPC is a **material safety data sheet (MSDS)**. The MSDS is a detailed profile of a single chemical or mixture of chemicals, provided by the manufacturer and/or supplier of a chemical (▲ Figure 27-5). The MSDS details a specific chemical's properties and all pertinent information about it. These chemical-specific data sheets are an important source of information for fire fighters when dealing with hazardous materials incidents. Later chapters will discuss MSDS in detail.

Each state has a **State Emergency Response Commission (SERC)**. The SERC is the liaison between local and state levels of authority. The SERC involves agencies such as the fire service, police services, and elected officials for the collection and dissemination of information relating to hazardous materials.

Fire Fighter Tips

You can't do it all! Different members of your department have different skill levels and capabilities. All personnel must use their collective talents and training at a hazardous materials incident.

Difference Between Hazardous Materials Incidents and Other Types of Emergencies

A critical distinction must be made between a hazardous materials event and other emergencies, such as a structure fire. Fire fighters cannot approach a hazardous materials incident with the same mindset used for a structure fire. Structural firefighting employs a fairly basic strategy, determined during the size-up phase, with the ultimate goal of extinguishing the fire. Hazardous materials incidents are more complicated. The strategies used must be thought out and planned in advance.

For example, most of the time, a hazardous materials emergency moves more slowly than a structure fire. If a rescue is required, or if the situation is imminently dangerous and requires rapid intervention, events may move quickly, but for the most part, a hazardous materials emergency requires time, forethought, and planning to ensure that the incident is handled safely and effectively.

The actions taken at hazardous materials incidents are largely dictated by the chemicals involved. In many cases, it takes considerable effort to identify them. It becomes much more difficult and time-consuming to handle an incident involving an unknown substance. Response objectives, the choice of personal protective equipment, and the type of decontamination are complicated decisions that depend on the chemical properties of the hazardous substance.

In the case study at the beginning of this chapter, there are some clues to the nature of the material involved, but not enough to indicate all the hazards involved. Is the material toxic or corrosive? What is the chemical name of the material involved? How will it respond when exposed to air, water, or other materials? There simply is not enough information available for you to enter the scene safely.

Fires are considerably more straightforward. Every fire will generate heat, smoke, flames, and toxic gases. Turnout gear does not change radically from fire to fire. The tactics and strategy change based on the type of building and degree of fire involvement, not because the actual chemical nature of the fire changes.

When approaching hazardous materials events, make a conscious effort to change your perspective. Slow down, think about the problem and available resources, and take well-considered actions to solve it.

Planning a Response

It is a mistake to assume that the response to a hazardous materials incident begins when the alarm sounds. In reality, the response begins with training, learning about the regulations and agencies involved, and finding out about potential hazards in your area. Response agencies also should conduct incident planning activities at target hazards and other potential problem areas throughout the jurisdiction (► Figure 27-6).

Preplanning activities enable agencies to develop logical and appropriate response procedures for anticipated incidents. Planning should focus on the real threats that exist in your community or adjacent communities you could be assisting.

Once the threats have been identified, agencies and fire departments must determine how they will respond. Some agencies establish parameters that guide their response to particular hazardous material incidents. Those parameters outline incident severity based on the nature of the chemical, the amount released, or the type of occupancy involved in the incident. The 2002 edition of NFPA 471, for example, offers guidelines for determining incident levels for response and training. (► Table 27-1) illustrates a planning guide for determining response levels.

Figure 27-6 Conduct preincident planning activities at target hazards throughout the jurisdiction.

Fire Fighter Tips

The goal of the fire fighter is to favorably change the outcome of the hazardous materials incident. This is accomplished by sound planning and by establishing safe and reasonable response objectives based on the level of training. Don't do it if you're not trained to do it!

Table 27-1 Planning Guide for Determining Incident Levels for Response and Training

Incident Conditions	Incident Level: One	Two	Three
Product indentifications	Placard not required, NFPA 0 or 1 all categories, all Class 9 and ORM-D	DOT placarded, NFPA 2 for any categories, PCBs without fire, EPA regulated waste	Class 2, Division 2.3—poisonous gases, Class 1 Division 1.1 and 1.2—explosives, organic peroxide, flammable solid, materials dangerous when wet, chlorine, fluorine, anhydrous ammonia, radioactive materials, NFPA 3 & 4 for any categories including special hazards, PCBs & fire, DOT inhalation hazard, EPA extremely hazardous substances, and cryogenics
Container size	Small [e.g., pail, drums, cylinders except 910-kg (1-ton), packages, bags]	Medium [e.g., 910-kg (1-ton) cylinders, portable containers, nurse tanks, multiple small packages	Large (e.g., tank cars, tank trucks, stationary tanks, hopper cars/trucks, multiple medium containers)
Fire/explosion potential	Low	Medium	High
Leak severity	No release or small release contained or confined with readily available resources	Release may not be controllable controllable without special resources	Release may not be controllable even with special resources
Life safety	No life-threatening situation from materials involved	Localized area, limited evacuation area	Large area, mass evacuation area
Environmental impact (potential)	Minimal	Moderate	Severe
Container integrity	Not damaged	Damaged but able to contain the contents to allow handling or transfer of product	Damaged to such an extent that catastrophic rupture is possible

Wrap-Up

Ready for Review

Chemicals, whether benign or hazardous, are ever-present in homes, industry, and business. Technological advancements in computers, space exploration, and automobile technology could not occur without chemicals. Medicines and medical care would not exist without chemicals; homes and workplaces would not function if it weren't for the thousands of chemicals used to create plastics, cleaning products, fertilizers, and motor fuels. Essentially, hazardous materials can be found almost everywhere, in manufacturing processes, transportation, and in some cases, illegal activities.

Despite all the uses and abuses of chemicals in society, significant hazardous materials events are relatively infrequent. The network of regulations that govern hazardous substances during use, storage, and transportation have created a safe, reliable system to manage even very dangerous chemicals. When incidents do occur, however, it is imperative that fire fighters recognize the presence of a chemical incident and understand what actions can be taken to handle the situation.

Being an effective hazardous materials responder boils down to being a good problem solver. Good problem solvers are aware of their surroundings, observant of a myriad of details, and confident in their ability to take appropriate actions under extreme circumstances.

The successful handling of hazardous materials incidents, like any other fire department function, depends on teamwork. The information in this and the following chapters on hazardous materials is directed at the individual fire fighter. But this information is not intended to create a group of individual hazardous materials responders—it is designed to educate a group that ultimately functions as a team.

Chief Concepts

- Personal safety is the number one priority at a hazardous materials incident!
- A hazardous material is a material that poses an unreasonable risk to the health and safety of operating emergency personnel, the public, and/or the environment if it is not properly controlled during handling, storage, manufacture, processing, packaging, use and disposal, or transportation.
- A hazardous waste is a substance that remains after a process or manufacturing plant has used some of the material so that it is no longer pure.
- The actions taken at hazardous materials incidents are largely dictated by the chemicals involved.
- The goal of the competencies at the awareness level shall be to recognize a potential hazardous materials emergency, isolate the area, and call for assistance. Awareness-level responders take protective actions.
- Fire fighters at the operational level are able to recognize a potential hazardous materials incident, isolate and deny entry to other responders and the public, and take defensive actions such as shutting off valves and protecting drains without having contact with the product. Operations-level responders act in a defensive fashion.
- Defensive actions may include protecting drains from the flow of the product, closing remote valves, evacuating civilians from affected areas or sheltering them in place, or isolating the area and calling a specialized hazardous materials team or hazardous materials clean-up contractor.
- Hazardous materials technicians are trained to enter heavily contaminated areas using the highest levels of chemical protection. Hazardous materials technicians take offensive actions.

Hot Terms

Awareness level Provides training that gives first responders the ability to recognize a potential hazardous materials emergency, to isolate the area, and to call for assistance; awareness-level responders take protective actions.

Department of Transportation (DOT) The federal Department of Transportation publicizes and enforces rules and regulations that relate to the transportation of many hazardous materials.

Emergency Planning and Community Right to Know Act (EPCRA) Requires a business that handles chemicals to report storage type, quantity, and storage methods to the fire department and the local emergency planning committee.

Environmental Protection Agency (EPA) Established in 1970, the federal Environmental Protection Agency ensures safe manufacturing, use, transportation, and disposal of hazardous substances.

Hazardous materials Any materials that pose an unreasonable risk of damage or injury to persons, property, or the environment if not properly controlled during handling, storage, manufacture, processing, packaging, use and disposal, or transportation.

Hazardous Materials Incident Commander (IC Hazardous Materials) This level of training is intended for those assuming command of a hazardous materials incident beyond the operations level. Individuals trained as incident commanders should have at least operations-level training and additional training specific to commanding a hazardous materials incident.

Hazardous waste A substance that remains after a process or manufacturing plant has used some of the material and it is no longer pure.

HAZWOPER Short for Hazardous Waste Operations and Emergency Response, this federal OSHA regulation governs hazardous materials waste sites and response training. Specifics can be found in 29 CFR, book number 29, part number 1910.120. Subsection (q) is specific to emergency response.

Local Emergency Planning Committee (LEPC) Committee comprised of members of industry, transportation, the public at large, media, and fire and police agencies. Gathers and disseminates information on hazardous materials stored in the community and ensures that there are adequate local resources to respond to a chemical event in the community.

Material Safety Data Sheet (MSDS) A form, provided by manufacturers and compounders (blenders) of chemicals, containing information about chemical composition, physical and chemical properties, health and safety hazards, emergency response, and waste disposal of the material.

National Fire Protection Association (NFPA) The association that develops and maintains nationally recognized minimum consensus standards on many areas of fire safety and specific standards on hazardous materials.

Operations level Responders trained to the operations level should be able to recognize a potential hazardous materials incident, isolate the incident, deny entry to other responders and the public, and take defensive actions such as shutting off valves and protecting drains without having contact with the product. Operations-level responders act in a defensive fashion.

Occupational Safety and Health Administration Agency (OSHA) The federal agency that regulates worker safety and in some cases, responder safety. OSHA is a part of the Department of Labor.

Specialist level Identified only in the HAZWOPER standard, this level receives more specialized training than a hazardous materials technician. Practically speaking, the two levels are not very different.

State Emergency Response Commission (SERC) The liaison between local and state levels that collects and disseminates information relating to hazardous materials; SERC involves agencies such as the fire service, police services, and elected officials.

Superfund Amendment Reauthorization Act (SARA) One of the first laws to affect how fire departments respond in a hazardous material emergency.

Target hazard Any occupancy type or facility that presents a high potential for loss of life or serious impact to the community resulting from fire, explosion, or chemical release.

Technician level At this level, responders are allowed to enter heavily contaminated areas using the highest levels of protection. This is the most aggressive level of training as identified in NFPA 472. Hazardous materials technicians take offensive actions.

Fire Fighter in Action

Your lieutenant has scheduled a walk-through of a new chrome-plating plant in your response area. The purpose is to learn the layout of the building, its safety features, and the hazardous materials that are used and stored. As you walk through the building, you consider the different types of hazardous materials that are stored in the plant.

1. What is a hazardous chemical?
   - **A.** A material that poses an unreasonable risk to the health of emergency personnel
   - **B.** A material that poses an unreasonable risk to the health of the public
   - **C.** A material that poses an unreasonable risk to the environment
   - **D.** All of the above

2. Hazardous materials can be found:
   - **A.** anywhere.
   - **B.** only in industrial and manufacturing areas.
   - **C.** only on highways and railroads.
   - **D.** only in hospitals and clinics.

You observe how the employees handle the chemicals and how they are stored. You watch as they process the material and how they dispose of it. You start to think about how you would respond to a chemical spill at the plant.

3. What level of training would you need to be able to take offensive actions at an incident at this plant?
   - **A.** Awareness
   - **B.** Operations
   - **C.** Technician
   - **D.** Specialist
   - **E.** Both C and D

4. A critical distinction between a hazardous materials incident and other emergencies is that:
   - **A.** hazardous materials incidents are more complicated.
   - **B.** a hazardous materials incident moves more slowly than a typical structure fire.
   - **C.** strategies must be thought out and planned in advance.
   - **D.** All of the above

www.FireFighter.jbpub.com

www.FireFighter.jbpub.com

Chapter Pretests

Interactivities

Hot Term Explorer

Web Links

Review Manual

FireLearn

Hazardous Materials: Properties and Effects

Chapter 28

NFPA 472 Standard

4.4.1 *Initiating Protective Actions.

(d) *Identify the general routes of entry for human exposure to hazardous materials for each hazard class.

5.2.2 Collecting Hazard and Response Information.

(8) Describe the properties and characteristics of the following:

(a) Alpha particles
(b) Beta particles
(c) Gamma rays
(d) Neutrons

5.2.3 *Predicting the Behavior of a Material and its Container.

Given an incident involving a single hazardous material, the first responder at the operational level should predict the likely behavior of the material and its container and also meet the following requirements:

(1) Given two examples of scenarios involving known hazardous materials, interpret the hazard and response information obtained from the current edition of the *Emergency Response Guidebook;* MSDS; CHEMTREC/CANUTEC/ SETIQ; local, state, and federal authorities; and shipper/ manufacturer contacts as follows:

(a) Match the following chemical and physical properties with their significance and impact on the behavior of the container and/or its contents:

i. Boiling point
ii. Chemical reactivity
iii. Corrosivity (pH)
iv. Flammable (explosive) range (LEL and UEL)
v. Flash point
vi. Ignition (autoignition) temperature
vii. Physical state (solid, liquid, gas)
viii. Specific gravity
ix. Toxic products of combustion
x. Vapor density
xi. Vapor pressure
xii. Water solubility
xiii. Radiation (ionizing and non-ionizing)

(b) Identify the differences between the following pairs of terms:

i. Exposure and hazard
ii. Exposure and contamination
iii. Contamination and secondary contamination
iv. Radioactive material exposure (internal and external) and radioactive contamination

Knowledge Objectives

After studying this chapter, you will be able to:

1. Describe:
   - State of matter
   - Physical and chemical change
   - Boiling point
   - Flash point
   - Ignition temperature
   - Flammable range
   - Vapor density
   - Vapor pressure
   - Specific gravity
   - Water solubility
   - Corrosivity
   - Toxic products of combustion
2. Describe alpha particles, beta particles, gamma rays, and neutrons.
3. Describe hazard, exposure, contamination, and secondary contamination.
4. Describe:
   - Nerve agents
   - Blister agents
   - Choking agents
   - Irritants
5. Describe the routes of exposure for humans.

Skills Objectives

There are no skills objectives for this chapter.

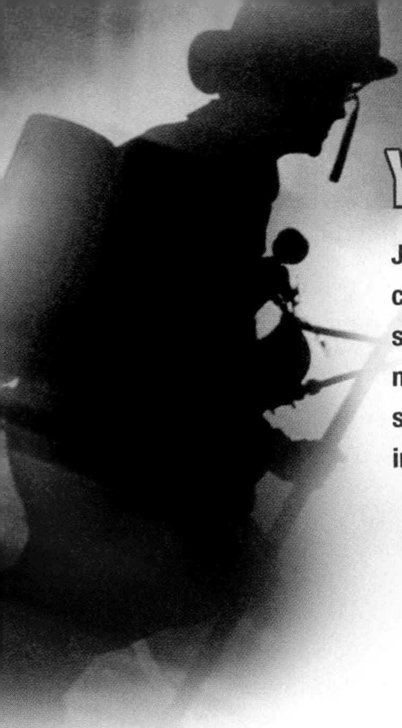

You Are the Fire Fighter

Just before midnight you receive a call regarding an explosion at a tree spraying and landscape company. Upon arrival, you find a detached storage shed with heavy structural damage and a small smoldering fire. A few minutes after you arrive, your captain orders you to shut down the nozzle because he is concerned about some leaking containers inside the shed. Additionally, some lower-floor residents of an adjacent multi-story apartment complex are complaining of eye irritation and asking about a strange odor in the air.

1. ***What would you want to know about the physical properties of the involved materials?***
2. ***Why might water be contraindicated for use on this fire?***
3. ***With the lower-floor residents of the apartment complaining of eye irritation, would you expect a vapor density less than one or greater than one?***

Characteristics of Hazardous Materials

To understand hazardous materials incidents, fire fighters must understand the physical properties of the chemicals involved. Physical properties are the characteristics of a chemical that are measurable, such as vapor density, flammability, corrosivity, water reactivity, etc. This does not mean that you need to be a chemist. In most cases, being an astute observer and correctly interpreting the visual clues presented will provide enough information. The basis for interpreting those visual clues lies in a solid grasp of some basic hazardous materials concepts. The first step in understanding the hazard of any chemical involves identifying the state of matter. The state of matter identifies the hazard as a solid, liquid, or gas (▼ **Figure 28-1 A–C**).

Figure 28-1 A–C The state of matter identifies the hazard as a solid, liquid, or gas. **A.** Solid. **B.** Liquid. **C.** Gas.

Why is this important? If you know the state of matter and physical properties of the chemical, you can predict what the substance will do if it escapes its container.

For example, it is vital to know if a released gas is heavier or lighter than air. In a trench rescue, a gas that is heavier than air will sink to the lowest point—posing a significant problem to the victim. It is also critical to determine if the released gas is flammable, potentially toxic, or if it could displace oxygen.

Based on the variety of potential hazards, you need to see the big picture before committing to a course of action. The bottom line is this: Think before you act—it could save your life or that of your team. This section identifies concepts and key terms to help you understand the hazards a chemical poses and helps you understand the information contained in material safety data sheets (MSDS) and other chemical reference sources.

Physical and Chemical Change

Chemicals exist in three physical states—solid, liquid, or gas. Chemicals can undergo a physical change when subjected to outside influences such as heat, cold, and pressure. When water is frozen, it undergoes a physical change from a liquid to a solid. The actual chemical make-up (H_2O) is still the same, but the state in which it exists is different.

Propane gas has the potential to undergo physical change based on how it is stored inside a vessel. Liquefied propane has a large expansion ratio. Expansion ratio is a description of the volume increase that occurs when a liquid material changes to a gas. For example, propane has an expansion ratio of 270 to 1. This means that for every 1 volume of liquid propane, there would be 270 times that amount of propane vapor. In other words, 1 gallon of liquid propane has the ability to vaporize to 270 gallons of propane vapor when released.

If the propane cylinder is exposed to heat, the liquid propane inside changes into gaseous propane, which increases the pressure inside the vessel. If this uncontrolled expansion takes place faster than the relief valve can vent, a catastrophic container failure could occur. The chemical make-up of propane has not changed; the substance has just physically changed state from liquid to gas.

Chemical change describes the ability of a chemical to undergo a change in its chemical make-up, usually with a release of some form of energy. The reactive nature of a chemical is influenced in many ways, such as mixing with another chemical or by the application of heat. Chemical change is different from physical change. Physical change is essentially a change in state; chemical change results in an alteration of the chemical nature of the material. Steel rusting or wood burning are examples of chemical change. When the chemical reaction is complete, the substance itself is not the same as it was. Think of a Mr. Potato Head Toy—taking off its arms and legs and changing them all around would be a physical change. Mr. Potato Head would still be the same toy—it would just look a bit different. A chemical change would be a transformation in which Mr. Potato Head changed from plastic to wood.

We can apply the concepts of physical and chemical change to our case study. Assume that the owner of the landscape company left some rags soaked with linseed oil wadded up in the corner of the shed. The rags ignited spontaneously and started a fire that ultimately caused the catastrophic failure of a small propane cylinder. The fire also caused the valve on a chlorine cylinder to fail, producing a release of chlorine gas.

Let's look closely at each event to see the impact of physical and chemical change. Linseed oil is an organic material that generates heat as it decomposes (chemical change). That heat, in the presence of oxygen in the surrounding air, ignited the rags, which in turn ignited other combustibles.

Propane is gas liquefied by pressure and temperature and stored in steel cylinders at –44°F. The surrounding heat caused the propane to expand (physical change) until it overwhelmed the ability of the container and the relief valve to handle the buildup of pressure. The cylinder ultimately exploded, demolished the shed, and damaged the chlorine cylinder. That cylinder leaked chlorine gas, which mixed with the moisture in the air and formed an acidic compound called hypochlorous acid (chemical change). The breeze carried the acid mist toward the apartment complex. Since the chlorine vapors are heavier than air, most of the residents complaining about eye irritation are located on the bottom floor of the complex.

Boiling Point

Boiling point is the temperature at which a liquid changes into a gas. The boiling point of water is 212°F. At this temperature, the water molecules have enough kinetic energy (energy in motion) to break free from the liquid and overcome the force of the surrounding atmospheric pressure. Standard atmospheric pressure at sea level is 14.7 pounds per square inch (psi). At sea level, there is 14.7 psi exerted on every surface of every object.

An example of boiling point can be found in popcorn. Popcorn pops because of water inside the kernel of corn. Water has an expansion ratio of 1700:1 and a boiling point of 212°F. A kernel of corn has no relief valve, so a rapid buildup of pressure will cause it to rupture. By heating the popcorn kernels, the small amount of water inside exceeds its boiling point, expands to 1700 times its original water volume, and the kernel pops. Unpopped kernels are those with insufficient water inside the kernel.

Flammable liquids with low boiling points are dangerous because of the potential to produce large volumes of flammable vapor when exposed to relatively low temperatures.

Flash Point, Ignition Temperature, and Flammable Range

When looking at the fire potential of a chemical, there are three important aspects to consider—flash point, ignition temperature, and flammable range. When it comes to combustion, materials must be in a gaseous or vapor state in order to burn; solids do not burn, and liquids do not burn—they must give off a gas or a vapor that must be ignited in order to sustain combustion. Think of a log in a fireplace. If you look closely, the fire is not directly on the log, it is slightly above the surface. The reason for this is that the wood has to be heated to the point where it produces enough "wood gas." The log, then, does not burn—the gas produced by heating the log is what starts the process of combustion.

Figure 28-2 Gasoline at –43°F will give off sufficient vapor to result in a flash point when an ignition source is present.

Table 28-1 Flammable Ranges of Common Gases

Hydrogen	4.0%–75%
Natural gas	5.0%–15%
Propane	2.5%–9.0%

Liquids behave the same way as the log. One way or another, there must be vapor production before there can be fire. **Flash point** is an expression of the temperature at which a liquid fuel gives off sufficient vapor that, when an ignition source is present, will result in a flash fire. The flash fire involves only the vapor phase of the liquid and will go out once the vapor fuel is consumed.

The flash point of gasoline is –43°F. When the temperature of gasoline is above –43°F, it gives off sufficient flammable vapors to support combustion (▲ Figure 28-2). If those vapors reach an ignition source, the vapors will ignite in a ball of fire. Diesel fuel has a much higher flash point of approximately 120°F. In either case, once the temperature of the liquid surpasses its flash point, the fuel will give off sufficient flammable vapors to support combustion.

Ignition temperature is the next important landmark after flash point. When a liquid fuel is heated beyond its ignition temperature, the substance will ignite without an external ignition source. Think of a pan full of cooking oil being heated on the stove. For illustration purposes, assume that the ignition temperature of the oil is 300°F. What would happen if the oil were heated past that point? Once the temperature exceeds the ignition temperature, the oil ignites; there is no need for an external ignition source. This is a common cause of stove fires.

Flammable range is another important concept. In broad terms, flammable range describes a mixture of fuel and air—it is an expression of a fuel/air mixture required for combustion, defined by upper and lower limits, and reflects an amount of vapor mixed with a volume of air. Gasoline again will serve as our example. The flammable range for gasoline vapors is 1.4% to 7.6%. The two percentages, called the **upper explosive limits (UEL)** (7.6%) and **lower explosive limits (LEL)** (1.4%), define the boundaries of a fuel/air mixture necessary for gasoline to burn properly. The upper explosive limit defines the amount of flammable vapor needed in order to keep a fire burning. The lower explosive limit defines the amount of vapor needed to cause an explosion. If a given gasoline/air mixture falls between the upper and lower explosive limits, and that mixture reaches an ignition source, there will be an explosive fire.

The concept of automobile carburetion is based on flammable range—the carburetor is the place where gasoline and air are mixed. When the mixture of gasoline vapors and air is right, the car runs smoothly. When there is too much fuel and not enough air, the mixture is considered to be too "rich." In the case of too much air and not enough fuel, the carburetion is called too "lean." In either case, optimum combustion is not achieved and the motor does not run correctly.

Understanding flammable range in the hazardous materials environment requires the same process. If gasoline is released from a rolled-over tanker, there is a potential of gasoline vapors and air mixing. If the mixture of vapors and air falls between the upper and lower explosive limits, and those vapors reach an ignition source, a flash fire will occur. As with carburetion, if there is too much or too little fuel, the mixture will not adequately support combustion.

In order to accurately assess the flammable range of a released vapor or gas, specialized combustible gas indicators must be used. These indicators are generally used by hazardous materials technicians and will help determine the presence of a dangerous flammable atmosphere. Some common flammable ranges are listed in (▲ Table 28-1).

Generally speaking, the wider the flammable range, the more dangerous the material. This is because the wider the flammable range, the more opportunity there is for an explosive mixture to find an ignition source.

Vapor Density

In addition to addressing the flammability of a particular vapor or gas, fire fighters must also determine if that vapor or gas is heavier or lighter than air. In essence, you must know the **vapor density** of the substance in question. Vapor density is the weight of an airborne concentration (vapor or gas) as compared to an equal volume of dry air (► Figure 28-3).

Basically, vapor density is a question of comparison—what will the gas or vapor do when it is released in the air? Will it collect in low spots in the topography or somewhere inside a building, or will it float upward into the ventilation system? These are important response considerations for hazardous materials incidents and may influence the severity of the incident.

You can find a chemical's vapor density by consulting a good reference source such as an MSDS. The vapor density will be expressed in numerical fashion: 1.2, 0.59, 4.0, etc. The vapor density of propane, for example, is 2.4. Air has a set value of 1.0. Gases like propane or chlorine are heavier than air and therefore have a vapor density value greater than 1.0. Substances such as acetylene, natural gas, and hydrogen are lighter than air and will have a value of less than 1.0.

In the absence of reliable reference sources in the field, there is a mnemonic you can use to remember a number of lighter than air gases—HA HA MINCE, which refers to:

H—Hydrogen
A—Ammonia
H—Helium
A—Acetylene
M—Methane
I—Illuminating gas (neon and hydrogen cyanide)
N—Nitrogen
C—Carbon monoxide
E—Ethylene

Although this list is not all-inclusive, most materials not on this list will be heavier than air. If you encounter a leaking propane cylinder, for example, just refer back to HA HA MINCE—propane is not on that list, so by default, you can assume that propane is heavier than air.

Figure 28-3 Vapor density.

Vapor Pressure

For our purpose, the definition of vapor pressure will pertain to liquids (▼ Figure 28-4). When liquids are held in a closed container, such as a 55-gallon drum, pressure will develop inside. All liquids, even water, will develop a certain amount of pressure in the airspace between the top of the liquid and the container.

The key point to understanding vapor pressure is this: The vapors released from the surface of the liquid must be contained to exert any pressure. Essentially, the liquid inside the container will vaporize until the molecules given off by the liquid reach equilibrium with the liquid itself. It is a balancing act between the liquid and the vapors—some molecules turn to vapor, others leave the vapor phase and return to the liquid.

Carbonated beverages illustrate this concept. Inside your basic cola drink is the formula for the soda and a certain amount of carbon dioxide (CO_2) molecules. It is the CO_2 that makes the bubbles in the soda. When the soda sits in a plastic bottle on the shelf, unopened, the balancing act of CO_2 is happening: CO_2 from the liquid becomes gaseous CO_2 above the liquid and back again. The pressure inside the bottle can be indirectly observed by feeling the rigidity of the plastic—until you open the bottle. Now the CO_2 is not in a closed vessel, and it escapes into the atmosphere. With the pressure released, the sides of the bottle can be indented easily in your fingers. Ultimately, the soda goes flat because all of the CO_2 has escaped. If you put the cap back tightly on the bottle before all of the CO_2 has escaped, after a short time the bottle will feel rigid again.

To that end, consider the technical definition of vapor pressure. The vapor pressure of a liquid is the pressure exerted by its vapor until the liquid and vapor are in equilibrium. Again, this process occurs inside a closed container and temperature has a direct influence on vapor pressure. For example, if heat impinges on a drum of acetone, the pressure above the liquid will increase and perhaps cause the drum to fail. If the temperature is dramatically reduced, the vapor pressure will drop. This is true for any liquid held inside a closed container.

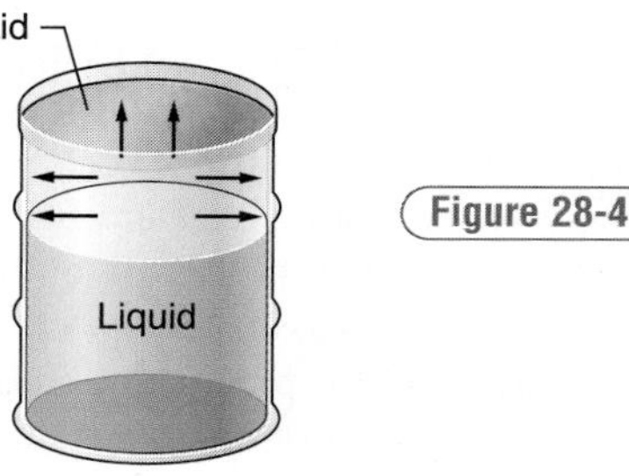

Figure 28-4 Vapor pressure.

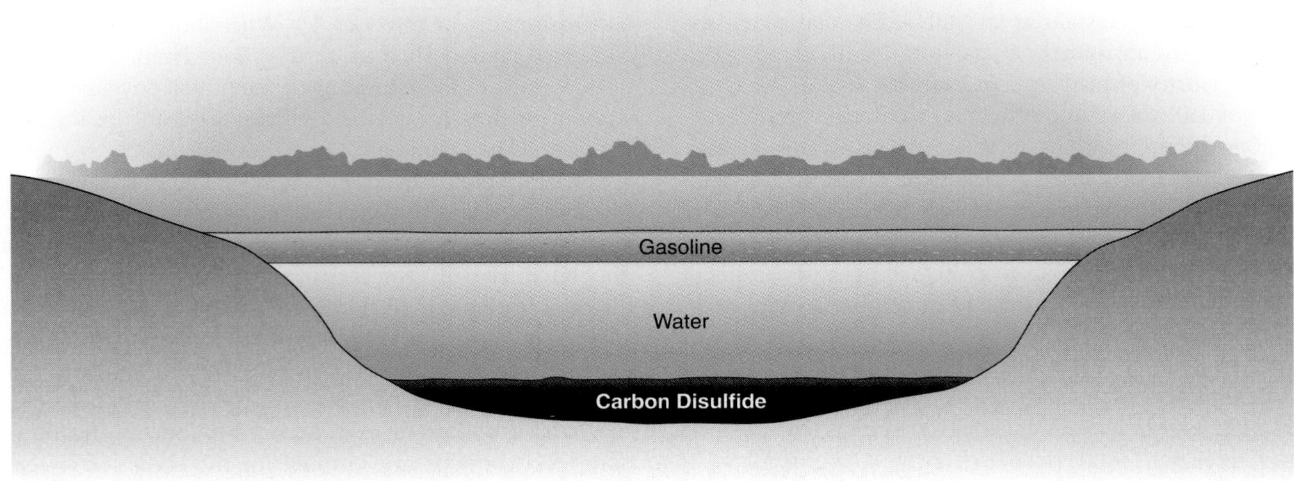

Figure 28-5 Gasoline will float on water while carbon disulfide will not.

What happens if the container is opened or spilled onto the ground to form a puddle? The liquid still has a vapor pressure but it is no longer confined to a container. In this case, we can conclude that liquids with high vapor pressures will evaporate much more quickly than liquids with low vapor pressures. Vapor pressure directly correlates to the speed at which a material will evaporate once it is released from its container.

For example, motor oil has a low vapor pressure. When it is released, it will stay on the ground a long time. Chemicals such as isopropyl alcohol or di-ethyl ether are the opposite. When either of these materials is released and collects on the ground, it will evaporate rapidly. If ambient air temperature or pavement temperatures are elevated, the evaporation rate will increase. Wind speed, shade, humidity, and the surface area of the spill also influence how fast the chemical will evaporate.

When consulting reference sources, vapor pressure may be expressed in pounds per square inch (psi), atmospheres (atm or torr), or millimeters of mercury (mm Hg) and is usually shown at a temperature of 20°C (68°F). For a frame of reference, use the following comparison:

14.7 psi = 1 atmosphere = 760 torr = 1mm Hg

The vapor pressure of water at room temperature is approximately 15 torr. Standard 40 weight motor oil, on the other hand, has a vapor pressure of approximately less than 0.1 mm Hg at 20 degrees centigrade—it is practically vaporless at room temperature. Isopropyl alcohol has a very high vapor pressure of 30 mm Hg at room temperature. Again, temperature has a direct correlation to vapor pressure.

Specific Gravity

Specific gravity is to liquids what vapor density is to gases and vapors. It is a comparison value. In this case, the comparison is between the weight of a liquid chemical and water. Specific gravity is the weight of a liquid as compared to water. Water is given the value of 1.0. Any material with a value less than 1.0 will float on water. Any material with a value greater than 1.0 will sink and remain below the surface of water. A comprehensive reference source will provide the specific gravity of the chemical in question.

Most flammable liquids will float on water (▲ Figure 28-5). Gasoline, diesel fuel, motor oil, and benzene are examples of liquids that float on water. Carbon disulfide, on the other hand, has a specific gravity of approximately 2.6. Consequently, if water is applied to a puddle of carbon disulfide, the water will cover it completely.

Water

When discussing the concept of specific gravity, it is also necessary to determine if a chemical will mix with water. Water is the predominate agent used to extinguish fire. When dealing with chemical emergencies, water may not always be the best and safest choice to mitigate the situation. It is a tremendously aggressive solvent and has the ability to react violently with certain other chemicals.

Concentrated sulfuric acid, metallic sodium, and magnesium are just a few examples of substances that will adversely react with water. If water is applied to burning magnesium, for example, the heat of the fire will break apart the water molecule, creating an explosive reaction. Adding water to sulfuric acid would be like throwing water on a pan full of hot oil—popping and spattering may occur.

On the other hand, water can be a friendly ally when attempting to mitigate a chemical emergency. Depending on the chemical involved, fog streams operated from handlines may knock down vapor clouds. Additionally, heavier-than-water flammable liquids can be extinguished by gently applying water to the surface of the liquid. Water solubility describes the ability of a substance to dissolve in water.

It is important to carefully consider the application of water in all cases involving spilled or released chemicals. Do not approach any problem with the "it is just water" mindset. The results could be something worse than the initial problem.

Corrosivity (pH)

Corrosivity is the ability of a material to cause damage (on contact) to skin, eyes, or other parts on the body. The technical definition, as written by the U.S. Department of Transportation, includes the language, "destruction or irreversible damage to living tissue at the site of contact." Such materials are often also damaging to clothing, rescue equipment, and other physical objects in the environment.

You should understand that corrosives are a complex group of chemicals and should not be taken lightly. There are tens of thousands of corrosive chemicals used in general industry, semiconductor manufacturing, and biotechnology that can be further broken down into two classes: acid and base.

There are number of technical definitions for acid and base, but the most common way to define them is by pH (▶ Figure 28-6). Think of pH as the 'potential of Hydrogen.' Essentially, pH is an expression of the concentration of hydrogen ions in a given substance. For our purposes, we will see pH as a measurement of corrosive strength, which can loosely translate to degree of hazard. In simple terms, we can use pH to judge how aggressive a particular corrosive might be. To judge that hazard, you must understand the pH scale.

Common acids like sulfuric, hydrochloric, phosphoric, nitric, and acetic (vinegar) will have pH values less than 7. Chemicals that are considered to be bases, like sodium and potassium hydroxide, sodium carbonate, and ammonium hydroxide, will have pH values greater than 7. The middle of the scale (7) is where a chemical is considered to be neutral—it is neither acidic nor basic. At this value, a chemical will not harm human tissue.

Generally, pH values of 2.5 or below and 12.5 and above are considered to be strong. What that means is that strong corrosives (acids and bases) will react more aggressively with metallic substances such as steel and iron; they will cause more damage to unprotected skin; they will react more adversely when contacting other chemicals; and they may react violently with water.

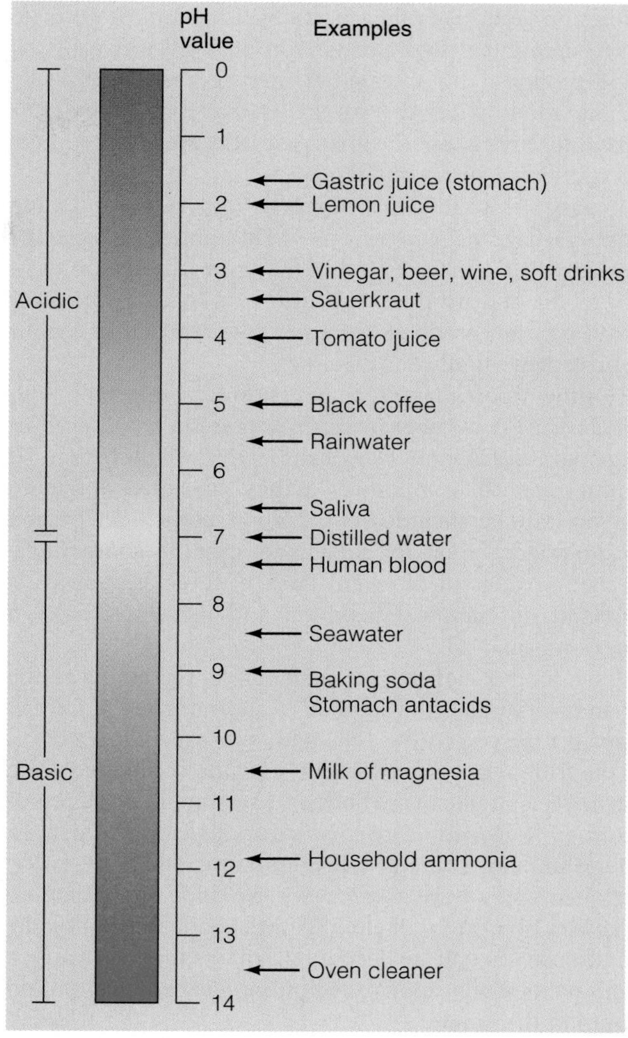

Figure 28-6 pH scale.

How do we determine pH in the field? One method is to use specialized pH test paper. You can also obtain this information from the MSDS, or call for specialized hazardous materials teams to determine the pH level of a chemical. Hazardous materials technicians have more specialized ways of determining pH and may be a useful resource when handling corrosive incidents.

At the operations level, it is critical to understand that corrosives may not be as straightforward as other types of chemical emergencies. The materials themselves are more complicated and therefore the tactics employed to deal with them could, in fact, be outside your scope.

Toxic Products of Combustion

Toxic products of combustion are the hazardous chemical compounds released when a material decomposes under heat. Remember that the process of combustion is a chemical reaction, and, like other reactions, is certain to generate a given amount of byproducts. You are well aware of the smoke produced by a structure fire, but have you really thought about what the smoke is made of or taken a moment to think about the toxic gases that are liberated during a residential structure fire?

An easy way to think about it is to apply a long-standing adage: "garbage in, garbage out." This phrase reflects the idea that whatever is involved in the fire will make up what is given off as a byproduct of the reaction. In short, the combustion reaction produces the toxic gases and other chemical substances found in the smoke.

Burning wood may seem like simple combustion, but consider this: A burning piece of a common wood used in residential construction, Douglass Fir, gives off more than 70 harmful chemical compounds. Other substances found in most fire smoke include soot, carbon monoxide, carbon dioxide, water vapor, formaldehyde, cyanide compounds, and many oxides of nitrogen. Each of these substances is unique in its chemical make-up and most are toxic to humans in small doses.

For example, carbon monoxide affects the ability of the human body to transport oxygen. In short, the red blood cell cannot get oxygen to the cells and, subsequently, a person will die from tissue asphyxiation. Cyanide compounds also affect oxygen uptake in the body and are found to be a prevalent cause of civilian death in structure fires. Formaldehyde is found in many plastics and resins and is one of the many components in smoke that causes eye and lung irritation. The oxides of nitrogen, including nitric oxide, nitrous oxide, and nitrogen dioxide are deep lung irritants that may cause a serious medical condition called **pulmonary edema**, or fluid buildup in the lungs.

As fire fighters, we should always consider the origin of any fire encountered. The conclusions reached may require evacuating affected citizens, dictate unique tactics (e.g., no water), or help you better understand and treat injuries to civilians or your own personnel.

Radiation

Most fire fighters have not been trained on the finer points of handling incidents involving radioactive materials and, therefore, have many misconceptions about radiation. **Radiation** is energy transmitted through space in the form of electromagnetic waves or energetic particles. Electromagnetic radiation (light or radio waves) has no mass or charge. Energetic particles have mass.

To begin, you should understand that you cannot escape radiation. It is all around you. During the course of your life you will receive a certain amount of radiation from the sun, the soil, or having an x-ray. Radiation is everywhere and has been around since the beginning of time. Our focus here, however, is not background radiation, but rather the occupational exposures encountered in the field. For the most part, the health hazards posed by radiation are a function of two factors. First, the amount of radiation absorbed by your body has a direct relationship to the degree of damage done. Secondly, the amount of exposure time will ultimately impact the extent of the injury.

The periodic table illustrates all the known elements that make up every known chemical compound. Those elements are made up of atoms. In the nucleus of those atoms are protons (positive [+] electrical charge) and neutrons (no electrical charge). Orbiting the nucleus are electrons (negative [-] electrical charge).

All atoms of any given element will have the same number of protons in the nucleus. Some elements have variations in the number of neutrons in the nucleus. Each variation in the number of neutrons results in a different isotope of the element. For most purposes, different isotopes of the same element can be considered to behave identically in chemical reactions. A **radioactive isotope** has an unstable configuration of protons and neutrons in the nucleus of the atom. Carbon-14, for example, is a radioactive isotope of a carbon atom. Additional examples include sulfur-35 and phosphorus-32, radioactive isotopes of sulfur and phosphorus atoms, respectively.

Radioactivity is the natural and spontaneous process by which unstable atoms (isotopes) of an element decay to a different state and emit or radiate excess energy in the form of particles or waves.

There are small radiation detectors available that can be worn on turnout gear. These detectors sound an alarm when dangerous levels are encountered and alert the fire fighters to leave the scene and call for more specialized assistance.

If the presence of radiation is suspected, steps must be taken to identify the type of radiation being emitted. In simple terms, radioactive isotopes give off energy from the nucleus of an unstable atom in an attempt to reach a stable state. Typically, you will encounter a combination of alpha, beta, and gamma radiation (▼ Figure 28-7). Each of these

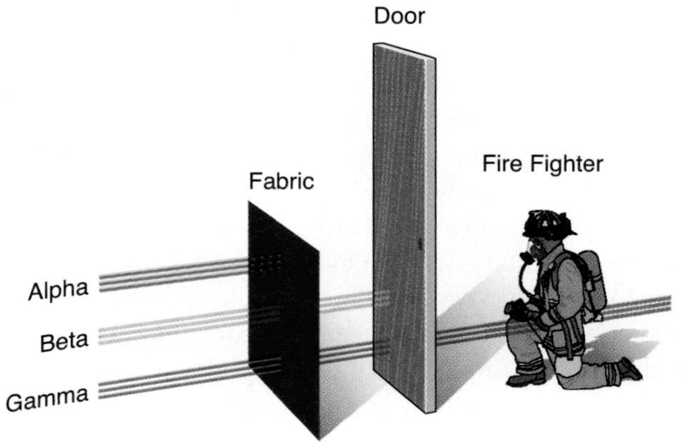

Figure 28-7 Alpha, beta, and gamma radiation.

forms of energy given off by a radioactive isotope will vary in intensity, and consequently determine your efforts to reduce the exposure potential.

Alpha Particles

As mentioned previously, radiation stems from the imbalance of protons and neutrons contained in the nucleus of an atom. Alpha radiation is a reflection of that instability. This form of radiation energy stems from an electrically charged particle given off by the nucleus of an unstable atom. **Alpha particles** have weight and mass and therefore cannot travel very far (less than a few centimeters) from the nucleus of the atom. For the purpose of comparison, alpha particles are like dust particles. Typical alpha emitters include americium (found in smoke detectors), polonium (identified in cigarette smoke), radium, radon, thorium, and uranium. You can protect yourself from alpha emitters by staying several feet away and by protecting your respiratory tract with a HEPA filter on a simple respirator or by using a SCBA. **HEPA (High Efficiency Particulate Air) filters** catch particles down to 0.3-micron size—much smaller than a typical dust particle. Although protection from exposure to alpha particles is offered through HEPA filters and proper PPE, it still poses a serious health risk if it enters into the body.

Beta Particles

Beta particles are more energetic than alpha particles and therefore pose a greater health hazard. Essentially, beta particles are like electrons, except that a beta particle is ejected from the nucleus of an unstable atom. Depending on the strength of the source, beta particles can travel several feet in the open air. Beta particles themselves are not radioactive; the radiation energy is generated by the speed at which the particles are emitted from the nucleus. Due to this, beta radiation can break chemical bonds at the molecular level and cause damage to living tissue. This breaking of chemical bonds creates an ion; therefore, beta particles are considered **ionizing radiation**. Ionizing radiation has the capability to cause changes in human cells. This damage may ultimately create a mutation of the cell and become the root cause of cancer. Examples of ionizing radiation are x-rays and gamma rays (discussed later).

Nonionizing radiation comes from electromagnetic waves, which are capable of causing a disturbance of activity at the atomic level but not enough to break the bonds and create an ion. Typical nonionizing waves include sound waves, radio waves, and microwaves.

Beta particles can redden and burn skin and also be inhaled. Beta particles that are inhaled can directly damage the cells of the human body. Most solid objects can stop beta particles, and your SCBA should provide adequate respiratory protection. Common beta emitters include tritium (luminous dials on gauges), iodine (medical treatment), and cesium.

Gamma Rays and Neutrons

Gamma radiation is the most energetic radiation fire fighters will encounter. Gamma radiation differs from alpha and beta in that it is not a particle ejected from the nucleus; rather, it is pure electromagnetic energy. It has no mass and no electrical charge and travels at the speed of light. Gamma rays can pass through thick solid objects (including the human body) very easily and generally follow the emission of a beta particle. If the nucleus still has too much energy after ejecting beta particles, a photon (packet of pure energy) is released.

Gamma radiation is ionizing radiation and can be deadly. Structural firefighting gear with SCBA will not protect you from gamma rays. If you are near the source, you will be exposed. Typical sources of gamma radiation come from cesium (cancer treatment and soil density testing at construction sites) and cobalt (medical instrument sterilization).

Neutrons are also penetrating particles found in the nucleus of the atom, and are removed through nuclear fusion or fission. Neutrons themselves are not radioactive; however, the exposure to neutrons can create radiation, such as gamma radiation.

Hazard Exposure and Contamination

Hazard and Exposure

According to NFPA 472, *Standard for Professional Competence of Responders to Hazardous Materials Incidents,* a **hazard** is defined as a material capable of posing an unreasonable risk to health, safety, or the environment; capable of causing harm. The same source identifies **exposure** as the process by which people, animals, the environment, and equipment are subjected to or come into contact with a hazardous material.

For our purposes, the two definitions are intertwined. When fire fighters respond to hazardous materials incidents, it is vital to determine the risk in order to reduce the potential for exposure. This is done by taking the time to understand the problem you are facing, including the physical properties of the chemical and the release scenario, and determining if you have the right training, equipment, and protective gear to positively influence the outcome of the incident. All of the information presented to this point has been directed at understanding the nature of the chemicals involved. Subsequent chapters will give you the tools necessary to fold that information into the big picture of handling a chemical emergency.

Contamination

When a chemical has been released and physically contacts people, the environment, animals, tools, etc., either intentionally or unintentionally, the residue is called **contamination**. In some cases, the process of removing those contaminants is complex and requires a significant effort to complete. In other cases, the contamination is easily removed and does not

Voices of Experience

"Different atmospheric conditions at this accident would have spurred us to different initial actions."

It was so cold on that January morning in central Canada that a mere 5 minutes outside transformed my turnout gear into 25 pounds of crackling cellophane. The cold on this particular morning also had the added complication of a thick blanket of "ice fog." The low visibility caused by the thick fog was the reason we found ourselves responding to a two-vehicle highway accident.

On arrival, we quickly determined that a semitrailer carrying a full load of a highly volatile petroleum base liquid had accidentally driven too close to another large vehicle. The two vehicles had rubbed the sides of their trailers together at about 30 miles per hour, causing about 20 feet of "tearing and creasing" along the side of the trailer containing the volatile liquid. At the scene, we noticed a puddle of product with about a 10-foot radius on the ground.

Different atmospheric conditions at this accident would have spurred us to different initial actions. In this case, it was so cold that the liquid was unable to generate any significant vapors that could cause immediate concerns about ignition outside the product's immediate area. Our captain recognized that the cold weather affected this product in specific ways and therefore decided on situation-specific actions and priorities that would provide reasonable safety for all responders and bystanders, prevent the leaking product from spreading, and work to safely offload the remaining product to a suitable replacement container. On a hot day, the liquid would have been actively producing dangerous vapors, changing our initial actions. In that case, our priorities would have involved removal of ignition sources through isolation of the area, as well as aggressive vapor suppression.

Changing factors, such as weather conditions and the environment immediately surrounding the affected area, will impact the properties and effects of hazardous materials. It is critical to gather as much information you can about the hazard to accurately determine your initial actions and priorities.

Ray Unrau
Saskatoon Fire and Protective Services
Saskatoon, Saskatchewan

pose much of a threat to the overall success of the incident. Chapter 33, Hazardous Materials: Decontamination Techniques, will cover the finer points of removing contamination and decontaminating the responders and their gear. For now, simply understand that when the chemical gets out of its container, it is possible that the fire fighters will need to organize a system to safely and efficiently remove it or reduce its physical hazard.

Secondary Contamination

Secondary contamination is when a person or object transfers the contamination or the source of contamination to another person or object by direct contact. In such cases, fire fighters may become contaminated and subsequently handle tools and equipment, touch door handles or other responders, and spread the contamination. The possibility of spreading contamination brings up a point that all fire fighters should understand: The cleaner you stay during the response, the less decontamination you will have to do later. Many fire fighters have the misperception that personal protective equipment is worn to enable you to contact the product. In fact, quite the opposite is true. Personal protective equipment, including specialized chemical protective gear, is worn to protect you in the event you cannot avoid product contact. The less contamination encountered or spread, the easier the decontamination will be.

Types of Hazardous Materials: Weapons of Mass Destruction

In recent years, weapons of mass destruction (WMD) have become a highly relevant topic for fire fighters. Weapons of mass destruction are a real threat in our country, and each fire fighter should have a basic knowledge of those potential threats. To that end, this section will briefly address the threats posed by nerve and blister agents, blood agents, and other potential hazards of choking agents and irritant materials.

In addition to these substances, keep in mind that WMD events can be complex, and the harm inflicted may be the result of several factors; nerve agents dispersed by explosions, radioactive contamination spread by the detonation of conventional explosives, (i.e. dirty bombs), etc. Generally speaking, however, the harm caused by a terrorism attack or any other hazardous materials incident can be broadly broken down into 6 categories, remembered by the mnemonic TRACEM:

Thermal: Heat created from intentional explosions or fires, or cold generated by cryogenic liquids

Radiological: Radioactive contamination from dirty bombs; alpha, beta, and gamma radiation

Asphyxiation: Oxygen deprivation caused by materials such as nitrogen; tissue asphyxiation from blood agents

Chemical: Injury and death caused by the intentional release of toxic industrial chemicals, nerve agents, vesicants, poisons, etc.

Etiological: Illness and death resulting form biohazards such as anthrax, plague, and smallpox; hazards posed by blood-borne pathogens

Mechanical: Property damage and injury caused by explosion, falling debris, shrapnel, firearms, and slips, trips, and falls

Nerve Agents

Nerve agents pose a significant threat to civilians and fire fighters alike. The main threat stems from the ability of a nerve agent to enter the body through the lungs or the skin and then systemically affect the function of the human body. Nerve agents, like many common pesticides, disrupt the central nervous system, possibly causing death or serious impairment.

Your central nervous system is laid out much like the electrical wiring in a house. The brain serves as the main panel and the wires, the nerves, go to muscles, organs, glands, and other critical areas of the body. When a nerve agent, such as sarin or VX, gets into the body, it affects the ability of the body to transact nerve impulses at the junction points between nerves and muscles, nerves and vital organs, and nerves and glands. Subsequently, those critical areas impacted will not function properly and associated bodily functions will be affected.

The signs and symptoms of nerve agent exposure can include pinpoint pupils, tearing of the eyes, twitching muscles, loss of bowel and bladder control, and slow or rapid heartbeat. Mild exposures may result in mild symptoms and include some or all of the symptoms listed above. Significant exposures may result in rapid death, serious impairment, or a more pronounced manifestation of symptoms. The bottom line is that an exposure to a nerve agent is not an automatic death sentence. The dose absorbed will have a direct correlation on the degree of damage done and the symptoms presented by the victim.

The important thing for you to remember is that recognition of the signs and symptoms of a nerve agent exposure is vital. Nerve agents in general are reported to have a fruity odor. In most cases, fire fighters will not have the specialized detection devices necessary to identify the presence of a nerve agent exposure; the symptoms presented by the victims should tip you off. To that end, you should commit the mnemonic SLUDGE to memory. It represents a brief summary of some of the more common signs and symptoms of nerve agent exposure.

S—Salivation
L—Lachrymation (tearing)
U—Urination
D—Defecation
G—Gastric disturbance
E—Emesis (vomiting)

Many times, nerve agents are incorrectly referred to as nerve gas. Nerve agents are liquids, not gases, and will each evaporate at different rates. Sarin, for example, is a water-like liquid and evaporates more easily than VX, a thick, oily

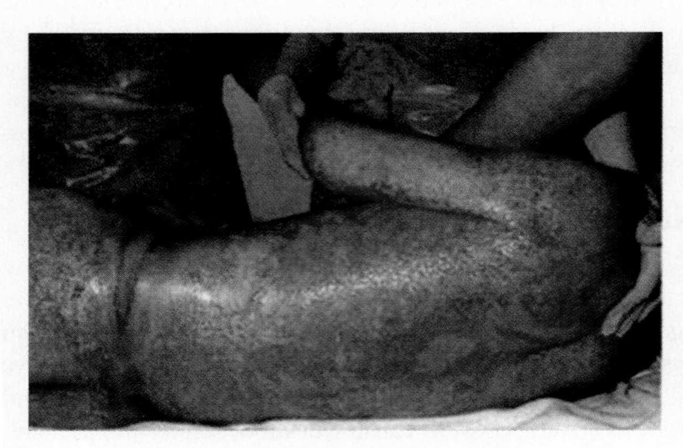

Figure 28-8 Blister agent exposure.

liquid. VX is the most toxic of all the nerve agents. Other nerve agents include soman, tabun, and the organophosphate and carbamate classes of pesticides.

Blister Agents

Blister agents, or vesicants, are classified as such because of their ability to cause blistering of the skin. Common blister agents include sulfur mustard and Lewisite and are unique in the way they interact with the human body.

Sulfur mustard was first used as a weapon in World War I. Chemical casualties suffered horrific injuries to the skin and lungs, and, in some cases, those exposures were fatal. Sulfur mustard is commonly referred to as mustard gas, but generally speaking, is found in a liquid state. It is colorless and odorless when pure, but when mixed with other substances or is otherwise impure, it can look brownish in color and smell like garlic or onions.

Exposures to sulfur mustard may not be immediately apparent. A victim may not experience pain or any other sign indicating contact with this agent. Redness and blistering may not appear for 2 to 24 hours after exposure. Once the blistering does occur, skin decontamination will not reduce the effects (▲ Figure 28-8).

On the other hand, if a sulfur mustard exposure is suspected, immediate skin decontamination is indicated. Early skin decontamination may reduce the effects of exposure.

Sulfur mustard produces injuries similar to second-degree and third-degree thermal burns. In mild skin exposures, complete recovery is common. Inhalation exposures are much more serious, and the damage to lung tissues could result in permanent respiratory function impairment. Additionally, sulfur mustard is very persistent in the environment when released. It can remain intact, sticking to the ground or other surfaces for up to several days.

Lewisite is another recognized vesicant and has many of the same characteristics of sulfur mustard. Lewisite, however, contains arsenic, and when absorbed by the skin may produce some symptoms specific to arsenic poisoning, such as vomiting and low blood pressure. Unlike sulfur mustard, Lewisite exposures to the skin will produce immediate pain. In the event a Lewisite exposure is suspected, immediate and aggressive skin decontamination is indicated. Like sulfur mustard, Lewisite does not occur naturally in the environment and is considered to be a warfare agent.

Cyanide

Cyanide compounds, including hydrogen cyanide and cyanogen chloride, prevent the body from using oxygen. Victims exposed to high levels of cyanide, for example, could be placed on supplemental oxygen but still not survive the exposure. The cause of death is not the lack of oxygen, it is the inability of the body to get oxygen to the cells.

The main route of exposure is through the lungs, but many cyanide compounds can also be absorbed through the skin. Hydrogen cyanide has an odor of bitter almonds. Sixty percent of the general population will not detect the presence of cyanide by smell. Cyanogen chloride, because of the chlorine atom within the chemical compound, has a much more irritating and pungent odor.

Some of the typical signs and symptoms of cyanide exposure include vomiting, dizziness, watery eyes, and deep and rapid breathing. High concentrations lead to convulsions, inability to breathe, loss of consciousness, and death.

Choking Agents

Choking agents are predominantly designed to incapacitate rather than kill. Death and serious injury are possible, but the nature of choking agents (extremely irritating odor) alerts potential victims of their presence and allows for escape. Some commonly recognized choking agents include chlorine, phosgene, and chloropicrin.

Each of these substances has a noticable odor. Chlorine smells like a swimming pool, while phosgene and chloropicrin have been reported to smell like freshly mown grass or hay. In either case, the odors are noticeable and, in many cases, strong enough to alert victims to the presence of an unusual situation.

In the event of a significant exposure, the main threat from these materials is pulmonary edema. When inhaled, choking agents damage sensitive lung tissue and cause fluid to be released by the injury. That fluid results in chemically induced pneumonia, which is fatal in many cases. Additionally, the choking agents act as skin irritants and can cause mild to moderate skin irritation and significant burning in high concentrations.

Irritants

Irritants, also known as riot control agents, are substances, such as mace, that can be dispersed to briefly incapacitate a person or groups of people. Chiefly, irritants cause pain and a burning sensation to exposed skin, eyes, and mucous membranes. The onset of symptoms occurs within seconds after contact and lasts several minutes to several hours, usually with no lasting effects. Irritants can be dispersed from canisters, hand-held sprayers, and grenades.

From a terrorism perspective, irritants may be employed to incapacitate rescuers or to drive a group of people into another area where a more dangerous substance can be released. Of all the groups of weapons of mass destruction, irritants pose the least amount of danger in terms of toxicity of all the potential agents a fire fighter may encounter. Exposed patients can be decontaminated with clean water, and the residual effects of the exposure are usually not significant.

How Harmful Substances Enter the Human Body

In order for a chemical or other harmful substance to injure you, it must first get into your body. Throughout your fire service career, you will be bombarded by harmful substances such as diesel exhaust, bloodborne pathogens, smoke, and accidentally and intentionally released chemicals. Fortunately, most of these exposures are not immediately deadly. Unfortunately, repetitive exposures to the materials listed above may have a negative health effect after a 20-year career. In order to protect yourself now and give yourself the best shot at a healthy retirement, it is important to understand some basic concepts about toxicology and how chemicals enter your body. Chemical substances can enter the human body in four ways (▼ Figure 28-9):

Inhalation: Through the lungs
Absorption: Permeating the skin
Ingestion: By way of the gastrointestinal tract
Injection: Entering cuts or other breaches in the skin

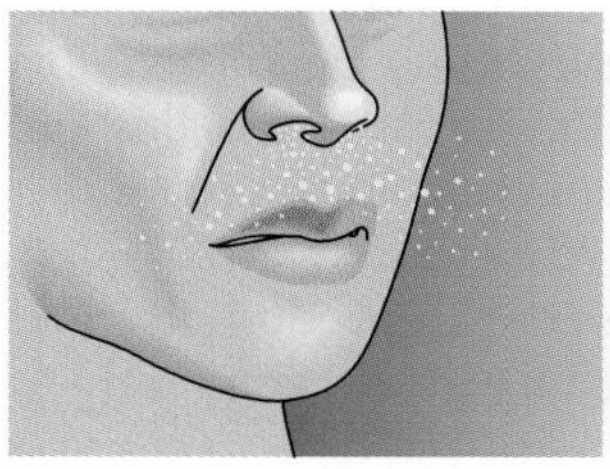
(a) Inhalation

(b) Absorption

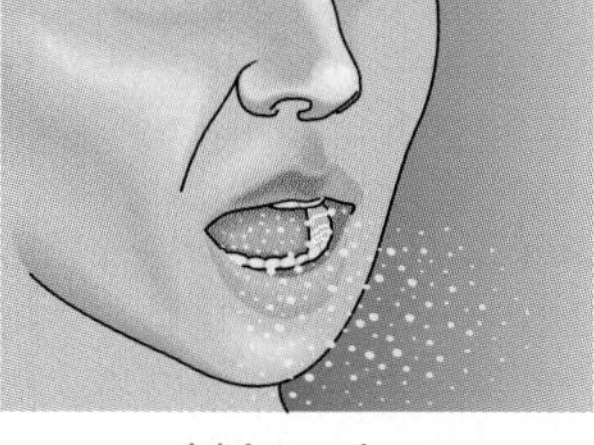
(c) Ingestion

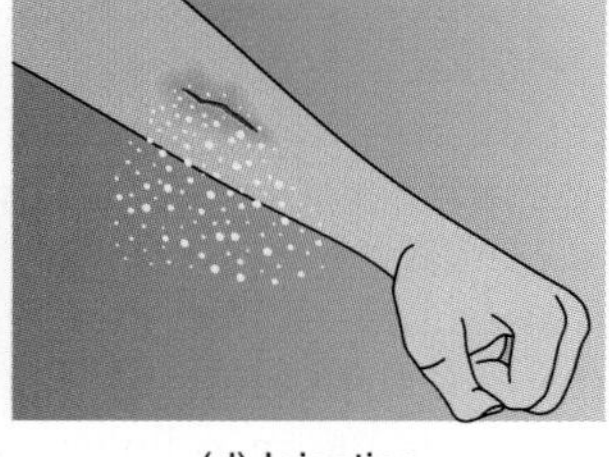
(d) Injection

Figure 28-9 Four ways a chemical substance can enter the body.

Toxicology is the study of the adverse effects of chemical or physical agents on living organisms. The following sections will discuss the potential routes of entry and methods you can employ to protect yourself.

Inhalation

Inhalation exposures occur when harmful substances are brought into the body through the respiratory system. The lungs are a direct point of access to the bloodstream and can quickly transfer an airborne substance into the bloodstream.

The entire respiratory system is vulnerable to attack from most weapons of mass destruction, corrosive materials such as chlorine and ammonia, solvent vapors such as gasoline and acetone, superheated air, and any other material finding its way into the air you breathe. In addition to gases and vapors, small particles of dust and smoke can lodge in sensitive lung tissue, causing substantial irritation. Essentially, everything you take into your lungs is made up of various sizes of matter.

Based on that, it is imperative that you always consider using respiratory protection when operating in airborne contamination. Fortunately, fire fighters have ready access to an excellent form of respiratory protection, the positive pressure open circuit SCBA, but there are other forms. Full and half-face air purifying respirators (APRs) or battery-powered air purifying respirators (PAPRs) offer a degree of protection if the chemical hazard is known and the appropriate filter canister is used (▼ Figure 28-10).

Respirators do not provide oxygen, however, so if the oxygen content of the work area is low, respirators are not a viable option. According to OSHA, any work environment containing less than 19.5% oxygen is considered to be oxygen deficient.

Respirators are lighter than SCBAs, more comfortable to wear, and usually allow longer work periods because they are not dependant on a limited source of breathing air.

Figure 28-10 APRs offer a degree of protection.

Table 28-2 Particle Sizes of Common Types of Respiratory Hazards

Fume	<1 micron
Smoke	<1 micron–1 micron
Dust	1 micron+
Fog	<40 microns
Mist	>40 microns

Table 28-3 Location of Respiratory Trapping by Particle Size

>7 microns	Nose
5–7 microns	Larynx
3–5 microns	Trachea and bronchi
2–3 microns	Bronchi
1–2.5 microns	Respiratory bronchioles
0.5–1 micron	Alveoli

SCBAs certainly offer a higher level of respiratory protection, but in the right circumstances, air-purifying respirators may provide a viable alternative.

When considering protection against airborne contamination, it is important to understand the origin and concentration of the contamination as well as the potential impact the contamination will have on oxygen levels in the area. In short, determining the appropriate type of respiratory protection must be accomplished by looking at the overall situation, including the nature of the contaminant. In some cases, the anticipated particle size of the contamination may determine the level of respiratory protection employed.

Anthrax spores offer an excellent example to illustrate this point. The typical range of weaponized anthrax spores vary in size from half a micron to 1 micron. Based on that, a typical full-face APR with a nuisance dust filter would not offer sufficient protection against that hazard. Anyone operating in an area contaminated with anthrax would wear an SCBA or, at minimum, an air-purifying respirator with HEPA filtration (protection from 0.3 micron particles and larger).

On the other hand, consider this scenario: A container of cryogenic nitrogen is leaking inside a poorly-ventilated storage room. The main threat is the cold vapor produced and the possibility of oxygen deficiency. In this case, SCBA is the appropriate type of respiratory protection based on the anticipated hazard. In many cases, hazardous materials technicians may need to respond; they will use air-monitoring devices to characterize the work area prior to entry. When this is done, the choices made on respiratory protection are based on more definitive information. ◄ Table 28-2 denotes the particle size of some common types of respiratory hazards compared to the average anthrax spore.

Particle size also determines where the inspired contamination will ultimately end up. Bigger particles that make up visible mists will be captured in the nose and upper airway, whereas smaller particles will work their way deeper into the lung. Gaseous substances such as carbon monoxide and hydrogen cyanide may diffuse into the bloodstream and affect entire body systems. ◄ Table 28-3 compares particle size and location in the respiratory tract to where those size particles are trapped.

Respiratory protection is one of the most important elements of personal protective equipment you can use. In all cases where airborne contamination is encountered, take the time to understand the nature of the threat, evaluate the respiratory protection you have, and decide if it provides adequate protection. When protecting your lungs, good enough is not an option. Later chapters will address the concept of chemical protective equipment and how respiratory protection fits into the big picture of remaining safe at chemical incidents.

Absorption

The skin is the largest organ in the body and susceptible to damage by a number of substances. In addition to serving as the body's protective shield against heat, light, and infection, the skin also plays a role in body temperature regulation, stores water and fat, and serves as a sensory center for painful and pleasant stimulation. The skin is an important organ and without it, human beings would not be able to survive.

When discussing chemical exposures, however, **absorption** is not limited only to the skin. Absorption is the process by which hazardous materials travel through body tissues until they reach the bloodstream. The eyes, nose, mouth, and, to a certain degree, the intestinal tract are also part of the equation. The eyes will absorb a high amount of liquid and vapor that come into contact with the sensitive tissues.

In the field, when you discuss the absorption potential of a certain chemical, do not limit your thinking only to the skin.

The skin functions as a shield for the body, but that shield can be pierced by a number of chemicals. Hydrocarbon-based substances, for example, can often be absorbed through the skin and ultimately enter the bloodstream. Aggressive solvents like methylene chloride (found in paint stripper) can be readily absorbed through the skin. The secondary hazard of this chemical occurs when the body attempts to metabolize the chemical after it is absorbed. A byproduct of that metabolization reaction is carbon monoxide, a cellular **asphyxiant**. Asphyxiants are substances that prevent the body from using oxygen, thus causing suffocation. In this scenario, the initial chemical is broken down to another substance that is a greater health hazard than the original chemical. Methylene chloride is also suspected to be a human **carcinogen**. A hydrocarbon is a substance consisting of carbon and hydrogen atoms, among other components. Examples include formaldehyde, benzene, hexane, acetone, propane, methane, and butane.

Absorption hazards are not limited to hydrocarbons. Hydrofluoric acid, for example, poses a significant threat to life when absorbed through the skin. This unique corrosive has the ability to bind with certain substances in the body (predominately calcium). Secondary health effects occurring after exposure can include muscular pain and potentially lethal cardiac arrhythmias.

In the field, you must constantly evaluate the possibility of chemical contact with your skin and eyes. In many cases, your structural firefighting turnout gear provides little or no protection against liquid chemicals. Consult the *North American Emergency Response Guidebook* (NAERG) or an MSDS for response guidance when deciding if turnout gear is appropriate for the hazard you are facing.

In the event your turnout gear does not offer adequate protection, you may have to increase your level of protection to include specialized chemical protective clothing. Keep in mind that the operations level of training is primarily intended to allow you to take defensive actions while avoiding contact with the released product. If you have to wear something other than turnout gear, perhaps you are exceeding your level of training! The use of self-contained breathing apparatus and proper protective firefighting gear can reduce many of the standard routes of exposure, but not always!

Ingestion

In addition to absorption through the skin, chemicals can also be brought into the body through the gastrointestinal tract, by the process of **ingestion**. The water, nutrients, and vitamins your body needs are predominately absorbed in this manner. This is an uncommon route of significant chemical exposure, except for fire fighters. For example, at a structure fire, fire fighters generally have an opportunity to rotate out of the building for rest and refreshment. Fire fighters may not take the time to wash up prior to eating or drinking. This leads to a high probability of spreading contamination from the hands to the food and subsequently to the intestinal tract. If you do not think about every situation where you might become exposed, you may put yourself in harm's way.

Injection

Chemicals brought into the body through open cuts and abrasions qualify as **injection** exposures. Protecting yourself from this route of exposure starts with realizing that you are going to work in a compromised state. Any cuts or open wounds should be addressed before reporting for duty. If they are significant, you may be excluded from operating in contaminated environments. Open wounds are a direct portal to the bloodstream and subsequently to muscles, organs, and other body systems. If a chemical substance contacts this open portal, the health effects could be immediate and pronounced. Remember, intact skin is a good protective shield. Do not go into battle if your shield is not up to the task.

Important Health and Safety Terminology

Chronic and Acute Health Effects

A **chronic health hazard** (chronic health effect) is an adverse health effect occurring after a long-term exposure to a substance. Chronic health effects occur after long-term exposures or by multiple short exposures over a shorter period of time. The exposures can occur by all routes of entry into the body and may result in such things as cancer, permanent loss of lung function, or repetitive skin rashes. For example, inhaling asbestos fibers for years without respiratory protection can result in a form of lung cancer called asbestosis.

Chemicals that pose a hazard to health after relatively short exposure periods are said to cause **acute health effects**. This exposure time frame is considered to be subchronic and may occur immediately after a single exposure or up to several days or weeks after an exposure. Essentially, acute exposures are "right now" exposures that produce some observable conditions such as eye irritation, coughing, dizziness, skin burns, etc.

Skin irritation and burning after a dermal exposure to sulfuric acid would be classified as an acute health effect. Other chemicals, such as formaldehyde, are capable of causing acute health effects. Among other things, formaldehyde is known as a sensitizer and capable of causing an immune system response when inhaled or absorbed through the skin. The reaction is similar to other allergic reactions in that an initial exposure produces mild health effects but subsequent exposures cause much more severe reactions such as breathing difficulties and skin irritation. OSHA defines a **sensitizer** as a chemical that causes a substantial proportion of exposed people or animals to develop an allergic reaction in normal tissue after repeated exposure to the chemical.

Convulsants

Chemicals classified as **convulsants** are capable of causing convulsions or seizures when absorbed by the body. Chemicals that fall into this classification include nerve agents such as sarin, soman, tabun, VX, and the organophosphate and carbamate classes of pesticides. These substances interfere with the central nervous system and disrupt normal transmission of neuromuscular impulses throughout the body. Pesticides capable of functioning as convulsants include parathion, aldicarb, diazinon, and fonofos.

When dealing with these materials, it is important to identify their presence and to avoid breathing vapors or allowing liquid to contact your skin. In some cases, a small exposure can be fatal.

Wrap-Up

Ready for Review

Hazardous materials incidents occur when highly toxic materials are released and the fire department is called to mitigate the hazard and bring the situation to a safe and timely conclusion. In most cases, chemicals released are not exotic or extremely toxic.

These efforts are based on correctly identifying the product and understanding the nature of the released material. If you do not understand what you're up against, it is difficult to make good decisions on how to solve the problem. You must be able to apply basic chemical concepts and terminology to a release scenario.

The art of hazardous materials response is like building a house—without a good foundation, the rest of the structure may be unable to stand up. Before moving to the next chapter, make sure you are comfortable with the terms and definitions and other information presented in this section.

Chief Concepts

- The methods of exposure and effects of a hazardous material incident can determine how to react to and mitigate the incident.
- All materials, whether in or outside a container, will react according to the physical properties of the material involved. The characteristics of a material and understanding how they will behave will assist the fire fighter.
- Physical change is a change in state.
- Avoid contamination whenever possible—it will reduce the likelihood of harmful exposures.
- Chemicals can enter the body four ways: inhalation, absorption, ingestion, and injection.
- Nerve agents affect the central nervous system and can be brought into the body by all routes of exposure.
- Blister agents cause damage to the skin and lung tissues.
- Blood agents interfere with the ability of the body to use oxygen.
- Choking agents may damage lung tissue, causing pulmonary edema.
- Irritants such as mace cause immediate irritation of the skin, eyes, and mucous membranes.
- HEPA filtration on APR cartridges and SCBA protects your lungs from small particulates such as smoke, dust, and fumes.
- Chronic health effects occur after years of exposure.

Hot Terms

Absorption (body) The process by which hazardous materials travel through body tissues until they reach the bloodstream.

Acid Materials within pH values less than 7.

Acute health effects Adverse health effects caused by relatively short exposure periods that produce observable conditions such as eye irritation, coughing, dizziness, skin burns, etc.

Alpha particle A type of radiation that quickly loses energy and can travel only 1″ and 2″ from its source. Clothing or a sheet of paper can stop this type of energy. This particle is not dangerous to plants, animals, or people unless the alpha-emitting substance has entered the body.

Asphyxiant A material that causes the victim to suffocate.

Base Materials with a pH value more than 7.

Beta particle A type of radiation that is capable of travelling 10′ to 15′. Heavier materials, such as metal and glass, can stop this type of energy.

Blister agents Chemicals that cause the skin to blister.

Blood agents Materials that, when absorbed by the body, interfere with the transfer of oxygen from the blood to the cells.

Boiling point The temperature at which the vapor pressure of a liquid equals the surrounding atmospheric pressure, and the liquid changes to a gas.

Carcinogen A cancer-causing agent.

Chemical change The ability of a chemical to undergo a change in its chemical make-up, usually with a release of some form of energy.

Chlorine A yellowish gas that is about 2.5 times heavier than air and slightly water-soluble. It has many industrial uses but also damages the lungs when inhaled (a choking agent).

Choking agent A chemical designed to inhibit breathing, and typically intended to incapacitate rather than kill.

Chronic health hazard An adverse health effect occurring after a long-term exposure to a substance.

Contamination The process of transferring a hazardous material from its source to people, animals, the environment, or equipment, all of which may act as a carrier.

Convulsant Chemicals capable of causing convulsions or seizures when absorbed by the body.

Corrosivity The ability of a material to cause damage (on contact) to skin, eyes, or other parts on the body.

Expansion ratio A description of the volume increase that occurs when a liquid material changes to a gas.

Exposure The process by which people, animals, the environment, and equipment are subjected to or come into contact with a hazardous material.

Flammable range The range of concentrations between the lower and upper flammable limits.

Flammable vapor Any substance that exists in the gaseous state at normal atmospheric temperature and pressure and is capable of being ignited and burned when mixed with the proper proportions of air, oxygen, or other oxidizers.

Flash point The minimum temperature at which a liquid or a solid releases sufficient vapor to form an ignitable mixture with the air.

Gamma radiation A type of radiation that can travel significant distances, penetrating most materials and passing through the body. Gamma radiation is the most destructive to the human body.

Hazard A material capable of posing an unreasonable risk to health, safety, or the environment; capable of causing harm.

HEPA (High Efficiency Particulate Air) filters Filters capable of catching particles down to 0.3-micron size—much smaller than a typical dust, or alpha radiation, particle.

Ignition temperature The minimum temperature at which a fuel, when heated, will ignite in air and continue to burn.

Ingestion Exposure to a hazardous material by swallowing.

Inhalation Exposure to a hazardous material by breathing it into the lungs.

Injection Hazardous materials entering cuts or other breaches in the skin.

Ionizing radiation Electromagnetic waves of such intensity that chemical bonds at the atomic level can be broken (creating an ion).

Irritants Substances such as mace that can be dispersed to briefly incapacitate a person or groups of people.

Lewisite Blister-forming agent that is an oily, colorless-to-dark brown liquid with an odor of geraniums.

Lower explosive limits (LEL) The minimum amount of gaseous fuel that must be present in the air mixture for the mixture to be flammable or explosive.

Nerve agents Toxic substances that attack the central nervous system in humans.

Neutrons Penetrating particles found in the nucleus of the atom that are removed through nuclear fusion or fission. Neutrons are not radioactive; however, exposure to neutrons can create radiation.

pH A measure of the acidity or base of a material; more technically, an expression of the concentration of hydrogen ions in the substance.

Physical change When a material changes its state of matter; for instance, from a liquid to a solid.

Physical properties Characteristics of a chemical that are measurable, such as vapor density, flammability, corrosivity, water reactivity, etc.

Pulmonary edema Fluid buildup in the lungs.

Radiation The combined process of emission, transmission, and absorption of energy traveling by electromagnetic wave propagation between a region of higher temperature and a region of lower temperature.

Radioactive isotope An atom with unequal numbers of protons and neutrons in the nucleus that emits radioactivity.

Radioactivity The spontaneous decay or disintegration of an unstable atomic nucleus accompanied by the emission of radiation.

Sarin A nerve agent that is primarily a vapor hazard.

Secondary contamination The process by which a contaminant is carried out of the hot zone and contaminates people, animals, the environment, or equipment.

Sensitizer A chemical that causes a large portion of people or animals to develop an allergic reaction after repeated exposure.

Specific gravity Specific gravity is the weight of a liquid as compared to water.

State of matter The physical state of a material—solid, liquid, or gas.

Sulfur mustard A clear, yellow, or amber oily liquid with a faint sweet odor or mustard or garlic that may be dispersed in an aerosol form. It causes blistering of exposed skin.

Toxic products of combustion Hazardous chemical compounds released when a material decomposes under heat.

Toxicology The study of the adverse effects of chemical or physical agents on living organisms.

TRACEM An acronym to help remember the effects and potential exposures to a hazardous materials incident—**T**hermal, **R**adiation, **A**sphyxiant, **C**hemical, **E**tiologic, and **M**echanical.

Upper explosive limits (UEL) Defines the boundaries of a fuel/air mixture necessary for a combustible material to burn properly.

Vapor The gas phase of a substance, particularly of those that are normally liquids or solids at ordinary temperatures.

Vapor density The weight of an airborne concentration (vapor or gas) as compared to an equal volume of dry air.

Vapor pressure The vapor pressure of a liquid is the pressure exerted by its vapor until the liquid and vapor are in equilibrium.

VX A nerve agent.

Water solubility Ability of a substance to dissolve in water.

Weapons of mass destruction (WMD) A weapon intended to cause mass casualties, damage, and chaos.

Fire Fighter in Action

It is a 70° F day with light winds, and your engine company is dispatched to a report of a hazardous materials leak at a farm. When you arrive on the scene, the farmer states that it appears someone was stealing anhydrous ammonia from his tank. Anhydrous ammonia is a commonly used fertilizer, but it is also a key ingredient in the production of methamphetamine, and as such, is a target for theft. The thief broke the valve on the tank, causing it to leak. Anhydrous ammonia is stored in a liquid state under pressure. The vapor density is .6 at 32° F, it is soluble in water, has a boiling point of –28.1° F, and has a specific gravity of .62 at 60° F. The pH is 11.6.

1. Based on the information from the farmer and your knowledge of hazardous materials, the fertilizer in the tank is:
   A. in a liquid state and will float in water.
   B. in a gas state and will be lighter than air.
   C. in a gas state and will be heaver than air
   D. in a liquid state and will sink in water.

2. Based on the information from the farmer and your knowledge of hazardous materials, the fertilizer is:
   A. acidic.
   B. basic.
   C. neutral.
   D. none of the above

You have used your Emergency Response Guidebook and have isolated the area. The hazardous material team arrives on the scene. One of the technicians states that the flammable range is 16% to 25% in air, but that the flash point and ignition temperature are so out of range that the anhydrous ammonia is not a fire hazard. In fact, it is considered a non-flammable gas. However, the technician also states that anhydrous ammonia poses a health risk. Anhydrous ammonia is corrosive to the skin, eyes, and lungs. It can be either inhaled or absorbed.

3. What is flash point?
   A. The amount of heat energy needed to raise one pound of water one degree Fahrenheit.
   B. The minimum temperature at which a liquid or a solid emits vapor sufficient to form an ignitable mixture with air near the surface of the liquid or the solid.
   C. The minimum temperature a substance should attain in order to ignite without an ignition source.
   D. The boundaries of a fuel/air mixture necessary for a combustible material to properly burn.

4. Anhydrous ammonia is corrosive to the skin, eyes, and lungs. Short-term exposure to this chemical will most likely cause:
   A. a chronic health hazard.
   B. an acute health effect.
   C. repetitive irritation to the skin, eyes, and lungs.
   D. convulsions.

Hazardous Materials: Recognizing and Identifying the Hazards

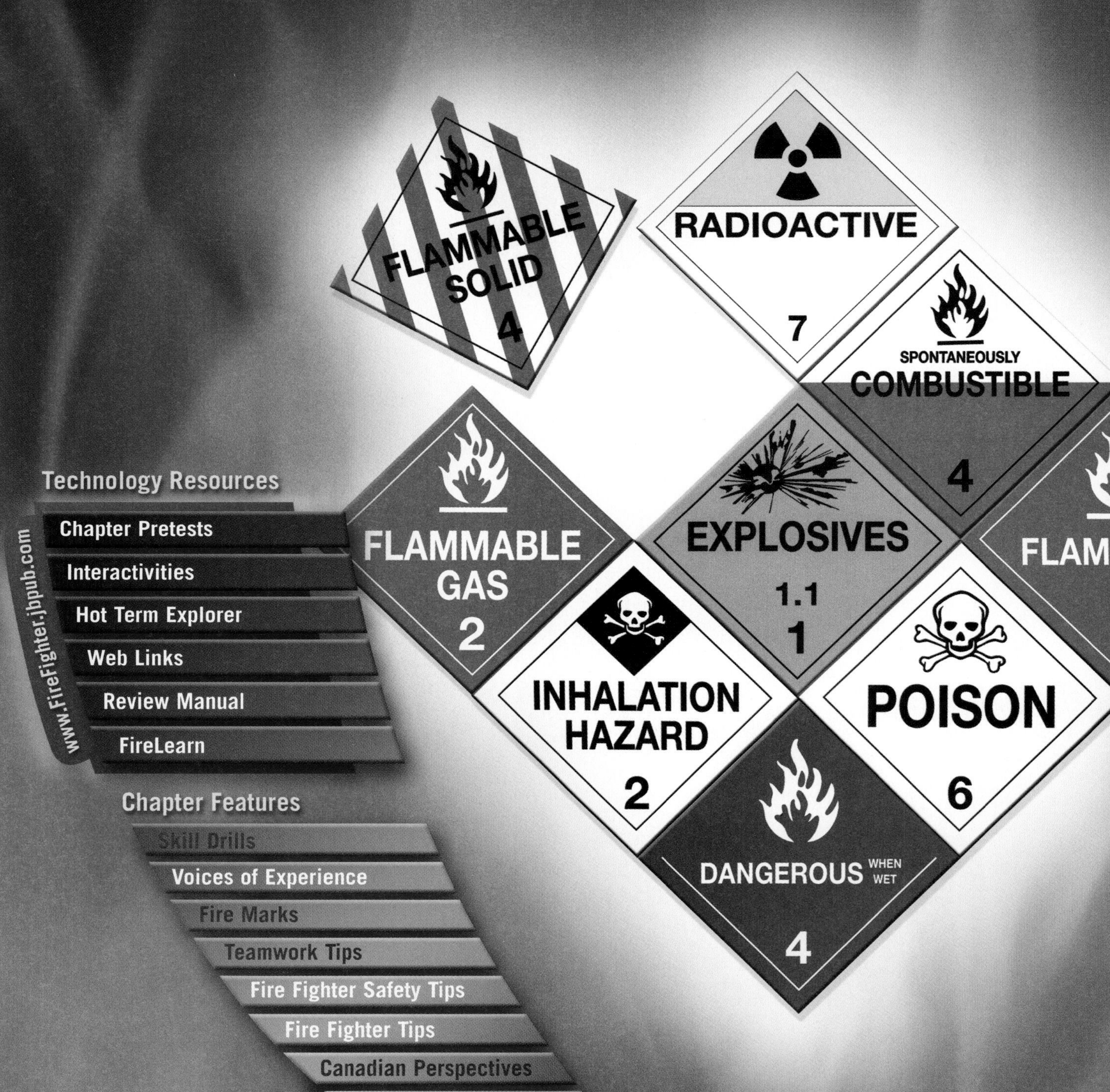

Technology Resources

www.FireFighter.jbpub.com

- Chapter Pretests
- Interactivities
- Hot Term Explorer
- Web Links
- Review Manual
- FireLearn

Chapter Features

- Skill Drills
- Voices of Experience
- Fire Marks
- Teamwork Tips
- Fire Fighter Safety Tips
- Fire Fighter Tips
- Canadian Perspectives
- Hot Terms
- Wrap-Up

Chapter 29

NFPA 472 Standard

Fire Fighter I and II

4.2 Competencies—Analyzing the Incident

4.2.1 Detecting the Presence of Hazardous Materials.

Given various facility or transportation situations, or both, with and without hazardous materials present, the first responder at the awareness level shall identify those situations where hazardous materials are present and also shall meet the following requirements:

(1) *Identify the definition of hazardous materials (or dangerous goods, in Canada).

(2) Identify the UN/DOT hazard classes and divisions of hazardous materials and identify common examples of materials in each hazard class or division.

(3) *Identify the primary hazards associated with each of the UN/DOT hazard classes and divisions of hazardous materials by hazard class or division.

(4) Identify the difference between hazardous materials incidents and other emergencies.

(5) Identify typical occupancies and locations in the community where hazardous materials are manufactured, transported, stored, used, or disposed of.

(6) Identify typical container shapes that can indicate the presence of hazardous materials.

(7) Identify facility and transportation markings and colors that indicate hazardous materials, including the following:

(a) Transportation markings, including UN/NA identification number marks, marine pollutant mark, elevated temperature (HOT) mark, commodity marking, and inhalation hazard mark

(b) NFPA 704, *Standard System for the Identification of the Hazards of Materials for Emergency Response,* markings

(c) Military hazardous materials markings

(d) Special hazard communication markings for each hazard class

(e) Pipeline markings

(f) Container markings

(8) Given an NFPA 704 marking, describe the significance of the colors, numbers, and special symbols.

(9) Identify U.S. and Canadian placards and labels that indicate hazardous materials.

(10) Identify the following basic information on material safety data sheets (MSDS) and shipping papers that indicates hazardous materials:

(a) Identify where to find MSDS.

(b) Identify entries on an MSDS that indicate the presence of hazardous materials.

(c) Identify the entries on shipping papers that indicate the presence of hazardous materials.

(d) Match the name of the shipping papers found in transportation (air, highway, rail, and water) with the mode of transportation.

(e) Identify the person responsible for having the shipping papers in each mode of transportation.

(f) Identify where the shipping papers are found in each mode of transportation.

(g) Identify where the papers can be found in an emergency in each mode of transportation.

(11) *Identify examples of clues (other than occupancy/location, container shape, markings/color, placards/labels, MSDS, and shipping papers) that use the senses of sight, sound, and odor to indicate hazardous materials.

(12) Describe the limitations of using the senses in determining the presence or absence of hazardous materials.

4.2.2 Surveying the Hazardous Materials Incident from a Safe Location.

Given examples of facility and transportation situations involving hazardous materials, the first responder at the awareness level shall identify the hazardous material(s) in each situation by name, UN/NA identification number, or type of placard applied, and also shall meet the following requirements:

(1) Identify difficulties encountered in determining the specific names of hazardous materials in both facilities and transportation.

(2) Identify sources for obtaining the names of, UN/NA identification numbers for, or types of placard associated with hazardous materials in transportation.

(3) Identify sources for obtaining the names of hazardous materials in a facility.

4.2.3* Collecting Hazard Information.

Given the identity of various hazardous materials (name, UN/NA identification number, or type of placard), the first responder at the awareness level shall identify the fire, explosion, and health hazard information for each material by using the current edition of the *Emergency Response Guidebook* and also shall meet the following requirements:

(1) *Identify the three methods for determining the guide page for a hazardous material.

(2) Identify the two general types of hazards found on each guide page.

Hazardous Materials: Recognizing and Identifying the Hazards

NFPA 472 Standard (continued)

Fire Fighter I and II

5.2.1* Surveying the Hazardous Materials Incident.

Given examples of both facility and transportation scenarios involving hazardous materials, the first responder at the operational level shall survey the incident to identify the containers and materials involved, determine whether hazardous materials have been released, and evaluate the surrounding conditions, and also shall meet the requirements in 5.2.1.1 through 5.2.1.6.

5.2.1.1* Given three examples each of liquid, gas, and solid hazardous materials, including various hazard classes, the first responder at the operational level shall identify the general shapes of containers in which the hazardous materials are typically found.

5.2.1.1 (A) Given examples of the following tank cars, the first responder at the operational level shall identify each tank car by type as follows:

(1) Cryogenic liquid tank cars
(2) High-pressure tube cars
(3) Nonpressure tank cars
(4) Pneumatically unloaded hopper cars
(5) Pressure tank cars

5.2.1.1 (B) Given examples of the following intermodal tanks, the first responder at the operational level shall identify each intermodal tank by type and identify at least one material and its hazard class that is typically found in each tank as follows:

(1) Nonpressure intermodal tanks, such as the following:
   (a) IM-101 (IMO Type 1 internationally) portable tank
   (b) IM-102 (IMO Type 2 internationally) portable tank
(2) Pressure intermodal tanks
(3) Specialized intermodal tanks, such as the following:
   (a) Cryogenic intermodal tanks
   (b) Tube modules

5.2.1.1 (C) Given examples of the following cargo tanks, the first responder at the operational level shall identify each cargo tank by type as follows:

(1) Nonpressure liquid tanks
(2) Low-pressure chemical tanks
(3) Corrosive liquid tanks
(4) High-pressure tanks
(5) Cryogenic liquid tanks
(6) Dry bulk cargo tanks
(7) Compressed gas tube trailers

5.2.1.1 (D) Given examples of the following tanks, the first responder at the operational level shall identify at least one material, and its hazard, that is typically found in each tank as follows:

(1) Nonpressure tank
(2) Pressure tank
(3) Cryogenic liquid tank

5.2.1.1 (E) Given examples of the following nonbulk packages, the first responder at the operational level shall identify each package by type as follows:

(1) Bags
(2) Carboys
(3) Cylinders
(4) Drums

5.2.1.1 (F) Given examples of the following radioactive material containers, the first responder at the operational level shall identify each container/package by type as follows:

(1) Type A
(2) Type B
(3) Industrial
(4) Excepted
(5) Strong, tight containers

5.2.1.2 Given examples of facility and transportation containers, the first responder at the operational level shall identify the markings that differentiate one container from another.

5.2.1.2 (A) Given examples of the following marked transport vehicles and their corresponding shipping papers, the first responder at the operational level shall identify the vehicle or tank identification marking as follows:

(1) Rail transport vehicles, including tank cars
(2) Intermodal equipment, including tank containers
(3) Highway transport vehicles, including cargo tanks

5.2.1.2 (B) Given examples of facility containers, the first responder at the operational level shall identify the markings indicating container size, product contained, and/or site identification numbers.

5.2.1.3 Given examples of facility and transportation situations involving hazardous materials, the first responder at the operational level shall identify the name(s) of the hazardous material(s) in each situation.

5.2.1.3 (A) The first responder at the operational level shall identify the following information on a pipeline marker:

(1) Product
(2) Owner
(3) Emergency telephone number

5.2.1.3 (B) Given a pesticide label, the first responder at the operational level shall identify each of the following pieces of information, and then match the piece of information to its significance in surveying the hazardous materials incident:

(1) Name of pesticide
(2) Signal word
(3) Pest control product (PCP) number (in Canada)
(4) Precautionary statement
(5) Hazard statement
(6) Active ingredient

5.2.1.3 (C) Given a label for a radioactive material, the first responder at the operational level shall identify vertical bars, contents, activity, and transport index.

5.2.2 Collecting Hazard and Response Information.

Given known hazardous materials, the first responder at the operational level shall collect hazard and response information using MSDS; CHEMTREC/CANUTEC/SETIQ; local, state, and federal authorities; and contacts with the shipper/manufacturer, and also shall meet the following requirements:

(1) Match the definitions associated with the UN/DOT hazard classes and divisions of hazardous materials, including refrigerated liquefied gases and cryogenic liquids, with the class or division.

(2) Identify two ways to obtain an MSDS in an emergency.

(3) Using an MSDS for a specified material, identify the following hazard and response information:

(a) Physical and chemical characteristics

(b) Physical hazards of the material

(c) Health hazards of the material

(d) Signs and symptoms of exposure

(e) Routes of entry

(f) Permissible exposure limits

(g) Responsible party contact

(h) Precautions for safe handling (including hygiene practices, protective measures, procedures for clean-up of spills or leaks)

(i) Applicable control measures, including personal protective equipment

(j) Emergency and first-aid procedures

(4) Identify the following:

(a) Type of assistance provided by CHEMTREC/CANUTEC/SETIQ and local, state, and federal authorities

(b) Procedure for contacting CHEMTREC/CANUTEC/SETIQ and local, state, and federal authorities

(c) Information to be furnished to CHEMTREC/CANUTEC/SETIQ and local, state, and federal authorities

(5) Identify two methods of contacting the manufacturer or shipper to obtain hazard and response information.

(6) Identify the type of assistance provided by local, state, and federal authorities with respect to criminal or terrorist activities involving hazardous materials.

(7) Identify the procedure for contacting local, state, and federal authorities as specified in the local emergency response plan (ERP) or the organization's standard operating procedures.

Other NFPA Standards

NFPA 704, *Standard System for the Identification of the Hazards of Materials for Emergency Response*

Knowledge Objectives

After studying this chapter, you will be able to:

- Describe occupancies that may contain hazardous materials.
- Describe how your senses can be used to detect the presence of hazardous materials.
- Describe specific containers and container shapes that might indicate hazardous materials.
- Describe tanks that could hold hazardous materials.
- Describe apparatus that can transport hazardous materials.
- Describe how to identify the product, owner, and emergency telephone number on a pipeline marker.
- Describe how to identify a placard and label.
- Describe how to use the *North American Emergency Response Guidebook.*
- Describe the NFPA 704 hazard identification system.
- Describe material safety data sheets (MSDS) and shipping papers.
- Describe CHEMTREC.

Skills Objectives

There are no skills objectives for this chapter.

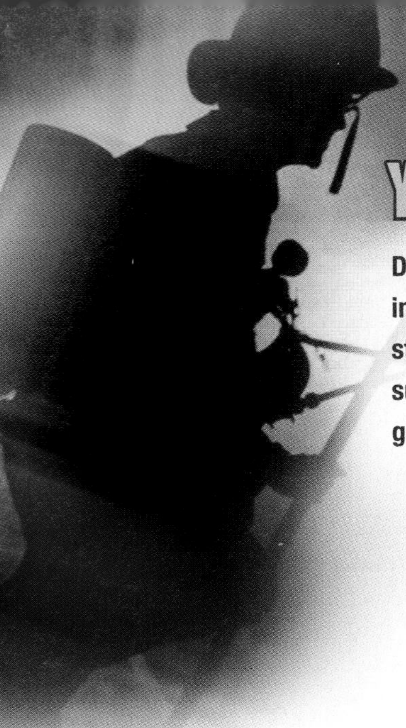

You Are the Fire Fighter

During an odor investigation in a light-industrial district, you are directed to a large metal building in the rear. The building has a large roll-up metal door and several standard doors. Several steel drums with red diamond-shaped labels are located next to the building. The maintenance supervisor says that he noticed an unusual odor when he walked by the drums and that the ground around them is wet.

1. ***What clues indicate the presence of hazardous material?***
2. ***What type of material might be stored in a steel drum?***
3. ***What do the labels on the drums signify?***

Introduction

Scene size-up is important in any emergency situation, but especially so in hazardous materials incidents. Situational awareness, or the ability to "read" the scene, is a critical skill. Fire fighters must be able to identify sensory clues to be useful and safe.

In hazardous materials incidents, it is not always possible to know the whole story before action is required. It may be possible, however, to identify the incident as involving hazardous materials based on information from the dispatcher, knowledge of the response area, and visual, auditory, or olfactory (odorous) clues. A leaking cargo trailer, colored placards on abandoned drums, and certain odors are all indications that hazardous materials may be present. Departmental standard operating procedures (SOP) and your level of training should guide any initial actions.

This chapter focuses on how to identify and obtain necessary information on hazardous materials incidents. Recognizing the potential for hazardous materials based on the containers involved or the available shipping information will help you take proper action at the scene.

Recognizing a Hazardous Material Incident

A hazardous material is any material that poses an unreasonable risk of damage or injury to persons, property, or the environment if it is not properly controlled during handling, storage, manufacture, processing, packaging, use and disposal, or transportation. Recognizing a hazardous materials event, and/or identifying the material(s) involved often requires some detective work. Take the time to look at the whole scene, to identify the critical visual indicators, and to fit them into what is known about the problem.

If dispatch information indicates that an incident involves hazardous materials, it may be a good idea to use binoculars and view the scene from a safe distance. Be sure to question anyone involved in the incident, such as the truck driver or the plant owner. Take time to scan the scene and interpret visual clues such as dead animals near the release, discolored pavement, dead grass, visible vapors or puddles, or labels that may help identify the presence of a hazardous material.

Occupancy and Location

Hazardous materials incidents are not confined to chemical facilities or nuclear power plants but can occur almost anywhere. Hazardous materials are stored in warehouses, hospitals, laboratories, industrial occupancies, residential garages, bowling alleys, home improvement centers, garden supply stores, restaurants, and a score of other facilities or businesses in your response area. So many different chemicals are used in daily life that you could encounter several types of hazardous materials as you respond to emergency situations. For example, a fire in a pesticide storage facility could release dangerous chemicals; a person exposed to a solvent while cleaning automotive parts could require emergency medical assistance; a rollover on the interstate could involve leaking diesel fuel. Any response could have a hazardous materials component.

The location and type of occupancy are two indicators for the presence of hazardous materials. The previous chapter outlined the basic reporting requirements for occupancies that use hazardous materials. By reporting the kinds and quantities of hazardous materials present to the local fire department, facilities that pose a significant threat to the community become part of a comprehensive community plan. These reports also alert fire departments to the need for a preincident plan for the site so that fire fighters are aware of the location and types of hazardous materials involved.

Senses

Another way to detect the presence of hazardous materials is to use your senses, although this must be done carefully to avoid exposure. Many hazardous materials can be recognized by smell, taste, or touch. Getting close enough to exercise these senses, however, may expose you to the hazard. The further you are from the incident when you notice a problem, the safer you will be.

Clues that are seen or heard provide warning information from a distance, enabling you to take precautionary steps. Vapor clouds at the scene, for example, are a signal to move away to safety; the sound of an alarm from a toxic gas cabinet is a warning to retreat. This chapter will focus on providing you with information on visual clues—such as containers, placards, labels, and shipping information—that signal possible hazardous materials incidents.

Containers

A **container** is any vessel or receptacle that holds material, including storage vessels, pipelines, and packaging. Often, container type, size, and material provide important clues about the nature of the substance inside. Fire fighters should not rely solely on the type of container, however, when making a determination about hazardous materials.

Red phosphorus from a drug lab, for example, might be found in an unmarked plastic jug. Acetone or other solvents may be stored in 55-gallon steel drums with two capped openings on the top. Sulfuric acid, at 97% concentration, is typically found in a polyethylene drum, but that drum could be colored black, red, white, or blue. The same sulfuric acid might also be found in a 1-gallon amber glass container. Hydrofluoric acid, on the other hand, is incompatible with silica (glass) and would be stored in a plastic container.

Container Type

Hazardous materials can be found in many different types of containers, ranging from 1-gallon glass containers to 5,000-gal steel storage tanks. Steel or polyethylene plastic drums, bags, high-pressure gas cylinders, railroad tank cars, plastic buckets, above-ground and underground storage tanks, truck tankers, and pipelines are all types of containers used with hazardous materials.

Some very recognizable chemical containers, such as 55-gallon drums and compressed gas cylinders, can be found in almost every commercial building. Materials in a cardboard drum are usually in solid form. Stainless steel containers hold particularly dangerous chemicals and cold liquids are kept in thermos-like **Dewar containers** designed to maintain the appropriate temperature.

> **Fire Fighter Tips**
>
> When you consider locations for possible hazardous materials incidents, do not limit your thinking to chemical plants. Explore your response district and you may be surprised at how many bulk diesel, oxygen, nitrogen, and high-volume corrosive storage tanks you will find.

Containers may be large or small. Smaller containers may be stored and shipped in larger containers. One way to distinguish containers is to divide them into two separate categories: bulk storage containers and nonbulk storage vessels.

Container Volume

Bulk storage containers, or large-volume containers, are defined by their internal capacity based on the following measures:

- Liquids: more than 119 gallons
- Solids: more than 882 pounds
- Gases: more than 882 pounds

Bulk storage containers include fixed tanks, large transportation tankers, totes, and intermodal tanks.

In general, bulk storage containers are found in occupancies that rely on and need to store large quantities of a particular chemical. Most manufacturing facilities have at least one bulk storage container. Often, these bulk storage containers are surrounded by a supplementary containment system to help control an accidental release. **Secondary containment** is an engineered method to control spilled or released product if the main containment vessel fails. A 5,000-gal vertical storage tank, for example, may be surrounded by a series of short walls that form a catch basin around the tank. The basin typically can hold the entire volume of the tank and accommodate water from hoselines or sprinkler systems in the event of fire. Many storage vessels, including 55-gal drums, may have secondary containment systems. If bulk storage containers in your response area have secondary containment systems, it will be easier to deal with large leaks in them.

Large-volume horizontal tanks are also common. When stored above ground, these tanks are referred to as ASTs (above-ground storage tanks); if they are placed underground, they are known as USTs (underground storage tanks). These tanks can hold a few hundred gallons to several thousand gallons of product and are usually made of aluminum, steel, or plastic. USTs and ASTs can be pressurized or nonpressurized. Nonpressurized horizontal tanks are usually steel or aluminum. Because it is difficult to relieve internal pressure on these tanks, they are dangerous when exposed to fire. Typically, they hold flammable or combustible materials such as gasoline, oil, or diesel fuel.

Table 29-1 Common Bulk Storage Vessels, Locations, and Contents

Tank Shape	Common Locations	Hazardous Materials Commonly Stored
Underground tanks	Residential, commercial	Fuel oil and combustible liquids
Covered floating roof tanks	Bulk terminals and storage	Highly volatile flammable liquids
Cone roof tanks	Bulk terminal and storage	Combustible liquids
Open floating roof tanks	Bulk terminal and storage	Flammable and combustible liquids
Dome roof tanks	Bulk terminal and storage	Combustible liquids
High-pressure horizontal tanks	Industrial storage and terminal	Flammable gases, chlorine, ammonia
High-pressure spherical tanks	Industrial storage and terminal	Liquid propane gas, liquid nitrogen gas
Cryogenic liquid storage tanks	Industrial and hospital storage	Oxygen, liquid nitrogen gas

Pressurized horizontal tanks have rounded ends and large vents or pressure-relief stacks. The most common above-ground pressurized tanks are liquid propane and liquid ammonia tanks, which can hold a few hundred gallons to several thousand gallons of product. These tanks also contain a small vapor space; 10% to 15% of total capacity can be vapor. Refer to the discussion of liquefied gases in the chapter on the properties and effects of hazardous materials for more information on these materials.

Another common bulk storage vessel is the **tote**. Totes are portable plastic tanks surrounded by a stainless steel web that adds both structural stability and protection. They can hold a few hundred gallons of product and may contain any type of chemical including flammable liquids, corrosives, food-grade liquids, or oxidizers.

Shipping and storing totes can be hazardous. They often are stacked atop one another and moved with a forklift. A mishap with the loading or moving process can compromise the container. Because totes have no secondary containment, any leak will create a large puddle. Additionally, the steel webbing around the tote makes leaks difficult to patch.

Intermodal tanks are both shipping and storage vehicles. They hold between 5,000 gallons and 6,000 gallons of product and can be either pressurized or nonpressurized. In most cases, an intermodal tank is shipped to a facility, where it is stored and used, and then returned to the shipper for refilling. Intermodal tanks can be shipped by all modes of transportation—air, sea, or land. These horizontal round tanks are surrounded by, or part of, a box-like steel framework for shipping. There are three basic types of intermodal (IM or IMO) tanks:

- IM-101 containers have a 6,000-gal capacity, with internal working pressures between 25 pounds per square inch (psi) and 100 psi. These containers typically carry mild corrosives, food-grade products, and flammable liquids.
- IM-102 containers have a 6,000-gal capacity, with internal working pressures between 14 psi and 30 psi. They primarily carry flammable liquids and corrosives.
- IMO Type 5 containers are high-pressure vessels with internal pressures of several hundred psi that carry liquefied gases like propane and butane.

▲ Table 29-1 summarizes the characteristics of several common bulk storage vessels.

Nonbulk Storage Vessels

Essentially, **nonbulk storage vessels** are all other types of containers. Nonbulk storage vessels can hold a few ounces to many gallons and include drums, bags, carboys, compressed gas cylinders, cryogenic containers, and more. Nonbulk storage vessels hold commonly used commercial and industrial chemicals such as solvents, industrial cleaners, and compounds. This section describes the most common nonbulk storage vessels.

Drums

Drums are easily recognizable, barrel-like containers. They are used to store a wide variety of substances, including food-grade materials, corrosives, flammable liquids, and grease. Drums may be constructed of low-carbon steel, polyethylene, cardboard, stainless steel, nickel, or other hybrid materials ► Figure 29-1. Generally, the nature of the material dictates the construction of the storage drum. Steel utility drums, for example, hold flammable liquids, cleaning fluids, oil, and other noncorrosive chemicals. Polyethylene drums are used for corrosives such as acids, bases, oxidizers, and other materials that cannot be stored in steel containers. Cardboard drums hold solid materials such as soap flakes, sodium hydroxide pellets, and food-grade materials. Stainless steel or other heavy-duty drums generally hold materials too aggressive for either plain steel or polyethylene.

Closed-head drums have a permanently attached lid with one or more small openings called **bungs**. Typically, these

Figure 29-1 Drums may be constructed of cardboard, polyethylene, or stainless steel.

openings are threaded holes sealed by removable caps that can only be removed by using a special wrench called a bung wrench. Closed-head drums usually have one 2″ bung and one 3/4″ bung. The larger hole is used to pump product from the drum, while the smaller bung functions as a vent.

An open-head drum has a removable lid fastened to the drum with a ring. The ring is tightened with a clasp or a threaded nut-and-bolt assembly.

Bags

Bags are commonly used to store solids and powders such as cement powder, sand, pesticides, soda ash, and slaked lime. Storage bags may be constructed of plastic, paper, or plastic-lined paper. Bags come in different sizes and weights, depending on their contents.

Pesticide bags must be labeled with specific information (► Figure 29-2). Fire fighters can learn a great deal from the label, including:

- Name of the product
- Statement of ingredients
- The total amount of product in the container
- The manufacturer's name and address
- The Environmental Protection Agengy (EPA) registration number, which provides proof that the product was registered with the EPA
- The EPA establishment number, which shows where the product was manufactured
- **Signal words** to indicate the relative toxicity of the material:
  - Danger—Poison: Highly toxic by all routes of entry
  - Danger: Severe eye damage or skin irritation
  - Warning: Moderately toxic
  - Caution: Minor toxicity and minor eye damage or skin irritation

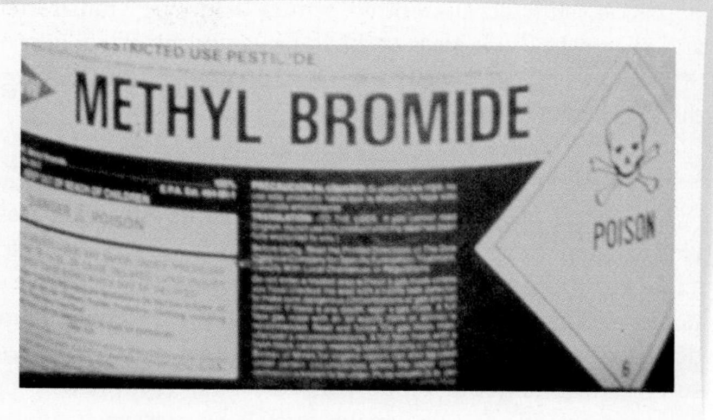

Figure 29-2 A pesticide bag must be labeled with specific information.

- Practical first aid treatment description
- Directions for use
- Agricultural use requirements
- Precautionary statements such as mixing directions or potential environmental hazards
- Storage and disposal information
- Classification statement on who may use the product

In addition, every pesticide label must have the statement, "Keep out of reach of children."

Carboys

Some corrosives and other types of chemicals are transported and stored in vessels called **carboys** (▼ Figure 29-3). A carboy is a glass, plastic, or steel container, that holds 5 gallons to 15 gallons of product. Glass carboys often have a protective wood or fiberglass box to help prevent breakage.

Figure 29-3 A carboy is used to transport and store corrosive chemicals.

Nitric, sulfuric, and other strong acids are transported and stored in thick glass carboys protected by a wooden or Styrofoam crate to shield the glass container from damage during normal shipping.

Cylinders

Several types of **cylinders** are used to hold liquids and gases. Uninsulated compressed gas cylinders are used to store gases such as nitrogen, argon, helium, and oxygen. They have a range of sizes and pressure readings. A typical oxygen cylinder used for medical purposes, for example, has a gas pressure reading of about 2,000 psi. Very large compressed gas cylinders found at a fixed facility may have pressure readings of 5,000 psi or greater.

The high pressures exerted by these cylinders are potentially dangerous to both fire fighters and others. If the cylinder is punctured or the valve assembly fails, the rapid release of compressed gas will turn the cylinder into a deadly missile. Additionally, if the cylinder is heated rapidly, it could explode with tremendous force, spewing product and metal fragments for long distances. Compressed gas cylinders do have pressure-relief valves, but those valves may not be sufficient to relieve the pressure created during a fast-growing fire. The result would be a catastrophic explosion.

A propane cylinder is another type of compressed gas cylinder. Propane cylinders have lower pressures (200 psi to 300 psi) and contain a liquefied gas. Liquefied gases such as propane are subject to the phenomenon called BLEVE, or boiling liquid expanding vapor explosion. For more information on BLEVEs, refer to Chapter 28, Hazardous Materials: Properties and Effects.

The low-pressure Dewar is another common cylinder type. Dewars are thermos-like vessels designed to hold **cryogenic liquids (cryogens)**—gaseous substances that have been chilled until they liquefy. Typical cryogens include oxygen, helium, hydrogen, argon, and nitrogen. Under normal conditions, each of these substances is a gas. A complex process turns them into liquids that can be stored in Dewar containers. Nitrogen, for example, becomes a liquid at –320°F (–160°C) and must be kept that cold to remain a liquid.

Fire Fighter Safety Tips

Keep in mind that hazardous materials can be found in standard tractor-trailers as well.

Cryogens pose a substantial threat if the Dewar fails to maintain the low temperature. Cryogens have large expansion ratios, even larger than the expansion ratio of propane. Cryogenic helium, for example, has an expansion ratio of approximately 750 to 1. If one volume of liquid helium is warmed to room temperature and vaporized in a totally enclosed container, it can generate a pressure of more than 14,500 psi. Therefore, cryogenic containers usually have two pressure-relief devices: a pressure-relief valve and a frangible (easily broken) disk.

Transporting Hazardous Materials

Although rail, air, and sea transport is used to deliver chemicals and other hazardous materials to their destinations, the most common method of transport is over land, by highway transportation vehicles. Even when another method of transport is used, vehicles often carry the shipments from the station, airport, or dock to the factory or plant. For this reason, fire fighters must become familiar with the types of chemical transport vehicles, or tankers, they might encounter during a traffic emergency.

One of the most common chemical tankers is the **MC-306 flammable liquid tanker**, also known as the DOT-406 or gasoline tanker (► Figure 29-4 A). It typically carries gasoline or other flammable and combustible materials. The oval-shaped tank is pulled by a diesel tractor and can carry between 6,000 gallons and 10,000 gallons. The MC-306 is nonpressurized, usually made of aluminum, and offloaded through valves at the bottom of the tank. It is a common highway sight and a reliable way to transport chemicals.

A similar vehicle is the **MC-307 chemical hauler**, which has a round or horseshoe-shaped tank, typically caries 6,000 gallons to 7,000 gallons (► Figure 29-4 B). The MC-307 is used to transport flammable liquids, mild corrosives, and poisons. Tanks that transport corrosives may have a rubber lining.

The **MC-312 corrosives tanker** is used for concentrated sulfuric and nitric acids and other corrosive substances (► Figure 29-4 C). This tanker has a smaller diameter than either the MC-306 or the MC-307 and is characterized by several reinforcing rings around the tank. These rings provide structural stability during transportation and in the event of a rollover. The inside of an MC-312 tanker operates at approximately 75 psi and holds approximately 6,000 gallons.

The **MC-331 pressure cargo tanker** carries materials like ammonia, propane, and butane. The tank has rounded ends, typical of a pressurized vessel, and is commonly constructed of steel with a single tank compartment (▼ Figure 29-4 D). The MC-331 operates at approximately 300 psi, and could be a significant explosion hazard if it accidentally rolls over or is threatened by fire. Fire fighters must use great care when dealing with this type of highway emergency.

The **MC-338 cryogenic tanker** operates much like the Dewar containers described earlier and carries many of the same substances. This low-pressure tanker relies on tank insulation to maintain the low temperatures required by the cryogens it carries (► Figure 29-4 E). A box-like structure containing the tank control valves is typically attached to the rear of the tanker. Special training is required to operate valves on this and any other tanker. An untrained individual who attempts to operate the valves may disrupt the normal operation of the tank, thereby compromising its ability to keep the liquefied gas cold and creating a potential explosion hazard.

Tube trailers carry compressed gases such as hydrogen, oxygen, helium, and methane (► Figure 29-4 F). Essentially, they are high-volume transportation vehicles comprised of several individual cylinders banded together and affixed to a trailer. The individual cylinders on the tube trailer are much like the smaller compressed gas cylinders previously discussed. These large-volume cylinders operate at 3,000 psi to 5,000 psi; one trailer may carry several different gases in individual tubes. Typically, there is a valve control box toward the rear of the trailer, and each individual cylinder has its own relief valve. These trailers can frequently be seen at construction sites or at facilities that use great quantities of these materials.

Fire Fighter Safety Tips

Cryogenic tankers have a relief valve near the valve control box. The vapor being vented from the tank comes out as small puffs of white "smoke" through this valve. Fire fighters should understand that this is a normal occurrence, not an emergency.

A

B

C

D

Figure 29-4 **A.** An MC-306 flammable liquid tanker. **B.** An MC-307 chemical hauler. **C.** An MC-312 corrosives tanker. **D.** An MC-331 pressure-cargo tanker.

Dry bulk cargo tanks also are commonly seen on the road and carry dry bulk goods such as powders, pellets, fertilizers, or grain (▼ Figure 29-4 G). These tanks are not pressurized, but may use pressure to offload product. Dry bulk cargo tanks are generally V-shaped with rounded sides that funnel the contents to the bottom-mounted valves.

E

F

G

Figure 29-4 E. An MC-338 cryogenic tanker. F. A tube trailer. G. A dry bulk cargo tank.

Railroad Transportation

Railroads move almost 2 million carloads of freight per year, with relatively few hazardous materials incidents. Railway tank cars carry volumes up to 30,000 gallons and have the potential to create large leaks or vapor clouds. Hazardous materials incidents involving railroad transportation, although relatively rare, are dangerous.

Fortunately, there are only three basic railcar configurations that fire fighters should recognize: nonpressurized, pressurized, and special use. Each has a distinct profile that can be recognized from a long distance. Additionally, railcars are usually labeled on both sides with the volume and maximum working pressure inside the tank. Dedicated haulers often have the chemical name clearly visible.

Nonpressurized (general service) railcars typically carry general industrial chemicals, consumer products such as corn syrup, flammable and combustible liquids, and mild corrosives. Nonpressurized railcars are easily identified by looking at the top of the car. Nonpressurized railcars will have visible valves and piping without a dome cover.

Pressurized railcars will have an enclosed dome on the top of the railcar. These cars transport materials such as propane, ammonia, ethylene oxide, and chlorine. Pressurized cars have internal working pressures ranging from 100 psi to 500 psi and are equipped with relief valves, similar to those on bulk storage tankers. Unfortunately, the high volumes carried in these cars can generate long-duration, high-pressure leaks that may be impossible to stop.

Special use railcars include boxcars, flat cars, cryogenic and corrosive tank cars, and high-pressure compressed gas tube cars (► Figure 29-5). In each case, the hazard will be unique to the particular railcar and its contents. Do not assume that only the chemical tank cars pose a threat; until you know what is in a particular car, assume it is a hazardous situation.

The train's engineer or conductor will have information about all the cars on that train and should be consulted during a railway emergency. Details about the shipping papers unique to railway transportation will be presented later in this chapter.

Pipelines

Of the various methods used to transport hazardous materials, the high volume pipeline is rarely involved in emergencies. In many areas, large-diameter pipelines transport natural gas, gasoline, diesel fuel, and other products from delivery terminals to distribution facilities. Pipelines are often buried underground, but may be above ground in remote areas. The pipeline right of way is an area, patch, or roadway that extends a certain number of feet on either side of the pipe itself. This area is maintained by the company that owns the pipeline. The company is also responsible for placing warning signs at regular intervals along the length of the pipeline.

Pipeline warning signs include a warning symbol, the pipeline owner's name, and an emergency contact phone number (▼ Figure 29-6). Pipeline emergencies are complicated events that require specially trained responders. If you suspect an emergency involving a pipeline, contact the owner of the line immediately. The company will dispatch a crew to assist with the incident.

Information about the pipe's contents and owner is also often found at the vent pipes. These inverted J-shaped tubes provide pressure relief or natural venting during maintenance and repairs. Vent pipes are clearly marked and are approximately 3′ above the ground.

Figure 29-5 Special use railcars can carry hazardous materials.

Figure 29-6 Information about the pipe's contents and owner is also often found.

Facility and Transportation Markings and Colorings

Labels, placards, and other markings on buildings, packages, boxes, and containers often enable fire fighters to identify a spilled or released chemical. When used correctly, marking systems indicate the presence of a hazardous material and provide clues about the substance. This section provides an introduction to various marking systems being used. It does not cover all the intricacies and requirements for every marking system; it will, however, acquaint you with the most common systems.

DOT System

The Department of Transportation (DOT) marking system is characterized by a system of labels and placards. The DOT's *North American Emergency Response Guidebook (NAERG)* is also part of this system and offers a certain amount of guidance for fire fighters operating at a hazardous materials incident.

Labels and Placards

Placards are diamond-shaped indicators (10¾″ on each side) that must be placed on all four sides of highway transport vehicles, railroad tank cars, and other forms of transportation carrying hazardous materials (▼ Figure 29-7). Labels are smaller versions (4″ diamond-shaped indicators) of placards, and are used on the four sides of individual boxes and smaller packages being transported (► Figure 29-8).

Placards and labels are intended to give fire fighters a general idea of the hazard inside a particular container. A placard may identify the broad hazard class (flammable, poison, corrosive) that a tanker contains, while the label on a box inside a delivery truck relates only to the potential hazard inside that package.

Figure 29-7 A placard.

Voices of Experience

"As we searched, the smell sometimes became fainter, but it was always present, and there was no sign of any container or spill anywhere in the facility."

Very early in my fire service career, my crew and I were dispatched late one evening to a smell of ammonia at a small industrial plant. Upon arrival at the plant's enclosed loading dock, I could immediately feel the ammonia in my eyes. The chief was already talking to the security staff when we joined the other crews at the loading dock. There was nothing visible that could have caused this incredibly strong smell of ammonia—there were no containers, vehicles, or equipment. In fact, the whole dock was wet—as if it had been hosed down at the end of the last shift.

Each crew was assigned areas of the facility to search for the source of the ammonia smell. As we searched, the smell sometimes became fainter, but it was always present, and there was no sign of any container or spill anywhere in the facility. We reassembled on the weathered, pitted concrete loading dock, still wet from its apparent cleaning. That's when it occurred to me to check the floor.

I reached down and touched the wet concrete with my glove and brought it up to my face. I immediately recognized the smell of ammonia... and nearly fell over. My partner urged me to tell the chief that I had found the source of the ammonia, but I couldn't speak and I was having trouble keeping my balance. So, in my place, my partner told the chief what I had discovered: a powerful, ammonia-based cleaning product had been used to wash down the concrete floors. I remember receiving a nod of appreciation and a pat on the shoulder, but what I really deserved was a stern lecture about proper procedures for handling hazardous materials. If we had been dealing with a chemical more dangerous than ammonia, I could have died on the spot.

I learned two things about hazardous materials that night: don't overlook the obvious, and don't take reckless chances. When I suspected that the ammonia smell was coming from the floor, my first step should have been to notify the chief or another fire fighter with training in identifying and handling hazardous materials. Only a trained specialist wearing the appropriate protective gear should conduct tests to confirm the presence or absence of a hazardous material.

Peter Sells
Toronto Fire Services
Toronto, Ontario

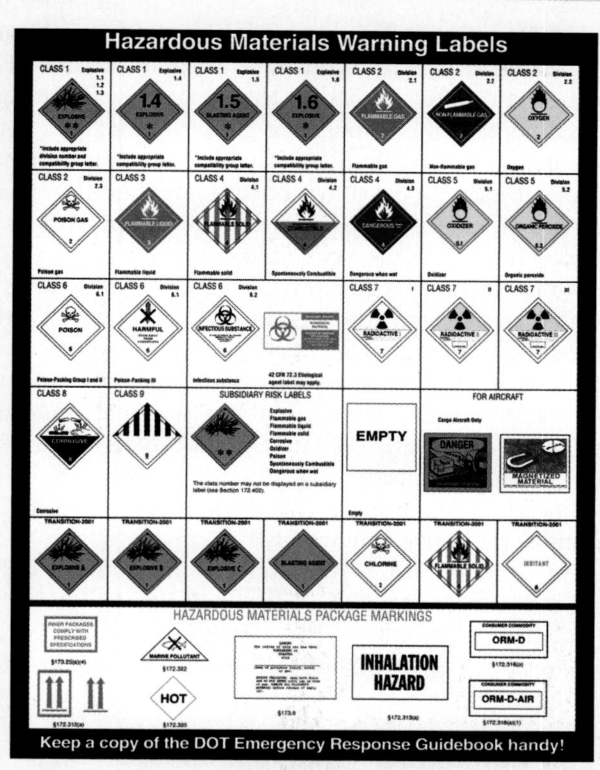

Figure 29-8 A DOT hazardous materials warning label chart.

The DOT's guidebook, the *NAERG,* can be used during the initial phase of a hazardous materials incident. The book organizes chemicals into nine basic hazard classes, or families, each of which exhibits similar properties. There also is a "Dangerous" placard, which indicates that more than one hazard class is contained in the same load. The DOT system is a broad-spectrum look at chemical hazards and a valuable resource to fire fighters.

The nine chemical families recognized in the *NAERG* include:

- Class 1—Explosives
- Class 2—Gases
- Class 3—Flammable combustible liquids
- Class 4—Flammable solids
- Class 5—Oxidizers
- Class 6—Poisons
- Class 7—Radioactive materials
- Class 8—Corrosives
- Class 9—Other regulated materials

Other Considerations

The DOT system does not require that all chemical shipments be marked with placards or labels. In most cases, the package or tank must contain a certain amount of hazardous material before a placard is required. For example, the "1,000 lb rule" applies to blasting agents, flammable and non-flammable gases, flammable/combustible liquids, flammable solids, air reactive solids, oxidizers and organic peroxides, poison solids, corrosives, and miscellaneous (class 9) materials. Placards are required for these materials only when the shipment weighs more than 1,000 lb.

Some chemicals are so hazardous that shipping any amount of them requires the use of labels or placards. These include Class 1.1, 1.2, or 1.3 explosives, poison gases, water-reactive solids, and high-level radioactive substances. A four-digit United Nations (UN) number may be required on some placards. This number identifies the specific material being shipped; a list of UN numbers is included in the *NAERG.*

> **Fire Fighter Tips**
>
> The fire fighter should always attempt to identify hazardous materials using more than one source. The *NAERG* is only one source and should not be used exclusively.

Using the *NAERG*

The *NAERG* is a preliminary action guide. It is useful during the initial 10 to 15 minutes of an incident, but it cannot be used to create a comprehensive action plan. When the *NAERG* refers to a small spill, it means a leak from one small package, or a small leak in a large container (up to a 55-gal drum), a small cylinder leak, or any small leak, even one in a large package. A large spill is a large leak or spill from a larger container or package, a spill from a number of small packages, or anything from a one-ton cylinder, tank truck, or rail car. The *NAERG* is divided into four colored sections: yellow, blue, orange, and green.

- **Yellow section**—Chemicals in this section are listed numerically by their four-digit UN identification number. Entry number 1017, for example, identifies chlorine. Use the yellow section when the UN number is known or can be identified. The entries include the name of the chemical and the emergency action guide number.
- **Blue section**—Chemicals in the blue section are listed alphabetically by name. The entry will include the emergency action guide number, the identification number. The same information, organized differently, is in both the blue and yellow sections.
- **Orange section**—This section contains the emergency action guides. Guide numbers are organized by general hazard class and indicate what basic emergency actions should be taken, based on hazard class.
- **Green section**—This section is organized numerically by UN identification number and provides the initial isolation distances for certain materials. Chemicals included in this section are highlighted in the blue or yellow sections.

Figure 29-9 Example of a placard using the NFPA 704 hazard identification system.

NFPA System

The DOT hazardous materials marking system is used when materials are being transported from one location to another. The National Fire Protection Association (NFPA) **NFPA 704 hazard identification system** is designed for fixed facility use. NFPA diamonds are found on the outside of buildings, on doorways to chemical storage areas, and on fixed storage tanks. Fire fighters can use the NFPA diamonds to determine a course of action at a hazardous material incident (▲ Figure 29-9).

The NFPA 704 hazard identification system uses a diamond-shaped symbol of any size, which is itself broken into four smaller diamonds, each representing a particular property or characteristic. The blue diamond (at the nine o'clock position) indicates the health hazard posed by the material. The top red diamond indicates flammability. The yellow diamond (at the three o'clock position) indicates reactivity. The bottom white diamond is used for special symbols and handling instructions.

The blue, red, and yellow diamonds will each contain a numerical rating of 0–4, with 0 being the least hazardous and 4 being the most hazardous. The white quadrant will not have a number but may contain special symbols. Among the symbols used are: a burning O (oxidizing capability), a three-bladed fan (radioactivity), and a W with a slash through it (water reactive). (► Table 29-2) provides a description of the numerical ratings in each category.

For more complete information on the NFPA 704 system, consult NFPA 704, *Standard for the Identification of the Fire Hazards of Materials for Emergency Response.*

HMIS Marking

Since 1983, the **Hazardous Materials Information System (HMIS)** hazard communication program has helped employers comply with the Hazard Communication Standard of the U.S. Occupational Safety and Health Administration (OSHA). The HMIS is similar to the NFPA diamond and uses a numerical hazard rating with similarly colored horizontal columns. The HMIS is more than just a label; it is a method used by employers to give their personnel necessary information to work safely around chemicals, and includes training materials to inform workers of chemical hazards in the workplace.

The HMIS is not required by law but is a voluntary system that employers choose to use to comply with OSHA's Hazard Communication Standard. In addition to describing the chemical hazards posed by a particular substance, the HMIS also provides guidance about personal protective equipment employees need to use to protect themselves from workplace hazards. Letters and icons, which are explained in the text, specify the different levels and combinations of protective equipment.

Fire fighters must understand the fundamental difference between NFPA 704 and the HMIS. NFPA 704 is intended for responders; the HMIS is intended for the employees of a facility. The HMIS is not a response information tool.

Other Reference Sources

Labels and placards may be helpful in identifying hazardous materials, but other sources of information are also avail-able. Among these reference sources are material safety data sheets (MSDS), shipping papers, and staffed, national resources such as CHEMTREC or the National Response Center. First-responding fire fighters should place a high priority on identifying the released material and finding a reliable source of information about the chemical and physical properties of the released substance. The most appropriate source will depend on the situation; use your best judgment in making your selection.

MSDS

A common source of information about a particular chemical is the **material safety data sheet (MSDS)** specific to that substance (► Figure 29-10). Essentially, an MSDS provides basic information about the chemical make-up of a substance, the potential hazards it presents, appropriate first aid in the event of an exposure, and other pertinent data for safe handling of the material. Generally, an MSDS will include:

- Physical and chemical characteristics
- Physical hazards of the material
- Health hazards of the material
- Signs and symptoms of exposure
- Routes of entry
- Permissible exposure limits
- Responsible party contact

Table 29-2 Hazard Levels in the NFPA Hazard Identification System

Flammability Hazards

4 Materials that will rapidly or completely vaporize at atmospheric pressure and normal ambient temperature, or that are readily dispersed in air and that will burn readily. Liquids with a flashpoint below 73°F (22°C) and a boiling point below 100°F (38°C).

3 Liquids and solids that can be ignited under almost all ambient temperature conditions. Liquids with a flashpoint below 73°F (22°C) and a boiling point above 100°F (38°C) or liquids with a flashpoint above 73°F (22°C) but not exceeding 100°F (38°C) and a boiling point below 100°F (38°C).

2 Materials that must be moderately heated or exposed to relatively high ambient temperatures before ignition can occur. Liquids with flashpoint above 100°F (38°C) but not exceeding 200°F (93°C).

1 Materials that must be preheated before ignition can occur. Liquids that have a flashpoint above 200°F (93°C).

0 Materials that will not burn.

Health Hazards

4 Materials that on very short exposure could cause death or major residual injury.

3 Materials that on short exposure could cause serious temporary or residual injury.

2 Materials that on intense or continued, but not chronic, exposure could cause incapacitation or possible residual injury.

1 Materials that on exposure would cause irritation but only minor residual injury.

0 Materials that on exposure under fire conditions would offer no hazard beyond that of ordinary combustible material.

Reactivity Hazards

4 Materials that in themselves are readily capable of detonation or of explosive decomposition or reaction at normal temperatures and pressures.

3 Materials that in themselves are capable of detonation or explosive decomposition or reaction but require a strong initiating source, or that must be heated under confinement before initiation, or that react explosively with water.

2 Materials that readily undergo violent chemical change at elevated temperatures and pressures, or that react violently with water, or that may form explosive mixtures with water.

1 Materials that in themselves are normally stable, but can become unstable at elevated temperatures and pressures.

0 Materials that in themselves are normally stable, even under fire exposure conditions, and are not reactive with water.

Special Hazards

ACID Acid

ALK Alkali

COR Corrosive

OX Oxidizer

~~W~~ Reacts with water

- Precautions for safe handling (including hygiene practices, protective measures, and procedures for cleaning up spills or leaks)
- Applicable control measures, including personal protective equipment
- Emergency and first aid procedures
- Appropriate waste disposal

When responding to a hazardous materials incident at a fixed facility such as a factory or plant, fire fighters should ask the site manager for an MSDS for the spilled material. All facilities that use or store chemicals are required, by law, to have an MSDS on file for each chemical used or stored in the facility. Although the MSDS is not a definitive response tool, it is a piece of the puzzle. Fire fighters should investigate as many sources as possible (preferably at least three) for emergency information about a released substance.

Shipping Papers

Shipping papers are required whenever materials are transported from one place to another. They include the names and addresses of the shipper and the receiver, identify the material being shipped, and specify the quantity and weight of each part of the shipment. Shipping papers for road and highway transportation are called **bills of lading** or **freight bills** and are located in the cab of the vehicle. Drivers transporting chemicals are required by law to have a set of shipping papers on their person or easily reachable inside the cab at all times.

A bill of lading may have additional information about a hazardous substance such as its packaging group designation. Packaging group designation is another way used by shippers to identify special handling requirements or hazards. Some DOT hazard classes require shippers to assign packaging groups based on the material's flash point and toxicity. A packaging group designation may signal that the material poses a greater hazard than similar materials in a hazard class. The three packaging group designations are:

- Packaging group I: high danger
- Packaging group II: medium danger
- Packaging group III: minor danger

MATERIAL SAFETY DATA SHEET

Syngenta Crop Protection, Inc.
Post Office Box 18300
Greensboro, NC 27419

In Case of Emergency, Call
1-800-888-8372

1. PRODUCT IDENTIFICATION

Product Name: **ENVOKE** Product No.: A9842A

EPA Signal Word: Caution

Active Ingredient(%): Trifloxysulfuron-Sodium (75.0%) CAS No.: 199119-58-9

Chemical Name: N-[(4,6-Dimethoxy-2-pyrimidinyl)amino]carbonyl-3-(2,2,2-trifluoro-ethoxy)-pyridin-2-sulfonamide sodium salt

Chemical Class: Sulfonylurea Herbicide

EPA Registration Number(s): 100-1132 **Section(s) Revised: 1**

2. COMPOSITION/INFORMATION ON INGREDIENTS

Material	OSHA PEL	ACGIH TLV	Other	NTP/IARC/OSHA Carcinogen
Surfactant	Not Established	Not Established	15 mg/m³ TWA (total dust)*	No
Sodium Sulfite	Not Established	Not Established	Not Established	IARC Group 3
Diatomaceous Earth	80 mg/m³/%SiO2 (20 mppcf) TWA	10 (inhalable);3 mg/m³ (respirable) TWA	6 mg/m³ TWA**	No
Crystalline Silica, Quartz	10 mg/m³/(%SiO2+2) (respirable dust)	0.1 mg/m³ (respirable silica)	Not Established	IARC Group 2A
Trifloxysulfuron-Sodium (75.0%)	Not Established	Not Established	Not Established	No

* recommended by manufacturer
** recommended by NIOSH

Ingredients not precisely identified are proprietary or non-hazardous. Values are not product specifications.

3. HAZARDS IDENTIFICATION

Symptoms of Acute Exposure
Presents slight hazard during normal handling.

Hazardous Decomposition Products
Can decompose at high temperatures forming toxic gases.

Physical Properties
Appearance: Light beige to brown granules
Odor: Not determined

Unusual Fire, Explosion and Reactivity Hazards
Tests have shown product does not explode when exposed to heat or mechanical shock.

During a fire, irritating and possibly toxic gases may be generated by thermal decomposition or combustion.

4. FIRST AID MEASURES

Product Name: **ENVOKE** Page: 1

Figure 29-10 Material safety data sheet.

Shipping papers for railroad transportation are called **waybills**; the list of every car on the train is called a **consist**. The conductor, engineer, or a designated member of the train crew will have a copy of both the waybill and the consist. On a marine vessel, shipping papers are called the **dangerous cargo manifest**. The manifest is generally kept in a tube-like container in the wheelhouse in the custody of the captain or master. For air transport, the **air bill** is the shipping paper; it is kept in the cockpit and is the pilot's responsibility. The responsible person in each situation should maintain the information about hazardous cargo and provide it in an emergency situation.

CHEMTREC

Located in Arlington, Virginia, and established by the Chemical Manufacturer's Association, the U.S. **Chemical Transportation Emergency Center (CHEMTREC)** is a clearinghouse of emergency response information. Essentially, this emergency call center is an information resource for fire fighters responding to chemical incidents. The toll-free number for CHEMTREC is 1-800-424-9300. When calling CHEMTREC, be sure to have the following basic information ready:

- Name of the caller and call back telephone number
- Location of the actual incident or problem
- Shipper or manufacturer of chemical (if known)
- Container type
- Rail car or vehicle markings or numbers
- The shipping carrier's name
- Recipient of material
- Local conditions and exact description of the situation

CHEMTREC is a free service that connects fire fighters with chemical manufacturers, chemists, and other product specialists who can help during a chemical incident. The Canadian equivalent of CHEMTREC is known as CANUTEC; the Mexican counterpart is SETIQ. Phone numbers for all of these resources can be found in the *NAERG*.

National Response Center

Whenever a significant hazardous incident occurs, the **National Response Center (NRC)** must be notified. The NRC is operated by the U.S. Coast Guard and serves as a central notification point, rather than a guidance center. Once the NRC is notified, it will alert the appropriate state and federal agencies. The NRC must be notified if any spilled material could enter a navigable waterway, through any method from a small spill into a local stream to a release at a sewerage treatment plant or a leak into an underground water table. The toll-free number is 1-800-424-8802.

Radiation

Fire fighters must be able to recognize the potential situations where radioactive materials might be encountered. Typical industries that routinely use radioactive materials include food testing labs, hospitals, medical research centers, biotechnology facilities, construction sites, and medical laboratories. For the most part, there will be some visual indicators (signs or placards) that indicate the presence of radioactive substances, but this is not always the case.

If you suspect a radiation incident at a fixed facility, you should initially ask for the radiation safety officer of the facility. This is the person responsible for the use, handling, and storage procedures for all the radioactive material at the site. This person likely will be a tremendous resource to you and will know exactly what is being used at the facility.

If the incident is not at a fixed site, the presence of radiation may never be apparent. Radioactive isotopes are not detected by sight, smell, taste, or any of the other senses. Therefore, if you have any suspicion that the incident involves radiation, it will be necessary to call a hazardous materials team or some other resource with radiation detection capabilities.

Significant incidents involving radiation are few and far between. This is not to say accidents cannot or will not happen, but regulations for using, storing, and transporting significant radioactive sources are so comprehensive that the entire process has become quite safe. Most of the incidents you may encounter will involve low-level radioactive sources and can be handled safely. These low-level sources are typically found in Type A packaging (▶ Figure 29-11). This packaging method is unique to radioactive substances and contains materials such as radiopharmaceuticals and other low-level emitters.

Type A packaging is characterized by having an inner containment vessel of glass, plastic, or metal and packaging materials made of polyethylene, rubber, or vermiculite. This type of packaging is designed to protect the contents from damage during normal shipping and handling. The key is to be able to suspect, recognize, and understand when and where you may encounter radioactive sources.

More dangerous radioactive sources might be found in Type B packaging (▶ Figure 29-12). This type of containment vessel contains materials such as spent radioactive waste and other high-level emitters. Type B packages are designed to protect the contents from greater exposure. The amount of protection is based on the potential severity of the hazard. Type B packages include small drums and heavily shielded casks weighing more than 100 metric tons.

Figure 29-11 Type A package.

Figure 29-12 Type B package.

Wrap-Up

Ready for Review

This chapter provided information on the visual indicators fire fighters can use to recognize a hazardous materials incident and identify the general characteristics of the released substance. It described the various containers used for storing chemicals and other hazardous materials, as well as methods of transporting these materials. It covered three marking systems for hazardous materials: the Department of Transportation's system of labels and placards and the corresponding *North American Emergency Response Guide,* the NFPA 704 hazard-identification system, and the Hazardous Materials Information System. Finally, this chapter also provided information on other resources that responders can use. Some information can be obtained from material safety data sheets and shipping papers; agencies such as CHEMTREC, CANUTEC, SETIQ, and the National Response Center are also helpful resources.

Chief Concepts

- Fire fighters should use all available resources to provide a greater degree of safety and improve their ability to mitigate the incident.
- Many resources are readily available and will provide initial guidance for handling the incident.
- Fire fighters should know where to obtain this initial information and how to best use it.
- Fire fighters should know how to obtain MSDS from various sources, including their department, the scene of the incident itself, or the manufacturer of the material.
- Fire fighters should be able to demonstrate proficiency in determining a proper guide to use when using the *NAERG.*
- Fire fighters should be able to name, understand, and locate the various types of shipping papers on various modes of transportation.

Hot Terms

Air bill The shipping papers on an airplane.

Bill of lading Shipping papers for roads and highways.

Bulk storage containers Large volume containers that have an internal volume greater than 119 gallons for liquids, greater than 882 pounds for solids, and a capacity of greater than 882 pounds for gases.

Bungs One or more small openings in closed-head drums.

Carboys A glass, plastic, or steel container, ranging in volume from 5 gallons to 15 gallons.

Chemical Transportation Emergency Center (CHEMTREC) A national call center for basic chemical information, established by the Chemical Manufacturer's Association.

Consist The list of every car on a train.

Container Any vessel or receptacle that holds material, including storage vessels, pipelines, and packaging.

Cryogenic liquids (cryogens) Gaseous substances that have been chilled to the point where they have liquefied; a liquid having a boiling point lower than –150°F (–101°C) at 14.7 psia (an absolute pressure of 101 kPa).

Cylinder A portable compressed-gas container.

Dangerous cargo manifest The shipping paper on a marine vessel, generally located in a tube-like container.

Department of Transportation (DOT) marking system A unique system of labels and placards that, in combination with the *North American Emergency Response Guide,* offers guidance for first responders operating at a hazardous materials incident.

Dewar containers Containers designed to preserve the temperature of the cold liquid held inside.

Drums Barrel-like containers built to DOT specification 5P (1A1).

Dry bulk cargo tanks Tanks designed to carry dry bulk goods such as powders, pellets, fertilizers, or grain; they are generally V-shaped with rounded sides that funnel toward the bottom.

Freight bills Shipping papers for roads and highways.

Wrap-Up

Hazardous materials Any materials or substances that pose an unreasonable risk of damage or injury to persons, property, or the environment if not properly controlled during handling, storage, manufacture, processing, packaging, use and disposal, or transportation.

Hazardous Materials Information System (HMIS) A color-coded marking system by which employers give their personnel the necessary information to work safely around chemicals.

Intermodal tanks Bulk containers that can be shipped by all modes of transportation—air, sea, or land.

Labels Smaller versions (4″ diamond-shaped markings) of placards, placed on four sides of individual boxes and smaller packages.

Material Safety Data Sheet (MSDS) A form, provided by manufacturers and compounders (blenders) of chemicals, containing information about chemical composition, physical and chemical properties, health and safety hazards, emergency response, and waste disposal of the material.

MC-306 flammable liquid tanker Commonly known as a gasoline tanker, this tanker typically carries gasoline or other flammable and combustible materials (also known as DOT-406).

MC-307 chemical hauler A tanker with a rounded or horseshoe-shaped tank.

MC-312 corrosives tanker A tanker that will often carry aggressive acids like concentrated sulfuric and nitric acid, with reinforcing rings along the side of the tank.

MC-331 pressure cargo tanker A tank commonly constructed of steel with rounded ends and a single open compartment inside; there are no baffles or other separations inside the tank.

MC-338 cryogenics tanker A low-pressure tanker designed to maintain the low temperature required by the cryogens it carries.

National Response Center (NRC) The U.S. Coast Guard maintained and staffed agency that should always be notified if any spilled material could possibly enter a navigable waterway.

NFPA 704 hazard identification system A hazardous materials marking system designed for fixed-facility use.

Nonbulk storage vessels Containers other than bulk storage containers.

***North American Emergency Response Guidebook* (NAERG)** The guidebook developed by the Department of Transportation to provide guidance for first responders operating at a hazardous materials incident in coordination with DOT's labels and placards marking system.

Pipeline A length of pipe, including pumps, valves, flanges, control devices, strainers, and/or similar equipment, for conveying fluids and gases.

Pipeline right of way An area, patch, or roadway that extends a certain number of feet on either side of the pipe itself and that may contain warning and informational signs about hazardous materials carried in the pipeline.

Placards Signage required to be placed on all four sides of highway transport vehicles, railroad tank cars, and other forms of hazardous materials transportation that identifies the hazardous contents of the vehicle, using a standardization system with $10\frac{3}{4}''$ × $10\frac{3}{4}''$ diamond-shaped indicators.

Secondary containment Any device or structure that prevents environmental contamination when the primary container or its appurtenances fail. Examples of secondary containment include dikes, curbing, and double-walled tanks.

Shipping papers A shipping order, bill of lading, manifest, or other shipping document serving a similar purpose and usually including the names and addresses of both the shipper and the receiver as well as a list of shipped materials with quantity and weight.

Signal words Information on a pesticide label that indicates the relative toxicity of the material.

Special use railcars Boxcars, flat cars, cryogenic and corrosive tank cars, and high-pressure compressed gas tube-type cars.

Totes Portable tanks, usually holding a few hundred gallons of product, characterized by a unique style of construction.

Tube trailers High-volume transportation devices made up of several individual compressed gas cylinders banded together and affixed to a trailer.

Vent pipes Inverted J-shaped tubes that allow for pressure relief or natural venting of the pipeline for maintenance and repairs.

Waybill Shipping papers for trains.

Fire Fighter in Action

It is a rainy afternoon when your engine company is dispatched to a motor vehicle accident involving a tanker truck. Upon arrival, you see the truck on its side, with a single tank compartment that has rounded ends. The tank appears to be a pressurized vessel.

1. What type of container is on the tanker truck?
A. MC-306
B. MC-307
C. MV-312
D. MC-331

2. Shipping papers for tanker trucks are known as:
A. bills of lading.
B. waybills.
C. the dangerous cargo manifest.
D. consists.

Upon investigation, you see a red placard with the number 1075. The pressurized tank does not seem to be leaking at this time.

3. In which section of the North American Emergency Response Guidebook should you look up the placard number?
A. Green section
B. Orange section
C. Yellow section
D. Blue section

You determine that you need to contact the Chemical Transportation Emergency Center (CHEMTREC) for additional information on the chemical contained in the tank.

4. What information do you need when calling CHEMTREC?
A. Location of incident and exact description of the incident
B. Shipper of manufacturer of chemical
C. Container type
D. All of the above

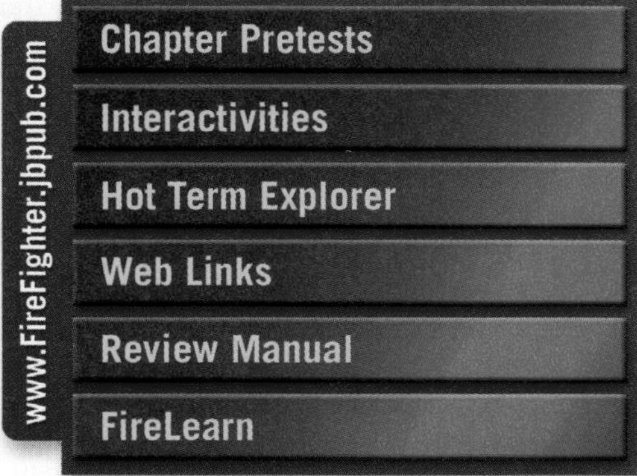

Hazardous Materials: Implementing a Response

Chapter 30

NFPA 472 Standard

4.4 Competencies—Implementing the Planned Response.

4.4.2 Initiating the Notification Process. Given either a facility or transportation scenario involving hazardous materials, regardless of the presence of criminal or terrorist activities, the first responder at the awareness level shall identify the initial notifications to be made and how to make them, consistent with the local emergency response plan or the organization's standard operating procedures.

5.1.2.1 The first responder at the operational level shall be able to perform the following tasks:

(2) Plan an initial response within the capabilities and competencies of available personnel, personal protective equipment, and control equipment by completing the following tasks:

(a) Describe the response objectives for hazardous materials incidents

(b) Describe the defensive options available for a given response objective

(c) Determine whether the personal protective equipment provided is appropriate for implementing each defensive option

(d) Identify the emergency decontamination procedures

5.2.4* Estimating the Potential Harm. The first responder at the operational level shall estimate the potential harm within the endangered area at a hazardous materials incident and also shall meet the following requirements:

(1) *Identify a resource for determining the size of an endangered area of a hazardous materials incident.

5.3 Competencies—Planning the Response.

5.3.2 Identifying Defensive Options. Given simulated facility and transportation hazardous materials problems, the first responder at the operational level shall identify the defensive options for each response objective and shall meet the following requirements:

(1) Identify the defensive options to accomplish a given response objective.

5.4.2* Initiating the Incident Management System. Given simulated facility and/or transportation hazardous materials incidents, the first responder at the operational level shall initiate the incident management system specified in the local emergency response plan and the organization's standard operating procedures and shall meet the following related requirements:

(1) Identify the role of the first responder at the operational level during hazardous materials incidents as specified in the local emergency response plan and the organization's standard operating procedures.

(2) Identify the levels of hazardous materials incidents as defined in the local emergency response plan.

(3) Identify the purpose, need, benefits, and elements of an incident management system at hazardous materials incidents.

(4) Identify the considerations for determining the location of the command post for a hazardous materials incident.

Knowledge Objectives

After studying this chapter, you will be able to:

- Describe how to contact the proper authorities.
- Describe how to plan an initial response.
- Describe how to estimate the size and scope of the incident.
- Describe how to identify a resource for determining the size of an endangered area.
- Describe resources available for determining the concentrations of a released hazardous material.
- Initiate an incident management system (IMS) for hazardous materials incidents.
- Identify considerations for determining the location of the command post.

Skills Objectives

There are no skills objectives for this chapter.

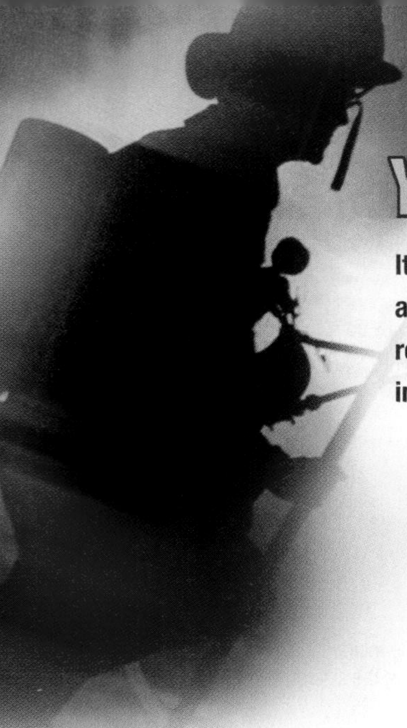

You Are the Fire Fighter

It is the middle of the afternoon. You arrive at the scene of a propane delivery truck with a noticeable leak directly under the main body of the vehicle. The vehicle is located in the middle of a residential area that is bordered by an old manufacturing-type mill building and row houses, all immediately adjacent to a river. You have limited resources initially and an active leak of propane.

1. ***Without defensive intervention, how would this incident progress on its own?***
2. ***What level of hazardous materials incident would this be?***
3. ***How would the damage to the container affect how the leak progressed and how the chemical affected the atmosphere?***

Plan an Initial Response

The first step after determining that there is a hazardous material present is to identify resources that can assist you in the process of reporting the hazardous materials incident. Fire fighters must follow their local standard operating procedures (SOP).

Who to Contact

Fire fighters should make a series of contacts after determining that there is a hazardous materials incident. Some of these notifications are related to operational concerns and some are for legal reasons. First and foremost, when a hazardous materials incident is detected, there should be an initial call for additional resources (► Figure 30-1). This should include support personnel, trained technicians, and technical specialists who will help identify the hazardous material and control the incident. There should be additional calls for decontamination personnel and equipment. Other notifications could include the Chemical Transportation Emergency Center (CHEMTREC), the National Response Center, local and state environmental agencies, and the local emergency planning commission. A predetermined list of contact names, agencies, and numbers should be established and maintained at the dispatch center.

It is important that no offensive action take place until the identity of the hazardous material involved is confirmed. It makes no sense to try to deal with a hazardous material until it is properly identified. This identification should be confirmed in a minimum of three references. For example, the hazardous material could first be identified by a placard. The placard is used to look up information in the *North American Emergency Response Guidebook* (NAERG). Then the shipping papers or Material Safety Data Sheets (MSDS) are checked to ensure that they list the same hazardous material as the placard. Then the National Institute for Occupational Safety and Health (NIOSH) references and CHEMTREC are consulted for a final confirmation of the identity of the hazardous material.

A variety of sources of information should be compared for consistency. If there are any variations in the information presented in the reference books, the information that reflects a more conservative course of action should be used. Protection and safety of fire fighters is always the first priority in any hazardous materials incident, and cannot be compromised for any reason. Therefore, if two references indicate that the use of water is acceptable and one reference cautions against it, the use of water should not be considered until further information is obtained. On-scene research should continue throughout the incident.

After the material is identified, an operations-level responder should perform only actions that do not involve contact

Figure 30-1 When a hazardous materials incident is detected, there should be an initial call for additional resources.

Fire Fighter Tips

If a hazardous material cannot be clearly identified, then the most conservative response strategy and tactics must be employed.

with the material. The responder must maintain full protective equipment during any activity, and must complete decontamination procedures prior to leaving any area where the hazardous material is present.

What to Report

When agencies are contacted, it is important that the information given is as clear, concise, and accurate as possible. An error in spelling, an incorrect measurement, a mispronunciation of a chemical name, or incorrect identification of a hazardous substance can be disastrous. The change or omission of just one letter in a chemical name could lead to incorrect identification. If a chemical is incorrectly identified, the proper response and safety procedures will not be carried out. This type of error can be deadly to both the responders and the public. The fire fighter should keep information as simple as possible, spell names that are complex or potentially confusing, and confirm that the receiver of the message has heard the correct information by repeating back what was heard.

In order for these agencies to plan, prepare, and begin assisting the fire fighters, they must have as much information as can be obtained. This will help ensure that every possible situation can be taken into account and planned for, in order to prevent any further injury, property loss, or environmental damage.

This information includes:

- The exact address and specific location of the leak or spill
- Identification of indicators and markers of hazardous materials
- All color or class information obtained from placards
- Four-digit UN/NA numbers
- Hazardous material identification obtained from shipping papers or MSDS and the potential quantity of hazardous material involved
- Description of container, including size, capacity, type, and shape
- Amount of chemical that could leak and amount that has leaked
- Exposures of people and the presence of special populations (children or elderly)
- The environment in the immediate area
- Current weather conditions, including wind direction and speed
- A contact or callback telephone number and two-way radio frequency or channel

Plan an Initial Response

When planning an initial hazardous materials incident response, the first priority is to consider the safety of the responding personnel. Responders are there to isolate, contain, and/or remedy the problem, not to become part of it. Proper incident planning will keep responders safe and provide a means to control the incident effectively, preventing further harm to persons or property.

Planning the response begins with the initial call for help. Information obtained during the call is used to determine the safest, most effective, and fastest route to the hazardous materials scene. Choose a route that approaches the scene from an upwind and upgrade direction, so that natural wind currents blow the hazardous material vapors away from arriving responders. A route that places the responders uphill of the site is also desired, so that a liquid hazardous material flows away from responders.

Responders need to know the type of material involved. Is the material a solid, liquid, or gas? Is it contained in a drum, a barrel, or a pressurized tank? Response to a spill of a solid hazardous material will differ from response to a liquid-release incident or vapor-release incident (▼ Figure 30-2). A solid is local and can be easily contained, whereas a released gas can be widespread and constantly moving, depending on the gas characteristics and weather conditions.

The characteristics of the affected area near the location of the spill or leak are important factors in planning the response to an incident. If an area is heavily populated, evacuation procedures and a decontamination process will have to be established very early in the incident. If the area is sparsely populated and rural, isolating the area from anyone trying to enter the location may be the top priority. A high-traffic area such as a major highway would necessitate immediate rerouting of traffic, especially during rush hours.

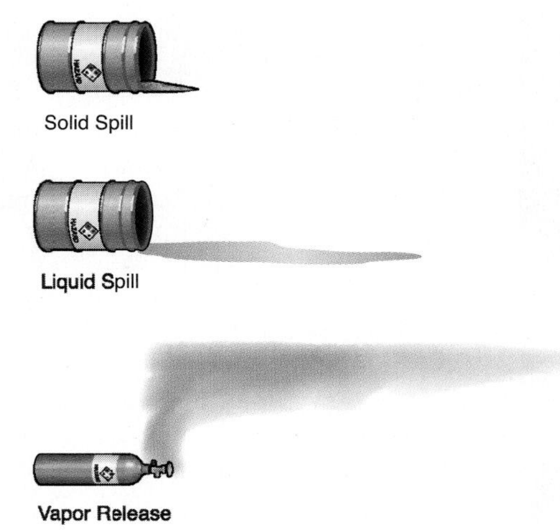

Figure 30-2 Response to a spill of a solid hazardous material will differ from response to a liquid-release incident or vapor-release incident.

Voices of Experience

"Hazardous materials incidents require much more than a single hazardous materials team."

Planning and implementing a hazardous materials response requires close cooperation between different agencies. My fire department consists of two stations that protect about 35,000 people, but we do not have a dedicated hazardous materials team. Through careful preincident planning and automatic dispatch, we rely on the hazardous materials team from a neighboring fire department. This mutual aid agreement enables fire departments to supply sufficient resources in a cost-effective manner, and is particularly beneficial in the case of specialized equipment and crews.

Many hazardous materials calls involve a spilled petroleum product. Mitigation of this type of incident often requires applying a blanket of foam to contain the combustible vapors until the source of the leak can be identified and fixed. When this type of incident occurs, the role of our department is to dispatch a crew of fire fighters who have completed specialized training in hazardous materials operations and the extinguishment of Class B fires utilizing foam equipment. They respond with a foam truck we fondly call "Mr. Bubbles," which carries 1500 gallons of flouroprotein foam concentrate and is equipped with a pump, supply hose, and special nozzles for applying foam. While this truck is used primarily to protect a large petroleum tank farm in our district, it fulfills a vital role in mitigating hazardous materials incidents involving spills from gasoline tankers.

Hazardous materials incidents require much more than a single hazardous materials team. "Mr. Bubbles" is a vital part of the response to hazardous materials incidents in our community. By working together with surrounding departments and by sharing resources, we are able to implement a prompt and effective response to hazardous materials incidents without excessive cost.

Dave Schottke
Fairfax City Volunteer Fire Department
Fairfax, Virginia

When responding to an incident, the more information that can be obtained the better. If information is unknown or is unconfirmed, then plan for the worst-case scenario. When planning for hazardous materials incidents, always plan for the largest margin of safety possible. (See the box 30-1: *Hazardous Materials Incident Levels,* later in this chapter.)

Response Objectives

At the operational level of training, all response objectives should be primarily defensive in purpose. Personnel are not actually in contact with the hazardous material. Some effective defense actions can be taken safely at a distance. **Defensive objectives** are ones that do not involve stopping the leak or release of a hazardous material. Defensive objectives are as follows:

- Isolate the area affected by the leak or spill, and evacuate victims who could become exposed to the hazardous material if the leak or spill were to progress.
- Control where the spill or release is spreading.
- Contain the spill to a specific area.

Defensive Actions

Some of the defensive actions that can be taken include diking and damming, absorbing or adsorbing material, stopping the flow remotely from a valve or shutoff, dilution or diverting material, and suppressing or dispersing vapor. These defensive actions are covered in detail in Chapter 32.

Proper Personal Protective Equipment

The determination of what personal protective equipment (PPE) is needed is based on the hazardous material involved, specific hazards, and the physical state of the material. At a minimum, fire fighters should wear full protective gear with no skin exposed, and use self-contained breathing apparatus (SCBA). This gear provides some respiratory protection, but standard structural firefighting PPE offers limited hazardous material protection depending on the type of hazardous material involved. Chapter 31 discusses proper PPE for a hazardous materials incident in detail.

Identify Emergency Decontamination Procedures

Even though there should be no intentional contact with the hazardous material involved, there must be a procedure or a plan in place to decontaminate any responder who accidentally becomes contaminated (► Figure 30-3). Victims removed from a contaminated zone must be decontaminated by personnel who have donned the appropriate protective gear and have been trained in the proper methods of decontamination. The methods of removal, or decontamination, are dictated by the hazardous material, the physical state of the material, and the hazards involved. Decontamination can be as simple as removing clothing and flushing material away with water or as complex as using drug therapy. Decontamination is covered in Chapter 33, Decontamination Techniques.

Figure 30-3 There must be a plan for decontamination.

Gauging the Potential Harm or Severity of the Incident

When attempting to evaluate a hazardous material event's potential to cause injury to people or destruction of property and the environment, responders need to consider factors such as the size of the container, the nature of the hazardous material involved, the amount released, and the area exposed. Based on the toxicity and the concentration of the hazardous material, how the incident might progress can be gauged. For example, the difference between a 20-pound propane cylinder and an 18,000-gallon propane tank will determine how extensive and lengthy an incident will become.

Resources for Determining the Size of the Incident

To determine the size of the incident, two resources should be used. The first is the *North American Emergency Response Guidebook,* published by the United States Department of Transportation. This reference book identifies and outlines predetermined evacuation distances and basic action plans, based on spill size estimates, for thousands of chemicals. The second is a computerized or hard-copy preincident plan. This includes reports submitted to the fire department and topographical mapping information.

Monitoring devices such as wind direction and weather forecasting equipment are critical resources for the incident commander (IC) in formulating response plans. Computer modeling programs such as ALOHA can plot the predicted movements of vapor clouds and plumes. The availability of

Box 30:1
Hazardous Materials Incident Levels

Level I

Lowest level of threat.

Small amount of hazardous material involved.

Can usually be handled by the local fire department.

Fire fighters must wear turnout gear and SCBA.

Example: a small gasoline spill from a motor vehicle accident.

Level II

A hazardous materials team is needed at this level.

Fire fighters only support the hazardous materials team.

Additional PPE required will be specialized and only carried by the hazardous materials team.

Civilian evacuations may be required.

Decontamination may need to be performed.

Example: a gasoline tanker has overturned in a tunnel and is spilling gasoline onto the highway.

Level III

The highest level of threat.

Large-scale evacuations may be needed.

Federal agencies will be called in.

Example: a ship in a highly populated harbor catches fire and begins to release chlorine vapors from its cargo area.

monitoring and portable detection devices will allow the IC to determine the hot, warm, and cold zones and the evacuation distances required.

Reporting the Size and Scope of the Incident

Reporting the estimated physical size of the area affected by a hazardous materials incident is accomplished by using information available at the scene. If a vehicle is transporting a known amount of material, an estimate of the size of the release can be made by subtracting the amount remaining in the container. For example, a gasoline tanker containing 9,000 gallons overturns, and 4,500 gallons remains in the tanker. An estimated 4,500 gallons of gasoline has thus spilled. The actual spill area can then be estimated in square feet. The affected area can be expressed in units as small as square feet or as large as square miles.

When it is initially unsafe to approach a vehicle and shipping papers are not immediately available, other methods must be used to determine the original amount of hazardous material and the remaining contained amount. Remember that the safety of responders is paramount to maintaining an effective response to any hazardous materials incident.

Determine the Concentration of a Released Hazardous Material

This information can be obtained from the MSDS. The Material Safety Data Sheet (MSDS) usually states the concentration of the hazardous material (► Figure 30-4). Litmus paper can be used to determine the concentration of an acid or a base by reporting the hazardous material's pH. To determine the concentration of a gaseous hazardous material, monitors are used to analyze the atmosphere from a safe distance. Once the concentration of the hazardous material is known, the IC can evaluate the incident response plan. A high concentration of an acid would call for a higher level of personal protection. It may also require the evacuation of civilians due to the concentration of the hazardous material in vaporized form.

Incident Management System

As we learned in Chapter 4, the IMS can be expanded to handle an incident of any size and complexity. Hazardous materials incidents can be complex, and local, state, and federal responders and agencies will be involved in many cases of long duration. The basic IMS system consists of five functions: command, operations, logistics, planning, and finance and administration.

In an incident involving hazardous materials, a special technical group develops under the Operations section. This special branch consists of some or all of the following positions:

- A second safety officer reporting directly to the hazardous materials team officer. This hazardous materials safety officer is responsible for the hazardous materials team's safety only.
- A hot zone entry team.
- A decontamination team.
- A backup team.
- A hazardous materials information research team.

The Command Post

The main hub of the IMS is the command post. The command post is the collection point for all information and resources. The command post must be located upwind and upgrade from the spill or leak in the cold zone to keep it from becoming contaminated or unsafe. If the command post, including those personnel housed there, were to become contaminated, the personnel inside the command post would no longer be able to control the operation. The officers in that location would become victims, the com-

MATERIAL SAFETY DATA SHEET

Syngenta Crop Protection, Inc.
Post Office Box 18300
Greensboro, NC 27419

In Case of Emergency, Call
1-800-888-8372

1. PRODUCT IDENTIFICATION

Product Name: **ENVOKE** Product No.: A9842A

EPA Signal Word: Caution

Active Ingredient(%): Trifloxysulfuron-Sodium (75.0%) CAS No.: 199119-58-9

Chemical Name: N-[(4,6-Dimethoxy-2-pyrimidinyl)amino]carbonyl-3-(2,2,2-trifluoro-ethoxy)-pyridin-2-sulfonamide sodium salt

Chemical Class: Sulfonylurea Herbicide

EPA Registration Number(s): 100-1132 **Section(s) Revised: 1**

2. COMPOSITION/INFORMATION ON INGREDIENTS

Material	OSHA PEL	ACGIH TLV	Other	NTP/IARC/OSHA Carcinogen
Surfactant	Not Established	Not Established	15 mg/m³ TWA (total dust)*	No
Sodium Sulfite	Not Established	Not Established	Not Established	IARC Group 3
Diatomaceous Earth	80 mg/m³/%SiO2 (20 mppcf) TWA	10 (inhalable);3 mg/m³ (respirable) TWA	6 mg/m³ TWA**	No
Crystalline Silica, Quartz	10 mg/m³/(%SiO2+2) (respirable dust)	0.1 mg/m³ (respirable silica)	Not Established	IARC Group 2A
Trifloxysulfuron-Sodium (75.0%)	Not Established	Not Established	Not Established	No

* recommended by manufacturer
** recommended by NIOSH

Ingredients not precisely identified are proprietary or non-hazardous. Values are not product specifications.

3. HAZARDS IDENTIFICATION

Symptoms of Acute Exposure
Presents slight hazard during normal handling.

Hazardous Decomposition Products
Can decompose at high temperatures forming toxic gases.

Physical Properties
Appearance: Light beige to brown granules
Odor: Not determined

Unusual Fire, Explosion and Reactivity Hazards
Tests have shown product does not explode when exposed to heat or mechanical shock.

During a fire, irritating and possibly toxic gases may be generated by thermal decomposition or combustion.

4. FIRST AID MEASURES

Product Name: **ENVOKE** Page: 1

Figure 30-4 MSDS contains critical information.

mand post would become a part of the hot zone, and all operations would have to be re-established elsewhere using a completely different pool of personnel, equipment, and supplies. The overall efficiency of command would be negatively affected, and control of the scene would be temporarily lost during transfer of command. Based on the potential threat or severity of the hazard, the command post could be as close as one block away or as far as miles away from the hot zone. When choosing a site for the command post, the maximum margin of safety must be used.

Wrap-Up

Ready for Review

Upon arrival at a hazardous materials incident, initiating the notification of agencies and resources who will join the response or help with technical support is essential. It is critical to report as much information as can be safely gathered. The initial response is planned with the safety of the responders as the first priority. Defensive actions such as absorbing and diking may occur during the response phase. Decontamination of responders and victims must also be considered during this phase.

Part of the early response is estimating the event's severity in terms of potential harm to people, property, and the environment. The incident management system provides a flexible way to adapt a response to an incident of any size and complexity. The main hub of IMS is the command post. It must be located in the cold zone to keep it from becoming contaminated.

Chief Concepts

- An important early notification to make is the request for additional response personnel, such as support personnel, trained technicians, and technical specialists.
- The approach to the incident should be from upwind, and from a direction that ensures that released liquids or vapors flow away from responders.
- Possible defensive actions include stopping the release with a valve or shutoff; absorbing, adsorbing, diking, damming, diverting, or diluting escaped material; and suppressing or dispersing vapor.
- The type of personal protective equipment required depends on the material involved and the nature of the incident; structural firefighting PPE provides no protection against hazardous materials.
- In a hazardous materials incident, a hazardous materials branch develops under the Operations sector in the incident management system. This branch includes a second safety office and a number of specialized operational teams.

Hot Terms

Defensive objectives Actions that do not involve the actual stopping of the leak or release of a hazardous material, or contact of responders with the material; these include preventing further injury and controlling or containing the spread of the hazardous material.

Material Safety Data Sheet (MSDS) A form, provided by manufacturers and compounders (blenders) of chemicals, containing information about chemical composition, physical and chemical properties, health and safety hazards, emergency response, and waste disposal of the material.

Fire Fighter in Action

It is 9:00 a.m. when your engine company is dispatched to the scene of a rollover tanker truck. Upon arrival, you find an overturned tanker truck leaking an unknown substance into a stream. You observe a red placard with the number 1993 on the truck.

1. Before an offensive action takes place, what is the minimum number of identification references that should be confirmed?
   - A. 1
   - B. 2
   - C. 3
   - D. 4

2. After a material is identified, what type of actions can an operational responder take?
   - A. No actions
   - B. Actions that do not involve contact with the material
   - C. Actions that involve contact with the material
   - D. Plug the leak

After consulting your *North American Emergency Response GuideBook,* you find that a red placard with the number 1993 could identify a number of different combustible liquids. The driver gives you the shipping papers and you discover the hazardous material is fuel oil. It appears that over 1,000 gallons of fuel has leaked into the creek.

3. At the operational level, what should your response objectives be?
   - A. Isolate the area affected by the spill and evacuate individuals who could become exposed to the hazardous material if the leak were to progress.
   - B. Control where the spill is spreading.
   - C. Contain the spill to a specific level.
   - D. All of the above

4. What level incident would you classify this truck rollover as?
   - A. Level I
   - B. Level II.
   - C. Level III.
   - D. Level IV

www.FireFighter.jbpub.com

www.FireFighter.jbpub.com

Chapter Pretests
Interactivities
Hot Term Explorer
Web Links
Review Manual
FireLearn

Hazardous Materials: Scene Safety and Control

Technology Resources

www.FireFighter.jbpub.com

- Chapter Pretests
- Interactivities
- Hot Term Explorer
- Web Links
- Review Manual
- FireLearn

Chapter Features

- Skill Drills
- Voices of Experience
- Fire Marks
- Teamwork Tips
- Fire Fighter Safety Tips
- Fire Fighter Tips
- Canadian Perspectives
- Hot Terms
- Wrap-Up

NFPA 472 Standard

4.4.1 *Initiating Protective Actions.

Given examples of facility and transportation hazardous materials incidents, the local emergency response plan, the organization's standard operating procedures, and the current edition of the *Emergency Response Guidebook,* first responders at the awareness level shall be able to identify the actions to be taken to protect themselves and others and to control access to the scene and shall also meet the following requirements:

(5) Given the name of a hazardous material, identify the recommended personal protective equipment from the following list:
- (a) Street clothing and work uniforms
- (b) Structural fire-fighting protective clothing
- (c) Positive-pressure self-contained breathing apparatus
- (d) Chemical-protective clothing and equipment

5.3.3 Determining Appropriateness of Personal Protective Equipment.

Given the name of the hazardous material involved and the anticipated type of exposure, the first responder at the operational level shall determine whether available personal protective equipment is appropriate for implementing a defensive option and also shall meet the following requirements:

(1) *Identify the respiratory protection required for a given defensive option and the following:
- (a) Identify the three types of respiratory protection and the advantages and limitations presented by the use of each at hazardous materials incidents.
- (b) Identify the required physical capabilities and limitations of personnel working in positive-pressure self-contained breathing apparatus (SCBA).

(2) Identify the personal protective clothing required for a given defensive option and the following:
- (a) Identify skin contact hazards encountered at hazardous materials incidents.
- (b) Identify the purpose, advantages, and limitations of the following levels of protective clothing at hazardous materials incidents:
  - i. Structural fire-fighting protective clothing
  - ii. High temperature-protective clothing
  - iii. Chemical-protective clothing
  - iv. Liquid splash-protective clothing
  - v. Vapor-protective clothing

5.4.1 Establishing and Enforcing Scene Control Procedures.

Given scenarios for facility and/or transportation hazardous materials incidents, the first responder at the operational level shall identify how to establish and enforce scene control including control zones, emergency decontamination, and communications and shall meet the following requirements:

(1) Identify the procedures for establishing scene control through control zones.

(2) Identify the criteria for determining the locations of the control zones at hazardous materials incidents.

5.4.3 Using Personal Protective Equipment.

The first responder at the operational level shall demonstrate the ability to don, work in, and doff the personal protective equipment provided by the authority having jurisdiction, and shall meet the following related requirements:

(1) Identify the importance of the buddy system in implementing the planned defensive options.

(2) Identify the importance of the back-up personnel in implementing the planned defensive options.

(3) Identify the safety precautions to be observed when approaching and working at hazardous materials incidents.

(4) Identify the symptoms of heat and cold stress.

(5) Identify the physical capabilities required for, and the limitations of, personnel working in the personal protective equipment as provided by the authority having jurisdiction.

Other NFPA Standards

NFPA 1991, *Standard on Vapor-Protective Ensembles for Hazardous Materials Emergencies*

Knowledge Objectives

After studying this chapter, you will be able to:
- Describe hazardous materials personal protective equipment.
- Identify the purpose, advantages, and limitations of: structural fire-fighting protective clothing, high temperature-protective clothing, chemical-protective clothing, liquid splash-protective clothing, and vapor-protective clothing.
- Discuss respiratory protection in a hazardous materials incident.
- Describe the levels of hazardous materials personal protective equipment.
- Identify skin-contact hazards encountered at hazardous materials incidents.
- Describe the safety precautions to be observed, including those for heat and cold stress, when approaching and working at hazardous materials incidents.
- Describe the physical capabilities required and limitations of personnel working in personal protective equipment.
- Describe techniques used to isolate hazard areas and deny entry.
- Describe the importance of the buddy system and back-up personnel.

Skills Objectives

There are no skills objectives for this chapter.

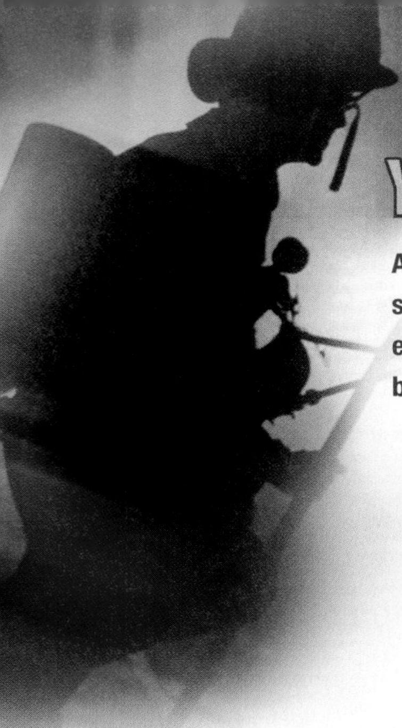

You Are the Fire Fighter

After local news media broadcast a report about picric acid, a potentially unstable, toxic explosive used in dyes, a local manufacturing company executive calls the station and says that an employee who saw the news report has brought a crystallized bottle of picric acid up to the business offices, which are adjacent to the employee cafeteria.

1. ***What additional information would you want to obtain to assess the hazard?***
2. ***What scene management techniques would you recommend for this situation?***
3. ***What level of protective clothing will be required for this situation?***

Introduction

Scene control is important at all emergencies; at a hazardous materials incident, however, scene control, site management, and personnel accountability are critical. The safe handling of the incident is often determined in the first five minutes, based on the actions of the fire fighters who are the initial responders.

Levels of Damage

The damage that a hazardous material will inflict on a human being depends upon the material's threshold limit value, the point at which the material begins to affect a person. The threshold limit value may be defined in several ways. The various definitions indicate the levels of protection that are required.

The **Threshold Limit Value/Short-Term Exposure Limit (TLV/STEL)** is the maximum concentration of material that a person can be exposed to in 15-minute intervals, up to four times a day, without experiencing irritation or chronic or irreversible tissue damage. There should be a minimum one-hour rest period between any exposures. The lower the TLV/STEL, the more toxic the substance.

The **Threshold Limit Value/Time Weighted Average (TLV/ TWA)** is the maximum concentration of material that a worker could be exposed to 8 hours a day, 40 hours a week, with no ill effects. As with the TLV/STEL, the lower the TLV/TWA, the more toxic the substance.

The **Threshold Limit Value/Ceiling (TLV/C)** is the maximum concentration of material that a worker should not be exposed to, even for an instant. Again, the lower the TLV/C, the more toxic the substance.

The **Threshold Limit Value/Skin** indicates that direct or airborne contact with a material could result in possible and significant exposure from absorption through the skin, mucous membranes, and eyes. This designation is intended to suggest that appropriate measures be taken to minimize skin absorption so that the TLV/TWA is not exceeded.

The **Permissible Exposure Limit (PEL)**, or **Recommended Exposure Level (REL)**, is comparable to TLV/TWA. These terms measure the maximum, time-weighted concentration of material to which 95% of healthy adults can be exposed without suffering any adverse effects over a 40-hour work week.

Immediate Danger to Life and Health (IDLH) is an atmospheric concentration of any toxic, corrosive, or asphyxiant substance that poses an immediate threat to life or could cause irreversible or delayed adverse health effects. There are three general IDLH atmospheres: toxic, flammable, and oxygen-deficient. Individuals exposed to atmospheric concentrations below the IDLH value could escape without experiencing irreversible damage to their health, even if their respiratory protection fails. Individuals who may be exposed to atmospheric concentrations at or above the IDLH value must use positive-pressure self-contained breathing apparatus (SCBA) or equivalent protection.

At most hazardous materials emergencies, fire fighters will not be able to measure concentrations of specific chemicals. Measuring concentrations of chemicals requires specific monitoring instruments and trained response personnel to interpret the results.

Using exposure guidelines helps to minimize the possibility that the emergency responder and the public will be exposed to hazardous materials or atmospheres. Once these exposure guidelines are understood, they can be applied at the scene of a hazardous materials emergency. For example, exposure guidelines can be used to define the three basic atmospheres at a hazardous materials emergency:

- Safe atmosphere—No harmful hazardous materials effects exist, so personnel can handle routine emergencies without specialized PPE.
- Unsafe atmosphere—A hazardous material that is no longer contained creates an unsafe condition or atmosphere. A person who is exposed to the material long enough will probably experience some form of acute or chronic injury.
- Dangerous atmosphere—These are environments where serious, irreversible injury or death may occur.

All exposure guidelines share a common goal: to ensure the safety and health of people exposed to hazardous material. As long as the exposure remains below these values, it is considered safe to the average healthy adult, based on current information. As additional research and studies on the toxic effects of various chemicals are completed, these values may be adjusted.

Personal Protective Equipment

Personal protective equipment (PPE) is the clothing and protection provided to shield or insulate a person from chemical, physical, and thermal hazards. It is essential that a fire fighter's PPE meet the standards of both the National Fire Protection Association (NFPA) and the United States Occupational Safety and Health Administration (OSHA), and that it be properly maintained and used.

Selecting the appropriate PPE for a hazardous materials incident is an important, potentially life-saving decision. Without the proper protective clothing and respiratory protection, fire fighters will be injured. Adequate PPE should protect the fire fighter's respiratory system, as well as the skin, eyes, face, hands, feet, head, body, and hearing. Every member of the responding team must know the shielding capabilities and limitations of their personal protective equipment. When product-specific chemical suits are unavailable, fire fighters must know what protection their regular firefighting clothing provides. Remember, protective clothing is your last line of defense, not your first.

Proper PPE

The selection of proper PPE depends on the activities a fire fighter is expected to perform. Standard structural firefighting gear may be suitable for support operations, but personnel involved in decontamination activities require chemical-protective clothing because of the possibility of contamination. The incident commander (IC) should approve the level of PPE required for a given activity. Fire fighters should operate only at the incident level that matches their knowledge, training, and equipment. If conditions indicate a need for a higher response level, additional personnel with the appropriate training and PPE should be summoned.

Hazardous Materials Personal Protective Equipment

Different levels of PPE may be required at a hazardous materials incident. This section reviews the protective qualities of various outfits, from the least protection to the greatest protection.

Clothing and Work Uniforms

Street clothing and work uniforms offer the least amount of protection in a hazardous materials emergency. Normal clothing may prevent a noncaustic powder from coming into direct contact with the skin, but it offers no more protection. A one-piece flame- or lightly-resistant coverall may enhance protection slightly.

Structural Firefighting Protective Clothing

The next level of protection is provided by structural firefighting protective equipment. This outfit includes a helmet, a bunker coat, bunker pants, boots, gloves, a hood, and a personal alert safety system (PASS) device. Standard firefighting turnout gear offers no chemical protection, although it does have some abrasion resistance and prevents direct skin contact. However, it easily absorbs materials, breaks down when exposed to chemicals, and does not provide adequate protec-

> **Fire Fighter Tips**
>
> Female fire fighters may have difficulty finding structural protective clothing that fits properly. Protective clothing designed for men does not adequately protect women because of differences in body shape. There may be gaps at the waist, wrists, and neck. A fire department should issue PPE that properly fits each fire fighter, regardless of gender, size, or shape.

tion from the harmful gases, vapors, liquids, or dusts that are encountered during hazardous materials incidents.

High Temperature-Protective Equipment

High temperature-protective equipment is PPE that protects the wearer during short-term exposures to high temperatures (► Figure 31-1). It allows the properly trained fire fighter to work in extreme fire conditions. It provides protection against high temperatures only and is not designed to protect the fire fighter from hazardous materials.

Figure 31-1 High temperature-protective equipment protects the wearer from high temperatures during a short exposure.

Chemical-Protective Clothing and Equipment

Chemical-protective clothing is designed to prevent chemicals from coming in contact with the body and may have varying degrees of resistance. Chemical-protective clothing material is any material or combination of materials used in an item of clothing to isolate parts of the wearer's body from contact with a hazardous chemical. Chemical resistance is the ability to resist chemical attack. Time, temperature, stress, and reagent are all factors that affect the chemical resistance of materials. Chemical-resistant materials are specifically designed to inhibit or resist the passage of chemicals into and through the material by the processes of penetration, permeation, or degradation.

Penetration is the flow or movement of a hazardous chemical through closures (like zippers), seams, porous materials, pinholes, or other imperfections in the material. Liquids are most likely to penetrate material, but solids (like asbestos) can also penetrate protective clothing materials.

Permeation is the process by which a hazardous chemical moves through a given material on the molecular level. Permeation differs from penetration in that permeation occurs through the material itself rather than through openings in the material.

Degradation is the physical destruction or decomposition of a clothing material due to chemical exposure, general use, or ambient conditions (like storage in sunlight). Degradation may be evidenced by visible signs such as charring, shrinking, swelling, color changes, or dissolving. Materials can also be tested for weight changes, loss of fabric tensile strength, and other properties to measure degradation.

Other performance requirements must be considered when selecting chemical-protective materials. These requirements include chemical resistance, flexibility, abrasion, temperature resistance, shelf life, and sizing criteria.

Chemical-protective clothing can be constructed as a single- or multi-piece garment. A single-piece garment completely encloses the wearer and is known as an encapsulated suit or acid suit. Encapsulated protective clothing offers full body protection from hostile environments and requires the use of air-supplied respiratory protection devices, such as

Figure 31-2 Liquid splash-protective clothing must be worn when there is the danger of chemical splashes.

Figure 31-3 Vapor-protective clothing retains body heat and increases the possibility of heat-related emergencies.

SCBA. A multi-piece garment works with the wearer's respiratory protection, an attached or detachable hood, gloves, and boots to protect against a specific hazard.

Various materials are used in both encapsulated and non-encapsulated protective clothing. The most common include butyl rubber, Tyvek, Saranex, polyvinyl chloride (PVC), and Vitron, either singly or in multiple layers of several materials. Special protective chemical clothing is adequate for some chemicals and useless for others; no single material provides satisfactory protection from all chemicals. Protective clothing materials must be compatible with the chemical substances involved and consistent with the manufacturer's instructions. The manufacturer's guidelines and recommendations should be consulted for material compatibility information.

Some materials are so caustic or corrosive that they could destroy SCBA after a short exposure. Personnel trained to the operations level should not be operating in encapsulated suits. Operational personnel can wear some nonencapsulated chemical protective clothing while providing support for those entering the hazardous materials area.

Liquid splash-protective clothing consists of several pieces of clothing and equipment that protects the skin and eyes from chemical splashes (▲ Figure 31-2). It does not provide total body protection from gases or vapors and should not be used for incidents involving liquids that emit vapors known to affect or be absorbed through the skin. **Vapor-protective clothing** must be used when hazardous vapors are present. Vapor-protective clothing traps the fire fighter's body heat, so fire fighters and support personnel must be aware of the possibility of heat-related emergencies (▲ Figure 31-3). SCBA or airline hose units must also be used for respiratory protection.

Fire fighters may wear liquid chemical splash-protective clothing over or under structural firefighting clothing in some situations. This provides limited chemical splash and thermal protection. Fire fighters trained to the operational level can wear liquid chemical splash-protective clothing when they are assigned to enter the initial site, to protect decontamination personnel, or to construct isolation barriers such as dikes, diversions, retention areas, or dams.

Respiratory Protection

Positive-Pressure Self-Contained Breathing Apparatus

Respiratory protection, usually provided by a self-contained breathing apparatus (SCBA), is an important separate level of protection at a hazardous materials incident. Using positive-pressure SCBA prevents both inhalation and ingestion,

Figure 31-4 A supplied air respirator is less bulky than an SCBA but is limited by the length and structural integrity of the air hose.

Figure 31-5 Air-purifying respirators can only be used where there is sufficient oxygen in the atmosphere.

two primary routes of exposure, and should be mandatory for fire service personnel. Although it is an excellent barrier in many hazardous environments, positive-pressure SCBA is not designed for and has never been tested or approved for use in atmospheres that may contain unknown hazards. Fire fighters must be aware of the limitations of SCBA in the hazardous atmosphere; review Chapter 2, Fire Fighter Qualifications and Safety, for more information.

Other Respiratory Protection

Supplied air respirators (SARs) have an external air source such as a compressor or storage cylinder. A hose connects the user to the air source and provides air to the face piece. SARs are useful during extended operations such as decontamination, clean-up, and remedial work. SARs may be less bulky and weigh less than SCBA, but the length of the air hose may limit movement (◄ Figure 31-4).

> **Fire Fighter Safety Tips**
>
> CBRN-certified SCBA refers to SCBA and APR that are safe to use during a Chemical, Biological, Radiological, or Nuclear incident. CBRN-certified SCBA and APR have gone through rigorous testing to ensure their integrity during such incidents.

Air-purifying respirators (APRs) are filtering devices that remove particulates and contaminants from the air. They should be worn only in atmospheres where the type and quantity of the contaminants are known and where there is sufficient oxygen for breathing (◄ Figure 31-5). APRs may be appropriate for operations involving volatile solids and for remedial clean-up and recovery operations where the type and concentration of contaminants is verifiable.

APRs use filtration devices to remove particulate matter, gases, or vapors from the atmosphere. These devices range from full-face piece, dual-cartridge masks with eye protection to half-mask, face piece-mounted cartridges with no eye protection. APRs do not have a separate source of air but filter and purify ambient air before it is inhaled. APRs used for gases or vapors are commonly equipped with an absorbent material that soaks up or reacts with the hazardous gas. Consequently, cartridge selection is based on the expected contaminants. Particle-removing respirators use a mechanical filter to separate the contaminants from the air. Both devices require that the ambient atmosphere contain a minimum of 19.5% oxygen.

APRs are easy to wear and offer no hose line restriction, but have some drawbacks. Because filtering cartridges are specific to expected contaminants, they will be ineffective if there is a sudden change in contaminant, endangering the lives of fire fighters. The air must be continually monitored for both the known substance and the ambient oxygen level throughout the incident. For these reasons, APRs should not be used at hazardous materials incidents until qualified personnel have tested the ambient atmosphere and determined that such devices can be used safely.

Chemical-Protective Clothing Ratings

Chemical-protective clothing is rated for its effectiveness in various ways. Testing measures whether chemicals will permeate the suit and, if so, how quickly and to what degree. This information can be obtained from the manufacturer. The Environmental Protection Agency (EPA) defines levels of protection, using an alphabetic system.

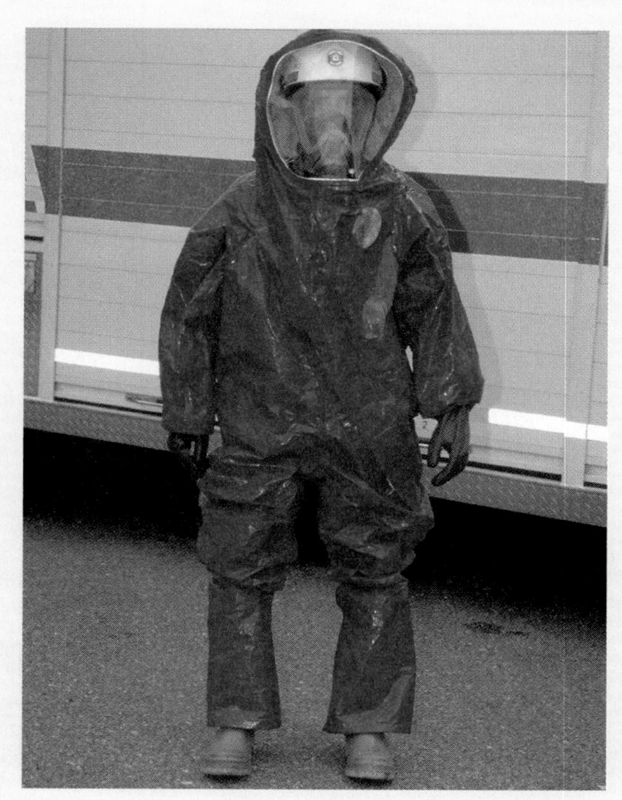

Figure 31-6 Level A protection envelops the wearer in a totally encapsulating suit.

Figure 31-7 Level B protection provides a high level of respiratory protection but less skin protection.

Level A

Level A protection consists of a heavy, encapsulating suit that envelops both the wearer and SCBA totally. It should be used when the hazardous material identified requires the highest level of protection for skin, eyes, and respiration. Level A protection is effective against vapors, gases, mists, and even dusts (▲ Figure 31-6).

Level A protection must meet the requirements for vapor-gas protection outlined in NFPA 1991, *Standard on Vapor-Protective Ensembles for Hazardous Materials Emergencies.* It requires open-circuit positive-pressure SCBA or an SAR for respiratory protection.

Recommended PPE of Level A includes:

- SCBA or SAR
- Fully encapsulating chemical-resistant suit
- Inner chemical-resistant gloves
- Chemical-resistant safety boots/shoes
- Two-way radio

Optional PPE of Level A includes:

- Coveralls
- Long cotton underwear
- Hard hat
- Disposable gloves and boot covers

Level B

Level B protection consists of chemical-protective clothing, boots, gloves, and SCBA (▲ Figure 31-7). It should be used when the type and atmospheric concentration of identified substances require a high level of respiratory protection but less skin protection. The type of gloves and boots worn depend on the identified chemical; wrists and ankles must be properly sealed to prevent splashed liquids from contacting skin.

Recommended PPE of Level B includes:

- SCBA or SAR
- Chemical-resistant clothing
- Inner and outer chemical-resistant gloves
- Chemical-resistant safety boots/shoes
- Hard hat
- Two-way radio

Optional PPE of Level B includes:

- Coveralls
- Disposable boot covers
- Face shield
- Long cotton underwear

Figure 31-8 Level C protection includes chemical-protective clothing and gloves as well as respiratory protection.

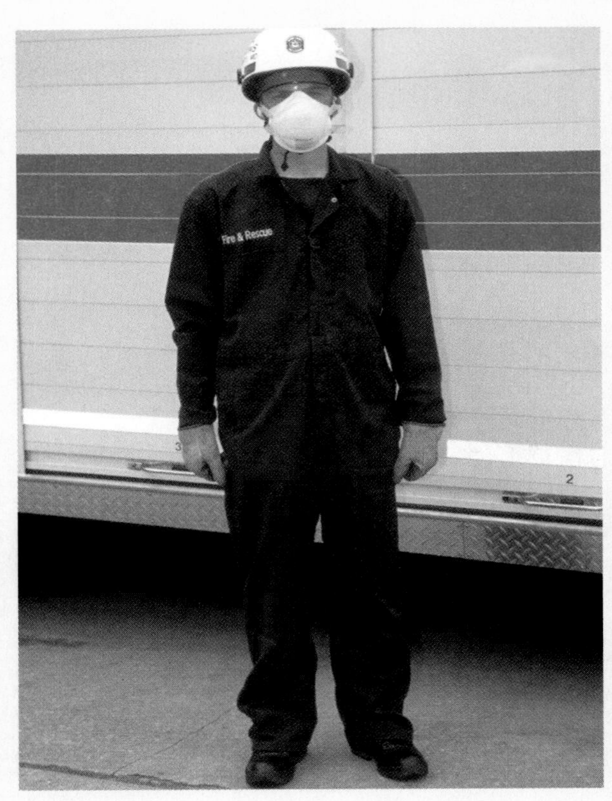

Figure 31-9 Level D protection is primarily a work uniform that includes coveralls and provides minimal protection.

Level C

Level C protection consists of standard work clothing plus chemical-protective clothing, chemically-resistant gloves, and a form of respiratory protection. Level C protection is appropriate when the type of airborne substance is known, its concentration is measured, the criteria for using APRs is met, and skin and eye exposure is unlikely (▲ Figure 31-8).

Level C respiratory protection can be provided with a filter or cartridge mask or an APR. The APR could be a half-face or a full-face mask that includes eye protection. Head protection and goggles are also required.

Recommended PPE of Level C includes:

- Full-face APR
- Chemical-resistant clothing
- Inner and outer chemical-resistant gloves
- Chemical-resistant safety boots/shoes
- Two-way radio
- Hard hat

Optional PPE of Level C includes:

- Coveralls
- Disposable boot covers
- Face shield
- Escape mask
- Long cotton underwear

Level D

Level D protection is the lowest level of protection—coveralls, work shoes, hard hats, gloves, and standard work clothing (▲ Figure 31-9). It should be used when the atmosphere contains no known hazard, and when work functions preclude splashes, immersion, or the potential for unexpected inhalation of or contact with hazardous levels of chemicals. Level D PPE should be used for nuisance contamination (like dust) only. It should not be worn on any site where respiratory or skin hazards exist.

Recommended PPE of Level D includes:

- Coveralls
- Safety boots/shoes
- Safety glasses or chemical-splash goggles
- Hard hat

Optional PPE of Level D includes:

- Gloves
- Escape mask
- Face shield

Skin Contact Hazards

The principal dangers of hazardous materials are toxicity, flammability, and reactivity. However, many hazardous materials have more than one dangerous characteristic. These materials can produce extremely dangerous and harmful effects on the unprotected or inadequately protected human body. The best action to take when dealing with an unknown chemical or with a chemical that has more than one dangerous characteristic is to assume the worst and to provide the largest safety margin possible.

The skin can absorb harmful toxins without any sensation to the skin itself. This means that fire fighters should not rely on pain or irritation as a warning sign of absorption. Some poisons are so concentrated that a few drops placed on the skin may result in death.

Skin absorption is enhanced by abrasions, cuts, heat, and moisture. This can create critical problems for fire fighters working at incidents that involve biological agents. Anyone with large open cuts, rashes, or abrasions should be prohibited from working in areas where they may be exposed to hazardous materials. Smaller cuts or abrasions should be covered with a nonporous dressing.

The rate of absorption can vary depending upon the body part that is exposed. For example, chemicals can be absorbed through the skin on the scalp much faster than through the forearm.

The high absorbency rate of the eyes make them one of the fastest means of exposure. This type of exposure may occur when a chemical is splashed directly into the eye, when the chemical is carried from a fire by toxic smoke particles, or when gases or vapors are absorbed through the eyes. Absorption of a chemical through the eyes also may be seen as an early warning signal for either PPE or SCBA failure.

Chemicals such as corrosives will immediately damage skin or body tissues upon contact. Acids have a strong affinity for moisture and can create significant skin and respiratory tract burns. However, acid injuries also create a clot-like barrier that blocks deep skin penetration. In contrast, alkaline materials dissolve the fats and lipids that make up skin tissue and change solid tissue into a soapy liquid. This process is similar to the way caustic chemical drain cleaners dissolve grease and other materials in sinks and drains. As a result, alkaline burns are often much deeper and more destructive than acid burns.

Safety Precautions

Fire fighters need to take several types of safety precautions during a hazardous materials incident. In addition to taking the standard safety precautions of using PPE and respirator protection, fire fighters must also take precautions against the dangers posed by temperature and stress.

Excessive Heat Disorders

Fire fighters operating in protective clothing should be aware of the signs and symptoms of heat exhaustion, heat stress, heat stroke, and dehydration. If the body is unable to disperse heat because it is covered by a closed, impermeable garment, serious short- and long-term medical situations could result. **Heat exhaustion** is a mild form of shock caused when the circulatory system begins to fail because the body is unable to dissipate excessive heat and becomes overheated. The body's core temperature rises, followed by weakness and profuse sweating. The fire fighter may become weak or dizzy and have episodes of blurred vision. Other signs and symptoms of heat exhaustion include: rapid, shallow breathing; weak pulse; cold, clammy skin; and, sometimes, loss of consciousness. An individual experiencing heat exhaustion, although not in an immediate life-threatening condition, should be removed at once from the source of heat, rehydrated with electrolyte solutions, and kept cool. If not properly treated, heat exhaustion may progress to heat stroke.

Heat stroke is a severe and potentially fatal condition resulting from the failure of the temperature-regulating capacity of the body. It is caused by exposure to the sun or high temperatures. Reduction or cessation of sweating is an early symptom. The body temperature can rise to 105°F (40.5°C) or higher, with rapid pulse, hot skin, headache, confusion, unconsciousness, and possible convulsions. Heat stroke is a true medical emergency that requires immediate transport to a medical facility.

The issue of dehydration should be addressed also. Fire fighters should be encouraged to drink 8 ounces to 16 ounces of water before donning any protective clothing. During rehabilitation, fire fighters should be required to rehydrate with 16 ounces of water for each SCBA tank used or consumed.

Cold Temperature Exposures

Fire fighters at hazardous materials incidents may be exposed to two types of cold temperatures: those caused by the materials and those caused by weather. Hazardous materials such as liquefied gases and cryogenic liquids expose fire fighters to the same low temperature hazards as those created by cold weather environments. Exposure to severe cold for even a short period of time may cause severe injury to body surfaces, particularly to the ears, the nose, the hands, and the feet.

Two factors in particular influence the development of cold injuries: temperature and wind speed. Because still air is a poor heat conductor, fire fighters working in low temperatures with little wind can endure these conditions for longer periods, assuming that their clothing remains dry. However, when low temperatures are combined with faster winds, wind chill occurs. As an example, if the temperature

Voices of Experience

"Upon arrival, I encountered two employees attempting to clean up the spilled sulfuric acid and the broken test tubes that had contained the chemical."

Many times, especially when responding to a research laboratory or other facility in which the use of hazardous materials is a daily operation, building occupants may get complacent.

While I was working on my fire department's hazardous materials response team, the dispatcher received an emergency call from a local laboratory indicating that several tubes of sulfuric acid had dropped to the floor and had spilled in the laboratory area. Upon arrival, I encountered two employees attempting to clean up the spilled sulfuric acid and the broken test tubes that had contained the chemical. The employees were not wearing proper protective clothing, and they did not have any respiratory protection.

As the first responder on the scene, I immediately set up an emergency decontamination area for the employees who may have been exposed to the chemical, and I established control zones. I requested additional resources and began the process for evacuation of the laboratory and shutdown of the HVAC system to prevent the spread of harmful vapors and gases. Although contained in the laboratory area, the spill had made contact with other products on the scene. The laboratory personnel had not noticed the additional contamination.

No one was seriously injured at this incident, but the importance of an emergency plan for those who work with hazardous materials cannot be stressed enough. Many times, employees who work with hazardous materials feel they can handle any emergency situation involving these products. This is simply not the case. Hazardous materials can cause serious injury to anyone, regardless of the person's level of comfort or experience with such chemicals.

Tracy E. Rickman
Rio Hondo College Fire Academy
Whittier, California

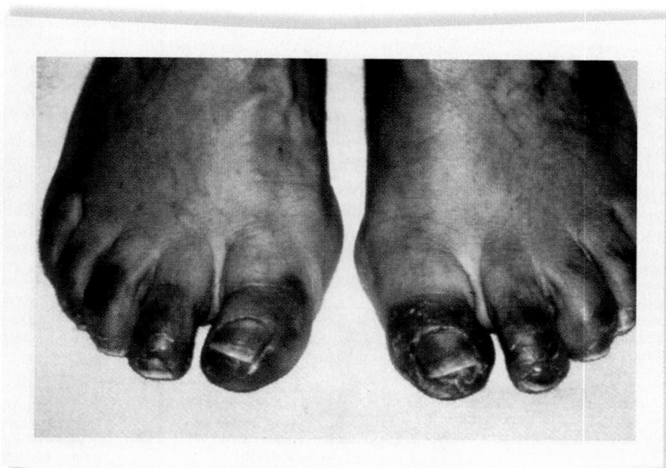

Figure 31-10 Trench foot can result when wet socks are worn at long-term emergencies in cool environments.

with no wind is 20°F, it will feel like -5°F when the wind speed is 15 mph.

Regardless of the temperature, fire fighters will perspire heavily in impermeable chemical-protective clothing. When this wet inner clothing is removed during the decontamination process, particularly in a cold environment, the body can cool rapidly. Wet clothing extracts heat from the body up to 240 times faster than dry clothing. It may also lead to hypothermia, a condition in which the body temperature falls below 95°F (35°C). Hypothermia is a true medical emergency.

Fire fighters must be aware of the dangers of frostbite, hypothermia, and impaired ability to work when temperatures are low, when the wind chill factor is low, or when they are working in wet clothing. All personnel should wear appropriate, layered clothing and should be able to warm themselves in heated shelters or vehicles such as buses.

The layer of clothing next to the skin, especially the socks, should be kept dry. Trench foot can result when wet socks are worn at long-term emergencies in cool (not cold) environments (▲ **Figure 31-10**). Fire fighters should carefully schedule their work and rest periods and monitor their physical working conditions. Warm shelters should be available for donning and doffing protective clothing.

Physical Capability Requirements

Hazardous materials response operations put a great deal of both physiological and psychological stress on responders. During the incident, they may be exposed to both chemical and physical hazards. They may face life-threatening emergencies, such as fire and explosions, or develop heat stress while wearing protective clothing or working under extreme temperatures.

For these reasons, every emergency response organization should have a health and safety management program. The components of a health and safety management system for hazardous materials responders are outlined in OSHA 1910.120, *Hazardous Waste Site Operations and Emergency Response* (HAZWOPER). Among the components of such a program are medical surveillance, personal protective equipment, and site safety practices and procedures.

A medical surveillance program is the cornerstone of an effective health and safety management system for responders. The two primary objectives of a medical surveillance program are to determine whether an individual can perform his or her assigned duties, including the use of personal protective clothing and equipment, and to detect any changes in body system functions caused by physical and/or chemical exposures.

Medical monitoring is the ongoing, systematic evaluation of response personnel who may experience adverse effects because of exposure to heat, cold, stress, or hazardous materials. Regular medical monitoring can quickly identify problems so they can be treated, thus preventing severe adverse effects and maintaining the optimal health and safety of on-scene personnel.

Fire fighters who wear SCBA should be examined by a physician at least once a year. The doctor may perform various tests such as a respiratory function test to measure lung capacity and lung function and to evaluate an individual's fitness for wearing SCBA and other respiratory protection devices.

Immediately before fire fighters don PPE to enter a hazardous materials environment, they should have their vital signs taken as part of the pre-entrance medical monitoring. This examination should also include an inspection of the individual's skin for rashes and open sores or wounds and an evaluation of the individual's mental status. Fire fighters at a hazardous materials incident should be alert, oriented to time and place, with clear speech, a normal gait, and the ability to respond appropriately. A recent medical history should be obtained, including: current weight; medications taken within the past 72 hours; alcohol consumption within the past 24 hours; any new medical treatment or diagnosis made within the past two weeks; and any symptoms of fever, nausea, vomiting, diarrhea, or cough within the past 72 hours.

This pre-entrance medical monitoring should be performed on all individuals who will wear chemical liquid splash- or vapor-protective clothing and execute hazardous materials operations. It should be completed in the hour before the fire fighter enters the hazardous environment.

Hazardous materials PPE puts significant stress on the body, particularly on the cardiovascular system. If a member of the hazardous materials team exhibits changes in gait, speech, or behavior, that team member should leave the environment for immediate decontamination, doffing of protective clothing, and assessment. Anyone who complains of chest pain, dizziness, shortness of breath, weakness, nausea, or headache should also undergo immediate decontamination, doffing of protective clothing, and assessment.

After the team members undergo decontamination, they should once again have all vital signs checked and monitored. Components of post-event medical monitoring should include history, vital signs, weight, skin evaluation, and mental status. Normal baseline values should be attained within 15 to 20 minutes after a person leaves the environment. Anyone who does not return to normal baseline entry levels should be treated and transported for follow-up monitoring.

During this post-event medical monitoring, vital signs should be checked every 5 to 10 minutes. The monitoring team should obtain information from medical control on latent reactions or symptoms and communicate this information to response personnel. If any of the symptoms appear, medical control should be contacted for direction and the affected individual should be prepared for possible transport to a medical facility.

Response Safety Procedures

One of the first things that first responders at the awareness level should do at a hazardous materials incident scene is to isolate the area and prevent anyone from entering the scene. These are basic protective actions that should be taken immediately. For the first several minutes, not even the first responders themselves should enter the area. During this time, first responders can gather necessary scene information and identify the materials involved. Any actions taken should follow the predetermined local response plan or guideline. For example, local and state officials, the local emergency planning committee, and other federal response agencies should be notified as required.

Figure 31-11 Isolate the area by using banner tape.

Any possible ignition sources should be eliminated, particularly if the incident involves the release or probable release of flammable materials. Whenever possible, electrical devices used in the immediate area of the hazard should be certified as safe by recognized organizations. Some flammable vapors are heavier than air and will travel along the ground to an ignition source.

The first step in gaining control of a hazardous materials incident is to isolate the problem and keep people away (▲ Figure 31-11). The incident commander (IC) cannot begin extended operations until the hazard area is identified and the perimeter is secured. Fire fighters can help to isolate the area by stretching banner tape across roads to divert traffic away from a spill and by coordinating their efforts with police, EMS, and other agencies.

Control Zones

Isolating the incident helps to limit the number of civilian and public service personnel who may be exposed to the hazardous materials. This can be done through a series of control zones around the incident site. Control zones are designated areas at a hazardous materials incident based upon safety and the degree of hazard (► Figure 31-12). Control zones should not be defined too narrowly. As the IC gets more information about the specifics of the chemical or material involved, the control zones can be reevaluated. If the incident is not as severe as first reported, the IC can readjust the control zones. It is always easier and safer to reduce the size of the control zones than to discover that they need to be larger.

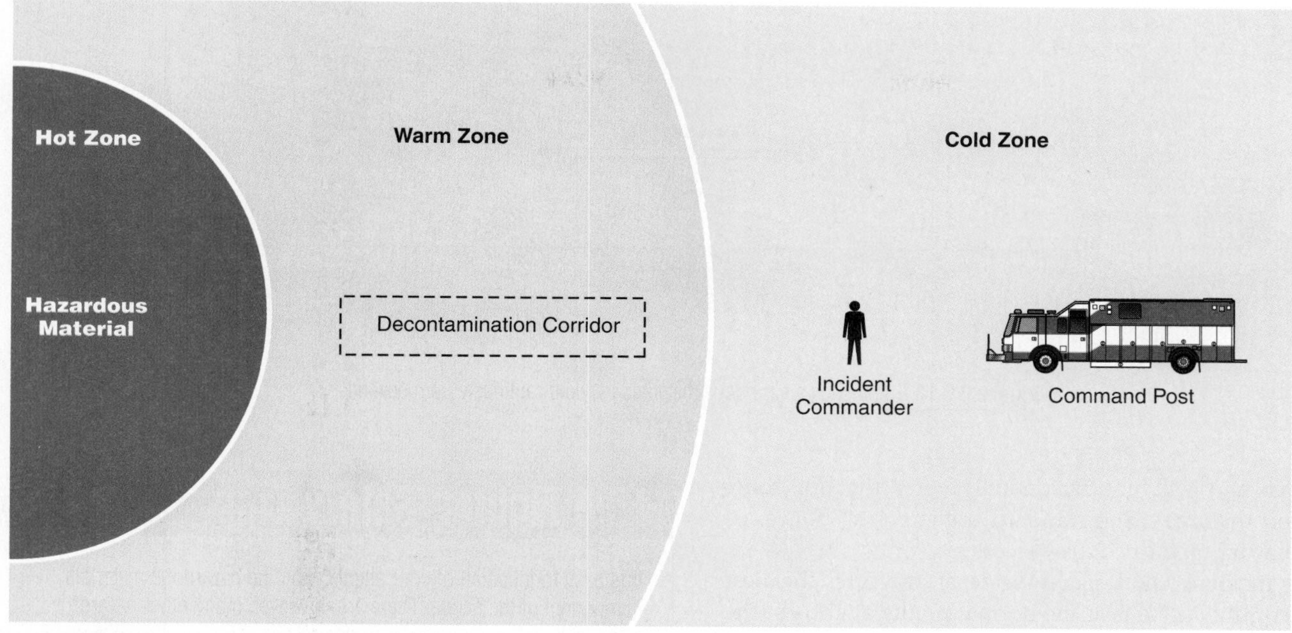

Figure 31-12 Control zones spread outward from the center of a hazardous materials incident.

Control zones are labeled as hot, warm, or cold. They may also be color-coded, such as red, yellow, and blue. The **hot zone** is the area immediately around and adjacent to the incident. Although the size of the hot zone will depend on the incident and the nature of the product, it should be large enough that adverse effects from the hazardous materials will not affect people outside the zone. An incident that involves a chemical gas will require a larger hot zone than one involving a liquid leak.

The hot zone should cover the incident site as well as the personnel and equipment necessary to control it. All personnel and equipment in the hot zone must be decontaminated when they leave the area, and personnel should be closely monitored. Access into the hot zone must be limited to those persons necessary to control the incident. Individuals must log in at the access control point, recording their entry and exit times. Only individuals trained to the technician level should be permitted in the hot zone.

The **warm zone** is the area where personnel and equipment are staged before entering and after leaving the hot zone. It contains control points for access corridors as well as the decontamination corridor. This area helps to reduce the spread of contamination. Only the minimal amount of personnel and equipment necessary to perform a task or support those operating in the hot zone should be permitted in the warm zone. Personnel trained to the operations level can work with moderate supervision in the warm zone.

Generally, personnel working in the warm zone can use PPE that is one level below the PPE used in the hot zone. If personnel in the hot zone are using Level A encapsulated protective suits, personnel in the warm zone might dress in Level B protection.

Beyond the warm zone is the **cold zone**. The cold zone is a safe area where personnel do not need to wear any special protective clothing for safe operation. Personnel staging, the command post, EMS providers, and an area for medical treatment after decontamination are all located in the cold zone.

Fire Fighter Safety Tips

The first step in establishing control zones is to immediately isolate the incident scene and deny entry to all personnel.

Isolation Techniques

Whenever possible, approach a hazardous materials incident cautiously from upwind (► Figure 31-13). Resist the urge to rush in; do not attempt to help others until the situation has been fully assessed. Without entering the immediate hazard area, isolate the scene so that the safety of people and the environment are assured. Keep people away from the scene and outside the safety perimeter. Only those directly involved in the emergency response operations should be permitted in the area, and even they must wait until control has been established and the situation assessed.

There are several ways to isolate the hazard area and create the control zones. The first step that a first responder should take is to identify the area verbally over the radio. This helps ensure that those who arrive later will remain in

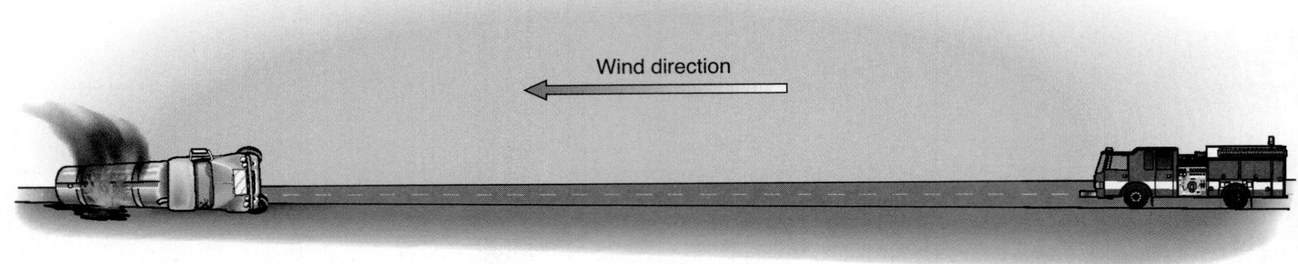

Figure 31-13 Approach a hazardous materials incident cautiously from upwind.

a safe area and will not accidentally enter the hot zone. Cordon off the area using standard traffic cones, police, or fire-line barrier tape to control access.

If the incident takes place inside a structure, the best place to control access is at the normal points of entry—the doors. Once the doors are secured so that no unauthorized personnel can enter the facility, appropriately trained emergency response crews can begin to isolate above and below the hazard.

The same concept applies to outdoor incidents. Secure the entry points and the area around the hazard. Begin by controlling intersections, on/off ramps, service roads, or other access routes to the scene.

Police officers should assist by diverting traffic at a safe distance outside the hazard area. They should block off streets, close intersections, or use their vehicles to redirect traffic. During a long-term incident, the highway department can set up traffic barriers. Crowd-control devices used by the police department can be used to identify the hazard zones and restrict access. Whatever devices are used to restrict access should not limit or inhibit rapid withdrawal from the area by personnel working inside the hot zone.

The Buddy System and Back-up Personnel

As discussed in Chapter 2, Fire Fighter Qualifications and Safety, the PASS and buddy systems are essential for fire fighter safety. On a hazardous materials incident, back-up personnel are also used to ensure the safety of emergency crews.

The **back-up personnel** remove those working in the hot zone if an emergency occurs. Back-up personnel must be in place and ready whenever personnel are operating in the hot zone.

All personnel must be briefed before approaching the hazard area or entering the hot zone. The decontamination team must be in place before anyone enters the hot zone. No one should enter the hot zone alone; at least two team members should suit up and enter the hot zone together. At least two additional team members should suit up and stand by to assist those who entered first. The IC should ensure that back-up personnel are prepared for immediate entry into the hot zone.

Team members should always remain within sight, sound, or touch of each other. Electronic communication devices may fail, or the individual needing rescue may not be physically able to summon assistance. Radios should be used for communications at hazardous materials incident sites. They cannot, however, replace visual or voice contact or be the only way to contact an individual working in the hot zone.

Fire Fighter Safety Tips

It is easy to focus all of your attention on the hazardous materials component of the scene. There are, however, many other hazards associated with emergency incident scenes. Watch out for downed power lines, jagged edges of metal, and tripping hazards. Do not get tunnel vision; consider all types of hazards.

Wrap-Up

Ready for Review

This chapter discussed the safety precautions that fire fighters must take at a hazardous materials incident. Personal protective equipment is the last line of defense and must be suited to both the type of incident and the activities the fire fighter is expected to perform. Various protective ensembles were described and the protection levels established by the Environmental Protection Agency were defined. The types of respiratory protection necessary at different incidents were also discussed.

All PPE has limitations that fire fighters should recognize. This chapter discussed the safety precautions that fire fighters should take when wearing PPE to prevent skin contact hazards, heat disorders, and cold injuries. In addition, this chapter reviewed the stresses that PPE places on the body and emphasized the importance of monitoring vital signs before, during, and after an incident.

Finally, this chapter outlined the safety procedures that fire fighters should take at an incident. It emphasized the need to isolate the incident and deny entry to protect people and the environment. It discussed the importance of establishing control zones and employing both the buddy system and back-up personnel to ensure the safety of individuals working at a hazardous materials incident.

Chief Concepts

- Proper personal protective equipment is critical to surviving a hazardous materials incident.
- The severity of the hazardous materials incident indicates the type of PPE to don.
- The EPA has defined four levels of hazardous materials PPE: Level A, Level B, Level C, and Level D.
- Awareness-level responders should ensure scene safety by setting up a barrier around the scene to keep out bystanders and those who are not properly trained or outfitted.
- The three levels of control around a hazardous materials incident are: cold zone, warm zone, and hot zone.
- Always use the buddy system.

Hot Terms

Air-purifying respirators (APRs) Devices worn to filter particulates and contaminants from the air.

Back-up personnel Individuals who remove or rescue those working in the hot zone in the event of an emergency.

Chemical-resistant materials Chemical-protective clothing that can maintain its integrity and protection qualities when it comes into contact with a hazardous material. They also resist penetration, permeation, and degradation.

Control zones Areas at a hazardous materials incident that are designated as hot, warm, or cold, based upon safety and the degree of hazard.

Cold zone A safe area at a hazardous materials incident for those agencies involved in the operations; the incident commander, command post, EMS providers, and other support functions necessary to control the incident should be located in the cold zone. This zone may also be referred to as the clean zone or the support zone.

Degradation The physical destruction or decomposition of clothing material due to chemical exposure, general use, or ambient conditions (like storage in sunlight).

Heat exhaustion A mild form of shock caused when the circulatory system begins to fail as a result of the body's inadequate effort to give off excessive heat.

Heat stroke A severe and sometimes fatal condition resulting from the failure of the temperature regulating capacity of the body. It is caused by prolonged exposure to the sun or high temperatures. Reduction or cessation of sweating is an early symptom; body temperature of 105°F or higher, rapid pulse, hot and dry skin, headache, confusion, unconsciousness, and convulsions may occur. Heat stroke is a true medical emergency requiring immediate transport to a medical facility.

High temperature-protective equipment Protective clothing designed to shield the wearer during short-term high temperature exposures.

Hot zone The area immediately surrounding a hazardous materials spill/incident site that is directly dangerous to life and health. All personnel working in the hot zone must wear complete, appropriate protective clothing and equipment. Entry requires approval by the IC or a designated sector officer. Complete back-up, rescue, and decontamination teams must be in place at the perimeter before operations begin.

Wrap-Up

Immediate Danger to Life and Health (IDLH) An atmospheric concentration of any toxic, corrosive, or asphyxiant substance that poses an immediate threat to life or could cause irreversible or delayed adverse health effects. There are three general IDLH atmospheres: toxic, flammable, and oxygen-deficient.

Level A protection PPE that provides protection against vapors, gases, mists, and even dusts—the highest level of protection; requires a totally encapsulating suit that includes SCBA.

Level B protection PPE used when the type and atmospheric concentration of substances have been identified. Generally requires a high level of respiratory protection but less skin protection: chemical-protective coveralls and clothing, chemical protection for shoes, gloves, and SCBA outside of a nonencapsulating chemical-protective suit.

Level C protection PPE used when the type of airborne substance is known, the concentration is measured, the criteria for using air-purifying respirators are met, and skin and eye exposure is unlikely. Consists of standard work clothing with the addition of chemical-protective clothing, chemically resistant gloves, and a form of respirator protection.

Level D protection PPE used when the atmosphere contains no known hazard, and work functions preclude splashes, immersion, or the potential for unexpected inhalation of or contact with hazardous levels of chemicals. It is primarily a work uniform that includes coveralls and affords minimal protection.

Liquid splash-protective clothing Several pieces of clothing and equipment designed to protect the skin and eyes from chemical splashes.

Penetration The flow or movement of a hazardous liquid (or solid) chemical through zippers, stitched seams, buttonholes, flaps, pinholes, or other imperfections in a material. Liquids and solids such as asbestos can penetrate chemical-protective clothing at several locations, including the face piece, exhalation valve, suit exhaust valve, and suit fasteners.

Permeation The process by which a hazardous liquid chemical moves through a given material on the molecular level. Permeation differs from penetration in that permeation occurs through the clothing material itself rather than through openings in the clothing material.

Permissible Exposure Limit (PEL) The established standard limit of exposure to a hazardous material. It is based on the maximum time-weighted concentration at which 95% of exposed, healthy adults suffer no adverse effects over a 40-hour work week.

Recommended Exposure Level (REL) The established standard limit of exposure to a hazardous material. It is based on the maximum time-weighted concentration at which 95% of exposed, healthy adults suffer no adverse effects over a 40-hour work week. Also knows as the Permissible Exposure Limit (PEL).

Supplied-air respirator (SAR) A respirator that gets its air through a hose from a remote source, such as a compressor or storage cylinder.

Threshold Limit Value/Ceiling (TLV/C) The maximum concentration of hazardous material that a worker should not be exposed to, even for an instant.

Threshold Limit Value/Short-Term Exposure Limit (TLV/STEL) The maximum concentration of hazardous material to which a worker can sustain a 15-minute exposure not more than four times daily with a 60-minute rest period required between each exposure. The lower the value, the more toxic the substance.

Threshold Limit Value/Skin Indicates that direct or airborne contact with a material could result in possible and significant exposure from absorption through the skin, mucous membranes, and eyes.

Threshold Limit Value/Time Weighted Average (TLV/ TWA) The airborne concentration of a material to which a worker can be exposed for 8 hours a day, 40 hours a week and not suffer any ill effects.

Vapor-protective clothing The garment portion of a chemical-protective ensemble designed and configured to protect the wearer against chemical vapors or gases.

Warm zone The area located between the hot zone and the cold zone at the incident. PPE is required. The decontamination corridor is located in the warm zone.

Fire Fighter in Action

It is 4:00 a.m. when your engine is dispatched to a report of a 2000-gallon PCB tank that is leaking at an electrical plant storage site. PCB is a toxic chemical that was banned in the U.S. in the late 1970s, but can still be found in many older electrical devices. This chemical is very harmful to the environment, and can cause skin irritations, liver damage, nausea, dizziness, and eye irritation. It appears that approximately 500 gallons of PCB has spilled onto the asphalt and is entering a storm drain. You and your crew must establish control zones to limit exposure to this dangerous chemical.

1. The warm zone is the area:
   A. immediately around and adjacent to the incident.
   B. where personnel and equipment are staged before they enter the hot zone.
   C. where personnel do not need to wear any special protective clothing for safe operation.
   D. where incident command is located.

2. An atmospheric concentration of any toxic, corrosive, or asphyxiant substance that poses an immediate threat to life or could cause irreversible or delayed adverse health effects known as:
   A. IDLH.
   B. TLV/TWA.
   C. PEL.
   D. TLV/STEL.

The hazardous materials team arrives on the scene. They prepare to don the appropriate PPE for the chemical. Appropriate PPE for this incident should be made of a chemical-resistant material designed to inhibit or resist the passage of chemicals into and through the material.

3. The process by which a hazardous chemical moves through a given material on the molecular level is called:
   A. penetration.
   B. degradation.
   C. permeation.
   D. none of the above

4. What level of PPE protection consists of fully encapsulating chemical-resistant suit, boots, gloves, and SCBA?
   A. Level A
   B. Level B
   C. Level C
   D. Level D

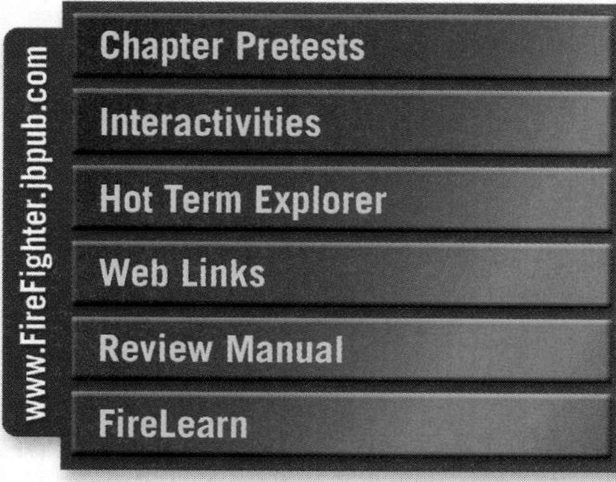

Hazardous Materials: Response Priorities and Actions

Technology Resources

www.FireFighter.jbpub.com

- Chapter Pretests
- Interactivities
- Hot Term Explorer
- Web Links
- Review Manual
- FireLearn

Chapter Features

- Skill Drills
- Voices of Experience
- Fire Marks
- Teamwork Tips
- Fire Fighter Safety Tips
- Fire Fighter Tips
- Canadian Perspectives
- Hot Terms
- Wrap-Up

Chapter 32

NFPA 472 Standard

5.4.4 Performing Defensive Control Actions. Given a plan of action for a hazardous materials incident within their capabilities, the first responder at the operational level shall demonstrate defensive control actions set out in the plan and shall meet the following related requirements:

(1) Using the type of fire-fighting foam or vapor suppressing agent and foam equipment furnished by the authority having jurisdiction, demonstrate the effective application of the fire-fighting foam(s) or vapor suppressing agent(s) on a spill or fire involving hazardous materials.

(2) Identify the characteristics and applicability of the following foams:

- (a) Protein
- (b) Fluoroprotein
- (c) Special purpose
  - i. Polar solvent alcohol-resistant concentrates
  - ii. Hazardous materials concentrates
- (d) Aqueous film-forming foam (AFFF)
- (e) High expansion

(3) Given the required tools and equipment, demonstrate how to perform the following defensive control activities:

- (a) Absorption
- (b) Damming
- (c) Diking
- (d) Dilution
- (e) Diversion
- (f) Retention
- (g) Vapor dispersion
- (h) Vapor suppression

(5) Describe the objectives and dangers of search and rescue missions at hazardous materials incidents

Knowledge Objectives

After studying this chapter, you will be able to:

- Discuss search and rescue missions at hazardous materials incidents.
- Discuss exposure protection.
- Discuss defensive control activities.

Skills Objectives

There are no skills objectives for this chapter.

You Are the Fire Fighter

On an extremely warm and humid summer day, a fire alarm and several telephone calls are received for an explosion and fire with a person burned at a local chemical manufacturing company. Upon your arrival, sprinklers have put out a fire that occurred when an "in the floor" mixing device exploded. Smoke is hanging in the air. The water from the sprinklers is running out of the building and is tinged with both a colored material and a powdery substance.

1. ***How would you protect the building occupants and public at this scene?***
2. ***What actions should you take to prevent further contamination from the hazardous materials?***

Risk-Benefit Assessment and Responder Safety

During the initial size-up at a hazardous materials incident, the first decision concerns personnel safety. The incident commander (IC) needs to decide whether any response would be effective to control and minimize the hazardous materials incident, or if it would unnecessarily risk harming fire fighters and responders with little or no benefit.

If an unknown hazardous material is covering a large area and already negatively affecting large numbers of residents, then the answer is to withdraw responding personnel for their safety. This decision goes against the basic nature of fire fighters, who are willing to take risks to help others. It will be difficult to accept the idea that no response action can be taken because the risk is too great or there are no living victims to save. This situation can be compounded if family members or friends are located in affected areas. However, if fire fighters or other responders expose themselves to unnecessary risk, injury, exposure, or contamination, it only complicates the problem.

Exposures

Exposures (hazardous materials) can be people, property, structures, or the environment that are subject to influence, damage, or injury. The amount of exposures that remain is determined by the location of the incident and how much progress has been made in the protection of those exposures. Incidents in urban areas will have more exposures, and will need more resources to protect those exposures.

Evacuation

Evacuating the exposed area is one of the first response priorities at the hazardous materials incident (► **Figure 32-1**).

The IC considers many factors before making decisions concerning evacuation. Evacuation has some significant risks involved even when properly planned. Once the evacuation order is given, it is the role of the fire fighter to assist in the physical evacuation of residents to a safe area. This might include traveling to homes and informing residents that they must leave their homes for a shelter.

Before an evacuation order is given, a safe area and suitable shelter are established. In many cases, schools are used for shelters. Depending on the time of day and the season, it may take a considerable amount of time to evacuate. The evacuation area should be located close enough to the exposure for the evacuation to be practical but far enough away from the incident to be safe. Temporary evacuation areas may be needed to shelter residents until evacuation sites or

Figure 32-1 Evacuating the exposed area is one of the first response priorities at the hazardous materials incident.

structures indicated in your community's emergency plan can be accessed and opened.

Transportation to temporary evacuation areas has to be arranged for all populations. The needs of handicapped and special needs persons should be considered as well. The initial areas that need to be evacuated can be determined from the *North American Emergency Response Guidebook* (NAERG). This guidebook provides the minimum basic guidelines for the initial response. The NAERG does not give complete detailed information for conditions that responders may encounter, such as weather extremes, road conditions, or actual conditions encountered at the scene.

The best way for the IC to determine an area of evacuation and how far to extend actual evacuation distances is to have hazardous materials technicians with detection devices monitoring both the concentrations of hazardous material and the direction and speed that the material is moving due to wind currents and terrain. Access to local weather conditions is invaluable in the decision-making process.

Another concern when considering evacuation is the potential for exposure to the hazardous material during the evacuation attempt. Personnel may have to go into or through the hazardous material to remove residents. Also, when the evacuated residents are removed from protected areas, they too may pass into or through the hazardous material. These people must be adequately protected from the hazardous material or they will become victims.

Shelter-in-place is a method of safeguarding people in a hazardous area by keeping them in a safe atmosphere, usually inside structures (► **Figure 32-2**). Local emergency planning should address local facilities that have large populations to evacuate, such as schools, churches, hospitals, nursing homes, and apartment complexes. When residents are sheltered in place, they remain indoors with windows and doors closed. All ventilation systems are turned off to prevent outside air from being drawn into the structure.

The toxicity of the hazardous material is a major factor in the decision whether to evacuate or to shelter in place. The more toxic the substance, the less likely a shelter-in-place strategy should be used. The expected duration of the incident is also a factor in determining whether sheltering in place is a viable option. The longer the incident, the more time the material has to enter or permeate into protected areas.

Figure 32-2 Shelter-in-place.

The ability to evacuate an area and staff to an adequate emergency shelter will be determined by the local resources available. There is no mathematical formula to calculate how many residents can be evacuated at any given time. In severe weather, evacuation can be challenging at best. Even in perfect circumstances, evacuation of residents will be a process that is measured in hours, not minutes. Local emergency planning for evacuations is key to a safe and successful emergency evacuation.

Search and Rescue

The protection of life is the first priority in any emergency response situation, and a hazardous materials incident is no exception. The search for and rescue of people involved in a hazardous materials event is the primary concern. However, this process is more complicated in a hazardous materials incident. In a structure fire, search and rescue is done quickly to remove victims from the smoke-filled environment. In a hazardous materials incident, all emergency response personnel (fire, law enforcement, emergency medical services) must first recognize and identify which hazardous materials may be present. After this is accomplished, the incident commander (IC) must slowly and cautiously

Fire Fighter Safety Tips

When evacuating resident populations, be sure that you are properly protected in case the wind shifts or the hazardous material engulfs you suddenly. Proper personal protection must be available to don before evacuation.

Fire Fighter Tips

When a fire fighter is injured in a hazardous materials incident, he or she is not the only fire fighter who has to leave the scene. One or more fire fighters are then needed to treat the injured fire fighter and/or transport that individual to medical care.

evaluate what conditions exist before sending in personnel to attempt a search and rescue.

After identifying the hazard, the IC must determine if entry into the hot zone by technicians to perform search and rescue can be accomplished safely. If the hazardous material is highly toxic, victims may not survive the exposure to the material. There would be no need for a search if a hazardous material exposure is not survivable. If it is determined that a search can be safely performed, rescue teams wearing proper personal protective equipment (PPE) may then enter the hot zone to retrieve victims. The victims are removed to the warm zone, where they can be decontaminated and given emergency medical attention in preparation for transport to a medical facility.

Figure 32-3 Foam should be gently applied.

Exposure Protection

Exposures can be protected in different ways. One way is to remove the exposure from the threat by evacuating residents. Another way is placing a barrier to prevent the hazardous materials from reaching the exposure. The hazardous material could also be neutralized chemically, such as treating an acid with a basic material to weaken the strength of the acid. Only a responder at the technician level should perform a neutralization technique.

Confinement and Containment

Confinement is the process of attempting to keep the hazardous material on the site or within the immediate area of the release. This can usually be done by damming or diking a material, or by confining vapors to a specific area.

Containment refers to actions that stop the hazardous material from leaking or escaping. Examples of containment include patching or plugging a breached container, or righting an overturned container. Only a responder at the technician level should patch or plug a container. When considering a containment response, certain factors must be evaluated, such as the maximum quantity of material that can be released, and what the likely duration of the incident would be without intervention.

Fire Extinguishment

Fires associated with a hazardous materials incident should be addressed with great caution. The principles of extinguishment should be dealt with in a systematic manner. First and foremost is to recognize and identify the hazardous material or materials involved. Some hazardous materials react violently if water is applied. This is important to know because water is the most commonly used firefighting agent. After identification is made and confirmed, the physical properties of the material, such as specific gravity, flash point, and flammable ranges, should be researched.

Most flammable and combustible liquid fires can be extinguished by the use of foam. (A complete discussion of foam appears in Chapter 16, Fire Hose, Nozzles, Streams, and Foam.) All of the foams listed below suppress the ignition of flammable vapors. Remember that as foam is applied, more volume will be added to the spill. There are several types of firefighting foams available:

- Aqueous film-forming foam (AFFF)
- Fluoroprotein
- Special purpose (polar solvent alcohol-resistant and hazardous materials concentrates)
- Protein
- High expansion

Foam should be gently applied or bounced off another adjacent object so that it flows down across the liquid and does not directly upset the burning surface (▲ Figure 32-3). Foam can also be applied in a rain-down method by directing the stream into the air over the material and letting foam gently fall as rain would. This is less effective when the fire is creating an intense thermal column. When the heat from the thermal column is high, the water content of the foam will turn to steam before it contacts the liquid surface. Foam can also be applied in the roll-on method by bouncing the stream directly into the front of the spill area and allowing it to gently push forward into the pool rather than directly "splashing" it in.

Defensive Control Activities

As fire fighters approach a hazardous materials incident, they should look for natural control points—areas in the terrain or places in a structure where materials such as liquids and powders or solids can be contained. Hazardous materials technicians should use monitoring equipment to determine where a material has traveled. Some natural barriers that might be used include doors to a room, doors to a building, and curb areas of roadways. A number of defensive control measures exist that operational personnel can use effectively.

Absorption

Absorption is the process of applying a material that will soak up and hold the hazardous material as a sponge holds water. This makes collection and disposal of a hazardous material in liquid form more manageable. This can be simply explained as adding a dry, granular, clay-based material or dry sand to a spill to help contain a spilled product. This should be done by distributing the absorbent material from a distance by shovels. The technique of absorption is difficult for operational personnel because it generally involves being in close proximity to the spill. It also involves the addition of material to a spilled product, which adds volume to the spill. The absorbent material can also react with certain hazardous substances. It is important to determine if an absorbent substance is compatible with the hazardous material before it is used. Tools such as shovels used for absorbent dispersal and pickup need to be nonsparking to prevent ignition of the spilled material. Fire departments need to plan for the proper disposal of contaminated materials collected from these types of incidents.

Diking, Damming, Diversion, and Retention

Controlling liquid spills can be difficult. Several techniques can be effective. **Diking** is the placement of materials to form a barrier that will keep a hazardous material in liquid form from entering an area, or hold the material in an area. **Damming** is used when liquid is flowing in a natural channel or depression and its progress can be stopped by blocking the channel. A **diversion** technique is used to redirect the flow of a liquid away from an area. Liquid hazardous materials spills can be controlled by using existing barriers such as curbs and channeling liquids away from storm drains that would allow them to leave the site. Diversion could also be used to protect sensitive environmental areas.

Retention is the process of creating an area to hold hazardous materials. For instance, this can be done by digging a depression and allowing material to pool in the depression. It is held there until a clean-up contractor can recover it. A hazardous material that is leaking in a room of an industrial facility might be safely approachable after proper recognition and identification by operational personnel wearing the proper level of PPE, who could close a door and contain the spill within one room. As another example, the public works department or an independent contractor might use a backhoe to create a retention area for runoff at a safe downhill distance from the release.

Dilution

Another protective action that can be taken from a safe distance is dilution. **Dilution** is the addition of water or another liquid to weaken the strength or concentration of a hazardous material. Dilution can only be used when the identity and properties of the hazardous material are known with certainty. One concern with dilution is that water applied to dilute a hazardous material increases the total volume and may overwhelm containment measures.

Vapor Dispersion and Suppression

When a hazardous material produces a vapor that collects in an area or increases in concentration, vapor dispersion or suppression is called for. **Vapor dispersion (hazardous materials)** is the process of lowering the concentration of vapors by spreading it out. Vapors can be dispersed with hose streams set on fog spray, large displacement fans, or heating and cooling systems. Before dispersing vapors, the consequences should be considered. If the vapors are highly flammable, an attempt at dispersing vapors can ignite them. The dispersed vapors can also contaminate areas outside the hot zone. Vapor dispersion with fans or a fog stream can be attempted only after the hazardous material is safely identified and in weather conditions that will promote the diversion of vapors away from any populated areas.

Vapor suppression is the technique of controlling fumes or vapors that are given off by certain materials, such as flammable liquid fuels. For example, gasoline is a material that gives off vapors. These vapors are dangerous because they are flammable. To control these vapors, a blanket of foam is

Fire Fighter Tips

Remember to take rehabilitation breaks throughout the hazardous materials incident. Participating in defensive operations takes a great deal of physical energy and mental concentration.

Voices of Experience

"High- and low-pressure cylinders of all sizes and containing many different products were launching like rockets from the facility."

When I arrived at an industrial gas manufacturing plant fire, the scene was something that most of us only see in the movies. High- and low-pressure cylinders of all sizes and containing many different products were launching like rockets from the facility. Many of them were exploding at the same time. Complicating matters, we had 150 to 200 citizen onlookers in the immediate vicinity, with shrapnel falling all around them. Protecting onlookers and emergency response personnel became our first priority. Even with five engines, three trucks, and two battalion chiefs initially dispatched, we knew this would be a challenge.

Arriving companies were immediately ordered to pull back a minimum of one block and begin evacuating the onlookers. To accomplish this, we used the public-address systems on our fire department vehicles to direct bystanders and residents to safe areas. Fire department commanders spoke with a lieutenant from the Sacramento Police Department and devised a plan to have law enforcement officers from all local agencies respond to the scene and evacuate everyone within half a mile of the incident. Within 20 minutes, the police department had between 35 and 45 officers on the scene, assisting with evacuation efforts.

After the incident, the command staff called for a situation and statistic report from our division commanders. This was when we learned some truly amazing details. The incident had been controlled in approximately 2.5 hours. Many buildings were damaged, but all were repairable. Not one piece of equipment suffered any damage. But the most amazing fact of all was that absolutely no civilians, law enforcement officers, or fire fighters were injured during the incident. Considering that the area surrounding the plant had been bombarded with rocketing cylinders and falling debris and shrapnel—some pieces hitting the ground several blocks away—this was quite a feat.

When faced with an extraordinary challenge, our companies did exactly what they were trained to do. Maintaining discipline regarding operating procedures is extremely important, and our companies did an excellent job of utilizing the incident command system and working with other emergency response personnel.

Lloyd Ogan
Sacramento Fire Department
Sacramento, California

placed on the surface of the liquid. The foam seals the surface of the liquid, prevents vapors from being given off, and thus reduces the danger of vapor ignition. Reducing the temperature of some hazardous materials will also suppress vapor formation.

Decision to Withdraw

After all factors have been considered, the IC may decide that the hazardous materials incident cannot be handled without unnecessary risks to all personnel. Either the resources are not available to mitigate the incident or there are no exposures to be saved. The IC may decide to withdraw to a safe distance, set a defensive perimeter, and wait for the additional resources to arrive—or to allow the hazardous materials incident to run its course. This is the case with certain chemicals, such as pesticides, which are consumed in a hot fire and therefore may be allowed to burn. The decision to extinguish a fire involving pesticides could result in runoff contamination of surrounding areas, and worsen the impact of the incident.

Recovery

The recovery phase of a hazardous materials incident occurs when the imminent danger to people, property, and the environment has passed or is controlled, and cleanup begins (► Figure 32-4). During the recovery phase, the appropriate state and federal agencies may become involved in cleaning up the site, determining the responsible party, and implementing cost-recovery methods. There is a transition between on-scene public safety responders and private-sector commercial cleanup companies. It must be stressed that the hazardous materials incident is not over at this point; as long as public safety personnel and responders are on scene, they should maintain a safety vigilance to ensure there are no hazards created or injuries incurred during this final phase.

Figure 32-4 Cleanup occurs during the recovery phase.

When to Terminate the Incident

The decision to terminate the hazardous materials incident is made by the IC. The recovery phase in large-scale incidents can go on for days, weeks, or months. Remember that our primary responsibility is for the safety of fire fighters. The recovery phase and cleanup period involves amounts of resources and equipment that are far beyond the capabilities of local responders. The ultimate goal of incident termination is to return the property or site of the incident to a preincident condition, then return the facility or mode of transportation to the responsible party. This often takes place long after fire fighters are released from the scene. The recovery phase includes completion of the records necessary for documenting the incident.

Wrap-Up

Ready for Review

The options for mitigating a hazardous materials incident include evacuation, confinement, containment, absorption, dilution, and dispersion. These options are acceptable practices that have been tried before. The concerns about evacuation versus sheltering in place are some of the most serious concerns that will face an IC. As difficult as the evacuation might be, a no-involvement decision may be even harder to make or accept. It is possible that the hazardous materials incident may progress and literally be over before you can intervene, or the incident could be so potentially deadly that fire fighters and responders should undertake no action. The concerns of life safety must weigh on the side of the fire fighters.

Chief Concepts

- The decision to evacuate exposed residents or shelter them in place involves multiple factors. The evacuation of resident populations will be complicated and time-consuming.
- Special foams can help in both prevention and extinguishment of fires associated with hazardous material releases. By aiding in vapor suppression, foam can also serve a more general safety function.
- Defensive actions generally aim to contain, redirect, or lower the concentration of the hazardous material involved, and to prevent its ignition.
- The recovery phase may last for days or months and aims to return the exposure area to its original condition.

Hot Terms

Absorption The process of applying a material that will soak up and hold a hazardous material in a sponge-like manner, for collection and disposal.

Confinement The process of keeping a hazardous material on the site or within the immediate area of the release.

Containment Actions relating to stopping a hazardous materials container from leaking, such as patching, plugging, or righting the container.

Damming A process used when liquid is flowing in a natural channel or depression and its progress can be stopped by blocking the channel.

Diking The placement of materials to form a barrier that will keep a hazardous material in liquid form from entering an area, or hold the material in an area.

Dilution The process of adding some substance, usually water, to a product to weaken its concentration.

Diversion Redirecting spilled or leaking material to an area where it will have less impact.

Exposures (hazardous materials) People, property, structures, or areas of the environment that are subject to influence, damage, or injury from hazardous materials.

Recovery phase The stage of a hazardous materials incident after imminent danger has passed, when cleanup and the return to normalcy have begun.

Retention The process of purposefully collecting hazardous materials in a defined area; may include creating that area, by digging, for instance.

Shelter-in-place A method of safeguarding people in a hazardous area by keeping them in a safe atmosphere, usually inside structures.

Vapor dispersion (hazardous materials) The process of lowering the concentration of vapors by spreading them out.

Vapor suppression The process of controlling vapors given off by hazardous materials, and therefore preventing vapor ignition, by covering the product with foam or other material or by reducing the temperature of the material.

Fire Fighter in Action

It is 11:00 a.m. on a snowy day in January. Your engine is dispatched to the scene of a rollover cargo truck on a nearby highway. Upon arrival, you find a cargo truck on its side with a granular substance spilling out of the trailer. You observe a placard with the number 1823 on the side of the truck and determine that the substance is caustic soda. The accident has occurred approximately 100' from an elementary school that is in session despite the snow.

1. While conducting the risk-benefit assessment, the first decision concerns:
   - A. the driver of the truck.
   - B. the school children.
   - C. personnel safety.
   - D. the environment.
2. Exposures in this incident include:
   - A. the staff and students at the school.
   - B. the school building itself.
   - C. the area immediately surrounding the truck.
   - D. all of the above

The incident commander determines that the best way to safeguard the staff and students at the school from exposure to the caustic soda is the shelter-in-place method.

3. Sheltering in place means to:
   - A. keep people in a safe atmosphere by keeping them inside structures.
   - B. move people to a safe area and suitable shelter.
   - C. purposely collect hazardous material in a defined area.
   - D. contain the hazardous material within the site or within the immediate area of release.

Cleaning up the caustic soda is likely to take some time, and will require the presence of public safety personnel and responders to ensure that no further hazards are created or injuries are incurred.

4. Once clean-up operations are completed, the decision to terminate this incident can only be made by:
   - A. the fire chief.
   - B. the hazardous materials team.
   - C. the safety officer.
   - D. the incident commander.

www.FireFighter.jbpub.com

Chapter Pretests
Interactivities
Hot Term Explorer
Web Links
Review Manual
FireLearn

www.FireFighter.jbpub.com

Hazardous Materials: Decontamination Techniques

NFPA 472 Standard

5.3.4* Identifying Emergency Decontamination Procedures. The first responder at the operational level shall identify emergency decontamination procedures and shall meet the following requirements:

(1) Identify ways that personnel, personal protective equipment, apparatus, tools, and equipment become contaminated.

(2) Describe how the potential for secondary contamination determines the need for emergency decontamination procedures.

(3) Identify the purpose of emergency decontamination procedures at hazardous materials incidents.

5.4.1 Establishing and Enforcing Scene Control Procedures. Given scenarios for facility and/or transportation hazardous materials incidents, the first responder at the operational level shall identify how to establish and enforce scene control including control zones, emergency decontamination, and communications and shall meet the following requirements:

(5)* Demonstrate the ability to perform emergency decontamination.

Knowledge Objectives

After studying this chapter, you will be able to:

- Describe how the potential for secondary contamination determines the need for emergency decontamination.
- Identify the types of decontamination.
- Identify emergency decontamination procedures.
- Identify where and how decontamination takes place.

Skills Objectives

After studying this chapter, you will be able to:

- Perform decontamination at a hazardous materials incident.

You Are the Fire Fighter

Your engine is dispatched to assist the hazardous materials team at the scene of a vehicle accident. A cargo van that was carrying several small containers of chlorine has overturned. Some of the containers ruptured in the roll over are leaking. The driver was unhurt in the accident and escaped without being exposed. The hazardous materials technicians are in the hot zone securing the remaining containers. Upon arrival your crew is assigned to set-up the decontamination corridor.

1. ***What method of decontamination would be most appropriate?***
2. ***Where should the decontamination corridor be constructed?***

What Is Decontamination?

Decontamination is the physical or chemical process of reducing and preventing the spread of hazardous materials by persons and equipment. Decontamination makes personnel, equipment, and supplies safe by removing or eliminating hazardous materials. Proper decontamination is essential at every hazardous materials incident to ensure the safety of personnel and property.

Secondary Contamination

Contamination is the process of transferring a hazardous material from its source to people, animals, the environment, or equipment, which may act as carriers of the contaminant. Secondary contamination (also known as cross contamination) occurs when a contaminated person or object comes into direct contact with another person or object. Secondary contamination may occur in several ways:

- A contaminated victim comes into physical contact with a fire fighter.
- A bystander or fire fighter comes into contact with a contaminated object from the hot zone.
- A decontaminated fire fighter re-enters the decontamination area and comes into contact with a contaminated fire fighter or object.

Because secondary contamination can occur in so many ways, control zones should be established, clearly marked, and enforced at hazardous materials incidents. The more people who are exposed, either directly or indirectly, to the contaminant, the larger the decontamination problem. A rapid, focused decontamination operation depends on an understanding of the contaminant and its chemical properties.

Types of Decontamination

Decontamination makes personnel, equipment, and supplies safe by removing or eliminating the hazardous materials. Government and industrial agencies are responsible for decontaminating the environment, while specialists decontaminate equipment such as collected tools. Fire fighters are responsible for establishing a decontamination corridor for the initial emergency response crews and victims. The decontamination corridor is a controlled area, usually within the warm zone, where decontamination procedures take place. Everyone who leaves the hot zone is required to pass through the decontamination corridor as they exit.

The major categories of decontamination are emergency decontamination, gross decontamination, formal decontamination, and fine decontamination. Each category provides a different level of decontamination.

Emergency Decontamination

Emergency decontamination is used in potentially life-threatening situations to rapidly remove most of the contaminants from an individual, regardless of a formal decontamination corridor. A more formal and detailed decontamination process may follow later. Emergency

Fire Fighter Safety Tips

Identify the hazardous material before beginning decontamination. Some hazardous materials are water-reactive and will require a special decontamination process.

Fire Fighter Safety Tips

The best method of decontamination for fire fighters and first responders is to avoid contact with the hazardous material. If you don't become contaminated, there's no need to decontaminate!

decontamination usually involves removing contaminated clothing and dousing the victim with quantities of water. If a decontamination corridor has not yet been established, isolate the exposed victims in a contained area and establish an emergency decontamination area. Do not allow the water runoff to flow into drains, streams, or ponds; try to divert it into an area where it can be treated and/or disposed of later. Do not delay decontamination; human life always comes first.

Gross Decontamination

Gross decontamination, like emergency decontamination, aims to significantly reduce the amount of surface contaminant by a continuous shower of water and removal of outer clothing. It differs from emergency decontamination, however, in that it is controlled through the decontamination corridor. Some hazardous materials teams use a large shower system for gross decontamination as the first part of a multistep decontamination process or line (► Figure 33-1).

The shower system works well if water is appropriate for removing the hazardous material. The victim or fire fighter, still fully clothed or wearing personal protective equipment (PPE) and self-contained breathing apparatus (SCBA), steps into a portable shower. High-pressure, low-volume water flow (or spray) is used to rinse off and dilute the contaminants. This step will remove most solid or liquid contaminants from outer clothing, PPE, and SCBA. Runoff water should be diverted to a safe area for treatment and/or disposal and should not be allowed to flow into drains, streams, and ponds.

After most of the exterior contaminants have been removed, the person will need to remove outer clothing before proceeding to the next step in the decontamination process. A fire fighter wearing fully encapsulated PPE should

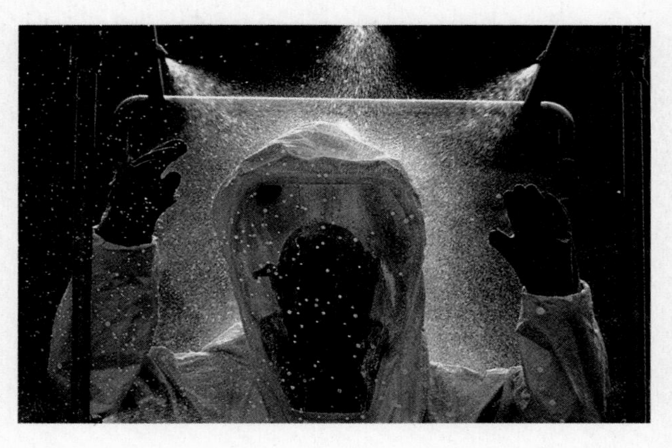

Figure 33-1 A shower system can be used for gross decontamination.

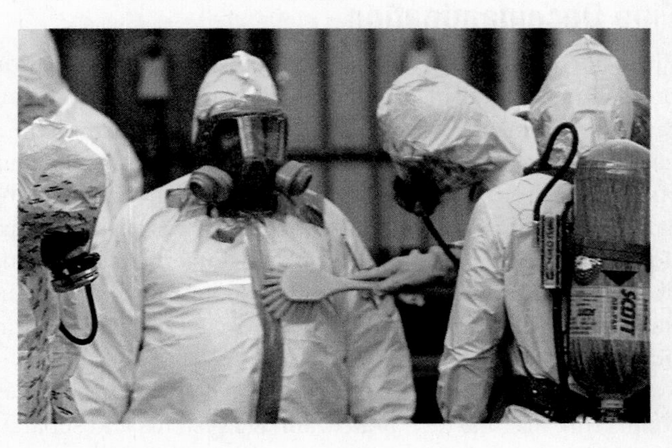

Figure 33-2 Formal decontamination is a more thorough process performed after gross decontamination.

remove the suit but continue to use SCBA to ensure adequate respiratory protection.

Formal Decontamination

Formal decontamination is performed after gross decontamination and is a more thorough cleaning process (▲ Figure 33-2). Formal decontamination may involve several stations or steps. Multiple personnel (the decontamination team) use brushes and/or swabs to scrub and wash the person or object to remove contaminants.

To clean a fire fighter's PPE, for example, the decontamination team would use water spray, long-handled scrub brushes, and a special cleaning solution. The type of cleaning solution used will depend on the type of hazardous material. Laundry detergent or soap flakes often are used, but in some situations special chemicals may be needed for formal decontamination.

Fire Fighter Tips

Because runoff water from gross decontamination is likely to contain contaminants, make some effort to contain it. Be aware of any obvious storm drains, streams, or ponds and divert contaminated runoff away from them if possible. However, do not delay decontamination of a victim to build structures to contain runoff; human life always takes precedence over environmental concerns.

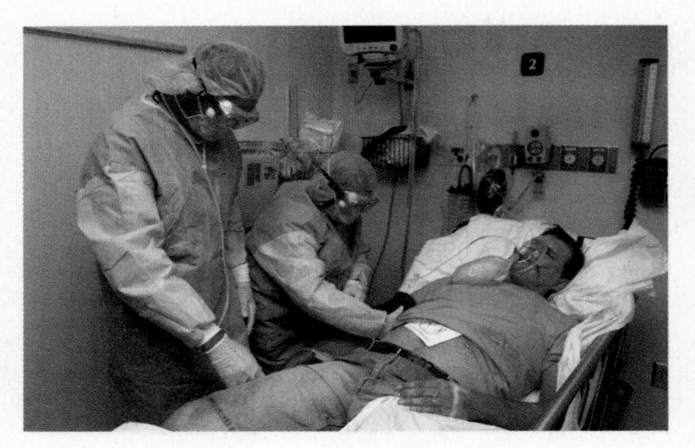

Figure 33-3 Fine decontamination is performed in a hospital, not in the field.

Figure 33-4 Rapid mass decontamination is used in incidents involving unknown agents and large groups of people.

Fine Decontamination

Fine decontamination is an advanced decontamination process performed in an isolated area of a hospital emergency department after gross and formal decontaminations have taken place in the field. It is performed by hospital emergency room personnel to further clean and evaluate the extent of hazardous material exposure. Fine decontamination includes cleaning eyes, ears, and fingernails; checking deep into the body's orifices; and possibly swabbing the nasal passages and mouth area to remove any hazardous particles (▲ Figure 33-3).

Fine decontamination usually takes place in isolated rooms that may have separate drainage systems and separate ventilation systems. Always notify the hospital in advance when transporting decontaminated hazardous materials victims to the emergency department for further treatment. This enables the hospital to prepare properly so that contaminants do not spread to medical workers or patients.

Rapid Mass Decontamination

Rapid mass decontamination is used in incidents involving unknown agents and large groups of people (► Figure 33-4). It takes place in the field and is a way of quickly performing gross decontamination on a large number of victims who have escaped from a hazardous materials incident.

Because water is a universal solvent, washing off as much of the contaminant as possible with a massive water spray is the best and quickest way to decontaminate a large group of people. Rapid mass decontamination requires a minimum of two fire apparatus placed side by side. Place fire nozzles on the side discharge outlets and use a spray pattern to douse the victims. An aerial ladder device can provide a complete overhead spray.

In rapid mass decontamination, the primary concern is to remove a hazardous material from a large number of victims. It can take place on any street, parking lot, or area where fire apparatus can be deployed with a continuous, uncontaminated water supply. Environmental concerns are secondary; there is usually not time to build structures to contain runoff.

These victims will need further decontamination and medical monitoring after they have been sprayed. When more emergency responders arrive, they can establish a decontamination corridor to clean and thoroughly decontaminate the victims before transporting them to a medical facility.

Methods of Decontamination

The previous descriptions of decontamination have focused on using water to dilute or wash away the hazardous material. However, some hazardous materials have chemical properties that may require different methods of decontamination. Other decontamination procedures that can be used include:

- Absorption
- Adsorption
- Dilution
- Disinfection
- Disposal
- Solidification
- Emulsification
- Vapor dispersion
- Removal
- Vacuuming

Absorption

In **absorption**, a spongy material is mixed with a liquid hazardous material and the contaminated mixture is collected and disposed of together. This technique is primarily used for decontaminating equipment and property and has limited application for decontaminating personnel.

Absorption minimizes surface areas of liquid spills, but is effective only on flat surfaces. Although absorbent materials such as soil and sawdust are inexpensive and readily available, they become hazardous themselves and must be disposed of properly. Most states have laws and regulations that dictate how to dispose of used absorbent materials.

Adsorption

The opposite of absorption is adsorption. In adsorption, the contaminant adheres to the surface of an added material, rather than combines with it as in absorption. Adsorbent materials, including sand and activated carbon, adhere to the outside of hazardous liquids or gases.

Dilution

Dilution uses plain water or a soap-and-water mixture to lower the concentration of a hazardous material while flushing it off of a contaminated person or object. Dilution is both fast and economical; water is a readily available solvent that usually generates no toxic fumes. The gross decontamination, formal decontamination, and mass rapid decontamination techniques described earlier all use dilution.

Fire fighters tend to use this method first because water is readily available from the tank on any engine or from a hydrant. However, before using water, consider whether the contaminant will react with water, be soluble in water, or spread to a larger area. The more water used, the more hazardous waste generated that must be disposed of safely.

Dilution also can be effective in removing some solids such as dusts and fibers; however, care must be used to prevent the spread of these contaminants to larger areas.

Disinfection

Disinfection is the process used to destroy recognized disease-carrying (pathogenic) microorganisms. Commercial disinfectants are packaged with a detailed brochure that describes the limitations and capabilities of the product. Fire fighters with medical research labs, hospitals, clinics, mortuaries, medical waste disposal facilities, blood banks, or universities in their response area should be familiar with the specific types of biological hazards present and the best disinfectants for each hazard. Take special precaution when responding to an emergency at a facility where bloodborne pathogens may be involved in an accident or spill.

Disposal

Disposal is a two-step removal process for items that cannot be properly decontaminated. First, the contaminated article is removed and isolated in a designated area. It is then packaged in a suitable container and transported to an approved facility, where it is either incinerated or buried in a hazardous waste landfill. Contaminated items such as disposable coveralls should be collected, bagged, and tagged. Contaminated tools and equipment should be placed in bags, barrels, or buckets.

Solidification

Solidification is the chemical process of treating a hazardous liquid to turn it into a solid material. This makes the material easier to handle, but does not change the inherent chemical properties of the hazardous material.

Emulsification

Emulsification is the process of neutralizing or reducing the harmful effects of a hazardous material by changing its chemical properties. This process can be used with some petroleum-based products. Although emulsification may change the chemical nature of the hazardous material, the resultant product still needs to be disposed of properly. Local regulations may apply to the use of emulsifier products.

The principal advantage of emulsification is that it renders the hazardous material less harmful, thereby decreasing the level of risk to the emergency responder. It can also help limit clean-up costs. However, the chemicals required for emulsification may be harmful to fire fighters. It also may take some time to determine which chemicals should be used and whether they are available.

Vapor Dispersion

Vapor dispersion is the process of separating and diminishing harmful vapors. A water spray is commonly used to disperse or move vapors away from certain areas or materials. Vapor dispersion should be used with extreme caution because the rapid introduction of a large volume of air can create dynamic results.

Removal

Removal as a mode of decontamination applies specifically to contaminated soil that can be taken away from the scene and disposed of properly. Removal may be necessary when highly toxic materials that cannot be rendered harmless are involved, when on-site treatment presents unacceptable risks to fire fighters, or when on-site treatment costs exceed disposal costs.

For example, it may be more cost-effective and less risky to pump the hazardous material into a drum and transport it to a hazardous waste dump than to neutralize a spill and decontaminate emergency equipment. Removing the contaminants reduces clean-up time and also limits the exposure risk to fire fighters. However, removal is not always the most economical decision and often requires the purchase of hazardous waste and transportation permits.

Vacuuming

Vacuuming is the removal of dusts, particles, and some liquids by sucking them up into a container. A filtering system prevents the contaminated material from re-circulating and re-entering the atmosphere. A special High-Efficiency Particulate Air (HEPA) vacuum cleaner is used to remove

VOICES OF EXPERIENCE

"The cold zone was enveloped in an ammonia vapor cloud, sending fire fighters scrambling for safety."

As a member of the Aurora Fire Department's hazardous materials team, I responded to an incident at a small, family-owned ice cream manufacturing facility. There was a leak in the plant's ammonia refrigeration system, and a vapor cloud had developed inside the plant.

A hot zone was established, taking into account the size of the vapor cloud and the wind direction. To ensure that harmful chemicals did not reach beyond the immediate vicinity, a decontamination corridor was set up in the warm zone, about 100′ from the building. The decontamination corridor incorporated a 3-step decontamination process, including rinse-off with soap and water, scrubbing, and removal of PPE. With the decontamination corridor in place, the hazardous materials team felt confident that the chemicals would be safely contained.

After communicating with the plant owner, the hazardous materials team pinpointed the location of the leak and prepared to enter the plant. Suddenly, the wind changed direction. The cold zone was enveloped in an ammonia vapor cloud, sending fire fighters scrambling for safety. After falling back and regrouping our resources, we immediately began reestablishing our operational zones. Setting up the new decontamination corridor was especially critical, given the fact that so many fire fighters and pieces of equipment had unexpectedly been exposed to the chemical. With the quick actions of the fire fighters on scene, all hazardous materials were contained.

The fire fighters at that incident learned a valuable lesson about the possibility of changing conditions. Having proper decontamination procedures in place is critical at every hazardous materials incident, but fire fighters must be able to quickly adapt these procedures to new circumstances. Otherwise, they risk losing control of the scene and exposing more personnel and equipment to the chemical than necessary.

John S. Lehman
Aurora Fire Department
Aurora, Illinois

hazardous dusts, powders, or fibers that are 0.3 microns or larger. HEPA filters allow air to pass through but capture the particulate matter in the air. HEPA filters must be replaced regularly to maintain their effectiveness.

The Decontamination Process

The decontamination process should take place in the decontamination corridor between the hot zone and the warm zone. Within this corridor, fire fighters must pass through several stations to complete the decontamination process. A clearly marked, easily seen, accessible entry point to the decontamination corridor should be established. If the decontamination operation occurs at night, the area should be well lit. All personnel leaving the hot zone must pass through the decontamination corridor.

The decontamination team must wear SCBA and a level of PPE equal to those being decontaminated. All decontamination team members must undergo decontamination before they leave the area.

Steps in Decontamination

Personnel leaving the hot zone should place any tools that they used in a tool drop area near the decontamination corridor. This tool drop area can be a container, a recovery drum, or a special tarp. The tools can then be used by another team entering the hot zone or collected for decontamination.

The fire fighter, still wearing full PPE and SCBA, steps into the decontamination corridor for gross decontamination. The first step is a portable shower, where high-pressure, low-volume water flow rinses off and dilutes the contaminants on the PPE.

The next step is formal decontamination. The decontamination team is responsible not only for scrubbing and swabbing the PPE worn by personnel but also for containing runoff. Formal decontamination involves one to three wash-and-rinse stations. At each station, one decontamination team member handles the scrubbing and another does the rinsing. Only one fire fighter is allowed in a station at a time. The decontamination team member who is scrubbing should pay special attention to the gloves, kneecaps, and boot bottoms.

After outwear is thoroughly scrubbed and rinsed, it can be removed along with chemical protective clothing. SCBA, however, should remain in place. The decontamination team unzips the PPE and peels back the suit, exposing the SCBA. The team should fold the PPE back, so that the contaminated side contacts only itself. If the procedure is done properly, the contaminated side of the suit will not touch either the interior of the suit or the person wearing it. If the fire fighter is wearing outer gloves, they can be removed now; inner gloves will be removed later. The fire fighter moves down the decontamination corridor to remove the remainder of the PPE, SCBA, and other support equipment.

When removing protective clothing, turn as much of the PPE inside-out as possible. This keeps any remaining trace contaminants inside the suit or glove. The fire fighter should be seated and allow the decontamination team members to remove PPE. Removing outside clothing properly is especially important when a fire fighter is wearing structural PPE because these clothes are porous and can absorb the hazardous material.

Remove the helmet, boots, and other support equipment along with the SCBA. The face shield is usually the last item removed because it provides an additional safeguard against trace contaminants. Place equipment on the contaminated side of the decontamination corridor area. Deposit the SCBA in a plastic bag or place it on a tarp for bagging later. Highly contaminated SCBA should be removed and isolated until it can undergo complete decontamination. Remove inner gloves, turn them inside-out, and sort them into individual containers for clean-up or disposal. Plastic bags can be used because they usually provide sufficient temporary protection from most materials. They should be sealed with tape and transported elsewhere for clean-up or disposal. Place the bags in a properly marked recovery drum when disposing of them.

After removing personal clothing, the fire fighter should wash the entire body. An overhead shower produces much better results than a hose line. Apply ample soap to all areas of the body, especially to the head and the groin. Liquid surgical soaps in plastic squeeze bottles give the best results; local hospitals may have a contact for obtaining prepackaged cleaning kits. Use small brushes and sponges for scrubbing the body surface. Afterward, bag and mark all cleaning items for disposal. The water runoff from this step should be controlled and contained, just as in previous steps.

After going through the decontamination corridor, the fire fighter should dry off using towels or sheets. Each towel can be used only once; it is then placed in bags on the contaminated side for disposal. Plastic garbage cans, lined with plastic bags are good storage containers.

Decontaminated personnel can now don clean clothes. Disposable cotton coveralls, hospital gowns, hospital booties, slippers, and flip-flops are inexpensive and easy to use. They

Fire Fighter Safety Tips

Contact lenses may trap contaminants under the lens. Anyone wearing contact lenses should remove them and flush the eyes thoroughly. If any discomfort or irritation remains, the eye should be evaluated.

can be prepackaged according to size and stored for easy access. Provisions should be made to bag clothing in the event uncontaminated clothing can be salvaged.

To perform decontamination at a hazardous materials incident, follow the steps in ▼ Skill Drill 33-1.

1. Drop any tools and equipment into a tool drum or onto a designated tarp. **(Step 1)**
2. Perform gross decontamination. **(Step 2)**
3. Perform formal decontamination. Wash and rinse the fire fighter one to three times. **(Step 3)**
4. Remove outer hazardous materials protective clothing and isolate PPE. **(Step 4)**
5. Remove SCBA. **(Step 5)**
6. Remove personal clothing. Bag and tag all personal clothing. **(Step 6)**
7. Shower and wash the body. Dry off the body and put on clean clothing. **(Step 7)**

Skill Drill

Decontamination

Drop any tools and equipment.

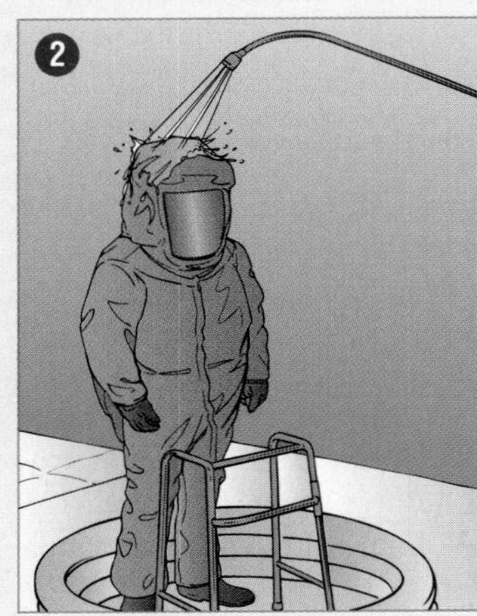

Perform gross decontamination.

Perform formal decontamination.

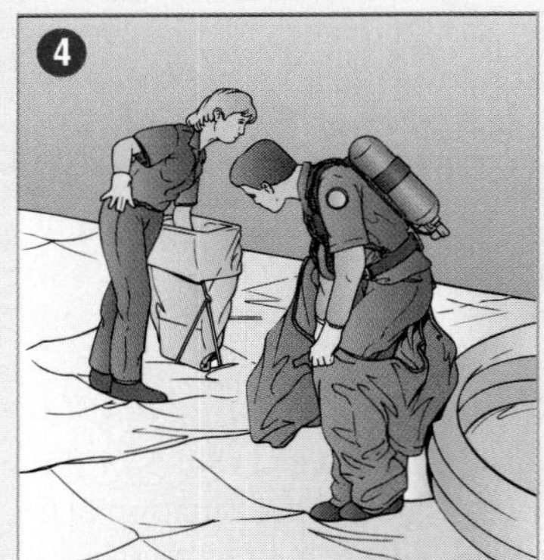

Remove outer hazardous materials protective clothing and isolate PPE.

Medical Follow-Up

After personnel are thoroughly decontaminated, they should proceed to a emergency medical station for a medical evaluation. Medical personnel will take vital signs for each person leaving decontamination and compare them to baseline data. Any open wounds or breaks in skin surfaces should be reported immediately to medical control. Unless advised otherwise by the medical command physician, all open wounds should be cleaned on the scene.

33-1 Skill Drill

5 Remove SCBA.

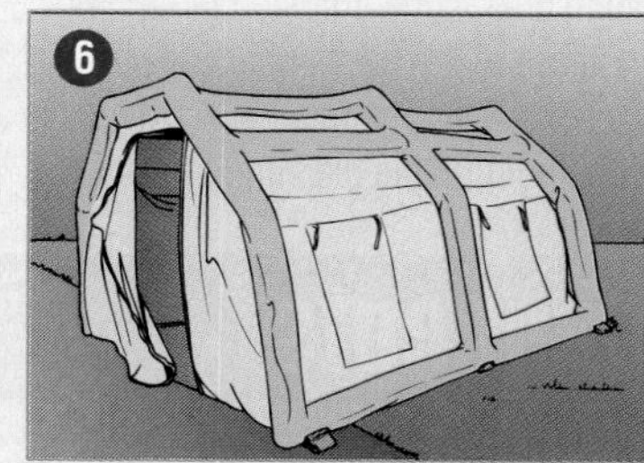

6 Remove personal clothing. Bag and tag all personal clothing.

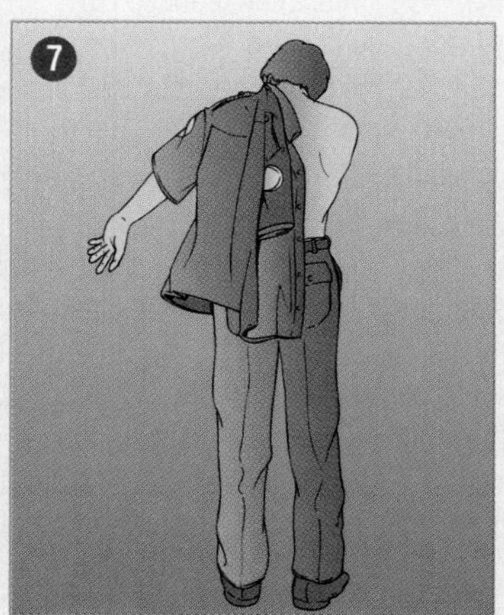

7 Shower and wash the body. Dry off the body and put on clean clothing.

Wrap-Up

Ready for Review

Controlling and minimizing contact with hazardous materials contaminants can reduce the need for decontamination, which is the process of removing contaminants from people, equipment, and the environment. Decontamination also involves containing and reducing the spread of contamination. The hazardous material must be separated from any person or object that came in contact with it. Everything and everyone that was in the hot zone or in contact with the hazardous material must be decontaminated.

The best field decontamination procedures concentrate on confining contaminants to a limited area. Establishing a designated decontamination corridor and decontamination area is the first step in limiting the spread of contaminants. Regardless of the type of decontamination required or the method used, decontamination is most effective when it is carried out by a trained team using multiple cleaning stations.

Chief Concepts

- Decontamination is the physical or chemical process of reducing and preventing the spread of hazardous materials from persons and equipment.
- Secondary contamination spreads contamination from the initial source to other people or objects via contaminated people or objects.
- The major categories of decontamination are emergency decontamination, gross decontamination, formal decontamination, and fine decontamination.
- The purpose of rapid mass decontamination is to perform gross decontamination quickly on a large number of victims.

Hot Terms

Absorption The process of applying a material that will soak up and hold a hazardous material in a sponge-like manner, for collection and disposal.

Adsorption The process of adding a material such as sand or activated carbon to a contaminant, which then adheres to the surface of the material for collection.

Contamination The process of transferring a hazardous material from its source to people, animals, the environment, or equipment, all of which may act as carriers.

Decontamination The physical or chemical process of removing any form of contaminant from a person, an object, or the environment.

Decontamination corridor A controlled area between the warm and hot zone through which all persons who entered the hot zone will exit for decontamination.

Decontamination team The team responsible for reducing and preventing the spread of contaminants from persons and equipment used at a hazardous materials incident. They establish the decontamination corridor and conduct all phases of decontamination.

Dilution The process of adding some substance, usually water, to a product to weaken its concentration.

Disinfection The process used to destroy recognized disease-carrying (pathogenic) microorganisms.

Disposal A two-step removal process for contaminated items that cannot be properly decontaminated; items are bagged and placed in appropriate containers for transport to a hazardous waste facility.

Emergency decontamination The process of getting the bulk of contaminants off a victim without regard for any form of containment; used in potentially life-threatening situations, without the formal establishment of a decontamination corridor.

Emulsification The process of changing the chemical properties of a hazardous material to reduce its harmful effects.

Fine decontamination An advanced decontamination process used by hospital and emergency room personnel to clean and to evaluate the extent of chemical exposure.

Formal decontamination A multistep process of carefully scrubbing and washing contaminants off a person or object, collecting runoff water, and collecting and properly handling all items that takes place after gross decontamination.

Gross decontamination The process of removing the outer clothing and generally flushing most contaminants from a victim; the initial phase of the decontamination process during which the amount of surface contaminant is significantly reduced.

Rapid mass decontamination A process of flushing a large number of victims with water for rapid decontamination in the field when the agent is unknown.

Removal A mode of decontamination that applies specifically to contaminated soil that is taken away from the scene.

Secondary contamination Contamination that occurs when a person or object comes into direct contact with a contaminated person or object, as might occur when a contaminant is carried out of the hot zone and comes into contact with people, animals, equipment, or the environment.

Solidification The process of chemically treating a hazardous liquid to turn it into a solid material, making the material easier to handle.

Vapor dispersion The process of separating and diminishing vapors, usually by using a water spray to disperse or move vapors away from certain areas or materials.

Vacuuming The process of cleaning up dusts, particles, and some liquids using a vacuum with HEPA filtration to prevent recontamination of the environment.

Fire Fighter in Action

It is Tuesday evening when your engine company is dispatched to a vehicle fire. Upon arrival, you find a fully involved older model sports car in a convenience store parking lot. The occupant of the vehicle has fled the scene. Your lieutenant tells your crew to pull a preconnected hand line and extinguish the fire. You extinguish the fire in the passenger compartment and force entry into the hatchback to conduct overhaul. During overhaul, you find evidence of a mobile methamphetamine lab. You report what you have found to your lieutenant. Lieutenant Gerth tells you to remain in your PPE and SCBA and calls for law enforcement and for a second engine to decontaminate you and your crew.

1. What types of decontamination should be used at this incident?
   A. Emergency and gross
   B. Gross and formal
   C. Formal and fine
   D. Fine and rapid mass

2. What method of decontamination will most likely be utilized at this incident?
   A. Removal
   B. Adsorption
   C. Dilution
   D. Emulsification

Engine 2 arrives on the scene and sets up a decontamination area. They finish scrubbing and rinsing your bunker gear, helmet, gloves, boots, and SCBA. You are now preparing to doff your personal protective equipment.

3. When removing protective clothing during the decontamination process, you should:
   A. drink plenty of water to stay hydrated.
   B. place the gear in a cardboard box for transport.
   C. turn as much of the PPE inside-out as possible.
   D. place gear in plastic bags and seal the bags for transport.
   E. both C and D

4. The decontamination team should be wearing what level of PPE?
   A. Full bunker gear and SCBA
   B. Class B hazardous materials suits
   C. Carbon monoxide monitors
   D. Splash and eye protection

www.FireFighter.jbpub.com

Terrorism Awareness

Technology Resources

www.FireFighter.jbpub.com

- Chapter Pretests
- Interactivities
- Hot Term Explorer
- Web Links
- Review Manual
- FireLearn

Chapter Features

- Skill Drills
- Voices of Experience
- Fire Marks
- Teamwork Tips
- Fire Fighter Safety Tips
- Fire Fighter Tips
- Canadian Perspectives
- Hot Terms
- Wrap-Up

Chapter 34

NFPA 1001 Standard

Fire Fighter I

NFPA 1001 contains no Fire Fighter I Job Performance Requirements for this chapter.

Fire Fighter II

NFPA 1001 contains no Fire Fighter II Job Performance Requirements for this chapter.

Additional NFPA Standards

NFPA 472 *Standard for Professional Competence of Responders to Hazardous Materials Incidents*

Knowledge Objectives

After studying this chapter, you will be able to:

- Describe the threat posed by terrorism.
- Identify potential terrorist targets in your jurisdiction.
- Describe the dangers posed by explosive devices.
- Describe the difference between chemical and biologic agents.
- Describe the dangers posed by radiological incidents.
- Describe the need for decontamination of exposed victims and response personnel.

Skills Objectives

There are no skills objectives for this chapter.

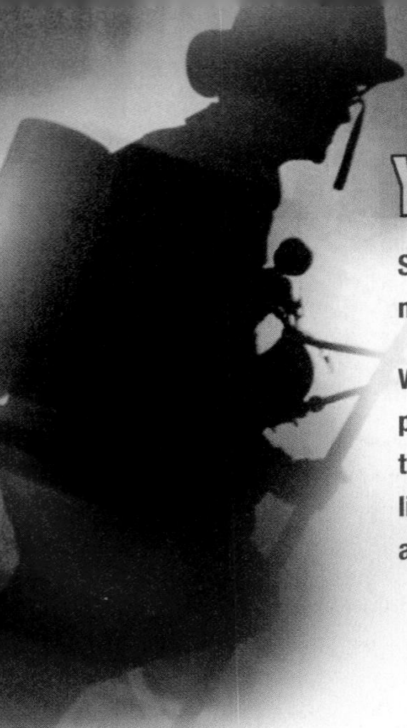

You Are the Fire Fighter

Shortly before noon one weekday morning, your engine company is dispatched to a local government office building. The dispatcher reports that an explosion has occurred.

When you arrive, you do not observe any indications of fire or damage to the structure, but people are running from the building. Many of them complain of difficulty breathing. One person says that a small package exploded in an upstairs office, showering nearby people with some type of liquid. This person also tells you that people in the office are having seizures and complaining of a burning sensation.

1. ***Is the information provided at this incident typical for most fires or explosions?***
2. ***What characteristics of this situation cause you concern?***
3. ***What is your primary concern as you enter the office building?***

What Is Terrorism?

Terrorism can be described as the unlawful use of violence or threats of violence to intimidate or coerce a government, the civilian population, or any segment thereof, in furtherance of political or social objectives. This broad definition can include a wide range of acts committed by different groups for different purposes.

The Federal Bureau of Investigation (FBI) classifies terrorism as either domestic or international. Domestic terrorism refers to acts that are committed within the United States by individuals or groups that operate entirely within the United States and are not influenced by any foreign interests. International terrorism includes any acts that transcend international boundaries.

Terrorism is a worldwide threat. Throughout the world, there were 1,106 incidents of international terrorism, resulting in the deaths of 2,494 innocent civilians in 2000. Although there were only 864 terrorist incidents recorded in 2001, the attack on the World Trade Center accounted for the majority of the 4,739 worldwide deaths[1] (► Figure 34-1).

Within the United States, there were 24 terrorist incidents between 1994 and 1999 and three additional crimes that were suspected to be terrorist in nature. Most of these incidents were classified as domestic terrorism. During the same period, law enforcement agencies prevented 47 terrorist incidents.

Fire Service Response to Terrorist Incidents

The fire service has a major role in protecting communities from terrorism. The fire service role includes emergency medical services (EMS), hazardous materials mitigation, technical rescue, and fire suppression. All of these functions may be needed when a terrorist incident occurs. Terrorism presents new challenges for the fire service. It also presents an unparalleled threat to the lives of fire fighters and emergency responders.

The terrorist threat requires fire fighters to work closely with local, state, and federal law enforcement agencies; emergency management agencies; allied health agencies; and

Figure 34-1 The September 11th attack on the World Trade Center accounted for the majority of the deaths caused by terrorists in 2001.

Fire Fighter Tips

The Code of Federal Regulations defines terrorism as: "the unlawful use of force and violence against persons or property to intimidate or coerce a government, the civilian population, or any segment thereof in furtherance of political or social objectives" (28 C.F.R. Section 0.85).

the military. It is critical that all of these agencies work together in a coordinated and cooperative manner. All emergency responders and law enforcement agencies must be prepared to face a wide range of potential situations.

The greatest threat posed by terrorists is the use of weapons of mass destruction (WMD), devices that are designed to cause maximum damage to property or people. These weapons include chemical, biologic, and radiological agents, as well as conventional weapons and explosives. Thousands of casualties could result from a WMD attack in an urban area. An incident of this magnitude could quickly overwhelm not only the largest and best-trained emergency response agencies but also the local healthcare system.

The fire service must adapt and be prepared for the threat of terrorist attacks by exploring new approaches and technologies to manage WMD incidents. One of the highest priorities is to improve the ability of first responders to identify and mitigate releases of chemical, biologic, and radiological agents.

Potential Targets and Tactics

Terrorists are usually motivated by a cause and choose targets they believe will help them achieve their goals and objectives. Many terrorist incidents aim to instill fear and panic among the general population and to disrupt daily ways of life. In other cases, they choose a symbolic target, such as a place of worship, a foreign embassy, a monument, or a prominent government building. Sometimes the objective is sabotage, to destroy or disable a facility that is significant to the terrorist cause. The ultimate goal could be to cause economic turmoil by interfering with transportation, trade, or commerce.

Terrorists choose a method of attack they think will make the desired statement or achieve the maximum results. They may vary their methods or change them over time. Explosive devices have been used in thousands of terrorist attacks; in recent years, however, there has been a significant increase in the number of suicide bombings.

Many terrorist incidents in the 1980s involved the taking of hostages on hijacked aircraft or cruise ships. Sometimes, only a few people were held; other times, hundreds of people were taken hostage. Diplomats, journalists, and athletes were targeted in several incidents, and the terrorists often offered to release their hostages in exchange for the release of imprisoned individuals allied with the terrorist cause. More recent terrorist actions have endangered thousands of lives with no opportunity to bargain.

Terrorism can occur in any community. A rural ski lodge, tucked away in the mountains, may be attacked by an environmental group that is upset with plans for expansion. A small retailer in an upscale suburban community could become the target of an animal rights group that objects to the sale of fur coats. An anti-abortion group might plant a bomb at a local community health clinic. There are many causes with supporters who range from peaceful, nonviolent organizations to fanatical fringe groups.

It is often possible to anticipate likely targets and potential attacks. Law enforcement agencies routinely gather intelligence about terrorist groups, threats, and potential targets. In many cases this information is shared with fire departments so that preincident plans can be developed for possible targets and scenarios. Even if no specific threats have been made, certain types of occupancies are known to be potential targets and preincident planning for those locations should include the possibility of a terrorist attack as well as an accidental fire.

News accounts of terrorist incidents abroad can help keep fire fighters current with trends in terrorist tactics. They can provide useful information about situations that could occur in your jurisdiction in the future. A variety of Internet resources can also help you to keep abreast of current threats. Periodic e-mail list servers, available through federal response agencies, provide useful information. By becoming familiar with potential targets and current tactics, emergency responders can plan appropriate strategies and tactics for potential attacks.

Ecoterrorism

Ecoterrorism refers to illegal acts committed by groups supporting environmental or related causes. Examples include spiking trees to sabotage logging operations, vandalizing a university research laboratory that is conducting experiments on animals, or firebombing a store that sells fur coats. Several incidents of domestic ecoterrorism have been attributed to special interest groups such as the Earth Liberation Front (ELF) and the Animal Liberation Front (ALF).

Infrastructure Targets

Terrorists might strike bridges, tunnels, or subways in an attempt to disrupt transportation and inflict a large number of casualties (► Figure 34-2). They could also attack the public water supply or try to disable the electrical power distribution system, telephones, or the Internet. Disruption of a community's 9-1-1 system or public safety radio network would have a very direct impact on emergency response agencies.

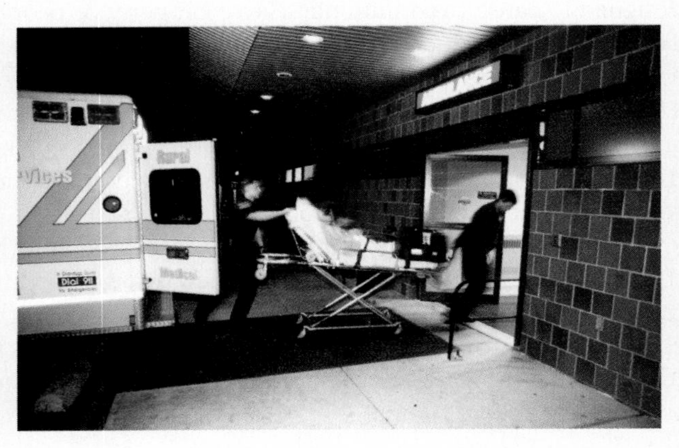

Figure 34-2 Subways, airports, bridges, and hospitals are all vulnerable to attack by terrorists who seek to interrupt a nation's infrastructure.

Figure 34-3 Terrorists might attempt to destroy visible national icons.

Symbolic Targets

Monuments such as the Lincoln Memorial, Washington Monument, or Mount Rushmore may be targeted by groups who want to attack symbols of national pride and accomplishment (◄ Figure 34-3). Foreign embassies and institutions might be attacked by groups promoting revolution within those countries or protesting their international policies. Religious institutions or other visible icons are potential

Fire Marks

The Earth Liberation Front claimed credit for arson fires that destroyed several mountaintop buildings at a ski resort in Vail, Colorado in 1998. The group was trying to prevent the development of additional ski trails on the mountain.

Fire Fighter Tips

Potential Terrorist Targets

Ecoterrorism Targets

Controversial development projects
Environmentally sensitive areas
Research facilities

Infrastructure Targets

Bridges and tunnels
Emergency facilities
Hospitals
Oil refineries
Pipelines
Power plants
Railroads
Telecommunications systems
Water reservoirs and treatment plants

Symbolic Targets

Embassies
Government buildings
Military bases
National monuments
Places of worship

Civilian Targets

Arenas and stadiums
Airports and railroad stations
Mass-transit systems
Schools and universities
Shopping malls
Theme parks

Cyberterrorism Targets

Banking and finance computer systems
Business computer systems
Court computer systems
Government computer systems
Law-enforcement computer systems
Military computer systems

Agroterrorism Targets

Crops
Feed storage
Grain elevators
Livestock and poultry

Fire Marks

Major Terrorist Incidents in the United States

September 1984 Dalles, Oregon

To influence local elections, a religious sect spread salmonella on salad bars in four restaurants, resulting in 750 cases of salmonella poisoning.

February 1993 New York, New York

A large explosive device was detonated in a van parked in the underground garage of the World Trade Center. Six workers were killed and more than a thousand people were injured.

April 1995 Oklahoma City, Oklahoma

The Alfred E. Murrah Federal Building was demolished by a truck bomb that also killed 167 people. (An NFPA Fire Investigations report is available on this incident.)

1978 to 1995 United States

Over a period of 17 years, the Unabomber mailed at least 16 packages containing explosives to university professors, corporate executives, and other targeted individuals. These attacks killed 3 individuals and injured 23 others.

July 1996 Atlanta, Georgia

A pipe bomb exploded in the Centennial Olympic Park, killing one person and injuring 111 people.

January 1997 Atlanta, Georgia

Following the bombing of an abortion clinic in suburban Atlanta, a secondary device exploded, wounding several emergency responders. A month later, another secondary device was found and disarmed at the scene of a bombing at a gay nightclub in Atlanta.

January 1998 Birmingham, Alabama

A bomb killed a police officer who was providing security at an abortion clinic.

October 1998 Vail, Colorado

Arson destroyed eight buildings at a ski resort. An extremist environmental group opposed to expansion of the resort claimed responsibility.

September 11, 2001 New York, New York/Washington D.C./Pennsylvania

Terrorists hijacked four commercial jets. Two were flown into the World Trade Center, one struck the Pentagon, and the fourth crashed into a field in rural Pennsylvania. More than 3,000 people died in the various incidents.

Fall 2001 United States

Five people died after letters containing anthrax virus were sent to various locations in the eastern United States.

targets of hate groups. By targeting these symbols, terrorist groups seek to make people aware of their demands and create a sense of fear in the public.

Civilian Targets

Terrorists who attack civilian targets such as shopping malls, schools, or stadiums indiscriminately kill or injure the maximum number of potential victims (▼Figure 34-4). Their goal is to create fear in every member of society and to make citizens feel vulnerable in their daily lives. Letter bombs or letters that contain a biologic agent have a similar effect.

Cyberterrorism

Groups might engage in cyberterrorism by electronically attacking government or private computer systems. Several attempts have been made to disrupt the Internet or to attack government computer systems and other critical networks.

Figure 34-4 By attacking a civilian target like a crowded stadium, terrorists might make citizens feel vulnerable in their everyday lives.

Figure 34-5 Agroterrorism might affect our food supply.

Canadian Perspectives

Terrorist Incidents in Canada

October 1963 to 1970 Montreal, Quebec

The Front de Liberation Québecois (FLQ) conducted both kidnapping and bombing campaigns. More than 85 bombs were planted, ultimately claiming six lives.

October 1982 Toronto, Ontario

A van with hundreds of kilograms of explosives was used to attack an industrial target, resulting in 10 injuries. The same group later firebombed several video stores.

1985 Vancouver, British Columbia

An Air India flight exploded in mid-air killing all 329 passengers on board. On the same day, a suitcase bomb exploded at Narita Airport in Tokyo, killing two baggage handlers. Both flights originated in Vancouver, British Columbia.

1991 Toronto, Ontario

Members of a Pakistani religious sect were arrested as they tried to enter Canada from the United States with explosives. Their planned targets included a Hindu temple, a movie theatre, and an East Indian restaurant in Toronto.

1999 Ottawa, Ontario and Montreal, Quebec

Kurdish expatriates conducted violent demonstrations against the Turkish government in Ottawa and Montreal. In Ottawa, a police officer's clothes were set on fire with a Molotov cocktail. In Montreal, a police officer lost an eye after being hit with a rock.

Agroterrorism

Agroterrorism could include the use of chemical or biologic agents to attack the agricultural industry or the food supply (◄Figure 34-5). The introduction of a disease such as foot-and-mouth to the livestock population could result in billions of dollars in losses to the food industry.

Agents and Devices

Terrorists can use several different kinds of weapons in an attack. Bombings are the most frequent terrorist acts, but fire fighters also must be aware of other potential weapons. By shooting into a crowd at a shopping mall or train station, a single terrorist with an automatic weapon could cause devastating carnage. A biologic agent that is only visible through a microscope could cause thousands to become ill and die. A computer virus that attacks the banking industry could cause tremendous economic losses. Planning should consider the full range of possibilities.

Explosives and Incendiary Devices

According to the FBI's Bomb Data Center, there has been a substantial increase in the number of bombings in the United States over recent years. Between 1987 and 1997, the FBI recorded 23,613 bombing incidents, resulting in 448 deaths and 4,170 injuries. Incendiary devices account for 20% to 25% of all bombing incidents in the United States.

Groups or individuals have used explosives to further a cause; to intimidate a co-worker or former spouse; to take revenge; or simply to experiment with a recipe found in a book or on the Internet.

Each year thousands of pounds of explosives are stolen from construction sites, mines, military facilities, and other locations. How much of this material makes its way to criminals and terrorists is not known (▼ Figure 34-6). Terrorists can also use commonly available materials, such as a mixture of ammonium nitrate fertilizer and fuel oil (ANFO), to create their own blasting agents.

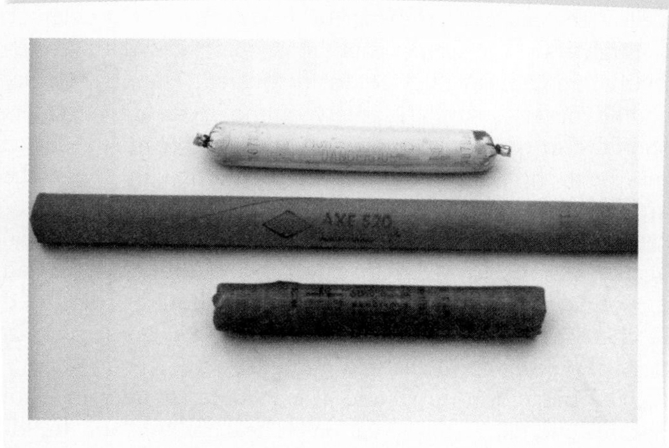

Figure 34-6 Every year, thousands of pounds of explosives are stolen.

Fire Fighter Safety Tips

A terrorist event may be designed to target emergency responders as well as civilians. Explosives do not discriminate. Anyone nearby can be severely injured or killed.

An **improvised explosive device** (IED) is any explosive device that is fabricated in an improvised manner. An IED could be contained in almost any type of package from a letter bomb to a truckload of explosives. The Unabomber constructed at least 16 bombs that were delivered in small packages through the postal service. The bombings of the World Trade Center in 1993 and the Alfred E. Murrah Federal Building in 1995 both involved delivery vehicles loaded with ANFO and detonated by a simple timer (◄ Figure 34-7).

Pipe Bombs

The most common improvised explosive device is the **pipe bomb**. A pipe bomb is simply a length of pipe filled with an explosive substance and rigged with some type of detonator (▼ Figure 34-8). Most pipe bombs are simple devices, made with black powder or smokeless powder and ignited by a hobby fuse. More sophisticated pipe bombs may use a variety of chemicals and incorporate electronic timers, mercury switches, vibration switches, photocells, or remote control detonators as triggers.

Pipe bombs are sometimes packed with nails or other objects so they will inflict as much injury as possible on anyone in the vicinity. A chemical or biologic agent or radiological material could be added to a pipe bomb to create a

Figure 34-7 The Alfred E. Murrah Federal Building in Oklahoma City was destroyed by a truck bomb.

Figure 34-8 Pipe bombs can come in many shapes and sizes.

much more complicated and dangerous incident. Experts can only speculate at the number of casualties that could result from such a weapon.

Secondary Devices

Emergency responders must realize that terrorists may have placed a **secondary device** in the area where an initial event has occurred. These devices are intended to explode some time after the initial device explodes. Secondary devices are designed to kill or injure emergency responders, law enforcement personnel, spectators, or news reporters. Terrorists may use this tactic to attack the best-trained and most experienced investigators and emergency responders, or simply to increase the levels of fear and chaos following an attack.

The use of secondary devices is a common tactic in incidents abroad and has occurred at a few incidents in North America. In 1998, a bomb exploded outside a Georgia abortion clinic. About an hour later, a second explosion injured seven people, including two emergency responders. A similar secondary device was discovered approximately one month later at the scene of a bombing in a nearby community. Responders were able to disable this device before it detonated.

Working with Other Agencies

Joint training with local, state, and federal agencies charged with handling incidents involving explosive devices should occur on a routine basis. Among these agencies are local and state police; the FBI; the Bureau of Alcohol, Tobacco, and Firearms; and military explosive ordnance disposal (EOD) units. In some jurisdictions, the fire department is responsible for handling explosive devices. Designated department members should be given the necessary training and equipment.

Potentially Explosive Devices

Fire fighters should consult their department's Standard Operating Procedures (SOP) for specific policies on bombing incidents. Fire department units responding to an incident that involves a potentially explosive device—one that has not yet exploded—should move all civilians from the area and establish a perimeter at a safe distance. At no time should fire fighters handle a potential explosive device unless they have received special training. Trained EOD personnel should assess the device and render it inoperative.

While waiting for the properly trained personnel to arrive, the fire department should establish an initial command post and a staging area (► Figure 34-9). Because there may be secondary devices, the staging area should be at least 3,000′ from the incident site. In this type of incident, a joint command structure, commonly referred to as a unified command, should be established to coordinate the actions of all agencies.

If the bomb disposal team decides to disarm the device, an emergency action plan must be developed in case there is an accidental detonation. The Incident Commander (IC) will work in concert with law enforcement and EOD personnel to determine a safe perimeter where fire fighters and emergency medical personnel should be staged (► Figure 34-10).

In some cases, a **forward staging area** will be established with a rapid intervention team standing by to provide immediate assistance to the bomb disposal team if something goes wrong. Other fire fighters and emergency medical personnel would remain in a remote staging area.

Fire Fighter Safety Tips

Radio waves can trigger electric blasting caps, initiating an explosion. Radio transmitters should not be used near a suspicious device.

Actions Following an Explosion

Unless the cause of an explosion is known to be accidental, fire fighters at the scene should always consider the possibility that an explosive device was detonated. The first priority should be to ensure the safety of the scene. Fire fighters should also consider the possibility that a secondary device may be in the vicinity. Responders should quickly survey the area for any suspicious bags, packages, or other items.

It is also possible that chemical, biologic, or radiological agents may be involved in a terrorist bombing. Qualified personnel with monitoring instruments should be assigned to check the area for potential contaminants. These precautions should be implemented immediately.

The initial size-up should also include an assessment of hazards and dangerous situations. The stability of any building involved in the explosion must be evaluated before anyone is permitted to enter. Entering an unstable area without

Figure 34-9 If the fire department is first to arrive at the scene of an explosion, it should establish an initial command post.

proper training and equipment may complicate rescue and recovery efforts.

Chemical Agents

Chemical agents have the potential to kill or injure great numbers of people. Several chemical weapons, including phosgene, chlorine, and mustard agents, were used in World War I, resulting in thousands of battlefield deaths and permanent injuries. Chemical weapons were also used during the Iran–Iraq War (1980 to 1988). They were also used by the Iraqi government against the minority Kurdish people during the same period. Although international agreements prohibit the use of chemical and biologic agents on the battlefield, there is great concern that chemical weapons could be used by terrorists or combatants.

In 1995, the religious cult Aum Shinrikyo released a nerve agent, **sarin** gas, into the Tokyo subway system. This attack resulted in 12 deaths and over 1,000 injuries. Many experts believe that this attack was ineffective, because there were relatively few casualties even though thousands of subway riders were potentially exposed. This attack instilled fear in a large population. A previous sarin attack in Matsumoto, Japan killed seven people and injured approximately 300 others.

The basic instructions for making chemical weapons are readily available through a variety of sources, including the Internet, books, and other publications. Additionally, the required chemicals can be obtained fairly easily. Many of the chemicals classified as WMD are routinely used in legitimate industrial processes. For instance, **chlorine** gas is used in swimming pools, at water treatment facilities, and in industry ▼ Figure 34-11. However, chlorine is classified as a choking agent, because inhalation causes severe pulmonary damage.

ATF	VEHICLE DESCRIPTION	MAXIMUM EXPLOSIVES CAPACITY	LETHAL AIR BLAST RANGE	MINIMUM EVACUATION DISTANCE	FALLING GLASS HAZARD
	COMPACT SEDAN	500 Pounds 227 Kilos *(In Trunk)*	100 Feet 30 Meters	1,500 Feet 457 Meters	1,250 Feet 381 Meters
	FULL SIZE SEDAN	1,000 Pounds 455 Kilos *(In Trunk)*	125 Feet 38 Meters	1,750 Feet 534 Meters	1,750 Feet 534 Meters
	PASSENGER VAN OR CARGO VAN	4,000 Pounds 1,818 Kilos	200 Feet 61 Meters	2,750 Feet 838 Meters	2,750 Feet 838 Meters
	SMALL BOX VAN *(14 FT BOX)*	10,000 Pounds 4,545 Kilos	300 Feet 91 Meters	3,750 Feet 1,143 Meters	3,750 Feet 1,143 Meters
	BOX VAN OR WATER/FUEL TRUCK	30,000 Pounds 13,636 Kilos	450 Feet 137 Meters	6,500 Feet 1,982 Meters	6,500 Feet 1,982 Meters
	SEMI-TRAILER	60,000 Pounds 27,273 Kilos	600 Feet 183 Meters	7,000 Feet 2,134 Meters	7,000 Feet 2,134 Meters

ATF I 5400.1 (01-99)

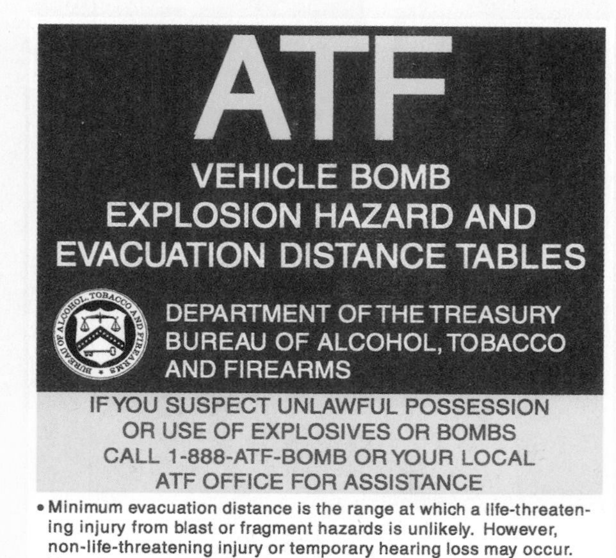

Figure 34-10 The Terrorist Bomb Threat Stand-Off Card helps the IC to determine a safe staging area.

Figure 34-11 Although chlorine is regularly used in swimming pools and water treatment facilities, it is also classified as a choking agent that can be used in a terrorist attack.

Cyanide compounds such as hydrogen cyanide (HCN) are used in the production of paper and synthetic textiles as well as in photography and printing. These chemicals are stored and transported in various sized containers, ranging from small containers and pressurized cylinders to railroad tank cars.

Chemical weapons can be disseminated in several ways. Simply releasing chlorine gas from a storage tank in an unguarded rail yard could cause thousands of injuries and deaths. A chemical agent can be added to an explosive device. Crop-dusting aircraft, truck-mounted spraying units, or hand-operated pump tanks could be used to disperse an agent over a wide area (► Figure 34-12).

Figure 34-12 Crop-dusting equipment can be used to distribute chemical agents.

Protection from Chemical Agents

Air flow is the most common dispersal method for chemical agents. Agents released outdoors will follow wind currents; agents released inside will be dispersed through a building's air circulation system. Some chemical agents have distinctive odors, if there is any unusual odor at an emergency scene, fire fighters must use full personal protective equipment (PPE) including self-contained breathing apparatus (SCBA) (► Figure 34-13). In addition, the area should be monitored with the appropriate detection devices. Remember that it is never sufficient to rely strictly on odor to determine the presence of a chemical agent.

Figure 34-13 If an unusual odor is reported at the scene, a fire fighter must use full PPE gear and a SCBA device.

Nerve Agents

Nerve agents are toxic substances that attack the central nervous system. They were first developed in Germany before World War II. Today, several countries maintain stockpiles of these agents. Nerve agents are similar to some pesticides, but they are extremely toxic. Nerve agents can be 100 to 1,000 times more toxic than similar pesticides. Exposure to these agents can result in injury or death within minutes.

In their normal states, common nerve agents are liquids (► Figure 34-14). As liquids, these agents are not likely to contaminate large numbers of people because direct contact with the agent is required. To be an effective weapon, the liquid must be dispersed in aerosol form, or broken down into fine droplets so that it can be inhaled or absorbed by the skin.

Pouring a liquid nerve agent onto the floor of a crowded building would probably not immediately contaminate large numbers of people; however, it would produce fear and panic. The effectiveness of the agent would depend on how long it remains in the liquid state and how widely it is dispersed throughout the building. Sarin, the most volatile nerve agent, evaporates at the same rate as water and is not considered persistent. The most stable nerve agent, V-agent or VX, is considered persistent because it takes several days or weeks to evaporate. Common nerve agents, their methods of contamination, and specific characteristics are shown in (► Table 34-1).

When a person is exposed to a nerve agent, symptoms will become evident within minutes. The symptoms include: pinpoint pupils, runny nose, drooling, difficulty breathing, tearing, twitching, diarrhea, convulsions or seizures, and loss of consciousness. Several mnemonics are used to help

Figure 34-14 In their normal states, nerve agents are liquids. They must be dispersed in aerosol form to be inhaled or absorbed by the skin.

Table 34-1 Common Nerve Agents

Nerve Agent	Method of Contamination	Characteristics
Tabun (GA)	Skin contact Inhalation	Disables the chemical connections between nerves and target organs
Soman (GD)	Skin contact Inhalation	Odor of camphor
Sarin (GB)	Inhalation	Evaporates quickly
V-agent (VX)	Skin contact	Oily liquid that can persist for weeks

Table 34-2 Symptoms of Nerve Agent Exposure

S – salivation (drooling)
L – lacrimation (tearing)
U – urination
D – defecation
G – gastric upset (upset stomach, vomiting)
E – emesis (vomiting)

you remember these symptoms. The mnemonic used most often by the fire service is SLUDGE (► Table 34-2).

The United States military has developed a nerve agent antidote kit, which contains two medications, atropine and pralidoxime chloride (2-PAM). These antidotes can be quickly injected into a person who has been contaminated by a nerve agent. The Mark 1 Nerve Agent Antidote Kit has been issued to many fire department hazardous materials teams (► Figure 34-15).

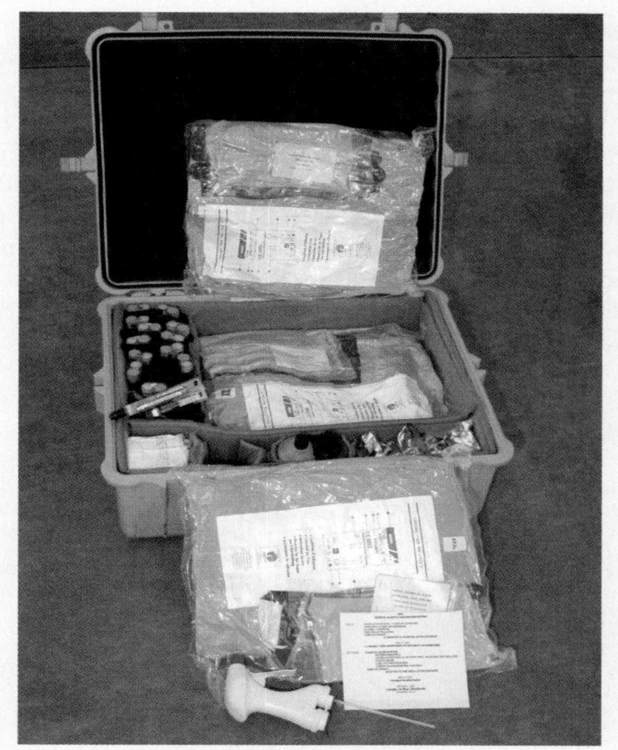

Figure 34-15 Mark 1 Nerve Agent Antidote Kit.

Blister Agents

Two chemical agents are generally used as blistering agents. Contact with either of these chemicals will cause the skin to blister.

Sulfur mustard (H) is a clear, yellow, or amber, oily liquid with a faint sweet odor of mustard or garlic. It vaporizes slowly at temperate climates and may be dispersed in an aerosol form.

Lewisite (L) is an oily, colorless-to-dark-brown liquid with an odor of geraniums.

Blister agents produce painful burns and blisters with even minimal exposure (► Figure 34-16). The major difference between these two blister agents is that lewisite causes pain

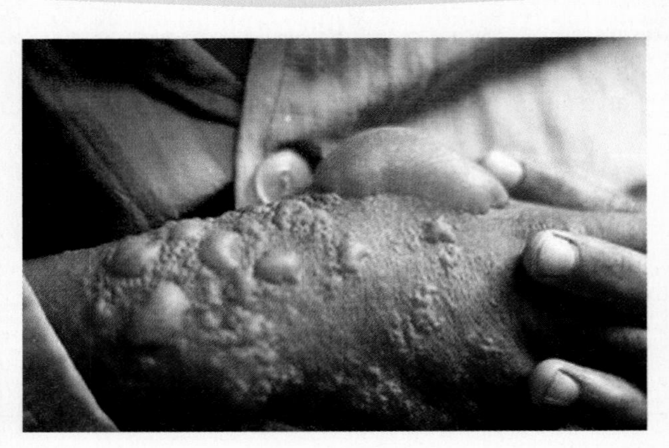

Figure 34-16 Typical effects of a blistering agent.

immediately upon contact with the skin, while the signs and symptoms from exposure to mustard gas may not appear for several hours. The patient will complain of burning at the site of the exposure, the skin will redden, and blisters will appear. The eyes may also itch, burn, and turn red. Inhalation of mustard gas produces significant respiratory damage.

Choking Agents

The two chemicals that are likely to be used as **choking agents** are **phosgene** (CG) and chlorine (CL). Both of these agents were used extensively as weapons in World War I, and both have several industrial uses. They cause severe pulmonary damage and asphyxia.

These choking agents are heavier than air, so they tend to settle in low areas. Subways, basements, and sewers are prime areas for these agents to accumulate. Fire fighters wearing appropriate PPE should quickly evacuate people from such areas when the possibility of phosgene or chlorine exposure exists.

Exposure to high concentrations of phosgene or chlorine will immediately irritate the eyes, nose, and upper airway. Within hours, the exposed individual will begin to develop pulmonary edema (fluid in the lungs). Those exposed to lower concentrations may not have any initial symptoms but can still experience respiratory damage.

Figure 34-17 Someone who has been exposed to phosgene or chlorine should be moved from the contaminated area and the skin flushed with water.

Although neither agent is absorbed through the skin, they can both cause skin burns on contact. **Decontamination** consists of removing exposed individuals from the area and flushing the skin with water (▲ Figure 34-17).

Blood Agents

Blood agents are highly toxic poisons that can cause death within minutes of exposure. The two most common blood agents are both cyanide compounds: hydrogen cyanide (AC) and cyanogen chloride (CK). Both of these chemicals are commonly used in industry.

Cyanide can be inhaled or ingested. Cyanide gas is typically associated with the death penalty and the gas chamber. A liquid cyanide mixed with fruit punch was used in the mass suicide of 913 members of a religious cult in Guyana in 1978.

The few symptoms associated with cyanide exposure occur quickly. Those exposed to the gas will begin gasping for air, and if enough agent is inhaled, the skin may begin to appear red. Seizures are also possible. A complete and rapid evacuation of the area is the most appropriate course of action.

Biologic Agents

Biologic agents are organisms that cause diseases and attack the body. Biologic agents include bacteria, viruses, and toxins. Some of these organisms, like anthrax, can live in the ground for years, while others are rendered harmless after being exposed to sunlight for only a short period of time. The highest potential for infection is through inhalation, although some biologic agents can be absorbed, injected through the skin, or ingested. The effects of a biologic agent depend on the specific organism or toxin, the dose, and the route of entry.

Some of these diseases, such as smallpox and pneumonic plague, are contagious and can be passed from person to person. Doctors are very concerned about the use of contagious diseases as weapons, because the resulting epidemic could overwhelm the healthcare system. There is growing concern that a terrorist group will use one of these agents against a civilian population.

Anthrax

Anthrax is an infectious disease caused by the bacteria *Bacillus anthracis,* which is typically found around farms and infects livestock. For use as a weapon, the bacteria is cultured to develop anthrax spores. The spores, in powdered form, can then be dispersed in a variety of ways (► Figure 34-18). Approximately 8,000 to 10,000 spores are typically required to cause an anthrax infection. Spores infecting the skin cause cutaneous anthrax; ingested spores cause gastrointestinal anthrax; and inhaled spores cause inhalational anthrax.

In 2001, four letters containing anthrax were mailed to locations in New York City; Boca Raton, Florida; and Washington D.C. Five people died after being exposed to the contents of these letters, including two postal workers who were exposed as the letters passed through postal sorting centers. Several major government buildings had to be shut down for months to be decontaminated.

Because these incidents followed shortly after the events of September 11, 2001, they caused tremendous public concern. Emergency personnel had to respond to thousands of incidents involving suspicious packages and citizens who believed that they might have been exposed to anthrax. Each of these situations had to be evaluated.

A 1970 publication from the World Health Organization is often used to show the dangers of anthrax. According to this publication, 110 pounds of anthrax sprayed over an urban center with 500,000 residents could cause 220,000 deaths.

Smallpox

Smallpox is a highly infectious disease caused by the virus variola. The disease kills approximately 30% of those infected. Although smallpox was once routinely encountered throughout the world, by 1980 it was successfully eradicated as a public health threat through the use of an extremely effective vaccine. Some nations maintained cultures of the disease for research purposes. International terrorist groups may have acquired the virus.

The smallpox virus could be dispersed over a wide area in an aerosol form. Infecting a small number of people could lead to a rapid spread of the disease throughout a targeted population because smallpox is highly contagious. The disease is easily spread by direct contact, droplet, and airborne transmission. Patients are considered highly infectious and should be quarantined until the last scab has fallen off (▼ Figure 34-19).

By 1980, when smallpox was eradicated, the worldwide vaccination program was discontinued. Today millions of people have never been vaccinated and millions more have reduced immunity because decades have passed since their last immunization.

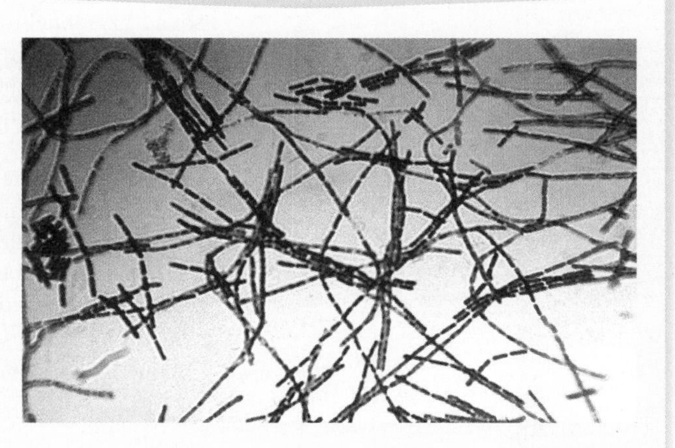

Figure 34-18 Anthrax spores can be dispersed in a variety of ways.

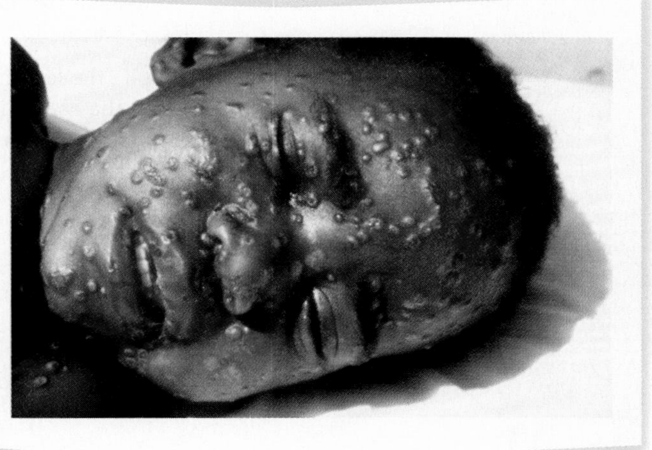

Figure 34-19 Smallpox is a highly contagious disease with a mortality rate of approximately 30%. *(Courtesy of Centers for Disease Control and Prevention)*

VOICES OF EXPERIENCE

"Because radio signals from police or fire department portable radio units could trigger the device, we had to await the return of the bomb squad for updates."

My engine company had been dispatched to assist police at one of the dozen high-rise buildings in our district. Outside the structure, police officers informed us that a suspicious package—a possible explosive device—was located on one of the upper floors.

I met with the Baltimore County Police Department's precinct commanders and bomb squad personnel to develop a plan and determine how fire department units could support their efforts. The bomb squad determined that the device could possibly be a pipe bomb. The safest course of action would have been to re-deploy our units and personnel approximately 1500′ away from the structure. However, should the device detonate, it would take us too long to reach the building and make our way to the upper floors where police were operating. The police and fire department commanders decided to stage fire department resources, both suppression and emergency medical personnel, three floors below the device's location. Working closely with bomb squad members, it had been determined that, considering the size of the possible explosive, the chosen area would be safe in the event of a detonation. Fire fighters took a variety of equipment to the staging area, including high-rise packs, self-contained breathing apparatus, forcible entry tools, and medical equipment.

Tense moments ensued as the two bomb squad members ascended the staircase. Because radio signals from police or fire department portable radio units could trigger the device, we had to await the return of the bomb squad for updates. After examining the suspicious package, the bomb squad determined that it was harmless. The threat of an explosive device in the building turned out to be a hoax, and the incident was safely handled through the cooperative efforts of both the police department and the fire department.

Despite the fact that the suspicious package in this incident did not pose any real danger in the end, it is important to remember that emergency personnel must treat all threats of terrorism with seriousness and professionalism. With so many hoax devices and false threats, fire fighters can easily become complacent. Yet an individual with the worst of intentions and a connection to the Internet can easily fashion a device or spread a lethal chemical that can terrorize a small town or a large city.

Dennis R. Krebs
Baltimore County Fire Department
Towson, Maryland

Plague

Plague is caused by the bacterium *Yersinia pestis,* which is commonly found on rodents. There are three forms of plague: bubonic, septicemic, and pneumonic. The bacterium is most often transmitted to humans by fleas that feed on infected animals and then bite humans. Those who are bitten generally develop bubonic plague, which attacks the lymph nodes (► Figure 34-20). Pneumonic plague can be contracted by inhaling the bacterium.

The plague bacteria can survive for weeks in water, moist soil, or grains. It has been cultured for distribution as a weapon in aerosol form. Inhalation of the aerosol form would put the target population at risk for pneumonic plague.

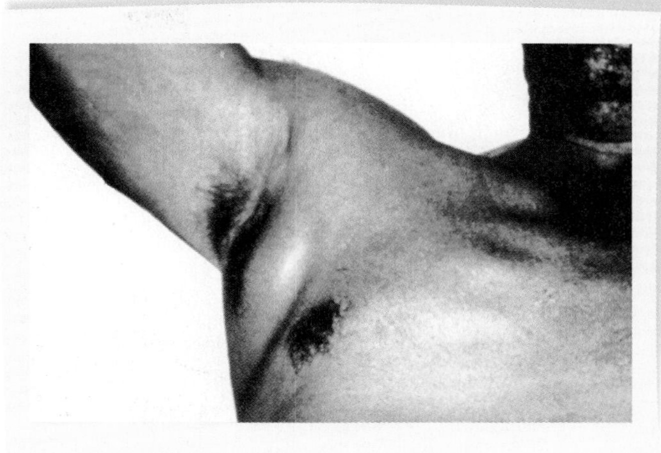

Figure 34-20 A bubo, one of the symptoms of plague, is a swollen, painful lymph node.

Dealing with Biologic Agents

It is unlikely that fire fighters or emergency responders would even recognize that a biologic weapon had been released in their communities. The time period between the actual infection and the appearance of symptoms is known as the **incubation period**. The symptoms of a biologic agent attack would typically manifest over a period of days after the exposure. If left untreated, many of these diseases will cause death in a large proportion of those infected.

The Centers for Disease Control and Prevention (CDC) and area hospitals would typically be the first to recognize the situation, after identifying significant numbers of people arriving at their facilities with similar symptoms. Multiple medical calls with patients exhibiting similar symptoms could provide a clue, especially if the location is considered to be a potential target.

Most experts believe that a biologic weapon would probably be spread by a device similar to a garden sprayer or crop-dusting plane. After the release of a biologic weapon, residents would begin experiencing flu-like symptoms, although the symptoms may not appear for 2 to 10 days.

Fire fighters or emergency responders treating people who may have been exposed to a biologic agent should follow the recommendations of their departments regarding **universal precautions** (protective measures for use in dealing with objects, blood, body fluids, or other potential exposure risks of communicable diseases). These measures include wearing gloves, masks, eye protection, and surgical gowns when treating patients.

Emergency responders who exhibit flu-like symptoms after a potential terrorist incident should seek medical care immediately. It is important that fire fighters report any possible exposure to department officials. Postincident actions may include medical screening, testing, or vaccinations.

Radiological Agents

Although unlikely, the possibility that terrorists could detonate a nuclear weapon is a threat that cannot be ignored. Several countries possess nuclear weapons and a number of others are attempting to develop or acquire them. The only limiting factor is the ability to acquire weapons-grade fuel, which is needed to cause a nuclear explosion. Some experts fear that terrorists could obtain and detonate a **suitcase nuke**, a nuclear weapon that is small enough to fit into a small package.

A more probable scenario is the use of **radiological agents** by terrorists. Terrorists could employ a variety of alternative approaches to disperse radiological agents and contaminate an area with radioactive materials. This threat is very different from a nuclear detonation and much more likely to occur.

The Dirty Bomb

The **radiation dispersal device** (RDD) or "dirty bomb" is an area of serious concern. An RDD is described as "any device that causes the purposeful dissemination of radioactive material across an area without a nuclear detonation."[2] Packing radioactive material around a conventional explosive device could contaminate a wide area. The size of the affected area would depend on the amount of radioactive material and the power of the explosive device.

To limit this threat, radioactive materials, even in small amounts, must be kept secure and protected. Radioactive materials are widely used in industry and health care, particularly with x-ray machines. The necessary materials could be obtained from medical clinics, nuclear power plants, university research facilities, and some industrial complexes (► Figure 34-21). These materials are also routinely transported in a variety of large and small containers. Terrorists could construct a dirty bomb with a small quantity of stolen radioactive material.

Security experts are also concerned that a large explosive device, such as a truck bomb, could be detonated near a nuclear power plant. Such an explosion could damage the reactor containment vessel sufficiently to release radioactive material. A large aircraft also could be used to damage a

Figure 34-21 The materials needed to build a radiation dispersal device can be obtained from a number of places, including a nuclear power plant.

Fire Fighter Safety Tips

Casualty Management After Detonation of a Nuclear Weapon

- Do not enter an area with possible contamination until radiation levels have been measured and found to be safe.
- Enter the area only to save lives; radiation levels will be very high.
- Wear PPE and SCBA.
- All personnel and equipment must be decontaminated after leaving the contaminated area. Because radioactive dust collects on clothing, all clothing should be removed and discarded after leaving the area. Failure to do so will result in continued radiation exposure to yourself and others.
- Wash thoroughly with lukewarm water as soon as possible after leaving the scene, even if you have been through a decontamination process.
- Do not eat, drink, or smoke while exposed to potential radioactive dust or smoke.
- Use radiation monitoring devices to map the location of areas with high radioactivity. Monitoring devices should be wrapped in plastic bags to prevent contamination.
- Vehicles should be washed before they leave the scene. The only exception is emergency units transporting critical victims.
- Life-threatening injuries are more serious than radioactive contamination. Treat such injuries before decontaminating the individual.
- Alert local hospitals that they may receive radioactive-contaminated victims.

nuclear reactor much as aircraft were used to destroy the World Trade Center.

Radiation

Radioactive materials release energy in the form of electromagnetic waves or energy particles. Radiation cannot be detected by the normal senses of smell and taste. Several instruments can detect the presence of radiation and measure dose rates. Fire fighters should become familiar with the radiation detectors used by their departments. There are three types of radiation: alpha particles, beta particles, and gamma rays.

Alpha particles quickly lose their energy and can travel only 1" to 2" from their source. Clothing or a sheet of paper can stop this type of energy. However, if ingested or inhaled, alpha particles can damage a number of internal organs.

Beta particles are more powerful, capable of traveling 10' to 15'. Heavier materials such as metal, plastic, and glass can stop this type of energy. Beta radiation can be harmful to both the skin and eyes, and ingestion or inhalation will damage internal organs.

Gamma rays can travel significant distances, penetrate most materials, and pass through the body. Gamma radiation is the most destructive to the human body. The only materials with sufficient mass to stop gamma radiation are concrete, earth, and dense metals such as lead.

Effects of Radiation

Symptoms of low-level radioactive exposure might include nausea and vomiting. Exposure to high levels of radiation can cause bone marrow destruction, nerve and digestive system damage, and radioactive skin burns. An extreme exposure may cause death rapidly; however, it often takes a considerable amount of time before the signs and symptoms of radiation poisoning become obvious. As the effects progress over the years, a prolonged death due to leukemia or carcinoma is likely.

A contaminated person will have radioactive materials on the exposed skin and clothing. In severe cases, the contamination could penetrate to the internal organs of the body. Decontamination procedures are required to remove the radioactive materials. Decontamination must be complete and thorough; the situation could escalate rapidly if decontamination procedures are inadequate. The contamination could quickly spread to ambulances, hospitals, and other locations where contaminated individuals or items are transported.

Radioactive materials expose rescuers to radiation injury as well as continue to affect the exposed person. Anyone who comes in contact with a contaminated person must be protected by appropriate PPE and shielding, depending on the particular contaminant. Rescuers and medical personnel will have to be decontaminated after handling a contaminated person.

A person can be exposed to radiation without coming into direct contact with a radioactive material through inhalation, skin absorption, or ingestion. A person can also be injured by exposure to radiation, without being contaminated by the radioactive material itself. Someone who has been exposed to radiation, but has not been contaminated by radioactive materials, requires no special handling.

There are three ways to limit exposure to radioactivity. Keep the time of the exposure as short as possible. Stay as far

Figure 34-22 A personal dosimeter measures the amount of radioactive exposure received by an individual.

away from the source of the radiation as possible, and use shielding to limit the amount of radiation absorbed by the body. Firefighting PPE and breathing apparatus will generally provide adequate shielding against alpha particles; breathing apparatus is needed to provide protection against beta radiation. Heavy shielding is needed to provide protection from sources emitting gamma radiation.

If radioactive contamination is suspected, everyone who enters the area should be equipped with a **personal dosimeter** to measure the amount of radioactive exposure (▲ Figure 34-22). Rescue attempts should be made as quickly as possible to reduce exposure time.

Operations

Responding to a terrorist incident puts fire fighters and other emergency personnel at risk. Responders must ensure their own safety at every incident; however, a terrorist incident has an extra dimension of risk. Because the terrorist's objective is to cause as much harm as possible, emergency responders are just as likely to be targets as ordinary civilians.

In many cases, the first units to respond will not be dispatched for a known WMD or terrorist incident. The initial dispatch might be for an explosion, for a possible hazardous materials incident, for a single person with difficulty breathing, or for multiple sick patients with similar symptoms. Emergency responders will usually not know that a terrorist incident has occurred until personnel on the scene begin to piece together information gained by their own observations and from witnesses.

If appropriate precautions are not taken, the initial responders may find themselves in the middle of a dangerous situation before they realize what has happened. Initial responders should be aware of any factors that suggest the possibility of a terrorist incident and immediately implement appropriate procedures. The possibility of a terrorist incident should be considered when responding to any location that has been identified as a potential terrorist target. It could be difficult to determine the true nature of the situation until a scene size-up is conducted.

Fire Marks

In 1993, the first units to respond to the World Trade Center thought a transformer explosion had occurred somewhere beneath the huge complex. As responders explored the smoke-filled interior, they realized that a large explosion had occurred in the underground parking area. As they assembled the information, they determined that terrorists had detonated a truck bomb, causing tremendous destruction and filling the twin towers with smoke.

At the same location in 2001, many of the first responders saw a plane crash into one of the towers and believed that they were responding to a terrible accident. They only realized the true nature of the situation when a second plane crashed into the other tower.

Initial Actions

Fire fighters should approach a known or potential terrorist incident just as they would a hazardous materials incident. The possibility that chemical, biologic, or radiological agents are involved cannot be ruled out until a hazardous materials team checks the area with detection and monitoring instruments. Apparatus and personnel should, if possible, approach the scene from a position that is uphill and upwind. Emergency responders should don PPE, including SCBA. Additional arriving units should be staged an appropriate distance away from the incident.

The first units to arrive should establish an outer perimeter to control access to and from the scene. They should deny access to all but emergency responders, and they should prevent potentially contaminated individuals from leaving the area before they have been decontaminated. The perimeter must completely surround the affected area, to keep people who were not initially involved from becoming additional victims.

Incident Command should be established in a safe location, which could be as far as 3,000′ from the actual incident scene. The command post must be outside the area of possible contamination and beyond the distance where a secondary device may be planted. The initial priority should be

Teamwork Tips

A hazardous materials team should respond to any potential or suspected terrorist incident. The hazardous materials personnel should test for chemical, biologic, or radiological contaminants using appropriate test instruments.

to determine the nature of the situation, the types of hazards that could be encountered, and the magnitude of the problems that will have to be faced.

An initial reconnaissance or "recon" team should be sent out to quickly examine the involved area and to determine how many people are involved. Proper use of PPE, including SCBA, is essential for the recon team, and the initial survey must be conducted very cautiously, although as rapidly as possible. The possibility that chemical, biologic, or radiological agents are involved cannot be ruled out until qualified personnel with instruments and detection devices have surveyed the area. Responders should take special care not to touch any liquids or solids or to walk through pools of liquid. Emergency responders who become contaminated must not leave the area until they have been decontaminated.

A process of elimination may be required to determine the nature of the situation. Occupants and witnesses should be asked if they observed any unusual packages or detected any strange odors, mists, or sprays. A large number of casualties, who have no outward signs of trauma, could indicate a possible chemical agent exposure. Possible symptoms include trouble breathing, skin irritations, or seizures.

A visible vapor cloud would also be a strong indicator of a chemical release, which could be an accident or a terrorist attack. Responders should look out for dead or dying animals, insects, or plant life as indications of a chemical agent release.

When approaching the scene of an explosion, responders should consider the possibility of a terrorist bombing incident and recall the potential for secondary explosive devices. Responders should note any suspicious packages and notify the IC. Fire fighters who have not been specially trained should never approach a suspicious object. Explosive ordnance disposal personnel should examine any suspicious articles and disable them.

Interagency Coordination

If a terrorist incident is suspected, the IC should consult immediately with local law enforcement officials. If a mass casualty situation is evident, the IC should notify area hospitals and activate the local medical response system. Local technical rescue teams should be requested to evaluate structural damage and initiate rescue operations.

State emergency management officials should be notified as soon as possible. This will help ensure a quick response by both state and federal resources to a major incident. Large-scale search-and-rescue incidents could require the response of Urban Search and Rescue (USAR) task forces through the Federal Emergency Management Agency (FEMA). Medical response teams, such as Disaster Medical Assistance Teams (DMAT) may be needed for incidents involving large numbers of people.

An Emergency Operations Center (EOC) can coordinate the actions of all involved agencies in a large-scale incident, particularly if terrorism is involved. The EOC is usually set up in a predetermined remote location and is staffed by experienced command and staff personnel (► Figure 34-23). The IC, who is at the scene of the incident, should provide detailed situation reports to the EOC and request additional resources as needed.

Fire fighters must remember that a terrorist incident is also a crime scene. To avoid destroying important evidence that could lead to a conviction of those responsible, responders should not disturb the scene any more than is necessary. Where possible, law enforcement personnel should be consulted prior to overhaul and before any material is removed from the scene. Fire fighters should also realize that one or more terrorists could be among the injured. Be alert

Fire Fighter Tips

Additional Resources for Terrorist Incidents

Resources	Action
Air Evacuation	Air transportation of injured
Air Supply Unit	Refill SCBA cylinders
Bomb Squad	Render explosive devices safe
Emergency Medical Services	Emergency medical care
Hazardous Materials Unit	Neutralize hazardous substances, including chemical agents
Health Department	Technical advice on chemical and biologic agents
Local Law Enforcement	Law enforcement, perimeter control, and evidence recovery
Structural Collapse Teams	Search and stabilize collapsed structures

Fire Fighter Tips

Emergency responders at a terrorist event may later be called to testify in court. Make mental notes if you must disturb or move any evidence to rescue someone in immediate danger. Document this information as soon as possible.

Figure 34-23 An EOC is set up in a predetermined location for large-scale incidents.

Figure 34-24 Mass decontamination procedures may be required to handle a large group of contaminated victims.

for threatening behavior, and aware of anyone who seems determined to leave the scene.

Decontamination

Everyone who is exposed to chemical, biologic, or radiological agents must be decontaminated to remove or neutralize any chemicals or substances on bodies or clothing. Decontamination should occur as soon as possible to prevent further absorption of a contaminant and to reduce the possibility of spreading the contamination. Equipment will also have to be decontaminated before it leaves the scene.

The perimeter must fully surround the area of known or suspected contamination. Every effort must be made to avoid contaminating any additional areas, particularly hospitals and medical facilities. Qualified personnel must monitor the perimeter with instruments or detection devices to ensure that contaminants are not spread.

Standard decontamination procedures usually involve a series of stations. At each station, clothing and protective equipment are removed and the individual is cleaned. Some contaminants require only soap and water, but chemical and biologic agents may require special neutralizing solutions.

A terrorist incident that results in a large number of casualties may require a different decontamination process. Special procedures for **mass decontamination** have been devised for incidents involving hundreds or thousands of people. These procedures use master stream devices from engine companies and aerial apparatus to create high-volume, low-pressure showers. This allows large numbers of people to be decontaminated rapidly (► Figure 34-24).

Emergency responders who enter a contaminated area must be decontaminated before they can leave the area (► Figure 34-25). Contaminated clothing and equipment must remain inside the perimeter. Depending on the involved agent, it may be necessary to maintain this perimeter for an extended period, until the entire incident scene can be decontaminated. It could take weeks or months to fully decontaminate some incident scenes.

Figure 34-25 All emergency responders and victims who have been exposed to chemical, radiological, or biologic agents must be decontaminated.

Mass Casualties

A terrorist or WMD incident may result in a large number of casualties. Some scenarios could involve hundreds or thousands of injuries. Special mass casualty plans are essential to manage this type of situation, which would quickly overwhelm the normal capabilities of most emergency response systems. Mass casualty plans usually involve resources from multiple agencies to handle large numbers of patients efficiently.

The basic principles of mass casualty operations are described in Chapter 24, *Emergency Medical Care*. A terrorist incident could involve several additional complications and considerations, including the possibility of contamination by chemical, biologic or radiological agents. The mass casualty plan must be expanded to address these problems.

If contamination is suspected, the plan must ensure that it does not spread beyond a defined perimeter. In some cases, patients may be decontaminated as they are moved to the triage and treatment areas (▶ Figure 34-26). This will keep these areas free of contaminants. In other cases, the triage and treatment areas may be considered contaminated zones. Patients would be decontaminated before they are transported from the scene.

During the early stages of an incident, it may be difficult to determine what agent was used so that appropriate treatment and decontamination procedures can be instituted. It may be necessary to quarantine exposed individuals within an area that is assumed to be contaminated until laboratory test results are evaluated. The strategic plan for this situation must assume that any personnel who provide assistance or treatment to patients during this period will also become contaminated, as will any equipment or vehicles used in the contaminated area.

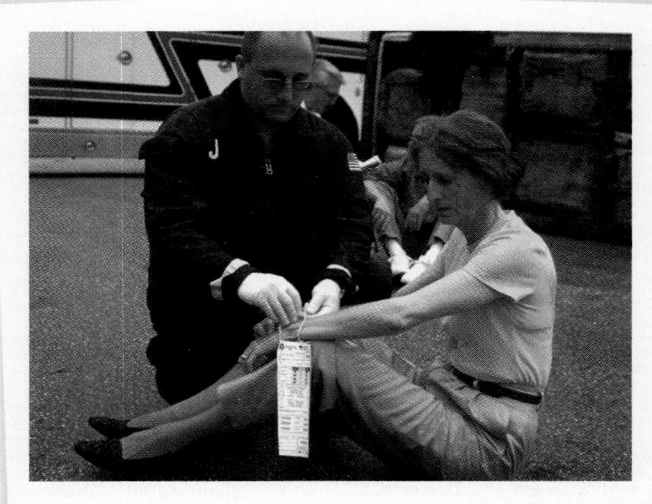

Figure 34-26 Patients should be decontaminated before they are placed in triage and treatment areas.

Additional Resources

A terrorist incident will likely result in a massive response by local, state, and federal government agencies. The FBI is the lead agency for crisis management, and FEMA is the lead agency for consequence management during a terrorist incident. Several federal agencies have specific responsibilities regarding terrorist incidents.

Funding for programs to train and equip local emergency responders has increased significantly since September 11, 2001. A Cabinet-level Department of Homeland Security has been established in the United States to coordinate these programs.

The Homeland Security Advisory System is designed to provide quick, comprehensive information concerning the potential threat of terrorist attacks or threat levels. Threat conditions can apply nationally, regionally, by industry, or by specific target. You should be aware of the current Homeland Security Threat Level at all times when on duty. Implement listed protective measures as directed in your department and recommended listed actions to the public.

Low Condition (Green)

This condition is declared when there is a low risk of terrorist attacks. Agencies should consider the following general measures in addition to any agency-specific protective measures they develop and implement:

1. Refine and exercise pre-planned protective measures.
2. Ensure that personnel receive proper training on the Homeland Security Advisory System and on specific pre-planned departmental or agency protective measures.
3. Institute a process to ensure that all facilities and regulated sectors are regularly assessed for vulnerability to terrorist attacks, and that all reasonable measures are taken to mitigate these vulnerabilities.

Advise the public to:

- Develop a family emergency plan, including family meeting locations in the event of an attack, and a contact point, such as the home of a relative in a distant state.
- Keep recommended immunizations up to date.
- Know how to turn off power, gas, and water service to homes or businesses. Keep tools available to do so.
- Know what hazardous materials are stored in homes or businesses and how to properly dispose of unneeded chemicals.
- Support the efforts of local emergency responders (fire fighters, law enforcement officers, and emergency medical service personnel).
- Know the natural hazards that are prevalent in your area, and what measures can be taken to mitigate these hazards.
- Maintain a first-aid kit.

Guarded Condition (Blue)

This condition is declared when there is a general risk of terrorist attacks. In addition to taking the protective measures listed under the previous Threat Condition, agencies should consider the following general measures as well as the agency-specific protective measures:

1. Check communications with designated emergency response or command locations.

2. Review and update emergency response procedures.
3. Provide the public with any information that would strengthen its ability to act appropriately.

Advise the public to:

- Continue normal activities, but be watchful for suspicious activities. Report criminal activity to local law enforcement.
- Review family emergency plans, including family meeting locations and distant points of contact in the event of an attack.
- Be familiar with the local natural and technological (man-made) hazards in your community and know what measures can be taken to mitigate these hazards.
- Maintain a first-aid kit.
- Increase individual or family emergency preparedness through training, maintaining good physical fitness and health, and storing a three-day supply of food, water, and emergency supplies, such as medications.
- Monitor local and national news for terrorist alerts.

Elevated Condition (Yellow)

An Elevated Condition Threat Alert is declared when there is a significant risk of terrorist attacks. In addition to the protective measures listed under the previous threat conditions, agencies should consider the following general measures as well as the protective measures that they will develop and implement:

1. Increase surveillance of critical locations.
2. Coordinate emergency plans as appropriate with nearby jurisdictions.
3. Address whether the precise characteristics of the threat require the further refinement of pre-planned protective measures.
4. Implement, as appropriate, contingency and emergency response plans.

Advise the public to:

- Continue normal activities, but report suspicious activities to local law enforcement agencies.
- Become trained/certified in first aid or emergency response protocol. Advise the public of local training programs.
- Maintain a first-aid kit.
- Become active in your local Neighborhood Crime Watch program. Report suspicious activities to local law enforcement.
- Network with your family, neighbors, and community for mutual support during a disaster or terrorist attack. Update previously established family emergency plans, including a distant point of contact. Update preparations to shelter in place, such as preparing of materials to seal openings to your home and shut down HVAC systems.
- Know the critical facilities and specific hazards located in your community, and report suspicious activities at or near these sites.
- Increase individual or family emergency preparedness through training, maintaining good physical fitness and health, and storing a three-day supply of food, water, and emergency supplies, such as medications.
- Monitor media reports concerning terrorism situations.

High Condition (Orange)

A High Condition Threat Alert is declared when there is a high risk of terrorist attacks. In addition to the protective measures listed under the previous Threat Conditions, agencies should consider the following general measures as well as agency-specific protective measures:

1. Coordinate necessary security efforts with Federal, State, and local law enforcement agencies or any National Guard or other appropriate armed forces organizations.
2. Take additional precautions at public events and consider alternative venues or even cancellation.
3. Prepare to execute contingency procedures, such as moving to an alternate site or dispersing a workforce.
4. Restrict threatened facility access to essential personnel only.

Advise the public to:

- Expect some delays, baggage searches, and restrictions as a result of heightened security at public buildings and facilities.
- Continue to monitor world and local events as well as local government threat advisories.
- Report suspicious activities at or near critical facilities to local law enforcement agencies by calling 9-1-1.
- Avoid leaving unattended packages or briefcases in public areas.
- Inventory and organize emergency supply kits (a three-day food and water supply) and discuss emergency plans with family members. Re-evaluate the meeting location based on any specific threat, and ensure that a mechanism to get to the location exists. Identify a distant point of contact for your family. Ensure that supplies for sheltering in place are maintained. Maintain a first-aid kit.
- Consider taking reasonable personal security precautions. Be alert to your surroundings, avoid placing yourself in a vulnerable situation, and monitor the activities of your children.
- Maintain close contact with your family and neighbors to ensure their safety and emotional welfare.

Severe Condition (Red)

Declaring a Severe Condition reflects a severe risk of terrorist attacks. Under most circumstances, the protective measures for a Severe Condition are not intended to be sustained for extended periods of time. In addition to the protective measures listed under the previous Threat Conditions, agencies should consider the following general measures as well as the agency-specific protective measures:

1. Increase or redirect personnel to address critical emergency needs.
2. Assign emergency response personnel and pre-position and mobilize specially trained teams or resources.
3. Monitor, redirect, or constrain transportation systems.
4. Close public and government facilities.

Advise the public to:

- Report suspicious activities and call 9-1-1 for immediate response.
- Expect delays, searches of purses and bags, and restricted access to public buildings.
- Expect traffic delays and restrictions.
- Take personal security precautions (such as avoiding target facilities) to avoid becoming a victim of a crime or a terrorist attack.
- Avoid crowded public areas and gatherings.
- Do not travel into areas affected by an attack.
- Keep emergency supplies accessible and your automobile's fuel tank full.
- Maintain a first-aid kit.
- Be prepared to evacuate your home or to shelter in place on the order of local authorities. Have supplies on hand for this purpose, including a three-day supply of water, food, and medication, and an identified distant point of contact.
- Be suspicious of persons taking photographs of critical facilities, asking detailed questions about physical security, or who are dressed inappropriately for weather conditions (suicide bombers). Report these incidents immediately to law enforcement.
- Closely monitor news reports and Emergency Alert System (EAS) radio/TV stations.
- Assist neighbors who may need help.
- Avoid passing unsubstantiated information.

Source: Homeland Security Presidential Directive #3, Washington, DC; March 12, 2002; and Homeland Security Advisory System Recommendations for Individuals, Families, Neighborhoods, Schools, and Businesses, The American Red Cross, www.redcross.org, March 2003.

Several Internet resources provide additional information. The U.S. Army Medical Research Institute of Chemical Defense (http://chemdef.apgea.army.mil/), the U.S. Army Medical Research Institute of Infectious Diseases (http://www.usamriid.army.mil/), and the National Homeland Security Knowledgebase (http://www.twotigersonline.com/resources.html) provide useful information and training materials.

Wrap-Up

Ready for Review

The fire service is responsible for responding to terrorist incidents. Fire fighters who respond to these incidents must follow standard operating procedures and safety precautions. Training in the field of terrorism and WMD is available from several sources.

Fire fighters should become familiar with potential terrorist targets in the local area and with the tactics, devices, and agents used by terrorists. Terrorists may use explosives and incendiary devices, chemical agents, biologic agents, nuclear weapons, and radiation dispersal devices. Fire fighters should understand the basic characteristics of these agents and devices and become familiar with the signs and symptoms of exposure.

Fire fighters must take special precautions to ensure their own safety at terrorist incidents. They should anticipate the release of a chemical agent, biologic agent, or radiological device. They should maintain a safe distance, remaining uphill and upwind until the situation is evaluated. Complete and appropriate PPE, including SCBA, must be used by anyone who could come in contact with any of these agents; serious health effects may result if such equipment is not properly used.

Emergency responders should remember that terrorists intend to cause as much harm as possible. Responders should always be aware of the potential for secondary devices at terrorist incidents. These devices may be planted to kill or injure emergency responders and law enforcement personnel.

A terrorist incident involving chemical, biologic, or radiological materials will require the implementation of appropriate decontamination procedures. All contaminated people, equipment, and vehicles must be decontaminated before leaving the scene. Decontamination may be complicated by large numbers of exposed individuals.

Terrorist incidents have the potential for massive numbers of casualties. Emergency response personnel will have to implement mass casualty plans to assemble the necessary resources and ensure that care is appropriately directed.

If a terrorist attack occurs, several agencies and organizations will become involved. Federal, state, and local agencies will perform designated functions and provide a variety of services to reinforce local resources.

Chief Concepts

- The possibility of secondary devices must be considered at all terrorist incidents.
- Fire fighters should become familiar with potential terrorist targets in their response area.
- Most chemical agents produce symptoms almost immediately upon exposure.
- Biological agents usually produce flu-like symptoms following an incubation period of 2 to 10 days.
- When the presence of radioactive material is suspected, fire fighters should limit their exposure time.
- Fire fighters should establish a staging area a safe distance from the scene of a terrorist incident and follow the direction of the IC.
- PPE will provide the fire fighter with limited protection from most agents used by terrorists.

Hot Terms

Agroterrorism The intentional act of using chemical or biologic agents against the agricultural industry or food supply.

Alpha particles A type of radiation that quickly loses energy and can travel only 1″ to 2″ from its source. Clothing or a sheet of paper can stop this type of energy.

ANFO An explosive material containing ammonium nitrate fertilizer and fuel oil.

Anthrax An infectious disease spread by the bacteria *Bacillus anthracis*; typically found around farms, infecting livestock.

Beta particles A type of radiation that is capable of traveling 10′ to 15′. Heavier materials such as metal, plastic and glass can stop this type of energy.

Biologic agents Disease-causing bacteria, viruses, and other agents that attack the human body.

Blister agents Chemicals that cause the skin to blister.

Chlorine A yellowish gas that is about 2.5 times heavier than air and slightly water-soluble. It has many industrial uses but also damages the lungs when inhaled (a choking agent).

Choking agent A chemical designed to inhibit breathing, and typically intended to incapacitate rather than kill.

Cyanide A highly toxic chemical agent that attacks the circulatory system.

Wrap-Up

Cyberterrorism The intentional act of electronically attacking government or private computer systems.

Decontamination The physical or chemical process of removing any form of contaminant from a person, an object, or the environment.

Ecoterrorism Terrorism directed against causes that radical environmentalists think would damage the earth or its creatures.

Forward staging area A strategically placed area, close to the incident site, where personnel and equipment can be held in readiness for rapid response to an emergency event.

Gamma radiation A type of radiation that can travel significant distances, penetrating most materials and passing through the body. Gamma radiation is the most destructive to the human body.

Improvised explosive device An explosive or incendiary device that is fabricated in an improvised manner.

Incubation period Time period between the initial infection by an organism and the development of symptoms.

Lewisite Blister-forming agent that is an oily, colorless-to-dark brown liquid with an odor of geraniums.

Mark 1 Nerve Agent Antidote Kit A military kit containing antidotes that can be administered to victims of a nerve agent attack.

Mass decontamination Special procedures for incidents that involve large numbers of people. Master stream devices from engine companies and aerial apparatus provide high-volume, low-pressure showers so that large numbers of people can be decontaminated rapidly.

Nerve agents Toxic substances that attack the central nervous system in humans.

Personal dosimeters Devices that measure the amount of radioactive exposure to an individual.

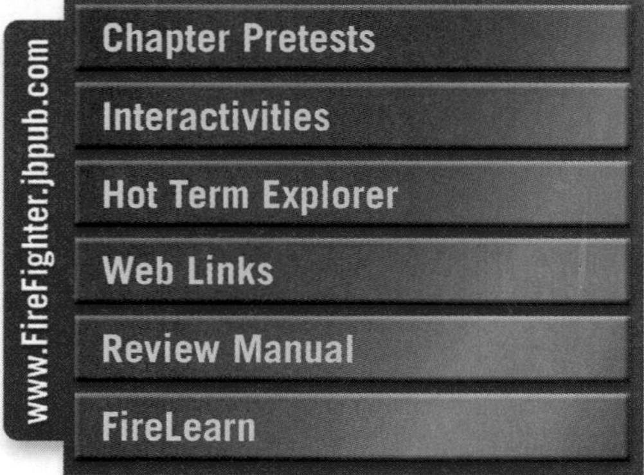

Phosgene A chemical agent that causes severe pulmonary damage. A by-product of incomplete combustion.

Pipe bomb A device created by filling a section of pipe with an explosive material.

Plague An infectious disease caused by the bacterium *Yersinia pestis*; commonly found on rodents.

Radiation dispersal device Any device that causes the purposeful dissemination of radioactive material without a nuclear detonation; a dirty bomb.

Radiological agents Materials that emit radioactivity.

Sarin A nerve agent that is primarily a vapor hazard.

Secondary device An explosive device designed to injure emergency responders who have responded to an initial event.

Smallpox A highly infectious disease caused by the virus variola.

Soman A nerve gas that is both a contact and a vapor hazard that has the odor of camphor.

Suitcase nuke A nuclear explosive device that is small enough to fit in a suitcase.

Sulfur mustard A clear, yellow, or amber oily liquid with a faint sweet odor of mustard or garlic that may be dispersed in an aerosol form. It causes blistering of exposed skin.

Tabun A nerve gas that is both a contact and a vapor hazard that operates by disabling the chemical connection between nerve and target organs.

Triage The process of sorting victims based on the severity of injury and medical needs to establish treatment and transportation priorities.

Universal precautions Procedures for infection control that treat blood and certain bodily fluids as capable of transmitting bloodborne diseases.

V-agent A nerve agent, principally a contact hazard; an oily liquid that can persist for several weeks.

Weapons of mass destruction (WMD) A weapon intended to cause mass casualties, damage, and chaos.

Fire Fighter in Action

You are on duty when the threat level for possible terrorist activity is raised to "Orange." Captain Thomas reviews the department's Standard Operating Procedure for Potential Terrorist Incidents with the crew and tells you to check all of the medical supplies and hazardous materials equipment that are carried on your apparatus.

During the afternoon rush hour, your engine company and an ambulance respond to a report of an unconscious person at a subway station. En route, the dispatcher advises Captain Thomas of additional information indicating that there may be several patients feeling ill at this location.

1. Which of the following factors would cause you to suspect that this is a terrorist incident?
   - **A.** Terrorist threat level is Orange.
   - **B.** Subway station during rush hour.
   - **C.** Multiple patients with similar symptoms suggest some type of exposure to a toxin or irritant could be involved.
   - **D.** There is no indication of an accidental cause.
   - **E.** All of the above
2. The subway station is a confusing scene with many passengers entering and exiting. You see several individuals sitting on the sidewalk close to the entrance. They look disoriented and confused. Concerned passers-by appear to be providing first aid. Your first priority is to:
   - **A.** Don your PPE and SCBA and enter the subway station to search for additional victims.
   - **B.** Begin to assess the patients and provide emergency medical care.
   - **C.** Establish a perimeter around the scene and prevent anyone from entering the subway station.
   - **D.** Set up a PPV fan to blow fresh air into the subway entrance.
   - **E.** Set up an exhaust fan to draw air out of the subway entrance.
3. A witness tells you that several individuals came out of the subway station complaining of burning eyes, a choking sensation, nausea, and dizziness. Some of them have already left the scene, while others are unable to walk without assistance. The witness does not have any information about the situation inside the station. What actions should be taken now?
   - **A.** Establish Incident Command and request additional units.
   - **B.** Request police assistance to secure the perimeter and control traffic.
   - **C.** Advise the Subway Authority to prevent the discharge of any additional passengers at this station.
   - **D.** Move patients away from the subway entrance and establish a treatment area.
   - **E.** All of the above
4. As HazMat units arrive, one of the first assignments is to enter the subway station to evacuate anyone who is still inside and determine if there are any additional patients. The crew that is assigned to enter the subway station should wear:
   - **A.** Latex gloves and facemasks
   - **B.** Firefighting PPE and SCBA
   - **C.** Level B hazardous materials ensembles
   - **D.** Level A hazardous materials ensembles
   - **E.** Nomex coveralls
5. Two additional patients who are coughing and have extremely red and watering eyes exit from the subway station. The patients report that they observed a canister on the floor of the station. A fine white, powdered residue surrounded the canister. The Incident Commander should call for the following actions:
   - **A.** Order an engine company to begin decontaminating the patients.
   - **B.** Order ambulances to transport all of the patients to a nearby hospital immediately.
   - **C.** Send the crew back into the subway station to retrieve the canister and bring it outside for close examination.
   - **D.** Assign a hazardous materials team to attempt to identify the substance.
   - **E.** Implement the procedures for a suspected anthrax incident.

Fire Prevention and Public Education

Chapter 35

NFPA 1001 Standard

Fire Fighter I

5.5 *Prevention, Preparedness, and Maintenance.* This duty involves performing activities that reduce the loss of life and property due to fire through hazard identification, inspection, education, and response readiness, according to the following job performance requirements.

5.5.1 Perform a fire safety survey in a private dwelling, given survey forms and procedures, so that fire and life-safety hazards are identified, recommendations for their correction are made to the occupant, and unresolved issues are referred to the proper authority.

5.5.1 (A) *Requisite Knowledge.* Organizational policy and procedures, common causes of fire and their prevention, the importance of a fire safety survey and public education programs to fire department public relations and the community, and referral procedures.

5.5.1 (B) *Requisite Skills.* The ability to complete forms, recognize hazards, match findings to preapproved recommendations, and effectively communicate findings to occupants or referrals.

5.5.2 Present fire safety information to station visitors or small groups, given prepared materials, so that all information is presented, the information is accurate, and questions are answered or referred.

5.5.2 (A) *Requisite Knowledge.* Parts of informational materials and how to use them, basic presentation skills, and departmental standard operating procedures for giving fire station tours.

5.5.2 (B) *Requisite Skills.* The ability to document presentations and to use prepared materials.

Fire Fighter II

6.5 *Prevention, Preparedness, and Maintenance.* This duty involves performing activities related to reducing the loss of life and property due to fire through hazard identification, inspection, and response readiness, according to the following job performance requirements.

Additional NFPA Standards

NFPA 1, *Uniform Fire Code*

NFPA 101, *Life Safety Code®*

Knowledge Objectives

After studying this chapter, you will be able to:

- Assist during a fire safety survey of a residential occupancy.
- Conduct a fire station tour.

Skills Objectives

There are no skills objectives for this chapter.

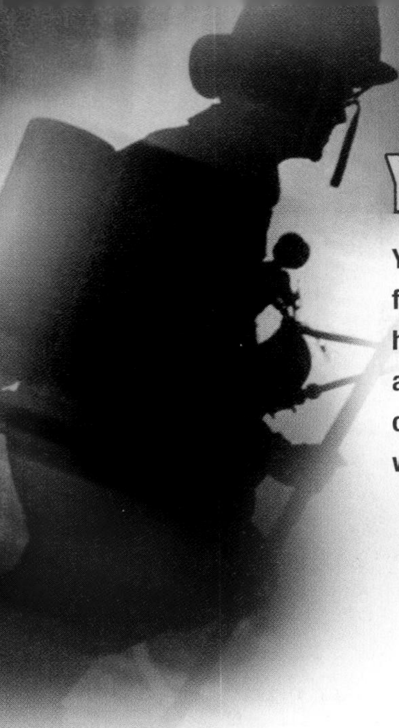

You Are the Fire Fighter

Your captain has asked you to conduct a one-hour fire station tour for a group of senior citizens from a local retirement community. There will be about 20 members of the group and the visit has been scheduled for early this afternoon. Captain Dickerson suggests showing the group around the station, including the apparatus, for about 30 minutes, then taking them into the classroom to provide a presentation on fire prevention and the 9-1-1 service. Other fire fighters will be available to assist you, but you must arrange and deliver the presentation.

1. ***What fire prevention topics are appropriate for this age group?***
2. ***What fire prevention tips would be beneficial for senior citizens?***
3. ***What written materials would be most helpful for this group?***
4. ***What types of apparatus would be most interesting to these citizens?***

Introduction

Fire prevention is one of the most important nonfirefighting activities performed by fire fighters. Most fires are caused by unsafe or careless acts, equipment failure, arson, or acts of nature. Many fires can be prevented by teaching safe behaviors and eliminating unsafe conditions. This will ultimately save lives and reduce the losses caused by fires.

Fire prevention duties for beginning fire fighters usually include performing fire safety inspections in residential occupancies and conducting fire station tours. This chapter will cover the information necessary to accomplish these tasks.

What Is Fire Prevention?

Fire prevention includes a range of activities that are intended to prevent the outbreak of fires or to limit the consequences if a fire does occur. These activities include enacting and enforcing fire codes, conducting property inspections, and presenting public fire safety education programs. Fire cause investigation is also included in the overall scope of fire prevention activities. All of these activities have the objective of limiting life loss, injuries, and property damage. Your highest priority as a fire fighter should be to prevent fires.

Every fire fighter should work to prevent fires and to educate the public about fire risks and hazards, even if a separate bureau or division within the department has primary responsibility for fire prevention. Some departments share the responsibility for conducting inspections and enforcing safety codes with the building department or another agency. Fire Fighter I and Fire Fighter II personnel may have limited responsibilities for fire prevention, but should understand the objectives of fire prevention, the delineation of responsibilities in the department, and their role in fire prevention.

Fire Fighter Tips

Fires that are prevented do not cause any deaths, injuries, or property loss. Most fires are caused by actions or situations that could have been prevented.

Enactment of Fire Codes

Preventing fires and protecting lives and property in the event of a fire are important community concerns and responsibilities. **Fire codes**—regulations that have been legally adopted by a governmental body with the authority to pass laws and enforce safety regulations—are enacted to ensure a minimum level of fire safety in the home and workplace environments. The fire code is enforced through a legal process similar to the way speed limits and other traffic regulations are enforced. A fire code can be adopted by a local government or by a state or province. In some cases, one basic set of regulations applies throughout a state, and local jurisdictions can adopt more stringent regulations.

One fire code, the *National Fire Codes,* is a set of documents produced by the National Fire Protection Association (NFPA) to address a wide range of issues relating to fire and safety. Many jurisdictions have adopted or incorporated these documents into their local codes and enforce them just as any other government regulation.

Most communities adopt and enforce a full set of codes that cover a range of areas and that are designed to establish basic health and safety standards. Generally, these codes include a building code, a fire code, an electrical code, a plumbing code, and other local rules and regulations. These codes and regulations must be coordinated to avoid conflicting requirements and to ensure a safe community. For example, the electrical code includes requirements that are intended to prevent electrical fires as well as requirements that reduce the risk of accidental electrocution.

The fire code generally includes regulations designed to prevent fires from occurring, to eliminate fire hazards, to protect lives, and to limit fire losses. Fire department personnel are usually involved in the development of fire codes. Every fire fighter needs to understand the effects of these codes and their involvement in establishing public safety requirements.

Fire codes are closely related to building codes. Building codes establish safety requirements that apply to the construction of new buildings and almost always include some fire protection requirements. When a new building is constructed, the building code sets the basic requirements for the type of construction, allowable heights and areas, distances between firewalls, and the building materials that can be used. In addition, the building code usually includes requirements for safe exits and determines when fire alarm systems, automatic sprinklers, and standpipes are required in new buildings. The knowledge you gained from Chapter 6, Building Construction, is closely related to building codes.

Fire codes apply to all buildings, new or old, and to many different situations related to fire risks and hazardous conditions. Fire codes require unobstructed, unlocked exits, properly maintained sprinkler systems, and fire alarm systems in operating condition. They may limit the number of people who can be present in a public assembly occupancy. Fire codes also include regulations for dispensing flammable liquids, for storing them in approved storage containers, and for prohibiting their use or storage in some types of buildings. Regulations that prohibit parking in a fire lane or too close to a hydrant are usually found in the fire code. All of these requirements are related to fire safety and protecting the community.

Inspection and Code Enforcement

After a jurisdiction adopts a fire code, it must be able to enforce it. Citizens have a legal obligation to comply with the rules and regulations. Regular inspections ensure compliance, note any violations, and require their correction. If violations are not corrected, a series of increasing penalties can be imposed, beginning with a verbal warning or a written notice. An individual who fails to correct the violation within a specified time can be fined. In some cases, a jail sentence can be imposed for very serious violations.

Fire codes usually specify the types of occupancies that can be inspected and the frequency of the inspections. For example, some fire codes require all schools to be inspected twice a year. In other cases, the schedule of inspections is discretionary.

The agency responsible for performing inspections and enforcing codes is usually named in the fire code and may vary in different communities. Sometimes the responsibility belongs to specific individuals or positions within the fire department, such as fire inspectors or fire marshals. Other fire departments train all fire fighters to conduct inspections and take code enforcement actions. In many communities, building inspectors or special code enforcement personnel are given this responsibility. Generally, newly assigned fire fighters are not directly involved in code enforcement activities, but you should understand how code enforcement is handled in your community.

In many cases, the fire code does not apply to the interior of a private dwelling, and the fire department cannot require an occupant to submit to an inspection. The occupant can request a voluntary inspection to identify fire hazards, and many fire departments offer this service. This voluntary inspection is called a home fire safety survey, and the legal requirements of the fire code are not applied. Voluntary inspections of private dwellings are one way that fire departments educate citizens about the risks of fire in their homes.

Every fire fighter should know how to conduct a home fire safety survey in a private dwelling. This includes knowing what to look for during the survey, how to contact the occupant, and how to communicate the survey findings. This information is presented in the section on conducting a fire safety survey in a private dwelling.

Public Fire Safety Education

The goals of public fire safety education are to help people understand how to prevent fires from occurring and to teach them how to react in an appropriate manner if a fire does occur. Making people aware of common fire risks and hazards and providing information about reducing or eliminating them can prevent many fires. Public education also teaches techniques to reduce the risks of death or injury in the event of a fire. Public fire safety education programs include:

- Learn Not to Burn®
- Stop, Drop, and Roll (► **Figure 35-1**).
- Change Your Clock—Change Your Battery
- Fire safety for babysitters
- Fire safety for seniors
- Wildland fire prevention programs

Most fire safety programs are presented to groups such as school classes, scout troops, church groups, senior citizen groups, civic organizations, hospital staff, or employees of different businesses. Presentations are often made at community events and celebrations, as well as during Fire Prevention Week.

One common fire safety education activity is a fire station tour. Children and adults enjoy the opportunity to tour the local fire station. A fire station tour is also an excellent opportunity to promote fire prevention. Every fire fighter must understand how to conduct an effective fire station tour. The skills and knowledge needed are presented in the section on conducting a fire station tour.

Figure 35-1 Stop, Drop, and Roll teaches young children what to do if their clothing catches fire.

Fire Cause Determination

Fire cause determination is the process of trying to establish the cause of a fire through careful investigation and analysis of available evidence. Fire cause determination is a part of fire prevention because finding the causes of fires can help prevent similar fires from occurring. Fire cause determination is equally important for both accidental and intentional fires. The fire fighter's role in fire cause determination is covered in Chapter 37, Fire Cause Determination.

Conducting a Fire Safety Survey in a Private Dwelling

Many fire departments offer to conduct fire safety surveys in private dwellings. A home fire safety survey helps identify fire and life safety hazards and provides the occupants with recommendations on making their home safer.

Residential fire safety surveys cannot be conducted without the occupant's permission. Usually, the survey is in response to a request by the occupant for an appointment. Fire departments may use public service announcements on radio and television to increase public awareness of the need for a fire-safe home and the availability of a fire safety survey. In many communities, the fire department offers to check home smoke alarms and to conduct a home fire safety survey at the same time. Some fire departments will install a smoke alarm free of charge in any dwelling that does not already have one.

A home fire safety survey is a joint effort by the fire department and the occupant to prevent fires and reduce fire fatalities. If the occupant accompanies you during the survey, you can point out hazards, explain the reasons for making different recommendations, and answer questions as they arise. Many fire departments provide educational materials that address safety issues and can be left with the occupant.

Getting Started

During a home inspection, be sure to present a neat, professional image (▼ Figure 35-2). Identify yourself by name, title, and organization. Inform the occupant of the purpose of your visit and the survey format. Always remember that you are a guest in a private home, and respect the privacy of the people who live there. You are there to help teach this person how to protect their family and property from fire.

During the survey, you will need to consider several types of hazards and issues. Concentrate on the hazard cat-

Figure 35-2 Present a neat, professional image during a home inspection.

Figure 35-3 Ensure that smoke alarms are properly installed, maintained, and fully operational during a home inspection.

egories that most often cause residential fires. These include cooking equipment, heating equipment, electrical wiring and appliances, smoking materials, candles and other open flames, and children playing with fire. Some of the hazards you encounter will depend on the local climate. For example, wood-burning stoves and fireplaces are more common in colder climates. Other hazards may be related to the age of the occupants. The fire hazards in a residence with young children, occupants with a physical or mental disability, and those in a residence occupied only by senior citizens would be quite different.

Always look for fire protection equipment such as smoke alarms, fire extinguishers, and residential fire sprinkler systems. Make sure that any fire alarm or suppression equipment in the home is properly installed, maintained, and fully operational (▲ Figure 35-3). If the occupant has no fire extinguishers, recommend the appropriate type and advise the homeowner on where to place it.

Some departments address additional safety issues during a home survey, such as swimming pool safety, slip-and-fall hazards, and the value of carbon monoxide detectors. Mentioning these additional safety items to the occupant can be very helpful.

Conduct the survey in a systematic fashion for both the inside and the outside of the home. Some fire fighters begin by inspecting the outside of the house, while others prefer to start with the interior. Many fire inspectors will start at the top of a building and systematically work their way down to ensure that all areas are covered. Inside the house, systematically inspect each room for fire hazards. Each section will have its own hazards.

Outside Hazards

On the exterior of a residential occupancy, you need to check several items. Make sure that the house number is

Figure 35-4 The home's address should be clearly visible.

clearly visible from the street so that emergency responders can identify the correct house quickly (▲ Figure 35-4). Look for accumulated trash that could present a fire hazard or obstruct an exit. Note any flammable materials located close to the house or cars that are parked so close that a vehicle fire could spread to the structure. Point out shrubs and vegetation that need to be trimmed or removed. For example, in an urban-wildland interface zone, there should be no dead vegetation within 30′ of a house.

Locate any chimneys and try to determine the condition of the mortar. Ask the occupant if the chimney has been cleaned within the past year, and recommend a cleaning if it has not.

Fire Fighter Safety Tips

Security bars and double-cylinder door locks are great deterrents for burglars, but can be deadly for occupants trying to escape from a house fire. They can also waste precious time if fire fighters must enter to rescue the occupants or control a fire. The occupants should always be able to release security bars and open locked doors from the inside.

Voices of Experience

“No matter how young your audience is a station tour should always educate them on preventing fires and what to do in case of a fire.”

When giving station tours, make sure that you adjust your message based upon the audience. Adults are likely to be interested in technical information: response times, staffing levels, and the department's organization. School-age children are more interested in seeing the engines, the “Jaws of Life”, SCBA, and the other firefighting tools. Pre-school children will be most interested in the lights, sirens, the big steering wheel, and the pump panel with all the gauges.

Likewise, the fire prevention message must also be adjusted. Pre-school children need to learn to never play with matches; stop, drop, and roll; and get low and go. School-age children need to learn about exit drills in the home and dialing 9-1-1. It is also important to review lessons like stop, drop, and roll with school-age children. Adults need to learn about exit drills in the home, checking smoke alarms, and other fire prevention methods.

Remember to always speak at the audience's level. The younger they are, the more animated your delivery style must be to keep their attention. No matter how young your audience is a station tour should always educate them on preventing fires and what to do in case of a fire.

David Hall
Springfield Missouri Fire Department
Rogersville, Missouri

Inside Hazards

During the interior inspection, explain why different situations are considered potential fire risks and hazards. A person who understands why an overloaded electrical circuit is a fire hazard will be more likely to avoid overloading circuits in the future. Your goal should be to educate the occupant, not simply to point out fire hazards.

As you walk through the house, help the occupant identify alternate escape routes that can be used in case of fire. Look for a primary and an alternate exit from each room and discuss them with the occupant. Be sure to mention the importance of home fire exit drills involving all family members.

Smoke Alarms

Studies show that the presence of working smoke alarms on each level of a house reduces the risk of death from fire by 50%. Take the time to verify and test all smoke alarms. Some fire departments keep donated smoke alarms available and install one in any dwelling that lacks this important protection. Give the residents a copy of the NFPA fact sheet on smoke alarms and request that they read it.

More information about the proper maintenance of smoke alarms is presented in Chapter 36, Fire Protection, Detection, and Suppression Systems.

These important smoke alarm tips should be shared with the home's occupants:

- Smoke alarms should be installed in or near every sleeping room of the home.
- Smoke alarms should be mounted on the ceiling, at least 4″ from a wall, or high on the wall, 4″ to 12″ below the ceiling.
- Smoke alarms should not be located near windows, exterior doors, or duct vents that could interfere with their operation.
- Only qualified electricians should install or replace AC-powered smoke alarms.
- Smoke alarms should be tested at least once a month by using the "test button" on the device (► Figure 35-5).
- Smoke alarms should be dusted and vacuumed regularly.
- Smoke alarm batteries should be replaced once a year, or as soon as the intermittent "chirp" warning that the battery is low sounds. Remind homeowners to "change the battery when you change your clock."

NOTE: Some smoke alarms are equipped with special batteries that last for 10 years.

Bedrooms

The most common causes of fires in bedrooms are defective wiring, improper use of heating devices, improper use of candles, children playing with matches, and smoking in bed. Smoking in bed causes only 5.4% of fires, but these fires are responsible for 23% of civilian fire deaths. Fire fighters conducting home safety surveys should stress the fire hazard that is associated with smoking in bed. In houses with children, it is equally important to discuss the risk of children playing with smoking materials.

Kitchens

Kitchen fires are responsible for 22% of residential fires and usually result from cooking accidents. Fires are often caused by leaving cooking food on the stove unattended and by faulty electric appliances. In the kitchen, look for signs of frayed wires on appliances, for flammable towels or potholders stored too close to the stove, and for flammable cooking oils stored near the stove. Nothing that can be ignited, such as a plastic food container or paper packaging, should ever be placed on a cooking surface, even if the burners are turned off and the surface is cold. Many fires have resulted from someone turning on the wrong burner and igniting something that was left on the stovetop.

An approved ABC-rated fire extinguisher should be located in the kitchen. Check to be sure the extinguisher is visible and properly charged (► Figure 35-6).

Additional recommendations for kitchen safety include:

- Do not leave anything cooking unattended on a stove.
- Keep all flammable materials, cleaning supplies, cooking oils, and aerosols away from the stove.
- Do not place anything that could ignite on a cooking surface, even when it is turned off and cold.
- Do not place towel racks near the stove.
- Store smoking marerials out of the reach of children.
- Do not overload electrical outlets or extension cords.
- Keep electric cords properly maintained; replace any frayed cords.

Figure 35-5 Smoke alarms should be tested at least monthly using the test button.

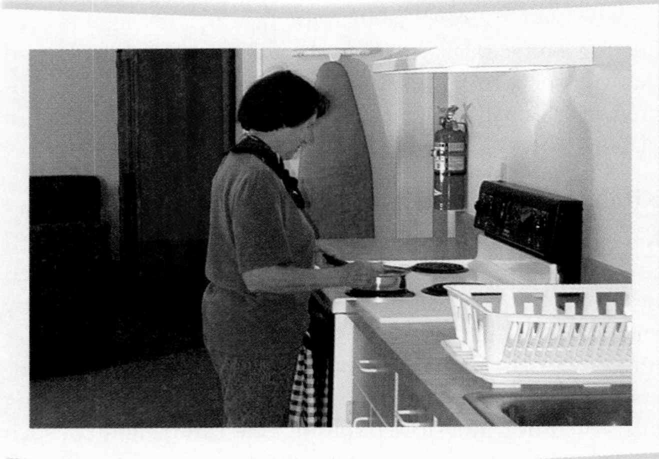

Figure 35-6 An approved ABC-rated fire extinguisher should be located in the kitchen.

Living Rooms

This room category includes formal living rooms, family rooms, and entertainment rooms. The primary causes of fires in these areas are smoking and electrical equipment. Look for indications of careless smoking or improperly installed home entertainment equipment, such as multiple devices plugged into a light-duty extension cord. Also look for overloaded electrical outlets and extension cords in home offices with computers, printers, scanners, fax machines, and other devices.

If the room contains a fireplace or a wood stove, ensure that no flammable or easily ignited materials are stored nearby. A fireplace should have a screen to keep sparks and hot embers from escaping. If solid fuels are used, the chimney or flue pipe should be professionally cleaned at least annually.

Garages, Basements, and Storage Areas

Explain the importance of good housekeeping and the need to clear accumulated junk out of garages, basements, and storage areas. These areas are often filled with large quantities of unneeded materials that can easily ignite, such as old newspapers and trash. Furnaces and hot water heaters that operate with an open flame are often located in a basement or garage. They could become ignition sources if flammable materials are stored too close to them.

Storage of gasoline and other flammable substances is a major concern, because an open flame or pilot light can easily ignite leaking vapors. Gasoline and other flammable liquids should be stored only in approved containers and in outside storage areas or outbuildings. Propane tanks, such as those used in gas grills, also should be kept in outside storage areas or outbuildings. Small quantities of flammable and combustible liquids (such as paint, thinners, varnishes, and cleaning fluids) should be stored in closed metal containers away from heat sources. Oily or greasy rags should also be stored in closed metal containers.

A fully-charged fire extinguisher (2A:10B:C classification) should be recommended for basements and garages.

Closing Review

During the home fire safety survey, listen carefully to any questions from the occupant. This will enable you to address any special concerns or correct any inaccurate information. At the end of the survey, complete the inspection form and give a copy to the occupants. Carefully review the findings and describe the steps that need to be taken to eliminate any identified hazards. If possible, talk to the occupants and take the time to answer their questions fully. Emphasize the importance of smoke alarms, home exit plans, and fire drills.

After you have completed a home fire safety survey or any other type of inspection, you must file your report according to the standard operating procedures of your department. If you identified hazards that require further action or follow-up, discuss them with your company officer or a designated representative from the code enforcement office.

Conducting a home fire safety survey is a complex, but important, task. Initially, you may accompany your company officer, a more experienced fire fighter, or code enforcement personnel to observe how it is done. You will work with a more experienced partner until you are ready to conduct home surveys on your own. Remember, a home fire safety survey is an important step in helping occupants prevent fires and the resultant loss of lives and property.

Conducting Fire Station Tours

Most people have a natural curiosity about fire. This curiosity often extends to fire apparatus, fire stations, and fire fighters. Fire station tours present a unique opportunity to help people learn how the fire department operates and how they can help prevent fires.

Most fire departments have a set format for conducting fire station tours. Your station commander is in charge of all activities and will teach you how to conduct tours in your station. Some tours start in the dayroom or in a meeting room with a general overview of the fire station, followed by a tour of the fire apparatus. A tour could also start with an inspection of the apparatus and end in a meeting room with a question-and-answer session and information about the department and fire prevention techniques. Showing a fire prevention video or a video about the operation of your department is a good way to start or finish a tour and stimulate questions from your audience.

In preparing for a fire station tour, remember that you will represent your department to the tour group. Your pro-

Figure 35-7 Young children always like to see fire fighters.

Figure 35-8 Always leave visitors with printed fire prevention materials.

fessional appearance and presentation will help project a positive image of the department to the community. One of the first things you should cover in your welcome is to tell the visitors what to expect and what they should do if the station receives an alarm during the tour. Visitors should not wander around the station while fire fighters respond to an emergency call. This is particularly important if the tour guide also might have to leave in response to a call.

The tour format will vary depending on the age and interests of the group, which could be comprised of small children, teenagers, adults, or senior citizens. The tour plan, including your presentation and fire prevention messages, should be designed to meet the needs of each group. The weather also can be an important factor if you are planning any outdoor demonstrations.

Young children like to see action, so squirting some water or raising an aerial ladder can be a real crowd pleaser (▲ Figure 35-7). Fire prevention messages for young people include learning what to do if their clothes catch fire, learning how to call 9-1-1, the importance of home fire drills, and the dangers of playing with smoking materials.

Teenagers are ready for lessons that they can apply in everyday life. Many teenagers work as babysitters, so prevention messages might be geared toward cooking safety and evacuating children in the event of a fire. Teenagers are also a high-risk group for motor vehicle collisions, so messages about seat belts, safe driving, the dangers of drinking and driving, and what to do if they see or hear an emergency vehicle while driving are appropriate. Some teenagers might be interested in becoming fire fighters or joining a fire fighter explorer group. This age group can be a good source of recruits for both career and volunteer departments.

Adults are probably more interested in home fire safety, while senior citizens are often more interested in the EMS services available. Remind smokers not to smoke in bed or throw burning cigarettes into wastepaper baskets. Seniors may forget to turn off burners or may leave cooking food unattended. They also may be worried about their ability to move quickly if a fire starts in their home. Tailor your prevention messages to the concerns of the age group you are addressing.

Try to leave every tour group with both a message and materials. First, leave them with a simple prevention or safety message that expands their awareness of fire and safety and helps them to deal with potential emergency situations (▲ Figure 35-8). Second, give them printed fire prevention materials. These materials could range from coloring books for young children to a home fire prevention checklist for adults.

Wrap-Up

Ready for Review

This chapter described the importance of fire prevention to reduce the loss of lives and property. It described your role as a new fire fighter in fire prevention. Fire prevention includes enacting and enforcing the fire code and inspecting properties to identify fire hazards. It also includes fire cause determination. Educating the public about fire safety is an important component of fire prevention, and this chapter reviewed various types of public fire safety education.

This chapter described the importance of conducting home fire safety surveys in private dwellings. It outlined the basic steps of this task, which every fire fighter should learn to perform. It described the common hazards that you should look for in each part of a dwelling and stressed the importance of conducting inspections in a systematic fashion using an inspection form. This chapter also pointed out the importance of conducting a review with the building occupant to discuss your findings and answer any questions the occupant might have.

Conducting fire station tours is another activity that can promote fire prevention and increase public awareness about the role of the fire department. This chapter covered the types of activities that could be included in a fire station tour. It stressed the importance of gearing the tour and the fire prevention message to the age and interest of the audience. Finally, it emphasized the importance of providing fire prevention materials that visitors can take home and study.

Chief Concepts

- Fire prevention is an important responsibility of every fire fighter.
- Fire prevention consists of four components: enactment of fire codes; inspection and fire code enforcement; public fire safety education; and fire cause determination.
- Conducting fire safety surveys in private dwellings is an effective strategy to reduce fire deaths and property loss in your community.
- Conducting fire station tours is a way to acquaint people with your department and provides an opportunity for public fire safety education.

Hot Terms

Fire code A set of legally adopted rules and regulations designed to prevent fires and protect lives and property in the event of a fire.

Fire prevention Activities conducted to prevent fires and protect lives and property in the event of a fire. Fire prevention activities include the enactment and enforcement of fire codes, the inspection of properties, the presentation of fire safety education programs, and the investigation of the causes of fires.

Fire Fighter in Action

Your fire company has been assigned to conduct voluntary inspections of private dwellings in a residential area. Each member of your company will be conducting separate inspections to maximize the number of inspections that can be done during the time allotted. You have reviewed the checklist used by your department for inspecting private dwellings.

1. What device should you look for during your inspection that can reduce the chance of fire deaths in the home by up to 50%?
   - **A.** 10 lb, dry chemical ABC fire extinguisher
   - **B.** Home security system
   - **C.** Working smoke alarm
   - **D.** Well maintained heating system
2. Which behavior is responsible for 22% of all fire deaths in private dwellings?
   - **A.** Excessive consumption of alcoholic beverages
   - **B.** Improper cooking practices
   - **C.** Children playing with matches
   - **D.** Smoking
3. What should the last part of the fire inspection be?
   - **A.** Inspection of the basement or garage
   - **B.** A review of your findings with the occupant
   - **C.** Inspection of the outside
   - **D.** Inspection of the fire extinguishers
4. What should you always give to tour group members at the end of the tour?
   - **A.** A smoke alarm
   - **B.** A fire extinguisher
   - **C.** Fire prevention materials
   - **D.** A fire hat

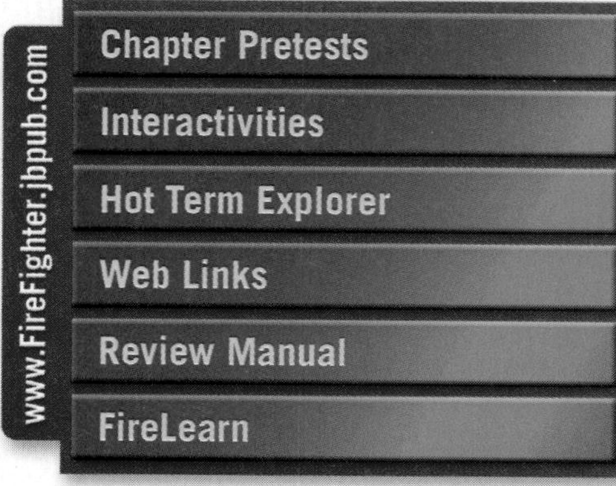

Fire Detection, Protection, and Suppression Systems

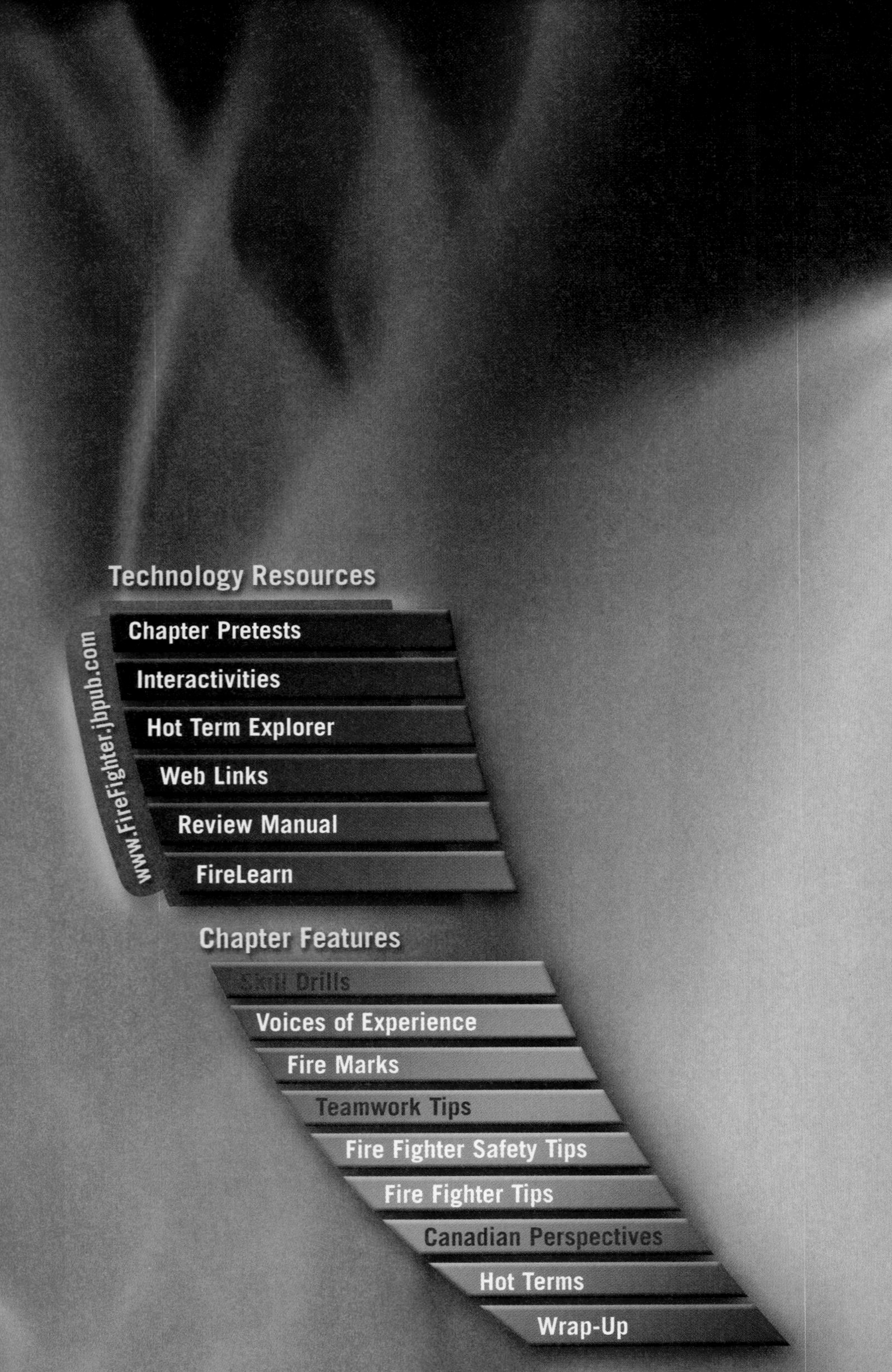

Chapter 36

NFPA 1001 Standard

Fire Fighter I

5.3.14 (A) *Requisite Knowledge.* The purpose of property conservation and its value to the public, methods used to protect property, types of and uses for salvage covers, operations at properties protected with automatic sprinklers, how to stop the flow of water from an automatic sprinkler head, identification of the main control valve on an automatic sprinkler system, and forcible entry related to salvage.

5.3.14 (B) *Requisite Skills.* The ability to cluster furniture; deploy covering materials; roll and fold salvage covers for reuse; construct water chutes and catch-alls; remove water; cover building openings, including doors, windows, floor openings, and roof openings; separate, remove, and relocate charred material to a safe location while protecting the area of origin for cause determination; stop the flow of water from a sprinkler with sprinkler wedges or stoppers; and operate a main control valve on an automatic sprinkler system.

Fire Fighter II

6.5.1 (A) *Requisite Knowledge.* The sources of water supply for fire protection; the fundamentals of fire suppression and detection systems; common symbols used in diagramming construction features, utilities, hazards, and fire protection systems; departmental requirements for a preincident survey and form completion; and the importance of accurate diagrams.

Knowledge Objectives

After studying this chapter, you will be able to:

- Explain why all fire fighters should have a basic understanding of fire protection systems.
- Describe the basic components and functions of a fire alarm system.
- Describe the basic types of fire alarm initiation devices and where each type is most suitable.
- Describe the fire department's role in resetting fire alarms.
- Explain the different ways that fire alarms may be transmitted to the fire department.
- Identify the four different types of sprinkler heads.
- Identify the different styles of indicating valves.
- Describe the operation and application of the following types of automatic sprinkler systems:
    - Wet-pipe system
    - Dry-pipe system
    - Preaction system
    - Deluge system
- Describe when and how water is shut off to a building's sprinkler system and how to stop water at a single sprinkler head.
- Describe the differences between commercial and residential sprinkler systems.
- Identify the three types of standpipes and the differences among them.
- Describe two problems that fire fighters could encounter when using a standpipe in a high-rise.
- Identify the hazards that specialized extinguishing systems can pose to responding fire fighters.

Skills Objectives

There are no skills objectives for this chapter.

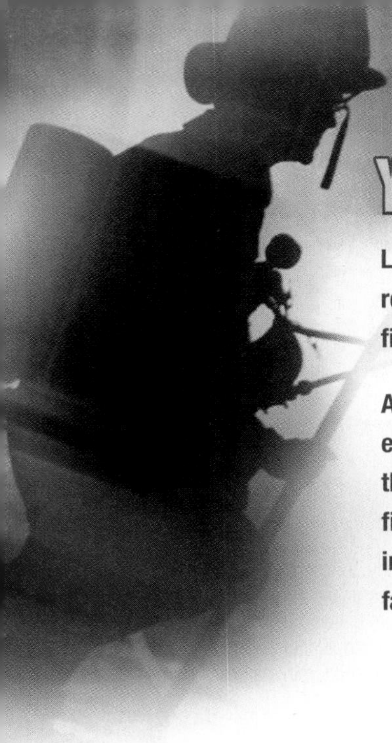

You Are the Fire Fighter

Late one night, you are on a second-due engine to a large, unoccupied industrial building in response to a reported fire alarm sounding. The building has a sprinkler system and a complete fire alarm system. Previous calls have all been false alarms.

As you arrive on scene, the first-in company reports that they have nothing showing and are entering the building to determine the source of the alarm. A water-motor gong sounds around the corner as you dismount the engine. Your officer orders a line to be connected to the building's fire department connection and tells you to prepare a hand line and get ready to enter the building. Your partner grumbles, "Why are we wasting all this time and energy pulling hose for another false alarm?"

1. ***Do you think that this call is another false alarm?***
2. ***Is there any need to pull an attack line into a building that is protected by a sprinkler system?***
3. ***What is the likelihood of encountering a serious fire in a building that has a full sprinkler and alarm system?***
4. ***If there is a fire, when should the water supply to the sprinkler system be shut off?***
5. ***What actions can your crew take to lessen any water damage if the sprinkler system was set off by accident?***

Introduction

Today, fire prevention and building codes require that most new structures have some sort of fire protection system installed. This makes it more important than ever for the line fire fighter to have a working knowledge of these systems, which include fire alarm systems and automatic fire detection and suppression systems. Understanding how these systems operate is important for the fire fighter's safety and necessary to provide effective customer service to the public.

From a safety standpoint, fire fighters need to understand the operations and limitations of fire detection and suppression systems. A building with a fire protection system will have very different working conditions during a fire than an unprotected building. Fire fighters need to know how to fight a fire in a building with a working fire suppression system, and how to shut down the system after the fire is extinguished.

From a customer service standpoint, fire fighters who understand how fire protection systems work can help dispell misconceptions about these systems and advise building owners and occupants after an alarm is sounded. Most people have no idea of how the fire prevention or detection systems in their building work, and there are often more false alarms in buildings with fire protection systems than actual fires. Fire fighters can help the owners/occupants determine what activated the system, how they can prevent future false alarms, and what needs to be done to restore a system to service. This chapter discusses the basic design and operation of fire alarm, fire detection, and fire suppression systems, each of which could be the subject of an entire book. It is not necessary to understand all of the design criteria for a fire protection system, only how the system is intended to function. Fire protection systems have fairly standardized design requirements across North America; most areas follow the applicable NFPA standards. Unfortunately, local fire prevention and building codes may require different types of systems for different buildings, and these requirements still vary considerably.

Fire Alarm and Detection Systems

Practically all new construction includes some sort of fire alarm and detection system. In most cases, both fire detection and fire alarm components are integrated in a single system. A fire detection system recognizes when a fire is occurring and activates the fire alarm system, which alerts building occupants and, in some cases, the fire department. Some fire detection systems also automatically activate fire suppression systems to control the fire.

Fire alarm and detection systems range from simple, single-station smoke alarms for private homes, to complex fire detection and control systems for high-rise buildings. Many fire alarm and detection systems in large buildings also control other systems to help protect occupants and control the spread of fire and smoke. Although these systems can be complex, they generally have the same basic components.

Fire Marks

Single-station smoke alarms have helped reduce death and injury rates caused by residential fires since they were first introduced in the 1970s. They provide valuable life-safety protection at a very low cost. Single-station smoke alarms are simple, reliable life savers.

Residential Fire Alarm Systems

Current building and fire prevention codes usually require the installation of a fire detection and alarm system in all residential dwelling units. The most common type of residential fire alarm system is a **single-station smoke alarm** (► Figure 36-1). This life-safety device includes both a smoke detection device and an audible alarm within a single unit, quickly alerting occupants when a fire occurs. Millions of them have been installed in private dwellings and apartments.

Smoke alarms can be battery-powered or hard-wired to a 110-volt electrical system. Most building codes require hard-wired, AC-powered smoke alarms in all newly constructed dwellings; battery-powered units are popular for existing occupancies. The major concern with a battery-powered smoke alarm is ensuring that the battery is replaced on a regular basis. The newest battery-powered units are available with batteries that will last for 10 years.

The most up-to-date codes require new homes to have a smoke alarm in every bedroom and on every floor level. They also require a battery backup in the event of a power failure. In newer installations, all of the smoke alarms in a dwelling are interconnected so they will all sound if one is activated. For example, if the basement smoke alarm is activated, alarms will sound on all levels to alert the occupants.

Many home fire alarm systems are part of security systems. Like commercial systems, these home fire alarm/security systems have an alarm control panel and require a pass code to set or reset the system. These systems may also be monitored by a central station.

Ionization versus Photoelectric Smoke Detectors

Two types of fire detection devices may be used in a smoke alarm to detect combustion. Ionization detectors are triggered by the invisible products of combustion. Photoelectric detectors are triggered by the visible products of combustion.

Ionization smoke detectors work on the principle that burning materials release many different products of combustion, including electrically charged microscopic particles. An ionization detector senses the presence of these invisible charged particles (ions).

An ionization smoke detector has a very small amount of radioactive material inside a chamber. The radioactive material releases charged particles into the chamber, and a small electric current flows between two plates (► Figure 36-2).

Figure 36-1 The most common residential fire alarm system today is a single-station smoke alarm.

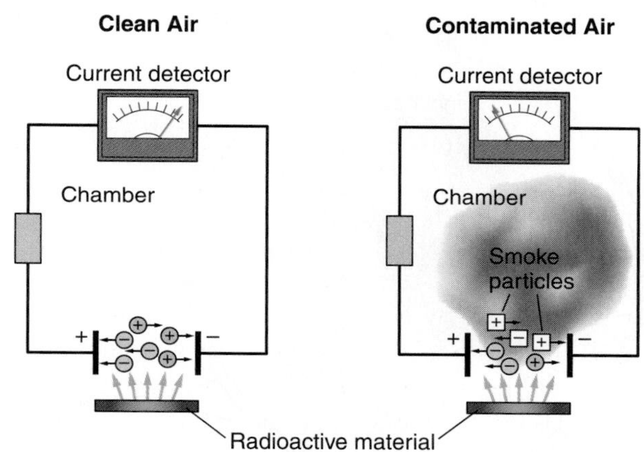

Figure 36-2 An ionization smoke alarm.

When smoke particles enter the chamber, they neutralize the charged particles and interrupt the current flow. The detector senses the interruption and activates the alarm.

Photoelectric smoke detectors use a light beam and a photocell to detect larger visible particles of smoke (► Figure 36-3). They operate by reflecting the light beam either away from or onto the photocell, depending on the design of the device. When visible particles of smoke pass through the light beam, they reflect the beam onto or prevent it from striking the photocell, activating the alarm.

Ionization smoke detectors are more common and less expensive than photoelectric smoke detectors. Ionization smoke detectors react more quickly than photoelectric smoke detectors to fast-burning fires, such as a fire in a wastepaper basket, which may produce little visible smoke

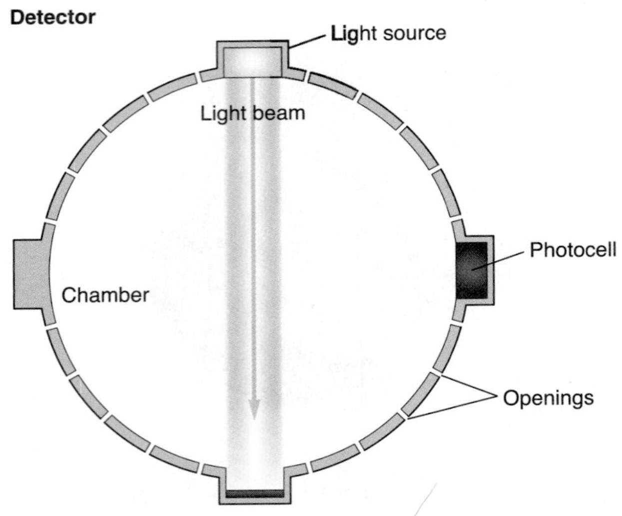

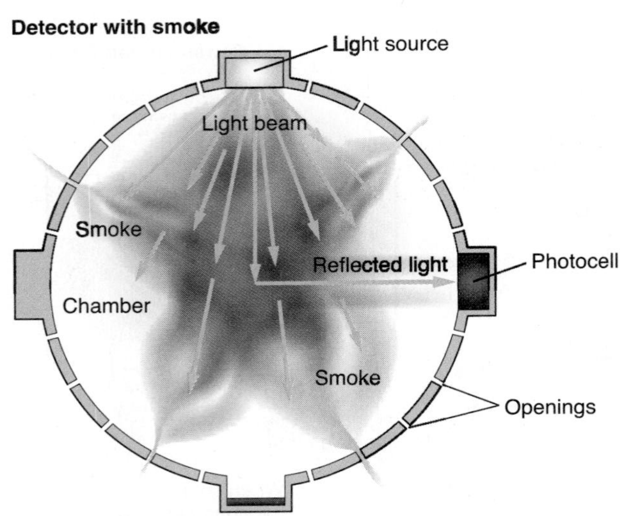

Figure 36-3 A photoelectric smoke alarm.

Figure 36-4 Ionization smoke detectors are slightly quicker to react than photoelectric smoke detectors to a fast-burning fire, such as a fire in a wastepaper basket, which may produce little visible smoke.

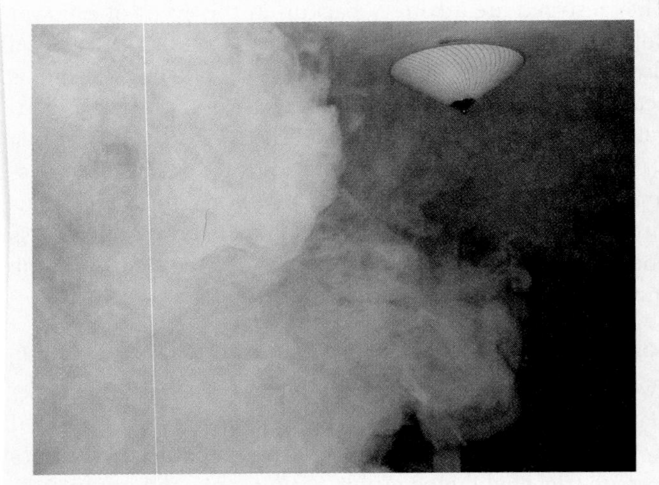

Figure 36-5 Photoelectric smoke alarms are slightly quicker to react to a slow-burning, or smoldering fire, such as a cigarette in furniture, which usually produces a large quantity of visible smoke.

► Figure 36-4. However, fumes and steam from common activities such as cooking and showering may trigger unwanted alarms.

Photoelectric smoke detectors are more responsive to a slow-burning or smoldering fire, such as one caused by a cigarette caught in a couch, which usually produces a large quantity of visible smoke (► Figure 36-5). Photoelectric smoke detectors are less prone to false alarms from regular cooking fumes and shower steam.

Both ionization and photoelectric smoke detectors are acceptable life-safety devices, and in most cases, they are interchangeable. Some fire codes require that photoelectric smoke detectors be used within a certain distance of cooking appliances or bathrooms, to help prevent unwanted alarms.

Combination ionization/photoelectric smoke alarms are also available. These alarms will quickly react to both fast-burning and smoldering fires. They are not suitable, however, for use near kitchens or bathrooms, because they are prone to the same unwanted alarms as regular ionization smoke detectors. (► Table 36-1) indicates the general features and recommended applications for each type of smoke detector.

Table 36-1 Types of Smoke Detectors

Type of Detector	Features	Application
Ionization	Uses radioactive material within the device to detect invisible products of combustion.	Used to detect fires that do not produce large quantities of smoke in their early states. React quickly to fast-burning fires. Inappropriate for use near cooking appliances or showers.
Photoelectric	Uses a light beam to detect the presence of visible particles of smoke.	Used to detect fires that produce visible smoke. React quickly to slow-burning, smoldering fires.

Most ionization and photoelectric smoke alarms look very similar to each other. The only sure way to identify the type of alarm is to read the label, which is often on the back of the case. An ionization alarm must have a label or engraving stating that it contains radioactive material.

Fire Alarm System Components

A fire alarm system has three basic components—an alarm initiation device, an alarm notification device, and a control panel. The **alarm initiation device** is either an automatic or manually operated device that, when activated, causes the system to indicate an alarm. The **alarm notification device** is generally an audible device, often accompanied by a visual device, that alerts the building occupants when the system is activated. But the most important part of any fire alarm system is the control panel, which links the initiation device to the notification device and performs other essential functions.

A single-station smoke alarm in a residential dwelling has the same basic components as a fire alarm system in a multistory office building. A single-station smoke alarm includes a smoke detector and an audible notification device, plus a control and power supply component.

Fire Alarm System Control Panels

Most installed fire alarm systems in buildings other than dwellings have several alarm initiation devices in different areas and use both audible and visible devices to notify the occupants of an alarm. The **fire alarm control panel** serves as the "brain" of the system, linking the activation devices to the notification devices.

The control panel manages and monitors the proper operation of the system. It can indicate the source of an alarm, so that responding fire personnel will know what activated the alarm and where the initial activation occurred. The control panel also manages the primary power supply and provides a backup power supply for the system. It may perform additional functions, such as notifying the fire department or central station monitoring company when the alarm system is activated, and may interface with other systems and facilities.

Figure 36-6 Most modern fire alarm control panels indicate the zone where an alarm was initiated.

Control panels vary greatly depending on the age of the system and the manufacturer. For example, an older system may simply indicate that an alarm has been activated, while a newer system may indicate a specific zone within the building (▲ Figure 36-6). The most modern panels actually specify the exact location of the activated initiation device.

Fire alarm control panels are used to silence the alarm and reset the system. These panels should always be locked; many newer systems require the use of a password before the alarm can be silenced or reset. Alarms should not be silenced or reset until the activation source has been found and checked by fire fighters to ensure that the situation is under control. In some systems, the individual alarm activation devices must be reset individually before the entire system can be reset; other systems reset themselves after the problem is resolved.

Many buildings have an additional display panel in a separate location, usually near the front door of the building. This is called a **remote annunciator** (► Figure 36-7). The remote annunciator enables fire fighters to ascertain the type and location of the activated alarm device as they enter the building.

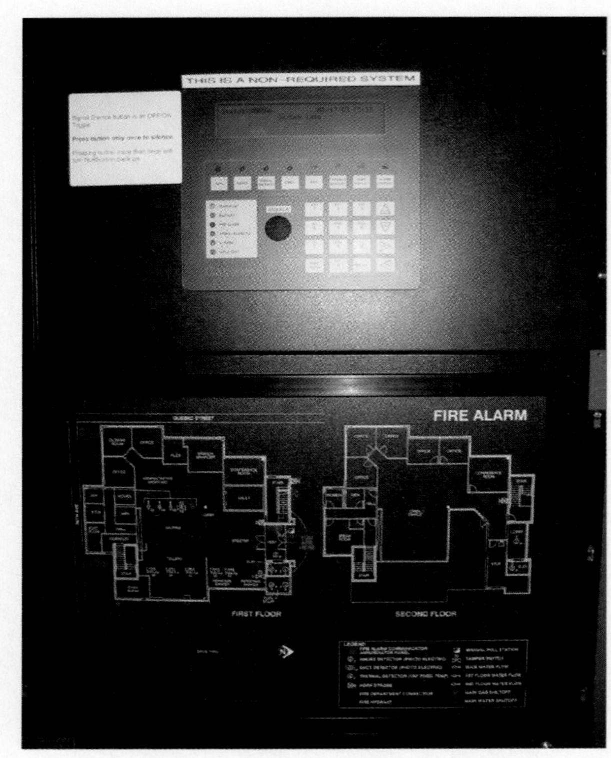

Figure 36-7 A remote annunciator allows fire fighters to quickly determine the type and location of the activated alarm device.

Figure 36-8 Fire prevention codes require that alarm sytems have a backup power supply, which is automatically activated when the normal electrical power is interrupted.

The fire alarm control panel should also monitor the condition of the entire alarm system to detect any faults. Faults within the system are indicated by a trouble alarm, which shows that a component of the system is not operating properly and requires attention. Trouble alarms do not activate the building's fire alarm, but should make an audible sound and illuminate a special light at the alarm control panel. It may also transmit a notification to a remote service location.

A fire alarm system is usually powered by a 110-volt line, even though the system's appliances may use a lower voltage. There are, however, some older alarm systems that require 110 volts for all components.

In addition to the normal power supply, fire prevention codes call for a backup power supply for all alarm systems (► Figure 36-8). In some systems, a battery in the fire alarm control panel will automatically activate when the external power is interrupted. Fire codes will specify how long the system must be able to function on the battery backup. In large buildings, the backup power supply could be an emergency generator or a secondary power source. If either the main power supply or the backup power source fails, the trouble alarm should sound.

Depending on the building's size and floor plan, an activated alarm may sound throughout the building or only in particular areas. In high-rise buildings, fire alarm systems are often programmed to alert only the occupants on the same floor as the activated alarm as well as those on the floors immediately above and below. Alarms on the remaining floors can be manually activated from the system control panel. Some systems have a public address feature, so that a fire department officer can provide specific instructions or information for occupants in different areas.

The control panel in a large building may be programmed to perform several additional functions. For example, it may automatically shut down or change the operation of air handling systems, recall elevators to the ground floor, and unlock stairwell doors so that a person in an exit stairway can reenter an occupied floor.

Alarm Initiating Devices

Alarm initiation devices are the components that activate a fire alarm system. Manual initiation devices require human activation; automatic devices function without human intervention. Manual fire alarm boxes are the most common type of alarm initiating devices that require human activation. Automatic initiation devices include various types of heat and smoke detectors and other devices that automatically recognize the evidence of a fire.

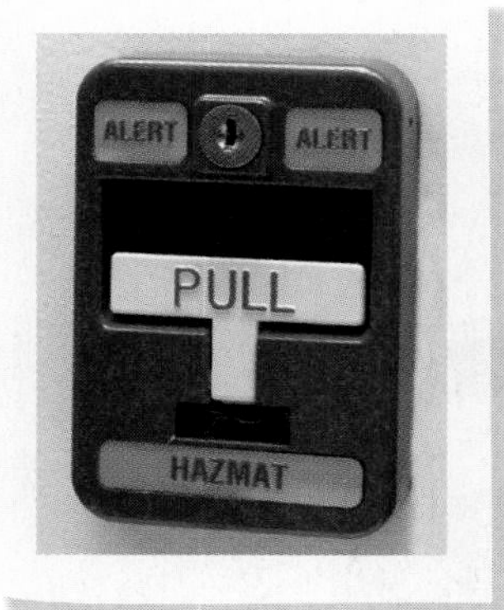

Figure 36-9 There are several different types of manual fire alarm boxes, also known as manual pull-stations.

Manual Initiation Devices

Manual initiation devices are designed so that building occupants can activate the fire alarm system if they discover a fire in the building. Many older alarm systems could only be activated manually. The primary manual initiation device is the manual fire alarm box, or manual pull-station (▲ Figure 36-9). These stations have a switch that either opens or closes an electrical circuit to activate the alarm. Pull-stations come in various sizes and designs, depending on the manufacturer. They can be either single-action or double-action devices.

Single-action pull-stations require a person to pull down a lever, toggle, or handle to activate the alarm. The alarm sounds as soon as the pull-station is activated. Double-action pull-stations require a person to perform two steps before the alarm will activate. They are designed to reduce the number of false alarms caused by accidental or intentional pulling of the alarm. The person must move a flap, lift a cover, or break a piece of glass to reach the alarm activation device. Designs that use glass are no longer in favor, because the glass must be replaced each time the alarm is activated and because the broken glass poses the risk of injury.

Once activated, a manual pull-station should stay in the "activated" position until it is reset. This enables responding fire fighters to determine which pull-station initiated the alarm. Resetting the pull-station requires a special key, screwdriver, or Allen wrench. In many systems, the pull-station must be reset before the building alarm can be reset at the alarm control panel.

A variation on the double-action pull-station, designed to prevent malicious false alarms, is covered with a piece of clear plastic (► Figure 36-10). These covers are often used in areas where malicious false alarms frequently occur, such as high schools and college dormitories. The plastic cover must be opened before the pull-station can be activated. Lifting the cover triggers a loud tamper alarm at that specific location, but does not activate the fire alarm system. Snapping the cover back into place resets the tamper alarm. The intent is that a person planning to initiate a false alarm will drop the cover and run when the tamper alarm sounds. In most cases, the pull-station tamper alarm is not connected to the fire alarm system in any way.

Figure 36-10 A variation on the double-action pull-station, designed to prevent malicious false alarms, has a clear plastic cover and a separate tamper alarm.

Even automatic fire alarm systems can be manually activated by pressing a button or flipping a switch on the alarm system control panel. In large buildings, particularly high-rise buildings, the control panel has separate manual activation switches for different areas or on particular floors.

Automatic Initiating Devices

Automatic initiating devices are designed to function without human intervention and will activate the alarm system when they detect evidence of a fire. These systems can be programmed to transmit the alarm to the fire department, even if the building is unoccupied, and to perform other functions when a detector is activated.

Automatic initiating devices can use several different types of detectors. Some detectors are activated by smoke or by the invisible products of combustion, and others react to heat, the light produced by an open flame, or specific gases.

Smoke Detectors

A smoke detector is designed to sense the presence of smoke. Among fire fighters, the term "smoke detector" generally refers to a sensing device that is part of a fire alarm system. This type of device is commonly found in school, hospital, business, and commercial occupancies with fire alarm systems (► Figure 36-11).

Smoke detectors come in different designs and styles for different applications. The most common smoke detectors are ionization and photoelectric detectors, which operate in the same way that residential smoke alarms do. However, smoke detectors used in commercial fire alarm systems are usually more sophisticated and more expensive than residential smoke alarms.

Each detection device is rated to protect a certain floor area, so in large areas the detectors are often placed in a grid pattern. Newer smoke detectors also have a visual indicator, such as a steady or flashing light, that indicates when the device has been activated.

A beam detector is a type of photoelectric smoke detector used to protect large open areas such as churches, auditoriums, airport terminals, and indoor sports arenas. In these facilities, it would be difficult or costly to install large numbers of individual smoke detectors, but a single beam detector could be used (► Figure 36-12).

A beam detector has two components: a sending unit that projects a narrow beam of light across the open area and a receiving unit that measures the intensity of the light when the beam strikes the receiver. When smoke interrupts the light beam, the receiver detects a drop in the light intensity and activates the fire alarm system.

Most photoelectric beam detectors are set to respond to a certain obscuration rate, meaning percentage of the light that is blocked. If the light is completely blocked, such as when a solid object is moved across the beam, the trouble alarm will sound, but the fire alarm will not be activated.

Smoke detectors are usually powered by a low voltage circuit and send a signal to the fire alarm control panel when they are activated. Both ionization and photoelectric smoke detectors are self-restoring. After the smoke condition clears, the alarm system can be reset at the control panel.

Figure 36-11 Commercial ionization smoke detector.

Figure 36-12 Beam detectors are used in large open spaces.

Heat Detectors

Heat detectors are also common automatic alarm initiation devices. Heat detectors can provide property protection, but cannot provide reliable life safety protection because they do not react quickly enough to incipient fires. They are generally used in situations where smoke alarms cannot be used, such as dusty environments and areas that experience extreme cold or heat. Heat detectors are often installed in unheated areas, such as attics and storage rooms, as well as in boiler rooms and manufacturing areas.

Heat detectors are generally very reliable and less prone to false alarms than smoke alarms. You may come across heat detectors that were installed 30 or more years ago and are still in service. However, older units have no visual trigger that tells which device was activated, so tracking down the cause of an alarm may be very difficult. Newer models have an indicator light that shows which device was activated.

There are several types of heat detectors, each designed for specific situations and applications. Single-station heat alarms are sometimes installed in unoccupied areas of buildings that do not have fire alarm systems, such as attics or storage rooms. Spot detectors are individual units that can be spaced throughout an occupancy; each detector covers a specific floor area. Spot detectors are usually installed in light commercial and residential settings; the units may be in individual rooms or spaced at intervals along the ceiling in larger areas.

Line detectors use wire or tubing strung along the ceiling of large open areas to detect an increase in heat. An increase in temperature anywhere along the line will activate the detector. Line detectors are found in churches, warehouses, and industrial or manufacturing applications.

Heat detectors can be designed to operate at a fixed temperature or to react to a rapid increase in temperature. Either fixed-temperature or rate-of-rise devices can be configured as spot or line detectors.

Fixed Temperature Heat Detectors

Fixed-temperature heat detectors, as the name implies, are designed to operate at a preset temperature (► **Figure 36-13**). A typical temperature for a light-hazard occupancy, such as an office building, would be 135°F (57°C). Fixed-temperature detectors usually use a metal alloy that will melt at the preset temperature. The melting alloy releases a lever-and-spring mechanism, to open or close a switch. Most fixed-temperature heat detectors must be replaced after they have been activated, even if the activation was accidental.

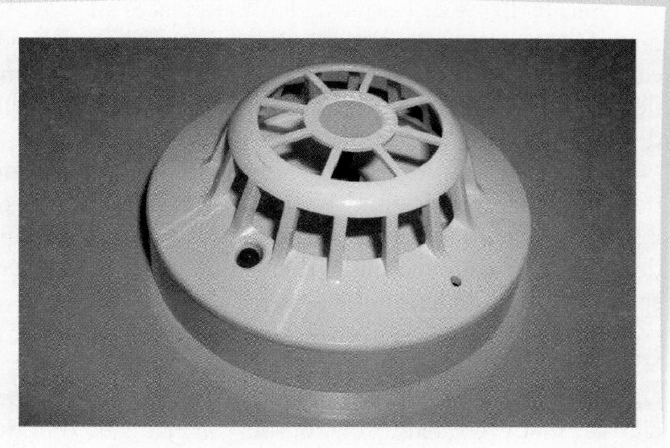

Figure 36-13 A fixed-temperature heat detector initiates an alarm at a preset temperature.

Rate-of-Rise Heat Detectors

Rate-of-rise heat detectors will activate if the temperature of the surrounding air rises more than a set amount in a given period of time. A typical rating might be "greater than 12°F in one minute." If the temperature increase is less than this rate, the rate-of-rise heat detector will not activate. An increase greater than this rate will activate the detector and set off the fire alarm. Rate-of-rise heat detectors should not be located in areas that normally experience rapid changes in temperature, such as near garage doors in heated parking areas.

Some rate-of-rise heat detectors have a bimetallic strip made of two metals that respond differently to heat. A rapid increase in temperature causes the strip to bend unevenly, which opens or closes a switch. Another rate-of-rise heat detector uses an air chamber and diaphragm mechanism. As air in the chamber heats up, the pressure increases. Gradual increases in pressure are released through a small hole, but a rapid increase in pressure will press upon the diaphragm and activate the alarm. Most rate-of-rise heat detectors are self-restoring, and do not need to be replaced after an activation unless they were directly exposed to a fire.

Rate-of-rise heat detectors generally respond faster to most fires than fixed-temperature heat detectors. However, a slow burning fire, such as a smoldering couch, may not activate a rate-of-rise heat detector until the fire is well-established.

Combination rate-of-rise and fixed-temperature heat detectors are available. These devices balance the faster response of the rate-of-rise detector with the reliability of the fixed-temperature heat detector.

Line Heat Detectors

Line heat detectors use wires or a sealed tube to sense heat. One wire-type line detector has two wires inside, separated by an insulating material. When heat melts the insulation, the wires short out and activate the alarm. The damaged section of insulation must be replaced with a new piece after activation.

Figure 36-14 Flame detectors are specialized devices that detect the electromagnetic light waves produced by a flame.

Another wire-type detector measures changes in the electrical resistance of a single wire as it heats up. This device is self-restoring and does not need to be replaced after activation unless it is directly exposed to a fire.

The tube-type line heat detector has a sealed metal tube filled with air or a nonflammable gas. When the tube is heated, the internal pressure increases and activates the alarm. Like the single-wire line heat detector, this device is self-restoring and does not need to be replaced after activation, unless it is directly exposed to a fire.

Flame Detectors

Flame detectors are specialized devices that detect the electromagnetic light waves produced by a flame (▲ **Figure 36-14**). These devices can quickly recognize even a very small fire.

Typically flame detectors are found in places such as aircraft hangars or specialized industrial settings in which early detection and rapid reaction to a fire is critical. Flame detectors are also used in explosion suppression systems to detect and suppress an explosion as it is occurring.

Flame detectors are complicated and expensive; in addition, other infrared or ultraviolet sources such as the sun or a welding operation can set off an unwanted alarm. Flame detectors that combine infrared and ultraviolet sensors are sometimes used to lessen the chances of a false alarm.

Gas Detectors

Gas detectors are calibrated to detect the presence of a specific gas that is created by combustion or that is used in the facility. Depending on the system, a gas detector may be programmed to activate either the building's fire alarm system or a separate alarm. Gas detectors are specialized instruments that need regular calibration. They are usually found only in specific commercial or industrial applications.

Air Sampling Detectors

Air sampling detectors continuously capture air samples and measure the concentrations of specific gases or products of combustion. These devices draw in the air samples through a sampling unit and analyze it using an ionization or photoelectric smoke detector (► Figure 36-15). Air sampling detectors are often installed in the return air ducts of large buildings. They will sound an alarm and shut down the air handling system if they detect smoke.

More complex systems are sometimes installed in special hazard areas to draw air samples from rooms, enclosed spaces, or equipment cabinets. The samples pass through gas analyzers that can identify smoke particles, products of combustion, and concentrations of other gases associated with a dangerous condition. Air sampling detectors are most often used in areas that contain valuable contents or sensitive equipment where it is important to detect problems early.

Alarm Initiation by Fire Suppression Systems

Other fire protection systems in a building may be used to activate the fire alarm system. Automatic sprinkler systems are usually connected to the fire alarm system, and will activate the alarm if there is a water flow (► Figure 36-16). This system not only alerts building occupants and the fire department to a possible fire, it also ensures that someone is aware water is flowing, in case of an accidental discharge. Any other fire extinguishing systems in a building, such as those found in kitchens or containing halogenated agents, should also be tied into the building's fire alarm.

False, Unwanted, and Nuisance Alarms

As the number of fire detection and alarm systems increases, so does the number of false, unwanted, or nuisance alarms. Knowing how to handle these alarms is just as important as knowing how to deal with an actual fire. Fire department personnel who are informed about fire alarm and detection systems can advise building owners and operators when these systems experience problems.

The term "false alarm" is generally used by the public to describe all fire alarm activations that are not associated with a true emergency. In reality, there are three distinct types of false alarms: malicious false alarms, unwanted alarms, and nuisance alarms. It is important to distinguish among the three types in determining the root cause of the fire alarm activation. The problem must be recognized before it can be rectified.

Regardless of the cause, all three types of false alarms have the same results. They waste fire department resources and may delay legitimate responses. Frequent false alarms at

Figure 36-15 Air sampling detector.

Figure 36-16 Automatic sprinkler systems use an electric flow switch to activate the building's fire alarm system.

the same site can desensitize building occupants to the alarm system so that they may not respond appropriately to a real emergency.

Malicious False Alarms

Malicious false alarms are caused by individuals who deliberately activate a fire alarm when there is no fire, causing a disturbance. Manual fire alarm boxes are popular targets for pranksters. A malicious false alarm is an illegal act.

Unwanted Alarms

An unwanted alarm occurs when an alarm system is activated by a condition that is not really an emergency. For example, a smoke alarm placed too close to a kitchen may be triggered by normal cooking activities. An unwanted alarm could also occur if a person who is smoking unknowingly stands under a smoke detector. The smoke detector does its job—recognizing smoke and activating the alarm system—but there is no real emergency.

Nuisance Alarms

Nuisance alarms are caused by improper functioning of an alarm system or one of its components. Alarm systems must be properly maintained on a continual basis. A mechanical failure or a lack of maintenance that causes an alarm system to activate when there is no emergency could also fail to activate the system when there is a real fire.

Preventing Unwanted and Nuisance Alarms

Systems that experience unwanted and nuisance alarms should be examined to determine the cause and correct the problem. Many unwanted alarms could be avoided by relocating a detector or using a different type of detector in a particular location. Proper design, installation, and system maintenance are essential to prevent nuisance alarms.

If an alarm system is activated whenever it rains, water is probably leaking into the wiring. Detectors can respond to a buildup of dust, dirt, or other debris by becoming more sensitive and needlessly going off. That is why smoke alarms and smoke detectors must be cleaned periodically.

Several different methods can be used to reduce unwanted and nuisance alarms caused by smoke detection systems. In a cross-zoned system, the activation of a single smoke detector will not sound the fire alarm, although it will usually set off a trouble alarm. A second smoke detector must be activated before the actual fire alarm will sound.

In a verification system, there is a delay of 30 to 60 seconds between activation and notification. During this time, the system may show a trouble or "prealarm" condition at the system control panel. After the preset interval, the system rechecks the detector. If the condition has cleared, the system returns to normal. If the detector is still sensing smoke, then the fire alarm is activated. Both cross-zoned and verification systems are designed to prevent brief exposures to common occurrences such as cigarette smoke from activating the alarm system.

Alarm Notification Appliances

Audible and visual alarm notification devices such as bells, horns, and electronic speakers produce an audible signal when the fire alarm is activated. Some systems also incorporate visual alerting devices. These audible and visual alarms alert occupants of a building to a fire.

Older systems used various sounds as notification devices. However, this often caused confusion over whether the sound was actually an alarm. More recent fire prevention codes have adopted a standardized audio pattern, called the temporal-3 pattern, that must be produced by any audio device used as a fire alarm. Even single-station smoke alarms designed for residential occupancies now use this sound pattern. This enables people to recognize a fire alarm immediately.

Some public buildings also play a recorded evacuation announcement in conjunction with the temporal-3 pattern. The recorded message is played through the fire alarm speakers and provides safe evacuation instructions (▼ Figure 36-17). In facilities such as airport terminals, this announcement is recorded in multiple languages. This type of system may include a public address feature that fire department or building security personnel can use. This feature may be used to provide specific instructions, information about the situation, or notice when the alarm condition is terminated.

Figure 36-17 A speaker fire alarm notification device.

Figure 36-18 This alarm notification device has both a loud horn and a high-intensity strobe light.

Many new fire alarm systems incorporate visual notification devices such as high-intensity strobe lights or other types of flashing lights as well as audio devices (▲ Figure 36-18). Visual devices alert hearing-impaired occupants to a fire alarm and are very useful in noisy environments where an audible alarm might not be heard.

Other Fire Alarm Functions

In addition to alerting occupants and summoning the fire department, fire alarm systems may also control other building functions, such as air handling systems, fire doors, and elevators. To control smoke movement through the building, the system may shut down or start up air handling systems. Fire doors that are normally held open by magnets may be released to compartmentalize the building and confine the fire to a specific area. Doors allowing re-entry from exit stairways into occupied areas may be unlocked. Elevators will be summoned to a predetermined floor, usually the main lobby, so they can be used by fire crews.

Responding fire personnel must understand which building functions are being controlled by the fire alarm, for both safety and fire suppression reasons. This information should be gathered during pre-incident planning surveys and should also be available in printed form or on a graphic display at the control panel location.

Table 36-2 Categories of Alarm Annunciation Systems

Category	Description
Non-coded alarm	No information is given on what device was activated or where it is located.
Zoned non-coded alarm	Alarm system control panel indicates the zone in the building that was the source of the alarm. It may also indicate the specific device that was activated.
Zoned coded alarm	The system indicates over the audible warning device which zone has been activated. This type of system is often used in hospitals, where it is not feasible to evacuate the entire facility.
Master-coded alarm	The system is zoned and coded. The audible warning devices are also used for other emergency-related functions.

Fire Alarm Annunciation Systems

Some fire alarm systems give little information at the alarm control panel; others will specify exactly which initiation device activated the fire alarm. The systems can be further subdivided based on whether they are zoned or coded systems.

In a **zoned system**, the alarm control panel will indicate where in the building the alarm was activated. Almost all alarm systems are now zoned to some extent. Only the most rudimentary alarm systems give no information at the alarm control panel about where the alarm was initiated. In a **coded system**, the zone is identified not only at the alarm control panel but also through the audio notification device. (▲ Table 36-2) shows how, using these two variables, systems can be broken down into four categories: **non-coded alarm**, **zoned non-coded alarm**, **zoned coded alarm**, and **master-coded alarm**.

Non-Coded Alarm System

In a non-coded alarm system, the control panel has no information indicating where in the building the fire alarm was activated. The alarm typically sounds a bell or horn. Fire department personnel must search the entire building to find which initiation device was activated. This type of system is generally found only in older, small buildings.

Zoned Non-Coded Alarm System

This is the most common type of system, particularly in newer buildings. The building is divided into multiple zones, often by floor or by wing. The alarm control panel indicates in which zone the activated device is located. It may also indicate the type of device that was activated. Responding

Table 36-3 Fire Department Notification Systems

Type of System	Description
Local Alarm	The fire alarm system sounds an alarm only in the building where it was activated. No signal is sent out of the building. Someone must call the fire department to respond.
Remote Station	The fire alarm system sounds an alarm in the building and transmits a signal to a remote location. The signal may go directly to the fire department or to another location where someone is responsible for calling the fire department.
Auxiliary System	The fire alarm system sounds an alarm in the building and transmits a signal to the fire department via a public alarm box system.
Proprietary System	The fire alarm system sounds an alarm in the building and transmits a signal to a monitoring location owned and operated by the facility's owner. Depending upon the nature of the alarm and arrangements with the local fire department, facility personnel may respond and investigate, or the alarm may be immediately retransmitted to the fire department. These facilities are monitored 24 hours a day.
Central Station	The fire alarm system sounds an alarm in the building and transmits a signal to an off-premises alarm monitoring facility. The off-premises monitoring facility is then responsible for notifying the fire department to respond.

personnel can go directly to that part of the building to search for the problem and check the activated device.

Many zoned non-coded alarm systems have an individual indicator light for each zone. When a device in that area is activated, the indicator light goes on. Computerized alarm systems also may use "addressable devices." In these systems, each individual initiation device—whether it is a smoke detector, heat detector, or pull-station—has its own unique identifier. When the device is activated, the identifier is indicated on a display or print-out at the control panel. Responding personnel know exactly which device or devices have been activated.

Zoned Coded Alarm

In addition to having all the features of a zoned alarm system, a zoned coded alarm system will also indicate which zone has been activated over the announcement system. Hospitals often use this type of system, because it is not possible to evacuate all staff and patients for every fire alarm. The audible notification devices give a numbered code that indicates which zone was activated. A code list tells building personnel which zone is in an alarm condition and which areas must be evacuated.

More modern systems in these occupancies use speakers as the alarm notification devices. This enables a voice message indicating the nature and location of the alarm to accompany the audible alarm signal.

Master-Coded Alarm

In a master-coded alarm system, the audible notification devices for fire alarms also are used for other purposes. For example, a school may use the same bell to announce a change in classes, to signal a fire alarm, to summon the janitor, or to make other notifications. Most of these systems have been replaced by modern speaker systems that use the temporal-3 pattern fire alarm signal and have public address capabilities. This type of system is not often installed in new buildings.

Fire Department Notification

The fire department should always be notified when a fire alarm system is activated. In some cases, a person must make a telephone call to the fire department. Or, the fire alarm system can be connected directly to the fire department or to a remote location where someone on duty calls the fire department. As shown in (▲ **Table 36-3**), fire alarm systems can be classified in five categories, based on how the fire department is notified of an alarm.

Local Alarm System

A local alarm system does not notify the fire department. The alarm sounds only in the building to notify the occupants. Buildings with this type of system should have notices posted requesting occupants to call the fire department and report the alarm after they exit (► **Figure 36-19**).

Remote Station System

A remote station system sends a signal directly to the fire department or to another monitoring location via a telephone line or a radio signal. This type of direct notification can only be installed in jurisdictions where the fire department is equipped to handle direct alarms. If the signal goes to a monitoring location, that site must be continually staffed by someone who will call the fire department.

Figure 36-19 Buildings with a local alarm system should post notices requesting occupants to call the fire department and report the alarm after they exit.

Auxiliary System

Auxiliary systems can be used in jurisdictions with a public fire alarm box system. A building's fire alarm system is tied into a master alarm box located outside the building. When the alarm activates, it trips the master box, which transmits the alarm directly to the fire department communications center.

Proprietary System

In a **proprietary system**, the building's fire alarms are connected directly to a monitoring site that is owned and operated by the building owner. Proprietary systems are often installed at facilities with multiple buildings belonging to the same owner, such as universities or industrial complexes. Each building is connected to a monitoring site on the premises, usually the security center, which is staffed at all times (▶ Figure 36-20). When an alarm sounds, the staff at the monitoring site report the alarm to the fire department, usually by telephone or a direct line.

Central Station

A **central station** is a third-party, off-site monitoring facility that monitors multiple alarm systems. Individual building owners contract and pay the central station to monitor their facilities (▶ Figure 36-21). An activated alarm transmits a signal to the central station by telephone or radio. Personnel at the central station then notify the appropriate fire department of the fire alarm. The central station facility may be located in the same city as the facility or in a different part of the country.

Usually, building alarms are connected to the central station through leased or standard telephone lines. However, the use of either cellular telephone frequencies or radio frequencies is becoming more common. Cellular or radio connections may be used to back up regular telephone lines in case they fail; in remote areas without telephone lines, they may be the primary transmission method.

Figure 36-20 In a proprietary system, fire alarms from several buildings are connected to a single monitoring site owned and operated by the buildings' owner.

Figure 36-21 A central station monitors alarm systems at many locations.

Fire Suppression Systems

Fire suppression systems include automatic sprinkler systems, standpipe systems, and specialized extinguishing systems such as dry chemical systems. Construction for most new commercial buildings incorporates at least one of these systems, and single-family dwellings are being built that include a residential sprinkler system.

Understanding how these systems work is important because they can affect fire behavior. In addition, fire fighters should know how to interface with the system and how to shut down a system to prevent unnecessary damage.

Automatic Sprinkler Systems

The most common type of fire suppression system is the automatic sprinkler system. Automatic sprinklers are reliable and effective, with a history of more than 100 years of successfully controlling fires. When properly installed and maintained, automatic sprinkler systems can help control fires and protect lives and property.

Unfortunately, the general public doesn't have an accurate understanding of how automatic sprinklers work. In movies and on television, when one sprinkler head is activated, the entire system begins to discharge water. This inaccurate portrayal of how automatic sprinkler systems operate has made people hesitant to install automatic sprinklers, fearing that they will cause unnecessary water damage.

The reality is quite different. In most automatic sprinkler systems, the sprinkler heads open one at a time as they are heated to their operating temperature. Usually, only one or two sprinkler heads open before the fire is controlled.

The basic operating principles of an automatic sprinkler system are simple. A system of water pipes is installed throughout a building to deliver water to every area where a fire might occur. Depending on the design and occupancy of the building, the pipes may be above or below the ceiling.

Automatic sprinkler heads are located along the system of pipes. Each sprinkler head covers a particular floor area. A fire in that area will activate the sprinkler head, which discharges water on the fire. It is like having a fire fighter in every room with a charged hose line, just waiting for a fire.

One of the major advantages of a sprinkler system is that it can function as both a fire detection system and a fire suppression system. An activated sprinkler head not only discharges water on the fire, it also triggers a **water-motor gong**, a flow alarm that signifies there is water flowing in the system. In addition, the system prompts electric flow switches to activate the building's fire alarm system, notifying the fire department and occupants. The system is so effective that by the time fire fighters arrive, the sprinklers have often already extinguished the fire.

Automatic Sprinkler System Components

The overall design of automatic sprinkler systems can be complex, especially in large buildings. However, even the largest systems have just four major components: the automatic sprinkler heads, piping, control valves, and a water supply, which may or may not include a fire pump (▼ Figure 36-22). Most sprinkler systems are directly connected to a public water supply system and do not need electricity to function.

Automatic Sprinkler Heads

Automatic sprinkler heads, commonly referred to as sprinkler heads or just heads, are the working ends of a sprinkler system. In most systems, the heads serve two functions: they activate the sprinkler system and they apply water to the fire. Sprinkler heads are composed of a body,

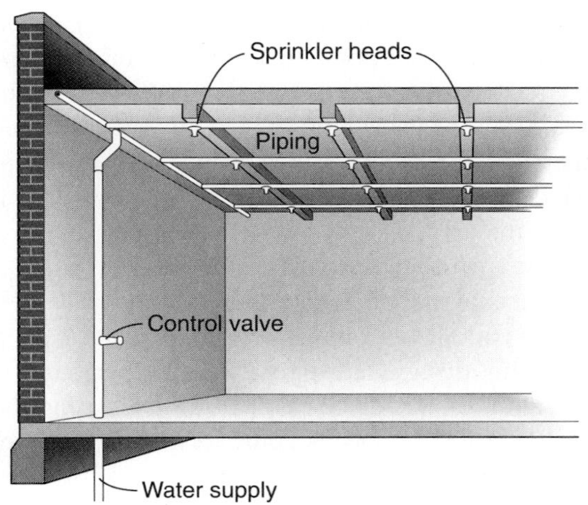

Figure 36-22 The basic components of an automatic sprinkler system include sprinkler heads, piping, control valves, and a water supply.

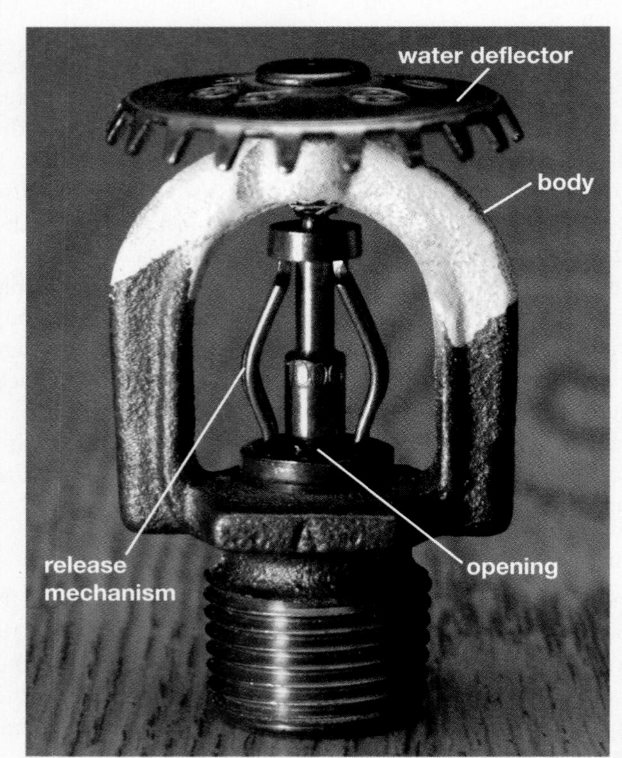

Figure 36-23 Automatic sprinkler heads have a body with an opening, a release mechanism, and a water deflector.

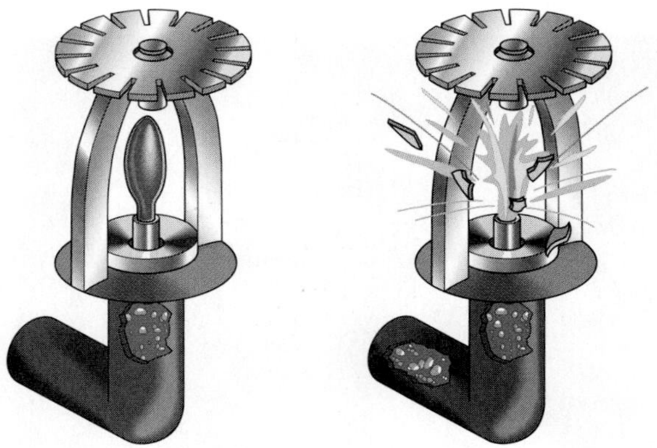

Figure 36-24 Automatic sprinkler heads are designed to activate at a wide range of temperatures and may react slowly or quickly, depending on these settings.

Figure 36-25 Fusible link sprinkler heads use two pieces of metal linked together by an alloy such as solder.

Figure 36-26 Frangible bulb sprinkler heads activate when the liquid in the bulb expands and breaks the glass.

which includes the orifice (opening); a release mechanism that holds a cap in place over the orifice; and a deflector that directs the water in a spray pattern (▲ Figure 36-23). Standard sprinkler heads have a ½″ orifice, but several other sizes are available for special applications.

Although sprinkler heads come in several styles, they are all categorized according to the type of release mechanism and the intended mounting position—upright, pendant, or horizontal. They are also rated according to their release temperature. The release mechanisms hold the cap in place until the release temperature is reached. At that point, the mechanism is released, and the water pushes the cap out of the way as it discharges onto the fire (► Figure 36-24).

Fusible link sprinkler heads use a metal alloy, such as solder, that melts at a specific temperature (► Figure 36-25). The alloy links two pieces of metal that keep the cap in place. When the designated operating temperature is reached, the solder melts and the link breaks, releasing the cap. Fusible link sprinkler heads come in a wide range of styles and temperature ratings.

Frangible bulb sprinkler heads use a glass bulb filled with glycerin or alcohol to hold the cap in place (► Figure 36-26). The bulb also contains a small air bubble. As the bulb is heated, the liquid absorbs the air bubble and

Figure 36-27 Chemical-pellet sprinkler heads have a plunger mechanism that holds the cap in place.

Figure 36-28 An early suppression fast response (ESFR) sprinkler head.

expands until it breaks the glass, releasing the cap. The volume and composition of the liquid and the size of the air bubble determine the temperature at which the head activates, as well as how quickly it responds.

Chemical-pellet sprinkler heads use a plunger mechanism and a small chemical pellet to hold the cap in place (▲ Figure 36-27). The chemical pellet will liquefy at a preset temperature. When the pellet melts, the liquid compresses the plunger, releasing the cap and allowing water to flow.

Special Sprinkler Heads

Sprinkler heads can also be designed for special applications such as covering large areas or discharging the water in extra-large droplets or as a fine mist. Some sprinkler heads have protective coatings to help prevent corrosion. Builders and installers should consider these characteristics when designing the system and selecting appropriate heads. It is important to ensure that the proper heads are installed and any replacement heads are of the same type.

Sprinkler heads that are intended for residential occupancies are manufactured with a release mechanism that provides a fast response. Residential heads usually have smaller orifices and release a limited flow of water, because they are used in small rooms with a limited fire load.

Early suppression fast response (ESFR) sprinkler heads have improved heat collectors to speed up the response and ensure rapid release (▲ Figure 36-28). They are used in large warehouses and distribution facilities where early fire suppression is important. These heads often have large orifices to discharge water onto a fire.

Figure 36-29 A deluge sprinkler head has no release mechanism.

Deluge Heads

Deluge heads are easily identifiable, because they have no cap or release mechanism. The orifice is always open (▲ Figure 36-29). Deluge heads are only used in deluge sprinkler systems, which are covered later in this chapter.

Temperature Ratings

Sprinkler heads are rated according to their release temperature. A typical rating for sprinkler heads in a light hazard occupancy, such as an office building, would be 165°F (74°C). Sprinkler heads that are used in areas with warmer ambient air temperatures would have higher ratings. The rating should be stamped on the body of the sprinkler head. Frangible bulb sprinkler heads use a color-coding system to identify the temperature rating (► Table 36-4). Some fusible link and chemical-pellet sprinklers also use this system.

The temperature rating on a sprinkler head must match the anticipated ambient air temperatures. If the rating is too low for the ambient air temperature, accidental alarms may occur. Conversely, if the rating is too high, the system will be slow to react to a fire, and the fire will be able to establish itself and grow before the system activates.

Spare heads that match those used in the system should always be available on-site. Usually the spare heads are kept in a clearly marked box near the main control valve (► Figure 36-30). Having spare heads handy enables sprinkler systems to be returned to full service quickly, whether they were set off by a fire or by accident.

Table 36-4 Temperature Rating Determined by Color of Sprinkler Head

Maximum Ceiling Temperature (°F)	Temperature Rating (°F)	Color Code	Glass Bulb Colors
100	135 to 170	uncolored or black	orange or red
150	175 to 225	white	yellow or green
225	250 to 300	blue	blue
300	325 to 375	red	purple
375	400 to 475	green	black
475	500 to 575	orange	black
625	650	orange	black

Figure 36-30 Spare sprinkler heads should be kept in a special box near the main sprinkler system valve.

Mounting Position

Sprinkler heads with different mounting positions are not interchangeable. Each mounting position has deflectors specifically designed to produce an effective water stream down or out toward the fire. Each automatic sprinkler head is designed to be mounted in one of three positions (► Figure 36-31 A–C):

A

B

C

Figure 36-31 Each sprinkler head model is designed to be mounted in **A.** upright, **B.** pendant, or **C.** sidewall.

- Pendant sprinkler heads are designed to be mounted on the underside of the sprinkler piping, hanging down toward the room. Pendant heads are commonly marked SSP, which stands for Standard Spray Pendant.
- Upright sprinkler heads are designed to be mounted on top of the supply piping as the name suggests. Upright heads are usually marked SSU, for Standard Spray Upright.
- Sidewall sprinkler heads are designed for horizontal mounting, projecting out from a wall.

Old Style Versus New Style Sprinkler Heads

Up until the early 1950s, deflectors in both pendant and upright sprinkler heads directed part of the water stream up toward the ceiling. It was believed that this helped cool the area and extinguish the fire. Sprinkler heads with this design are called old style sprinklers. There are still many old style heads in service today.

Automatic sprinklers manufactured after the mid-1950s deflect the entire water stream down to the fire. These types of heads are referred to as new style heads or standard spray heads. New style heads can replace old style heads but the reverse is not true. Due to different coverage patterns, old style heads should not be used to replace any new style heads.

Sprinkler Piping

Sprinkler piping, the network of pipes that delivers water to the sprinkler heads, includes the main water supply lines, risers, feeder lines, and branch lines. Sprinkler pipes are usually made of steel, but other metals can be used (► Figure 36-32). Plastic pipe is sometimes used in residential sprinkler systems.

Sprinkler system designers use piping schedules or hydraulic calculations to determine the size of pipe and the layout of the "grid." Most new systems are designed using computer software. Near the main control valve, pipes have a large diameter; as the pipes approach the sprinkler heads, the diameter generally decreases.

Figure 36-32 Most new sprinkler systems use steel pipes.

Valves

A sprinkler system includes several different valves such as the main water supply control valve, the alarm valve, and other, smaller valves used for testing and service. Many large systems have zone valves so the water supply to different areas can be shut down without turning off the entire system. All of the valves play a critical role in the design and function of the system.

Figure 36-33 An OS&Y valve is often used to control the flow of water into a sprinkler system.

Figure 36-34 A post indicator valve (PIV) is used to open or close an underground valve.

Water Supply Control Valves

Every sprinkler system must have at least one main control valve that allows water to enter the system. This water supply control valve must be of the "indicating" type, meaning that the position of the valve itself indicates whether it is open or closed. Two examples are the outside stem and yoke valve (OS&Y), and the post indicator valve (PIV).

The OS&Y valve has a stem that moves in and out as the valve is opened or closed (▲ Figure 36-33). If the stem is out, the valve is open. OS&Y valves are often found in a mechanical room in the building, where water to supply the sprinkler system enters the building. In warmer climates, OS&Y valves may be found outside.

The PIV has an indicator that reads either open or shut depending on its position (► Figure 36-34). A PIV is usually located in an open area outside of the building and controls an underground valve. Opening or closing a PIV requires a wrench, which is usually attached to the side of the valve.

A wall post indicator valve (WPIV) is similar to a PIV but is designed to be mounted on the outside wall of a building (► Figure 36-35).

The main control valve, whether it is an OS&Y valve or a PIV, should be locked in the open position. This ensures that the water supply to the sprinkler system is never shut off unless the proper people are notified that the system is out of service. It is critically important that the sprinkler system is always charged with water and ready to operate if needed.

An alternative to locking the valves open is equipping them with tamper switches (► Figure 36-36). These devices monitor the position of the valve. If someone opens or closes the valve, the tamper switch sends a signal to the fire alarm control panel indicating a change in valve position. If the change has not been authorized, the cause of the signal can be investigated and the problem can be corrected.

Main Sprinkler System Valves

The type of main sprinkler system valve used depends on the type of sprinkler system installed. Options include an alarm valve, a dry-pipe valve, or a deluge valve. These valves are usually installed on the main riser, above the water supply control valve.

The primary functions of an alarm valve are to signal an alarm when a sprinkler head is activated and to prevent nuisance alarms caused by pressure variations and surges in the water supply to the system. The alarm valve has a clapper mechanism that remains in the closed position until a sprinkler head opens. The closed clapper prevents water from flowing out of the system and back into the public water mains when water pressure drops.

When a sprinkler head is activated, the clapper opens fully and allows water to flow freely through the system. The open clapper also allows water to flow to the water-motor gong, sounding an alarm. Electrical flow switches activate connections to external alarm systems.

Figure 36-35 A wall post indicator valve (WPIV) controls the flow of water from an underground pipe into a sprinkler system.

Figure 36-36 A tamper switch activates an alarm if someone attempts to close a valve that should remain open.

Without a properly functioning alarm valve, sprinkler system flow alarms would occur frequently. The normal pressure changes and surges in a public water system will not open the clapper. This prevents water from flowing to the water-motor gong or tripping the electrical water flow switches.

In dry-pipe and deluge systems, the main valve functions both as an alarm valve and as a dam, holding back the water until the sprinkler system is activated. When the system is activated, the valve opens fully so water can enter the sprinkler piping. Both dry-pipe and deluge systems are described later in this chapter.

Additional Valves

Sprinkler systems are equipped with various other control valves. Several smaller valves are usually located near the main control valve, with others located elsewhere in the building. These smaller valves include drain valves, test valves, and connections to alarm devices. All of these valves should be properly labeled.

In larger facilities, the sprinkler system may be divided into zones. Each zone has a valve that controls the flow of water to that particular zone. This design makes maintenance easy and also is valuable when a fire occurs. After the fire is extinguished, water flow to the affected area can be shut off so that the activated heads can be replaced. Fire protection in the rest of the building is unaffected by the shut-down.

Water Supplies

The water used in an automatic sprinkler system may come from a municipal water system, from on-site storage tanks, or from static water sources such as storage ponds or rivers. Whatever the source, the water supply must be able to handle the demand of the sprinkler system, as well as the needs of the fire department in the event of a fire.

The preferred water source for a sprinkler system is a municipal water supply, if one is available. If the municipal supply cannot meet the water pressure and volume requirements of the sprinkler system, alternative supplies must be provided.

Fire pumps are used when the water comes from a static source. They may also be used to boost the pressure in some sprinkler systems, particularly for tall buildings ► Figure 36-37. Because most municipal water supply systems do not provide enough pressure to control a fire on upper floors of high-rise buildings, fire pumps will turn on automatically when the sprinkler system activates or when the pressure drops to a pre-set level. In high-rise buildings, a series of fire pumps on upper floors may be needed to provide adequate pressure.

A large industrial complex could have more than one water source, such as a municipal system and a back-up storage tank ► Figure 36-38. Multiple fire pumps can provide water to the sprinkler and standpipe systems in different areas through underground pipes. Private hydrants may also be connected to the same underground system.

Figure 36-37 In tall buildings, a fire pump may be needed to maintain appropriate pressure in the sprinkler system.

Figure 36-38 An elevated storage tank ensures that there will be sufficient water and adequate pressure to fight a fire.

Each sprinkler system should also have a fire department connection (FDC). This connection allows the department's engine to pump water into the sprinkler system (▶ Figure 36-39). The FDC is used as either a supplement or the main source of water to the sprinkler system if the regular supply is interrupted or a fire pump fails.

The FDC usually has two or more 2½″ female couplings or one large-diameter hose coupling mounted on an outside wall or placed near the building. It ties directly into the sprinkler system after the main control valve or alarm valve. Each fire department should have a standard procedure for first-arriving companies. The procedure should specify how to connect to the FDC and when to charge the system.

In large facilities, a single FDC may be used to deliver water to all fire protection systems in the complex. The water from the FDC flows into the private underground water mains instead of into each system. Water pumped into this type of FDC should come from a source that does not service the complex, such as a public hydrant on a different grid.

Water Flow Alarms

All sprinkler systems should be equipped with a method for sounding an alarm whenever there is water flowing in the pipes. This is important both in cases of an actual fire, and in cases of accidental activation. Without these alarms, the occupants or the fire department might not be aware of the sprinkler activation. If a building is unoccupied the sprinkler system could continue to discharge water long after a fire is extinguished, resulting in excessive water damage.

Most systems incorporate a mechanical flow alarm called a water-motor gong (▶ Figure 36-40). When the sprinkler system is activated and the main alarm valve opens, some water is fed through a pipe to a water-powered gong located on the outside of the building. This alerts people outside the building that there is water flowing. This type of alarm will function even if there is no electricity.

Accidental soundings of water-motor gongs are rare. If a water-motor gong is sounding, water is probably flowing from the sprinkler system somewhere in the building. Fire companies who arrive and hear the distinctive sound of a water-motor gong know that there is a fire or that something else is causing the sprinkler system to flow water.

Most modern sprinkler systems also are connected to the building's fire alarm system by either an electric flow switch or a pressure switch. This connection will trigger the alarm to alert the building's occupants. A monitored system also will notify the fire department. Unlike water-motor gongs, flow and pressure switches can be accidentally triggered by water pressure surges in the system. To reduce the risk of accidental activations, these devices usually have a time delay before they will sound an alarm.

Figure 36-39 A fire department connection delivers additional water and boosts the pressure in a sprinkler system.

Figure 36-40 A water-motor gong sounds when water is flowing in a sprinkler system.

Types of Automatic Sprinkler Systems

Automatic sprinkler systems are divided into four categories, depending on what type of sprinkler head is used, and how the system is designed to activate. The four categories of sprinkler systems are:

- Wet sprinkler systems
- Dry sprinkler systems
- Preaction sprinkler systems
- Deluge sprinkler systems

Many buildings may use the same type of system to protect the entire facility; it is not uncommon, however, to see two or three systems combined in one building. Some facilities use a wet sprinkler system to protect most of the structure, but will have a dry sprinkler or preaction system in a specific area. In many cases, a dry sprinkler or preaction system will branch off from the wet sprinkler system.

Wet sprinkler systems

A wet sprinkler system is the most common and the least expensive type of automatic sprinkler system. As the name implies, the piping in a wet system is always filled with water. As a sprinkler head activates, water is immediately discharged onto the fire. The major drawback to a wet sprinkler system is that it cannot be used in areas where temperatures drop below freezing. They will also flow water if a sprinkler head is accidentally opened or a leak occurs in the piping.

If only a small, unheated area needs to be protected, two options are available. A dry-pendant sprinkler head can be used in very small areas, such as walk-in freezers (▶ Figure 36-41). The bottom part of a dry-pendant head, which resembles a standard sprinkler head, is mounted inside the freezer. The head has an elongated neck, usually 6" to 18" long, that extends up and connects to the wet sprinkler piping in the heated area above the freezer. The vertical neck section is filled with air and capped at each end. The top cap prevents water from entering the lower section where it would freeze. The bottom cap acts just like the cap on a standard sprinkler head. When the head is activated and the lower cap drops out, a device inside the neck releases the upper cap, so water can flow down. The entire dry-pendant head assembly must be replaced after it has been activated.

Larger unheated areas, such as a loading dock, can be protected with an antifreeze loop. An antifreeze loop is a small section of the wet sprinkler system that is filled with glycol or glycerin instead of water. A check valve separates the antifreeze loop from the rest of the sprinkler system. When a sprinkler head in the unheated area is activated, the antifreeze sprays out first, followed by water.

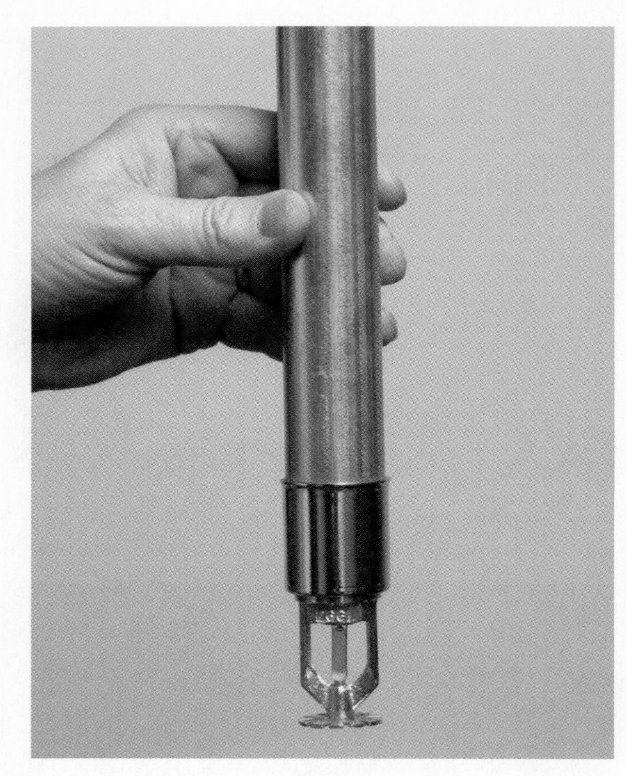

Figure 36-41 A dry-pendant sprinkler head can be used to protect a freezer box.

Dry Sprinkler Systems

A **dry sprinkler system** operates much like a wet sprinkler system, except that the pipes are filled with pressurized air instead of water. A dry-pipe valve keeps water from entering the pipes until the air pressure is released. Dry systems are used in large facilities that may experience below-freezing temperatures, such as unheated warehouses or garages.

The air pressure is set high enough to hold a clapper inside the dry-pipe valve in the closed position (▶ Figure 36-42). When a sprinkler head opens, the air escapes. As the air pressure drops, the water pressure on the other side of the clapper forces it open and water flows into the pipes. When the water reaches the open sprinkler head, it is discharged onto the fire.

Dry sprinkler systems do not eliminate the risk of water damage from accidental activation. If a sprinkler head breaks, the air pressure will drop and water will flow, just as in a wet sprinkler system.

The clapper assembly inside most dry-pipe valves works on a pressure differential. The system or air side of the clapper has a larger surface area than the supply or wet side. In this way, a lower air pressure can hold back a higher water pressure. A small compressor is used to maintain the air pressure in the system.

Dry sprinkler systems should have an air pressure alarm to alert building personnel if the air pressure drops. This could mean that the compressor is not working, or that there is an air leak in the system. If the air pressure in the system is too low, the clapper will open, and the system will fill with water. At that point, the system would essentially be a wet sprinkler system, which could freeze in low temperatures. The system would have to be drained and reset to prevent the pipes from freezing.

Dry sprinkler systems must be drained after every activation so the dry-pipe valve can be reset. The clapper also must be reset, and the air pressure must be restored before the water is turned back on.

Accelerators and Exhausters

One problem encountered in dry sprinkler systems is the delay between the activation of a sprinkler head and the actual flow of water out of the head. The pressurized air that fills the system must escape through the open head before the water can flow. For personal safety and property protection reasons, any delay longer than 90 seconds is unacceptable. Large systems, however, can take several minutes to empty of air and refill with water. To compensate for this problem, two additional devices are used: accelerators and exhausters.

An **accelerator** is installed at the dry-pipe valve. The rapid drop in air pressure caused by an open sprinkler head triggers the accelerator, which allows air pressure to flow to the supply side of the clapper valve. This quickly eliminates the pressure differential, opening the dry-pipe valve and

Fire Fighter Tips

Some dry-pipe valves will not open if there is water above the **clapper valve**, a mechanical device that allows the water to flow in only one direction. Unless the main valve is designed to operate as either a wet- or dry-pipe valve, a dry system that has been filled with water should be immediately drained and reset for dry-pipe operation.

allowing the water pressure to force the remaining air out of the piping.

An **exhauster** is installed on the system side of the dry-pipe valve, often at a remote location in the building. Like an accelerator, the exhauster monitors the air pressure in the piping. If it detects a drop in pressure, it opens a large-diameter portal, so the air in the pipes can escape. The exhauster closes when it detects water, diverting the flow to the open sprinkler heads. Large systems may have multiple exhausters located in different sections of the piping.

Preaction Sprinkler Systems

A **preaction sprinkler system** is similar to a dry sprinkler system with one key difference. In a preaction sprinkler system, a secondary device, such as a smoke detector or a manual-pull alarm, must be activated before water is released into the sprinkler piping. When the system is filled with water, it functions as a wet sprinkler system.

A preaction system uses a deluge valve instead of a dry-pipe valve. The deluge valve will not open until it receives a signal that the secondary device has been activated. Because a detection system usually will activate more quickly than a sprinkler system, water in a preaction system will generally reach the sprinklers before a head is activated.

The primary advantage of a preaction sprinkler system is in preventing accidental water discharges. If a sprinkler head is accidentally broken or the pipe is damaged, the deluge valve will prevent water from entering the system. This makes preaction sprinkler systems well-suited for locations where water damage is a major concern, such as computer rooms, libraries, and museums.

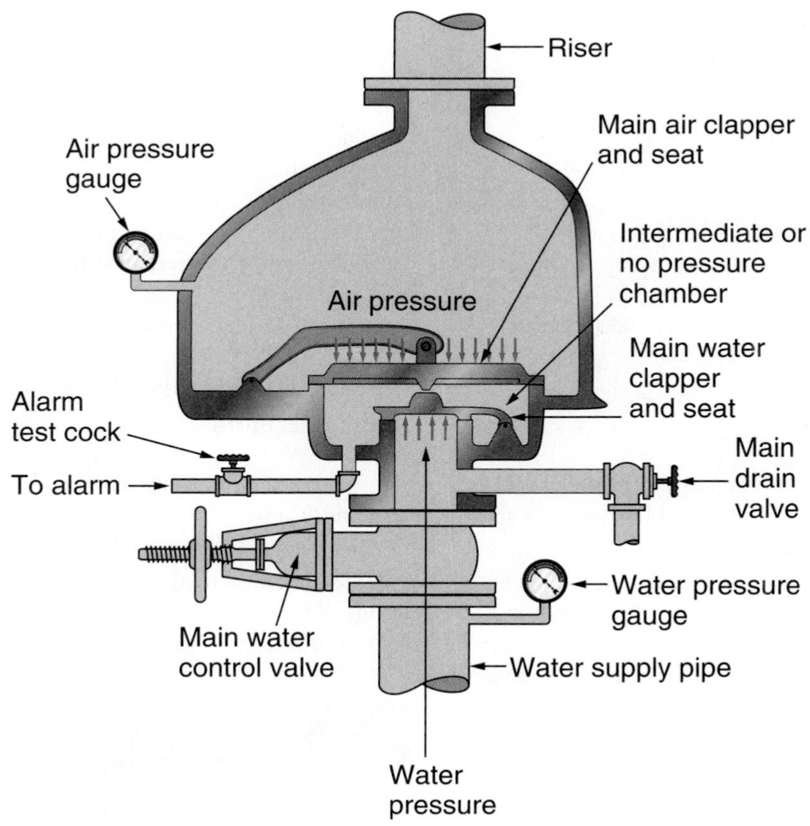

Figure 36-42 Water pressure on one side of the dry-pipe valve is balanced by air pressure on the other side.

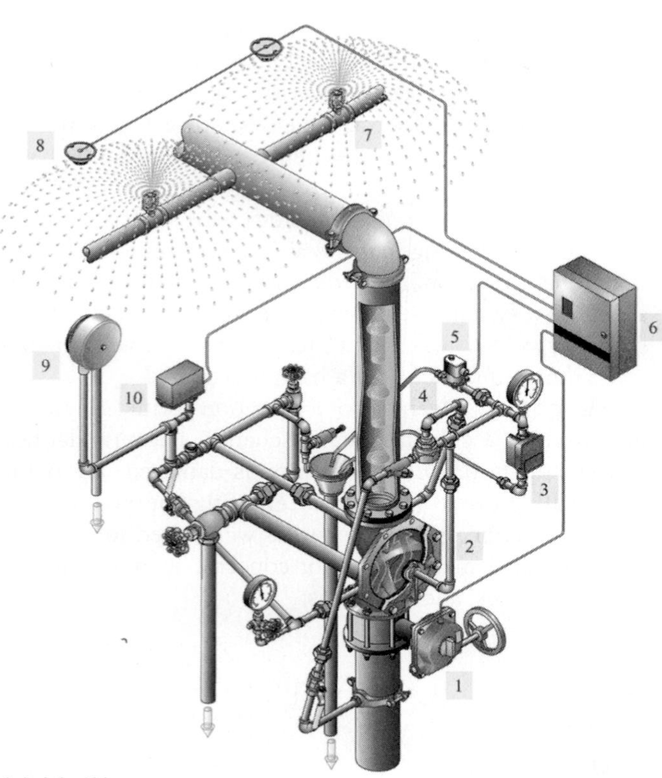

Figure 36-43 Water flows from all of the heads in a deluge system as soon as the system is activated.

Deluge Sprinkler Systems

A **deluge sprinkler system** is a type of dry sprinkler system in which water flows from all of the sprinkler heads as soon as the system is activated (▲ Figure 36-43). A deluge system does not have closed heads that open individually at the activation temperature; all of the heads in a deluge system are always open.

Deluge systems can be activated in three ways. A detection system can release the deluge valve when a detector is activated. The deluge system can also be connected to a separate pilot system of air-filled pipes with closed sprinkler heads. When a head on the pilot line is activated, the air pressure drops, opening the deluge valve. Most deluge valves can be released manually as well.

Deluge systems are only used in special applications such as aircraft hangars or industrial processes, where rapid fire suppression is critical. In some cases, foam concentrate added to the water will discharge a foam blanket over the hazard. Deluge systems are also used for special hazard applications, such as liquid propane gas loading stations. In these situations, a heavy deluge of water is needed to protect exposures from a large fire that occurs very rapidly.

Shutting Down Sprinkler Systems

Responding fire companies should know where and how to shut down any automatic sprinkler system if needed. If the system is accidentally activated, it should be shut down as soon as possible to avoid excessive water damage.

In an actual fire, the order to shut down the sprinkler system should come only from the Incident Commander. Generally, the sprinkler system should not be shut down until the fire is completely extinguished. However, if the system is damaged and is wastefully discharging water, shutting down the sprinkler system could make more water available to fight the fire manually.

When the system is shut down, a fire fighter with a portable radio should stand by at the control valve, in case it has to be reopened quickly.

In most cases, the entire sprinkler system can be shut down by closing the main control valve (OS&Y or PIV). In a zoned system, the affected zone can be shut down by closing the appropriate valve. The rest of the system can then remain operational while the activated heads are being replaced.

Placing a wooden wedge or a commercial sprinkler stopper into the sprinkler head can quickly stop the flow of water (► Figure 36-44), although this will not work with all types of heads. This shuts off the flow of water until the control valve can be located and shut down. After the main valve is closed, opening the drain valve allows the system to drain so that the activated head can be replaced.

Residential Sprinkler Systems

Residential sprinkler systems are relatively new, but many homes now being built include them. Some communities require residential sprinkler systems in every newly constructed home.

The design and theory of residential sprinkler systems are similar to commercial systems, but there are significant variations. The primary objective of a residential sprinkler system is life safety, so the sprinkler heads are designed to

Fire Fighter Safety Tips

Do not shut down the water to a sprinkler system without orders from the Incident Commander. The fire must be completely out and hose lines must be available should the fire re-ignite. A fire fighter should be stationed at the valve with a portable radio, ready to reopen the valve if necessary.

Figure 36-44 A sprinkler stopper can be used to stop the flow of water from a sprinkler head that has been activated.

Figure 36-45 Special sprinkler heads are used in residential systems. They open more quickly than commercial heads and discharge less water.

respond quickly (► Figure 36-45). A fast response helps protect the occupants and control incipient-stage fires.

Residential systems typically use smaller piping and sprinkler heads with smaller orifices and less water discharge. To control costs, plastic pipe may be used instead of metal. These systems protect high-risk areas such as rooms and corridors; small areas such as closets and bathrooms may not be covered.

Residential systems are usually wet sprinkler systems, although some dry-pipe systems exist. They are usually connected to the domestic water supply instead of a public water main. Local regulations may not require a residential sprinkler system to have a fire department connection or to be tied into a separate fire alarm system, but usually do require a water flow alarm.

The responding fire companies may not be familiar with every home that has a sprinkler system, but fire fighters should know how these systems work and how to shut down a system that has been activated. Usually there is a main shut-off valve for the sprinklers, near the location where water enters the house. If the house has a basement, the shut-off valve may be located with other utility controls. If a separate shut-off for the sprinkler system cannot be found, the main shut-off for the house can be used to shut down the system.

Standpipe Systems

A standpipe system is a network of pipes and outlets for fire hoses built into a structure to provide water for firefighting purposes. Standpipe systems are usually used in high-rise buildings, although they are found in many other structures as well. At set intervals throughout the building, there will be a valve where fire fighters can connect a hose to the standpipe (► Figure 36-46).

Figure 36-46 Standpipe outlets allow fire hoses to be connected inside a building.

Standpipes are found in buildings with and without sprinkler systems. In many newer buildings, sprinklers and standpipes are combined in one system. Older buildings more commonly have separate sprinkler and standpipe systems.

The three categories of standpipes—Class I, Class II, and Class III—are defined by their intended use.

Class I Standpipe

A Class I standpipe is designed for use by fire department personnel only. Each outlet has a $2^1/_2$" male coupling and a valve to open the water supply after the hose is connected (► Figure 36-47). Often, the connection is located inside a cabinet, which may or may not be locked. Responding fire personnel carry the hose into the building with them, usually in some sort of roll, bag, or backpack. A Class I standpipe system must be able to supply an adequate volume of water with sufficient pressure to operate fire department attack lines.

Figure 36-47 A Class I standpipe provides water for fire department hose lines.

Class II Standpipes

A Class II standpipe is designed for use by the building occupants. The outlets are generally equipped with a length of $1^1/_2$" single-jacket hose preconnected to the system (► Figure 36-48). These systems are intended to enable occupants to attack a fire before the fire department arrives, but their safety and effectiveness is questionable. Most building occupants are not trained to attack fires safely. If a fire cannot be controlled with a regular fire extinguisher, it is usually safer for the occupants to evacuate the building and call the fire department. Class II standpipes may be useful at facilities such as refineries and military bases, where workers are trained as an in-house fire brigade.

Responding fire personnel should not use Class II standpipes for fighting a fire. The hose and nozzles may be of inferior quality and the water flow may not be adequate to control a fire. Class II standpipe outlets are frequently connected to the domestic water piping system in the building rather than an outside main or a separate system. Instead of using equipment that may not be reliable or adequate, fire fighters should always use department issued equipment.

Class III Standpipes

A Class III standpipe has the features of both Class I and Class II standpipes in a single system. They have $2^1/_2$" outlets for fire department use as well as smaller outlets with attached hoses for occupant use. The occupant hoses have been removed, either intentionally or by vandalism, in many facilities, so the system basically becomes a Class I system. Fire fighters should use only the $2^1/_2$" outlets, even if they are using an adapter to connect a smaller hose. The $1^1/_2$" outlets may have pressure-reducing devices to limit the flow and pressure for use by untrained civilians.

Water Flow in Standpipe Systems

Standpipes are designed to deliver a minimum amount of water at a particular pressure to each floor. The design requirements depend on the code requirements in effect when the building was constructed. The actual flow also depends on the water supply, as well as on the condition of the piping system and fire pumps.

Flow-restriction devices (► Figure 36-49) or pressure-reducing valves (► Figure 36-50) are often installed at the outlets to limit the pressure and flow. A vertical column of water, such as the water in a standpipe riser, exerts a backpressure (also called head pressure). In a tall building, this backpressure can be hundreds of pounds per square inch at lower floor levels. If a hose line is connected to an outlet without a flow restrictor or a pressure-reducing valve, the water pressure could rupture the hose and excessive nozzle pressure could make the line difficult or dangerous to handle. Building and

Fire Fighter Safety Tips

Fire fighters should never use the hose lines on a Class II standpipe to attack a fire. These hoses may not be well-maintained or able to withstand the pressure. Valves and nozzles may not function adequately. Always use your department's hose lines and a reliable source of water.

Figure 36-48 A Class II standpipe is intended to be used by building occupants to attack incipient-stage fires.

Figure 36-49 A flow-restriction device on a standpipe outlet can cause problems for fire fighters.

fire codes limit the height of a single riser and may also require the installation of pressure-reducing valves on lower floors.

If they are not properly installed and maintained, these devices can cause problems for fire fighters. An improperly adjusted pressure-reducing valve could severely restrict the flow to a hose line. A flow restriction device could also limit the flow of water to fight a fire.

The flow and pressure capabilities of a standpipe system should be determined during pre-incident planning. Many standpipe systems deliver water at a pressure of only 65 psi (pounds per square inch) at the top of the building. The combination fog-and-straight-stream nozzles used by many fire departments are designed to operate at 100 psi. As a result, many fire departments use low-pressure combination nozzles for fighting fires in high-rise buildings or require the use of only smooth-bore nozzles when operating from a standpipe system.

Pre-incident plans for high-rise buildings should include an evaluation of the building's standpipe system and a determination of the anticipated flows and pressures. This information should be used to make decisions about the appropriate nozzles and tactics for those buildings.

Fire Fighter Safety Tips

When using a Class III standpipe, always connect your hose line to the $2^1/_2$" outlet to ensure that you get as much water as possible. The smaller $1^1/_2$" outlet may be equipped with a pressure-reducing device.

Figure 36-50 A pressure-reducing valve on a standpipe outlet may be necessary on lower floors to avoid problems caused by backpressure.

Engine companies that respond to buildings equipped with standpipes should carry a kit that includes the appropriate hose and nozzle, a spanner wrench, and any required adapters. This kit should also include tools to adjust the settings of pressure-reducing valves or to remove restrictors that are obstructing flows.

Water Supplies

Both standpipe systems and sprinkler systems are supplied with water in essentially the same way. Many wet standpipe systems in modern buildings are connected to a public water supply with an electric or diesel fire pump to provide additional pressure. Many of these systems also have a water storage tank as a back-up supply. In these systems the FDC on the outside of the building can be used to increase the

Voices of Experience

"Because the historic building was not regularly occupied, its first line of defense against a fire was its fire detection system and automatic alarm."

The tones sounded at 5:24 a.m. on a cold and dark January morning. "Engine 403, Truck 403, respond for a fire alarm activation at 10386 Main Street." Ninety seconds later, having donned our personal protective clothing, six seat belt buckles clicked and the engine and truck pulled out of the station.

We arrived at a historic building in the old downtown section of the city. The original part of the wood frame two-story building was constructed in 1812. When the building was given to the city in the early 1970s, it was set up to house a local historical office and a museum.

We sized-up the building and found no sign of fire. Entry was gained by using the lockbox. Inside, the annunciator panel showed the activation of one smoke detector. Further checking indicated no signs of fire or smoke in the building, and the alarm system was reset.

As we were leaving the scene, I considered how much this building meant to the city's residents, and what a devastating loss it would be if the building were destroyed by fire. Because the historic building was not regularly occupied, its first line of defense against a fire was its fire detection system and automatic alarm. Without that system in place, if a fire started in this old wood building, we would have little chance of saving it. A fire would grow until someone passing by noticed smoke or flames.

While fire detection and suppression systems do sometimes generate false alarms, they enable us to respond to a fire while it is small enough to be extinguished easily. Fire detection and suppression systems help us in our mission of saving lives and property.

Jennifer Schottke
Fairfax City Volunteer Fire Department
Fairfax, Virginia

flow, boost the pressure, or obtain water from an alternative source.

Dry standpipe systems are found in many older buildings. If freezing weather is a problem, such as in open parking structures, bridges, and tunnels, dry standpipe systems are still acceptable. Most dry standpipe systems do not have a permanent connection to a water supply, so the FDC must be used to pump water into the system. If there is a fire in a building with dry standpipes, connecting the hose lines to the FDC and charging the system with water is a high priority.

Some dry standpipe systems are connected to a water supply through a dry-pipe or deluge valve, similar to a sprinkler system. Opening an outlet valve or tripping a switch next to the outlet releases water into the standpipes in these systems.

High-rise buildings often have complex systems of risers, storage tanks, and fire pumps to deliver the needed flows to upper floors. The details of these systems should be obtained during pre-incident planning surveys. Department procedures should dictate how responding units will supply the standpipes with water as well as how crews should use the standpipes inside.

Specialized Extinguishing Systems

Automatic sprinkler systems are used to protect whole buildings, or at least major sections of buildings. But in certain situations, more specialized extinguishing systems are needed. Specialized extinguishing systems are often used in areas where water would not be an acceptable extinguishing agent (► **Figure 36-51**). For example, water is not the agent of choice for areas containing sensitive electronic equipment or contents such as computers, valuable books, or documents. Water is also incompatible with materials such as flammable liquids or water-reactive chemicals. Areas where these materials are stored or used may have a separate extinguishing system.

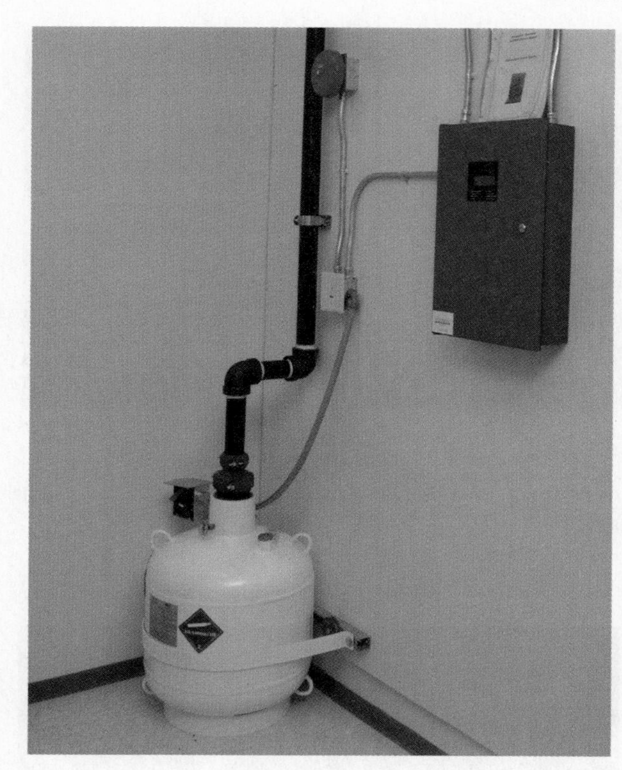

Figure 36-51 Special extinguishing systems are used in areas where water would not be effective or desirable.

Dry Chemical and Wet Chemical Extinguishing Systems

Dry chemical and wet chemical extinguishing systems are the most common specialized agent systems. In commercial kitchens, they are used to protect the cooking areas and exhaust systems. Many self-service gas stations have dry chemical systems that protect the dispensing areas. These systems are often installed inside buildings to protect areas where flammable liquids are stored or used. Both dry chemical and wet chemical extinguishing systems are similar in basic design and arrangement.

Dry chemical extinguishing systems use the same types of finely powdered agents as dry chemical fire extinguishers (► **Figure 36-52**). The agent is kept in self-pressurized tanks or in tanks with an external cartridge of carbon dioxide or nitrogen that provides pressure when the system is activated.

Figure 36-52 Dry chemical extinguishing systems are installed at many self-service gasoline filling stations.

Figure 36-53 Wet chemical extinguishing systems are used in most new commercial kitchens.

Figure 36-54 Fusible links can be used to activate a special agent extinguishing system.

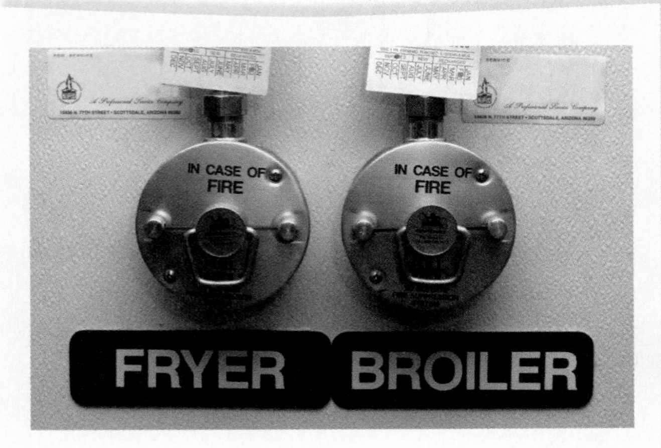

Figure 36-55 Most special extinguishing systems can also be manually activated.

Wet chemical extinguishing systems are used in most new commercial kitchens (◄ Figure 36-53). These systems use a proprietary liquid extinguishing agent, which is much more effective on vegetable oils than the dry chemicals used in older kitchen systems. Wet chemical systems are also easier to clean up after a discharge, so the kitchen can resume operations more quickly after the system has discharged.

Wet chemical extinguishing agents are not compatible with normal all-purpose dry chemical extinguishing agents. Only wet agents or B:C-rated dry chemical extinguishers should be used where these systems are installed.

Fusible links or other automatic initiation devices are placed above the target hazard to activate both dry chemical and wet extinguishing agent systems (◄ Figure 36-54). A manual discharge button is also provided so that workers can activate the system if they discover a fire (◄ Figure 36-55). Open nozzles are located over the target areas to discharge the agent directly onto a fire. When the system is activated, the extinguishing agent flows out of all the nozzles.

Many kitchen systems discharge agent into the ductwork above the exhaust hood, as well as onto the cooking surface. This helps prevent a fire from igniting any grease build-up inside the ductwork and spreading throughout the system. Although the ductwork should be cleaned regularly, it is not unusual for a kitchen fire to extend into the exhaust system.

Most dry and wet chemical extinguishing systems are tied into the building's fire alarm system. Kitchen extinguishing systems should also shut down the gas or electricity to the cooking appliances and exhaust fans.

Clean Agent Extinguishing Systems

Clean agent extinguishing systems are often installed in areas where computers or sensitive electronic equipment are used, or where valuable documents are stored. These agents are nonconductive and leave no residue. Halogenated agents or carbon dioxide are generally used because they will extinguish a fire without causing significant damage to the contents.

Clean agent systems operate by discharging a gaseous agent into the atmosphere at a concentration that will extinguish a fire. Smoke detectors or heat detectors installed in these areas activate the system. A manual discharge button is also provided with most installations. Discharge is usually delayed 30 to 60 seconds after the detector is activated to allow workers in the area to evacuate.

Fire Fighter Safety Tips

Halon and clean agents are considered non-toxic. However, you must use SCBA at all times when working in rooms where these agents have discharged.

Fire Fighter Safety Tips

Carbon dioxide displaces the oxygen in a room and creates a dangerous situation. Wear SCBA at all times when entering these rooms!

During this delay (the pre-alarm period), an abort switch can be used to stop the discharge. In some systems, the abort button must be pressed until the detection system is reset; releasing the abort button too soon causes the system to discharge.

If there is a fire, the clean agent system should be completely discharged before fire fighters arrive. Whether the fire was successfully extinguished or not, fire fighters entering the area should use self-contained breathing apparatus (SCBA) until the area has been properly ventilated. Although these agents are not considered immediately dangerous to life and health, it is better to avoid any unnecessary exposure to them. Toxic products or by-products of combustion could be present in the atmosphere, or the oxygen level could be reduced.

Clean agent systems should be tied to the building's fire alarm system and indicated as a zone on the control panel. This alerts fire fighters that they are responding to a situation where a clean agent has discharged. If the system has a preprogrammed delay, the pre-alarm should activate the building's fire alarm system.

Until the 1990s, **Halon 1301** was the agent of choice for protecting areas such as computer rooms, telecommunications rooms, and other sensitive areas. Halon 1301 is a nontoxic, odorless, colorless gas that leaves no residue. It is very effective at extinguishing fires because it interrupts the chemical reaction of combustion. However, this agent is damaging to the environment and its production has been terminated. Alternative agents have been developed for use in new systems; they are also replacing Halon 1301 in many existing systems. A stockpile of Halon 1301 is still available to recharge existing systems when there is no suitable substitute agent.

Figure 36-56 Carbon dioxide extinguishes a fire by displacing the oxygen in the room and smothering the fire.

Carbon Dioxide Extinguishing Systems

Carbon dioxide extinguishing systems are similar in design to clean agent systems. The primary difference is that carbon dioxide extinguishes a fire by displacing the oxygen in the room and smothering the fire. Large quantities of carbon dioxide are required, because the area must be totally flooded to extinguish a fire (▲ **Figure 36-56**).

Carbon dioxide systems are designed to protect a single room or a series of rooms. They usually have the same series of pre-alarms and abort buttons as Halon systems. Because discharge creates an oxygen-deficient atmosphere in the room, it is immediately dangerous to life. Any occupant who is still in the room when the agent is discharged is likely to be rendered unconscious and asphyxiated. Fire fighters responding to a carbon dioxide extinguishing system discharge must use SCBA protection until the area is fully vented.

Carbon dioxide extinguishing systems should be connected to the building's fire alarm system. Responding fire fighters should see that a carbon dioxide system discharge has been activated. Using this knowledge, they can deal with the situation safely.

Wrap-Up

Ready for Review

This chapter covered the basic operation and principles of fire alarm, fire detection, and fire suppression systems. New fire fighters should be familiar with these basic principles. Understanding the basic operating principles of each system enables you to deal with fire protection systems as a professional, whether they are activated by a fire or by accident.

Fire departments are not in the business of servicing fire alarm and suppression systems. However, when fire fighters show up on the scene of an activated alarm or sprinkler system, the public expects a professional level of both knowledge and action. Every fire department should have a clear policy stating the actions crews should take in resetting systems.

The operation—both proper and improper—of fire suppression systems can affect fire fighter safety. Crews must have an understanding of the systems in their jurisdiction and how to fight a fire in a protected building. Any department that has high-rise buildings within its jurisdiction must include training for fighting fires in these buildings. Fire fighters must have a clear understanding of the standpipe and sprinkler systems in these buildings. The available flows and pressures in these buildings must be determined to ensure that the appropriate attack lines and nozzles are selected.

By understanding residential alarm and suppression systems, fire fighters can help educate the public and promote residential fire safety.

Chief Concepts

- All fire fighters, not just officers and inspectors, need to have a general understanding of how fire protection systems function.
- Fire fighters will respond to more false alarms in their careers than actual fires; knowing how to handle false alarms is critical for providing customer service and preventing future false alarms.
- Although it is not the fire fighter's job to maintain fire protection systems, fire fighters should be able to advise the public on how to restore systems to service.
- Regardless of how sophisticated a fire protection system is, a serious fire can still occur.
- It is every fire fighter's job to try to limit water damage from the activation of a sprinkler system.
- Fire fighters must understand the potential shortcomings of using a standpipe system to prevent injuries and fatalities.
- Home fires cause the most deaths from fire each year in the United States. Fire fighters should encourage the use of fire protection systems in all dwellings.

Hot Terms

Accelerator A device that accelerates the removal of the air from a dry-pipe or preaction sprinkler system.

Air sampling detector A system that captures a sample of air from a room or enclosed space and passes it through a smoke detection or gas analysis device.

Alarm initiation device An automatic or manually operated device in a fire alarm system that, when activated, causes the system to indicate an alarm condition.

Alarm notification device An audible and/or visual device in a fire alarm system that makes occupants or other persons aware of an alarm condition.

Alarm valve This valve signals an alarm when a sprinkler head is activated and prevents nuisance alarms caused by pressure variations.

Automatic sprinkler heads The working ends of a sprinkler system. They serve to activate the system and to apply water to the fire.

Automatic sprinkler system A system of pipes filled with water under pressure that discharges water immediately when a sprinkler head opens.

Auxiliary system A fire alarm system that sounds an alarm in the building and transmits a signal to the fire department via a public alarm box system.

Beam detector A smoke detection device that projects a narrow beam of light across a large open area from a sending unit to a receiving unit. When the beam is interrupted by smoke, the receiver detects a reduction in light transmission and activates the fire alarm.

Bimetallic strip A device with components made from two distinct metals that respond differently to heat. When heated, the metals will bend or change shape.

Carbon dioxide extinguishing system A system designed to protect a single room or series of rooms by flooding the area with carbon dioxide.

Central station An off-premises facility that monitors alarm systems and is responsible for notifying the fire department of an alarm. These facilities may be geographically located some distance from the protected building(s).

Chemical-pellet sprinkler head A sprinkler head activated by a chemical pellet that liquefies at a preset temperature.

Clapper valve A mechanical device installed within a piping system that allows water to flow in only one direction.

Class I standpipe A standpipe system designed for use by fire department personnel only. Each outlet should have a valve to control the flow of water and a 2½" male coupling for fire hose.

Class II standpipe A standpipe system designed for use by occupants of a building only. Each outlet is generally equipped with a length of 1½" single-jacket hose and a nozzle, preconnected to the system.

Class III standpipe A combination system that has features of both Class I and Class II standpipes.

Coded system A fire alarm system design that divides a building or facility into zones and has audible notification devices that can be used to identify the area where an alarm originated.

Cross-zoned system A fire alarm system that requires activation of two separate detection devices before initiating an alarm condition. If a single detection device is activated, the alarm control panel will usually show a problem or trouble condition.

Deluge head A sprinkler head that has no release mechanism; the orifice is always open.

Deluge sprinkler system A sprinkler system in which all sprinkler heads are open. When an initiating device, such as a smoke detector or heat detector, is activated, the deluge valve opens and water discharges from all of the open sprinkler heads simultaneously.

Deluge valve A valve assembly designed to release water into a sprinkler system when an external initiation device is activated.

Double-action pull-station A manual fire alarm activation device that takes two steps to activate the alarm. The person must push in a flap, lift a cover, or break a piece of glass before activating the alarm.

Dry chemical extinguishing system An automatic fire extinguishing system that discharges a dry chemical agent.

Dry-pipe valve The valve assembly on a dry sprinkler system that prevents water from entering the system until the air pressure is released.

Dry sprinkler system A sprinkler system in which the pipes are normally filled with compressed air. When a sprinkler head is activated, it releases the air from the system, which opens a valve so the pipes can fill with water.

Early suppression fast response (ESFR) sprinkler head A sprinkler head designed to react quickly and suppress a fire in its early stages.

Exhauster A device that accelerates the removal of the air from a dry-pipe or preaction sprinkler system.

False alarm The activation of a fire alarm system when there is no fire or emergency condition.

Fire alarm control panel That component in a fire alarm system that controls the functions of the entire system.

Fire department connection (FDC) A fire hose connection through which the fire department can pump water into a sprinkler system or standpipe system.

Fixed-temperature heat detector A sensing device that responds when its operating element is heated to a predetermined temperature.

Flame detector A sensing device that detects the radiant energy emitted by a flame.

Flow switch An electrical switch that is activated by water moving through a pipe on a sprinkler system.

Frangible bulb sprinkler head A sprinkler head with a frangible bulb

Fusible link sprinkler head A sprinkler head with an activation mechanism that incorporates two pieces of metal held together by low-melting-point solder. When the solder melts, it releases the link and water begins to flow.

Wrap-Up

Gas detector A device that detects and/or measures the concentration of dangerous gases.

Halon 1301 A liquefied gas extinguishing agent that puts out the fire by chemically interrupting the combustion reaction between fuel and oxygen. Halon agents leave no residue.

Heat detector A fire alarm device that detects either abnormally high temperatures or rate-of-rise in temperature, or both.

Ionization smoke detector A device containing a small amount of radioactive material that ionizes the air between two charged electrodes to sense the presence of smoke particles.

Line detector Wire or tubing that can be strung along the ceiling of large open areas to detect an increase in heat.

Local alarm system A fire alarm system that sounds an alarm only in the building where it was activated. No signal is sent out of the building.

Malicious false alarm A fire alarm signal when there is no fire, usually initiated by individuals who wish to cause a disturbance.

Manual pull-station A device with a switch that either opens or closes a circuit, activating the fire alarm.

Master-coded alarm An alarm system in which audible notification devices can be used for multiple purposes, not just for the fire alarm.

Non-coded alarm An alarm system that provides no information at the alarm control panel indicating where the activated alarm is located.

Nuisance alarm A fire alarm signal caused by malfunction or improper operation of a fire alarm system or component.

Obscuration rate A measure of the percentage of light transmission that is blocked between a sender and a receiver unit.

Outside stem and yoke (OS&Y) valve A sprinkler control valve with a valve stem that moves in and out as the valve is opened or closed.

Pendant sprinkler head A sprinkler head designed to be mounted on the underside of sprinkler piping so the water stream is directed down.

Photoelectric smoke detector A device to detect visible products of combustion using a light source and a photosensitive sensor.

Post indicator valve (PIV) A sprinkler control valve with an indicator that reads either open or shut depending on its position.

Preaction sprinkler system A dry sprinkler system that uses a deluge valve instead of a dry-pipe valve and requires activation of a secondary device before the pipes fill with water.

Proprietary system A fire alarm system that transmits a signal to a monitoring location owned and operated by the facility's owner.

Rate-of-rise heat detector A device that responds when the temperature rises at a rate that exceeds a predetermined value.

Remote annunciator A secondary fire alarm control panel in a different location than the main alarm panel, usually near the front door of a building.

Remote station system A fire alarm system that sounds an alarm in the building and transmits a signal to the fire department or an off-premise monitoring location.

Residential sprinkler system A sprinkler system designed to protect dwelling units.

Sidewall sprinkler head A sprinkler that is mounted on a wall and discharges water horizontally into a room.

Single-action pull-station A manual fire alarm activation device that takes a single step, such as moving a lever, toggle, or handle, to activate the alarm.

Single-station smoke alarm A single device that both detects visible or invisible products of combustion and sounds an alarm.

Smoke detector A device that detects smoke and sends a signal to a fire alarm control panel.

Spot detector Single heat-detector devices, spaced throughout an area.

Sprinkler piping The network of piping in a sprinkler system that delivers water to the sprinkler heads.

Standpipe system A system of pipes and hose outlet valves used to deliver water to various parts of a building for fighting fires.

Tamper switch A switch on a sprinkler valve that transmits a signal to the fire alarm control panel if the normal position of the valve is changed.

Temporal-3 pattern A standard fire alarm audible signal for alerting occupants of a building.

Unwanted alarm A fire alarm signal caused by a device reacting properly to a condition that is not a true fire emergency.

Upright sprinkler head A sprinkler head designed to be installed on top of the supply piping and usually marked SSU (Standard Spray Upright).

Verification system A fire alarm system that does not immediately initiate an alarm condition when a smoke detector activates. The system will wait a preset interval, generally 30 to 60 seconds, before checking the detector again. If the condition is clear, the system returns to normal status. If the detector is still sensing smoke, the system activates the fire alarm.

Wall post indicator valve (WPIV) A sprinkler control valve that is mounted on the outside wall of a building. The position of the indicator tells whether the valve is open or shut.

Water-motor gong An audible alarm notification device that is powered by water moving through the sprinkler system.

Wet chemical extinguishing systems An extinguishing system that discharges a proprietary liquid extinguishing agent.

Wet sprinkler system A sprinkler system in which the pipes are normally filled with water.

Zoned coded alarm A fire alarm system that indicates which zone was activated both on the alarm control panel and through a coded audio signal.

Zoned non-coded alarm A fire alarm system that indicates the activated zone on the alarm control panel.

Zoned system A fire alarm system design that divides a building or facility into zones so the area where an alarm originated can be identified.

Fire Fighter in Action

You are on Engine Company 12 responding to an automatic fire alarm at a 12-story office building. The pre-incident plan for this building shows a Class III standpipe system and automatic sprinklers only in the underground parking levels. Smoke detectors are located in the elevator lobby on each floor and in the return air duct to the air handling system in the mechanical penthouse. Combination fixed-temperature and rate-of-rise heat detectors are installed in the corridors on each floor.

Your department's written standard operating procedures for high-rise buildings state that the first arriving engine company should report to the main lobby and check the annunciator panel. The ladder company also comes into the building. The second-due engine company has hooked up to a hydrant and is connecting dry hose lines to the fire department connection.

A security guard meets you at the annunciator panel. He is sure that something is wrong with the alarm system, because several zones have activated. He silenced the alarm and pressed the reset button, but it keeps reactivating. He thinks the problem is in the alarm control panel.

Your lieutenant looks at the panel and calls the Battalion Chief on his portable radio, stating that he believes there is an actual fire, probably on the 10th floor. Your crew prepares to go upstairs as the Battalion Chief calls for a second alarm response.

On the annunciator panel, red lights are flashing for the following zones:
- **10th floor elevator lobby smoke detector**
- **10th floor heat detectors**
- **11th floor elevator lobby smoke detector**
- **11th floor heat detectors**
- **Penthouse air-return smoke detector**

1. Why do you think your lieutenant reported a probable ongoing fire?
   - **A.** It is standard operating procedure when a smoke detector is activated.
   - **B.** It is standard operating procedure when a heat detector is activated.
   - **C.** It is standard operating procedure when a smoke detector in an air return duct is activated.
   - **D.** The fact that multiple alarms of different types have been activated in the same area of the building suggests that there is a real fire.

2. The information provided by the security guard suggests that the smoke detection system in this building is:
   - **A.** Cross-zoned
   - **B.** A verification system
   - **C.** Zoned-coded
   - **D.** None of the above

3. There are two fire department connections on the front of the building. They look identical, except that one is labeled "Auto Spklr" and the other is labeled "Dry Standpipe." Which connection should the engine company use for the hose lines?
   - **A.** Either connection—it doesn't matter which one is used.
   - **B.** Both connections—if there is a fire both will be needed.
   - **C.** Only the "Dry Standpipe" connection
   - **D.** Only the "Auto Spklr" connection

4. When should the engine company charge the lines with water?
   - **A.** Only when requested by the crews inside
   - **B.** Only when directed by the Incident Commander
   - **C.** As soon as the lines are connected
   - **D.** Only if the system does not provide enough pressure for the crews attacking the fire

5. What should your crew do when you reach the 10th floor?
   A. Connect your hose line to a 2 1/2" standpipe outlet in the stairway.
   B. Utilize the preconnected 1 1/2" hose in the cabinet in the corridor on the 9th floor and extend it up the stairs.
   C. Remove the preconnected hose from the hose cabinet and attach your hose to that outlet.
   D. Work with the ladder company to extend a hose line over the aerial ladder and into the building through a window.

On a previous call to the same building, you could hear the water-motor gong ringing when your engine company arrived. On the annunciator panel, the alarm light was flashing for the zone marked "water flow—parking garage." The security guard said that a delivery truck tried to enter the underground garage and struck one of the sprinkler heads.

6. You are told to go to the control valve in the mechanical room and stand by to shut the system down. Your lieutenant will call you by radio when he wants you to stop the flow of water. What should you look for in the mechanical room?
   A. A post indicator valve
   B. A deluge valve
   C. A pressure-reducing valve
   D. An OS&Y valve
7. After you have shut down the system, your lieutenant explains that the underground parking garage has a dry sprinkler system, because it is unheated and could freeze in the winter. What will you need to do before leaving the scene?
   A. Completely drain the sprinkler system, replace the damaged sprinkler head, reset the dry-pipe valve, and open the OS&Y valve.
   B. Replace the damaged sprinkler head with a dry pendant head and reopen the OS&Y valve.
   C. Replace the sprinkler head, fill the pipes with an antifreeze solution, reset the dry-pipe valve, and open the PIV.
   D. Reset the dry-pipe valve, replace the sprinkler head, reopen the OS&Y, and drain the water that is still in the system.
8. You are conducting a fire station tour for a group of schoolchildren. Their teacher asks you to explain how a smoke detector works. What should you include in your explanation?
   A. The difference between a smoke detector and a smoke alarm
   B. The difference between ionization and photoelectric smoke detectors
   C. The importance of regularly checking smoke alarms, changing batteries, and locating smoke alarms in or near all sleeping areas
   D. All of the above

www.FireFighter.jbpub.com

Chapter Pretests
Interactivities
Hot Term Explorer
Web Links
Review Manual
FireLearn

Fire Cause Determination

Chapter 37

NFPA 1001 Standard

Fire Fighter I

5.3.8 Extinguish fires in exterior Class A materials, given fires in stacked or piled and small unattached structures or storage containers that can be fought from the exterior, attack lines, hand tools and master stream devices, and an assignment, so that exposures are protected, the spread of fire is stopped, collapse hazards are avoided, water application is effective, the fire is extinguished, and signs of the origin area(s) and arson are preserved.

5.3.8 (A) *Requisite Knowledge.* Types of attack lines and water streams appropriate for attacking stacked, piled materials and outdoor fires; dangers—such as collapse—associated with stacked and piled materials; various extinguishing agents and their effects on different material configurations; tools and methods to use in breaking up various types of materials; the difficulties related to complete extinguishment of stacked and piled materials; water application methods for exposure protection and fire extinguishment; dangers such as exposure to toxic or hazardous materials associated with storage building and container fires; obvious signs of origin and cause; and techniques for the preservation of fire cause evidence.

5.3.8 (B) *Requisite Skills.* The ability to recognize inherent hazards related to the material's configuration, operate handlines or master streams, evaluate for complete extinguishment, operate hose lines and other water application devices, evaluate and modify water application for maximum penetration, search for and expose hidden fires, assess patterns of origin determination, and evaluate for complete extinguishment.

5.3.13 Overhaul a fire scene, given personal protective equipment, attack line, hand tools, a flashlight, and an assignment, so that structural integrity is not compromised, all hidden fires are discovered, fire cause evidence is preserved, and the fire is extinguished.

5.3.13 (A) *Requisite Knowledge.* Types of fire attack lines and water application devices most effective for overhaul, water application methods for extinguishments that limit water damage, types of tools and methods used to expose hidden fire, dangers associated with overhaul, obvious signs of area of origin or signs of arson, and reasons for protection of fire scene.

5.3.13 (B) *Requisite Skills.* The ability to deploy and operate an attack line; remove flooring, ceiling, and wall components to expose void spaces without compromising structural integrity; apply water for maximum effectiveness; expose and extinguish hidden fires in walls, ceilings, and subfloor spaces; recognize and preserve obvious signs of area of origin and arson; and evaluate for complete extinguishment.

5.3.14 (B) *Requisite Skills.* The ability to cluster furniture; deploy covering materials; roll and fold salvage covers for reuse; construct water chutes and catch-alls; remove water; cover building openings, including doors, windows, floor openings, and roof openings; separate, remove, and relocate charred material to a safe location while protecting the area of origin for cause determination; stop the flow of water from a sprinkler with sprinkler wedges or stoppers; and operate a main control valve on an automatic sprinkler system.

Fire Fighter II

6.3.4 Protect evidence of fire cause and origin, given a flashlight and overhaul tools, so that the evidence is noted and protected from further disturbance until investigators can arrive on scene.

6.3.4 (A) *Requisite Knowledge.* Methods to assess origin and cause; types of evidence; means to protect various types of evidence; the role and relationship of Fire Fighter II, criminal investigators, and insurance investigators in fire investigations; and the effects and problems associated with removing property or evidence from the scene.

6.3.4 (B) *Requisite Knowledge.* The ability to locate the fire's origin area, recognize possible causes, and protect the evidence.

Other NFPA Standards

NFPA 921, *Guide to Fire and Explosion Investigations, 2001 Edition*

Knowledge Objectives

After studying this chapter, you will be able to:

- Describe the role and relationship of the fire fighter to criminal investigators and insurance investigators.
- Differentiate accidental fires from incendiary fires.
- Describe the point of origin.
- Define the chain of custody.
- Describe demonstrative, direct, and circumstantial evidence.
- Describe techniques for preserving fire cause evidence.
- Describe the observations fire fighters should make during fireground operations.
- Describe the steps needed to secure a property.
- Explain the importance of protecting a fire scene.

Skills Objectives

There are no skills objectives for this chapter.

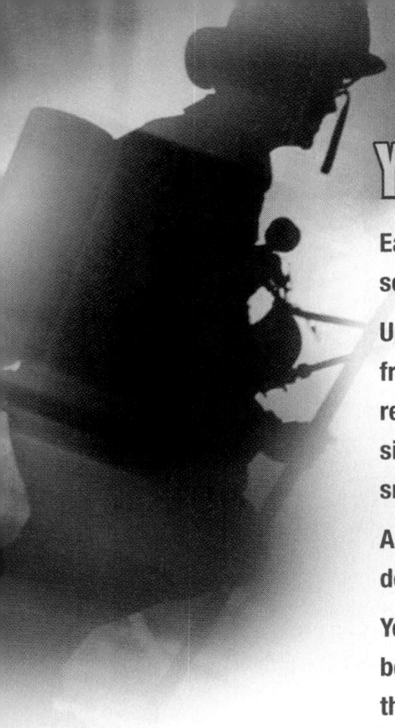

You Are the Fire Fighter

Early one summer morning, your engine company is dispatched for an automatic fire alarm sounding at a pool hall in an older section of the downtown district.

Upon arrival, you see that the parking lot is empty except for two cars in a remote corner, away from the building. You also notice three bystanders across the street. You overhear the Captain reporting, "On the scene of a one-story, wood-frame structure; light smoke showing on the west side of the structure. Advancing a line to the front door for an interior attack." You can smell the smoke, but you don't see any flames.

As you advance the hose line to the front door, you stumble over a small pile of smoldering debris. The front door is closed and locked, but the lock mechanism is damaged and loose.

Your experience tells you that a fire investigator will be asking you questions about this fire. You begin making mental notes and thinking about other suspicious items to look for while working this fire.

1. ***Why will the fire investigator ask you about your observations?***
2. ***What observations have you made so far, and what should you be watching for at each stage of the suppression sequence?***
3. ***How can you protect evidence that has been found and/or might be found at this fire?***

Introduction

Fire fighters usually arrive at a fire scene before a trained fire investigator arrives. This means that fire fighters are able to observe important signs and patterns that the investigators can use in determining how and where the fire started. By identifying and preserving possible evidence, as well as recalling and reporting objective findings, fire fighters provide essential assistance to fire investigators. In some cases, the observations and actions of fire fighters could be significant in apprehending and convicting arsonists.

Fire departments determine the causes of fires so they can take steps to prevent future fires. For example, a fire department might develop an education program to reduce accidental fires, such as those caused by an unattended candle left burning. A series of fires could point to a product defect, such as a design error in a chimney flue or the improper installation of chimney flues. Identifying fires that were intentionally set could lead to the arrest of the person responsible.

Fire fighters must understand the basic principles of fire investigation and participate in this important departmental responsibility. Fire cause determination is often difficult, because important evidence can be consumed by the fire or destroyed during fire suppression operations or during salvage and overhaul. Investigators must rely on fire fighters to observe, capture, and retain information, as well as to preserve evidence until it can be examined. Evidence that is lost can never be replaced.

Who Conducts Fire Investigations?

In most jurisdictions, the chief of the fire department has a legal responsibility to determine the causes of fires. In large fire departments, the fire chief usually delegates this responsibility to a special unit staffed with trained investigators. In smaller jurisdictions, one or two fire department members are trained to conduct investigations. Sometimes a special unit conducts fire investigations for several area fire departments, under an established policy that determines when they should be called to conduct an investigation. Only in very small communities will the fire chief personally examine the scene to determine the cause.

Many fire departments automatically dispatch an investigator to all working structure fires as well as to any other fire where the incident commander (IC) suspects something unusual. Other jurisdictions may have policies that require an investigator only when the damage exceeds a predetermined level or when there are injuries or fatalities. In other departments, the IC may be expected to conduct a preliminary investigation and decide whether an investigator is needed. If the cause of the fire is evident and accidental, the IC would be responsible for gathering the information and filing the necessary reports. If the cause cannot be determined or appears to be suspicious, the IC would summon a fire investigator.

When a fire investigator is not on the scene while the fire is being extinguished or overhauled, fire fighters must be particularly careful to preserve evidence. They must serve as the eyes and ears of the investigator, and pass on their observations. In some cases, fire fighters may need to be interviewed or to prepare a written report documenting their observations. This is part of the fire fighter's role and responsibility in the fire investigation process.

Fire cause investigation should not be confused with a criminal investigation of **arson**, which is the malicious burning of property with a criminal intent. The objective of a fire cause investigation is to identify the cause of the fire; the investigator should never have a preconceived idea about what started the fire. The evidence and the facts determine whether the fire was started deliberately or resulted from an accidental cause. Only if the fire investigator determines that arson is the cause should a criminal investigation begin to find the person responsible.

Law Enforcement Authority

An arson investigation must determine not only the cause and origin of the fire, but also who was responsible for starting it and what sequence of events led up to it. Determination of cause and origin is a small but important step in the overall investigation. To prove that a crime took place, investigators must establish a link from the fire to the cause and origin and eliminate any other possible cause.

Whether the fire investigators have police powers and can conduct a criminal investigation depends on state and local laws. In some jurisdictions, fire investigators may have police powers during fire investigations. In other areas, police officers are trained as fire investigators. Sometimes fire investigators determine the cause and origin, and turn the investigation over to a law enforcement agency if the fire is determined to be intentionally set. In other jurisdictions, fire department and police department personnel work together throughout the investigation.

Investigation Assistance

Most fire departments and government agencies have limited time, personnel, and resources for fire investigations. A state fire marshal or similar authority may have an investigations unit that concentrates on major incidents and supports local investigators on smaller fires. Because these investigators cannot always reach the fire scene quickly, the local fire department must be prepared to conduct a thorough preliminary investigation and to protect the scene and preserve evidence.

Federal resources are also available for major investigations. The U.S. Bureau of Alcohol, Tobacco and Firearms (ATF) has individual agents who can assist local jurisdictions with fire investigations. The ATF also has response teams across the country that respond to large-scale incidents. Each of these teams has about 15 agents and equipment that ranges from simple tools for digging out a fire scene to laser-surveying devices for fully documenting the scene.

Fire Fighter Tips

Fire investigations involve much more than simply determining what caused the fire, where it started, and whether it was accidental or incendiary (intentionally set). Many fire investigations also examine many other aspects of a fire incident.

If the fire results in injuries or fatalities, the investigation often will examine every factor that could have contributed to or prevented those losses. In a fire that causes millions of dollars in damage, a whole range of factors could contribute to the large loss. Investigators could examine the construction and contents of the building to determine whether fire and building codes were followed, if the built-in fire protection systems functioned properly, or what lessons could be applied to amending codes or inspection procedures. A fire investigation can teach many lessons, even if the specific cause and origin are never determined.

The National Fire Protection Association (NFPA) investigates major incidents around the world to identify factors that contributed to an unusual number of deaths or widespread property damage. The NFPA uses this information to amend its codes and standards to help avoid future catastrophes.

Insurance companies often investigate fires to determine the validity of a claim or to identify factors that might help prevent future fires. The cost of an investigation is more than offset by the savings the company would realize by identifying a fraudulent claim. Some insurance companies have their own investigators, while others retain independent investigators. These outside investigators often have valuable experience and can provide technical support to determine the cause of a fire.

Causes of Fires

Every fire has a cause, which the fire investigator tries to uncover. A cause and origin investigation determines where, why, and how the fire originated. Some fires have simple causes that are easily identified and understood; others result from a complex set of circumstances that must be examined carefully to determine what actually happened. In some cases, the cause of a fire will never be determined with absolute certainty.

Basically, every fire has a starting point where ignition occurs and fuel begins to burn. This location is the **point of origin** of the fire. At the point of origin, an ignition source comes into contact with a fuel supply, such as a lighted match touching a piece of paper. The cause of the fire is the particular set of circumstances that brought the ignition source into contact with the fuel.

Fires result when a **competent ignition source** and a fuel come together for long enough to ignite. A competent ignition source must have enough heat energy to ignite the fuel and must be in contact with the fuel until the fuel reaches its ignition temperature, which could be a fraction of a second to hours, days, or weeks.

A fire can be caused by an act or by an omission. Igniting a piece of paper with a match is an act. Leaving a pot of grease unattended on a hot stove is considered an omission. The cause of a fire can also be classified as either incendiary or accidental. Arson, which is the malicious burning of one's own or another's property with a criminal intent, is an **incendiary** cause. **Accidental** fires do not involve a criminal or malicious intent, even if they are caused by human error or carelessness. Falling asleep with a lit cigarette would be considered an accidental cause.

Investigators should always consider a fire to have an **undetermined** cause until the specific cause is established. The evidence from both incendiary and accidental fires can be very similar. Untrained individuals should not attempt to categorize fires as either incendiary or accidental. Fire fighters are responsible for helping to identify and preserve possible evidence for the fire investigator to examine.

Fire Cause Statistics

According to the NFPA, there were 1.8 million fire incidents reported in the United States in 1999, the latest year for which statistics are available. These fires caused 3,570 fatalities and over $10 billion in direct property damage.

Most structure fires occur in residential occupancies; home fires represent 73% of all structure fires. The 10 leading causes of fires in the home, averaged over the years 1994 to 1998, are shown in ▼ Table 37-1.

Accidental Fire Causes

Accidental fires have hundreds of possible causes and involve multiple factors and circumstances. The most important reason for investigating and determining the causes of accidental fires is to prevent future fires. To reduce the number of fires, efforts must concentrate on the most frequent causes and those involving the greatest risks of death, injury, and property damage.

Most fires, fire deaths, and injuries occur in residential occupancies. The most commonly reported accidental causes of fire in these occupancies involve smoking, cooking, heating equipment, and electrical equipment. These statistics provide a foundation for fire prevention and public education efforts. Additional analysis can provide more specific information. For instance, fires caused by electrical equipment can be divided into four groups: those caused by worn-out or defective equipment, by improper use of approved equipment, by defective installations, and by other accidents. A proper, thorough investigation of an electrical fire would identify and classify the specific cause within one of these groups.

For example, worn-out or defective electrical equipment would include deteriorating 50-year-old wiring circuits or a computer circuit board with an internal defect. Other worn-out equipment that might still be in use includes electrical motors, switches, appliances, and extension cords. Properly used and maintained equipment that has been tested and listed by a recognized laboratory rarely causes a fire, but should be replaced when it wears out.

Placing a portable heater too close to flammable materials or using a toaster oven to heat a container of flammable glue are examples of the improper use of electrical equipment. Defective installations are those not acceptable under electrical codes or printed instructions, such as using a light-duty extension cord to connect a heavy-duty appliance to a wall outlet.

Table 37-1 Leading Causes of Fires in the Home

Major Cause	Percentage of Total Total Fires	Ranking	Percentage of Civilian Deaths	Ranking	Percentage of Property Damage	Ranking
Cooking equipment	23%	1	9%	5	9%	5
Heating equipment	15%	2	13%	3	13%	3
Incendiary	12%	3	16%	2	18%	1
Other equipment	11%	4	8%	7	12%	4
Electrical	9%	5	10%	4	14%	2
Appliance	7%	6	4%	9	6%	6
Smoking materials	5%	7	23%	1	6%	7
Open flame	5%	8	3%	10	6%	8
Child playing	5%	9	8%	6	6%	8
Exposure to other hostile fire	4%	10	1%	11	5%	9

Source: *The U.S. Fire Problem Overview Report, Leading Causes and Other Patterns and Trends, June 2001* Marty Ahrens, NFPA Fire Analysis and Research Division

Some electrical fires result from an accidental misuse or oversight, such as unintentionally leaving a cooking appliance turned on. Natural events can also cause accidental electrical fires, such as when a tree falls on a wire. A fire caused by an electrical overload or a short circuit can start wherever there is electricity, such as in the electrical panel, fuses, fuse boxes, circuit breakers, wiring, and appliances.

Sometimes it may be difficult or impossible to identify the specific source of ignition because the fire destroyed every trace of evidence. Investigators should classify these fires as having an undetermined cause, rather than use a "best guess." For example, low temperature ignition can occur when wood is subjected to low heat, such as that generated by steam pipes or incandescent light bulbs, for a long period of time. Gradually, the wood can deteriorate and eventually ignite. If the fire destroys the building, the point of origin would be difficult to identify.

Incendiary Fire Causes

People may set fires for several reasons and in many different ways. The same type of cause and origin investigation is needed to identify where and how the fire started. With incendiary fires, it is particularly important to rule out possible accidental causes to prove beyond doubt that the fire was deliberately set.

A fire caused by arson requires a second phase of investigation to identify the person responsible. All of the evidence relating to the cause of the fire must be handled in a way that ensures it would be admissible as evidence in a criminal trial. A trained, qualified fire investigator should always be called to determine the cause of any fire that may have been deliberately set.

Arson and the factors that could indicate an incendiary fire cause are discussed in more detail later in this chapter.

Determining the Cause and Origin of a Fire

A systematic analysis is needed to determine the cause and origin of a fire. The investigator must determine where the fire started and how it was ignited. The investigator must look at the situation objectively to be sure that the evidence is convincing and fully explains the situation. If there is more than one possible explanation for the observations, each possibility must be considered. The cause cannot be determined with absolute certainty until all alternative explanations have been ruled out.

Identifying the Point of Origin

One of the first steps in a fire investigation is identifying the point of origin. At this location, the investigator can look for clues indicating the specific cause of the fire.

The investigation process usually begins with an examination of the building's exterior. The investigator should look for indications that the fire originated outside the building before looking inside, and should size-up the building to identify important information. The overall size, construction, layout, and occupancy of the building should be noted, as well as the extent of damage that is visible from the exterior. The investigator should look for any openings that might have created drafts that influenced the fire spread and should examine the condition of outside utilities, such as the electrical power connection and gas meter.

The search for indications to the point of origin continues inside the building, beginning with the area of lightest damage to the area of heaviest damage. This is probably the area that was burning for the longest time; areas with less damage were probably not as involved in the fire, or were involved for shorter periods.

The **depth of char** can be used to help determine how long a fire burned in a particular location (▼ Figure 37-1). The depth of char is related to how long the surface of a material was exposed to the fire, from the time of ignition to the time of extinguishment. An area that burned only for a few minutes would have mostly surface damage and a shallow depth of char. Materials that burned for longer periods will show evidence of charring deeper into their cores. Charring is usually deepest at the point of origin; however, flammable liquids and combustible materials in other locations can also leave heavy charring. The area with the deepest char is not necessarily the point of origin.

Burn patterns and smoke residue can be helpful in identifying the area of origin, but again are not conclusive. Burn patterns and damage will often spread outward from the room or area where the damage is most severe. Because heat rises, the flow of heated gases from a fire will almost always be up and out from the point of origin. This upward, outward flow can usually be recognized, even when all of the building's contents were involved in the fire. Often the point of origin is found directly below the most damaged area on the ceiling, where the heat of the fire was most concentrated.

Figure 37-1 Depth of char.

Figure 37-2 Often the point of a V-pattern is near or at the point of origin.

Figure 37-3 Fire investigators systematically dig through the debris of a fire to search for evidence.

A charred V-pattern on a wall indicates that fire spread up and out from something at the base of the V. The best place to start looking for a specific fire cause in a room with a V-pattern on the wall, a pile of charred debris at the base of the V, and minimal damage to the other contents is in the pile of debris (▲ Figure 37-2).

An experienced fire investigator knows that many factors can influence burn patterns, including ventilation, fire suppression efforts, and the burning materials themselves. A V-pattern on a wall could indicate that there was an easily ignited and intensely burning fuel source at that location. It might also indicate that something fell from a higher level and burned on the floor.

An inverted V-pattern on a wall could indicate that a flammable liquid was used along the base of the wall to set the fire intentionally. A fire burning across a wide area at the floor level can funnel into a thermal column as it rises, creating the inverted V-pattern.

Once the investigator identifies the exact or approximate location of the point of origin, the search for indications of a specific cause can begin. The investigator must determine what happened at that location to cause the fire. This involves identifying both the source of ignition and the fuels that were involved. The investigator must ultimately determine how the source of ignition and fuel came together, either accidentally or intentionally.

Digging Out

After the investigator identifies the area of origin, fire fighters could be asked to assist in digging out the fire scene. "Digging out" is a term used to describe the process of carefully looking for evidence within the debris. Sometimes the entire fire scene must be closely examined to determine the cause of the fire and gather evidence.

The fire investigator will take extensive photographs of the fire scene as it first appears. Then, he or she will begin to remove and inspect the debris, layer by layer, from the top of the pile down to the bottom. The type of evidence uncovered and its location in the layers of debris can provide important indications of how the fire originated and progressed.

Removing and inspecting the layers of debris enables the investigator to determine the sequence in which items burned, whether an item burned from the top down or from the bottom up, and how long it burned. Did the fire start at a low point and burn up or did burning items fall down from above and ignite combustible materials below? Did the fire spread along the ceiling or along the floor? Is there a residue of rags containing a flammable liquid under the furniture? Why are papers that should have been in a metal filing cabinet stacked on the floor and partially burned?

Systematically digging through the debris often can uncover the exact point of origin and cause of both accidental and deliberate fires (▲ Figure 37-3). If circumstances or eyewitness accounts indicate a deliberate fire, the investigator may request that fire fighters help examine the entire area. Common search methods include the "grid" search (also known as the "double strip" search) and the "strip" search. The investigator will explain generally what to look for, how to search, and what to do with any potential evidence.

When you find possible evidence, stop so the investigator can examine it in place. It is the investigator's job to document, photograph, and remove any potential evidence, whether or not it supports the suspected cause.

To determine the cause and origin of a fire, the investigator must evaluate all potential causes for the fire. The process of eliminating possible alternative causes and documenting the reasons for rejecting them is as important as properly documenting the ultimate cause of the fire.

For example, if the fire started in the kitchen, potential causes could include:

- The stove
- All of the electrical appliances
- Light fixtures
- Smoking materials
- Cleaning supplies
- Other factors, such as candles

Evidence

Evidence refers to all of the information gathered and used by an investigator in determining the cause of a fire. Evidence can be used in a legal process to establish a fact or prove a point. To be admissible in court, evidence must be gathered and processed under strict procedures.

Physical evidence consists of items that can be observed, photographed, measured, collected, examined in a laboratory, and presented in court to prove or demonstrate a point (► Figure 37-4). Fire investigators can gather physical evidence, such as a burn pattern on a wall or an empty gasoline can, at the fire scene to explain how the fire started or to document how it burned.

Trace or transfer evidence is a minute quantity of physical evidence that is conveyed from one place to another. For example, a suspect's clothing may contain the residue of the same flammable liquid found at the scene of a fire.

Demonstrative evidence is anything that can be used to validate a theory or to show how something could have occurred. To demonstrate how a fire could spread, for example, an investigator may use a computer model of the burned building.

Evidence used in court can be considered either direct or circumstantial. **Direct evidence** includes facts that can be observed or reported first hand. Testimony from an eyewitness who saw a person actually ignite a fire or a videotape from a security camera showing the person starting the fire are examples of direct evidence. Direct evidence is rare in arson cases, but eyewitnesses often can describe the circumstances of an accidental fire.

Circumstantial evidence is information that can be used to prove a theory, based on facts that were observed first hand. For example, an investigation might show that gasoline from a container found at the scene was used to start a fire. Two different witnesses testify that the suspect purchased a container of gasoline before the fire and walked away from the fire scene without the container a few minutes before the fire department arrived. Such circumstantial evidence clearly places the suspect at the fire site with an ignition source at the time the fire started. Fire investigators must often work with circumstantial evidence to connect an arson suspect to a fire.

Figure 37-4 Physical evidence at a fire scene.

Preservation of Evidence

Fire fighters have a responsibility to preserve evidence that could indicate the cause or point of origin of a fire. Fire fighters who discover something that could be evidence while digging out a fire scene or performing other activities should leave it in place, make sure that no one interferes with it or the surrounding area, and notify a company officer or fire investigator immediately.

Evidence is most often found during the salvage and overhaul phases of a fire. Salvage and overhaul should always be performed carefully and can often be delayed until an investigator has examined the scene. Do not move debris any more than is absolutely necessary and never discard debris until the investigator gives approval. The fire investigator must decide whether the evidence is relevant, not the fire fighter.

Fire fighters at the scene are not in the position to decide whether the evidence they find will be admissible in court and therefore worthy of preservation; that is the fire investigator's decision. Your responsibility is to make sure that potential evidence is not destroyed or lost. Too much evidence is better than too little, so no piece of potential evidence should be considered insignificant.

If evidence could be damaged or destroyed during fire suppression activities, cover it with a salvage cover or some other type of protection, such as a garbage can. Use barrier tape to keep others from accidentally walking through evidence. Before moving an object to protect it from damage, be sure that witnesses are present, that a location sketch is drawn, and that a photograph is taken.

Fire Fighter Tips

Gasoline-powered tools and equipment, including saws and generators, are often used during firefighting operations. Gasoline from the equipment could contaminate the area and lead to an erroneous assumption that **accelerants** (materials used to initiate or increase the spread of fire) were used in the fire. Limit the use of gasoline-powered equipment inside the structure, particularly in the area of origin. At a minimum, refuel such equipment outside the investigation area to prevent spilled fuel contamination.

Fire Fighter Tips

To avoid contaminating evidence, fire investigators always wash their tools between taking samples. This ensures that material from one piece of evidence will not be unintentionally transferred to another piece. Investigators also change their gloves each time they take a sample of evidence and place evidence only in absolutely clean containers. Before entering the fire scene, investigators often wash their boots to keep from transporting any contamination into the fire scene.

Evidence should not be **contaminated**, or altered from its original state, in any way. Fire investigators use special containers to store evidence and prevent contamination from any other products.

Chain of Custody

To be admissible in a court of law, physical evidence must be handled according to certain prescribed standards. Because the cause of the fire may not be known when evidence is first collected, all evidence should be handled according to the same procedure. In an incendiary or arson fire, evidence relating to the cause of the fire will probably have to be presented in court. The same is true for accidental fires, because lawsuits might be filed to claim or recover damages.

Chain of custody (also known as chain of evidence or chain of possession) is a legal term that describes the process of maintaining continuous possession and control of the evidence from the time it is discovered until it is presented in court. Every step in the capture, movement, storage, and examination of the evidence must be properly documented. For example, if a gasoline can is found in the debris of a suspected arson fire, documentation must record the person who found it, where and when it was found and under what circumstances. Photographs should be taken to show where it was found and what condition it was in. In court, the investigator must be able to show that the gas can presented is the specific can that was found.

The person who takes initial possession of the evidence must keep it under his or her personal control until it is turned over to another official. Each successive transfer of possession must be recorded. If evidence is stored, documentation must indicate where and when it was placed in storage, whether the storage location was secure, and when it was removed. Often, evidence is maintained in a secured evidence locker to ensure that only authorized personnel have access to it.

Everyone who had possession of the evidence must be able to attest that it has not been contaminated, damaged, or changed. If evidence is examined in a laboratory, the laboratory tests must be documented. The documentation for chain of custody must establish that the evidence was never out of the control of the responsible agency and that no one could have tampered with it.

Fire fighters are frequently the first link in the chain of custody. The fire fighter's responsibility in protecting the integrity of the chain is relatively simple: report everything to a senior officer and disturb nothing needlessly. The individual who finds the evidence should remain with it until it is turned over to a company officer or to the fire investigator (► **Figure 37-5**). Remember that you could be called as a witness to state that you are the fire fighter who discovered this particular piece of evidence at the specific location pictured.

The fire investigator's standard operating procedures for collecting and processing evidence generally include the following steps:

- Take photographs of each piece of evidence as it is found and collected. If possible, photograph the item as it was found, before it is moved or disturbed.
- On the fire scene, sketch, mark, and label the location of the evidence. Sketch the scene as near to scale as possible.
- Place evidence in appropriate containers to ensure safety and prevent contamination. Unused paint cans with lids that automatically seal when closed are the best containers for transporting evidence. Glass mason jars sealed with a sturdy sealing tape are appropriate for transporting smaller quantities of materials. Plastic containers and plastic bags should not be used for evidence containing petroleum products since it may deteriorate the plastic. Paper bags can be used for dry clothing or metal articles, matches, or papers. Soak up small quantities of liquids with either a cellulose sponge or cotton batting. Protect partially burned paper and ash by placing them between layers of glass (assuming that small sheets or panes of glass are available at the scene).
- Tag all evidence at the fire scene. Evidence being transported to the laboratory should include a label with the date, time, location, discoverer's name, and witnesses' names.
- Record the time the evidence was found, where it was found, and the name of the person who found it. Keep a record of each person who handled the evidence.

Figure 37-5 Evidence should remain where you find it until you can turn it over to your company officer or the fire investigator.

Figure 37-6 Jesting remarks and jokes should never be made at the scene, because your comments could be overheard by the media.

- Keep a constant watch on the evidence until it can be stored in a secure location. Evidence that must be moved temporarily should be put in a secure place accessible only to authorized personnel.
- Preserve the chain of custody in handling all the evidence. A broken chain of custody may result in a court ruling that the evidence is inadmissible.

Only one person should be responsible for collecting and taking custody of all evidence at a fire scene, no matter who discovers it. If someone other than the assigned evidence collector must seize the evidence, that person must photograph, mark, and contain the evidence properly and turn it over to the evidence collector as soon as possible. The evidence collector must also document all evidence that is collected, including the date, time, and location of discovery, the name of the finder, known or suspected nature of the evidence, and the evidence number from the evidence tag or label. A log of all photographs taken should also be recorded at the scene.

Witnesses

Although fire fighters do not interview witnesses, you can identify potential witnesses to the investigator. People who were on the scene when fire fighters arrived could have invaluable information about the fire. If a fire fighter learns something that might be related to the cause of the fire, this information should be passed on to a company officer or to the fire investigator.

Interviews with witnesses should be conducted by the fire investigator or by a police officer. If the fire investigator is not on the scene or does not have the opportunity to interview the witness, obtain the person's name, address, and telephone number and give it to the investigator. A witness who leaves the scene without providing this information could be difficult or impossible to locate later.

Fire fighters have a primary responsibility to save lives and property. Until the fire is under control, fire fighters must concentrate on fighting the fire, not investigating the cause. However, fire fighters should pay attention to the situation and make mental notes about any observations. Fire fighters must tell the investigator about any odd or unusual happenings. Information, suspicions, or theories about the fire should be shared only with the fire investigator and only in private.

Do not make statements of accusation, personal opinion, or probable cause to anyone other than the investigator. Comments that are overheard by the property owner, the occupant, a news reporter, or a bystander can impede the efforts of the fire investigator to obtain complete and accurate information. A witness trying to be helpful might report an overheard comment as a personal observation. In this way, inaccurate information can generate a rumor that becomes a theory and turns into a reported "fact" as it passes from person to person.

Never make jesting remarks or jokes at the scene (▲ Figure 37-6). Careless, unauthorized, or premature remarks could embarrass the fire department. Statements to news reporters about the fire's cause should be made only by an official spokesperson after the fire investigator and ranking fire officer have agreed to their accuracy and validity. Until then, "The fire is under investigation" is a sufficient reply to any questions concerning the cause of the fire.

Fire Fighter Tips

Do not speculate on the cause of the fire with or in front of bystanders. Speculation that is reported or printed tends to be treated as fact.

Voices of Experience

> **"Uncovering the cause of the fire, whether accidental or incendiary, is critical in helping fire departments prevent future fires."**

My department was dispatched to a fire at a local nightclub at approximately 0230 hours. The building was moderately involved with fire on arrival of fire units. Fire was quickly knocked down using several handlines. During the extinguishment, command was notified that there was an odor of kerosene near the side door.

Following good investigative technique, we began eliminating causes for the fire. This included a thorough check of the building wiring, which was not to code but showed no signs of failure (this structure was in an area of the county protected by our department, but not routinely inspected). Discarded smoking materials in trash cans was also eliminated as were other accidental and natural causes. The weather was clear and warm.

Deductive investigation led the investigation team to the side door area where the odor of kerosene was reported during the fire. We found that the bottom third of the side door was charred, and appeared to have flammable liquid splash patterns on it. The concrete floor showed spalling and splash patterns. We called for the arson dog, and as we worked the scene, the dog alerted near the side door. Samples were taken of the door and floor area for lab analysis.

During the investigation, the nightclub's owner was interviewed. He stated that earlier in the day, he had thrown one of his regular patrons out of the establishment for causing a disturbance. This was just one of several disturbances this person had created over the past several weeks, and the owner informed the patron that he was permanently barred from the nightclub. The patron stormed off, shouting at the owner, "If I can't be here, nobody can be here!"

While completing the investigation, the owner told us that the patron was nearby. We approached the patron to conduct an interview. During the interview, the patron was asked by the investigator if the dog could approach him. The patron agreed to allow the dog near him and to sniff his clothing. The arson dog was trained to sit down when it detects the presence of a flammable liquid. Sure enough, when the dog sniffed at the patron's pant leg and boots, it immediately sat down. The patron was arrested and his pants and boots were then taken as evidence. Lab tests would later confirm that the patron's clothing contained residue of the same kerosene that was used to start the fire.

The fire scene gives us many clues about the cause of the fire, from burn patterns on the walls and floors, unusual or out of place odors, to layers of debris on the floor. Uncovering the cause of the fire, whether accidental or incendiary, is critical in helping fire departments prevent future fires. As such, fire fighters should make careful observations during all phases of fireground operations.

Larry Morabito
Wayne County Community College
Detroit, Michigan

Observations During Fireground Operations

Although a fire fighter's primary concern is saving lives and property, you will make observations and gather information as you perform your duties. What you observe could be significant in the subsequent investigation of the incident. The following sections will help you identify specific signs, patterns, and evidence during various fireground operations, from dispatch and response to fire attack and overhaul.

Dispatch and Response

During dispatch and response, form a mental image of the scene you expect to encounter. Note the time of day, weather conditions, and route obstructions. As the incident progresses, pay attention to things that do not match your expectations because they could help fire investigators determine the origin and cause of the fire.

Time of Day

Time of day and type of occupancy can indicate the number and type of people at an incident. Offices, stores, and other business places are filled with people during the day and mostly empty at night. Restaurants and nightclubs are likely to be crowded after dark. Residential occupancies are more likely to be occupied at night and on weekends and holidays.

As victims evacuate the building, fire fighters may be able to note any that stand out, either because they are fully dressed when everyone else is in pajamas, or because their behavior and demeanor is quite different.

Weather Conditions

Note whether the day is hot, cold, cloudy, or clear, and whether conditions in the burning structure match the weather. On a cold day, windows should be closed; on a hot day, the furnace should be off.

Lightning, heavy snow, ice, flooding, fog, or other hazardous conditions can help cover an arsonist's activities because they delay the fire department's arrival and make a fire fighter's job more difficult. Because wind direction and velocity help determine the natural path of fire spread, being aware of these conditions will help you recognize when a fire behaves in an unnatural way.

Route Obstructions

Unusual traffic patterns or barriers blocking the route to the scene may be early indications of a suspicious fire. Be sure to note these and any other obstructions, such as barricades, felled trees, downed cables, or trash containers that cause delays.

Fire Fighter Safety Tips

Safety is always the first and most important priority on the fireground.

Arrival and Size-Up

Size-up operations can provide valuable information for fire investigators. Pay attention to the fire conditions, building characteristics, and vehicles and people at or leaving the scene.

Description of the Fire

Compare the dispatcher's description with the actual fire conditions. If the fire has intensified dramatically in a short time, an accelerant could have been used. Note whether flames are visible or only smoke. Also observe the quantity, color, and source of the smoke. Does the fire appear to be burning in one place, or in multiple locations?

Vehicles and People on the Scene

The appearance and behavior of people and vehicles at the scene of a fire can provide valuable clues. Did you notice a suspicious car as you arrived? Do you recognize any bystanders from other fire scenes? Note the attitude and dress of the owner and/or occupants of the building, as well as other individuals at the scene of a fire. Normally, the owner or occupants of a building would be distressed and wearing clothing appropriate to the time of day.

Anyone who seems out of place at a fire or someone who has been observed at several fires in various locations should be reported to the investigator. Some arsonists are emotionally disturbed individuals who receive personal satisfaction in watching a "working" fire. Sometimes investigators will photograph the crowd at a fire scene, particularly if there is a series of similar fires, and look for the same faces at different incidents.

Most people at a fire scene are serious and intent on watching the drama unfold (► Figure 37-7). Someone who is talking loudly, laughing, or making light of the situation should be considered suspicious. Fire investigators may also want to know about someone who eagerly volunteered multiple theories or too much information.

Unusual Items or Conditions

Always note any unusual items or conditions about the property, such as whether windows and doors are open or closed.

Fire Fighter Tips

Photographs—whether taken by film or digital cameras—can document conditions before the arrival of fire investigators. However, specific procedures must be followed for the photographs to be admissible in court as evidence. Follow your department's SOP regarding picture taking.

Figure 37-7 Most persons at a fire scene are intent on watching fire fighters at work.

Arsonists commonly draw the shades or cover windows and doors with blankets to delay the discovery of a fire. Gasoline cans, forcible entry tools, and a damaged hydrant or sprinkler connection might all suggest an intentionally set fire.

Entry

As you prepare to enter the burning structure, look for evidence of any prior entry, such as shoeprints leading into or out of the structure or tracks from vehicle tires. Note whether the windows and doors are intact, whether they are locked or unlocked, and whether there are any unusual barriers limiting access to the structure. Also note any signs of forced entry by others. If you see evidence of forced entry, ask yourself if the forced entry likely occurred before the fire, or if it could have been caused by fire fighters gaining access to the structure.

Forced entry could leave impressions of tools on the windows and doors. Cut or torn edges of wood, metal, or glass also may indicate forced entry. Investigators might be able to determine whether the glass was been broken by heat or by mechanical means. Look for signs of a burglary; the fire may have been set to destroy evidence of another crime.

Search and Rescue

As you enter the building to perform search and rescue or interior fire suppression activities, consider the location and extent of the fire. Separate or seemingly unconnected fires may indicate multiple points of origin and are often found at arson scenes. The location and condition of the building's contents may also provide clues. Unusual building contents or conditions—such as barriers in doors and windows, the absence of inventory from a warehouse, or insurance papers or deeds left out on a desk—should be noted.

The team responsible for shutting off electric power to the building should note whether the circuit breakers were on or off when they arrived. The location of any people found in the building should also be noted.

Ventilation

The ventilation crew should note whether the windows and doors were open or closed, locked or unlocked. They should also note the color and quantity of the smoke, as well as the presence of any unusual odors.

Color of Smoke

The color of smoke often indicates what is burning. Unusual smoke might indicate that additional fuel was added to the mix.

Unusual Odors

Self-contained breathing apparatus (SCBA) protects fire fighters from hazardous fumes and toxic odors. However, sometimes an odor is so strong that it can be detected en route or linger after the fire has been extinguished. Cooking fires and overheated light ballasts produce distinctive, identifiable odors familiar to experienced fire fighters. Other common odors familiar to fire fighters include gasoline, kerosene, paint thinner, lacquers, turpentine, linseed oil, furniture polish, or natural gas.

Odors often linger in the soil under a building without a basement, particularly if an accelerant has been used and if the ground is wet. Concrete, brick, and plaster will all retain vapors after a fire has been extinguished.

Effects of Ventilation on Burn Patterns

Fire department ventilation operations can dramatically influence the behavior of a fire and alter burn patterns. For example, if the ventilation crew opens a window or makes an opening in the roof, the heat, smoke, and fire are likely to move toward this opening, creating new burn patterns on the walls, floors, and ceilings. Relay such information to the fire investigator so that he or she can correctly interpret these patterns.

Suppression

Fire behavior, the presence of **incendiary devices** (materials used to start a fire), obstacles encountered during fire suppression operations, and charring and burn patterns are among the factors that might help to determine the origin and cause of the blaze.

Behavior of Fire

During the fire attack, observe the behavior of the fire and how it reacts when an extinguishing agent is applied. Look for unusual flame colors, sounds, or reactions. For example,

Fire Fighter Safety Tips

If you encounter an incendiary device that has not ignited, notify your supervisor and others in the area immediately! Do not make any attempt to move it or disable it.

most flammable liquids will float, continue to burn, and spread the fire when water is applied. Rekindles in the same area or a flare-up when water is applied could indicate the presence of an accelerant.

Incendiary Devices, Trailers, and Accelerants

While fighting the fire, be aware of streamers or trailers (combustible materials placed to spread the fire). Look for combustible materials like wood, paper, or rags in unusual locations. Note containers of flammable or combustible liquids normally not found in the type of occupancy. Very intense heat or rapid fire spread might indicate the use of an accelerant to increase the fire spread.

Incendiary devices can include unusual items in unlikely places, such as a packet of matches tied to a bundle of combustible fibers or attached to a mechanical device. Often, incendiary devices fail to ignite or burn out without igniting other materials. Try to extinguish the fire without unduly disturbing any suspicious contents.

Condition of Fire Alarm or Suppression Systems

If the building is equipped with a fire alarm or fire suppression system, fire fighters should examine it to see whether the system operated properly or was disabled. If the fire suppression system failed to work, the problem could be poor maintenance or deliberate tampering. Notify the fire inspector of any findings.

Obstacles

An arsonist may place obstacles to hinder the efforts of fire fighters. Note whether furniture was moved to block entry. Arsonists also may prop open fire doors, pull down plaster to expose the wood structure, or punch holes in walls and ceilings to increase the rate of fire spread.

Contents

Fire fighters involved in interior fire suppression, like the members of the search and rescue team, should make note of anything unusual about the contents of the building. The absence of personal items in a residence may indicate that they were removed and the fire was intentionally set. Empty boxes in a warehouse may belong there, or they may indicate that the valuable contents were removed. If only one item or a particular stock of products burns, and there is no reasonable explanation for the fire, arson could be suspected. The fire investigator will consider these factors with other evidence before reaching any conclusion.

Fire Marks

Michigan v. Tyler

Because of this court decision, fire investigators may remain at a fire for a reasonable time without a search warrant to determine the cause of the fire. Care should be taken to only secure evidence that is in plain view. Searching for evidence in drawers and closets where there is no reasonable likelihood of evidence would violate the spirit of this decision.

Normally, once fire investigators leave the premise, a search warrant should be obtained prior to reentry. Failure to do so may prevent the use in court of any evidence obtained.

Charring and Burn Patterns

Charring in unusual places—like open floor space away from any likely accidental ignition source—could indicate that the fire was deliberately set. Char on the underside of doors or on the underside of a low horizontal surface, such as a tabletop, could indicate that there was a pool of flammable liquid.

Overhaul

During overhaul, the smoke and steam should begin to dissipate, enabling both fire fighters and fire investigators to get a better look at the surroundings. Fire fighters should continue to look for indications of the signs, patterns, and evidence previously discussed while conducting overhaul in a way that allows evidence to be identified and preserved. The overhaul process, if not done carefully, can quickly destroy valuable evidence.

If possible, the investigator should take a good look at an area before overhaul begins. The investigator can often quickly identify potential evidence, and can help direct or guide the overhaul operation so that it is properly preserved. Evidence located during overhaul should be left where it is found, untouched and undisturbed, until the investigator examines it. Evidence that must be removed from the scene should be properly identified, documented, photographed, packaged, and placed in a secure location.

Fire suppression personnel and investigators must work as a team to ensure that the fire is completely extinguished and properly overhauled while continually searching for and preserving signs, patterns, and evidence. Taking photographs during this phase of the fire operation is a good idea.

Be careful not to destroy evidence during overhaul (► Figure 37-8). Avoid throwing materials into a pile. Use low-velocity hose streams to avoid breaking up and scattering debris needlessly. Thermal imaging devices can be used to find hot spots without tearing apart the interior structure.

Watch for evidence that was shielded from the fire and is lying beneath burned debris. For example, a wall clock may have fallen during the fire and been covered with debris. If the clock stopped approximately when the fire broke out, it could be an important piece of evidence.

A fire investigator will always try to determine whether the building contents were changed or removed prior to a fire. If all of a business' computers are missing, the fire might have been lit to conceal a theft. Missing portraits or other sentimental items might have been removed by the owner before setting the fire.

Wildland and vehicle fires also should be investigated. The evidence at these fires may be somewhat different from the evidence at a building fire. An accidental wildland fire might have started from an untended campfire, while intentional wildland fires can be started from a cigarette or a book of matches.

A vehicle fire can start from an accidental electrical short or from a combination of gasoline and matches placed in the passenger compartment. Signs of spilled fuel on the ground around a vehicle may be evidence of an intentional fire. Do not move the car without documenting this spilled fuel; the evidence will be lost.

Injuries and Fatalities

Any fire that results in an injury or fatality must be thoroughly investigated and the fire scene documented. A search and rescue operation should never be compromised, but it is important to document the location and position of any victims, especially in relation to the fire and the exits.

Clothing removed from any victim should be preserved as evidence. It may contain traces of flammable liquids, and burn patterns can indicate the fire flow. If the clothing is removed in the ambulance or at the hospital, personnel should be assigned to collect it and keep it as intact as possible.

Figure 37-8 Be aware of the need to preserve evidence during overhaul.

Document what may be lying under the victim's body after it is removed. This often is a protected area and may reveal important evidence.

Securing and Transferring the Property

Fire departments have the legal authority to maintain control of a building after a fire to determine the cause of the fire. Maintaining site integrity is critical to the fire investigation. The building and premises must be properly secured and guarded until the fire investigator has finished gathering evidence and documenting the fire scene. Otherwise, any efforts to determine the cause of a malicious or incendiary fire, no matter how efficient or complete, will be wasted.

Fire fighters must be aware of any state or local laws pertaining to right of access by the owners or occupants while the property is under the control of the fire department. The fire department should keep a log of the name of anyone who enters the property, times of entry and departure, and a description of any items taken from the scene. A fire officer or a fire fighter should accompany anyone who enters the premises for any reason until the scene is released.

If a fire investigator is not immediately available, the premises should be guarded and maintained under the control of the fire department until the investigation takes place and all evidence is collected. In the interim, take the following steps:

- Suspend salvage and overhaul, and secure the scene. Keep nonessential personnel out of the area. Deny entry to all unauthorized and unnecessary persons.
- Photograph the fire scene extensively. Start from the area of least damage and work toward the area of possi-

ble origin. Take several pictures of the point of origin from various angles. Photograph any incendiary devices on the premises exactly where they were found.

- If weather, traffic, or other factors could destroy the evidence, take steps to preserve it in the best way possible. Protect tire tracks or footprints by placing boxes over them to prevent dust accumulation. Use barricades to block off the area to further traffic. Rope off areas surrounding plants, trailers, and devices, and post a guard.

The property should be secured by cordoning off the area with fire- or police-line tape. A member of the fire department or law enforcement agency should remain at the scene to ensure that no unauthorized persons cross the line. As previously mentioned, the fire department has the authority to deny access to a building for as long as necessary to conduct a thorough investigation and ensure the safety of the public.

Before leaving the scene, make sure that the building is properly secured, and no hazards to public safety exist. Shut off all utilities and seal any openings in the roof to prevent additional water damage. Board up and secure windows and doors to prevent unauthorized entry.

Fire departments can secure and protect the premises in several ways. Lock and guard gates if necessary, rope off dangerous areas and mark them with signs. Most fire departments have contracts with local companies that provide 24-hour board-up services, and sometimes hire private security guards to secure a property after a fire (▶ Figure 37-9).

Figure 37-9 Some fire departments have contracts with local companies that provide 24-hour board-up service.

Eventually, when fire department operations are over, the property will be returned to the owner. This should not be done, however, until the investigation is complete and all evidence is collected. The fire department's authority ends when the property is formally released to the property owner. Afterward, the investigator might need a search warrant or written consent to search, and unguarded evidence could be contaminated.

Incendiary Fires

The term incendiary fires refers to all fires that were deliberately started for malicious or criminal intent. In many jurisdictions, arson has a narrower, specific legal definition. This chapter uses arson to mean the malicious burning of property with criminal intent.

Fire fighter must be aware of factors that could indicate an intentionally set fire, to report any observations to a company officer or fire investigator, and to help protect evidence.

Indications of Arson

Arson fires have several distinct, recognizable patterns or indications. For example, a deliberate fire might have multiple points of origin or multiple simultaneous fires. An arsonist will ignite fires in different areas to create a large fire as quickly as possible and involve the entire building. Arsonists also use trailers made from combustible materials, such as paper, rags, clothing, curtains, kerosene-soaked rope or other fuels, to spread a fire. Trailers often leave distinctive char and/or burn patterns that investigators can used to trace the spread of a fire.

An **incendiary device** is a device or mechanism, such as candles, timers, or electrical heaters that is used to start a fire or explosion. Many incendiary devices leave evidence, such as metal parts, electrical components, or mechanical devices, at the point of origin. Often an arsonist will use more than one incendiary device; sometimes, a failed incendiary device is found at the fire scene. If trailers were used, investigators may be able to trace the burn patterns back to an incendiary device.

Evidence of a flammable liquid often indicates an incendiary fire, but does not necessarily establish arson as the cause because there could be an accidental explanation for a flammable liquid spill. To prove arson, the fire investigator must determine that a flammable liquid was used and that there is no explanation other than arson for the presence of the flammable liquid.

Extensive burn damage on a floor's surface could indicate that a flammable liquid was poured and ignited. For example, a wood floor might have a pool-shaped burn pattern, or floor tiles could be discolored or blistered with an irregular burn pattern. On concrete and masonry floors, a flammable liquid pattern is usually irregular with various shades of gray to black. Cracked and pitted concrete may also suggest flammable liquid use. Liquids flow to the lowest level possible, so the evidence of pooled liquids could be visible in corners and along the base of walls, where there might also be low levels of charring.

Sometimes the first indications of a possible arson fire are entirely circumstantial. A particular fire could fit into a pattern, such as a series of fires in the same area, at about the same time on the same days of the week. There might be a series of fires in the same type of business or in properties with the same owner. A fire fighter who notices a pattern or a set of similar circumstances should immediately report the observation to a company officer or a fire investigator.

Arsonists

Arsonists fall into several categories, with various explanations for their behavior. The fire service has identified two groups who are responsible for a large number of fires: pyromaniacs and juvenile fire-setters. Many other arsonists start fires for a wide range of motives.

Pyromaniacs

A **pyromaniac** is a pathological fire-setter. Most are adult males, often loners. They are usually introverted, polite but timid, and have difficulty relating to other people. The fires set by pyromaniacs have the following characteristics:

- Fires are set in easily accessible locations, such as immediately inside entrances, on basement stairs, in trash bins, or on porches.
- Fires are set in structures such as occupied residences of all types, barns, and vacant buildings.
- Accelerants are rarely used. The pyromaniac is impulsive, so materials readily at hand are used.
- Each pyromaniac usually has a pattern, setting fires at the same time of day or night, using the same method, and in similar locations.

Juvenile Fire-Setters

Juvenile fire-setters are usually divided into three groups according to age: 8 years old and under, 9 to 12 years old (preadolescent) and 13 to 17 years old (adolescent). Preadolescent and adolescent fire-setters often exhibit the same personality traits as adult pyromaniacs: introverted, difficulty with interpersonal relationships, and extreme politeness when questioned.

Children under 8 years old are seldom criminally motivated when they set fires; they usually are just curious and experimenting. Children of this age do not really understand the danger of fire. They usually set fires in or near their homes: in attics, basements, bathrooms, garages, closets, or in nearby fields or vacant lots. They start fires with matches or by sticking combustible material into equipment such as electric heaters that provide an ignition source. The remains of matches, matchboxes, or matchbooks are often found at the point of origin.

Preadolescent fire-starters do not venture far from home, but this group does set most of the fires that involve schools and churches. Preadolescents have motivations other than idle curiosity. The motivations of boys range from spite to revenge and disruptive behavior. They will set fires in closets, classrooms, and wastebaskets, and may try burning hallway bulletin boards. Girls usually set less aggressive fires and are motivated by a need for attention or in response to a particular stress, such as a test they don't want to take.

The preadolescent usually does not use elaborate trailers or incendiary devices. They will use common, available accelerants such as gasoline, kerosene, or lighter fluid. The preadolescent often uses whatever materials are at hand and on site, including trash container contents, loose papers, and rags. The fire will show a lack of planning, preparation, and sophistication, particularly if it was a group effort.

Preadolescents may use fire to cover vandalism and theft. The items taken from the scene are typically personal, rather than especially valuable. For example, the child may take athletic equipment, which is later displayed to classmates, or may reclaim an item that had been confiscated by a teacher.

The fires set by adolescents are similar to those set by adults. Adolescents have better access to transportation and can travel farther, so they have access to a wider variety of buildings. They also have many of the same motivations of adult fire-setters, such as revenge or attempts to hide larceny, and they often use accelerants. Two-thirds of fires set in vacant buildings are set by adolescents, but no class of property is exempt from harm. Often a great amount of vandalism at the scene will be a clue that the fire was set by an adolescent.

Arsonist Motives

An **arsonist** is someone who deliberately sets a fire with criminal intent. Arsonists set all types of fires and have a range of motives. Male arsonists are usually motivated by profit, revenge, or vanity, and often use accelerants. In the past, female fire-setters were seldom motivated by profit and seldom used accelerants, but this seems to be changing.

There are six common motives listed in NFPA 921, *Guide to Fire and Explosion Investigations:*

1. Vandalism
2. Excitement
3. Revenge
4. Crime concealment
5. Profit
6. Extremism

Arsonists who are motivated by excitement are relatively easy to apprehend because they usually have some identifiable connection with their targets, such as a failing student to a school or an employee to a business. One category of excitement fires includes fires started by a would-be hero. The would-be hero can be either a discoverer or an assister. The discoverer sets the fire, "discovers" it, and turns in the alarm. The assister sets the fire, sometimes discovers it and reports it, and is always on hand to help fire fighters pull hose, raise ladders, and rescue people.

An unfortunate category of excitement arsonists is closely related to the would-be hero. These are fire fighters and would-be fire fighters (persons who have tried to become fire fighters and failed) who intentionally start fires. Some just want the opportunity to operate a particular piece of apparatus or equipment; others want to get some practice. Some have even started fires in nearby jurisdictions in an attempt to harm another fire department's reputation.

Fire Fighter Tips

Arson Facts

- Arson is often called a "young man's crime;" 54% of those arrested for arson are under the age of 18.
- Only about 2% of all incendiary or suspected incendiary fires lead to conviction.
- Arson is the leading cause of property damage in the United States, resulting in $1.3 billion of property damage annually.
- One out of eight fire fatalities is due to a fire started by arson.

SOURCE: U.S. Arson Trends and Patterns, NFPA

Fire Fighter Tips

Serial Arson

There are three types of serial (repetitive) arson identified by NFPA 921:

1. **Serial arson** involves an offender who sets three or more fires, with a cooling-off period between fires.
2. **Spree arson** involves an arsonist who sets three or more fires at separate locations with no emotional cooling-off period between fires.
3. **Mass arson** involves an offender who sets three or more fires at the same site or location during a limited period of time.

SOURCE: NFPA 921, *Guide to Fire and Explosion Investigations,* 2001 Edition, Section 19.4.8.2

Arsonists—continued.

Revenge arsonists are often very careful, and make detailed plans for setting the fire and escaping detection afterwards. Arsonists motivated by revenge seldom consider the extent of damage fire can inflict. Their intent is to harm a particular person or group to get revenge for a perceived injustice.

Some individuals with extreme beliefs will set fires to aid their cause. These fires are prompted by political, religious, and racial differences. Churches, abortion clinics, and government buildings have been targets of arsonists in the extremism category.

Arsonists also may set fires to conceal a crime. The fire is used to destroy evidence of another crime such as burglary, vandalism, fraud, or murder. Fires have also been set to divert attention while another crime is being committed or during attempted escapes from jails, hospitals, and other institutions.

Arson for profit occurs because the arsonist will benefit from the fire either directly or indirectly. Fires that directly benefit the arsonist often involve businesses. The business owner might set the fire or arrange to have it set, to gain an insurance settlement that can be used to cover other financial needs. An increasing problem connected with this arson for profit is fraud. For example, a person may buy and insure a property, only to burn it for the sole purpose of defrauding an insurance company.

The other type of profit arsonist sets a fire to benefit from another person's loss. Included in this motive category are:

- Insurance agents who want to sell more insurance to the victim's neighbors
- Contractors who want to secure a contract for rebuilding or wrecking or to get a salvage
- Competitors who want to drive the victim out of business
- Owners of adjoining property who would like to buy the property and expand their own holdings
- Owners of nearby property who want to prevent an occupancy change, such as the conversion of a building to a drug rehabilitation center

Wrap-Up

Ready for Review

This chapter covered the information fire fighters need to assist in identifying and preserving signs, patterns, and evidence that can be used to determine the origin and cause of a fire. Fire cause determination helps fulfill the fire department's mission of preventing fires. A fire's cause is considered undetermined until it is officially identified as either accidental or incendiary.

Fire departments use fire cause information to prevent further accidental fires and prevent people from intentionally starting fires. Fire investigation must identify the point of origin and use physical evidence to determine the cause of the blaze. Fire investigators are responsible for finding and preserving both direct and circumstantial evidence and for interviewing witnesses. A chain of custody must be established and maintained for all evidence, starting at the time of discovery. Fire fighters can preserve evidence by leaving it where it is found, covering it, or placing something over it.

Fire fighters can help identify patterns, signs, unusual activities, or conditions, and preserve any materials that might be valuable evidence. Throughout the fire suppression sequence, fire fighters should note the time of day; people leaving the scene; factors related to the fire; devices that might be used for burglary or arson; whether doors and windows are locked, unlocked, open, or closed; the condition of building utilities; signs of illegal activity; the condition of the building contents; any unusual odors or smoke conditions; the condition of fire detection, notification, and suppression systems, and the reaction of the fire during suppression.

Fire fighters and fire investigators must work together to ensure that the fire is completely extinguished and that evidence is identified and preserved. The fire scene must be properly secured and turned over to the fire investigator or to law enforcement personnel. The fire fighters on the scene will need to be the eyes and ears of the investigator until the investigator arrives. Their observations can be passed on to the investigator through an interview or through written reports, photographs, and proper evidence handling.

Chief Concepts

- Preserving evidence assists fire fighters with the primary goal of preventing loss of lives and property loss.
- Fires are caused by either incendiary or accidental causes.
- Basic fire investigation includes locating the point of origin, determining the fuel used, and identifying the ignition source.
- Fire investigation should be performed by one of the following: trained fire department investigators, the fire marshal's office, insurance company investigators, or a law enforcement agency.
- Physical evidence must be preserved by maintaining an unbroken chain of custody.
- The fire fighter's role in fire investigation is to identify and preserve possible evidence until it can be turned over to a trained fire investigator.
- The fire fighter's role in identifying and preserving evidence continues throughout the fire suppression sequence, and includes the following factors:
  - Time of day, weather
  - People leaving the scene
  - Extent of fire, number of locations of fire
  - Security of the building
  - Signs of property break-in
  - Vehicles or people in the area
  - Indications of unusual fire situations
  - Unusual color of smoke
  - Position of windows and roof
  - The reaction of the fire during initial attack
  - Abnormal behavior of fire
  - Condition of the building contents
  - Need to coordinate overhaul and evidence preservation activities
  - Need to transfer responsibility of the property from fire suppression personnel to fire investigators
  - Need to secure the property

Wrap-Up

Hot Terms

Accelerants Materials, usually flammable liquids, used to initiate or increase the spread of fire.

Accidental Fire cause classification that includes fires with a proven cause that does not involve a deliberate human act.

Arson The malicious burning of one's own or another's property with a criminal intent.

Arsonist A person who deliberately sets a fire to destroy property with criminal intent.

Chain of custody A legal term used to describe the paperwork or documentation describing the movement, storage, and custody of evidence, such as the record of possession of a gas can from a fire scene.

Circumstantial evidence The means by which alleged facts are proven by deduction or inference from other facts that were observed first hand.

Competent ignition source A competent ignition source is one that can ignite a fuel under the existing conditions at the time of the fire. It must have sufficient heat and be in proximity to the fuel for sufficient time to ignite the fuel.

Contaminated A term used to describe evidence that may have been altered from its original state.

Demonstrative evidence Term used to describe materials used to demonstrate a theory or explain an event.

Depth of char The thickness of the layer of a material that has been consumed by a fire. The depth of char on wood can be used to help determine the duration of a fire.

Direct evidence Evidence that is reported first hand, such as statements from an eyewitness who saw or heard something.

Incendiary device A device or mechanism used to start a fire or explosion.

Incendiary fires Intentionally set fires.

Mass arson Involves an offender who sets three or more fires at the same site or location during a limited period of time.

Physical evidence Items that can be observed, photographed, measured, collected, examined in a laboratory, and presented in court to prove or demonstrate a point.

Point of origin The exact location where a heat source and a fuel come in contact with each other and a fire begins.

Pyromaniac A pathological fire-setter.

Serial arson A series of fires set by the same offender, with a cooling-off period between fires.

Spree arson A series of fires started by an arsonist who sets three or more fires at separate locations with no emotional cooling-off period between fires.

Trace (transfer) evidence Evidence of a minute quantity that is conveyed from one place to another.

Trailers Combustible material, such as rolled rags, blankets, and newspapers or flammable liquid, used to spread fire from one point or area to other points or areas, often used in conjunction with an incendiary device.

Undetermined Cause classification that includes fires for which the cause has not or cannot be proven.

Fire Fighter in Action

It is 2:00 a.m. on a chilly Saturday morning when your engine is dispatched to a report of a local high school on fire. Upon arrival, you find a fully involved dumpster underneath an overhang of the school. The fire has spread through the overhang into the attic of the school. You note that the dumpster has been moved from its designated area behind the school. The lieutenant tells you and your crew to extinguish the dumpster fire.

1. What observation would you make about the time of this incident?
   A. The time is not a factor.
   B. The school should be vacant at this time.
   C. School should be in session at this time.
   D. None of the above

As you advance the hose line, you notice an empty gasoline can on the ground several feet away from the dumpster. This could be a critical piece of evidence for the fire investigation.

2. What type of evidence consists of items that can be observed, photographed, measured, collected, examined in a laboratory, and presented in court?
   A. Physical
   B. Trace or transfer
   C. Demonstrative
   D. Circumstantial

3. If a fire investigator is not immediately available, what action(s) can you take to secure the property?
   A. Suspend salvage and overhaul operations.
   B. Photograph the scene extensively.
   C. Preserve evidence from damaging traffic, weather, or other factors.
   D. All of the above

4. As you consider the unusual circumstances of this fire, you wonder if the cause is incendiary. Who should you share your observations with?
   A. The school principal
   B. Other fire fighters
   C. The fire investigator
   D. The local news media

www.FireFighter.jbpub.com

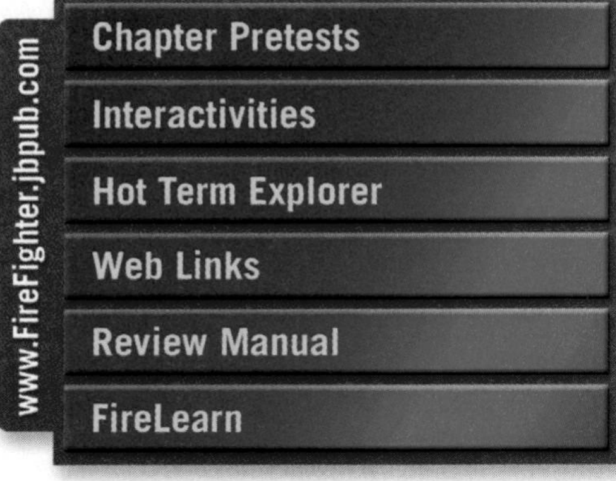

NFPA 1001, *Standard for Fire Fighter Professional Qualifications*, 2002 Edition

Chapter 1: Administration

1.1 Scope. This standard identifies the minimum job performance requirements for career and volunteer fire fighters whose duties are primarily structural in nature.

1.2 Purpose. The purpose of this standard is to specify the minimum job performance requirements for fire fighters. It is not the intent of the standard to restrict any jurisdiction from exceeding these requirements.

1.3 General.

1.3.1 The job performance requirements shall be accomplished in accordance with the requirements of the authority having jurisdiction and NFPA 1500, *Standard on Fire Department Occupational Safety and Health Program.*

1.3.2* It is not required for the job performance requirements to be mastered in the order they appear. The authority having jurisdiction shall establish instructional priority and the training program content to prepare individuals to meet the job performance requirements of this standard.

1.3.3* Performance of each requirement of this standard shall be evaluated by individuals approved by the authority having jurisdiction.

1.3.4 The entrance requirements in Chapter 4 of this standard shall be met prior to beginning training at the Fire Fighter I level.

1.3.5* Prior to being certified at the Fire Fighter I level, the fire fighter candidate shall meet the general knowledge and skills requirements and the job performance requirements of Chapter 5.

1.3.6 Prior to being certified at the Fire Fighter II level, the Fire Fighter I shall meet the general knowledge and skills requirements and the job performance requirements of Chapter 6.

1.3.7 Wherever in this standard the terms *rules, regulations, procedures, supplies, apparatus,* or *equipment* are referred to, it is implied that they are those of the authority having jurisdiction.

1.4 Units. In this standard, values for measurement are followed by an equivalent in SI units, but only the first stated value shall be regarded as the requirement. Equivalent values in SI units shall not be considered as the requirement, as these values can be approximate. (*See Table 1.4.*)

Table 1.4 SI Conversions

Quantity	U.S. Unit/ Symbol	SI Unit/ Symbol	Conversion Factor
Length	inch (in.)	millimeter (mm)	1 in. = 25.4 mm
	foot (ft)	meter (m)	1 ft = 0.305 m
Area	square foot (ft^2)	square meter (m^2)	1 ft^2 = 0.0929 m^2

Chapter 2: Referenced Publications

2.1 General. The documents or portions thereof listed in this chapter are referenced within this standard and shall be considered part of the requirements of this document.

2.2 NFPA Publications. National Fire Protection Association, 1 Batterymarch Park, P.O. Box 9101, Quincy, MA 02169-7471.

NFPA 472, *Standard for Professional Competence of Responders to Hazardous Materials Incidents,* 2002 edition.
NFPA 1500, *Standard on Fire Department Occupational Safety and Health Program,* 2002 edition.
NFPA 1582, *Standard on Medical Requirements for Fire Fighters and Information for Fire Department Physicians,* 2000 edition.

2.3 Other Publications. (Reserved)

Chapter 3: Definitions

3.1* General. The definitions contained in this chapter shall apply to the terms used in this standard. Where terms are not included, common usage of the terms shall apply.

3.2 NFPA Official Definitions.

3.2.1* Approved. Acceptable to the authority having jurisdiction.

Appendix A

3.2.2* Authority Having Jurisdiction (AHJ). The organization, office, or individual responsible for approving equipment, materials, an installation, or a procedure.

3.2.3* Listed. Equipment, materials, or services included in a list published by an organization that is acceptable to the authority having jurisdiction and concerned with evaluation of products or services, that maintains periodic inspection of production of listed equipment or materials or periodic evaluation of services, and whose listing states that either the equipment, material, or service meets appropriate designated standards or has been tested and found suitable for a specified purpose.

3.3 General Definitions.

3.3.1 Fire Department. An organization providing rescue, fire suppression, and related activities. The term "fire department" shall include any public, governmental, private, industrial, or military organization engaging in this type of activity.

3.3.2 Fire Fighter Candidate. The person who has fulfilled the entrance requirements of Chapter 4 of this standard but has not met the job performance requirements for Fire Fighter I.

3.3.3 Fire Fighter I. The person, at the first level of progression as defined in Chapter 5, who has demonstrated the knowledge and skills to function as an integral member of a firefighting team under direct supervision in hazardous conditions.

3.3.4* Fire Fighter II. The person, at the second level of progression as defined in Chapter 6, who has demonstrated the skills and depth of knowledge to function under general supervision.

3.3.5 Job Performance Requirement (JPR). A statement that describes a specific job task, lists the items necessary to complete the task, and defines measurable or observable outcomes and evaluation areas for the specific task.

3.3.6 Personal Protective Clothing. The full complement of garments fire fighters are normally required to wear while on emergency scene including turnout coat, protective trousers, fire fighting boots, fire fighting gloves, a protective hood, and a helmet with eye protection.

3.3.7 Personal Protective Equipment. Consists of full personal protective clothing, plus a self-contained breathing apparatus (SCBA) and a personal alert safety system (PASS) device.

3.3.8 Procedure. The series of actions, conducted in an approved manner and sequence, designed to achieve an intended outcome.

3.3.9 Requisite Knowledge. Fundamental knowledge one must have in order to perform a specific task.

3.3.10 Requisite Skills. The essential skills one must have in order to perform a specific task.

3.3.11 Structural Fire Fighting. The activities of rescue, fire suppression, and property conservation in buildings, enclosed structures, aircraft interiors, vehicles, vessels, aircraft, or like properties that are involved in a fire or emergency situation. [**1500**:3.3]

3.3.12 Task. A specific job behavior or activity.

3.3.13 Team. Two or more individuals who have been assigned a common task and are in proximity to and in direct communications with each other, coordinate their activities as a work group, and support the safety of one another.

Chapter 4: Entrance Requirements

4.1 General. Prior to entering training to meet the requirements of Chapters 5 and 6 of this standard, the candidate shall meet the following requirements:

(1) Minimum educational requirements established by the authority having jurisdiction
(2) Age requirements established by the authority having jurisdiction
(3)* Medical requirements of NFPA 1582, *Standard on Medical Requirements for Fire Fighters and Information for Fire Department Physicians*

4.2 Fitness Requirements. Physical fitness requirements for entry-level personnel shall be developed and validated by the authority having jurisdiction.

Chapter 4: Entrance Requirements (continued)

4.3* Emergency Medical Care. Minimum emergency medical care performance capabilities for entry level personnel shall be developed and validated by the authority having jurisdiction to include infection control, CPR, bleeding control, and shock management.

Chapter 5: Fire Fighter I

5.1 General.

5.1.1 For certification at Level I, the fire fighter candidate shall meet the general knowledge requirements in 5.1.1.1, the general skill requirements in 5.1.1.2, and the job performance requirements defined in Sections 5.2 through 5.5 of this standard and the requirements defined in Chapter 4, Competencies for the First Responder at the Awareness Level, of NFPA 472, *Standard for Professional Competence of Responders to Hazardous Materials Incidents.*

5.1.1.1 General Knowledge Requirements. The organization of the fire department; the role of the Fire Fighter I in the organization; the mission of fire service; the fire department's standard operating procedures and rules and regulations as they apply to the Fire Fighter I; the role of other agencies as they relate to the fire department; aspects of the fire department's member assistance program; the critical aspects of NFPA 1500, *Standard on Fire Department Occupational Safety and Health Program,* as they apply to the Fire Fighter I; knot types and usage; the difference between life safety and utility rope; reasons for placing rope out of service; the types of knots to use for given tools, ropes, or situations; hoisting methods for tools and equipment; and using rope to support response activities.

5.1.1.2 General Skill Requirements. The ability to don personal protective clothing within one minute; doff personal protective clothing and prepare for reuse; hoist tools and equipment using ropes and the correct knot; tie a bowline, clove hitch, figure eight on a bight, half hitch, becket or sheet bend, and safety knots; and locate information in departmental documents and standard or code materials.

5.2 Fire Department Communications. This duty involves initiating responses, receiving telephone calls, and using fire department communications equipment to correctly relay verbal or written information, according to the following job performance requirements.

5.2.1* Initiate the response to a reported emergency, given the report of an emergency, fire department standard operating procedures, and communications equipment, so that all necessary information is obtained, communications equipment is operated correctly, and the information is promptly and accurately relayed to the dispatch center.

(A) *Requisite Knowledge:* Procedures for reporting an emergency, departmental standard operating procedures for taking and receiving alarms, radio codes or procedures, and information needs of dispatch center.

(B) *Requisite Skills:* The ability to operate fire department communications equipment, relay information, and record information.

5.2.2 Receive a business or personal telephone call, given afire department business phone, so that procedures for answering the phone are used and the caller's information is relayed.

(A) *Requisite Knowledge:* Fire department procedures for answering nonemergency telephone calls.

(B) *Requisite Skills:* The ability to operate fire station telephone and intercom equipment.

5.2.3 Transmit and receive messages via the fire department radio, given a fire department radio and operating procedures, so that the information is accurate, complete, clear, and relayed within the time established by the AHJ.

(A) *Requisite Knowledge:* Departmental radio procedures and etiquette for routine traffic, emergency traffic, and emergency evacuation signals.

(B) *Requisite Skills:* The ability to operate radio equipment and discriminate between routine and emergency traffic.

5.3 Fireground Operations. This duty involves performing activities necessary to ensure life safety, fire control, and property conservation, according to the following job performance requirements.

5.3.1* Use SCBA during emergency operations, given SCBA and other personal protective equipment, so that the SCBA is correctly donned and activated within one minute, the SCBA is correctly worn, controlled breathing techniques are used, emergency procedures are enacted if the SCBA fails, all low-air warnings are recognized, respiratory protection is not intentionally compromised, and hazardous areas are exited prior to air depletion.

(A) *Requisite Knowledge:* Conditions that require respiratory protection, uses and limitations of SCBA, components of SCBA, donning procedures, breathing techniques, indications for and emergency procedures used with SCBA, and physical requirements of the SCBA wearer.

(B) *Requisite Skills:* The ability to control breathing, replace SCBA air cylinders, use SCBA to exit through restricted passages, initiate and complete emergency procedures in the event of SCBA failure or air depletion, and complete donning procedures.

5.3.2* Respond on apparatus to an emergency scene, given personal protective clothing and other necessary personal protective equipment, so that the apparatus is correctly mounted and dismounted, seat belts are used while the vehicle is in motion, and other personal protective equipment is correctly used.

(A) *Requisite Knowledge:* Mounting and dismounting procedures for riding fire apparatus; hazards and ways to avoid hazards associated with riding apparatus; prohibited practices; types of department personal protective equipment and the means for usage.

(B) *Requisite Skills:* The ability to use each piece of provided safety equipment.

5.3.3* Operate in established work areas at emergency scenes, given protective equipment, traffic and scene control devices, structure fire and roadway emergency scenes, traffic hazards and downed electrical wires, so that procedures are followed, protective equipment is worn, protected work areas are established as directed using traffic and scene control devices, and the fire fighter performs assigned tasks only in established, protected work areas.

(A) *Requisite Knowledge:* Potential hazards involved in operating on emergency scenes including vehicle traffic, utilities, and environmental conditions; proper procedures for dismounting apparatus in traffic; procedures for safe operation at emergency scenes; and the protective equipment available for members' safety on emergency scenes and work zone designations.

(B) *Requisite Skills:* The ability to use PPC, the deployment of traffic and scene control devices, dismount apparatus and operate in the protected work areas as directed.

5.3.4* Force entry into a structure, given personal protective equipment, tools, and an assignment, so that the tools are used as designed, the barrier is removed, and the opening is in a safe condition and ready for entry.

(A) *Requisite Knowledge:* Basic construction of typical doors, windows, and walls within the department's community or service area; operation of doors, windows, and locks; and the dangers associated with forcing entry through doors, windows, and walls.

(B) *Requisite Skills:* The ability to transport and operate hand and power tools and to force entry through doors, windows, and walls using assorted methods and tools.

5.3.5* Exit a hazardous area as a team, given vision-obscured conditions, so that a safe haven is found before exhausting the air supply, others are not endangered, and the team integrity is maintained.

(A) *Requisite Knowledge:* Personnel accountability systems, communication procedures, emergency evacuation methods, what constitutes a safe haven, elements that create or indicate a hazard, and emergency procedures for loss of air supply.

(B) *Requisite Skills:* The ability to operate as a team member in vision-obscured conditions, locate and follow a guideline, conserve air supply, and evaluate areas for hazards and identify a safe haven.

5.3.6* Set up ground ladders, given single and extension ladders, an assignment, and team members if needed, so that hazards are assessed, the ladder is stable, the angle is correct for climbing, extension ladders are extended to the necessary height with the fly locked, the top is placed against a reliable structural component, and the assignment is accomplished.

(A) *Requisite Knowledge:* Parts of a ladder, hazards associated with setting up ladders, what constitutes a stable foundation for ladder placement, different angles for various tasks, safety limits to the degree of angulation, and what constitutes a reliable structural component for top placement.

(B) *Requisite Skills:* The ability to carry ladders, raise ladders, extend ladders and lock flies, determine that a wall and roof will support the ladder, judge extension ladder height requirements, and place the ladder to avoid obvious hazards.

5.3.7* Attack a passenger vehicle fire operating as a member of a team, given personal protective equipment, attack line, and hand tools, so that hazards are avoided, leaking flammable liquids are identified and controlled, protection from flash fires is maintained, all vehicle compartments are over-hauled, and the fire is extinguished.

(A) *Requisite Knowledge:* Principles of fire streams as they relate to fighting automobile fires; precautions to be followed when advancing hose lines toward an automobile; observable results that a fire stream has been properly applied; identifying alternative fuels and the hazards associated with them; dangerous conditions created during an automobile fire; common types of accidents or injuries related to fighting automobile fires and how to avoid them; how to access locked passenger, trunk, and engine compartments; and methods for overhauling an automobile.

Chapter 5: Fire Fighter I (continued)

(B) *Requisite Skills:* The ability to identify automobile fuel type; assess and control fuel leaks; open, close, and adjust the flow and pattern on nozzles; apply water for maximum effectiveness while maintaining flash fire protection; advance $1^{1}/_{2}$-in. (38-mm) or larger diameter attack lines; and expose hidden fires by opening all automobile compartments.

5.3.8* Extinguish fires in exterior Class A materials, given fires in stacked or piled and small unattached structures or storage containers that can be fought from the exterior, attack lines, hand tools and master stream devices, and an assignment, so that exposures are protected, the spread of fire is stopped, collapse hazards are avoided, water application is effective, the fire is extinguished, and signs of the origin area(s) and arson are preserved.

(A) *Requisite Knowledge:* Types of attack lines and water streams appropriate for attacking stacked, piled materials and outdoor fires; dangers—such as collapse—associated with stacked and piled materials; various extinguishing agents and their effect on different material configurations; tools and methods to use in breaking up various types of materials; the difficulties related to complete extinguishment of stacked and piled materials; water application methods for exposure protection and fire extinguishment; dangers such as exposure to toxic or hazardous materials associated with storage building and container fires; obvious signs of origin and cause; and techniques for the preservation of fire cause evidence.

(B) *Requisite Skills:* The ability to recognize inherent hazards related to the material's configuration, operate hand lines or master streams, break up material using hand tools and water streams, evaluate for complete extinguishment, operate hose lines and other water application devices, evaluate and modify water application for maximum penetration, search for and expose hidden fires, assess patterns for origin determination, and evaluate for complete extinguishment.

5.3.9 Conduct a search and rescue in a structure operating as a member of a team, given an assignment, obscured vision conditions, personal protective equipment, a flashlight, forcible entry tools, hose lines, and ladders when necessary, so that ladders are correctly placed when used, all assigned areas are searched, all victims are located and removed, team integrity is maintained, and team members' safety—including respiratory protection—is not compromised.

(A) *Requisite Knowledge:* Use of forcible entry tools during rescue operations, ladder operations for rescue, psychological effects of operating in obscured conditions and ways to manage them, methods to determine if an area is tenable, primary and secondary search techniques, team members' roles and goals, methods to use and indicators of finding victims, victim removal methods (including various carries), and considerations related to respiratory protection.

(B) *Requisite Skills:* The ability to use SCBA to exit through restricted passages, set up and use different types of ladders for various types of rescue operations, rescue a fire fighter with functioning respiratory protection, rescue a fire fighter whose respiratory protection is not functioning, rescue a person who has norespiratory protection, and assess areas to determine tenability.

5.3.10* Attack an interior structure fire operating as a member of a team, given an attack line, ladders when needed, personal protective equipment, tools, and an assignment, so that team integrity is maintained, the attack line is deployed for advancement, ladders are correctly placed when used, access is gained into the fire area, effective water application practices are used, the fire is approached correctly, attack techniques facilitate suppression given the level of the fire, hidden fires are located and controlled, the correct body posture is maintained, hazards are recognized and managed, and the fire is brought under control.

(A) *Requisite Knowledge:* Principles of fire streams; types, design, operation, nozzle pressure effects, and flow capabilities of nozzles; precautions to be followed when advancing hose lines to a fire; observable results that a fire stream has been properly applied; dangerous building conditions created by fire; principles of exposure protection; potential long-term consequences of exposure to products of combustion; physical states of matter in which fuels are found; common types of accidents or injuries and their causes; and the application of each size and type of attack line, the role of the backup team in fire attack situations, attack and control techniques for grade level and above and below grade levels, and exposing hidden fires.

(B) *Requisite Skills:* The ability to prevent water hammers when shutting down nozzles; open, close, and adjust nozzle flow and patterns; apply water using direct, indirect, and combination attacks; advance charged and uncharged $1^{1}/_{2}$-in. (38-mm) diameter or larger hose lines up ladders and up and down interior and exterior stairways; extend hose lines; replace burst hose sections; operate charged hose lines of $1^{1}/_{2}$-in. (38-mm) diameter or larger while secured to a ground ladder; couple and uncouple various hand line connections; carry hose; attack fires at grade level and above and below grade levels; and locate and suppress interior wall and subfloor fires.

5.3.11 Perform horizontal ventilation on a structure operating as part of a team, given an assignment, personal protective equipment, ventilation tools, equipment, and ladders, so that the ventilation openings are free of obstructions, tools are used as designed, ladders are correctly placed, ventilation devices are correctly placed, and the structure is cleared of smoke.

(A) *Requisite Knowledge:* The principles, advantages, limitations, and effects of horizontal, mechanical, and hydraulic ventilation; safety considerations when venting a structure; fire behavior in a structure; the products of combustion found in a structure fire; the signs, causes, effects, and prevention of backdrafts; and the relationship of oxygen concentration to life safety and fire growth.

(B) *Requisite Skills:* The ability to transport and operate ventilation tools and equipment and ladders and to use safe procedures for breaking window and door glass and removing obstructions.

5.3.12 Perform vertical ventilation on a structure as part of a team, given an assignment, personal protective equipment, ground and roof ladders, and tools, so that ladders are positioned for ventilation, a specified opening is created, all ventilation barriers are removed, structural integrity is not compromised, products of combustion are released from the structure, and the team retreats from the area when ventilation is accomplished.

(A) *Requisite Knowledge:* The methods of heat transfer; the principles of thermal layering within a structure on fire; the techniques and safety precautions for venting flat roofs, pitched roofs, and basements; basic indicators of potential collapse or roof failure; the effects of construction type and elapsed time under fire conditions on structural integrity; and the advantages and disadvantages of vertical and trench/strip ventilation.

(B) *Requisite Skills:* The ability to transport and operate ventilation tools and equipment; hoist ventilation tools to a roof; cut roofing and flooring materials to vent flat roofs, pitched roofs, and basements; sound a roof for integrity; clear an opening with hand tools; select, carry, deploy, and secure ground ladders for ventilation activities; deploy roof ladders on pitched roofs while secured to a ground ladder; and carry ventilation-related tools and equipment while ascending and descending ladders.

5.3.13 Overhaul a fire scene, given personal protective equipment, attack line, hand tools, a flashlight, and an assignment, so that structural integrity is not compromised, all hidden fires are discovered, fire cause evidence is preserved, and the fire is extinguished.

(A) *Requisite Knowledge:* Types of fire attack lines and water application devices most effective for overhaul, water application methods for extinguishment that limit water damage, types of tools and methods used to expose hidden fire, dangers associated with overhaul, obvious signs of area of origin or signs of arson, and reasons for protection of fire scene.

(B) *Requisite Skills:* The ability to deploy and operate an attack line; remove flooring, ceiling, and wall components to expose void spaces without compromising structural integrity; apply water for maximum effectiveness; expose and extinguish hidden fires in walls, ceilings, and subfloor spaces; recognize and preserve obvious signs of area of origin and arson; and evaluate for complete extinguishment.

5.3.14 Conserve property as a member of a team, given salvage tools and equipment and an assignment, so that the building and its contents are protected from further damage.

(A) *Requisite Knowledge:* The purpose of property conservation and its value to the public, methods used to protect property, types of and uses for salvage covers, operations at properties protected with automatic sprinklers, how to stop the flow of water from an automatic sprinkler head, identification of the main control valve on an automatic sprinkler system, and forcible entry issues related to salvage.

(B) *Requisite Skills:* The ability to cluster furniture; deploy covering materials; roll and fold salvage covers for reuse; construct water chutes and catch-alls; remove water; cover building openings, including doors, windows, floor openings, and roof openings; separate, remove, and relocate charred material to a safe location while protecting the area of origin for cause determination; stop the flow of water from a sprinkler with sprinkler wedges or stoppers; and operate a main control valve on an automatic sprinkler system.

5.3.15* Connect a fire department pumper to a water supply as a member of a team, given supply or intake hose, hose tools, and a fire hydrant or static water source, so that connections are tight and water flow is unobstructed.

(A) *Requisite Knowledge:* Loading and off-loading procedures for mobile water supply apparatus; fire hydrant operation; and suitable static water supply sources, procedures, and protocol for connecting to various water sources.

(B) *Requisite Skills:* The ability to hand lay a supply hose, connect and place hard suction hose for drafting operations, deploy portable water tanks as well as the equipment necessary to transfer water between and draft from them, make hydrant-to-pumper hose connections for forward and reverse lays, connect supply hose to a hydrant, and fully open and close the hydrant.

Chapter 5: Fire Fighter I (continued)

5.3.16* Extinguish incipient Class A, Class B, and Class C fires, given a selection of portable fire extinguishers, so that the correct extinguisher is chosen, the fire is completely extinguished, and correct extinguisher-handling techniques are followed.

(A) *Requisite Knowledge:* The classifications of fire; the types of, rating systems for, and risks associated with each class of fire; and the operating methods of, and limitations of portable extinguishers.

(B) *Requisite Skills:* The ability to operate portable fire extinguishers, approach fire with portable fire extinguishers, select an appropriate extinguisher based on the size and type of fire, and safely carry portable fire extinguishers.

5.3.17 Illuminate the emergency scene, given fire service electrical equipment and an assignment, so that designated areas are illuminated and all equipment is operated within the manufacturer's listed safety precautions.

(A) *Requisite Knowledge:* Safety principles and practices, power supply capacity and limitations, and light deployment methods.

(B) *Requisite Skills:* The ability to operate department power supply and lighting equipment, deploy cords and connectors, reset ground-fault interrupter (GFI) devices, and locate lights for best effect.

5.3.18 Turn off building utilities, given tools and an assignment, so that the assignment is safely completed.

(A) *Requisite Knowledge:* Properties, principles, and safety concerns for electricity, gas, and water systems; utility disconnect methods and associated dangers; and use of required safety equipment.

(B) *Requisite Skills:* The ability to identify utility control devices, operate control valves or switches, and assess for related hazards.

5.3.19* Combat a ground cover fire operating as a member of a team, given protective clothing, SCBA if needed, hose lines, extinguishers or hand tools, and an assignment, so that threats to property are reported, threats to personal safety are recognized, retreat is quickly accomplished when warranted, and the assignment is completed.

(A) *Requisite Knowledge:* Types of ground cover fires, parts of ground cover fires, methods to contain or suppress, and safety principles and practices.

(B) *Requisite Skills:* The ability to determine exposure threats based on fire spread potential, protect exposures, construct a fire line or extinguish with hand tools, maintain integrity of established fire lines, and suppress ground cover fires using water.

5.4 Rescue Operations. This duty involves no requirements for Fire Fighter I.

5.5 Prevention, Preparedness, and Maintenance. This duty involves performing activities that reduce the loss of life and property due to fire through hazard identification, inspection, education, and response readiness, according to the following job performance requirements.

5.5.1 Perform a fire safety survey in a private dwelling, given survey forms and procedures, so that fire and life-safety hazards are identified, recommendations for their correction are made to the occupant, and unresolved issues are referred to the proper authority.

(A) *Requisite Knowledge:* Organizational policy and procedures, common causes of fire and their prevention, the importance of a fire safety survey and public fire education programs to fire department public relations and the community, and referral procedures.

(B) *Requisite Skills:* The ability to complete forms, recognize hazards, match findings to preapproved recommendations, and effectively communicate findings to occupants or referrals.

5.5.2* Present fire safety information to station visitors or small groups, given prepared materials, so that all information is presented, the information is accurate, and questions are answered or referred.

(A) *Requisite Knowledge:* Parts of informational materials and how to use them, basic presentation skills, and departmental standard operating procedures for giving fire station tours.

(B) *Requisite Skills:* The ability to document presentations and to use prepared materials.

5.5.3 Clean and check ladders, ventilation equipment, self-contained breathing apparatus (SCBA), ropes, salvage equipment, and hand tools, given cleaning tools, cleaning supplies, and an assignment, so that equipment is clean and maintained according to manufacturer's or departmental guidelines, maintenance is recorded, and equipment is placed in a ready state or reported otherwise.

(A) *Requisite Knowledge:* Types of cleaning methods for various tools and equipment, correct use of cleaning solvents, and manufacturer's or departmental guidelines for cleaning equipment and tools.

(B) *Requisite Skills:* The ability to select correct tools for various parts and pieces of equipment, follow guidelines, and complete recording and reporting procedures.

5.5.4 Clean, inspect, and return fire hose to service, given washing equipment, water, detergent, tools, and replacement gaskets, so that damage is noted and corrected, the hose is clean, and the equipment is placed in a ready state for service.

(A) *Requisite Knowledge:* Departmental procedures for noting a defective hose and removing it from service, cleaning methods, and hose rolls and loads.

(B) *Requisite Skills:* The ability to clean different types of hose; operate hose washing and drying equipment; mark defective hose; and replace coupling gaskets, roll hose, and reload hose.

Chapter 6: Fire Fighter II

6.1 General.

6.1.1 For certification at Level II, the Fire Fighter I shall meet the general knowledge requirements in 6.1.1.1, the general skill requirements in 6.1.1.2, and the job performance requirements defined in Sections 6.2 through 6.5 of this standard and the requirements defined in Chapter 5, Competencies for the First Responder at the Operational Level, of NFPA 472, *Standard for Professional Competence of Responders to Hazardous Materials Incidents.*

6.1.1.1 General Knowledge Requirements. Responsibilities of the Fire Fighter II in assuming and transferring command within an incident management system, performing assigned duties in conformance with applicable NFPA and other safety regulations and authority having jurisdiction procedures, and the role of a Fire Fighter II within the organization.

6.1.1.2 General Skill Requirements. The ability to determine the need for command, organize and coordinate an incident management system until command is transferred, and function within an assigned role in the incident management system.

6.2 Fire Department Communications. This duty involves performing activities related to initiating and reporting responses, according to the following job performance requirements.

6.2.1 Complete a basic incident report, given the report forms, guidelines, and information, so that all pertinent information is recorded, the information is accurate, and the report is complete.

(A) *Requisite Knowledge:* Content requirements for basic incident reports, the purpose and usefulness of accurate reports, consequences of inaccurate reports, how to obtain necessary information, and required coding procedures.

(B) *Requisite Skills:* The ability to determine necessary codes, proof reports, and operate fire department computers or other equipment necessary to complete reports.

6.2.2* Communicate the need for team assistance, given fire department communications equipment, standard operating procedures (SOPs), and a team, so that the supervisor is consistently informed of team needs, departmental SOPs are followed, and the assignment is accomplished safely.

(A) *Requisite Knowledge:* SOPs for alarm assignments and fire department radio communication procedures.

(B) *Requisite Skills:* The ability to operate fire department communications equipment.

6.3 Fireground Operations. This duty involves performing activities necessary to insure life safety, fire control, and property conservation, according to the following job performance requirements.

6.3.1* Extinguish an ignitable liquid fire, operating as a member of a team, given an assignment, an attack line, personal protective equipment, a foam proportioning device, a nozzle, foam concentrates, and a water supply, so that the correct type of foam concentrate is selected for the given fuel and conditions, a properly proportioned foam stream is applied to the surface of the fuel to create and maintain a foam blanket, fire is extinguished, reignition is prevented, team protection is maintained with a foam stream, and the hazard is faced until retreat to safe haven is reached.

(A) *Requisite Knowledge:* Methods by which foam prevents or controls a hazard; principles by which foam is generated; causes for poor foam generation and corrective measures; difference between hydrocarbon and polar solvent fuels and the concentrates that work on each; the characteristics, uses, and limitations of fire fighting foams; the advantages and disadvantages of using fog nozzles versus foam nozzles for foam application; foam stream application techniques; hazards associated with foam usage; and methods to reduce or avoid hazards.

(B) *Requisite Skills:* The ability to prepare a foam concentrate supply for use, assemble foam stream components, master various foam application techniques, and approach and retreat from spills as part of a coordinated team.

Chapter 6: Fire Fighter II (continued)

6.3.2* Coordinate an interior attack line for team's accomplishment of an assignment in a structure fire, given attack lines, personnel, personal protective equipment, and tools, so that crew integrity is established; attack techniques are selected for the given level of the fire (for example, attic, grade level, upper levels, or basement); attack techniques are communicated to the attack teams; constant team coordination is maintained; fire growth and development is continuously evaluated; search, rescue, and ventilation requirements are communicated or managed; hazards are reported to the attack teams; and incident command is apprised of changing conditions.

(A) *Requisite Knowledge:* Selection of the nozzle and hose for fire attack given different fire situations; selection of adapters and appliances to be used for specific fire ground situations; dangerous building conditions created by fire and fire suppression activities; indicators of building collapse; the effects of fire and fire suppression activities on wood, masonry (brick, block, stone), cast iron, steel, reinforced concrete, gypsum wall board, glass, and plaster on lath; search and rescue and ventilation procedures; indicators of structural instability; suppression approaches and practices for various types of structural fires; and the association between specific tools and special forcible entry needs.

(B) *Requisite Skills:* The ability to assemble a team, choose attack techniques for various levels of a fire (e.g., attic, grade level, upper levels, or basement), evaluate and forecast a fire's growth and development, select tools for forcible entry, incorporate search and rescue procedures and ventilation procedures in the completion of the attack team efforts, and determine developing hazardous building or fire conditions.

6.3.3* Control a flammable gas cylinder fire operating as a member of a team, given an assignment, a cylinder outside of a structure, an attack line, personal protective equipment, and tools, so that crew integrity is maintained, contents are identified, safe havens are identified prior to advancing, open valves are closed, flames are not extinguished unless the leaking gas is eliminated, the cylinder is cooled, cylinder integrity is evaluated, hazardous conditions are recognized and acted upon, and the cylinder is faced during approach and retreat.

(A) *Requisite Knowledge:* Characteristics of pressurized flammable gases, elements of a gas cylinder, effects of heat and pressure on closed cylinders, boiling liquid expanding vapor explosion (BLEVE) signs and effects, methods for identifying contents, how to identify safe havens before approaching flammable gas cylinder fires, water stream usage and demands for pressurized cylinder fires, what to do if the fire is prematurely extinguished, valve types and their operation, alternative actions related to various hazards and when to retreat.

(B) *Requisite Skills:* The ability to execute effective advances and retreats, apply various techniques for water application, assess cylinder integrity and changing cylinder conditions, operate control valves, choose effective procedures when conditions change.

6.3.4* Protect evidence of fire cause and origin, given a flashlight and overhaul tools, so that the evidence is noted and protected from further disturbance until investigators can arrive on the scene.

(A) *Requisite Knowledge:* Methods to assess origin and cause; types of evidence; means to protect various types of evidence; the role and relationship of Fire Fighter IIs, criminal investigators, and insurance investigators in fire investigations; and the effects and problems associated with removing property or evidence from the scene.

(B) *Requisite Skills:* The ability to locate the fire's origin area, recognize possible causes, and protect the evidence.

6.4 Rescue Operations. This duty involves performing activities related to accessing and disentangling victims from motor vehicle accidents and helping special rescue teams, according to the following job performance requirements.

6.4.1* Extricate a victim entrapped in a motor vehicle as part of a team, given stabilization and extrication tools, so that the vehicle is stabilized, the victim disentangled without further injury, and hazards are managed.

(A) *Requisite Knowledge:* The fire department's role at a vehicle accident, points of strength and weakness in auto body construction, dangers associated with vehicle components and systems, the uses and limitations of hand and power extrication equipment, and safety procedures when using various types of extrication equipment.

(B) *Requisite Skills:* The ability to operate hand and power tools used for forcible entry and rescue as designed; use cribbing and shoring material; and choose and apply appropriate techniques for moving or removing vehicle roofs, doors, windshields, windows, steering wheels or columns, and the dashboard.

6.4.2* Assist rescue operation teams, given standard operating procedures, necessary rescue equipment, and an assignment, so that procedures are followed, rescue items are recognized and retrieved in the time as prescribed by the AHJ, and the assignment is completed.

(A) *Requisite Knowledge:* The fire fighter's role at a special rescue operation, the hazards associated with special rescue operations, types and uses for rescue tools, and rescue practices and goals.

(B) *Requisite Skills:* The ability to identify and retrieve various types of rescue tools, establish public barriers, and assist rescue teams as a member of the team when assigned.

6.5 Prevention, Preparedness, and Maintenance. This duty involves performing activities related to reducing the loss of life and property due to fire through hazard identification, inspection, and response readiness, according to the following job performance requirements.

6.5.1* Prepare a preincident survey, given forms, necessary tools, and an assignment, so that all required occupancy information is recorded, items of concern are noted, and accurate sketches or diagrams are prepared.

(A) *Requisite Knowledge:* The sources of water supply for fire protection; the fundamentals of fire suppression and detection systems; common symbols used in diagramming construction features, utilities, hazards, and fire protection systems; departmental requirements for a preincident survey and form completion; and the importance of accurate diagrams.

(B) *Requisite Skills:* The ability to identify the components of fire suppression and detection systems; sketch the site, buildings, and special features; detect hazards and special considerations to include in the preincident sketch; and complete all related departmental forms.

6.5.2 Maintain power plants, power tools, and lighting equipment, given tools and manufacturers' instructions, so that equipment is clean and maintained according to manufacturer and departmental guidelines, maintenance is recorded, and equipment is placed in a ready state or reported otherwise.

(A) *Requisite Knowledge:* Types of cleaning methods, correct use of cleaning solvents, manufacturer and departmental guidelines for maintaining equipment and its documentation, and problem-reporting practices.

(B) *Requisite Skills:* The ability to select correct tools; follow guidelines; complete recording and reporting procedures; and operate power plants, power tools, and lighting equipment.

6.5.3 Perform an annual service test on fire hose, given a pump, a marking device, pressure gauges, a timer, record sheets, and related equipment, so that procedures are followed, the condition of the hose is evaluated, any damaged hose is removed from service, and the results are recorded.

(A)* *Requisite Knowledge:* Procedures for safely conducting hose service testing, indicators that dictate any hose be removed from service, and recording procedures for hose test results.

(B) *Requisite Skills:* The ability to operate hose testing equipment and nozzles and to record results.

6.5.4* Test the operability of and flow from a fire hydrant, given a Pitot tube, pressure gauge, and other necessary tools, so that the readiness of the hydrant is assured and the flow of water from the hydrant can be calculated and recorded.

(A) *Requisite Knowledge:* How water flow is reduced by hydrant obstructions; direction of hydrant outlets to suitability of use; the effect of mechanical damage, rust, corrosion, failure to open the hydrant fully, and susceptibility to freezing; and the meaning of the terms *static, residual,* and *flow pressure*.

(B) *Requisite Skills:* The ability to operate a pressurized hydrant, use a Pitot tube and pressure gauges, detect damage, and record results of test.

NFPA 1001, *Standard for Fire Fighter Professional Qualifications*, 2002 Edition, Correlation Guide

NFPA 1001, *Standard for Fire Fighter Professional Qualifications*, 2002 Edition	Corresponding Textbook Chapter

Chapter 4: Entrance Requirements

4.1	General Objective
4.2	General Objective
4.3	24

Chapter 5: Fire Fighter I

5.1	General Objective
5.1.1	1
5.1.1.1	1, 9
5.1.1.2	1, 2, 9
5.2	3
5.2.1	3
5.2.1A	3
5.2.1B	3
5.2.2	3
5.2.2A	3
5.2.2B	3
5.2.3	3
5.2.3A	3
5.2.3B	3
5.3	General Objective
5.3.1	2
5.3.1A	2
5.3.1B	2, 17
5.3.2	10
5.3.2A	10
5.3.2B	10
5.3.3	10
5.3.3A	10
5.3.3B	10
5.3.4	8, 11
5.3.4A	11
5.3.4B	11
5.3.5	17
5.3.5A	17

NFPA 1001, *Standard for Fire Fighter Professional Qualifications*, 2002 Edition	Corresponding Textbook Chapter
5.3.5B	17
5.3.6	12
5.3.6A	12
5.3.6B	12
5.3.7	21
5.3.7A	21, 25
5.3.7B	21
5.3.8	21, 37
5.3.8A	21, 37
5.3.8B	21, 37
5.3.9	13
5.3.9A	13
5.3.9B	13
5.3.10	21
5.3.10A	5, 16, 18, 21
5.3.10B	16, 21
5.3.11	14
5.3.11A	5, 14
5.3.11B	14
5.3.12	14
5.3.12A	5, 6, 14
5.3.12B	14
5.3.13	18, 37
5.3.13A	8, 18, 37
5.3.13B	18, 37
5.3.14	18
5.3.14A	11, 18, 36
5.3.14B	18, 36, 37
5.3.15	15, 16
5.3.15A	15, 16
5.3.15B	15, 16
5.3.16	7
5.3.16A	7
5.3.16B	7
5.3.17	18

Appendix B

NFPA 1001, *Standard for Fire Fighter Professional Qualifications,* 2002 Edition	Corresponding Textbook Chapter
5.3.17A	18
5.3.17B	18
5.3.18	10
5.3.18A	10
5.3.18B	10
5.3.19	20
5.3.19A	20
5.3.19B	20
5.4	General Objective
5.5	35
5.5.1	35
5.5.1A	1, 35
5.5.1B	35
5.5.2	35
5.5.2A	35
5.5.2B	35
5.5.3	2, 8, 9, 12
5.5.3A	8
5.5.3B	8
5.5.4	16
5.5.4A	16
5.5.4B	16

Chapter 6: Fire Fighter II

NFPA 1001, *Standard for Fire Fighter Professional Qualifications,* 2002 Edition	Corresponding Textbook Chapter
6.1	General Objective
6.1.1	1, 2
6.1.1.1	1, 4
6.1.1.2	4
6.2	3
6.2.1	3
6.2.1A	3
6.2.1B	3
6.2.2	1, 3
6.2.2A	3
6.2.2B	3
6.3	21
6.3.1	16
6.3.1A	16
6.3.1B	16
6.3.2	14, 21
6.3.2A	11, 14, 21
6.3.2B	14, 21
6.3.3	21
6.3.3A	21
6.3.3B	21
6.3.4	37
6.3.4A	37
6.3.4B	37
6.4	25, 26
6.4.1	25
6.4.1A	25
6.4.1B	25
6.4.2	8, 9, 26
6.4.2A	8, 26
6.4.2B	8, 26
6.5	22, 35
6.5.1	22
6.5.1A	22, 36
6.5.1B	22
6.5.2	8, 18
6.5.2A	8, 18
6.5.2B	8, 18
6.5.3	16
6.5.3A	16
6.5.3B	16
6.5.4	15
6.5.4A	15
6.5.4B	15

Glossary

9-1-1 dispatcher/telecommunicator From the communications center, the dispatcher takes the calls from the public, sends appropriate units to the scene, assists callers with treatment instructions until the medic unit arrives, and assists the incident commander with needed resources.

A posts Vertical support members that form the sides of the windshield.

A tool A cutting tool with a pry bar built into the cutting part of the tool.

Abrasion Loss of skin as a result of a body part being rubbed.or scraped across a rough or hard surface.

Absorption The process of applying a material that will soak up and hold a hazardous material in a sponge-like manner, for collection and disposal.

Absorption (body) The process by which hazardous materials travel through body tissues until they reach the bloodstream.

Accelerants Materials, usually flammable liquids, used to initiate or increase the spread of fire.

Accelerator A device that accelerates the removal of the air from a dry-pipe or preaction sprinkler system.

Accidental Fire cause classification that includes fires with a proven cause that does not involve a deliberate human act.

Accordion hose load A method of loading hose on a vehicle whose appearance resembles accordion sections. It is achieved by standing the hose on its edge, then placing the next fold on its edge and so on.

Accountability system A method of accounting for all personnel at an emergency incident and ensuring that only personnel with specific assignments are permitted to work within the various zones.

Acid Materials within pH values less than 7.

Acquired immunodeficiency syndrome (AIDS) An immune disorder caused by HIV, resulting in an increased vulnerability to infections and to certain rare cancers.

Activity logging system Device that keeps a detailed record of every incident and activity that occurs.

Acute health effects Adverse health effects caused by relatively short exposure periods that produce observable conditions such as eye irritation, coughing, dizziness, skin burns, etc.

Adaptor A device that joins hose couplings of the same type, such as male to male or female to female.

Adjustable gallonage fog nozzle A nozzle that allows the operator to select a desired flow from several settings.

Adsorption The process of adding a material such as sand or activated carbon to a contaminant, which then adheres to the surface of the material for collection.

Advanced life support (ALS) Advanced lifesaving procedures, such as cardiac monitoring, administration of IV fluids and medications, and use of advanced airway adjuncts.

Adz The prying part of the Halligan tool.

Adze A hand tool constructed of a thin arched blade set at right angles to the handle. It is used to chop brush for clearing a fire line, or to mop up a wildland fire.

Aeration Inducting air into the foam solution, which expands and finishes the foam.

Aerial fuel Fuels more than six feet off the ground, usually part of or attached to trees.

Aerial ladder A power-operated ladder permanently mounted on a piece of apparatus.

Agroterrorism The intentional act of using chemical or biologic agents against the agricultural industry or food supply.

Air bill The shipping papers on an airplane.

Air cylinder The component of the SCBA that stores the compressed air supply.

Air line The hose through which air flows, either within an SCBA or from an outside source to a supplied air respirator.

Air management The way in which an individual utilizes a limited air supply to be sure it will last long enough to enter a hazardous area, accomplish needed tasks, and return safely.

Air sampling detector A system that captures a sample of air from a room or enclosed space and passes it through a smoke detection or gas analysis device.

Aircraft/crash rescue fire fighter Individuals with specialized training in aircraft fires, extrication, and extinguishing agents; they wear special types of turnout gear and respond in fire apparatus that protect them from high-temperature fires.

Air-purifying respirators (APRs) Devices worn to filter particulates and contaminants from the air.

Airway The passages from the openings of the mouth and nose to the air sacs in the lungs through which air enters and leaves the lungs.

Airway obstruction Partial or complete obstruction of the respiratory passages resulting from blockage by food, small objects, or vomitus.

Alarm initiation device An automatic or manually operated device in a fire alarm system that, when activated, causes the system to indicate an alarm condition.

Alarm notification device An audible and/or visual device in a fire alarm system that makes occupants or other persons aware of an alarm condition.

Alarm valve This valve signals an alarm when a sprinkler head is activated and prevents nuisance alarms caused by pressure variations.

Alcohol resistant foam A synthetic foam that resists mixing with polar solvents (alcohol type liquids).

Alpha particles A type of radiation that quickly loses energy and can travel only 1″ to 2″ from its source. Clothing or a sheet of paper can stop this type of energy. This particle is not dangerous to plants, animals, or people unless the alpha-emitting substance has entered the body.

Alternative-powered vehicles Vehicles that are powered by compressed natural gas.

Alveoli The air sacs of the lungs where the exchange of oxygen and carbon dioxide takes place.

Ammonium phosphate An extinguishing agent used in dry chemical fire extinguishers that can be used on Class A, B, and C fires.

Anaphylactic shock Severe shock caused by an allergic reaction to food, medicine, or insect stings.

ANFO An explosive material containing ammonium nitrate fertilizer and fuel oil.

Annealed The process of forming standard glass.

Anthrax An infectious disease spread by the bacteria *Bacillus anthracis*; typically found around farms, infecting livestock.

Aqueous film-forming foam (AFFF) A water-based extinguishing agent used on Class B fires that forms a foam layer over the liquid and stops the production of flammable vapors.

Arched roof A rounded roof usually associated with a bow-truss.

Area of origin The room or area where a fire began.

Arson The malicious burning of one's own or another's property with a criminal intent.

Arsonist A person who deliberately sets a fire to destroy property with criminal intent.

Arterial bleeding Serious bleeding from an artery in which blood frequently pulses or spurts from an open wound.

Asphyxiant A material that causes the victim to suffocate.

Assistant or division chief Midlevel chief who often has a functional area of responsibility, such as training, and answers directly to the fire chief.

Atrium Either of the two upper chambers of the heart.

Attack engine The engine from which the attack lines have been pulled.

Attack hose (attack line) The hose that delivers water from a fire pump to the fire. Attack hoses range in size from 1″ to 2½″.

Automatic adjusting fog nozzle A nozzle that can deliver a wide range of water stream flows. It operates by an internal spring loaded piston.

Automatic Location Identification (ALI) An enhanced 9-1-1 service feature that displays where the call originated or where the phone service is billed.

Automatic Number Identification (ANI) An enhanced 9-1-1 service feature that shows the calling party's phone number on the display screen at the telecommunicator's terminal.

Automatic sprinkler heads The working ends of a sprinkler system. They serve to activate the system and to apply water to the fire.

Automatic sprinkler system A system of pipes filled with water under pressure that discharges water immediately when a sprinkler head opens.

Auxiliary system A fire alarm system that sounds an alarm in the building and transmits a signal to the fire department via a public alarm box system.

Avulsion An injury in which a piece of skin is either torn completely loose from all of its attachments or is left hanging by a flap.

Awareness level Provides training that gives first responders the ability to recognize a potential hazardous materials emergency, to isolate the area, and to call for assistance; awareness-level responders take protective actions.

Awareness level (TRI) The first level of rescue training provided to all responders with an emphasis on recognizing the hazards, securing the scene, and calling for appropriate assistance. There is no actual use of rescue skills.

Awning windows Windows that have one large or two medium panels operated by a hand crank from the corner of the window.

B posts Vertical support members located between the front and rear doors.

Backdraft The sudden explosive ignition of fire gases when oxygen is introduced into a superheated space previously deprived of oxygen.

Backfiring A planned operation to remove fuel from a wildland fire by burning out large selected areas. This should be done only under the close supervision of experienced, authorized fire officers.

Backpack The harness of the SCBA that supports the components worn by a fire fighter.

Backpack pump extinguisher A portable fire extinguisher consisting of a four-gallon to eight-gallon water tank worn on the user's back and featuring a hand-powered piston pump for discharging the water.

Back-up personnel Individuals who remove or rescue those working in the hot zone in the event of an emergency.

Ball valves Valves used on nozzles, gated wyes, and engine discharge gates. Made up of a ball with a hole in the middle of the ball.

Balloon-frame construction An older type of wood frame construction in which the wall studs extend vertically from the basement of a structure to the roof without any fire stops.

Bam-bam tool A tool with a case-hardened screw, which is secured in the keyway of a lock, to remove the keyway from the lock.

Bangor ladder A ladder equipped with tormentor poles or staypoles that stabilize the ladder during raising and lowering operations.

Banked Covering a fire to ensure low burning.

Bankshot (bank down) method A method that applies the stream onto a nearby object, such as a wall, instead of directly aiming at the fire.

Base Materials with a pH value more than 7.

Base station Radios are permanently mounted in a building, such as a fire station, communications center, or remote transmitter site.

Basic life support (BLS) Noninvasive emergency lifesaving care that is used to treat airway obstruction, respiratory arrest, or cardiac arrest.

Batch mixing Pouring foam concentrate directly into the fire apparatus water tank, mixing a large amount at one time.

Battalion chief Usually the first level of fire chief; also called district chief. These chiefs are often in charge of running calls and supervising multiple stations or districts within a city. A battalion chief is usually the officer in charge of a single-alarm working fire.

Battering ram A tool made of hardened steel with handles on the sides used to force doors and to breach walls. Larger versions are used by up to four people; smaller versions are made for one or two people.

Beam One of the two main structural pieces running the entire length of each ladder or ladder section. The beams support the rungs.

Beam detector A smoke detection device that projects a narrow beam of light across a large open area from a sending unit to a receiving unit. When the beam is interrupted by smoke, the receiver detects a reduction in light transmission and activates the fire alarm.

Bearing wall A wall that is designed to support the weight of a floor or roof.

Bed section The lowest and widest section of an extension ladder. The fly sections of the ladder extend from the bed section.

Bends Knots used to join two ropes together.

Beta particles A type of radiation that is capable of traveling 10′ to 15′. Heavier materials such as metal, plastic, and glass can stop this type of energy.

Bight A U-shape created by bending a rope with the two sides parallel.

Bill of lading Shipping papers for roads and highways.

Bimetallic strip A device with components made from two distinct metals that respond differently to heat. When heated, the metals will bend or change shape.

Biologic agents Disease-causing bacteria, viruses, and other agents that attack the human body.

Bite A small opening made to enable better tool access in forcible entry.

Black An area that has already been burned.

Blanking (blinding) The absolute closure of a pipe, line, or duct by the fastening of a solid plate that completely covers the base and is capable of withstanding the maximum pressure of the pipe, line, or duct with no leakage beyond the plate.

Blister agents Chemicals that cause the skin to blister.

Block creel construction Rope constructed without knots or splices in the yarns, ply yarns, strands, braids, or rope.

Blood agents Materials that, when absorbed by the body, interfere with the transfer of oxygen from the blood to the cells.

Blood pressure The pressure of the circulating blood against the walls of the arteries.

Body substance isolation (BSI) An infection control concept that treats all bodily fluids as potentially infectious.

Boiling point The temperature at which the vapor pressure of a liquid equals the surrounding atmospheric pressure, and the liquid changes to a gas.

Boiling-liquid, expanding-vapor explosion (BLEVE) An explosion that can occur when a tank containing a volatile liquid is heated.

Bolt cutter A cutting tool used to cut through thick metal objects, such as bolts, locks, and wire fences.

Booster hose (booster lines) A rigid hose that is 3/4" or 1" in diameter. This hose only delivers 30 to 60 gpm, but can do so at high pressures. It is used for small outdoor fires.
Bowstring trusses Trusses that are curved on the top and straight on the bottom.
Box-end wrench A hand tool used to tighten or loosen bolts. The end is enclosed, as opposed to an open-end wrench. Each wrench is a specific size and most have ratchets for easier use.
Brachial artery The major vessel in the upper extremity that supplies blood to the arm.
Braided rope Rope constructed by intertwining strands in the same way that hair is braided.
Branch A supervisory level established to manage the span of control above the division, group, or sector level; usually applied to operations or logistics functions.
Branch director Officer in charge of all resources operating within a specified branch, responsible to the next higher level in the incident organization (either a Section Chief or the Incident Commander).
Breakaway nozzle A nozzle with a tip that can be separated from the shut-off valve.
Bresnan distributor nozzle A device that can be placed in confined spaces. The nozzle spins, spreading water over a large area.
Bronchi The two main branches of the windpipe that lead into the right and left lungs. Within the lungs, they branch into smaller airways.
Bruise Injury caused by a blunt object striking the body and crushing the tissue beneath the skin; also called a contusion.
Buddy system A system in which two fire fighters always work as a team for safety purposes.
Bulk storage containers Large volume containers that have an internal volume greater than 119 gallons for liquids, greater than 882 pounds for solids, and a capacity of greater than 882 pounds for gases.
Bulkhead The wall that separates the engine compartment from the passenger compartment, formally known as a firewall.
Bungs One or more small openings in closed-head drums.
Bunker coat The protective coat worn by a fire fighter for interior structural firefighting; also called a turnout coat.
Bunker pants The protective trousers worn by a fire fighter for interior structural firefighting; also called turnout pants.
Butt Often called the heel or base, the butt is the end of the ladder that is placed against the ground when the ladder is raised.
Butt plate (footpad) An alternative to a simple butt spur; a swiveling plate with both a spur and a cleat or pad that is attached to the butt of the ladder.
Butt spurs The metal spikes attached to the butt of a ladder. The spurs help prevent the butt from slipping out of position.
Butterfly valves Valves that are found on the large pump intake valve where the hard or soft suction hose connects.
C posts Vertical support members located behind the rear doors.
Call box System of telephones connected by phone lines, radio equipment, or cellular technology to communicate with a communications center or fire department.
Capillaries The smallest blood vessels that connect small arteries and small veins. Capillary walls serve as the membrane to exchange oxygen and carbon dioxide.
Capillary bleeding Bleeding from the capillaries in which blood oozes from the open wound.
Captain The second rank of promotion, between the lieutenant and battalion chief. Captains are responsible for a fire company and for coordinating the activities of that company among the other shifts.
Carabiner A piece of metal hardware used extensively in rope rescue operations. It is generally an oval-shaped device with a spring-loaded clip that can be used for connecting together pieces of rope, webbing, or other hardware. Also referred to as a "snap link."
Carbon dioxide (CO_2) extinguisher A fire extinguisher that uses carbon dioxide gas as the extinguishing agent.
Carbon dioxide extinguishing system A system designed to protect a single room or series of rooms by flooding the area with carbon dioxide.
Carbon monoxide A toxic gas produced through incomplete combustion.
Carboys A glass, plastic, or steel container, ranging in volume from 5 gallons to 15 gallons.
Carcinogen A cancer-causing agent.
Cardiac arrest A sudden ceasing of heart function.
Cardiogenic shock Shock resulting from inadequate functioning of the heart.
Cardiopulmonary resuscitation (CPR) The artificial circulation of the blood and movement of air into and out of the lungs in a pulseless, nonbreathing victim.
Carotid artery The major artery that supplies blood to the head and brain.
Carpenter's handsaw A saw designed for cutting wood.
Carryall A piece of heavy canvas with handles, which can be used to tote debris, ash, embers, and burning materials out of a structure.
Cartridge/cylinder fire extinguisher A fire extinguisher that has the expellant gas in a separate container from the extinguishing agent storage container. The storage container is pressurized by a mechanical action that releases the expellant gas.
Cascade system An apparatus consisting of multiple tanks used to store compressed air and fill SCBA cylinders.
Case-hardened steel A process that uses carbon and nitrogen to harden the outer core of a steel component, while the inner core remains soft. Case-hardened steel can be cut only with specialized tools.
Casement windows Windows in a steel or wood frame that open away from the building via a crank mechanism.
Ceiling hook A tool with a long wooden or fiberglass pole that has a metal point with a spur at right angles at one end. It can be used to probe ceilings and pull down plaster lath material.
Cellar nozzles Nozzles used to fight fires in cellars and other inaccessible places. These devices work by spreading water in a wide pattern as the nozzle is lowered by a hole into the cellar.
Central station An off-premises facility that monitors alarm systems and is responsible for notifying the fire department of an alarm. These facilities may be geographically located some distance from the protected building(s).
Chain of command A rank structure, from the fire fighter through the fire chief, for managing a fire department and fireground operations.
Chain of custody A legal term used to describe the paperwork or documentation describing the movement, storage, and custody of evidence, such as the record of possession of a gas can from a fire scene.
Chain saw A power saw that uses the rotating movement of a chain equipped with sharpened cutting edges. Typically used to cut through wood.
Chase Open space within walls for wires and pipes.
Chemical change The ability of a chemical to undergo a change in its chemical make-up, usually with a release of some form of energy.
Chemical energy Energy created or released by the combination or decomposition of chemical compounds.
Chemical Transportation Emergency Center (CHEMTREC) A national call center for basic chemical information, established by the Chemical Manufacturer's Association.
Chemical-pellet sprinkler head A sprinkler head activated by a chemical pellet that liquefies at a preset temperature.
Chemical-resistant materials Chemical-protective clothing that can maintain its integrity and protection qualities when it comes into contact with a hazardous material. They also resist penetration, permeation, and degradation.
Chief of the department The top position in the fire department. The fire chief has ultimate responsibility for the fire department and usually answers directly to the mayor or designated public official.
Chief's trumpet An obsolete amplification device that enabled a chief officer to give orders to fire fighters during an emergency; precursor to a bullhorn and portable radios.
Chisel A metal tool with one sharpened end used to break apart material in conjunction with a hammer, mallet, or sledgehammer.

Chlorine A yellowish gas that is about 2.5 times heavier than air and slightly water-soluble. It has many industrial uses but also damages the lungs when inhaled (a choking agent).
Choking agent A chemical designed to inhibit breathing, and typically intended to incapacitate rather than kill.
Chronic health hazard An adverse health effect occurring after a long-term exposure to a substance.
Churning Recirculation of exhausted air that is drawn back into a negative-pressure fan in a circular motion.
Circulatory system The heart and blood vessels, which together are responsible for the continuous flow of blood throughout the body.
Circumstantial evidence The means by which alleged facts are proven by deduction or inference from other facts that were observed first hand.
Clapper valve A mechanical device installed within a piping system that allows water to flow in only one direction.
Class I standpipe A standpipe system designed for use by fire department personnel only. Each outlet should have a valve to control the flow of water and a $2^1/_2''$ male coupling for fire hose.
Class II standpipe A standpipe system designed for use by occupants of a building only. Each outlet is generally equipped with a length of $1^1/_2''$ single-jacket hose and a nozzle, preconnected to the system.
Class III standpipe A combination system that has features of both Class I and Class II standpipes.
Class A fires Fires involving ordinary combustible materials, such as wood, cloth, paper, rubber, and many plastics.
Class A foam A firefighting foam for use on a particular type of fire. Class A refers to fires involving ordinary combustible materials such as wood, paper, and textiles.
Class B fires Fires involving flammable and combustible liquids, oils, greases, tars, oil-based paints, lacquers, and flammable gases.
Class B foam A firefighting foam for a particular type of fire. Class B refers to fires involving flammable liquids.
Class C fires Fires that involve energized electrical equipment where the electrical conductivity of the extinguishing media is of importance.
Class D fires Fires involving combustible metals such as magnesium, titanium, zirconium, sodium, and potassium.
Class K fires Fires involving combustible cooking media such as vegetable oils, animal oils, and fats.
Claw The forked end of a tool.
Claw bar A tool with a pointed claw-hook on one end and a forked- or flat-chisel pry on the other end. It is often used for forcible entry.
Clean agent Volatile or gaseous fire extinguishing agent that does not leave a residue when it evaporates. Also known as halogenated agents.
Clemens hook A multipurpose tool that can be used for several forcible entry and ventilation applications because of its unique head design.
Closed wound Injury in which soft-tissue damage occurs beneath the skin even though there is no break in the surface of the skin.
Closed-circuit breathing apparatus SCBA designed to recycle the user's exhaled air. The system removes carbon dioxide and generates fresh oxygen.
Closet hook A type of pike pole intended for use in tight spaces, commonly 2' to 4' in length.
Cockloft The concealed space between the top floor ceiling and the roof of a building.
Coded system A fire alarm system design that divides a building or facility into zones and has audible notification devices that can be used to identify the area where an alarm originated.
Cold zone A safe area at a hazardous materials incident for those agencies involved in the operations; the incident commander, command post, EMS providers, and other support functions necessary to control the incident should be located in the cold zone. This zone may also be referred to as the clean zone or the support zone.
Cold zone (TRI) A safe area for those agencies involved in the operations; the IC, command post, EMS providers, and other support functions necessary to control the incident should be located in the cold zone.
Combination attack A type of attack employing both the direct and indirect attack methods.
Combination EMS system In this system, the fire department provides medical first response, while another agency transports the patient to the hospital emergency room.
Combination ladder A ladder that converts from a straight ladder to a step ladder configuration (A-frame) or from an extension ladder to a step ladder configuration.
Combustibility Determines whether or not a material will burn.
Combustion A chemical process of oxidation that occurs at a rate fast enough to produce heat and usually light in the form of either a glow or flames.
Come along A hand-operated tool used for dragging or lifting heavy objects. Sometimes known as lever blocks.
Command The first component of the IMS system. This is the only position in the IMS system that must always be staffed.
Command post The location at the scene of an emergency where the Incident Commander is located and where command, coordination, control, and communications are centralized.
Command staff Staff positions established to assume responsibility for key activities in the incident management system; individuals at this level report directly to the IC. Command staff include the Safety Officer, Public Information Officer, and Liaison Officer.
Communications center Facility that receives emergency or non-emergency reports from citizens. Many communications centers are responsible for dispatching fire department units as well.
Company officer Usually a lieutenant or captain in charge of a team of fire fighters, both on scene and at the station. The company officer is responsible for firefighting strategy, safety of personnel, and the overall activities of the fire fighters on their apparatus.
Competent ignition source A competent ignition source is one that can ignite a fuel under the existing conditions at the time of the fire. It must have sufficient heat and be in proximity to the fuel for sufficient time to ignite the fuel.
Compressed air foam systems (CAFS) Combines foam concentrate, water, and compressed air to produce a foam that can stick to both vegetation and structures.
Compressor A mechanical device that increases the pressure and decreases the volume of atmospheric air, used to refill SCBA cylinders.
Computer-Aided Dispatch (CAD) Computer-based, automated systems used by telecommunicators to obtain and assess dispatch information. The system recommends the type of response required.
Conduction Heat transfer to another body or within a body by direct contact.
Confined space A space with limited or restricted access that is not meant for continuous occupancy, such as a manhole, well, or tank.
Confinement The process of keeping a hazardous material on the site or within the immediate area of the release.
Conflagration A large fire, often involving multiple structures.
Congestive heart failure (CHF) Heart disease characterized by breathlessness, fluid retention in the lungs, and generalized swelling of the body.
Consensus document A code document developed through agreement between people representing different organizations and interests. NFPA codes and standards are consensus documents.
Consist The list of every car on a train.
Container Any vessel or receptacle that holds material, including storage vessels, pipelines, and packaging.
Containment Actions relating to stopping a hazardous materials container from leaking, such as patching, plugging, or righting the container.
Contaminated A term used to describe evidence that may have been altered from its original state.
Contamination The process of transferring a hazardous material from its source to people, animals, the environment, or equipment, all of which may act as carriers.
Control zones Areas at a hazardous materials incident that are designated as hot, warm, or cold, based upon safety and the degree of hazard.
Convection Heat transfer by circulation within a medium such as a gas or a liquid.
Conventional vehicle A vehicle that uses an internal combustion engine and is fueled with either gasoline or diesel fuel.
Convulsant Chemicals capable of causing convulsions or seizures when absorbed by the body.

Coping saw A saw designed to cut curves in wood.
Corrosivity The ability of a material to cause damage (on contact) to skin, eyes, or other parts on the body.
Coupling One set or a pair of connection devices attached to a fire hose that allows the hose to be interconnected to additional lengths of hose.
Crew An organized group of fire fighters under the leadership of a company officer, crew leader, or other designated official.
Cribbing Short lengths of robust and usually hardwood timber (4″ × 4″ and 18″ to 24″ long) that are used in a variety of ways, usually in the stabilization of vehicles.
Critical incident stress debriefing (CISD) A confidential group discussion among those who served at a traumatic incident to address emotional, psychological, and stressful issues; usually occurs within 24 to 72 hours of the incident.
Critical Incident Stress Management (CISM) A system to help personnel deal with critical incident stress in a positive manner; its aim is to promote long-term mental and emotional health after a critical incident.
Cross-zoned system A fire alarm system that requires activation of two separate detection devices before initiating an alarm condition. If a single detection device is activated, the alarm control panel will usually show a problem or trouble condition.
Crowbar A straight bar made of steel or iron with a forked-like chisel on the working end suitable for performing forcible entry.
Cryogenic liquids (cryogens) Gaseous substances that have been chilled to the point where they have liquefied; a liquid having a boiling point lower than –150°F (–101°C) at 14.7 psia (an absolute pressure of 101 kPa).
Curtain walls Nonbearing walls used to separate the inside and outside of the building, but not part of the support structure for the building.
Curved roofs Roofs that have a curved shape.
Cutting tools Tools that are designed to cut into metal or wood.
Cutting torch A torch that produces a high temperature flame capable of heating metal to its melting point, thereby cutting through an object. Because of the high temperatures (5,700ºF) that these torches produce, the operator must be specially trained before using this tool.
Cyanide A highly toxic chemical agent that attacks the circulatory system.
Cyberterrorism The intentional act of electronically attacking government or private computer systems.
Cylinder A portable compressed-gas container.
Cylinder (fire extinguisher) The body of the fire extinguisher where the extinguishing agent is stored.
Cylindrical locks The most common fixed lock in use today. The locks and handles are placed into predrilled holes in the doors. One side of the door will usually have a key-in-the-knob lock, and the other will have a keyway, a button, or some other type of locking/unlocking device.
Damming A process used when liquid is flowing in a natural channel or depression and its progress can be stopped by blocking the channel.
Dangerous cargo manifest The shipping paper on a marine vessel, generally located in a tube-like container.
Dead bolts Surface or interior mounted locks on or in a door with a bolt that provide additional security.
Dead load The weight of a building; the dead load consists of the weight of all materials of construction incorporated into a building, including but not limited to walls, floors, roofs, ceilings, stairways, built-in partitions, finishes, cladding, and other similarly incorporated architectural and structural items, as well as fixed service equipment, including the weight of cranes.
Deadbolt A secondary locking device used to secure a door in its frame.
Decapitation Occurs when the head is separated from the rest of the body.
Decay phase The phase of fire development where the fire has consumed either the available fuel or oxygen and is starting to die down.
Deck gun Apparatus-mounted master stream device intended to flow large amounts of water directly onto a fire or exposed building.
Decontamination The physical or chemical process of removing any form of contaminant from a person, an object, or the environment.
Decontamination corridor A controlled area between the warm and hot zone through which all persons who entered the hot zone will exit for decontamination.
Decontamination team The team responsible for reducing and preventing the spread of contaminants from persons and equipment used at a hazardous materials incident. They establish the decontamination corridor and conduct all phases of decontamination.
Defend-in-place A strategy where the victims are protected from the fire without relocation.
Defensive attack Exterior fire suppression operations directed at protecting exposures.
Defensive objectives Actions that do not involve the actual stopping of the leak or release of a hazardous material, or contact of responders with the material; these include preventing further injury and controlling or containing the spread of the hazardous material.
Degradation The physical destruction or decomposition of clothing material due to chemical exposure, general use, or ambient conditions (like storage in sunlight).
Dehydration A state in which fluid losses are greater than fluid intake into the body, leading to shock and even death if untreated.
Deluge head A sprinkler head that has no release mechanism; the orifice is always open.
Deluge sprinkler system A sprinkler system in which all sprinkler heads are open. When an initiating device, such as a smoke detector or heat detector, is activated, the deluge valve opens and water discharges from all of the open sprinkler heads simultaneously.
Deluge valve A valve assembly designed to release water into a sprinkler system when an external initiation device is activated.
Demonstrative evidence Term used to describe materials used to demonstrate a theory or explain an event.
Department of Transportation (DOT) The federal Department of Transportation publicizes and enforces rules and regulations that relate to the transportation of many hazardous materials.
Department of Transportation (DOT) marking system A unique system of labels and placards that, in combination with the *North American Emergency Response Guide,* offers guidance for first responders operating at a hazardous materials incident.
Dependent lividity The red or purple color that appears on the parts of the victim's body closest to the ground. It is caused by blood seeping into the tissues on the dependent, or lower, part of the person's body.
Depressions Indentations felt on a kernmantle rope that indicate damage to the interior, or kern, of the rope.
Depth of char The thickness of the layer of a material that has been consumed by a fire. The depth of char on wood can be used to help determine the duration of a fire.
Designated incident facilities Assigned locations where specific functions are always performed.
Dewar containers Containers designed to preserve the temperature of the cold liquid held inside.
Diaphragm A muscular dome that separates the chest from the abdominal cavity. Contraction of the diaphragm and the chest wall muscles brings air into the lungs; relaxation expels air from the lungs.
Diking The placement of materials to form a barrier that will keep a hazardous material in liquid form from entering an area, or hold the material in an area.
Dilution The process of adding some substance, usually water, to a product to weaken its concentration.
Direct attack Firefighting operations involving the application of extinguishing agents directly onto the burning fuel.
Direct attack (wildland and ground fire) A method of fire attack mounted by containing and extinguishing the fire at its burning edge.
Direct evidence Evidence that is reported first hand, such as statements from an eyewitness who saw or heard something.
Direct line Telephone that connects two predetermined points.
Discipline The guidelines that a department sets for fire fighters to work within.
Disinfection The process used to destroy recognized disease-carrying (pathogenic) microorganisms.
Dispatch A summons to fire department units to respond to an emergency. Also known as alerting, dispatch is performed by the telecommunicator at the communications center.
Disposal A two-step removal process for contaminated items that cannot be properly decontaminated; items are bagged and placed in appropriate containers for transport to a hazardous waste facility.
Distributors Relatively small-diameter underground pipes that deliver water to local users within a neighborhood.
Diversion Redirecting spilled or leaking material to an area where it will have less impact.

Division An organizational level within IMS that divides an incident into geographic areas of operational responsibility.
Division of labor Breaking down an incident or task into smaller, more manageable tasks and assigning personnel to complete those tasks.
Division supervisor The officer in charge of all resources operating within a specified division, responsible to the next higher level in the incident organization, and the point-of-contact for the division within the organization.
Do Not Resuscitate (DNR) Written documentation giving orders to medical personnel to not attempt resuscitation in the event of cardiac arrest.
Doff To take off an item of clothing or equipment.
Dogs (also referred to as pawls, ladder locks, and rung locks) A mechanical locking device used to secure the fly section(s) of a ladder after they have been extended.
Don To put on an item of clothing or equipment.
Door An entryway; the primary choice for forcing entry into a vehicle or structure.
Double female connection A hose adaptor that is equipped with two female connectors. It allows two hoses with male couplings to be connected together.
Double jacket hose A hose constructed with two layers of woven fibers.
Double male connection A hose adaptor that is equipped with two male connectors. It allows two hoses with female couplings to be connected together.
Double-action pull-station A manual fire alarm activation device that takes two steps to activate the alarm. The person must push in a flap, lift a cover, or break a piece of glass before activating the alarm.
Double-female adaptor Used to join two male couplings.
Double-hung windows Windows that have two movable sashes that can go up and down.
Double-male adaptor Used to join two female hose couplings.
Double-pane glass Window design that traps air or inert gas between two pieces of glass to help insulate a house.
Drafting sites Location where an engine can draft water directly from a static source.
Dressing A bandage.
Driver/operator Often called an engineer, this person operates the fire apparatus. The driver/operator is responsible for getting the apparatus safely to the scene, setting up, and operating the apparatus on scene.
Drums Barrel-like containers built to DOT specification 5P (1A1).
Dry bulk cargo tanks Tanks designed to carry dry bulk goods such as powders, pellets, fertilizers, or grain; they are generally V-shaped with rounded sides that funnel toward the bottom.
Dry chemical extinguishing system An automatic fire extinguishing system that discharges a dry chemical agent.
Dry chemical fire extinguisher An extinguisher that uses a mixture of finely divided solid particles to extinguish fires. The agent is usually sodium bicarbonate-, potassium bicarbonate-, or ammonium phosphate-based, with additives to provide resistance to packing and moisture absorption and to promote proper flow characteristics.
Dry hydrant An arrangement of pipe that is permanently connected to allow a fire department engine to draft water from a static source.
Dry powder extinguishing agent Extinguishing agent used in putting out Class D fires. The common dry powder extinguishing agents include sodium chloride and graphite-based powders.
Dry powder fire extinguisher A fire extinguisher that uses an extinguishing agent in powder or granular form, designed to extinguish Class D combustible metal fires by crusting, smothering, or heat-transferring means.
Dry sprinkler system A sprinkler system in which the pipes are normally filled with compressed air. When a sprinkler head is activated, it releases the air from the system, which opens a valve so the pipes can fill with water.
Dry-barrel hydrant A type of hydrant used in areas subject to freezing weather. The valve that allows water to flow into the hydrant is located underground and the barrel of the hydrant is normally dry.
Dry-pipe valve The valve assembly on a dry sprinkler system that prevents water from entering the system until the air pressure is released.
Drywall hook A specialized version of a pike pole that can remove drywall more effectively because of its hook design.
Duck-billed lock breaker A tool with a point that can be inserted in the shackles of a padlock. As the point is driven further into the lock, it gets larger and forces the shackles apart until they break.
Dump valve A large valve installed on a mobile water supply apparatus to allow the contents of the tank to be discharged quickly.
Duplex channel A radio system that uses two frequencies per channel. One frequency transmits and the other receives messages. The system uses a repeater site to transmit messages over a greater distance than a simplex system.
Dutchman A term used for a short fold placed in a hose when loading it into the bed. This fold prevents the coupling from turning in the hose bed.
Dynamic rope A rope generally made out of synthetic materials that is designed to be elastic and stretch when loaded. Used often by mountain climbers.
Early suppression fast response (ESFR) sprinkler head A sprinkler head designed to react quickly and suppress a fire in its early stages.
Ecoterrorism Terrorism directed against causes that radical environmentalists think would damage the earth or its creatures.
Edge roller A device used to prevent damage to a rope from jagged edges or friction.
Egress A method of exiting from an area or a building.
Ejectors Electrical fans used in negative-pressure ventilation.
Electrical energy Heat produced by electricity.
Electric-powered vehicles A vehicle propelled by an electric motor that is powered by batteries.
Electrolytes Certain salts and other chemicals that are dissolved in body fluids and cells.
Elevated master stream Nozzle mounted on the end of an aerial device capable of delivering large amounts of water onto a fire or exposed building from an elevated position.
Elevated water storage tower An above-ground water storage tank that is designed to maintain pressure on a water distribution system.
Elevation pressure The amount of pressure created by gravity.
Emergency by-pass mode Operating mode that allows an SCBA to be used even if part of the regulator fails to function properly.
Emergency decontamination The process of getting the bulk of contaminants off a victim without regard for any form of containment; used in potentially life-threatening situations, without the formal establishment of a decontamination corridor.
Emergency incident rehabilitation A function on the emergency scene that cares for the well-being of the fire fighters. It includes physical assessment, revitalization, medical evaluation and treatment, and regular monitoring of vital signs.
Emergency medical services (EMS) company EMS companies may be made up of medical units and first-response vehicles. They respond to and assist in the transport of medical and trauma patients to medical facilities. They often have medications, defibrillators, and paramedics who can stabilize a critical patient.
Emergency medical services (EMS) personnel EMS personnel are those responsible for administering prehospital care to people who are sick and injured. Prehospital calls make up the majority of responses in most fire departments and EMS personnel are cross-trained as fire fighters.
Emergency Medical Technician (EMT)–Basic EMT-Basics account for most of the EMS providers in the country. EMT-Basics have training in basic emergency care skills, including oxygen therapy, bleeding control, cardiopulmonary resuscitation (CPR), automated external defibrillation, use of basic airway devices, and assisting patients with certain medications.
Emergency Medical Technician (EMT)–Intermediate EMT-Intermediate personnel can perform limited procedures that usually fall between those provided by an EMT- Basic and those provided by an EMT-Paramedic, including intravenous (IV) therapy, interpretation of cardiac rhythms, defibrillation, and airway intubation.
Emergency Medical Technician (EMT)–Paramedic An EMT-Paramedic has the highest level of training in EMS, including cardiac monitoring, administering drugs, inserting advanced airways, manual defibrillation, and other advanced assessment and treatment skills.
Emergency Planning and Community Right to Know Act (EPCRA) Requires a business that handles chemicals to report storage type, quantity, and storage methods to the fire department and the local emergency planning committee.
Emergency traffic An urgent message, such as a call for help or evacuation, transmitted over a radio that takes precedence over all normal radio traffic.

Employee assistance program (EAP) Program adopted by many departments for fire fighters to receive confidential help with problems such as substance abuse, stress, depression, or burn out that can affect their work performance.
Emulsification The process of changing the chemical properties of a hazardous material to reduce its harmful effects.
Endothermic reactions Reactions that absorb heat or require heat to be added.
Engine company Engine companies are responsible for securing a water source, deploying hose lines, conducting search-and-rescue operations, and putting water on the fire.
Entrapment A condition in which a victim is trapped by debris, soil, or other material and is unable to extricate himself or herself.
Environmental Protection Agency (EPA) Established in 1970, the federal Environmental Protection Agency ensures safe manufacturing, use, transportation, and disposal of hazardous substances.
Esophagus The tube through which food passes. It starts at the throat and ends at the stomach.
Evacuation signal Warn all personnel to pull back to a safe location.
Exhauster A device that accelerates the removal of the air from a dry-pipe or preaction sprinkler system.
Exothermic reactions Reactions that result in the release of energy in the form of heat.
Expansion ratio A description of the volume increase that occurs when a liquid material changes to a gas.
Exposure Any person or property that may be endangered by flames, smoke, gases, heat, or runoff from a fire.
Exposure (hazardous materials) The process by which people, animals, the environment, and equipment are subjected to or come into contact with a hazardous material.
Exposures (hazardous materials) People, property, structures, or areas of the environment that are subject to influence, damage, or injury from hazardous materials.
Extension Fire that moves into areas not originally involved, including walls, ceilings, and attic spaces; also the movement of fire into uninvolved areas of a structure.
Extension ladder An adjustable-length, multiple-section ladder.
Exterior wall A wall—often made of wood, brick, metal, or masonry—that makes up the outer perimeter of a building. Exterior walls are often load-bearing.
External chest compressions A means of applying artificial circulation by applying rhythmic pressure and relaxation on the lower half of the sternum.
Extinguishing agent Material used to stop the combustion process; extinguishing agents may include liquids, gases, dry chemical compounds, and dry powder compounds.
Extra hazard locations Occupancies where the total amounts of Class A combustibles and Class B flammables are greater than expected in occupancies classed as ordinary (moderate) hazard.
Face piece Component of SCBA that fits over the face.
False alarm The activation of a fire alarm system when there is no fire or emergency condition.
Federal Communications Commission (FCC) United States federal regulatory authority that oversees radio communications.
Femoral artery The principal artery of the thigh.
Film-forming fluoroprotein (FFFP) A water-based extinguishing agent used on Class B fires that forms a foam layer over the liquid and stops the production of flammable vapors.
Finance/Administration Section The command-level section of IMS responsible for all costs and financial aspects of the incident, as well as any legal issues that arise.
Fine decontamination An advanced decontamination process used by hospital and emergency room personnel to clean and to evaluate the extent of chemical exposure.
Fine fuel Fuel that ignites and burns easily, such as dried twigs, leaves, needles, grass, moss, and light brush.
Finger A narrow point of fire caused by a shift in wind or a change in topography.
Fire alarm annunciator panel Part of the fire alarm system that indicates the source of an alarm within a building.
Fire alarm control panel That component in a fire alarm system that controls the functions of the entire system.
Fire and life safety education specialist This person deals with the public on education, fire safety, and juvenile fire safety programs.
Fire apparatus maintenance personnel The people who repair and service the fire and EMS vehicles so that they are always ready to respond to emergencies.
Fire code A set of legally adopted rules and regulations designed to prevent fires and protect lives and property in the event of a fire.
Fire department connection (FDC) A fire hose connection through which the fire department can pump water into a sprinkler system or standpipe system.
Fire department EMS system In this system, the fire department provides both the medical first response and transportation of the patient to the hospital emergency room.
Fire enclosures Fire-rated assemblies used to enclose vertical openings such as stairwells, elevator shafts, and chases for building utilities.
Fire fighter This person is tasked with anything from hose line placement to extinguishing fires. Generally, the fire fighter is not responsible for any command functions or supervising other personnel.
Fire helmet Protective head covering worn by fire fighters to protect the head from falling objects, blunt trauma, and heat.
Fire hook A tool used to pull down burning structures.
Fire hydraulics The physical science of how water flows through a pipe or hose.
Fire load The weight of combustibles in a fire area or on a floor in buildings and structures, including either contents or building parts, or both.
Fire mark Historically, an identifying symbol on a building to let fire fighters know that the building was insured by a company that would pay them for extinguishing the fire.
Fire marshal/fire inspector/fire investigator These people inspect businesses and enforce laws that deal with public safety and fire codes. Fire investigators may respond to fire scenes to help incident commanders investigate the cause of a fire. Investigators may have full police powers of arrest and deal directly with investigations and arrests.
Fire partitions Interior walls extending from the floor to the underside of the floor above.
Fire plug A valve installed to control water accessed from wooden pipes.
Fire point (flame point) The lowest temperature at which a substance releases enough vapors to ignite and sustain combustion.
Fire police Fire police protect fire fighters by controlling traffic and securing the scene from public access. Many fire police are sworn peace officers as well as fire fighters.
Fire prevention Activities conducted to prevent fires and protect lives and property in the event of a fire. Fire prevention activities include the enactment and enforcement of fire codes, the inspection of properties, the presentation of fire safety education programs, and the investigation of the causes of fires.
Fire protection engineer The person in this position has received a college degree in one of several engineering disciplines. He or she is responsible for reviewing plans and working with building owners to ensure that the design of and systems for fire detection and suppression will meet code and function as needed.
Fire shelter A special tent-like device that protects fire fighters from the radiant heat of fire for a short period of time.
Fire tetrahedron A geometric shape used to depict the four components required for a fire to occur: fuel, oxygen, heat, and chemical chain reactions.
Fire wall A wall with a fire-resistive rating and structural stability that separates buildings or subdivides a building to prevent the spread of fire.
Fire wardens Individuals designated to enforce fire regulations in colonial America.
Fire window A window or glass block assembly with a fire-resistive rating.
Fireground Command System (FCS) An incident management system developed in the 1970s for day-to-day fire department incidents (generally handled with fewer than 25 units or companies).
Fire-resistive construction Buildings that have structural components of noncombustible materials with a specified fire resistance. Materials can include concrete, steel beams, and masonry block walls. Type I building construction, as defined in NFPA 220, *Types of Building Construction.*
FIRESCOPE (**FI**re **RES**ources of **C**alifornia **O**rganized for **P**otential **E**mergencies) An organization of agencies established in the early 1970s to develop a standardized system for managing fire resources at large-scale incidents such as wildland fires.
First responder The first trained individual to arrive at the scene of an emergency to provide initial medical assistance.

Fixed gallonage fog nozzle A nozzle delivers a set number of gallons per minute that the nozzle was designed for, no matter what pressure is applied to the nozzle.
Fixed-temperature heat detector A sensing device that responds when its operating element is heated to a predetermined temperature.
Flame detector A sensing device that detects the radiant energy emitted by a flame.
Flammability limits (explosive limits) The upper and lower concentration limits (at a specified temperature and pressure) of a flammable gas or vapor in air that can be ignited, expressed as a percentage of fuel by volume.
Flammable range The range of concentrations between the lower and upper flammable limits.
Flammable vapor Any substance that exists in the gaseous state at normal atmospheric temperature and pressure and is capable of being ignited and burned when mixed with the proper proportions of air, oxygen, or other oxidizers.
Flash point The minimum temperature at which a liquid or solid releases sufficient vapor to form an ignitable mixture with the air.
Flashover The condition where all combustibles in a room or confined space have been heated to the point at which they release vapors that will support combustion, causing all combustibles to ignite simultaneously.
Flat bar A specialized type of prying tool made of flat steel with prying ends suitable for performing forcible entry.
Flat hose load A method of putting a hose on a vehicle in which the hose is laid flat and stacked on of top the previous section.
Flat roofs Horizontal roofs often found on commercial or industrial occupancies.
Flat-head axe A tool that has a head with an axe on one side and a flat head on the opposite side.
Floodlight A light that can illuminate a broad area.
Floor runner A piece of canvas or plastic material, usually 3′ to 4′ wide and in various lengths, used to protect flooring from dropped debris and/or dirt from shoes and boots.
Flow pressure The amount of pressure created by moving water.
Flow switch An electrical switch that is activated by water moving through a pipe on a sprinkler system.
Fluoroprotein foam A blended organic and synthetic foam that is made from animal byproducts and synthetic surfactants.
Fly section A section of an extension ladder that is raised or extended from the base section or from another fly section. Some extension ladders have more than one fly section.
Foam concentrate Foam that is in its raw state.
Foam eductor A device placed in the hose line that draws foam concentrate from a container and introduces it into the fire stream.
Foam injector A device installed on a fire pump that meters foam by pumping or injecting it into the firestream.
Foam proportioner A device that meters and introduces the foam concentrate into the fire stream at a proper rate.
Foam solution Foam that has been mixed with water.
Fog stream nozzle Device placed at the end of a fire hose that separates water into fine droplets to aid in heat absorption.
Folding ladder A ladder that collapses by bringing the two beams together for portability. Unfolded, the folding ladder is narrow and used for access to attic scuttle holes and confined areas.
Forcible entry Techniques used by fire fighters to gain entry into structures, vehicles, aircraft, or other areas when normal means of entry are locked or blocked.
Formal decontamination A multistep process of carefully scrubbing and washing contaminants off a person or object, collecting runoff water, and collecting and properly handling all items that takes place after gross decontamination.
Forward lay A method of laying a supply line where the line starts at the water source and ends at the attack engine.
Forward staging area A strategically placed area, close to the incident site, where personnel and equipment can be held in readiness for rapid response to an emergency event.
Four-way hydrant valve A specialized type of valve that can be placed on a hydrant that allows another engine to increase the supply pressure without interrupting flow.
Frangible bulb sprinkler head A sprinkler head with a frangible bulb.
Freelancing Dangerous practice of acting independently of command instructions.
Freight bills Shipping papers for roads and highways.
Fresno ladder A narrow, two-section extension ladder that has no halyard. Because of its limited length, it can be extended manually.
Friction loss The reduction in pressure due to the water being in contact with the side of the hose. This contact requires force to overcome the drag the wall of the hose creates.
Frostbite Damage to tissues as the result of exposure to cold; frozen or partially frozen body parts.
Fuel All combustible materials. The actual material that is being consumed by a fire, allowing the fire to take place.
Fuel compactness How tightly packed fuels are.
Fuel continuity The relative closeness of wildland fuels, a factor in a fire's ability to spread from one area of fuel to another.
Fuel moisture The amount of moisture present in a fuel, a factor in how readily the fuel will ignite and burn.
Fuel volume The amount of fuel present in a given area.
Fully developed phase The phase of fire development where the fire is free-burning and consuming much of the fuel.
Fully-encapsulated suits A protective suit that completely covers the fire fighter, including the breathing apparatus, and does not let any vapor or fluids enter the suit.
Fusible link sprinkler head A sprinkler head with an activation mechanism that incorporates two pieces of metal held together by low-melting-point solder. When the solder melts, it releases the link and water begins to flow.
Gamma radiation A type of radiation that can travel significant distances, penetrating most materials and passing through the body. Gamma radiation is the most destructive to the human body. Gamma rays are best stopped or shielded against by dense material, such as depleted uranium, lead, water, concrete, or iron.
Gas One of the three phases of matter. A substance that will expand indefinitely and assume the shape of the container that holds it.
Gas detector A device that detects and/or measures the concentration of dangerous gases.
Gastric distention Inflation of the stomach caused when excessive pressures are used during artificial ventilation and air is directed into the stomach rather than the lungs.
Gate valves Valves found on hydrants and sprinkler systems.
Gated wye A valved device that splits a single hose into two separate hoses, allowing each hose to be turned on and off independently.
Generator An engine-powered device that provides electricity.
Glass blocks Thick pieces of glass similar to bricks or tiles.
Glazed Transparent glass.
Glucose The source of energy for the body. One of the basic sugars, it is the primary fuel, along with oxygen.
Grade The level at which the ground intersects the foundation of a structure.
Gravity feed system A water distribution system that depends on gravity to provide the required pressure. The system storage is usually located at a higher elevation than the end users.
Green An area of unburned fuels.
Gripping pliers A hand tool with a pincer-like working end that can be used to bend wire or hold smaller objects.
Gross decontamination The process of removing the outer clothing and generally flushing most contaminants from a victim; the initial phase of the decontamination process during which the amount of surface contaminant is significantly reduced.
Ground cover fire A fire that burns loose debris on the surface of the ground.
Ground duff Partly decomposed organic material on a forest floor; a type of light fuel.
Group An organization level within IMS that divides an incident according to functional areas of operation.
Group supervisor The officer in charge of all resources operating within a specified group, responsible to the next higher level in the incident organization, and the point-of-contact for the group within the organization.
Growth phase The phase of fire development where the fire is spreading beyond the point of origin and beginning to involve other fuels in the immediate area.
Guideline A rope used for orientation when inside a structure when there is low or no visibility; it is attached to a fixed object outside the hazardous area.

Guides Strips of metal or wood that serve to guide a fly section during extension. Channels or slots in the bed or fly section may also serve as guides.

Gusset plates The connecting plate made of wood or lightweight metal used in trusses.

Gypsum A naturally occurring material composed of calcium sulfate and water molecules

Gypsum board The generic name for a family of sheet products consisting of a noncombustible core primarily of gypsum with paper surfacing.

Hacksaw A cutting tool designed for use on metal. Different blades can be used for cutting different types of metal.

Halligan tool A prying tool that incorporates a pick and a fork, designed for use in the fire service. Sometimes known as a Hooligan tool.

Halogenated extinguishing agent A liquefied gas extinguishing agent that puts out fires by chemically interrupting the combustion reaction between the fuel and oxygen.

Halogenated-agent extinguisher An extinguisher that uses halogenated extinguishing agents.

Halon 1211 Bromochlorodifluoromethane ($CBrClF_2$), a halogenated agent that is effective on Class A, B, and C fires.

Halon 1301 A liquefied gas extinguishing agent that puts out the fire by chemically interrupting the combustion reaction between fuel and oxygen. Halon agents leave no residue.

Halyard The rope or cable used to extend or hoist the fly section(s) of an extension ladder.

Hammer A striking tool.

Hand light Small, portable light carried by fire fighters to improve visibility at emergency scenes, often powered by rechargeable batteries.

Handle The grip used for holding and carrying a portable fire extinguisher.

Handline nozzles Used on hoses ranging from 1½" to 2½" hose lines, usually flow between 90 and 350 gallons per minute.

Handsaw A manually powered saw designed to cut different types of materials. Examples include hacksaws, carpenter's handsaws, keyhole saws, and coping saws.

Hard suction hose A hose designed to prevent collapse under vacuum conditions so that it can be used for drafting water from below the pump (lakes, rivers, wells, or sea water, etc.).

Hardware The parts of a door or window that enable it to be locked or opened.

Harness A piece of equipment worn by a rescuer that can be attached to a life safety rope.

Hazard A material capable of posing an unreasonable risk to health, safety, or the environment; capable of causing harm.

Hazardous materials Materials or substances that pose an unreasonable risk of damage or injury to persons, property, or the environment if not properly controlled during handling, storage, manufacture, processing, packaging, use and disposal, or transportation.

Hazardous materials company Hazardous materials companies respond to and control scenes where hazardous materials spilled or leaked. Responders wear special suits and are trained to deal with most chemicals.

Hazardous Materials incident commander (IC hazardous materials) This level of training is intended for those assuming command of a hazardous materials incident beyond the operations level. Individuals trained as incident commanders should have at least operations-level training and additional training specific to commanding a hazardous materials incident.

Hazardous Materials Information System (HMIS) A color-coded marking system by which employers give their personnel the necessary information to work safely around chemicals.

Hazardous materials technician Fire fighters who receive training in chemical identification, leak control, decontamination, and clean-up procedures. They would be called to handle the threat of weapons of mass destruction.

Hazardous waste A substance that remains after a process or manufacturing plant has used some of the material and it is no longer pure.

HAZWOPER Short for Hazardous Waste Operations and Emergency Response, this federal OSHA regulation governs hazardous materials waste sites and response training. Specifics can be found in 29 CFR, book number 29, part number 1910.120. Subsection (q) is specific to emergency response.

Head of the fire The main or running edge of a fire. The part of the fire that spreads with the greatest speed.

Head tilt-chin lift technique Opening the airway by tilting the victim's head backward and lifting the chin forward, bringing the entire lower jaw with it.

Heat detector A fire alarm device that detects either abnormally high temperatures or rate-of-rise in temperature, or both.

Heat exhaustion A mild form of shock caused when the circulatory system begins to fail as a result of the body's inadequate effort to give off excessive heat.

Heat sensor label A piece of heat-sensitive material on each section of a ladder that identifies when the ladder has been exposed to high heat conditions.

Heat stroke A severe and sometimes fatal condition resulting from the failure of the temperature regulating capacity of the body. It is caused by prolonged exposure to the sun or high temperatures. Reduction or cessation of sweating is an early symptom; body temperature of 105°F or higher, rapid pulse, hot and dry skin, headache, confusion, unconsciousness, and convulsions may occur. Heat stroke is a true medical emergency requiring immediate transport to a medical facility.

Heavy fuels Fuels of a large diameter, such as large brush, heavy timber, snags, stumps, branches, and dead timber on the ground; these ignite and are consumed more slowly than light fuels.

Heavy timber construction Buildings constructed with noncombustible or limited-combustible exterior walls and interior walls and floors made of large dimension combustible materials. Also known as Type IV building construction.

Heel of the fire The side opposite the head of the fire, often close to the area of origin.

Heimlich maneuver A series of manual thrusts to the abdomen to relieve upper airway obstruction.

HEPA (High Efficiency Particulate Air) filters Filters capable of catching particles down to 0.3-micron size—much smaller than a typical dust, or alpha radiation, particle.

Hepatitis B virus (HBV) A virus that causes inflammation of the liver.

Hepatitis C virus (HCV) A virus that causes inflammation of the liver. It is transmitted through blood.

Higbee indicators An indicator on the male and female threaded couplings that indicate where the threads start. These indicators should be aligned before starting to thread the couplings together.

High temperature-protective equipment Protective clothing designed to shield the wearer during short-term high temperature exposures.

High-angle operations A rope rescue operation where the angle of the slope is greater than 45°; rescuers depend on life safety rope rather than a fixed support surface such as the ground.

High-pressure lift airbags An extrication tool, consisting of airbags, filler hoses, air regulators, control valves, and a supply of compressed air.

Hitches Knots that attach to or wrap around an object.

Hollow-core A door made of panels that are honeycombed inside, creating an inexpensive and lightweight design.

Horizontal evacuation Moving occupants from a dangerous area to a safe area on the same floor level.

Horizontal ventilation The process of making openings so that smoke, heat, and gases can escape horizontally from a building through openings such as doors and windows.

Horizontal-sliding windows Windows that slide open horizontally.

Horn The tapered discharge nozzle of a carbon dioxide-type fire extinguisher.

Horseshoe hose load A method of loading hose where the hose is laid into the bed along the three walls of the bed, resembling a horseshoe.

Hose appliance Any device used in conjunction with fire hose for the purpose of delivering water.

Hose clamp A device used to compress a fire hose to stop water flow.

Hose jacket A device used to stop a leak in a fire hose or to join hoses that have damaged couplings.

Hose liner (hose inner jacket) The inside portion of a hose that is in contact with the flowing water.

Hose roller A device that is placed on the edge of a roof and is used to protect hose as it is hoisted up and over the roof edge.

Hot zone The area immediately surrounding a hazardous materials spill/incident site that is directly dangerous to life and health. All personnel working in the hot zone must wear complete, appropriate protective clothing and equipment. Entry requires approval by the IC or a designated sector officer. Complete back-up, rescue, and decontamination teams must be in place at the perimeter before operations begin.

Human immunodeficiency virus (HIV) The virus that causes AIDS.

Hux bar A multipurpose tool that can be used for several forcible entry and ventilation applications because of its unique design. Also can be used as a hydrant wrench.

HVAC system Heating, ventilation, and air-conditioning system in large buildings.
Hybrid vehicle A vehicle that uses battery-powered electric motors and a gasoline-powered engine.
Hydrant wrench A hand tool is used to operate the valves on a hydrant; may also be used as a spanner wrench. Some are plain wrenches, and others have a ratchet feature.
Hydraulic shears A lightweight hand-operated tool that can produce up to 10,000 pounds of cutting force.
Hydraulic spreader A lightweight hand-operated tool that can produce up to 10,000 pounds of prying and spreading force.
Hydraulic ventilation Ventilation that relies upon the movement of air caused by a fog stream.
Hydrostatic testing Periodic testing of an extinguisher to verify it has sufficient strength to withstand internal pressures.
Hypothermia A condition in which the internal body temperature falls below 95°F after prolonged exposure to cool or freezing temperatures.
Hypoxia A state of inadequate oxygenation of the blood and tissue.
I-beam A ladder beam constructed of one continuous piece of I-shaped metal or fiberglass to which the rungs are attached.
Ignition phase The phase of fire development where the fire is limited to the immediate point of origin.
Ignition point The minimum temperature at which a substance will burn.
Ignition temperature The minimum temperature at which a fuel, when heated, will ignite in air and continue to burn.
Immediate Danger to Life and Health (IDLH) An atmospheric concentration of any toxic, corrosive, or asphyxiant substance that poses an immediate threat to life or could cause irreversible or delayed adverse health effects. There are three general IDLH atmospheres: toxic, flammable, and oxygen-deficient.
Improvised explosive device An explosive or incendiary device that is fabricated in an improvised manner.
IMS General Staff The chiefs of each of the four major sections of IMS: Operations, Planning, Logistics, and Finance/Administration.
Incendiary device A device or mechanism used to start a fire or explosion.
Incendiary fires Intentionally set fires.
Incident Action Plan (IAP) The objectives for the overall incident strategy, tactics, risk management, and member safety that are developed by the IC. Incident Action Plans are updated throughout the incident.
Incident Command System (ICS) The first standard system for organizing large, multi-jurisdictional and multi-agency incidents involving more than 25 resources or operating units; eventually developed into the Incident Management System (IMS).
Incident Commander The person in charge of the incident site who is responsible for all decisions relating to the management of the incident.
Incident Management System (IMS) The combination of facilities, equipment, personnel, procedures, and communications under a standard organizational structure to manage assigned resources effectively to accomplish stated objectives for an incident. Also known as Incident Command System (ICS).
Incipient The initial stage of a fire.
Incomplete combustion A burning process in which the fuel is not completely consumed, usually due to a limited supply of oxygen.
Incubation period Time period between the initial infection by an organism and the development of symptoms.
Indirect application of water Using a solid object such as a wall or ceiling to break apart a stream of water, causing it to create more surface area on the water droplets and thereby absorb more heat.
Indirect attack Firefighting operations involving the application of extinguishing agents to reduce the buildup of heat released from a fire without applying the agent directly onto the burning fuel.
Indirect attack (wildland and ground fires) A method of attack in which the control line is located along natural fuel breaks, favorable breaks in the topography, or at considerable distance from the fire, and the intervening fuel is burned out.
Information management Fire fighters or civilians who take care of all the computer and networking systems that a fire department needs to operate.
Ingestion Exposure to a hazardous material by swallowing.
Inhalation Exposure to a hazardous material by breathing it into the lungs.
Injection Hazardous materials entering cuts or other breaches in the skin.
Integrated communications The ability of all appropriate personnel at the emergency scene to communicate with their supervisor and their subordinates.
Interior attack The assignment of a team of fire fighters to enter a structure and attempt fire suppression.
Interior finish Any coating or veneer applied as a finish to a bulkhead, structural insulation, or overhead, including the visible finish, all intermediate materials, and all application materials and adhesives.
Interior wall A wall inside a building that divides a large space into smaller areas. Also known as a partition.
Intermodal tanks Bulk containers that can be shipped by all modes of transportation—air, sea, or land.
Inverter A device that converts the direct current from an apparatus electrical system into alternating current.
Ionization smoke detector A device containing a small amount of radioactive material that ionizes the air between two charged electrodes to sense the presence of smoke particles.
Ionizing radiation Electromagnetic waves of such intensity that chemical bonds at the atomic level can be broken (creating an ion).
Irons A combination tool, normally the Halligan tool and the flat-head axe.
Irritants Substances such as mace that can be dispersed to briefly incapacitate a person or groups of people.
Island An unburned area surrounded by fire.
J tool A tool that is designed to fit between double doors equipped with panic bars.
Jalousie windows Windows made of small slats of tempered glass that overlap each other when closed. Often found in trailers and mobile homes, jalousie windows are held by a metal frame and operated by a small hand wheel or crank found in the corner of the window.
Jamb The part of a doorway that secures the door to the studs in a building.
Jaw-thrust technique Opening the airway by bringing the victim's jaw forward without extending the neck.
Junction box A device that attaches to an electrical cord to provide additional outlets.
K tool Used to remove lock cylinders from structural doors so the locking mechanism can be unlocked.
Kelly tool A steel bar with two main features: a large pick and a large chisel or fork.
Kerf cut A cut that is only the width and depth of the saw blade. It is used to inspect cockloft spaces from the roof.
Kernmantle rope Rope made of two parts—the kern (interior component) and the mantle (the outside sheath).
Kevlar® Strong, synthetic material used in the construction of protective clothing and equipment.
Keyhole saw A saw designed to cut keyholes in wood.
Knots A fastening made by tying together lengths of rope or webbing in a prescribed way, used for a variety of purposes.
Labels Smaller versions (4″ diamond-shaped markings) of placards, placed on four sides of individual boxes and smaller packages.
Laceration An irregular cut or tear through the skin.
Ladder belt A belt specifically designed to secure a fire fighter to a ladder or elevated surface.
Ladder gin An A-shaped structure formed with two ladder sections. It can be used as a makeshift lift when raising a trapped person. One form of the device is called an A-frame hoist.
Ladder halyards Rope used on extension ladders to raise a fly section.
Ladder pipe Nozzle attached to the end of a straight ladder truck designed to provide large volumes of water from an elevated position.
Laminated glass Also known as safety glass; the lamination process places a thin layer of plastic between two layers of glass, so that the glass does not shatter and fall apart when broken.
Laminated windshield glass Type of window glazing that incorporates a sheeting material that stops the glass from breaking into individual shards. Commonly used in windshields.

Laminated wood Pieces of wood that are glued together.
Large Diameter Hose (LDH) Hose in the 4″, 5″, and 6″ range.
Larynx A structure composed of cartilage in the neck that guards the entrance to the windpipe and functions as the organ of voice; also called the voice box.
Latch The part of the door lock that secures into the jamb.
Laths Thin strips of wood used to make the supporting structure for roof tiles.
Leap-frogging A fire spread from one floor to the other through exterior windows (auto-exposure).
Level A protection PPE that provides protection against vapors, gases, mists, and even dusts—the highest level of protection; requires a totally encapsulating suit that includes SCBA.
Level B protection PPE used when the type and atmospheric concentration of substances have been identified. Generally requires a high level of respiratory protection but less skin protection: chemical-protective coveralls and clothing, chemical protection for shoes, gloves, and SCBA outside of a nonencapsulating chemical-protective suit.
Level C protection PPE used when the type of airborne substance is known, the concentration is measured, the criteria for using air-purifying respirators are met, and skin and eye exposure is unlikely. Consists of standard work clothing with the addition of chemical-protective clothing, chemically resistant gloves, and a form of respirator protection.
Level D protection PPE used when the atmosphere contains no known hazard, and work functions preclude splashes, immersion, or the potential for unexpected inhalation of or contact with hazardous levels of chemicals. It is primarily a work uniform that includes coveralls and affords minimal protection.
Lewisite Blister-forming agent that is an oily, colorless-to-dark brown liquid with an odor of geraniums.
Liaison Officer The position within IMS that establishes a point of contact with outside agency representatives.
Lieutenant A company officer who is usually responsible for a single fire company on a single shift; the first in line of company officers.
Lifeline A rope secured to a fire fighter that enables him or her to retrace his or her steps out of a structure.
Life safety rope Rope used solely for the purpose of supporting people during firefighting, rescue, other emergency operations, or during training exercises.
Light hazard locations Occupancies where the total amount of combustible materials is less than expected in an ordinary hazard location.
Light-emitting diode (LED) An electronic semiconductor that emits a single-color light when activated.
Lightweight construction The use of small dimension members such as as 2′ × 4′ or 2′ × 6′ wood assemblies as structural supports in buildings.
Line detector Wire or tubing that can be strung along the ceiling of large open areas to detect an increase in heat.
Liquid One of the three phases of matter. A nongaseous substance that is composed of molecules that move and flow freely and assumes the shape of the container that holds it.
Liquid splash-protective clothing Several pieces of clothing and equipment designed to protect the skin and eyes from chemical splashes.
Live load The weight of the building contents.
Load-bearing wall A wall that supports structural members or upper floors of a building.
Loaded-stream extinguisher A water-based fire extinguisher that uses an alkali metal salt as a freezing point depressant.
Lob A method of discharging extinguishing agent in an arc to avoid splashing or spreading the burning fuel.
Local alarm system A fire alarm system that sounds an alarm only in the building where it was activated. No signal is sent out of the building.
Local Emergency Planning Committee (LEPC) Committee comprised of members of industry, transportation, the public at large, media, and fire and police agencies. Gathers and disseminates information on hazardous materials stored in the community and ensures that there are adequate local resources to respond to a chemical event in the community.
Lock body The part of a padlock that holds the main locking mechanisms and secures the shackles.
Locking mechanism A device that locks an extinguisher's trigger to prevent accidental discharge.
Lockout/tagout systems Lockout and tagout are methods of ensuring that electricity and other utilities have been shut down and switches are "locked" so that they cannot be switched on, in order to prevent flow of power or gases into the area where rescue is being conducted.
Logistics Section The section within IMS responsible for providing facilities, services, and materials for the incident.
Logistics Section Chief The General Staff position responsible for directing the logistics function; generally assigned on complex, resource-intensive, or long-duration incidents.
Loop A piece of rope formed into a circle.
Louver cut A cut that is made using power saws and axes to cut along and between roof supports so that the sections created can be tilted into the opening.
Low volume nozzles Nozzles that flow 40 gallons per minute or less.
Low-angle operations A rope rescue operation on a mildly sloping surface (less than 45°) or flat land where fire fighters are dependent on the ground for their primary support, and the rope system is a secondary means of support.
Lower explosive limits (LEL) The minimum amount of gaseous fuel that must be present in the air mixture for the mixture to be flammable or explosive.
Lungs The organs that supply the body with oxygen and eliminate carbon dioxide from the blood.
Malicious false alarm A fire alarm signal when there is no fire, usually initiated by individuals who wish to cause a disturbance.
Mallet A short-handled hammer.
Manual pull-station A device with a switch that either opens or closes a circuit, activating the fire alarm.
Manufactured (mobile) homes A factory-assembled structure or structures transportable in one or more sections that is built on a permanent chassis and designed to be used as a dwelling without a permanent foundation when connected to the required utilities, including the plumbing, heating, air-conditioning, and electric systems contained therein.
Mark 1 Nerve Agent Antidote Kit A military kit containing antidotes that can be administered to victims of a nerve agent attack.
Masonry Built-up unit of construction or combination of materials such as brick, clay tiles, or stone set in mortar.
Mass arson Involves an offender who sets three or more fires at the same site or location during a limited period of time.
Mass decontamination Special procedures for incidents that involve large numbers of people. Master stream devices from engine companies and aerial apparatus provide high-volume, low-pressure showers so that large numbers of people can be decontaminated rapidly.
Mass-casualty incident An emergency situation involving more than one victim, and which can place such great demand on equipment or personnel that the system is stretched to its limit or beyond.
Master stream device A large capacity nozzle that can be supplied by two or more hose lines or fixed piping. Can flow in excess of 300 gallons per minute. Includes deck guns and portable ground monitors.
Master stream nozzles A nozzle used on deck guns, portable monitors, and ladder pipes that flow more than 350 gallons per minute.
Master-coded alarm An alarm system in which audible notification devices can be used for multiple purposes, not just for the fire alarm.
Material Safety Data Sheet (MSDS) A form, provided by manufacturers and compounders (blenders) of chemicals, containing information about chemical composition, physical and chemical properties, health and safety hazards, emergency response, and waste disposal of the material.
Maul A specialized striking tool, weighing six pounds or more, with an axe on one end and a sledgehammer on the other end.
Mayday Code that indicates that a fire fighter is lost, missing, or requires immediate assistance.
MC-306 flammable liquid tanker Commonly known as a gasoline tanker, this tanker typically carries gasoline or other flammable and combustible materials (also known as DOT-406).
MC-307 chemical hauler A tanker with a rounded or horseshoe-shaped tank.
MC-312 corrosives tanker A tanker that will often carry aggressive acids like concentrated sulfuric and nitric acid, with reinforcing rings along the side of the tank.
MC-331 pressure cargo tanker A tank commonly constructed of steel with rounded ends and a single open compartment inside; there are no baffles or other separations inside the tank.

MC-338 cryogenics tanker A low-pressure tanker designed to maintain the low temperature required by the cryogens it carries.
McLeod tool A hand tool used for constructing fire lines and overhauling wildland fires. One side of the head consists of a five-toothed to seven-toothed fire rake and the other side is a hoe.
Mechanical energy Heat energy created by friction.
Mechanical saw Usually powered by electric motors or gasoline engines. The three primary types are chain saws, rotary saws, and reciprocating saws.
Mechanical ventilation Ventilation that uses mechanical devices to move air.
Medical control Physician instructions that are given directly by radio or indirectly by protocol/guidelines as authorized by the medical director of the service.
Medical director The physician providing direction for patient care activities in the prehospital setting.
Medium diameter hose (MDH) Hose of $2^1/_2''$ or $3''$ size.
Mildew A condition that can occur on hose if it is stored wet. Mildew can damage the jacket of a hose.
Mobile data terminals (MDT) Technology that allows fire fighters to receive information in the apparatus or at the station.
Mobile radio Two-way radio that is permanently mounted in a fire apparatus.
Mobile water supply apparatus A fire department vehicle that is designed to transport water to the scene of a fire; often called a tanker.
Mortise locks Door locks with both a latch and a bolt built into the same mechanism; the two locking devices operate independently of each other. This is a common lock found in places such as hotel rooms.
Mouth-to-mask ventilation device A piece of equipment that consists of a mask, a one-way valve, and a mouthpiece. Rescue breathing is performed by breathing into the mouthpiece after placing the mask over the victim's mouth and nose.
Multipurpose dry chemical extinguishers Extinguishers rated to fight Class A, B, and C fires.
Multipurpose hook A long pole with a wooden or fiberglass handle and a metal hook on one end used for pulling.
Municipal water system A water distribution system that is designed to deliver potable water to end-users for domestic, industrial, and fire protection purposes.
Mushrooming The process that occurs when rising smoke, heat, and gases encounter a horizontal barrier such as a ceiling and begin to move out and back down.
National Fire Incident Reporting System (NFIRS) System used by fire departments to report and maintain computerized records of fires and other fire department incidents in a uniform manner.
National Fire Protection Association (NFPA) The association that develops and maintains nationally recognized minimum consensus standards on many areas of fire safety and specific standards on hazardous materials.
National Institute for Occupational Safety and Health (NIOSH) A U.S. Federal agency responsible for research and development on occupational safety and health issues.
National Response Center (NRC) The U.S. Coast Guard maintained and staffed agency that should always be notified if any spilled material could possibly enter a navigable waterway.
Natural ventilation Ventilation that relies upon the natural movement of heated smoke and wind currents.
Negative-pressure ventilation Ventilation that relies upon electric fans to pull or draw the air from a structure or area.
Nerve agents Toxic substances that attack the central nervous system in humans.
Neutrons Penetrating particles found in the nucleus of the atom that are removed through nuclear fusion or fission. Neutrons are not radioactive; however, exposure to neutrons can create radiation.
NFPA 704 hazard identification system A hazardous materials marking system designed for fixed-facility use.
Nomex® A fire-resistant synthetic material used in the construction of personal protective equipment for fire fighting.
Nonbearing wall A wall that does not support a ceiling or structural member, but simply divides a space. See interior wall and partition.
Nonbulk storage vessels Containers other than bulk storage containers.
Non-coded alarm An alarm system that provides no information at the alarm control panel indicating where the activated alarm is located.
Noncombustible construction Buildings where the structural members are of non-combustible materials without fire resistance. Also know as Type II building construction.
Normal operating pressure The observed static pressure in a water distribution system during a period of normal demand.
***North American Emergency Response Guidebook* (NAERG)** The guidebook developed by the Department of Transportation to provide guidance for first responders operating at
a hazardous materials incident in coordination with DOT's labels and placards marking system.
Nose cups An insert inside the face piece of an SCBA that fits over the user's mouth and nose.
Nozzle (fire extinguisher) The discharge orifice of a portable fire extinguisher.
Nozzle shut off Device that enables the person at the nozzle to start or stop the flow of water.
Nozzles Attachments to the discharge end of attack hoses to give fire streams shape and direction.
Nuisance alarm A fire alarm signal caused by malfunction or improper operation of a fire alarm system or component.
Obscuration rate A measure of the percentage of light transmission that is blocked between a sender and a receiver unit.
Occupancy The purpose for which a building or other structure, or part thereof, is used or intended to be used.
Occupational Safety and Health Administration Agency (OSHA) The federal agency that regulates worker safety and in some cases, responder safety. OSHA is a part of the Department of Labor.
Offensive attack An advance into the fire building by fire fighters with hose lines or other extinguishing agents to overpower the fire.
One-person rope A rope rated to carry the weight of a single person (300 lbs).
Open wound Injury that breaks the skin or mucous membrane.
Open-circuit breathing apparatus SCBA in which the exhaled air is released into the atmosphere and is not reused.
Open-end wrench A hand tool used to tighten or loosen bolts. The end is open, as opposed to a box-end wrench. Each wrench is a specific size.
Operating lever The handle of a door that turns the latch to open it.
Operations level (TRI) The technical rescue training level geared toward working in the warm zone of an incident, which allows responders to directly assist those conducting the rescue operation and to use certain rescue skills and procedures.
Operations level Responders trained to the operations level should be able to recognize a potential hazardous materials incident, isolate the incident, deny entry to other responders and the public, and take defensive actions such as shutting off valves and protecting drains without having contact with the product. Operations-level responders act in a defensive fashion.
Operations Section The section within IMS responsible for all tactical operations at the incident.
Operations Section Chief The general staff position responsible for managing all operations activities; usually assigned when complex incidents involve more than 20 single resources or when the IC cannot be involved in the details of tactical operations.
Ordinary construction Buildings in which the exterior walls are noncombustible or limited-combustible, but the interior floors and walls are made of combustible materials. Also known as Type III building construction.
Ordinary hazard location Occupancies that contain more Class A and Class B materials than are found in light hazard locations.
Outside stem and yoke (OS&Y) valve A sprinkler control valve with a valve stem that moves in and out as the valve is opened or closed.
Overhaul Examination of all areas of the building and contents involved in a fire to ensure that the fire is completely extinguished.
Oxidation A chemical reaction initiated by combining an element with oxygen, resulting in the form of the element or one of its compounds.
Oxidizing agent A substance that will release oxygen or act in the same manner as oxygen in a chemical reaction.

Oxygen deficiency Any atmosphere where the oxygen level is below 19.5%. Low oxygen levels can have serious effects on people, including adverse reactions such as poor judgment and lack of muscle control.
Packaging Preparing the victim for movement as a unit, often accomplished with a long spine board or similar device.
Padlocks The most common lock on the market today, built to provide regular-duty or heavy-duty service. Several types of locking mechanism are available, including a keyway, combination wheels, or combination dials.
Parallel chord trusses A truss in which the top and bottom chords are parallel.
Parapet walls Walls on a flat roof that extend above the roof line.
Partition A wall that does not support a ceiling or structural member, but simply divides a space. See interior or nonbearing wall.
Party walls Walls constructed on the line between two properties.
P-A-S-S Acronym used for operating a portable fire extinguisher: Pull pin, Aim nozzle, Squeeze trigger, Sweep across burning fuel.
Passing command Option that can be used by the first-arriving company officer to direct the next arriving unit to assume command.
Pathogens Microorganisms capable of causing disease.
PBI® A fire-retardant synthetic material used in the construction of personal protective equipment.
Peak cut A ventilation opening that runs along the top of a peaked roof.
Pendant sprinkler head A sprinkler head designed to be mounted on the underside of sprinkler piping so the water stream is directed down.
Penetration The flow or movement of a hazardous liquid (or solid) chemical through zippers, stitched seams, buttonholes, flaps, pinholes, or other imperfections in a material. Liquids and solids such as asbestos can penetrate chemical-protective clothing at several locations, including the face piece, exhalation valve, suit exhaust valve, and suit fasteners.
Permeation The process by which a hazardous liquid chemical moves through a given material on the molecular level. Permeation differs from penetration in that permeation occurs through the clothing material itself rather than through openings in the clothing material.
Permissible Exposure Limit (PEL) The established standard limit of exposure to a hazardous material. It is based on the maximum time-weighted concentration at which 95% of exposed, healthy adults suffer no adverse effects over a 40-hour work week.
Personal alert safety system (PASS) Device worn by a fire fighter that sounds an alarm if the fire fighter is motionless for a period of time.
Personal dosimeters Devices that measure the amount of radioactive exposure to an individual.
Personal escape rope An emergency use rope designed to carry the weight of only one person and to be used only once.
Personal flotation device (PFD) Also commonly known as a life vest; it allows the body to float in water.
Personal protective equipment (PPE) Gear worn by fire fighters that includes helmet, gloves, hood, coat, pants, SCBA, and boots. The personal protective equipment provides a thermal barrier for fire fighters against intense heat.
Personnel Accountability Report (PAR) A report confirming that all members of a company are present.
Personnel accountability system A method of tracking the identity, assignment, and location of fire fighters operating at an incident scene.
Personnel accountability tag (PAT) Identification card used to track the location of a fire fighter on an emergency incident.
pH A measure of the acidity or base of a material; more technically, an expression of the concentration of hydrogen ions in the substance.
Phosgene A chemical agent that causes severe pulmonary damage. A by-product of incomplete combustion.
Photoelectric smoke detector A device to detect visible products of combustion using a light source and a photosensitive sensor.
Physical change When a material changes its state of matter; for instance, from a liquid to a solid.
Physical evidence Items that can be observed, photographed, measured, collected, examined in a laboratory, and presented in court to prove or demonstrate a point.
Physical properties Characteristics of a chemical that are measurable, such as vapor density, flammability, corrosivity, water reactivity, etc.
Pick The pointed end of a pick axe that can be used to make a hole or bite in a door, floor, or wall.
Pick-head axe A tool that has a head with an axe on one side and a pointed end, or "pick" on the opposite side.
Piercing nozzle A nozzle that can be driven through sheet metal or other material to deliver a water stream to that area.
Pike pole A pole with a sharp point, or pike, on one end coupled with a hook. Used to make openings in ceilings, walls, etc.
Pipe bomb A device created by filling a section of pipe with an explosive material.
Pipe wrench A wrench having one fixed grip and one moveable grip that can be adjusted to fit securely around pipes and other tubular objects.
Pipeline A length of pipe, including pumps, valves, flanges, control devices, strainers, and/or similar equipment, for conveying fluids and gases.
Pipeline right of way An area, patch, or roadway that extends a certain number of feet on either side of the pipe itself and that may contain warning and informational signs about hazardous materials carried in the pipeline.
Pitched chord truss Type of truss typically used to support a sloping roof.
Pitched roof A roof with sloping or inclined surfaces.
Pitot gauge A type of gauge that is used to measure the velocity pressure of water that is being discharged from an opening. Used to determine the flow of water from a hydrant.
Placards Signage required to be placed on all four sides of highway transport vehicles, railroad tank cars, and other forms of hazardous materials transportation that identifies the hazardous contents of the vehicle, using a standardized system with $10^3/_4$" × $10^3/_4$" diamond-shaped indicators.
Plague An infectious disease caused by the bacterium *Yersinia pestis*; commonly found on rodents.
Planning Section The section within IMS responsible for the collection, evaluation, and dissemination of tactical information related to the incident and for preparation and documentation of incident management plans.
Planning Section Chief The general staff position responsible for planning functions; assigned when the IC needs assistance in managing information.
Plasma The fluid part of the blood that carries blood cells, transports nutrients, and removes cellular waste materials.
Plaster hook A long pole with a pointed head and two retractable cutting blades on the side.
Plate glass A type of glass with additional strength so it can be formed in larger sheets, but will still shatter upon impact.
Platelets Microscopic disc-shaped elements in the blood that are essential to the process of blood clot formation, the mechanism that stops bleeding.
Platform frame Type of vehicle frame resembling a ladder type made up of two parallel rails joined by a series of cross members. Typically used for luxury vehicles, sport utility vehicles, and all types of trucks.
Platform-frame construction Construction technique for building the frame of the structure one floor at a time. Each floor has a top and bottom plate that acts as a fire stop.
Plume The column of hot gases, flames, and smoke that rises above a fire, also called a convection column, thermal updraft, or thermal column.
Pocket A deep indentation of unburned fuel along the fire's perimeter, often between a finger and the head of the fire.
Point of origin The exact location where a heat source and a fuel come in contact with each other and a fire begins.
Polar solvent One of the water-soluble flammable liquids such as alcohol, acetone, ester, and ketone.
Policies Formal statements that provide guidelines for present and future actions; policies often require personnel to make judgments.
Pompier ladder (scaling ladder) A lightweight, single beam ladder.
Portable ladder Ladder carried on fire apparatus, but designed to be removed from the apparatus and deployed by fire fighters where needed.
Portable monitor A master stream appliance designed to be set up and then left to operate unattended. This device is designed to flow large amounts of water onto a fire or exposed building from a ground-level position.
Portable radio A battery-operated, hand-held transceiver.
Portable tanks Folding or collapsible tanks that are used at the fire scene to hold water for drafting.
Positive-pressure ventilation Ventilation that relies upon fans to push or force clean air into a structure.
Post indicator valve (PIV) A sprinkler control valve with an indicator that reads either open or shut depending on its position.

Posts One of the vertical support members of a vehicle that holds up the roof and forms the upright columns of the passenger compartment.
Pounds per square inch (psi) Standard unit used in measuring pressure.
Preaction sprinkler system A dry sprinkler system that uses a deluge valve instead of a dry-pipe valve and requires activation of a secondary device before the pipes fill with water.
Preincident plan A written document resulting from the gathering of general and detailed information to be used by public emergency response agencies and private industry for determining the response to reasonable anticipated emergency incidents at a specific facility.
Preincident survey The process used to gather information to develop a preincident plan.
Premixed foam Commonly used in fire extinguishers, fire foam extinguishers are charged with compressed air or nitrogen, which discharges the foam solution through an attached aerating nozzle.
Pressure gauge A device that measures and displays pressure readings. In an SCBA, the pressure gauges indicate the quantity of breathing air that is available at any time.
Pressure indicator A gauge on a pressurized portable fire extinguisher that indicates the internal pressure of the expellant.
Pressure points Points where a blood vessel lies near a bone; pressure can be applied to these points to help control bleeding.
Primary cut The main ventilation opening made in a roof to allow smoke, heat, and gases to escape.
Primary feeder The largest diameter pipes in a water distribution system, carrying the greatest amounts of water.
Primary search An initial search conducted to determine if there are victims who must be rescued.
Private water system A privately owned water system operated separately from the municipal water system.
Products of combustion Heat, smoke, and toxic gases.
Projected windows Also called factory windows. Usually found in older warehouse or commercial buildings. They project inward or outward on an upper hinge.
Proprietary system A fire alarm system that transmits a signal to a monitoring location owned and operated by the facility's owner.
Protection plates Reinforcing material placed on a ladder at chaffing and contact points to prevent damage from friction and contact with other surfaces.
Protective hood A part of a fire fighter's PPE designed to be worn over the head and under the helmet to provide thermal protection for the neck and ears.
Protein foam An organic foam that is made from animal byproducts.
Pry axe A specially designed hand axe that serves multiple purposes. Similar to a Halligan bar, it can be used to pry, cut, and force doors, windows, and many other types of objects. Also called a multipurpose axe.
Pry bar A specialized prying tool made of a hardened steel rod with a tapered end that can be inserted into a small area.
Public Information Officer The position within IMS responsible for gathering and releasing incident information to the media and other appropriate agencies.
Public safety answering point (PSAP) The community's emergency response communications center.
Pulaski axe A hand tool that combines an adze and an axe for brush removal.
Pulley A small, grooved wheel through which the halyard runs. The pulley is used to change the direction of the halyard pull, so that a downward pull on the halyard creates an upward force on the fly section(s).
Pulmonary edema Fluid buildup in the lungs.
Pulse The wave of pressure created by the heart as it contracts and forces blood out into the major arteries.
Pump tank extinguishers Nonpressurized, manually operated water extinguishers, usually have a nozzle at the end of a short hose.
Pump tank water-type fire extinguisher A nonpressurized portable water fire extinguisher. Discharge pressure is provided by a hand-operated double-acting piston pump.
Puncture wounds A wound resulting from a bullet, knife, ice pick, splinter, or any other pointed object.
Pyrolysis The chemical decomposition of a compound into one or more other substances by heat alone; pyrolysis often precedes combustion.
Pyromaniac A pathological fire-setter.
Rabbet A type of door frame that has the stop for the door cut into the frame.
Rabbet tool Hydraulic spreading tool designed to pry open doors that swing inward.
Radial artery The major artery in the forearm; it is palpable at the wrist on the thumb side.
Radiation The combined process of emission, transmission, and absorption of energy traveling by electromagnetic wave propagation between a region of higher temperature and a region of lower temperature.
Radiation dispersal device Any device that causes the purposeful dissemination of radioactive material without a nuclear detonation; a dirty bomb.
Radio repeater system A radio system that automatically retransmits a radio signal on a different frequency.
Radioactive isotope An atom with unequal numbers of protons and neutrons in the nucleus that emits radioactivity.
Radioactivity The spontaneous decay or disintegration of an unstable atomic nucleus accompanied by the emission of radiation.
Radiological agents Materials that emit radioactivity.
Rafters Solid structural components that support a roof.
Rail The top or bottom piece of a trussed-beam assembly used in the construction of a trussed ladder. The term rail is also sometimes used to describe the top and bottom surfaces of an I-beam ladder. Each beam will have two rails.
Rapid intervention company/crew (RIC) A minimum of two fully equipped personnel on site, in a ready state, for immediate rescue of injured or trapped fire fighters. In some departments, this is also known as a Rapid Intervention Team.
Rapid mass decontamination A process of flushing a large number of victims with water for rapid decontamination in the field when the agent is unknown.
Rapid oxidation Chemical process that occurs when a fuel is combined with oxygen, resulting in the formation of ash or other waste products and the release of energy as heat and light.
Rate-of-rise heat detector A device that responds when the temperature rises at a rate that exceeds a predetermined value.
Rear of the fire The side opposite the head of the fire, also called the heel of the fire.
Reciprocating saw Powered by electric or battery motors, a reciprocating saw's blade moves back and forth.
Recommended Exposure Level (REL) The established standard limit of exposure to a hazardous material. It is based on the maximum time-weighted concentration at which 95% of exposed, healthy adults suffer no adverse effects over a 40-hour work week. Also knows as the Permissible Exposure Limit (PEL).
Reconnaissance report The inspection and exploration of a specific area in order to gather information for the incident commander.
Recovery phase The stage of a hazardous materials incident after imminent danger has passed, when cleanup and the return to normalcy have begun.
Red blood cells Cells that carry oxygen to the body's tissues.
Reducer A device that can join two hoses of different sizes.
Regulations Rules, usually issued by a government or other legally authorized agency, that dictate how something must be done; regulations are often developed to implement a law.
Rehabilitate To restore to a condition of health or to a state of useful and constructive activity.
Rehabilitation A systematic process to provide periods of rest and recovery for emergency workers during an incident; usually conducted in a designated area away from the hazardous area.
Rekindle A situation where a fire, which was thought to be completely extinguished, reignites.
Relative humidity The ratio of the amount of water vapor present in the air compared to the maximum amount the air can hold at a given temperature.
Remote annunciator A secondary fire alarm control panel in a different location than the main alarm panel, usually near the front door of a building.
Remote station system A fire alarm system that sounds an alarm in the building and transmits a signal to the fire department or an off-premise monitoring location.
Removal A mode of decontamination that applies specifically to contaminated soil that is taken away from the scene.

Rescue Those activities directed at locating endangered persons at an emergency incident, removing those persons from danger, treating the injured, and providing for transport to an appropriate health care facility.
Rescue breathing Artificial means of breathing for a victim.
Rescue company Rescue companies usually are tasked with rescue of victims from fires, confined spaces, trenches, and high-angle situations.
Rescue lift airbags An extrication tool, consisting of airbags, filler hoses, air regulators, control valves, and a supply of compressed air. Used to lift vehicles involved in crashes.
Reservoir A water storage facility.
Residential sprinkler system A sprinkler system designed to protect dwelling units.
Residual pressure The pressure remaining in a water distribution system while water is flowing. The residual pressure indicates how much more water is potentially avaiable.
Resource management A standard system of assigning and keeping track of the resources involved in the incident.
Respirator A protective device used to provide safe breathing air to a user in a hostile or dangerous atmosphere.
Respiratory arrest Sudden stoppage of breathing.
Response Activities that occur in preparation for an emergency and continue until the arrival of emergency apparatus at the scene.
Retention The process of purposefully collecting hazardous materials in a defined area; may include creating that area, by digging, for instance.
Reverse lay A method of laying a supply line where the line starts at the attack engine and ends at the water source.
Rigor mortis The temporary stiffening of muscles that occurs several hours after death.
Rim locks Surface or interior mounted locks on or in a door with a bolt that provide additional security.
Risk-benefit analysis The process of weighing predicted risks against potential benefits and making decisions based on the outcome of that analysis.
Rocker lug (rocker pins) Fittings on threaded couplings that aid in coupling the hoses.
Rollover (flameover) The condition where unburned products of combustion from a fire have accumulated in the ceiling layer of gas to a sufficient concentration (i.e., at or above the lower flammable limit) that they ignite momentarily.
Roof covering The material or assembly that makes up the weather-resistant surface of a roof.
Roof decking The rigid component of a roof covering.
Roof hooks The spring-loaded, retractable, curved metal pieces that allow the tip of a roof ladder to be secured to the peak of a pitched roof. The hooks fold outward from each beam at the top of a roof ladder.
Roof ladder (hook ladder) A straight ladder equipped with retractable hooks so that the ladder can be secured to the peak of a pitched roof. Once secured, the ladder lies flat against the surface of the roof, providing secure footing for fire fighters.
Roofman's hook A long pole with a solid metal hook used for pulling.
Rope bag A bag used to protect and store rope so that the rope can be easily and rapidly deployed without kinking.
Rope record A record for each piece of rope that includes a history of when the rope was placed in service, when it was inspected, when and how it was used, and what types of loads were placed on it.
Rotary saw Powered by electric motors or gasoline engines, a rotary saw uses a large rotating blade to cut through material. The blades can be changed depending upon the material that is being cut.
Round turn A piece of rope looped to form a complete circle with the two ends parallel.
Rubber-covered hose (rubber-jacket hose) Hose whose outside covering is made of rubber, said to be more resistant to damage.
Run cards Cards used to determine a predesignated response to an emergency.
Rung A ladder crosspiece that provides a climbing step for the user. The rung transfers the weight of the user out to the beams of the ladder or back to a center beam on an I-beam ladder.
Running end The part of a rope used for lifting or hoisting.
Safe haven A temporary place of refuge to await rescue.
Safety knot A knot used to secure the leftover working end of the rope. Also known as an overhand knot or keep knot.
Safety Officer The position within IMS responsible for identifying and evaluating hazardous or unsafe conditions at the scene of the incident. Safety officers have the authority to stop any activity that is deemed unsafe.
Salvage Removing or protecting property that could be damaged during firefighting or overhaul operations.
Salvage cover Large square or rectangular sheets made of heavy canvas or plastic material spread over furniture and other items to protect them from water run-off and falling debris.
San Francisco hook A multipurpose tool that can be used for several forcible entry and ventilation applications because of its unique design, which includes a built-in gas shut-off and directional slot.
Saponification The process of converting the fatty acids in cooking oils or fats to soap or foam.
Sarin A nerve agent that is primarily a vapor hazard.
SCBA Harness Part of SCBA that allows fire fighters to wear it as a "backpack."
SCBA regulators Part of the SCBA that reduces the high pressure in the cylinder to a usable lower pressure and controls the flow of air to the user.
Screwdriver A tool used for turning screws.
SCUBA dive rescue technician A responder trained to handle water rescues and emergencies, including recovery and search procedures, in both water and under-ice situations. (SCUBA stands for self-contained underwater breathing apparatus.)
Scupper An opening through which water can be removed from a building.
Search The process of looking for victims who are in danger.
Search and rescue The process of searching a building for a victim and extricating the victim from the building.
Search rope A guide rope used by fire fighters that allows them to maintain contact with a fixed point.
Seat of the fire The main area of the fire.
Seatbelt cutter A specialized cutting device that cuts through seatbelts.
Secondary collapse A collapse that occurs following the primary collapse. This can occur in trench, excavation, and structural collapses.
Secondary containment Any device or structure that prevents environmental contamination when the primary container or its appurtenances fail. Examples of secondary containment include dikes, curbing, and double-walled tanks.
Secondary contamination Contamination that occurs when a person or object comes into direct contact with a contaminated person or object, as might occur when a contaminant is carried out of the hot zone and comes into contact with people, animals, equipment, or the environment.
Secondary cut An additional ventilation opening to create a larger opening, or to limit fire spread.
Secondary device An explosive device designed to injure emergency responders who have responded to an initial event.
Secondary feeder Smaller diameter pipes that connect the primary feeders to the distributors.
Secondary loss Property damage that occurs due to smoke, water, or other measures taken to extinguish the fire.
Secondary search A more thorough search undertaken after the fire is under control. This search is done to ensure that there are no victims still trapped inside of the building.
Sector Alternate terminology used for an organizational level defined by either a geographic or functional assignment; comparable to a division or group.
Self-contained breathing apparatus (SCBA) Respirator with independent air supply used by fire fighters to enter toxic and otherwise dangerous atmospheres.
Self-contained underwater breathing apparatus (SCUBA) Respirator with independent air supply used by underwater divers.
Self-expelling A fire extinguisher in which the agents have sufficient vapor pressure at normal operating temperatures to expel themselves.
Self-rescue The activity of a fire fighter using techniques and tools to remove himself from a hazardous situation.
Sensitizer A chemical that causes a large portion of people or animals to develop an allergic reaction after repeated exposure.
Serial arson A series of fires set by the same offender, with a cooling-off period between fires.
Severe Acute Respiratory Syndrome (SARS) A viral respiratory illness caused by a coronavirus.
Shackles The U-shaped part of a padlock that runs through a hasp and secures back into the lock body.

Shelter-in-place A method of safeguarding people in a hazardous area by keeping them in a safe atmosphere, usually inside structures.
Shipping papers A shipping order, bill of lading, manifest, or other shipping document serving a similar purpose and usually including the names and addresses of both the shipper and the receiver as well as a list of shipped materials with quantity and weight.
Shock A state of collapse of the cardiovascular system; the state of inadequate delivery of blood to the organs of the body.
Shock load An instantaneous load that places a rope under extreme tension, such as when a falling load is suddenly stopped when the rope becomes taut.
Shoring A method of supporting a trench wall or building components such as walls, floors, or ceilings using either hydraulic, pneumatic, or wood shoring systems; it is used to prevent collapse.
Shut-off valve Any valve that can be used to shut down water flow to a water user or system.
Siamese A device that allows two hoses to be connected together and flow into a single hose.
Sidewall sprinkler head A sprinkler that is mounted on a wall and discharges water horizontally into a room.
Signal words Information on a pesticide label that indicates the relative toxicity of the material.
Simplex channel Radio system that uses one frequency to transmit and receive all messages.
Single resource An individual vehicle and the personnel that arrive on that unit.
Single-action pull-station A manual fire alarm activation device that takes a single step, such as moving a lever, toggle, or handle, to activate the alarm.
Single-station smoke alarm A single device that both detects visible or invisible products of combustion and sounds an alarm.
Size-up The ongoing observation and evaluation of factors that are used to develop objectives, strategy, and tactics for fire suppression.
Slash The leftovers of a logging operation; includes large and small pieces of logs, branches, bark, stumps, and other vegetative debris.
Sledgehammer A long, heavy hammer that requires the use of both hands.
Small Diameter Hose (SDH) Hose in the 1″ to 2″ range.
Smallpox A highly infectious disease caused by the virus variola.
Smoke An airborne particulate product of incomplete combustion suspended in gases, vapors, or solid and liquid aerosols.
Smoke detector A device that detects smoke and sends a signal to a fire alarm control panel.
Smoke inversion Smoke hanging low to the ground due to the cold air.
Smoke particles Airborne solid material consisting of ash and unburned or partially burned fuel released by a fire.
Smooth bore nozzle Nozzle that produces a straight stream that is a solid column of water.
Smooth bore tip A nozzle device that is a smooth tube, used to deliver a solid column of water.
Sniffing position Position where the head and chin are thrust slightly forward.
Socket wrench A wrench that fits over a nut or bolt and uses the ratchet action of an attached handle to tighten or loosen the nut or bolt.
Soft suction hose A large diameter hose that is designed to be connected to the large port on a hydrant (steamer connection) and into the engine.
Solid One of three phases of matter. A substance that has three dimensions and is firm in substance.
Solid beam A ladder beam constructed of a solid rectangular piece of material, typically wood, to which the ladder rungs are attached.
Solid-core A door design that consists of wood filler pieces inside the door. This creates a stronger door that may be fire-rated.
Solidification The process of chemically treating a hazardous liquid to turn it into a solid material, making the material easier to handle.
Solid stream A stream made by using a smooth bore nozzle to produce a penetrating stream of water.
Soman A nerve gas that is both a contact and a vapor hazard that has the odor of camphor.
Sounding The process of striking a roof with a tool to determine if the roof is solid enough to support the weight of a fire fighter.
Spalling Chipping or pitting of concrete or masonry surfaces.
Span of control The number of people that single person supervises. The maximum number of people that one person effectively is about five.
Spanner wrench A type of wrench used in coupling or uncoupling hoses by turning the rocker lugs on the connections.
Special use railcars Boxcars, flat cars, cryogenic and corrosive tank cars, and high-pressure compressed gas tube-type cars.
Specialist level Identified only in the HAZWOPER standard, this level receives more specialized training than a hazardous materials technician. Practically speaking, the two levels are not very different.
Specific gravity Specific gravity is the weight of a liquid as compared to water.
Split hose bed A hose bed arranged such that a 1″-long supply line can be laid out, or two supply lines can be laid out.
Split hose lay A scenario where the attack engine will lay a supply line from an intersection to the fire, and the supply engine will lay a supply line from the hose left by the attack engine to the water source.
Spoil pile The pile of dirt that has been removed from an excavation, which may be unstable and prone to collapse.
Spot detector Single heat-detector devices, spaced throughout an area.
Spot fire A new fire that starts outside areas of the main fire; usually caused by flying embers and sparks.
Spotlight A light designed to project a narrow, concentrated beam of light.
Spree arson A series of fires started by an arsonist who sets three or more fires at separate locations with no emotional cooling-off period between fires.
Spring-loaded center punch A spring-loaded punch used to break automobile glass.
Sprinkler piping The network of piping in a sprinkler system that delivers water to the sprinkler heads.
Sprinkler stop A mechanical device inserted between the deflector and the orifice of a sprinkler head to stop the flow of water.
Sprinkler system An automatic fire protection system designed to turn on sprinklers if a fire occurs.
Sprinkler wedge A piece of wedge-shaped wood placed between the deflector and the orifice of a sprinkler head to stop the flow of water.
Squelch An electric circuit designed to cut off weak radio transmissions that are only capable of generating noise.
Stack effect The vertical movement of smoke and products of combustion within a high-rise building caused by temperature differentials between the interior and exterior.
Staging area A prearranged, strategically placed area where support personnel, vehicles, and other equipment can be held in an organized state of readiness for use during an emergency.
Standard operating procedures (SOPs) Written rules, policies, regulations, and procedures enforced to structure the normal operations of most fire departments.
Standing orders Written documents, signed by the system's medical director, that outline specific directions, permissions, and sometimes prohibitions regarding patient care; also called protocols.
Standing part The part of a rope between the working end and the running end.
Standpipe system An arrangement of piping, valves, and hose connections installed in a structure to deliver water for fire hoses.
State Emergency Response Commission (SERC) The liaison between local and state levels that collects and disseminates information relating to hazardous materials; SERC involves agencies such as the fire service, police services, and elected officials.
State of matter The physical state of a material—solid, liquid, or gas.
Static pressure The pressure in a water pipe when there is no water flowing.
Static rope A rope generally made out of synthetic material that stretches very little under load.
Static water source A water source such as a pond, river, stream, or other body of water that is not under pressure.
Static water supply A water supply that is not under pressure, such as a pond, lake, or stream.
Staypole (also known as a tormentor) A long piece of metal attached to the top of the bed section of an extension ladder and used to help stabilize the ladder during raising and lowering. The pole attaches to a swivel point and has a spur on the other end. One pole is attached to each beam of long (40′ or longer) extension ladders.

Steamer port The large diameter port on a hydrant.
Steel joists A lightweight steel truss used as a horizontal beam.
Step blocks Specialized cribbing assemblies made out of wood or plastic blocks in a step configuration. These are usually used to stabilize vehicles.
Sternum The breastbone.
Stop A piece of material that prevents the fly section(s) of a ladder from overextending and collapsing the ladder.
Stored-pressure extinguisher A fire extinguisher in which both the extinguishing agent and the expellant gas are kept in a single container; generally equipped with a pressure indicator or gauge.
Stored-pressure water-type fire extinguisher A fire extinguisher in which water or a water-based extinguishing agent is stored under pressure.
Storz-type coupling A hose coupling that has the property of being both the male and female coupling. It is connected by engaging the lugs and turning the coupling a one-third turn.
Straight stream A stream made by using an adjustable nozzle that is set to provide a straight stream of water.
Strike team Five units of the same resource category, such as engines or ambulances, with a leader.
Strike team leader The person in charge of a strike team, responsible to the next higher level in the incident organization, and the point-of-contact for the strike team within the organization.
Striking tools Tools designed to strike other tools or objects such as walls, doors, or floors.
Subsurface fuels Partially decomposed matter that lies beneath the ground, such as roots, moss, duff, and decomposed stumps.
Suitcase nuke A nuclear explosive device that is small enough to fit in a suitcase.
Sulfur mustard A clear, yellow, or amber oily liquid with a faint sweet odor of mustard or garlic that may be dispersed in an aerosol form. It causes blistering of exposed skin.
Superfund Amendment Reauthorization Act (SARA) One of the first laws to affect how fire departments respond in a hazardous material emergency.
Supplied air breathing apparatus (SABA) Emergency breathing systems similar to SCBA, but which utilize an airline running from the rescuers to a fixed air supply located outside of the confined space.
Supplied-air respirator (SAR) A respirator that gets its air through a hose from a remote source, such as a compressor or storage cylinder.
Supply hose (supply line) The hose used to deliver water from a source to a fire pump.
Surface fuels Fuels that are close to the surface of the ground, such as grass, leaves, twigs, needles, small trees, logging slash, and low brush; also called ground fuels.
Surface to mass ratio (STMR) The surface area of a material in proportion to the mass of that material.
Surface to volume ratio The surface area of a liquid in proportion to the mass.
Sweep (roll-on) method Foam applying method that is done by sweeping the stream just in front of the target.
Tabun A nerve gas that is both a contact and a vapor hazard that operates by disabling the chemical connection between nerve and target organs.
Talk around channel A simplex channel used for on-site communications
Tamper seal A retaining device that breaks when the locking mechanism is released.
Tamper switch A switch on a sprinkler valve that transmits a signal to the fire alarm control panel if the normal position of the valve is changed.
Tanker shuttle A method of transporting water from a source to a fire scene using a number of mobile water supply apparatus.
Target hazard Any occupancy type or facility that presents a high potential for loss of life or serious impact to the community resulting from fire, explosion, or chemical release.
Task force Any combination of single resources assembled for a particular tactical need; has common communications and a leader.
Task force leader The person in charge of a task force, responsible to the next higher level in the incident organization, and the point-of-contact for the task force within the organization.
TDD/TTY/Text Phone system User devices that allow speech- and/or hearing-impaired citizens to communicate over a telephone system. TDD stands for Telecommunications Device for the Deaf; TTY stands for teletype, and Text Phones visually display text. The displayed text is the equivalent of a verbal conversation between two hearing persons.
Technical rescue incident (TRI) A complex rescue incident involving vehicles or machinery, water or ice, rope techniques, a trench or excavation collapse, confined spaces, a structural collapse, an SAR, or hazardous materials, and which requires specially trained personnel and special equipment.
Technical rescue team A group of rescuers specially trained in the various disciplines of technical rescue.
Technical rescue technician Someone trained in special rescue techniques for incidents involving structural collapse, trench rescue, swiftwater rescue, confined space rescue, and other unusual rescue situations.
Technician level At this level, responders are allowed to enter heavily contaminated areas using the highest levels of chemical protection. This is the most aggressive level of training as identified in NFPA 472. Hazardous materials technicians take offensive actions.
Technician level (TRI) The training level that provides a high level of competency in the various disciplines of technical or hazardous materials rescue for rescuers who will be directly involved in the rescue operation itself.
Telecommunicator A trained individual responsible for answering requests for emergency and nonemergency assistance from citizens. This individual assesses the need for response and alerts responders to the incident.
Telephone interrogation Phase in a 9-1-1 call when a telecommunicator asks questions to obtain vital information such as the location of the emergency.
Tempered glass A type of safety glass that is heat-treated so that it will break into small pieces that are not as dangerous.
Temporal-3 pattern A standard fire alarm audible signal for alerting occupants of a building.
Ten-codes System of predetermined, coded messages, such as "What is your 10-20?" used by responders over the radio.
Thermal column A cylindrical area above a fire in which heated air and gases rise and travel upward.
Thermal conductivity Describes how quickly a material will conduct heat.
Thermal imaging devices Electronic devices that detect differences in temperature based on infrared energy and then generate images based on that data. Commonly used in obscured environments to locate victims.
Thermal layering The stratification, or heat layers, that occur in a room as a result of a fire.
Thermoplastic material Plastic material capable of being repeatedly softened by heating and hardened by cooling and, that in the softened state, can be repeatedly shaped by molding or forming.
Thermoset material Plastic material that, after having been cured by heat or other means, is substantially infusible and cannot be softened and formed.
Threaded hose couplings A type of coupling that requires a male and female fitting to be screwed together.
Threshold Limit Value/Ceiling (TLV/C) The maximum concentration of hazardous material that a worker should not be exposed to, even for an instant.
Threshold Limit Value/Short-Term Exposure Limit (TLV/STEL) The maximum concentration of hazardous material to which a worker can sustain a 15-minute exposure not more than four times daily with a 60-minute rest period required between each exposure. The lower the value, the more toxic the substance.
Threshold Limit Value/Skin Indicates that direct or airborne contact with a material could result in possible and significant exposure from absorption through the skin, mucous membranes, and eyes.
Threshold Limit Value/Time Weighted Average (TLV/ TWA) The airborne concentration of a material to which a worker can be exposed for 8 hours a day, 40 hours a week and not suffer any ill effects.
Tie rod A metal rod that runs from one beam of the ladder to the other to keep the beams from separating. Tie rods are typically found in wood ladders.
Time marks Status updates provided to the communications center every 10 to 20 minutes. This update should include the type of operation, the progress of the incident, the anticipated actions, and the need for additional resources.
Tip The very top of the ladder.
Topography The features of the earth's surface; changes in land elevation and the position of natural and manmade features.
Totes Portable tanks, usually holding a few hundred gallons of product, characterized by a unique style of construction.
Toxic products of combustion Hazardous chemical compounds released when a material decomposes under heat.
Toxicology The study of the adverse effects of chemical or physical agents on living organisms.

Trace (transfer) evidence Evidence of a minute quantity that is conveyed from one place to another.

TRACEM An acronym to help remember the effects and potential exposures to a hazardous materials incident—**T**hermal, **R**adiation, **A**sphyxiant, **C**hemical, **E**tiologic, and **M**echanical.

Trachea The windpipe.

Trailers Combustible material, such as rolled rags, blankets, and newspapers or flammable liquid, used to spread fire from one point or area to other points or areas, often used in conjunction with an incendiary device.

Training officer Responsible for updating the training of current employees and for training new fire fighters in the current techniques of firefighting and EMS.

Transfer of command Reassignment of command authority and responsibility from one individual to another.

Trench cut A cut that is made from bearing wall to bearing wall to prevent horizontal fire spread in a building.

Triage The process of sorting victims based on the severity of injury and medical needs to establish treatment and transportation priorities.

Triangular cut A triangle-shaped ventilation cut in the roof decking that is made using saws or axes.

Trigger The button or lever used to discharge the agent from a portable fire extinguisher.

Triple layer load A hose loading method that utilizes folding the hose back onto itself to reduce the overall length to one-third before loading in the bed. This load method reduces deployment distances.

Truck company Truck companies specialize in forcible entry, ventilation, roof operations, search-and-rescue operations above the fire, and deployment of ground ladders.

Trunking system A radio system that uses a shared bank of frequencies to make the most efficient use of radio resources.

Truss A collection of lightweight structural components joined in a triangular configuration that can be used to support either floors or roofs.

Truss block A piece of wood or metal that ties the two rails of a trussed beam ladder together and serves as the attachment point for the rungs.

Trussed beam A ladder beam constructed of top and bottom rails joined by truss blocks that tie the rails together and support the rungs.

Tube trailers High-volume transportation devices made up of several individual compressed gas cylinders banded together and affixed to a trailer.

Turnout coat Protective coat that is part of a protective clothing ensemble for structural firefighting; also called a bunker coat.

Turnout gear Defined as helmet, hood, coat, pants, boots, gloves, personal alert safety system, and self-contained breathing apparatus. Sometimes referred to as personal protective equipment.

Turnout pants Protective trousers that are part of a protective clothing ensemble for structural firefighting; also called bunker pants.

Twisted rope Rope constructed of fibers twisted into strands, which are then twisted together.

Two-in/two-out rule A safety procedure that requires a minimum of two personnel to enter a hazardous area and a minimum of two back-up personnel to remain outside the hazardous area during the initial stages of an incident.

Two-person rope A rope rated to carry the weight of two people (600 lbs).

Two-way radio A portable communication device used by fire fighters. Every firefighting team should carry at least one radio to communicate distress, progress, changes in fire conditions, and other pertinent information.

Type I construction Buildings with structural members made of noncombustible materials that have a specified fire resistance.

Type II construction Buildings with structural members made of noncombustible materials without fire resistance.

Type III construction Buildings with the exterior walls made of noncombustible or limited-combustible materials, but interior floors and walls made of combustible materials.

Type IV construction Buildings constructed with noncombustible or limited-combustible exterior walls, and interior walls and floors made of large dimension combustible materials.

Type V construction Buildings with exterior walls, interior walls, floors, and roof structures made of wood.

Underwriters' Laboratories Inc. (UL) The U.S. organization that tests and certifies that fire extinguishers (among many other products) meet established standards. The Canadian equivalent is Underwriters' Laboratories of Canada (ULC).

Undetermined Cause classification that includes fires for which the cause has not or cannot be proven.

Unibody (Unit body) Most commonly used frame construction in vehicles. The base unit is made of formed sheet metal and structural components then added to the base to form the passenger compartment. Subframes are attached to each end. This type of construction does away with rail-beams used in platform framed vehicles.

Unified command IMS option that allows representatives from multiple jurisdictions and/or agencies to share command authority and responsibility, working together as a "joint" incident command team.

Unity of command A characteristic of the IMS structure that has each individual reporting to a single supervisor and everyone reporting to the IC directly or through the chain of command.

Universal precautions Procedures for infection control that treat blood and certain bodily fluids as capable of transmitting bloodborne diseases.

Unlocking device A key way, combination wheel, or combination dial.

Unwanted alarm A fire alarm signal caused by a device reacting properly to a condition that is not a true fire emergency.

Upper explosive limits (UEL) Defines the boundaries of a fuel/air mixture necessary for a combustible material to burn properly.

Upright sprinkler head A sprinkler head designed to be installed on top of the supply piping and usually marked SSU (Standard Spray Upright).

Utility rope Rope used for securing objects, for hoisting equipment, or for securing a scene to prevent bystanders from being injured. It is never to be used in life safety operations.

Vacuuming The process of cleaning up dusts, particles, and some liquids using a vacuum with HEPA filtration to prevent recontamination of the environment.

V-agent A nerve agent, principally a contact hazard; an oily liquid that can persist for several weeks.

Vapor The gas phase of a substance, particularly of those that are normally liquids or solids at ordinary temperatures.

Vapor density The weight of an airborne concentration (vapor or gas) as compared to an equal volume of dry air.

Vapor dispersion The process of separating and diminishing vapors, usually by using a water spray to disperse or move vapors away from certain areas or materials.

Vapor dispersion (hazardous materials) The process of lowering the concentration of vapors by spreading them out.

Vapor pressure The vapor pressure of a liquid is the pressure exerted by its vapor until the liquid and vapor are in equilibrium.

Vapor suppression The process of controlling vapors given off by hazardous materials, and therefore preventing vapor ignition, by covering the product with foam or other material or by reducing the temperature of the material.

Vapor-protective clothing The garment portion of a chemical-protective ensemble designed and configured to protect the wearer against chemical vapors or gases.

Venous bleeding External bleeding from a vein, characterized by steady flow; the bleeding may be profuse and life-threatening.

Vent pipes Inverted J-shaped tubes that allow for pressure relief or natural venting of the pipeline for maintenance and repairs.

Ventilation The process of removing smoke, heat, and toxic gases from a burning structure and replacing them with clean air.

Ventricle Either of the two lower chambers of the heart.

Verification system A fire alarm system that does not immediately initiate an alarm condition when a smoke detector activates. The system will wait a preset interval, generally 30 to 60 seconds, before checking the detector again. If the condition is clear, the system returns to normal status. If the detector is still sensing smoke, the system activates the fire alarm.

Vertical ventilation The process of making openings so that the smoke, heat, and gases can escape vertically from a structure.

Voice recording system Recording devices or computer equipment connected to telephone lines and radio equipment in a communications center to record telephone calls and radio traffic.

Volatility The ready ability of a substance to produce combustible vapors.

VX A nerve agent.

Wall post indicator valve (WPIV) A sprinkler control valve that is mounted on the outside wall of a building. The position of the indicator tells whether the valve is open or shut.

Warm zone The area located between the hot zone and the cold zone at an incident. PPE is required. The decontamination corridor is located in the warm zone.

Water catch-all A salvage cover folded to form a container to hold water until it can be removed.

Water chute A salvage cover folded to direct water flow out of a building or away from sensitive items or areas.

Water curtain nozzles Nozzles used to deliver a flat screen of water to form a protective sheet of water.

Water hammer An event that occurs when flowing water is suddenly stopped; the velocity force of the moving water is transferred to everything it is in contact with. These can be tremendous forces that can damage equipment and cause injury.

Water main The generic term for any underground water pipe.

Water solubility Ability of a substance to dissolve in water.

Water supply A source of water.

Water thief A device that has a 2½″ inlet and a 2½″ outlet in addition to two 1½″ outlets. The device is used to supply many hoses from one source.

Water vacuum A device similar to a household vacuum cleaner, but with the ability to pick up liquids. Used to remove water from buildings.

Water-motor gong An audible alarm notification device that is powered by water moving through the sprinkler system.

Waybill Shipping papers for trains.

Weapons of mass destruction (WMD) A weapon intended to cause mass casualties, damage, and chaos.

Wedges Used to snug loose cribbing under the load or when using lift airbags to fill the void between the crib and the object as it is raised.

Wet chemical extinguisher A fire extinguisher for use on Class K fires that contains wet chemical extinguishing agents.

Wet chemical extinguishing agent An extinguishing agent for Class K fires; commonly uses solutions of water and potassium acetate, potassium carbonate, potassium citrate, or any combination thereof.

Wet chemical extinguishing systems An extinguishing system that discharges a proprietary liquid extinguishing agent.

Wet sprinkler system A sprinkler system in which the pipes are normally filled with water.

Wet-barrel hydrant A hydrant used in areas that are not susceptible to freezing. The barrel of the hydrant is normally filled with water.

Wetting-agent water-type extinguisher An extinguisher that expels water combined with a chemical or chemicals to reduce its surface tension.

Wheeled fire extinguisher A portable fire extinguisher equipped with a carriage and wheels intended to be transported to the fire by one person.

White blood cells Blood cells that play a role in the body's immune defense mechanisms against infection.

Wilderness search and rescue (SAR) The process of locating and removing a victim from the wilderness.

Wildland Land in an uncultivated natural state that is covered by timber, woodland, brush, or grass.

Wildland/brush company Wildland/brush companies are dispatched to woods and brush fires where larger engines cannot gain access. Wildland/brush companies have four-wheel drive vehicles and special firefighting equipment.

Wildland fire An unplanned and uncontrolled fire burning in vegetative fuels.

Wildland urban interface An area where undeveloped land with vegetative fuels is mixed with manmade structures.

Wire cutters Tool to cut into small gauge wire and fences.

Wired glass Glass made by molding glass around a special wire mesh.

Wood panels Thin sheets of wood glued together.

Wood trusses Assemblies of small pieces of wood or wood and metal.

Wooden beams Load-bearing members assembled from individual wood components.

Wood-frame construction Buildings with exterior walls, interior walls, floors, and roof made of combustible wood material. Type V building construction, as defined in NFPA 220, *Types of Building Construction.*

Working end The part of the rope used for forming the knot.

Wye A device used to split a single hose into two separate lines.

Xiphoid process The flexible cartilage at the lower end of the sternum (breastbone), a key landmark in the administration of CPR and the Heimlich maneuver.

Zoned coded alarm A fire alarm system that indicates which zone was activated both on the alarm control panel and through a coded audio signal.

Zoned non-coded alarm A fire alarm system that indicates the activated zone on the alarm control panel.

Zoned system A fire alarm system design that divides a building or facility into zones so the area where an alarm originated can be identified.

Index

C

D

E

G

J

K

L

O

P

Q

R

S

U

V

W

X

Y

Z

·e Fighter in Action © :hael Heller/911 Pictures ·ices of Experience © Corbis

hapter 1
·pener Courtesy of Aurora :gional Fire Museum; 1-2 © ·nicago Historical Society #ICHi- 2808; 1-11 Courtesy of AAOS; 1- 3 © Craig Jackson/In the Dark ·notography; 1-14 Courtesy of FPA

hapter 2
·pener © Nancy ·reifenhagen/911 Pictures; 2-2 ·ourtesy of NFPA; 2-3 © Tom ·arter/911 Pictures; 2-24 Courtesy f Draeger Safety; 2-33 Courtesy f Micmac Fire and Safety Ltd

hapter 3
·pener Courtesy of Captain David ·ackson, Saginaw Township Fire ·epartment; 3-9 Courtesy of ·Hamish Reid/Mistrale

Chapter 4
·Opener © Michael Heller/911 Pictures; 4-1 Courtesy of the USDA Forest Service; 4-2, 4-17 © Keith D. Cullom; 4-3 © Dan Myers; 4-8 Courtesy of Captain David Jackson, Saginaw Township Fire Department

Chapter 5
Opener © Glen E. Ellman

Chapter 6
Opener Courtesy of Dr. Edwin P. Ewing, Jr./The Centers for Disease Control and Prevention; 6-1, 6-5 © AbleStock; 6-2 Courtesy of Captain David Jackson, Saginaw Township Fire Department; 6-6 © John Foxx/Alamy Images; 6-7 Courtesy of Universal Fire Shield; 6-9 © Ken Hammon/USDA; 6-10 Courtesy of NFPA; 6-13 © Justin Kase/Alamy Images; 6-14 © Calvin College. Used with permission. Photograph by Luke Robinson; 6-15A: © AbleStock; B: Courtesy of Pacific Northwest National Laboratory; C: Courtesy of Doyal Dum/US Army Corps of Engineers; D: © AbleStock; 6-16A: © Paul Springett/Up the Resolution/Alamy Images; B: Ron Chapple/Thinkstock/ Alamy Images; C: TH Photo/Alamy Images

Chapter 7
7-1, 7-18, 7-19 Courtesy of Amerex Corporation; 7-6 Courtesy of Tyco Fire and Building Products; 7-7 © Linda Gheen; 7-8 Courtesy of Vent-A-Hood Company; 7-15 © James Smalley/Index Stock Imagery, Inc; 7-SD3, 7-SD7 Courtesy of Berta A. Daniels

Chapter 8
8-12, 8-17, 8-30 Courtesy of Berta A. Daniels; 8-22 © Dan Myers

Chapter 9
9-9 Courtesy of Yale Cordage; 9-10 Courtesy of Michael Nevins/US Army Corps of Engineers; Detail courtesy of John Symonds; 9-19 Courtesy of Huy Richard Mach/*Loudoun Times-Mirror*; 9-20 Courtesy of Captain David Jackson, Saginaw Township Fire Department; 9-23 © AbleStock; 9-24 © Michael Heller/911 Pictures

Chapter 10
Opener © David French/911 Pictures; 10-5 © Warren Gretz/NREL; 10-7 Courtesy of Berta A. Daniels; 10-10 © Glen E. Ellman; 10-18 © Gregory Price, *Lewiston Sun Journal*/AP Photo

Chapter 11
11-8 © AbleStock; 11-13 Courtesy of Horton Automatics; 11-14A: Courtesy of St. Cloud Overhead Door Company; B: © Photodisc/Creatas; 11-16, 11-19, 11-20 Used with permission of Andersen Corporation; 11-21 Courtesy of Potomac Communications Group, Inc; 11-23, 11-26, 11-27 Courtesy of IR Safety and Security—Americas; 11-29 © Ed Hancock/NREL

Chapter 12
12-2 Courtesy of NASA-AMES Disaster Assistance and Rescue Team; 12-6, 12-7, 12-26 Courtesy of Dennis Wetherhold, Jr; 12-8, 12-10, 12-13, 12-14 Courtesy of Duo-Safety Ladder Corporation; 12-20 Courtesy of Underwriters Laboratories, Inc

Chapter 13
13-1 Courtesy of Donald M. Colarusso, ALLHANDSFIRE.COM; 13-2, 13-12 Courtesy of District Chief Chris E. Mickal/New Orleans Fire Department, Photo Unit; 13-14, 13-SD13 Courtesy of AAOS

Chapter 14
Opener, 14-9 © Glen E. Ellman; 14-4 © Keith D. Cullom; 14-10, 14-11 Courtesy of District Chief Chris E. Mickal/New Orleans Fire Department, Photo Unit; 14-14 Courtesy of Super Vacuum Mfg. Co., Inc; 14-19, 14-22, 14-23, 14-30 Courtesy of Captain David Jackson, Saginaw Township Fire Department; 14-20 © 2003 Gene Blevins/CFPA; 14-25 Courtesy of Four Seasons Solar Products

Chapter 15
15-2 © Dan Myers; 15-3 © Tom Carroll/Index Stock Imagery, Inc; 15-7, 15-9 Courtesy of American AVK Company; 15-8, 15-10 Courtesy of Captain David Jackson, Saginaw Township Fire Department; 15-15 Courtesy of District Chief Chris E. Mickal/New Orleans Fire Department, Photo Unit; 15-17 Courtesy of Tom Fields, Douglas Forest Protective Association; 15-18 Courtesy of Dennis Wetherhold, Jr
15-19 Courtesy of Steven Townsend/Code 3 Images; 15-20 © Jim Smalley

Chapter 16
Opener © Chris Mickal/911 Pictures; 16-11, 16-12, 16-30B Courtesy of Akron Brass Company; 16-14, 16-15, 16-16 Courtesy of Berta A. Daniels; 16-29 Courtesy of Flamefighter Corporation; 16-31 Courtesy of POK of North America, Inc

Chapter 17
Opener © Glen E. Ellman; 17-1 Courtesy of District Chief Chris E. Mickal/New Orleans Fire Department, Photo Unit; 17-2 Courtesy of NFPA

Chapter 18
Opener © Glen E. Ellman; 18-3 Courtesy of Tyco Fire and Building Products; 18-6 Courtesy of Cascade Fire Equipment Company; 18-7 Courtesy of Berta A. Daniels; 18-8 Courtesy of GFE Manufacturing, Inc

Chapter 19
19-7 © Stan Honda/AFP Photo/Getty Images; 19-8 Courtesy of Michael Rieger/FEMA News Photo; 19-11 © Roland Gosselin/911 Pictures; 19-15 © Photodisc/Creatas; 19-16 © Mark Adams/SuperStock, Inc; 19-17 © Brandon Jacob/ FourAlarmPhotos.com; 19-19 Courtesy of AAOS

Chapter 20
Opener Courtesy of John Hutmacher/USFS; 20-3 © Karen Wattenmaker Photography; 20-4 Courtesy of Anchor Industries, Inc

Chapter 21
Opener Courtesy of District Chief Chris E. Mickal/New Orleans Fire Department, Photo Unit; 21-7 © Kurt Hegre, *The Fresno Bee*/AP Photo

Chapter 22
Opener Courtesy of Dennis Wetherhold, Jr; 22-7 © John Foxx/Alamy Images; 22-8 © Joe Sohm/Alamy Images; 22-9 © Ken Hammon/USDA; 22-10 Courtesy of APA - The Engineered Wood Association; 22-12 Courtesy of Underwriters Laboratories Inc © Underwriters Laboratories Inc ®; 22-13 © NREL/DOE; 22-16 © Jim Smalley; 22-17 © Keith D. Cullom; 22-19 © Fred Prouser/Reuters; 22-20 © Sami Sarkis/Photodisc/Getty Images; 22-21 © Corbis/PictureQuest; 22-23 Courtesy of Brad Emerson/US Army Corps of Engineers; 22-24 © Arthur S. Aubry/Photodisc/Getty Images; 22-25 Courtesy of Image*After

Chapter 23
Opener © Eddie M. Sperling; 23-1 © Spencer Grant/PhotoEdit; 23-3, 23-4, 23-5, 23-6, 23-7, 23-9, 23-11, 23-12, 23-13 Courtesy of AAOS; 23-8 © Tony Freeman/ PhotoEdit/ PictureQuest

Chapter 24
Opener © Bruce Preston/911 Pictures; 24-1, 24-2, 24-3, 24-4, 24-5, 24-6, 24-7, 24-8, 24-9, 24-10, 24-11, 24-12, 24-13, 24-14, 24-15, 24-16, 24-17, 24-18, 24-19, 24-20, 24-21, 24-22, 24-23, 24-24, 24-25, 24-26, 24-27, 24-28, 24-29, 24-30, 24-31, 24-32, 24-33, 24-34, 24-35, 24-36, 24-37, 24-38, 24-39, 24-40, 24-41, 24-SD03, 24-SD05, 24-SD12, 24-SD13, 24-SD14 Courtesy of AAOS

...25
... © Craig Jackson/In the ...hotography; 25-1 © Warren Gretz/NREL; 25-3, 25-12 © Dan Myers; 25-4 © SHOUT/ Alamy Images

Chapter 26
26-5 Courtesy of Lara Shane/ FEMA Photo News; 26-8 © Keith D. Cullom; 26-9 Courtesy of Captain David Jackson, Saginaw Township Fire Department; 26-10 Courtesy of Dave Gatley/FEMA; 26-11 © Nate Rawner; 26-13 Courtesy of Richard E. Avallone, Elevator Safety Training Programs Inc

Chapter 27
27-1 © Gary Stelzer, *Middletown Journal*/AP Photo; 27-3 © Mark E. Gibson/Corbis

Chapter 28
Opener © Joaquim Pol Martinez/911 Pictures; 28-8 Courtesy of Dr. Saeed Keshavarz/ RCCI (Research Center of Chemical Injuries)/Iran; 28-10 Courtesy of Survivair

Chapter 29
Opener Placards courtesy of ICC The Compliance Center, Inc; 29-1, 29-2 © USDA; 29-3 Courtesy of EMD Chemicals, Inc; 29-4A, B, and G: Courtesy of Polar Tank Trailer LLC, C: Courtesy of National Tank Truck Carriers Association, D and F: Courtesy of Jack B. Kelley, Inc, E Courtesy of Jack B. Kelley, Inc, and Applied LNG Technologies; 29-5 © Mark Gibson/Index Stock Imagery, Inc/Alamy Images; 29-6 Courtesy of Panhandle Eastern Pipe Line Company, photograph provided by TSI; 29-7, 29-8 Courtesy of AAOS; 29-9 Courtesy of NFPA; 29-10 Courtesy of Syngenta Crop Protection, Inc; 29-11, 29-12 Courtesy of Sandia National Laboratories

Chapter 30
Opener © Ed Kreiser Photography; 30-1 © Gary Stelzer, *Middletown Journal*/AP Photo; 30-3 © Bill Karrow, *Dunkirk Observer*/AP Photo; 30-4 Courtesy of Syngenta Crop Protection, Inc

Chapter 31
Opener Courtesy of the Tempe, Arizona Fire Department; 31-1 © Photodisc; 31-2, 31-3 Courtesy of Lakeland Industries, Inc; 31-5 Courtesy of Survivair; 31-6, 31-9 Courtesy of AAOS; 31-7 Courtesy of US Army Corps of Engineers; 31-8 Courtesy of the United States Fire Administration; 31-10 © NMSB/Custom Medical Stock Photo; 31-11 © Jim Ruymen/Reuters

Chapter 32
Opener Courtesy of Donald M. Colarusso, ALLHANDSFIRE.COM; 32-1 © Glen E. Ellman; 32-2 © Steve Mason/Photodisc/Getty Images; 32-4 © Cory Morse, *The State News*/AP Photo

Chapter 33
Opener © Chris Mickal/911 Pictures; 33-1 © Patrick Schneider, *The Charlotte Observer*/AP Photo; 33-2 © Tom Mihalek/AP Photo/Corbis; 33-4 © Jim Mone/AP Photo

Chapter 34
Opener Courtesy of Andrea Booher/FEMA Photo News; 34-1 © Todd Hollis/AP Photo; 34-2A: © Steve Allen/Brand X Pictures/Alamy Images; B: © AbleStock; C: © Arthur S. Aubry/Photodisc/Getty Images; D: © Craig Jackson/In the Dark Photography; 34-3 © James P. Blair/Photodisc/Getty Images; 34-4 © Photodisc/Getty Images; 34-5 © Larry Rana/USDA; 34-7 Courtesy of FEMA News Photo; 34-8, 34-9, 34-25 Courtesy of Captain David Jackson, Saginaw Township Fire Department; 34-10 Courtesy of the Department of Defense; 34-11 © Andrew Lambert/Leslie Garland Picture Library/Alamy Images; 34-12 © Tim McCabe/USDA; 34-14, 34-21 © AbleStock; 34-16 Courtesy of Dr. Saeed Keshavarz/ RCCI (Research Center of Chemical Injuries)/Iran; 34-17 © AP Photo/Corbis; 34-18, 34-19, 34-20 Courtesy of the Centers for Disease Control and Prevention; 34-22 Courtesy of Atomex Scientific and Production Enterprise; 34-24 © Jime Mone/AP Photo; 34-26 Courtesy of AAOS

Chapter 35
Opener, 35-1 Courtesy of NFPA; 35-3 Courtesy of Nampa Fire Department, Nampa, Idaho; 35-4 © Corbis; 35-5 Courtesy of BRK Brands, Inc; 35-6 Courtesy of Pacific Northwest National Laboratory; 35-7 Courtesy of Captain David Jackson, Saginaw Township Fire Department

Chapter 36
Opener, 36-27, 36-29, 36-43 Courtesy of Tyco Fire and Building Products; 36-1 Courtesy of BRK Brands, Inc; 36-5 Courtesy of Kevin Reed; 36-6, 36-8 Courtesy of Honeywell/Gamewell; 36-7 Courtesy of Aaron Fire & Safety and Honeywell/Gamewell; 36-9, 36-17 Courtesy of Honeywell/Fire-Lite Alarms; 36-10 Courtesy of STI-USA; 36-12 Courtesy of Fire Fighting Enterprises Ltd; 36-13 Courtesy o... Firetronics Pte Ltd; 36-21 Courtes... of www.acimonitoring.com, Doug Beaulieu; 36-33, 36-48 Courtesy o... Ralph G. Johnson (nyail.com/fsd); 36-45 © AbleStock; 36-47 Courtesy of NFPA; 36-49, 36-50 © 2003 Kidde Fire Fighting Inc. All rights reserved. Reprinted with permission; 36-52 © Michael Heller/911 Pictures; 36-56 Courtesy of Chemetron Fire Systems

Chapter 37
Opener © Doug McSchooler, *The Indianapolis Star*/AP Photo; 37-1, 37-2, 37-4, Courtesy of Charles B. Hughes/Unified Investigations & Sciences, Inc; 37-3 Courtesy of Captain David Jackson, Saginaw Township Fire Department; 37-6 © Nate Rawner; 37-7 Courtesy of Rescue Resources, Inc; 37-8 © Joe Giblin/AP Photo; 37-9 © Steve Hamblin/Alamy Images

Unless otherwise indicated, photographs have been supplied by Jones and Bartlett Publishers and the Maryland Institute of Emergency Medical Services System. Illustrations for this text have been created by Precision Graphics, Rolin Graphics, and Studio Montage.

Works Cited

Chapter 34

1. Dugan, Laura. *Summary of Emergency Response and Research Institute Terrorism Statistics: 2000 and 2001*. Online. http//www.emergency.com/2002/terroris00-01.pdf. 24 May 2002
2. Jarrett, Colonel David G. *Medical Management of Radiological Casualties Handbook*, First Edition. Online. http://www.afrri.usuhs.mil/www/outreach/pdf/radiologicalhandbooksp99-2.pdf. 24 May 2002